G000128318

# FOREST SAMPLING DESK REFERENCE

# FOREST SAMPLING DESK REFERENCE

*Evert W. Johnson*

CRC Press
Boca Raton London New York Washington, D.C.

**Library of Congress Cataloging-in-Publication Data**

Johnson, Evert W.
Forest sampling desk reference / Evert W. Johnson.
p. cm.
Includes bibliographical references.
ISBN 0-8493-0058-5 (alk. paper)
1. Forests and forestry—Statistical methods—Handbooks, manuals, etc. 2. Sampling (Statistics)—Handbooks, manuals, etc. I. Title.

SD387.S73 J65 2000
634.9′07′27—dc21 99-086777



Direct all inquiries to CRC Press LLC, 2000 N.W. Corporate Blvd., Boca Raton, Florida 33431.


International Standard Book Number 0-8493-0058-5
Library of Congress Card Number 99-086777
Printed in the United States of America 1 2 3 4 5 6 7 8 9 0
Printed on acid-free paper

# Preface

This book is written for students and for forest practitioners. It is a further development of the class notes used in a two-course sequence in forest sampling in the Auburn University School of Forestry. Certain chapters were used as text material and were well received by the students and their instructors, which gave rise to the idea that a book based on that material might be useful.

The material covered is basic but is further developed than is the case with standard works in forest mensuration. The book describes the derivations of many of the statistical expressions used in forestry. These derivations can be bypassed, but they explain why the expressions take their utilitarian form. At times, to the average student and/or practitioner, it seems as if those expressions were handed down from above, engraved on tablets of stone. From my own student experience I could never understand why, in the calculations of a sample standard deviation, the sum of squared residuals is divided by $n - 1$ instead of $n$. Why subtract 1? Why not some other number? Standard forest mensuration texts rarely explain this. It is hoped that derivations in this book will fill this gap.

Standard forest mensuration books describe many of the sampling designs that can be used in forest inventory operations, but rarely explain their limitations — for example, why systematic sampling cannot yield a valid estimate of the population variance. *Forest Sampling Desk Reference* provides a door through which interested readers can become aware of the limitations and why they exist.

The book is basic. It does not describe sophisticated techniques that are used by specialists. To follow the derivations, some knowledge of differential and integral calculus and matrix algebra is needed. However, the mathematical backgrounds of most forestry students should be adequate for comprehension of the material. The examples given are deliberately made simple so that the student can concentrate on the statistical aspects.

# The Author

E. W. Johnson was born April 6, 1921 in Astoria, New York. In December, 1950, he married Janice E. Hesse of Rutland, Vermont. They had three children; two, a son and a daughter, survive. Dr. Johnson went to public schools in New York and Connecticut. He was graduated from Roger Ludlowe High School in Fairfield, Connecticut in 1939 and from the University of New Hampshire with a B.S. in forestry in 1943. He was in the Army in World War II, spending about 18 months in the Mountain Infantry (6 months in the Aleutian Islands) and 18 months in the Field Artillery (Solomon Islands, Philippine Islands, and Japan).

He took advantage of the G.I. Bill and obtained a Master of Forestry degree from Yale in 1947 and in 1957 the Ph.D. from the New York State College of Forestry at Syracuse University. The subject matter for the Ph.D. was forest measurements and photogrammetry. This educational experience was augmented by short courses and seminars at institutions such as Colorado State University, University of Michigan, University of Georgia, and the Georgia Institute of Technology.

From 1947 to 1950 he was employed by the Sable Mountain Corporation, a consulting forestry firm, working primarily in large-scale timber inventory operations in which aerial photographs and sampling were heavily involved. This work was in Ontario, Maine, New Hampshire, Vermont, Virginia, North Carolina, South Carolina, and Florida. In 1948 he became a member of the firm.

During this consulting work he injured his back, leaving him with a paralyzed right leg which forced him to turn to teaching and research. In the fall of 1956 he joined the forestry faculty in the Alabama Polytechnic Institute (now Auburn University) as an instructor. In 1985 he retired from Auburn as an emeritus full professor. His teaching assignments were varied but his main responsibility was in the forest mensuration–photogrammetry area. His research work paralleled his teaching responsibilities. In addition to these activities he was involved in the planning and execution of a number of short courses in forest photogrammetry, 3P sampling, etc. These courses were taught at a number of locations other than Auburn, one at the Marshall Space Flight Center in Huntsville, Alabama.

From 1958 to 1960 he served as associate director of the Society of American Foresters "Study of Education in Forestry and Related Fields of Natural Resource Management," which resulted in the book *Forestry Education in America* (the Dana–Johnson Report), which had a very significant impact on forestry curricula across the United States.

Within the Society of American Foresters he is a 50-year (golden) member and a fellow. In addition he is an inductee into the Alabama Foresters' Hall of Fame. For 2 years in the 1970s he served as National Chairman of the Society of American Foresters Inventory Working Group.

He is a member of the Society of the Sigma Xi (science honorary), Alpha Zeta (agriculture honorary), Xi Sigma Pi (forestry honorary), Phi Sigma (biology honorary), and Gamma Sigma Delta (agriculture honorary). In addition he is a member of the American Society of Photogrammetry and Remote Sensing Society. His social fraternity is Alpha Gamma Rho.

Dr. Johnson is the author of 18 peer-reviewed journal papers (in the *Journal of Forestry, Photogrammetric Engineering and Remote Sensing,* and the *Scientific Monthly*), Alabama Agricultural Experiment Station Bulletins, and eight Auburn University Forestry Department series publications.

# Acknowledgments

Thanks must be extended to Dr. Emmett Thompson, Dean of Forestry at Auburn University, who provided time and secretarial service to the author. Drs. Glenn Glover, Greg Somers, and Ralph Meldahl, biometricians at Auburn University, and Dr. Marian Eriksson, biometrician at Texas A&M University, provided technical assistance and encouragement. Drs. Glover, Somers, and Meldahl used chapters of the book in their classes. Chief among the transcribers of the written material to computer disks were Bessie Zachery and Linda Kerr, who labored with the mathematical expressions. I especially wish to thank Dr. Ralph Meldahl for the assistance and encouragement he gave me in preparing the manuscript for publication. Finally, but not least, I wish to thank my wife Janice for supporting me in this endeavor.

**Evert W. Johnson**

# Table of Contents

**Chapter 6 Measures of Dispersion**



**Chapter 7 Probability**

**Chapter 8 Probability Distributions**



**Chapter 9 Multidimensional Probability Distributions**



**Chapter 10 Discrete Probability Distributions**



**Chapter 11 Continuous Probability Distributions**

## Chapter 12 Sampling Theory



## Chapter 13 Estimation of Parameters



## Chapter 14 Fundamentals of Hypothesis Testing

**Chapter 17 Cluster Sampling**

**Chapter 20 Double Sampling**

**Chapter 21 Sampling with Unequal Probabilities**

# 1 Introduction to Sampling

## 1.1 INTRODUCTION

Persons engaged in the management or utilization of forest resources are required to make decisions of many kinds regarding those resources. For example, timber managers must decide whether a certain slash pine plantation, damaged by ice, should be clear-cut and a new stand established or if the plantation should be permitted to recover as much as possible and carried to its normal rotation age; or wood procurement foresters must decide what they would be willing to pay for the timber on a certain tract of land; or state wildlife managers must decide whether or not the deer herd in a certain portion of the state is large enough to survive hunting pressure during a specified open season. If such decisions are to be made in an intelligent manner, they must be made on the basis of sound information about the resource in question. Much of this information can only be obtained by assessment or inventory of the resource. Preferably such assessments or inventories would be based on *all* the individual organisms or items involved, but only rarely is this possible. The costs, in terms of money and time, for complete inventories are almost always excessive. Consequently, the decisions have to be made using knowledge based on an evaluation of only a small portion of the whole. When this occurs, *sampling* is involved.

Often many sampling options are available. The validity and strength of the results vary from option to option. In addition, the costs, in terms of money and time, also vary. It is desirable, therefore, that the decision maker have an idea of the nature of the available procedures, their strengths and weaknesses, and their relative costs. The procedure used in any specific case should be one that yields the needed information, at the desired level of precision, at the lowest cost in money and time. It should be recognized at the outset that inventories are not, by themselves, productive. They produce income only for those doing the work. Consequently, it is rational to use procedures that produce the greatest benefit: cost ratios. This book is written to provide background for land managers so that they can apply this philosophy to their inventory operations.

## 1.2 SOME BASIC CONCEPTS AND TERMINOLOGY

A *universe* is a complete set of objects, each of which is referred to as an *element*. In set terminology it would be called a *universal set*. If a universe contains a definite number of individual objects or elements, it is a *finite universe* (e.g., the set of "forties" in a specified section or the set of logs purchased by a certain sawmill on a certain day or during a certain week). On the other hand, if the set of elements is so large that the number of elements cannot be counted, it is an *infinite universe*. If each element is a distinct entity and the elements can be put on a one-to-one relationship with the set of all integers, it is said that the universe is *countably* infinite. On the other hand, if the individual elements of the universe cannot be isolated, as in the case of the set of points making up a line segment or the surface of a piece of land, the universe is *noncountably* infinite.

Each element in a universe possesses one or more characteristics or properties (e.g., each of the logs mentioned above comes from a tree of a certain species, has a certain small-end inside bark diameter, a certain length, a certain cubic foot volume, a board foot volume in terms of the International 1/4-in. log rule, a certain weight, etc. Each of these is a characteristic.). When such a characteristic is the same for all elements in the specified universe, it is referred to as a *constant* (e.g., if all the logs were from trees of the same species). When the characteristic varies from element to element, it is called a *variable* (e.g., the weights of the logs). Variables may be of several types. When the values of the variables can be expressed in a numerically ordered manner, the variable is a *measurement variable*. If the measurement variable can be described only in terms of whole numbers, it is a *discrete*, *discontinuous*, or *meristic* variable (e.g., the counts of filled-out seeds in individual pine cones). On the other hand, if the variable can take on any value between two limits, it is a *continuous* variable (e.g., the basal areas attained by a tree during a specified time span). In some cases an element either possesses or does not possess a certain characteristic. For example, a tree is either a loblolly pine or it is not a loblolly pine. This type of characteristic is known as an *attribute*. Since neither measurement nor counting is involved in the assessment of this type of characteristic, it is *not* a measurement variable. Last, it may be necessary in some cases to identify the order in which something has occurred or to indicate a ranking among a number of individuals which is nonquantitative or where the quantitative criteria are loose. For example, one might be interested in the order in which tree species begin to show fall colors or one might describe woods roads following heavy rains as passable for all types of vehicles, passable for high-clearance vehicles such as pickups, passable for four-wheel-drive vehicles, or impassable. In either case a numerical rank could be assigned to each condition. Variables handled in this manner are *ranked variables*, which, like attributes, are not measurement variables.

The variable that is of primary concern to the person requiring the information is referred to as the *variable of interest* or the *response variable*. An individual value of the variable of interest is known as a *variate* (e.g., if the cubic foot volume of logs is the variable of interest, the volume of an individual log would be a variate). The set of variates for a given variable of interest obtained from all elements of a specified universe is known as a *population* (e.g., the set of cubic foot volumes of logs processed by a certain mill on a certain day). Since each element of a universe usually possesses a number of different attributes, any one of which could be the variable of interest, each universe usually has associated with it a number of populations. (For example, each of the infinite number of points making up the surface of a piece of land has an altitude; a slope; an aspect; is in some cover category, such as grassland, timberland, water, road; etc.; and falls inside the boundary of some soil type. The set of values associated with each of these attributes is a separate population.)

If it is assumed that there is only one variate of the variable of interest associated with each element of a universe, it can be said that finite universes give rise to *finite populations* and infinite universes give rise to *infinite populations*. However, an *infinite population* can be derived from a finite universe, even one with only one element, if the variable of interest is the *measured* value of some characteristic. For example, when the least reading on a measuring device is very small, such as on the horizontal circle of a theodolite, it often is impossible to obtain consistent measurements of the desired dimension. Consequently, the use of the instrument can give rise to a set of individual measurement values. Collectively, these values form a population. From a conceptual point of view, there is no limit to the number of measurements that could be made. Consequently, the set could be considered to be infinite in size.

Each population possesses a set of characteristics that are known as *parameters*. Among these might be the total of all the variates, the mean of all the variates, the range of values of the variates, or the way the individual variates are distributed along a number line. When decision makers desire information about some variable, such as the total volume of timber on a tract of land or the average volume per acre on the tract, what they really want is a knowledge of these population parameters.

Usually this information is not available and must be obtained through an inventory or assessment operation. When every element of the specified universe is measured or otherwise evaluated, the process is known as a *complete enumeration*. In the case of the timber on a given piece of land, the process would be called a *100% cruise*. Such complete enumerations usually are excessively costly in time and money. Consequently, recourse is taken to *sampling*, where only a portion of the entire universe is evaluated. The sample is made up of *sampling units*, each of which contains one or more elements of the universe, depending on how the sampling operation has been designed. For the purpose of sampling it is necessary to have a listing or a mapping of all the sampling units that are available for sampling. Such a list or map is referred to as the *sampling frame*.

There are many ways by which the sample may be drawn from the frame. A number of these sampling procedures are described in this book.

The evaluation of the elements in the sample gives rise to a set of variates of the variable of interest. This set is, of course, a subset of the values in the corresponding population. The total of these values or their mean or their range or their distribution along a number line, to mention but a few of the possibilities, can be determined for the sample in the same way as for the population. Each of these sample-derived values is a *statistic*. Sample statistics correspond to and are *estimates* of the population parameters. This is the key concept in sampling. By evaluating a small portion of an entire universe, one can arrive at estimates of population parameters. These estimates can then be used by the decision maker. The soundness of the decisions in such cases is heavily dependent on the quality of the estimates. It is the responsibility of the information supplier to provide the highest-quality estimates possible given the constraints on time and money imposed. Suppliers can meet this responsibility only if they have a sound understanding of the principles and procedures of scientific sampling.

## 1.3 REASONS WHY SAMPLING MAY BE PREFERRED OVER COMPLETE ENUMERATION

It must be admitted at the outset that sampling does not provide the decision maker with perfect information, yet sampling usually is used rather than complete enumeration. Primary among the reasons for preferring sampling are the following:

1. It may be impossible to obtain a complete enumeration. For example, if the evaluation process is destructive, as when testing wooden trusses by applying loads until the trusses break, or when cutting open tree seeds to determine soundness. In either case, a complete enumeration would lead to destruction of the universe. It also should be obvious that a complete enumeration of an infinite population would not be possible.
2. A complete enumeration may be too expensive. Insofar as budgets are concerned, inventories are cost items and usually the funds available for such activities are limited. Only rarely are such funds sufficient for a complete enumeration.
3. A complete enumeration may require more time than is available, It is not uncommon in timber management operations that information about the resource has to be obtained within a very narrow time span. Under such circumstances sampling provides the only solution to the problem.
4. In some cases sampling may actually yield a more reliable final value than would a complete enumeration. Since the time requirement for sampling is less than that for a complete inventory, more time can be taken to obtain each individual evaluation, and this usually results in better or stronger individual values. Furthermore, the cost of sampling is less than that of a complete evaluation and the saving can be used to improve the instrumentation.

## 1.4 SOME FURTHER TERMINOLOGY

As is common in the English language, the word *statistic* has more than one meaning. In Section 1.2 it is used as a general name for a sample characteristic in the same way as *parameter* is used for a population characteristic. In nontechnical colloquial speech the word is often used as a descriptive term for an individual piece of information, especially when referring to a person killed or injured in an automobile accident. Such a person becomes a "statistic" in the collection of accident casualty figures. This is an improper use of the word. Technically, the fact of the person's death or injury is a *datum*, an individual piece of information which usually is numerical. The plural of datum is *data*, a collection of pieces of information, a commonly encountered term. It should be emphasized that the word *data* should never be used in a singular sense. Unfortunately, this rule is often ignored.

The word *statistics* is very similar to "statistic" but differs from it considerably in meaning. Actually, there are two ways of using the word. In the first, "statistics" or, preferably, *descriptive statistics*, is used as a general term for collections or listings of numerical data such as the tabular material in U.S. Forest Service publications showing pulpwood production or amounts of commercial forestland by counties or states. These data usually have been processed to a certain extent so that they appear in the form of means or totals. Only rarely would the term *statistics*, in this sense, be used with unprocessed data. The data themselves shown in the collections or lists may have been obtained from either complete enumerations (as in the case of census data) or by sampling. The method of acquisition has no effect on the meaning of the word when used in this context.

The second, and probably more important, use of the word *statistics* is as the name of the scientific discipline that is concerned with the theory and procedures associated with the acquisition, summarization, analysis, and interpretation of numerical data obtained using sampling. This discipline or scientific field involves such things as estimating population characteristics, the testing of hypotheses involving differences between sets of experimental data, the description of relationships between variables, and the monitoring of industrial processes. Each of these has evolved, to a certain extent, into a discipline or specialization in its own right. These subdivisions might be referred to, respectively, as "parameter estimation," "hypothesis testing," "regression analysis," and "quality control." "Time series analysis," used heavily by economists seeking trends, is a subdivision of regression analysis. All of these are interrelated but in each the emphasis is different. Because of the heavy use of hypothesis testing in research, statistics has become practically synonymous with hypothesis testing. As a consequence, courses and textbooks dealing with statistics usually are strongly oriented toward this subdivision of the field. Regression analysis has so many applications in research and elsewhere that it, too, is stressed in courses and texts. Parameter estimation, on the other hand, is rarely used in research and usually is relegated to a minor place in the statistics picture. This is unfortunate because it is of primary importance to *managers* because of its utility in inventory operations. As the broad field of statistics has developed, the term *sampling* has been attached to the parameter estimation segment of the field in the same way that statistics has been attached to hypothesis testing. Quality control is essentially a subfield under parameter estimation since it is concerned with dimensional (in some sense) characteristics of manufactured products. If these characteristics change as the manufacturing process goes on, it means that adjustments must be made in the process so that the characteristics of the individual items of the product fall within certain specified limits. Control of manufacturing processes using these techniques is so important that this subfield of parameter estimation has become a separate area of specialization with its own literature and instructional packages.

Because of the sheer size of the parent field this splitting of statistics into subfields is defensible. It is a rare individual who is competent in all aspects of statistics. Most research workers use statistical methods as tools in their work and usually learn only enough to be able to do that work. The same situation occurs with management personnel. It is logical, therefore, to set up curricula and other educational programs so that persons entering professional fields are prepared to use statistical tools appropriate to their fields. In the general area of natural resource *management*, including forestry, fisheries, wildlife management, watershed management, and recreational land management, the stress in the statistical area is on *inventory*, not research, and the statistical training of people in these fields should reflect this fact.

# 2 Accuracy and Precision of Data

## 2.1 ACCURACY OF DATA

In Section 1.4 it was explained that the word *data* is the plural of *datum* and that each datum is an individual piece of information, usually in the form of a number. A set of such items of information is known as a *data set*. In a sampling or statistical context each datum usually is a *variate*, an individual value of a variable of interest. Thus, a population, which is composed of variates, is a data set. In most cases the magnitudes of the variates making up the population are not known. Nonetheless, they are real quantities. The purpose of the data-gathering step of any inventory is the determination of some or all of these values.

When an attempt is made to arrive at the value of a variate, the result may be exactly correct or it may be, to a certain extent, off the mark. The *accuracy* of such an evaluation refers to the closeness of the measure or otherwise assessed value to the true value. The closer the assessed value is to the true value, the more accurate the evaluation. If for some reason the assessed values tend to be consistently too high or too low, they are *biased*. Bias can be introduced into the evaluation process through faults in the instrumentation, improper techniques, or human error. The presence of bias is often difficult or impossible to detect, and consequently the accuracy of a set of observations is difficult or impossible to ascertain.

## 2.2 PRECISION

If repeated evaluations are made of the same element, the same result may or may not be obtained each time. If the same result is obtained each time, the value obtained is said to be *precise*. As the degree of repeatability decreases, the assessments become *less precise*. Consequently, *accuracy* and *precision* refer to different things. Accuracy is concerned with the nearness of evaluations to the true values, whereas precision is concerned with the repeatability of evaluations. It is possible for a set of evaluations to be biased and, consequently, not accurate, yet be highly precise because repeating the assessment process yields essentially the same results. Conversely, the individual values in a set of repeated measurements may be too large in some cases and too low in others. Individually they are inaccurate. However, if *on the average*, across all the measurements in the set, they were correct it would be said that *compensating errors* or *random errors* have occurred and that on the average the set of measurements is accurate. The precision of the measurements, of course, would be less than perfect.

## 2.3 ACCURACY AND PRECISION WITH DISCRETE VARIABLES

When the variable of interest is *discrete*, the value of the variate is determined by *counting*. For example, if the variable of interest is the number of cones on longleaf pine trees, as is the case of the data forming the basis of Table 3.1, each value in the data set would be obtained by counting the cones on the associated tree. In this context a cone either is or is not present and its condition (broken, poorly developed, etc.) is not a factor. Consequently, the counts produce only *whole* numbers. If no human errors are involved, the counts should be exactly correct, that is, *accurate*. Furthermore, if repeated counts were made on each tree, presumably the same results would be obtained each time. In such a case the counts would be *precise*. On the other hand, if human errors were to occur, the individual counts would be inaccurate, the degree of accuracy depending on how far a specific count was from the correct value. The effect of the human errors on precision would depend on the nature of the errors. If they were such that repeated counts yielded the same but inaccurate results, the counts would be biased but still precise. If, on the other hand, the errors resulted in varying counts, the precision would be reduced. Usually both accuracy and precision are high when the variable of interest is discrete. However, much depends on the magnitudes of the counts. When these are relatively low, as in the case of some cone counts, accuracy and precision should be high. If, however, the variable of interest is the number of needles on the trees, the situation would be considerably different. Accurate counts would be difficult to obtain and the likelihood of repeated counts yielding the same results is not high. Consequently, both accuracy and precision would be relatively low in spite of the fact that the data are discrete.

## 2.4 ACCURACY AND PRECISION WITH CONTINUOUS VARIABLES

If the variable of interest is *continuous*, each variate is obtained by *measurement*. In essence, the true value of each variate occupies a position along a number line and the measurement process is used to determine that position. Fundamentally, measurement involves the counting of the number of line segments, of some specified size, which make up the space between the origin of the number line and the position of the variate. The length of the line *segments* used in this counting process is determined by the instrument or procedure used in the measurement process and is known as the *least reading* of that instrument or procedure. For example, the smallest graduation on a weighing scale or balance might represent a pound. If so, the least reading of the scale or balance is 1 lb. A more sensitive scale, such as a laboratory balance, might have a least reading of 0.001 g. The smaller the least reading the more accurate a measurement is apt to be, provided no bias exists or human or other errors are present. However, exactly correct values are rarely obtained because only rarely will the distance between the origin of the number of line and the variate be such that can be divided into segments corresponding to the least reading without leaving a remainder. Consequently, absolutely accurate values are rarely obtained when the variable of interest is continuous, even if no human or other errors are present. On the other hand, the level of accuracy can be controlled to a considerable extent by the choice of measurement device or procedure.

Perfect precision, in the case of a continuous variable, cannot be attained. To have perfect precision, the repeated measurements would have to agree when the least reading is an infinitesimal. However, with a continuous variable, it is possible to have an indication of the level of precision that has been attained. For example, if the distance between two points is measured and recorded as 362.7 ft, the *implied precision* of the measurement is ±0.05 ft. In other words, it implies that repeated measurements would fall between 362.65 and 362.75 ft. Furthermore, *it implies that the true distance lies within the indicated interval.*

If procedural and human errors are not a factor, the implied precision usually is equal to plus or minus half the least reading of the measuring device. The rationale for this assumption stems from the nature of measurement. It was pointed out earlier that when a measurement is made it involves counting the number of least reading units lying in the interval between the origin of a number line and the position of the variate in question. When this is done, a remainder usually is left. If the remainder is less than half the least reading, it is ignored and the count is left unchanged.

On the other hand, if the remainder is more than half the least reading, the count is increased by one unit. If the remainder is exactly a half least reading unit, certain rules must be applied to see if it is to be ignored or the count increased by 1. These rules are discussed in Section 3.3.4. Thus, the last unit added to the count is approximate. The actual value may be up to one-half a least reading unit more than is indicated or it may be up to one-half unit smaller than is indicated.

In some cases using the least reading to imply a precision is misleading. For example, tree heights can be measured on vertical aerial photographs using at least two different procedures. One of these makes use of a quantity known as differential parallax. Average humans can detect a differential parallax if it is larger than 0.001 in. Several instruments are available for measuring differential parallaxes. Several have least readings of 0.01 mm. A measurement made with one of these instruments might be recorded as 1.07 mm which implies a precision of ±0.005 mm.

Assuming an average human as the person making the measurement, this implied precision is too good. Since the *human's* least reading is 0.001 in. or 0.0254 mm, the recorded value should be rounded back from 1.07 to 1.1 mm. This implies a precision of ±0.05 mm which is more nearly correct. Implying a precision greater than actually attained is bad practice and should be discouraged whenever possible.

## 2.5 SIGNIFICANT FIGURES

When a measurement is made, the result is a count of some unit of measure plus a remainder. The count might be a number such as 50,962. The 2 in the number is approximate since the remainder is absorbed into it. However, the count is as exact as can be attained with the unit used. Each of the digits in the number is therefore considered *significant*. If the unit used in the counting process were 0.001 g, the measurement would be recorded as 50.962 g. The measurement also could be recorded in terms of kilos, in which case the number would be 0.050962 kg. It also could be shown in terms of metric tons, 0.000050962 metric tons. Regardless of the units used to describe the count, the count remained the same. In spite of the change of recorded units, the significant figures remained the same. The zeros to the *left* of the 5 are not significant because they had nothing to do with the count. They merely are used to show the position of the decimal point.

In some cases the counting unit used in a measurement is a multiple of some smaller unit. This often is the case with timber inventories where the volume is expressed in thousands of board feet rather than individual board feet. For example, an inventory might indicate a volume of 1,206,000 bd. ft. Since the counting unit was 1000 bd. ft, the count actually was only 1206. Consequently, only these figures are significant. The three zeros to the *right* of the 6 are used only to establish the position of the decimal point.

As can be seen, zeros in a measurement value may or may not be significant. If they appear to the *left* of the first nonzero digit in the number, they are there only to locate the decimal point and, consequently, are not significant. If they appear to the *right* of the last nonzero digit, the situation is ambiguous. To determine which, if any, of such zeros are significant one would have to know the units used in the counting process. Only if a zero appeared in the position appropriate for the counting unit would it be significant. An internal zero, such as the one in the number 1206, resulted from the counting process and would be significant.

The system of expressing numbers called *standard* or *engineering notation* isolates the portion of a number made up of significant figures, places the decimal point to the right of the first significant digit, and then multiplies the number by 10 raised to the appropriate power. For example, using the weight values used earlier, 50.962 g would be shown 5.0962 * 10 g; 0.050962 kg would appear as $5.0962 * 10^{-3}$ kg; and 0.000050962 metric tons would be $5.0962 * 10^{-5}$ metric tons. The board foot volume used earlier would be shown as $1.206 * 10^{6}$ bd. ft. Much use was made of scientific notation when the slide rule was the primary computing device used by scientists and engineers. Since the slide rule could be read only to three decimal places, numbers were routinely rounded to three significant figures and converted to scientific notation so as to keep track of the decimal point. At the present time, the slide rule has been supplanted by electronic calculators and computers, and since such devices commonly can be read to eight or ten decimal places the need for scientific notation has declined. However, even the modern calculators resort to scientific notation when a number is too large to be displayed in conventional form. In such a case, up to eight or ten significant figures can be used so the attainable precision is much greater with calculators than with the slide rule.

## 2.6 ROUNDING OFF

When a number contains more significant figures than are desired or can be used, the number can be *rounded off* to the point that it expresses the count in terms of a larger unit which is some multiple of 10 larger than the original counting unit. In this process the portion of the number to the right of the new right end position becomes a remainder. If this remainder is less than half the new counting unit, it is ignored and the right end digit is left unchanged. If, on the other hand, the remainder is more than half the new counting unit, the right end digit is increased to be next higher value. If the remainder is exactly half the new counting unit, recourse must be made to one of the rules described in Section 3.3.4. For example, if 608.6 g is to be rounded to the nearest gram, the new right end position would be that occupied by the 8. The remainder is 0.6, which is greater than half a gram. Consequently, the 8 is increased to a 9 and the rounded off measurement would be 609 g. In another example, 50,422 lb are to be rounded to the nearest thousand pounds. The zero would be the new right end digit. The remainder, 422 lb, is less than half of 1000 lb, so the zero is unchanged and the rounded off measurement would be 50,000 lb. Note that in this case the first zero in the number is significant. The remaining three are not.

## 2.7 COMPUTATIONS USING APPROXIMATE NUMBERS

In many cases in sampling and statistical operations the data values are not simple measurements such as those discussed up to this point. Instead they are *derived* or *computed* values resulting from some sort of computation routine. Computations may be, and usually are, based on measurements as well as other numerical values such as trigonometric functions or physical constants. Both the measurement and nonmeasurement values entering a computation routine may be, and usually are, approximations. When such values are used in a computation, the resulting value also is an approximation. The degree of approximation must be understood by workers in this field because it often is much greater than what might be anticipated.

To obtain an understanding of the problem, let us consider a situation where the area of a rectangular piece of land is to be determined. Assume that a tape graduated in feet was used to measure the dimensions and that these were 98 × 123 ft. The commonly used procedure would be to multiply the two values to obtain the area:

$$98 * 123 = 12{,}054 \text{ ft}^2$$

However, neither 98 nor 123 is exact. The 98 stands for a dimension which might be as small as 97.5 ft or as large as 98.5 ft while the 123 stands for one which might be as small as 122.5 feet or as large as 123.5 ft. If the extreme low values were used in the computation, the area would be:

$$97.5 * 122.5 = 11{,}943.75 \text{ ft}^2$$

which is 110.25 ft$^2$ less than was obtained in the first computation. If the extreme high values are used, the area would be

$$98.5 * 123.5 = 12{,}164.75 \text{ ft}^2$$

which is 110.75 $ft^2$ greater than the first result. Consequently, the area could be anything between 11,943.75 and 12,164.75 $ft^2$. Rounding both back, to the point where the figures are significant, results in an area of 12,000 $ft^2$, which has two significant figures.

The rules controlling the multiplication or division of approximate numbers are

1. Determine the number of significant figures in the number with the fewest significant figures. Round off all the other numbers until they have one more significant figure than the one with the fewest.
2. Multiply or divide in the usual manner.
3. Round off the product or quotient to the number of significant figures possessed by the original number with the fewest.

This process is demonstrated in the following example. Let us assume that the area of a piece of land has been determined to be 11,860 acres (contains four significant figures) and the average volume of timber has been estimated to be 6900 bd. ft/acre (contains two significant figures). The total volume of timber on the tract is then a product of these two values. Before multiplying, however, the area value must be rounded to three significant figures (one more than the number in the volume per acre value). The computation would then be

$$11{,}900 * 6900 = 82{,}110{,}000$$

which should be rounded off to two significant figures, producing

82,000,000 bd. ft on the tract

The rules controlling the addition or subtraction of approximate numbers are as follows:

1. Arrange the numbers in a column so that their decimal points are lined up.
2. Find the column farthest *toward the left* which contains a nonsignificant figure.
3. Round off all the other numbers until their right-most significant figure is in the column found in step 2.
4. Add or subtract in the normal manner.
5. Round off the sum or difference so that it has the same number of significant figures as has the item which had its right-most significant figure farthest to the left when the numbers were arranged in a column.

An example involving this process is described as follows. Assume that the total area in a certain ownership is needed by the owner. The ownership consists of three widely scattered parcels. The area of the first, 362.52 acres, was determined from a transit and tape survey and all five digits are significant. The area of the second, 226 acres, was obtained from a map using a polar planimeter. This value contains three significant figures. The last parcel had an area of 1620 acres which had been obtained from a compass-and-pace survey. The value contains three significant figures. Carrying out the rules yields the following:

1. Arrange values in column form

```
 362.52
 226.
1620.
```

2. Determine the column with the left-most nonsignificant figure. The zero in 1620 is not significant, so this column is the one immediately to the left of the decimal point.
3. Rounding off the values and adding

    363.
    226.
    1620.
    2209

4. Rounding off to three significant figures, the number of significant figures in 1620 yields a total of 2210 acres.

An example involving subtraction might be where the total land area in a certain county has been determined using survey data to be 1,362,520 acres. This value contains six significant figures. The area of land in the category "Forest" has been found, using a sampling procedure, to be 622,000 acres. This value contains three significant figures. The area in "Non-Forest" would be obtained by subtraction. The procedure, following the rules, would be:

1. Arrange values in column form

    1,362,520.
    622,000.

2. Determine the column with the left-most nonsignificant figure. Since the left-most zero in 622,000 is not significant, the desired column is the fourth to the left of the decimal point.
3. Rounding off the values and subtracting

    1,363,000
    – 622,000
    741,000

4. Rounding off the difference in this case is not necessary because it already contains the correct number of significant figures. Consequently, the area in the Non-Forest category is 741,000 acres.

# 3 Data Arrangement

## 3.1 INTRODUCTION

The manner in which the individual data items are arranged is of significance for two reasons. First, by appropriate arrangement the person attempting to interpret the data may obtain at least a subjective impression of the existing situation. Second, the manner of arrangement controls to a certain extent the computational procedures which are to be used in the analysis of the data. As a matter of fact, the data may have to be put into specific arrangements before certain parameters or statistics can be determined. Consequently, attention must be paid to how the data are arranged.

## 3.2 METHODS OF ARRANGEMENT

Data that have not been subjected to any form of arrangement are referred to as *raw data*. The term usually implies that the data are in the order in which they were obtained in the field. An example of raw data is shown in column 1 of Table 3.1. This set of data is very small yet a person probably would have difficulty obtaining a reasonable impression of the cone production by simply looking at the data set. If the set were larger, and in most cases the sets are much larger, visualizing the cone production situation by inspection of the raw data would be impossible.

The problem of visualization is eased to some extent if the data are arranged in ascending or descending order. When this is done, the data are in the form of an *array*, as in column 2 of Table 3.1. Arranging the data in this way facilitates visualization to the extent that the lowest and highest values of the variable are identified and the range of values in the set is easily obtained. In the case of the cone counts in Table 3.1, the range of counts is from zero to 72 cones per tree. Even this little gain of information may be of value in some situations.

In the majority of cases data sets are large and the listing of the individual values, in either raw data or array form, rarely yields benefits that are commensurate with the time and effort needed to get the job done. It usually is preferable to condense the data in some way to reduce the work and also to improve the likelihood of visualizing the situation. As a first step in condensation, one might list the counts of the values that have occurred. Such a listing is known as a *frequency table*. The cone count data are shown in this form in column 3 in Table 3.1. When data are arranged in this manner, the number of times a value occurs is referred to as its *frequency*. Arranging the data in this form will condense them considerably if the values appear repeatedly. In the case of the cone count data the repeating was minimal, and as a result the frequency table does not make visualization appreciably easier than does the array.

It should be noted that in the development of the frequency table described above only the values actually appearing in the data set are shown. If the remaining intervening numbers were inserted into the list and were shown as having zero frequency, the resulting frequency table would show the *frequency distribution*. In the case of the cone count data, the frequency distribution based on all the possible counts from zero to 72 (column 4 in Table 3.1) would be clumsier and less

**TABLE 3.1**
**Data Arrangements That May Be Used with a Set of Cone Counts from Longleaf Pine Trees**

| (1) | (2) | (3) | | (4) | | | | | | (5) | | |
|---|---|---|---|---|---|---|---|---|---|---|---|---|
| Raw Data | Array | Simple Frequency Tables | | Frequency Distribution | | | | | | Frequency Distribution | | |
| Y[a] | Y | Y | f[b] | Y | f | Y | f | Y | f | Y | Class Limit | f |
| 28 | 0 | 0 | 1 | 0 | 1 | 25 | 1 | 50 | 0 | 4 | 0–8 | 3 |
| 41 | 5 | 5 | 1 | 1 | 0 | 26 | 0 | 51 | 1 | 13 | 9–17 | 1 |
| 36 | 7 | 7 | 1 | 2 | 0 | 27 | 0 | 52 | 2 | 22 | 18–26 | 4 |
| 17 | 17 | 17 | 1 | 3 | 0 | 28 | 2 | 53 | 0 | 31 | 27–35 | 3 |
| 52 | 21 | 21 | 2 | 4 | 0 | 29 | 0 | 54 | 1 | 40 | 36–44 | 5 |
| 25 | 21 | 24 | 1 | 5 | 1 | 30 | 0 | 55 | 1 | 49 | 45–53 | 4 |
| 54 | 24 | 25 | 1 | 6 | 0 | 31 | 0 | 56 | 0 | 58 | 54–62 | 2 |
| 21 | 25 | 28 | 2 | 7 | 1 | 32 | 0 | 57 | 0 | 67 | 63–71 | 0 |
| 43 | 28 | 35 | 1 | 8 | 0 | 33 | 0 | 58 | 0 | 76 | 72–80 | 1 |
| 52 | 28 | 36 | 1 | 9 | 0 | 34 | 0 | 59 | 0 | | | |
| 7 | 35 | 37 | 1 | 10 | 0 | 35 | 1 | 60 | 0 | | | |
| 40 | 36 | 40 | 1 | 11 | 0 | 36 | 1 | 61 | 0 | | | |
| 51 | 37 | 41 | 1 | 12 | 0 | 37 | 1 | 62 | 0 | | | |
| 72 | 40 | 43 | 1 | 13 | 0 | 38 | 0 | 63 | 0 | | | |
| 24 | 41 | 46 | 1 | 14 | 0 | 39 | 0 | 64 | 0 | | | |
| 28 | 43 | 51 | 1 | 15 | 0 | 40 | 1 | 65 | 0 | | | |
| 37 | 46 | 52 | 2 | 16 | 0 | 41 | 1 | 66 | 0 | | | |
| 21 | 51 | 54 | 1 | 17 | 1 | 42 | 0 | 67 | 0 | | | |
| 46 | 52 | 55 | 1 | 18 | 0 | 43 | 1 | 68 | 0 | | | |
| 55 | 52 | 72 | 1 | 19 | 0 | 44 | 0 | 69 | 0 | | | |
| 0 | 54 | | | 20 | 0 | 45 | 0 | 70 | 0 | | | |
| 35 | 55 | | | 21 | 2 | 46 | 1 | 71 | 0 | | | |
| 5 | 72 | | | 22 | 0 | 47 | 0 | 72 | 2 | | | |
| | | | | 23 | 0 | 48 | 0 | | | | | |
| | | | | 24 | 1 | 49 | 0 | | | | | |

[a] $Y$ is used as a symbol for the variable of interest. In this case it represents the cone count.
[b] $f$ is used as a symbol for the count or frequency of occurrence of the variate shown.

effective than either the array or the simple frequency table because of the very large number of values with zero occurrence. However, if the individual values were lumped or aggregated into *classes*, the situation could be considerably improved. This has been done in column 5 in Table 3.1 with each class including nine of the values used in column 4. The compactness of this arrangement of the data is clearly evident. This compactness is accomplished, however, only at the expense of detailed knowledge. For example, we know that four trees had cone counts ranging from 18 to 26 but we no longer know exactly what those counts were. In many cases this loss of detailed information is acceptable, and, consequently, the frequency distribution is a very common way of arranging and displaying data. In forestry a *stand table* is a frequency distribution of trees by dbh classes* or by dbh and tree height classes. Such tables are required in many silvicultural, mensurational, economic, and other operations carried out by professional foresters.

* dbh is the acronym for "diameter at breast height."

## 3.3 RULES CONTROLLING FREQUENCY DISTRIBUTIONS

### 3.3.1 Number of Classes

In the frequency distribution shown in column 4 in Table 3.1 there were 73 classes, many of which were unoccupied. Visualization of the situation using this arrangement is difficult because of the large number of classes. It would be preferable to reduce the number of classes by using broader classes. This was done in column 5 in Table 3.1. Here nine classes are used, each consisting of nine of the original classes. The result is a much more useful arrangement of the data.

A number of rules of thumb have been devised which provide the designer of an inventory with an idea of how many classes should be used. One, which might be said to occupy a middle ground, states that no fewer than 8 classes and *no more than 20 classes* should be used. If fewer than 8 are used, the lumping or aggregating becomes excessive and too much information is lost. On the other hand, if more than 20 classes are used, the arrangement begins to take on the character of an array and comprehension of the set becomes difficult. With 9 classes, the frequency distribution in column 5 in Table 3.1 meets the requirements of this rule of thumb. It must be emphasized that this rule, like any of the others like it, is only a guide and if, for some reason, it must be violated, the resulting distribution usually is acceptable.

### 3.3.2 Class Width or Interval

Associated with the number of classes is the span of values which are to be included in each class. The primary rule in this area is that all the classes should have the same width, provided that the results obtained are reasonable. In most cases this provision can be met and the rule obeyed. Failure to comply with the rule can easily lead to incorrect visualization of the situations and to problems when computations are to be made using the data in the frequency distribution.

The number of classes and their widths are related. In any given case, if the number of classes is increased, the class width would be reduced. The reverse would occur if the number of classes is reduced. It is up to the designer to arrive at an appropriate balance between the two factors. Guiding the designer should be a knowledge of what the range of data values would be. For example, if dbh is the variable of interest and the trees involved are old-growth coastal Douglas firs, the range of dbh values might be as much as 100 in. In such a case, the use of 5-in. dbh classes would be appropriate because 20 5-in. classes would span the 100-in. range and 20 is an acceptable number under the number-of-classes rule. On the other hand, if the trees involved are second-growth slash pines in central Florida, the largest tree probably would not have a dbh in excess of 20 in. and the use of a 1- or 2-in. dbh class width would be appropriate.

There are some situations where unequal class widths have been used. They are most common in social science and medical contexts. For example, the incidence of a certain disease might be very high in children under 1 year of age but would then taper off until it was virtually zero at ages beyond 20 years. There is no reason equal width classes could not be used in a frequency distribution showing the incidence of this disease. However, in some cases, perhaps to make the distribution more compact or to avoid empty classes or classes with very low counts, persons using or displaying such data will aggregate certain of the classes. The resulting frequency distribution might appear as the one in Table 3.2.

Another situation that might give rise to what might appear to be, or be considered to be, unequal class widths is when only certain values of the variable of interest are involved. For example, if a tally is made, by fuel tank size, of the trucks operated by a certain organization, the frequency distribution might appear as in Table 3.3. Actually, even though the tank sizes used would be unevenly spaced along a number line, there is no ambiguity. There is only one tank size associated with each class. In essence, what has been done is simply to leave out the empty classes in a regular

**TABLE 3.2**
**Hypothetical Distribution of Cases of a Childhood Disease by Age in a Certain Population Segment**

| Age (years) | Class Limits (years) | *f* |
|---|---|---|
| 0.5 | 0–1 | 103 |
| 2.0 | 1–3 | 69 |
| 4.5 | 3–6 | 42 |
| 8.0 | 6–10 | 16 |
| 12.5 | 10–15 | 9 |
| 17.5 | 15–20 | 2 |

frequency distribution. Table 3.3 is, thus, a simple frequency table with equal class widths, similar to column 3 in Table 3.1.

**TABLE 3.3**
**Frequency Distribution of Fuel Tank Sizes in Trucks Operated by the XYZ Agency**

| Tank Size (gal) | *f* |
|---|---|
| 18 | 6 |
| 20 | 13 |
| 30 | 9 |
| 100 | 4 |

In some cases the temptation is great to violate the constant-class-width rule when the data values scatter out at the ends of the distribution. For example, one might set the upper limit of the lowest class relatively high and then fail to set any lower limit to that class. This would permit the gathering into this class of a number of items which properly should belong to classes below the established lowest class. This might be done in an attempt to meet the number-of-classes rule while still using convenient class midpoints. The same situation might occur at the upper end of the distribution, but there it would be the upper end of the class that would be undefined. Such classes with undefined widths or limits are called *open-ended* classes. Use of such classes should be avoided because of the ambiguities that can result from their use. If computations are to be made using the frequency distribution data, open-ended classes may lead to distortions of the computed values.

### 3.3.3 Class Midpoints

A *class midpoint* or *class mark* is the numerical value at the center of the range of values associated with a class. For example, in column 5 in Table 3.1, 31 is the center value in the range from 27 to 35 and consequently is the class midpoint. Class midpoints are used to identify the classes; furthermore, it is assumed, in most analyses of frequency distribution data, that every item in a class possesses the value of the midpoint of that class.

Considerable thought should be given to the choice of class midpoints. If the data are *discrete,* the class midpoints must be *whole numbers*. One must consider this requirement concurrently with the others controlling the construction of a frequency distribution. An example uses the cone count data in Table 3.1. If a class width of 10 units had been chosen, the class ranges would have been:

**TABLE 3.4**
**Frequency Distribution of Cone Counts When Trees Possessing Fewer Than 20 Cones Are Ignored**

| Cone Count | Class Limits | *f* |
|---|---|---|
| 24 | 20–28 | 6 |
| 33 | 29–37 | 3 |
| 42 | 38–46 | 4 |
| 51 | 47–55 | 5 |
| 60 | 56–64 | 0 |
| 69 | 65–73 | 1 |

0–9; 10–19; 20–29; 30–39; 40–49; 50–59; 60–69; and 70–79. The number-of-classes rule would have been met, but the class midpoints would not have been whole numbers: 4.5, 14.5, 24.5, etc. If any class width *described by an even number* had been used, the class midpoints *would not have been whole numbers*. When the class width is expressed by an *odd* number, however, there is no problem, as can be seen in column 5 where a class width of 9 units was used.

A further factor that should be considered when designing a frequency distribution is that in some cases there is a lower limit to the magnitude of the variable of interest which will be recognized. In the case of the data in Table 3.1, the lower limit was zero, indicating a complete absence of cones on the tree. In other cases the lower limit might be a value other than zero. For example, the trees bearing fewer than 20 cones might be of no interest to the decision maker. In such a case the data relating to trees bearing fewer than 20 cones would be disregarded and not included in the frequency distribution. When this occurs, the resulting set of values is referred to as *amputated data*. the lower limit of the acceptable values becomes the beginning point for the development of the classes in the frequency distribution (e.g., see Table 3.4).

When the variable of interest is *continuous*, the problem of defining class midpoints becomes more complex. In some cases, depending on the nature of the data, the midpoints might be whole numbers. This is the case when the variable of interest is dbh. The dbh values in a frequency distribution are always shown as whole numbers. In other cases, however, the data may be such that it would be appropriate to use fractional values. For example, if the variable of interest is the growth in dbh over a period of 5 years, the increments probably would be measured to the nearest 50th or 100th in., and the range of values might be from essentially 0 to 2 in. Using whole numbers as class midpoints would be ridiculous. What probably would be done is to divide the total range into 20 classes, each 0.1 in. in width as shown in Table 3.5. The class midpoints in this case are commonly encountered quantities which can be visualized without difficulty. Furthermore, if computations are to be made using the data from the frequency distribution, the simplicity of the midpoint values would be beneficial. The use of fractional values in this situation is proper.

It should be noted in Table 3.5 that a gap exists between zero and the lower limit of the first class. Such a gap will occur whenever the class width is an *odd* value. In most cases the presence of the gap is not a problem because no data values fall in it. However, if the range of values in the first class must include zero the class width would have to be an *even* value to adhere to the equal-class-width rule. The class midpoints in such a case would be *odd* numbers. This can be seen in Table 3.6, which involves the data used in Table 3.5.

### 3.3.4 Class Limits

When the data are *discrete*, no ambiguity exists with respect to the limits of the classes. This, however, is not true when the data are *continuous*. It can be seen in Tables 3.5 and 3.6 that the

**TABLE 3.5**
**Frequency Distribution of 5-Year dbh Growth Increments**

| Growth Increment (in.) | Class Limits (in.) | f | Growth Increment (in.) | Class Limits (in.) | f |
|---|---|---|---|---|---|
| 0.1 | 0.05–0.15 | 1 | 1.1 | 1.05–1.15 | 10 |
| 0.2 | 0.15–0.25 | 3 | 1.2 | 1.15–1.25 | 8 |
| 0.3 | 0.25–0.35 | 4 | 1.3 | 1.25–1.35 | 8 |
| 0.4 | 0.35–0.45 | 7 | 1.4 | 1.35–1.45 | 6 |
| 0.5 | 0.45–0.55 | 12 | 1.5 | 1.45–1.55 | 4 |
| 0.6 | 0.55–0.65 | 13 | 1.6 | 1.55–1.65 | 0 |
| 0.7 | 0.65–0.75 | 11 | 1.7 | 1.65–1.75 | 2 |
| 0.8 | 0.75–0.85 | 16 | 1.8 | 1.75–1.85 | 0 |
| 0.9 | 0.85–0.95 | 15 | 1.9 | 1.85–1.95 | 0 |
| 1.0 | 0.95–1.05 | 12 | 2.0 | 1.95–2.05 | 1 |

lower limit of one class coincides with the upper limit of the class immediately below the first. For example, consider the situation when a tree had a dbh growth increment of exactly 0.45 in. and the frequency distribution was designed as is the one in Table 3.5. To which class would this tree be assigned? A number of options exist for resolving this problem:

1. *Toss a coin* after assigning the 0.4-in. class to one face and the 0.5-in. class to the other. If the coin is "fair," the heads side having precisely the same probability of occurrence as the tails side, the assignment of borderline values would presumably balance out over the *long run*. There is no guarantee that balancing out *would occur in any given situation*.
2. Use a rule similar to the following: assign to the 0.5-in. class all trees with growth increments *equal to or greater than 0.45 in. and less than 0.55 in.* In symbolic form this would be: 0.45 in. ≤ 0.5-in. class < 0.55 in.

This option has the virtue of being applicable regardless of the magnitude of the least reading of the instrument being used. Furthermore, it leaves no gaps between classes and does not rely on the laws of probability to provide for averaging out of the upward and downward assignments.

3. Use a rule that requires the assignment of the value to the adjoining class which has an *odd* number for a class midpoint. This option is known as "favoring the odd." An alternative would be to "favor the even." For example, if the odd is being favored and the situation is that of Table 3.5, a value of 0.45 would be assigned to the 0.5-in. class. This procedure would be applicable in the case of Table 3.5, where class midpoints are

**TABLE 3.6**
**Frequency Distribution of 5-Year dbh Growth Increments**

| Growth Increment (in.) | Class Limits (in.) | f | Growth Increment (in.) | Class Limits (in.) | f |
|---|---|---|---|---|---|
| 0.1 | 0.00–0.20 | 3 | 1.1 | 1.00–1.20 | 19 |
| 0.3 | 0.20–0.40 | 10 | 1.3 | 1.20–1.40 | 14 |
| 0.5 | 0.40–0.60 | 24 | 1.5 | 1.40–1.60 | 5 |
| 0.7 | 0.60–0.80 | 27 | 1.7 | 1.60–1.80 | 2 |
| 0.9 | 0.80–1.00 | 28 | 1.9 | 1.80–2.00 | 1 |

both odd and even, but could not be used in the case of Table 3.6, where all the class midpoints are odd.

When a rule of this type is used it is assumed that the number of items assigned to the higher class would be balanced by the number assigned to the lower class. This would only occur when a very large number of such assignments are made. Usually the number of assignments on a specific job would be insufficient. Consequently, one can only say that *in the long run* balance would be achieved. This would require a *constant policy* whether the odd or the even is to be favored.

4. Use a rule similar to the following: assign to the 0.5-in. class all trees with growth increments falling in the range from 0.45 to 0.54 in., to the 0.6-in. class trees in the range from 0.55 to 0.64 in., etc. This procedure leaves gaps along the number line representing the values of the variable of interest. This is of no significance if the gap is the size of the least reading of the measuring instrument. If, however, the measurements are made using least readings smaller than the gap, anomalies will occur when the resulting values fall on the midpoint of the gap. If this occurs, one would have to fall back on either option 1, 2, or 3 to resolve the problem.

## 3.4 CUMULATIVE FREQUENCY DISTRIBUTIONS

In some cases it is desirable to know the number of items above or below certain levels. When such listings are developed, as in Tables 3.7 and 3.8, the results are known as *cumulative frequency distributions*. Furthermore, when, as in Table 3.7, the sums of the items *above* the specified levels are listed, the values in the second column are referred to as *more-than frequencies*. Conversely, when the sums of the items *below* the specified levels are listed, as in Table 3.8, the values are known as *less-than frequencies*. In either case, the sums are referred to as *cumulative frequencies*.

It should be noted that in column 2 in Table 3.7 there is no way to establish a level below the 4-cone class and, consequently, the frequencies shown *equal* or exceed the stated levels. If simple more-than frequencies were desired, the distribution would be as shown in column 3.

**TABLE 3.7**
**Cumulative Frequency Distribution of the Number of Longleaf Pine Trees Producing Cones Equal to or above the Indicated Levels**

| (1) Level of Production (No. of cones) | (2) Number of Trees with Cone Counts Equal to or Greater Than the Level Indicated | (3) Number of Trees with Cone Counts Exceeding the Indicated Level |
|---|---|---|
| 4 | 23 | 20 |
| 13 | 20 | 19 |
| 22 | 19 | 15 |
| 31 | 15 | 12 |
| 40 | 12 | 7 |
| 49 | 7 | 3 |
| 58 | 3 | 1 |
| 67 | 1 | 1 |
| 76 | 1 | 0 |
| 85 | 0 | 0 |

**TABLE 3.8**
**Cumulative Frequency Distribution of the Number of Longleaf Pine Trees Producing Cones below the Indicated Levels**

| (1)<br>Level of Production (No. of cones) | (2)<br>Number of Trees with Cone Counts Equal to or Less Than the Level Indicated | (3)<br>Number of Trees with Cone Counts Less Than the Indicated Level |
|---|---|---|
| 4 | 3 | 0 |
| 13 | 4 | 3 |
| 22 | 8 | 4 |
| 31 | 11 | 8 |
| 40 | 16 | 11 |
| 49 | 20 | 16 |
| 58 | 22 | 20 |
| 67 | 22 | 22 |
| 76 | 23 | 22 |
| 85 | 23 | 23 |

## 3.5 RELATIVE FREQUENCY DISTRIBUTIONS

If the proportion of total count of items in the data set is determined for each class, the resulting set of proportions forms a *relative frequency distribution*. Table 3.9 shows the relative distribution of the cone counts using the same classes as in column 5 in Table 3.1.

**TABLE 3.9**
**Relative Frequency Distribution of Cone Production Levels by a Set of Longleaf Pine Trees**

| Level of Production (No. of cones) | Class Limits | Relative Frequency |
|---|---|---|
| 4 | 0–8 | 3/23 = 0.1304 |
| 13 | 9–17 | 1/23 = 0.0435 |
| 22 | 18–26 | 4/23 = 0.1739 |
| 31 | 27–35 | 3/23 = 0.1304 |
| 40 | 36–44 | 5/23 = 0.2174 |
| 49 | 45–53 | 4/23 = 0.1739 |
| 58 | 54–62 | 2/23 = 0.0870 |
| 67 | 63–71 | 0/23 = 0.0000 |
| 76 | 72–80 | 1/23 = 0.0435 |

In the same manner, a cumulative frequency distribution can be converted to a *cumulative relative frequency distribution*. Table 3.10 is the relative equivalent of column 2 of Table 3.8.

## 3.6 GRAPHICAL PRESENTATIONS OF FREQUENCY DISTRIBUTIONS

If the frequencies or counts in a frequency distribution or the relative frequencies in a relative frequency distribution are plotted against the class midpoints and then the plotted points are connected by straight lines, as in Figure 3.1, the resulting figure is called a *frequency polygon*. If,

**TABLE 3.10**
**Cumulative Relative Frequency Distribution of the Number of Longleaf Pine Trees with Cone Counts Less Than or Equal to the Indicated Levels**

| Level of Production (No. of cones) | Relative Frequency of Trees with Cone Counts Less Than or Equal to the Level Indicated |
|---|---|
| 4 | 3/23 = 0.1304 |
| 13 | 4/23 = 0.1739 |
| 22 | 8/23 = 0.3478 |
| 31 | 11/23 = 0.4783 |
| 40 | 16/23 = 0.6957 |
| 49 | 20/23 = 0.8696 |
| 58 | 22/23 = 0.9565 |
| 67 | 22/23 = 0.9565 |
| 76 | 23/23 = 1.0000 |

instead of connecting the plotted points, a bar graph is constructed *so that the bars touch*, as in Figure 3.2, the resulting figure is called a *histogram*. If the figure is constructed so that the bars do not touch, it is referred to as a *bar graph* (Figure 3.3). According to some workers, the term *histogram* should be used only with *continuous* variables. The argument used for this position is that the touching bars indicate continuity. However, as will be seen in Chapter 8, the use of touching bars with a discrete variable has certain advantages and, consequently, in this book the term *histogram* will be used with both kinds of variables.

The frequency polygon of a cumulative frequency distribution, as is shown in Figures 3.4 and 3.5, is an *ogive*. When the distribution is of the "less-than" type the ogive characteristically rises monotonically as the magnitudes of the class midpoints increase. The reverse occurs when the distribution is of the "more-than" type.

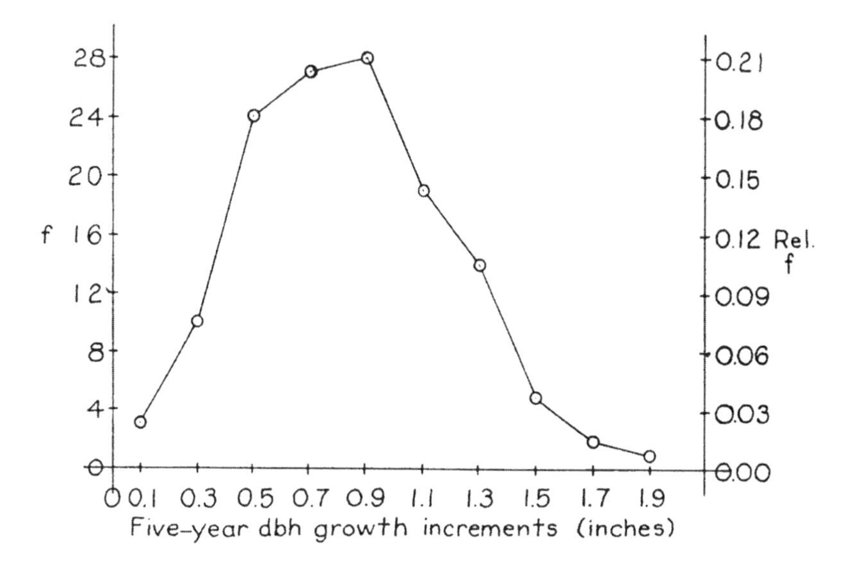


**FIGURE 3.1** Frequency polygon showing the distribution of 5-year dbh growth increments, using the classes in Table 3.6.

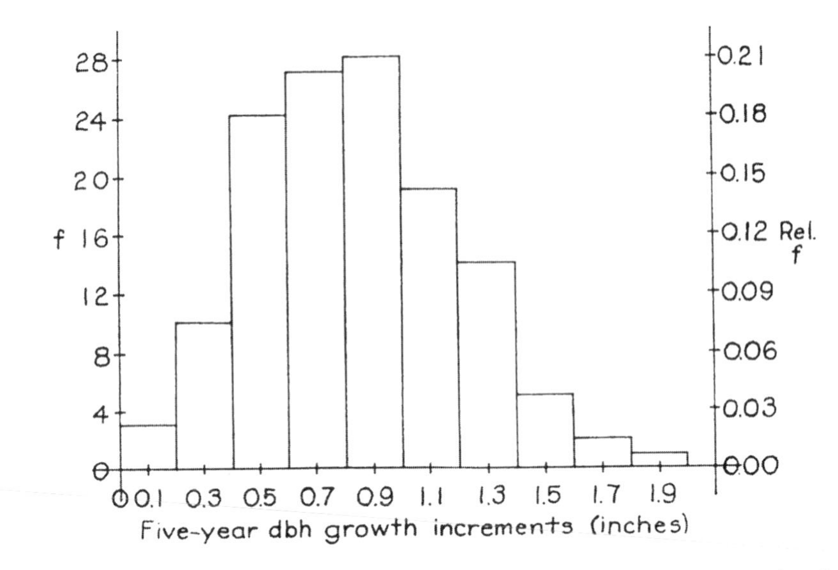


**FIGURE 3.2** Histogram showing the distribution of 5-year dbh growth increments, using the classes in Table 3.6.

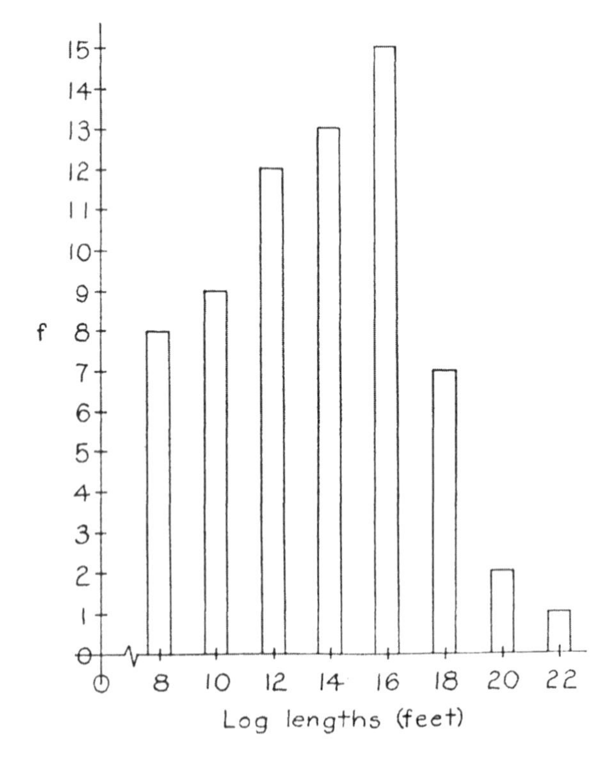


**FIGURE 3.3** Bar graph showing the distribution of standard log lengths in a hypothetical set of pine logs. Note that the horizontal axis has been broken to make the diagram more compact.

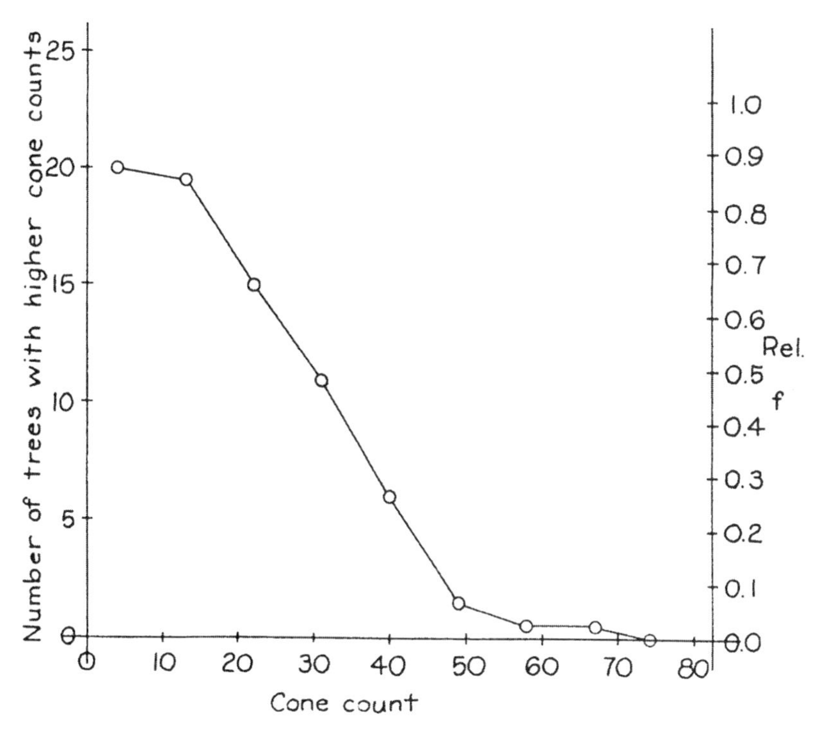


**FIGURE 3.4** Ogive of the more-than cumulative frequency distribution of cone counts shown in column 3 of Table 3.7.

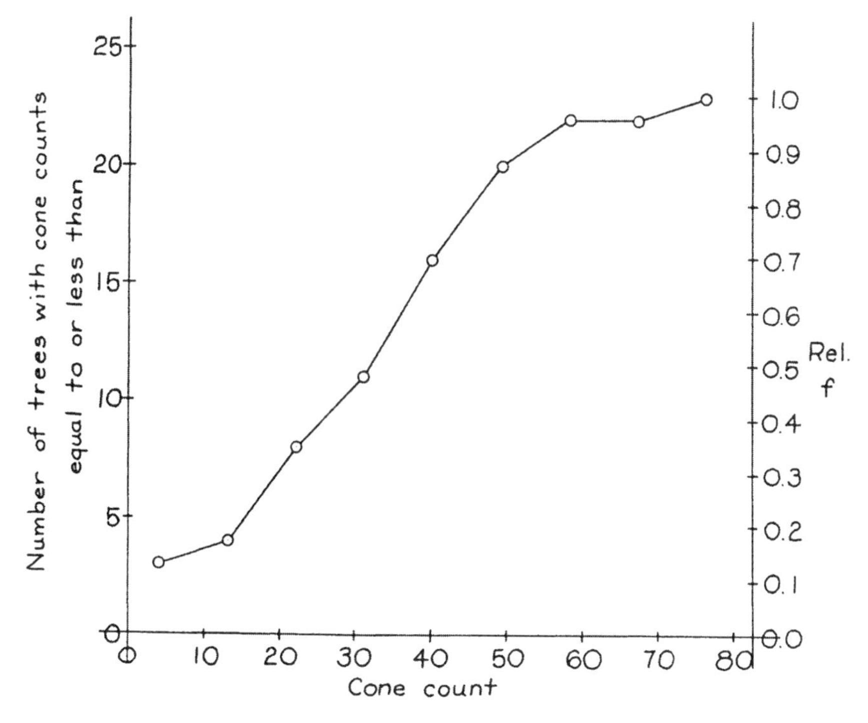


**FIGURE 3.5** Ogive of the "less-than or equal-to" cumulative frequency distribution of cone counts shown in column 2 of Table 3.8.

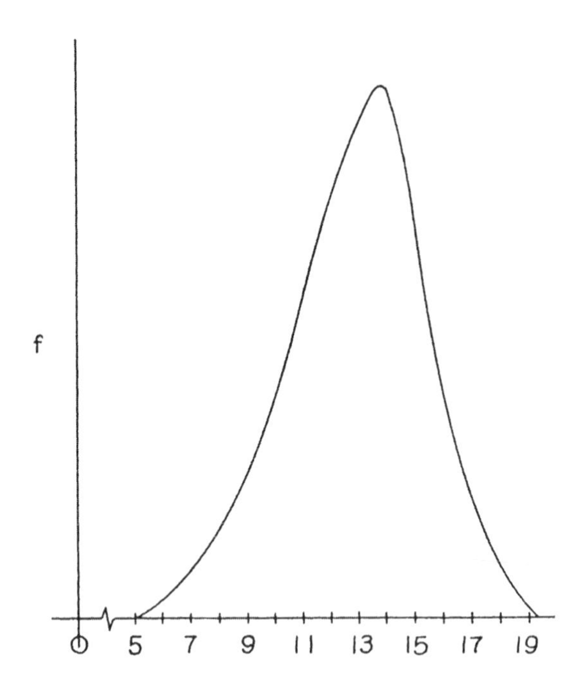


**FIGURE 3.6** Frequency curve showing the distribution of values of a hypothetical continuous variable of interest.

As can be seen in Figures 3.1, 3.2, 3.4, and 3.5, the graphical representations of relative frequency distributions are essentially the same as those of the parent distributions. The plotted points maintain the same relative positions regardless of whether the distribution is or is not a relative frequency distribution.

If the variable of interest is continuous and the class interval is made smaller and smaller while the number of data items is increased, the frequency polygon or the tops of the histogram bars will approach a smooth curve, as is shown in Figure 3.6. This curve is known as a *frequency curve*. In the limit, as the class widths become infinitesimals, the curve becomes smooth. The algebraic expression that produces that curve can then be used to describe the distribution. Such an expression is known as a *frequency function* or *frequency distribution function*.

Graphical presentations are of great value when one needs to visualize a situation and the information relating to that situation is in the form of a frequency distribution.

## 3.7 DESCRIPTIVE TERMS USED FOR FREQUENCY DISTRIBUTIONS

Graphical presentations of frequency distributions provide pictures of the way the values occur in the distribution. The shape formed by the frequency polygon usually can be loosely described in terms of one of the following descriptive terms.

1. *Mound shaped*, as in Figures 3.1 and 3.2, where the frequencies are low at the ends of the distribution but high near its center. If the mound can be divided into two parts by a vertical line and the two are mirror images of each other, as in Figure 3.7A, the distribution is *symmetrical*. If the two halves are not mirror images of each other, as in Figure 3.7B, the distribution is *skewed*. If the peak of the skewed distribution is to the right of the center as in Figure 3.7B, it is said that the distribution is skewed toward the *left* because skewness actually refers to the way the frequencies taper off toward the ends. Skewness toward the left means that the frequencies taper off more slowly to the

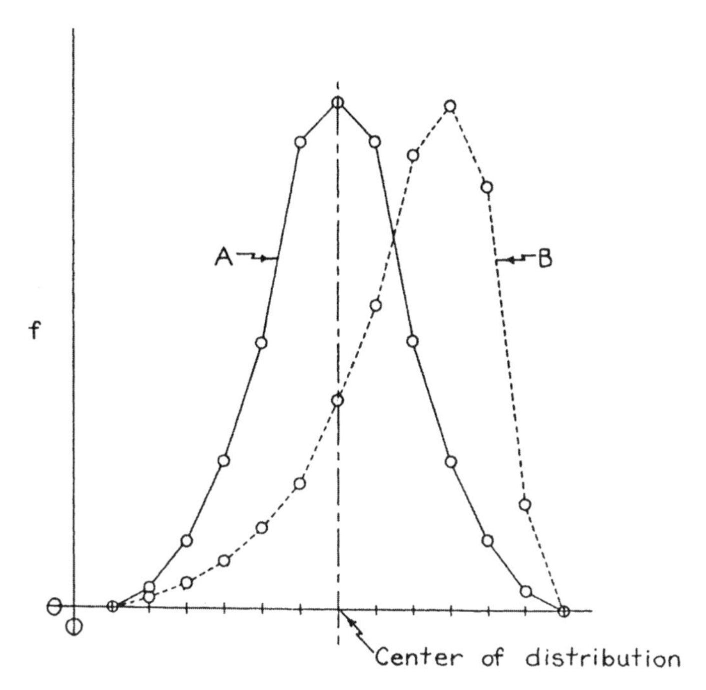


**FIGURE 3.7** Symmetrical (A) and skewed (B) mound-shaped frequency distributions; the skewness in B is toward the left.

left than they do to the right. The reverse occurs when the skewness is toward the right. Mound-shaped distributions are common in biology. For example, the distribution of dbh values in even-aged forest stands is mound shaped and usually skewness is present.

2. *J shaped*, as in Figure 3.8A, where the frequencies are low at the lower end of the distribution but increase at an increasing rate toward the upper end. When the reverse

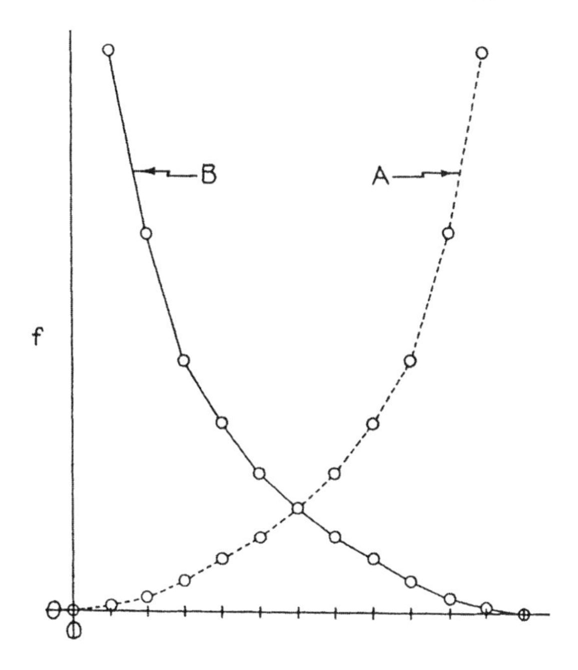


**FIGURE 3.8** J-shaped (A) and reversed J-shaped (B) frequency distributions.

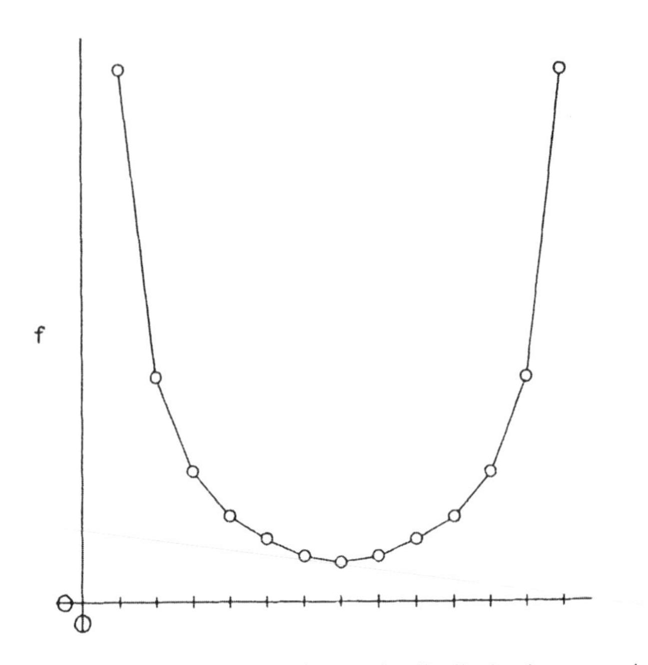


**FIGURE 3.9** U-shaped frequency distribution. In this case the distribution is symmetrical.

occurs, as in Figure 3.8B, the distribution has a *reversed J shape*. The distribution of dbh values in all-aged forest stands characteristically has a reversed J shape.

3. *U shaped*, as in Figure 3.9, is the reverse of the mound shape. The frequencies are high at both ends of the distribution but low in the center. Such distributions, like the mound-shaped ones may be either symmetrical or skewed. U-shaped distributions, symmetrical or skewed, are not common.
4. *S shaped or sigmoid*, as in Figure 3.10, where the frequencies are low at one end of the distribution, then increase at an increasing rate, and then increase at a decreasing rate.

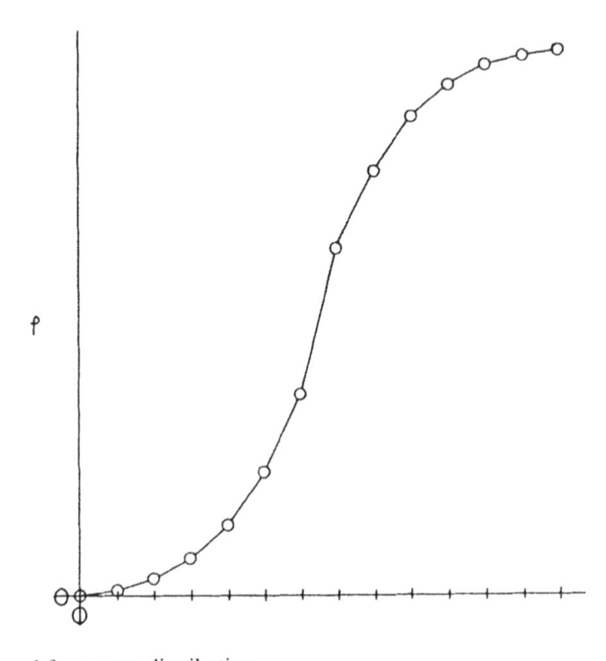


**FIGURE 3.10** S-shaped frequency distribution.

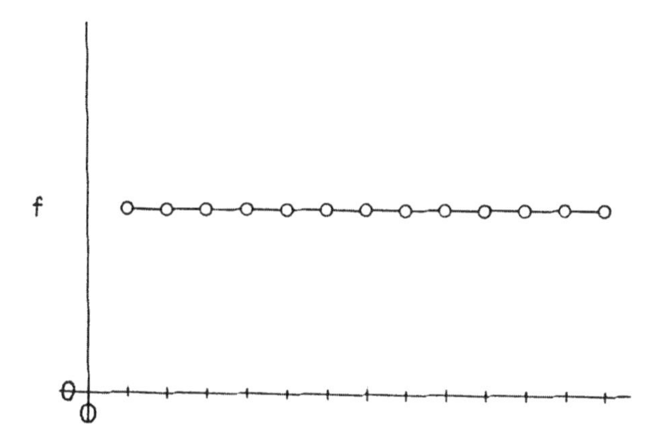


**FIGURE 3.11** Rectilinear frequency distribution.

The increase may be toward the right, as in Figure 3.10, but the reverse may occur, in which case it is referred to as a *reversed* S *shape*. Ogives commonly are S shaped.

5. *Rectilinear*, as in Figure 3.11, where all the classes contain the same number of items.

# 4 Symbology

## 4.1 INTRODUCTION

In the following chapters a great deal of use will be made of symbols in both the text and in algebraic expressions. Some of these are standard algebraic symbols, but many more have special meanings within a sampling or statistical context. Unfortunately, statistical symbology is not standard and considerable variation is encountered in the literature. In recent years the situation has improved considerably and it is probable that standardization will eventually be achieved. An attempt will be made in this text to remain in the mainstream of current thought in this matter, but where it appears that understanding would be facilitated by modifications they will be made.

## 4.2 IDENTIFICATION OF VARIABLES

Traditionally, capital $X$ and $Y$ have been used as symbols for variables, with $X$ probably being used more often than $Y$. However, in some statistical work, e.g., regression analysis, which is discussed in Chapter 15, more than one variable is involved. In these cases, $Y$ is reserved for the variable of interest while the other variable(s) are identified by $X$, which is subscripted if more than one variable is involved. Since $Y$ represents the variable of interest in that context, it will be used in that manner throughout this book. A capital letter, such as $X$ or $Y$, represents the *entire set* of values of the variable. If a lowercase letter is used, it means that an *individual variate* is being represented. Subscripting is used to identify the specific variates.

## 4.3 SUBSCRIPTING

The use of subscripts is common in statistical operations. They are used to identify individual items in data sets and also to identify subsets of larger assemblages of values. For example, a population of variates might consist of the following values: 3, 9, 11, 14, 19, and 23. These values could be identified as follows:

| Symbol | Value |
|---|---|
| $y_1$ | 3 |
| $y_2$ | 9 |
| $y_3$ | 11 |
| $y_4$ | 14 |
| $y_5$ | 19 |
| $y_6$ | 23 |

If two different samples (subsets), each with three items, were drawn from this population, the values in these samples could be identified as follows:

| Sample 1 | | Sample 2 | |
|---|---|---|---|
| Symbol | Value | Symbol | Value |
| $y_{11}$ | 9 | $y_{21}$ | 3 |
| $y_{12}$ | 14 | $y_{22}$ | 9 |
| $y_{13}$ | 23 | $y_{23}$ | 19 |

where the first subscript identifies the sample and the second the item within the sample.

When the subscripting is generalized, the lowercase letters *i*, *j*, *k*, and *l* are most often used. When only one set is involved the letter *i* usually has been the one used to represent the individual items in the set, as $y_i$, which would be interpreted as the *i*th item in the set. If two or more data sets are involved, as with the two samples shown above, the letter *i* would be used to identify the set. For example, in the case of the two samples, $y_{ij}$ would be interpreted as the *j*th item in the *i*th sample. If subsets within subsets are involved, and this does occur in sampling and statistical work, the subscripting would be extended appropriately. For example, $Y_{ijk}$ would be interpreted as the *k*th item in the *j*th subset of the *i*th set.

## 4.4 SUMMATION SYMBOLOGY

The Greek letter Σ, which is the uppercase *sigma* and corresponds to *S* in the English alphabet, is used to show that the values in a set are to be *summed*. In complete form the symbol is

$$\sum_{i=a}^{b} y_i$$

which shows that the values of *Y* bearing subscripts from *a* to *b* are to be added together. In the case of the population of values used earlier, the sum of all the values would be

$$\sum_{i=1}^{6} y_i = 3 + 9 + 11 + 14 + 19 + 23 = 79$$

Notice that in this case the symbol *a* is equal to 1 and *b* is equal to 6. If only the last three values in the set were to be summed, the expression would be

$$\sum_{i=4}^{6} y_i = 14 + 19 + 23 = 56$$

*a* in this case is 4, indicating that the summing should begin with the fourth item in the set; *b* is equal to 6, indicating that the summing should end when the sixth item had been processed. It is possible to have a situation such as the following:

$$\sum_{i=2}^{5} y_i = 9 + 11 + 14 + 19 = 53$$

where the summing began with the second item and terminated with the fifth.

In most statistical work the entire set would be involved and, consequently, $a$ would be equal to 1. When this is the case, the symbol may be simplified to:

$$\sum_{i}^{b} y_i$$

In many cases there is no question about the $i$ and it, too, may be eliminated, leaving

$$\sum^{b} y_i$$

When the entire set is to be summed the $b$ may be left off and the $y$ replaced by $Y$, leaving

$$\sum Y$$

In this text any or all of these forms may be encountered. In the early chapters the complete symbol will be used to make clear precisely what is to be done. In later chapters, where the summation may be a portion of a complex algebraic expression, the simplified forms may be used.

It should be pointed out that the summing process is not limited to individual data items. In some cases the values to be summed are values obtained by arithmetic or algebraic operations, such as

$$\sum_{i=a}^{b} (y_i + C)^2 = \sum_{i=a}^{b} (y_i^2 + 2Cy_i + C^2)$$

This expression also could be written as

$$\sum_{i=a}^{b} y_i^2 + \sum_{i=a}^{b} 2Cy_i + \sum_{i=a}^{b} C^2$$

In this expression, 2 is a constant and may be factored out of the second term, yielding

$$\sum_{i=a}^{b} y_i + 2\sum_{i=a}^{b} Cy_i + \sum_{i=a}^{b} C^2$$

If $C$ is also assumed to be a constant, it can be factored out of the second term, in the same manner as was the 2:

$$\sum_{i=a}^{b} y_i + 2C\sum_{i=a}^{b} y_i + \sum_{i=a}^{b} C^2$$

If $C$ is a constant and $a$ is equal to 1, the expression becomes

$$\sum_{i=1}^{b} y_i + 2C\sum_{i=1}^{b} y_i + bC^2$$

since summing $b$ values of the constant $C^2$ is equivalent to multiplying $C^2$ by $b$.

This type of algebraic manipulation occurs fairly frequently in the explanations and derivations of statistical procedures and formulae. It behooves the student, therefore, to be aware of what can and what cannot be done with summation expressions.

## 4.5 DISTINGUISHING BETWEEN POPULATIONS AND SAMPLE VALUES

As was brought out in Chapter 1, population parameters are population characteristics. An example of such a parameter is the arithmetic mean of all the values in the population. When decisions must be made about the universes from which the populations are derived, the decision maker would prefer to base these decisions on absolute knowledge; in other words, on the population parameters. However, parameters are rarely known and recourse must be made to estimates of the parameters. These estimates are based on samples and, in actuality, are the sample characteristics, called sample statistics, that correspond to the parameters. The computational procedures used to obtain the sample statistics are sometimes identical to, but usually somewhat different from, those used to obtain the parameters. Consequently, distinctions must be made between parameters and statistics. Usually *parameters* are identified by, or have associated with them, *lowercase Greek* letters or *uppercase Roman* letters. Sample *statistics*, on the other hand, usually are identified by *lowercase Roman* letters augmented, in some cases, by special symbols. For example, the arithmetic mean of a population is indicated by the Greek letter $\mu$, which is the lowercase *mu*. The symbol for the arithmetic mean of a sample is $\bar{y}$, if the variable is symbolized by $Y$, or $\bar{x}$, if the variable is symbolized by $X$. The straight bar across the top of the letter is almost universally accepted as the symbol of the arithmetic mean of a sample. The verbal equivalent to $\bar{y}$ is "bar $y$" or "$y$ bar" and for $\bar{x}$ "bar $x$" or "$x$ bar." When statistics other than the mean are used as estimators, they usually are identified as such by an upside-down $\vee$ over the letter symbolizing the statistic. For example, the symbol for an estimated proportion is $\hat{p}$. The technical name for the upside-down $\vee$ is *caret* but usually is called a "hat." Thus, $\hat{p}$ would be referred to as "$p$ hat."

It is essential that the meanings of the symbols be understood. The mean of a population is different from that of a sample, population size is different from sample size, and so it goes. Failure to distinguish between symbols will lead to confusion and to an inability to understand what is being done.

# 5 Averages

## 5.1 INTRODUCTION

A major problem facing persons working with masses of data is visualizing the situation represented by those data. Arranging the data into arrays or frequency distributions and the construction of frequency polygons, histograms, or bar graphs help immensely, but in many cases it would be advantageous to be able to collapse the set of data items into a single representative value. A number of such single-value representations have been developed. Collectively they are *averages* or *measures of central tendency*. Each of these averages has its own characteristics, and situations exist where each is the appropriate one to use.

## 5.2 ASSESSING AVERAGES

Since a multiplicity of averages exist, it is necessary to have a way of choosing among them and assessing their strengths and weaknesses. The following criteria form a framework for such evaluations (Waugh, 1952).

1. The average should be defined in such a manner that if it is determined repeatedly using the same data set the same result is obtained each time. Preferably the definition should be stated in the form of an algebraic expression.
2. The average should be easily understood and interpreted.
3. The average should be easy to compute, but this is less important than being easy to understand.
4. The average should depend on every item in the data set. It should typify the entire set.
5. The average should not be influenced greatly by any single value in the set.
6. The average should be "stable" in a statistical sense. In other words, if a number of samples of a given size are drawn from a given population, the averages of the individual samples should be reasonably similar. This criterion has its roots in the concept that sample statistics are used as estimators of population parameters and, if such an estimator is highly erratic, estimates made with it are apt to be unsatisfactory.
7. The average should be of such a nature that it can be used in statistical operations beyond its own determination. At this stage in the text, it is difficult to explain what is meant by this statement. However, it can be said that much of the remainder of this book is concerned with statistical operations within which averages are embedded.

## 5.3 THE ARITHMETIC MEAN

The most commonly used and best known of the averages is the *arithmetic mean*, also known simply as the *mean* or the *average*. It can be computed regardless of the manner in which the data are arranged, but the computational procedures are dependent on the way the data are arranged.

### 5.3.1 Ungrouped Data

In its basic form, the arithmetic mean is computed simply by summing the individual values and then dividing the sum by the count of items summed. The data may be in random (raw) order or in the form of an array. Expressed algebraically, if the *entire population* is involved, the expression is

$$\mu = \frac{\sum_{i=1}^{N} y_i}{N} \tag{5.1}$$

where $\mu$ = arithmetic mean of the population
$N$ = the total number of variates in the population, otherwise known as the *population size*

If the dataset is from a *sample*, the algebraic expression is:

$$\bar{y} = \frac{\sum_{i=1}^{n} y_i}{n} \tag{5.2}$$

where $\bar{y}$ = arithmetic mean of the sample
$n$ = the number of variates in the sample, otherwise known as the *sample size*

For example, using the cone count data in either columns 1 or 2 in Table 3.1, the arithmetic mean is

$$\bar{y} = 790/23 = 34.34782609 \text{ cones per tree}$$

It should be noted that one of the variates in the data set is a zero. In spite of the fact that the zero contributes nothing to the sum of variates, it *is* a variate and it should be counted as a variate. Failure to do this will lead to incorrect means and to distorted visualizations of the data set.

When the arithmetic mean is to be used as a component in further computations, it should be computed to as many decimal places as the computing device will permit and then used *without any rounding back*. Rounding back almost invariably introduces so-called *rounding errors*, which distort the values being computed. This will be discussed further in the sections of the book where the procedures subject to this problem are described.

Rounding back of the mean should be carried out, however, when it itself is the end product of the computation process. This rounding should be done so that the *appropriate precision of the mean* is indicated. Thus, the amount of rounding depends on the situation. In the case of the cone count in Table 3.1, the variable of interest was *discrete*. If it can be assumed that the counts were *free of error*, the mean would be shown as computed, 34.34782609, because the values are exact and no approximations are involved. If, on the other hand, one cannot assume that the counts are free of error, the computed value of the mean should be appropriately rounded back. Assessing the error of the counts may be very difficult. When this is the case, a rule of thumb which may be used is to round back the mean to the nearest $^1/_{10}$ of the counting unit. This, of course, is only an approximation of what actually should be done, but it usually offers the only feasible solution to the problem. Using this rule with the cone count data would yield a mean of 34.3 cones per tree.

When the variable whose mean is being computed is *continuous*, the usually accepted procedure is to round back to *one significant figure more than what was present in the data*. It is difficult to reconcile this accepted procedure with the rules governing significant figures that were described in Section 2.7. Under those rules the computed value of the mean should be rounded back to the *same* number of significant figures that were in the original data. An argument for the usual procedure, however, might be developed as follows. The divisor in the formula for the mean is a count and, thus, is made up of discrete units so no approximation is involved. Furthermore, the count is accurate. Consequently, the divisor contributed no error to the computation. All the error in the mean, therefore, can be attributed to the sum of measured values. Let us assume that in a certain case the variable of interest was the weight of individual animals of a certain species. Further assume that 31 animals were weighed and the weights were recorded to the nearest whole gram. If the sum of the weights was 46,392 g, the mean would be

$$\bar{y} = 46{,}392/31 = 1496.516129 \text{ g per animal}$$

Since the actual weights had been rounded back to the nearest whole gram, each of the weights could be as much as a half gram higher or a half gram lower than what was recorded. It is likely that in the summing process considerable canceling of high and low value takes place, but the extent of this would not be known. Consequently, the true sum probably would not be precisely equal to the computed values. If it is assumed that the range of error could be from a half gram high to a half gram low, a mean could be computed using each of these two extremes. By using the high value, the mean would be

$$\bar{y}_{\text{high}} = 46{,}392.5/31 = 1496.532258 \text{ g per animal}$$

By using the low value, the mean would be

$$\bar{y}_{\text{low}} = 46{,}391.5/31 = 1496.500000 \text{ g per animal}$$

Rounding back the three means *until they are equal* results in the composite mean, 1496.5 g, which contains one more significant figure than did the original data values. This agrees with the usual procedure.

### 5.3.2 Grouped Data

When the data are in the form of a frequency distribution, it is necessary to include the class frequencies in the computation of the arithmetic mean. Thus, when the *entire population* is involved, the arithmetic mean is

$$\mu = \frac{\sum_{i=1}^{M} (f_i C_i)}{\sum_{i=1}^{M} f_i} \tag{5.3}$$

where $M$ = total number of *classes* in the *population*
$C_i$ = the class midpoint value for the $i$th class
$f_i$ = the frequency or count of items in the $i$th class

When the data are from a *sample*, the algebraic expression for the arithmetic mean is

$$\bar{y} = \frac{\sum_{i=1}^{m}(f_i C_i)}{\sum_{i=1}^{m} f_i} \tag{5.4}$$

A lowercase $m$ is used in this expression because the span of classes encountered in the sampling operation may be less than that actually present in the population. In the majority of cases $m$ is equal to $M$, but since it is rarely known if this actually is the case, the lowercase $m$ is the more appropriate of the two.

The arithmetic mean of the cone count data, shown in frequency distribution form in column 4 in Table 3.1, is computed as follows:

$$\bar{y} = \frac{(1*0)+(0*1)+(0*2)+\cdots+(1*72)}{1+0+0+\cdots+1}$$

$$= 790/23 = 34.34782609 \text{ cones per tree}$$

Since no lumping of the data was done in this case, the mean of the frequency distribution is precisely the same as that for the ungrouped data and the rounding rules described earlier for discrete data are applicable. When data are lumped, however, as in column 5 in Table 3.1, the values entering the computations are less precise and exact agreement between the mean of the frequency distribution and that of the ungrouped data rarely occurs. This is demonstrated when the mean of the frequency distribution in column 5 is computed:

$$\bar{y} = \frac{(3*4)+(1*13)+(4*22)+\cdots+(1*72)}{3+1+4+\cdots+1}$$

$$= 794/23 = 34.52173913 \text{ cones per tree}$$

The difference between this mean and that of the ungrouped data is only 0.1739130404 cones per tree. Such a small difference is typical. Class width has no direct effect on the mean. The source of the difference between the means is differences between the class midpoints and the means of the items within those classes. This can be seen from the following argument.

The mean of the values in the $i$th class is

$$\bar{y}_i = \frac{\sum_{j=1}^{f_i} y_{ij}}{f_i}$$

The weighted mean of the class means (Section 5.3.3) is

$$\bar{y}_w = \frac{\sum_{i=1}^{m} f_i \bar{y}_i}{\sum_{i=1}^{m} f_i}$$

Substituting for $\bar{y}_i$ in the expression for $\bar{y}_w$:

$$\bar{y}_w = \frac{\sum_{i=1}^{m} f_i \left( \frac{\sum_{j=1}^{f_i} y_{ij}}{f_i} \right)}{\sum_{i=1}^{m} f_i} = \frac{\sum_{i=1}^{m} \sum_{j=1}^{f_i} y_{ij}}{\sum_{j=1}^{m} f_i}$$

Since

$$\sum_{i=1}^{m} \sum_{j=1}^{f_i} y_{ij} = \sum_{i=1}^{n} y_i \quad \text{and} \quad \sum_{i=1}^{m} f_i = n$$

then

$$\bar{y}_w = \frac{\sum_{i=1}^{n} y_i}{n} = y$$

Thus, the correct value for the mean of the frequency distribution is obtained when the class *means* are used in the computations. Since $\bar{y}$ can occupy any position in the span of values making up the $i$th class, $C_i$ rarely equals $\bar{y}_i$. Consequently,

$$C_i = \bar{y}_i + \varepsilon_i$$

where $\varepsilon_i$ = error of $C_i$

If $C_i$ is used in the computation of the mean, it is the equivalent of substituting $C_i - \varepsilon_i$ for $\bar{y}_i$ in Equation 5.5.

$$\bar{y}_w = \frac{\sum_{i=1}^{m} f_i (C_i - \varepsilon_i)}{\sum_{i=1}^{m} f_i} = \frac{\sum_{i=1}^{m} f_i C_i - \sum_{i=1}^{m} f_i \varepsilon_i}{\sum_{i=1}^{m} f_i}$$

Thus, the value of $\bar{y}_w$, or $\bar{y}$, would be correct only when

$$\sum_{i=1}^{m} f_i \varepsilon_i$$

equals zero. The more it differs from zero, the greater the difference between the mean of the frequency distribution and that of the ungrouped data. As can be seen, class width has no direct effect on the mean. However, when the classes are wide, the potential for large $\varepsilon_i$ values is greater than when the classes, are narrow. If the values in the classes are numerous and well distributed within the classes the $\varepsilon_i$ values will be small, regardless of class width.

When the mean of the frequency distribution of cone counts was computed earlier in this section, no attempt was made to round it back. Leaving it in its unmodified form is appropriate if the mean is to be used in further calculations, but, if the mean is the final product of the computations and is going to be used as the representative value for the data set, it is necessary to round back. This raises the question of significant figures. In the case of the frequency distribution in column 5 in Table 3.1, the class size or width is 9 cones. Thus, in the 4-cone class the actual number of cones on a tree may be as few as none at all or as many as 8. This type of situation occurs in all the classes. Consequently, considerable uncertainty exists regarding the cone counts. Using the lower limits of the classes, instead of the class midpoints, in the calculation of the mean yields the following value:

$$\bar{y}_{\text{low}} = \frac{(3*0)+(1*9)+(4*18)+\cdots+(1*72)}{3+1+4+\cdots+1}$$

$$= 702/23 = 30.52173913 \text{ cones per tree}$$

If the upper class limits are used in such a calculation, the following value is obtained:

$$\bar{y}_{\text{high}} = \frac{(3*8)+(1*17)+(4*26)+\cdots+(1*80)}{3+1+4+\cdots+1}$$

$$= 886/23 = 38.52173913 \text{ cones per tree}$$

Rounding back these three means until they are equal cannot be done in this case, indicating that the mean of the frequency distribution has *no* significant figures and, strictly speaking, is essentially meaningless. This problem would be compounded if uncertainty exists regarding the accuracy of the counts themselves. In many cases, of course, the class widths would be smaller than was the case in this example and rounding to the significant figures could be carried out. As far as can be ascertained, no general rules covering this situation have been formulated and recourse must be made to the procedure used above if a technically correct value is to be obtained. It should be pointed out that most people working with frequency distributions ignore the problem of significant figures and simply leave the mean as calculated or round to what they feel is appropriate. This, of course, is not proper procedure. When means are not properly rounded back, they should be used with caution.

The rules regarding the rounding off of arithmetic means of frequency distributions are, with one exception, equally applicable when the variable is continuous. The exception occurs when the class width is equal to unity. When the variable is discrete, as in the case of the distribution in column 4 in Table 3.1, the mean of the frequency distribution is based on precisely the same values that would be used if ungrouped data were used. Consequently, since no approximation are involved, all the digits in the computed mean are significant. When the variable is continuous, however, this situation cannot occur and rounding must be done, provided that the mean is the desired end product of the calculations.

### 5.3.3 Weighted Mean

There are situations where the variates entering the computation of an arithmetic mean are not inherently equal insofar as their strength or importance are concerned. When this occurs it is necessary to recognize this lack of equality in the computation process. This is done by *weighting* each variate according to its relative importance:

$$\bar{y}_w = \frac{\sum_{i=1}^{n} w_i y_i}{\sum_{i=1}^{n} w_i} \tag{5.5}$$

where $\bar{y}_w$ = weighted arithmetic mean
$w_i$ = weight of the $i$th variate

The resulting mean is a *weighted arithmetic mean.*

An example of this can be drawn from forest management. It is not uncommon in forest management that information is needed about the mean per acre or per hectare volume of timber on a piece of property. Furthermore, when a piece of property is under management it typically is divided into subdivisions called compartments. When forest inventories are made, the data usually are kept separate by compartments and compartment means and other statistics are computed. Rarely are the compartments of equal size so the simple averaging of the compartment means is not appropriate. Instead, each compartment mean, now a variate, must be weighted by the compartment size. This has been done in Table 5.1. If the weighting had not been done, the overall mean would have been 256.0 m³/ha. The error in this case is small because the range in weights is small. If compartment 10 had had 1375 ha instead of 375, the weighted mean would have been 296.1 m³/ha. The effect of the weight is very obvious in this case.

**TABLE 5.1**
**Computation of the Weighted Mean per Hectare Volume of Pine on the Property of a Hypothetical Landowner**

| Compartment No. | Compartment Area ($w_i$) | Mean Volume per Hectare ($y_i$), m³ | $w_i y_i$ |
|---|---|---|---|
| 1 | 450 | 180 | 81,000 |
| 2 | 510 | 300 | 153,000 |
| 3 | 400 | 216 | 86,400 |
| 4 | 300 | 324 | 97,200 |
| 5 | 500 | 228 | 114,000 |
| 6 | 325 | 264 | 85,800 |
| 7 | 425 | 204 | 86,700 |
| 8 | 480 | 312 | 149,760 |
| 9 | 525 | 300 | 157,500 |
| 10 | 375 | 228 | 85,500 |
| 11 | 550 | 156 | 85,800 |
| 12 | 350 | 360 | 126,000 |

$$\sum_{i=1}^{12} w_i y_i = 1,3( \qquad \sum_{i=1}^{12} w_i y_i = 1,308,660$$

$$\bar{y}_w = \frac{\sum_{i=1}^{12} w_i y_i}{\sum_{i=1}^{12} w_i} = \frac{1,308,660}{5190} = 252.150289 \text{ or } = 252.2 \text{ m}^3/\text{ha}$$

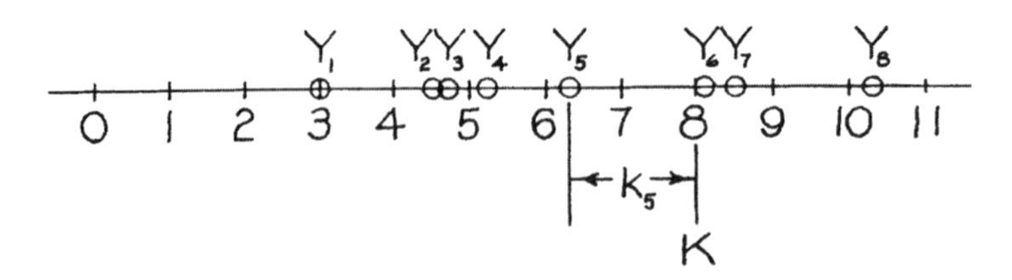


**FIGURE 5.1** Number line showing the position of several items of data, the arbitrarily chosen position of the representative value $K$, and the difference $k_I$.

### 5.3.4 Characteristics of the Arithmetic Mean — The Concept of Least Squares

The arithmetic mean is only one of a number of so-called averages that can be used to represent data sets. Which of these is the *most* representative single value? Various criteria can be proposed to answer this question, but the one that is generally considered to be of preeminent importance is referred to as the *least squares criterion*. This concept was originally developed by geodetic engineers for use in high-precision surveys, but it since has been adopted as a fundamental concept in many other disciplines, including statistics.

The concept of least squares probably can best be visualized with the aid of a number line along which are a number of points representing items of data (Figure 5.1). Which point along the line, already occupied or not, would identify the value that would be the most representative of the set of plotted values? One might arbitrarily choose some point along the line, label it $K$, and then determine the algebraic difference between each of the data values and $K$:

$$k_i = y_i - K$$

For each data value, there is a corresponding $k_i$ value. Each of these $k_i$ values is then squared and the squared values are summed:

$$\sum_{i=1}^{n} k_i^2$$

If now another $K$ is chosen and the process repeated, a second sum of squared differences would be obtained. In all likelihood it would be different from the first. As a matter of fact, if this process is repeated a number of times, beginning with a $K$ close to one end of the span of $y_i$ values, and using $K$ values at either regular or irregular intervals until the other end of the span of $y_i$ values is reached, the resulting sums of squared differences would follow a pattern similar to that shown in Figure 5.2. It can be seen from this figure that there is a $K$ value where the sum of squared differences is at a minimum. According to the least squares criterion, the $K$ at this point is the one that is most representative of the set.

The least squares value of $K$ can be determined using the minimization procedures of differential calculus, as follows.

Under the least squares criterion

$$\sum_{i=1}^{n} k_i^2$$

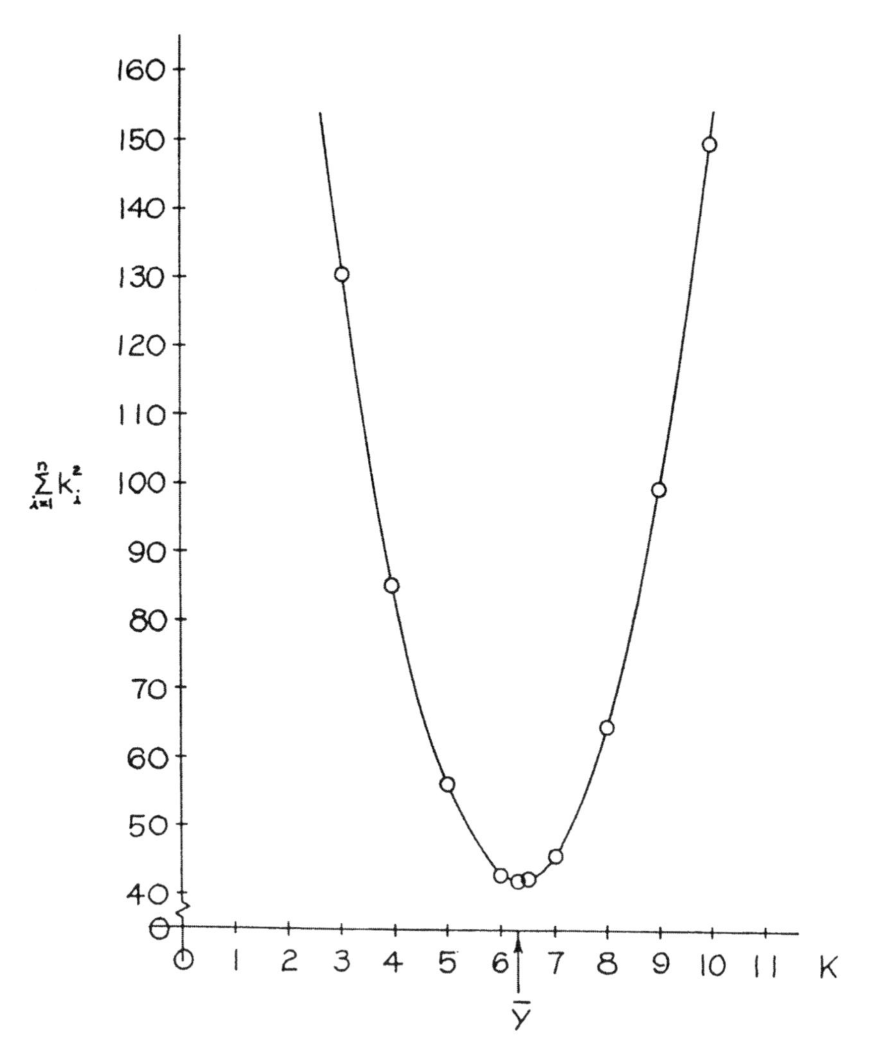


**FIGURE 5.2** Sums of squared differences $\left(\sum_{i=1}^{n} k_i^2\right)$ plotted against $K$.

must be at a minimum. Thus,

$$\frac{d\sum_{i=1}^{n} k_i^2}{dK} = \frac{d\sum_{i=1}^{n} (y_i - K)^2}{dK}$$

$$= 2\sum_{i=1}^{n} (K - y_i) = 2nK - 2\sum_{i=1}^{n} y$$

Setting the derivative equal to zero and solving for $K$:

$$2nK - 2\sum_{i=1}^{n} y_i = 0$$

$$2nK = 2\sum_{i=1}^{n} y_i$$

$$K = \frac{2\sum_{i=1}^{n} y_i}{2n} = \frac{\sum_{i=1}^{n} y_i}{n} = \bar{y}$$

Thus, the least squares value of $K$ occurs when it equals the arithmetic mean of the set and, consequently, *the arithmetic mean is considered to be the value most representative of the set under the least squares criterion.*

### 5.3.5 Errors and Residuals

In the previous discussion, the difference between a data item, $y_i$, and $K$ was identified by the symbol $k_i$ but was not given a special name. However, when the arithmetic mean of the set is substituted for $K$, the situation changes. Thus, when a *population* mean is involved:

$$\varepsilon_i = y_i - \mu \tag{5.6}$$

where $\varepsilon_i$ = the *error* of the $i$th variate.

It should be emphasized that $\varepsilon_i$ *does not represent a mistake or a blunder.* The term *error* simply indicates the algebraic difference between a variate and the mean of the population of which it is a member.

When a *sample* mean is involved:

$$e_i = y_i - \bar{y} \tag{5.7}$$

where $e_i$ = the *residual* of the $i$th variate

Only when the sample mean is equal to the population mean will the residual of a given variate be equal to its error. This rarely occurs.

### 5.3.6 Further Characteristics of the Arithmetic Mean

*The algebraic sum of the errors or residuals is equal to zero.* In the case of a population,

$$\sum_{i=1}^{N} \varepsilon_i = \sum_{i=1}^{N} (y_i - \mu) = 0 \tag{5.8}$$

and when a *sample* is involved

$$\sum_{i=1}^{n} e_i = \sum_{i=1}^{n} (y_i - \bar{y}) = 0 \tag{5.9}$$

The algebraic proof of this statement is straightforward. Assuming that the mean is of a sample,

$$\sum_{i=1}^{n}(y_i - \bar{y}) = \sum_{i=1}^{n} y_i - \sum_{i=1}^{n} \bar{y}$$

but

$$\sum_{i=1}^{n} \bar{y} = n\bar{y} = \frac{n\sum_{i=1}^{n} y_i}{n} = \sum_{i=1}^{n} y_i$$

Consequently,

$$\sum_{i=1}^{n}(y_i - \bar{y}) = \sum_{i=1}^{n} y_i - \sum_{i=1}^{n} y_i = 0$$

An arithmetic example can easily be found which would appear to invalidate this proof. If one computes the mean and the residuals of the cone count data in column 1 of Table 3.1 the results will be as shown in Table 5.2. The sum of residuals approaches, but does not equal, zero. The effect of rounding back the mean when further calculations are involved is evident in this table. The further the rounding back is carried out, the greater the discrepancy in the sum of residuals. This is the reason it was said in Sections 5.3.1 and 5.3.2 that the mean should be computed to as many decimal places (not significant figures) as possible, and no rounding back be done, if the mean is to be used for further calculations.

*The arithmetic mean is greatly affected by extreme values.* If all the items, except one, in a set of data are close to each other in value and the exception is considerably different, the arithmetic mean of the set can yield a distorted impression of the set. For example, assume that 11 students at a certain university were weighed and the weights were 169, 180, 168, 186, 173, 179, 166, 175, 182, 163, and 402 lb. The mean of the set is 194.8 lb. Of the 11 students, 10 had a weight less than the mean so a person having only the mean available on which to base an impression of the student "body" would obtain a false idea of the physical size of the students.

*A valid arithmetic mean of a set of arithmetic means can only be obtained if each of the means is appropriately weighted.* Assume a set of three means: 362, based on 15 items; 403, based on 25 items; and 296, based on 42 items. The unweighted mean of the set is 353.7, which is incorrect since in its computation the fact that each was based on a different number of data items was ignored. The 362 item was considered to be just as strong a value as the 296 item. The proper procedure is to weight the means using the size of the data set on which each mean was based as the weighting factor. When this is done, using Equation 5.5, the mean is 340.7. In this process the grand total of all the items in all three data sets was computed and then this total was divided by the total count across all three data sets.

*Miscellaneous characteristics.* Alone among the averages, the arithmetic mean can be determined without knowledge of the individual values in the set. If, for example, values are cumulated as obtained and a count of the observations is maintained, the arithmetic mean can be obtained by dividing the total of the values by the count. The arithmetic mean is the most statistically stable of the averages, changing relatively little from sample to sample drawn from the same population. If all four of the computed means (the arithmetic, geometric, harmonic, and quadratic means) are

**TABLE 5.2**
**Computation of the Sums of the Residuals of the Cone Count Data in Table 3.1**

| Tree No. | Cone Count | Residuals Using $\bar{y} = 34$ | Residuals Using $\bar{y} = 34.35$ | Residuals Using $\bar{y} = 34.3478$ | Residuals Using $\bar{y} = 34.347826$ |
|---|---|---|---|---|---|
| 1 | 28 | –6 | –6.35 | –6.3478 | –6.347826 |
| 2 | 41 | +7 | +6.65 | +6.6522 | +6.652174 |
| 3 | 36 | +2 | +1.65 | +1.6522 | +1.652174 |
| 4 | 17 | –17 | –17.35 | –17.3478 | –17.347826 |
| 5 | 52 | +18 | +17.65 | +17.6522 | +17.652174 |
| 6 | 25 | –9 | –9.35 | –9.3478 | –9.347826 |
| 7 | 54 | +20 | +19.65 | +19.6522 | +19.652174 |
| 8 | 21 | –13 | –13.35 | –13.3478 | –13.347826 |
| 9 | 43 | +9 | +8.65 | +8.6522 | +8.652174 |
| 10 | 52 | +18 | +17.65 | +17.6522 | +17.652174 |
| 11 | 7 | –27 | –27.35 | +27.3478 | +27.347826 |
| 12 | 40 | +6 | +5.65 | +5.6522 | +5.652174 |
| 13 | 51 | +17 | +16.65 | +16.6522 | +16.652174 |
| 14 | 72 | +38 | +37.65 | +37.6522 | +37.652174 |
| 15 | 24 | –10 | –10.35 | –10.3478 | –10.347826 |
| 16 | 28 | –6 | –6.35 | –6.3478 | –6.347826 |
| 17 | 37 | +3 | +2.65 | +2.6522 | +2.652174 |
| 18 | 21 | –13 | –13.35 | –13.3478 | –13.347826 |
| 19 | 46 | +12 | +11.65 | +11.6522 | +11.652174 |
| 20 | 55 | +21 | +20.65 | +20.6522 | +20.652174 |
| 21 | 0 | –34 | –34.35 | –34.3478 | –34.347826 |
| 22 | 5 | –29 | –29.35 | –29.3478 | –29.347826 |
| 23 | 35 | +1 | +0.65 | +0.6522 | +0.652174 |
| | | +8 | –0.05 | +0.0006 | +0.000002 |

Note the reduction in the magnitude of the differences from zero as the number of decimal places used in the mean increases.

computed using the same data set, the arithmetic mean will be somewhat larger than the geometric but smaller than the other two.

## 5.4 THE MEDIAN

The median is the value which occupies the *midpoint position* in a mass of data *when those data are arranged into an array.* It can be further described by stating that the number of data items larger than the median is equal to the number of data items smaller than the median. The implication is that the *count of items in the array* is an *odd* number. *A true median does not exist if the array contains an even number of items* because there is no unique midpoint position. It is possible in some cases to approximate the median when the count is even, but in others even an approximation cannot be obtained.

### 5.4.1 Ungrouped Data

When the array contains an *odd* number of items, the *median position* is determined using the formula (assuming the array to be from a sample):

$$\text{M.P.} = \frac{n+1}{2} \tag{5.10}$$

The *median value* is the value of the item occupying the median position. An example using the student weights used in Section 5.2.6 follows:

| *Position in array* | 1 | 2 | 3 | 4 | 5 | 6 | 7 | 8 | 9 | 10 | 11 |
|---|---|---|---|---|---|---|---|---|---|---|---|
| *Value* (lb) | 163 | 166 | 168 | 169 | 173 | 175 | 179 | 180 | 182 | 186 | 402 |

The median *position* is number 6 — M.P. = (11 + 1)/2 = 6 — and the median *value* is 175 lb.

When the array contains an *even* number of items, there is no median position and, consequently, no median exists. To approximate the median the *two positions* in the middle of the array are determined using the formula:

$$\text{M.P.} = \frac{n}{2} \tag{5.11}$$

This indicates the count from either end of the array. If the variable of interest is *continuous*, the median *value* is approximated by the arithmetic mean of the values occupying the two middle positions. For example, using the sample of student weights used previously but deleting the 186-lb item, the array would be as shown below:

| *Position in array* | 1 | 2 | 3 | 4 | 5 | 6 | 7 | 8 | 9 | 10 |
|---|---|---|---|---|---|---|---|---|---|---|
| *Value* (lb) | 163 | 166 | 168 | 169 | 173 | 175 | 179 | 180 | 182 | 402 |

The median *positions* in this case are those fifth from each end — M.P. = (10/2) = 5 — or positions 5 and 6. The approximation to the median is 175 lb — (173+175)/2 = 174. *Since the variable is continuous, a fractional value for the median is acceptable.*

When the variable of interest is *discrete*, an approximation of the median can be obtained *only* if the mean of the two middle values is a *whole* number. When the mean is not a whole number, the median cannot be approximated because a fractional discrete value has no meaning.

### 5.4.2 Grouped Data

In some instances a median is desired when the data are in the form of a frequency distribution. When this occurs, the work must be done under one or the other of two assumptions.

> *Assumption 1.* The data items in the classes are all actually equal to the midpoints of their respective classes.

Under this assumption, using the frequency distribution of dbh values in Table 5.3 as an example, it would be assumed that all the dbh values in the 12-in. class are exactly 12 in., all those in the 14-in. class are exactly 14 in., and so forth.

The determination of the median in a situation such as this first requires the determination of the median position or positions. This would be done using Equations 5.10 or 5.11. The class within which the median position is located would be determined by cumulating the class frequencies until the median position is reached. Since it is assumed that all values in a class are equal to the class midpoint, the median *value* is the value of the class midpoint of the class that contains the median *position*.

**TABLE 5.3**
**Frequency Distribution, by 2-in. Classes, of a Set of dbh Values**

| Class Midpoint (in.) | Class Limits (in.) | f | Cumulative f |
|---|---|---|---|
| 12 | 11.0–13.0 | 3 | 3 |
| 14 | 13.0–15.0 | 8 | 11 |
| 16 | 15.0–17.0 | 15 | 26 |
| 18 | 17.0–19.0 | 22 | 48 |
| 20 | 19.0–21.0 | 36 | 84 |
| 22 | 21.0–23.0 | 42 | 126 |
| 24 | 23.0–25.0 | 41 | 167 |
| 26 | 25.0–27.0 | 32 | 199 |
| 28 | 27.0–29.0 | 24 | 223 |
| 30 | 29.0–31.0 | 12 | 235 |
| 32 | 31.0–33.0 | 5 | 240 |
| 34 | 33.0–35.0 | 0 | 240 |
| 36 | 35.0–37.0 | 1 | 241 |

Using the dbh data in Table 5.3 as an example, the total count is 241, an odd number, so the median position is determined using equation 5.10. It is position 121 — M.P. = (241 + 1)/2 = 121 — which lies in the 22-in. class. Consequently, the median value is assumed to be 22 in.

If the count of all items had been *even*, the median would be obtained in the same way, *provided that the two middle positions were in the same class*. The arithmetic mean procedure for arriving at the approximation of the median would be in force, but since all the values in a class are assumed to be equal to the class midpoint, the midpoint would be used as the median value. When the two middle positions are not in the same class, *the two class midpoints are averaged*. If the variable is *continuous*, this approximation of the median is acceptable whether it is a whole or fractional number. If the variable is *discrete*, an acceptable approximate median would have to be a *whole* number. If the mean is not a whole number, there is no median.

*Assumption 2*. The values of the data items in any class are equally spaced within the class.

This assumption can best be visualized and assessed using a number line, as in Figure 5.3, which represents two of the classes of dbh values shown in Table 5.3. The 12-in. class contains three items and the 14-in. class 8 items. The values are assumed to be equally spaced through their

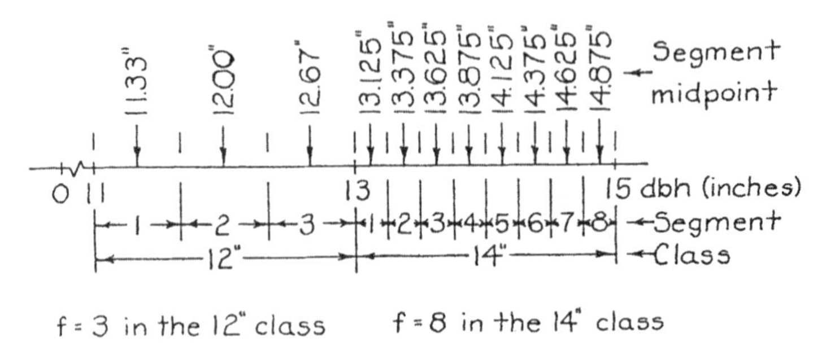


**FIGURE 5.3** Number line showing the assumed positions of the values in the first two classes of the frequency distribution of dbh values shown in Table 5.3.

respective classes. To find these positions the classes are divided into segments, the number of which corresponds to the class frequency. Thus, the 12-in. class is divided into three segments and the 14-in. class is divided into eight segments. The width of one of these segments is the class width divided by the frequency. Consequently, in the 12-in. class the segment widths are 0.67 in. (2 in./3 = 0.67 in.) while in the 14-in. class they are 0.25 in. (2 in./8 = 0.25 in.). The *midpoints* of the *segments* are the values assumed under Assumption 2. In the 12-in. class the first value is half a segment width from the left limit of the class and thus is 11.00 + 0.67/2 = 11.33 in. The second value is 0.67 in. to the right of the first or 11.33 + 0.67 = 12.00 in. The third is 0.67 in. to the right of the second, 12.00 + 0.67 = 12.67 in.. The same procedure yields the values for the items in the other classes.

If the variable is *continuous*, fractional values such as these shown in Figure 5.3 would provide no problem. However, if it is *discrete*, its values will have to be whole numbers and fractional values would be unacceptable. Since this procedure rarely yields whole numbers, it is reasonable to say that the second assumption should only be used when the variable is continuous.

The procedure for determining the class containing the median position is the same as that used under Assumption 1. That class is then subdivided into segments as described above and the segment(s) corresponding to the median position(s) identified. The midpoint(s) of the segment(s) would then be determined. If the total frequency is an odd number, the median will be the segment midpoint value for the median segment. If the total frequency is even, the median will be the mean of the midpoint values of the two median segments.

The median value for the dbh data in Table 5.3, determined under Assumption 2, is obtained in the following manner. The median *position* is obtained using Equation 5.10. It is position 121. By using the cumulative frequency column in Table 5.3, position 121 was found to be in the 22-in. class. Since 84 trees had dbh values smaller than the lower limit of the 22-in. class, the median position is coincident with the midpoint of segment 37 in the 22-in. class (121 – 84 = 37). Each segment in the 22-in. class is 2/42 = 0.0476190476 in. wide. The value corresponding to the midpoint of segment 37 is 21.0 + 37(0.0476190476) – (0.0476190476/2) = 22.73809524 in., which is the median *value*.

If the total frequency is an even number, it is possible that one of the middle positions will be in one class and the other in the adjacent class. When this occurs, the two *segment* midpoint values are averaged in the normal manner even though they come from different classes.

### 5.4.3 Characteristics of the Median

Unlike the arithmetic mean, the median is not a *computed* value. It simply is the value of the item occupying the central position in an array. This is an easily understood and unambiguous concept. It can even be extended to nonnumerical data for which some form of ranking can be devised. However, a true median exists only when an odd number of items are arranged into an array. With an even number of items or when the data are in the form of a frequency distribution, the median can only be approximated and in some cases even an approximation cannot be made. Consequently, it is not always available and it is not always unambiguous. This reduces its value considerably.

One of the criteria used to judge averages is that the average should not be subject to manipulation on the part of the person determining its value. This criterion is not met when the median of a frequency distribution is involved since the determination can be made under either one of two different assumptions regarding the distribution of the values within the classes in the distribution.

Since the median is not a computed value, but rather is a value occupying a position in an array, it is not based on all the values as is the arithmetic mean. For example, using the weight data from Sections 5.3.6 and 5.4.1, the median weight is 175 lb while the arithmetic mean of the weights is 194.8 lb. The 175-lb value provides a much better idea of the sizes of the individuals in the set than does the arithmetic mean. Because of the insensitivity to extreme values, the median is valuable when extreme values can distort an evaluation of a situation. It finds considerable use among economists and social scientists, particularly when income levels are under study. It is rarely used with biological data.

The statistical stability of the median is less than that of the arithmetic mean. In other words, if one were to draw a number of samples of a given size from a certain population, the medians of the samples would exhibit more variability than would the arithmetic means.

In Section 5.3.4 it was pointed out that the sum of squared $k_i$ values would be at a minimum if the $k_i$ values were obtained using the arithmetic mean. If the *absolute* magnitudes of the $k_i$ values are summed, rather than their squares, the minimum sum would be obtained using the median.

The statistical utility of the median is not great. It is sometimes used by statisticians in preference to the arithmetic mean when, as was described earlier, the sample set contains a nonrepresentative item or the distribution of the data set is highly skewed. However, the use of the median, even in these cases, is limited to the simple expression of the average because it cannot be subjected to further statistical analysis.

### 5.4.4 Fractiles

The median is one of a group of values that are referred to collectively as *fractiles*. A fractile is a *data value* occupying a position in an array that is some *fraction* or *proportion* of the total number of *positions* in the array. Thus, the median occupies the midpoint position in the array. If the median is excluded, one half or 50% of the items in the set are smaller than or equal to the median and one half or 50% are larger than or equal to the median. When the set is divided into quarters, the data items at the breaking points are *quartiles*. The median is the second quartile. The data items at the breaking points are *deciles* when the set is divided into tenths and *percentiles* when it is divided into 100ths.

Fractiles are rarely used in the interpretation of biological data, but in certain types of non-biological statistical operations they are used extensively. A familiar use is to indicate a person's standing in a set of examinees after an achievement test has been given.

## 5.5 THE MODE

Fundamentally, the *mode* is the value in the set that occurs most frequently. It is interpreted as being the value that is most typical of the set.

### 5.5.1 Determination of the Mode

Usually the mode is determined from a frequency distribution. In such a case the class containing the greatest number of items is the *modal class* and the midpoint value of that class is considered to be the *mode*. Thus, in the frequency distribution of dbh values in Table 5.3, the 22-in. class, with 42 items, is the modal class and the mode is 22 in. A more precise method of determining the mode, when the variable is continuous, requires the development of a frequency curve. The value of the abscissa corresponding to the highest point on the curve is the mode. To be done properly this requires the fitting of an algebraically defined curve to the plotted data points. This can be a difficult task. Consequently, it is rarely done.

### 5.5.2 Multimodality

When a frequency distribution has two or more classes that have the same frequency and when that frequency is greater than that of any of the remaining classes, the distribution is *multimodal* and it has no true mode. It may be said, in such a case, that the distribution has two, or more, modes. However, strictly speaking, there should be only one and, if multimodality exists, the mode is undefined. This is the case with the distribution of cone counts in column 4 in Table 3.1 where the 21 and 52 count classes each contains two trees. The occurrence of two or more modal classes may be an indication of an excessive number of classes. Resorting the data into fewer, but broader, classes may eliminate the multiplicity of modal classes. The cone count distribution in column 4 in Table 3.1 is a good example. When the classes were broadened and reduced in number to form the distribution in column 5, a definite mode was obtained. It should be recognized, however, that multimodality may be an inherent characteristic of the set such as when the distribution is U shaped. When this occurs, the practice of referring to more than one mode is acceptable.

### 5.5.3 Characteristics of the Mode

The concept of the mode is easily understood and, if multimodality is absent, it is relatively easy to determine. However, if multimodality is present, the situation becomes ambiguous.

Since it is possible to reduce or eliminate multiple modes, or even to shift the position of the mode, by restructuring the classes in the frequency distribution, the mode is not a rigidly determined average.

The fourth criterion used to judge averages requires that every item in the data set have some impact on the magnitude of the average. The mode does not meet this requirement. Neither the range nor the distribution of values in the data set has any effect on the mode if the modal class remains the same. This, of course, means that, in contrast to the arithmetic mean, the mode is not influenced by erratic extreme values. Consequently, where such values are present in a data set the mode may provide a better picture of the set than would the arithmetic mean.

The mode is a measure of central tendency, but often it is far from the center of the set of values. Indeed, it is possible for the mode to be at the very end of a set, as would be the case of a J shaped distribution, or near the end of a heavily skewed mound-shaped distribution. The fact that the mode is the most frequently occurring single value is important in cases such as this because it provides a better picture of the set than does any of the other averages.

It should be noted that, in the case of a *symmetrical mound-shaped* distribution, the arithmetic mean, the median, and the mode coincide. If skewness is present, the three will separate. Referenced to the arithmetic mean, both the median and the mode shift away from the direction of skewness (the side of the distribution where the frequencies trail off at the slower rate) with the median moving farther than the mode, the amount of shift increasing as the skewness increases. The measures of skewness are discussed in Section 8.12.

As has been previously stated, the mode provides the statistical worker or the decision maker with a single value condensation of its data set. In some ways it is a very good condensation. In some cases it is the best available. However, it cannot be used in, nor subjected to, statistical operations or analyses beyond its simple determination. Consequently, its utility is limited.

## 5.6 THE GEOMETRIC MEAN

The geometric mean is unfamiliar to most people, including many engaged in the inventory or other statistical assessment of natural resources. However, while certainly not as useful as the arithmetic mean, it has a role of some importance in certain aspects of the work.

### 5.6.1 Ungrouped Data

In its basic form the geometric mean is the *n*th root of the grand product of the data items*:

---

* The symbols shown are appropriate for a sample. In the case of a population, the symbol for the mean would be $G$. The symbol $\Pi$ is an uppercase *pi* and symbolizes a series of multiplications in the same sense that $\Sigma$ symbolizes a series of additions.

$$g = (y_1 * y_2 * y_3 * \cdots * y_n)^{1/n}$$
$$= \left( \prod_{i=1}^{n} y_i \right) \tag{5.12}$$

Assume a set of data containing the values 1, 2, 4, 8, 16, 32, and 64. The geometric mean would be

$$g = (1 * 2 * 4 * 8 * 16 * 32 * 64)^{1/7}$$
$$= (2{,}097{,}152)^{1/7} = 8.0$$

It should be noted that the presence of a zero in the data set will preclude the determination of the geometric mean.

An alternative computation procedure for the geometric mean requires the use of logarithms:

$$g = \text{antilog}\left( \frac{\sum_{i=1}^{n} (\log y_i)}{n} \right) \tag{5.13}$$

This procedure is demonstrated below using the data set used above:

| $y^i$ | $\log y^i$ |
|---|---|
| 1 | 0.00000000 |
| 2 | 0.30103000 |
| 4 | 0.60205999 |
| 8 | 0.90308999 |
| 16 | 1.20411998 |
| 32 | 1.50514998 |
| 64 | 1.80617997 |

$$\sum_{i=1}^{7} \log y_i = 6.32162991$$
$$g = \text{antilog } (6.32162991/7) = 8.0$$

Under most circumstances the procedure using logarithms is preferred. If the data set is small, the basic procedure, using Equation 5.12, can be carried out without difficulty. However, even with just a few items in the data set, the product of the series of multiplications grows very rapidly and is very likely to exceed the capacity of the computing device to store, use, or display the value unless scientific notation is used. When this is done, the product is rounded back *at each step* to the number of decimal places that can be maintained by the device. The cumulative effect of this rounding can become significant. If the logarithmic approach is used, the sum increases at a much slower rate than does the equivalent product in the primary procedure and the rounding effect is much smaller. Consequently, in most cases the logarithmic procedure is preferred.

### 5.6.2 Grouped Data

When the data are in the form of a frequency distribution, the geometric mean is computed in the following manner:

$$g = (C_1^{f_1} * C_2^{f_2} * C_3^{f_3} * \cdots * C_m^{f_m})^{1/n}$$

$$= \left(\prod_{i=1}^{m} C_i^{f_i}\right)^{1/n} \tag{5.14}$$

where $n = \sum_{i=1}^{m} f_i$

The logarithmic equivalent of this expression is

$$g = \text{antilog}\left[\frac{\sum_{i=1}^{m} f_i \log C_i}{\sum_{i=1}^{m} f_i}\right] \tag{5.15}$$

It should be noted that an *empty class* created no problem since the value of any number raised to the zero power is equal to 1. However, *if one of the class midpoints is zero*, the geometric mean cannot be computed.

### 5.6.3 Use of the Geometric Mean

The use of the geometric mean is appropriate when the values in the data set, arranged in an array, approximate the values of a geometric sequence. The data set used to demonstrate the computation process meets this requirement. The values in that set, the $Y$ values, are the terms in the geometric series:

$$\sum_{x=1}^{n} 2^{x_i}$$

where $x_i = i - 1$ and, thus,

$$y_i = 2^{x_i} \tag{5.16}$$

In cases such as this where the variable of interest is a function of another variable, the variable of interest is referred to as the *dependent* variable, whereas the variable whose values determine those of the dependent variable is the *independent* variable. Independent variables are symbolized using the uppercase $X$ while dependent variables are symbolized by $Y$. If more than one independent variable is involved, they are identified by subscripts such as $X_1$, $X_2$, etc. In Equation 5.16 the independent variable is the exponent $X$. Thus, when

$$\begin{array}{ll} x_1 = 0 & y_1 = 2^0 = 1 \\ x_2 = 1 & y_2 = 2^1 = 2 \\ x_3 = 2 & y_3 = 2^2 = 4 \\ x_4 = 3 & y_4 = 2^3 = 8 \\ x_5 = 3 & y_5 = 2^4 = 16 \\ x_6 = 5 & y_6 = 2^5 = 32 \\ x_7 = 6 & y_7 = 2^6 = 64 \end{array}$$

The function $y_i = 2^{x_i}$ can be graphed as shown in Figure 5.4. The arithmetic mean of the $x$ values is

$$\bar{x} = \frac{\sum_{i=1}^{n} x_i}{n} = \frac{21}{7} = 3.0$$

Substituting $\bar{x}$ for $x_i$ in Equation 5.16 yields

$$y = 2^{3.0} = 8.0$$

which is the value of the geometric mean of the $y$ values.

It can be shown that the curve of a function will pass through the point that has as its coordinates the means of the two variables. The curve in Figure 5.4 *does* pass through the point (3, 8). It cannot, however, be made to pass through the point (3, 18.1429), where 18.1429 is the arithmetic mean of the $y$ values. This is interpreted to mean that the geometric mean is appropriate in this case and the arithmetic mean is not.

Situations requiring the use of the geometric mean arise when the average growth over a period of time is needed for an individual organism, a population of organisms, or an investment. This is not an uncommon occurrence among foresters, wildlife managers, or other natural resources managers.

### 5.6.4 Characteristics of the Geometric Mean

The computation procedure for the geometric mean is clearly defined, permitting no personal judgment to enter the process. All persons using the same data would arrive at the same value for the mean, provided that the same number of decimal places were used at each stage of the operation. In spite of its computational unambiguity, the geometric mean is much more difficult to understand or to interpret than are the three previously described averages.

Like the arithmetic mean and unlike the median and the mode, the geometric mean is computed in such a manner that every number of the data set exerts an influence on its magnitude. Changing, adding, or deleting even a single data value will modify the geometric mean. Erratic extreme values do tend to make it less representative of the data set than it otherwise would be, but it is the least influenced by such values of any of the four computed means.

The geometric mean can be used in statistical operations beyond its simple determination. However, its utilization in such operations is limited.

Finally, the geometric mean of a set of data is always larger than the harmonic mean of the set but is smaller than either the arithmetic or quadratic means.

## 5.7 THE HARMONIC MEAN

Like the geometric mean, the harmonic mean is seldom used but there are times when it, rather than any other measure of central tendency, should be used.

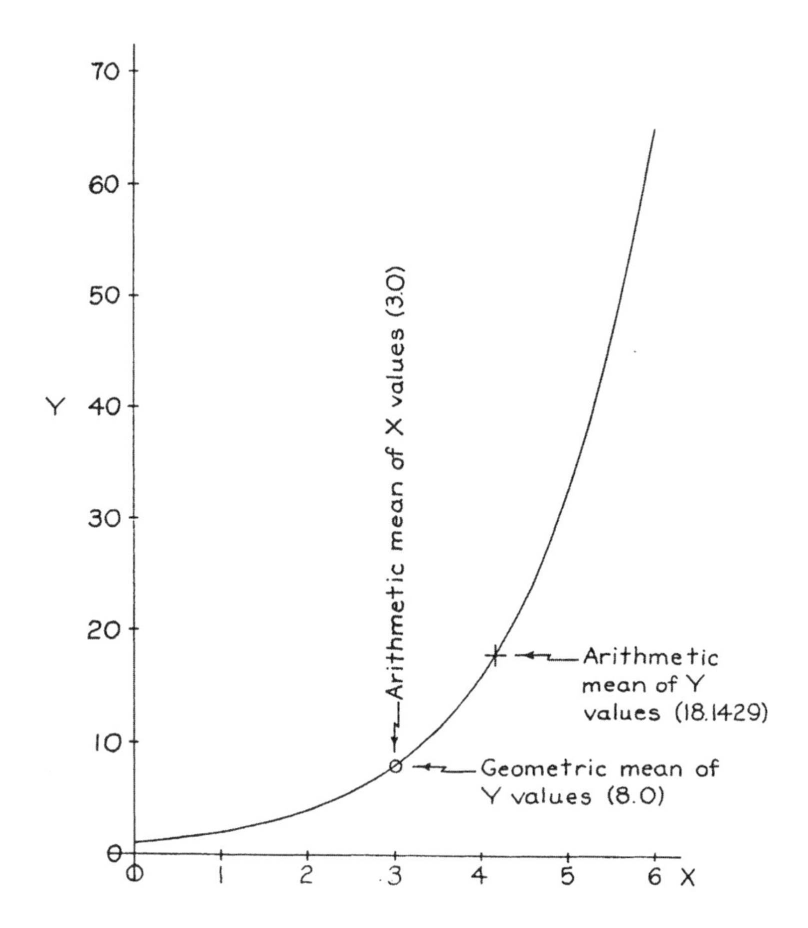


**FIGURE 5.4** Graphical representation of the experimental function $Y = 2^X$.

### 5.7.1 Ungrouped Data

The harmonic mean is defined as the reciprocal of the arithmetic mean of the reciprocals of the data values:

$$h = \frac{1}{\left[\frac{\sum_{i=1}^{n} (1/y_i)}{n}\right]} = \frac{n}{\sum_{i=1}^{n} (1/y_i)} \tag{5.17}$$

If a certain data set contains the following values (5, 3, 8, 7, 2, 3, 7, 4, 3), the harmonic mean of the set is

$$h = \frac{9}{1/5 + 1/3 + 1/8 + 1/7 + 1/2 + 1/3 + 1/7 + 1/4 + 1/3}$$

As in the case of the geometric mean, the presence of a zero in the data set will preclude the calculation of the harmonic mean. Since the reciprocal of zero is infinity, when a zero is included in a data set, the harmonic mean of that set automatically becomes equal to zero.

### 5.7.2 Grouped Data

When the data are grouped into a frequency distribution, the harmonic mean is computed as follows:

$$h = \frac{\sum_{i=1}^{m} f_i}{\sum_{i=1}^{m} [f_i(1/C_i)]} \tag{5.18}$$

With the previously used data set, the computation procedure is as follows:

| $C_i$ | $f_i$ | $1/C_i$ | $F_i(1/C_i)$ |
|---|---|---|---|
| 2 | 1 | 0.5000 | 0.5000 |
| 3 | 3 | 0.3333 | 1.0000 |
| 4 | 1 | 0.2500 | 0.2500 |
| 5 | 1 | 0.2000 | 0.2000 |
| 6 | 0 | 0.1667 | 0.0000 |
| 7 | 2 | 0.1429 | 0.2858 |
| 8 | 1 | 0.1250 | 0.1250 |

$$\sum_{i=1}^{m} [f_i(1/C_i)] = 2.3608$$

$$\sum_{i=1}^{m} f_i = 9$$

$$h = \frac{9}{2.3608} = 3.8124$$

Note that the set had an empty class and that this caused no difficulty. If, however, a class midpoint had been zero, the computation could not have been carried out.

### 5.7.3 Use of the Harmonic Mean

The harmonic mean is the appropriate average to use under certain circumstances when *rates* are being averaged. When a rate is stated, such as the price per unit of some commodity or the number of units of some item made by a worker in a certain period of time, the statement contains a variable (the price of the number of units) and a constant (the unit of the commodity or the unit of time). It is possible to reverse a rate statement so that the variable term becomes the constant and the constant term becomes the variable. For example, the first of the two examples given above could be restated to read, "the number of units of the commodity for a given price," while the second example would produce the statement, "the amount of time needed to make one unit." If the arithmetic mean of a set of rates is computed and then the rates are reversed as was done above and the arithmetic mean of the reversed rates is computed, it will be found that the two means do not correspond. For example, assume that three people are engaged in planting trees by hand. Their rates of production, on a per hour basis, are 80 trees for $A$, 92 trees for $B$, and 106 trees for $C$. The arithmetic mean of the three rates is

| | |
|---|---|
| *A* | 80 trees in 1 hour |
| *B* | 92 trees in 1 hour |
| *C* | 106 trees in 1 hour |
| All | 278 trees in 3 hours |

The average rate of production = $\bar{y}$ = 278/3 = 92.7 trees per hour.

If the tree rates had been stated in reverse form, they would state the time needed to plant one tree.

| | |
|---|---|
| *A* | 1/80 or 0.01250 hours |
| *B* | 1/92 or 0.01087 hours |
| *C* | 1/106 or 0.00943 hours |

The arithmetic mean of the time needed to plant one tree is

$$\bar{y} = \frac{0.01250 + 0.01087 + 0.00943}{3} = 0.01093 \text{ hours}$$

The average time needed to plant 1000 trees could be determined from either of the two average rates. By using the first, the time needed would be 1000/92.7 = 10.79 hours. By using the second, the time needed would be 1000 * 0.01093 = 10.93 hours. The two times are not the same. This is caused partially by the rounding-back process but it is primarily caused by the difference in the ways that the rates are stated.

Which of the two average times needed for planting should be used by the decision maker? There is no clear-cut answer to this question. However, the usual procedure is to use the value which was computed using the rate expression *in common use* in the organization. In forestry and the allied fields rates *per hour* or *per year* are common and such expressions usually are preferred over the reciprocal expressions *hours per* or *years per*. Consequently, the average time needed to plant 1000 trees that probably would be used is that based on the *per hour* rates.

If the available data in the above case had been those in the *hours per* category, the desired result would have been obtained by using the harmonic mean of these rates.

$$h = \frac{3}{1/0.01250 + 1/0.01087 + 1/0.00943}$$

$$= 0.01079$$

The average time needed to plant 1000 trees would be 1000 * 0.01079 = 10.79 hours. This agrees with the value obtained using the arithmetic mean of the *per hour* rates. Thus, the use of the harmonic mean is appropriate when the available rate data are in the form of reciprocals of the rate expressions commonly used.

### 5.7.4 Characteristics of the Harmonic Mean

The harmonic mean is one of the computed means and, in common with them, it is defined in rigid mathematical terms that permit no personal judgment to influence its value. It is not easily understood, nor is the problem of computing the averages of rates, with which is associated. Like the other computed means, the harmonic mean is dependent on every item in the data set. It is very sensitive to erratic extreme values, especially those that are very small. Finally, when the four computed means are determined for a given data set, the harmonic mean is always the smallest of the four.

## 5.8 THE QUADRATIC MEAN

The quadratic mean is rarely used as an average of a set of data. However, it is very frequently used, in the form of the standard deviation, as a measure of the variability or dispersion of the items in data sets about the arithmetic means of those sets. This will be discussed in considerable detail in Chapter 6.

### 5.8.1 UNGROUPED DATA

By definition, the quadratic mean is the square root of the arithmetic mean of the squares of the data values.

$$q = \sqrt{\frac{\sum_{i=1}^{n} y_i^2}{n}} \tag{5.19}$$

For example, the quadratic mean of the data set (5, 3, 8, 7, 2, 3, 7, 4, 3) is

$$q = \sqrt{\frac{5^2 + 3^2 + 8^2 + 7^2 + 2^2 + 3^2 + 7^2 + 4^2 + 3^2}{9}}$$

$$= 5.0990$$

### 5.8.2 GROUPED DATA

When the data are in the form of a frequency distribution, the formula for the quadratic mean is

$$q = \sqrt{\frac{\sum_{i=1}^{m} f_i C_i^2}{\sum_{i=1}^{m} f_i}} \tag{5.20}$$

By using the above data set arranged into a frequency distribution, the quadratic mean is

| $C_i$ | $f_i$ | $C_i^2$ | $F_iC_i^2$ |
|---|---|---|---|
| 2 | 1 | 4 | 4 |
| 3 | 3 | 9 | 27 |
| 4 | 1 | 16 | 16 |
| 5 | 1 | 25 | 25 |
| 6 | 0 | 36 | 0 |
| 7 | 2 | 49 | 98 |
| 8 | 1 | 64 | 64 |

$$\sum_{i=1}^{m} f_i C_i^2 = 234$$

$$q = \sqrt{\frac{234}{9}} = 5.0990$$

### 5.8.3 Use of the Quadratic Mean

Theoretically, the quadratic mean should be used when the squares of data items, rather than the data items themselves, are under study. This rarely occurs except when the variability of data is being assessed. The standard deviation, a very commonly used measure of data dispersion, is a quadratic mean.

### 5.8.4 Characteristics of the Quadratic Mean

The quadratic mean is one of the four computed means and, like the others, is rigidly defined by a mathematical formula. Personal judgment cannot enter into its determination. Consequently, everyone computing the quadratic mean of a certain data set will obtain the same result.

The meaning of the quadratic mean is very difficult to visualize or to express in words. The problem of meaning arises when the mean is in the form of a standard deviation. People familiar with standard deviations usually have an intuitive understanding of what they represent but, aside from describing how they are computed, have little success in describing what they are.

Like the other computed means, the quadratic mean is dependent on every value in the data set. Changing any single value will change the mean. Erratic extreme values will have a profound effect on the mean and can cause the mean to be, in some senses, not representative of the data set. Finally, among the computed means, all computed for a given data set, the quadratic mean always has the greatest magnitude.

# 6 Measures of Dispersion

## 6.1 INTRODUCTION

Data sets, especially when large, are difficult to comprehend. It is for this reason that averages of such sets are determined. They are single values which, in one sense or another, represent the sets from which they are derived. It is far easier to work with one number than with an entire set. However, it is often found that the average alone provides only a portion of the information needed to visualize the set. For example, consider the situation in Table 6.1. There are two stands of trees, A and B, the mean dbhs of which are identical, 15.0 in. However, it is obvious from the frequency distributions that the stands are not alike. Stand A contains trees whose diameters range all the way from 10 to 24 in. while Stand B is made up of trees which are almost all the same size, ranging only from 13 to 17 in. To complete the representation of a data set begun by the use of an average, it is necessary to have some way of showing, or providing a measure of, the variability of the items in the set. Frequency distributions, such as those in Table 6.1, certainly can be used for this purpose. However, the distributions are often clumsy to use and, in some cases, difficult to interpret. For these reasons it is desirable to use a single value, analogous to the average, which can indicate the degree of *dispersion* or *variability* in the set. Expressions of this type are known as *measures of dispersion*.

## 6.2 THE RANGE

The simplest, and probably the most easily understood, measure of dispersion is the *range*. It is the numerical difference between the largest and the smallest values in the set.

$$R = y_L - y_S \tag{6.1}$$

where $y_L$ = largest number in the set
$y_S$ = smallest number in the set

The greater the range, the more variable is the set. In the case of the two stands in Table 6.1, the range of A is 24 – 10 = 14 in., while that of B is 17 – 13 = 4 in. These values indicate that A is more variable than B, which agrees with our previous assessment, which was based on the frequency distributions.

The range is most easily obtained when the data are in the form of an array or a frequency distribution, but it can be determined when the data are in raw form. In the latter case, the searching for the high and low values can become tedious but it can be done. Some authors are concerned that the true range cannot be determined when the data are in frequency distribution form because the exact maximum and minimum values are not known. The loss of accuracy caused by this is usually of little significance because the range, at best, is only a crude measure of the dispersion. The statistical utility of ranges is relatively limited but they are valuable for quick assessments.

**TABLE 6.1**
**Distributions of Diameters Breast High in Two Hypothetical Stands of Trees**

| dbh (in.) | Stand A | Stand B |
|---|---|---|
| 10 | 2 | 0 |
| 11 | 3 | 0 |
| 12 | 4 | 0 |
| 13 | 3 | 4 |
| 14 | 2 | 7 |
| 15 | 5 | 10 |
| 16 | 4 | 5 |
| 17 | 2 | 5 |
| 18 | 1 | 0 |
| 19 | 2 | 0 |
| 20 | 0 | 0 |
| 21 | 1 | 0 |
| 22 | 0 | 0 |
| 23 | 1 | 0 |
| 24 | 1 | 0 |

One of the reasons they have limited statistical value is that they are based only on the two extreme values in the set. The remainder of the set has no effect at all, which can lead to false impressions. For example, if Stand B in Table 6.1 had one more tree and it had a dbh of 24 in., the range would jump from 4 to 10 in. The bulk of the stand would be unchanged but the indicated variability is much greater than it was in the original situation.

## 6.3 THE INTERQUARTILE AND SEMI-INTERQUARTILE RANGES

The concept of the fractile was introduced in Section 5.4.4. Through the use of specified fractiles the problem of extreme values associated with the range can be substantially reduced or even eliminated. For example, if the range from the first to the third quartile were used to express the variability of the set, the extreme values of the set would not be involved. Presumably the resulting range value would be more representative of the set than would the range proper. The range from the first to third quartiles is the *interquartile range* and takes the following form when expressed algebraically:

$$I = y_{Q_3} - y_{Q_1} \tag{6.2}$$

where $y_{Q_1}$ = first quartile

$y_{Q_3}$ = third quartile

The interquartile range is seldom used. A more popular measure of dispersion is the *semi-interquartile range* which is equal to one half the interquartile range.

$$Q = I/2 = \frac{y_{Q_3} - y_{Q_1}}{2} \tag{6.3}$$

In the case of the stands in Table 6.1, the semiquartile ranges are determined as follows:
*For stand A:*

$$\text{Position of first quartile} = \frac{n+1}{4} = \frac{31+1}{4} = 8$$

$$\text{Position of third quartile} = \frac{3(n+1)}{4} = \frac{3(31+1)}{4} = 24$$

The eighth position in the set is in the 12-in. dbh class. Thus, $y_{Q_1} = 12$ in.
The 24th position in the set is in the 17-in. dbh class and thus,

$$y_{Q_3} = 17 \text{ in.}$$

$$Q_A = \frac{17-12}{2} = 2.5 \text{ in.}$$

Similarly for stand B.

Position of first quartile, 8, which is in the 14-in. dbh class
Position of third quartile, 24, which is in the 16-in. dbh class

$$Q_B = \frac{16-14}{2} = 1.0 \text{ in.}$$

Again, Stand A is shown to be more variable than Stand B.

The reason for using the semi-interquartile range rather than the interquartile range itself is that the former indicates the average spread of values on either side of the median, which is the second quartile. Since the semi-interquartile range is related to the median, it is the logical expression of dispersion when the median is used as the measure of central tendency. However, the semi-interquartile range, like the other ranges, cannot be used for deeper statistical analyses of the data. Consequently, it is used only when conditions preclude the use of more powerful expressions such as the variance or the standard deviation. It also should be noted that under certain circumstances the quartiles, like the median, can only be approximated. However, since the semi-interquartile range cannot be used in further statistical operations, this matter of approximation is of minor importance.

## 6.4 THE AVERAGE DEVIATION

A serious flaw associated with all of the measures of dispersion, which are based on the concept of the range, is that in their determination most of the values in the data set are ignored. Any measure that is determined using all the data items is inherently a better expression. Several of the measures of dispersion are of this type. All of these are based on the use of residuals or errors. As was discussed in Section 5.3.5, a *residual* is the difference between a variate and the arithmetic mean of the *sample,* while an *error* is the difference between a variate and the arithmetic mean of the *population.* Generally speaking, if the residuals associated with a population or a sample are for the most part large, there must be considerable variability in the set. If the residuals are for the most part small, the reverse is true. Consequently, it is logical to use them in expressions of variability.

A logical extension of the above argument is to use the mean of the residuals or errors. This, however, leads to meaningless results because these algebraic sums are always equal to zero, as was brought out in Section 5.3.6. To eliminate this problem the *absolute* rather than the signed values of the residuals or errors could be used in the computations. The resulting mean absolute residual or error is known as an *average deviation*, a *mean deviation*, or an *absolute deviation*. The algebraic expression representing this value, using the symbology appropriate to a sample, is

$$\text{A.D.} = \frac{\sum_{i=1}^{n} | y_i - \bar{y} |}{n} = \frac{\sum_{i=1}^{n} | e_i |}{n} \tag{6.4}$$

A simple example of the computation of an average deviation is shown below:

| $y_i$ | $y_i - \bar{y} = e_i$ | $\lvert e_i \rvert$ |
|---|---|---|
| 3 | 3 – 5.0 = –2.0 | 2.0 |
| 8 | 8 – 5.0 = 3.0 | 3.0 |
| 2 | 2 – 5.0 = –3.0 | 3.0 |
| 7 | 7 – 5.0 = 2.0 | 2.0 |
| 5 | 5 – 5.0 = 0.0 | 0.0 |
| 25 | | 10.0 |

$\bar{y}$ = 25/5 = 5.0
A.D. = 10.0/5 = 2.0

In the case of a frequency distribution the formula is

$$\text{A.D.} = \frac{\sum_{i=1}^{m} [f_i | c_i - y |]}{\sum_{i=1}^{m} f_i} \tag{6.5}$$

The computation of an average deviation of a frequency distribution is demonstrated below using the data for Stand B in Table 6.1.

| $C_i$ (in.) | $f_i$ | $f_iC_i$ (in.) | $C_i - \bar{y}$ (in.) | $f_i(C_i - \bar{y})$ (in.) | $f_i \lvert C_i - \bar{y} \rvert$ (in.) |
|---|---|---|---|---|---|
| 13 | 4 | 52 | 13 – 15.0 = –2.0 | –8.0 | 8.0 |
| 14 | 7 | 98 | 14 – 15.0 = –1.0 | –7.0 | 7.0 |
| 15 | 10 | 150 | 15 – 15.0 = 0.0 | 0.0 | 0.0 |
| 16 | 5 | 80 | 16 – 15.0 = 1.0 | 5.0 | 5.0 |
| 17 | 5 | 85 | 17 – 15.0 = 2.0 | 10.0 | 10.0 |
| | 31 | 465 | | | 30.0 |

$\bar{y}$ = 465/31 = 15.0 in.
A.D. = 30.0/31 = 0.97 in.

In the case of Stand A the average deviation, computed in the same manner, is 2.71 in.. These results are consistent with the evidence of the two frequency distributions.

The average deviation is easy to interpret. It simply indicates the average scatter of the data values about their arithmetic mean. Unfortunately, eliminating the negative signs on the residuals by using absolute values makes it impossible to use the average deviation in further statistical operations. As a simple expression of the scatter of the data, it is superior to range-based expressions because it makes use of all the values in the set. While certainly sensitive to erratic extreme values, it is much less so than is the range. On the other hand, it is probably more sensitive to such values than are the interquartile or semi-interquartile ranges.

The average deviation is rarely used. Its flaws are too great and advantages too small relative to the other measures of dispersion. However, its inclusion in a book of this type is justified because it provides a very good introduction to the variance, which without doubt is the most important and useful of any of the measures of dispersion.

## 6.5 THE VARIANCE

The basic concept of the average deviation, that of using the arithmetic mean of the residuals or errors as a measure of dispersion, is good. However, since it is based on absolute rather than signed values, its statistical utility is limited. This problem can be overcome by using the *squares*, rather than the absolute values, of the residuals or errors. The *arithmetic mean of the squared residuals or errors* is the *variance* of the set. In the case of a *sample*, the variance is obtained using the following expression*:

$$s^2 = \frac{\sum_{i=1}^{n} (y_i - \bar{y})^2}{n} = \frac{\sum_{i=1}^{n} e_i^2}{n} \tag{6.6}$$

The symbology in the case of the *population* is

$$\sigma^2 = \frac{\sum_{i=1}^{N} (y_i - \mu)^2}{N} = \frac{\sum_{i=1}^{N} \varepsilon_i^2}{N} \tag{6.7}$$

The variance of the five-item data set used earlier is computed as follows:

| $y_i$ | $y_i - \bar{y} = e_i$ | $e_i^2$ |
|---|---|---|
| 3 | 3 – 5.0 = –2.0 | 4 |
| 8 | 8 – 5.0 = 3.0 | 9 |
| 2 | 2 – 5.0 = –3.0 | 9 |
| 7 | 7 – 5.0 = 2.0 | 4 |
| 5 | 5 – 5.0 = 0.0 | 0 |
| 25 | | 26 |

$\bar{y} = 25/5 = 5.0$
$s^2 = 26/5 = 5.2$

* For the benefit of those readers who have had some background in statistics, it should be emphasized that Equation 6.6 is appropriate for *defining* the variance of a sample. At this point we are *not* considering the sample variance as an estimator of the population variance.

When the data are in the form of a frequency distribution, the formula for the variance of a sample is

$$s^2 = \frac{\sum_{i=1}^{m}[f_i(C_i - \bar{y})^2]}{\sum_{i=1}^{m} f_i} \tag{6.8}$$

while that for a population is

$$\sigma^2 = \frac{\sum_{i=1}^{M}[f_i(C_i - \mu)^2]}{\sum_{i=1}^{M} f_i} \tag{6.9}$$

The variance of the dbh data from Stand B in Table 6.1 is

| $C_i$ (in.) | $f_i$ | $f_iC_i$ (in.) | $C_i - \bar{y}$ (in.) | $(C_i - \bar{y})^2$ | $f_i(C_i - \bar{y})$ (in.) |
|---|---|---|---|---|---|
| 13 | 4 | 52 | –2.0 | 4.00 | 16.00 |
| 14 | 7 | 98 | –1.0 | 1.00 | 7.00 |
| 15 | 10 | 150 | 0.0 | 0.00 | 0.00 |
| 16 | 5 | 80 | 1.0 | 1.00 | 5.00 |
| 17 | 5 | 85 | 2.0 | 4.00 | 20.00 |
| | 31 | 465 | | | 48.00 |

$\bar{y} = 465/31 = 15.0$
$s^2 = 48/31 = 1.55$

It should be noted at this point that the *variance bears no label as to the units involved.* In the example above it would be incorrect to show the variance as 1.55 *in.*$^2$. The variance is a pure number and is never labeled.

The computation procedures shown above may be referred to as *primary* methods because they follow the definition exactly. However, they are inefficient for several reasons. They are tedious, clumsy to program for a computer, and considerable loss of precision can result from their use because of the rounding back of computed values as was discussed in Section 5.3.6. Fortunately, a more efficient procedure for computing the *sum of squared residuals* or *errors* has been devised. Its use simplifies operations on calculators and, consequently, the procedure is sometimes referred to as the *calculator* method. In the form appropriate for ungrouped data from a sample, the procedure is as follows:

$$\sum_{i=1}^{n} e_i^2 = \sum_{i=1}^{n} y_i^2 - \frac{\left(\sum_{i=1}^{n} y_i\right)^2}{n} \tag{6.10}$$

The derivation of this expression is straightforward:

$$\sum_{i=1}^{n} e_i^2 = \sum_{i=1}^{n} (y_i - \bar{y})^2$$

$$= \sum_{i=1}^{n} (y_i^2 - 2y_i\bar{y} + \bar{y}^2)$$

$$= \sum_{i=1}^{n} y^2 - 2\bar{y}\sum_{i=1}^{n} y + n\bar{y}^2$$

But

$$n\bar{y}^2 = \bar{y}(n\bar{y}) = \bar{y}n\frac{\left(\sum_{i=1}^{n} y_i\right)}{n} = \bar{y}\sum_{i=1}^{n} y_i$$

Therefore,

$$\sum_{i=1}^{n} e_i^2 = \sum_{i=1}^{n} y_i^2 - 2\bar{y}\sum_{i=1}^{n} y_i + \bar{y}\sum_{i=1}^{n} y_i$$

$$= \sum_{i=1}^{n} y_i^2 - \bar{y}\sum_{i=1}^{n} y_i$$

$$= \sum_{i=1}^{n} y_i^2 - \left(\frac{\sum_{i=1}^{n} y_i}{n}\right)\sum_{i=1}^{n} y_i$$

$$= \sum_{i=1}^{n} y_i^2 \frac{\left[\sum_{i}^{n} y_i\right]^2}{n}$$

It probably should be brought out at this point that statisticians usually use the term *sum of squares* rather than the complete terms "sum of squared residuals" or "sum of squared errors." Usually this causes no problem since the contexts of their statements will indicate whether residuals or errors are involved.

The sum of squares for the five-item data set used earlier is computed as follows when using the calculator method.

| $y_i$ | $y_i^2$ |
|---|---|
| 3 | 9 |
| 8 | 64 |
| 2 | 4 |
| 7 | 49 |
| 5 | 25 |
| 25 | 151 |

$$\sum^{n} e_i^2 = 151 - \frac{(25)^2}{5} = 151 - 125 = 26$$

The formula for the calculator method when sample data are in the form of a frequency distribution is

$$\sum_{i=1}^{n} e_i^2 = \sum_{i=1}^{m} (f_i C_i^2) - \frac{\left[\sum_{i=1}^{m} (f_i C_i)\right]^2}{\sum_{i=1}^{m} f_i} \tag{6.11}$$

The sum of squares of the dbh data for Stand B in Table 6.1, computed in this manner, is

| $C_i$ (in.) | $f_i$ | $f_iC_i$ | $C_i^2$ | $f_iC_i^2$ |
|---|---|---|---|---|
| 13 | 4 | 52 | 169 | 676 |
| 14 | 7 | 98 | 196 | 1372 |
| 15 | 10 | 150 | 225 | 2250 |
| 16 | 5 | 80 | 256 | 1280 |
| 17 | 5 | 85 | 289 | 1445 |
| | 31 | 465 | | 7023 |

$$\sum^{n} e_i^2 = 7023 - \frac{(465)^2}{31} = 7023 - 6975 = 48$$

When the variable is continuous but the data have been grouped to form a frequency distribution, a *grouping error* will occur, biasing the variance upward. The grouping error occurs because the class midpoints do not coincide with the class means. As can be seen in Figure 6.1, the class midpoints for the classes below the modal class are too far to the left, while those for the classes above the modal class are too far to the right. This means that, on the average, the residuals or errors used in the calculations are always a bit too large. The bias caused by the grouping error can be counteracted by the use of *Sheppard's correction*, provided that the distribution of the variable is such that, when plotted, both tails are asymptotic to the horizontal axis. The correction is the $(-w^2/12)$ term in the expression below:

$$s^2 = \frac{\sum_{i=1}^{m} [f_i (C_i - \bar{y})^2]}{\sum_{i=1}^{m} f_i} - \frac{w^2}{12} \tag{6.12}$$

where $w$ = the class width

Using the correction with the dbh data from Stand B in Table 6.1 yields the value:

$$s^2 = \frac{48}{31} - \frac{(1)^2}{12} = 1.548 - 0.083 = 1.465$$

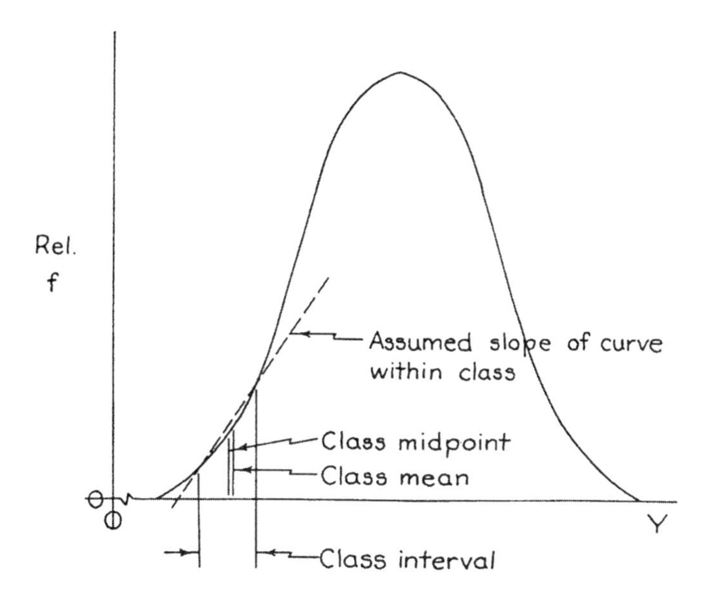


**FIGURE 6.1** The grouping error which occurs when the values of a continuous variable are grouped into a frequency distribution.

The use of the correction in this case is *not* appropriate because the distribution of dbh values has definite upper and lower limits. However, it does serve to show the computation procedure.

There is no question about the preeminence of the variance among the measures of dispersion. Although perhaps not quite as easily interpreted as the average deviation, it can be understood without difficulty. It simply is the average squared residual or the average squared error, depending on whether a sample or a population is involved. Practically all statistical operations involving the variability of the data, and this includes a very large proportion of all statistical operations, make use of the variance or a value derived from the variance. It is essential that a person engaged in any type of statistical work understand what a variance is and how it can be obtained.

## 6.6 THE STANDARD DEVIATION

The *standard deviation*, or the *root mean square*, is very easily computed. It simply is the square root of the variance:

$$s = \sqrt{s^2}, \text{ in the case of a sample} \tag{6.13}$$

or

$$\sigma = \sqrt{\sigma^2}, \text{ in the case of a population} \tag{6.14}$$

As can be seen, this makes it the quadratic mean of the residuals or errors. Through the square root process, the original units of measure are recovered. Consequently, the standard deviation can be labeled with respect to these units. For example, the variance of Stand B in Table 6.1 is 1.55. The standard deviation is

$$s = \sqrt{1.55} = 1.24 \text{ in.}$$

This demonstrates the primary value of the standard deviation — it is expressed in the same units as the data set. To some this means that it is more understandable than the variance. However, if one attempts to verbalize a meaning for the standard deviation, without simply stating how it is obtained, one will be forced to agree that the task is very difficult.

Like any square root, the standard deviation actually possesses two values, one positive and one negative. In most cases it is understood that the positive root is used. However, in some cases the idea of variability on both sides of the mean needs to be emphasized. In such cases it is appropriate to use the ± symbol, as in ±1.24 in.

## 6.7 THE COEFFICIENT OF VARIATION

There are times when the actual numerical value of a measure of dispersion is of less value than when that value is expressed as a proportion or percentage of the measure of central tendency being used. Several such *relative measures of dispersion* could be devised, but only one is in common use. This is the *coefficient of variation*, which is the ratio of the standard deviation to the arithmetic mean:

$$\text{C.V.} = s / \bar{y}, \text{ in the case of a sample} \tag{6.15}$$

or

$$\text{C.V.} = \sigma / \mu, \text{ in the case of a population} \tag{6.16}$$

The coefficient of variation of the dbh data from Stand B in Table 6.1 is 1.24/15.0 = 0.0827 or 8.27%. As in the case of the standard deviation, this can be shown with or without the ± symbol.

The value of a relative measure of dispersion, such as the coefficient of variation, is that it is a pure number. It has no relationship to any units of measure. As a consequence, when relative measures of dispersion are used it is possible to compare the variabilities of different variables. For example, one might wish to know if the heights of a certain set of trees varied relatively in the same manner as did the dbh of these trees. Assume that the mean height ($\bar{y}_H$) is 80 ft and the standard deviation of the heights ($s_H$) is 20 ft, while the mean dbh ($\bar{y}_D$) is 20 in. and the standard deviation of the dbh values ($s_D$) is 5 in. Do the two variables vary relatively in the same manner?

$$\text{C.V.}_H = s_H/\bar{y}_H = 20.80 = 0.25 \text{ or } 25\%$$

or

$$\text{C.V.}_D = s_D/\bar{y}_D = 5/20 = 0.25 \text{ or } 25\%$$

The evidence of the coefficients of variation indicates that they do indeed vary relatively in the same manner.

The coefficient of variation can also be used to describe a *general* situation. For example, assume that a number of sample plots of a certain size were established in an unmanaged stand of one of the southern pines. The stand was made up of trees of merchantable size. Further assume that the plots were cruised and the cubic foot volume of pine timber on each was determined. The mean, standard deviation, and coefficient of determination of the plot volumes were then computed. It is likely that the coefficient of variation would be in the neighborhood of 100% because such stands, over a broad range of conditions, tend to have coefficients of variation of about 100%. This is a *general* value which can be used in the absence of more specific information.

# 7 Probability

In Chapter 1 it was stated that the field of statistics is concerned with the acquisition, summarization, analysis, and interpretation of numerical data obtained using sampling. In the analysis and interpretation phases of the statistical process, estimates are made. When this is done, there must be some measure of the likelihood that the estimate is adequate. As a consequence, the concepts of probability permeate the entire structure of statistics, making it imperative that workers in the field have at least some knowledge of, and appreciation for, probability.

Probability is concerned primarily with the *quantification of likelihood*. For example, it might be said about someone setting up as a consulting forester that the person is *likely* to succeed if he or she goes into general practice but is *less likely* to do so if he or she specializes in tree diseases. Precisely how likely is the person to succeed as a general practitioner? How much less likely to succeed as a specialist? These questions require answers beyond the loose subjective terms used in the likelihood statement. The answers should be expressed numerically and these numbers should indicate the appropriate degrees of likelihood.

## 7.1 SOME BASIC CONCEPTS AND TERMS

As can be seen in the statement about the consultant-to-be, when one speaks of likelihood or probability it is implied that something is going to occur *in the future* and that the *results are uncertain*. Let us consider this in the context of a relatively simple situation. Assume that a litter of puppies contains three males and five females. Further, assume that a child will be blindfolded and told to choose one of the puppies. What is the probability that the puppy will be a male? The future action is the blind choice of a puppy. The variable of interest is the sex of the puppy. Last, the result of the choice is uncertain since the choice has not yet occurred. In many ways this is a typical problem in probability.

Like those in any subject-matter field, workers in the field of probability have developed a specialized terminology. In this terminology, the act of choosing a puppy, as was done above, constitutes an *experiment*. Experiments are actions taken to secure information about some specific variable of interest within some specific context. Experiments may be very simple, such as choosing a puppy, tossing a coin, or casting a die, but they may also be very complex as in the case of many chemical, physical, biological, or social experiments. It must be emphasized that, in the context of probability, an experiment, regardless of its internal complexity, is a *single* action or trial. For example, if a coin is tossed 100 times and each time it is recorded that a head or tail appeared, 100 experiments have been carried out.

Experiments, to have meaning in a probabilistic sense, must have at least two possible outcomes relative to the variable of interest. Each of these outcomes is an *event*. In the case of a coin being tossed, one event is the appearance of a head and the second is the appearance of a tail. In the case of the puppy problem, one event is the choice of a male and the other is the choice of a female. When one speaks of probability, the reference is to a specific event. In the case of the puppies it

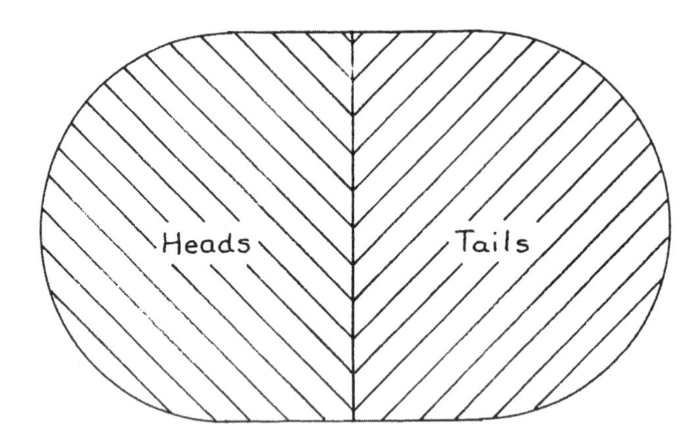


**FIGURE 7.1** Venn diagram showing the two disjoint sets generated by the tossing of a coin.

is the choice of a male. Such events are *events of interest*. The occurrence of the alternative event or events is usually interpreted as the *failure of occurrence of the event of interest*.

An event is *simple* when it cannot be decomposed into simpler events. In other words, if an event can be obtained in more than one way, as when the event of interest is the drawing of a king from a deck of playing cards, the event is *not* simple, because four kings are available. On the other hand, if the event of interest is the drawing of a specific king, such as the king of hearts, the event is simple because there is only one way by which the requirement of the event of interest can be met.

Events are *mutually exclusive* if the occurrence of one of the possible events precludes the occurrence of the other(s). This can be demonstrated with a coin. When it is tossed *either* a head *or* tail will be face up when the coin comes to rest. Both cannot occur simultaneously. In the context of sets, each belongs to a separate set and these sets are disjoint, as can be seen in the Venn diagram in Figure 7.1. Most experiments involving probability are based on mutually exclusive events. However, it is possible to devise experiments where this is not the case. For example, assume that a tract of land in the Alabama Piedmont contains a number of dead shortleaf pine trees, some of which had been killed by the littleleaf disease, some by the southern pine beetle, and some by a joint attack by both agents. This situation is shown in the form of a Venn diagram in Figure 7.2. If one of these dead trees were to be selected blindly, one could *not* say that it was killed by *either* the disease *or* the beetle because in some cases *both* were involved. Consequently, these events are not mutually exclusive.

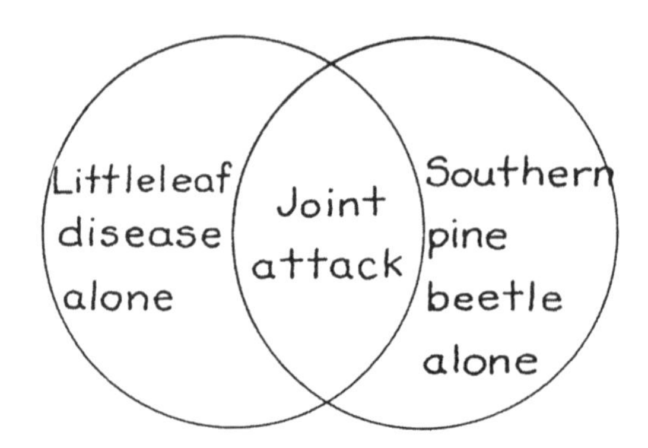


**FIGURE 7.2** Venn diagram showing the situation where some shortleaf pines were killed by littleleaf disease, some by southern pine beetles, and some by joint attack by both agents.

## 7.2 RELATIVE FREQUENCY

Let us assume that a coin was tossed 100 times and a record was kept of the number of heads which appeared. Each toss was an individual experiment. The event of interest was the appearance of a head. The ratio of the count or frequency of heads to the total number of tosses is the *relative frequency* of the occurrence of a head.

In general terms, relative frequency is the ratio of the frequency of occurrence of the event of interest to the total number of trials.

$$\mathrm{RF}(E) = N(E)\,/\,n \tag{7.1}$$

where $\mathrm{RF}(E)$ = relative frequency of the event of interest
$N(E)$ = count or frequency of occurrence of the event of interest
$n$ = total number of trials

If, out of the 100 tosses of the coin, heads appeared 55 times, the relative frequency of heads would be

$$\mathrm{RF\ (heads)} = 55/100 = 0.55 \text{ or } 55\%$$

It should be noted that a relative frequency can be expressed in fractional, decimal, or percent form. Of these, the fractional is inherently the most accurate since no approximations due to rounding back are involved.

When the event of interest *occurred in every trial*, $n(E) = n$ and the relative frequency is

$$\mathrm{RF}(E) = n/n = 1 \tag{7.2}$$

On the other hand, if the event of interest *never occurred*, $n(E) = 0$ and the relative frequency is

$$\mathrm{RF}(E) = 0/n = 0 \tag{7.3}$$

Since all other possible values of $n(E)$ are in the range $0 < n(E) < n$, a negative relative frequency is impossible, as is a relative frequency greater than 1. A *property* of relative frequencies which is of fundamental importance is

$$0 \le \mathrm{RF}(E) \le 1 \tag{7.4}$$

Equation 7.1 can be rewritten in the following form:

$$\mathrm{RF}(E) = \frac{n(E)}{n(E) + n(\tilde{E})}$$

where $n(\tilde{E})$ = count or frequency of the *failure* of occurrence of the event of interest; variants of the symbol of failure are $(\tilde{E})$ or $(\bar{E})$

The relative frequency of failures to obtain the event of interest is then

$$\mathrm{RF}(\tilde{E}) = \frac{n(\tilde{E})}{n(E) + n(\tilde{E})} \tag{7.5}$$

where $R\tilde{F}$ = relative frequency of the failure of occurrence of the event of interest

Summing the relative frequencies of success and failure yields

$$RF(E) + RF(\tilde{E}) = \frac{n(E)}{n(E) + n(\tilde{E})} + \frac{n(\tilde{E})}{n(E) + n(\tilde{E})} = \frac{n(E) + n(\tilde{E})}{n(E) + n(\tilde{E})} = 1 \tag{7.6}$$

provided the events are mutually exclusive.

In many cases more than two outcomes (occurrence and nonoccurrence of an event of interest) are possible. For example, consider a die (a die is one of a pair of dice). When it is tossed there are six possible outcomes. If it is tossed a number of times, say 100, and a record is kept of the occurrence of each face, the frequencies might be $n(1) = 15$, $n(2) = 10$, $n(3) = 20$, $n(4) = 25$, $n(5) = 10$, and $n(6) = 20$. A relative frequency can be obtained for each face:

RF (Face 1) = $n(1)/n$ = 15/100 RF (Face 4) = $n(4)/n$ = 25/100
RF (Face 2) = $n(2)/n$ = 10/100 RF (Face 5) = $n(5)/n$ = 10/100
RF (Face 3) = $n(3)/n$ = 20/100 RF (Face 6) = $n(6)/n$ = 20/100

which can be generalized to

$$RF(E_i) = n_i/n \tag{7.7}$$

where $RF(E_i)$ = relative frequency of the $i$th event
$n_i$ = count or frequency of event $i$

Summing the relative frequencies of all the possible outcomes when a die is cast yields

$$\frac{n(1)}{n} + \frac{n(2)}{n} + \frac{n(3)}{n} + \frac{n(4)}{n} + \frac{n(5)}{n} + \frac{n(6)}{n} = \frac{15}{100} + \frac{10}{100} + \frac{20}{100} + \frac{25}{100} + \frac{20}{100} = \frac{100}{100} = 1$$

because

$$n = n(1) + n(2) + n(3) + n(4) + n(5) + n(6)$$

This can be generalized to

$$\sum_{i=1}^{n}\left(\frac{n_i}{n}\right) = \sum_{i=1}^{n}[RF(E_i)] = 1 \tag{7.8}$$

provided all the events are mutually exclusive. This is known as the *special additive property* of relative frequencies. Equation 7.6 is a special case in which only two outcomes are involved.

When the possible outcomes of an experiment are not mutually exclusive, as in the case of the shortleaf pines killed by littleleaf disease, southern pine beetles, or by joint attack by both, Equation 7.8 would not be applicable. Assume that 500 of the dead trees were examined to determine the cause of death and it was found that 120 had been killed by the disease, 80 by the beetles, and the rest by a joint attack. The relative frequency of trees *bearing evidence of the disease* would be

$$RF\ (\text{disease}) = RF(E_1) = \frac{120 + 300}{500} = \frac{420}{500}$$

while the relative frequency of trees *showing evidence of beetle attack* would be

$$\text{RF (beetles)} = \text{RF}(E_2) = \frac{80+30}{500} = \frac{380}{500}$$

The sum of these two relative frequencies is

$$\text{RF}(E_1) + \text{RF}(E_2) = \frac{420}{500} + \frac{380}{500} = \frac{800}{500} \neq 1$$

This result was caused by the fact that the 300 trees that showed evidence of both disease and beetle attack were counted twice. Consequently, if the relative frequency of the trees killed by the joint attack is subtracted from the sum, the result will be unity.

$$\text{RF}(E_1) + \text{RF}(E_2) - \text{RF}(E_1 \textit{ and } E_2) = 1 \tag{7.9}$$

$$\frac{420}{500} + \frac{380}{500} - \frac{300}{500} = \frac{500}{500} = 1$$

When more than two events are involved, and they are not mutually exclusive, the situation becomes more complex. Assume that a third death-causing factor, fire, was added to the list of those events killing shortleaf pines. The situation then would be as shown in the Venn diagram in Figure 7.3. If one assumes that 500 trees were examined, the frequencies might be as follows:

| | |
|---|---|
| Littleleaf disease alone | 70 trees |
| Southern pine beetle alone | 50 |
| Fire alone | 10 |
| Littleleaf disease and southern pine beetle | 100 |
| Littleleaf disease and fire | 160 |
| Southern pine beetle and fire | 90 |
| All three together | 20 |

Using the reasoning behind Equation 7.9,

$\text{RF}(E_1) = \text{RF (disease)} = (70 + 100 + 160 + 20)/500 = 350/500$
$\text{RF}(E_2) = \text{RF (beetle)} = (50 + 100 + 90 + 20)/500 = 260/500$
$\text{RF}(E_3) = \text{RF (fire)} = (10 + 160 + 90 + 20)/500 = 280/500$

Because of the overlaps, the sum of these relative frequencies is 890/500, which is in excess of unity, violating the additive rule. To compensate for this surplus, the sum of the relative frequencies of the two-way joint events is subtracted from 890/500.

$\text{RF}(E_1 \text{ and } E_2) = \text{RF (disease and beetle)} = (100 + 20)/500 = 120/500$
$\text{RF}(E_1 \text{ and } E_3) = \text{RF (disease and fire)} = (160 + 20)/500 = 180/500$
$\text{RF}(E_2 \text{ and } E_3) = \text{RF (beetle and fire)} = (90 + 20)/500 = 110/500$

The sum of these joint relative frequencies is 410/500 which, when subtracted from 890/500, yields a deficit of 20/500, caused by the fact that the trees killed by all three agents were again counted more than once. This deficit is overcome by adding

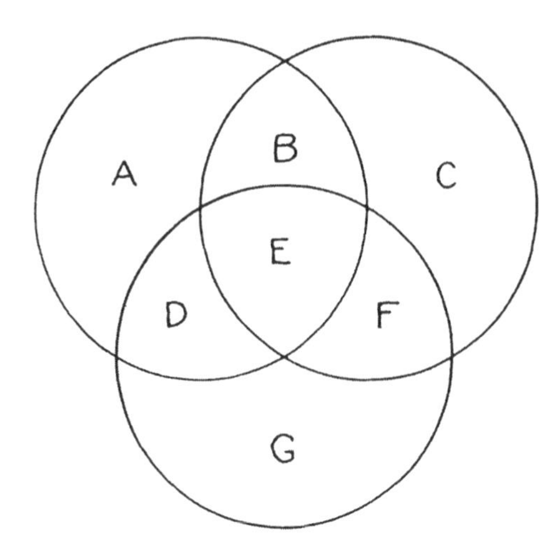


A Littleleaf disease alone
B Littleleaf disease and southern pine beetle
C Southern pine beetle alone
D Littleleaf disease and fire
E Littleleaf disease, southern pine beetle, and fire
F Southern pine beetle and fire
G Fire alone

**FIGURE 7.3** Venn diagram of the situation where some shortleaf pines were killed by the littleleaf disease, some by the southern pine beetle, some by fire, while the rest were killed by combined attacks by two or three of the lethal factors.

$$\text{RF}(E_1 \text{ and } E_2 \text{ and } E_3) = \text{RF (disease, beetle, and fire)} = 20/500$$

to the previous sum, which yields the grand sum of 500/500 = 1. In equation form

$$\frac{890}{500} - \frac{410}{500} + \frac{20}{500} = \frac{500}{500} = 1$$

Generalizing,

$$\begin{aligned}
&\sum_{i}^{M}[\text{RF}(E_i)] - \left\{\sum_{i=1}^{M-1}\sum_{j=i+1}^{M}[\text{RF}(E_i \text{ and } E_j)]\right\} \\
&\quad + \left\{\sum_{i=1}^{M-2}\sum_{j=i+1}^{M-1}\sum_{k=i+2}^{M}[\text{RF}(E_i \text{ and } E_j \text{ and } E_k)]\right\} \\
&\quad - \cdots + \left\{\sum_{i=1}^{M-(M-1)}\sum_{j=i+1}^{M-(M-2)}\sum_{k=i+2}^{M(M-3)}\cdots\sum_{\gamma=i+(M-1)}^{M}[\text{RF}(E_i \text{ and } E_j \text{ and } E_k \text{ and} \cdots E_\gamma)]\right\} = 1
\end{aligned} \tag{7.10}$$

where $M$ = number of different primary events.

Note that the signs preceding the terms in braces alternate, taking into account the pattern of surpluses and deficits which are generated by the process. Equation 7.10 expresses the *general additive property* of relative frequencies. It is considered general because it can be used whether or not the events are mutually exclusive. If the events are mutually exclusive, the joint terms are equal to zero and the general form reduces to the special form shown in Equation 7.8.

In some cases the occurrence of a number of *alternative* outcomes will satisfy the requirements of the event of interest. If the possible outcomes of an experiment are mutually exclusive, as in the case of the die, the relative frequency of the event of interest would be the sum of the relative frequencies of the individual outcomes that are acceptable.

$$\text{RF}(E_i \text{ or } E_j \text{ or } \ldots) = \text{RF}(E_i) + \text{RF}(E_j) + \cdots \tag{7.11}$$

This stems directly from the additive property of relative frequencies. An example of this can be drawn from the die-tossing problem used earlier. If the relative frequency of 2 and 3 *together* is desired:

$$\text{RF (2 or 3)} = (n(2)/n) + (n(3)/n) = 10/100 + 20/100 = 30/100$$

If the events are not mutually exclusive, the relative frequency of the combined events would be obtained using the appropriate portion of the general additive formula (Equation 7.10), such as

$$\text{RF}(E_i \text{ or } E_j) = \text{RF}(E_i) + \text{RF}(E_j) - \text{RF}(E_i \text{ and } E_j)$$

where $E_i$ and $E_j$ are different events out of a set of three or more possible events.

## 7.3 EMPIRICAL PROBABILITY

A probability is a numerical value expressing the likelihood that an event of interest will occur *in the future*. One way of arriving at such a value is to make a series of trials and then assume that the pattern of the past will be continued into the future. A classic example of this process is the determination of the probability of occurrence of a head when a coin is tossed. If a coin is *fair*, the probability of obtaining a head is exactly equal to that of obtaining a tail. Actual coins, however, are rarely, if ever, fair, so the question of probability of a head is not trivial.

The testing process involves tossing the coin a specified number of times, say, 100, and determining the relative frequency of the occurrence of a head. The process is then repeated and a new relative frequency computed. This relative frequency would be computed using the data from all 200 tosses. It is very likely that this second relative frequency will be different from that initially obtained. Then the coin would be tossed 100 more times and the relative frequency computed, this time on the basis of 300 tosses. Again, it is likely that the relative frequency would be different from the previous values. This process would be repeated, time after time, until the relative frequency values cease to change. This process is illustrated in Figure 7.4. When the relative frequency values cease changing, or change so little that the change is inconsequential, it is said that they have *stabilized* or have *converged on a limit*. In the case of the coin in Figure 7.4, the relative frequency stabilized at 0.46. This value is accepted as the best available estimate of the true probability of a head appearing with that specific coin. It is known as the *empirical* or *statistical* probability. It also is known as the *a posteriori* probability since it was determined *after* the trials were run.

It should be noted that, while the word *limit* is used to describe the stable relative frequency, the latter is not equivalent to a limit approached by a converging series. Instead, it is a *probabilistic*

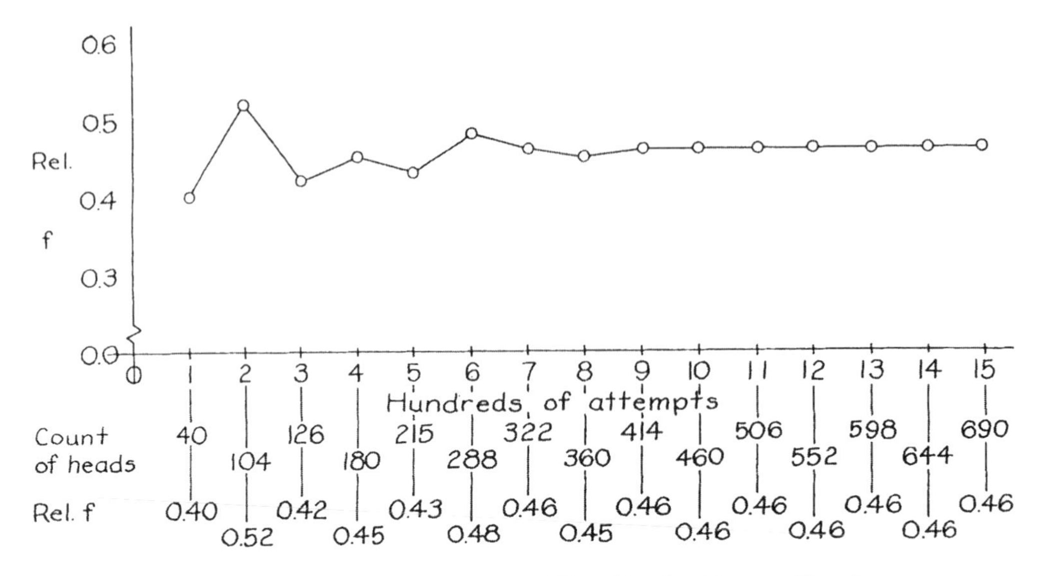


**FIGURE 7.4** The determination of the empirical probability of heads with a specific coin.

limit that can only be determined after a sufficiently large number of trials have been carried out. It is a consequence of the *law of large numbers,* which is discussed in Section 13.5.

From this description of the determination of an empirical probability, it can be seen that in actuality a probability is a limiting case of a relative frequency. The relative frequency is historical in nature. When it is used as a measure of likelihood, it becomes the probability of occurrence of a future event. Since an empirical probability is simply a special case of a relative frequency, it possesses the same properties as do all relative frequencies. The significance of this will become evident later.

## 7.4 CLASSICAL PROBABILITY

It was pointed out in the previous section that an empirical probability is the best available *estimate* of the *true* probability. True probabilities exist for all likelihood situations, but only under certain circumstances can they be determined with exactitude. A discussion of these circumstances will be returned after a discussion of some additional terms.

Closely related to the concept of the population (see Section 1.1) is that of the *sample space.* The sample space, $S$, is the *set of all possible events or outcomes* that could result from an experiment. Each item in this set is an *element of the sample space*, $s_i$.* Thus,

$$S = \{s_i\};\ i = 1, 2, \ldots, M$$

where $M$ is the total number of possible events or outcomes. Events or outcomes can be either *simple* or *compound* (see Section 7.1). Simple events can occur in only one way, whereas compound events can occur in more than one way. For example, the drawing of an ace of spades from a standard deck of playing cards is a simple event because it can only occur in one way — the drawing of the ace of spades. However, if the event of interest is the drawing of an ace, regardless of suit, the event would be compound since it can occur in any one of four ways. Sample spaces

* It should be noted that the term *element* was defined in Chapter 1 as being one of the objects making up the universe. Used in that sense with the puppy problem, an element is an individual puppy, not its sex. This confusion of meanings stems from the fact that the term *element* is used to identify a member of a universal set and both the universe and the sample space are universal sets.

can be developed in terms of either type of event. When the events are simple, the sample space is equivalent to the population and $M = N$, the size of the population. When the events are compound, however, the sample space cannot be considered the equivalent of the population since more than one element of the population would be associated with an element of the sample space. For example, consider a single die. The universe, $U$, is made up of the faces of the die.* The size of the universe, $N$, is equal to 6. If the variable of interest is the number appearing on the face, the population, $T$, is the set of numbers (variates):

$$T = \{1, 2, 3, 4, 5, 6\}$$

and the size of the population, $N$, is 6. The sample space corresponding to this population is made up of single events since this is only one way to obtain any one of the six numbers:

$$S = \{1, 2, 3, 4, 5, 6\}$$

and, thus, $M = N = 6$. If, on the other hand, the example is the litter of puppies, the universe is the set of puppies:

$$U = \{\text{Puppies } 1, 2, 3, 4, 5, 6, 7, 8\}$$

and $N = 8$. If the variable of interest is the sex of the puppies, the population would be

$$T = \{f, m, f, m, m, f, f, f\}$$

The sample space corresponding to this population would be made up of compound events since there is more than one way to draw a male or female:

$$S = \{f, m\}$$

In this case, $M = 2$.

In the case of the puppies, the event of interest could be the drawing of a specific puppy. In such a case the population and sample space would coincide:

$$S = \{\text{Puppy 1, Puppy 2, Puppy 3, Puppy 4, Puppy 5, Puppy 6, Puppy 7, Puppy 8}\}$$

Each puppy has a sex. Consequently, the simple events associated with sex are the sexes of the individual puppies.

$$\begin{aligned} S &= \{f_1, m_2, f_3, m_4, m_5, f_6, f_7, f_8\} \\ &= \{f_1, f_3, f_6, f_7, f_8\} \cup \{m_2, m_4, m_5\} \\ &= \{f \cup m\} \end{aligned}$$

Thus, regardless of whether the events are simple or compound, an element of a sample space is a subset of the sample space, and the sizes of the subsets depend on how the variable of interest is defined.

Events in populations and sample spaces are mutually exclusive, regardless of whether they are simple or compound. For example, consider the case of the shortleaf pine trees killed by littleleaf

* The terms *universe* and *universal set* should not be confused. Universal set refers to the complete set of whatever is under consideration. It can be a set of objects, which would make it synonymous with universe, but, on the other hand, it can be a set of variates, in which case it would be synonymous with *population* or *sample space*.

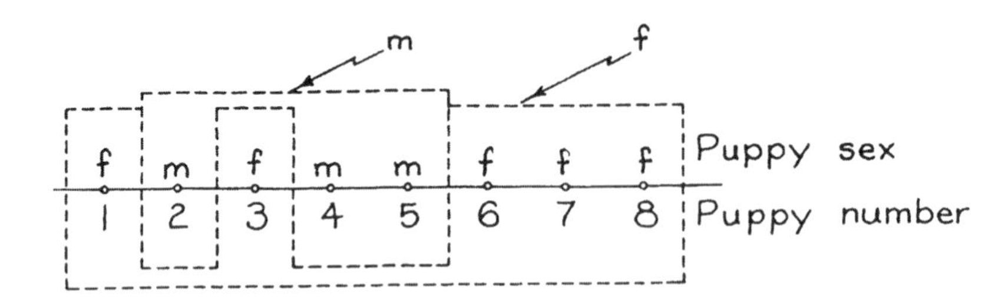


**FIGURE 7.5** Graphical representation of the sample space associated with the puppy experiment. The dashed lines delineate the two disjoint subsets, *m* and *f*, in the sample space.

disease, southern pine beetle, or both jointly, which was diagrammed in Figure 7.2. Events involving littleleaf and the beetle are not mutually exclusive since in some cases the two acted jointly. The sample space, however, contains three events which are mutually exclusive:

$$S = \{L\} \cup \{B\} \cup \{J\}$$

where $L$ = killed by littleleaf disease alone
$B$ = killed by southern pine beetle alone
$J$ = killed jointly

Thus, the sample space for any experiment consists of two or more disjoint subsets, each consisting of the simple events yielding a particular outcome:

$$S = \{s_1\} \cup \{s_2\} \cup \cdots \cup \{s_M\} = \bigcup_{j=1}^{M} s_j \tag{7.12}$$

where $s_j$ = subset $j$, corresponding to event $j$
$M$ = total number of events

In Figure 7.5 the presence of the two subsets, *m* and *f*, of the puppy experiment, is shown. The division of the sample space into its component disjoint subsets is referred to as the *partitioning* of the sample space and an individual subset is called a *partition*.

In this case the probability of choosing a male from the litter of puppies can be stated *exactly*. It is

$$P(E) = P(\text{male}) = 3/8$$

where $P(E)$ = probability of occurrence of the event of interest; *this symbology is appropriate regardless of the type of probability*

This probability is exact because (1) the sample size is finite and its size is known exactly and (2) the size of the subset associated with the event of interest is also known exactly. When these values are not known exactly, the probability can only be estimated, which is what is done when an empirical probability is determined. In the case of the coin being tested for fairness, both the sample space and the subset containing the events yielding heads were infinitely large and obtaining an exact probability would have been impossible.

In general terms, an exact probability is

$$P(E) = N(E)/N \tag{7.13}$$

where $N(E)$ = size of the subset of interest
$N$ = size of the sample space*

This is the *classical* definition of probability. It is obvious that it can be determined only when the sample space is *finite*. When a sample space is infinite or, if finite, its size is unknown, the definition cannot be used. Because of this, there has been a tendency in recent years to use the definition of empirical probability as the general definition and to consider the exact probability, as defined above, as a special case.

In any case, whether empirical or exact, a probability is a special case of a relative frequency and, consequently, must possess the same properties as do all relative frequencies. As a result, these properties, which were discussed in Section 7.2, are used as standards against which probability statements can be judged to see if they are valid. Because they are used in this manner, they are referred to as *axioms* or *laws* of probability.

The first law states that the probability of occurrence of a simple event cannot be negative nor can it exceed 1. In algebraic form, it is

$$0 \le P(E) \le 1 \tag{7.14}$$

The second law is known as the *general additive law*. It states that the sum of the probabilities of all the events that could appear as a result of an experiment is always equal to 1. In algebraic form it is

$$\begin{aligned} P(S) = {} & \sum_{i=1}^{M}[P(E_i)] - \left\{\sum_{i=1}^{M-1}\sum_{j=i+1}^{M}[P(E_i \text{ and } E_j)]\right\} \\ & + \left\{\sum_{i=1}^{M-2}\sum_{j=i+1}^{M-1}\sum_{k=i+2}^{M}[P(E_i \text{ and } E_j \text{ and } E_k)]\right\} \\ & - \cdots + \left\{\sum_{i=1}^{M-(M-1)}\sum_{j=i+1}^{M-(M-2)}\sum_{k=i+2}^{M-(M-3)}\cdots\sum_{\gamma=i+(M-1)}^{M}[P(E_i \text{ and } E_j \text{ and } E_k \text{ and} \cdots E_\gamma)]\right\} \\ = {} & 1 \end{aligned} \tag{7.15}$$

This is a *general* law and is appropriate whether the events are or are not mutually exclusive. In some cases some of the terms in Equation 7.15 that represent joint events may be equal to zero. This indicates that the subsets upon which they are based are empty. In other words, those joint events cannot occur. If *all* the subsets associated with joint events are empty, the primary events are mutually exclusive. When this occurs, the *general* law reduces to the *special* additive law, which is shown below in algebraic form:

$$P(S) = \sum_{i-1}^{M}[P(E_i)] = 1 \tag{7.16}$$

* The uppercase $N$ is used in these expressions because the entire sample space is involved. If only a part of the sample space is being used, as when an empirical probability is being determined, the appropriate symbol is lowercase $n$.

Special cases of the additive laws are as follows:

1. The sum of the probability of occurrence and the probability of nonoccurrence of the event of interest is, from the special additive law, equal to 1.

$$P(E) + P(\tilde{E}) = 1 \tag{7.17}$$

where $P(\tilde{E})$ = probability of nonoccurrence of the event of interest

$$= \frac{N - N(E)}{N} \tag{7.18}$$

2. The sum of the probabilities of two primary events when they are *not* mutually exclusive (as in the case shown in Figure 7.2) is, from the general additive law, equal to 1.

$$P(E_1) + P(E_2) - P(E_1 \text{ and } E_2) = 1 \tag{7.19}$$

where $P(E_1 \text{ and } E_2)$ = probability of occurrence of the joint event, $E_1$ *and* $E_2$

In addition to these laws, there is a law that is concerned with the multiplication of probabilities. It will be discussed in Section 7.6.2.

## 7.5 PROBABILITY OF ALTERNATIVE EVENTS

Situations occur where experiments have more than two possible outcomes and where two or more of these outcomes are acceptable as the event of interest. In other words, there are *alternative* events of interest. For example, assume that a box contains 50 marbles of which 10 are red, 14 are white, 7 are green, 12 are blue, and 7 are yellow. The sample space thus contains 50 items in five disjoint subsets. The probabilities of drawing from these subsets, in a single blind draw, are: $P(\text{red}) = 10/50$; $P(\text{white}) = 14/50$; $P(\text{green}) = 7/50$; $P(\text{blue}) = 12/50$; and $P(\text{yellow}) = 7/50$. If the event of interest is the drawing, while blindfolded, of *either* a red *or* a white marble, the sample space could be considered to contain only two, rather than five, disjoint subsets. The events in the first of these subsets would satisfy the requirements of the event of interest while the second subset would contain the remaining events. The model of the experiment would then be the same as Equation 7.17:

$$P(S) = P(E) + P(\tilde{E}) = 1$$

where $P(E)$ = probability of drawing a red or white marble
$P(\tilde{E})$ = probability of drawing a green, blue, or yellow marble
$= [P(\text{red}) + P(\text{white})] + [P(\text{green}) + P(\text{blue}) + P(\text{yellow})]$
$= 1$

$$P(S) = \left(\frac{10}{50} + \frac{14}{50}\right) + \left(\frac{7}{50} + \frac{12}{50} + \frac{7}{50}\right) = 1$$

$$= \frac{24}{50} + \frac{26}{50} = \frac{50}{50} = 1$$

Thus, when there are alternative mutually exclusive events that will satisfy the requirements of the event of interest, the probability of the occurrence of the event of interest is the *sum* of the probabilities of the individual alternative events:

$$P(E_i \text{ or } E_j \text{ or } \cdots) = P(E_i) + P(E_j) + \cdots \tag{7.20}$$

Notice that the key words in the statement describing the event of interest are "either … or." When these words occur, it means that alternative events are acceptable and the probability of occurrence of the event of interest is the sum of the probabilities of the individual alternative events.

When there are more than two possible outcomes of an experiment and those outcomes are not mutually exclusive, and alternative events will meet the requirements of the event of interest, it becomes necessary to use the general additive law (Equation 7.15) to model the experiment. Consider the situation, shown in Figure 7.3 and discussed in Section 7.2, where there are three different causes of mortality of shortleaf pines (littleleaf disease, southern pine beetle, and fire) which can act jointly as well as individually. Assume that it is necessary to determine the probability that littleleaf disease and/or fire caused the death of a blindly chosen tree. Further, assume that the relative frequencies used in the discussions in Section 7.2 are true probabilities. The model of the experiment would be

$$\begin{aligned} P(S) &= P(E) + P(\tilde{E}) = 1 \\ &= \{[P(E_1) + P(E_3)] - P(E_1 \text{ and } E_3)\} \\ &+ \{P(E_2) - [P(E_1 \text{ and } E_2) + P(E_2 \text{ and } E_3)] + P(E_1 \text{ and } E_2 \text{ and } E_3)\} = 1 \end{aligned}$$

where $E$ = littleleaf disease and/or fire
$E_1$ = littleleaf disease
$E_2$ = southern pine beetle
$E_3$ = fire

$$\begin{aligned} P(S) &= \left[\left(\frac{350}{500} + \frac{280}{500}\right) - \frac{180}{500}\right] + \left[\frac{260}{500} - \left(\frac{120}{500} + \frac{110}{500}\right) + \frac{20}{500}\right] = 1 \\ &= \frac{450}{500} + \frac{50}{500} = \frac{500}{500} = 1 \end{aligned}$$

and the probability of occurrence of the event of interest would be

$$P(E) = \frac{450}{500}$$

## 7.6 PROBABILITY OF MULTIPLE EVENTS

Up to this point, probability only in the context of a single action, such as making *one* choice of a puppy, *one* toss of a coin, or *one* choice of a marble has been considered. However, in the majority of cases where probability is involved, a number of separate actions are carried out and the desired probability is of an event of interest resulting from those multiple actions.

In Section 7.4 it was stated that the sample space was essentially equivalent to the population and that its size was the same as that of the population. The probability of occurrence of any given item in the sample space was equal to the reciprocal of the size of the sample space. When multiple draws are involved, it usually is assumed that a series of draws are being made from the sample space. In actuality, however, what is being done is the *drawing of a single item out of a new sample space. The concept of a single draw is basic.* The probability of the multiple event of interest is actually the reciprocal of the size of the new sample space. The composition and size of this new sample space is dependent on the method of sampling being used.

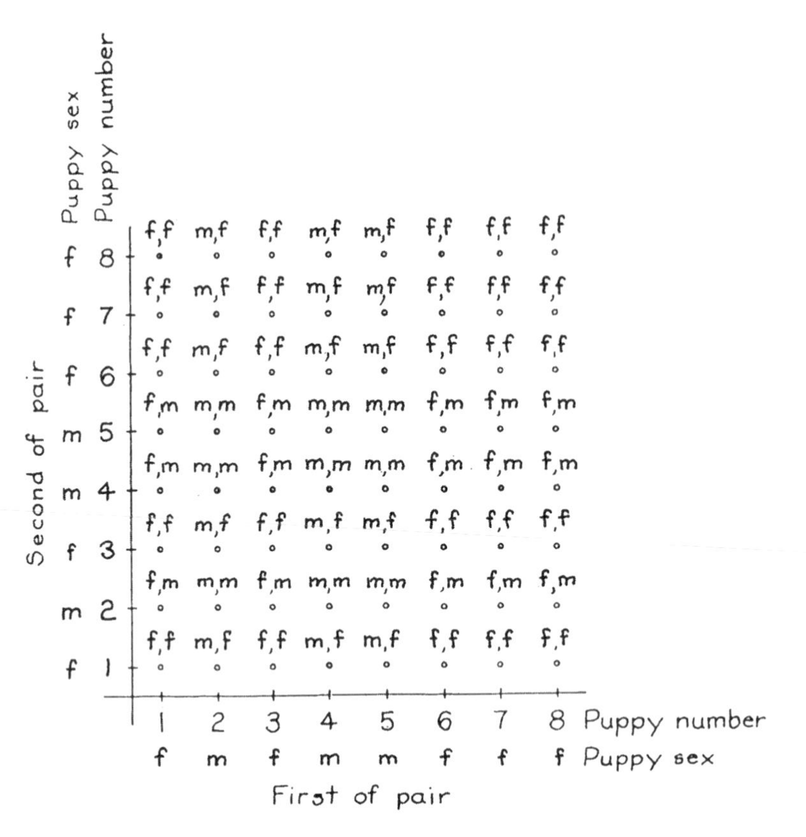


**FIGURE 7.6** Graphical representation of a two-dimensional sample space associated with the experiment in which two puppies are to be drawn from the litter of eight using sampling with replacement.

### 7.6.1 Sampling with and without Replacement

In the case of the original puppy problem, the sample space was as shown in Figure 7.5. It contained eight events, three yielding males and five yielding females. If the experiment was modified so that *two* puppies were to be chosen, the event of interest might be the occurrence of two males. If a puppy is chosen, its sex determined, and *then it is returned* to the box prior to the next draw, *sampling with replacement* is being used. Whenever multiple selections or draws are involved and each chosen item is returned or made available for the subsequent draw, the sampling is with replacement. If two puppies are to be chosen from the litter of eight, using sampling with replacement, the original eight-item sample space (Figure 7.5) is converted into the 64-item sample space shown in Figure 7.6. As can be seen from the figure, each event in this new sample space is a compound event. There is one such event for every possible outcome. Each puppy is paired with every other puppy and also with itself. Every pair has two permutations (arrangements) and both are present. In general form, the formula for the size of the sample space is

$$N_S = N_T^n \tag{7.21}$$

where $N_S$ = number of items or events in the new sample space
$N_T$ = number of items or events in the original sample space
$n$ = number of draws per experiment (i.e., *sample size*)

In the case of the puppy experiment:

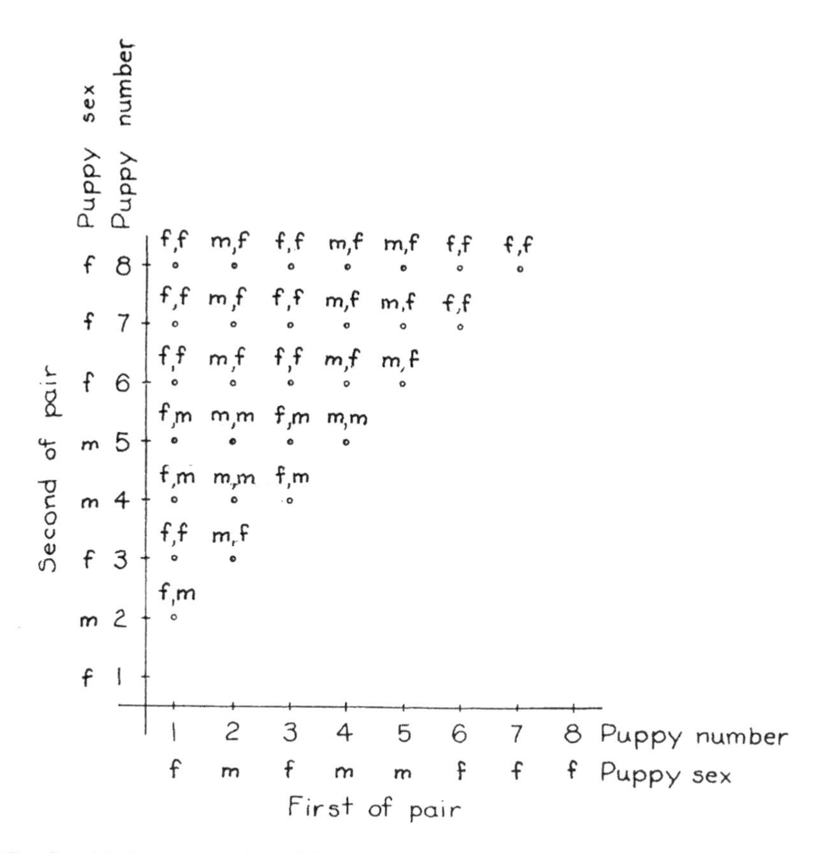


**FIGURE 7.7** Graphical representation of the sample space associated with the two-puppy experiment when the sampling was without replacement.

$$N_S = 8^2 = 64$$

The probability of drawing an individual item out of this new sample space is

$$P(e_i) = \frac{1}{N_S} = \frac{1}{64}$$

When an item is drawn and then is *not* made available for a subsequent draw, *sampling without replacement* is being used. If this was done in the puppy problem, the new sample space would be as shown in Figure 7.7. Since an item cannot be drawn more than once, it can only be paired once with each of the other items and it cannot be paired with itself. The size of this new sample space is*

$$N_S = \binom{N_T}{n} = \frac{N_T!}{n!(N_T - n)!} = \frac{8!}{2!(8-2)!} = 28 \tag{7.22}$$

* A *combination* is a subset, of a given size, of a set of items, each of which is identifiable in some manner. A primary problem in combinatorial analysis is the determination of the number of different combinations, or subsets, of a given size that can be obtained from the total set. In such problems the order of occurrence of the individuals within the subset is of no significance. The formula that will yield the number of combinations of size $r$ which can be obtained from a set of size $M$ is

$$\binom{M}{r} = \frac{M!}{r!(M-r)!}$$

and the probability of drawing an individual item from it is

$$P(s_i) = \frac{1}{N_S} = \frac{1}{28}$$

The effect of the method of sampling on the probability is obvious. When the sampling is without replacement, the probability of obtaining a specific event is increased substantially.

### 7.6.2 Probability in the Case of Independent Events

Assume a situation where a cage contains three geese and five ducks and a second cage contains six geese and four ducks. One bird is to be chosen, at random, from each of the two cages. What is the probability that both will be geese?

The probability of choosing a goose from the first cage is

$$P(E)_1 = P(\text{goose})_1 = \frac{3}{3+5} = \frac{3}{8}$$

while that of choosing a goose from the second cage is

$$P(E)_2 = P(\text{goose})_2 = \frac{6}{6+4} = \frac{6}{10}$$

It should be noted that, regardless of what occurs when the bird is chosen from the first cage, the probability of choosing a goose from the second cage remains unchanged. This means that the two actions and their corresponding results (events) are *independent* of one another. Independence is present whenever the carrying out of the process leading to one event has no effect on the probabilities associated with any of the other events involved in the problem.

In the case of the goose problem, the probability of choosing a goose out of each of the cages is the *product of the probabilities associated with the two draws.*

$$P(E)_0 = P(E)_1 * P(E)_2$$

$$P(2\text{ geese}) = P(\text{goose})_1 * P(\text{goose})_2$$

$$= \frac{3}{8} * \frac{6}{10} = \frac{18}{80}$$

The rationale behind this rule can be demonstrated as follows. Two cages of birds are two different universes, and consequently there are two different sample spaces.

$$S_1 = \{g_{11}, g_{12}, g_{13}\} \cup \{d_{11}, d_{12}, d_{13}, d_{14}, d_{15}\} = G_1 \cup D_1$$
$$S_2 = \{g_{21}, g_{22}, g_{23}, g_{24}, g_{25}, g_{26}\} \cup \{d_{21}, d_{22}, d_{23}, d_{24}\} = G_2 \cup D_2$$

where $g_{ij}$ = the $j$th goose in cage $i$
$d_{ij}$ = the $j$th duck in cage $i$

When a draw is to be made from $S_1$ and then another from $S_2$, it amounts to one draw from the composite sample space $S_c$, each element of which is a pair of birds, as is shown in Figure 7.8. Since each of the three geese in $S_1$ can be paired with any of the six geese in $S_2$, there are $3 * 6 =$

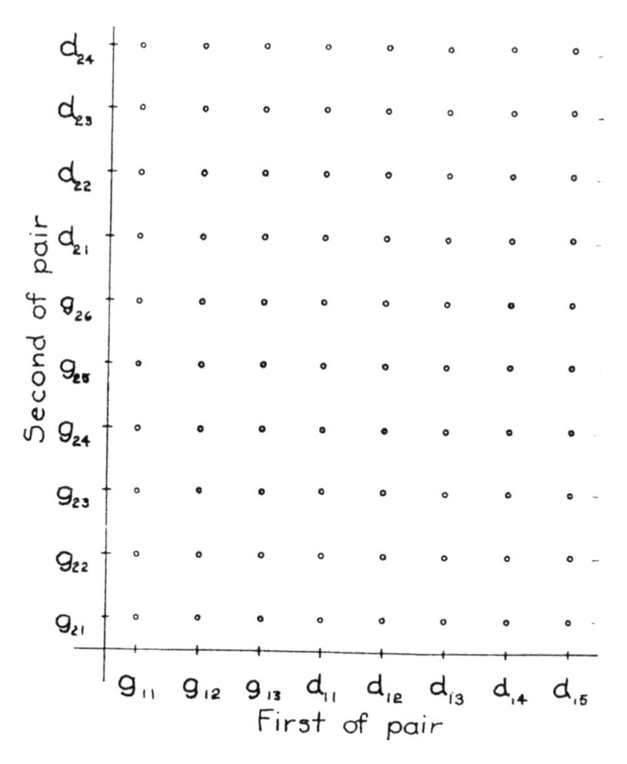


**FIGURE 7.8** Graphical representation of the composite sample space when one draw is to be made from a cage containing three geese and five ducks and the second from a cage containing six geese and four ducks.

18 possible pairs of geese. Likewise, each of the three geese in $S_1$ can be paired with any of the four ducks in $S_2$, which gives rise to 3 * 4 = 12 goose–duck pairs. In the same way, the six geese in $S_2$ can be paired with the five ducks in $S_1$, yielding 6 * 5 = 30 goose–duck pairs, and the four ducks in $S_2$ can be paired with the five ducks in $S_1$, yielding 4 * 5 = 20 pairs of ducks. Consequently, the composite sample space contains 18 + 12 + 30 + 20 = 80 pairs of birds. The probability of drawing any one of these pairs in a single blind draw out of the composite sample space is

$$P(e_i) = 1/80$$

Of these pairs, 18 are pairs of geese. Consequently, the probability of obtaining a pair of geese in a blind draw is

$$P(2 \text{ geese}) = N(E \text{ geese}) * P(s_i) = 18\ (1/80) = 18/80$$

or

$$= N(\text{ geese})/N_C = 18/80$$

which agrees with the previous result.

If more than two actions are involved, the grand product of the individual probabilities is the overall probability.

$$P(E)_0 = \prod_{i-1}^{M} P(E_i) \tag{7.23}$$

This expression describes the *multiplication law* of probability.

### 7.6.3 Dependent Events and Conditional Probability

Assume that a cage contains three swans, five geese, and eight ducks. Three birds are to be blindly chosen, using sampling *without* replacement. The event of interest is the drawing of a swan, a goose, and a duck, *in that order.*

The initial sample space contains three disjoint subsets:

$$S = W \cup G \cup D$$

where $W = \{w_1, w_2, w_3\}$, representing swans
$G = \{g_1, g_2, g_3, g_4, g_5\}$, representing geese
$D = \{d_1, d_2, d_3, d_4, d_5, d_6, d_7, d_8\}$, representing ducks

The sizes of the sample space and the subsets are

$$\begin{aligned} N_W &= 3 \\ N_G &= 5 \\ N_D &= 8 \\ N &= N_W + N_G + N_D = 3 + 5 + 8 = 16 \end{aligned}$$

The probabilities of selection for the three kinds of birds are

$$\begin{aligned} P(E_i) &= N_i/N \\ P(W) &= 3/16 \\ P(G) &= 5/16 \\ P(D) &= 8/16 \end{aligned}$$

Assume that the first draw is made and a swan is chosen, as would have to be the case if the requirements of the event of interest are to be met. The *composition of the sample space is now altered.* There are now only two swans, five geese, and eight ducks in the cage.

$$S' = W' \cup G' \cup D'$$

where $S'$ = sample space after the first draw
$W'$ = set of swans after the first draw
$G'$ = set of geese after the first draw
$D'$ = set of ducks after the first draw

and $N' = N'_W + N'_G + N'_D = 2 + 5 + 8 = 15$.

Consequently, the probabilities of selection in the second phase of the operation are not the same as they were in the beginning.

$$\begin{aligned} P(E_j|E_i) &= N'_j/N' \\ P(W|W) &= 2/15 \\ P(G|W) &= 5/15 \\ P(D|W) &= 8/15 \end{aligned}$$

This indicates that the probabilities following the draw are *dependent* on the results of that draw. As a consequence, these probabilities are known as *conditional* probabilities. The expression $P(W|W)$ refers to the probability of a swan being chosen after a swan has been chosen. In generalized form, the expression is $P(E_j|E_i)$. It represents the probability of the occurrence of $E_j$ after $E_i$ has occurred. Such a probability is the ratio of the size of the subset of interest, after the prior draw, *to the size of the altered or new sample space*. The symbology can be extended to situations where the event in question is to occur after any number of previous events. In such a case it would be

$$P(E_m|E_iE_jE_k \ldots)$$

Returning to the bird problem, assume that, as required, a goose was chosen in the second draw. Again the composition of the sample space was altered. It now contains two swans, four geese, and eight ducks.

$$S'' = W'' \cup G'' \cup D''$$

where $S''$ = sample space after the second draw
$W''$ = set of swans after the second draw
$G''$ = set of geese after the second draw
$D''$ = set of ducks after the second draw

and $N'' = N''_W + N''_G + N''_D = 2 + 4 + 8 = 14$.

The probabilities prior to the third draw are

$$P(E_k|E_iE_j) = N''_k/N''$$
$$P(W|WG) = 2/14$$
$$P(G|WG) = 4/14$$
$$P(D|WG) = 8/14$$

The overall probability of the selection of a swan, a goose, and a duck, in that order, is, according to the multiplication law:

$$P(E_iE_jE_k) = P(E_i) * P(E_j|E_i) * P(E_k|E_iE_j)$$

$$P(E_0) = (3/16) * (5/15) * (8/14) = 120/3360$$

Consider this situation using an approach similar to that used in the case of multiple independent events. In that case, a new composite sample space was constructed in which each element was a set of two simple events. Consequently, a single draw from the composite sample space was equivalent to a single draw from each of the two primary sample spaces. In the swan–goose–duck problem, the composite sample space consists of all possible *permutations** of the 16 birds.

$$N_c = {}_{16}P_3 = \frac{16!}{(16-3)!} = 16 * 15 * 14 = 3360$$

---

* A permutation is a specific arrangement of the items in a set. The number of permutations (arrangements) which can be obtained from a specific number of items ($n$), taken $r$ at a time, is

$${}_nP_r = \frac{n!}{(n-r)!}$$

It should be pointed out that none of the $r$ items in any of the subsets appears more than once and, consequently, no item appears more than once in any individual permutation.

This same result can be obtained using the *fundamental principle of counting* from combinatorial analysis.*

$$N_c = N * N' * N'' \\ = 16 * 15 * 14 = 3360$$

because there were $N = 16$ ways to make the first selection, $N' = 15$ ways to make the second, and $N'' = 14$ ways to make the third. Consequently, the probability of drawing any item out of the composite sample space, in a single blind draw, is

$$P(e_i) = 1/3360$$

Using the same procedure the number of ways by which the requirements of the event of interest can be satisfied is

$$s = s_W * s_G * s_D \\ = 3 * 5 * 8 = 120$$

These 120 ways of obtaining the event of interest form the subset of interest. This subset is shown in Table 7.1. The entire sample space cannot be shown because of a lack of space. Using these values, the probability of obtaining a swan, a goose, and a duck, in that order, is

$$P(E)_0 = P(E_iE_jE_k) = N(E) * P(s_i) = 120 * (1/3360) = 120/3360$$

or

$$P(E_iE_jE_k) = N(E)/N_c = 120/3360$$

which agrees with the probability previously obtained.

In the preceding discussion $P(E_j|E_i)$ and $P(E_k|E_iE_j)$ were described as conditional probabilities and the manner by which they are determined was described as

$$P(E_j|E_i) = N'_j/N'$$

and

$$P(E_k|E_iE_j) = N''_j/N''$$

In general, if the overall probability of a compound event, $P(E)_0$, is known, along with the probability of the prior event, the conditional probability can be computed using Equation 7.24:

$$\text{if} \quad P(E_i) * P(E_j|E_i) = P(E)_0 \\ P(E_j|E_i) = P(E)_0/P(E_i) \tag{7.24}$$

provided $P(E_i) > 0$. *This expression is the mathematical definition of conditional probability.*

---

* The fundamental principle of counting states that, if an event can happen in any one of $n_1$ ways, and if, when this has occurred, another event can happen in any one of $n_2$ ways, then the number of ways in which both events can happen, in the specified order, is $n_1 * n_2$. This rule can be expanded to any number of events.

**TABLE 7.1**
**The Permutations of Three Swans, Five Geese, and Eight Ducks That Would Satisfy the Requirement of a Swan (*S*), a Goose (*G*), and a Duck (*D*), in That Order**

| | | | | |
|---|---|---|---|---|
| $S_1G_1D_1$ | $S_1G_2D_1$ | $S_1G_3D_1$ | $S_1G_4D_1$ | $S_1G_5D_1$ |
| $S_1G_1D_2$ | $S_1G_2D_2$ | $S_1G_3D_2$ | $S_1G_4D_2$ | $S_1G_5D_2$ |
| $S_1G_1D_3$ | $S_1G_2D_3$ | $S_1G_3D_3$ | $S_1G_4D_3$ | $S_1G_5D_3$ |
| $S_1G_1D_4$ | $S_1G_2D_4$ | $S_1G_3D_4$ | $S_1G_4D_4$ | $S_1G_5D_4$ |
| $S_1G_1D_5$ | $S_1G_2D_5$ | $S_1G_3D_5$ | $S_1G_4D_5$ | $S_1G_5D_5$ |
| $S_1G_1D_6$ | $S_1G_2D_6$ | $S_1G_3D_6$ | $S_1G_4D_6$ | $S_1G_5D_6$ |
| $S_1G_1D_7$ | $S_1G_2D_7$ | $S_1G_3D_7$ | $S_1G_4D_7$ | $S_1G_5D_7$ |
| $S_1G_1D_8$ | $S_1G_2D_8$ | $S_1G_3D_8$ | $S_1G_4D_8$ | $S_1G_5D_8$ |
| $S_2G_1D_1$ | $S_2G_2D_1$ | $S_2G_3D_1$ | $S_2G_4D_1$ | $S_2G_5D_1$ |
| $S_2G_1D_2$ | $S_2G_2D_2$ | $S_2G_3D_2$ | $S_2G_4D_2$ | $S_2G_5D_2$ |
| $S_2G_1D_3$ | $S_2G_2D_3$ | $S_2G_3D_3$ | $S_2G_4D_3$ | $S_2G_5D_3$ |
| $S_2G_1D_4$ | $S_2G_2D_4$ | $S_2G_3D_4$ | $S_2G_4D_4$ | $S_2G_5D_4$ |
| $S_2G_1D_5$ | $S_2G_2D_5$ | $S_2G_3D_5$ | $S_2G_4D_5$ | $S_2G_5D_5$ |
| $S_2G_1D_6$ | $S_2G_2D_6$ | $S_2G_3D_6$ | $S_2G_4D_6$ | $S_2G_5D_6$ |
| $S_2G_1D_7$ | $S_2G_2D_7$ | $S_2G_3D_7$ | $S_2G_4D_7$ | $S_2G_5D_7$ |
| $S_2G_1D_8$ | $S_2G_2D_8$ | $S_2G_3D_8$ | $S_2G_4D_8$ | $S_2G_5D_8$ |
| $S_3G_1D_1$ | $S_3G_2D_1$ | $S_3G_3D_1$ | $S_3G_4D_1$ | $S_3G_5D_1$ |
| $S_3G_1D_2$ | $S_3G_2D_2$ | $S_3G_3D_2$ | $S_3G_4D_2$ | $S_3G_5D_2$ |
| $S_3G_1D_3$ | $S_3G_2D_3$ | $S_3G_3D_3$ | $S_3G_4D_3$ | $S_3G_5D_3$ |
| $S_3G_1D_4$ | $S_3G_2D_4$ | $S_3G_3D_4$ | $S_3G_4D_4$ | $S_3G_5D_4$ |
| $S_3G_1D_5$ | $S_3G_2D_5$ | $S_3G_3D_5$ | $S_3G_4D_5$ | $S_3G_5D_5$ |
| $S_3G_1D_6$ | $S_3G_2D_6$ | $S_3G_3D_6$ | $S_3G_4D_6$ | $S_3G_5D_6$ |
| $S_3G_1D_7$ | $S_3G_2D_7$ | $S_3G_3D_7$ | $S_3G_4D_7$ | $S_3G_5D_7$ |
| $S_3G_1D_8$ | $S_3G_2D_8$ | $S_3G_3D_8$ | $S_3G_4D_8$ | $S_3G_5D_8$ |

In the previous discussion the conditional probabilities were obtained for mutually exclusive events. When the events are not mutually exclusive, the basic concept of a conditional probability remains unchanged but the sample space situation has to be reconsidered. For example, assume that one of the three swans, one of the five geese, and two of the eight ducks are males and that the event of interest is the drawing of a male. It would be very simple in this case to determine the probability of drawing a male by disregarding the species factor and simply setting up a sample space containing two disjoint subsets, one containing the events yielding a male while the other contains the events yielding females.

$$P(E) = P(\text{male}) = N(M)/N = 4/16 = 1/4$$

However, the problem can be solved in another way, making use of conditional probability. Consider the Venn diagram in Figure 7.9. Each bird now has two attributes, a species and a sex. In a case such as this, a *single* draw will yield *two* attributes, a species and a sex. The probability of occurrence of one of these events, when the other event is specified, is a conditional probability.

$$P(E_j|E_i) = P(\text{male}|\text{swans})$$

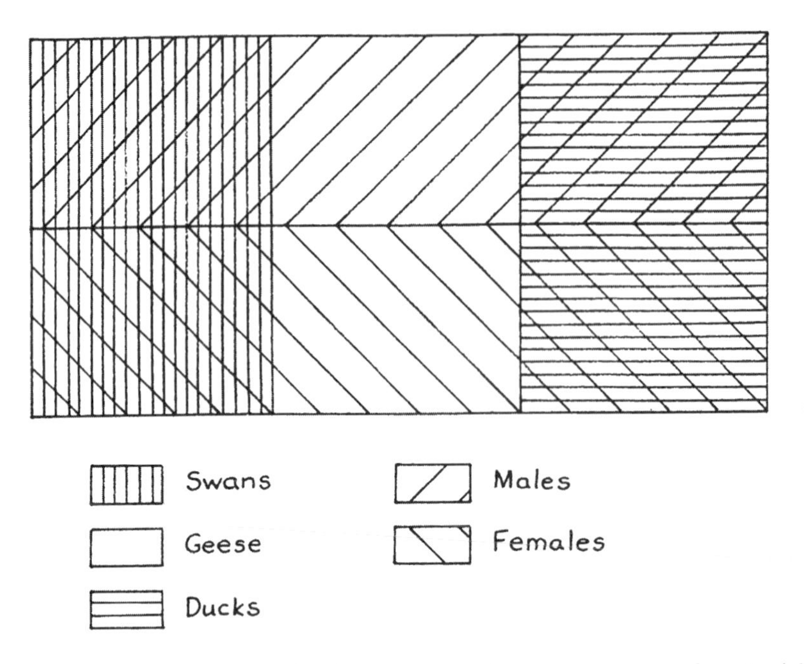


**FIGURE 7.9** Venn diagram showing the situation when a sample space is made up of events yielding swans, geese, and ducks upon which is superimposed the condition of sex.

In this case the sample space involved is the subset of the original sample space which contains *the events yielding a swan*. This subset contains three elements, $N_W = 3$. One of these three birds is a male. Consequently, the new sample space contains two disjoint subsets, one containing one event yielding a male while the other contains two events which yield females. Under these conditions the conditional probability of drawing a male swan is

$$P(M|W) = N_{MW}/N_W = 1/3$$

In the same manner the conditional probabilities of drawing a male goose or duck are

$$P(M|G) = N_{MG}/N_G = 1/5$$

and

$$P(M|D) = N_{MD}/N_D = 2/8 = 1/4$$

From the previous work

$$P(W) = 3/16, \quad P(G) = 5/16, \quad \text{and} \quad P(D) = 8/16 = 1/2$$

The product, $P(E_j|E_i) * P(E_i)$, as, for example,

$$P(M|W) * P(W) = (1/3) * (3/16) = 1/16$$

is the proportion of the original sample space included in the subset associated with the specified event, which, in this case, is the drawing of a male swan. When this is done for the other events involving a male, the results are

$$P(M|G) * P(G) = (1/15) * (5/16) = 1/16$$

and

$$P(M|D) * P(D) = (2/8) * (8/16) = 2/16 = 1/8$$

These proportions are associated with mutually exclusive events so they can be summed, using the special additive law, to arrive at the *total* probability of drawing a male when making a single blind draw from the cage of birds.

$$P(E) = P(E_m \mid E_i) * P(E_i) + P(E_m \mid E_j) * \{(E_j) + \cdots + P(E_m \mid E_n) * P(E_n)\}$$

$$P(M) = P(M \mid W) * P(W) + P(M \mid G) * P(G) + P(M \mid D) * P(D) \tag{7.25}$$

$$= (1/16) + (1/16) + (2/16) = 4/16 = 1/4$$

which agrees with the value originally computed.

A somewhat more realistic problem, which can be solved using this approach, might be where there are three suppliers of loblolly pine seedlings. All three obtain their seed from areas in which longleaf pine is present and, consequently, some cross-fertilization occurs forming Sonderegger pines. Let $B_1$ represent the first supplier, $B_2$ the second, and $B_3$ the third. $B_1$ supplies 20% of the needed seedlings, of which 1% is Sonderegger pines; $B_2$ supplies 30%, of which 2% is Sonderegger pines, and $B_3$ supplies 50%, of which 3% is Sonderegger pines. In this situation, what is the probability that a blindly chosen seedling will be Sonderegger pine?

$$P(B_1) = 20/100, \quad P(B_2) = 30/100, \quad P(B_3) = 50/100$$

$$P(E_S|B_1) = 1/100, \quad P(E_S|B_2) = 2/100, \quad P(E_S|B_3) = 3/100$$

where $E_S$ = the drawing of a Sonderegger pine.

$$P(E)_0 = P(E_S \mid B_1) * P(B_1) + P(E_S \mid B_2) * P(B_2) + P(E_S \mid B_3) * P(B_3)$$

$$= (1/100) * (20/100) + (2/100) * (30/100) + (3/100) * (50/100)$$

$$= 20/10000 + 60/10000 + 150/10000 = 230/10000 = 23/1000$$

The total probability of drawing a Sonderegger pine is $\{(E)_0 = 23/1000$, which also can be interpreted as meaning that out of each 1000 seedlings, 23 will be Sonderegger pines.

In the Sonderegger pine problem, the total size of the sample space was not specified but the proportions in the three disjoint subsets ($B_1$, $B_2$, and $B_3$) were given. Presumably these proportions were obtained by experimentation and were statistically stable. Consequently, the final result, the probability of drawing a Sonderegger pine, is a derived empirical probability. This, and the male bird problem, indicates that this procedure can be used with either exact or empirical probabilities.

### 7.6.4 Baysian Probability

By turning the Sonderegger pine problem around, what is the probability that a given seedling came from $B_1$ if the seedling is a Sonderegger pine?

The attempt is to find the proportion of the total sample space occupied by the intersection of the $B_1$ and the Sonderegger pine subsets. In Section 7.6.3 the proportion of the whole sample space that was occupied by the joint event "$B_1$, Sonderegger pine" was obtained by

$$P(E_S|B_1) * P(B_1) = 20/10000$$

The ratio of this to the total proportion *involving Sonderegger pine* is

$$P(B_1 \mid E_S) = \frac{P(E_S \mid B_1) * P(B_1)}{P(E_S \mid B_1) * P(B_1) + P(E_S \mid B_2) * P(B_2) + P(E_S \mid B_3) * P(B_3)}$$

This is the probability of drawing a seedling from $B_1$ from the set of Sonderegger pines.

$$P(B_1|E_S) = (20/10000)/(230/10000) = 2/23$$

In other words, there are two chances out of 23 that a Sonderegger pine came from $B_1$.

Generalizing, let $A$ represent events such as $E_S$ and let $B_1$ be the event of interest; then,

$$P(B_j \mid A) = \frac{P(A \mid B_j) * P(B_j)}{\sum_{i=1}^{M} [P(A \mid B_i) * P(B_i)]} \tag{7.26}$$

when $j = 1, 2, \ldots,$ or $M$.

This expression is called the *rule of Bayes* or *Bayes' theorem* and the probability $P(B_j|A)$ is the *Baysian* probability (see Roeder, 1979, and Swindel, 1972). It is applicable only when the $B$ events are mutually exclusive and one of them must occur.

As one can see from the previous discussion, a Baysian probability is simply the probability that the $B_j$ event will be drawn from the sample space defined by the presence of $A$. In some cases $A$ events are considered "effects" and $B$ events "causes." By using this terminology, the Baysian probability can be interpreted as the probability that a certain effect $A$ was caused by the specified $B$.

In the Sonderegger pine problem, the probabilities of the $B$ events (the suppliers) and the conditional probabilities of the occurrence of Sonderegger pines were known from the records and it was presumed that these values would continue to be applicable in the future. In many cases, however, information of this type is not available and in work done using Baysian probability recourse has had to be made to subjective assessments of the probabilities (see Section 7.7) and the results have been unsatisfactory. Consequently, Baysian probabilities have been used very little. However, some workers in the area of forest inventory (Issos et al., 1977; Ek and Issos, 1978; Burk and Ek, 1982; and Maxim and Huntington, 1983) have found that potentials exist for its use and it is probable that it will become a more commonly used statistical tool in the future.

### 7.6.5 Hypergeometric Probability

If the swan–goose–duck problem had not required the specific order that was specified, the compound probability would have been a *hypergeometric* probability. This type of probability is associated with multiple events when the sampling is without replacement and specific orders or arrangements of the items in the events are not required. In its simplest form the expression used to obtain the hypergeometric probability is

$$P(E)_0 = \frac{\binom{N_1}{C_1} * \binom{N_2}{n - C_1}}{\binom{N}{n}} \tag{7.27}$$

where $N$ = size of original sample space
$n$ = size of sample
$N_1$ = size of subset associated with characteristic No. 1
$N_2$ = size of subset associated with characteristic No. 2
$c_1$ = number of items with characteristic No. 1 required in event of interest
$c_2$ = number of items with characteristic No. 2 required in event of interest

This expression can be expanded to accommodate more than two characteristics, as can be seen when it is applied to the swan–goose–duck problem.

$$P(E)_0 = P(\text{swan goose duck})$$

$$= \binom{N_i}{C_i} * \binom{N_2}{C_2} * \left( n - \binom{N_3}{C_1} - C_2 \right) \Big/ \binom{N}{n}$$

$$= \binom{3}{1} * \binom{5}{1} * \left( 3 - \binom{8}{1} - 1 \right) \Big/ \binom{16}{3}$$

$$= \frac{3 * 5 * 8}{560} = \frac{120}{560}$$

In this case, so far as the requirements of the event of interest are concerned, the order of occurrence within the subsets is of no consequence. As a result, the sample space is much smaller than when all the different permutations (orders or arrangements of occurrence) have to be included. The sample space is shown in Figure 7.10. Note that it has 560 items and 120 of these meet the requirement that a swan, a goose, and a duck be chosen. These 120 items are inside the dashed lines within the sample space. It also should be noted that the probability of selection of any item in this set is

$$P(E_i) = 1/560$$

and the probability of the compound event is

$$P(E)_0 = N(E) * P(s_i) = 120 * (1/560) = 120/560$$

or

$$P(E)_0 = N(E)/N = 120/560$$

## 7.7 OBJECTIVE AND SUBJECTIVE PROBABILITIES

In the introduction to this chapter it was said that the subject matter area of probability is concerned with the quantification of likelihood. One can approach the problem in either one of two ways. In the first the probability value is based on numerical evidence and is determined using definite

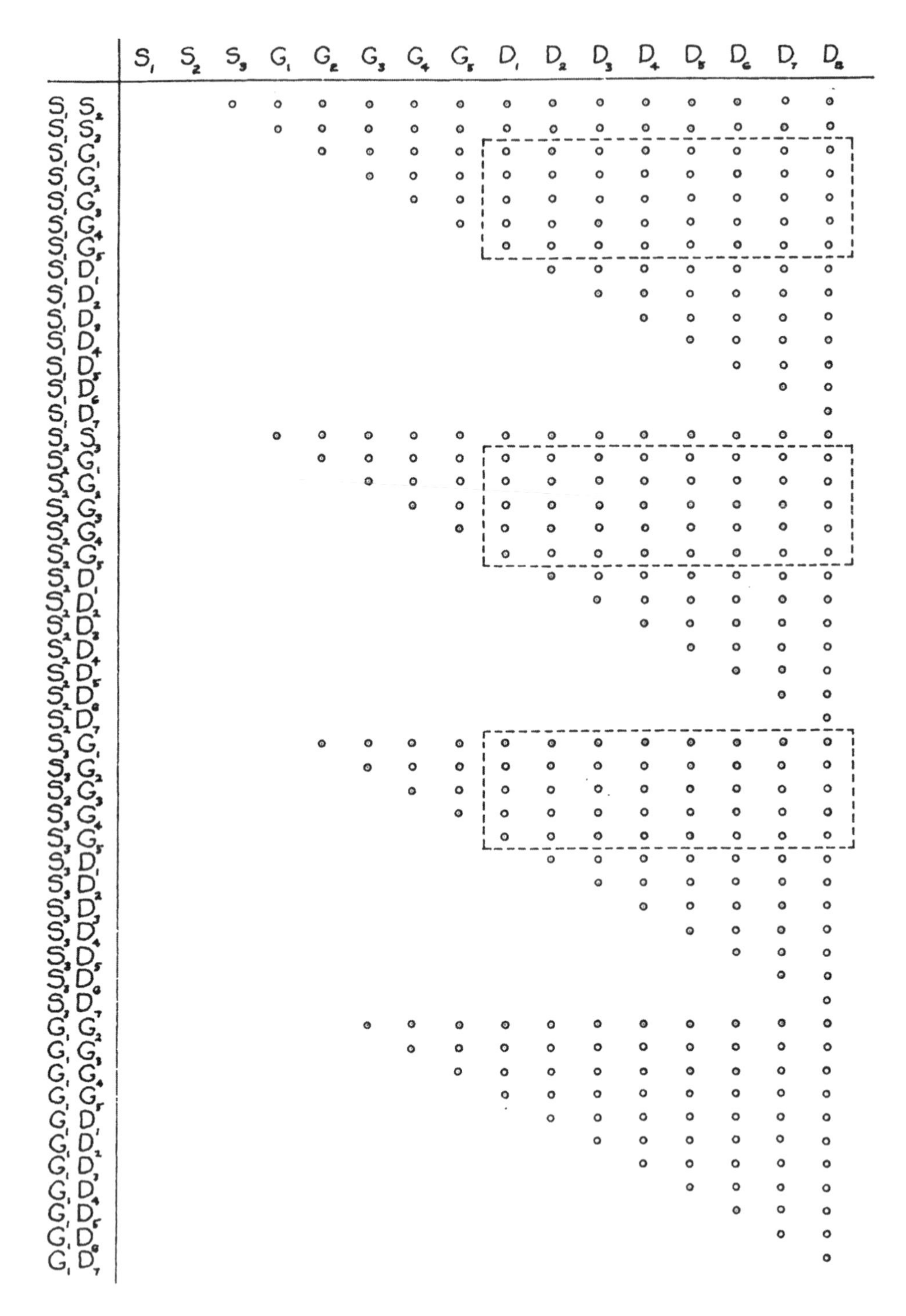

FIGURE 7.10 Graphical representation of the sample space associated with the swan–goose–duck problem when order of occurrence in the sample is not a factor.

| | $S_1$ | $S_2$ | $S_3$ | $G_1$ | $G_2$ | $G_3$ | $G_4$ | $G_5$ | $D_1$ | $D_2$ | $D_3$ | $D_4$ | $D_5$ | $D_6$ | $D_7$ | $D_8$ |
|---|---|---|---|---|---|---|---|---|---|---|---|---|---|---|---|---|
| $G_2 G_3$ | | | | | | | o | o | o | o | o | o | o | o | o | o |
| $G_2 G_4$ | | | | | | | | o | o | o | o | o | o | o | o | o |
| $G_2 G_5$ | | | | | | | | | o | o | o | o | o | o | o | o |
| $G_2 D_1$ | | | | | | | | | | o | o | o | o | o | o | o |
| $G_2 D_2$ | | | | | | | | | | | o | o | o | o | o | o |
| $G_2 D_3$ | | | | | | | | | | | | o | o | o | o | o |
| $G_2 D_4$ | | | | | | | | | | | | | o | o | o | o |
| $G_2 D_5$ | | | | | | | | | | | | | | o | o | o |
| $G_2 D_6$ | | | | | | | | | | | | | | | o | o |
| $G_2 D_7$ | | | | | | | | | | | | | | | | o |
| $G_3 G_4$ | | | | | | | | o | o | o | o | o | o | o | o | o |
| $G_3 G_5$ | | | | | | | | | o | o | o | o | o | o | o | o |
| $G_3 D_1$ | | | | | | | | | | o | o | o | o | o | o | o |
| $G_3 D_2$ | | | | | | | | | | | o | o | o | o | o | o |
| $G_3 D_3$ | | | | | | | | | | | | o | o | o | o | o |
| $G_3 D_4$ | | | | | | | | | | | | | o | o | o | o |
| $G_3 D_5$ | | | | | | | | | | | | | | o | o | o |
| $G_3 D_6$ | | | | | | | | | | | | | | | o | o |
| $G_3 D_7$ | | | | | | | | | | | | | | | | o |
| $G_4 G_5$ | | | | | | | | | o | o | o | o | o | o | o | o |
| $G_4 D_1$ | | | | | | | | | | o | o | o | o | o | o | o |
| $G_4 D_2$ | | | | | | | | | | | o | o | o | o | o | o |
| $G_4 D_3$ | | | | | | | | | | | | o | o | o | o | o |
| $G_4 D_4$ | | | | | | | | | | | | | o | o | o | o |
| $G_4 D_5$ | | | | | | | | | | | | | | o | o | o |
| $G_4 D_6$ | | | | | | | | | | | | | | | o | o |
| $G_4 D_7$ | | | | | | | | | | | | | | | | o |
| $G_5 D_1$ | | | | | | | | | | o | o | o | o | o | o | o |
| $G_5 D_2$ | | | | | | | | | | | o | o | o | o | o | o |
| $G_5 D_3$ | | | | | | | | | | | | o | o | o | o | o |
| $G_5 D_4$ | | | | | | | | | | | | | o | o | o | o |
| $G_5 D_5$ | | | | | | | | | | | | | | o | o | o |
| $G_5 D_6$ | | | | | | | | | | | | | | | o | o |
| $G_5 D_7$ | | | | | | | | | | | | | | | | o |
| $D_1 D_2$ | | | | | | | | | | | o | o | o | o | o | o |
| $D_1 D_3$ | | | | | | | | | | | | o | o | o | o | o |
| $D_1 D_4$ | | | | | | | | | | | | | o | o | o | o |
| $D_1 D_5$ | | | | | | | | | | | | | | o | o | o |
| $D_1 D_6$ | | | | | | | | | | | | | | | o | o |
| $D_1 D_7$ | | | | | | | | | | | | | | | | o |
| $D_2 D_3$ | | | | | | | | | | | | o | o | o | o | o |
| $D_2 D_4$ | | | | | | | | | | | | | o | o | o | o |
| $D_2 D_5$ | | | | | | | | | | | | | | o | o | o |
| $D_2 D_6$ | | | | | | | | | | | | | | | o | o |
| $D_2 D_7$ | | | | | | | | | | | | | | | | o |
| $D_3 D_4$ | | | | | | | | | | | | | o | o | o | o |
| $D_3 D_5$ | | | | | | | | | | | | | | o | o | o |
| $D_3 D_6$ | | | | | | | | | | | | | | | o | o |
| $D_3 D_7$ | | | | | | | | | | | | | | | | o |
| $D_4 D_5$ | | | | | | | | | | | | | | o | o | o |
| $D_4 D_6$ | | | | | | | | | | | | | | | o | o |
| $D_4 D_7$ | | | | | | | | | | | | | | | | o |
| $D_5 D_6$ | | | | | | | | | | | | | | | o | o |
| $D_5 D_7$ | | | | | | | | | | | | | | | | o |
| $D_6 D_7$ | | | | | | | | | | | | | | | | o |

**FIGURE 7.10** *Continued.*

mathematical processes. The discussions up to this point in this chapter have been concerned with probabilities of this type. When a probability is obtained in this manner any number of people, using the same information, will arrive at the same value. Human judgment, or lack thereof, is not a factor. Consequently, a probability obtained in this manner is considered to be an *objective* probability.

The alternative to the objective approach is to use personal knowledge, experience, and intuition to arrive at a value. In the case of the consultant-to-be in the introduction to this chapter, a person might make the statement, "In my opinion, the probability of his success as a general practitioner is 80% but as a specialist it is only 50%." They are *subjective* assessments of the situation and they may or may not be sound. The meanings of such statements are often unclear. In this case it could be assumed to mean that, in the observer's opinion, if the consultant-to-be were to make 100 attempts to set up as a general practitioner he or she would succeed 80 times, but if the person set up as a specialist in tree diseases he or she would succeed only 50 times.

Each person making an assessment of this type works from a unique experience background and consequently views the situation somewhat differently than would anyone else. This means that the process of assessment cannot be rigorously described. However, as was mentioned in the discussion of Baysian probability, in many cases this is the only way a probability value can be obtained because of time, money, or other constraints. As a result, it is not unusual to encounter subjectively derived probabilities. Persons using them should realize what they are and should interpret them accordingly.

# 8 Probability Distributions

## 8.1 RANDOM VARIABLES

In Chapter 1 it was stated that a *variable* was a characteristic of the elements of a specified universe which varied from element to element. If the original puppy experiment is viewed in terms of this statement, the litter is the universe, the individual puppies are the elements of this universe, and the variable of interest is the sex of the puppies. The sample space associated with the experiment is composed of a set of events yielding puppy sexes. These events are not numerical in nature. The same is true of the other experiments used as examples in Chapter 7. This does not mean that numerical events cannot occur. As a matter of fact, experiments involving numerical events are very common, and, in addition, since certain advantages are associated with the use of numerical events, nonnumerical events are often converted to numerical equivalents. For example, using the puppy experiment, each event yielding a male might be converted to the drawing of a "1" while each event yielding a female is converted to the drawing of a "0." When this is done, the sample space $S$ is converted to the *range space*, $R$, as in Table 8.1. As can be seen in Table 8.1, each event in $S$ has a corresponding value in $R$. These corresponding events are known as *equivalent events*.

Even when the events of an experiment involve numbers, it may be necessary to convert these numbers to new values which are more easily analyzed. For example, when a pair of fair dice are tossed, the fundamental event is the pair of faces facing upward when the dice come to rest. The original sample space $S$ is then made up of pairs of numbers. It is convenient to convert these pairs into single number values. If the numbers on the faces of each pair are summed, a single number is obtained. The set of these numbers is the range space R. This process is shown in Table 8.2.

In both these cases a *process* was used to arrive at the values in the range space. In the case of the puppies it involved the assignment of a number on the basis of the nonnumerical event:

$$y_i = \begin{cases} 1, & \text{if } s_i \text{ is a male} \\ 0, & \text{if } s_i \text{ is a female} \end{cases}$$

while in the dice problem

$$y_i = (s_1 + s_2)_i$$

where $s_1$ = number on face of first die
$s_2$ = number on face of second die

These processes are mathematical functions and can be generalized as follows:

$$y_i = Y(s_i) \tag{8.1}$$

**TABLE 8.1**
**The Conversion of the Sample Space Associated with the Puppy Experiment to the Equivalent Range Space**

| $S = \{s_i\}$ | $R(Y) = \{y_i\}$ |
|---|---|
| $s_1$ = F | $y_1 = 0$ |
| $s_2$ = M | $y_2 = 1$ |
| $s_3$ = F | $y_3 = 0$ |
| $s_4$ = M | $y_4 = 1$ |
| $s_5$ = M | $y_5 = 1$ |
| $s_6$ = F | $y_6 = 0$ |
| $s_7$ = F | $y_7 = 0$ |
| $s_8$ = F | $y_8 = 0$ |

The *function* $Y(s_i)$ is known as a *random variable*. It should be pointed out that even if no conversion to a numerical value is needed, it is assumed that the identity function, $y_i = Y(s_i) = s_i$, has been used and that the sample space has been converted to a range space. Furthermore, the usual practice, when numerical values are involved, is to ignore the original sample space $S$ and to sample directly from the range space.

Random variables receive their name from the fact that prior to a random draw from the sample space, one has no idea of the value that the random variable function will generate. The values obtained as repeated drawing is carried out will vary randomly as the events in the sample space are randomly chosen. It should be noted that an individual element in the range space, $y_i$, cannot vary. Its value is fixed by its equivalent event, $s_i$, in the sample space. Thus, one must distinguish between the random variable and the items in the range space.

The range space, like the sample space, can be partitioned into subsets containing like elements. In the case of the range space these elements are numbers. Thus, a given subset would contain all the events yielding a specific number. The number can be considered a subset label. *The set of such labels constitutes the values of the random variable.*

**TABLE 8.2**
**The Conversion of the Sample Space Associated with the Tossing of a Pair of Fair Dice to the Equivalent Range Space**

| $S = \{s_i\}$ | $R(Y) = \{y_i\}$ | $y_i$ | $N_i$ |
|---|---|---|---|
| 1,1 | 2 | 2 | 1 |
| 1,2: 2,1 | 3, 3 | 3 | 2 |
| 1,3: 2,2; 3,1 | 4, 4, 4 | 4 | 3 |
| 1,4; 2,3; 3,2; 4,1 | 5, 5, 5, 5 | 5 | 4 |
| 1,5; 2,4; 3,3; 4,2; 5,1 | 6, 6, 6, 6, 6 | 6 | 5 |
| 1,6; 2,5; 3,4; 4,3; 5,2; 6,1 | 7, 7, 7, 7, 7, 7 | 7 | 6 |
| 2,6; 3,5; 4,4; 5,3; 6,2 | 8, 8, 8, 8, 8 | 8 | 5 |
| 3,6; 4,5; 5,4; 6,3 | 9, 9, 9, 9 | 9 | 4 |
| 4,6; 5,5; 6,4 | 10, 10, 10 | 10 | 3 |
| 5,6; 6,5 | 11, 11 | 11 | 2 |
| 6,6 | 12 | 12 | 1 |

**TABLE 8.3**
**Probability Distribution Associated with the Puppy-Choosing Experiment**

| $i$ | 1 | 2 |
|---|---|---|
| $y_i$ | 0 | 1 |
| $N_i$ | 5 | 3 |
| $P(y_i)$ | $\frac{5}{8}$ | $\frac{3}{8}$ |

Random variables are usually symbolized by uppercase letters such as $X$, $Y$, or $Z$. Individual values of the random variable are symbolized by the lowercase letter corresponding to the symbol of the random variable. For example, as can be seen in Table 8.1, the random variable is symbolized by $Y$, an individual value would be symbolized by $y$ and an individual value in the range space is identified by subscript. This leads to some confusion since the *subsets* are identified also by the *same* lowercase letter and subscripts. For example, $y_i$ could refer to the $i$th item in the range space or to the $i$th subset. In the following discussions only rarely will the identifications of individual variates be needed and, consequently, the subscripting of the lowercase letters should be understood to refer to the subsets or partitions. If identification of the individual elements of the range space is needed, that fact will be brought out to reduce the confusion. Symbolically, the subscript $i$ will be reserved for individual variates and $i$ also will be used for the subsets. This is demonstrated in Table 8.2.

## 8.2 PROBABILITY DISTRIBUTIONS

An examination of $R$ in the puppy experiment (Table 8.1) will reveal that, like $S$, it consists of two disjoint subsets, one containing three 1s ($N(Y = 1)$ or ($N(1) = 3$) while the other contains five 0s ($N(Y = 0)$ or ($N(0) = 5$). The probability of drawing a 1 in a single blind draw from $R$ is then:

$$P(Y = 1) \quad \text{or} \quad P(1) = N(Y = 1)/N \quad \text{or} \quad N(1)/N = 3/8$$

while the probability of drawing a 0 is

$$P(Y = 0) \quad \text{or} \quad P(0) = N(Y = 0)/N \quad \text{or} \quad N(0)/N = 5/8$$

Table 8.3 shows these possible values of the random variable and their probabilities of occurrence. By using the same procedure, Table 8.4 was constructed for the problem involving the tossing of two fair dice. A table of this sort is a tabular representation of a *probability distribution*. In a

**TABLE 8.4**
**Probability Distribution Associated with the Tossing of a Pair of Fair Dice**

| $i$ | 1 | 2 | 3 | 4 | 5 | 6 | 7 | 8 | 9 | 10 | 11 |
|---|---|---|---|---|---|---|---|---|---|---|---|
| $y_i$ | 2 | 3 | 4 | 5 | 6 | 7 | 8 | 9 | 10 | 11 | 12 |
| $N_i$ | 1 | 2 | 3 | 4 | 5 | 6 | 5 | 4 | 3 | 2 | 1 |
| $P(y_i)$ | $\frac{1}{36}$ | $\frac{2}{36}$ | $\frac{3}{36}$ | $\frac{4}{36}$ | $\frac{5}{36}$ | $\frac{6}{36}$ | $\frac{5}{36}$ | $\frac{4}{36}$ | $\frac{3}{36}$ | $\frac{2}{36}$ | $\frac{1}{36}$ |

probability distribution each value of the random variable has associated with it the probability of it occurring in a single blind draw from its range space.

If a functional relationship can be established between the value of the random variable and the *size of the subset* of the range space associated with that value, the function is called a *frequency distribution function* or, simply, a *frequency function*. This is an extension of the meaning of these terms which were originally described, in the context of continuous frequency distributions, in Section 3.5.

$$N_i \text{ or } N(y_i) = \phi(y_i) \tag{8.2}$$

Since it is customary to assume that every range space contains an infinite number of cells, the usual method of presentation recognizes the existence of the empty cells:

$$N_i \text{ or } N(y_i) = \begin{cases} \phi(y_i), & \text{when } y_i = \ldots; \quad i = 1, 2, 3, \ldots M \\ 0, & \text{elsewhere} \end{cases} \tag{8.3}$$

where $M$ = total number of $Y$ values (this is *not* the same as the number of items in the sample or range spaces; the latter are symbolized by $N$)
$i$ = 1, corresponds to the lowest possible value of $Y$

Thus, in the case of the pair of fair dice $M$ = 11, $N$ = 36, and the frequency distribution is

$$N_i \text{ or } N(y_i) = \begin{cases} (y_i - 1), & \text{when } 2 \leq y_i 7 \\ (13 - y_i), & \text{when } 8 \leq y_i \leq 12 \\ 0, & \text{elsewhere} \end{cases}$$

The division of each $N_i$ value by the total count of events in the range space ($N$) converts the $N_i$ value to a probability.

$$P(y_i) = N_i / N = \begin{cases} \phi(y_i)/N, & \text{when } y_i = \ldots; \quad i = 1, 2, 3, \ldots M \\ 0, & \text{elsewhere} \end{cases} \tag{8.4}$$

This function is a *probability distribution function*, or, simply, a *distribution function*. The probability distribution function for the pair of fair dice is

$$P(y_i) = \begin{cases} (y_i - 1)/36, & \text{when } 2 \leq y_i 7 \\ (13 - y_i)/36, & \text{when } 8 \leq y_i \leq 12 \\ 0, & \text{elsewhere} \end{cases}$$

The relationship between the values generated by the two functions is evident in Figure 8.1.

According to the laws of probability, probability distribution functions must meet the following criteria:

$$0 \leq P(y_i) \leq 1 \tag{8.5}$$

for all values of Y, and

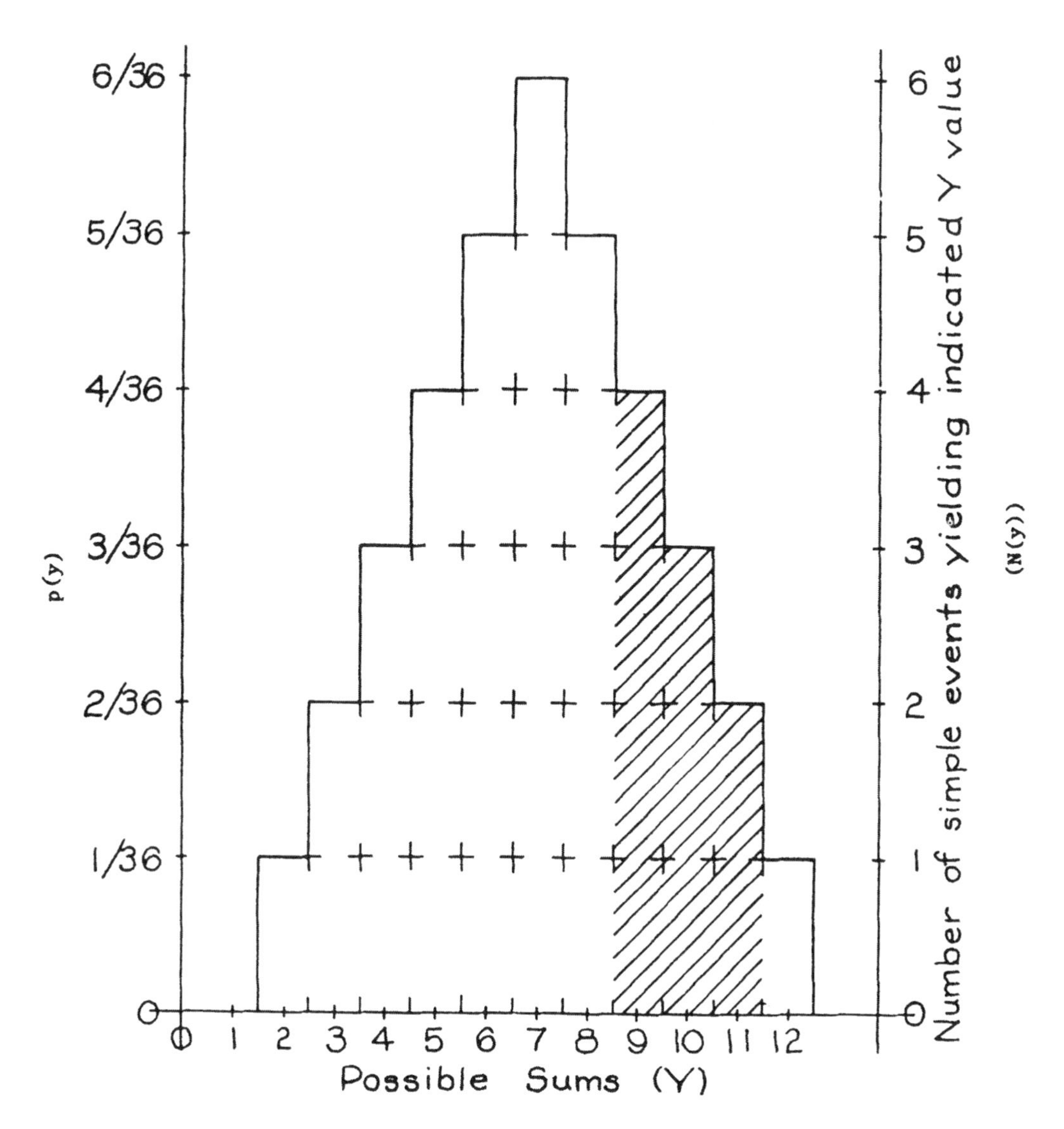


**FIGURE 8.1** Histogram of the discrete probability distribution associated with a pair of fair dice. Each bar has been divided into blocks representing individual events. The cross-hatching shows the portion of the area under the upper boundary of the histogram which is associated with the events 9, 10, and 11.

$$\sum_{i=1}^{M} P(y_i) = 1 \tag{8.6}$$

The probability distribution associated with the tossing of the pair of fair dice meets these requirements. As can be seen in Table 8.4 and Figure 8.1, no $Y$ value is negative nor does any value exceed 1. Furthermore, if the individual probabilities are summed:

$$P(Y = 2) + P(Y = 3) + \cdots + P(Y = 12)$$

$$\frac{1}{36} + \frac{2}{36} + \frac{3}{36} + \frac{4}{36} + \frac{5}{36} + \frac{6}{36} + \frac{5}{36} + \frac{4}{36} + \frac{3}{36} + \frac{2}{36} + \frac{1}{36} = \frac{36}{36} = 1$$

**TABLE 8.5**
**Empirical Relative Frequency Distribution Associated with the Tossing of a Single Die of Unknown Fairness**

| $i$ | 1 | 2 | 3 | 4 | 5 | 6 |
|---|---|---|---|---|---|---|
| $y_i$ | 1 | 2 | 3 | 4 | 5 | 5 |
| $RF(Y_i)$ | $\frac{n_1}{n}$ | $\frac{n_2}{n}$ | $\frac{n_3}{n}$ | $\frac{n_4}{n}$ | $\frac{n_5}{n}$ | $\frac{n_6}{n}$ |

Consequently, the second requirement is also met. It should be noted that each term in this expression represents the probability of occurrence of a specific value of the random variable, that the expression is linear, that all terms are positive, and the sum is equal to unity. This is characteristic of all probability distributions.

## 8.3 PROBABILITY DISTRIBUTION AND EMPIRICAL RELATIVE FREQUENCY DISTRIBUTIONS

In Section 3.4 a relative frequency distribution was described as the equivalent of a frequency distribution in which proportions of the total count of observations were used in place of the actual class counts or frequencies (Table 3.9 and Figures 3.1 and 3.2). When a situation arises involving probability and the probabilities cannot be deduced, as would be the case if the fairness of a die was not known, a series of trials can be carried out and observations made of the relative occurrence of the possible outcomes. The result is a relative frequency distribution in which the values of the random variable take the place of class midpoints. Such a distribution is known as an *empirical relative frequency distribution*. Table 8.5 is an example of such a distribution. It shows the relative frequency of occurrence of each face when a die has been tossed $n$ times.

$$RF(y = 1) = n_1/n;\ RF(y = 2) = n_2/n;\ RF(y = 3) = n_3/n;$$
$$RF(y = 4) = n_4/n;\ RF(y = 5) = n_5/n;\ RF(y = 6) = n_6/n$$

where $n = \sum_{i=1}^{6} n_i$ = total number of tosses

Since, by definition, a probability is a relative frequency, a probability distribution is a special case of a relative frequency distribution. It is special because it can be defined exactly, either from complete knowledge of the composition of the sample or range spaces, as in the case of the puppy experiment, or through the use of the probability distribution function. When the relative frequencies must be obtained through the use of repeated trials, as in the case of the die of unknown fairness, the resulting distribution is, as was explained earlier, an empirical relative frequency distribution — not a probability distribution. If the repeated trials had continued until stability of the relative frequencies had been obtained, the distribution could logically be called an *empirical probability distribution*, but such a term is not in use in the literature. Distributions based on incomplete knowledge can only be considered approximations of the true underlying probability distributions. Thus, *probability* distributions are associated with populations and, in a certain sense, are parameters, whereas empirical relative frequency distributions are associated with samples and could be considered statistics.

In real-life situations when sampling is used to obtain estimates of population parameters, it usually is *assumed* that the relative frequencies of the variates of the population of interest are distributed in a manner *similar to* some known probability distribution. Rarely is the actual probability distribution known. Consequently, persons carrying out such work must decide which known probability distribution most nearly matches what is known of the actual distribution. This decision is made using evidence such as empirical relative frequency distributions obtained from previous similar experiments or data obtained in the course of the current operation. When a probability distribution is used in this way, it is considered to be a *model* of the underlying true distribution. It may be a very good model, but it also may be very poor. In any case, it is being used as a representation of the true distribution because its characteristics are known and it is assumed that the unknown true distribution has similar characteristics. Needless to say, the matching of an appropriate probability distribution to a real-life situation, known as "choosing a model," is of critical importance.

## 8.4 DISCRETE PROBABILITY DISTRIBUTIONS

When the random variable can only assume a finite or countably infinite number of values, it is a discrete random variable and its probability distribution is referred to as a *discrete probability distribution*. The probability distributions associated with both the puppy and dice problems are of this type. In the case of discrete probability distributions, the probabilities associated with the individual values of the random variable are known as *point probabilities* and functions, such as Equation 8.4, when defined for $1 = 1, 2, \ldots M$, as *point probability functions*.

When a discrete probability distribution is shown as a *histogram*, with the adjacent bars touching, the result is a figure similar to that in Figure 8.1, which shows the point probabilities associated with the possible outcomes when a pair of fair dice are tossed. Note that the bar widths ($\Delta y$) are constant and that the bar heights ($N_i$) indicate the number of outcomes that yield the indicated sum ($y_i$). Consequently, each event in the range space is represented by a rectangle or block in a bar. Each of these blocks is $\Delta y$ units wide and one unit high. The probability of obtaining any given $y$ is the ratio of the count of blocks in the bar of that $Y$ value to the total count of blocks in all the bars:

$$P(y_i) = \frac{\text{Block count in bar } i}{\text{Total block count}} = N_i \Big/ \sum_{i=1}^{M} N_i \tag{8.7}$$

Since the blocks have the same area, the probability can be stated in terms of *relative area:*

$$P(y_i) = \frac{\Delta y * N_i}{\sum_{i=1}^{M} (\Delta y * N_i)} \tag{8.8}$$

For example, in the problem involving the two fair dice, the probability of obtaining a 9 is

$$P(Y = 9) = \frac{4}{36}$$

If more than one outcome is acceptable, as, for example, 9, 10, or 11, as shown in Figure 8.1, the probability of success is obtained using the special additive law:

$$P(9 \leq Y \leq 11) = \frac{\text{Area in bars 9, 10, and 11}}{\text{Total area}} = \frac{4+3+2}{36} = \frac{9}{36}$$

This expression can be generalized in the following manner:

$$P(y_a \leq Y \leq y_b) = \frac{\sum_{i=a}^{b} (\Delta y * N_i)}{\sum_{i=1}^{M} (\Delta y * N_i)} = \sum_{i=a}^{b} P(y_i) \tag{8.9}$$

where $y_a$ = smallest acceptable $Y$ value
$y_b$ = largest acceptable $Y$ value

Thus, in a properly drawn histogram, the *probability of the occurrence of an event, or a set of alternative events, can be obtained by using the relative area under the boundary of the histogram, associated with that (those) event(s).* This is of slight value insofar as discrete probability distributions are concerned, but the *concept* is of critical importance when probabilities are to be obtained from continuous probability distributions.

## 8.5 CONTINUOUS PROBABILITY DISTRIBUTIONS

When a random variable is *continuous,* it takes on *all* the infinite number of values lying between two specified points on the appropriate number line. Each of these values, as is the case when the variable is discrete, is a label for a subset or partition of the range space and, consequently, has a probability of occurrence. The distribution of these probabilities is a *continuous probability distribution.*

When a histogram is made of a continuous probability distribution, the bars of the histogram are infinitesimally narrow and the tops of the bars form a smoothly curving line, as in Figure 8.2. The algebraic expression that describes this line is the frequency distribution function (Equation 8.2):

$$N(y) = \phi(y)$$

Since $N(y)$ represents the size of a subset in the sample or range space, it *cannot be a negative quantity.* Consequently, *a distribution curve cannot dip below* the *horizontal axis in the range from the smallest possible value to the largest possible value of* Y. If portions of the curve generated by the function being used to describe the distribution lie below the horizontal axis (i.e., the $N$ values computed using the function become negative in certain regions), the usable portion of the curve is defined and the frequency function is set equal to zero for all other values of $Y$. Thus,

$$N(y) = \begin{cases} \phi(y_i), & \text{when } (y_A \leq Y \leq y_B) \\ 0, & \text{elsewhere} \end{cases} \tag{8.10}$$

where $y_A$ = smallest possible value of $Y$
$y_B$ = largest possible value of $Y$

In this manner the first law of probability is met.

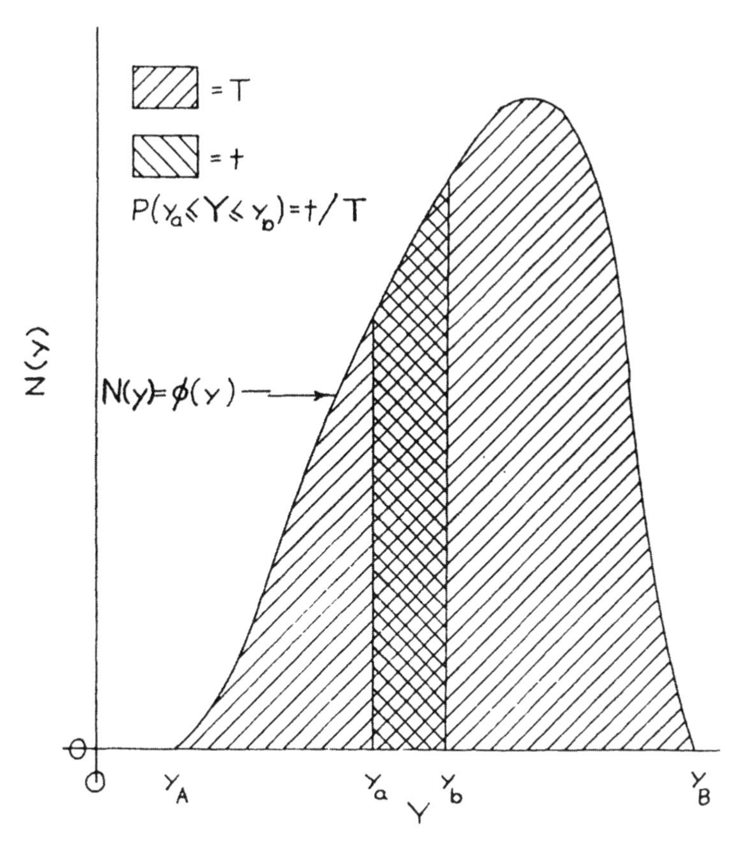


**FIGURE 8.2** Graph of the locus of a probability distribution function showing how the probability of occurrence of an event yielding a $Y$ value between, and including, $y_a$ and $y_b$ is determined.

The identification of an individual value of the random variable is, of course, impossible, but, conceptually, $N(y)$ represents the height in the histogram of the bar associated with that event. The size of the area under the curve associated with the event is

$$\text{Area} = N(y)\Delta y = \phi(y)\Delta y$$

where $\Delta y$ represents the infinitesimally narrow width of the bar. $N(y)\Delta y$ is, of course, essentially equal to zero. The sum of these infinitesimally small areas is

$$T = \int_{-\infty}^{\infty} N(y)dy = \int_{y_A}^{y_B} \phi(y)dY \tag{8.11}$$

where $T$ = total area under the curve from $y_A$ to $y_B$.

The probability of occurrence of the specified $Y$ value is then the ratio:

$$P(y) = N(y)\frac{\Delta y}{T} = \phi(y)\frac{\Delta y}{T}$$

This expression, *less the* $\Delta$y *term*, is the *probability density function* or *pdf:*

$$f(y) = \begin{cases} N(y)/T, & \text{when } y_A \le Y \le y_B \\ 0, & \text{elsewhere} \end{cases} \tag{8.12}$$

*Note that the pdf does not express a probability.* It simply is the frequency distribution function modified by the scalar $1/T$ so that the area under the curve from $y_A$ to $y_B$ is equal to unity, thus meeting the requirements of the special additive law:

$$P(-\infty < Y < +\infty) = \frac{1}{T}\int_{-\infty}^{\infty} N(y)dy = \frac{1}{T}\int_{y_A}^{y_B} \phi(y)dY = \frac{T}{T} = 1 = \int_{-\infty}^{\infty} f(y)dy = 1 \tag{8.13}$$

The area under the curve associated with the alternative events of interest (Figure 8.2) is

$$t = \int_{y_a}^{y_b} N(y)dy = \int_{y_a}^{y_b} \phi(y)dY$$

where $t$ = area under the curve when $Y_a \le Y \le Y_b$
$y_a$ = smallest acceptable value of $Y$
$y_b$ = largest acceptable value of $Y$

Following the reasoning used in Section 8.4, the ratio $t/T$ is the probability of occurrence of an event yielding a $Y$ value between, and including, $y_a$ and $y_b$:

$$P(y_a \le Y \le y_b) = \frac{t}{T} \tag{8.14}$$

This expression is the continuous variable equivalent to Equation 8.8. In terms of the pdf, the expression is

$$P(y_a \le Y \le y_b) = \int_{y_a}^{y_b} f(Y)dy \tag{8.15}$$

As can be seen, the use of the pdf simplifies the process of determining probabilities. Consequently, the pdf, rather than the frequency distribution function, is used to describe continuous probability distributions.

## 8.6 CUMULATIVE PROBABILITY DISTRIBUTIONS

If, in Equation 8.9, $y_a$ is set equal to $y_1 = y_A$, the smallest or lowest value of $Y$, and the probabilities are computed as $b$ is increased, in unit increments, from 1 to $M$, where $y_M = y_B$, the largest value of $Y$, the resulting set of probabilities constitute a *discrete cumulative probability distribution.*

$$P(y_1 \le Y \le y_i) = \frac{\sum_{i=1}^{b}(\Delta y * N_i)}{\sum_{i=1}^{M}(\Delta y * N_i)} = \sum_{i=1}^{b} P(y_i) \tag{8.16}$$

**TABLE 8.6**
**Discrete Cumulative Probability Distribution Associated with the Tossing of a Pair of Fair Dice**

| $i$ | 1 | 2 | 3 | 4 | 5 | 6 | 7 | 8 | 9 | 10 | 11 |
|---|---|---|---|---|---|---|---|---|---|---|---|
| $y_i$ | 2 | 3 | 4 | 5 | 6 | 7 | 8 | 9 | 10 | 11 | 12 |
| $y_i * P(y_i)$ | $\frac{1}{36}$ | $\frac{3}{36}$ | $\frac{6}{36}$ | $\frac{10}{36}$ | $\frac{15}{36}$ | $\frac{21}{36}$ | $\frac{26}{36}$ | $\frac{30}{36}$ | $\frac{33}{36}$ | $\frac{35}{36}$ | $\frac{36}{36}$ |

when $j = 1, 2, \ldots, M$ and $b = 1, 2, \ldots, M$. Equation 8.16 is a *cumulative probability distribution function* or simply a *cumulative distribution function,* which is abbreviated to *cdf.* Table 8.6 is an example of such a distribution. It shows the cumulative probabilities associated with the tossing of a pair of fair dice. It is equivalent to a "less than or equal to" cumulative relative frequency distribution as shown in Table 3.10 and Figure 3.5. The distribution is shown in graphical form in Figure 8.3. As can be seen from Table 8.6, the cumulative probabilities cannot exceed 1 and, since negative probabilities are not possible, the cumulative probabilities increase monotonically as the number of acceptable alternative $Y$ values increase.

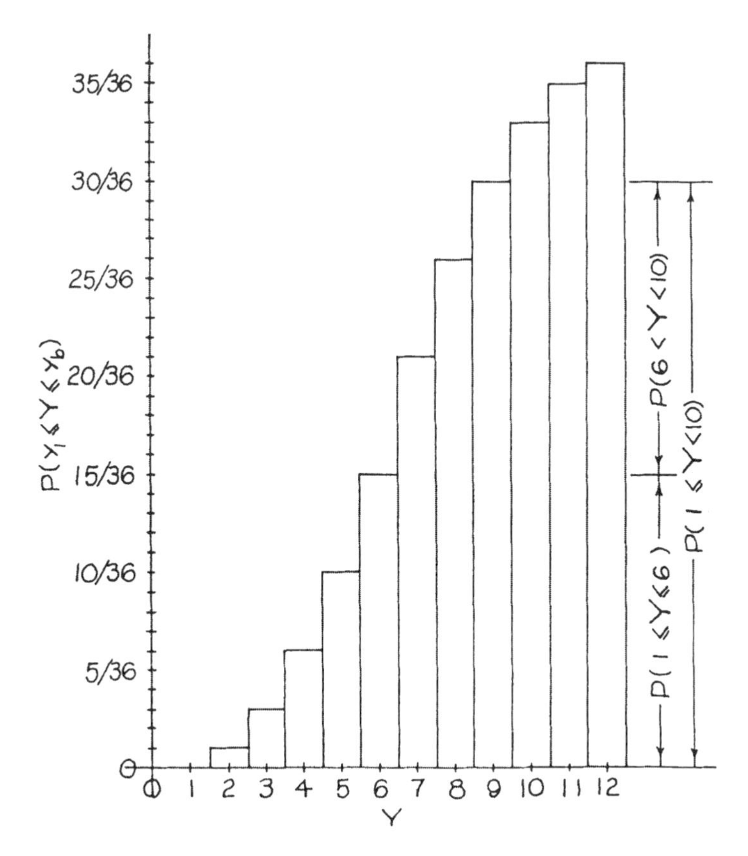


**FIGURE 8.3** Graphical representation of the discrete cumulative probability distribution associated with the tossing of a pair of fair dice.

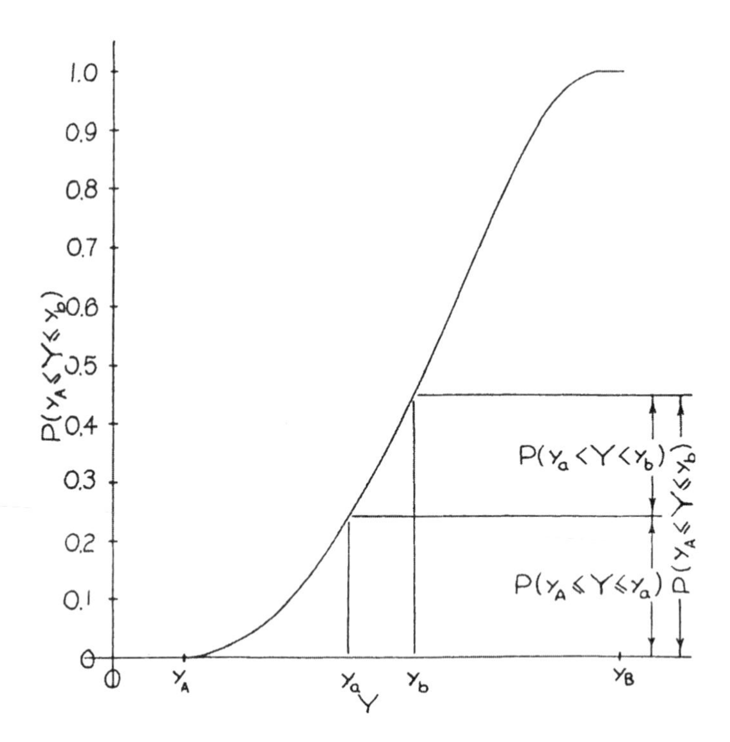


**FIGURE 8.4** Cumulative continuous probability distribution developed from Figure 8.2.

In much the same manner, but using Equation 8.17 and infinitesimal rather than unit increments, a *continuous cumulative probability distribution* can be developed.

$$F(Y) = P(y_A \leq Y \leq y_b) = \int_{y_A}^{y_b} f(Y)dy \tag{8.17}$$

Equation 8.17 is a cumulative probability distribution function, but functions of this type are usually referred to as *cumulative density functions* and are symbolized by cdf, as in the case of the discrete distribution. Figure 8.4 is the graphical representation of the cumulative probability distribution developed from the probability distribution shown in Figure 8.2.

The values labeled $y_b$ in the preceding discussion are the upper ends of the intervals ($y_A \leq Y \leq y_b$). Consequently, they identify fractions of the entire distribution. For this reason they are referred to as *fractiles*, *quantiles*, or percentage points. Certain of these have special meanings, as was brought out in Section 5.3.4. There is a probability associated with each fractile. Tables showing the cumulated probabilities by fractiles or the fractiles associated with specified levels of cumulated probability have been developed for many of the known distributions. Several of these tables are included in the Appendix and references will be made to them in the remainder of this book. The symbolism used to identify a specific fractile usually takes the following form:

$$y_b = I_C(\text{Pars}) \tag{8.18}$$

where $y_b$ = desired fractile
$I$ = a symbol identifying the distribution (e.g., $N$ for the normal, $X^2$ for the chi square, or $\beta$ for the beta distributions)
$C$ = the stated level of cumulated probability

Pars = the parameters controlling the characteristics of the distribution (e.g., in the case of the normal, the mean, $\mu$, and the variance, $\sigma^2$, or the standard deviation, $\sigma$)

The primary value of cumulative probability distributions is that through their use the process of computing probabilities can be simplified. The principles of this process are most easily demonstrated using a discrete variable, but they apply with equal force to continuous distributions. Using the cumulative probability distribution for the pair of fair dice (Table 8.6 and Figure 8.3), one can determine the probability of occurrence of a $Y$ value *between* 6 and 10 in the following manner:

$$P(6 < Y < 10) = P(1 \le Y < 10) - P(1 \le Y \le 6)$$
$$= \left(\frac{30}{36}\right) - \left(\frac{15}{36}\right) = \frac{15}{36} \tag{8.19}$$

If the end point values in the range of acceptable events are to be included, the problem is solved as follows:

$$P(6 \le Y \le 10) = P(1 \le Y < 10) - P(1 \le Y \le 6)$$
$$= \left(\frac{33}{36}\right) - \left(\frac{10}{36}\right) = \frac{23}{36} \tag{8.20}$$

The validity of these two answers can be checked by counting the blocks in the bars in Figure 8.1. In the case of Equation 8.19, the count of blocks up to, but *not* including, $Y = 10$, is 30, while the count up to *and* including $Y = 6$ is 15. Thus, the number of blocks between these two limits is 30 – 15 or 15 and the probability of obtaining a $Y$ in that range is 15/36.

Essentially the same process is used when the random variable is continuous (Figure 8.4).

$$P(y_a < Y < y_b) = P(y_A \le Y \le y_b) - P(y_A \le Y \le y_a) \tag{8.21}$$

Since the bar widths are infinitesimally narrow in a histogram of a continuous probability distribution, it makes no difference if the end point values are or are not assumed to be included.

## 8.7 THE MEAN OF A RANDOM VARIABLE

The arithmetic mean of the values of a random variable is a weighted mean (Section 5.2.3), where the weights are the probabilities of occurrence of the values of the random variable. When the variable is discrete,

$$\mu_Y = \frac{\sum_{i=1}^{M}[y_i * P(y_i)]}{\sum_{i=1}^{M} P(y_i)}$$

However, according to the special additive law,

$$\sum_{i=1}^{M} P(y_i) = 1$$

Consequently, the expression can be simplified to

$$\mu_Y = \sum_{i=1}^{M} [y_i * P(y_i)] \tag{8.22}$$

When the probability of occurrence is constant for all values of the random variable, Equation 8.22 reduces to the basic expression for an arithmetic mean. The probability of a specific $Y$ value being drawn, in such a case, is $1/M$ and Equation 8.22 becomes

$$\mu_Y = \sum_{i=1}^{M} \left[ y_i * \frac{1}{M} \right] = \frac{1}{M} \sum_{i=1}^{M} y_i \tag{8.23}$$

which is the conventional expression for the arithmetic mean. For example, consider the case of the single fair die. There are six faces to the die and, consequently, $M = 6$. The probability of any particular face appearing is $P(y_i) = 1/M = 1/6$. The mean is, therefore,

$$\mu_Y = \frac{1}{6} \sum_{i=1}^{6} y_i = \frac{1}{6} * 21 = 3.5$$

In the case of the pair of fair dice the probabilities are not constant. In such a case the computation process is as follows:

| ***i*** | = | **1** | **2** | **3** | **4** | **5** | **6** | **7** | **8** | **9** | **10** | **11** |
|---|---|---|---|---|---|---|---|---|---|---|---|---|
| $y_i$ | = | 2 | 3 | 4 | 5 | 6 | 7 | 8 | 9 | 10 | 11 | 12 |
| $P(y_i)$ | = | $\frac{1}{36}$ | $\frac{2}{36}$ | $\frac{3}{36}$ | $\frac{4}{36}$ | $\frac{5}{36}$ | $\frac{6}{36}$ | $\frac{5}{36}$ | $\frac{4}{36}$ | $\frac{3}{36}$ | $\frac{2}{36}$ | $\frac{1}{36}$ |
| $y_i * P(y_i)$ | = | $\frac{2}{36}$ | $\frac{6}{36}$ | $\frac{12}{36}$ | $\frac{20}{36}$ | $\frac{30}{36}$ | $\frac{42}{36}$ | $\frac{40}{36}$ | $\frac{36}{36}$ | $\frac{30}{36}$ | $\frac{22}{36}$ | $\frac{12}{36}$ |

$$\mu_Y = \sum_{i=1}^{11} [y_i * P(y_i)] = \frac{252}{36} = 7$$

When the random variable is continuous, the expression for the mean is

$$\mu_Y = \int_{-\infty}^{\infty} y * f(y)dy \tag{8.24}$$

For example, assume a random variable $0 \leq Y \leq 6$, with the probability density function $f(y) = y/18$.

$$\mu_Y = \int_0^6 y * \left( \frac{y}{18} \right) dy = \frac{1}{18} \int_0^6 y^2 dy = \frac{1}{18} * \frac{y^3}{3} \Bigg| \begin{matrix} 6 = \frac{6^3}{54} = \frac{216}{54} = 4 \\ 0 = \frac{0^3}{54} = \frac{0}{54} = \frac{0}{4} \end{matrix}$$

$$= 4$$

## 8.8 THE FUNCTION OF A RANDOM VARIABLE

When a variable, say, $W$, is obtained through a functional relationship with a random variable as, for example, $W = H(Y)$, that derived variable is a *function of the random variable*. Thus, if the random variable, $Y$, is discrete, its range space is

$$R(Y_i) = \{y_1 = Y(s_1),\ y_2 = Y(s_2),\ y_3 = (s_3),\ \ldots,\ y_N = Y(s_N)\}$$

and the range space of the function of $Y$ would be

$$R(w_i) = \{w_1 = g(y_1),\ w_2 = g(y_2),\ w_3 = g(y_3),\ \ldots,\ w_N = g(y_N)\}$$

The one-to-one relationship between the elements of the two range spaces is obvious. When the two range spaces are partitioned, there is a subset of $W$ values corresponding to each subset of $Y$ values and the sizes of these corresponding subsets are equal, provided that each value of $Y$ gives rise to a unique value of $W$. If this condition is met, the probability distribution of $W$ follows that of $Y$ with the probability of each $W$ value being that of its corresponding $Y$ value. If, on the other hand, individual values of $W$ can be associated with more than one $Y$ value, as when $W = \sin Y$, the probability distribution of $W$ will be different from that of $Y$. If $Y$ is discrete, the general procedure used to develop the probability distribution of $W$ is to sum the probabilities of occurrence of all the values of $Y$, which give rise to a given value of $W$. Thus,

$$P(W = w_i) = P(y_{i1}) + P(y_{i2}) + P(y_{i3}) + \cdots \tag{8.25}$$

where $y_{i1}$ = first value of $Y$ giving rise to $w$
$y_{i2}$ = second such value
$y_{i3}$ = third such value, etc.

This is done for each value of $W$. Each such sum is the point probability of a value of $W$, and the list of these probabilities forms the probability distribution function for $W$.

When $Y$ and $W$ are continuous variables, the pdf of $W$ is obtained through the use of the cumulative distribution function of $Y$. For example, if $W = 4Y^2$, $0 < Y < 10$, and the frequency function of $Y$ is

$$\phi(y) = \begin{cases} y^2, & 0 < Y < 10 \\ 0, & \text{elsewhere} \end{cases}$$

Then, the total area under the frequency distribution curve is

$$T = \int_{-\infty}^{\infty} \phi(y)dy = \int_0^{10} y^2 dy = \frac{y^3}{3}\bigg|_0^{10} = \frac{1000}{3}$$

and the pdf of $Y$ is

$$f(y) = \frac{1}{T}\phi(y) = \begin{cases} \dfrac{3y^2}{1000}, & 0 < Y < 10 \\ 0, & \text{elsewhere} \end{cases}$$

The cdf of $Y$ is

$$F(y) = P(Y \leq y) = \frac{3}{1000} T = \int_0^y y^2 dy = \frac{y^3}{1000}\Bigg|_0^y = \frac{y^3}{1000}$$

Converting this to the cdf of $W$ requires the substitution of $w$ equivalents of $y$ and $dy$. Since $W = 4Y^2$,

$$Y = \left(\frac{w}{4}\right)^{1/2} = \frac{w^{1/2}}{2}$$

and since $0 < Y < 10$, $0 < W < 400$. In addition, since

$$\frac{dy}{dw} = \frac{1}{2}\frac{d}{dw}(w^{1/2}) = \frac{w^{-1/2}}{4}, \quad dy = \left(\frac{w^{-1/2}}{4}\right)dw$$

Thus,

$$\begin{aligned} G(w) = P(W \leq w) &= \frac{3}{1000}\int_0^w \left(\frac{w^{1/2}}{2}\right)^2 \left(\frac{w^{-1/2}}{4}\right)dw \\ &= \frac{3}{16,000}\int_0^w w^{1/2} dw \\ &= \frac{3}{16,000} * w^{3/2} * \frac{2}{3}\Bigg|_0^w = \frac{w^{3/2}}{8000} \end{aligned}$$

The derivative of the cdf of a random variable is its pdf. Thus,

$$\frac{d}{dw}\left(\frac{w^{3/2}}{8000}\right) = \frac{1}{8000} * \frac{3}{2} * w^{1/2}$$

and

$$g(w) = \begin{cases} \dfrac{3w^{1/2}}{16,000}; & 0 < W < 400 \end{cases}$$

In the above case, $W$ was an *increasing* monotonic function of $Y$. In other words, as the values of $Y$ increased in magnitude, those of $W$ also increased. If $W$ had been a *decreasing* monotonic function of $Y$, as, for example, $W = -4Y^2$, the derivation of $f(w)$ would have proceeded as follows:

$$y = -\left(\frac{w}{4}\right)^{1/2} = -\frac{w^{1/2}}{2}$$

$$\frac{dy}{dw} = -\frac{1}{4}w^{-1/2}$$

and

$$dy = \left(-\frac{w^{-1/2}}{4}\right)dw$$

Then,

$$G(w) = P(W \le w) = \frac{3}{1000}\int_0^w \left(-\frac{w^{1/2}}{2}\right)^2 \left(-\frac{w^{-1/2}}{4}\right)dw$$

$$= -\frac{3}{1000} * \left(+\frac{1}{4}\right) * \left(-\frac{1}{4}\right) * \int_0^w w^{1/2}dw$$

$$= -\frac{3}{16,000} * w^{3/2} * \frac{2}{3}\bigg|_0^w = -\frac{w^{3/2}}{8000}$$

This cumulative probability is negative, which is not permitted by the first law of probability. The source of the negative sign is the differential

$$\left(-\frac{w^{-1/2}}{4}\right)dw$$

which replaces dy in the integral when $F(y)$ is converted to $F(w)$. The differential is introduced into the integral to convert the increments from being in terms of $Y$ to being in terms of $W$. Thus, this is a conversion of units process and the sign has no significance. Consequently, if one uses the *absolute value* of the differential, the problem is resolved.

This transformation of the pdf of a random variable to that of a function of the random variable can be generalized as

$$g(w) = f(y)\left|\frac{dy}{dw}\right| \tag{8.26}$$

By using the above example,

$$f(y) = 3\frac{y^2}{1000}, \quad 0 < Y < 10 \quad \text{and} \quad \left|\frac{dy}{dw}\right| = \frac{w^{-1/2}}{4}$$

Substituting,

$$g(w) = \frac{3y^2}{1000}\left(\frac{w^{-1/2}}{4}\right)$$

Substituting $w^{-1/2}/2$ for $y$,

$$g(w) = \frac{3\left(\frac{w^{1/2}}{2}\right)^2}{1000}\left(\frac{w^{-1/2}}{4}\right) = \begin{cases} \frac{3w^{1/2}}{16,000}, & 0 < W < 400 \\ 0, & \text{elsewhere} \end{cases}$$

The proof, assuming a monotonically increasing or decreasing function, is as follows:

$$G(w) = P(W \le w) = P[H(Y) \le w] = P[Y \le H^{-1}(w)] = F[H^{-1}(w)]$$

where $H^{-1}(w)$ is the inverse of the function $H(y)$. By differentiating with respect to w, using the chain rule,

$$\frac{dG(w)}{dw} = \frac{dG(w)}{dy}\frac{dy}{dw} = \frac{dF[H^{-1}(w)]}{dy}\frac{d}{d}$$

Since $y - H^{-1}(w)$, $F[H^{-1}(w)] = F(y)$, and

$$\frac{dG(w)}{dw} = \frac{dF(y)}{dy}\frac{dy}{dx}$$

$$= f(y)\frac{dy}{dx}$$

and since $dG(w)/dw = g(w)$,

$$g(w) = f(y)\left|\frac{dy}{dx}\right|$$

In the above cases the function of the random variable was monotonically increasing or decreasing. If this condition of monotonicity is not met, the above-described procedure for arriving at the pdf is not appropriate. For example, if $W = 4Y^2$, as before, but the range of values of $Y$ is $-10 < Y < 10$, the graph of the relationship would show an upward-opening parabola passing through the origin of the coordinate system. Consequently, a given value of $W$ could arise from two different values of $Y$. Then,

$$\phi(y) = \begin{cases} y^2, & -10 < Y < 10 \\ 0, & \text{elsewhere} \end{cases}$$

$$T = \int_{-10}^{10} y^2 dy = \frac{y^3}{3}\Bigg|_{-10}^{10} = \frac{2000}{3}$$

$$f(y) = \begin{cases} \dfrac{3y^2}{2000}, & -10 < Y < 10 \\ 0, & \text{elsewhere} \end{cases}$$

$$F(y) = \frac{3}{2000}\int_{-10}^{10} y^2 dy = \frac{y^3}{2000}\Bigg|_{-10}^{y} = \frac{y^3}{2000} + \frac{1}{2}$$

$$G(w) = P(W \le w) = P(4Y^2 \le w) = P\left(-\frac{\sqrt{w}}{2} \le Y \le \frac{\sqrt{w}}{2}\right)$$

$$= P\left(Y \le \frac{\sqrt{w}}{2}\right) - P\left(Y \le \frac{\sqrt{w}}{2}\right)$$

$$= F\left(\frac{\sqrt{w}}{2}\right) - F\left(-\frac{\sqrt{w}}{2}\right)$$

$$g(w) = \frac{dG(w)}{dw} = \frac{dF\left(\frac{\sqrt{w}}{2}\right)}{dw} - \frac{dF\left(-\frac{\sqrt{w}}{2}\right)}{dw}$$

$$= f\left(\frac{\sqrt{w}}{2}\right)\frac{dy}{dw} + f\left(-\frac{\sqrt{w}}{2}\right)\frac{dy}{dw}$$

Since

$$\frac{d\left(\frac{\sqrt{w}}{2}\right)}{dw} = \frac{1}{4\sqrt{w}}$$

and

$$\frac{d\left(-\frac{\sqrt{w}}{2}\right)}{dw} = -\frac{1}{4\sqrt{w}}$$

then

$$g(w) = \frac{f\left(\frac{\sqrt{w}}{2}\right)}{4\sqrt{w}} - \frac{f\left(-\frac{\sqrt{w}}{2}\right)}{4\sqrt{w}}$$

$$= \frac{1}{4w^{1/2}} f\left(\frac{\sqrt{w}}{2}\right) + f\left(-\frac{\sqrt{w}}{2}\right) \tag{8.27}$$

## 8.9 THE MEAN OF A FUNCTION OF A RANDOM VARIABLE

If $W$ is a function of the random variable $Y$, such that $W = g(Y)$, the probability distribution of $Y$ can be used to obtain the mean of $W$, because the probability of each $y$ value can be used as the weight of the $w$ value computed using the $y$ value. Consequently, if $Y$ is discrete,

$$\mu_W = \sum_{i=1}^{M} [W * P(y_i)] \tag{8.28}$$

For example, if

$$W = 3y^2 = 2$$

and the values of $Y$ are the values obtained when tossing a pair of fair dice (Table 8.2), the computation would be as follows:

| $i$ | = | **1** | **2** | **3** | **4** | **5** | **6** | **7** | **8** | **9** | **10** | **11** |
|---|---|---|---|---|---|---|---|---|---|---|---|---|
| $y_i$ | = | 2 | 3 | 4 | 5 | 6 | 7 | 8 | 9 | 10 | 11 | 12 |
| $P(y_i)$ | = | $\frac{1}{36}$ | $\frac{2}{36}$ | $\frac{3}{36}$ | $\frac{4}{36}$ | $\frac{5}{36}$ | $\frac{6}{36}$ | $\frac{5}{36}$ | $\frac{4}{36}$ | $\frac{3}{36}$ | $\frac{2}{36}$ | $\frac{1}{36}$ |
| $w_i$ | = | 14 | 29 | 50 | 77 | 110 | 149 | 194 | 245 | 302 | 365 | 434 |
| $w_i * P(y_i)$ | = | $\frac{14}{36}$ | $\frac{58}{36}$ | $\frac{150}{36}$ | $\frac{308}{36}$ | $\frac{550}{36}$ | $\frac{894}{36}$ | $\frac{970}{36}$ | $\frac{980}{36}$ | $\frac{906}{36}$ | $\frac{730}{36}$ | $\frac{434}{36}$ |

$$\mu_w = \frac{5994}{36} = 166.5$$

When the random variable, $Y$, is continuous, the expression for the mean of the set of $W$ values is

$$\mu_W = \int_{-\infty}^{\infty} [W * f(y)]dy \tag{8.29}$$

For example, if the functional relationship between the derived variable $W$ and the predicting variable $Y$ is

$$W = 3y + 5, \quad 5 \leq Y \leq 10, \quad \text{and} \quad f(y) = \frac{y}{18}$$

when $5 \leq Y \leq 10$, and 0 elsewhere,

$$\mu_W = \int_5^{10} \left[ (3y+5)\left(\frac{y}{18}\right) \right] dy$$

$$= \frac{1}{18} \int_5^{10} (3y^2 + 5y)dy$$

$$= \frac{1}{18} \left[ \int_5^{10} 3y^2 dy + \int_5^{10} 5ydy \right]$$

$$= \frac{1}{18} \left[ \left( \frac{3Y^3}{3} \middle| \begin{matrix} 10 = 10^3 = 1000 \\ 5 = 5^3 = \frac{125}{875} \end{matrix} \right) + \left( \frac{5Y^2}{2} \middle| \begin{matrix} 10 = 5(10)^2 = 250 \\ 5 = 5(5)^2 = \frac{62.5}{187.5} \end{matrix} \right) \right]$$

$$= \frac{1}{18}(875 + 187.5) = 59.03$$

## 8.10 EXPECTATION

The mean of a probability distribution is usually referred to as the *expected value* of the random variable. Thus,

$$E(Y) = \mu_Y$$

where $E(Y)$ = expected value of $Y$.

This terminology arises from the idea that, if a series of blind draws are made from a range space, the average of the values drawn would tend to be close to the mean of the entire distribution. In other words, *prior to the start of the drawing process*, the *anticipated* mean of the drawn values is the mean of the distribution. The term *expected value* is never used as a substitute for the term *arithmetic mean* in the case of samples or sets of numbers that are not variates from a probability distribution. It should be pointed out, however, that when a set of values have been generated through a function of a random variable, as was done in Section 8.8, that set is distributed as was the random variable and the mean of that distribution is an expected value. Thus, where $W$ is a function of $Y$, and $Y$ is a discrete variable,

$$E(W) = \sum_{i=1}^{\infty} [w_i * P(y_i)] \tag{8.30}$$

and when the variable is continuous,

$$E(W) = \int_{-\infty}^{\infty} W * f(y)dy \tag{8.31}$$

Expected values have a number of properties which should be known by persons engaged in sampling operations. However, since the concept of the expected value is applicable in the case of multidimensional random variables, which are discussed in Chapter 9, the listing and explanation of the properties will be delayed until Section 9.6.

## 8.11 THE VARIANCE OF A RANDOM VARIABLE

By definition, the variance of a set of numbers is the arithmetic mean of the squared differences between the individual numbers in the set and the arithmetic mean of the set. In the case of a random variable, $Y$, its range space forms a population and an individual difference is an *error*, which is symbolized by $\varepsilon$. Thus,

$$\varepsilon_i = y_i - \mu_Y = y_i - E(Y)$$

Obviously, $\varepsilon_i$ and $\varepsilon_i^2$ are functions of $Y$. Consequently, the variance of $Y$ is, according to the reasoning in Section 8.10,

$$\sigma_Y^2 = E(\varepsilon^2) = V(Y) = \sum_{i=1}^{\infty} [\varepsilon_i^2 P(y_i)] \tag{8.32}$$

if the random variable is discrete, or

$$\sigma_Y^2 = \int_{-\infty}^{\infty} \varepsilon^2 f(y)dy \tag{8.33}$$

if the random variable is continuous. It should be noted that $V(Y)$ is currently the most commonly encountered symbol for the variance of a random variable.

In the case of two fair dice, the sum of the values appearing on the two dice constitutes a discrete random variable. This variable has a mean of 7.0 and its variance is computed as follows:

| i | = | 1 | 2 | 3 | 4 | 5 | 6 | 7 | 8 | 9 | 10 | 11 |
|---|---|---|---|---|---|---|---|---|---|---|---|---|
| $y_i$ | = | 2 | 3 | 4 | 5 | 6 | 7 | 8 | 9 | 10 | 11 | 12 |
| $N_i$ | = | 1 | 2 | 3 | 4 | 5 | 6 | 5 | 4 | 3 | 2 | 1 |
| $\varepsilon_i$ | = | –5 | –4 | –3 | –2 | –1 | 0 | 1 | 2 | 3 | 4 | 5 |
| $\varepsilon_i^2$ | = | 25 | 16 | 9 | 4 | 1 | 0 | 1 | 4 | 9 | 16 | 25 |
| $P(y_i)$ | = | $\frac{1}{36}$ | $\frac{2}{36}$ | $\frac{3}{36}$ | $\frac{4}{36}$ | $\frac{5}{36}$ | $\frac{6}{36}$ | $\frac{5}{36}$ | $\frac{4}{36}$ | $\frac{3}{36}$ | $\frac{2}{36}$ | $\frac{1}{36}$ |

$$\sum_{i=1}^{11}[\varepsilon_i^2 P(y_i)] = \frac{[25+32+27+16+5+0+5+16+27+32+25]}{36} = \frac{210}{36}$$

$$V(Y) = \frac{210}{36} = 5.83$$

As an example of a situation when the random variable is continuous, assume the random variable $Y$ which can take on the values $0 \leq Y \leq 6$. Further assume that $\mu_Y = 4$ and that $f(y) = Y/18$; then,

$$V(Y) = \int_0^6 \left[(y-4)^2\left(\frac{y}{18}\right)\right]dy$$

$$= \frac{1}{18}\int_0^6 [y^2 - 8y + 16)y]dy$$

$$= \frac{1}{18}\int_0^6 (y^3 - 8y^2 + 16y)dy$$

$$= \frac{1}{18}\left[\int_0^6 y^3 dy - 8\int_0^6 y^2 dy + 16\int_0^6 ydy\right]$$

$$= \frac{1}{18}[324 - 8(72) + 16(18) = \frac{36}{18} = 2.0$$

If $Y$ is discrete and its range space is finite with size $N$, and the probability is constant across all values of $Y$ so that $P(y) = 1/N$, Equation 8.33 reduces to

$$V(y) = \sum_{i=1}^{N}\left[\varepsilon_i^2\left(\frac{1}{N}\right)\right] = \frac{1}{N}\sum_{i=1}^{N}\varepsilon_i^2 \tag{8.34}$$

which is equivalent to Equation 6.7. This is the situation when a single fair die is tossed. Since there are six faces on the die, $N = 6$ and $P(y) = 1/6$.

$$V(Y) = \frac{1}{6}\sum_{i=1}^{6}\varepsilon_i^2 = \frac{1}{6}(17.5) = 2.92$$

The value of

$$\sum_{i=1}^{6} \varepsilon_i^2$$

is obtained as follows:

| $i$ | = | 1 | 2 | 3 | 4 | 5 | 6 |
|---|---|---|---|---|---|---|---|
| $y_i$ | = | 1 | 2 | 3 | 4 | 5 | 6 |
| $N_i$ | = | 1 | 1 | 1 | 1 | 1 | 1 |
| $\varepsilon_i$ | = | –2.5 | –1.5 | –0.5 | 0.5 | 1.5 | 2.5 |
| $\varepsilon_i^2$ | = | 6.25 | 2.25 | 0.25 | 0.25 | 2.25 | 6.25 |
| $\sum_{i=1}^{6} \varepsilon_i^2$ | = | 17.5 | | | | | |

Assuming a constant probability, $P(Y) = 1/N$, the "calculator" method of arriving at the sum of squared errors and the variance (Equation 6.10) is

$$V(Y) = \frac{\sum_{i=1}^{N} y_i^2 - \frac{\left(\sum_{i=1}^{N} y_i\right)^2}{N}}{N}$$

which can be written

$$V(Y) = \frac{\sum_{i=1}^{N} y_i^2}{N} - \frac{\left(\sum_{i=1}^{N} y_i\right)^2}{N^2}$$

In this expression

$$\sum_{i=1}^{N} \frac{y^2}{N}$$

is the expected value of $Y^2$ and

$$\sum_{i=1}^{N} \frac{y_i}{N}$$

is the expected value of $Y$. Consequently, the expression can be written:

$$V(Y) = E(Y)^2 - [E(Y)]^2 \tag{8.35}$$

In the next section it will be seen that Equation 8.35 is one of a set of general expressions that are applicable whether a variable is discrete or continuous and whether the probability of occurrence is constant or variable.

Like expected values, variances of random variables have a number of properties that must be recognized by workers in the fields of probability and statistics. The discussion of these properties has been centered in Section 9.7 so that it might include properties arising from multidimensional random variables, which form the subject matter of Chapter 9.

## 8.12 MOMENTS, SKEWNESS, AND KURTOSIS

The expected values of powers of the differences between the values of a random variable and a constant are known as *moments* about that constant. Thus, the $r$th moment about $k$ for random variable $Y$ is

$$\text{Moment}_r = E(Y-k)^r$$
$$= \sum_{i=1}^{M} (y_i - k)^r P(y_i) \quad \text{or} \quad \int_{-\infty}^{\infty} (y-k)^r f(y)dy \tag{8.36}$$

as $r = 0, 1, 2, \ldots$ .

When $k = 0$, the *moments about zero*, for a discrete random variable, are

$$\mu_0' = E(Y^0) = \sum_{i=1}^{M} (y_i - 0)^0 P(y_i) = 1, \quad \text{the 0th moment}$$

$$\mu_1' = E(Y^1) = \sum_{i=1}^{M} (y_i - 0)^1 P(y_i), \quad \text{the 1st moment}$$

$$\mu_2' = E(Y^2) = \sum_{i=1}^{M} (y_i - 0)^2 P(y_i), \quad \text{the 2nd moment}$$

and so forth. In the case of a continuous random variable, the process would be the same except that integration would be used rather than summation. As can be seen, the arithmetic mean of the distribution is the first moment about zero.

When $k$ is equal to the arithmetic mean or expected value of the distribution, the values raised to the various powers are errors, $\varepsilon = y - \mu$, as were described in Section 5.3.5. The *moments about the mean* or the *central moments* are then (again, using a discrete random variable as an example)

$$\mu_0 = E(\varepsilon^0) = \sum_{i=1}^{M} \varepsilon_1^0 P(y_i) = 1, \quad \text{the 0th moment}$$

$$\mu_1 = E(\varepsilon^1) = \sum_{i=1}^{M} \varepsilon_1^1 P(y_i) = 0, \quad \text{the 1st moment}$$

$$\mu_2 = E(\varepsilon^2) = \sum_{i=1}^{M} \varepsilon_1^2 P(y_i) = \sigma_Y^2, \quad \text{the 2nd moment}$$

$$\mu_3 = E(\varepsilon^3) = \sum_{i=1}^{M} \varepsilon_1^3 P(y_i), \quad \text{the 3rd moment}$$

$$\mu_4 = E(\varepsilon^4) = \sum_{i=1}^{M} \varepsilon_1^4 P(y_i), \quad \text{the 4th moment}$$

As can be seen, the second moment about the mean is the variance of the probability distribution.

As will be shown in Section 8.13, equivalent values can be obtained for infinite discrete and continuous distributions. The description of the procedures used to obtain them will be delayed until then to prevent confusion. It should be pointed out, however, that the following discussion of the moments and their uses is applicable whether the distributions are finite or infinite, discrete or continuous, or the probabilities are constant or variable.

In all frequency and probability distributions, the values of the first three moments are as shown. Furthermore, the first, third, fifth, etc. moments about the mean are equal to zero *if the distribution is symmetrical*. This occurs because the errors, when raised to the odd powers, retain their signs and the positive and negative values cancel each other. This provides a means by which the symmetry of a distribution can be assessed. Although any of the odd moments could be used for this purpose, the standard procedure uses only the third. As has been stated, if the third moment is equal to zero, the distribution is symmetrical. If the moment is positive (i.e., $\mu_3 > 0$), it indicates that the sum of the cubed errors to the right of the mean is greater than that to the left of the mean and it is said that the distribution is *positively skewed* or *skewed to the right*. On the other hand, if the moment is negative (i.e., $\mu_3 < 0$) the pattern is reversed, and it is said that the distribution is *negatively skewed* or *skewed to the left*.

When the third moment is used in this way, the presence of skewness can be detected and its direction determined. However, on the basis of the moment one cannot make any meaningful statements regarding the *amount* of skewness. There is no basis for judging the skewness to be slight or great. This is caused by the fact that the moment is a function of both the skewness and the magnitude of the values making up the distribution. All the moments share this dependence on the magnitudes of the values of the random variable. To eliminate this effect, each error prior to the computation of the moment is converted to a dimensionless value by being divided by the standard deviation, which is the quadratic mean of the set. The moments are then computed using these relative values:

$$\alpha_r = \sum_{i=1}^{M} \left(\frac{\varepsilon_i}{\sigma}\right)^2 P(y_i) \quad \text{or} \quad \int_{-\infty}^{\infty} \left(\frac{\varepsilon}{\sigma}\right)^2 f(y)dy \tag{8.37}$$

and thus, in the case of the discrete random variable:

$$\alpha_1 = \sum_{i=1}^{M} \left(\frac{\varepsilon_i}{\sigma}\right)^1 P(y_i) = \frac{1}{\sigma} \sum_{i=1}^{M} \varepsilon P(y_i) = \frac{\mu_1}{\sigma} = \frac{0}{\sigma} = 0$$

$$\alpha_2 = \sum_{i=1}^{M} \left(\frac{\varepsilon_i}{\sigma}\right)^2 P(y_i) = \frac{1}{\sigma^2} \sum_{i=1}^{M} \varepsilon_i^2 P(y_i) = \frac{\mu_2}{\sigma^2} = \frac{\sigma^2}{\sigma^2} = 1$$

$$\alpha_3 = \sum_{i=1}^{M}\left(\frac{\varepsilon_i}{\sigma}\right)^3 P(y_i) = \frac{1}{\sigma^3}\sum_{i=1}^{M}\varepsilon_i^2 P(y_i) = \frac{\mu_3}{\sigma^3}$$

and

$$\alpha_4 = \sum_{i=1}^{M}\left(\frac{\varepsilon_i}{\sigma}\right)^4 P(y_i) = \frac{1}{\sigma^4}\sum_{i=1}^{M}\varepsilon_i^4 P(y_i) = \frac{\mu_4}{\sigma^4}$$

When the moments are in this form, it is possible to compare different distributions and to establish standards against which distributions can be judged.

$\alpha_3$ is the *moment coefficient of skewness*.* As in the case of the third moment proper, when $\alpha_3 = 0$ the distribution is symmetrical, when $\alpha_3 > 0$ the distribution is positively skewed, and when $\alpha_3 < 0$ the distribution is negatively skewed. For example, in the case of the normal distribution shown in Figure 8.5, which is symmetrical, $\alpha_3 = 0$. On the other hand, the value of $\alpha_3$ for the gamma distribution shown in Figure 8.6 is 1.414, indicating positive skewing. In this case, the tail on the right side of the peak descends more slowly than does the tail on the left side of the peak. This is characteristic of positive skewing.

Simply because two distributions are mound shaped and have the same mean, variance, and moment coefficient of skewness does not mean that they are the same. They may have different shapes. One may be relatively low and flat, while the other may have a tall narrow mound. In other words, they possess different degrees of peakedness or *kurtosis*. Because of its central position in the fields of probability and statistics, the normal distribution has been arbitrarily chosen as the standard against which kurtosis can be judged. Distributions having flatter curves or profiles than the normal are called *platykurtic*, while those with curves or profiles that have narrower, more-peaked mounds than the normal are called *leptokurtic*. The normal curve itself is called *mesokurtic*. Examples of these curves are shown in Figure 8.5. The measure used to assess kurtosis is $\alpha_4$ and consequently it is known as the *moment coefficient of kurtosis*. In the case of the normal distribution $\alpha_4 = 3$. When $\alpha_4 > 3$, the curve or profile is leptokurtic, and when $\alpha_4 < 3$, it is platykurtic. These results are caused by the fact that as a curve or profile becomes more peaked the percentage of small errors increases, causing the variance and the fourth moment to decrease. However, the moment decreases at a faster rate than does the variance and consequently the $\alpha_4$ value increases. This difference in rates of decrease can be seen in the following demonstration.

| $i$ | = $\varepsilon_i$ | $\varepsilon_i^2$ | $\frac{(\varepsilon_i^2 - \varepsilon_{i-1}^2)}{\varepsilon_i^2}$ | $\varepsilon^4$ | $\frac{(\varepsilon_i^4 - \varepsilon_{i-1}^4)}{\varepsilon_i^4}$ |
|---|---|---|---|---|---|
| 1 | 5 | 25 | | 625 | |
| 2 | 4 | 16 | 0.3600 | 256 | 0.5904 |
| 3 | 3 | 9 | 0.4375 | 81 | 0.6836 |
| 4 | 2 | 4 | 0.5556 | 16 | 0.8025 |
| 5 | 1 | 1 | 0.7500 | 1 | 0.9375 |

As the squared terms decrease, the percentage decrease is smaller than the equivalent decrease of the fourth power terms. As the curves or profiles flatten, the $\alpha_4$ values become smaller. When the

* In some publications the symbols used for the coefficients of skewness and kurtosis are $\sqrt{\beta_1}$ *or* $\sqrt{\beta_1(Y)} = \alpha_3$ and $\beta_2$ *or* $\beta_2(Y) = \alpha_4$. The coefficients of kurtosis may be further modified so that $\beta_2 = \alpha_4 - 3$ in order that it be equal to zero when the distribution is coincident with the normal. For simplicity, the basic $\alpha$ system will be used in this book.

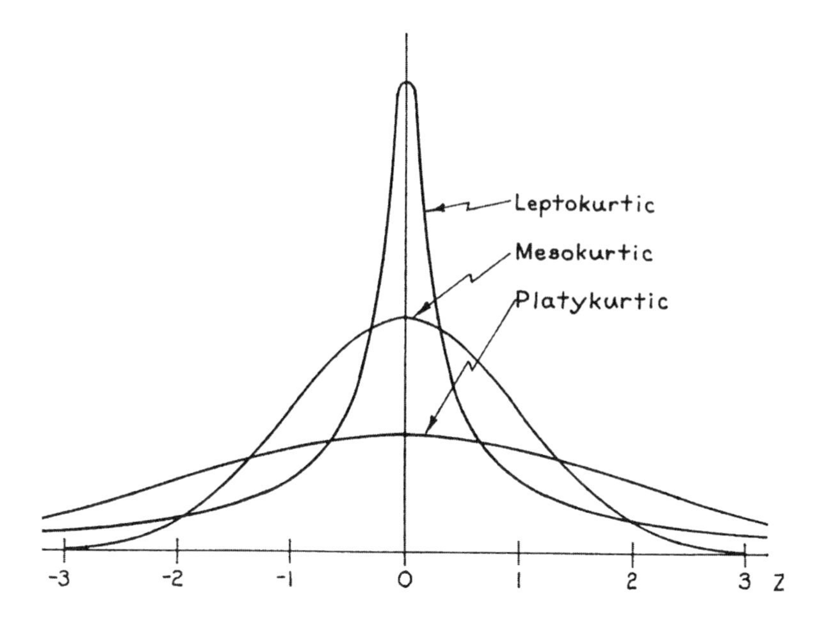


**FIGURE 8.5** Examples of different degrees of kurtosis. The horizontal scale is in terms of units equal in size to the standard deviations of the three distributions.

distribution curve or profile becomes a straight horizontal line, the curve form for a rectangular or uniform distribution $\alpha_4 = 1.8$. If $\alpha_4$ is less than 1.8, the distribution is U shaped. Finally, in the opposite direction, if the distribution becomes more and more peaked, the coefficient will become larger and larger until it approaches infinity. In the limiting situation, however, where the curve is, in one sense, a vertical spike or, in another sense, a single point, the coefficient of kurtosis becomes equal to 0/0 which is indeterminate.

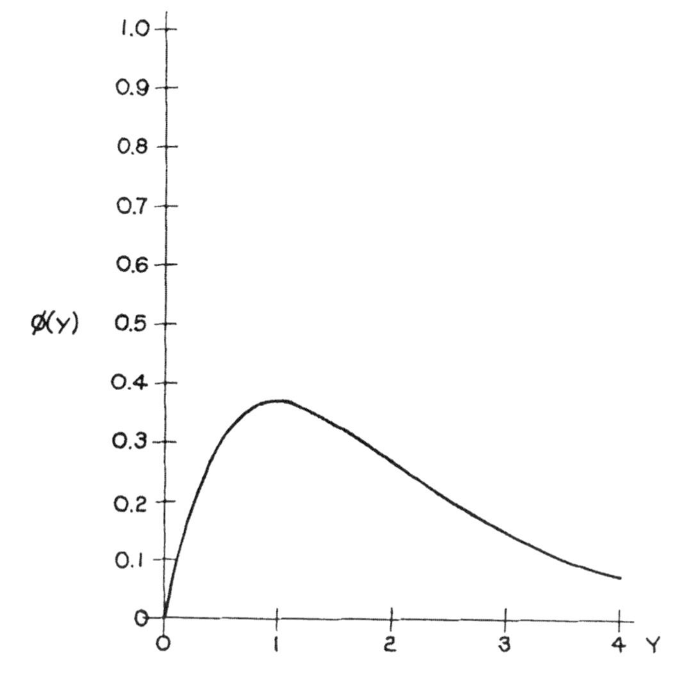


**FIGURE 8.6** Gamma distribution with the parameters $r = 2$ and $\alpha = 1$, showing positive skewness.

## 8.13 MOMENT-GENERATING FUNCTIONS

In the previous section the concept of moments was introduced and it was shown how the third and fourth moments about the mean can be used to assess skewness and kurtosis. The determination of the moments using Equation 8.37, in either its discrete or continuous form, may be difficult. It has been found that an alternative procedure, making use of an exponential transformation* of the point probability function or the pdf is preferred. The transformed function is known as the *moment-generating function* or mgf. In the case of a point probability function the mgf is:

$$M_Y(t) = \sum_{i=1}^{\infty} e_i^{ty} P(y_i) \tag{8.38}$$

and in the case of a pdf it is

$$M_Y(t) = \int_{-\infty}^{\infty} e^{ty} f(y)dy \tag{8.39}$$

In Section 8.10 the *expected value* of the values in a probability distribution was defined as the arithmetic mean of the distribution. An examination of the two mgf expressions shown above will reveal that in both cases $Y$ has been transformed to $e^{ty}$ but the probabilities of the occurrence of the individual $Y$ values have not been changed. Consequently, these are probability distributions of $e^{tY}$ values, and the means of these distributions are equal to the mfg values:

$$M_Y(t) = E(e^{tY}) \tag{8.40}$$

The MacLaurin series expansion of $e^{tY}$ is

$$e^{tY} = 1 + tY + \frac{(tY)^2}{2!} + \frac{(tY)^3}{3!} + \cdots + \frac{(tY)^n}{n!} + \cdots$$

Therefore, assuming convergence of the series,

$$M_Y(t) = E(e^{tY}) = E\left[1 + tY + \frac{(tY)^2}{2!} + \frac{(tY)^3}{3!} + \cdots + \frac{(tY)^n}{n!} + \cdots\right]$$

$$= 1 + tE(Y) + \frac{t^2E(Y^2)}{2!} + \frac{t^3E(Y^3)}{3!} + \cdots + \frac{t^nE(Y^n)}{n!} + \cdots$$

Then

* Transformations are often used by workers in the field of probability and statistics to simplify computations or to make it possible to meet the requirements underlying analytical procedures. A very simple example of how a transformation can simplify a computational procedure is in the determination of the root of a number using logarithms. For example, the seventh root of a number is found by dividing the logarithm of the number by 7 and then finding the antilogarithm of the quotient. In this case, the number was *transformed* into its logarithm and the computations were made with the transformed value. Finally, the result of the computation was transformed back into its original form to yield the desired answer. Entire populations and/or probability distributions can be converted into different forms, which may be of value to the workers involved. An interesting example of such a conversion is the transformation of a rectangular or uniform distribution into a Cauchy distribution, which is described in Hald (1952a).

$$\frac{dM}{dt}Y(t) = M'(t) = E(Y) + tE(Y^2) + \frac{t^2E(Y^3)}{2!} + \cdots + \frac{t^{n-1}E(Y^n)}{(n-1)!}$$

If $t$ is set equal to zero

$$M'(0) = E(Y) = \mu_1'$$

The second derivative of the mgf is

$$M''(t) = E(Y^2) + tE(Y^3) + \frac{t^2E(Y^4)}{2!} + \cdots + \frac{t^{n-2}E(Y^n)}{(n-2)!} + \cdots$$

and when $t$ is set equal to zero

$$M''(0)E(Y^2) = \mu_2'$$

Similarly,

$$M'''(0) = E(Y^3) = \mu_3'$$

$$M''''(0) = E(Y^4) = \mu_4'$$

and

$$M^r(0) = E(Y^r) = \mu_r'$$

Returning to the concept of the error,

$$\varepsilon_j = y_j - \mu_1'$$

The expected value of $\varepsilon$ is

$$E(\varepsilon) = E(Y - \mu) = E(Y) - \mu_1'$$

and since $\mu_1 = E(Y)$

$$E(\varepsilon) = E(Y) - E(Y) = 0 \tag{8.41}$$

This is also the arithmetic mean of the errors, which is the first moment about the mean. The expected value of $\varepsilon^2$ is

$$E(\varepsilon^2) = E(Y - \mu_1')^2 = E(Y^2 - 2\mu_1' + \mu_1'^2)$$

$$= E(Y^2) - 2\mu_1'(Y) + \mu_1'^2$$

and since $\mu = E(Y)$,

$$
\begin{aligned}
E(\varepsilon^2) &= E(Y^2) - 2[E(Y)]^2 + [E(Y)]^2 \\
&= E(Y^2) - [E(Y)]^2
\end{aligned} \tag{8.42}
$$

which is Equation 8.36, the calculator expression for the variance of the distribution. It also is the second moment about the mean. In the same manner,

$$
\begin{aligned}
\mu^3 &= E(\varepsilon^3) = E(Y - \mu_1')^3 \\
&= E(Y^3) - 3E(Y^2)E(Y) + 2[E(Y)]^3
\end{aligned} \tag{8.43}
$$

and

$$
\begin{aligned}
\mu^4 &= E(\varepsilon^4) = E(Y - \mu_1')^4 \\
&= E(Y^4) - 4E(Y^3)E(Y) + 6E(Y^2)[E(Y)]^2 - 3[E(Y)]^4
\end{aligned} \tag{8.44}
$$

In spite of the fact that the process of arriving at the moment-generating functions and their first, second, third, and fourth derivatives is not easy, it often is easier than trying to arrive at the expected values directly.

There are several properties of moment-generating functions that bear on subjects which will be discussed in later chapters. These are

1. Given the random variable $X$ which has the mgf, $M_x$. Then, let $Y = \alpha + \beta X$:

$$M_y(t) = e^{\alpha t} M_x(\beta t) \tag{8.45}$$

The proof:

$$
\begin{aligned}
M_y(t) &= E(e^{Yt}) = E[e^{(\alpha + \beta X)t}] \\
&= e^{\alpha t} E(e^{\beta x t}) \\
&= e^{\alpha t} M_x(\beta t)
\end{aligned}
$$

2. Given two random variables, $X$ and $Y$. If $M_x(t) = M_y(t)$ for all values of $T$, then $X$ and $Y$ have the same probability distribution. The proof is beyond the scope of this presentation.
3. Given two independent random variables, $X$ and $Y$. Let $Z = X + Y$. Then,

$$M_z(t) = M_x(t) * M_y(t) \tag{8.46}$$

The proof:

$$
\begin{aligned}
M_z(t) &= E(e^{zt}) = E[e^{(x+Y)t}] \\
&= E(e^{Xt} e^{Yt}) = E(e^{Xt})E(e^{Yt}) \\
&= M_X(t) * M_Y(t)
\end{aligned}
$$

Finally, there are certain probability distributions, including the normal, the gamma, and the Poisson, that possess a characteristic which is known as a *reproductive property.* When two or more random variables, distributed as one of these probability distributions, are added, the distribution of the sums will be of the same type as that possessed by the original random variables. When this is the case, the mgf of the distribution of sums will be equal to the product of the mgf of the original random variables (see Equation 8.46). When the mgf of the distribution of sums is used to generate the moments about the mean and, subsequently, the parameters of the distribution, those parameters will be equal to the sums of the corresponding parameters of the distributions of the original random variables. For example, if $X$ and $Y$ are random variables, both of which are normally distributed and their parameters are ($\mu_X$, $\sigma_X^2$) and ($\mu_Y$, $\sigma_Y^2$), respectively, the distribution of $Z = X + Y$ will also be normal, and the parameters of the distribution of $Z$ will be $\mu_z = \mu_X + \mu_Y$ and $\sigma_X^2 = \sigma_X^2 + \sigma_Y^2$. The proof of the existence of the reproductive property must be carried out on a distribution-by-distribution basis. Readers interested in such proofs are referred to Meyer (1970).

# 9 Multidimensional Probability Distributions

## 9.1 MULTIDIMENSIONAL RANDOM VARIABLES

The probability distribution associated with the tossing of a pair of fair dice was described in Section 8.2 and shown in Table 8.4 and Figure 8.1. The random variable of interest in that case was the sum of the values on the upfacing sides of the two dice following the toss. This may appear simple, and in many respects it is simple, but it actually is an example of a rather complex situation. The values of the random variable, the $y$ values, upon which the distribution is based, are sums of values obtained from *two different sources*, the two dice. *Each* of these dice had associated with it a sample space, a definition of the random variable ($x_i = X(s_i) = s_i$), a range space, and a probability distribution such as that shown in Table 9.1. Let us refer to the values associated with the individual dice as *primary* variables and their distributions as *primary* distributions. In the discussion in Section 8.2, the tossing of the two dice was construed to be a single act and, consequently, an individual experiment. Actually, however, a draw was made from *each* of the two primary range spaces yielding a *joint event*, the appearance of two $x$ values ($x_1$, $x_2$), where $x_1$ is the value from the first die and $x_2$ is from the second die.* This *pair* of $x$ values is a *two-dimensional random variable* or *random vector*. This concept can be extended to any number ($M$) of primary variables giving rise to an *M-dimensional* or *multidimensional random variable*.

It must be emphasized that the random variable is the *pair* (or larger set in the case of multidimensional random variables), *not some value derived from the pair* (such as its sum). Furthermore, the values of the pair arise from the functions defining the primary random variables and, consequently, are real number values. If the values making up a pair (or large set) are finite or countably infinite the variable is *discrete*. For example, if the variable can be represented as ($x_{1i}$, $x_{2j}$), where $i = 1, 2, 3, \ldots, L_1$, and $j = 1, 2, 3, \ldots, L_2$, the variable is discrete. If, on the other hand, the values in the pair can assume *any* value within certain ranges, the variable is *continuous*. For example, if $X_1$ can take on all values in the range $A$ to $B$ and $X_2$ can take on all the values in the range $C$ to $D$, the variable is continuous. In a graphical sense, the values of $X_2$ can be plotted against the values of $X_1$ in a two-dimensional coordinate system (Figure 9.1). If the variable is discrete, the plotted points will be distinct from one another (Figure 9.1A), whereas if the variable is continuous the plotted points will form a surface within the indicated ranges (Figure 9.1B). It is possible for a multidimensional random variable to be made up of one or more discrete primary

* It should be noted that in this chapter the subscripting will be critical and confusion can develop if the scheme is not clear. In the case of $X$ the subscripts refer to primary random variables. In the case of $X$ the first subscript refers to a primary random variable while the second refers to a value of that random variable. In the case of $N$, $N_R$ indicates that it refers to the size of the composite range space: if subscripted otherwise the first subscript refers to a primary random variable and the second, if present, to a subset or partition of the range space of that primary random variable. $L$ refers to the number of values which a primary random variable can assume. Consequently, its subscript refers to that primary random variable.

**TABLE 9.1**
**The Sample Space, Range Space, and a Probability Distribution Associated with the Tossing of a Single Fair Die When the Random Variable $x_i = X(s_i) = s_i$**

| $i$ | 1 | 2 | 3 | 4 | 5 | 6 | |
|---|---|---|---|---|---|---|---|
| $s_i$ | 1 | 2 | 3 | 4 | 5 | 6 | (the sample space) |
| $x_i$ | 1 | 2 | 3 | 4 | 5 | 6 | (the range space) |
| $P(x_i)$ | $\frac{1}{6}$ | $\frac{1}{6}$ | $\frac{1}{6}$ | $\frac{1}{6}$ | $\frac{1}{6}$ | $\frac{1}{6}$ | (the probability distribution) |

variables and one or more continuous primary variables. If a certain two-dimensional random variable is of this type, its graphical presentation would consist of a series of parallel continuous lines crossing the coordinate system (Figure 9.1C). The positions of the lines would correspond to the values of the discrete primary variable.

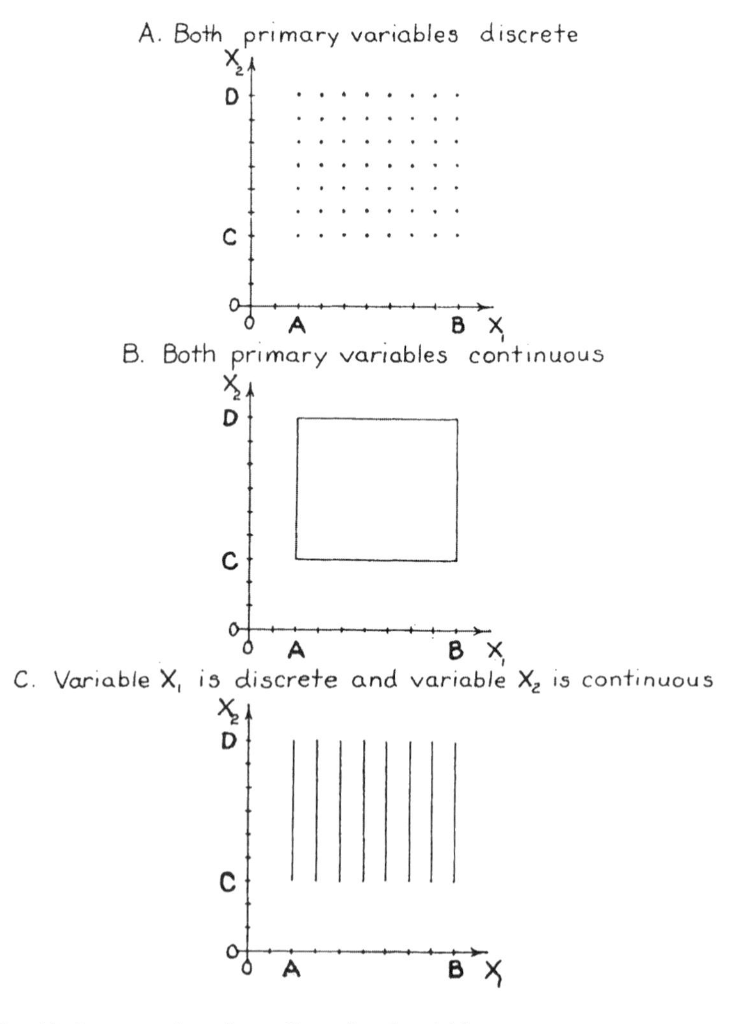


**FIGURE 9.1** Graphical presentation of two-dimensional variables.

To be meaningful, multidimensional random variables must be made up of primary variables that are in some manner linked together. In the case of the two fair dice, the two values making up the two-dimensional random variable were obtained following a specific toss of the two dice. The link between the two values was that the dice had been tossed at the *same time*. Let us assume another situation. Consider a large number of logs on the deck of a certain sawmill. Each of these logs has a small end diameter and length. $X_1$ can be defined as the diameter and $X_2$ as the length. Thus, a pair of values $(X_1, X_2)$ can be obtained from each log. The link between the values is that they were obtained from the *same object*. Time and object links are probably the most common but other types are possible. The important thing to remember is that there must be some rationale for the linkage.

## 9.2 JOINT PROBABILITY DISTRIBUTIONS

Obviously, multidimensional random variables are generated by multiple events and the probabilities of occurrence of specific multidimensional random variables are controlled by the rules governing the probabilities of multiple events. These rules are discussed in Section 7.6. Each act giving rise to a pair or larger set of events is an individual experiment. Thus, an experiment involves two or more draws from one or more specified primary sample spaces. Each draw yields an event. The set of events generated by the experiment constitutes a single multiple event. The set of all possible results of such an experiment constitutes a composite sample space, which is the sample space associated with the multidimensional random variable. Applying the functions that convert the primary events in the multiple events to their numerical equivalents converts the composite sample space to the range space of the multidimensional random variable. The size of this range space is dependent on the nature of the experiment. If the primary random variables are discrete and independent and all possible combinations are acceptable,

$$N_R = N_1 * N_2 * N_3 * \dots * N_M \tag{9.1}$$

where $N_R$ = number of items in the composite range space
$N_i$ = number of items in the $i$th primary range space, i = 1, 2, 3, …, $M$
$M$ = number of primary random variables that are involved

This, of course, reduces to Equation 7.19 if multiple draws are to be made from one primary sample space using sampling with replacement. If the sampling from this one primary sample space is to be without replacement, Equation 7.20 would be appropriate. In many cases, however, the size of the composite sample space is determined by the nature of the problem rather than by computation using these equations.

As can be seen in Figure 9.1, each pairing or larger set in the composite space is the result of the joint occurrence of two or more simple events. Each of these joint events is unique and the probability of its occurrence is equal to the reciprocal of the size of the composite sample space. When the primary events making up the joint events are converted to their equivalent random variables, which are real numbers, it is possible that specific multidimensional random variables will occur more than once. When this occurs the range space can be partitioned in the normal manner and the probability of occurrence of each specific multidimensional random variable is simply the ratio of its frequency of occurrence to the total size of the range space. This is demonstrated in Table 9.2. In this case the universe consists of 120 loblolly pine trees. Two variable characteristics of these trees are of interest, dbh $(X_1)$ and total height $(X_2)$. Each of these is a random variable;

$$x_{1i} = \begin{cases} 1 \text{ in.,} & \text{if } 0.5 \le \text{dbh} < 1.5 \text{ in.} \\ 2 \text{ in.,} & \text{if } 1.5 \le \text{dbh} < 2.5 \text{ in.} \\ 3 \text{ in.,} & \text{if } 2.5 \le \text{dbh} < 3.5 \text{ in.} \\ \vdots & \\ 50 \text{ in.,} & \text{if } 49.5 \le \text{dbh} < 50.5 \text{ in.}^* \\ 0, & \text{elsewhere} \end{cases}$$

$$x_{2j} = \begin{cases} 10 \text{ ft,} & \text{if } 5.0 \le \text{height} < 15.0 \text{ ft} \\ 20 \text{ ft,} & \text{if } 15.0 \le \text{height} < 25.0 \text{ ft} \\ 30 \text{ ft,} & \text{if } 25.0 \le \text{height} < 35.0 \text{ ft} \\ \vdots & \\ 200 \text{ ft,} & \text{if } 195.0 \le \text{height} < 205.0^* \\ 0, & \text{elsewhere} \end{cases}$$

To obtain a pair, a tree is blindly chosen and its dbh and total height are measured. If, for example, the dbh is 10.7 in. and the total height is 62 ft, the two-dimensional random variable is (11 in., 60 ft).

Since the two primary variables are linked through the individual trees and the dbh of one tree cannot be paired with the total height of another tree, the composite or joint sample space for the two-dimensional random variable can contain only 120 events, as is shown in Table 9.2A. Part B of Table 9.2 is the equivalent range space. An examination of the range space will reveal that many of the pairings occur more than once. Consequently, using the methods described in Section 8.2, the range space can be partitioned and the subset sizes determined, giving rise to a *joint frequency distribution*. This has been done in Part C of Table 9.2. The probabilities of the pairings are shown in Part D. These probabilities form the probability distribution of the two-dimensional random variable. Such a distribution is known as a *joint* or *multidimensional probability distribution* of the primary random variables.

## 9.3 MARGINAL PROBABILITIES

In Table 9.2D, the column labeled $P(X_1)$ contains the probabilities of occurrence of the dbh, or $X_1$, values, disregarding tree height:

$$P(X_1 = x_{1i}) = \sum_{j=1}^{L_2} [P(x_{1i}, x_{2j})] \tag{9.2}$$

where $L_2$ = count of all possible values of $X_2$

$x_{1i}$ = the $i$th value of $X_1$

Similarly, the row labeled $P(X_2)$ contains the probabilities of the occurrence of the tree height, or $X_2$, values, disregarding dbh:

* It is expected that no tree would have a dbh in excess of 50.5 in. nor a total height in excess of 205.0 ft.

**TABLE 9.2**
**The Composite Sample Space, Range Space, and Probabilities Associated with the Two-Dimensional Random Variable (dbh, total height) for 120 Loblolly Pine Trees**

**A. Sample Space**

| | | | | | | | | | |
|---|---|---|---|---|---|---|---|---|---|
| $(d_1, h_1)$ | $(d_2, h_2)$ | $(d_3, h_3)$ | $(d_4, h_4)$ | $(d_5, h_5)$ | $(d_6, h_6)$ | $(d_7, h_7)$ | $(d_8, h_8)$ | $(d_9, h_9)$ | $(d_{10}, h_{10})$ |
| $(d_{11}, h_{11})$ | $(d_{12}, h_{12})$ | $(d_{13}, h_{13})$ | $(d_{14}, h_{14})$ | $(d_{15}, h_{15})$ | $(d_{16}, h_{16})$ | $(d_{17}, h_{17})$ | $(d_{18}, h_{18})$ | $(d_{19}, h_{19})$ | $(d_{20}, h_{20})$ |
| $(d_{21}, h_{21})$ | $(d_{22}, h_{22})$ | $(d_{23}, h_{23})$ | $(d_{24}, h_{24})$ | $(d_{25}, h_{25})$ | $(d_{26}, h_{26})$ | $(d_{27}, h_{27})$ | $(d_{28}, h_{28})$ | $(d_{29}, h_{29})$ | $(d_{30}, h_{30})$ |
| $(d_{31}, h_{31})$ | $(d_{32}, h_{32})$ | $(d_{33}, h_{33})$ | $(d_{34}, h_{34})$ | $(d_{35}, h_{35})$ | $(d_{36}, h_{36})$ | $(d_{37}, h_{37})$ | $(d_{38}, h_{38})$ | $(d_{39}, h_{39})$ | $(d_{40}, h_{40})$ |
| $(d_{41}, h_{41})$ | $(d_{42}, h_{42})$ | $(d_{43}, h_{43})$ | $(d_{44}, h_{44})$ | $(d_{45}, h_{45})$ | $(d_{46}, h_{46})$ | $(d_{47}, h_{47})$ | $(d_{48}, h_{48})$ | $(d_{49}, h_{49})$ | $(d_{50}, h_{50})$ |
| $(d_{51}, h_{51})$ | $(d_{52}, h_{52})$ | $(d_{53}, h_{53})$ | $(d_{54}, h_{54})$ | $(d_{55}, h_{55})$ | $(d_{56}, h_{56})$ | $(d_{57}, h_{57})$ | $(d_{58}, h_{58})$ | $(d_{59}, h_{59})$ | $(d_{60}, h_{60})$ |
| $(d_{61}, h_{61})$ | $(d_{62}, h_{62})$ | $(d_{63}, h_{63})$ | $(d_{64}, h_{64})$ | $(d_{65}, h_{65})$ | $(d_{66}, h_{66})$ | $(d_{67}, h_{67})$ | $(d_{68}, h_{68})$ | $(d_{69}, h_{69})$ | $(d_{70}, h_{70})$ |
| $(d_{71}, h_{71})$ | $(d_{72}, h_{72})$ | $(d_{73}, h_{73})$ | $(d_{74}, h_{74})$ | $(d_{75}, h_{75})$ | $(d_{76}, h_{76})$ | $(d_{77}, h_{77})$ | $(d_{78}, h_{78})$ | $(d_{79}, h_{79})$ | $(d_{80}, h_{80})$ |
| $(d_{81}, h_{81})$ | $(d_{82}, h_{82})$ | $(d_{83}, h_{83})$ | $(d_{84}, h_{84})$ | $(d_{85}, h_{85})$ | $(d_{86}, h_{86})$ | $(d_{87}, h_{87})$ | $(d_{88}, h_{88})$ | $(d_{89}, h_{89})$ | $(d_{90}, h_{90})$ |
| $(d_{91}, h_{91})$ | $(d_{92}, h_{92})$ | $(d_{93}, h_{93})$ | $(d_{94}, h_{94})$ | $(d_{95}, h_{95})$ | $(d_{96}, h_{96})$ | $(d_{97}, h_{97})$ | $(d_{98}, h_{98})$ | $(d_{99}, h_{99})$ | $(d_{100}, h_{100})$ |
| $(d_{101}, h_{101})$ | $(d_{102}, h_{102})$ | $(d_{103}, h_{103})$ | $(d_{104}, h_{104})$ | $(d_{105}, h_{105})$ | $(d_{106}, h_{106})$ | $(d_{107}, h_{107})$ | $(d_{108}, h_{108})$ | $(d_{109}, h_{109})$ | $(d_{110}, h_{110})$ |
| $(d_{111}, h_{111})$ | $(d_{112}, h_{112})$ | $(d_{113}, h_{113})$ | $(d_{114}, h_{114})$ | $(d_{115}, h_{115})$ | $(d_{116}, h_{116})$ | $(d_{117}, h_{117})$ | $(d_{118}, h_{118})$ | $(d_{119}, h_{119})$ | $(d_{120}, h_{120})$ |

**B. Range Space**

| | | | | | | | | | |
|---|---|---|---|---|---|---|---|---|---|
| (8,30) | (8,30) | (8,40) | (8,40) | (8,40) | (8,40) | (8,50) | (8,50) | (8,50) | (8,60) |
| (9,30) | (9,40) | (9,40) | (9,40) | (9,40) | (9,50) | (9,50) | (9,50) | (9,50) | (9,50) |
| (9,60) | (9,60) | (9,60) | (10,40) | (10,50) | (10,50) | (10,50) | (10,50) | (10,50) | (10,50) |
| (10,50) | (10,60) | (10,60) | (10,60) | (10,60) | (10,60) | (10,60) | (10,70) | (10,70) | (10,70) |
| (10,70) | (11,40) | (11,40) | (11,50) | (11,50) | (11,50) | (11,50) | (11,50) | (11,60) | (11,60) |
| (11,60) | (11,60) | (11,60) | (11,60) | (11,60) | (11,70) | (11,70) | (11,70) | (11,70) | (11,70) |
| (12,40) | (12,50) | (12,50) | (12,50) | (12,50) | (12,60) | (12,60) | (12,60) | (12,60) | (12,60) |
| (12,60) | (12,60) | (12,60) | (12,60) | (12,70) | (12,70) | (12,70) | (12,70) | (12,70) | (12,70) |
| (12,70) | (12,70) | (13,50) | (13,50) | (13,60) | (13,60) | (13,60) | (13,60) | (13,60) | (13,60) |
| (13,60) | (13,70) | (13,70) | (13,70) | (13,70) | (13,70) | (13,70) | (13,70) | (13,80) | (14,50) |
| (14,60) | (14,60) | (14,60) | (14,70) | (14,70) | (14,70) | (14,70) | (14,70) | (14,80) | (14,80) |
| (14,80) | (14,80) | (15,60) | (15,70) | (15,80) | (15,80) | (15,90) | (16,70) | (16,80) | (16,90) |

**C. Subset Sizes, $N(X_1, X_2)$**

| | $X_2$ = Total Height, ft | | | | | | | |
|---|---|---|---|---|---|---|---|---|
| $X_1$ = dbh, in. | 30 | 40 | 50 | 60 | 70 | 80 | 90 | Σ |
| 8 | 2 | 4 | 3 | 1 | | | | 10 |
| 9 | 1 | 4 | 5 | 3 | | | | 13 |
| 10 | | 1 | 7 | 6 | 4 | | | 18 |
| 11 | | 2 | 5 | 7 | 5 | | | 19 |
| 12 | | 1 | 4 | 9 | 8 | | | 22 |
| 13 | | | 2 | 7 | 7 | 1 | | 17 |
| 14 | | | 1 | 3 | 5 | 4 | | 13 |
| 15 | | | | 1 | 1 | 2 | 1 | 5 |
| 16 | | | | | 1 | 1 | 1 | 3 |
| Σ | 3 | 12 | 27 | 37 | 31 | 8 | 2 | 120 |

**D. Probabilities, $P(X_1, X_2)$**

| | $X_2$ = Total Height, ft | | | | | | | |
|---|---|---|---|---|---|---|---|---|
| $X_1$ = dbh, in. | 30 | 40 | 50 | 60 | 70 | 80 | 90 | $P(x_1)$ |
| 8 | $\frac{2}{120}$ | $\frac{4}{120}$ | $\frac{3}{120}$ | $\frac{1}{120}$ | | | | $\frac{10}{120}$ |
| 9 | $\frac{1}{120}$ | $\frac{4}{120}$ | $\frac{5}{120}$ | $\frac{3}{120}$ | | | | $\frac{13}{120}$ |

**TABLE 9.2** *(Continued)*
**The Composite Sample Space, Range Space, and Probabilities Associated with the Two-Dimensional Random Variable (dbh, total height) for 120 Loblolly Pine Trees**

| | $X_2$ = Total Height, ft | | | | | | | |
|---|---|---|---|---|---|---|---|---|
| $X_1$ = dbh, in. | 30 | 40 | 50 | 60 | 70 | 80 | 90 | $P(x_1)$ |
| 10 | | $\frac{1}{120}$ | $\frac{7}{120}$ | $\frac{6}{120}$ | $\frac{4}{120}$ | | | $\frac{18}{120}$ |
| 11 | | $\frac{2}{120}$ | $\frac{5}{120}$ | $\frac{7}{120}$ | $\frac{5}{120}$ | | | $\frac{19}{120}$ |
| 12 | | $\frac{1}{120}$ | $\frac{4}{120}$ | $\frac{9}{120}$ | $\frac{8}{120}$ | | | $\frac{22}{120}$ |
| 13 | | | $\frac{2}{120}$ | $\frac{7}{120}$ | $\frac{7}{120}$ | $\frac{1}{120}$ | | $\frac{17}{120}$ |
| 14 | | | $\frac{1}{120}$ | $\frac{3}{120}$ | $\frac{5}{120}$ | $\frac{4}{120}$ | | $\frac{13}{120}$ |
| 15 | | | | $\frac{1}{120}$ | $\frac{1}{120}$ | $\frac{2}{120}$ | $\frac{1}{120}$ | $\frac{5}{120}$ |
| 16 | | | | | $\frac{1}{120}$ | $\frac{1}{120}$ | $\frac{1}{120}$ | $\frac{3}{120}$ |
| $P(x_2)$ | $\frac{3}{120}$ | $\frac{12}{120}$ | $\frac{27}{120}$ | $\frac{37}{120}$ | $\frac{31}{120}$ | $\frac{8}{120}$ | $\frac{2}{120}$ | |

$$P(X_2 = x_{2j}) = P(x_{2j}) = \sum_{i=1}^{L_1} [P(x_{1i}, x_{2j})] \tag{9.3}$$

where $L_1$ = count of all possible values of $X_1$
$x_{2j}$ = the $j$th value of $X_2$

These probabilities are those of the primary variables and from their probability distribution. Because of their locations along the edges or margins of the table, they are referred to as *marginal probabilities*. The set of such probabilities associated with one of the primary variables is the *marginal probability distribution* of that variable and the function that describes it is a *marginal probability function*. For example, the marginal probability function for $X_1$ in a two-dimensional discrete random variable is

$$P(x_{1i}) = \begin{cases} \sum_{j=1}^{L_2} [P(x_{11}, x_{2j})] = P(x_{11}), & \text{when } X_1 = x_{11} \\ \sum_{j=1}^{L_2} [P(x_{12}, x_{2j})] = P(x_{12}), & \text{when } X_1 = x_{12} \\ \vdots & \\ \sum_{j=1}^{L_2} [P(x_{1L_1}, x_{2j})] = P(x_{1L_1}), & \text{when } X_1 = x_{1L_1} \\ 0, & \text{elsewhere} \end{cases} \tag{9.4}$$

while for $X_2$ it is

$$P(x_{2j}) = \begin{cases} \sum_{i=1}^{L_1} [P(x_{1i}, x_{21})] = P(x_{21}), & \text{when } X_2 = x_{21} \\ \sum_{i=1}^{L_1} [P(x_{1i}, x_{22})] = P(x_{22}), & \text{when } X_2 = x_{22} \\ \vdots & \\ \sum_{i=1}^{L_1} [P(x_{1i}, x_{2L_2})] = P(x_{2L_2}), & \text{when } X_2 = x_{2L_2} \\ 0, & \text{elsewhere} \end{cases} \tag{9.5}$$

## 9.4 CONDITIONAL PROBABILITIES

The multiplication law of probability (Equation 7.23) states that the probability of occurrence of a multiple event is equal to the product of the probabilites of the events making up the multiple event. The probability of occurrence of a specific value for the first item in the joint event ($X_1$) is simply the marginal probability associated with that event. The probability of specific values for the second ($X_2$) and subsequent items, however, is dependent on the nature of the primary variables. If they are *independent*, as in the case of the values on a pair of fair dice, the probability of occurrence of a specific combination is simply the product of the marginal probabilties of the primary values, as is shown in Table 9.3.

$$\begin{aligned} P(x_{1i}, x_{2j}) &= P(x_{1i}) * P(x_{2j}) \\ &= \left(\frac{1}{6}\right)\left(\frac{1}{6}\right) = \left(\frac{1}{36}\right) \end{aligned} \tag{9.6}$$

where $x_{1i}$ = desired face value of first die
$x_{2j}$ = desired face value of second die

Equation 9.6 defines the term *independence* as it refers to the primary variables making up a two-dimensional random variable. *If the probability of occurrence of a specific pairing is not equal to the product of the marginal probabilities of the two primary variables, the two primary variables are not independent.* This concept extends to all multidimensional random variables. As would be expected when the occurrence of one event has no effect on the probability of occurrence of the other events, the probabilities of the joint events are constant.

If the primary random variables are not independent, as in the case of the 120 loblolly pines, the probability of occurrence of a specific value of the first primary variable is, as before, the marginal probability of that value. The probabilities of specific values of the remaining primary variables are, however, dependent on the values obtained for the earlier primary variables. Consequently, the probability of a joint event involving dependent primary variables is the product of the marginal probability of the first primary event and the conditional probabilities of the subsequent primary events, as in Equation 7.24:

$$P(x_1, x_2, x_3, \ldots, x_M) = P(x_1) * P(x_2 \mid x_1) * P(x_3 \mid x_1 x_2) * \cdots * P(x_M \mid x_1 x_2 x_3 \ldots x_{M-1}) \tag{9.7}$$

**TABLE 9.3**
**The Probabilities Associated with the Tossing of a Pair of Fair Dice**

| $X_1$ = Face Value of First Die | $X_2$ = Face Value of Second Die | | | | | | |
|---|---|---|---|---|---|---|---|
| | 1 | 2 | 3 | 4 | 5 | 6 | $P(x_1)$ |
| 1 | $\frac{1}{36}$ | $\frac{1}{36}$ | $\frac{1}{36}$ | $\frac{1}{36}$ | $\frac{1}{36}$ | $\frac{1}{36}$ | $\frac{1}{6}$ |
| 2 | $\frac{1}{36}$ | $\frac{1}{36}$ | $\frac{1}{36}$ | $\frac{1}{36}$ | $\frac{1}{36}$ | $\frac{1}{36}$ | $\frac{1}{6}$ |
| 3 | $\frac{1}{36}$ | $\frac{1}{36}$ | $\frac{1}{36}$ | $\frac{1}{36}$ | $\frac{1}{36}$ | $\frac{1}{36}$ | $\frac{1}{6}$ |
| 4 | $\frac{1}{36}$ | $\frac{1}{36}$ | $\frac{1}{36}$ | $\frac{1}{36}$ | $\frac{1}{36}$ | $\frac{1}{36}$ | $\frac{1}{6}$ |
| 5 | $\frac{1}{36}$ | $\frac{1}{36}$ | $\frac{1}{36}$ | $\frac{1}{36}$ | $\frac{1}{36}$ | $\frac{1}{36}$ | $\frac{1}{6}$ |
| 6 | $\frac{1}{36}$ | $\frac{1}{36}$ | $\frac{1}{36}$ | $\frac{1}{36}$ | $\frac{1}{36}$ | $\frac{1}{36}$ | $\frac{1}{6}$ |
| $P(x_2)$ | $\frac{1}{6}$ | $\frac{1}{6}$ | $\frac{1}{6}$ | $\frac{1}{6}$ | $\frac{1}{6}$ | $\frac{1}{6}$ | |

For example, in the case of the 120 loblolly pines, if the desired joint random variable is (8 in. dbh, 50 ft height), the probability is

$$P(8,50) = \left(\frac{10}{120}\right) * \left(\frac{3}{10}\right) = \frac{30}{1200} = \frac{3}{120} = \frac{1}{40}$$

since $P(8 \text{ in.}) = 10/120$ and $P(50 \text{ ft}|8 \text{ in.}) = 3/10$. The conditional probability used in this computation was obtained by considering only the ten trees having dbh values of 8 in. The heights of those trees formed a range space which was partitioned into the four classes whose subset sizes are listed in the 8-in. dbh row in Table 9.2C. The probabilities associated with the four height classes were the ratios of the subset sizes to the size of the range space:

$$P(30 \text{ ft} \mid 8 \text{ in.}) = \frac{2}{10}$$

$$P(40 \text{ ft} \mid 8 \text{ in.}) = \frac{4}{10}$$

$$P(50 \text{ ft} \mid 8 \text{ in.}) = \frac{3}{10}$$

$$P(60 \text{ ft} \mid 8 \text{ in.}) = \frac{1}{10}$$

Generalizing, in the context of a two-dimensional discrete random variable,

$$P(x_{2j} \mid x_{1i}) = N_{2ij} \Big/ \sum^{L_2} N_{2ij} \qquad \begin{array}{l} i = 1,\ 2,\ 3,\ \ldots,\ L_1 \\ j = 1,\ 2,\ 3,\ \ldots,\ L_2 \end{array} \tag{9.8}$$

where $N_{2ij}$ = size of the subset or partition of its composite range space associated with $X_2 = x_{2j}$ when $X_1 = x_{1i}$

## 9.5 JOINT PROBABILITY DISTRIBUTION FUNCTIONS

The set of conditional probabilities associated with a specific value of $X_1$ forms a *conditional probability distribution* and the function that describes that distribution is a *conditional probability distribution function*. If the primary variables are discrete, the function takes on the following form:

$$P(x_2, x_{1i}) = \begin{cases} P(x_{21} \mid x_{1i}) = \dfrac{N_{21i}}{T_i}, & \text{when } X_2 = x_{21} \text{ and } X_1 = x_{1i} \\ P(x_{22} \mid x_{1i}) = \dfrac{N_{22i}}{T_i}, & \text{when } X_2 = x_{22} \text{ and } X_1 = x_{1i} \\ \vdots & \\ P(x_{2j} \mid x_{1i}) = \dfrac{N_{2ji}}{T_i}, & \text{when } X_2 = x_{2i} \text{ and } X_1 = x_{1i} \\ 0, & \text{otherwise} \end{cases} \tag{9.9}$$

where

$$T_i = \sum_{j=1}^{L_2} N_{2ij}$$

When the primary variables are all discrete the multidimensional random variable is discrete and its *joint probability distribution function* or *joint probability function* is typical of discrete distributions. For example, in the case of the 120 loblolly pines:

$$P(x_1, x_2) = \begin{cases} \dfrac{2}{120}, & \text{when } X_1 = 8 \text{ in. and } X_2 = 30 \text{ ft} \\ \dfrac{4}{120}, & \text{when } X_1 = 8 \text{ in. and } X_2 = 40 \text{ ft} \\ \vdots & \\ \dfrac{1}{120}, & \text{when } X_1 = 16 \text{ in. and } X_2 = 90 \text{ ft} \\ 0, & \text{otherwise} \end{cases} \tag{9.10}$$

In the case of a continuous multidimensional random variable, the joint frequency distribution function is

$$N(x_1, x_2, x_3, \ldots, x_M) = \begin{cases} \phi(x_1, x_2, x_3, \ldots, x_M) & \text{when} \quad x_{1A} \leq X_1 \leq x_{1B} \\ & \phantom{\text{when}} \quad x_{2A} \leq X_2 \leq x_{2B} \\ & \phantom{\text{when}} \quad x_{3A} \leq X_3 \leq x_{3B} \\ & \phantom{\text{when}} \quad \vdots \\ & \phantom{\text{when}} \quad x_{MA} \leq X_M \leq x_{MB} \\ 0, & \text{elsewhere} \end{cases} \tag{9.11}$$

where $M$ = number of primary random variables; $A$ identifies the smallest possible value of the primary random variables for which it is a subscript and $B$ identifies the largest possible value of the primary random variable; and the *joint probability density function* (joint pdf) is

$$f(z) = f(x_1, x_2, x_3, \ldots, x_M) = \frac{N(x_1, x_2, x_3, \ldots, x_M)}{T}$$

$$= \begin{cases} \dfrac{\phi(x_1, x_2, x_3, \ldots, x_M)}{T} & \text{when} \quad x_{1A} \leq X_1 \leq x_{1B} \\ & \phantom{\text{when}} \quad x_{2A} \leq X_2 \leq x_{2B} \\ & \phantom{\text{when}} \quad x_{3A} \leq X_3 \leq x_{3B} \\ & \phantom{\text{when}} \quad \vdots \\ & \phantom{\text{when}} \quad x_{MA} \leq X_M \leq x_{MB} \\ 0, & \text{elsewhere} \end{cases} \tag{9.12}$$

where

$$T = \int_{x_{MA}}^{x_{MB}} \cdots \int_{x_{3A}}^{x_{3B}} \int_{x_{2A}}^{x_{2B}} \int_{x_{1A}}^{x_{1B}} \phi(x_1, x_2, x_3, \ldots, x_M) dx_1 dx_2 dx_3 \ldots dx_M$$

As can be seen by comparing Equations 8.10 and 9.11 and Equations 8.12 and 9.12, the joint or multidimensional frequency distribution and probability density functions are simply logical extensions of the concepts introduced in Section 8.5. The joint frequency distribution function, in the case of a continuous two-dimensional random variable, describes a surface within a three-dimensional rectangular coordinate system. The horizontal axes are occupied by the primary variables, $X_1$ and $X_2$, as in Figure 9.1B, while the vertical axis is occupied by the subset size, $N(x_1, x_2)$. The total volume under this surface is equal to $T$. Thus, in the case of the joint pdf, the vertical dimension is equal to $N(x_1, x_2)/T$ and the total volume under the surface is equal to unity or 100%. Probabilities of specific joint random variables are, of course, infinitesimally small and, consequently, ranges of values of the joint random variable must be specified to obtain meaningful probabilities. For example, with a two-dimensional random variable a probability would be computed as follows:

$$P(x_{1a} \leq X_1 \leq x_{1b},\ x_{2a} \leq X_2 \leq x_{2b}) = \int_{x_{2a}}^{x_{2b}} \int_{x_{1a}}^{x_{1b}} f(z) dx_1 dx_2 \tag{9.13}$$

where $x_{1a}$ = minimum acceptable value of $X_1$
$x_{1b}$ = maximum acceptable value of $X_1$
$x_{2a}$ = minimum acceptable value of $X_2$
$x_{2b}$ = maximum acceptable value of $X_2$

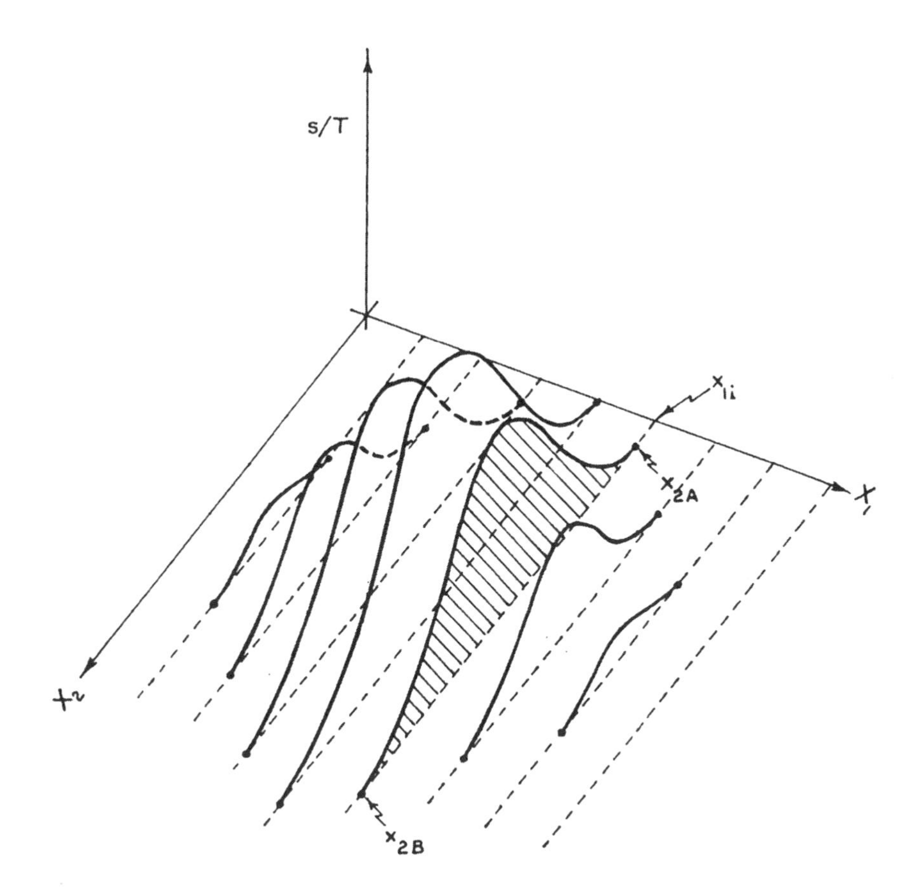


**FIGURE 9.2** Profiles through the solid representing the probability distribution of the two-dimensional random variable $(Y_1, Y_2)$. These profiles represent the conditional probability distributions of $Y_2$ at specific values of $Y_1$.

In the previous discussion of discrete multidimensional random variables, it was brought out that the marginal probability of a specific value of one of the primary variables was the sum of the probabilities of events yielding that value (Equations 9.2 and 9.3). In the case of continuous multidimensional random variables the basic concept of a marginal probability is still valid. Figure 9.2 is a graphic representation of the probability distribution of the two-dimensional random variable $(X_1, X_2)$. The profiles across the solid represent the probability distributions of $X_2$ at specific values of $X_1$. The volume under such a profile is a measure of the probability of occurrence of the specific $X_1$ with which the profile is associated.

$$P(X_1 = X_{1i}) = \Delta X_1 \int_{x_2iA}^{x_2iB} f(z)dx_2 \tag{9.14}$$

where $\Delta X_1$ = infinitesimally small increment from one value of $X_1$ to the next adjacent value
$x_{2iA}$ = minimum possible value of $X_2$ when $X_1 = x_{1i}$
$x_{2iB}$ = maximum possible value of $X_2$ when $X_1 = x_{1i}$

Since $\Delta X_1$ is essentially equal to zero, $P(X_1 = x_{1i})$ is infinitesimally small. However, if $\Delta X_1$ is considered to be simply a constant, the area under the curve is representative of the marginal probability of the specific $X_1$. The function describing the distribution of such marginal probabilities is the *marginal density function*:

$$f(x_1) = \begin{cases} \int_{x_{2A}}^{x_{2B}} f(z)dx_2, & \text{when } x_{1A} \leq X_1 \leq x_{1B} \\ 0, & \text{elsewhere} \end{cases} \tag{9.15}$$

There is, of course, a marginal density function for each of the primary variables.

The probabilities associated with the individual values of $X_2$, when $X_1$ is held constant at a specific value, are conditional probabilities, comparable with the discrete conditional probabilities discussed earlier (Equations 9.7 through 9.9). Thus, the function that describes a profile such as these in Figure 9.2 is a *conditional probability density function*:

$$f(x_2 \mid x_{1i}) = \begin{cases} f(z), & \text{when } X_1 = x_{1i} \text{ and } x_{2A} \leq X_2 \leq x_{2B} \\ 0, & \text{elsewhere} \end{cases} \tag{9.16}$$

The actual determination of a joint pdf is made using the multiplication law. If the primary random variables are *independent*, the joint pdf is

$$\begin{aligned} f(z) &= f(x_1, x_2, x_3, \ldots, x_M) \\ &= f(x_1) * f(x_2) * f(x_3) * \ldots * f(x_M) \end{aligned} \tag{9.17}$$

If, on the other hand, the primary variables are *not independent*, the expression is

$$f(z) = f(x_1) * f(x_2 | x_1) * f(x_3 | x_1 x_2) * \cdots f(x_M | x_1 x_2 x_3 \cdots x_{M-1}) \tag{9.18}$$

assuming that the subscripts refer to the order in which the primary variables appear.

Equations 9.17 and 9.18 provide a basis for another way of defining the marginal and conditional probability density functions. Referring to Equation 9.17 and assuming a two-dimensional random variable, $(Y_1, Y_2)$, the marginal density function of $X_1$ is

$$f(x_1) = \frac{f(z)}{f(x_2)} = \frac{f(x_1, x_2)}{f(x_2)} \tag{9.19}$$

while that for $X_2$ is

$$f(x_2) = \frac{f(z)}{f(x_1)} = \frac{f(x_1, x_2)}{f(x_1)}$$

If the variables are not independent, Equation 9.18 indicates that

$$f(x_1) = \frac{f(z)}{f(x_2 \mid x_1)} = \frac{f(x_1, x_2)}{f(x_2 \mid x_1)}$$

and

$$f(x_2) = \frac{f(z)}{f(x_1 \mid x_2)} = \frac{f(x_1, x_2)}{f(x_1 \mid x_2)}$$

As can be seen, the marginal density functions are the pdfs of the primary variables acting alone, *regardless of the independence or lack thereof between the variables.* On the other hand, if dependence exists between the variables, the conditional density functions are

$$f(x_2 \mid x_1) = \frac{f(z)}{f(x_1)} = \frac{f(x_1, x_2)}{f(x_1)}$$

and

$$f(x_1 \mid x_2) = \frac{f(z)}{f(x_2)} = \frac{f(x_1, x_2)}{f(x_2)} \tag{9.20}$$

The cumulative density function, cdf, is

$$\begin{aligned} F(z) &= F(X_1 \le x_{1b},\ X_2 \le x_{2b},\ X_3 \le x_{3b},\ \ldots,\ X_M \le x_{Mb}) \\ &= \int_{-\infty}^{x_{Mb}} \cdots \int_{-\infty}^{x_{3b}} \int_{-\infty}^{x_{2b}} \int_{-\infty}^{x_{1b}} f(z)dx_1 dx_2 dx_3 \ldots dx_M \end{aligned} \tag{9.21}$$

Like all probability distributions, joint probability distributions, discrete or continuous, must meet the requirements of the several laws of probability. Thus, for discrete distributions,

$$0 \le P(x_1, x_2, \ldots, x_M) \le 1$$

for all combinations of $X$ values, and

$$\sum_{i=1}^{\infty} \sum_{j=1}^{\infty} \cdots \sum_{k=1}^{\infty} P(x_{1i}, x_{2j}, \ldots, x_{Mk}) = 1$$

In the case of continuous distributions,

$$0 \le P(z) \le 1$$

for all values of the primary variables within the stated ranges, and

$$\int_{-\infty}^{\infty} f(z)dz = 1$$

which can also be written:

$$\frac{1}{T} \int_{-\infty}^{\infty} \cdots \int_{-\infty}^{\infty} \int_{-\infty}^{\infty} \phi(x_1, x_2, \ldots, x_k)dx_M \ldots dx_2 dx_1 = 1$$

Similarly, the laws of probability apply to the marginal and conditional probability distributions. For example, in the case of the 120 loblolly pine trees:

$0 \leq P(x_1) \leq 1,$ for all values of $X_1$ (see the column labeled $P(x_1)$ in Table 9.2D)

$0 \leq P(x_2) \leq 1,$ for all values of $X_2$ (see the row labeled $P(x_2)$ in Table 9.2D)

$$\sum_{i=1}^{9} P(x_{1i}) = \frac{10}{120} + \frac{13}{120} + \frac{18}{120} + \frac{19}{120} + \frac{22}{120} + \frac{17}{120} + \frac{13}{120} + \frac{5}{120} + \frac{3}{120} = 1$$

$$\sum_{j=1}^{7} P(x_{2j}) = \frac{3}{120} + \frac{12}{120} + \frac{27}{120} + \frac{37}{120} + \frac{31}{120} + \frac{8}{120} + \frac{2}{120} = 1$$

$$0 \leq P(x_{2j} \mid x_{1i}) \leq 1 \quad \text{and} \quad \sum_{j=1}^{n} (Px_{2j} \mid x_{1i}) = 1$$

which can be seen in Table 9.4.

## 9.6 FUNCTIONS OF MULTIDIMENSIONAL RANDOM VARIABLES

In Section 8.8 the concept of a function of a random variable was introduced. This was done within the context of a one-dimensional random variable. The concept, however, extends as well to

**TABLE 9.4**
**Conditional Probabilities of $X_2$ (total height) by Values of $X_1$ (dbh) for the 120 Loblolly Pine Trees Used in Table 9.2**

| | $P(X_2 \mid x_{1i})$ $X_2$ = Total Height, ft | | | | | | | |
|---|---|---|---|---|---|---|---|---|
| $X_1$ = dbh, in. | 30 | 40 | 50 | 60 | 70 | 80 | 90 | $\Sigma(X_2 \mid X_1)$ |
| 8 | $\frac{2}{10}$ | $\frac{4}{10}$ | $\frac{3}{10}$ | $\frac{1}{10}$ | | | | 1 |
| 9 | $\frac{1}{13}$ | $\frac{4}{13}$ | $\frac{5}{13}$ | $\frac{3}{13}$ | | | | 1 |
| 10 | | $\frac{1}{18}$ | $\frac{7}{18}$ | $\frac{6}{18}$ | $\frac{4}{18}$ | | | 1 |
| 11 | | $\frac{2}{19}$ | $\frac{5}{19}$ | $\frac{7}{19}$ | $\frac{5}{19}$ | | | 1 |
| 12 | | $\frac{1}{22}$ | $\frac{4}{22}$ | $\frac{9}{22}$ | $\frac{8}{22}$ | | | 1 |
| 13 | | | $\frac{2}{17}$ | $\frac{7}{17}$ | $\frac{7}{17}$ | $\frac{1}{17}$ | | 1 |
| 14 | | | $\frac{1}{13}$ | $\frac{3}{13}$ | $\frac{5}{13}$ | $\frac{4}{13}$ | | 1 |
| 15 | | | | $\frac{1}{5}$ | $\frac{1}{5}$ | $\frac{2}{5}$ | $\frac{1}{5}$ | 1 |
| 16 | | | | | $\frac{1}{3}$ | $\frac{1}{3}$ | $\frac{1}{3}$ | 1 |

functions of multidimensional variables. For example, assume the two-dimensional random variable $(X_1, X_2)$ and let $Y = X_1 + X_2$, $x_{1A} \le x_1 \le x_{1B}$ and $x_{2A} \le x_2 \le x_{2B}$. Obviously, $Y$ is a one-dimensional function of $(X_1, X_2)$ and is itself a random variable.

Assume further that $Y_1$ and $Y_2$ are discrete, independent of one another, and that their marginal probability distributions are as shown in Table 9.5A. In such a case, the point probability of a specific pair $(x_{1i}, x_{2j})$ is, according to Equation 9.6,

$$P(x_{1i}, x_{2j}) = P(x_{1i}) * P(x_{2j})$$

$$= \frac{N(x_{1i})}{T} * \frac{N(x_{2j})}{T}$$

For example, if $x_{1i} = 3$ and $x_{2j} = 2$,

$$P(3,2) = \left(\frac{4}{15}\right) * \left(\frac{3}{15}\right) = \frac{12}{225}$$

Now consider $Y = X_1 + X_2$. If each pair of $X$ values gave rise to a *unique* $Y$ value, the point probability of that $Y$ value could be equal to the point probability of the pair. However, when $Y = X_1 + X_2$, as can be seen in Table 9.5B, the individual $Y$ values can be obtained from more than one pair of $X$ values. For example, a $Y$ value of 3 can arise when $(X_1 = 0, X_2 = 3)$, $(X_1 = 1, X_2 = 2)$, $(X_1 = 2, X_2 = 1)$, or $(X_1 = 3, X_2 = 0)$. Consequently, the point probability of $Y = 3$ is the sum of the probabilities of these four pairs:

$$P(Y = 3) = P(0,3) + P(1,2) + P(2,1) + P(3,0)$$

$$= \left(\frac{4}{225}\right) + \left(\frac{6}{225}\right) + \left(\frac{6}{225}\right) + \left(\frac{4}{225}\right) = \frac{20}{225}$$

This is typical of functions of multidimensional random variables. Rarely will the joint probability distribution function, whether discrete or continuous, be applicable to the function of the multidimensional variable.

When the multidimensional random variable is discrete and the ranges of values of the primary variables are short, the point probability distribution can be developed by constructing and then partitioning the range space as is done in Table 9.5C. This approach rapidly becomes impractical as the ranges of values of the primary variables increase in size. In some cases algebraic procedures may be used to obtain the probability distribution function. One of these involves the development of the cumulative probability function and from it derivation of the probability distribution function. Another makes use of "probability-generating functions." The following discussion will be limited to the use of the first of these approaches in a single, limited situation. For further information on both approaches reference is made to Hoel et al. (1971).

Consider the discrete two-dimensional random variable $(X_1X_2)$ and the function $Y = X_1 + X_2$. Let $X_1$ and $X_2$ be independent with the marginal probabilities $P(x_1)$ and $P(x_2)$, $x_1 = 0, 1, 2, \ldots, x_{1B}$ and $x_2 = 0, 1, 2, \ldots, x_{2B}$. The subscript $B$ indicates the maximum possible values of the variable. If a plot is made of $X_2$ against $X_1$, each point represents an individual pairing of the primary variables. Since the probabilities of individual $X_1$ and $X_2$ values are not constant, each point has associated with it a column whose height indicates the relative frequency (point probability) of the pair:

$$\frac{N(x_{1i}, x_{2j})}{T} = P(x_{1i}, x_{2j}) = P(x_{1i}) * P(x_{2j})$$

**TABLE 9.5A**
**Range Space for the Discrete Two-Dimensional Random Variable $(X_1, X_2)$, When $x_1 = 0, 1, 2, 3, 4$, and $x_2 = 0, 1, 2, 3, 4$ [the values in parentheses are the subset sizes, $N(x_1, x_2)$]**

| $X_2$ | $P(x_2)$ | | | | | |
|---|---|---|---|---|---|---|
| 4 | $\frac{5}{15}$ | (5) | (10) | (15) | (20) | (25) |
| 3 | $\frac{4}{15}$ | (4) | (8) | (12) | (16) | (20) |
| 2 | $\frac{3}{15}$ | (3) | (6) | (9) | (12) | (15) |
| 1 | $\frac{2}{15}$ | (2) | (4) | (6) | (8) | (10) |
| 0 | $\frac{1}{15}$ | (1) | (2) | (3) | (4) | (5) |
| $X_1 =$ | | 0 | 1 | 2 | 3 | 4 |
| $P(x_1) =$ | | $\frac{1}{15}$ | $\frac{2}{15}$ | $\frac{3}{15}$ | $\frac{4}{15}$ | $\frac{5}{15}$ |

**TABLE 9.5B**
**Values of $Y = X_1 + X_2$ (in parentheses)**

| $X_2$ | | | | | |
|---|---|---|---|---|---|
| 4 | (4) | (5) | (6) | (7) | (8) |
| 3 | (3) | (4) | (5) | (6) | (7) |
| 2 | (2) | (3) | (4) | (5) | (6) |
| 1 | (1) | (2) | (3) | (4) | (5) |
| 0 | (0) | (1) | (2) | (3) | (4) |
| $X_1 =$ | 0 | 1 | 2 | 3 | 4 |

**TABLE 9.5C**
**Subset Sizes, Point Probabilites, and Cumulative Probabilities for $Y$**

| $Y =$ | **0** | **1** | **2** | **3** | **4** | **5** | **6** | **7** | **8** |
|---|---|---|---|---|---|---|---|---|---|
| | 1 | 2 | 3 | 4 | 5 | 10 | 15 | 20 | 25 |
| | | 2 | 4 | 6 | 8 | 12 | 16 | 20 | |
| | | | 3 | 6 | 9 | 12 | 15 | | |
| | | | | 4 | 8 | 10 | | | |
| | | | | | 5 | | | | |
| $N(Y) =$ | 1 | 4 | 10 | 20 | 35 | 44 | 46 | 40 | 25 |
| $P(Y) =$ | $\frac{1}{225}$ | $\frac{4}{225}$ | $\frac{10}{225}$ | $\frac{20}{225}$ | $\frac{35}{225}$ | $\frac{44}{225}$ | $\frac{46}{225}$ | $\frac{40}{225}$ | $\frac{25}{225}$ |
| $P(Y \le y_k) =$ | $\frac{1}{225}$ | $\frac{5}{225}$ | $\frac{15}{225}$ | $\frac{35}{225}$ | $\frac{70}{225}$ | $\frac{114}{225}$ | $\frac{160}{225}$ | $\frac{200}{225}$ | $\frac{225}{225}$ |

This situation is demonstrated in Table 9.5A. Each point on the plot also has a $Y$ value associated with it. If $Y$ is held constant at some value, such as $y_k$, the points associated with the pairs of $X_1$ and $X_2$ values which sum to $y_k$ form a diagonal row across the plot, as can be seen in Table 9.5B. The point probability of $y_k$ is the sum of the point probabilities of the $(X_1, X_2)$ value in the diagonal row. The probability of occurrence of a $Y$ value less than $y_k$ is the sum of the point probabilities of the $(X_1, X_2)$ values associated with the points lying to the left of the diagonal row while the probability of occurrence of a $Y$ value greater than $y_k$ is the sum of the point probabilities of the $(X_1, X_2)$ values whose points are to the right of the diagonal row.

If $y_k$ is incremented in unit increments from 0 to $x_{1B} + x_{2B}$, and the point probabilities cumulated, the result is the cumulative probability distribution of $Y$. If this distribution is to be described algebraically, it must be remembered that $X_1$ can take on all values from 0 to $x_{1B}$ but $X_2$ is constrained since $Y = X_1 + X_2$ and, thus, $X_2 = Y - X_1$; then,

$$P(Y \leq y_i) = \sum_{x_1=0}^{y_k} \sum_{x_2=0}^{y_k - x_1} P(X_1) * P(X_2)$$

If $P(Y - X_1)$ is substituted for $P(X_2)$, the second summation must be modified to recognize that $Y$ rather than $X_2$ is being incremented. Since a unit increment in $X_2$ results in a unit increment in $Y_1$, the scale factor between the two is equal to unity, and the summation can begin with $Y = 0$. Furthermore, since $X_2$ cannot take on a negative value, the probability of a negative $Y - X_1$ value would be equal to zero. Consequently, $Y$ can be incremented until it equals $y_k$ and the function becomes

$$P(Y \leq y_k) = \sum_{x_1=0}^{y_k} \sum_{y=0}^{y_k} P(x_1) * P(y - x_1)$$

Reversing the order of summation,

$$P(Y \leq y_k) = \sum_{y=0}^{y_k} \sum_{x_1=0}^{y_k} P(x_1) * P(y - x_1) \tag{9.22}$$

This is the cumulative probability distribution function for $Y = X_1 + X_2$. Cumulative probability distribution functions cumulate the probabilities of individual values of the variable of interest, in this case, $Y$. The inner summation provides these individual probabilities and thus is the point probability function:

$$P(y) = \begin{cases} \sum_{x_1=0}^{y_k} P(x_1) * P(y - x_1), & \begin{array}{l} x_1 = 0, 1, 2, \ldots, x_{1B} \\ y = 0, 1, 2, \ldots, x_{1B} + x_{2B} \end{array} \\ 0, & \text{elsewhere} \end{cases} \tag{9.23}$$

For example, when $Y = 3$, assuming the situations in Table 9.5:

$$P(Y=3) = P(X_1 = 0) * P(Y - X_1 = 3 - 0 = 3 = X_2) = \left(\frac{1}{15}\right)\left(\frac{4}{15}\right) = \frac{4}{225}$$

$$= P(X_1 = 1) * P(Y - X_1 = 3 - 1 = 2 = X_2) = \left(\frac{2}{15}\right)\left(\frac{3}{15}\right) = \frac{6}{225}$$

$$= P(X_1 = 2) * P(Y - X_1 = 3 - 2 = 1 = X_2) = \left(\frac{3}{15}\right)\left(\frac{2}{15}\right) = \frac{6}{225}$$

$$= P(X_1 = 3) * P(Y - X_1 = 3 - 3 = 0 = X_2) = \left(\frac{4}{15}\right)\left(\frac{1}{15}\right) = \frac{4}{225}$$

$$\frac{20}{225}$$

When the multidimensional random variable is continuous, the procedure used to arrive at the pdf of a function of that variable is analagous to that used when the random variable is discrete. Assume that $X_1$ and $X_2$ are continuous independent random variables whose pdfs are $g(x_1)$, $-\infty \leq X_1 \leq \infty$, and $h(x_2)$, $-\infty \leq X_2 \leq \infty$, respectively. The joint pdf of $X_1$ and $X_2$ is then

$$f(x_1, x_2) = g(x_1)h(x_2)$$

and the cdf is

$$F(x_1, x_2) = \int_{-\infty}^{\infty}\int_{-\infty}^{\infty} f(x_1 x_2)dx_2 dx_1$$

Further assume that $Y = K_1(X_1, X_2)$ and that we desire to determine its pdf. To do this it is necessary to develop a second function of $(X_1, X_2)$, say, $Z = K_2(X_1, X_2)$. To keep mathematical complications at a minimum, this auxilliary function should be kept as simple as possible. $(Y, Z)$ is then a multidimensional random variable with its own pdf and cdf. To obtain this cdf, say, $L(y, z)$, one must first determine the values of $X_1$ and $X_2$ in terms of $(Y, Z)$:

$$x_1 = L_1(y, z) \quad \text{and} \quad x_2 = L_2(y, z)$$

$L_1(y, z)$ and $L_2(y, z)$ are then substituted for $x_1$ and $x_2$ in $F(x_1 x_2)$. Since the increments of integration may change in this process a transformation term must be introduced into the expression. The cdf of $(Y, Z)$ is then

$$M(y, z) = \int_{-\infty}^{\infty}\int_{-\infty}^{\infty} f[L_1(y, z),\ L_2(y, z)] \mid J \mid dz dy \tag{9.24}$$

where

$$J(y, z) = \begin{vmatrix} \dfrac{\partial x_1}{\partial y} & \dfrac{\partial x_1}{\partial z} \\ \dfrac{\partial x_2}{\partial y} & \dfrac{\partial x_2}{\partial z} \end{vmatrix} = \frac{\partial x_1}{\partial y} * \frac{\partial x_2}{\partial z} - \frac{\partial x_2}{\partial y} * \frac{\partial x_1}{\partial z} = \frac{\partial(x_1, x_2)}{\partial(y, z)} \tag{9.25}$$

$J(y, z)$ is referred to as the *Jacobian* of the transformation from $(x_1, x_2)$ to $(y, z)$. In this case the original random variable is two dimensional and thus the Jacobian is a $2 \times 2$ determinant. If the original random variable had been $n$-dimensional, the Jacobian would have been an $n \times n$ determinant. As was brought out in Section 8.8, the transformation process may produce a negative probability since the differential obtained in the process may be negative. The Jacobian is the equivalent of that differential and may be negative. Consequently, the absolute value of the Jacobian is used in the expression for the cdf.

Since the cdf is the integral of the pdf, the joint pdf of $(Y, Z)$ is

$$m(y,z) = f[L_1(y,z),\ L_2(y,z)]\,|J| \tag{9.26}$$

Then, the pdf of $Y$ is its marginal density function (see Equation 9.15):

$$p(y) = \int_{-\infty}^{\infty} f[L_1(y,z),\ L_2(y,z)]\,|J|\,dz \tag{9.27}$$

In order for the preceding development to be valid, the following conditions must be met:

1. The equation $y = K_1(x_1, x_2)$ and $Z = K_2(x_1, x_2)$ must be such that unique solutions for $x_1$ and $x_2$, in terms of $y$ and $z$, can be obtained.
2. The partial derivatives used in the Jacobian must exist and be continuous.

Furthermore, as will be seen in the following discussion, it is imperative that the limits of integration be properly identified. In some cases it will be necessary to break the total range of integration into segments. The process of determining these partial ranges can be tricky and great care must be taken or serious errors can be introduced into the pdf.

### 9.6.1 The PDF of a Sum

Assume that $X_1$ and $X_2$ are independent random variables with the following pdfs:

$$g(x_1) = 5e^{-5x}1, \quad x_1 \geq 0$$

$$h(x_2) = 3e^{-3x}1, \quad x_2 \geq 0$$

Let $Y = X_1 + X_2$ and $Z = X_2$. Then $y = x_1 + x_2$ and $z = x_2$. Consequently, $x_1 = L_1(y, z) = y - z$ and $x_2 = L_2(y, z) = z$. Then,

$$\frac{\partial x_1}{\partial y} = 1, \quad \frac{\partial x_1}{\partial z} = -1, \quad \frac{\partial x_2}{\partial y} = 0, \quad \text{and} \quad \frac{\partial x_2}{\partial z} = 1$$

and

$$J(y,z) = \begin{vmatrix} 1 & -1 \\ 0 & 1 \end{vmatrix} = 1*1 - 0*(-1) = 1$$

It should be noted that *in the case of a sum*, regardless of the pdfs of the primary variables, *the absolute value of the Jacobian is always equal to one.*

The pdf of $Y$ is then

$$p(y) = \int_{-\infty}^{\infty} g(y-z)h(z)\,|1|\,dz, \quad y \geq 0$$

Since $X_1 \geq 0$ and $X_2 \geq 0$, $Y$ must be >0. Consequently, the range of integration must be modified. Since $(Y - Z) = X_1$, $(Y - Z) \geq 0$, and since $Z = X_2$, $Z \geq 0$. Furthermore, since $(Y - Z) \geq 0$, $Y \geq Z$. Combining these inequalities yields $0 \leq Z \leq y$. As a result, the range of integration for Z is from 0 to $y$:

$$p(y) = \int_0^y 5e^{-5(y-z)} * 3e^{-3z}\,dz$$

$$= 15\int_0^y e^{-5y+5z-3z}\,dz$$

$$= 15e^{-5y}\int_0^y e^{2z}\,dz$$

$$= 15e^{-5y} * \frac{e^{2z}}{z}\Big|_0^y$$

$$= 15e^{-5y} * \frac{e^{2y}}{z} - \frac{1}{2}$$

$$= \frac{15}{2}(e^{-3y} - e^{-5y}), \quad y \geq 0$$

The validity of this pdf can be checked by integrating it with respect to $y$:

$$\int_0^{\infty} p(y)dy = \frac{15}{2}\left[\int_0^{\infty} e^{-3y}dy - \int_0^{\infty} e^{-5y}dy\right]$$

$$= \frac{15}{2}\left[\left(e^{-3y}\Big|_0^{\infty}\right) - \left(-e^{-5y}\Big|_0^{\infty}\right)\right]$$

$$= \frac{15}{2}\left(\frac{2}{15}\right) = 1$$

In the above case the problem of the limits of integration was easily solved. In the following example, the problem is more complex. Assume that the pdfs of $X_1$ and $X_2$ are, respectively,

$$g(x_1) = \frac{2x_1}{75}; \quad 5 \leq x_1 \leq 10, \ 0 \ \text{elsewhere}$$

$$h(x_2) = \frac{x_2}{150}; \quad 10 \leq x_2 \leq 20, \ 0 \ \text{elsewhere}$$

$$\text{Then} \quad Y = X_1 + X_2, \ Z = X_2, \ \text{and thus}$$

$$Y = X_1 + X_2 = X_1 + Z$$

$$Z = X_2$$

$$X_1 = Y - Z$$

$$\text{and} \quad X_2 = Z$$

Consequently, $5 \le (Y - Z) \le 10$ and $10 \le Z \le 20$. Furthermore, $(5 + 10) \le Y \le (10 + 20)$ or $15 \le Y \le 30$. Then,

$$5 \le (Y - Z) \quad \text{and} \quad (Y - Z) \le 10$$

$$Z \le (Y - 5) \qquad\qquad Y - 10 \le Z$$

As a result, $(Y - 10) \le Z \le (Y - 5)$. However, since $Z$ cannot assume values less than 10, $(Y - 10)$ must be greater than or equal to 10. Consequently, $Y \ge 20$. Furthermore, since $Z$ cannot assume values greater than 20, $(Y - 5)$ must be less than or equal to 20 and, thus, $Y \le 25$. This means that *when* $(Y - 10) \le Z \le (Y - 5)$, $20 \le Y \le 25$. Since the total range of $Y$ is $15 \le Y \le 30$, this means that there are at least three segments to the pdf of $Y$: that when $15 \le Y \le 20$, when $20 \le Y \le 25$, and when $25 \le Y \le 30$.

Each of the above ranges of $y$ applies to a separate pdf. These can be labeled $p(y_A)$, $p(y_B)$, and $p(y_C)$, respectively. Then each of these is integrated across its appropriate range of values of $Y$ and these integrals are summed; the sum should equal unity:

$$\int_{15}^{20} p(y_A)dy + \int_{20}^{25} p(y_B)dy + \int_{25}^{30} p(y_C)dy = 1$$

To obtain these pdfs, the pdf of $(Y, Z)$ is integrated across the ranges of $Z$ corresponding to the ranges of $Y$. We already know that for $p(y_B)$ the integration range is $(Y - 10) \le Z \le (Y - 5)$. When $15 \le Y \le 20$, the range of integration of $Z$ must begin with 10 since that is the minimum value of $Z$. Since $Y = X_1 + Z$, $Z = Y - X_1$. The range of $X_1$ is $5 \le X_1 \le 10$. This means that $X_1$ cannot be less than 5 and $Z$ cannot be more than $(Y - 5)$. Consequently, for $p(y_A)$ the range of integration is $10 \le Z \le (Y - 5)$. When $25 \le Y \le 30$, the upper limit of the range of integration must be 20 since that is the maximum value of $Z$. Since $X_1$ cannot exceed 10, $Z$ cannot be less than $(Y - 10)$ and thus the range of integration for $p(y_C)$ is $(Y - 10) \le Z \le 20$.

The three pdfs are then

$$p(y_A) = \int_{10}^{(y-5)} g(y-z)h(z)dz = \int_{10}^{(y-5)} \left[\frac{2(y-z)}{75} * \frac{1}{150}\right] dz = \frac{1}{5625}\int_{10}^{(y-5)} (yz - z^2)dz$$

$$= \frac{1}{33,750}(y^3 - 375y + 2250)$$

$$p(y_B) = \frac{1}{5625}\int_{(y-10)}^{(y-5)} (yz - z^2)dz = \frac{1}{33,750}(225y - 1750)$$

$$p(y_C) = \frac{1}{5625}\int_{(y-10)}^{20} (yz - z^2)dz = \frac{1}{33,750}(1500y - y^3 - 18,000)$$

The integrals of these pdfs, using the appropriate ranges of $y$, sum to 1.

If $Y$ is the sum of more than two random variables, the determination of $p(y)$ follows the preceding procedure but requires $M - 1$ cycles. For example, if $Y = X_1 + X_2 + X_3$, $Z$ would be defined as $X_2 + X_3$, and

$$p(y) = \int_{-\infty}^{\infty} g(y-z)h(z)dz$$

However, $Z$ itself is now a sum of two random variables and, consequently, its pdf must be determined before the computation of $p(y)$ can proceed. To this end, let $W = X_3$ and then

$$h(z) = \int_{-\infty}^{\infty} k(z-w)l(w)dw$$

Now $p(y)$ can be determined.

### 9.6.2 The pdf of a Product

Let $X_1$ and $X_2$ be independent random variables with the pdfs

$$g(x_1) = \frac{2x_1}{75}; \quad 5 \leq X_1 \leq 0, \ 0, \text{ elsewhere}$$

$$\text{and} \quad h(x_2) = \frac{x_2}{150}; \quad 10 \leq X_2 \leq 0, \ 0, \text{ elsewhere}$$

which were used in the preceding discussion. Then let $Y = X_1 * X_2$ and $Z = X_2$. Consequently, $Y = X_1X_2$, $Z = X_2$, $X_1 = L_1(y, z) = y/z$, and $X_2 = L_2(y, z) = Z$. The partial derivatives are

$$\frac{\partial x_1}{\partial y} = z, \quad \frac{\partial x_1}{\partial z} = yz, \quad \frac{\partial x_2}{\partial y} = 0, \quad \text{and} \quad \frac{\partial x_2}{\partial z} = 1$$

The Jacobian is

$$J(y,z) = \begin{vmatrix} z^{-1} & -yz^{-2} \\ 0 & 1 \end{vmatrix} = z^{-1}$$

The pdf of $Y$ is

$$p(y) = \int_{-\infty}^{\infty} g\left(\frac{y}{z}\right)h(z)\,|\,J\,|\,dz$$

$$= \int_{-\infty}^{\infty} \left[\frac{2}{75}\left(\frac{y}{z}\right) * \left(\frac{z}{150}\right) * \left(\frac{1}{z}\right)\right]dz$$

$$= \frac{1}{5625}\int_{-\infty}^{\infty}\left(\frac{y}{z}\right)dz$$

At this point it is necessary to determine the range of integration. Since

$$\left(\frac{Y}{Z}\right) = X_1, \quad 5 \le \left(\frac{Y}{Z}\right) \le 10$$

and since $Z = X_2$, $10 \le Z \le 20$. In addition, $(5 * 10) \le Y \le (10 * 20)$ or $50 \le Y \le 200$. Then,

$$5 \le \left(\frac{Y}{Z}\right) \quad \text{and} \quad \left(\frac{Y}{Z}\right) \le 10$$

$$z \le \left(\frac{Y}{5}\right) \qquad \left(\frac{Y}{10}\right) \le Z$$

Consequently,

$$\left(\frac{Y}{10}\right) \le Z \le \left(\frac{Y}{5}\right)$$

Since $10 \le Z$, $(Y/10)$ must be $\ge 10$, which means that $Y \ge 100$, and since $Z \le 20$, $(Y/5)$ must be $\le 20$, which means that $Y \le 100$. Thus, when $(Y/10) \le Y \le (Y/5) 100 \le Y \le 100$, indicating no range at all. This means that $p(y)$ involves two subdivisions, $p(y_A)$ when $50 \le Y \le 100$ and $p(y_B)$ when $100 \le Y \le 200$. The problem now is to determine the ranges of $Z$ which correspond to these ranges of $Y$.

Consider that $10 \le Z \le 20$ and $(Y/10) \le Z \le (Y/5)$. Setting $Z = 10$, its minimum, yields $(Y/10) \le 10$ or $Y \le 100$ and $10 \le (Y/5)$ or $50 \le Y$. Setting $Z = 20$, its maximum, yields $(Y/10) \le 20$ or $Y \le 200$ and $20 \le (Y/5)$ or $100 \le Y$. The range of $Z$ that matches $50 \le Y \le 100$ is then $10 \le Z \le (Y/5)$ and that which matches $100 \le Y \le 200$ is $(Y/10) \le Z \le 20$. Consequently,

$$p(y_A) = \frac{1}{5625}\int_{10}^{y/5}\left(\frac{y}{z}\right)dz = \frac{y}{5625}(\ln y - \ln 50)$$

$$\text{and} \quad p(y_B) = \frac{1}{5625}\int_{y/10}^{20}\left(\frac{y}{z}\right)dz = \frac{y}{5625}(\ln 200 - \ln y)$$

This can be checked by integrating each of these expressions across its appropriate range of $y$ values.

$$\int_{50}^{100} p(y_A)dy + \int_{100}^{200} p(y_B)dy = 1$$

### 9.6.3 The PDF of a Quotient

Assume the same primary random variables, $X_1$ and $X_2$, used in the previous discussion and let $Y = X_1/X_2$ and $Z = X_2$. Then, $Y = X_1/X_2$, $Z = X_2$, $X_1 = L_1(y, z) = YZ$, and $X_2 = L_2(y, z) = Z$. The partial derivatives are

$$\frac{\partial x_1}{\partial y} = z, \quad \frac{\partial x_1}{\partial z} = y, \quad \frac{\partial x_2}{\partial y} = 0, \quad \text{and} \quad \frac{\partial x_2}{\partial z} = 1$$

The Jacobian is

$$J(y,z) = \begin{vmatrix} z & y \\ 0 & 1 \end{vmatrix} = z$$

The pdf of $Y$ is

$$p(y) = \int_{-\infty}^{\infty} g(yz)h(z)\,|\,J\,|\,dz$$

$$= \int_{-\infty}^{\infty} \frac{2}{75}(yz) * \frac{z}{150} * zdz$$

$$= \frac{1}{5625}\int_{-\infty}^{\infty} yz^3 dz$$

Since $YZ = X$, $5 \leq YZ \leq 10$, and since $Z = X_2$, $10 \leq Z \leq 20$. Furthermore, $(5/20) \leq Y \leq (10/10)$ or $(1/4) \leq Y \leq 1$. Then,

$$5 \leq YZ \quad \text{and} \quad YZ \leq 10$$

$$\frac{5}{Y} \leq Z \qquad Z \leq \frac{10}{Y}$$

As a result, $(5/Y) \leq Z\ (10/Y)$. Since $10 \leq Z$, $(5/Y)$ must be $\geq 10$ and since $Z \leq 20$, $10/7$ must be $\leq 20$. Then $1/2 \leq Y \leq 1/2$, indicating that $Y$ has no range and that $p(y)$ is made up of two subdivisions, $p(y_A)$, when $1/4 \leq Y \leq 1/2$, and $p(y_B)$, when $1/2 \leq Y \leq 1$. Setting $Z = 10$, its minimum, yields $5/Y \leq 10$ or $1/2 \leq Y$ and $10 \leq 10/Y$ or $Y \leq 1$. Setting $Z = 20$, its maximum, yields $5/Y \leq 20$ or $1/4 \leq Y$ and $20 \leq 10/Y$ or $Y \leq 1/2$. Then, the range of $Z$ matching $1/4 \leq Y \leq 1/2$ is $5/Y \leq Z \leq 20$ and that matching $1/2 \leq Y \leq 1$ is $10 \leq Z \leq 10/y$. The pdf of $Y$ then made up of

$$p(y_A) = \frac{1}{5625}\int_{10}^{10/y} yz^3 dz = \frac{2500}{5625}(y^{-3} - y)$$

$$\text{and} \quad p(y_B) = \frac{1}{5625}\int_{5/y}^{20} yz^3 dz = \frac{1}{5625}\left(40{,}000y - \frac{625}{y}y^{-3}\right)$$

This can be confirmed using

$$\int_{1/4}^{1/2} p(y_A)dy + \int_{1/2}^{1} p(y_B)dy = 1$$

## 9.7 EXPECTATION OF FUNCTIONS OF MULTIDIMENSIONAL RANDOM VARIABLES

It was brought out in Section 8.10 that the expected value of a random variable is the weighted mean of the values of the variable when the weights are the probabilities of occurrence of those values. Multidimensional random variables, as such, do not possess means, but functions of multidimensional variables, being one-dimensional random variables, do have means and, consequently, expected values. The procedures used to obtain such expected values will be based on one or more of the following properties of expected values. In this discussion it will be assumed that the random variables have finite expectations.

1. The expected value of a constant is equal to the constant.

$$E(K) = K$$

where $K$ = constant

The proof:
If $Y = K$ and $Y$ is assumed to be a discrete variable,

$$E(Y) = \sum_{i=1}^{\infty} [KP(y_1)]$$

$$= K\sum_{i=1}^{\infty} [P(y_1)]$$

Since $\Sigma_{i=1}^{\infty}$ $[P(y_i)] = 1$,

$$E(Y) = K$$

Similarly, if it is assumed that $Y$ is a continuous variable

$$E(Y) = \int_{-\infty}^{\infty} Kf(y)dY$$

$$= K\int_{-\infty}^{\infty} f(y)dY$$

Since $\int_{-\infty}^{\infty} f(y)dY = 1$,

$$E(Y) = K$$

2. The expected value of the product of a constant and a random variable is equal to the product of the constant and the expected value of the random variable.

$$E(KY) = KE(Y)$$

The proof:
If $Y$ is a discrete variable,

$$E(KY) = \sum_{i=1}^{\infty} [Ky_iP(y_i)]$$

$$= K\sum_{i=1}^{\infty} [y_iP(y_i)]$$

Since $\Sigma_{i=1}^{\infty} [y_i P(y_i)] = \mu_Y = E(Y)$,

$$E(KY) = KE(Y)$$

and if $Y$ is a continuous variable,

$$E(KY) = \int_{-\infty}^{\infty} KYf(y)dY$$

$$= K\int_{-\infty}^{\infty} Yf(y)dY$$

Since $\int_{-\infty}^{\infty} yf(y)dy = \mu_Y = E(Y)$,

$$E(KY) = KE(Y)$$

3. The expected value of the sum of two random variables, each of which is a function of a multidimensional random variable, is equal to the sum of the expected values of the two random variables. Assuming that the multidimensional random variable is two dimensional,

$$E(V + W) = E(V) + E(W)$$

where $V = g_1(X_1, X_2)$
$W = g_2(X_1, X_2)$

The proof:
If $X_1$ and $X_2$ are discrete variables:

$$E(V+W) = \sum_{i=1}^{\infty}\sum_{j=1}^{\infty}[(V_{ij} + W_{ij})P(x_{1i}, x_{2j})]$$

$$= \sum_{i=1}^{\infty}\sum_{j=1}^{\infty}[v_{ij}P(x_{1i}, x_{2j})] + \sum_{i=1}^{\infty}\sum_{j=1}^{\infty}[w_{ij}P(x_{1i}, x_{2j})]$$

$$= E(V) + E(W)$$

For example, assume a pair of fair dice, one red and one white. Let $X_1$ equal the value appearing on the red die and $X_2$ equal the white die. $(X_1, X_2)$ is the two-dimensional random variable. Now let $V = X_1 + X_2$ and $W = X_1 - X_2$. The computations leading to $E(V)$ are

| | $V = X_1 + X_2$ | | | | | |
|---|---|---|---|---|---|---|
| $X_1 =$ | $X_2 =$ 1 | 2 | 3 | 4 | 5 | 6 |
| 1 | 2 | 3 | 4 | 5 | 6 | 7 |
| 2 | 3 | 4 | 5 | 6 | 7 | 8 |
| 3 | 4 | 5 | 6 | 7 | 8 | 9 |
| 4 | 5 | 6 | 7 | 8 | 9 | 10 |
| 5 | 6 | 7 | 8 | 9 | 10 | 11 |
| 6 | 7 | 8 | 9 | 10 | 11 | 12 |

| | | | | | | | | | | | | |
|---|---|---|---|---|---|---|---|---|---|---|---|---|
| $k$ | = | 1 | 2 | 3 | 4 | 5 | 6 | 7 | 8 | 9 | 10 | 11 |
| $w_k$ | = | 2 | 3 | 4 | 5 | 6 | 7 | 8 | 9 | 10 | 11 | 12 |
| $N_k$ | = | 1 | 2 | 3 | 4 | 5 | 6 | 5 | 4 | 3 | 2 | 1 |
| $P(w_k)$ | = | $\frac{1}{36}$ | $\frac{2}{36}$ | $\frac{3}{36}$ | $\frac{4}{36}$ | $\frac{5}{36}$ | $\frac{6}{36}$ | $\frac{5}{36}$ | $\frac{4}{36}$ | $\frac{3}{36}$ | $\frac{2}{36}$ | $\frac{1}{36}$ |

$$E(V) = \frac{[2(1)+3(2)+4(3)+5(4)+6(5)+7(6)+8(5)+9(4)+10(3)+11(2)+12(1)]}{36}$$

$$= 7.0$$

The computations leading to $E(W)$ are

| $X_1 =$ | $X_2 =$ 1 | 2 | 3 | 4 | 5 | 6 |
|---|---|---|---|---|---|---|
| | $w = X_1 - X_2$ | | | | | |
| 1 | 0 | –1 | –2 | –3 | –4 | –5 |
| 2 | 1 | 0 | –1 | –2 | –3 | –4 |
| 3 | 2 | 1 | 0 | –1 | –2 | –3 |
| 4 | 3 | 2 | 1 | 0 | –1 | –2 |
| 5 | 4 | 3 | 2 | 1 | 0 | –1 |
| 6 | 5 | 4 | 3 | 2 | 1 | 0 |

| | | | | | | | | | | | | |
|---|---|---|---|---|---|---|---|---|---|---|---|---|
| $k$ | = | 1 | 2 | 3 | 4 | 5 | 6 | 7 | 8 | 9 | 10 | 11 |
| $w_k$ | = | –5 | –4 | –3 | –2 | –1 | 0 | 1 | 2 | 3 | 4 | 5 |
| $N_k$ | = | 1 | 2 | 3 | 4 | 5 | 6 | 5 | 4 | 3 | 2 | 1 |
| $P(w_k)$ | = | $\frac{1}{36}$ | $\frac{2}{36}$ | $\frac{3}{36}$ | $\frac{4}{36}$ | $\frac{5}{36}$ | $\frac{6}{36}$ | $\frac{5}{36}$ | $\frac{4}{36}$ | $\frac{3}{36}$ | $\frac{2}{36}$ | $\frac{1}{36}$ |

$$E(V) = \frac{[(-5)(1)+(-4)(2)+(-3)(3)+(-2)(4)+(-1)(5)+(0)(6)+(1)(5)+(2)(4)+(3)(3)+(4)(2)+(5)(1)]}{36}$$

$$= 0.0$$

The computations leading the $E(V + W)$ are as follows. The numbers in the parentheses are the subset sizes. $N_k$ is the sum of the subset sizes associated with $(V + W)_k$.

| $V =$ | $W =$ –5 | –4 | –3 | –2 | –1 | 0 | 1 | 2 | 3 | 4 | 5 |
|---|---|---|---|---|---|---|---|---|---|---|---|
| | $V + W =$ | | | | | | | | | | |
| 2 | –3 | –2 | –1 | 0 | 1 | 2 | 3 | 4 | 5 | 6 | 7 |
| | (1) | (2) | (3) | (4) | (5) | (6) | (5) | (4) | (3) | (2) | (1) |
| 3 | –2 | –1 | 0 | 1 | 2 | 3 | 4 | 5 | 6 | 7 | 8 |
| | (2) | (4) | (6) | (8) | (10) | (12) | (10) | (8) | (6) | (4) | (2) |
| 4 | –1 | 0 | 1 | 2 | 3 | 4 | 5 | 6 | 7 | 8 | 9 |
| | (3) | (6) | (9) | (12) | (15) | (18) | (15) | (12) | (9) | (6) | (3) |
| 5 | 0 | 1 | 2 | 3 | 4 | 5 | 6 | 7 | 8 | 9 | 10 |
| | (4) | (8) | (12) | (16) | (20) | (24) | (20) | (16) | (12) | (8) | (4) |

| | | V + W = | | | | | | | | | | |
|---|---|---|---|---|---|---|---|---|---|---|---|---|
| **V =** | **W =** | **–5** | **–4** | **–3** | **–2** | **–1** | **0** | **1** | **2** | **3** | **4** | **5** |
| 6 | | 1 | 2 | 3 | 4 | 5 | 6 | 7 | 8 | 9 | 10 | 11 |
| | | (5) | (10) | (15) | (20) | (25) | (30) | (25) | (20) | (15) | (10) | (5) |
| 7 | | 2 | 3 | 4 | 5 | 6 | 7 | 8 | 9 | 10 | 11 | 12 |
| | | (6) | (12) | (18) | (24) | (30) | (36) | (30) | (24) | (18) | (12) | (6) |
| 8 | | 3 | 4 | 5 | 6 | 7 | 8 | 9 | 10 | 11 | 12 | 13 |
| | | (5) | (10) | (15) | (20) | (25) | (30) | (25) | (20) | (15) | (10) | (5) |
| 9 | | 4 | 5 | 6 | 7 | 8 | 9 | 10 | 11 | 12 | 13 | 14 |
| | | (4) | (8) | (12) | (16) | (20) | (24) | (20) | (15) | (12) | (5) | (4) |
| 10 | | 5 | 6 | 7 | 8 | 9 | 10 | 11 | 12 | 13 | 14 | 15 |
| | | (3) | (6) | (9) | (12) | (15) | (18) | (15) | (12) | (9) | (6) | (3) |
| 11 | | 6 | 7 | 8 | 9 | 10 | 11 | 12 | 13 | 14 | 15 | 16 |
| | | (2) | (4) | (6) | (8) | (10) | (12) | (10) | (8) | (6) | (4) | (2) |
| 12 | | 7 | 8 | 9 | 10 | 11 | 12 | 13 | 14 | 15 | 16 | 17 |
| | | (1) | (2) | (3) | (4) | (5) | (6) | (5) | (4) | (3) | (2) | (1) |
| $k$ | = | 1 | 2 | 3 | 4 | 5 | 6 | 7 | 8 | 9 | 10 | 11 |
| $(v+w)_k$ | = | –3 | –2 | –1 | 0 | 1 | 2 | 3 | 4 | 5 | 6 | 7 |
| $N_k$ | = | 1 | 4 | 10 | 20 | 35 | 56 | 80 | 104 | 125 | 140 | 146 |
| $P(v+w)_k$ | = | $\frac{1}{1296}$ | $\frac{4}{1296}$ | $\frac{10}{1296}$ | $\frac{20}{1296}$ | $\frac{35}{1296}$ | $\frac{56}{1296}$ | $\frac{80}{1296}$ | $\frac{104}{1296}$ | $\frac{125}{1296}$ | $\frac{140}{1296}$ | $\frac{146}{1296}$ |

$$E(V) = \frac{[(-5)(1)+(-4)(2)+(-3)(3)+(-2)(4)+(-1)(5)+(0)(6)+(1)(5)+(2)(4)+(3)(3)+(4)(2)+(5)(1)]}{36}$$

$$= 0.0$$

| | | | | | | | | | | |
|---|---|---|---|---|---|---|---|---|---|---|
| $k$ | = 12 | 13 | 14 | 15 | 16 | 17 | 18 | 19 | 20 | 21 |
| $(v+w)_k$ | = 8 | 9 | 10 | 11 | 12 | 13 | 14 | 15 | 16 | 17 |
| $N_k$ | = 140 | 125 | 104 | 80 | 56 | 35 | 20 | 10 | 4 | 1 |
| $P(v+w)_k$ | = $\frac{140}{1296}$ | $\frac{125}{1296}$ | $\frac{104}{1296}$ | $\frac{80}{1296}$ | $\frac{56}{1296}$ | $\frac{35}{1296}$ | $\frac{20}{1296}$ | $\frac{10}{1296}$ | $\frac{4}{1296}$ | $\frac{1}{1296}$ |

$$E(V) = [(-3)(1) + (2)(-4) + (-1)(10) + (0)(20) + (1)(35) + (2)(56) + (3)(80) + (4)(104) + (5)(125) + (6)(140) + (7)(146) + (8)(140) + (9)(125) + (10)(104) + (11)(80) + (12)(56) + (13)(35) + (14)(20) + (15)(10) + (16)(4) + (17)(1)]/1296$$

$$= \frac{9072}{1296}$$

$$= 7.0$$

Thus, $E(V + W) = 7.0 = E(V) + E(W) = 7.0 + 0.0$.

The perfect symmetry of the probability distribution which is present in this example is not usually encountered. A more typical situation would be seen if $U = X_1X_2$ is substituted for $W = X_1 - X_2$. The expected value of this sum is 19.25. The computations verifying this value are left to the student.

The proof of Property 3, if $X_1$ and $X_2$ are continuous variables, is as follows:

$$E(V+W) = \int_{-\infty}^{\infty}\int_{-\infty}^{\infty}(v+w)f(z)dx_1dx_2$$

$$= \int_{-\infty}^{\infty}\int_{-\infty}^{\infty}vf(z)dx_1dx_2 + \int_{-\infty}^{\infty}\int_{-\infty}^{\infty}wf(z)dx_1dx_2$$

$$= E(V) + E(W)$$

It should be noted that the pdf, or point probability function, of the multidimensional variable was used to weight the values of the functions $V$ and $W$ in the computations leading to the expected values in spite of the fact that the values of $v$, $w$, or $v + w$, might not be unique. This, however, is not a contradiction since probabilities are additive and since there is a one-to-one relationship between the elements of the range space of the derived values and the range space of the multidimensional random variable. If several sets of the multidimensional variable produce the same value of the derived value, the probability of that value is the sum of the probabilities of the sets producing that value. For example, in the range space for $v$, the value 5 appears four times. The probability of a 5 is then

$$P(V=5) = P(X_1=1,\ X_2=4) + P(X_1=2,\ X_2=3) + P(X_1=3,\ X_2=2) + P(X_1=1,\ X_2=4)$$

$$= \frac{1}{36} + \frac{1}{36} + \frac{1}{36} + \frac{1}{36}$$

$$= \frac{4}{36}$$

which would be the weight of $V = 5$ in the computations of $E(A)$. Thus, the weighting needed to obtain the expected values is accomplished without knowledge of the actual probability distributions of the functions of the multidimensional random variable. This argument applies to the situations covered in the remainder of this discussion of the properties of expected values.

4. The expected value of the sum of any two random variables is equal to the sum of the expected values of the two random variables.

$$E(X_1 + X_2) = E(X_1) + E(X_2)$$

where $X_1$ = a random variable and
$X_2$ = another random variable

It is obvious that the sum $(X_1 + X_2)$ is a function of the two-dimensional random variable $(X_1, X_2)$. Furthermore, the joint probability distribution function or the pdf of $(X_1, X_2)$ is the product of the joint marginal probability function of one of the primary variables and the conditional probability function of the other. In the case of a discrete two-dimensional random variable,

$$P(x_{1i}, x_{2j}) = \begin{cases} P(x_{1i}) * P(x_{2j} \mid x_{1i}), & \text{when } i = 1, 2, \ldots, N_1 \\ & \text{and } j = 1, 2, \ldots, N_2 \\ 0, & \text{elsewhere} \end{cases}$$

while in the case of a continuous variable it would be

$$f(x_1, x_2) = \begin{cases} f(x_1) * f(x_2 \mid x_{1i}), & \text{when } x_{1A} \le x_1 \le x_{1B} \\ & \text{and } x_{2A} \le x_2 \le x_{2B} \\ 0, & \text{elsewhere} \end{cases}$$

If $X_1$ and $X_2$ are independent of one another, the conditional probability terms in these expressions reduce to

$$P(x_{2j}|x_{1i}) = P(x_{2j}) \quad \text{and} \quad f(x_2|x_1) = f(x_2)$$

The expected value of $(X_1 + X_2)$ is then

$$E(X_1 + X_2) = \sum_{i=1}^{\infty} \sum_{j=1}^{\infty} [(x_{1i} + x_{2j})P(x_{1i}, x_{2j})]$$

or

$$E(X_1, X_2) = \int_{-\infty}^{\infty} \int_{-\infty}^{\infty} (x_1 + x_2) f(z) dx_1 dx_2$$

$$= \int_{-\infty}^{\infty} \int_{-\infty}^{\infty} (x_1) f(z) dx_1 dx_2 + \int_{-\infty}^{\infty} \int_{-\infty}^{\infty} (x_2) f(z) dx_1 dx_2$$

$$= E(X_1) + E(X_2)$$

The example used for Property 3 applies to Property 4 if $X_1$ is substituted for $V$ and $X_2$ for $W$.

5. The expected value of the sum of any number of random variables is equal to the sum of the expected values of the individual random variables

$$E(X_1 + X_2 + \cdots + X_k) = E(X_1) + E(X_2) + \cdots + E(X_k)$$

The proof follows that of Property 4.

6. The expected value of the product of the two primary variables in a two-dimensional random variable depends on the degree of relationship existing between the two primary variables. To discuss this it is necessary to introduce the concept of *covariance*. By definition, covariance is

$$\text{COV}(X_1 X_2) = \sigma_{X_1 X_2} = E\{[X_1 - E(X_1)][X_2 - E(X_2)\} \tag{9.28}$$

$$= \text{E}[\text{X}_1 X_2 - X_1 E(X_2) - X_2 E(X_1) + E(X_1)E(X_2)]$$

$$= E(X_1 X_2) - E(X_1)E(X_2) - E(X_2)E(X_1) + E(X_1)E(X_2) \tag{9.29}$$

$$= E(X_1 X_2) - E(X_1)E(X_2)$$

Consequently, the expected value of the product of the primary variables of the two-dimensional random variable is

$$E(X_1X_2) = E(X_1)E(X_2) + \text{COV}(X_1X_2) \tag{9.30}$$

When every variate of one of the primary variables is free to pair with every variate of the second primary variable, the covariance is equal to zero and it is said that the two primary variables are *independent*. This can be shown as follows. According to Equation 9.29,

$$\text{COV}(X_1X_2) = E(X_1X_2) - E(X_1)E(X_2)$$

$$= \frac{\sum_{i=1}^{N(X_1)}\sum_{j=1}^{N(X_2)} x_{1i}x_{2j}}{N(X_1)N(X_2)} - \left[\frac{\sum_{i=1}^{N(X_1)} x_{1i}}{N(X_1)}\right]\left[\frac{\sum_{j=1}^{N(X_2)} x_{2j}}{N(X_2)}\right]$$

where $N(X_1) = N_{X_1}$
$N(X_2) = N_{X_2}$

Then,

$$\text{COV}(X_1X_2) = \frac{1}{N(X_1)N(X_2)}(x_{11}x_{21} + x_{11}x_{22} + x_{11}x_{23} + \cdots$$

$$+x_{11}x_{2N(X_2)} + x_{12}x_{21} + x_{12}x_{22}$$

$$+x_{12}x_{23} + \cdots + x_{12}x_{2N(X_2)} + \cdots$$

$$+x_{1N(X_1)}x_{2N(X_2)}) - \frac{1}{N(X_1)N(X_2)}$$

$$(x_{11}x_{21} + x_{11}x_{22} + x_{11}x_{23} + \cdots$$

$$+x_{11}x_{2N(X_2)} + x_{12}x_{21} + x_{12}x_{22}$$

$$+x_{12}x_{2N(X_2)} + x_{12}x_{21} + x_{12}x_{22}$$

$$+x_{12}x_{23} + \cdots + x_{12}x_{2N(X_2)} + \cdots$$

$$+x_{1N(X_1)}x_{2N(X_2)})$$

Since the second term is identical to the first,

$$\text{COV} = \frac{1}{N(X_1)N(X_2)}(0) = 0$$

Thus, when the variates of the two primary variables are free to pair, the expected product of the two primary variables is

$$E(X_1X_2) = E(X_1)(X_2)$$

which is the same as Equation 9.6.

As an example of this, if the two-dimensional random variable is discrete, referring back to the pair of fair dice with $X_1$ representing the values appearing on the red die and $X_2$ those on the white die, the two-dimensional random variable is $(X_1X_2)$, and the computations for $X_1X_2$ and $E(X_1, X_2)$ are

| | $X_2 =$ | | | | | |
|---|---|---|---|---|---|---|
| $X_1 =$ | **1** | **2** | **3** | **4** | **5** | **6** |
| 1 | 1 | 2 | 3 | 4 | 5 | 6 |
| 2 | 2 | 4 | 6 | 8 | 10 | 12 |
| 3 | 3 | 6 | 9 | 12 | 15 | 18 |
| 4 | 4 | 8 | 12 | 16 | 20 | 24 |
| 5 | 5 | 10 | 15 | 20 | 25 | 30 |
| 6 | 6 | 12 | 18 | 24 | 30 | 36 |

| $k$ | = | 1 | 2 | 3 | 4 | 5 | 6 | 7 | 8 | 9 | 10 | 11 |
|---|---|---|---|---|---|---|---|---|---|---|---|---|
| $(x_1x_2)_k$ | = | 1 | 2 | 3 | 4 | 5 | 6 | 8 | 9 | 10 | 12 | 15 |
| $N_k$ | = | 1 | 2 | 2 | 3 | 2 | 4 | 2 | 1 | 2 | 4 | 2 |
| $P(x_1x_2)_k$ | = | $\frac{1}{36}$ | $\frac{2}{36}$ | $\frac{2}{36}$ | $\frac{3}{36}$ | $\frac{2}{36}$ | $\frac{4}{36}$ | $\frac{2}{36}$ | $\frac{1}{36}$ | $\frac{2}{36}$ | $\frac{4}{36}$ | $\frac{2}{36}$ |

| $k$ | = | 12 | 13 | 14 | 15 | 16 | 17 | 18 |
|---|---|---|---|---|---|---|---|---|
| $(x_1x_2)_k$ | = | 16 | 18 | 20 | 24 | 25 | 30 | 36 |
| $N_k$ | = | 1 | 2 | 2 | 2 | 1 | 2 | 1 |
| $P(x_1x_2)_k$ | = | $\frac{1}{36}$ | $\frac{2}{36}$ | $\frac{2}{36}$ | $\frac{2}{36}$ | $\frac{1}{36}$ | $\frac{2}{36}$ | $\frac{1}{36}$ |

$$\begin{aligned} E(X_1X_2) &= [1(1) + 2(2) + 3(2) + 4(3) + 5(2) + 6(4) + 8(2) + 9(1) + 10(2) \\ &\quad + 12(4) + 15(2) + 16(1) + 18(2) + 20(2) + 24(2) + 25(1) + 30(2) + 36(1)]/36 \\ &= \frac{441}{36} = 12.25 \end{aligned}$$

Since $E(X_1) = [1(1) + 2(1) + 3(1) + 4(1) + 5(1) + 6(1)]/6 = 21.0 = 3.5$ and $E(X_2)$ also equals 3.5,

$$E(X_1X_2) = 12.25 = E(X_1)E(X_2) = 3.5(3.5)$$

In the case of a continuous two-dimensional random variable, the proof follows the same pattern.

$$E(X_1X_2) = \int_{-\infty}^{\infty}\int_{-\infty}^{\infty}(x_1x_2)f(z)dx_1dx_2$$

Since the probability of the joint event ($X_1$, $X_2$) is the product of the probabilities of occurrence of the individual values of $X_1$ and $X_2$,

$$E(X_1X_2) = \int_{-\infty}^{\infty}\int_{-\infty}^{\infty}(x_1x_2)f(x_1)f(x_2)dx_1dx_2$$

$$= \int_{-\infty}^{\infty} x_1f(x_1)dx_1\int_{-\infty}^{\infty} x_2f(x_2)dx_2$$

$$= E(X_1)E(X_2)$$

When the variates of the first primary variable are not free to pair with any of the variates of the second primary variable, or when the drawing of the $i$th sampling unit yields the pair ($x_{1i}$, $x_{2i}$), the covariance may or may not be equal to zero and must be considered in the determination of the expected value of the product of the two primary variables of a two-dimensional random variable. When the pairing is constrained in this manner,

$$N(X_1) = N(X_2) = N$$

and $N$ is the size of the defined universe. Then returning to Equation 9.29,

$$\text{COV}(X_1X_2) = E(X_1X_2) - E(X_1)E(X_2)$$

$$= \frac{\sum_{i=1}^{N}(x_1x_2)_i}{N} - \left(\frac{\sum_{i=1}^{N}x_{1i}}{N}\right)\left(\frac{\sum_{i=1}^{N}x_{2i}}{N}\right)$$

$$= \frac{1}{N}(x_{11}x_{21} + x_{12}x_{22} + \cdots + x_{1N}x_{2N}) \tag{9.31}$$

$$- \frac{1}{N^2}(x_{11}x_{21} + x_{11}x_{22} + x_{11}x_{23} + \cdots$$

$$+x_{11}x_{2N} + x_{12}x_{21} + x_{12}x_{22} + x_{12}x_{23} + \cdots$$

$$+x_{12}x_{2N} + \cdots + x_{1N}x_{2N})$$

In this case the difference between the two terms is not necessarily equal to zero. If the two terms are not in balance, covariance exists. For example, assume that the range space of $X_1$ consists of the set

$$X_1 = \{x_{11} = 1, x_{12} = 2, x_{13} = 3, x_{14} = 4, x_{15} = 5\}$$

and the range space of $X_2$ is

$$X_2 = \{x_{21} = 1, x_{22} = 2, x_{23} = 3, x_{24} = 4, x_{25} = 5\}$$

Then the second term of Equation 9.31 consists of the following products:

| $X_2$ = | $X_1$ = 1 | 2 | 3 | 4 | 5 |
|---|---|---|---|---|---|
| 1 | 1 * 1 = 1 | 1 * 2 = 2 | 1 * 3 = 3 | 1 * 4 = 4 | 1 * 5 = 5 |
| 2 | 2 * 1 = 2 | 2 * 2 = 4 | 2 * 3 = 6 | 2 * 4 = 8 | 2 * 5 = 10 |
| 3 | 3 * 1 = 3 | 3 * 2 = 6 | 3 * 3 = 9 | 3 * 4 = 12 | 3 * 5 = 15 |
| 4 | 4 * 1 = 4 | 4 * 2 = 8 | 4 * 3 = 12 | 4 * 4 = 16 | 4 * 5 = 20 |
| 5 | 5 * 1 = 5 | 5 * 2 = 10 | 5 * 3 = 15 | 5 * 4 = 20 | 5 * 5 = 25 |

The first term of Equation 9.31 consists of the diagonal: 1 * 1 = 1, 2 * 2 = 4, 3 * 3 = 9, 4 * 4 = 16, 5 * 5 = 25. Then

$$\begin{aligned}\text{COV}(XY) &= \frac{1}{5}(1+4+9+16+25) + \frac{1}{5^2}(1+2+3+\cdots+25) \\ &= \frac{55}{5} - \frac{225}{25} \\ &= 2\end{aligned}$$

The expected value of the product is then

$$E(X_1X_2) = E(X_1)E(X_2) + \text{COV}(X_1X_2)$$

An example uses the data from the 120 loblolly pine trees in Tables 9.2 and 9.4. Using the fundamental method,

$$\begin{aligned}E(X_1X_2) &= \frac{\sum_{i=1}^{120}(x_{1i}x_{2i})}{120} \\ &= \frac{83810}{120} \\ &= 698.41667\end{aligned}$$

The expected values of $X_1$ and $X_2$ are:

$$E(X_1) = \frac{\sum_{i=1}^{120}(x_{1i})}{120} = \frac{1376}{120} = 11.46667$$

and

$$E(X_2) = \frac{\sum_{i=1}^{120}(x_{2i})}{120} = \frac{7130}{120} = 59.41667$$

The product is

$$E(X_1)E(X_2) = 11.46667(59.41667) = 681.31113$$

which is not equal to $E(X_1X_2) = 698.41667$. The difference,

$$E(X_1X_2) - E(X_1)E(X_2) = 698.41667 - 681.31113 = 17.10554$$

is the covariance. This can be confirmed by computing the covariance in the fundamental manner:

$$\begin{aligned} \mathrm{COV}(X_1X_2) &= \frac{\sum_{i=1}^{120}[x_{1i} - E(X_1)][x_{2i} - E(X_2)]}{120} \\ &= \frac{2052.66667}{120} \\ &= 17.10556 \end{aligned}$$

The difference is due to rounding error.

## 9.8 CORRELATION

In Section 9.4 and in Item 6 of Section 9.7 reference is made to the independence of the primary variables making up a two-dimensional random variable. Independence is defined by Equation 9.6 and the concept is discussed in Item 6 of Section 9.7. When the primary variables are not independent, it often is necessary to make an assessment of the *strength* of the relationship. The measure used is the *correlation coefficient*. The rationale underlying the measure is introduced in Section 9.10 and is expanded in Chapter 15. Consequently, the following discussion is limited to a simple description of the measure and to certain of its properties.

The correlation coefficient is defined as

$$\rho_{X_1X_2} = \rho(X_1X_2) = \frac{\mathrm{COV}(X_1X_2)}{\sqrt{V(X_1)V(X_2)}} \tag{9.32}$$

As can be seen, the correlation coefficient involves two random variables. It is not applicable when more than two such variables are involved. Furthermore, it is applicable only when the regression of the mean (Section 9.10) is linear.

The correlation coefficient can take on values only in the range from –1 to +1. The sign simply indicates the direction of slope of the regression of the mean. If the values of one variable increase as the values of the other also increase, the sign is positive, whereas if the values of one variable decrease while the other increases, the sign is negative. The sign of the expression, therefore, is not a factor insofar as the *strength* of the relationship between the variables is concerned. When the absolute value of $\rho$ is 1, the two variables are *perfectly correlated*, meaning that if one chooses a value of one of the variables, the value of the second can be determined exactly. In other words, the two variables are in perfect step with one another. If $\rho$ is equal to 0, the reverse is true. The two variables are said to be *uncorrelated* and a knowledge of a value of one variable provides no information whatsoever with respect to a corresponding value of the other. Absolute values of $\rho$ lying between 1 and 0 provide a measure of the degree of correlation between the two variables with values of $\rho$ near 1 indicating strong but not perfect relationship while values near 0 indicate that relationships exist but that they are weak.

Obviously, when the covariance is equal to 0 the value of the correlation coefficient is also 0, indicating no correlation between the variables. One should not interpret this as proof that the variables are independent. It must be remembered that independence exists only if Equation 9.6 is appropriate. Examples can be found where the covariance is equal to 0 but the joint pdf is not equal to the product of the marginal pdfs of the two variables. For example, assume the two-dimensional random variables $(X_1, X_2)$ which can take on only the following values: (–1, 0), (0, 1), (1, 0), and (0, 1). The subset sizes, $N(x_1x_2)$ are

| | $X_2 =$ | | |
|---|---|---|---|
| $X_1 =$ | **–1** | **0** | **1** |
| –1 | 0 | 1 | 0 |
| 0 | 1 | 0 | 1 |
| 1 | 0 | 1 | 0 |

Since the total size of the range space, $T$, is equal to 4, the probabilities of occurrence are

| | $X_2 =$ | | | |
|---|---|---|---|---|
| $X_1 =$ | **–1** | **0** | **1** | $P(x_1)$ |
| –1 | 0 | $\frac{1}{4}$ | 0 | $\frac{1}{4}$ |
| 0 | $\frac{1}{4}$ | 0 | $\frac{1}{4}$ | $\frac{2}{4}$ |
| 1 | 0 | $\frac{1}{4}$ | 0 | $\frac{1}{4}$ |
| $P(x_2)$ | $\frac{1}{4}$ | $\frac{2}{4}$ | $\frac{1}{4}$ | |

The bottom row contains the marginal probabilities of $X_2$ while the column farthest to the right contains the marginal probabilities of $X_1$. If the two variables were independent $P(x_1, x_2)$ would be equal to the product of $P(x_1)$ and $P(x_2)$ as shown below:

| | $X_2 =$ | | |
|---|---|---|---|
| $X_1 =$ | **–1** | **0** | **1** |
| –1 | $\frac{1}{16}$ | $\frac{2}{16}$ | $\frac{1}{16}$ |
| 0 | $\frac{2}{16}$ | $\frac{4}{16}$ | $\frac{2}{16}$ |
| 1 | $\frac{1}{16}$ | $\frac{2}{16}$ | $\frac{1}{16}$ |

Obviously, $P(x_1x_2) \neq P(x_1)P(x_2)$. Yet, the covariance is equal to zero:

$$\text{COV}(X_1X_2) = E(X_1X_2) - E(X_1)E(X_2)$$

$$= \frac{(-1*0)+(0*1)+(1*0)+(0*-1)}{4} - \frac{(-1)+0+1+0}{4} * \frac{0+1+0-1}{4}$$

$$= \frac{0}{4} - \frac{0}{4} * \frac{0}{4} = 0$$

## 9.9 VARIANCES OF FUNCTIONS OF MULTIDIMENSIONAL RANDOM VARIABLES

The concept of and the basic computational procedures required for the variance of a random variable were introduced in Section 8.11. Functions of multidimensional random variables are one-dimensional random variables which have probability distributions and expected values. They also possess variances, as is brought out in the following discussion of the properties of variances of random variables.

1. The variance of the sum of a random variable and a constant ($K$) is the variance of the random variable.

$$V(Y + K) = V(Y)$$

The proof:

$$V(Y+K) = E[(Y+K) - E(Y+K)]^2$$
$$= E[(Y+K) - E(Y) - K]^2$$
$$= E[Y - E(Y)]^2$$
$$= V(Y)$$

The property is logical since the variance is a measure of the *scatter* of the values of the random variable about the arithmetic mean of the set of values. If a constant is added to each of the values of the random variable, it merely shifts the entire set laterally along the number line without affecting the positions of the values relative to one another or relative to the mean of the set. This is demonstrated in Figure 9.3.

2. The variance of the product of a random variable and a constant is the product of the square of the constant and the variance of the random variable.

$$V(KY) = K^2V(Y)$$

The proof:

$$V(KY) = E(KY)^2 - [E(KY)]^2$$
$$= K^2E(Y^2) - K^2[E(Y)]^2$$
$$= K^2\{E(Y^2) - [E(Y)]^2\}$$
$$= K^2V(Y)$$

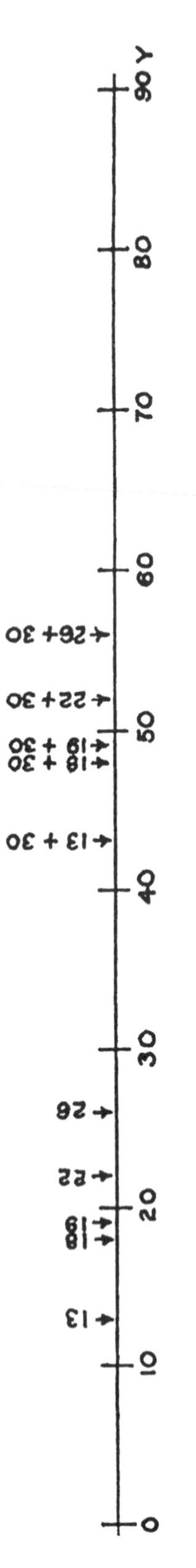


**FIGURE 9.3** Number line showing the scatter of a set of five $Y$-values and the scatter of the same set of $Y$-values to each of which a constant, $K = 30$, has been added.

This property indicates that the variance of a random variable cannot be expanded simply by multiplication by a constant. For example, assume a situation where a tract of land has been subdivided into 1/5-acre plots which are nonoverlapping and between which there are no gaps. Further, assume that the mean volume of pine timber on these plots is $\mu_Y = E(Y) = 1000$ bd. ft/plot and that the variance is $\sigma_Y^2 = V(Y) = 1{,}000{,}000$. If one wishes to put these values on a per *acre* basis, one would, in essence, multiply the volume on each plot by $K = 5$. The mean volume per acre would then be, according to the second property of expected values,

$$\mu_{KY} = E(KY) = KE(Y) = 5(1000) = 5000 \text{ bd. ft/acre}$$

The variance of the per acre values would not be simply $KV(Y)$. It must be

$$\sigma_{KY}^2 = V(KY) = K^2V(Y) = 5^2(1{,}000{,}000) = 25{,}000{,}000$$

Since this is a commonly encountered situation in timber inventory operations, people engaged in such work must be aware of this property of the variance and carry out their computations accordingly.

3. The variance of the sum of the two primary random variables of a two-dimensional random variable is the sum of the variances and the covariance of the two primary variables.

$$V(X_1 + X_2) = V(X_1) + V(X_2) + 2\text{COV}(X_1X_2)$$

The proof:

$$V(X_1 + X_2) = E(X_1 + X_2)^2 - [E(X_1 + X_2)]^2$$

According to Property 4 of expected values (Section 9.7)

$$E(X_1 + X_2) = E(X_1) + E(X_2)$$

Consequently,

$$V(X_1 + X_2) = E(X_1 + X_2)^2 - [E(X_1) + E(X_2)]^2$$

Expanding both binominals:

$$\begin{aligned}
V(X_1 + X_2) &= E(X_1^2 + 2X_1X_2 + X_2^2) - \{[E(X_1)]^2 + 2E(X_1)E(X_2) + [E(X_2)]^2\} \\
&= E(X_1^2) + 2E(X_1X_2) + E(X_2^2) - [E(X_1)]^2 - 2E(X_1)E(X_2) - [E(X_2)]^2 \\
&= \{E(X_1^2 - [E(X_1)]^2\} + \{E(X_2^2) - [E(X_2)]^2\} + [2E(X_1X_2) - 2E(X_1)E(X_2)] \\
&= V(X_1) + V(X_2) = 2[E(X_1X_2) - E(X_1)E(X_2)] \\
&= V(X_1) + V(X_2) + 2\,\text{COV}(X_1X_2)
\end{aligned}$$

If the two primary random variables are independent of one another, their covariance is equal to zero and the expression reduces to

$$V(X_1 + X_2) = V(X_1) + V(X_2)$$

4. The variance of the difference between the two primary random variables of a two-dimensional random variable is

$$V(X_1 - X_2) = V(X_1) + V(X_2) - 2\,\mathrm{COV}(X_1X_2)$$

The proof follows that of Property 3.

5. The variance of the sum of $n$ primary random variables of a multidimensional random variable is the sum of the variances and covariances of the $n$ primary variables.

$$\begin{aligned} V(X_1 + X_2 + \cdots + X_n) = V(X_1) + V(X_2) + \cdots + V(X_n) \\ +2[\mathrm{COV}(X_1X_2) + \mathrm{COV}(X_1X_3) + \cdots \\ +\mathrm{COV}(X_{n-1}X_n)] \end{aligned}$$

The proof follows that of the third property. The number of covariance terms in the expression depends on which of the set of random variables are interrelated. If all of the random variables are independent of one another, the expression becomes

$$V(X_1 + X_2 + \cdots + X_n) = V(X_1) + V(X_2) + \cdots + V(X_n)$$

6. The variance of the product of two independent random variables is

$$V(X_1X_2) = [E(X_1)]^2V(X_2) + [E(X_2)]^2V(X_1)$$

The proof:

$$x_{1i} = E(X_1) + \varepsilon_{1i}$$

$$x_{2j} = E(X_2) + \varepsilon_{2j}$$

Since $\varepsilon_{1i}$ and $\varepsilon_{2j}$ are independent, then

$$\begin{aligned} x_{1i}x_{2j} &= [E(X_1) + \varepsilon_{1i}][E(X_2) + \varepsilon_{2j}] \\ &= E(X_1)E(X_2) + E(X_1)\varepsilon_{2j} + E(X_2)\varepsilon_{1i} + \varepsilon_{1i}\varepsilon_{2j} \end{aligned}$$

The error of the product, $\varepsilon_p$, is

$$\varepsilon_{pij} = x_{1i}x_{2j} - E(X_1)E(X_2)$$

and

$$x_{1i}x_{2j} = \varepsilon_{pij} + E(X_1)E(X_2)$$

Then, substituting

$$\varepsilon_{pij} + E(X_1)E(X_2) = E(X_1)E(X_2) + E(X_1)\varepsilon_{2j} + E(X_2)\varepsilon_{1i} + \varepsilon_{1i}\varepsilon_{2j}$$

$$\varepsilon_{pij} = E(X_1)\varepsilon_{2j} + E(X_2)\varepsilon_{1i} + \varepsilon_{1i}\varepsilon_{2j}$$

$$\varepsilon^2_{pij} = [E(X_1)\varepsilon_{2j} + E(X_2)\varepsilon_{1i} + \varepsilon_{1i}\varepsilon_{2j}]^2$$

$$= [E(X_1)]^2\varepsilon^2_{2j} + [E(X_2)]^2\varepsilon^2_{1i} + (\varepsilon_{1i}\varepsilon_{2j})^2$$

$$+2[E(X_1)\varepsilon_{2j}E(X_2)\varepsilon_{1i} + E(X_1)\varepsilon_{2j}\varepsilon_{1i}\varepsilon_{2j} + E(X_1)\varepsilon_{1i}\varepsilon_{2j}\varepsilon_{1i}]$$

The expected value of $\varepsilon^2_{ij}$ is then

$$E(\varepsilon^2_p) = E\{[E(X_1)]^2\varepsilon^2_2\} + E\{[E(X_2)]^2\varepsilon^2_1\} + E(\varepsilon_1\varepsilon_2)^2$$

$$+2\{E[E(X_1)\varepsilon_2 E(X_2)\varepsilon_1] + E[E(X_1)\varepsilon_2\varepsilon_1\varepsilon_2]$$

$$+E[E(X_2)\varepsilon_1\varepsilon_2\varepsilon_1]\}$$

$$= [E(X_1)]^2 E(\varepsilon^2_2) + [E(X_2)]^2 E(\varepsilon^2_1) + E(\varepsilon_1\varepsilon_2)^2$$

$$+2[E(X_1)E(X_2)E(\varepsilon_1\varepsilon_2) + E(X_1)E(\varepsilon_2)E(\varepsilon_1\varepsilon_2)$$

$$+E(X_2)E(\varepsilon_1)E(\varepsilon_1\varepsilon_2)$$

$$= [E(X_1)]^2 V(X_2) + [E(X_2)]^2 V(X_1) + E(\varepsilon_1\varepsilon_2)^2$$

$$+2[E(X_1)E(X_2)\mathrm{COV}(X_1X_2) + E(X_1)E(X_2)\mathrm{COV}(X_1X_2)$$

$$+E(X_1)E(X_2)\mathrm{COV}(X_1X_2)$$

Since $X_1$ and $X_2$ are independent, $\mathrm{COV}(X_1X_2) = 0$, and

$$E(\varepsilon^2_p) = V(X_1X_2) = [E(X_1)]^2 V(X_2) + [E(X_2)]^2 V(X_1) + E(\varepsilon_1\varepsilon_2)^2$$

Continuing,

$$E(\varepsilon_1\varepsilon_2)^2 = E\{[X_1 - E(X_1)][X_2 - E(X_2)]\}^2$$

$$= E[X_1X_2 - X_1E(X_2) - X_2E(X_1) + E(X_1)E(X_2)]^2$$

$$= E\{X^2_1X^2_2 + X^2_1[E(X_2)]^2 + X^2_2[E(X_1)]^2 + [E(X_1)]^2[E(X_2)]^2$$

$$-2[X^2_1X_2E(X_2) + X_1X^2_2E(X_1) - X_1X_2E(X_1)E(X_2)$$

$$-X_1X_2E(X_1)E(X_2) + X_1E(X_1)[E(X_2)]^2$$

$$+X_2[E(X_1)]^2[E(X_2)]\}$$

$$= E(X_1^2X_2^2) + E(X_1^2)[E(X_2)]^2 + E(X_2^2)[E(X_1)]^2$$

$$+[E(X_1)]^2[E(X_2)]^2 - 2\{E(X_1^2)[E(X_2)]^2 + E(X_2^2)[E(X_1)]^2$$

$$-[E(X_1)]^2[E(X_2)]^2 - [E(X_1)]^2[E(X_2)]^2 + [E(X_1)]^2[E(X_2)]^2$$

$$+[E(X_1)]^2[E(X_2)]^2\}$$

$$= E(X_1^2X_2^2) + E(X_1^2)[E(X_2)]^2 + E(X_2^2)[E(X_1)]^2$$

$$+[E(X_1)]^2[E(X_2)]^2 - 2E(X_1^2)[E(X_2)]^2 - 2E(X_2^2)[E(X_1)]^2$$

$$= E(X_1^2X_2^2) + [E(X_1)]^2[E(X_2)]^2 - E(X_1^2)[E(X_2)]^2$$

$$-E(X_2^2[E(X_1)]^2$$

This quantity is usually negligible and, consequently, is ususally ignored. Thus,

$$V(X_1X_2) = [E(X_1)]^2V(X_2) + [E(X_2)]^2V(X_1)$$

## 9.10 THE REGRESSION OF THE MEAN

In Figure 9.2, each value of $X_1$ has associated with it a distribution of $X_2$ values. Such a distribution has a mean, or expected value, $E(X_2|X_1)$, and a variance, $V(X_2|X_1)$. If the means or expected values of these distributions are plotted against their associated $X_1$ values, a curve is generated which is known as the *regression of the mean*. Examples of such regressions are shown in Figure 9.4. If the process is reversed so that the means of the distributions of $X_1$ values are plotted against their associated $X_2$ values, a second regression of the mean is generated which may or may not be similar in shape to the first. Regressions of the mean provide a source of information about bivariate probability distributions which, when used with the expected values and variance, helps describe those distributions.

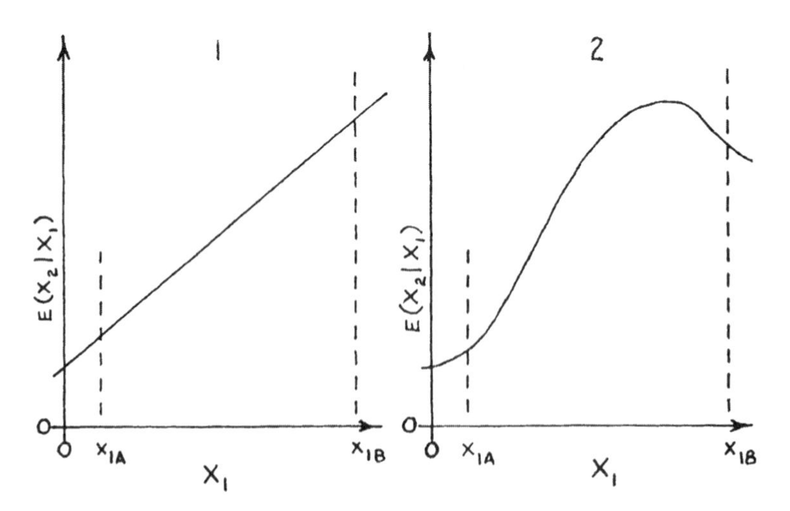


**FIGURE 9.4** Examples of regressions of the mean. Example 1 shows a linear relationship between $E(X_2|X_1)$ and $X_1$ while Example 2 shows a nonlinear relationship.

As can be seen in Figure 9.4, regressions of the mean may be linear or nonlinear. The following discussion is limited to situations where the regression is linear. Reference is made to Chapter 15 for information relating to nonlinear regressions. If the relationship is linear, the algebraic expression corresponding to the curve of $E(X_2|X_1)$ on $X_1$ is

$$E(X_2 \mid X_1) = \beta_0 + \beta_1 X_1 \tag{9.33}$$

where $\beta_0$ = the value of $E(X_2|X_1)$ when $X_1$ is equal to zero
$\beta_1$ = regression coefficient, which indicates the change in $E(X_2|X_1)$ per unit change in $X_1$

In Chapter 15 it will be shown that

$$\beta_0 = E(X_2) - \left[\frac{\text{COV}(X_1X_2)}{V(X_1)}\right]E(X_1) = E(X_2) - \beta_1 E(X_1) \tag{9.34}$$

and

$$\beta_1 = \frac{\text{COV}(X_1X_2)}{V(X_1)} \tag{9.35}$$

An alternative to Equation 9.33 makes use of the correlation coefficient, $\rho$, and is of value under certain circumstances:

$$E(X_2 \mid X_1) = E(X_2) + \left\{\frac{\rho\sqrt{V(X_2)}}{\sqrt{V(X_1)}}[X_1 - E(X_1 0)]\right\} \tag{9.36}$$

As was brought out in Item 6 in Section 9.7, when the two primary variables of a two-dimensional random variable are independent, their covariance equals 0. In such a case $\beta_1$ would be equal to 0 and $\beta_0$ to $E(X_2)$, indicating that the mean of the distributions of $X_2$ values associated with the $X_1$ values would all be equal to $E(X_2)$, producing a straight horizontal line on the graph. If the order of the primary variables is reversed so that $E(X_2|X_1)$ is plotted against $X_2$, the regression coefficient again would be equal to zero because the covariance would still be equal to zero. Again, the plot would produce a straight line. If the two regression lines are plotted on the same sheet of coordinate paper, as in Figure 9.5, they would cross at $[E(X_1), E(X_2)]$, forming a right angle. This, as was pointed out in Section 9.8, cannot be interpreted as indicating that the two variables are independent. They are, however, *uncorrelated*. $\rho$ would be equal to zero. As the correlation between the variables increases the angle between the two regression lines decreases, as is shown in Figures 9.5 and 9.6. When the relationship between the variable is perfect, when $\rho = 1$, the lines become superimposed. In such a case, the value of $X_2$ associated with a given value of $X_1$, or *vice versa*, could be determined without error by using the regression equation.

When the variances, $V(X_2|X_1)$, for all values of $X_1$ are all equal, it is said that they are *homogeneous* or that *homoscedasticity* exists. If they differ they are *heterogeneous* and *heteroscedasticity* is present. It should be evident that if the regression line is horizontal, i.e., $\beta_1 = 0$, and the variances of $X_2$ are homogeneous for all values of $X_1$,

$$V(X_2 \mid X_1) = V(X_2) \tag{9.37}$$

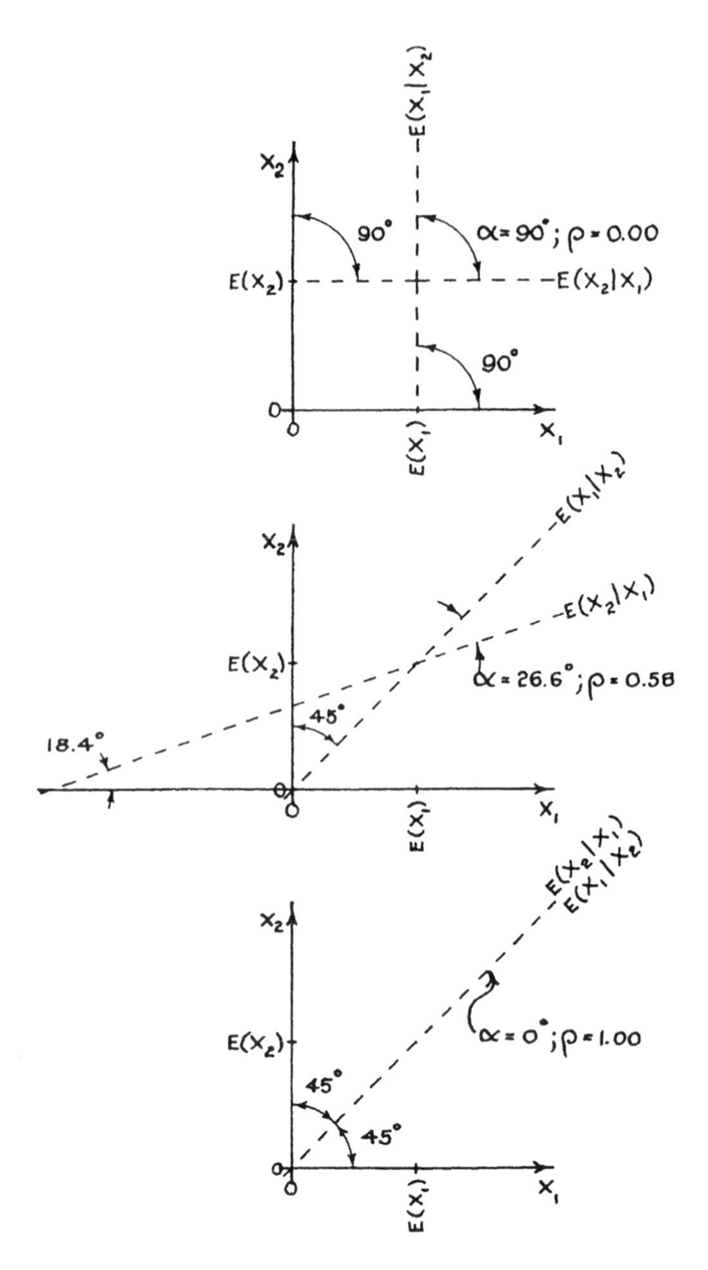


**FIGURE 9.5** Plots of regressions of the mean showing the angles found between the lines under different levels of correlation.

If the variances are homogeneous and the regression line is not horizontal, i.e., $\beta_1 \neq 0$,

$$V(X_2 \mid X_1) < V(X_2) \tag{9.38}$$

because $V(X_2)$ contains two sources of variability. One of these is the variability of the $E(X_2|X_1)$ values with respect to $E(X_2)$ while the other is the variability of the individual variates about their respective $E(X_2|X_1)$ values. $V(X_2|X_1)$ is based only on the second of these two sources of variability and consequently is smaller than $V(X_2)$. If the variances $V(X_2|X_1)$ are heterogeneous and $\beta_1 \neq 0$, the variances associated with some values of $X_1$ may be greater than $V(X_2)$ but, *on the average*, Equation

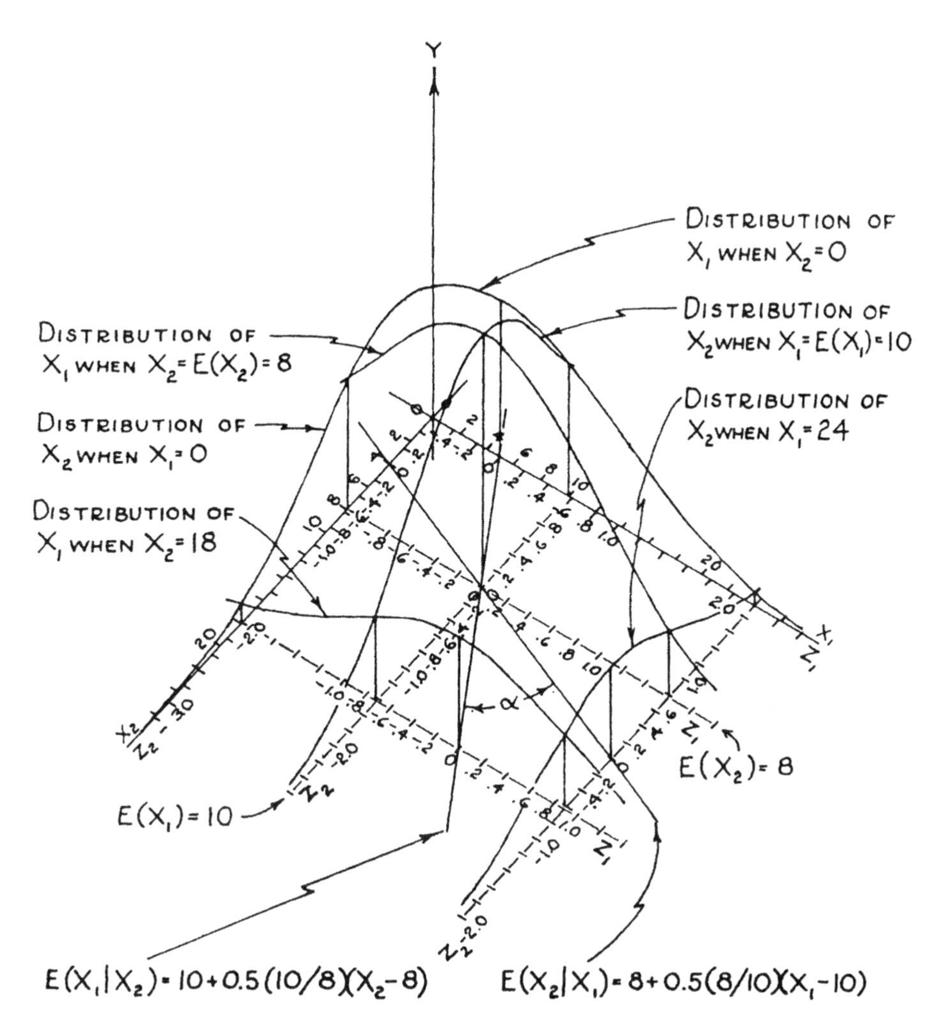


**FIGURE 9.6** Diagram of the two regressions of the mean associated with the bivariate random variable $(X_1, X_2)$. Both primary variables are normally distributed. The parameters of their distributions are $E(X_1) = 10$, $V(X_1) = 100$, and $E(X_2) = 8$, $V(X_2) = 64$. The correlation coefficient $\rho = 0.5$. The conditional variances are homogeneous, with $V(X_2|X_1) = 48$ and $V(X_1|X_2) = 75$. The conditional distributions of the two primary variables are represented by profiles at $X_1 = 0$, 10, and 24 and $X_2 = 0$, 8, and 18. The reason for the $Z$ scales is brought out in Section 11.2.

9.38 is still appropriate. In other words, if one were to pool, across the entire two-dimensional distribution, the squared differences between the variates and their associated regression line values $E(X_2|X_1)$, the mean squared difference, known as the *variance from regression*, $V(X_2 \cdot X_1)$, would be less than if the difference had been computed using the overall expected value $E(X_2)$.

The difference between $V(X_2 \cdot X_1)$ and $V(X_2)$ forms the basis for the correlation coefficient. The ratio of the two variances, $V(X_2 \cdot X_1)/V(X_2)$ provides a relative measure of the amount of variability of $X_2$ that remains after the effect of the relationship between $X_2$ and $X_1$ has been removed. If this relationship is such that $E(X_2|X_1)$ is constant for all values of $X_1$, it will equal $E(X_2)$, $V(X_2 \cdot X_1)$ will equal $V(X_2)$ and $V(X_2 \cdot X_1)/V(X_2)$ will equal unity. In other words, a relationship between $X_2$ and $X_1$ does not exist. On the other hand, if $V(X_2 \cdot X_1)$ is equal to 0, it means that the values of all the variates in the $X_2$ range space fall *on* the regression line. There are no errors. Consequently, the ratio is equal to zero and the relationship is perfect. Thus, the ratio can change in value from 0 to 1 with the relationship between the variables becoming weaker as the ratio increases. Since it is intellectually pleasing to have a measure of some quantity increase as the quantity increases, the ratio is reversed, through the use of its complement, giving rise to the *coefficient of determination*:

$$\rho^2 = 1 - \frac{V(X_2 \cdot X_1)}{V(X_2)} \quad \text{or} \quad \frac{V(X_2) - V(X_2 \cdot X_1)}{V(X_2)} \tag{9.39}$$

The coefficient $\rho^2$ increases as the relationship strengthens and ranges in value from 0 to 1. The square root of $\rho^2$ is the *correlation coefficient*:

$$\rho = \sqrt{1 - \frac{V(X_2 \cdot X_1)}{V(X_2)}} \tag{9.40}$$

The difference between Equations 9.32 and 9.39 is bridged algebraically in Chapter 15.

The coefficient of determination, $\rho^2$, has two roots, one positive and one negative. The sign of the root has no significance insofar as the strength of the relationship between the variables is concerned. The sign is used, however, to indicate the direction of slope of the regression line. The positive root is assigned to regressions where the values of $X_2$ increase as those of $X_1$ increase. The negative root is assigned in cases when the values of $X_2$ decrease as those of $X_1$ increase.

It should be noted that the coefficient of determination is a true proportion and can be expressed as a percent. The correlation coefficient, on the other hand, is the square root of a proportion and cannot be expressed as a percent. In light of this, it is difficult to rationalize the preferred status given the correlation coefficient over that of the coefficient of determination.

Earlier it was stated that the variance from regression, $V(X_2 \cdot X_1)$ is less than the variance from the fixed mean, $V(X_2)$. The magnitude of the difference is a function of $\rho$, as can be seen in Equation 9.41. From Equation 9.39,

$$\rho^2 = \frac{V(X_2) - V(X_2 \cdot X_1)}{V(X_2)}$$

$$V(X^2)\rho^2 = V(X_2) - V(X_2 \cdot X_1) \tag{9.41}$$

$$V(X_2 \cdot X_1) = V(X_2) - V(X_1)\rho^2 = V(X_2)(1 - \rho^2)$$

Thus, if, as in Figure 9.6, $V(X_2) = 64$ and $\rho = 0.5$, the magnitude of the variance from regression would be

$$V(X_2 \cdot X_1) = 64(1 - 0.5^2) = 64(0.75) = 48$$

It should be noted that the reduction is the same, in a relative sense, for both primary variables.

# 10 Discrete Probability Distributions

## 10.1 INTRODUCTION

In real-life situations one is almost never able to set up a range space and determine the sizes of the subsets so as to compute the probabilities needed for a probability distribution. This is true with discrete random variables and the situation becomes impossible if the random variable is continuous. How then is it possible to put the concepts discussed in the previous chapter to work? This question cannot be answered completely at this stage, but it can be said that at the heart of the procedure an assumption is made by the person responsible for the work that the real, but unknown, distribution is sufficiently similar to a theoretical distribution of known characteristics so that the theoretical distribution can be used as a substitute for the real one. The responsible person should have some idea of the nature of the real distribution and also a knowledge of the available theoretical distributions so that a rational matching can be made. The closer the match, the better the results. This chapter describes a number of discrete theoretical distributions. The next chapter is devoted to a number of continuous distributions. Neither is complete. However, with a knowledge of the distributions described one should be able to cope with virtually all problems in forest and natural resource inventory that require a probability distribution. For a more comprehensive coverage of probability distributions, see Johnson and Kotz (1969, 1970), Patel et al. (1976), and Hastings and Peacock (1974).

## 10.2 THE RECTANGULAR PROBABILITY DISTRIBUTION ($D{:}M, y_A, y_B$)

The simplest probability distribution is one where every value of the random variable has the same probability of being drawn. A distribution of this type is known as a *rectangular*, *rectilinear*, or *uniform* probability distribution and may be either discrete or continuous and, if discrete, can be either finite or countably infinite.* An example of a finite discrete rectangular probability distribution is that generated when a fair die is tossed one time. The values of the random variable, $Y$, are the values appearing on the faces of the die. Each of the six faces has the same probability of occurrence, as can be seen in Table 10.1. When a rectangular probability distribution, whether discrete or continuous, is shown as a histogram, the tops of the bars form a straight horizontal line, as in Figure 10.1. Hence, the name "rectangular," or "uniform."

* Simply stated, if the events making up a sample or range space are *discrete*, they are *countable*. If the total count of such events is less than infinity, it is said that the set is *finite*. On the other hand, if the count is infinitely large, it is said that the set is *countably infinite*. If the events are not discrete, as in the case of a continuous distribution, the events are *noncountably infinite*.

**TABLE 10.1**
**The Rectangular Probability Distribution Associated with the Tossing of a Single Fair Die**

| $i$ | 1 | 2 | 3 | 4 | 5 | 6 |
|---|---|---|---|---|---|---|
| $y_i$ | 1 | 2 | 3 | 4 | 5 | 6 |
| $P(y_i)$ | 1/6 | 1/6 | 1/6 | 1/6 | 1/6 | 1/6 |

The frequency function for a discrete random variable, having a rectangular probability distribution, is

$$N_j = \begin{cases} \frac{N}{M}; & j = 1, 2, 3, \ldots, M; \; y_1 = y_A; \; y_M = y_B \\ 0, & \text{elsewhere} \end{cases} \tag{10.1}$$

where $N$ = total number of nonempty cells in the range space, i.e., cells that contain $Y$ values corresponding to events in the associated sample space
$M = (y_B - y_A + h)/h$ = total number of possible values of $Y$
$h$ = magnitude of the difference between two adjacent $Y$ values

The point probability function, using Equation 8.4, is

$$P(y_i) = \begin{cases} \frac{N_i}{N} = \left(\frac{N}{M}\right) * \left(\frac{1}{N}\right) = \frac{1}{M}; & i = 1, 2, 3, \ldots, M; \; y_1 = y_A; \; y_M = y_B \\ 0, & \text{elsewhere} \end{cases} \tag{10.2}$$

Thus, the distribution has three parameters, the scale parameter $M$ and the two location parameters $y_A$ and $y_B$. The cumulative probability function is

$$P(y_A \leq Y \leq y_i) = \sum_{i=1}^{b} \left(\frac{1}{M}\right) = \frac{b}{M} \qquad i = 1, 2, 3, \ldots, M \tag{10.3}$$

where $b = (y_i - y_A + h)/h$.

The mean and variance of the rectangular distribution can be computed using the standard method (Equations 8.30 and 8.32). However, simplified expressions, which are modifications of the formulae derived from the continuous versions of the rectangular distribution (see Section 11.1), are available. These are

$$E(Y) = \frac{(y_Z + y_B)}{2} \tag{10.4}$$

and

$$V(Y) = \frac{(M^2 - 1)}{12} \tag{10.5}$$

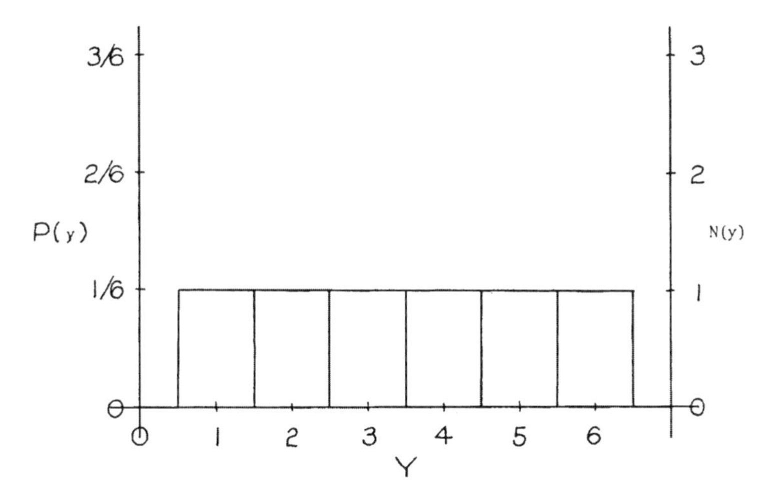


**FIGURE 10.1** Histogram of the rectangular probability distribution associated with the tossing of a single fair die.

Thus, in the case of the single fair die,

$$E(Y) = \frac{(1+6)}{2} = 3.5$$

and

$$V(Y) = \frac{(6^2 - 1)}{12} = \frac{35}{12} = 2.92$$

since

$$h = 1 \quad \text{and} \quad M = \frac{(6-1+1)}{1} = 6$$

The rectangular distribution is symmetrical. Consequently, the moment coefficient of skewness is equal to zero. The magnitude of the coefficient of kurtosis is dependent on $M$, the number of $Y$ values. When $M = 2$,

$$\mu_2 = \frac{[(y_1 - \mu_Y)^2 + (y_2 - \mu_Y)^2]}{2}$$

Since there are only two values involved,

$$(y_1 - \mu_Y)^2 = (y_2 - \mu_Y)^2 \quad \text{and} \quad \mu_2 = (y_1 - \mu_Y)^2 = \sigma_Y^2$$

Then

$$\mu_4 = \frac{[(y_1 - \mu_y)^4 + (y_2 - \mu_y)^4]}{2}$$

By using the same argument as above,

$$\mu_4 = (y_1 - \mu_y)^4$$

Then

$$\alpha_4 = \frac{(y_1 - \mu_y)^4}{(y_1 - \mu_y)^4} = 1$$

As $M$ increases, the magnitude of $\alpha_4$ increases until it reaches a maximum of 1.8 when $M \to \infty$. This is shown in Section 11.1. The rate of closure of the $\alpha_4$ values on 1.8 is quite rapid, significantly so that when $M = 7$, the coefficient has already reached 1.7.

Obviously, the rectangular distributions have no mode.

The utility of the rectangular probability distribution in the field of forestry is difficult to assess. In some instances random compass directions are involved, in which case the random variable could be degrees of azimuth from north. Such a random variable would have a rectangular probability distribution. The distribution is used at the decision points in certain simulators where theory indicates equal probabilities or where theory or field evidence fails to support the use of any other distribution.

## 10.3 THE BINOMIAL PROBABILITY DISTRIBUTION (*B:n, p*)

Assume a situation where a certain wooded house lot contains ten loblolly pine trees, of which three show evidence of southern fusiform rust. If a tree is chosen blindly from the set of ten trees, the probability of choosing one with rust ($E = R$) is

$$P(E) = \frac{N(R)}{N} = \frac{3}{10} = 0.3 = p \tag{10.6}$$

where $E$ = occurrence of a tree with rust and the probability of choosing a rust-free tree ($\tilde{E} = F$) is:

$$P(\tilde{E}) = P(F) = \frac{N(F)}{N} = \frac{(10-3)}{10} = \frac{7}{10} = 0.7 = q \tag{10.7}$$

In a case such as this, only two events are possible. When a draw is made, one either obtains a tree with rust or a tree without rust. Traditionally, one refers to the occurrence of interest as a "success" and to its nonoccurrence as a "failure." If more than one draw is made, the result is a set of successes or failures. A number of possible mixes can occur. For example, if three draws are to be made from the set of ten loblolly pines, and the sampling is to be *with* replacement, the possible results are

$R, R, R$
$R, R, F; R, F, R; F, R, R$
$R, F, F; F, R, F; F, F, R$
$F, F, F$

where $R$ represents a tree with rust and $F$ a tree free of rust. If the choosing of a tree with symptoms of rust is considered a "success," it can be seen that it is possible to obtain either 3, 2, 1, or no successes in the three draws.

**TABLE 10.2**
**The Binomial Probability Distribution Associated with the Occurrence of Southern Fusiform Rust in a Stand of Loblolly Pines When 30% of the Trees Show Evidence of the Rust and Blind Draws Are to Be Made Using Sampling with Replacement**

| $i$ | 1 | 2 | 3 | 4 |
|---|---|---|---|---|
| $y_i$ | 0 | 1 | 2 | 3 |
| $p(y_i)$ | 0.343 | 0.441 | 0.189 | 0.027 |

Define the random variable, $Y$, in these terms:

$Y_1 = 3$, when 3 trees with rust are chosen
$Y_2 = 2$, when 2 trees with rust are chosen
$Y_3 = 1$, when 1 tree with rust is chosen
$Y_4 = 0$, when no trees with rust are chosen

As can be seen, $M$, the number of possible values of $Y$, is equal to $n + 1$, where $n$ is the size of the sample (i.e., the number of draws).

Symbolically, the probabilities associated with these results are

$$p * p * p = p^3$$
$$p * p * q + p * q * p + q * p * p = 3\,pq^2$$
$$p * q * q + q * p * q + q * q * p = 3\,pq^2$$
$$q * q * q = q^3$$

Substituting the numerical values for $p$ and $q$ yields

$$P(Y = 3) = p^3 = (0.3)^3 = 0.027$$
$$P(Y = 2) = 3\,p^2q = 3(0.3)^2(0.7) = 0.189$$
$$P(Y = 1) = 3\,pq^2 = 3(0.3)(0.7)^2 = 0.441$$
$$P(Y = 0) = q^3 = (0.7)^3 = 0.343$$

which is a probability distribution, as shown in Table 10.2. The sum of these probabilities is

$$\sum_{i=1}^{M} p(y_1) = p^3 + 3p^2q + 3pq^2 + q^3$$

$$= 0.27 + 0.187 + 0.441 + 0.343 = 1$$

which indicates that the special additive law has been obeyed.

According to the special additive law, when only one draw is made

$$p + q = 1$$

which can be rewritten

$$(p + q)^1 = 1^1 = 1$$

When more than one draw is made, using sampling *with* replacement, the expression becomes

$$(p + q)^n = 1^n = 1 \tag{10.8}$$

Expanding this binomial yields an algebraic expression which corresponds exactly with the situation in the loblolly pine example:

$$(p+q)^3 = p^3 + 3p^2q + 3pq^2 + q^3 = 1$$

Each term of the expansion represents the probability of occurrence of a certain mix of successes and failures. A probability distribution that follows this pattern is a *binomial* distribution.*

As can be seen from the previous development, each term in the expansion represents the probability of occurrence of a specific number of successes in a set of draws or, in other words, the probability of a specific $Y$ value. The appropriate term in the expansion can be identified by the exponent of $p$. For example, if one wanted the probability of obtaining two trees with rust when three draws are made, one would locate the term in which the exponent of $p$ was 2. This is the second term. Thus,

$$p(Y = 2) = 3p^2q$$

Such an individual term of the expansion is a *point binomial probability* or, simply, a *point probability*. Each of the coefficients of the terms in the expansion indicates the number of ways by which the mix of successes and failures associated with the term can occur. For example, in the case of the loblolly pine problem, there was only one way by which $Y = 3$ could be obtained ($R, R, R$). $Y = 2$ could be obtained in three ways ($R, R, F$; $R, F, R$; and $F, R, R$), as could $Y = 1$ ($R, F, F$; $F, R, F$; and $F, F, R$). Like $Y = 3$, $Y = 0$ can be obtained in only one way ($F, F, F$). Consequently, the coefficient of a term is a count of the combinations yielding the indicated results, and any individual term or point probability can be obtained using the point probability function:

$$P(y_i) = \begin{cases} \binom{n}{y_i} p^{y_i} q^{n-y_i}, & y = 0,\ 1,\ 2,\ \ldots,\ n; \\ & n \text{ is a positive integer;} \\ & p > 0;\ q > 0 \\ 0, & \text{elsewhere} \end{cases} \tag{10.9}$$

For example, assume that *five* draws are to be made from the set of ten loblolly pines. What is the probability that two of the chosen trees show evidence of rust:

$$P(Y = 2) = \binom{5}{2} p^2 q^3 = \frac{5!}{2!(5-3)!}(0.3)^2(0.7)^3 = 0.3087$$

When $n$ is not large, the coefficients for the terms in the expansion of the binomial can be obtained from "Pascal's triangle," a portion of which is shown in Table 10.3. Note the pattern of values in the triangle. Each number, with the exception of those along the outer bounds of the

* Such distributions are also known as *Bernoulli* distributions after James Bernoulli, who first described them in the early 18th century.

**TABLE 10.3**
**Pascal's Triangle, Showing the Coefficients of the Terms When the Binomial is Expanded**

| *n* | Coefficients |
|---|---|
| 1 | 1 1 |
| 2 | 1 2 1 |
| 3 | 1 3 3 1 |
| 4 | 1 4 6 4 1 |
| 5 | 1 5 10 10 5 1 |
| 6 | 1 6 15 20 15 6 1 |
| 7 | 1 7 21 35 35 21 7 1 |
| 8 | 1 8 28 56 70 56 28 8 1 |
| 9 | 1 9 36 84 126 126 84 36 9 1 |
| 10 | 1 10 45 120 210 256 210 120 45 10 1 |

figure, is the sum of the two values above it. The triangle can be expanded indefinitely, but when only specific coefficients are needed it is preferable to use Equation 10.9.

In this discussion of the binomial distribution no attempt has been made to define the sample or range spaces or the frequency function. None of these is necessary for the utilization of the binomial distribution, but an explanation of how they fit in may be of value.

It was brought out in Chapter 7 that, as soon as multiple draws and compound events are involved, the original or primary sample space must be converted to the appropriate derived or secondary sample space. In the case of the binomial distribution, the primary sample space is that upon which the $p$ and $q$ values are based. Assume a situation where the universe in question contains three trucks, one of which is a Ford while the remaining two are Chevrolets. If the variable of interest is the manufacturer of the truck, the primary sample space is

$$S_p = \{F\} \cup \{C_1, C_2\}$$

where $F$ = Ford
$C_1$ = First Chevrolet
$C_2$ = Second Chevrolet

Furthermore,

$N_P$ = size of the primary space = 3
$N_F$ = size of the subset containing events yielding Fords = 1
$N_C$ = size of the subset containing events yielding Chevrolets = 2

Then,

$$P(\text{Ford}) = \frac{N_F}{N_P} = \frac{1}{3} = p$$

$$P(\text{Chevrolet}) = \frac{N_C}{N_P} = \frac{2}{3} = q$$

If two draws are to be made from this primary sample space, using sampling with replacement, a derived sample space is generated in which each event consists of two results:

$$S_D = \{F,F\} \cup \{F, C_1; F, C_2; C_1, F; C_2, F\} \\ \cup \{C_1, C_1; C_1, C_2; C_2, C_1; C_2, C_2\}$$

As can be seen, the derived sample space contains

$$N_D = 9 \text{ events}$$

If the random variable, $Y$, is defined as the number of Fords in the event, the random variable can take on any one of the three values:

$y_1 = 2$, when two Fords are in the event
$y_2 = 1$, when one Ford is in the event
$y_3 = 0$, when no Fords are in the event

The derived sample space can be converted into the range space:

$$R_D = \{2, 1, 1, 1, 1, 0, 0, 0, 0\}$$

The subset sizes are

$N_1 = 1$
$N_2 = 4$
$N_3 = 4$

and the probabilities of occurrence of the three possible outcomes are

$$P(Y = 2) = \frac{N_1}{N_D} = \frac{1}{9}$$

$$P(Y = 1) = \frac{N_2}{N_D} = \frac{4}{9}$$

$$P(Y = 0) = \frac{N_3}{N_D} = \frac{4}{9}$$

which is the probability distribution.

Shifting now to the procedures usually associated with a problem of this type, if the *primary* event of interest is the drawing of a Ford,

$$p = \frac{N_F}{N_P} = \frac{1}{3} \quad \text{and} \quad q = \frac{N_C}{N_P} = \frac{2}{3}$$

Then, if two draws are made from the primary sample space, using sampling with replacement,

$$(p + q)^n = (p + q)^2 = p^2 + 2pq + q^2$$

$$= \left(\frac{1}{3}\right)^2 + 2\left(\frac{1}{3}\right)\left(\frac{2}{3}\right) + \left(\frac{2}{3}\right)^2$$

$$= \frac{1}{9} + \frac{4}{9} + \frac{4}{9} = 1$$

which agrees with the previous development, demonstrating that the probabilities associated with a binomial distribution are based on single draws from the appropriate sample space. By using Equation 10.9,

$$P(y_i) = \binom{n}{y_i} p^{y_i} q^{n-y_i}$$

$$= \binom{n}{y_i} \left(\frac{N_F}{N_P}\right)^{y_i} \left(\frac{N_C}{N_P}\right)^{n-y_i}$$

Since

$$\left(\frac{N_F}{N_P}\right)^{y_i} \left(\frac{N_C}{N_P}\right)^{n-y_i} = \frac{[(N_F)^{y_i}(N_C)^{n-y_i}]}{N_p^n}$$

$$P(y_i) = \frac{1}{N_p^n}\left[\binom{n}{y_i}(N_F)^{y_i}(N_C)^{n-y_i}\right]$$

Consequently, the frequency function for the binomial distribution is

$$N_i = \begin{cases} \binom{n}{y_i}(N_S)^{y_i}(N_f)^{n-y_i}, & y = 0,\ 1,\ 2,\ \ldots,\ n \\ 0, & \text{elsewhere} \end{cases} \tag{10.10}$$

where $N_s$ = size of the subset in the primary sample space that contains the events yielding successes

$N_f$ = size of the subset in the primary sample space that contains the events yielding failures

and the size of the derived sample space is

$$N_D = N_p^n \tag{10.11}$$

When one cumulates the point probabilities as $Y$ increases from zero, in order to construct the cumulative binomial probability distribution it is convenient to reverse the order of the terms in the expansion; e.g., using the loblolly pine problem involving three draws,

$$(q + p)^3 = q^3 + 3q^2p + 3qp^2 + p^3 = 1$$

The cumulative probability distribution is then

$$\begin{aligned} P(Y = 0) &= q^3 && = 0.343 \\ P(0 \le Y \le 1) &= q^3 + 3q^2p && = 0.784 \\ P(0 \le Y \le 2) &= q^3 + 3q^2p + 3qp^2 && = 0.973 \\ P(0 \le Y \le 2) &= q^3 + 3q^2p + 3qp^2 + q^3 && = 1.000 \end{aligned}$$

and the cumulative distribution function is

$$P(0 \le Y \le y_b) = \sum_{i=1}^{b} \binom{n}{y_i} q^{n-y_i} p^{y_i} \tag{10.12}$$

where $b = 1, 2, 3, \ldots, M$; and $M = n + 1$. As $n$ increases, the computation needed to obtain cumulative probabilities becomes excessively burdensome. Table A.1 in the Appendix is one of many that have been developed to ease this problem. It can be used, for example, to solve problems of the following type. Assume that 15 trees are to be drawn from the set of loblolly pines. Remember that the drawing is *with* replacement and, consequently, one can draw 15 times from a set of 10. What is the probability of obtaining 5, 6, 7, or 8 trees with rust?

$$p(5 \le Y \le 8) = P(0 \le Y \le 8) - P(0 \le Y \le 4)$$

From Table A.1 in the Appendix:

$$P(0 \le Y \le 8) = 0.9848$$
$$P(0 \le Y \le 4) = 0.5155$$
$$P(5 \le Y \le 8) = 0.9848 - 0.5155 = 0.4693$$

This result can be confirmed by computing the individual probabilities of 5, 6, 7, and 8, using Equation 10.9, and then summing those probabilities.

The moment-generating function for the binomial distribution is:

$$\begin{aligned} M_Y(t) = E(e^{tY}) &= \sum_{j=1}^{M} e^{tY_i} * P(y_i) \\ &= \sum_{i=1}^{M} \left[ e^{tY} \binom{n}{y_i} p^{y_i} q^{n-y_i} \right] \\ &= e^{t(0)} \frac{n!}{0!(n-0)!} p^0 q^{n-0} + e^{t(1)} \frac{n!}{1!(n-1)!} p^1 q^{n-1} \\ &\quad + e^{t(2)} \frac{n!}{2!(n-2)!} p^2 q^{n-2} + \cdots \\ &\quad + e^{t(n)} \frac{n!}{n!(n-n)!} p^n q^{n-n} \\ &= [p(e^t) + q]^n \end{aligned} \tag{10.13}$$

From the mgf the moments about zero can be computed (see Section 8.13):

$$\begin{aligned} \mu_1' &= E(Y) = np \\ \mu_2' &= E(Y^2) = np + n(n-1)p^2 \\ \mu_3' &= E(Y^3) = np + 3n(n-1)p^2 + n(n-1)(n-2)p^3 \\ \mu_4' &= E(Y^4) = np + 7n(n-1)p^2 + 6n(n-1)(n-2)p^3 + n(n-1)(n-2)(n-3)p^4 \end{aligned} \tag{10.14}$$

Thus, the mean or expected value is given by Equation 10.14, and the variance is

$$\begin{aligned} V(Y) &= E(Y^2) - [E(Y)]^2 \\ &= np + n(n-1)p^2 - (np)^2 \\ &= npq \end{aligned} \tag{10.15}$$

When a binomial distribution is shown in the form of a histogram, as in Figures 10.2 and 10.3, it is evident that it is mound shaped. The moment coefficient of skewness is

$$\begin{aligned} \alpha_3 &= \frac{\mu_3}{\sigma^3} \\ &= \{E(Y^3) - 3E(Y^2)E(Y) + [E(Y)]^3\} * [V(Y)]^{-3/2} \\ &= \{np + 3n(n-1)p^2 + n(n-1)(n-2)p^3 - 3[np + n(n-1)p^2]np + 2(np)^3\}\left(\frac{1}{npq}\right)^{3/2} \\ &= \frac{(q-p)}{(npq)^{1/2}} \end{aligned} \tag{10.16}$$

Thus, when samples of size 3 are drawn from the set of ten loblolly pieces, the distribution is as shown in Figure 10.2. The moment coefficient of skew is

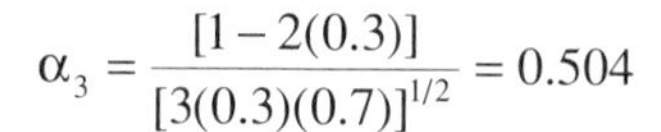

$$\alpha_3 = \frac{[1 - 2(0.3)]}{[3(0.3)(0.7)]^{1/2}} = 0.504$$

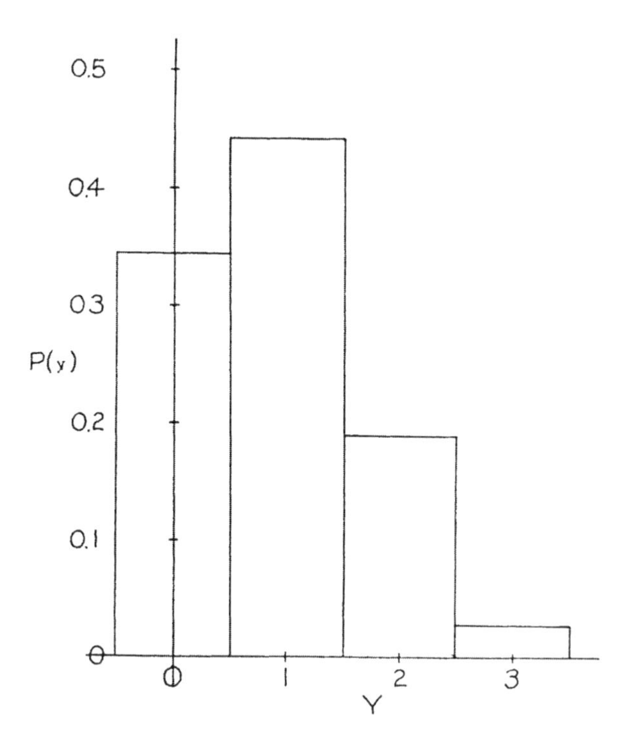


**FIGURE 10.2** Histogram of the binomial probability distribution associated with the drawing of fusiform rust–infected loblolly pine trees when 30% of the stand is infected and when $n = 3$.

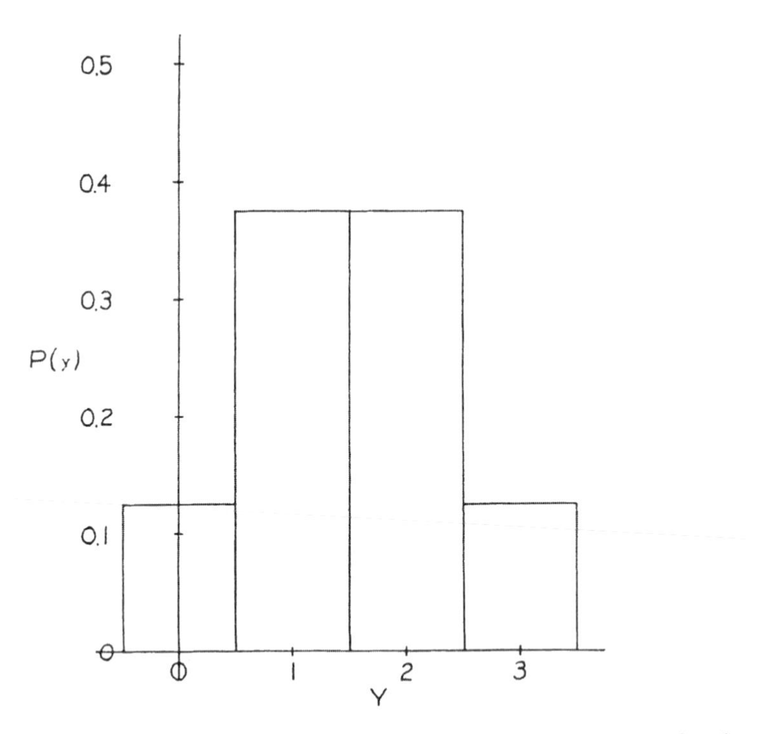


**FIGURE 10.3** Histogram of the binomial probability distributions which result when $n = 3$ and $p = 0.5$.

The skew is positive or to the right. This is evident in Figure 10.2. When $p = q = 0.5$, as in Figure 10.3,

$$\alpha_3 = \frac{[1 - 2(0.5)]}{[n(0.5)(0.5)]^{1/2}} = 0$$

indicating symmetry. When $p > 0.5$ the skew is to the left, and when $p < 0.5$ it is to the right. The latter is demonstrated in Figure 10.2. If $p$ is held constant at some value other than 0.5 and $n$ is increased, the distribution tends to become more symmetrical. In the limit, when $n \to \infty$, $\alpha_3 \to 0$, and the distribution becomes symmetrical, regardless of the magnitude of $p$.

The moment coefficient of kurtosis is

$$\begin{aligned}
\alpha_4 &= \frac{\mu_4}{\sigma^4} \\
&= \{E(Y^4) - 4E(Y^3)E(Y) + 6E(Y^2)[E(Y)]^2 - 3[E(Y)]^4\} * [V(Y)]^{-2} \\
&= \{np + 7n(n-1)p^2 + 6n(n-1)(n-2)p^3 + n(n-1)(n-2)(n-3)p^4 \\
&\quad - 4[np + 3n(n-1)p^2 + n(n-1)(n-2)p^3]np \\
&\quad + 6[np + n(n-1)p^2](np)^2 - 3(np)^4\} * \left(\frac{1}{npq}\right)^2 \\
&= 3 + \frac{(1 - 6pq)}{npq}
\end{aligned} \tag{10.17}$$

In the case of the distribution in Figure 10.2 the coefficient is

$$\alpha_4 = 3 + \frac{[1 - 6(0.3)(0.7)]}{3(0.3)(0.7)} = 2.5873$$

indicating a certain degree of platykurtosis. In the case of the distribution in Figure 10.3, the coefficient is

$$\alpha_4 = 3 + \frac{[1 - 6(0.5)(0.5)]}{3(0.5)(0.5)} = 2.3333$$

also indicating platykurtosis. When both $n$ and $p$ are small, e.g., $n = 3$ and $p = 0.01$, the distribution becomes leptokurtic:

$$\alpha_4 = 3 + \frac{[1 - 6(0.1)(0.99)]}{3(0.1)(0.99)} = 34.6700$$

Regardless of the magnitude of $p$, however, as $n$ increases the distribution moves closer to the normal. As $n \to \infty$ the second term in Equation 10.17 approaches zero and $\alpha_4 \to 3$, the value for the normal. Some evidence of the shift toward the normal as $n$ increases can be seen by comparing Figures 10.2 and 10.4.

The importance of the binomial distribution rests primarily on the fact that many other probability distributions are derived from it.

## 10.4 THE HYPERGEOMETRIC PROBABILITY DISTRIBUTION ($H$:$N$, $N_s$, $n$)

If, in the loblolly pine problem, the sampling had been *without* replacement, the same outcomes ($Y$ values) would have occurred but, because the primary sample space was being altered as the drawing proceeded, the probabilities would have been hypergeometric (Section 7.6.5) rather than binomial and the probability distribution would be *hypergeometric*. The point probability function in the case of the hypergeometric distribution is

$$P(y_i) = \begin{cases} \dfrac{\dbinom{N_S}{y_i}\dbinom{N - N_S}{n - y_i}}{\dbinom{n}{N}}, & 0 \le y \le n;\ 0 \le y \le N_S; \\ & 0 \le n - y \le N - N_S; \\ & N_S < N;\ n < N_S \\ 0, & \text{elsewhere} \end{cases} \tag{10.18}$$

Equation 10.18 is the equivalent of Equation 7.25, with $y_i$ replacing $C_1$ and $(n - y_i)$ replacing $C_2$. The probability distribution in the case of the loblolly pine problem is shown in Table 10.4. The individual computations of the values in that table are as follows:

$$\binom{N}{n} = \frac{10!}{3!(10-3)!} = 120$$

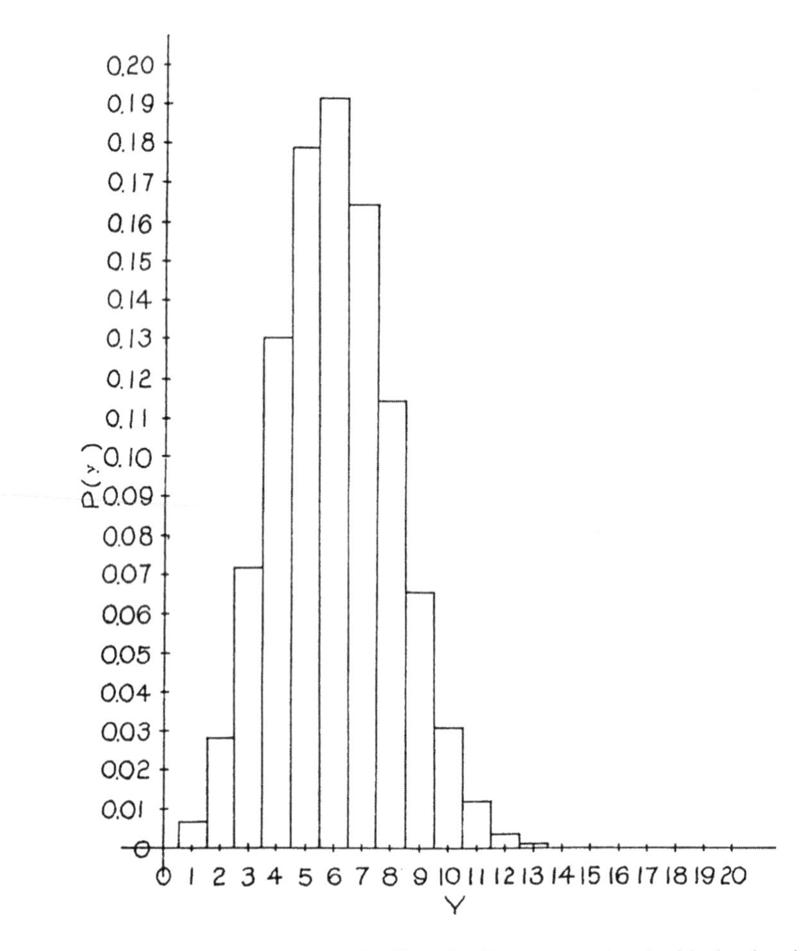


**FIGURE 10.4** Histogram of the binomial probability distribution associated with the drawing of fusiform rust–infected loblolly pine trees when 30% of the stand is infected and when $n = 20$.

**TABLE 10.4**
**The Hypergeometric Probability Distribution Associated with the Occurrence of Southern Fusiform Rust in a Stand of Ten Loblolly Pines in Which 30% of the Trees Show Evidence of the Rust and Three Blind Draws Are to Be Made Using Sampling without Replacement**

| $i$ | 1 | 2 | 3 | 4 |
|---|---|---|---|---|
| $y_i$ | 0 | 1 | 2 | 3 |
| $p(y_i)$ | 0.292 | 0.525 | 0.175 | 0.008 |

When

| $Y =$ | $\binom{N_S}{y_i} =$ | $\binom{N - N_S}{n - y_i} =$ | and $P(y_i) =$ |
|---|---|---|---|
| 0 | $\frac{3!}{0!(3-0)!} = 1$ | $\frac{7!}{3!(7-3)!} = 35$ | $\frac{1(35)}{120} = 0.29$ |
| 1 | $\frac{3!}{1!(3-1)!} = 3$ | $\frac{7!}{2!(7-2)!} = 21$ | $\frac{3(21)}{120} = 0.52$ |
| 2 | $\frac{3!}{2!(3-2)!} = 3$ | $\frac{7!}{1!(7-1)!} = 7$ | $\frac{3(7)}{120} = 0.17$ |
| 3 | $\frac{3!}{3!(3-3)!} = 1$ | $\frac{7!}{0!(7-0)!} = 1$ | $\frac{1(1)}{120} = 0.00$ |

It should be noted that the sum of the numerators of these point probability expressions is equal to 120, demonstrating that the hypergeometric distribution obeys the special additive law of probability.

A comparison of Tables 10.2 and 10.4 will reveal that the method of sampling had an effect on the point probabilities. However, as the relative size of the sample, or the *sampling fraction*, $n/N$, decreases, the differences between the two distributions become smaller and smaller. As a matter of fact, in the limit, as $N \to \infty$ and $n$ is held constant, the hypergeometric distribution converges to the binomial. The rate of convergence is quite rapid. This is demonstrated in Table 10.5. For the distributions in Table 10.5 the size of the sample space, $N$, was increased from that originally used, 10, to 1000; the proportion of the trees with rust was kept constant at 0.3 so that $N_1 = 300$; and the sample size, $n$, was held at 3. As can be seen, the differences between the hypergeometric and binomial distributions are extremely small. It has been found that it is usually safe to use the binomial as an approximation of the hypergeometric distribution when $n$ is 1/10 or less of $N$.

**TABLE 10.5**
**Comparison of the Hypergeometric to the Binomial Probability Distribution When $N$ = 1000, $n$ =3, and $p = N_1/N = 0.3$**

| $i$ | 1 | 2 | 3 | 4 |
|---|---|---|---|---|
| Hypergeometric $P(y_i)$ | 0.342558 | 0.441694 | 0.188936 | 0.026811 |
| Binonomial $p(y_i)$ | 0.343 | 0.441 | 0.189 | 0.027 |

The cumulative probability distribution function for the hypergeometric distribution is

$$P(0 \leq Y \leq y_b) = \binom{N}{n}^{-1} \sum_{i=1}^{b} \binom{N_S}{y_i} \binom{N - N_S}{n - y_i} \tag{10.19}$$

where $b = 1, 2, 3, \ldots, M$; and $M = n + 1$. When sample sizes are small relative to the size of the sample or range spaces, $n \leq N/10$, the hypergeometric distribution differs only slightly from the equivalent binomial distribution and, consequently, Table A.1 in the Appendix can be used to obtain satisfactory cumulative probabilities.

The moment-generating function for the hypergeometric distribution is

$$M_Y(t) = E(e^{tY})$$

$$= \left[\frac{(N-n)!(N-N_S)!}{N!}\right] F(-n, -N_S;\ N - N_S - N + 1;\ e^t) \qquad (10.20)$$

where the $F$ function is a *hypergeometric function*. An explanation of such functions is beyond the scope of this book. Reference is made to Johnson and Kotz (1969) or Kendall and Stuart (1963). The computation of the moments about zero using this function is complicated, lengthy, and tedious. The moments themselves are complex. Consequently, no attempt will be made to show or use them.

The expected value is

$$E(Y) = \frac{nN_S}{N} \qquad (10.21)$$

Since $(N_S/N) = p$, the probability of success in a single random draw, Equation 10.21 can be written $E(Y) = np$, which is Equation 10.14, the expected value in the case of the binomial. Thus, the binomial and hypergeometric distributions have the same means or expected values.

In the case of the original loblolly pine problem, with samples of size 3 and sampling without replacement,

$$E(Y) = \frac{3(3)}{10} = 0.9$$

This can be confirmed using Equation 8.21.

The variance of a hypergeometric distribution is

$$V(Y) = \left[\frac{(N-n)}{(N-1)}\right]\left(\frac{nN_S}{N}\right)\left(1 - \frac{N_S}{N}\right)$$

which can be written

$$V(Y) = \left(\frac{N-n}{N-1}\right)npq \qquad (10.22)$$

provided one realizes that $p$ and $q$ refer to the initial situation, prior to any sampling. The term

$$\left[\frac{(N-n)}{(N-1)}\right]$$

is known as the *finite population correction factor* or fpc. It appears whenever corrections to computations must be made to compensate for the change in the size and character of the range space as sampling without replacement proceeds. If one compares the variances of the binomial and hypergeometric distributions (Equations 10.15 and 10.22), one can see that the only difference involves the fpc. As the sampling fraction, $n/N$, increases, the fpc becomes smaller; and when $n = N$, the fpc and, consequently, the variance equal zero. This reduction in the size of the variance when one samples without replacement, as compared with when the sampling is with replacement, is of profound importance in sampling operations.

Applying Equation 10.22 to the loblolly pine problem yields

$$V(Y) = \left[\frac{(10-3)}{(10-1)}\right](3)\left(\frac{3}{10}\right)\left[1-\left(\frac{3}{10}\right)\right] = 0.49$$

which can be confirmed using Equation 8.32.

The moment coefficient of skewness is

$$\alpha_3 = \frac{\left[\left(1-\frac{N_S}{N}\right)-\left(\frac{N_S}{N}\right)\right]}{\left[\left(\frac{N-n}{N-1}\right)\frac{nN_S}{N}\left(1-\frac{N_S}{N}\right)\right]^{1/2}}\left(\frac{N-2n}{N-2}\right) \tag{10.23}$$

which can be rewritten:

$$\alpha_3 = \frac{(q-p)}{\left[\left(\frac{N-n}{N-1}\right)npq\right]^{1/2}}\left(\frac{N-2n}{N-2}\right) = \frac{(q-p)}{(npq)^{1/2}}\left[\left(\frac{N-1}{N-n}\right)^{1/2}\left(\frac{N-2n}{N-2}\right)\right]$$

In this form the relationship between the coefficients of skew for the binomial and hypergeometric distribution can be seen without difficulty. Again the fpc appears as a factor. As the sampling fraction increases, the magnitude of $\alpha_3$ decreases. When $n = N/2$, the

$$\left[\frac{(N-2n)}{(N-2)}\right]$$

term and, consequently, $\alpha_3$ equal zero, indicating symmetry, regardless of the magnitudes of $p$ and $q$. When $n > N/2$, the

$$\left[\frac{(N-2n)}{(N-2)}\right]$$

term becomes negative, and the direction of skew normally associated with the relative magnitude of $p$ and $q$ becomes reversed. Finally, when $n = N$, $(N - n) = 0$, and $\alpha_3$ becomes indeterminate.

It is obvious from the revised version of Equation 10.23 that the distribution is symmetrical when $p = q = 0.5$. When $p < 0.5$ the skew is negative, and when $p > 0.5$ it is positive. In the case of the loblolly pine problem,

$$\alpha_3 = \left\{\frac{(0.7-0.3)}{\left[\left(\frac{10-3}{10-1}\right)(3)(0.3)(0.7)\right]^{1/2}}\right\}\left[\frac{10-2(3)}{10-2}\right] = 0.2857$$

The skew is toward the right.

The moment coefficient of kurtosis is

$$\alpha_4 = \frac{3(N-1)(N+6)}{(n-2)(N-3)} + \left[\frac{(N-1)N(N+1)}{(N+n)(N-2)(N-3)}\right] * \left\{1 - \frac{6N}{(N+1)}\left[pq + n\frac{(N-n)}{N^2}\right]\right\} * \frac{1}{npq} \tag{10.24}$$

All three parameters have an effect on the skewness and kurtosis of the hypergeometric distribution. If $N$ and $n$ are held constant, as the magnitude of $p$ shifts either way from 0.5, the distribution becomes skewed, and the greater the shift from 0.5, the greater the skew. This skewing causes the mound to become more and more peaked. Consequently, as $p$ shifts farther and farther away from 0.5, the greater the magnitude of $\alpha_4$. As the sampling fraction increases, the degree of skewing is reduced and concomitantly the degree of kurtosis also changes. Finally, as was mentioned earlier, when $N \to \infty$ and $n$ is held constant, the distribution converges to the binomial.

## 10.5 THE POISSON PROBABILITY DISTRIBUTION ($p{:}\lambda$)

In the upland portion of the Lower Coastal Plain of Alabama, above the relatively wet areas along the watercourses of all sizes, is a forest containing an enormously large number of trees. Scattered through this forest are a few spruce pines that have managed to survive outside of their usual habitat in the stream bottoms. If a universe is defined as consisting of all the trees in this region and the event of interest is the choosing of a spruce pine, the probability of doing so would be very small because the size of the subset of interest relative to that of the entire sample space is very small ($N_{SP} \to 0$):

$$P(\text{spruce pine}) = \frac{N_{SP}}{N} = p \to 0$$

If repeated draws are to be made from the range space associated with this situation, using sampling with replacement, the distribution of the results, the $Y$ values, would be binomial. These $Y$ values would be discrete: 0, 1, 2, …, $n$. Theoretically, the sampling could be continued indefinitely and, consequently, the number of Y values could become countably infinite. When $N \to \infty$ and $p \to 0$, a special case of the binomial distribution is generated, which is known as the *Poisson distribution*.*

According to Equation 10.14, the expected value of a binomial distribution is

$$E(Y) = np$$

If $\lambda$ is defined as the expected value, $np$, then $p = \lambda/n$ and $q = 1 - \lambda n$. Substituting these values into the expression for the point binomial probability (Equation 10.9) yields

$$P(y_i) = \binom{n}{y_i} p^{y_i} q^{n-y_i}$$

$$= \frac{n!}{[y_i!(n-y_i)!]}\left(\frac{\lambda}{n}\right)^{y_i}\left[1 - \left(\frac{\lambda}{n}\right)\right]^{n-y_i}$$

$$= \left(\frac{1}{y_i}!\right)[n(n-1)(n-2)\cdots(n-y_i+1)]\left(\frac{\lambda}{n}\right)^{y_i}\left[1 - \left(\frac{\lambda}{n}\right)\right]^{n-y_i}$$

* After S. D. Poisson, who first described the distribution in 1837.

Since

$$\left(\frac{\lambda}{n}\right)^{y_i} = \frac{\lambda^{y_i}}{n^{y_i}}$$

and there are $y_i$ terms within the first set of brackets,

$$P(y_i) = \left(\frac{1}{y_i}!\right)\left[\left(\frac{n}{n}\right)*\frac{(n-1)}{n}*\frac{(n-2)}{n}*\cdots*\frac{(n-y_i=1)}{n}\right]*\lambda^{y_i}*\left[1-\left(\frac{\lambda}{n}\right)\right]^{n-y_i}$$

$$=\left(\frac{1}{y_i!}\right)\left\{\left[1*\left(\frac{1}{n}\right)\right]*\left[1-\left(\frac{2}{n}\right)\right]*\cdots*\left[1-\left(\frac{y_i}{n}\right)+\left(\frac{1}{n}\right)\right]\right\}*\lambda^{y_i}*\left[1-\left(\frac{\lambda}{n}\right)\right]^{n-y_i}$$

However, the

$$\lim_{n\to\infty}\left\{1*\left(\frac{1}{n}\right)*\left[1-\left(\frac{2}{n}\right)\right]*\cdots*\left[1-\left(\frac{y_i}{n}\right)+\left(\frac{1}{n}\right)\right]\right\}\to 1$$

Therefore,

$$p(y_i) = \left(\frac{1}{y_i!}\right)\lambda^{y_i}\left[1-\left(\frac{\lambda}{n}\right)\right]^{n-y_i}$$

Furthermore,

$$\left[1-\left(\frac{\lambda}{n}\right)\right]^{n-y_i} = \left[1-\left(\frac{\lambda}{n}\right)\right]^{n}*\left[1-\left(\frac{\lambda}{n}\right)\right]^{-y_i}$$

$$=\left\{\left[1-\left(\frac{\lambda}{n}\right)\right]^{(n/\lambda)-\lambda}*\left[1-\left(\frac{\lambda}{n}\right)\right]\right\}^{-y_i}$$

Let $\lambda/n = -X$. Then

$$\left[1-\left(\frac{\lambda}{n}\right)\right]^{n/\lambda} = \left[1-\left(\frac{\lambda}{n}\right)\right]^{-1/X} = \frac{1}{(1+X)^{1/X}}$$

Since

$$X = \frac{-\lambda}{n},\quad \lim_{n\to\infty} X \to 0$$

and, by definition,

$$\lim_{X\to 0}(1+X)^{1/X} \to e = 2.718281828\ \ldots$$

the base for Naperian or natural logarithms. Thus, the

$$\lim_{n\to 0}\left[\frac{1}{(1-X)^{1/X}}\right]\to\frac{1}{e}$$

and

$$P(y_i)=\left(\frac{1}{y_i!}\right)\lambda^{y_i}\left(\frac{1}{e}\right)^{\lambda}\left[1-\left(\frac{\lambda}{n}\right)\right]^{-y_i}$$

but the

$$\lim_{n\to\infty}\left[1-\left(\frac{\lambda}{n}\right)\right]^{-y_i}\to 1$$

therefore

$$P(y_i)=\begin{cases}\dfrac{\lambda^{y_i}}{(y_i!e^{\lambda})}, & \lambda\geq 0;\ Y=0,\ 1,\ 2,\ \ldots \\ 0, & \text{elsewhere}\end{cases} \tag{10.25}$$

which is the expression for the point probability in the case of a Poisson distribution. To illustrate the use of this expression, assume that the probability of selecting a spruce pine in a single blind draw from the set of trees on a specific ownership on the Lower Coastal Plain in Alabama is 0.0001. Further assume that 100 blind draws are to be made using sampling with replacement. What is the probability that three of the trees chosen will be spruce pines? In this case, $\lambda = np = 100(0.0001) = 0.01$, the number of spruce pines expected to be drawn in 100 blind draws, and

$$P(Y=3)=\frac{\lambda^y}{y!e^{\lambda}}=\frac{(0.01)^3}{3!(2.718281828)^{0.01}}=0.000001650$$

The moment-generating function is

$$\begin{aligned}M_Y(t)&=\sum_{i=1}^{n+1}e_i^{ty}P(y_i)\\&=\frac{e^{t(0)}\lambda^0}{0!e^0}+\frac{e^{t(1)}\lambda^1}{1!e^1}+\frac{e^{t(2)\lambda^2}}{2!e^2}+\cdots+\frac{e^{t(n)}\lambda^n}{n!e^n}\\&=e^{\lambda}(e^t-1)\end{aligned} \tag{10.26}$$

By using this, the moments about zero are

$$\begin{aligned}
\mu_1' &= E(Y) = \lambda \text{ or } np \\
\mu_2' &= E(Y^2) = \lambda + \lambda^2 \\
\mu_3' &= E(Y^3) = \lambda + 3\lambda^2 + \lambda^3 \\
\mu_4' &= E(Y^4) = \lambda + 7\lambda^2 + 6\lambda^3 + \lambda^4
\end{aligned} \tag{10.27}$$

The mean or expected value of a Poisson distribution is then equal to $\lambda$ or $np$. This result is to be expected since the Poisson is a special case of the binomial.

The moments about the mean are

$$\begin{aligned}
\mu_2 &= V(Y) = \lambda \text{ or } np \\
\mu_3 &= \lambda \\
\mu_4 &= \lambda + 3\lambda^2
\end{aligned} \tag{10.28}$$

Since $E(Y) = V(Y) = \lambda$, the Poisson probability distribution has only one parameter: $P(\lambda)$ or $P(np)$.

The coefficient of skewness is

$$\alpha_3 = \frac{1}{\lambda} \tag{10.29}$$

Regardless of the value of $\lambda$, which cannot be negative, the coefficient of skew will be positive indicating a skew to the right. This can be seen from the values in Table 10.5. When $\lambda \to 0$, $\alpha_3 \to +\infty$, and the distribution has a reversed J shape. Conversely, when $\lambda \to +\infty$, $\alpha_3 \to 0$, and the distribution approaches symmetry.

As one would suspect, when the distribution can change shape as much as this, the amount of kurtosis also changes. The moment coefficient of kurtosis is

$$\alpha_4 = 3 + \frac{1}{\lambda} \tag{10.30}$$

Since $\alpha_4 \geq 3$ in all cases, the Poisson distribution is inherently leptokurtic. As $\lambda$ increases, the degree of peakedness becomes less and in the limit, as $\lambda \to \infty$, $\alpha_4 \to 3$, and the Poisson converges to the normal.

The mode is the largest value of $Y$ less than $\lambda$. However, when $\lambda$ is an integer, the distribution becomes bimodal with modes at $\lambda - 1$ and $\lambda$. This can be seen in Table 10.6 where the modes are underlined.

The cumulative probability distribution function is derived from Equation 8.16:

$$P(0 \leq Y \leq y_b) = \sum_{i=1}^{b} \left( \frac{\lambda^{Y_i}}{y_i! e^{\lambda}} \right) \tag{10.31}$$

where $b = 1, 2, 3, \ldots, M$; and $M = n + 1$. In the case of the spruce pine in the Lower Coastal Plain of Alabama, the cumulative probability distributions for values of $Y$ from 0 to 7 are shown in Table 10.7. By using this table, the probability of occurrence of 3, 4, 5, 6, or 7 spruce pines in 100 blind draws is

**TABLE 10.6**
**Point Poisson Probabilities Showing the Skewness Inherent in Poisson Distribution**

| | $\lambda$ | | | | | | |
|---|---|---|---|---|---|---|---|
| $Y$ | 0.5 | 1.0 | 3.0 | 5.0 | 7.0 | 14.0 | 15.0 |
| 0 | 0.607 | 0.368 | 0.050 | 0.007 | 0.001 | 0.000 | 0.000 |
| 1 | 0.303 | 0.368 | 0.149 | 0.034 | 0.006 | 0.000 | 0.000 |
| 2 | 0.076 | 0.184 | 0.224 | 0.084 | 0.022 | 0.000 | 0.000 |
| 3 | 0.013 | 0.061 | 0.224 | 0.140 | 0.052 | 0.000 | 0.000 |
| 4 | 0.002 | 0.015 | 0.168 | 0.176 | 0.091 | 0.001 | 0.001 |
| 5 | | 0.003 | 0.101 | 0.176 | 0.128 | 0.004 | 0.002 |
| 6 | | 0.001 | 0.050 | 0.146 | 0.149 | 0.009 | 0.005 |
| 7 | | | 0.022 | 0.104 | 0.149 | 0.017 | 0.010 |
| 8 | | | 0.008 | 0.065 | 0.130 | 0.030 | 0.019 |
| 9 | | | 0.003 | 0.036 | 0.101 | 0.047 | 0.032 |
| 10 | | | 0.001 | 0.018 | 0.071 | 0.066 | 0.049 |
| 11 | | | | 0.008 | 0.045 | 0.084 | 0.066 |
| 12 | | | | 0.003 | 0.026 | 0.098 | 0.083 |
| 13 | | | | 0.001 | 0.014 | 0.106 | 0.096 |
| 14 | | | | 0.001 | 0.007 | 0.106 | 0.102 |
| 15 | | | | | 0.003 | 0.099 | 0.102 |
| 16 | | | | | 0.001 | 0.087 | 0.096 |
| 17 | | | | | 0.001 | 0.071 | 0.085 |
| 18 | | | | | | 0.055 | 0.071 |

*Note:* The underlined values are the modes in each distribution.

$$
\begin{aligned}
P(3 \le Y \le 7) &= P(0 \le Y \le 7) - P(0 \le Y \le 2) \\
&= 0.9999999999 - 0.9999998345 \\
&= 0.0000001654
\end{aligned}
$$

The task of computing the cumulative probabilities using Equation 10.31 is tedious, although the advent of programmable calculators has made the task much less onerous. Because of the tediousness of the calculations, recourse has been made to tabulations such as Table A.2 in the

**TABLE 10.7**
**Point and Cumulative Probabilities Associated with the Occurrence of Spruce Pines in Samples of Size 100 When $p = 0.0001$ and $\lambda = 0.01$**

| Point Probabilities | Cumulative Probabilities |
|---|---|
| $P(Y = 0) = (0.01)^0/0!e^{0.01} = 0.9900498338$ | $P(Y = 0) = 0.9900498338$ |
| $P(Y = 1) = (0.01)^1/1!e^{0.01} = 0.0099004983$ | $P(0 \le Y \le 1) = 0.9999503321$ |
| $P(Y = 2) = (0.01)^2/2!e^{0.01} = 0.0000495024$ | $P(0 \le Y \le 2) = 0.9999998345$ |
| $P(Y = 3) = (0.01)^3/3!e^{0.01} = 0.0000001605$ | $P(0 \le Y \le 3) = 0.9999999995$ |
| $P(Y = 4) = (0.01)^4/4!e^{0.01} = 0.0000000004$ | $P(0 \le Y \le 4) = 0.9999999999$ |
| $P(Y = 5) = (0.01)^5/5!e^{0.01} < 0.0000000001$ | $P(0 \le Y \le 5) > 0.9999999999$ |
| $P(Y = 6) = (0.01)^6/6!e^{0.01} < 0.0000000001$ | $P(0 \le Y \le 6) > 0.9999999999$ |
| $P(Y = 7) = (0.01)^7/7!e^{0.01} < 0.0000000001$ | $P(0 \le Y \le 7) > 0.9999999999$ |

Appendix. These are used in the same manner as was done with Table 10.7 to solve the spruce pine problem. Unfortunately, because of space limitations, these tables are abbreviated and, consequently, cannot be used in the present case. When the situation permits their use, however, they greatly speed up the probability determination process.

The Poisson distribution meets the requirements for a probability distribution which are imposed by the several laws of probability. The primary sample space, like that of the binomial distribution, is made up of two types of mutually exclusive events, "successes" and "failures" of occurrence of the event of interest. While $p$ approaches zero, it can never be negative. The parameter $\lambda$ is the product of two positive values, $n$ and $p$, and, consequently, cannot be negative. Finally, the sum of the point probabilities is equal to unity, as is required by the special additive law:

$$P(S) = \sum_{y=0}^{\infty}[P(y)] = \sum_{y=0}^{\infty}\left(\frac{\lambda^y}{y!e^{\lambda}}\right) = e^{-\lambda}\sum_{y=0}^{\infty}\left(\frac{\lambda^y}{y!}\right)$$

$$= e^{-\lambda}\left[\frac{\lambda^0}{0!} + \frac{\lambda^1}{1!} + \frac{\lambda^2}{2!} + \frac{\lambda^3}{3!} + \cdots + \frac{\lambda^{k-1}}{(k-1)!} + \cdots\right]$$

Since

$$\lim_{k\to\infty}\left[1 + \lambda + \frac{\lambda^2}{2!} + \frac{\lambda^3}{3!} + \cdots + \frac{\lambda^{k-1}}{(k-1)!} + \cdots\right] \to e^{\lambda}$$

$$P(S) = e^{-\lambda} * e^{\lambda} = 1$$

When $n$ is relatively large and $p$ is relatively small, it often is easier to compute probabilities using the Poisson rather than the binomial expressions, even though the problem is actually binomial in nature. This can be appreciated if one considers the computational problem involved with factorials, or even partial factorials, of large numbers and with the raising of numbers to high powers. For example, if $n = 1000$, $p = 0.001$, and $Y = 20$, the probability of the occurrence of $y$, using the binomial is

$$P(Y = 20) = \binom{n}{y}p^y q^{n-1} = \frac{10,000 * 999 * 998 * \cdots * 981}{20!}(0.001)^{20}(0.999)^{(980)}$$

which would indeed be difficult to compute. Even with a calculator or a computer it would be difficult because of the problem of the loss of significant figures. This value could be approximated using the Poisson expression for point probability:

$$\lambda = np = 1000(0.001) = 1$$

$$P(Y = 20) = \frac{\lambda^Y}{Y!e^{\lambda}} = \frac{1^{20}}{20!e^1} = \frac{1}{20!(2.718281828)}$$

which would be much easier to compute. Approximations of the binomial using the Poisson are considered to be acceptable, in most cases, if $n \geq 20$ and $p \geq 0.05$. When $n \geq 100$ and $\lambda \leq 10$, the approximations are usually considered to be excellent. In the case of the spruce pines, the binomial probability is

$$P(Y = 3) = \binom{100}{3}(0.001)^3(0.9999)^{97} = 0.0000001601$$

while the Poisson probability, obtained from Table 10.7, is

$$P(Y = 3) = 0.0000001650$$

Since the problem is properly a binomial probability problem, the correct probability is that obtained using the binomial expression. Thus, the error incurred by using the Poisson is +2.97%.

When occurrences of some event take place randomly or stochastically along some continuum, such as a timescale or on a surface or within a three-dimensional space, the occurrences are known as *random processes*. For example, assume a sawmill–planing mill operation which sustains 52 lost-time accidents in an average year. An accident of this sort, which could be anything from a cut finger to a serious injury, is a random process because it can occur at any time. In the context of probability, the sample space is made up of two types of events, the occurrences of accidents and the nonoccurrences of accidents. This is classically binomial. However, since an accident can occur at any instant of time, the sample space is infinitely large and the probability of occurrence of an accident at any instant of time is infinitesimally small. This means that if it is necessary to determine the probability of occurrence of some number of accidents, say, two, in some time span, say, 4 days, it is impossible to use the binomial distribution. It is possible, however, to arrive at a satisfactory approximation of the probability using the Poisson distribution.

To illustrate this approximation, consider the milling operation mentioned above. Assume that each workweek consists of 5 8-hour days. If this is the case, the accident rate is

$$p = \frac{52 \text{ accidents}}{(52 \text{ weeks} * 5 \text{ days})} = \frac{52}{260} = \frac{1}{5} \text{ or } 0.20 \text{ per day}$$

In other words, if the timescale is divided into 1-day segments, the probability, $p$, of occurrence of an accident in a given day is 0.20. The binomial probability of two accidents in a given 4-day period is

$$P(Y = 2 \text{ in } 4 \text{ days}) = \binom{4}{2}(0.20)^2(0.80)^2 = 0.1536$$

*provided that only one accident can occur on a given day.* This is, however, an unrealistic assumption. It is entirely possible to have two, or even more, accidents occur on a single day. This invalidates the probability computed above. To reduce the likelihood of multiple accidents in a time segment, the time segment can be shortened. For example, if the time segments were reduced to half-days,

$$p = \frac{52 \text{ accidents}}{(52 \text{ weeks} * 10 \text{ half-days})} = \frac{52}{520} = \frac{1}{10} \text{ or } 0.10 \text{ per half-day}$$

and the binomial probability would be

$$P(Y = 2 \text{ in } 4 \text{ days}) = \binom{8}{2}(0.10)^2(0.90)^6 = 0.1488$$

Again the possibility of multiple accidents would be present, which would invalidate the result. Cutting the time segments to quarter-days would yield the following results:

$$p = \frac{52 \text{ accidents}}{(52 \text{ weeks} * 20 \text{ quarter-days})} = \frac{52}{1040} = \frac{1}{20} \text{ or } 0.05 \text{ per quarter-day}$$

and the binomial probability

$$P(Y = 2 \text{ in 4 days}) = \binom{16}{2}(0.050)^2(0.95)^{14} = 0.1463$$

As can be seen, the binomial probabilities change as the time segments become shorter. Furthermore, at each level the possibility of multiple occurrences in a time segment is present, invalidating the computed probabilities. However, it should be noted that, in the above process, the value of λ, *the number of occurrences of the event of interest expected to occur in the specified time segment*, remains constant:

| Time Segment | $p$ | λ |
|---|---|---|
| 1 day | 0.20 | 4 * 0.20 = 0.80 |
| 1/2 day | 0.10 | 8 * 0.10 = 0.80 |
| 1/4 day | 0.05 | 16 * 0.05 = 0.80 |

As a matter of fact, *regardless of how short the time segment is made, the value of λ remains constant*. Consequently, it becomes feasible to use the Poisson distribution to obtain the desired probability:

$$P(Y = 2 \text{ in 4 days}) = \frac{\lambda_y}{y!\,e^{\lambda}} = \frac{0.80^2}{2!\,e^{0.80}} = 0.1438$$

The use of the Poisson distribution in situations such as this is appropriate provided that the following assumptions are valid:

1. The value of $p$ for a specific time (or distance or volume of space) segment is known or can be estimated with suitable precision (e.g., from historical records, as was the case with the milling operation).
2. The probability of multiple successes in any time (or distance or volume of space) segment is negligible.
3. The value of $p$ is independent of previous results (i.e., conditional probability is not present).

In the accident problem the Poisson distribution was used as a substitute for the binomial because the conditions of the problem did not permit the computation of probabilities using the binomial. Actually, the accident problem is but one of many where the Poisson is the appropriate model. Random processes that produce results which are appropriately modeled by the Poisson distribution are called *Poisson processes*. To be a Poisson process the following requirements must be met:

1. The process yields results of the binomial type. That is, they can be described as the occurrence of the event of interest or the failure of that event to occur.
2. The events occur randomly or stochastically along a continuum such as a timescale, a number line, or in a volume of space.
3. The occurrence of an event of interest is independent of any other occurrence of the event of interest. This should hold true if the continuum is segmented and the events occur within the same segment or in separate segments.
4. The continuum is such that it is possible, at least theoretically, to subdivide any segment into an infinite number of subdivisions and that it is possible for the event of interest to occur in any one of this infinite set of subdivisions.
5. The probability of two or more occurrences of the event of interest in one of the infinite set of segment subdivisions is negligible.
6. The probability of occurrence of an event of interest in a segment of the continuum is proportional to the length of the segment, regardless of where the segments are located along the continuum.

## 10.6 THE GEOMETRIC PROBABILITY DISTRIBUTION ($G:p$)

Return to the situation where the stand of ten loblolly pine trees contain three trees infected with southern fusiform rust. What is the probability that, if blind draws were made using sampling with replacement, it would require a certain number of draws, say, three, before a rust-infected tree would be drawn? As has been the case with the distributions previously discussed, each draw results in either the occurrence or the nonoccurrence of the event of interest. The probability of occurrence of the event of interest is constant from draw to draw, and the result of any draw has no effect on any subsequent draw. In this case, however, the *number of draws* is not specified. Instead, the random variable $Y$ is the count of draws needed to obtain the event of interest.

The probability that three draws would be needed before a tree with rust would be drawn can easily be computed using the multiplication law (Equation 7.21):

$$P(Y = 3) = q * q * p = (0.7)(0.7)(0.3) = 0.147$$

As can be seen, this problem is simply one of multiple independent events where the first two events involve failing to obtain a tree with rust and the third involves drawing a tree with rust. This expression can be generalized:

$$P(Y = m) = q^{m-1}p \tag{10.32}$$

where $m$ = the specified number of draws or trials needed to obtain the event of interest.

If this expression is used with a specified $p$ and as $m = 1, 2, 3, \ldots$, the resulting set of probabilities constitutes a *geometric* probability distribution with the point probability function:

$$P(Y = M) = \begin{cases} pq^{m-1}, & Y = 1,\ 2,\ 3,\ \ldots;\ 0 < p \leq 1;\ q = 1 - p \\ 0, & \text{elsewhere} \end{cases} \tag{10.33}$$

Since $q = 1 - p$, the distribution has only one parameter: $p$.

A portion of the geometric distribution associated with the loblolly pine problem is shown in Table 10.8. As can be seen, as the value of $Y$ increases, the probability of having to make that many draws decreases.

If the problem involves the probability of needing *at most m* draws or trials, it can be solved in the following manner, making use of the special addition law:

**TABLE 10.8**
**Geometric Probability Distribution When $p$ = 0.30**

| $Y$ | $P(y)$ |
|---|---|
| $m = 1$ | $p = 0.3000$ |
| 2 | $qp = 0.2100$ |
| 3 | $q^2p = 0.1470$ |
| 4 | $q^3p = 0.1029$ |
| 5 | $q^4p = 0.0720$ |
| 6 | $q^5p = 0.0504$ |
| 7 | $q^6p = 0.0353$ |
| 8 | $q^7p = 0.0247$ |
| 9 | $q^8p = 0.0173$ |
| 10 | $q^9p = 0.0121$ |

$$
\begin{aligned}
P(Y \leq m) &= p\sum_{c=1}^{m} q^{c-1} = p[1 + q + q^2 + \cdots + q^{m-1}] \\
&= p + pq + pq^2 + \cdots + pq^{m-1}
\end{aligned} \tag{10.34}
$$

This, of course, is the cumulative probability function. The probability of obtaining the event of interest in the first draw is $p$. The probability that it will occur in the second draw is $pq$, in the third $pq^2$, and so forth. Since they are alternative events their probabilities are added. The above expression describes the situation correctly. However, it usually is easier to use the following alternative procedure, which involves the *failure* to achieve the event of interest in the specified number of trials.

$$
\begin{aligned}
P(Y > m) &= p\sum_{c=m+1}^{\infty} q^{c-1} \\
&= p[q^m + q^{m+1} + q^{m+2} + \cdots + q^{m+n} + \cdots] \\
&= pq^m(1 + q + q^2 + \cdots + q^{m+n} + \cdots) \\
&= pq^m\left(\frac{1}{p}\right) = q^m
\end{aligned}
$$

Since

$$
\lim_{n\to\infty}[1 + q + q^2 + \cdots + q^{n-1} + \cdots] \to \frac{1}{(1-q)} = \frac{1}{p}
$$

and

$$
P(Y \leq m) = 1 - q^m \tag{10.35}
$$

which is the alternative form of the cumulative probability function.

For example, the probability that *no more* than five draws would be needed to obtain a loblolly pine tree infected with rust is

$$
P(Y \leq 5) = 1 - (0.7)^5 = 0.83193
$$

This can be confirmed using Equation 10.34.

Without proof, the moment-generating function for the geometric probability distribution is

$$M_Y(t) = E(e^{tY}) = \frac{p}{(e^{-t} - q)} \tag{10.36}$$

and the expected value is

$$E(Y) = \frac{1}{p} \tag{10.37}$$

Thus, the expected value of the geometric distribution when $p = 0.30$, as in the case of the loblolly pines, is $1/0.30 = 3.3333$, which means that, on the average, it requires 3.3333 draws to obtain a tree with rust.

The moments about the mean are

$$\begin{aligned} \mu_2 &= V(Y) = \frac{q}{p^2} \\ \mu_3 &= \frac{q(2-p)}{p^3} \\ \mu_4 &= \left(\frac{9q^2}{p^4}\right) + \left(\frac{q}{p^2}\right) \end{aligned} \tag{10.38}$$

and the coefficients of skewness and kurtosis are

$$\alpha_3 = \frac{(2-p)}{9} \tag{10.39}$$

and

$$\alpha_4 = 9 + \frac{p^2}{q} \tag{10.40}$$

Since $p$ and $q$ are always positive and since $p$ is always equal to or less than 1, the coefficient of skew is always positive indicating skew to the right. In actuality, histograms of geometric distributions always have reversed J shapes. The distribution can never take on a mound shape and, consequently, the coefficient of kurtosis is essentially meaningless. However, it cannot take on a value less than $q$ but when $p \to q$, $q \to 0$, and the coefficient $\to \infty$.

Obviously, if the distribution has a reversed J, the mode will be at the minimum value of $Y$, which is 1.

To meet the requirement of the special additive law of probability, the probabilities making up the distribution must sum to 1. This requirement is met by the geometric distribution:

$$\sum_{y=1}^{\infty} P(y) = \sum_{y=1}^{\infty} (q^{y-1}p) = p\sum_{y=1}^{\infty} q^{y-1}$$

$$= p(1 + q + q^2 + q^3 + \cdots + q^{n-1} + \cdots)$$

Since

$$\lim_{n \to \infty}(1 + q + q^2 + q^3 + \cdots + q^{n-1} + \cdots) \to \frac{1}{(1-q)}$$

$$\sum_{y=1}^{\infty} P(y) = p\left[\frac{1}{(1-q)}\right] = p\left(\frac{1}{p}\right) = 1$$

The geometric distribution has been used as a model for diameter distributions in all-aged stands of trees. This was first proposed and used by de Liocourt in 1898. When the ratio of the counts between adjacent diameter classes remains constant as one progresses from the smallest to the largest diameters, the use of the geometric distribution is appropriate. This is referred to as the *law of de Liocourt* (Meyer, 1952). The geometric distribution is the discrete equivalent of the exponential distribution (Section 11.4), which possesses this same characteristic. In recent years the exponential has been used rather than the geometric for such modeling because of the possibility of changing class widths.

## 10.7 THE NEGATIVE BINOMIAL AND PASCAL PROBABILITY DISTRIBUTIONS (*NB*:*m*, *p* and *C*:*m*, *p*)

If the question asked in the loblolly pine problem had been, "What is the probability that three trees with rust would be chosen in five draws?" it could not have been answered using the geometric distribution. However, it obviously is a further extension of the ideas behind the geometric distribution.

If the sampling was done blindly, with replacement, the probability of choosing a tree with rust would be constant for all draws and the result of any draw would have no effect on the results of another draw. This sampling could be continued indefinitely until the desired number of trees had been chosen. The number of draws needed to achieve this goal is the random variable $Y$, which can take on only whole-number values that are equal to or greater than $m$, the desired number of occurrences of the event of interest. If $k$ is the number of trials and if $Y$ can equal $k$ only if the event of interest occurs at the $k$th trial and if $m - 1$ events of interest have already occurred somewhere in the previous $k - 1$ trials, the probability of $Y$ equaling $k$ is, according to the multiplication law, the product of the probability of success at the $k$th trial, which is $p$, and the probability of having had $m - 1$ successes in the previous $k - 1$ trials:

$$P(Y = k) = p\left[\binom{k-1}{m-1} p^{m-1} q^{(k-1)-(m-1)}\right]$$

$$= p^m\left[\binom{k-1}{m-1} q^{(k-m)}\right]$$

$$= \frac{p^m}{q^m}\left[\binom{k-1}{m-1} q^k\right]$$

This forms the basis for the point probability function for the *Pascal* probability distribution.*

* After Blaise Pascal, an eminent French mathematician and religious philosopher of the 17th century.

$$P(Y = k) = \begin{cases} \dfrac{p^m}{q^m}\dbinom{k-1}{m-1}q^k, & y = m,\ m+1,\ M+2,\ \ldots; \\ & m = 1,\ 2,\ 3,\ \ldots; \\ & 0 \le p \le 1;\ q = 1 - p \\ 0, & \text{elsewhere} \end{cases} \tag{10.41}$$

Applying this to the question raised at the beginning of this discussion ($m = 3$, $k = 8$, $p = 0.3$):

$$P(5) = \left[\frac{(0.3)}{(0.7)}\right]^3 \left[\frac{7!}{2!}(7-2)!\right](0.7)^8 = 0.9530$$

Thus, the probability of obtaining three trees with rust in eight random draws, with replacement, is 0.14464.

If $m$ is set equal to 1, Equation 10.41 becomes

$$P(Y = K) = \left(\frac{p}{q}\right)\binom{k-1}{0}q^k = pq^{k-1}$$

which is the point probability function for the geometric distribution (Equation 10.32). Thus, the geometric distribution is a special case of the Pascal.

Looking at the reverse side of this case, the number of *failures* incurred in the $k$ trials which resulted in the $m$ successes is $y = k - m$. Thus, $k = m + y$, $k - 1 = m + y - 1$, and $(m + y - 1) - (m - 1) = y$. Consequently,

$$\frac{p^m}{q^m}\binom{k-1}{m-1}q^k \quad \text{becomes} \quad \frac{p^m}{q^m}\binom{m+y-1}{y}q^{m+y} = p^m\binom{m+y-1}{y}q^y$$

which forms the basis for the point probability function for the *negative binomial* probability distribution:

$$P(Y = y) = \begin{cases} p^m\dbinom{m+y-1}{y}q^y; & Y = 0,\ 1,\ 2,\ 3,\ \ldots; \\ & m = 1,\ 2,\ 3,\ \ldots; \\ & 0 \le p \le 1;\ q = 1 - p \\ 0, & \text{elsewhere} \end{cases} \tag{10.42}$$

Applying this to the loblolly pine problem, the probability of five failures in eight trials is

$$P(2) = (0.3)^3\left[\frac{7!}{2!}(7-2)!\right](0.7)^5 = 0.09530$$

Obviously, the two distributions are closely related. The difference lies in the variable of interest. In the case of the Pascal the variable of interest is the number of draws or trials needed to achieve a specified number of successes, whereas in the case of the negative binomial the variable of interest is the number of failures incurred prior to achieving the specified number of successes. Thus, the

Pascal variable is equal to the sum of the negative binomial variable and $m$, the specified number of successes. The two distributions become identical when this relationship is recognized. For this reason, some writers (e.g., Meyer, 1970) consider them identical. Johnson and Kotz (1969) distinguish between the two by indicating that $m$ can take on noninteger values in the case of the negative binomial but only integer values in the case of the Pascal. Other writers (e.g., Hastings and Peacock, 1974) list them as separate distributions.

The cumulative probability in the case of the Pascal is the probability that *no more than k trials or draws* would need to be made to obtain $m$ occurrences of the event of interest. Conventionally, this would be

$$P(Y \leq k) = \left(\frac{p}{q}\right)^m \sum_{a=m}^{k} \binom{a-1}{m-1} q^a \tag{10.43}$$

For example, assume that one desires to know the probability that no more than eight draws ($k = 8$) would have to be made from the stand of loblolly pines to obtain three trees with rust ($m = 3$).

$$P(Y = 3) = \left(\frac{0.3}{0.7}\right)^3 \binom{2}{2} (0.7)^3 = 0.02700$$

$$P(Y = 4) = \left(\frac{0.3}{0.7}\right)^3 \binom{3}{2} (0.7)^4 = 0.05670$$

$$P(Y = 5) = \left(\frac{0.3}{0.7}\right)^3 \binom{4}{2} (0.7)^5 = 0.07938$$

$$P(Y = 6) = \left(\frac{0.3}{0.7}\right)^3 \binom{5}{2} (0.7)^6 = 0.09261$$

$$P(Y = 7) = \left(\frac{0.3}{0.7}\right)^3 \binom{6}{2} (0.7)^7 = 0.09724$$

$$P(Y = 8) = \left(\frac{0.3}{0.7}\right)^3 \binom{7}{2} (0.7)^8 = 0.09530$$

$$P(Y \leq 8) \qquad = 0.44823$$

The probability that no more than eight draws would be needed is 0.44823.

This is a relatively laborious process. It can often be expedited as follows. If the probability of obtaining *exactly a* successes in $k$ trials is, from the binomial,

$$P(a \text{ in } k) = \binom{k}{a} p^a q^{k-a} = \left(\frac{p}{q}\right)^a \binom{k}{a} q^k$$

and if the probability of failure to meet the requirement of $m$ successes in $k$ trials is

$$P(\widetilde{m \text{ in }} k) = \left(\frac{p}{q}\right)^a \sum_{a=0}^{m-1} \binom{k}{a} q_k$$

the probability of needing no more than $k$ trials is

$$P(Y \le k) = 1 - \left(\frac{p}{q}\right)^a \sum_{a=0}^{m-1} \binom{k}{a} q_k \tag{10.44}$$

Thus, if $k = 8$ and $m = 3$,

$$P(Y \le 8) = 1 - \left(\frac{0.3}{0.7}\right)^0 (0.7)^8 - \left(\frac{0.3}{0.7}\right)^1 (0.7)^8 - \left(\frac{0.3}{0.7}\right)^2 (0.7)^8$$

$$= 1 - 0.05765 - 0.19765 - 0.29647 = 0.44823$$

The reduction in work can be significant.

Since the Pascal and negative binomial distribution are identical except for their positions along a number line, the cumulative probability functions are similar. Thus, in the case of the negative binomial,

$$P(Y \le y) = p^m \sum_{a=0}^{y} \binom{m+a-1}{a} q^a \tag{10.45}$$

Thus,

$$P(Y = 0) = (0.3)^3 \binom{2}{0} (0.7)^0 = 0.02700$$

$$P(Y = 1) = (0.3)^3 \binom{3}{1} (0.7)^1 = 0.05670$$

$$P(Y = 2) = (0.3)^3 \binom{4}{2} (0.7)^2 = 0.07938$$

$$P(Y = 3) = (0.3)^3 \binom{5}{3} (0.7)^3 = 0.09261$$

$$P(Y = 4) = (0.3)^3 \binom{6}{4} (0.7)^4 = 0.09724$$

$$P(Y = 5) = (0.3)^3 \binom{7}{5} (0.7)^5 = 0.09530$$

$$P(Y \le 5) = \overline{0.44823}$$

Equation 10.43 can be used with the Pascal by simply substituting $m + y$ for $k$:

$$P(Y \le y) = 1 - \left(\frac{p}{q}\right)^a \sum_{a=0}^{m-1} \binom{m+y}{a} q^{m+y} \tag{10.46}$$

The moment-generating function for the Pascal distribution, without proof, is

$$M_Y(t) = \frac{p^m e^{tm}}{(1 - qe^t)^m} \tag{10.47}$$

while that for the negative binomial is, again without proof,

$$M_Y(t) = p^m(1 - qe^t)^m \tag{10.48}$$

Since the two distributions are identical except for their positions along the number line, only the means or expected values are different. These could be computed classically using Equation 8.21 or the mgf. However, in the case of the Pascal it is simpler to recognize that:

$Q_1$ = the number of trials needed to obtain the first occurrence of the event of interest
$Q_2$ = the number of trials needed to obtain the second occurrence of the event of interest
⋮
$Q_m$ = the number of trials needed to obtain the $m$th occurrence of the event of interest after the $(m - 1)$ occurrence has taken place

Each of these $Q$ values is subject to random variation and all are geometrically distributed. Since all of these distributions are based on the same parent sample space, they all have the same expected value, $1/p$ (see Equation 10.35) and variance, $q/p^2$ (see Equation 10.38). Furthermore, since there are $m$ distributions, the composite mean or expected value of the Pascal is

$$E(Y) = m\left(\frac{1}{p}\right) = \frac{m}{p} \tag{10.49}$$

and the composite variance is

$$V(Y) = m\left(\frac{q}{p^2}\right) = \frac{mq}{p^2} \tag{10.50}$$

The expected value of the Pascal distribution is interpreted as being the number of trials needed, on the average, to obtain the desired number of occurrences of the event of interest. For example, in the case of the loblolly pines, the average number of draws which would have to be made to obtain three trees with rust is 3(1/0.3) = 10.

In the case of the negative binomial, the expected value, computed using the mgf, is

$$E(Y) = \frac{mq}{p} \tag{10.51}$$

Consequently, the expected number of failures prior to achieving the third success in the loblolly pine problem is 3(0.7)(0.3) = 7. This is $m = 3$ units less than the expected number of trials obtained using Equation 10.49, which points up the difference between the two distributions.

As has been stated, the variance and higher moments about the mean of the two distributions are identical. Thus, the variance of the negative binomial is

$$V(Y) = \frac{mq}{p^2} \tag{10.52}$$

while the moment coefficients of skewness and kurtosis for both distributions are:

$$\alpha_3 = (1 + 9)(mq)^{-1/2} \tag{10.53}$$

and

$$\alpha_4 = 3 + \left(\frac{6}{m}\right) + (p^2 mq) \tag{10.54}$$

The distributions are unimodal and mound shaped. Since neither $m$ nor $q$ can assume negative values, $\alpha_3$ is always positive, indicating a positive skew. Furthermore, because $m$ and $q$ cannot assume negative values, $\alpha_4$ can never become less than 3, which means that the distributions are inherently leptokurtic. If $m$ is held constant and $q \to 0$, the mean of the Pascal approaches $m$ and the mean of the negative binomial approaches zero. In both cases the variances approach zero and the coefficients of skewness and kurtosis approach positive infinity. The distributions would then be spikelike, centered on their respective means. If $m$ is held constant and $q \to 1$, the mean and variances of both distributions approach zero and the coefficients of skewness and kurtosis approach positive infinity, regardless of the magnitude of $q$. Finally, when $M \to \infty$, the means and variances of both distributions approach positive infinity, $\alpha_3 \to 0$, and $\alpha_4 \to 3$. In other words, when $m \to \infty$, the distributions converge to the normal.

As one would expect, the relationships that exist between the two distributions and the binomial all rest on a sample space that is made up of occurrences and nonoccurrences of an event of interest. Fundamental to all three distributions is the probability of occurrence of that event of interest, $p$. As a consequence of their common base, it is possible to obtain probabilities of Pascal and negative binomial outcomes from the binomial probabilities listed in Table A.1 in the Appendix. For example, assume that someone desires to know the probability of having to make more than ten draws from the set of loblolly pines to obtain four trees with rust. In such a case, the binomial distribution would be the set of probabilities of $R$ successes in $n = 10$ draws when $p = 0.3$ and the Pascal distribution would be the set of probabilities of $Y$ draws needed to obtain $m = 4$ trees with rust when $p = 0.3$. The binomial distribution would be based on the expansion

$$(p+q)^{10} = p^{10} + \binom{10}{9}p^9 q + \binom{10}{8}p^8 q^2 + \binom{10}{7}p^7 q^3 + \binom{10}{6}p^6 q^4$$

$$+ \binom{10}{5}p^5 q^5 + \binom{10}{4}p^4 q^6 + \binom{10}{3}p^3 q^7 + \binom{10}{2}p^2 q^8$$

$$+ \binom{10}{1}pq^9 + q^{10} = 1$$

Let $A$ include the first seven terms in the expansion and $B$ the last four terms. As can be seen, $P(R \geq m)$ is the sum of the terms in $A$ while $P(R \leq m)$ is the sum of the terms in $B$. Furthermore, the sum of the terms in $A$ is $P(Y \leq n)$ and the sum in $B$ is $P(Y \geq n)$. Consequently,

$$P(Y \leq n) = P(R \geq m) \quad \text{and} \quad P(Y > N) = P(R < m)$$

From these relationships the derived probability is

$$P(Y > 10) = P(R < 3) = \sum_{k=0}^{3} \binom{10}{k} p^k q^{10-k}$$

which, from Table A.1 in the Appendix, is 0.6496.

Since $p, q, m, k,$ and $y$ can take on only positive values, no point probability for either distribution can be negative. The remaining arguments proving that the two distributions meet the laws of probability are somewhat more complex.

The generalized form of the binomial theorem is

$$(1+q)^m = 1 + mq + \frac{m(m-1)}{2!}q^2 + \frac{m(m-1)(m-2)}{3!}q^3 + \cdots + \frac{m(m-1)\cdots(m-n+1)}{n!}q^n + \cdots + q^m = \sum_{a=0}^{m}\binom{m}{a}q^a \tag{10.55}$$

If $q$ is negative and $m$ is positive, the expression becomes

$$(1-q)^m = 1 - mq + \frac{m(m-1)}{2!}q^2 - \frac{m(m-1)(m-2)}{3!}q^3 + \cdots = \sum_{a=0}^{m}\binom{m}{a}q^a \tag{10.56}$$

This is an alternating series containing $m + 1$ terms. If both $q$ and $m$ are negative,

$$\begin{aligned}(1-q)^{-m} &= \binom{-m}{0}(-q)^0 + \binom{-m}{1}(-q)^1 + \binom{-m}{2}(-q)^2 + \binom{-m}{3}(-q)^3 + \cdots \\ &= 1(1) + \frac{(-m)}{1}(-q) + \frac{(-m)[-(m+1)]}{2!}(-q)^2 \\ &\quad + \frac{(-m)[-(m+1)][-(m+2)]}{3!}(-q)^3 + \cdots \\ &= 1 + mq + \frac{m(m+1)}{(2)(1)}q^2 + \frac{m(m+1)(m+2)}{(3)(2)(1)}q^3 + \cdots \\ &= \sum_{a=0}^{\infty}\binom{-m}{a}(-q)^a\end{aligned} \tag{10.57}$$

which is a positive series with an infinite number of terms. In this case, the numerator

$$\binom{-m}{a}$$

is an *ascending* factorial in which each term is negative. For example, if $m = 3$,

$$
\begin{aligned}
(1-q)^{-3} &= \binom{-3}{0}(-q)^0 + \binom{-3}{1}(-q)^1 + \binom{-3}{2}(-q)^2 + \binom{-3}{3}(-q)^3 + \cdots \\
&= 1 + (-3)(-q) + \frac{(-3)(-4)}{(2)(1)}(-q)^2 + \frac{(-3)(-4)(-5)}{(3)(2)(1)}(-q)^3 \\
&= 1 + 3q + 6q^2 + 10q^3 + \cdots
\end{aligned}
$$

It has been shown by Feller (1957) that

$$
\binom{-m}{y} = (-1)^y \binom{m+y-1}{y} \tag{10.58}
$$

Dividing both sides by $(-1)^y$ yields

$$
(-1)^{-y}\binom{-m}{y} = \binom{m+y+1}{y}
$$

Substituting this in Equation 10.40

$$
\begin{aligned}
P(Y=y) &= (-1)^{-y}\binom{-m}{y} p^m q^y \\
&= \binom{-m}{y} p^m \left[\frac{q}{(-1)}\right]^y \\
&= p^m \binom{-m}{y}(-q)^y
\end{aligned} \tag{10.59}
$$

Substituting this in Equation 10.43 yields

$$
P(Y \le y) = p^m \sum_{a=0}^{y} \binom{-m}{a}(-q)^a \tag{10.60}
$$

and, since

$$
(1-q)^{-m} = \sum_{a=0}^{\infty} \binom{-m}{a}(-q)^a
$$

if $y$ is set equal to positive infinity,

$$
P(Y \le \infty) = p^m (1-q)^{-m} = \frac{p^m}{p^m} = 1
$$

indicating that the sum of the probabilities in the negative binomial distribution is equal to 1 and that the special additive law has been observed. The proof is equally applicable to the Pascal.

Foresters and forest scientists are often faced with the problem of describing, in a mathematical sense, the frequency distribution of counts of occurrences of some event of interest per some unit of time or space. For example, the frequencies of quadrats of a certain size bearing 0, 1, 2, 3, …, trees of a certain species that are less than 4.5 ft in height, or the frequencies of leaves bearing 0, 1, 2, 3, …, egg masses of a certain insect. When the relative frequencies of these counts are matched by the probabilities of a probability distribution, that probability distribution can be used to describe the frequency distribution.

If the event of interest (e.g., the occurrence of a seedling or an egg mass) occurs in a random manner over the space or time continuum, the likelihood of any one sampling unit (e.g., a quadrat or leaf) including an event of interest is constant across all the sampling units. When this occurs, the expected frequency of occurrence of the event of interest on each sampling unit is equal to the mean of the population. Under these circumstances, the occurrence of the event of interest is a Poisson process, and the relative frequency distribution will take on the characteristics of the Poisson probability distribution, including having a variance that is equal to its mean.

If the event of interest is not randomly distributed over the space or time continuum, the likelihood of any one sampling unit including an event of interest is not constant. Consequently, the occurrence of the event of interest is not a Poisson process, and the Poisson is not an appropriate model for the relative frequency distribution. This lack of randomness can arise from many sources, some of which may be obvious while others are not. An example of an obvious relationship is the clustering of seedlings around the mother tree. When clustering (or nonrandomness) occurs, the variance of the relative frequency distribution tends to increase relative to the mean and it is said that "overdispersion" or "contagion" exists. Probability distributions that can be used to describe relative frequency distributions of this type are classed as *contagious* distributions. The negative binomial is one of these contagious distributions and has been used to describe the distributions of seedlings (e.g., Ker, 1954) and counts of different stages of forest insects (e.g., Waters, 1955).

Since the relative frequency distributions mentioned above are strongly influenced by the size and configuration of the sampling units, the use of contagious probability distributions in the study of sampling unit efficiency has been extensive (Ker, 1954; Smith and Ker, 1957; Waters and Henson, 1959). The negative binomial has been widely used in these studies.

The formulae in common use in these studies are derived from Anscombe (1949), and they are somewhat different from those used in the preceding material. This can be confusing. The following development ties together the two systems and should reduce the level of confusion.

According to Anscombe, the parameters of the negative binomial distribution are the mean, $\mu$, and an exponent which corresponds to $m$. The variance, according to Anscombe, is

$$V(Y) = \mu + \frac{\mu^2}{m} \tag{10.61}$$

Substituting $mq/p$ for $\mu$

$$\begin{aligned} V(Y) &= \left[\frac{mg}{p}\right] + \left(\frac{mg}{p}\right)^2\left(\frac{1}{m}\right) \\ &= \left(\frac{mq}{p}\right)\left[1 + \left(\frac{mq}{mp}\right)\right] \\ &= \left(\frac{mq}{p}\right)\left(1 + \frac{q}{p}\right) \\ &= \left(\frac{mq}{p}\right)\left(\frac{1}{p}\right) \\ &= \frac{mq}{p^2} \end{aligned}$$

which is Equation 10.50.

The expected frequency of zeros:

$$\begin{aligned} \phi_0 &= N\left[1+\left(\frac{\mu}{m}\right)\right]^{-m} \\ &= N\left[1+\left(\frac{mq}{p}\right)\left(\frac{1}{m}\right)\right]^{-m} \\ &= N\left[1+\left(\frac{q}{p}\right)\right]^{-m} \\ &= N\left(\frac{1}{p}\right)^{-m} \\ &= Np^m \end{aligned} \tag{10.62}$$

and the expected frequency of any positive count, $y$, is

$$\begin{aligned} \phi_y &= \phi_0\binom{m+y-1}{y}\left(\frac{\mu}{\mu+m}\right)^y \\ &= Np^m\binom{m+y-1}{y}\left[\frac{\left(\frac{mq}{p}\right)}{\left(\frac{mq}{p}\right)+m}\right]^y \end{aligned} \tag{10.63}$$

but

$$\left(\frac{mq}{p}\right)+m = \frac{m}{p}$$

and thus

$$\left[\frac{\left(\frac{mq}{p}\right)}{\left(\frac{mq}{p}\right)+m}\right]^y = \left[\frac{mq}{p}\left(\frac{p}{m}\right)\right]^y = q^y$$

and

$$\phi_y = N\left[p^m\binom{m+y-1}{y}q^y\right]$$

This should make the correlating of formulae somewhat easier when reading the literature dealing with the use of the negative binomial.

## 10.8 THE MULTINOMIAL PROBABILITY DISTRIBUTION ($M$:$p_1$, $p_2$, ...)

With the exception of the rectangular, all the previously discussed probability distributions have been based on a sample space made up of occurrences and nonoccurrences of a single event of interest. In other words, in the case of the loblolly pines, one *succeeded* in drawing a tree with rust or one *failed* to draw such a tree. No other outcomes were possible. Suppose, however, that a stand of trees was made up of loblolly, shortleaf, and longleaf pines, and the event of interest was the drawing of a specific mix of the three species in $n$ draws, when the sampling was with replacement. This would require the isolation of a specific term in the expansion of a trinomial.

$$(p_1 + p_2 + p_3)^n = 1 \tag{10.64}$$

where $p_1 = p$(loblolly pine)
$p_2 = p$(shortleaf pine)
$p_3 = p$(longleaf pine)

Assume that $p_1 = 0.2$, $p_2 = 0.3$, and $p_3 = 0.5$ and that three draws are to be made. The expansion is

$$\begin{aligned}(p_1 + p_2 + p_3)^3 &= p_1^3 + p_2^3 + p_3^3 + 3p_1^2 p_3 + 3p_1 p_2^2 \\ &+ 3p_1 p_3^2 + 3p_2^2 p_3 + 3p_2 p_3^2 + 6p_1 p_2 p_3\end{aligned} \tag{10.65}$$

Each of the terms in the expansion represents the probability of occurrence of a specific mix of species and, consequently, the expansion forms the probability distribution shown in Table 10.9. This is an example of a *multinomial* probability distribution.

Multinomial probability distributions arise when the sample space contains two or more mutually exclusive sets of events; and multiple draws, using sampling with replacement, are to be made from the sample space. The binomial distribution is obviously one of the multinomial distributions

**TABLE 10.9**
**The Trinomial Distribution Associated with the Occurrence of Species Mixes When Three Blind Draws Are Made from a Stand of Trees Made up of 20% Loblolly (L), 30% Shortleaf (S), and 50% Longleaf (LL) Pine**

| Species Mix | | P(Species Mix) | |
|---|---|---|---|
| (L, L, L) | or ($Y_1 = 3, Y_2 = 0, Y_3 = 0$) | $p_1^3 = (0.2)^3$ | = 0.008 |
| (S, S, S) | or ($Y_1 = 0, Y_2 = 3, Y_3 = 0$) | $p_2^3 = (0.3)^3$ | = 0.027 |
| (LL, LL, LL) | or ($Y_1 = 0, Y_2 = 0, Y_3 = 3$) | $p_3^3 = (0.5)^3$ | = 0.125 |
| (L, L. S) | or ($Y_1 = 2, Y_2 = 1, Y_3 = 0$) | $3p_1^2p_2 = 3(0.2)^2(0.3)$ | = 0.036 |
| (L, L, LL) | or ($Y_1 = 2, Y_2 = 0, Y_3 = 1$) | $3p_1^2p_3 = 3(0.2)^2(0.5)$ | = 0.060 |
| (L, S, S) | or ($Y_1 = 1, Y_2 = 2, Y_3 = 0$) | $3p^1p_2^2 = 3(0.2)(0.3)^2$ | = 0.054 |
| (L, LL, LL) | or ($Y_1 = 1, Y_2 = 0, Y_3 = 2$) | $3p^1p_3^2 = 3(0.2)(0.5)^2$ | = 0.150 |
| (S, S, LL) | or ($Y_1 = 0, Y_2 = 2, Y_3 = 1$) | $3p_2^2p_3 = 3(0.3)^2(0.5)$ | = 0.135 |
| (S, LL, LL) | or ($Y_1 = 0, Y_2 = 1, Y_3 = 2$) | $3p_2p_3^2 = 3(0.3)(0.5)^2$ | = 0.225 |
| (L, S, LL) | or ($Y_1 = 1, Y_2 = 1, Y_3 = 1$) | $6p_1p_2p_3 = 6(0.2)(0.3)(0.5)$ | = 0.180 |
| | | | 1.000 |

and many of the characteristics of binomial distributions are common with multinomial distributions in general.

The point probability function for a multinomial distribution is

$$P(Y_1 = n_1,\ Y_2 = n_2,\ \ldots,\ Y_k = n_k) = \frac{n!}{n_1!n_2!\ldots n_k!} p_1^n 1 p_2^n 2 \ldots p_k^n k \qquad (10.66)$$

where $Y_i$ = number of times simple event $i$ occurs in the sample (e.g., the number of times a loblolly pine appears in a sample)

$n_i$ = the specified number of times event $i$ is to occur in the desired mix of events

Thus, if the probability of drawing a sample of size 3 from the stand of mixed pines which would contain two loblolly pines and one longleaf pine is

$$P(Y_1 = 2,\ Y_2 = 0,\ Y_2 = 1) = \frac{3!}{2!0!1!}(0.2)^2(0.3)^0(0.5)^1 = 0.060$$

which agrees with the corresponding term in Table 10.9.

The expected number of occurrences of a specific event in a sample of size $n$ is

$$E(Y_i) = np_i \qquad (10.67)$$

This arises from the fact that, when only one of the possible simple events is considered the event of interest (event $i$), the sample space which actually contains several mutually exclusive subsets is restructured into two subsets, the first containing the event of interest and the second all the other events. Thus, when a draw is made from the sample space, either one *succeeds* in obtaining the event of interest or *fails* to do so — a classical binomial situation. Consequently, Equation 10.14, the expression for the expected value for a binomial distribution, is appropriate. In the same manner,

$$V(Y_i) = np(1 - p) \qquad (10.68)$$

# 11 Continuous Probability Distributions

## 11.1 THE RECTANGULAR PROBABILITY DISTRIBUTION ($R$:$y_A$, $y_B$)

The discrete form of this distribution is described in Section 10.2. The continuous form is basically similar. The probability of occurrence is constant for all values of $Y$ in the interval $y_A$ to $y_B$. When graphed, the pdf produces a straight horizontal line running from $y_A$ to $y_B$ at the distance of $f(y)$ from the horizontal axis and the cdf produces a curve of constant slope rising from $F(y_A)$ to $F(y_B)$.

In the case of a continuous random variable, $N(y)$ is undefined, since the interval between adjacent $Y$ values is infinitesimally small. However, since it is a constant and serves only as a scalar, it can be set equal to unity. Then

$$T = \int_{y_A}^{y^B} N(y)dy = h\int_{y_A}^{y^B} dy = N(y)(y_B - y_A) = y_B - y_A$$

and the pdf is

$$f(y) = \frac{N(y)}{T} = \begin{cases} \dfrac{1}{(y_B - y_A)}, & y_A \leq Y \leq y_B \\ 0, & \text{elsewhere} \end{cases} \tag{11.1}$$

Note that the pdf is a constant and is a function of the length of the interval between $y_A$ and $y_B$. The cumulative probability function is

$$\begin{aligned} F(y) &= P(y_A \leq Y \leq y_b) \\ &= \frac{1}{y_B - y_A}\int_{y_A}^{y^B} dy = \frac{y_b - y_A}{y_B - y_A} \end{aligned} \tag{11.2}$$

as $y_b$ takes all values from $y_A$ to $y_B$.

The moments about zero are obtained using Equation 8.35:

$$\mu_1' = E(Y) = \int_{y_A}^{y_B} y\left[\frac{1}{(y_B - y_A)}\right]dy = \frac{1}{y_B - y_A}\left(\frac{y^2}{2}\right)\Bigg|_{y_A}^{y_B} = \frac{y_B^2 - y_A^2}{2(y_B - y_A)}$$

$$\mu_2' = E(Y^2) = \frac{y_B^3 - Y_A^3}{3(y_B - y_A)}$$

$$\mu_3' = E(Y^3) = \frac{y_B^4 - Y_A^4}{4(y_B - y_A)}$$

$$\mu_4' = E(Y^4) = \frac{y_B^5 - Y_A^5}{5(y_B - y_A)}$$

Thus, the mean or expected value is

$$E(Y) = \frac{y_B^2 - y_A^2}{2(y_B - y_A)} = \frac{(y_B + y_A)(y_B - y_A)}{2(y_B - y_A)} = \frac{y_B + y_A}{2} \tag{11.3}$$

The variance is

$$V(Y) = E(Y^2) - [E(Y)]^2 = \frac{y_B^3 - y_A^3}{3(y_B - y_A)} - \left(\frac{y_B - y_A}{2}\right)^2 = \frac{(y_B - y_A)^2}{12} \tag{11.4}$$

The moment coefficient of skew is equal to zero since the distribution is symmetrical. The coefficient of kurtosis is, after much tedious algebra,

$$\alpha_4 = \{E(Y^4) - 4E(Y^3)E(Y) + 6E(Y^2)[E(Y)]^2 - 3[E(Y)]^4\} * [V(Y)]^{-2} \tag{11.5}$$

As in the case of the discrete form of the distribution, there is no mode.

The utility of the rectangular probability distribution was discussed in Section 10.2.

## 11.2 THE NORMAL PROBABILITY DISTRIBUTION ($N$:$\mu$, $\sigma$)

Assume a primary sample space which contains two mutually exclusive subsets. When samples of size $n$ are drawn from this space, using sampling with replacement, the probabilities of occurrence of the compound events generated by this sampling process are distributed binomially. If the probability of occurrence of the primary event of interest is 0.5, the binomial distribution is mound shaped and symmetrical (Section 10.3). Assume now that the sample size is increased indefinitely so that $n \to \infty$. In such a case the lateral extent of the distribution, if shown in histogram form, would be to infinity in both directions from the distribution mean [$E(y) = np$]. If the bar widths of the histogram were reduced to infinitesimals, the tops of the bars would form a smooth bell-shaped *continuous* curving line. This line is the trace of the frequency distribution function of the *normal* or *Gaussian** probability distribution, which is, without doubt, the most important probability distribution in existence.

* After Carl Friedrich Gauss (1777–1855), who first described the distribution. Gauss is considered to be one of the foremost mathematicians of all time, ranking with Archimedes and Newton.

The frequency distribution function for the normal distribution is derived from the binomial in the following manner. In terms of the binomial the probability of obtaining $y$ occurrences of the event of interest ($Y = y$) in a sample of size $n$ is (see Equation 10.9):

$$P(y) = \binom{n}{y} p^y q^{n-y}$$

and since it has been specified that $p = 0.5$ or $^1/_2$,

$$P(y) = \binom{n}{y}\left(\frac{1}{2}\right)^y \left(\frac{1}{2}\right)^{n-y}$$

$$= \binom{n}{y}\left(\frac{1}{2}\right)^n = \left[\frac{n!}{y!(n-y)!}\right]\left(\frac{1}{2}\right)^n$$

The frequency of the event of interest is

$$N_y = N_D \left(\frac{1}{2}\right)^n \left[\frac{n!}{y!(n-y)!}\right]$$

where $N_D = N_P^n$ = size of the derived sample space
$N_P$ = size of the primary sample space

The size of the subset of interest, when $Y = m + 1$ is

$$N_y + 1 = N_D \left(\frac{1}{2}\right)^n \left\{\frac{n!}{(y+1)![n-(y+1)]!}\right\}$$

$$= N_D \left(\frac{1}{2}\right)^n \left[\frac{n!}{(y+1)!(n-y-1)!}\right]$$

As can be seen, in both the $N_Y$ and $N_{y-1}$ expressions the $N_D(^1/_2)^n$ terms are the same, and the difference between the two frequencies is due entirely to the difference in the binomial coefficients. Thus,

$$\left[\frac{n!}{y!(n-y)!}\right] = \frac{n(n-1)(n-2)\cdots(n-y+1)}{y(y-1)(y-2)\cdots(2)(1)}$$

and

$$\left[\frac{n!}{(y+1)!(n-y-1)!}\right]=\frac{n(n-1)(n-2)\cdots[n-(y+1)+2][n-(y+1)+1]}{(y+1)(y)(y-1)(y-2)\cdots(2)(1)}$$

$$=\frac{n(n-1)(n-2)\cdots(n-y+1)(n-y)}{(y+1)(y)(y-1)(y-2)\cdots(2)(1)}$$

$$=\frac{(n-y)}{(y+1)}\left[\frac{n(n-1)(n-2)\cdots(n-y+1)}{y(y-1)(y-2)\cdots(2)(1)}\right]$$

$$=\frac{(n-y)}{(y+1)}\left[\frac{n!}{y!(n-y)!}\right]$$

Therefore,

$$\left[\frac{(n-y)}{(y+1)}\right]$$

is a multiplier which can be used to obtain the frequency of an event of interest from the frequency of the event immediately preceding that event of interest.

$$s_{y+1}=s_y\left[\frac{(n-y)}{(y+1)}\right]$$

When the multiplier is greater than 1, $N_{y+1}$ is greater than $N_y$. When it is equal to 1, $N_{y+1}$ and $N_y$ are equal, and when it is less than 1, $N_{y+1}$ is smaller than $N_y$. If this is done systematically, term by term in the binomial expansion, it will be seen that the subset sizes increase to a maximum and then decrease. If $n$ is an even number, and since $n + 1$ outcomes* are possible, the maximum subset size will be associated with one specific outcome and that outcome is also the mean of the distribution. This can be demonstrated as follows. Let $n = 4$, then $Y$ can take on the values 0, 1, 2, 3, and 4. In this case $\mu_y = np4(1/2) = 2$.

Computing the values of the multiplier yields

$$\frac{4-0}{0+1}=\frac{4}{1}>1,\quad N_1>N_0$$

$$\frac{4-1}{1+1}=\frac{3}{2}>1,\quad N_2>N_1$$

$$\frac{4-2}{2+1}=\frac{2}{3}<1,\quad N_3<N_2$$

$$\frac{4-3}{3+1}=\frac{1}{4}<1,\quad N_4<N_3$$

As can be seen, $N_2$ is greater than either $N_1$ or $N_3$ and is equal to $\mu$. If $n$ is an odd number, the maximum subset size is associated with the two adjacent outcomes at the center of the array of $Y$ values and the mean of the distribution is the mean of these two outcomes. Let $n = 5$. Then $Y$ can take on the values 0, 1, 2, 3, 4, and 5. $\mu = 5(1/2) = 2.5$.

* Since $Y$ can assume the value 0 as well as all the values in the set {1, 2, 3, …, $n$}, the total number of possible outcomes is $n + 1$.

$$\frac{5-0}{0+1} = \frac{5}{1} > 1, \quad N_1 > N_0$$

$$\frac{5-1}{1+1} = \frac{4}{2} > 1, \quad N_2 > N_1$$

$$\frac{5-2}{2+1} = \frac{3}{3} = 1, \quad N_3 = N_2$$

$$\frac{5-3}{3+1} = \frac{2}{4} < 1, \quad N_4 < N_3$$

$$\frac{5-4}{4+1} = \frac{1}{5} < 1, \quad N_5 < N_4$$

In this case, the maximum subset sizes occurred when $Y = 2$ and $Y = 3$ and the mean of these two $Y$ values is 2.5. In either case, *the maximum probability of occurrence was associated with the mean of the distribution.*

In the remaining portion of this derivation it will be assumed that $n$ is infinitely large and yet is an even number. This is being done to keep the algebra as simple as possible. Assuming $n$ to be odd would lead to the same final result but the work would be considerably more complex. Let $n = 2k$. Consequently, $\mu = np = 2k(^1/_2) = k$, the value at the center of the array of $Y$ values. $k$ has the maximum probability of occurrence, the greatest frequency, and in a histogram its ordinate would be the highest.

$$N_k = N_D\left(\frac{1}{2}\right)^{2k}\left[\frac{(2k)!}{k!(2k-k)!}\right]$$

$$= N_D\left(\frac{1}{2}\right)^{2k}\left[\frac{(2k)!}{k!k!}\right]$$

The subset size or frequency of any specific $Y$ value is

$$S_{k+z} = N_D\left(\frac{1}{2}\right)^n\left\{\frac{(2k)!}{(k+z)![2k-(k+z)]!}\right\} \quad \text{where } z = |y-k|, \; y = 0, 1, 2, \ldots, n$$

$$= N_D\left(\frac{1}{2}\right)^n \frac{(2k)!}{(k+z)!(k-z)!}$$

and the ratio of $S_{k+z}$ to $S_k$ is

$$\frac{N_z}{N_k} = \frac{N_D\left(\frac{1}{2}\right)^n\left[\frac{(2k)!}{(k+z)!(k-z)!}\right]}{N_D\left(\frac{1}{2}\right)^n\left[\frac{(2k)!}{k!k!}\right]}$$

$$= \frac{k!k!}{(k+z)!(k-z)!}$$

$$= \frac{k(k-1)(k-2)(k-3)\cdots(k-z+1)}{(k+1)(k+2)(k+3)\cdots[k+(z-1)](k+z)}$$

Multiplying both numerator and denominator of the ratio by $1/k$:

$$\frac{N_z}{N_k} = \frac{(1)\left(1-\frac{1}{k}\right)\left(1-\frac{2}{k}\right)\left(1-\frac{3}{k}\right)\cdots\left[1-\frac{(z-1)}{k}\right]}{\left(1+\frac{1}{k}\right)\left(1+\frac{2}{k}\right)\left(1+\frac{3}{k}\right)\cdots\left[1+\frac{(z-1)}{k}\right]\left(1+\frac{z}{k}\right)}$$

The natural logarithm of this expression is

$$\begin{aligned}
\ln\left(\frac{N_{k+z}}{N_k}\right) & \ln\left\{\left(1-\frac{1}{k}\right)\left(1-\frac{2}{k}\right)\left(1-\frac{3}{k}\right)\cdots\left[1-\frac{(z-1)}{k}\right]\right\} \\
& -\ln\left\{\left(1-\frac{1}{k}\right)\left(1+\frac{2}{k}\right)\left(1+\frac{3}{k}\right)\cdots\left[1+\frac{(z-1)}{k}\right]\left(1+\frac{z}{k}\right)\right\} \\
& = \ln\left(1-\frac{1}{k}\right)+\ln\left(1-\frac{2}{k}\right)+\ln\left(1-\frac{3}{k}\right)+\cdots+\ln\left[1-\frac{(z-1)}{k}\right] \\
& -\ln\left(1+\frac{1}{k}\right)-\ln\left(1+\frac{2}{k}\right)-\ln\left(1+\frac{3}{k}\right)-\cdots \\
& -\ln\left[1+\frac{(z-1)}{k}\right]-\ln\left(1+\frac{z}{k}\right)
\end{aligned}$$

Since

$$\ln(1+\psi) = \psi - \frac{\psi^2}{2} + \frac{\psi^3}{3} - \frac{\psi^4}{4} + \cdots$$

and

$$\ln(1-\psi) = -\psi - \frac{\psi^2}{2} - \frac{\psi^3}{3} - \frac{\psi^4}{4} - \cdots$$

and, since both of these series converge very rapidly to their respective limits, the first term of the appropriate series can be substituted for each term in the logarithmic form of the ratio without undue loss of precision:

$$\begin{aligned}
\ln\left(\frac{N_{k+z}}{N_k}\right) & = -\frac{1}{k}-\frac{2}{k}-\frac{3}{k}-\cdots-\frac{(z-1)}{k}-\frac{1}{k}-\frac{2}{k}-\frac{3}{k}-\cdots-\frac{(z-1)}{k}-\frac{z}{k} \\
& = \frac{-2}{k}[1+2+3+\cdots+(z-1)]-\frac{z}{k} \\
& = \frac{-z(z-1)}{k}-\frac{z}{k}=\frac{-z^2}{k}
\end{aligned}$$

$$\frac{N_{k+z}}{N_k} = \text{anti}\ln\left(\frac{-z^2}{k}\right) = e^{-z^2/k}$$

Thus, the magnitude of $N_{k+z}$ relative to $N_k$ is established. Since $\sigma_Y^2$, or simply

$$\sigma^2 = npq = 2k\left(\frac{1}{2}\right)\left(\frac{1}{2}\right) = \frac{k}{2}$$

and consequently $k = 2\sigma^2$, then $N_{k+z} = N_k e^{-z^2/2\sigma^2}$.

In this expression $N_k$ is the standard from which the ordinates of all other $Y$ values can be determined. It is, of course, an infinitesimal and its actual value cannot be computed. However, if it is set equal to 1, the ordinates of all the $Y$ values become proportions of 1 which preserves the curve and also makes it possible to scale the curve to any desired height if a graphical plot is desired. Thus, when $N_k = 1$, the expression becomes

$$N_{k+z} = e^{-z^2/2\sigma^2}$$

Earlier in this development, $z$ was defined as $|Y - k|$. Since $k = \mu$, $z$ can be interpreted as the distance, along the $y$-axis of the histogram, from $\mu$ to the $Y$ value of interest. Because of the symmetry of the distribution, the heights of the ordinate at $Y = k + z$ and that at $k - z$ are equal. Furthermore, because of the symmetry, it is possible to redefine $z$ as $y - k$ or $y - \mu$ and the expression becomes $z = Y - \mu$; thus,

$$N_y = e^{-(y-\mu)^2/2\sigma^2} \quad \text{or} \quad e^{-1/2[(y-\mu)^2/\sigma^2]} \tag{11.6}$$

which is the frequency distribution function for the *normal* distribution. If $\mu$ and $\sigma^2$ are specified, the curve of the function is defined exactly.

At this point it is well to remember that one is still working with a binomial distribution of $Y$ values, each of which is the number of occurrences of an event of interest in an infinitely large sample. Since negative occurrences are impossible, $Y$ cannot be negative. Consequently, the $Y$ values range from zero to infinity. The mean or expected value of this distribution is $\mu = np$, which, because $n$ is infinitely large, cannot be defined. The same is true of the variance, $\sigma^2 = npq$. This might lead one to think that the distribution has little or no application. This is not true. What has been done is to derive a curve that can be used to describe nonbinomial distributions of random variables which do have definite means and variances. Furthermore, since the curve extends to infinity in both directions from the mean, no matter where the mean is placed on the horizontal axis of the graph, the curve will extend to $Y$ values which are negative. It is in this latter context that the discussion continues.

Following the reasoning behind Equation 8.11, the total area under the normal curve is

$$T = P(-\infty \leq Y \leq \infty) = \int_{-\infty}^{\infty} e^{-1/2[(y-\mu)^2/\sigma^2]} dy$$

There is no known function for which $e^{-1/2[(y-\mu)^2/\sigma^2]}$ is a derivative. Consequently, the integration must be done indirectly. Let

$$z = \frac{(y-\mu)}{\sigma}$$

When this is done, the element of area, $dy$, must be converted to its equivalent in terms of $z$.

$$\frac{dz}{dy} = \frac{d\left[\frac{(y-\mu)}{\sigma}\right]}{dy}$$

$$= \frac{1}{\sigma}\frac{d}{dy}(Y-\mu) = \frac{1}{\sigma}$$

$$\text{and} \quad dz = f'(y)dy = \frac{1}{\sigma dy}$$

$$\sigma dz = dy$$

Substituting,

$$T = \int_{-\infty}^{\infty} e^{-z^2/2}\sigma dz = \sigma\int_{-\infty}^{\infty} e^{-z^2/2}dz$$

$$T^2 = \sigma^2\int_{-\infty}^{\infty} e^{-z^2/2}dz\int_{-\infty}^{\infty} e^{-s^2/2}ds = \sigma^2\int_{-\infty}^{\infty}\int_{-\infty}^{\infty} e^{-(s^2+z^2)/2}dsdz$$

Making use of polar coordinates to solve this integral, let $s = r\cos\alpha$ and $z = r\sin\alpha$. Then $s^2 + z^2 = r^2$.

As $s$ and $z$ vary from $-\infty$ to $+\infty$, $r$ varies from 0 to $\infty$ and $\alpha$ varies from 0 to $2\pi$ radians. The element of area $dsdt$ becomes $rdrd\alpha$ and thus,*

$$T^2 = \sigma^2\int_0^{2\pi}\int_0^{\infty} e^{-r^2/2}rdrd\alpha$$

Since

$$\frac{d(-e^{-r^2/2})}{dr} = -e^{-r^2/2}\left(-\frac{1}{2}\right)(2)(r) = e^{-r^2/2}r$$

$$\int_0^{\infty} e^{-r^2/2}rdr = -e^{-r^2/2}\Big|_0^{\infty} = 0-(-1) = 1$$

$$T^2 = \sigma^2\int_0^{2\pi} d\alpha = \sigma^2\alpha\Big|_0^{2\pi} = \sigma^2(2\pi - 0) = \sigma^2 2\pi$$

$$T = \int_{-\infty}^{\infty} e^{-1/2[(y-\mu)^2/\sigma^2]}dy = \sigma\sqrt{2\pi} \qquad (11.7)$$

and the probability density function, or pdf, of the *normal* distribution, is

$$f(y) = \frac{N_y}{T} = \frac{1}{\sigma\sqrt{2\pi}}e^{-1/2[(y-\mu)^2/\sigma^2]}, \quad -\infty < Y < \infty \qquad (11.8)$$

* For an explanation of the extra $r$ in $rdrd\alpha$, see Randolph (1967), p. 469.

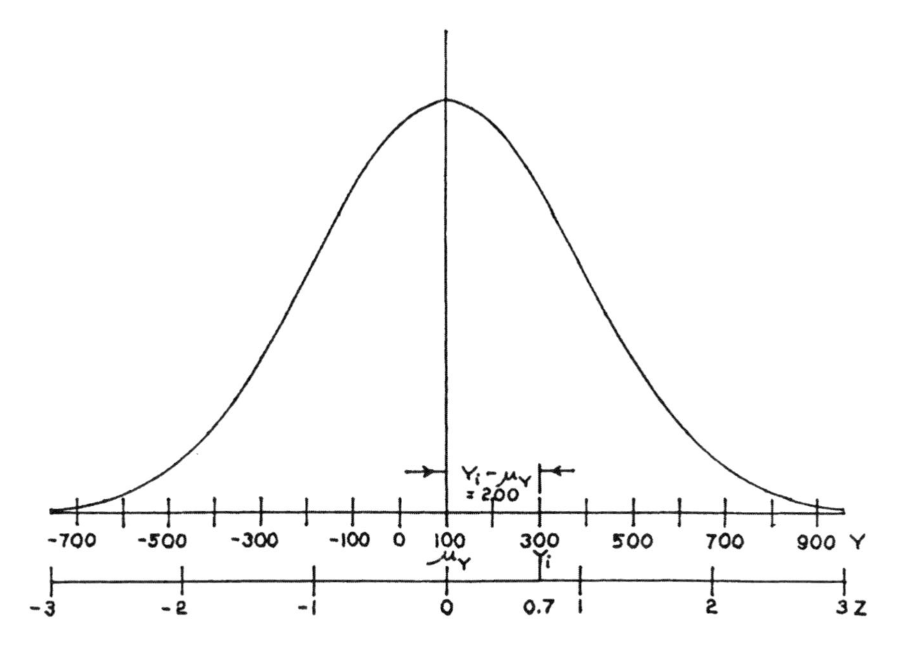


**FIGURE 11.1** The conversion of a specific normal distribution, with the parameters $I(Y) = 100$ and $V(Y) = 286$, into the standard form.

The cumulative probability distribution function is

$$F(y) = P(-\infty \leq Y \leq y_b) = \frac{1}{\sigma\sqrt{2\pi}} \int_{-\infty}^{y_b} e^{-1/2[(y-\mu)^2/\sigma^2]} dy \tag{11.9}$$

Unfortunately, due to the technical difficulties mentioned earlier, these cumulative probabilities cannot be obtained using this expression. Instead, recourse must be made to the methods of numerical integration such as Simpson's rule. Fortunately, this can be done so as to obtain excellent approximations, provided that the bar widths, equivalent to $dY$, are kept sufficiently narrow.

In Equations 11.6 through 11.9, the controlling variable is actually the error $\varepsilon$, $\varepsilon = y - \mu$, which indicates the distance, along the horizontal axis of the histogram, from the $Y$ value to the mean, as is shown in Figure 11.1. However, as far as the exponent of $e$ is concerned, $\varepsilon$ actually is used in the form of a proportion of the standard deviation:

$$Z = \frac{\varepsilon}{\sigma} = \frac{(Y - \mu)}{\sigma} \tag{11.10}$$

Consequently, as can be seen in Figure 11.1, it is possible to show the position of $y$ in terms of either the conventional $Y$ scale or a scale made up of standard deviation units. These units are called *standard deviates* and are symbolized by the letter $Z$. When $Y = \mu$, $\varepsilon = 0$, $Z = 0$. The index or zero point on the $Z$ scale is therefore equivalent to $\mu$ on the $Y$ scale. As can be seen, any continuous probability distribution which can be described using the normal curve, regardless of the magnitudes of its mean and variance, can be converted to the $Z$ form where the mean is equal to zero and the standard deviation (and variance) is equal to 1. Thus, the $Z$ form is known as the *standard* form of the normal. In terms of the standard form, the frequency function is

$$N_z = e^{-z^2/2} \tag{11.11}$$

which has been tabulated (see Table A.3 in the Appendix) while the pdf is

$$f(z) = \frac{N_z}{T} = \frac{1}{\sqrt{2\pi}} e^{-z^2/2}, \quad -\infty \leq Z \leq \infty \tag{11.12}$$

and the cumulative probability distribution function is

$$F(z) = P(-\infty \leq Z \leq z_b) = \frac{1}{\sqrt{2\pi}} \int_{-\infty}^{z_b} e^{-z^2/2} dz \tag{11.13}$$

The general procedure used to determine probabilities when continuous probability distributions are involved was described in Section 8.6. Its essence was stated by Equation 8.21:

$$P(y_a \leq Y \leq y_b) = P(y_A \leq Y \leq y_b) - P(y_A \leq Y \leq y_a)$$

In the case of the standardized normal, this becomes

$$P(z_a \leq Z \leq z_b) = P(-\infty \leq Z \leq z_b) - P(-\infty \leq Z \leq z_a)$$

Since the standard form of the normal distribution is invariant, it is possible to avoid using the numerical integration equivalent of Equation 11.13 in the determination of the above probabilities. Instead one can use tables, such as Table A.4 in the Appendix, to obtain values for $P(-\infty \leq Z \leq z_b)$ or $P(0 \leq Z \leq z_b)$ for specified fractiles. Table A.4 in the Appendix is of the latter type. It is used in the following manner.

$$P(0 \leq Z \leq 1.43) = 0.4236, \text{ see Figure 11.2a}$$
$$P(0 \leq Z \leq 0.36) = 0.1406, \text{ see Figure 11.2b}$$

and

$$\begin{aligned} P(0.36 \leq Z \leq 1.43) &= P(0 \leq Z \leq 1.43) - P(0 \leq Z \leq 0.36) \\ &= 0.4236 - 0.1406 \\ &= 0.2830, \text{ see Figure 11.2c} \end{aligned}$$

Because of the symmetry of the distribution,

$$P(-2.31 \leq Z \leq 0) = P(0 \leq Z \leq 2.31) = 0.4896, \text{ see Figure 11.2d and e}$$

and

$$\begin{aligned} P(-2.31 \leq Z \leq 0.36) &= P(-2.31 \leq Z \leq 0) + P(0 \leq Z \leq 0.36) \\ &= 0.4896 + 0.1406 \\ &= 0.6302, \text{ see Figure 11.2f} \end{aligned}$$

Finally, if

$$P(-0.87 \leq Z \leq 0) = P(0 \leq Z \leq 0.87) = 0.3078, \text{ see Figure 11.2g and h}$$

and

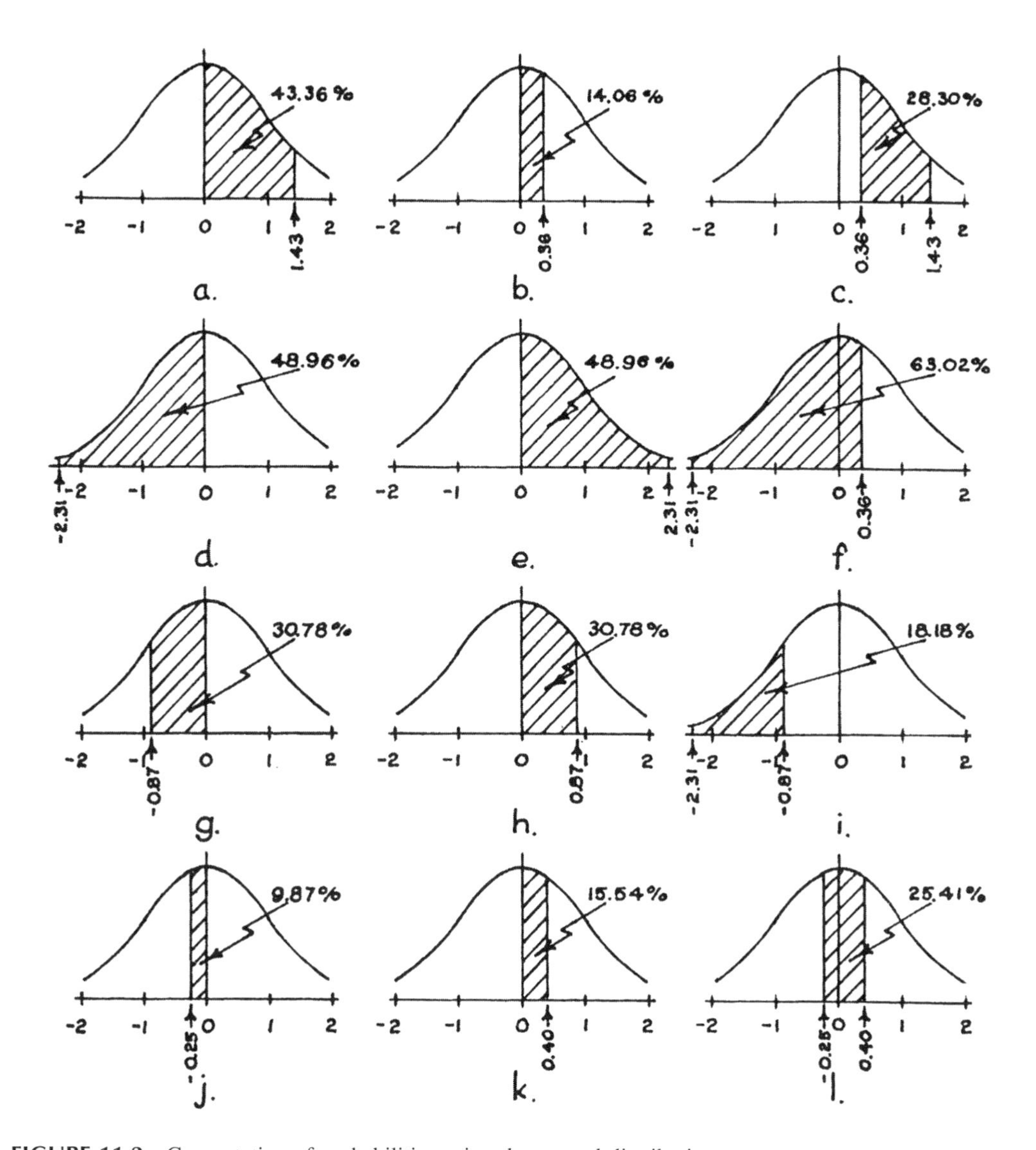


**FIGURE 11.2** Computation of probabilities using the normal distribution.

$$
\begin{aligned}
P(-2.31 \le Z \le -0.87) &= P(-2.31 \le Z \le 0) - P(-0.87 \le Z \le 0) \\
&= 0.4896 - 0.3078 \\
&= 0.1818, \text{ see Figure 11.2i}
\end{aligned}
$$

When the critical values are in terms of $Y$ rather than $Z$, it is necessary to convert the $Y$ values to $Z$ values using Equation 11.10 and then proceeding as was done above. For example, if $\mu = 100$ and $\sigma = 80$ and the desired probability is $P(80 \le Y \le 132)$, it would be necessary first to convert the $y_a$ and $y_b$ values to their equivalent $Z$ values (see Figures 11.2j and k):

$$z_a = \frac{(y_a - \mu)}{\sigma} = \frac{(80 - 100)}{80} = -\frac{20}{80} = -0.25$$

$$z_b = \frac{(y_b - \mu)}{\sigma} = \frac{(132 - 100)}{80} = \frac{32}{80} = 0.40$$

Then

$$P(-0.25 \leq Z \leq 0) = P(0 \leq Z \leq 0.25) = 0.0987$$
$$P(0 \leq Z \leq 0.40) = 0.1554$$

and

$$\begin{aligned} P(-0.25 \leq Z \leq 0.40) &= P(-0.25 \leq Z \leq 0) + P(0 \leq Z \leq 0.40) \\ &= 0.0987 + 0.1554 \\ &= 0.2541, \text{ see Figure 11.21} \end{aligned}$$

The normal distribution, in either the specific or standard form, is a typical continuous probability distribution as described in Section 8.5. It meets the requirements of the first law of probability, that all the probabilities are in the range 0 to +1 and the special additive law which specifies that the probabilities must sum to 1:

$$P(-\infty \leq Y \leq \infty) = \frac{1}{\sigma\sqrt{2\pi}} \int_{-\infty}^{\infty} e^{-1/2[(y-\mu)^2/\sigma^2]} dy$$

Since, from Equation 11.7,

$$\int_{-\infty}^{\infty} e^{-1/2[(y-\mu)^2/\sigma^2]} dy = \sigma\sqrt{2\pi}$$

$$P(-\infty \leq Y \leq \infty) = \frac{\sigma\sqrt{2\pi}}{\sigma\sqrt{2\pi}} = 1 \tag{11.14}$$

Without proof, the moment-generating function is

$$M_Y(t) = e^{\mu t + [(\sigma^2 t^2)/2]} \tag{11.15}$$

The first derivative of the mgf is

$$\begin{aligned} \frac{d}{dt} M_Y(t) &= \frac{d}{dt}[e^{\mu t} e^{(\sigma^2 t^t)/z}] \\ &= e^{\mu t + [(\sigma^2 t^2)/2]} \left( \frac{\sigma^2}{2} \right)(2t) + e^{[(\sigma^2 t^2)/2] + \mu t} \mu \end{aligned}$$

Setting $t = 0$ and solving for $E(Y)$ yields

$$\mu_1' = E(Y) = e^0 \left( \frac{\sigma^2}{2} \right)(0) + e^0 \mu = \mu \tag{11.16}$$

This can be confirmed using the fundamental approach with Equation 8.23:

$$E(Y) = \frac{1}{\sigma\sqrt{2\pi}} \int_{-\infty}^{\infty} y e^{-1/2[(y-\mu)^2/\sigma^2]} dy$$

Converting this to the standard form requires

$$z = \frac{(y-\mu)}{\sigma}$$

Then $y = z\sigma + \mu$; $dy/dz = \sigma$; and $dy = \sigma dz$ and

$$E(Y) = \frac{1}{\sigma\sqrt{2\pi}} \int_{-\infty}^{\infty} (z\sigma + \mu) e^{-z^2/2} \sigma dz$$

$$= \frac{1}{\sqrt{2\pi}} \int_{-\infty}^{\infty} (z\sigma + \mu) e^{-z^2/2} dz$$

$$= \frac{\sigma}{\sqrt{2\pi}} \int_{-\infty}^{\infty} z e^{-z^2/2} dz = \frac{\mu}{\sqrt{2\pi}} \int_{-\infty}^{\infty} e^{-z^2/2} dz$$

Since the integrand in the first of these integrals changes sign in the indicated interval, $-\infty < z < \infty$, the function is discontinuous and the integral must be treated as the sum of two integrals:

$$\frac{\sigma}{\sqrt{2\pi}} \int_{-\infty}^{\infty} z e^{-z^2/2} dz = \frac{\sigma}{\sqrt{2\pi}} \int_{-\infty}^{\infty} z e^{-z^2/2} dz + \frac{\sigma}{\sqrt{2\pi}} \int_{0}^{\infty} z e^{-z^2/2} d$$

By implicit integration, since

$$\frac{d}{dz} - e^{-z^2/2} = -e^{-z^2/2}\left(-\frac{1}{2}\right)(2z) = z e^{-z^2/2}$$

$$\int_{-\infty}^{0} z e^{-z^2/2} dz = -e^{-z^2/2}\Big|_{-\infty}^{0} = 0 - 0 = 0$$

and

$$\int_{0}^{\infty} z e^{-z^2/2} dz = -e^{-z^2/2}\Big|_{0}^{\infty} = 0 - 0 = 0$$

Therefore,

$$\frac{\sigma}{\sqrt{2\pi}} \int_{-\infty}^{\infty} z e^{-z^2/2} dz = 0$$

and

$$E(Y) = \frac{\mu}{\sqrt{2\pi}} \int_{-\infty}^{\infty} e^{-z^2/2} dz = \frac{\mu}{\sqrt{2\pi}} * \sqrt{2\pi} = \mu$$

which confirms Equation 11.16.

The variance can be obtained through the method of moments or by Equation 8.34. By using the mgf,

$$\frac{d^2}{dt^2}M_Y(t) = \sigma^2\{e^{\mu t+[(\sigma^2 t^2)/2]} + te^{\mu t+[(\sigma^2 t^2)/2]}(\mu + \sigma^2 t)\} + \mu\{e^{[(\sigma^2 t^2)/2]+\mu t}[(\sigma^2 t + \mu]\}$$

Setting $t = 0$ yields

$$\mu_2' = E(Y^2) = \sigma^2(e^0 + 0) + \mu[e^0(0) + e^0(\mu)]$$
$$= \sigma^2 + \mu^2$$

The variance then is

$$\begin{aligned} V(Y) &= E(Y^2) - [E(Y)]^2 \\ &= \sigma^2 + \mu^2 - \mu^2 \\ &= \sigma^2 \end{aligned} \tag{11.17}$$

It is possible to use the mgf to obtain the third and forth moments about zero and then to use them to obtain the coefficients of skewness and kurtosis. However, it is simpler to recognize, from the earlier discussion, that the normal distribution is symmetrical and that

$$\alpha_3 = 0 \tag{11.18}$$

It can be shown (Kendall and Stuart, 1958), that when $k$ (see Section 8.12) is even,

$$\mu_k = \frac{k!\sigma^k}{2^{k/2}\left(\frac{k}{2}\right)!} \tag{11.19}$$

Thus, the fourth moment is

$$\mu_4 = \frac{4!}{2^{4/2}\left(\frac{4}{2}\right)!} = \frac{4!}{4(2)!} = 3$$

Since the variance of the normal, when in standard form, is equal to unity, the moment coefficient of kurtosis is

$$\alpha_4 = \frac{\mu_4}{\sigma^4} = 3 \tag{11.20}$$

This is the basis for the standard used to express the degree of kurtosis possessed by a distribution.

The importance of the normal distribution stems from the fact that many random occurrences tend to have distribution patterns that coincide with, or are very similar to, the normal. This phenomenon is sufficiently common that much of the edifice of statistical evaluation of sample-based estimates and statistical testing of hypotheses in research operations is based on the use of the normal. As will be described in more detail in Chapter 13, even when a real-life distribution is not normal, the normal is the appropriate model for the distribution of all possible means or sums of samples, of a given size, drawn from the real distribution. This is a fact of profound importance when considering the importance of the normal distribution.

It is not uncommon in forestry research or management that frequency distributions, especially diameter distributions, are involved and must be described in mathematical terms. This occurs in growth and yield studies, simulations of stand development, harvesting simulations, and other similar operations. Often the pdfs of probability distributions are used for this purpose. The pdfs not only describe the frequency distribution in algebraic terms but since the probability distributions have expressions for expected values, variances, modes, and degrees of skewness and kurtosis, the descriptions are far more complete than if simply an algebraic expression was used to indicate the shape of the curve. When the pdf is used in this manner, the determination of probabilities often is not involved. Consequently, it is possible to consider the probability distributions in the classical sense, in which they are used to arrive at probabilities, and also in a descriptive sense where probability is not involved.

The normal distribution has been used with some success as a model of frequency distributions which are mound shaped. However, it is inflexible, having only one shape and often the matches are poor. This can be seen if one considers the stand table for an even-aged stand. The histogram would be mound shaped, but it is likely to be skewed to some extent. The normal is symmetrical and if used as the model of the stand would yield misleading impressions. Growth and/or yield studies based on such a model would be biased. Because of this inflexibility, the normal is rarely used as a model of diameter distributions or whenever skewness and/or nonnormal kurtosis is known to be present.

In addition to serving as a model for actual frequency distributions, the normal, in some cases, can be used as a substitute for another probability distribution in the same way as the Poisson can be used as a substitute for the binomial, as was described in Section 10.5. The Poisson is an appropriate substitute for the binomial when $n$ is large and $p$ is small. When $n$ is large and $p$ is in the neighborhood of 0.5, the Poisson is not appropriate but the normal can be used. For example, assume a binomial distribution where $n = 10$ and $p = 0.5$. Admittedly, $n$ is not large in this case, but the example will serve as a demonstration. The event of interest is the occurrence of 3, 4, or 5 successes.

Thus, using the binomial,

$$P(Y = 3,\ 4,\ \text{or}\ 5) = \binom{10}{3}(0.5)^3(0.5)^7 + \binom{10}{4}(0.5)^4(0.5)^6 + \binom{10}{5}(0.5)^5(0.5)^5$$

$$= 0.568359375$$

Figure 11.3 shows the distribution. The bars representing the event of interest are crosshatched. The area in the crosshatched bars relative to the total area of the figure is equal to the probability of occurrence of 3, 4, or 5. The solution to the problem using the normal as a substitute for the binomial is as follows. $\mu = np = 10(0.5) = 5$, $\sigma = npq = 10(0.5)(0.5) = 2.5$, and $\sigma = 1.58113883$. Using these data the normal curve has been computed and superimposed on the histogram in Figure 10.3. It should be remembered that relative area is the key to obtaining probabilities from continuous distributions. To obtain an area under the normal curve that would correspond to the crosshatched area in the histogram of the binomial the limits of integration must be at the *outer edges* of the crosshatched bars. Thus,

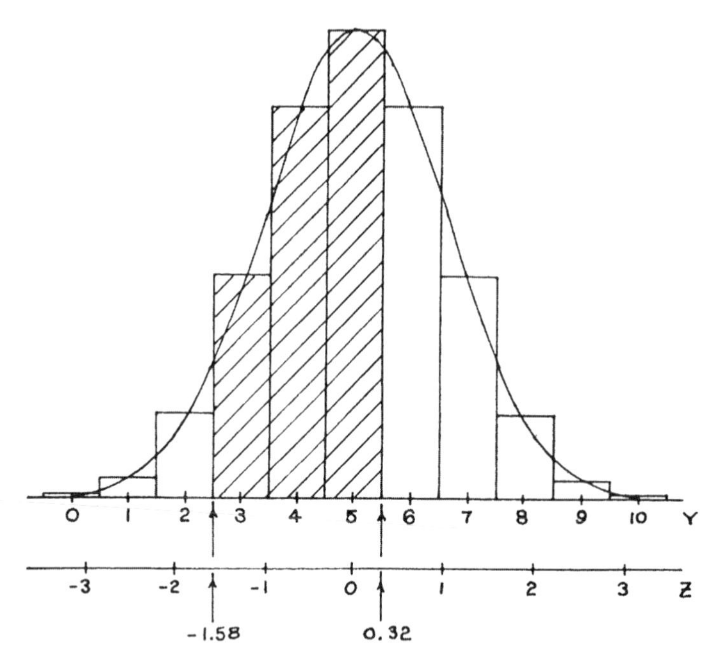


**FIGURE 11.3** Histogram of the binomial distribution when $n = 10$ and $p = q = 0.5$. Superimposed on the histogram is the curve of the corresponding normal distribution. The crosshatched bars represent the event of interest, the occurrence of 3, 4, or 5 successes.

$$z_a, \text{ when } y = 2.5, \quad \frac{(2.5 - 5)}{1.5113883} = -1.58$$

$$z_b, \text{ when } y = 5.5, \quad \frac{(5.5 - 5)}{1.5113883} = 0.32$$

Areas, from Table A.4 in the Appendix, are

| | |
|---|---|
| From $z_a = -1.58$ to $Z = 0.00$, | 0.4429 |
| From $z_b = 0.00$ to $Z = 0.32$, | 0.1255 |
| $P(y = 3, 4, \text{ or } 5) =$ | 0.5684 |

As can be seen the approximation is very good. Even if $p$ is not equal to 0.5, the approximation is often acceptable. For example, assume $n = 10$ and $p = 0.4$ and that the event of interest is the occurrence of 3, 4, or 5 successes.

Using the binomial,

$$P(y = 3, 4, \text{ or } 5) = \binom{10}{3}(0.4)^3(0.6)^7 + \binom{10}{4}(0.4)^4(0.6)^6 + \binom{10}{5}(0.4)^5(0.6)^5$$

$$= 0.6664716288$$

In this case, $\mu = np = 10(0.4) = 4$, $\sigma^2 = npq = 10(0.4)(0.6) = 2.4$, and $\sigma = 1.54919338$. Using the normal,

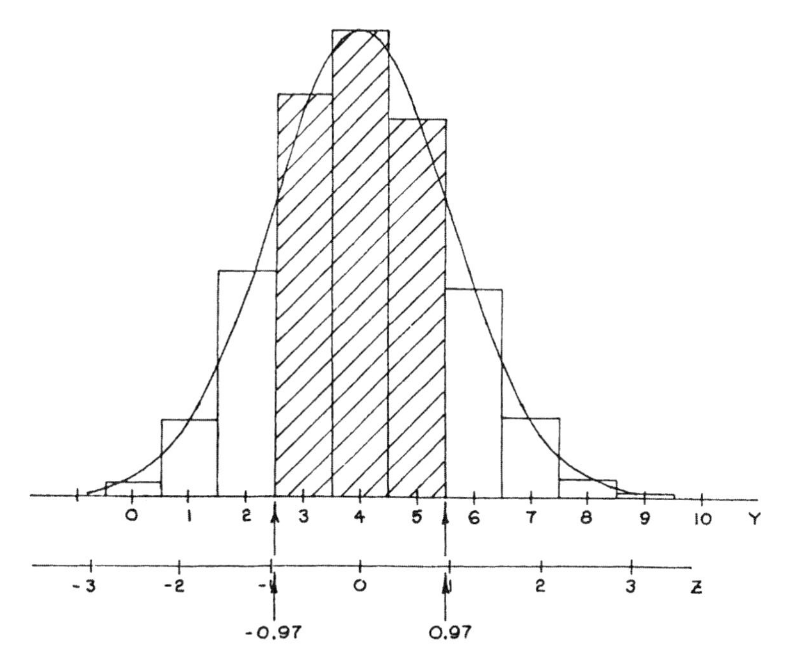


**FIGURE 11.4** Histogram of the binomial distribution if $p = 0.4$ instead of 0.5. Note the skewing to the right. In spite of this, probabilities obtained using the normal are reasonably close to their correct values.

$$z_a, \text{ when } y = 2.5, = \frac{(2.5 - 4)}{1.549193338} = -0.97$$

$$z_b, \text{ when } y = 5.5, = \frac{(5.5 - 4)}{1.549193338} = 0.97$$

Areas from Table A.4 in the Appendix:

| | |
|---|---|
| From $z_a = -0.97$ to $Z = 0.0$, | 0.3340 |
| From $z_b = 0.00$ to $Z = 0.97$, | 0.3340 |
| $P(y = 3, 4, \text{ or } 5) =$ | 0.6680 |

Again the approximation is quite good, in spite of the fact that the binomial is noticeably skewed, as can be seen in Figure 11.4.

It should be noted that in the approximation process the integration was made from 2.5 to 5.5 rather than from 3 to 5. As was brought out earlier, this is done to obtain corresponding relative areas. The use of the extra half bar width at each end of the integration interval is known as the *continuity correction,* which must be used whenever the normal is used as a substitute for the binomial or whenever any continuous distribution is used as a substitute for a discrete distribution. This could be shown formally as

$$z_a = \frac{\left[\left(y_a - \frac{1}{2}\right) - \mu_y\right]}{\sigma_y}$$

$$z_b = \frac{\left[\left(y_b + \frac{1}{2}\right) - \mu_y\right]}{\sigma_y}$$

## 11.3 THE GAMMA PROBABILITY DISTRIBUTION ($\Gamma$:*r, a*)

As was brought out in Chapter 10 and in the discussion of the normal distribution in Section 11.2, the binomial distribution forms the basis for a number of other distributions. In somewhat the same manner, a number of distributions are based on a mathematical expression known as the *gamma function*:

$$\Gamma(r) = \int_0^\infty y^{r-1} e^{-y} dy \tag{11.21}$$

In this expression $r$ is a parameter that controls the shape of the curve of the integrand. The gamma function has the property that, if the integration is carried out, the resulting expression contains an integral that is identical to the original except that the exponent of $y$ has been reduced by 1:

$$\Gamma(r) = (r-1)\int_0^\infty y^{r-2} e^{-y} dy = (r-1)\Gamma(r-1)$$

When this integration is carried out, the expression becomes

$$\Gamma(r) = (r-1)(r-2)\int_0^\infty y^{r-3} e^{-y} dy = ((r-1)(r-2)\Gamma(r-2)$$

This process will continue until the exponent of $y$ equals zero. At this point

$$\Gamma(r) = (r-1)(r-2)\cdots 2\int_0^\infty y^0 e^{-y} dy$$

$$= (r-1)(r-2)\cdots 2\int_0^\infty e^{-y} dy$$

Since

$$\int_0^\infty e^{-y} dy = -e^{-y} \Bigg|\begin{matrix} \infty = \frac{-1}{e^\infty} = 0 \\ 0 = \frac{+1}{e^0} = 1 \end{matrix} = 1 \tag{11.22}$$

$$\Gamma(r) = (r-1)(r-2)\cdots(2)(1) = (r-1)! \tag{11.23}$$

It should be evident that Equation 11.22 appears, and the cycling ceases, only when the exponent of $y$ assumes the value zero. Such a termination of the cycling will always occur when $r$ is a positive integer. When $r$ is not an integer, the cycling cannot be terminated normally. However, when the fractional portion of $r$ is exactly equal to ½ (e.g., $r = 1.5, 2.5, 3.5$, etc.) the cycling can be terminated as shown below. Let $r = n/2$, where $n$ is an odd whole number. Then

$$\Gamma(r) = \Gamma\left(\frac{n}{2}\right) = \left[\left(\frac{n}{2}\right) - 1\right] * \left[\left(\frac{n}{2}\right) - 2\right] * \cdots * \left(\frac{5}{2}\right) * \left(\frac{3}{2}\right) * \left(\frac{1}{2}\right) * \Gamma\left(\frac{1}{2}\right)$$

To complete the process it is necessary to determine the value of $\Gamma(1/2)$. This must be done indirectly, following the pattern of argument used to find the total area under the normal curve (see Equation 11.7).

$$\Gamma\left(\frac{1}{2}\right) = \int_0^\infty y^{-1/2} e^{-y} dy$$

Let $\mu^2/2 = y$. Then $dy = udu$ and

$$\Gamma\left(\frac{1}{2}\right) = \int_0^\infty \left(\frac{\mu^2}{2}\right)^{-1/2} e^{-u^2/2} udu = \int_0^\infty \left(\frac{2}{u^2}\right)^{1/2} e^{-u^2/2} udu$$

$$= \sqrt{2} \int_0^\infty e^{-u^2/2} du$$

Let

$$A = \int_0^\infty e^{-u^2/2} du$$

and

$$\int_0^\infty e^{-v^2/2} dv = \int_0^\infty e^{-u^2/2} du$$

Then

$$A^2 = \int_0^\infty e^{-u^2/2} du \int_0^\infty e^{-v^2/2} dv = \int_0^\infty \int_0^\infty e^{-(u^2+v^2)/2} dvdu$$

Converting to polar coordinates, let $u = w \sin \alpha$ and $v = w \cos \alpha$, then $u^2 + v^2 = w^2$. As $u$ and $v$ vary from 0 to $\infty$, $w$ varies from 0 to $\infty$ and, since $u$ and $v$ are always positive and only the first quadrant is involved, $\alpha$ varies from 0 to $\pi/2$ radians. Furthermore, the element of area $dvdu$ becomes $wdwd\alpha$.* Thus,

$$A^2 = \int_0^{\pi/2} \int_0^\infty e^{-w^2/2} wdwd\alpha$$

Since

* For an explanation of the extra $r$ in rdrdα, see Randolph (1967), p. 469.

$$\int_0^\infty e^{-w^2/2} w\,dw = -e^{-w^2/2} \Bigg| \begin{matrix} \infty = -\dfrac{1}{e^{\infty^2/2}} = 0 \\ 0 = +\dfrac{1}{e^{0^2/2}} = 1 \end{matrix} = 1$$

$$A^2 = \int_0^{\pi/2} d\alpha = \alpha \Bigg| \begin{matrix} \pi/2 = \pi/2 \\ 0 = 0 \end{matrix} = \frac{\pi}{2}$$

and

$$A = \sqrt{\frac{\pi}{2}}$$

Then

$$\Gamma\left(\frac{1}{2}\right) = \sqrt{2}A = \sqrt{2}\sqrt{\frac{\pi}{2}} = \sqrt{\pi}$$

permitting the termination of the cycling.

When the fractional portion of $r$ is something other than 0.5 recourse must be made to methods of numerical integration if the numerical value of the gamma function is needed. Table A.5 in the Appendix provides the values of $\Gamma(r)$ for values of $r$ ranging from 1000 to 2000 in 0.005 increments. For example, assume $r = 5.165$.

$$\begin{aligned} \Gamma(5.165) &= (5.165 - 1)! \\ &= (4.165)(3.165)(2.165)(1.165)\Gamma(1.165) \end{aligned}$$

From Table A.5 in the Appendix, $\Gamma(1.165) = 0.9282347$. Thus,

$$\Gamma(5.165) = (4.165)(3.165)(2.165)(1.165)(0.9282347) = 30.862$$

The integrand in the gamma function

$$w(y) = y^{r-1}e^{-y}, \quad r > 0$$

is actually a special case of a more general two-parameter function which is the frequency function of the *gamma* probability distribution. This function is

$$N(y) = \begin{cases} y^{r-1}e^{-ay}, & 0 < Y < \infty, \; r > 0, \; a > 0 \\ 0, & \text{elsewhere} \end{cases} \tag{11.24}$$

The parameter $r$ controls the fundamental *shape* of the curve of the function while parameter $a$ controls the *scale*. The effects of changes in the two parameters can be seen in Figure 11.5.

The total area under the curve of the frequency function is

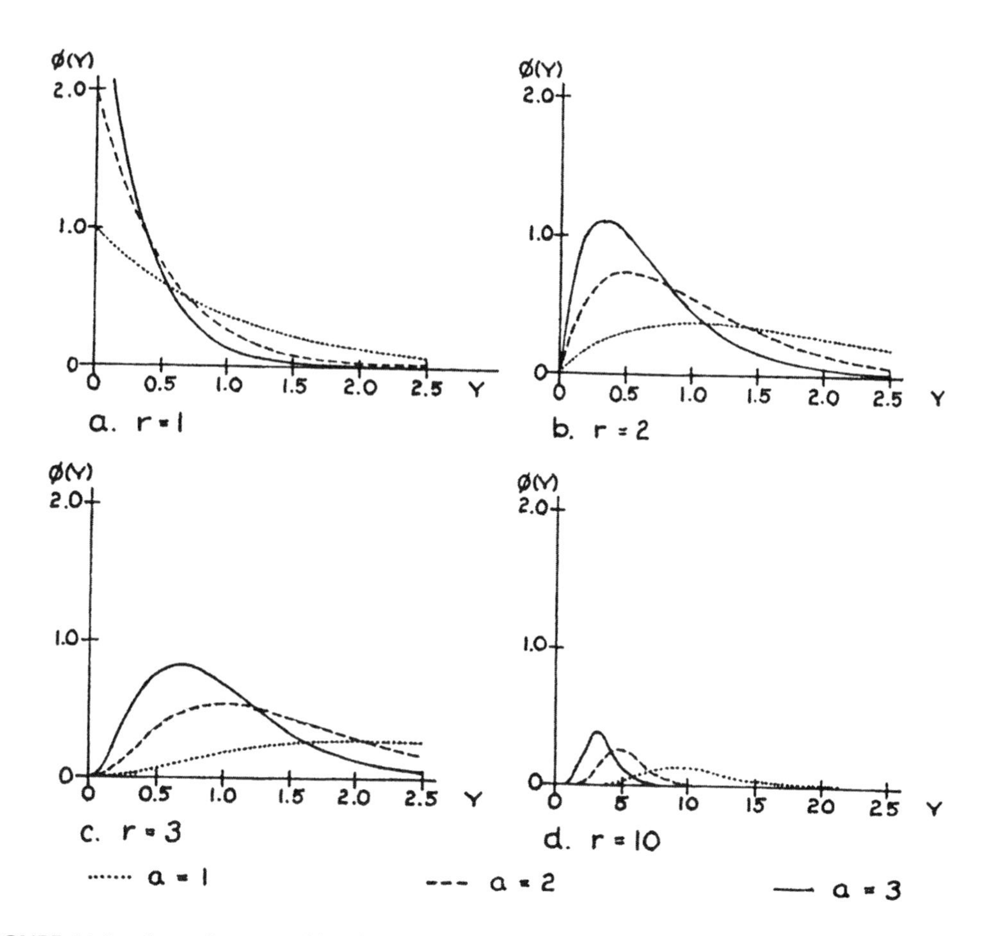


**FIGURE 11.5** Curve forms resulting from the use of different values of the parameters $r$ and $a$ in the pdf of the gamma distribution. Note the reduced scale of the horizontal scale in $d$ compared with the others.

$$T = \int_0^\infty y^{r-1} e^{-ay}\, dy$$

It should be remembered that it may not be possible to obtain a value for $T$ if $r$ is not a positive integer. Assuming that $r$ is a positive integer and integrating by parts, let $u = y^{r-1}$ and $dv = c^{-ay}dy$. Then

$$v = \int e^{-ay}\, dy = -a^{-1} e^{-ay}$$

$$du = (r-1) y^{r-2}\, dy$$

$$T = uv\Big|_0^\infty - \int_0^\infty v\, du = \frac{(r-1)}{a} \int_0^\infty y^{r-2} e^{-ay}\, dy$$

Obviously cycling will take place until the exponent of $y$ is equal to zero. Then

$$T = \frac{(r-1)!}{a^r} = \frac{\Gamma(r)}{a^r} \tag{11.25}$$

Thus, the pdf of the gamma distribution, $\Gamma(y{:}r, a)$, is

$$f(y) = \frac{N(y)}{T} = \begin{cases} \dfrac{a^r y^{r-1} e^{-ay}}{\Gamma(r)}, & 0 < Y < \infty, \; r > 0, \; a > 0 \\ 0, & \text{elsewhere} \end{cases} \tag{11.26}$$

This is a valid pdf. Since neither $Y$, $r$, nor $a$ can assume negative values within their indicated ranges, the pdf cannot take on negative values, thus meeting the requirements of the first law of probability. Furthermore, since the integral of the pdf, as $Y$ goes from 0 to $\infty$, is equal to unity, the additive law of probability is also met.

The cdf of the gamma distribution is

$$F(y) = P(0 \leq Y \leq y_b) = \frac{a^r}{\Gamma(r)} \int_0^{y_b} y^{r-1} e^{-ay} dy$$

Again, the integral has no simple solution except when $r$ is a positive integer. When integrated by parts, using $u = y^{r-1}$ and $dv = d^{-ay}dy$,

$$F(y) = \frac{a^{r-1}}{(r-2)!} \int_0^{y_b} y^{r-2} e^{-ay} dy - \frac{(ay)^{r-1} e^{-ay}}{(r-1)!}$$

Continuing the cycling $r - 1$ times, until the exponent of $y$ becomes zero, yields

$$F(y) = 1 - \sum_{k=1}^{r-1} \left[ \frac{(ay_b)^k}{k! e^{ay} b} \right], \quad y_b > 0 \tag{11.27}$$

As an example, assume that $r = 4$, then cycling $r - 1 = 3$ times and remembering that

$$\int_0^{y_b} y^{1-1} e^{-ay} dy = \int_0^{y_b} e^{-ay} dy = -\frac{1}{ae^{ay}} \bigg|_0^{y_b} = -\frac{1}{ae^{ay}} + \frac{1}{a}$$

then

$$f(y) = 1 - \frac{a^3 y_b^3}{3! e^{ay} b} - \frac{a^2 y_b^2}{2! e^{ay} b} - \frac{ay_b}{e^{ay} b} - \frac{1}{e^{ay} b}$$

The terms in this expression are mathematically equivalent to the point probabilities that $r$ would have if it had a Poisson distribution with the parameter $\lambda = ay_b$.

$$\begin{aligned} F(y) &= 1 - \frac{\lambda^3}{3! e^{\lambda}} - \frac{\lambda^2}{2! e^{\lambda}} - \frac{\lambda^1}{1! e^{\lambda}} - \frac{\lambda^0}{0! 1e} \\ &= 1 - \sum_{k=0}^{r-1} \left( \frac{\lambda^k}{k! e^{\lambda}} \right) \end{aligned} \tag{11.28}$$

Consequently, the tabulated cumulative Poisson probabilities in Table A.2 in the Appendix can be used to obtain the cumulative probabilities of the gamma distribution. For example, continuing the situation where $r = 4$, assume that $a = 2$ and $y_b = 2$. Then, $\lambda = ay_b = 4$. Table A.2 is entered with $y = r = 4$ and $\lambda = 4$. The tabular value is 0.433 and thus

$$F(Y \leq 2) = 1 - 0.433 = 0.567$$

This can be confirmed, within rounding error, by direct computation.

$$\begin{aligned} F(Y \leq 2) &= 1 - \frac{4^3}{3!e^4} - \frac{4^2}{2!e^4} - \frac{4}{e^4} - \frac{1}{e^4} \\ &= 1 - 0.195 - 0.147 - 0.073 - 0.018 \\ &= 0.567 \end{aligned}$$

It should be pointed out that although the cumulative Poisson probabilities are used in this operation, one should not assume that $r$ is the variable of interest. $r$ has a definite value and does not change. The variable is $y_b$ and the cumulative probabilities change as $y_b$ is changed.

The moment-generating function for the gamma distribution is

$$\begin{aligned} M_Y(t) &= \frac{a^r}{\Gamma(r)} \int_0^\infty e^{ty} y^{r-1} e^{-ay} dy \\ &= \frac{a^r}{\Gamma(r)} \int_0^\infty y^{r-1} e^{-y(a-t)} dy \end{aligned}$$

Setting $u = y(a - t)$, then $du/dy = (a - t)$ and $dy = du/(a - t)$. Substitute these values into the equation:

$$\begin{aligned} M_Y(t) &= \frac{a^r}{\Gamma(r)} \int_0^\infty \frac{\mu^{r-1}}{(a-t)} e^{-u} (a-t)^{-1} du \\ &= \frac{a^r}{\Gamma(r)(a-t)^{r-1}(a-t)} \int_0^\infty u^{r-1} e^{-u} du \\ &= \frac{a^r \Gamma(r)}{(a-t)^r \Gamma(r)} = \frac{a^r}{(a-t)^r} \end{aligned} \tag{11.29}$$

The moments about zero are then

$$\begin{aligned} \mu_1' &= E(Y) = \frac{r}{a} \\ \mu_2' &= E(Y^2) = \frac{r(r+1)}{a^2} \\ \mu_3' &= E(Y^3) = \frac{r(r+1)(r+2)}{a^3} \\ \mu_4' &= E(Y^4) = \frac{r(r+1)(r+2)(r+3)}{a^4} \end{aligned} \tag{11.30}$$

and the variance is

$$V(Y) = E(Y^2) - [E(Y)]^2 = \left[\frac{r(r+1)}{a^2}\right] - \left(\frac{r}{a}\right)^2 = \frac{r}{a^2} \tag{11.31}$$

The moment coefficients of skewness and kurtosis are similarly derived:

$$\alpha_3 = \frac{2}{\sqrt{r}} \tag{11.32}$$

$$\alpha_4 = 3 + \frac{6}{r} \tag{11.33}$$

It already has been stated that the parameter $r$ controls the shape of the curve of the gamma distribution and that the parameter $a$ is a scaling factor. These effects are demonstrated in Figure 11.5. The curves are positively skewed or skewed to the right, which is what is indicated by $\alpha_3$. There is no way by which $\alpha_3$ can become negative if $r$ is a positive value. As $r$ increases, the degree of skewing decreases until, in the limit, as $r$ approaches infinity, the distribution approaches symmetry.

It is difficult to assess kurtosis when looking at heavily skewed curves such as those in Figure 11.5. However, $\alpha_4$ indicates that the curves of gamma distributions are leptokurtic or relatively narrower or more peaked than the curve of the normal distribution. Furthermore, $\alpha_4$ indicates that, as $r$ increases, the degree of excess peakedness decreases and in the limit, as $r$ approaches infinity, the gamma distribution approaches the same shape as that of the normal distribution.

One of the most significant characteristics of the gamma distribution is its relationship to the normal. If $X$ is a continuous random variable that is normally distributed, $N(0, \sigma_X^2)$, and $Y$ is a second random variable defined as $Y = X^2$, $Y$ will have a gamma distribution. To demonstrate the validity of this statement it first will be necessary to establish the relationship between the pdfs of the $X$ and $Y$ regardless of how $X$ is distributed. Let $F(x)$ be the cdf of $X$ and $F(y)$ be that of $Y$. Furthermore, define $F(y) = 0$ when $y_b \leq 0$. Then, when $Y > 0$,

$$\begin{aligned} F(y) &= P(Y \leq y_b) = P(X^2 \leq y_b) \\ &= P(-\sqrt{y_b} \leq X \leq \sqrt{y_b}) \\ &= F(X \leq \sqrt{y_b}) - F[X \leq (-\sqrt{y_b})] \end{aligned}$$

Differentiating this expression with respect to $y_b$, using the chain rule, and remembering that the derivative of $F(x)$ with respect to $x_b$ is $f(x)$, yields

$$f(y) = \frac{1}{2\sqrt{y_b}}[f(X = \sqrt{y_b}) + f(X = -\sqrt{y_b})]$$

Therefore,

$$f(y) = \begin{cases} \dfrac{1}{2\sqrt{y_b}}[f(X = \sqrt{y_b}) + f(X = -\sqrt{y_b})], & y_b > 0 \\ 0, & \text{elsewhere} \end{cases} \tag{11.34}$$

Now, if $X$ is normally distributed, $N(0, \sigma_X^2)$,

$$f(x) = \frac{1}{\sigma_X\sqrt{2\pi}} e^{-(x-\mu_X)^2/2\sigma_x^2}, \quad -\infty < X < \infty$$

$$= \frac{1}{\sigma_X\sqrt{2\pi}} e^{-x^2/2\sigma_X^2}$$

the pdf of $x_2$ when $\mu_X = 0$. Substituting this pdf in Equation 11.34, and recognizing that $y_b = x_b^2$

$$f(y) = \frac{1}{2\sqrt{y_b}}\left[\frac{1}{\sigma_X\sqrt{2\pi}} e^{-y_b/2\sigma_X^2} + \frac{1}{\sigma_X\sqrt{2\pi}} e^{-y-b/2\sigma_X^2}\right]$$

$$= \frac{1}{\sqrt{y_b}}\left(\frac{1}{\sigma_X\sqrt{2\pi}} e^{-y_b/2\sigma_X^2}\right) \tag{11.35}$$

$$= \begin{cases} \dfrac{1}{\sigma_X\sqrt{2\pi y_b}} e^{-y_b/2\sigma_X^2}, & y_b > 0 \\ 0, & \text{elsewhere} \end{cases}$$

This same result can be obtained using the gamma distribution when $r = 1/2$ and $a = 1/2\sigma_x^2$.

$$f(y) = \frac{a^r}{\Gamma(r)} y_b^{r-1} e^{-ay_b} = \frac{\left(\dfrac{1}{2\sigma_X^2}\right)^{1/2}}{\Gamma\left(\dfrac{1}{2}\right)} y_b^{(1/2)-1} e^{-y_b/2\sigma_X^2}$$

Since

$$\Gamma\left(\frac{1}{2}\right) = \sqrt{\pi} \quad \text{and} \quad y_b^{(1/2)-1} = \frac{1}{\sqrt{y_b}}$$

$$f(y) = \begin{cases} \dfrac{1}{\sigma_X\sqrt{2\pi y_b}} e^{-y_b/2\sigma_X^2}, & y_b > 0 \\ 0, & \text{elsewhere} \end{cases} \tag{11.36}$$

Thus, if $X$ is normally distributed with parameters $(0, \sigma_X^2)$, $X^2$ will be distributed as gamma with parameters $r = 1/2$ and $a = 1/2\sigma^2$.

Because of its flexibility, the gamma distribution has been used, in both cumulative and noncumulative form, to model diameter distributions (e.g., Nelson, 1964). However, its flexibility is less than that of Johnson's $S_B$, the beta, or the Weibull (Hafley and Schreuder, 1977) and, consequently, its use as a diameter distribution model has been minimal. Its primary importance rests with the distributions that are derived from it.

## 11.4 THE EXPONENTIAL PROBABILITY DISTRIBUTION (*E:a*)

If parameter $r$ in the gamma distribution is set equal to 1, the resulting special case of the gamma distribution is known as the *exponential, negative exponential,* or the *reversed* or *inverse J-shaped* distribution.

When $r$ equals 1, the exponent of $y$ equals zero and the frequency function becomes

$$N(y) = \begin{cases} e^{-ay}, & Y > 0, \ a > 0 \\ 0, & \text{elsewhere} \end{cases} \tag{11.37}$$

The total area under the curve is

$$T = \int_0^{\infty} e^{-ay} dy = a^{-1} \tag{11.38}$$

and, thus, the pdf of the exponential distribution is

$$f(y) = \frac{N(y)}{T} = \begin{cases} ae^{-ay}, & Y > 0, \ a > 0 \\ 0, & \text{elsewhere} \end{cases} \tag{11.39}$$

Again, as in the case of the gamma distribution, this is a proper pdf, meeting the laws of probability.

The cdf is

$$\begin{aligned} F(y) &= \int_0^{y_b} ae^{-ay} dy \\ &= 1 - e^{-ay}b \end{aligned} \tag{11.40}$$

Since the exponential is a special case of the gamma distribution, the expected value is

$$E(Y) = \frac{r}{a}$$

but, since $r = 1$,

$$E(y) = a^{-1} \tag{11.41}$$

Similarly, the variance of the exponential distribution is

$$V(y) = a^{-2} \tag{11.42}$$

It should be noted that the standard deviation equals the mean.

As is shown in Figure 11.5a, the curves associated with the exponential distribution have reversed *J* shapes with their vertical extent increasing as $a$ increases. Since they are not mound shaped, the moment coefficients of skew and kurtosis have little significance.

The exponential distribution has been used to describe mathematically the distribution of diameters of trees in all-aged or all-sized stands. An example of this is shown in Figure 11.6. This use of the exponential was first suggested by Leak (1965) and his suggestion has been accepted by a number of workers concerned with the simulation of all-aged stands. Moser (1976) describes how stand density constraints could be applied in such simulations.

The exponential distribution is an appropriate model for diameter distributions in all-aged stands when the ratio of class frequencies in adjacent classes ($h$) remains constant or nearly constant as one moves from class to class across the range of diameters (Meyer, 1952). In other words, when

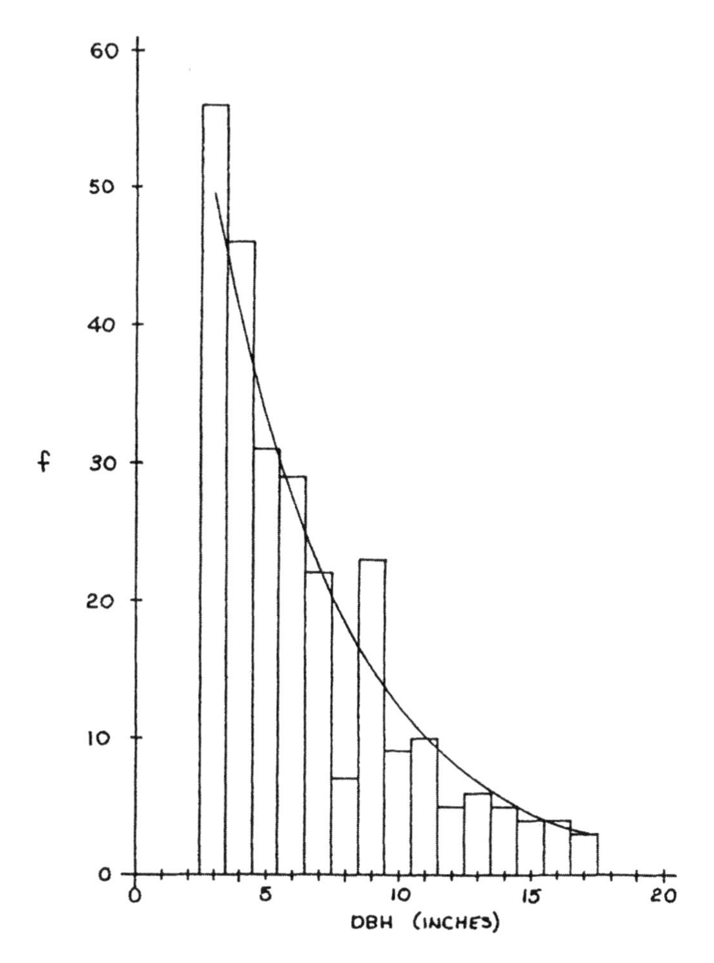


**FIGURE 11.6** Histogram of the per acre stand table for a hypothetical all-aged stand on which is superimposed the curve of an exponential probability distribution with parameter $a = 0.2$.

$$h = \frac{N_i}{N_{i+1}}, \quad i = 1, 2, \ldots, M \tag{11.43}$$

where $N_i$ = number of trees in the $i$th dbh class
$M$ = largest dbh class

Like the geometric distribution (Section 10.6), the exponential distribution is said to have "no memory." This property is exploited by engineers in certain situations when studying or modeling the reliability, under stress, of components in a system (Meyer, 1970).

## 11.5 THE CHI-SQUARE PROBABILITY DISTRIBUTION ($\chi^2$:$k$)

"Chi-square" is the name of a certain statistical value, which is defined as the sum of the squares of $k$ independent variables, $W$, which are normally distributed with parameters (0, 1).

$$\chi^2 = \sum_i^k w_i^2, \quad 0 < \chi^2 \leq \infty \tag{11.44}$$

Chi-square has a gamma distribution with parameters $r = k/2$ and $a = 1/2$. This specific case of the gamma is referred to as the *chi-square distribution*. Variables whose distributions follow this pattern are said to have a chi-square distribution.

Since $k$ is discrete, there is an infinitely large set of chi-square distributions. The shapes of distribution curves change as $k$ changes in much the same manner as do the curves in Figure 11.5.

The pdf of the chi-square distribution is

$$f(y) = \begin{cases} \dfrac{1}{2^{k/2}\Gamma\left(\dfrac{k}{2}\right)} y^{(k/2)-1} e^{-y/2}, & Y > 0 \\ 0, & \text{elsewhere} \end{cases} \tag{11.45}$$

The random variable $Y$ is said to have a *chi-square distribution with* k *degrees of freedom*. This is denoted by $\chi^2_b(Y; k, 2k)$. The meaning of the term "degrees of freedom" is explained in Section 12.11. It must be remembered that variables other than $\chi^2$ can be distributed in this manner and that any of these, including $\chi^2$, can be substituted for $Y$.

The cdf is

$$F(y) = P(Y \leq y_b) = \frac{1}{2^{k/2}\Gamma\left(\dfrac{k}{2}\right)} \int_0^{y_b} y^{(k/2)-1} e^{-y/2} dy \tag{11.46}$$

This expression, however, is of utility only when $k$ is an *even* number. As is typical of gamma distributions, the integration of the pdf involves recursion, the repeated appearance of the integral in the integration with the only change being the exponent of $y$, which is reduced by 1 after each cycle. The cycling will terminate only when the exponent becomes equal to zero. This will occur in the case of chi-square only when $k$ is an *even* number. Tables of fractiles of the $\chi^2$ distribution, such as that in Table A.6 in the Appendix, have been compiled for statistical workers. Because of the problem with odd values of $k$, these tables result from numerical integration of the areas under the distribution curves, rather than from the use of the cdf. To obtain a probability using Table A.6 in the Appendix, the column heads indicate the level of probability, the rows are by values of $k$, the degrees of freedom, while the tabular values are the fractiles or values of $y_b$ or $\chi^2$. Thus, when $k = 3$,

$$P(Y < y_b) = 0.95 \text{ when } \chi^2_b(k = 3) = 7.81$$

$$P(Y < y_b) = 0.01 \text{ when } \chi^2_b(k = 3) = 0.11$$

and

$$P(0.11 \leq y_b \leq 7.81) = 0.95 - 0.01 = 0.94 \text{ or } 94\%$$

Since the chi-square distributions are gamma distributions, their expected values and variances can be obtained directly from Equations 11.30 and 11.31

$$E(Y) = \frac{r}{a} = \frac{\frac{k}{2}}{\frac{1}{2}} = k \tag{11.47}$$

$$V(Y) = \frac{r}{a^2} = \frac{\frac{k}{2}}{\left(\frac{1}{2}\right)^2} = 2k \tag{11.48}$$

Thus, the expected value of a chi-square distribution is equal to the number of degrees of freedom and the variance to twice the number of degrees of freedom.

The moment coefficients of skewness and kurtosis are obtained from Equations 11.32 and 11.33.

$$\alpha_3 = \frac{2}{\sqrt{r}} = \frac{2}{\sqrt{\frac{k}{2}}} \tag{11.49}$$

As $k$ increases, the distribution curves become more and more symmetrical and in the limit, as $k$ approaches infinity, $\alpha_3$ approaches zero, and the distribution becomes symmetric.

$$\alpha_4 = 3 + \frac{6}{r} = 3 + \frac{6}{\frac{k}{2}} = 3 + \frac{12}{k} \tag{11.50}$$

As $k$ increases the curve becomes less peaked. In the limit, as $k$ approaches infinity, $\alpha_4$ approaches 3 and the distribution converges on the normal.

It was shown in Section 11.3 that a relationship exists between the normal and gamma distributions (Equations 11.34 through 11.36). If, in that context, the distribution of random variable $X$ is in standardized form, $\mu_x$ will be equal to zero and $\sigma_x^2$ will be equal to 1. Under these conditions $r$ can be defined as $k/2$, with $k = 1$, and $a$ defined as $^1/_2$. Then, if $Y = X^2$, Equation 11.36 can be written

$$f(y) = \begin{cases} \frac{1}{\sqrt{2\pi y_b}} e^{-y_b/2}, & y_b > 0 \\ 0, & \text{elsewhere} \end{cases} \tag{11.51}$$

Thus, if $X$ is normally distributed, $Y = X^2$ is distributed as chi-square with one degree of freedom. This relationship is exploited by statisticians in a number of ways (e.g., see Section 12.11).

To put this in perspective, assume a continuous variable, $Y$, that is normally distributed with parameters $\mu$ and $\sigma^2$. As in Section 5.2.5, let $\varepsilon_i$ be the *error* associated with $y_i$ (see Equation 5.6):

$$\varepsilon_i = y_i - \mu$$

Since $Y$ is normally distributed and $\mu$ is a constant, $\varepsilon$ must also be normally distributed. This distribution of $\varepsilon$ values can be converted to standard form (Section 11.3) using Equation 11.10:

$$z_i = \frac{\varepsilon_i}{\sigma}$$

When this is done, $Z$ is distributed normally with a mean of zero and a variance equal to unity.

Assume now that all possible samples of size $n$ are drawn from the population of $Y$ values. This, of course, is impossible but the concept is valid. The number of samples would be infinitely large. If, in each sample the $Y$ values were converted to their equivalent $Z$ values and the $Z$ values were squared and then summed, that sum would be a $\chi^2$ value.

$$\chi^2 = \sum_{i=1}^{n} z_i^2$$

The infinitely large number of $\chi^2$ values so generated would have a gamma distribution with parameters $r = n/2$ and $a = 1/2$. Thus, this version of the gamma is known as the chi-square distribution.

The chi-square distribution may be used as a model of a real-life distribution, in the same manner as any probability distribution, but one usually finds it being used in the context of hypothesis testing rather than as a model. In sampling, however, it plays an important role since it can be used to see if the distribution of values in a sample is such that a proposed model is appropriate or if the sample distribution is such that it can or cannot be considered representative of a population assumed to have a certain distribution. For example, assume an all-aged stand in which the frequencies in the dbh classes are as shown in Figure 11.6. Further assume that a forester, for one reason or another, hypothesizes (in other words, is of the opinion) that this stand is one of a set of stands which, on the average, have stand tables in which the frequencies in the dbh classes are exponentially distributed (Section 11.4) with the parameter $a = 0.2$. To test the hypothesis, recourse is made to the *goodness of fit* test, which makes use of the chi-square distribution. This test is described in Chapter 14, which is concerned with a number of hypothesis tests that may be encountered by persons engaged in sampling operations.

## 11.6 THE BETA PROBABILITY DISTRIBUTION ($\beta$:$r_1$, $r_2$)

Consider the two-dimensional random variable ($X_1$, $X_2$) in which the primary random variables are independent and both have gamma distributions with the shape parameters $r_1$ and $r_2$, respectively. The scale parameters of both primary distributions are equal, so that $a_1 = a_2 = a$. Since the primary variables are independent, the regressions of the mean are linear, paralleling the coordinate axes, and intersect at [$\mu(X_1)$, $\mu(X_2)$]. Furthermore, $0 \leq X_1$ and $X_2 < \infty$. The joint pdf is the product of the two marginal pdfs:

$$f(x_1, x_2) = \left[\frac{a^{r_1}}{\Gamma(r_1)} x_1^{r_1-1} e^{-ax_1}\right] * \left[\frac{a^{r_2}}{\Gamma(r_2)} x_2^{r_2-1} e^{-ax_2}\right]$$

$$= \begin{cases} \dfrac{a^{r_1+r_2}}{\Gamma(r_1)\Gamma(r_2)} x_1^{r_1-1} x_2^{r_2-1} e^{-za(x_1+x_2)}, & 0 \leq X_1 \text{ and } X_2 < \infty; \\ & a > r_1 > 0; \text{ and } r_2 > 0 \\ 0, & \text{elsewhere} \end{cases} \tag{11.52}$$

and the cdf is

$$F(x_1, x_2) = P(X_1 \leq x_{1b};\ X_2 \leq x_{2b})$$

$$= \frac{a^{r_1+r_2}}{\Gamma(r_1)\Gamma(r_2)} \int_0^{x_{1b}} \int_0^{x_{2b}} x_1^{r_1-1} x_2^{r_2-1} e^{-2a(x_1+x_2)} dx_1 dx_2 \tag{11.53}$$

This pdf and cdf are for the special case of the *bivariate gamma* distribution when the primary random variables are independent. Since the primary random variables are constrained to take on only positive values, the values of $(X_1, X_2)$ occupy only the upper-right quadrant of the coordinate system. The shape of the surface is determined by the values of the $r$ parameters but, except when $r_1 = r_2 = 1$, it is asymmetrically mound shaped. When $r_1 = r_2 = 1$, the distribution is exponential and the surface slopes downward from a maximum on the vertical axis as both $X_1$ and $X_2$ increase in magnitude.

Now consider the random variable $Z = X_1 + X_2$. It is a function of $(X_1, X_2)$ and there is a value of $Z$ for each pairing of the primary variables. The pdf of $Z$ is obtained using Equation 9.26.

$$
\begin{aligned}
f(z) &= \frac{a^{r_1+r_2}}{\Gamma(r_1)\Gamma(r_2)} \int_0^z x_1^{r_1-1}(z-x_1)^{r_2-1} e^{-a[x_1+(z=x_1)}dx_1 \\
&= \frac{a^{r_1+r_2}e^{-az}}{\Gamma(r_1)\Gamma(r_2)} \int_0^z x_1^{r_1-1}(z-x_1)^{r_2-1} dx_1
\end{aligned}
$$

Let $ZY = X_1$. Since $dzy/dx_1 = zdy/dx_1$, $dx_1 = zdy$. Since $ZY = X_1$ each $Y$ unit is $1/Z$ of an $X_1$ unit. Consequently, the integration is from 0 to $z(1/z) = 1$:

$$
\begin{aligned}
f(z) &= \frac{a^{r_1+r_2}e^{-az}}{\Gamma(r_1)\Gamma(r_2)} \int_0^1 (zy)^{r_1-1}(z-zy)^{r_2-1} zdy \\
&= \frac{a^{r_1+r_2}e^{-az}z}{\Gamma(r_1)\Gamma(r_2)} \int_0^1 z^{r_1-1}y^{r_1-1}z^{r_2-1}(1-y)^{r_2-1} dy \\
&= \frac{a^{r_1+r_2}e^{-az}zz^{r_1-1}z^{r_2-1}}{\Gamma(r_1)\Gamma(r_2)} \int_0^1 y^{r_1-1}(1-y)^{r_2-1} dy \\
&= \left[\frac{a^{r_1+r_2}\int_0^1 y^{r_1-1}(1-y)^{r_2-1} dy}{\Gamma(r_1)\Gamma(r_2)}\right] z^{r_1+r_2-1}e^{-az}
\end{aligned}
\tag{11.54}
$$

The frequency function of $Z$ in the above expression is

$$N(z) = z^{r_1+r_2=1}e^{-az}$$

and

$$T = \int_0^\infty z^{r_1+r_2-1}e^{-az}dz = \frac{\Gamma(r_1+r_2)}{a^{r_1+r_2}}$$

Consequently,

$$f(z) = \frac{N(z)}{T} = \left[\frac{a^{r_1+r_2}}{\Gamma(r_1+r_2)}\right] z^{r_1+r_2-1}e^{-az} \tag{11.55}$$

Setting Equations 11.54 and 11.55 equal to one another:

$$\left[\frac{a^{r_1+r_2}\int_0^1 y^{r_1-1}(1-y)^{r_2-1}dy}{\Gamma(r_1)\Gamma(r_2)}\right]z^{r_1+r_2-1}e^{-az} = \left[\frac{a^{r_1+r_2}}{\Gamma(r_1+r_2)}\right]z^{r_1+r_2-1}e^{-az} \tag{11.56}$$

$$\int_0^1 y^{r_1-1}(1-y)^{r_2-1}dy = \frac{\Gamma(r_1)\Gamma(r_2)}{\Gamma(r_1+r_2)}$$

Substituting this into Equation 11.54 yields the pdf:

$$f(z) = \begin{cases} \left[\dfrac{a^{r_1+r_2}}{\Gamma(r_1+r_2)}\right]z^{r_1+r_2-1}e^{-az}, & 0 \le Z \le 1,\ r_1 > 0,\ r_2 > 0, \text{ and } a > 0 \\ 0, & \text{elsewhere} \end{cases} \tag{11.57}$$

Thus, the pdf of $Z = X_1 + X_2$ is for a gamma distribution with the parameters $(r_1 + r_2, a)$.

If the expression under the integral in Equation 11.54 is considered to be the frequency function of $Y$,

$$N(y) = y^{r_1-1}(1-y)^{r_2-1}, \quad 0 < Y \le 1$$

then, from Equation 11.56,

$$T = \frac{\Gamma(r_1)\Gamma(r_2)}{\Gamma(r_1+r_2)}$$

and

$$f(y) = \frac{N(y)}{T} = \begin{cases} \left[\dfrac{\Gamma(r_1+r_2)}{\Gamma(r_1)\Gamma(r_2)}\right]y^{r_1-1}(1-y)^{r_2-1}, & 0 < Y \le 1,\ r_1 > 0, \text{ and } r_2 > 0 \\ 0, & \text{elsewhere} \end{cases} \tag{11.58}$$

$Y$ has a *beta* probability distribution with parameters $r_2$ and $r_2$. It is called a beta distribution because the expression

$$B(r_1, r_2) = \int_0^1 y^{r_1-1}(1-y)^{r_2-1}dy = \frac{\Gamma(r_1)\Gamma(r_2)}{\Gamma(r_1+r_2)} \tag{11.59}$$

is known as the *beta function*. Since $r_1$ and $r_2$ can vary, there is a family of beta distributions. Some are shown in Figure 11.7.

When the integration in Equation 11.59 is carried out from zero to some value of $Y$ other than 1 (i.e., to $y_b$), the function is known as the *incomplete beta function*:

$$B_{y_b}(r_1, r_2) = \int_0^{y_b} y^{r_1-1}(1-y)^{r_2-1}dy; \quad 0 \le Y < 1 \tag{11.60}$$

This, of course, is equivalent to obtaining the subset size $N(0 \le Y \le y_b)$ and the ratio of this subset size to the total size of the range space, $T$, is the cumulative probability,

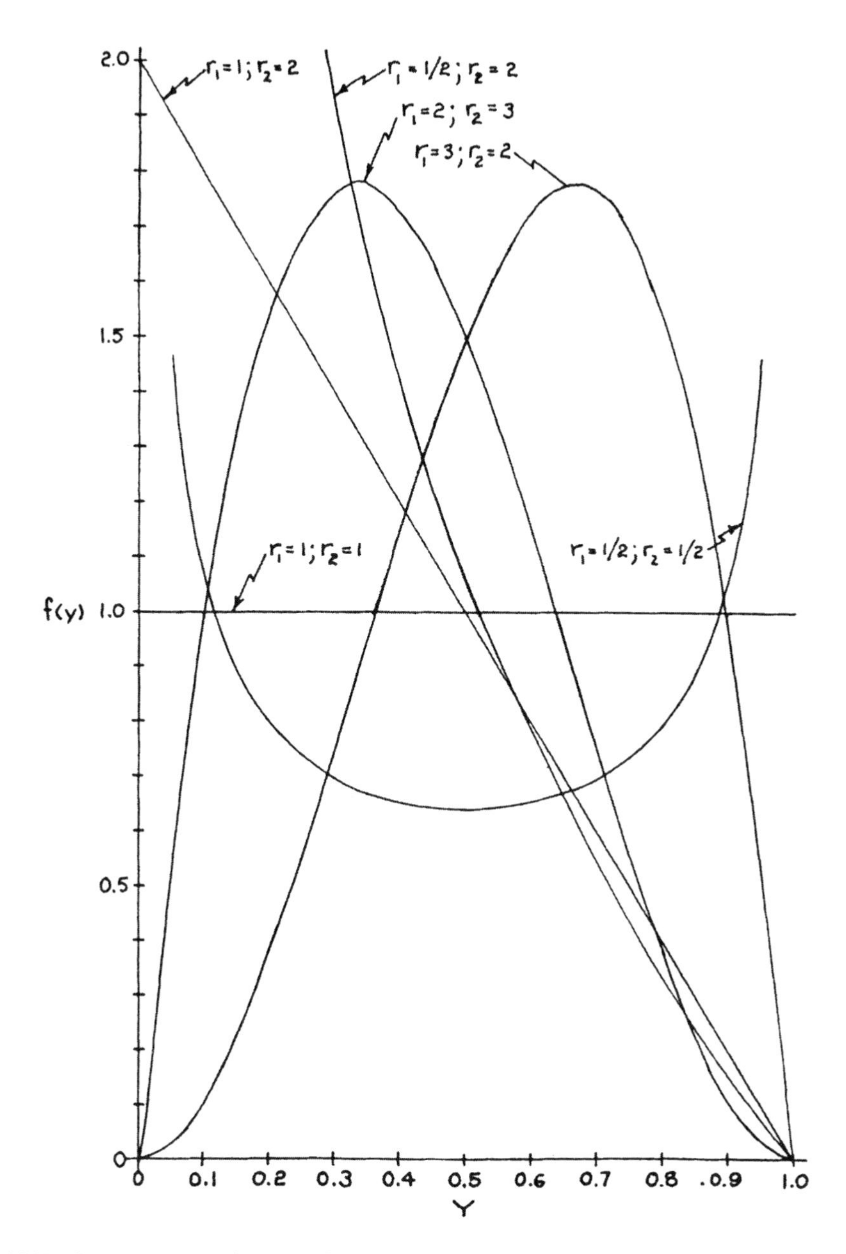


**FIGURE 11.7** Some representative beta distributions.

$$P(0 \le Y \le y_b) = F(Y \le y_b) = \frac{N(0 \le Y \le y_b)}{T}$$

$$= \frac{B_{y_b}(r_1, r_2)}{B(r_1, r_2)} \tag{11.61}$$

$$F(y) = I_{y_b}(r_1, r_2)$$

Cumulative probabilities can be computed using these equations but, because of the recursions (cycling) involved, the computations are tedious and time-consuming. Tables have been prepared that may or may not simplify the problems. Reference is made to Beyer (1968, pp. 252–265). These tables provide the fractiles or *percentage points* for probability levels 0.50, 0.25, 0.10, 0.05, 0.025, 0.01, and 0.005, for different values of $r_1$ and $r_2$. Table A.7 in the Appendix is from this set, being that for a probability level of 0.05. As an example of the use of these tables, if one needs to know the $y_b$ value at which $I_{y_b} = 0.05$ and $r_1 = 3$ and $r_2 = 5$, the value is obtained from Table A.7 using $v_1 = 2r_2 = 2(5) = 10$ and $v_2 = 2r_1 = 2(3) = 6$. Then $y_b = 0.12876$. If a probability level is needed for which there is no table, one must interpolate between the tabular values that are available. Straight-line interpolation is not appropriate and may introduce significant errors. The reverse process, in which the $I_{y_b} = 0.05$ for a specific $y_b$ value is to be obtained, is difficult to determine using the table and recourse may have to be made to numerical integration methods. Reference is made to Johnson and Kotz (1970, pp. 48–51).

It can be shown that the *k*th moment about zero is

$$u'_k = E(Y^k) = \frac{\Gamma(r_1 + r_2)\Gamma(r_1 + k)}{\Gamma(r_1 + r_2 + k)\Gamma(r_1)} \tag{11.62}$$

The derivation of this expression is too lengthy for inclusion here. From it the following expected values are obtained:

$$\begin{aligned}
\mu'_i &= E(Y) = u_Y = \frac{(r_1 + r_2 - 1)!(r_1)!}{(r_1 + r_2)!(r_1 - 1)!} = \frac{r_1}{(r_1 + r_2)} \\
\mu'_2 &= E(Y^2) = \frac{(r_1 + r_2 - 1)!(r_1 - 1)!}{(r_1 + r_2 + 1)!(r_1 - 1)!} = \frac{(r_1 + 1)r_1}{(r_1 + r_2 + 1)(r_1 + r_2)} \\
\mu'_3 &= E(Y^3) = \frac{(r_1 + r_2 - 1)!(r_1 - 2)!}{(r_1 + r_2 - 1)!(r_1 - 1)!} = \frac{(r_1 + 2)(r_1 + 1)r_1}{(r_1 + r_2 + 2)(r_1 + r_2 + 1)(r_1 + r_2)} \\
\text{and}\quad \mu'_4 &= E(Y^4) = \frac{(r_1 + r_2 - 1)!(r_1 - 3)!}{(r_1 + r_2 + 3)!(r_1 - 1)!} = \frac{(r_1 + 3)(r_1 + 2)(r_1 + 1)r_1}{(r_1 + r_2 + 3)(r_1 + r_2 + 2)(r_1 + r_2 + 1)(r_1 + r_2)}
\end{aligned} \tag{11.63}$$

Thus, Equation 11.63 yields the expected value or mean of the distribution. The variance is

$$V(Y) = E(Y^2) - [E(Y)]^2 = \frac{r_1 r_2}{(r_1 + r_2 + 1)(r_1 + r_2)^2} \tag{11.64}$$

It should be noted that the expected value, *E*(*Y*), remains constant, regardless of the magnitude of $r_1$ and $r_2$, if the ratio *r*/*r* remains constant. For example,

$$E(Y) = \frac{r_1}{(r_1 + r_2)} = \frac{3r_1}{3r_1 + 3r_2} = \frac{50r_1}{50r_1 + 50r_2}, \quad \text{etc.}$$

The variance, however, does not behave in this manner. When the ratio is held constant and the values of $r_1$ and $r_2$ are increased indefinitely, the variance tends toward zero. For example, let $r_1/r_2 = 2$ and setting $r_1 = 4$ and $r_2 = 2$:

$$V(Y) = \frac{r_1 r_2}{(r_1 + r_2 + 1)(r_1 + r_2)^2} = \frac{4(2)}{7(36)} = \frac{1}{7}(0.2222222)$$

Now, set $r_1 = 100$ and $r_2 = 50$:

$$V(Y) = \frac{100(500)}{151(22,500)} = \frac{1}{151}(0.2222222)$$

and now, setting $r_1 = 1000$ and $r_2 = 500$,

$$V(Y) = \frac{1000(500)}{1501(2,250,000)} = \frac{1}{1501}(0.2222222)$$

If $r_1 \to \infty$ and $r_2 \to \infty/2$

$$V(Y) = \frac{1}{\infty}(0.2222222) \to 0$$

The third and fourth moments about the mean are obtained using Equations 8.43 and 8.44. Hence the moment coefficient of skew is (bypassing much algebra):

$$\alpha_3 = \frac{2(r_2 - r_1)(r_1 + r_2 + 1)^{1/2}}{(r_1 + r_2 + 2)(r_1 r_2)^{1/2}}$$

When $r_1 = r_2$, $(r_2 - r_1) = 0$, and the coefficient equals zero, indicating symmetry. When $r_1 > r_2$, the coefficient will be negative and the skew will be to the left. Conversely, when $r_1 < r_2$, the coefficient will be positive and the skew will be to the right.

The moment of kurtosis is (again bypassing much algebra)

$$\alpha_4 = \left[\frac{6(r_1 + r_2 + 1)(r_1 + r_2)^2 + 3(r_1 + r_2 + 1)(r_1 + r_2 - 6)r_1 r_2}{r_1 r_2 (r_1 + r_2 + 2)(r_1 + r_2 + 3)}\right] \tag{11.66}$$

When symmetric, the distribution is platykurtic for all values of $r = r_1 = r_2$. When $r = 1$, the distribution is rectilinear and $\alpha_4 = -1.2$. As the magnitude of $r$ increases, the curve becomes mound shaped and increasingly peaked. $\alpha_4$ increases but at a decreasing rate. In the limit, as $r \to \infty$, the coefficient $\to 0$. Some of this trend can be seen in Table 11.1. Thus, the beta converges to the standardized normal as $r \to \infty$. This occurs in spite of the fact that the range of $Y$ values is only from 0 to 1. As was mentioned earlier, if $r_1/r_2$ is constant and $r_1$ and $r_2 \to \infty$, the variance $\to 0$. Thus, there are an infinite number of standard deviates on either side of the mean, even if the number line is finite in length.

The mode of the distribution is obtained using the methods of differential calculus. It is

$$y(\text{mode}) = \frac{(r_1 - 1)}{(r_1 + r_2 - 1)} \tag{11.67}$$

When $r_1 < 1$ and $r_2 < 1$, the distribution becomes *U* shaped and Equation 11.57 locates the *antimode* or minimum value point. When the product $(r_1 - 1)(r_2 - 1) < 0$ the distribution has a *J* or reversed-*J* shape and no mode or antimode exists.

The beta distribution is extremely flexible insofar as shape is concerned, as can be seen in Figure 11.7. When $r_1 = r_2$ the distribution is symmetrical and, as the magnitude of $r = r_1 = r_2$

**TABLE 11.1**
**Values of the Moment Coefficient of Kurtosis in the Case of the Beta Distribution When $r_1 = r_2$**

| $r = r_1 = r_2$ | $\alpha_4$ |
|---|---|
| 1 | –1.2000000 |
| 3 | –0.6666667 |
| 5 | –0.4615385 |
| 7 | –0.3529412 |
| 9 | –0.2857143 |
| 11 | –0.2400000 |
| 13 | –0.2068966 |
| 15 | –0.1818182 |
| 20 | –0.1431813 |
| 25 | –0.1132075 |
| 30 | –0.0952381 |
| 35 | –0.0821918 |
| 50 | –0.0582524 |
| 70 | –0.0419580 |
| 90 | –0.0327869 |
| 100 | –0.0295567 |

*Note:* Note the approach of the shape of the distribution to the normal as the value of $r$ increases.

increases, the degree of platykurtosis decreases until the normal is approached. When $r_1 > r_2$ the distribution is skewed to the left and when $r_1 < r_2$ the skew is to the right. When $r_1 = r_2 = 1$, the distribution is rectilinear. When $r_1 = r_2 < 1$, the distribution is *U* shaped. When $r_1 \leq 1$ and $r_2 > 1$, the distribution has a reversed *J* shape which becomes *J*-shaped when $r_1 > 1$ and $r_2 \leq 1$. When $r_1 = 1$ and $r_2 = 2$, the distribution curve is a straight line running from (0,2) to (1,0). The pattern is reversed when $r_1 = 2$ and $r_2 = 1$. Hafley and Schreuder (1977), in a study of the performance of a number of probability distributions as models of diameter distributions, ranked the beta second only to the Johnson $S_B$ in degree of flexibility.

The beta distribution is particularly valuable when the random variable is a proportion or percent because the range of $Y$ values is from zero to one. It has been used in this manner to describe the proportion of stem counts or basal areas occurring in stem diameter classes. This has been done in the context of growth and yield studies (Clutter and Bennett, 1965; McGee and Della-Bianca, 1967; Clutter and Shephard, 1968; Lenhart and Clutter, 1971; Zöhrer, 1970; Sassaman and Barrett, 1974; van Laar and Geldenhuys, 1975).

## 11.7 THE BIVARIATE NORMAL PROBABILITY DISTRIBUTION ($N_1{:}\mu_1, \sigma_1; N_2{:}\mu_2, \sigma_2$)

When the two primary variables in a two-dimensional random variable are both normally distributed, the distribution of the two-dimensional random variable is known as a *bivariate normal distribution*. As was brought out in Section 9.5, the joint pdf of a two-dimensional distribution is the product of the marginal pdf of the first primary random variable and the conditional pdf of the second primary random variable (Equation 9.18). Assume two primary random variables, $Y_1$ and $Y_2$. Both are normally distributed with the parameters: $-\infty < \mu_1 < \infty$, $-\infty < \mu_2 < \infty$, $\sigma_1^2 > 0$; $\sigma_2^2 > 0$. Further, assume that the two primary variables are correlated, $-1 \leq \rho \leq 1$, their regressions of the mean are linear, and the variances about the regression lines are homogeneous. In such a case the conditional distributions of the two primary random variables are both normal with the following parameters:

$$\mu_{1|2} = \mu_1 + \rho\left(\frac{\sigma_1}{\sigma_2}\right)(Y_2 - \mu_2);\quad \sigma^2_{1|2} = \sigma_1^2(1-\rho^2)$$

$$\mu_{2|1} = \mu_2 + \rho\left(\frac{\sigma_2}{\sigma_1}\right)(Y_1 - \mu_1);\quad \sigma^2_{2|1} = \sigma_1^2(1-\rho^2)$$

Reference is made to Equation 9.35 for these means and Equation 9.36 for the variances.

If one assumes $Y_1$ to be the first of the pair of primary random variables, the joint pdf of the bivariate normal distribution is the product of the marginal pdf of $Y_1$ and the conditional pdf of $Y_2$:

$$\begin{aligned} f(y_1y_2) &= f(y_1) * f(y_2 \mid y_1) \\ &= \left[\frac{1}{\sigma_1\sqrt{2\pi}} e^{-(y_1-\mu_1)^2/2\sigma_1^2}\right] * \left\{\frac{1}{\sqrt{\sigma_2^2(1-\rho^2)2\pi}} e^{[y_2-\mu_2-\rho(\sigma_2/\sigma_1)(y_1-\mu_1)]^2/2[\sigma_2^2(1-\rho)]}\right\} \quad (11.68) \\ &= \frac{1}{2\pi\sigma_1\sigma_2\sqrt{1-\rho^2}} * \exp\left\{-\frac{1}{2(1-\rho^2)}\left[\left(\frac{y_1-\mu_1}{\sigma_1}\right)^2 + \left(\frac{y_2-\mu_2}{\sigma_2}\right)^2 - \frac{2\rho(y_1-\mu_1)(Y_2-\mu_2)}{\sigma_1\sigma_2}\right]\right\} \\ & -\infty < Y_1;\ Y_2 < \infty;\ -1 \le \rho \le 1 \end{aligned}$$

The algebraic process leading to Equation 11.68 is lengthy but reasonably straightforward if one remembers that $\rho = \mathrm{COV}(Y_1Y_2)/\sqrt{V(Y_1)V(Y_2)} = \sigma_1\sigma_2/\sqrt{\sigma_1^2\sigma_2^2}$ (see Equation 9.32).

If the correlation coefficient $\rho = 0$, the conditional pdf of the second primary variable reduces to the marginal pdf of that variable. When this occurs, the joint pdf becomes the product of the two marginal pdfs and the requirements for independence between the primary variables, as set forth in Equation 9.6, are met. Thus, in the case of the bivariate normal distribution, when the correlation coefficient is equal to zero the primary variables are independent. This is not generally the case with bivariate distributions.

When the primary random variables are independent, Equation 11.68 becomes

$$f(y_1y_2) = \frac{1}{2\pi\sigma_1\sigma_2} * \exp\left\{-\left[\frac{1}{2}\left(\frac{y_1-\mu_1}{\sigma_1}\right)^2 + \left(\frac{y_2-\mu_2}{\sigma_{y_2}}\right)^2\right]\right\} \quad (11.69)$$

Finally, if the marginal distributions are put into standard form by setting

$$\left(\frac{y_1-\mu_1}{\sigma_1}\right) = z_1 \quad \text{and} \quad \left(\frac{y_2-\mu_2}{\sigma_2}\right) = z_2$$

(see Equation 11.10) Equation 11.68 becomes

$$f(z_1z_2) = \frac{1}{2\pi(1-\rho^2)} * \exp\left[-\frac{1}{2(1-\rho^2)}(z_1^2 + z_2^2 - 2\rho z_1 z_2)\right] \quad (11.70)$$

and Equation 11.69 becomes

$$f(z_1z_2) = \frac{1}{2\pi} * \exp\left[-\frac{1}{2}(z_1^2 + z_2^2)\right] \tag{11.71}$$

The distribution described by Equation 11.71 is known as the *standard bivariate normal distribution*. In some cases, however, the title is given to distributions described by Equation 11.70. Care should be taken when encountering the name to avoid confusion.

The cdf of the bivariate normal distribution is obtained classically (Equation 9.21):

$$F(y_1y_2) = P(Y_1 \leq y_{1b},\ Y_2 \leq y_{2b} \mid y_{1b}) \frac{1}{2\pi\sigma_1\sigma_2\sqrt{1-\rho^2}}$$

$$\int_{-\infty}^{y_{1b}} \int_{-\infty}^{y_{2b}} \exp\left\{-\frac{1}{2(1-\rho^2)}\left[\left(\frac{y_1-\mu_1}{\sigma_1}\right)^2 + \left(\frac{y_2-\mu_2}{\sigma_2}\right)^2 - \frac{2\rho(y_1-\mu_1)(y_2-\mu_2)}{\sigma_1\sigma_2}\right]\right\} dy_2 dy_1 \tag{11.72}$$

As was the case with the univariate normal, this integral does not exist and the equation must be solved using numerical methods. This has been done using Equation 11.71 and a table can be found in Owen (1962). Pearson (1931) also prepared a table but cumulated the probabilities from the positive rather than the negative extreme. As with any continuous distribution, the probabilities of occurrence are for intervals along the coordinate axes, $P(y_{1a} \leq Y_1 \leq y_{1b};\ y_{2a} \leq Y_2 \leq y_{2b})$. In a graphical sense, these intervals form a rectangle in the plane of the coordinate axes. Vertical lines, extending upward from the corners of the rectangle to the surface defined by the pdf, define a column whose volume, relative to the total volume under the surface, is equal to the desired probability.

Regardless of which pdf is used, the surface that is generated is dome shaped with the maximum height at $(\mu_{Y1}, \mu_{Y2})$. If the surface is sliced by a vertical plane parallel to either of the coordinate axes, the intersection of the slicing plane and the pdf surface forms a normal curve. This can be confirmed if one holds one of the primary variables constant (e.g., let $Y_1 = a$) in Equation 11.69.

$$f(y_1y_2) = \frac{1}{2\pi\sigma_1\sigma_2} * \exp\left\{-\frac{1}{2}\left[\left(\frac{a-\mu_1}{\sigma_1}\right)^2 + \left(\frac{y_2-\mu_2}{\sigma_2}\right)^2\right]\right\}$$

$$= \left\{\frac{1}{\sigma_1\sqrt{2\pi}} * \exp\left[-\frac{1}{2}\left(\frac{a-\mu_{Y1}}{\sigma_1}\right)^2\right]\right\} * \left\{\frac{1}{\sigma_2\sqrt{2\pi}} * \exp\left[-\frac{1}{2}\left(\frac{y_2-\mu_2}{\sigma_{y_2}}\right)^2\right]\right\}$$

The first term of this expression is now a constant and acts as a scalar. The second term is the marginal pdf of $Y_2$, which is normally distributed. Thus, the line of intersection is a normal curve whose vertical scale has been modified. It can be shown that this property of normality is possessed by the distribution even when $\rho \neq 0$ (Hald, 1952).

Consider the term

$$\left[\left(\frac{y_1-\mu_1}{\sigma_1}\right)^2 + \left(\frac{y_2-\mu_2}{\sigma_2}\right)^2\right]$$

in Equation 11.69. If it remains constant,

$$\left[\left(\frac{y_1-\mu_1}{\sigma_1}\right)^2 + \left(\frac{y_2-\mu_2}{\sigma_2}\right)^2\right] = k \tag{11.73}$$

$f(y_1y_2)$ also remains constant:

$$f(y_1y_2) = \frac{1}{2\pi\sigma_1\sigma_2} * \exp\left(-\frac{k}{2}\right)$$

The set of values of the random variable $(Y_1Y_2)$ which would make the term constant, when plotted on the $Y_1Y_2$ coordinate system, forms an ellipse. Setting $C^2 = k$,

$$\left(\frac{y_1 - \mu_1}{c\sigma_1}\right)^2 + \left(\frac{y_2 - \mu_2}{c\sigma_2}\right)^2 = 1 \tag{11.74}$$

which is the equation of an ellipse whose center is at $(\mu_1, \mu_2)$ and whose semiaxes are $c\sigma_1$ and $c\sigma_2$. If $k$, or $c^2$, is reduced in magnitude, a smaller ellipse is formed since the semiaxes are reduced. Furthermore, as $k$ is reduced, the $f(y_1y_2)$ value is increased. At the limit, when $k = 0$, the ellipse degenerates to a dimensionless point at $(\mu_1, \mu_2)$ and the magnitude of $f(y_1y_2)$ is at the maximum, $1/2\pi\sigma_1\sigma_2$. The set of points, on and within the boundary of the original ellipse, is associated with $f(y_1y_2)$ values equal to or greater than the $f(y_1y_2)$ value associated with the original $k$.

In Section 11.5, $\chi^2$ was defined as the sum of squared standardized errors. Thus, for *any* value of the random variable $(Y_1Y_2)$,

$$\left(\frac{y_1 - \mu_1}{\sigma_1}\right)^2 + \left(\frac{y_2 - \mu_2}{\sigma_2}\right)^2 = \chi^2 \tag{11.75}$$

In other words, each of the infinitely large set of points making up the plane of the $Y_1Y_2$ coordinate system has associated with it a $\chi^2$ value. It was also brought out in Section 11.5 that if the errors are normally distributed, as they are in this case, the $\chi^2$ values are distributed as $\chi^2$ with two degrees of freedom.

Since Equations 11.73 and 11.75 are equivalent, a specified value of $\chi^2$ can be used in place of $k$ to define the ellipse. The statement then can be made that all points on or within the boundary of the ellipse so defined have $\chi^2$ values less than the specified $\chi^2$. The probability of drawing one of these points is equivalent to $\rho(\chi^2 \leq k)$. This probability can be determined using Table A.6 in the Appendix. For example, according to Table A.6, 5% of $\chi^2_2$ values are equal to or less than 0.103. Substituting this for $k$ in Equation 11.73:

$$\left(\frac{y_1 - \mu_1}{c\sigma_1}\right)^2 + \left(\frac{y_2 - \mu_2}{c\sigma_2}\right)^2 = 0.103$$

$$\left(\frac{y_1 - \mu_1}{\sigma_1}\right)^2 + \left(\frac{y_2 - \mu_2}{\sigma_2}\right)^2 = \frac{0.103}{c} = 1$$

Then,

$$c^2 = 0.103$$

$$c = \sqrt{0.103} = 0.32094$$

and the ellipse has the semiaxes $0.32094c_{Y1}$ and $0.32094\sigma_{Y2}$.

The use of $\chi^2$ provides an alternative approach to the determination of probabilities using the bivariate normal distribution. However, one must realize that the probabilities associated with this approach are inherently different from those obtained using the cdf approach. With the conventional approach, the volume of a rectangular column, which can be anywhere within the coordinate system, is related to the total volume under the surface to obtain the probability. In the case of the ellipse approach, the probability obtained is that a certain value of the random variable is within a given ellipse which is centered on the distribution mean. Care should be taken to avoid confusion when computing or interpreting a probability involving the bivariate normal distribution.

## 11.8 THE WEIBULL PROBABILITY DISTRIBUTION ($W{:}r, a, y_a$)

Assume a random variable, $X$, which is exponentially distributed

$$f(x) = e^{-x}; \quad 0 \le X \le \infty$$

Let $Y = aX^{1/r}$. Then, $X = (Y/a)^r$. If $c$ is a specific value of $X$, the corresponding value of $Y$ is $c^{1/r}$. Then,

$$P(0 \le X \le c^{1/r}) = \int_0^{c^{1/r}} e^{-x}dx$$

since $y = ax^{1/r}$, $dy/dx = ar^{-1}x^{-(r-1)/r}$, and $dx = a^{-1}rx^{(r-1)/r}dy$.

Substituting $(y/a)^r$ for $x$ in the differential yields

$$dx = a^{-1}r\left(\frac{y}{a}\right)^{(r-1)} dy$$

Thus,

$$P(0 \le Y \le c) = \int_0^{c} a^{-1}r\left(\frac{y}{a}\right)^{r-1} e^{-(y/a)^2} dy \tag{11.76}$$

which is the cumulative density function for $Y$. The pdf of $Y$ is

$$f(y) = \begin{cases} a^{-1}r\left(\dfrac{y}{a}\right)^{r-1} e^{-(y/a)^2}, & 0 \le Y \le \infty,\ a > 0,\ r > 0 \\ 0, & \text{elsewhere} \end{cases} \tag{11.77}$$

In this form the minimum value of $Y$, $y_A$, is located at zero. To provide flexibility to the location of the distribution along a number line, the pdf is written

$$f(y) = \begin{cases} a^{-1}r\left[\dfrac{(y - y_A)}{a}\right]^{r-1} e^{-[(y-y_A)/a]^r}, & y_A \le Y \le \infty,\ 0 \le y_A,\ a > 0,\ r > 0 \\ 0, & \text{elsewhere} \end{cases} \tag{11.78}$$

$Y$ has a *Weibull* probability distribution.* Weibull probability distributions have three parameters. $r$ is the *shape* parameter which controls the basic shape of the distribution curve, as can be seen

* The distribution is named after Waloddi Weibull, a Swedish physicist, who first described it in 1939.

in Figure 11.8. When $r = 1$, the pdf reduces to the exponential and the curve is a reversed *J* with a definite *y*-intercept. When $r < 1$, the pdf produces a reversed-*J* hyperbolic curve which is asymptotic to both axes. When $r > 1$, the curve becomes mound shaped. When $r = 3.6$, the curve is essentially symmetrical and is similar to, but not coincident with, the normal curve. When $1 < r < 3.6$, the curve is positively skewed and when $r > 3.6$ the curve is negatively skewed.*

The parameter $a$ controls the *scale*. As can be seen from Figure 11.8, as $a$ increases, the curve broadens and flattens.

As was brought out earlier, the *position of the distribution along a number line* is controlled by the parameter $y_A$. In many cases $y_A$ is set equal to zero, in which case the pdf reduces to Equation 11.77.

When in *standard form*, $a$ is set equal to 1 and $y_A$ to zero. The pdf then becomes

$$f(y) = \begin{cases} ry^{r-1}e^{-y^r}, & 0 \le Y \le \infty,\ r > 0 \\ 0, & \text{elsewhere} \end{cases} \tag{11.79}$$

Equation 11.76 is the cumulative density function for $Y$. It usually does not appear in this form. Instead it is shown as Equation 11.80 below.

$$P(0 \le Y \le c) = \int_0^c a^{-1}r\left(\frac{y}{a}\right)^{r-1} e^{-(y/a)^2}\,dy$$

Since

$$\frac{d}{dy}[-e^{-(y/a)^r}] = -e^{-(y/a)^r}\left[-r\left(\frac{y}{a}\right)^{r-1} a^{-a}\right]$$

$$= a^{-1}r\left(\frac{y}{a}\right)^{r-1} e^{-(y/a)^r}$$

then

$$P(0 \le Y \le c) = -e^{-(y/a)^r}\Bigg|\begin{matrix} c = -e^{-(y/a)^r} \\ 0 = +e^{-(y/a)^r} \end{matrix} = 1$$

$$F(y) = 1 - e^{-(c/a)^r} \tag{11.80}$$

The mean or expected value of $Y$, assuming $y_A = 0$, is obtained classically using Equation 8.31:

$$E(Y) = \int_0^\infty y\left[a^{-1}r\left(\frac{y}{a}\right)^{r-1} e^{-(y/a)^r}\right]dy$$

* It should be brought out here that the symbology usually used in forestry is that originally used by Dubey (1967). It was adopted by Bailey and Dell (1973), in their paper which did much to call the distribution to the attention of American foresters. In this symbol system, $a$ is the location parameter, $b$ is the scale parameter, and $c$ is the shape parameter. In engineering references (e.g., Meyer, 1970, or Miller and Freund, 1977) $\alpha$ is usually the scale parameter and $\beta$ the shape parameter. The location parameter apparently is rarely used. It also should be noted that in the engineering literature the scale parameter is equivalent to $b^{-c}$ in the Dubey system or to $a^{-r}$ in the system used in this book.

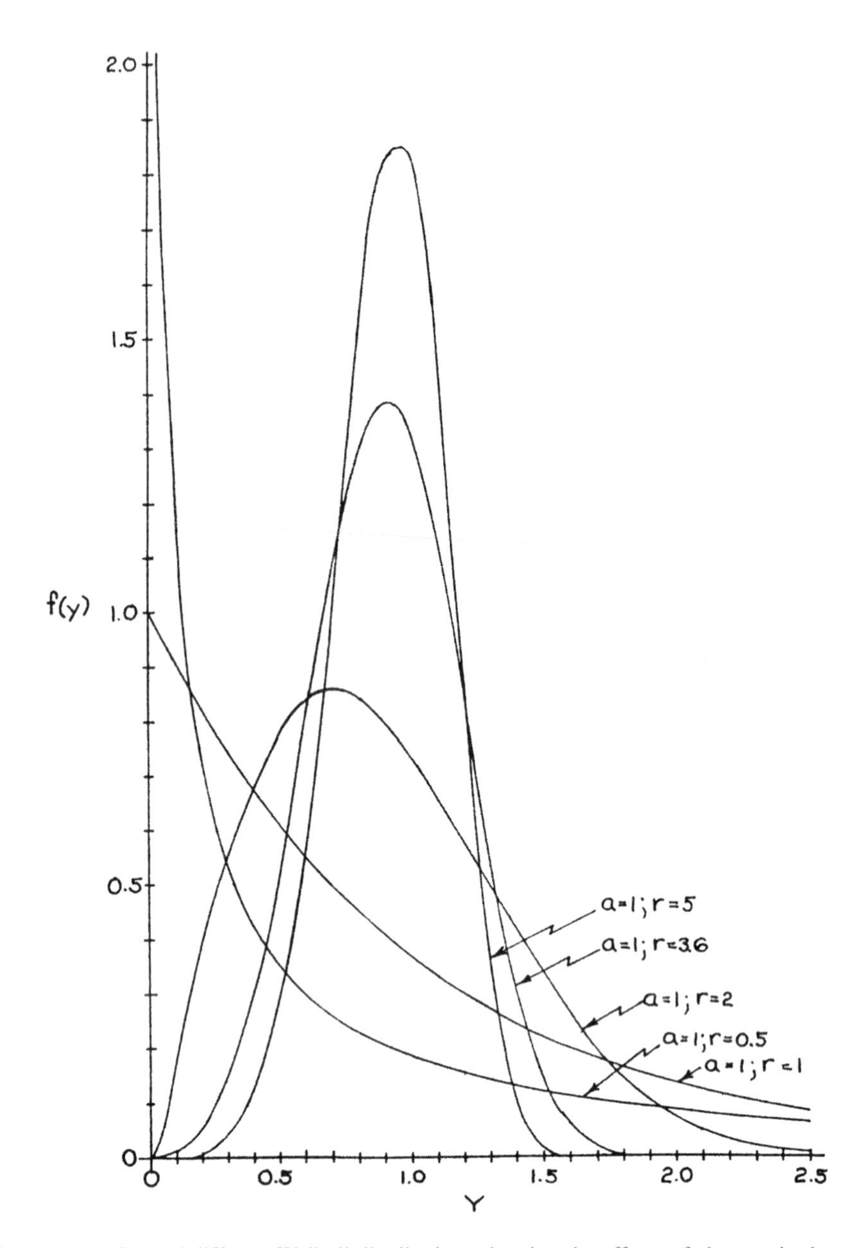


**FIGURE 11.8** (A) Several different Weibull distributions showing the effects of changes in the parameter $r$ when $a$ is held constant at 1. (B) The Weibull distribution when $a = 2$.

Letting $w = (y/a)^r$, then $dy = r^{-1}(y/a)^{1-r}adw$. Furthermore, since $(y/a) = w^{1/r}$, $dy = r^{-1}(w^{1/r})^{1-r}adw = r^{-1}w^{(1/r)-1}adw$. Thus,

$$E(Y) = ra\int_0^\infty we^{-w}r^{-1}w^{(1/r)-1}dw$$

$$= a\int_0^\infty w^{1/r}e^{-w}dw$$

$w^{1/r}$ can be written $w^{[(1/r)+1]-1}$. Consequently,

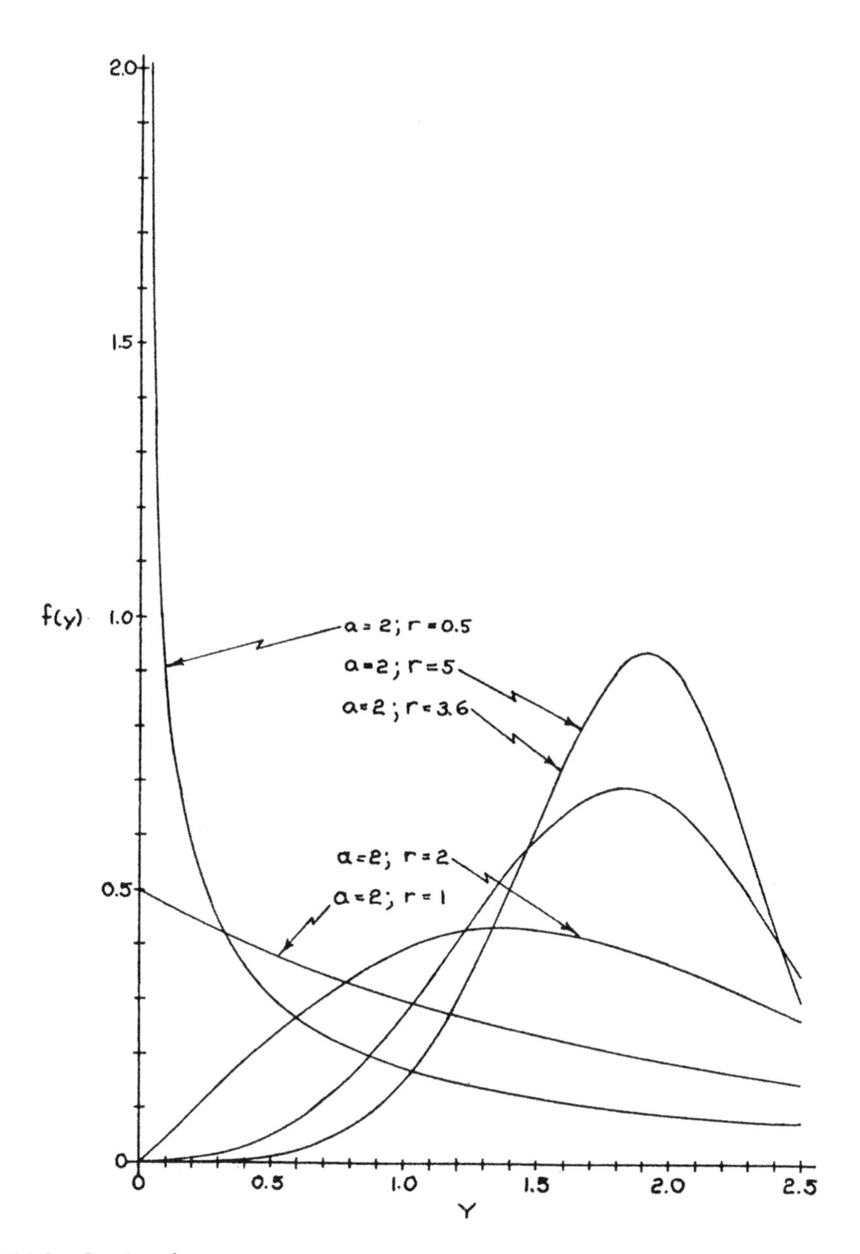


**FIGURE 11.8** *Continued.*

$$E(Y) = a\int_0^{\infty} w^{[(1/r)+1]-1} e^{-w} dw$$

which, by definition (see Equation 11.21), is equal to

$$\mu_1' = E(Y) = a\Gamma\left[\left(\frac{1}{r}\right)+1\right] \tag{11.81}$$

Thus, if $a = 1$ and $r = 2$,

$$E(Y) = \Gamma(1.5) = 0.88622$$

The derivations of the following expected values follow the pattern used for $E(Y)$:

$$\mu_2' = E(Y^2) = a^2\Gamma\left[\left(\frac{2}{r}\right)+1\right]$$

$$\mu_3' = E(Y^3) = a^3\Gamma\left[\left(\frac{3}{r}\right)+1\right]$$

$$\mu_4' = E(Y^4) = a^4\Gamma\left[\left(\frac{4}{r}\right)+1\right]$$

The variance of the Weibull distribution is obtained using Equation 8.42:

$$V(Y) = E(Y^2) - [E(Y)]^2 = a\Gamma\left[\left(\frac{2}{r}\right)+1\right] - \left\{a\Gamma\left[\left(\frac{1}{r}\right)+1\right]\right\}^2 \tag{11.82}$$

For example, if $a = 1$ and $r = 2$,

$$V(Y) = \Gamma(2) - [\Gamma(1.5)]^2 = 1 - (0.88622)^2 = 0.21461$$

The moment coefficient of skew is obtained using Equations 8.36 and 8.43:

$$\alpha_3\{E(Y^3) - 3E(Y^2)E(Y) + 2[E(Y)]^3\} * [V(Y)]^{-3/2} \tag{11.83}$$

It was stated earlier that when $r = 3.6$ the Weibull approximates the normal distribution. If $a$ is set equal to 1 and $r$ equal to 3.6 in Equation 11.83, $\alpha_3$ equals zero (within rounding error) and indicates that the distribution is essentially symmetrical. When $1 \leq r < 3.6$, $\alpha_3$ is positive, indicating a skew to the right. When $r > 3.6$, $\alpha_3$ is negative and the skew is to the left.

The moment coefficient of kurtosis is computed using Equations 8.36 and 8.44:

$$\alpha_4 = \{E(Y^4) - 4E(Y^3)E(Y) + 6E(Y^2)[E(Y)]^2 - 3[E(Y)]^4\} * [V(Y)]^{-2} \tag{11.84}$$

Substituting $a = 1$ and $r = 3.6$ into this expression yields a value of $\alpha_4$ of approximately 2.615, which is less than 3, indicating that the Weibull, when symmetrical, is somewhat platykurtic. The lack of coincidence with the normal is evident. Compounding the lack of coincidence is that the Weibull has a definite point of origin on the number line, $Y_A$, while the normal is asymptotic to the number line with its point of origin at negative infinity.

The mode of the Weibull is obtained using the methods of differential calculus.

$$\frac{d}{dy}\left[a^{-1}r\left(\frac{y}{a}\right)^{r-1} e^{-(y/a)^2}\right] = a^{-1}r\frac{d}{dy}\left[\left(\frac{y}{a}\right)^{r-1} e^{-(y/a)^r}\right]$$

$$= a^{-1}r\left[e^{-(y/a)^r}(r-1)\left(\frac{y}{a}\right)^{r-2} a^{-1} - e^{-(y/a)^r} r\left(\frac{y}{a}\right)^{2r-2_a-1}\right]$$

Setting the derivative equal to zero and dividing both sides of the equation by $a^{-1}r$ yields

$$e^{-(y/a)^r}(r-1)\left(\frac{y}{a}\right)^{r-2}a^{-1} - e^{-(y/a)^r}r\left(\frac{y}{a}\right)^{2r-2}a^{-1} = 0$$

$$e^{-(y/a)^r}(r-1)\left(\frac{y}{a}\right)^{r-2}a^{-1} = e^{-(y/a)^r}r\left(\frac{y}{a}\right)^{2r-2}a^{-1}$$

Multiplying both sides by $(y/a)^2e^{(y/a)^r}a$ yields:

$$(r-1)\left(\frac{y}{a}\right)^r = r\left(\frac{y}{a}\right)^{2r}$$

$$\frac{(r-1)}{r} = \left(\frac{y}{a}\right)^r$$

$$\left[\frac{(r-1)}{r}\right]^{1/r} = \frac{y}{a} \tag{11.85}$$

$$a\left[\frac{(r-1)}{r}\right]^{1/r} = y$$

which is the mode.

If the position parameter, $y_A$, has a value other than zero, the expression for the mode is

$$\text{Mode} = a\left[\frac{(r-1)}{r}\right]^{1/r} + y_A \tag{11.86}$$

Thus, when $a = 2$, $r = 3$, and $y_A = 10$, the mode is

$$2\left[\frac{(3-1)}{3}\right]^{1/3} + 10 = 11.747$$

As $r$ approaches infinity, $1/r$, $2/r$, $3/r$, and $4/r$, in Equations 11.81 through 11.84, all approach zero and, consequently, the gamma terms all approach unity. In turn, this means that $E(y) = a$, $E(y^2) = a^2$, $E(y^3) = a^3$, and $E(y^4) = a^4$. Thus, when $r \to \infty$ and $y_A = 0$:

$$E(y) = a$$

$$V(y) = a^2 - a^2 = 0$$

$$\alpha_3 = \frac{(a^3 - 3a^3 + 2a^3)}{(0)^{3/2}} = \frac{0}{0}$$

$$\alpha_4 = \frac{(a^4 - 4a^4 + 6a^4 - 3a^4)}{(0)^2} = \frac{0}{0}$$

$$\text{Mode} = a\left[\frac{\infty}{\infty}\right]^{1/\infty} = a(1)^0 = a$$

Under these circumstances the distribution has no variance and is centered on $Y = a$. In short, the curve of the distribution forms a spike at $a$. If $y_A$ has not equaled zero, the spike would have been at $y_A + a$.

In recent years the Weibull has become the most widely used distribution for modeling diameter distributions and distributions of basal area by diameter classes (Bailey and Dell, 1973; Bailey, 1974; Smalley and Bailey, 1974a, b; Dell et al., 1979; Feduccia et al., 1979; Hyink and Moser, 1979; 1983; Avery and Burkhart, 1983; Clutter et al., 1983). When $r < 1$, the distribution has a reversed-*J* shape and can be used to model all-aged or all-sized stands. When $r > 1$ the distribution is mound shaped and is often an appropriate model for even-aged stands such as plantations. The fact that the distribution is flexible and can be made to be skewed either positively or negatively and can be either lepto- or platykurtic makes it particularly valuable for describing diameter distributions in mathematical terms. In addition, the Weibull's cdf has been used to model mortality in growth and yield studies (Shifley et al., 1982). Industrial engineers, including forest engineers, use the Weibull in the context of reliability theory (Meyer, 1970; Miller and Freund, 1977). It even has been used by geophysicists to predict the tensile strength–volume relationship of snow (Sommerfeld, 1974).

# 12 Sampling Theory

## 12.1 INTRODUCTION

It was brought out in Chapter 1 that sampling is used to arrive at estimates of the characteristics of the populations of interest and that this was done because the cost in money, time, and/or effort to make complete enumerations or evaluations is usually excessive. This implies that such estimates can be made, which, in turn, implies that relationships exist between the populations and the samples drawn from them. Sampling theory is concerned with these interrelationships.

## 12.2 SAMPLING DESIGN

In order for samples to be useful they must be representative of the population being sampled. The methods used to ensure representativeness are included in the field of *sampling design*. When the sampling is being done within the context of research, the end result is usually the evaluation of a hypothesis (e.g., the application of nitrogen fertilizer increases yield) and the sampling methods are included in the field of *statistical* or *experimental* design. When the sampling is being used to arrive at estimates of the quantity of some item (e.g., volume of standing timber on a certain tract) the sampling methods are included in the field of *inventory* design. If the sampling is being used to determine if some industrial operation (e.g., sawing 1-in. boards out of a log or cant) is functioning as it should, the sampling methods are included in the field of *quality control*. Regardless of the ultimate purpose of the sample, the sample should be representative of the population or incorrect inferences will be made relative to that population.

## 12.3 SAMPLING, SAMPLING UNITS, AND SAMPLING FRAMES

Sampling units are the units chosen for selection in the sampling process. An individual sampling unit may consist of a single element of the universe or it may be made up of several such elements. When more than one element is involved, the sampling unit is known as a *cluster*. Both single and multielement sampling units are common. In either case, the universe of interest must be rigidly defined so that confusion does not arise when the sampling process begins. For example, if one is interested in the volume of pine sawtimber on a certain ownership and a timber cruise is to be made so that the volume can be estimated, what is the universe of interest? What is an individual element of that universe? Is it an individual pine tree? Or is it a patch of ground? Both could be used in the cruising process, but the sampling methods would be considerably different.

Once a decision has been made regarding the nature of the sampling unit, a *listing* or a *mapping* of *all the available sampling units* is made. This list or map is known as the *sampling frame*. The sampling is made from this frame. It would seem that the sampling frame and the universe would be the same. This may or may not be the case. The universe is made up of individual elements, whereas the sampling frame is made up of sampling units that may contain more than one element.

Furthermore, for various reasons some of the elements of the universe may not be available for sampling and thus would be excluded from the frame. For example, if a telephone survey is to be made of the inhabitants of a certain town, only those inhabitants having telephones could be reached. The ones without telephones are inhabitants of the town and, thus, elements of the universe but, since they are not available for interviewing, they cannot be included in the frame. This can lead to biased estimates, and, consequently, most theoretical work in sampling assumes that the frame and the universe are coincident. One must remember, however, that in practice this often is not the case and that, when coincidence is not present, that fact should be recognized when interpretations are made of the sample-based data.

The sample consists of a certain number of sampling units drawn from the frame. The *sample size*, $n$, is the *number of sampling units* drawn in this process. The way they are drawn constitutes the *sampling design*. There are many designs that can be used, each with advantages and disadvantages insofar as statistical theory and actual execution are concerned.

## 12.4 RANDOM SAMPLING AND RANDOM NUMBERS

In the discussion of probability in Chapter 7 the term *blind draw* was used repeatedly. It implied that the probability of selection was constant for all events in the sample space (e.g., one was just as likely to draw female number 3 as male number 1 from the box of puppies or that no one of the six faces of a die was more likely to appear than any of the others). This, technically speaking, is *random sampling with constant probability.*

In some situations the probability of selection is not constant. *Provided that the probability of selection in known for each sampling unit in the frame and the sampling is done in such a manner that this probability is observed*, the sampling is still random. For example, assume a situation where a tree is to be chosen for measurement if a randomly chosen point falls inside a circle centered on the tree. If all trees had equal-sized circles, a randomly chosen point would be just as likely to fall inside the circle of one tree as it would to fall inside any other circle. This would be random sampling with equal probability. However, if the sizes of the circles varied, the probabilities would no longer be constant. A tree with a small circle would be less likely to be chosen than one with a large circle. This would be *random sampling with variable probability.* There are a number of sampling designs used in forestry that make use of varying probabilities. Usually they are quite efficient and, consequently, desirable.

Technically speaking, random sampling occurs when the probability of obtaining a certain sample is the product of the probabilities of occurrence of the individuals included in the sample:

$$P(y_1, y_2, \ldots, y_n) = P(y_1) * P(y_2) * \cdots * P(y_n) \tag{12.1}$$

A problem in random sampling is the procedure to be used to ensure randomness. A number of ways of doing this have been devised, such as drawing slips of paper from a hat or drawing marbles from a bowl, but the one using *random numbers* is the one commonly used in statistical operations.

A set of random numbers (or pseudo-random numbers) is a tabular arrangement of the ten digits (0, 1, 2, …, 9) which are in random order. See Figure 12.1 and Table A.13 in the Appendix. The frequency distribution of these ten digits in the table, which can extend over many pages, is rectangular, which means that the probabilities of occurrence of the several digits are constant. Furthermore, if one were to pair adjacent digits, the frequency distribution of the pairs also would be rectangular. Theoretically, the frequency distributions arising from all possible sized groupings would be rectangular. This, of course, is impossible for any finite table but when the table is large, e.g., Tippett (1959), one probably could use groups as large as five or six and the not seriously invalidate the assumption of constant probability. With smaller tables the practicable maximum

(XVIII) Random Sample Numbers

| 1 2 3 4 | 5 6 7 8 | 9 10 11 12 | 13 14 15 16 | 17 18 19 20 | 21 22 23 24 | 25 26 27 28 | 29 30 31 32 |
|---|---|---|---|---|---|---|---|
| 6926 | 6841 | 4032 | 4025 | 9833 | 1514 | 2955 | 1764 |
| 4396 | 6952 | 9432 | 8935 | 2900 | 5640 | 2024 | 0022 |
| 9897 | 1918 | 7330 | 4389 | 1172 | 9372 | 0338 | 3534 |
| 7971 | 0005 | 9635 | 0284 | 4681 | 5835 | 2125 | 2276 |
| 0708 | 8720 | 8330 | 3547 | 5469 | 0274 | 9689 | 9542 |
| 0788 | 5662 | 2505 | 4538 | 7724 | 2790 | 7140 | 7736 |
| 3022 | 8212 | 0627 | 3817 | 4169 | 0324 | 0691 | 3151 |
| 6139 | 4767 | 7495 | 2463 | 5666 | 3904 | 4133 | 1607 |
| 8362 | 0592 | 8557 | 6498 | 8123 | 3227 | 2187 | 4169 |
| 0037 | 7052 | 1868 | 4960 | 6094 | 3140 | 5074 | 0365 |
| 1426 | 8963 | 9355 | 6556 | 7959 | 6826 | 9065 | 6001 |
| 9434 | 2162 | 5312 | 3837 | 7374 | 9385 | 5820 | 5405 |
| 6902 | 6370 | 4174 | 6856 | 0784 | 7070 | 1091 | 1750 |
| 5664 | 1297 | 8435 | 3568 | 2703 | 7405 | 3649 | 5165 |
| 3739 | 6252 | 6610 | 1895 | 6986 | 1877 | 7971 | 5683 |
| 0648 | 6603 | I 3840 | 5091 | 1960 | 8340 | 6950 | 6826 |
| 0412 | 9671 | 5677 | 9403 | 6879 | 4711 | 2555 | 0994 |
| 3869 | 0242 | 3601 | 6670 | 3329 | 6267 | 2455 | 1289 |
| 7455 | 9279 | 5691 | 0975 | 8569 | 5735 | 7742 | 5108 |
| 3492 | 0743 | 2336 | 8269 | 2375 | 9325 | 4486 | 2769 |
| 1559 | 7356 | 6096 | 3325 | 9386 | 1130 | 5833 | 8990 |
| 3196 | 8721 | 6487 | 6260 | 6280 | 8669 | 0623 | 0423 |
| 3877 | 7201 | 8336 | 4831 | 9621 | 8034 | 3918 | 4636 |
| 2212 | 4715 | 4087 | 4660 | 7476 | 2437 | 7963 | 1894 |
| 2301 | 1103 | 8953 | 6227 | 1612 | 9529 | 5218 | 8799 |
| 6339 | 4972 | 9172 | 0120 | 7356 | 7814 | 5451 | 4741 |
| 4019 | 3113 | 9937 | 2588 | 9626 | 8241 | 7300 | 0797 |
| 2738 | 1091 | 6068 | 9615 | 3524 | 4512 | 2093 | 2169 |
| 6584 | 6800 | 8963 | 5544 | 0158 | 5742 | 2013 | 1115 |
| 5784 | 0386 | 0399 | 7480 | 0824 | 4486 | 0294 | 2278 |
| 0107 | 8400 | 0186 | 6154 | 0285 | 3516 | 7354 | 5337 |
| 4113 | 9440 | 4896 | 7167 | 6892 | 2309 | 6175 | 0729 |
| 7925 | 8050 | 9962 | 8730 | 1695 | 2418 | 6185 | 2695 |
| 3896 | 5092 | 2633 | 0325 | 7299 | 1536 | 9004 | 0849 |
| 1949 | 4754 | 5976 | 2587 | 9616 | 1336 | 7689 | 7785 |
| 5108 | 4137 | 2906 | 5244 | 7986 | 3488 | 4263 | 2087 |
| 3300 | 5475 | 1435 | 1165 | 5176 | 2450 | 3972 | 7724 |
| 6354 | 0621 | 3508 | 1350 | 2475 | 8345 | 2012 | 8624 |
| 3344 | 0490 | 4634 | 7022 | 5663 | 1353 | 3175 | 7880 |
| 1405 | 8777 | 6581 | 3960 | 9691 | 4935 | 3611 | 3323 |
| 2362 | 0940 | 1947 | 8003 | 3988 | 6973 | 2905 | 8851 |
| 7789 | 1643 | 8306 | 4174 | 1865 | 7991 | 1724 | 0937 |
| 2454 | 6605 | 3716 | 3022 | 8666 | 0015 | 2884 | 3944 |
| 8473 | 4583 | 1162 | 9973 | 9658 | 0214 | 2359 | 6522 |
| 1772 | 1082 | 4945 | 2143 | 8250 | 2741 | 2044 | 1869 |
| 3507 | 3956 | 5104 | 4558 | 3388 | 8146 | 2718 | 8927 |
| 2719 | 9473 | 4072 | 9721 | 7505 | 0788 | 5440 | 9793 |
| 8100 | 0726 | 0018 | 0692 | 1774 | 4598 | 8198 | 2478 |
| 4049 | 8678 | 8251 | 7077 | 6876 | 9631 | 3738 | 0723 |
| 1856 | 3172 | 3203 | 1438 | 6045 | 7867 | 9996 | 2848 |

**FIGURE 12.1** Random number table showing the randomly located entry point (I) at column 10 and row 16 and the path taken within the table. The table is from Tippett (1959) and is the 18th page of a 26-page table. The circled numbers are those of the students included in the sample. (From Tippett, L., *Random Sampling Numbers,* Cambridge University Press, London, 1959. With permission.)

size of the groups is correspondingly smaller. With a one-page table, it probably would be best to use groups of no more than two digits.

If one wished to make a random selection of, say, 10 students out of a class of 125, the procedure would be as follows. The universe is the set of 125 students. Since only 1 student would be chosen per draw, the sampling unit would be removed from the frame since it consists only of the sampling units available for sampling. Each student would be assigned a number corresponding to the position of the student's name on the roll (in the frame). Since 125 students are in the class (assuming all to be present), the identification numbers will have to contain three digits: 001, 002, 003, …, 125.

These numbers correspond to the numbers in the random number table when those numbers are in groups of three digits.

Prior to using the table one must *randomly* choose an entry point and have a plan of movement within the table from that entry point. One can move in any direction, up, down, to the right, to the left, or diagonally. One also should include in the plan of movement what should be done when an edge is reached. If a plan of this sort is not in place prior to the drawing, one might move knowingly in a direction so as to guarantee the selection of a specific sampling unit.

If we assume a multipage table, such as Tippett's (1959), the first step leading to the location of the entry point involves the random election of a page. To do this the page numbers are entered separately on slips of paper which are then placed in a container and thoroughly stirred up. A blind draw is made from the container for the page. In this way the table in Figure 12.1 was chosen. The second and third steps are similar except that they are used to identify the column and row. In the case of the student problem (see Figure 12.1), the entry point turned out to be page 18, column 10, and row 16.

Usually the entry point locates the first digit in the group. Thus, in the student problem, the first group is the number 840. This number does not fall in the range of numbers in the frame, 001 to 125, and consequently is passed by. The sampler then moves to the next group, following the predetermined plan of movement. In Figure 12.1, this movement is downward until the bottom is reached, at which time a shift to the right of three digits will be made, and then the movement is upward. When a number is encountered that lies in the desired range, that individual is included in the sample. Thus, the 10 students chosen to be in the sample are 012, 018, 032, 066, 072, 087, 096, 104, and 116.

The preceding procedure is appropriate when the sampling frame is of reasonable size. When the frame is very large, the procedures may have to be modified but fundamentally they are similar to that described above.

It should be brought out that many computer routines are available that will generate random numbers. Even some of the more powerful handheld calculators have this capability. In some cases these generators are general in nature and can be used in a multiplicity of contexts. In others, however, the programs are designed to provide random number sampling tools for specific sampling designs.

## 12.5 SAMPLING FRAMES AND INFINITE POPULATIONS

As can be seen from the preceding discussion, each sampling unit in the sampling frame is identifiable and can be singled out for evaluation. For this to be possible, the sampling frame must be of finite size and must be made up of distinct units. How then does one sample a population that is infinitely large?

This situation occurs whenever sampling is done along a continuum — a line, a surface, or within a multidimensional space. One way this problem can be overcome is to superimpose a grid, or the equivalent of a grid, over the continuum and to consider only the points at the grid intersections. For example, a tract of land has a surface that is made up of an infinite number of points. Each point possesses a multitude of characteristics, any one of which could be of interest: elevation above sea level, soil type, vegetation cover, etc. Thus, if one wished to determine the proportion of the tract in one of these categories, one could superimpose a grid over a detailed map of the area and randomly locate a sample of grid intersections which would then be evaluated with respect to the characteristic of interest either on the map or in the field. Another possibility when sampling an area is to establish a point near the center of the area and to use randomly chosen directions and distances from the centerpoint to locate the sampling point. This still would result in a finite frame since the directions would be established in terms of some unit of angular measure, such as the degree or grad, while the distances would be laid off in terms of some linear unit such as the foot, link, etc. There is, however, a major flaw with this procedure. As one moves away from the central point, the spacing of the points increases and, consequently, all points do not represent the same amount of ground area. This could introduce a bias into the sampling operation. In any case, conversion of a continuous population to a finite discrete sampling frame is common practice in forest sampling.

When a grid is used to define a sampling frame, the sample would consist of randomly chosen grid intersections. To do this one would use random numbers to obtain coordinates. For example, if one desires to locate a sample of 150 points on the map in Figure 12.2, in order to determine the proportion in forest, one would draw 150 numbers in the range 01 to 70, using sampling with replacement, to obtain the $x$-coordinates. Then, 150 $y$-coordinate values would be obtained in the range from 01 to 99. These would be paired with the $x$-coordinates in the order of draw. If the sampling is to be with replacement, the repetition of a specific pair of coordinates would be of no significance. If the sampling is to be without replacement, the repetition of a pair of coordinates would call for an additional draw. This could be done by drawing both an extra $x$- and an extra $y$-coordinate or by keeping one, either $x$ or $y$, and drawing a new companion coordinate. This has been done in Figure 12.2. The 150 points so drawn are shown as dots.

## 12.6 THE SAMPLE SIZE AND THE SAMPLING FRACTION

It has already been brought out that the *sample size*, usually symbolized by $n$, is the number of sampling units included in the sample. The ratio of the sample size to the total size of the *sampling frame*, $n/N$, is the *sampling fraction*. The sampling fraction is sometimes referred to as the *sampling intensity*.

It is common practice among foresters to refer to the timber cruise intensity as the *percent cruise. This may or may not be equivalent to the sampling fraction or intensity.* The percent cruise refers to the percentage of the tract area that actually was included inside sample plots that were cruised. Thus, percent cruise and sampling intensity are the same if the frame is made up of sample plots that do not overlap and are so shaped that no gaps exist between plots. This rarely is the case. Consequently, one must use the two terms with care. They usually do not mean the same thing. For example, assume that the 150 points located on the map in Figure 12.2 were used as centers of $^1/_5$ acre plots. The total area cruised in such a case would be 150(0.2) = 30.0 acres. The total area in the tract is 4,557 acres. Consequently, the percent cruise is (30/4557)100 = 0.658%. On the other hand, there are, by count, 4988 grid intersections inside the tract boundaries and the sampling fraction is 150/4998 = 0.03007 or 3.007%.

## 12.7 SAMPLING DISTRIBUTIONS

Let us assume a universe that is made up of elements possessing a number of variable characteristics. One of these characteristics is the variable of interest. The set of variates of this variable forms the sample space. Each of the variates is converted to a numerical value of the random variable and the set of these numerical values forms the range space or population. This population possesses a number of parameters — its mean or expected value, its variance, its standard deviation, its range, its mode, its median, etc. Parameters provide concise information about the population and sampling is used to obtain estimates of them.

Assume that sampling from this universe is to be carried out. Further assume that the sampling units are made up of single elements of the universe and, consequently, the sampling frame and universe are coincident. If a sample of size $n$ (any specific size) is drawn from the frame using a specified sampling procedure (e.g., sampling with replacement or sampling without replacement), a set of sample statistics, corresponding to the population parameters, can be obtained. Thus, there is a sample mean, a sample variance, etc. If a second sample of the same size is then drawn from the frame, following the same statistical procedure, a second set of statistics would be obtained. If this process of obtaining samples of the specified size, using the specified procedure, is continued

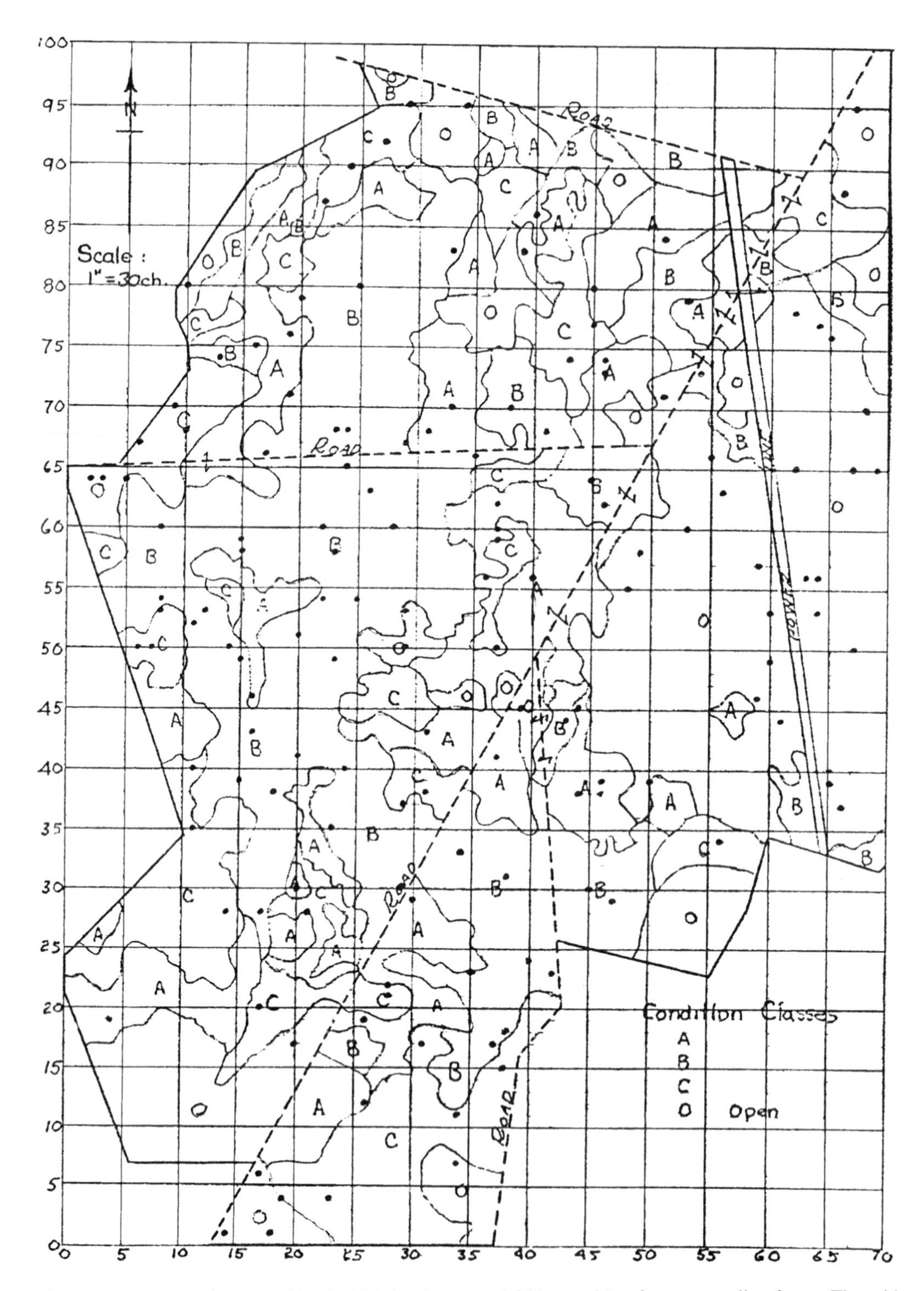


**FIGURE 12.2** Map of a tract of land which has been overlaid by a grid to form a sampling frame. The grid intersections form the sampling units. The positions of the 150 randomly chosen points are shown by heavy dots.

**TABLE 12.1**
**The Set of All Possible Samples of Size 2 Drawn from a Parent Population Containing the Values 3, 5, 9, 12, 13, and 15, with the Parameters $\mu_Y = 9.5$ and $\sigma_Y^2 = 18.5833$**

| Sampling with Replacement | | | | | | Sampling without Replacement | | | | |
|---|---|---|---|---|---|---|---|---|---|---|
| **The Samples** | | | | | | | | | | |
| (3,3) | (3,5) | (3,9) | (3,12) | (3,13) | (3,15) | (3,5) | (3,9) | (3,12) | (3,13) | (3,15) |
| (5,3) | (5,5) | (5,9) | (5,12) | (5,13) | (5,15) | | (5,9) | (5,12) | (5,13) | (5,15) |
| (9,3) | (9,5) | (9,9) | (9,12) | (9,13) | (9,15) | | | (9,12) | (9,13) | (9,15) |
| (12,3) | (12,5) | (12,9) | (12,12) | (12,13) | (12,15) | | | | (12,13) | (12,15) |
| (13,3) | (13,5) | (13,9) | (13,12) | (13,13) | (13,15) | | | | | (13,15) |
| (15,3) | (15,5) | (15,9) | (15,12) | (15,13) | (15,15) | | | | | |
| **Sampling Distributions of Means** | | | | | | | | | | |
| 3.0 | 4.0 | 6.0 | 7.5 | 8.0 | 9.0 | 4.0 | 6.0 | 7.5 | 8.0 | 9.0 |
| 4.0 | 5.0 | 7.0 | 8.5 | 9.0 | 10.0 | | 7.0 | 8.5 | 9.0 | 10.0 |
| 6.0 | 7.0 | 9.0 | 10.5 | 11.0 | 12.0 | | | 10.5 | 11.0 | 12.0 |
| 7.5 | 8.5 | 10.5 | 12.0 | 12.5 | 13.5 | | | | 12.5 | 13.5 |
| 8.0 | 9.0 | 11.0 | 12.5 | 13.0 | 14.0 | | | | | 14.0 |
| 9.0 | 10.0 | 12.0 | 13.5 | 14.0 | 15.0 | | | | | |
| $N = 36$ | $\mu_{\bar{y}} = 9.5$ | $\sigma_{\bar{y}}^2 = 9.2917$ | | | | $N = 15$ | $\mu_{\bar{y}} = 9.5$ | $\sigma_{\bar{y}}^2 = 7.433$ | | |
| **Sampling Distributions of Uncorrected Variances** | | | | | | | | | | |
| 0.00 | 1.00 | 9.00 | 20.25 | 25.00 | 36.00 | 1.00 | 9.00 | 20.25 | 25.00 | 36.00 |
| 1.00 | 0.00 | 4.00 | 12.25 | 16.00 | 25.00 | | 4.00 | 12.25 | 16.00 | 25.00 |
| 9.00 | 4.00 | 0.00 | 2.25 | 4.00 | 9.00 | | | 2.25 | 4.00 | 9.00 |
| 20.25 | 12.25 | 2.25 | 0.00 | 0.25 | 2.25 | | | | 0.25 | 2.25 |
| 25.00 | 16.00 | 4.00 | 0.25 | 0.00 | 1.00 | | | | | 1.00 |
| 36.00 | 25.00 | 9.00 | 2.25 | 1.00 | 0.00 | | | | | |
| $\mu_{s^2} = 9.2917$ | $\sigma_{s^2}^2 = 111.9045$ | | | | | $\mu_{s^2} = 11.15$ | $\sigma_{s^2}^2 = 113.565$ | | | |
| **Sampling Distributions of Corrected Variances** | | | | | | | | | | |
| 0.00 | 2.00 | 18.00 | 40.50 | 50.00 | 72.00 | 1.67 | 15.00 | 33.75 | 41.67 | 60.00 |
| 2.00 | 0.00 | 8.00 | 24.50 | 32.00 | 50.00 | | 6.67 | 20.42 | 26.67 | 41.67 |
| 18.00 | 8.00 | 0.00 | 4.50 | 8.00 | 18.00 | | | 3.75 | 6.67 | 15.00 |
| 40.50 | 24.50 | 4.50 | 0.00 | 0.50 | 4.50 | | | | 0.42 | 3.75 |
| 50.00 | 32.00 | 8.00 | 0.50 | 0.00 | 2.00 | | | | | 1.67 |
| 72.00 | 50.00 | 18.00 | 4.50 | 2.00 | 0.00 | | | | | |
| $\mu_{s^2} = 18.5833$ | $\sigma_{s^2}^2 = 447.6181$ | | | | | $\mu_{s^2} = 18.5833$ | $\sigma_{s^2}^2 = 315.4496$ | | | |

until all possible samples have been obtained, there will be a set of sample means, a set of sample variances, a set of sample standard deviations, etc., as is shown in Table 12.1. Each of these sets is a *sampling distribution.* Thus, in Table 12.1 there are shown two sampling distributions of means — two sampling distributions of uncorrected variances, and two sampling distributions of corrected variances. The reason for listing two of each is to show the difference caused by the method of sampling. The matter of correction of the variances is discussed in Section 12.12.

Each sampling distribution is a complete set of variates of some function of a random variable. Consequently, it is a range space or population and possesses its own parameters. Thus, in Table 12.1, when the sampling was with replacement

$$\mu_{\bar{y}} = E(\bar{y}) = 9.5$$

$$\sigma^2_{\bar{y}} = V(\bar{y}) = 9.2917$$

$$\mu_{s^2} \text{ (uncorrected)} = 9.2917$$

$$\sigma^2_{s^2} \text{ (uncorrected)} = 111.9045$$

$$\mu_{s^2} \text{ (corrected)} = 18.5833$$

$$\sigma^2_{s^2} \text{ (corrected)} = 447.6181$$

and when the sampling was without replacement

$$\mu_{\bar{y}} = E(\bar{y}) = 9.5$$

$$\sigma^2_{\bar{y}} = V(\bar{y}) = 7.4333$$

$$\mu_{s^2} \text{ (uncorrected)} = 11.15$$

$$\sigma^2_{s^2} \text{ (uncorrected)} = 113.565$$

$$\mu_{s^2} \text{ (corrected)} = 18.5833$$

$$\sigma^2_{s^2} \text{ (corrected)} = 315.4496$$

If the sample size is changed, a different set of sampling distributions would be generated, each with its own set of parameters. Thus, a given parent population can give rise to a large number of different sampling distributions. Consequently, one must identify a given sampling distribution not only in terms of the statistic but also in terms of the size of the sample.

## 12.8 STANDARD ERRORS

The standard deviations of sampling distributions are referred to as *standard errors*. Thus, the standard deviation of a sampling distribution of means is known as the *standard error of the mean* and is symbolized as $\sigma_{\bar{y}}$; the standard deviation of a sampling distribution of variances is known as the *standard error of the variance*, symbolized as $\sigma_s$; etc.

## 12.9 BIAS

Each variate in a sampling distribution is an *estimate* of its equivalent population parameter. For example, a sample mean is an estimate of the population mean:

$$\bar{y} \rightarrow \mu_Y$$

and the sample variance is an estimate of the population variance

$$s_Y^2 \rightarrow \sigma_Y^2$$

The arrows indicate estimation. *When the mean of the sampling distribution of a statistic is equal to its equivalent parameter, it is said that the estimator is unbiased.* In other words, if an estimator, *on the average, equals that being estimated, it is unbiased. If not, it is biased.* Bias can creep into many estimators from outside sources such as faulty instrumentation or technique, but some estimators are inherently biased. This will be explored in some detail in the following sections.

## 12.10 PRECISION

A rule of thumb that can be used to arrive at precision levels for a statistic and its standard error is as follows (Sokol and Rohlf, 1969). Divide the standard error of the statistic by 3, note the number of decimal places to the first nonzero digit of the quotient, give the statistic that number of decimal places, and, finally, give the standard error of the statistic one more decimal place than was given the statistic itself. For example, if the statistic is the sample mean, $\bar{y}$, and it is computed to be 10.1362589, and the standard error of the mean, $s_{\bar{y}}$, is computed to be 2.6295623,

$$\frac{s_{\bar{y}}}{3} = \frac{2.6295623}{3} = 0.8765208$$

The first nonzero digit in the quotient is in the first position to the right of the decimal point. Consequently, the mean would be written with one decimal place to the right of the decimal point,

$$\bar{y} = 10.1$$

and the standard error of the mean would be written with two decimal places to the right of the mean,

$$s_{\bar{y}} = 2.63$$

## 12.11 SAMPLING DISTRIBUTION OF MEANS

The sections headed "Means" in Table 12.1 contain sampling distributions of means of samples of size $n = 2$ drawn from the specified parent population using first sampling with replacement and then sampling without replacement. The parent population is finite with $N = 6$. When the sampling is with replacement, the size of the sampling distribution is

$$M = N^n = 6^2 = 36$$

whereas when the sampling is without replacement the size of the sampling distribution is

$$M = \binom{N}{n} = \frac{N!}{n!(N-n)!} = \frac{6!}{2!(6-2)!} = \frac{6*5}{2*1} = 15$$

Obviously, if the parent population is infinitely large, the sampling distribution, regardless of sample size or method of sampling, is also infinitely large.

A definite set of relationships exists between the parameters of the sampling distribution of means and the parameters of the parent population. They are

1. $E(\bar{y}) = \mu_{\bar{Y}} = \mu_Y$ (12.2)

The proof, which is applicable whether the sampling is with or without replacement, provided that the difference in the meaning of $M$ is recognized:

$$\mu_Y = \left(\frac{1}{N}\right)\sum_{i=1}^{N} y_i$$

and

$$\mu_{\bar{y}} = \left(\frac{1}{M}\right)\sum_{i=1}^{M} \bar{y}_i$$

$$= \left(\frac{1}{M}\right)\sum_{i=1}^{M}\left[\left(\frac{1}{n}\right)\sum_{j=1}^{n} y_j\right]_i$$

$$= \left(\frac{1}{M}\right)\sum_{i=1}^{M}\sum_{j=1}^{n} y_{ij}$$

Each $Y = y_i$ value appears $Mn/N$ times. Since $M$ = number of samples and $n$ = sample size, $Mn$ = the total number of appearances of $Y$ values, and, since the frequency of occurrence is constant across all the $Y$ values, $Mn/N$ = the frequency of occurrence of an individual $Y$ value. Then

$$\sum_{i=1}^{M}\sum_{j=1}^{n} y_{ij} = \sum_{i=1}^{N}\left(\frac{Mn}{N}\right) y_i = \left(\frac{Mn}{N}\right)\sum_{i=1}^{N} y_i$$

and

$$\mu_{\bar{y}} = \left(\frac{1}{Mn}\right)\sum_{i=1}^{M}\sum_{j=1}^{n} y_{ij} = \frac{1}{Mn}\left(\frac{Mn}{N}\right)\sum_{i=1}^{N} y_i$$

$$= \left(\frac{1}{N}\right)\sum_{i=1}^{N} y_i = \mu_Y$$

For example, from Table 12.1, when the sampling is with replacement:

$$\mu_{\bar{y}} = \left[\frac{1}{(6^2)(2)}\right]\left[\frac{(6^2)(2)}{6}\right](57)$$

$$= \left(\frac{1}{6}\right)(57) = 9.5 = \mu_Y$$

and, when the sampling is without replacement:

$$\mu_{\bar{y}} = \left[\frac{1}{\binom{6}{2}(2)}\right]\left[\frac{\binom{6}{2}(2)}{6}\right](57)$$

$$= \left(\frac{1}{6}\right)(57) = 9.5 = \mu_Y$$

Thus, the sample mean is an unbiased estimator of the population mean.

2. When the sampling is with replacement:

$$V(\bar{y}) = \sigma_{\bar{y}}^2 = \frac{\sigma_Y^2}{n} \tag{12.3}$$

The proof:

$$\sigma_{\bar{y}}^2 = E[\bar{y} - E(\bar{y})]^2 = E[\bar{y} - \mu_Y]^2 = E\left(\left(\frac{\sum_{i=1}^{n} y_i}{n}\right) - \mu_y\right)^2$$

Multiply $\mu_Y$ by $n/n$. Then,

$$\sigma_{\bar{y}}^2 = \frac{1}{n^2}\left(\left(\sum_{i=1}^{n} y_i\right) - n\mu_Y\right)^2$$

$$= \frac{1}{n^2} E[(y_1 - \mu_Y) + (y_2 - \mu_Y) + (y_3 + \mu_Y) + \cdots + (y_n - \mu_Y)]^2$$

$$= \frac{1}{n^2} E[(y_1 - \mu_Y)^2 + (y_2 - \mu_Y)^2 + (y_3 + \mu_Y)^2 + \cdots + (y_n - \mu_Y)^2$$

$$+ 2(y_1 - \mu_Y)(y_2 - \mu_Y) + 2(y_1 + \mu_Y)(y_3 - \mu_Y) + \cdots$$

$$+ 2(y_1 - \mu_Y)(y_n - \mu_Y) + \cdots + 2(y_{n-1} - \mu_Y)(y_n - \mu_Y)]$$

$$= \frac{1}{n^2}\{E(y_1 - \mu_Y)^2 + E(y_2 - \mu_Y)^2 + E(y_3 - \mu_Y)^2 + \cdots + E(y_n - \mu_Y)^2$$

$$+ 2E[(y_1 - \mu_Y)(y_2 - \mu_Y)] + 2E[(y_1 - \mu_Y)(y_3 - \mu_Y)] + \cdots$$

$$+ 2E[(y_1 - \mu_Y)(y_n - \mu_Y)] + \cdots + 2E[(y_{n-1} - \mu_Y)(y_n - \mu_Y)]\}$$

$$= \frac{1}{n^2}\{\sigma_Y^2 + \sigma_Y^2 + \sigma_Y^2 + \cdots + \sigma_Y^2 + 2E[(y_1 - \mu_Y)(y_2 - \mu_Y)]$$

$$+ 2E[(y_1 - \mu_Y)(y_3 - \mu_Y)] + \cdots + 2E[(y_1 - \mu_Y)(y_n - \mu_Y)]$$

$$+ \cdots + 2E[(y_{n-1} - \mu_Y)(y_n - \mu_Y)]\}$$

$$= \frac{1}{n^2}\left[n\sigma_Y^2 + \sum_{i=1}^{n(n-1)} \text{cross-products}\right] \tag{12.4}$$

Consider any individual cross-product. Let $i$ be the subscript of $y$ in the first term and $j$ in the second term.

$$E[(y_1 - \mu_Y)(y_j - \mu_Y)] = (y_i - \mu_Y)E(y_j - \mu_Y)$$

since, once the first draw is made, $y_i$ is known and thus becomes a constant. Then

$$\begin{aligned}(y_i - \mu_Y)E(y_j - \mu_Y) &= (y_i - \mu_Y)[E(y_j) - \mu_Y] \\ &= (y_i - \mu_Y)(\mu_Y - \mu_Y) \\ &= (y_i - \mu_Y)(0) = 0\end{aligned}$$

Thus, the sum of the cross-product is equal to zero and

$$\sigma_{\bar{y}}^2 = \frac{1}{n^2}(n\sigma_Y^2) = \frac{\sigma_Y^2}{n}$$

This can be demonstrated using the data from Table 12.1. Consider the sampling distribution of means which was obtained using sampling with replacement:

$$\sigma_Y^2 = 18.5833$$

$$\sigma_{\bar{y}}^2 = 9.2917$$

$$\frac{\sigma_Y^2}{n} = \frac{18.5833}{2} = 9.29165 = \sigma_{\bar{y}}^2$$

3. When the sampling is without replacement,

$$V(\bar{y}) = \sigma_{\bar{y}}^2 = \frac{\sigma_Y^2}{n}\left(\frac{N-n}{N-1}\right) \tag{12.5}$$

The proof follows that in item 2 up to Equation 12.4:

$$\sigma_{\bar{y}}^2 = \frac{1}{n^2}\left[n\sigma_Y^2 + \sum_{i=1}^{n(n-1)} \text{cross-products}\right]$$

and

$$E[(y_i - \mu_Y)(y_j - \mu_Y)] = (y_i - \mu_Y)[E(y_j) - \mu_Y]$$

Now, however, $E(y_j)$ must be corrected for the fact that $y_i$ has been drawn from the range space and not replaced. To do this it will be necessary to first compute the population total, then subtract $y_i$ from that total, and finally divide the difference by $N - 1$ to arrive at the corrected mean. Thus,

$$(y_1 - \mu_Y)[E(y_j) - \mu_Y] = (y_i - \mu_Y)\left[\left(\frac{N\mu_Y - y_i)}{N-1}\right) - \mu_Y\right]$$

$$= (y_i - \mu_Y)\left[\frac{N\mu_Y - y_i - (N-1)\mu_Y}{N-1}\right]$$

$$= (y_i - \mu_Y)\left(\frac{N\mu_Y - y_i - N\mu_Y + \mu_Y}{N-1}\right)$$

$$= (y_i - \mu_Y)\left(\frac{-y_i + \mu_Y}{N-1}\right)$$

$$= -\left[\frac{(y_i - \mu_Y)(y_i - \mu_Y)}{N-1}\right]$$

$$= -\frac{(y_i - \mu_Y)^2}{N-1}$$

By definition, $E(y_i - \mu_Y)^2 = \mu_Y^2$. Thus,

$$(y_i - \mu_Y)[E(y_j) - \mu_Y] = -\frac{\sigma_Y^2}{N-1}$$

There are $n(n - 1)$ cross-products. To obtain the total across all cross-products, the expected value must be multiplied by $n(n - 1)$:

$$E\left[\sum_{i=1}^{n(n-1)} \text{cross-products}\right] = \frac{-n(n-1)\sigma_Y^2}{N-1}$$

Substituting this back into Equation 12.3,

$$\sigma_{\bar{Y}}^2 = \frac{1}{n^2}\left[n\sigma_Y^2 - \frac{n(n-1)\sigma_Y^2}{N-1}\right]$$

$$= \frac{\sigma_Y^2}{n} - \frac{(n-1)\sigma_Y^2}{n(N-1)}$$

$$= \frac{\sigma_Y^2}{n}\left[1 - \frac{(n-1)}{(N-1)}\right]$$

$$= \frac{\sigma_Y^2}{n}\left(\frac{N-1-n+1}{N-1}\right)$$

$$= \frac{\sigma_Y^2}{n}\left(\frac{N-n}{N-1}\right)$$

This can be demonstrated using the sampling distribution of means, obtained by sampling without replacement, which is shown in Table 12.1.

$$\sigma_Y^2 = 18.5833$$

$$\sigma_Y^2 = 7.4333$$

$$\frac{\sigma_Y^2}{n}\left(\frac{N-n}{N-1}\right) = \frac{18.5833}{2}\left(\frac{6-2}{6-1}\right) = 7.4333 = \sigma_{\bar{y}}^2$$

As was brought out in Section 10.4, the

$$\left(\frac{N-n}{N-1}\right)$$

term is a *finite population correction factor* (fpc) which has to be applied to a variance if the sampling is without replacement.

The actual probability distribution function of a sampling distribution of means is dependent on that of the parent population. However, sampling distributions of means possess a characteristic described by the *central limit theorem*. This theorem states that sampling distributions of means, if based on infinitely large parent populations, tend to be distributed as the normal and that this tendency increases in strength as the sample size, $n$, increases. Stated more formally, if $\bar{y}_i$ is one value in the sampling distribution, which is based on samples of size $n$ drawn from an infinitely large population with a finite mean, $\mu_Y$, and finite variance, $\mu_Y^2$, then

$$z_i = \frac{\bar{y}_i - \mu_Y}{\frac{\sigma_Y}{n}} \tag{12.6}$$

is the corresponding value of a random variable whose pdf approaches the standard normal as $n$ approaches infinity. If the distribution of the parent population is normal, the distribution of the sampling distribution is also normal regardless of the magnitude of $n$, but even if the parent population is not normally distributed the sampling distribution will approach normality as the sample size increases. The proof of this is beyond the scope of this presentation. Meyer (1970) offers a partial proof of the situation described above. Actually, the central limit theorem is broader than has been stated, applying to sums as well as means and also to situations where the individual values making up the sums or means are drawn from different parent populations with different means, variances, and distributions.

Even when sample sizes are quite small, the approximation to the normal is surprisingly close. Demonstrating this using complete sets of data is practically impossible because of the sheer size of the range spaces. However, sampling can provide an idea of the rate of closure. Schumacher and Chapman (1948) drew 550 random samples of size 5 from a uniform population of the digits 1, 2, 3, 4, 5, 6, 7, 8, 9, and 0, using sampling with replacement, and computed the sample means. The distribution of these means is shown in histogram form in Figure 12.3A. Superimposed on the histogram is the equivalent normal curve. As can be seen, the histogram indicates a mound-shaped distribution which follows the normal curve fairly closely. One must realize that the histogram is based on a sample of only 550 out of a sampling distribution of means containing $10^5$ values. If more than 550 means had been used, the picture would have been clearer. In Figure 12.3B is a similar histogram, but in this case the samples were of size 10. As can be seen, the approximation to the normal is closer than it is in Figure 12.3A. Again, one must realize that the histogram is based only on 550 means, not the $10^{10}$ in the sampling distribution.

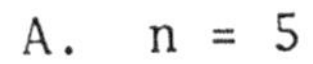


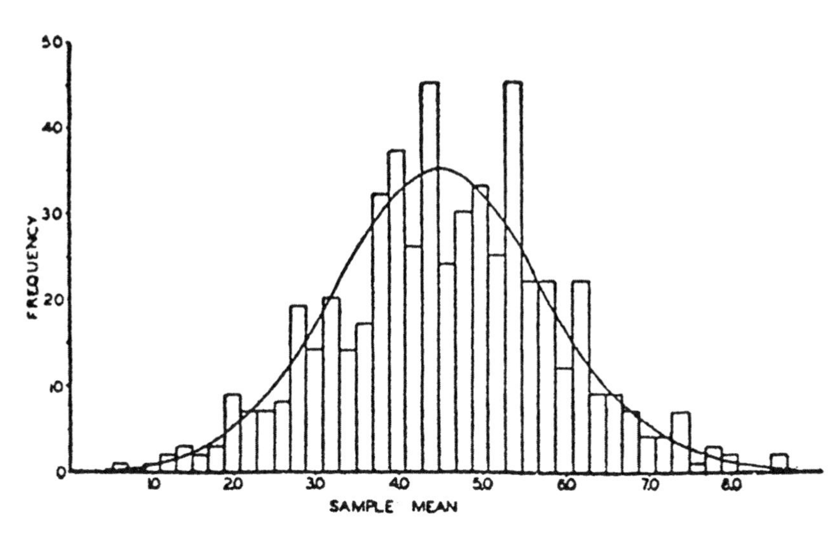


B. n = 10

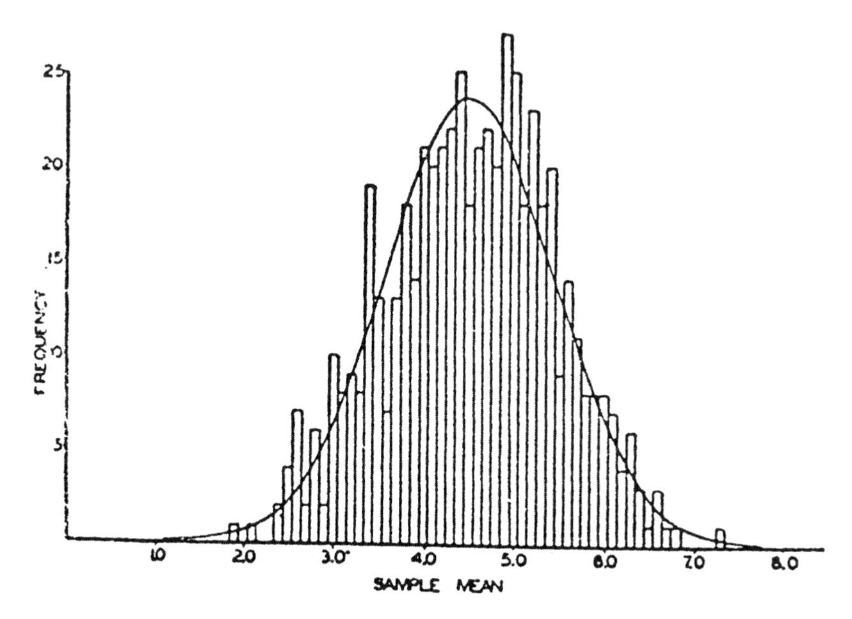


**FIGURE 12.3** Histograms showing the distributions of 500 sample means based on samples of size $n$ drawn from a parent population containing the variates 1, 2, 3, 4, 5, 6, 7, 8, 9, and 0 using sampling with replacement. (From Schumacher, F.X. and Chapman, R.A., Duke University School of Forestry Bulletin 7, 1948. With permission.)

When the parent population is finite and the sampling is without replacement, the central limit theorem is not applicable. However, it has been found that even under these conditions the sampling distribution tends to the normal and the approximation is good for large sample sizes ($n > 30$), provided that the sampling fraction is small.

Regardless of the type of sampling and the type of distribution of parent population, it usually is assumed that, if the sample size exceeds 30, the normal can be used as a model for the sampling distribution of means. However, even when $n$ is less than 30, the approximation in the middle of the distribution usually is quite good. The divergence from the normal is greatest in the tails. Consequently, if probability statements regarding the sample means are needed and they refer to ranges near the middle of the distributions, the normal usually can be used without incurring serious error.

## 12.12 SAMPLING DISTRIBUTION OF VARIANCES

Table 12.1 contains two sampling distributions of variances, one obtained using sampling with replacement and one using sampling without replacement. The means or expected values of these distributions are, respectively, 9.2917 and 11.15. The variance of the parent population, $\sigma_Y^2$, is 18.5833. Obviously, the expected values do not match the parameter and consequently *the sample variances are biased estimators* of the population variance:

$$s_y^2 = \left[\frac{\sum_{i=1}^{n}(y_i - \bar{y})^2}{n}\right] \xrightarrow{\text{BIASED}} \sigma_Y^2 \tag{12.7}$$

The source of the bias lies with the fact that a *residual* (e.g., $y_i - \bar{y}$) is not the same thing as an *error* (e.g., $y_i - \mu_Y$). A sample variance is the mean of a set of squared residuals while the population variance is the mean of the complete set of squared errors. An individual error is an unbiased estimator of the population variance because, on the average, across the entire population, the squared errors will equal the population variance. This must occur since, by definition, the population variance is the expected value of the squared errors. If errors could be obtained from a sample and a sample variance computed using those errors, it would be unbiased.

$$s_\varepsilon^2 = \left[\frac{\sum_{i=1}^{n}(y_i - \mu_Y)^2}{n}\right] \xrightarrow{\text{UNBIASED}} \sigma_Y^2 \tag{12.8}$$

Unfortunately, this requires a knowledge of $\mu_Y$, which is rarely known. Consequently, the best that can be done is to compute the sample variance using residuals.

Reference is made to Section 5.2.4, in which the concept of "least squares" is discussed. At that time it was brought out that the sum of squared differences between a set of values and some constant would be at a minimum when the constant is the arithmetic mean of the set (Figure 5.2). If a sample is drawn from a population and the differences between the values in the sample and the *population* mean are computed, squared, and summed, that sum will be *larger* than if the *sample* mean had been used. This always will be the case except when the population and sample means are equal. Thus, on the average, the expected value of the sample variance is always less than the population variance.

It has been found, however, that this bias can be removed by dividing the sum of squared residuals by $n - 1$ rather than by $n$. Thus,

$$s_Y^2 = \left[\frac{\sum_{i=1}^{n}(y_i - \bar{y})^2}{n-1}\right] \xrightarrow{\text{UNBIASED}} \sigma_Y^2 \tag{12.9}$$

provided that the sampling was with replacement. The proof is as follows:

Assume a random sample of size $n$ drawn from a population whose mean, $\mu_Y$, is unknown. Then, $\bar{y} \xrightarrow{\text{UNBIASED}} \mu_Y$,

$\varepsilon_i = y_i - \mu_Y$, the unknown error of $y_i$,
$\bar{\varepsilon} = \bar{y} - \mu_Y$, the unknown error of $\bar{y}$,
$\mu_Y = y_i - \varepsilon_i$, and $\mu_Y = \bar{y} - \bar{\varepsilon}$

Thus, $y_i - \varepsilon_i = \bar{y} - \bar{\varepsilon}$ and $y_i - \bar{y} = \varepsilon_i - \bar{\varepsilon}$. Then,

$$s_Y^2 = \sum_{i=1}^{n}\frac{(y_i - \bar{y})^2}{n} = \sum_{i=1}^{n}\frac{(\varepsilon_i - \bar{\varepsilon})^2}{n}$$

The expected value of the sum of squared residuals is

$$E\left[\sum_{i=1}^{n}(y_i - \bar{y})^2\right]$$

which can be modified to

$$E\left[\sum_{i=1}^{n}(y_i - \mu_Y + \mu_Y - \bar{y})^2\right] = E\left\{\sum_{i=1}^{n}[(y_i - \mu_Y) - (\bar{y} - \mu_Y)]^2\right\}$$

Then, substituting $\varepsilon_i$ for $(y_i - \mu_Y$ and $\bar{\varepsilon}$ for $(\bar{y} - \mu_Y)$,

$$E\left[\sum_{i=1}^{n}(y_i - \bar{y})^2\right] = E\left[\sum_{i=1}^{n}(\varepsilon_i - \bar{\varepsilon})^2\right]$$

$$= E[(\varepsilon_i^2 + \bar{\varepsilon}^2 - 2\varepsilon_1\bar{\varepsilon}) + (\varepsilon_2^2 + \bar{\varepsilon}^2 - 2\varepsilon_2\bar{\varepsilon}) + \cdots$$

$$+ (\varepsilon_n^2 + \bar{\varepsilon}^2 - 2\varepsilon_n\bar{\varepsilon})$$

$$= E\left[\sum_{i=1}^{n}\varepsilon_i^2 + n\bar{\varepsilon}^2 - 2\bar{\varepsilon}\sum_{i=1}^{n}\varepsilon_i\right]$$

Since

$$\varepsilon_i = y_i - \mu_Y, \quad \sum_{i=1}^{n}\varepsilon_i = \sum_{i=1}^{n}y_i - n\mu_Y, \text{ and } E\left[\sum_{i=1}^{n}(y_i - \bar{y})^2\right] = E\left[\sum_{i=1}^{n}\varepsilon_i^2 + n\bar{\varepsilon}^2 - 2\bar{\varepsilon}\left(\sum_{i=1}^{n}y_i - n\mu_Y\right)\right]$$

Multiplying $\Sigma_{i=1}^{n} y_i$ by $n/n$,

$$E\left[\sum_{i=1}^{n}(y_i-\bar{y})^2\right]=E\left[\sum_{i=1}^{n}\varepsilon_i^2+n\bar{\varepsilon}^2-2\bar{\varepsilon}\left(n\sum_{i=1}^{n}\frac{y_i}{n}-n\mu_Y\right)\right]$$

$$=E\left[\sum_{i=1}^{n}\varepsilon_i^2+n\bar{\varepsilon}^2-2\varepsilon n(\bar{y}-\mu_Y)\right]$$

$$=E\left[\sum_{i=1}^{n}\varepsilon_i^2+n\bar{\varepsilon}^2-2\bar{\varepsilon}^2 n\right]$$

$$=E\left[\sum_{i=1}^{n}\varepsilon_i^2-n\bar{\varepsilon}^2\right] \tag{12.10}$$

$$=E[\varepsilon_i^2+\varepsilon_2^2+\cdots+\varepsilon_n^2-n\bar{\varepsilon}^2]$$

$$=E(\varepsilon_i^2)+E(\varepsilon_2^2)+\cdots+E(\varepsilon_n^2)-nE(\bar{\varepsilon}^2)$$

$$=\sigma_Y^2+\sigma_Y^2+\cdots+\sigma_y^2-n\sigma_{\bar{y}}^2$$

$$=n\sigma_Y^2-n\frac{\sigma_Y^2}{n}$$

Then

$$E\left[\sum_{i=1}^{n}(y_i-\bar{y})^2\right]=\sigma_Y^2(n-1)$$

and

$$\sigma_Y^2=\frac{E\left[\sum_{i=1}^{n}(y_i-\bar{y})^2\right]}{(n-1)}$$

Consequently,

$$s_Y^2=\left[\frac{\sum_{i=1}^{n}(y_i-\bar{y})^2}{n-1}\right]\xrightarrow{\text{UNBIASED}}\sigma_Y^2$$

and thus,

$$E(s_Y^2)=\mu_{s^2}=\sigma_Y^2 \tag{12.11}$$

When the sampling is without replacement there are two sources of bias, one of which is corrected by dividing the sum of squared residuals by $(n - 1)$ rather than $n$, while the second is corrected by using a finite population correction factor.

$$s_Y^2\left(\frac{N-1}{N}\right) = \frac{\sum_{i=1}^{n}(y_i - \bar{y})^2}{n-1}\left(\frac{N-1}{N}\right) \xrightarrow{\text{UNBIASED}} \sigma_Y^2 \tag{12.12}$$

The proof follows that used above up to Equation 12.10

$$E\left[\sum_{i=1}^{n}(y_i - \bar{y})^2\right] = \sigma_Y^2 + \sigma_Y^2 + \cdots + \sigma_Y^2 - n\sigma_Y^2$$

In the case of sampling without replacement,

$$\sigma_{\bar{y}}^2 = \frac{\sigma_Y^2}{n}\left(\frac{N-n}{N-1}\right)$$

(see Equation 12.5). Thus,

$$E\left[\sum_{i=1}^{n}(y_i - \bar{y})^2\right] = n\sigma_Y^2 - n\frac{\sigma_Y^2}{n}\left(\frac{N-n}{N-1}\right)$$

$$= \sigma_Y^2\left[n - \left(\frac{N-n}{N-1}\right)\right]$$

$$= \sigma_Y^2\left(\frac{nN - n - N + n}{N-1}\right)$$

$$= \sigma_Y^2\left(\frac{nN - N}{N-1}\right)$$

$$= \sigma_Y^2\left[N\left(\frac{n-1}{N-1}\right)\right]$$

$$= (n-1)\sigma_Y^2\left(\frac{N}{N-1}\right)$$

and

$$\sigma_Y^2\left(\frac{N}{N-1}\right) = \frac{E\left[\sum_{i=1}^{n}(y_i - \bar{y})^2\right]}{(n-1)}$$

Consequently,

$$s_Y^2 = \frac{\sum_{i=1}^{n}(y_i - \bar{y})^2}{n-1} \xrightarrow{\text{UNBIASED}} \sigma_Y^2\left(\frac{N}{N-1}\right)$$

and

$$s_Y^2\left(\frac{N-1}{N}\right) = \frac{\sum_{i=1}^{n}(y_i - \bar{y})^2}{n-1}\left(\frac{N-1}{N}\right) \xrightarrow{\text{UNBIASED}} \sigma_Y^2$$

Again,

$$E\left[s_Y^2\left(\frac{N-1}{N}\right)\right] = \mu_{s^2} = \sigma_Y^2 \tag{12.13}$$

If the values in the samples in Table 12.1 are used in Equations 12.9 and 12.12 the sampling distributions of variances are modified to those shown under the heading "corrected variances." Thus, when the sampling is with replacement

$$\mu_{s^2} = E[s_{y^2}] = 18.5833 = \sigma_{y^2}$$

and when the sampling is without replacement

$$\mu_{s^2} = E\left[s_Y^2\left(\frac{N-1}{N}\right)\right] = 18.5853 \cong 18.5853 = \sigma_Y^2$$

The difference in the latter case is caused by rounding errors.

It is obvious that confusion can develop with respect to what is meant by the term *sample variance* or the symbol $s_Y^2$. Furthermore, use is made of all these forms. Strictly speaking, the basic formula for the variance in which the divisor is $n$ should never be used when the estimation of a population variance is involved. However, when $n$ is "large" (large is usually associated with $n >$ 30), the correction becomes, for most workers, negligible and can be ignored. Similarly, the finite population correction is dependent on the size of the sampling fraction, $n/N$. When the sampling fraction is small, the correction becomes small and can be ignored. Consequently, the uncorrected form of the sample variance is used as the standard in some statistical references. This is typically the case with those oriented toward opinion surveys where the populations are very large and the sampling fractions are small. When the corrections are ignored, it is said that the investigator is operating under *large sample theory*. When corrections are used, the work is being done under *small sample theory* or *exact sampling theory*.

Cochran (1963) and some other writers in the field of sampling theory have approached the problem of finite population correction in a different manner. They do not apply the fpc to the corrected sample but, instead, apply it to the population variance:

$$s_Y^2 = \frac{\sum_{i=1}^{n}(y_i - \bar{y})^2}{(n-1)} \rightarrow S^2 = \sigma_Y^2\left(\frac{N}{N-1}\right) \tag{12.14}$$

The advantage of this approach is that the formulation and computation of the sample variance is always the same. However, if one encounters the symbol $S^2$ without prior knowledge of its meaning, it can cause problems.

Returning to the correction for bias, let us consider the residuals $(y_i - \bar{y})$, $i = 1, 2, \ldots, n$. Each is a linear function of the $n$ random variables $Y_1, Y_2, \ldots Y_n$. (Note that $Y_1, Y_2, \ldots Y_n$ are identified as random variables. This is done because one is referring to samples in general, not to a specific sample. Consequently, the values of $Y_1, Y_2, \ldots$, etc. are subject to random change from sample to sample.)

$$\begin{aligned}(y_i - \bar{y}) &= y_i - \left(\frac{Y_1}{n} + \frac{Y_2}{n} + \cdots + \frac{Y_i}{n} + \cdots + \frac{Y_n}{n}\right) \\ &= \frac{[-Y_1 - Y_2 - \cdots + y_i(n-1) - \cdots - Y_n]}{n}\end{aligned} \tag{12.15}$$

Furthermore

$$\sum_{i=1}^{n} (y_i - \bar{y}) = 0 \tag{12.16}$$

Consequently, the residuals are not mutually independent, even if the values of the random variable $Y$ are independent. They are constrained in that their sum must equal zero. It is said that they are *linearly dependent*. The number of variables (residuals) in the set, less the number of linear relations or constraints between them, is known as the number of *degrees of freedom* (often symbolized by d.f.). Thus, in the case of a sample of size $n$, there are $n - 1$ degrees of freedom because of the one constraint that the residuals sum to zero. In other words, because of the constraint, $n - 1$ of the variables (residuals) can take on any numerical values but the remaining variable must have a magnitude such that the sum of the residuals equals zero. As will be seen later, other constraints can be imposed with concomitant losses of degrees of freedom. It so happens that the divisor needed in the computation of the sample variance to obtain an unbiased estimator is equal to the degrees of freedom. Because of the great importance of the variance in expressing the precision of sample estimates, the use of the term *degrees of freedom* is common in the statistical literature.

The variance of the sampling distribution of variances is much less readily determined than the mean because it is a function of the distribution of the parent population. Without proof, since the proof is beyond the scope of this book, the variance of the variance is (Lass and Gottlieb, 1971, pp. 220–221):

$$V(s_Y^2) = \sigma_{s^2}^2 = \frac{\mu_4}{n} - \frac{(\sigma_Y^2)^2}{n}\left(\frac{n-3}{n-1}\right) \tag{12.17}$$

where $\mu_4$ is the fourth moment about the mean of the parent population. When $n$ is large, say, $n > 30$, the term

$$\left(\frac{n-3}{n-1}\right)$$

approaches unity and often is disregarded, leading to the approximation

$$V(s_Y^2) \cong \frac{\mu_4 - (\sigma_Y^2)^2}{n} \tag{12.18}$$

In the case of the sampling distribution of corrected variances, based on sampling with replacement, in Table 12.1:

$$\sigma_{s^2}^2 = \frac{549.8958}{2} - \frac{(18.5833)^2}{2}\left(\frac{2-3}{2-1}\right) = 447.6174$$

which agrees, except for rounding error, with the variance obtained using the values making up the sampling distributions.

It should be noted that Equations 12.17 and 12.18 are general expressions but they cannot be used unless the distribution of the parent population is known so that the needed moments about the mean ($\mu_2 = \sigma_Y^2$ and $\mu_4$) can be determined. If the distribution of the parent population is not known, the needed moments must be estimated from a sample using the method of computation described in Section 8.12. Relative frequencies, of course, must be substituted for the probabilities. The sample must be large enough to obtain good estimates of these relative frequencies. Consequently, depending on the situation, sample size could reach into the hundreds. Failure to obtain good estimates of the moments will, in turn, distort the estimate of $V(s_Y^2)$.

The distributional pattern of sampling distributions of variances is variable, being a function of the distribution of the parent population. However, variances are arithmetic means and, consequently, according to the central limit theorem, when the sample size is sufficiently large their distribution will tend to assume the shape of the normal. Usually the convergence to the normal is considered to be sufficient for making probability statements using the normal as the model when $n > 30$. When $n$ is less than 30, the situation is usually not clear and, if probability statements are needed, the only recourse is to use Chebyshev's theorem or the Camp–Meidell extension of that theorem (see Section 13.6).

If, however, the parent population is distributed normally with mean $\mu_Y$ and variance $\sigma_Y^2$, it is known that the sampling distribution of variances will be distributed as chi-square based on $n - 1$ degrees of freedom. Thus, according to Equation 12.9,

$$s_Y^2 = \frac{\sum_{i=1}^{n}(y_i - \bar{y})^2}{(n-1)}$$

Multiplying by $\sigma_Y^2/\sigma_Y^2$ does not change the value of $s_Y^2$.

$$s_Y^2 = \frac{\sigma_Y^2}{\sigma_Y^2}\frac{\sum_{i=1}^{n}(y_i - \bar{y})^2}{(n-1)} = \frac{\sigma_Y^2}{n-1}\sum_{i=1}^{n}\left(\frac{y_i - \bar{y}}{\sigma}\right)^2$$

$$= \frac{\sigma_Y^2}{n-1}\left[\left(\frac{y_1 - \bar{y}}{\sigma_Y}\right)^2 + \left(\frac{y_2 - \bar{Y}}{\sigma_Y}\right)^2 + \cdots + \left(\frac{y_n - \bar{y}}{\sigma_Y}\right)^2\right]$$

The residuals have now been put in standardized form. In the above format the symbols are appropriate for a specific sample. However, the wish is to generalize the situation as was done in the discussion of degrees of freedom. Consequently, $Y_1$ represents the random variable which fills the first position in the sample, $Y_2$ the second, etc. All of these random variables are coincident with $Y$ and consequently have precisely the same distributions. Thus,

$$s_Y^2 = \frac{\sigma_Y^2}{n-1}\left[\left(\frac{Y_i - \bar{Y}}{\sigma_Y}\right)^2 + \left(\frac{Y_2 - \bar{Y}}{\sigma_Y}\right)^2 + \cdots + \left(\frac{Y_n - \bar{Y}}{\sigma_Y}\right)^2\right]$$

$\bar{y}$ also applies to a specific sample. $\bar{Y}$ is the random variable of which $\bar{y}$ is an individual value. Since $Y_1, Y_2, \ldots, Y_n$ are identically distributed as $Y$,

$$E(\bar{Y}) = E\left[\frac{(Y_1 + Y_2 + \cdots + Y_n}{n}\right] = \left[\frac{E(Y_1) + E(Y_2) + \cdots + E(Y_n)}{n}\right]$$

$$= \frac{\mu_Y + \mu_Y + \cdots + \mu_Y}{n} = \frac{n\mu_Y}{n} = \mu_Y$$

Then,

$$s_Y^2 = \frac{\sigma_Y^2}{n-1}\left[\left(\frac{Y_i - \mu_Y}{\sigma_Y}\right)^2 + \left(\frac{Y_2 - \mu_Y}{\sigma_Y}\right)^2 + \cdots + \left(\frac{Y_n - \mu_Y}{\sigma_Y}\right)^2\right] \tag{12.19}$$

Since $Y_1, Y_2, \ldots, Y_n$ all have the distribution $N(\mu_Y, \sigma_Y^2)$ and since $\mu_Y$ is a constant, $(Y_1 - \mu_Y)$, $(Y_2 - \mu_Y)$, …, $(Y_n - \mu_Y)$ all have the distribution $N(0, \sigma_Y^2)$. Division by $\sigma_Y$ puts these random variables into standard form so that they have distribution $N(0, 1)$.

According to Equation 11.34, if $U$ is a continuous random variable with the pdf $f(u)$, and $W = U^2$, the pdf of $W$ is

$$f(w) = \frac{1}{2\sqrt{w}}[f(\sqrt{w}) + f(-\sqrt{w})]$$

Applying this to the problem at hand, let

$$Z_i = \left(\frac{Y_i - \mu_Y}{\sigma_Y}\right), \quad i = 1, 2, \ldots, n$$

The distribution of $Z_i$ is then $N(0, 1)$ and the pdf of

$$W_i = Z_i^2 = \left(\frac{Y_i - \mu_Y}{\sigma_Y}\right)^2$$

is

$$f(w) = \frac{1}{2z}\left(\frac{1}{\sqrt{2\pi}}e^{-z^2/2} + \frac{1}{\sqrt{2\pi}}e^{-z^2/2}\right)$$

$$= \frac{1}{2z}\left(\frac{1}{\sqrt{2\pi}}\right)2e^{-z^2/2} \tag{12.20}$$

$$= \frac{1}{\sqrt{2\pi w}}e^{-w/2}$$

The pdf of the chi-square distribution (Equation 11.45) is

$$f(w) = \begin{cases} \dfrac{1}{2^{k/2}\Gamma\left(\dfrac{k}{2}\right)} w^{(k/2)-1}e^{-w/2}, & w > 0 \\ 0, & \text{elsewhere} \end{cases}$$

By setting $k = 1$

$$f(w) = \frac{1}{2^{1/2}\Gamma\left(\dfrac{1}{2}\right)} w^{-(1/2)}e^{-w/2}$$

and since $\Gamma(1/2) = \sqrt{\pi}$

$$f(w) = \frac{1}{\sqrt{2\pi w}}e^{-w/2} \tag{12.21}$$

Obviously Equations 12.20 and 12.21 are identical and, thus, the distribution of each

$$\left(\frac{Y_i - \mu_Y}{\sigma_Y}\right)^2$$

is chi-square based on one degree of freedom.

Reference is made to Equation 8.46, which states that if $Z$ is the sum of $n$ independent random variables the moment-generating function of $Z$ will be equal to the product of the moment-generating functions of the $n$ independent random variables. Applying this, let

$$W_i = \left(\frac{Y_i - \mu_Y}{\sigma_Y}\right)^2$$

Consequently, all $W_i$ are distributed as $\chi^2(1)$. The mgf for the gamma distribution (Equation 11.29) is

$$M_y(t) = \frac{a^r}{(a-t)^r}$$

When $a = 1/2$ and $r = k/2$, the gamma becomes the chi-square distribution with the mgf

$$M_y(t) = \left[\frac{\frac{1}{2}}{\left(\frac{1}{2}\right) - t}\right]^{k/2} = \left[\frac{1}{2}\left(\frac{2}{1-2t}\right)\right]^{k/2} = (1-2t)^{-k/2} \tag{12.22}$$

Thus, when there is one degree of freedom, $k = 1$, and

$$M_y(t) = (1-2t)^{-1/2} \tag{12.23}$$

Consequently, the mgf for $W_i$ is as Equation 12.23, and the mgf for $\Sigma_{i=1}^{n} W_i$ is

$$M_W(t) = n(1-2t)^{-1/2} \tag{12.24}$$

indicating that $\Sigma_{i=1}^{n} W_i$ is distributed as $\chi^2$ based on $k = n$ degrees of freedom. This is appropriate only when errors, not residuals, are used, since with errors there is no constraint on the freedom of the variable to vary. The algebraic sum of the errors making up the sample does not need to equal zero. When residuals are used, however, the algebraic sum must equal zero and consequently only $n - 1$ of the random variables $Y_i$ are free to vary. The remaining value is therefore determined by the preceding values. This means that only $n - 1$ degrees of freedom are available and the distribution of $\Sigma_{i=1}^{n} W_i$ is distributed as $\chi^2(k)$, where $k = n - 1$. Consequently, when the parent population is normally distributed, sample variances are distributed as $\chi^2(k)$. Thus, Equation 12.19 can be written

$$s_Y^2 = \frac{\sigma_Y^2 \chi^2(k)}{k} \tag{12.25}$$

Furthermore, the fractiles of the $\chi^2(k)$ distribution, through the transformation

$$\chi(k) = s_Y^2\left(\frac{k}{\sigma_Y^2}\right) \tag{12.26}$$

can be used to obtain probabilities of occurrence within the sampling distribution of variances:

$$\begin{aligned} P(s_Y^2 \le s_b^2) &= P\left[s_Y^2 \le \left(\frac{\sigma_Y^2}{k}\right)\chi^2(k)\right] \\ &= P\left[s_Y^2\left(\frac{k}{\sigma_Y^2}\right) \le \chi^2(k)\right] \end{aligned} \tag{12.27}$$

## 12.13 SAMPLING DISTRIBUTION OF STANDARD DEVIATIONS

Standard deviations are functions of their corresponding variances and consequently, according to Section 8.8, are distributed in the same manner as those variances. In other words, the probability of occurrence of a specific variance and the probability of occurrence of its positive square root, the standard deviation, are identical and the methods described in Section 12.11 to obtain probabilities of variances can be used to obtain probabilities of the corresponding standard deviations.

Even though the sample standard deviations are distributed in the same manner as their corresponding variances, the mean or expected value of the standard deviation is *not* equal to the square root of the expected value of the variance. As a result, the individual sample standard deviations are biased estimators of the standard deviation of the parent population. This can be seen if the sampling distributions of variances in Table 12.1 are converted to sampling distributions of standard deviations and their expected values computed. When the sampling was with replacement, $E(s) = 3.4177$, and when the sampling was without replacement, $E(s) = 3.7444$. Neither of these is equal to the square root of $E(s^2)$, which is 4.3108. This bias arises from the fact that the square root of $E(s^2)$ is a *quadratic* mean (see Section 5.7),

$$\sqrt{E(s^2)} = \sqrt{\frac{\sum_{i=1}^{N} s_i^2}{N}} = \sqrt{\sigma_Y^2} = \sigma_Y \tag{12.28}$$

while $E(s)$ is an *arithmetic* mean

$$E(s) = \frac{\sum_{i=1}^{N} s_i}{N} \neq \sqrt{\frac{\sum_{i=1}^{N} s_i^2}{N}} \tag{12.29}$$

Arithmetic means are always smaller than their corresponding quadratic means (see Section 5.7.4) and consequently the bias of the standard deviation is always negative.

When the parent population is normally distributed, the expected value of the standard deviation can be computed as follows:

$$E(s) = \sigma_Y \sqrt{\frac{2}{k}} \left\{ \frac{\left[\frac{(k-1)}{2}\right]!}{\left[\frac{(k-2)}{2}\right]!} \right\} \tag{12.30}$$

The derivation of this expression is as follows. From Equation 12.25,

$$s_Y = \sigma_Y \sqrt{\frac{\chi^2}{k}}$$

Then

$$\frac{s_Y}{\sigma_Y} = \sqrt{\frac{\chi^2}{k}}$$

According to Equation 11.45,

$$f(\chi^2) = \frac{1}{2^{k/2}\Gamma\left(\frac{k}{2}\right)} (\chi^2)^{(k/2)-1} e^{-\chi^2/2}; \quad \chi^2 > 0$$

Let $\chi^w = (\chi^2)^{w/2}$. Then

$$E(\chi^w) = \frac{1}{2^{k/2}\Gamma\left(\frac{k}{2}\right)}\int_0^\infty \chi^w (\chi^2)^{(k/2)-1} e^{-\chi^2/2} d(\chi^2)$$

Since $\chi^w(\chi^2)^{(k/2)-1} = (\chi^2)^{w/2}(\chi^2)^{(k/2)-1} = (\chi^2)^{[(w+k)/2]-1}$,

$$E(\chi^w) = \frac{1}{2^{k/2}\Gamma\left(\frac{k}{2}\right)}\int_0^\infty (\chi^2)^{[(w+k)/2]-1} e^{-X^2/2} d(\chi^2)$$

Then

$$\int_0^\infty (\chi^2)^{[(w+k)/2]-1} e^{-\chi^2/2} d(\chi^2) = \Gamma\frac{\left[\frac{(w+k)}{2}\right]}{\left(\frac{1}{2}\right)^{(w+k)/2}} = \Gamma\left[\frac{(w+k)}{2}\right]2^{(w+k)/2}$$

Then

$$E(\chi^2) = \frac{\Gamma\left[\frac{(w+k)}{2}\right]2^{(w+k)/2}}{2^{k/2}\Gamma\left(\frac{k}{2}\right)} = 2^{w/2}\frac{\Gamma\left[\frac{(w+k)}{2}\right]}{\Gamma\left(\frac{k}{2}\right)} \tag{12.31}$$

Then, when $w = 1$,

$$E(\chi) = \frac{2^{1/2}\Gamma\left[\frac{(1+k)}{2}\right]}{\Gamma\left(\frac{k}{2}\right)} \tag{12.32}$$

According to Equation 11.23,

$$\Gamma(r) = (r-1)!$$

Then

$$\Gamma\left[\frac{(1+k)}{2}\right] = \left\{\left[\frac{(1+k)}{2}\right]-1\right\}! = \left[\frac{(k-1)}{2}\right]! \tag{12.33}$$

and

$$\Gamma\left(\frac{k}{2}\right) = \left[\left(\frac{k}{2}\right)-1\right]! = \left[\frac{(k-2)}{2}\right]! \tag{12.34}$$

Consequently,

$$E(\chi) = \sqrt{2}\left\{\frac{\left[\frac{(k-1)}{2}\right]!}{\left[\frac{(k-2)}{2}\right]!}\right\} \tag{12.35}$$

Then, from Equation 12.25,

$$E(s_Y) = \frac{\sigma_Y}{\sqrt{k}} E(\chi)$$

$$= \sigma\sqrt{\frac{2}{k}}\left\{\frac{\left[\frac{(k-1)}{2}\right]!}{\left[\frac{(k-2)}{2}\right]!}\right\}$$

which is Equation 12.30.

This can be approximated by

$$E(s_Y) \cong \sigma_Y\sqrt{1 - \frac{1}{2k}} \tag{12.36}$$

or

$$E(s_Y) \cong \sigma_Y \tag{12.37}$$

For example, if $k = 6$, according to Equation 12.30,

$$E(s_Y) = 0.9594\sigma_Y$$

while according to Equation 12.36,

$$E(s_Y) \cong 0.9574\sigma_Y$$

As $k \to \infty$, $E(s_Y) \to \sigma_Y$, and, consequently, Equation 12.36 is appropriate when $k$ is large, say, $k > 30$.

When the distribution of the parent population is known but is not normal, the expected value of the standard deviation must be obtained using the methods of Section 8.9. If the distribution of the parent population is not known, there is no general theorem upon which one can base a determination of the expected value.

The variance of the standard deviation can be obtained using Equation 12.38, provided that the parent population is normally distributed.

$$V(s_Y) = \sigma_Y^2 - \frac{2\sigma_Y^2}{k}\left\{\frac{\left[\frac{(k-1)}{2}\right]!}{\left[\frac{(k-2)}{2}\right]!}\right\}^2 \tag{12.38}$$

The derivation of Equation 12.38 is as follows. Returning to Equation 12.31, if $w = 2$,

$$E(\chi^2) = \frac{2\Gamma\left[\frac{(2+k)}{2}\right]}{\Gamma\left(\frac{k}{2}\right)} = \frac{2\left(\frac{k}{2}\right)!}{\left[\frac{(k-2)}{2}\right]!} = k \tag{12.39}$$

Then

$$V(\chi) = E(\chi^2) - [E(\chi)]^2$$

$$= k - \left(\sqrt{2}\left\{\frac{\left[\frac{(k-1)}{2}\right]!}{\left[\frac{(k-2)}{2}\right]!}\right\}\right)^2 \tag{12.40}$$

$$= k - 2\left\{\frac{\left[\frac{(k-1)}{2}\right]!}{\left[\frac{(k-2)}{2}\right]!}\right\}^2$$

Then, since (see Equation 12.25)

$$s_Y = \sigma_Y \sqrt{\frac{\chi^2}{k}}$$

$$= \sigma_Y \frac{\chi}{\sqrt{k}}$$

the variance of $s_Y$ is

$$V(s_Y) = V\left(\sigma_Y \frac{\chi}{\sqrt{k}}\right)$$

Since $\sigma_Y$ and $k$ are constants (see Section 9.9, Item 2),

$$V(s_Y) = \frac{\sigma_Y^2}{k} V(\chi)$$

$$= \frac{\sigma_Y^2}{k}\left[k - 2\left\{\frac{\left[\frac{(k-1)}{2}\right]!}{\left[\frac{(k-2)}{2}\right]!}\right\}^2\right]$$

$$= \sigma_Y^2 - \frac{2\sigma_Y^2}{k}\left\{\frac{\left[\frac{(k-1)}{2}\right]!}{\left[\frac{(k-2)}{2}\right]!}\right\}^2$$

which is Equation 12.38. As in the case of the expected value, an approximation is available.

$$V(s_Y) \cong \frac{\sigma_Y^2}{2k} \tag{12.41}$$

Thus, if $k = 6$, according to Equation 12.38

$$V(s_Y) = 0.0796\sigma_Y^2$$

while according to Equation 12.41, $V(s_Y) = 0.0833\sigma_Y^2$. As $k \to \infty$, $V(s_Y) \to 0$, or in other words, as $k \to \infty$, $s_Y$ varies less and less from $\sigma_Y$.

As in the case of the expected value, if the parent population is known but not normal, the variance of the standard error can be obtained using the methods of Section 8.9. When the distribution of the parent population is not known, there is no known way of arriving at $V(s_Y)$.

## 12.14 SAMPLING DISTRIBUTION OF PROPORTIONS

Assume a universe made up of elements, each of which possesses or does not possess a certain characteristic. For example, the set of eight puppies in Chapter 7, each of which was a male or was not a male; or the set of ten loblolly pines in Chapter 10, each of which was infected with fusiform rust or was not so infected; or the set of grid intersections on the map in Figure 12.2, each of which occurs in a forested area or does not so occur. The variable of interest in each of these cases is of the binomial type. An evaluation of an element in any of these universes will yield a result which is either "yes" or "no" with respect to the variable of interest. These outcomes are converted to the random variable, $Y$, using the function

$$y = \begin{cases} 1, & \text{if "yes"} \\ 0, & \text{if "no"} \end{cases}$$

This produces a range space or parent population that is made up only of zeros and ones. The expected value or mean of this population is

$$E(Y) = \mu_Y = \sum_{i=1}^{N} \frac{y_i}{N}$$

Since $Y$ can only take on values of 0 or 1, $\mu_Y$ is the *proportion* of the parent population possessing the characteristic of interest. Thus,

$$\mu_Y = p \tag{12.42}$$

It should be noted that since $p$ is a proportion, it can assume values only in the range $0 \leq p \leq 1$.

The variance of the parent population is

$$\sigma_Y^2 = \sum_{i=1}^{N} \frac{(y_i - \mu_Y)^2}{N} = \left(\frac{1}{N}\right)\sum_{i=1}^{N}(y_i^2 + \mu_Y^2 - 2y_i\mu_Y)$$

$$= \left(\frac{1}{N}\right)\left[\sum_{i=1}^{N} y_i^2 + N\mu_Y^2 - 2\mu_Y \sum_{i=1}^{N} y_i\right]$$

Since $Y$ can only take on the values 0 and 1, $\Sigma_{i=1}^{N} y_i = N\mu_Y$. Thus,

$$\sigma_Y^2 = \left(\frac{1}{N}\right)\left[\sum_{i=1}^{N} y_i^2 + N\mu_Y^2 - 2N\mu_Y^2\right] = \left(\frac{1}{N}\right)\left[\sum_{i=1}^{N} y_i^2 - N\mu_Y^2\right]$$

Since $Y$ can only take on the values 0 and 1, $Y^2$, likewise, can only take on the values 0 and 1. Therefore, $\Sigma_{i=1}^{N} y_i^2 = \Sigma_{i=1}^{N} y_i = N\mu_Y$, and

$$\sigma_Y^2 = \frac{N\mu_Y - N\mu_Y^2}{N} = p = p^2 = p(1-p)$$

and, since $q = 1 - p$

$$\sigma_Y^2 = pq \tag{12.43}$$

As can be seen, the parent population is distributed as the binomial when $n = 1$ (see Section 10.3). As an example, consider a fleet of ten log trucks, six of which are diesel powered and four of which are gasoline powered. Let the random variable, $Y$, be

$$y = \begin{cases} 1, & \text{if diesel} \\ 0, & \text{if gasoline} \end{cases}$$

Then,

| **Truck (*i*)** | $Y = y_i$ | $y_i - \mu_Y$ | $(y_i - \mu_Y)^2$ |
|---|---|---|---|
| 1 | 1 | 0.4 | 0.16 |
| 2 | 1 | 0.4 | 0.16 |
| 3 | 0 | –0.6 | 0.36 |
| 4 | 1 | 0.4 | 0.16 |
| 5 | 0 | –0.6 | 0.36 |
| 6 | 0 | –0.6 | 0.36 |
| 7 | 1 | 0.4 | 0.16 |
| 8 | 1 | 0.4 | 0.16 |
| 9 | 0 | –0.6 | 0.36 |
| 10 | 1 | 0.4 | 0.16 |

$$\sum_{i=1}^{N} y_i = 6$$

$$\sum_{i=1}^{N} (y_i - \mu_Y) = 0.0$$

$$\sum_{i=1}^{N} (y_i - \mu_Y)^2 = 2.40$$

$$\mu_Y = \frac{6}{10} = 0.6 = p \qquad q = (1-p) = 0.4$$

$$\sigma_Y^2 = \frac{2.40}{10} = 0.24 = pq$$

Now, assume that all possible samples of size $n$ are drawn from a parent population of the type described above. As was the case with the population mean, *each sample mean is a proportion.*

$$\bar{y}_i = \sum_{j=1}^{n} \frac{y_{ij}}{n} = \hat{p}_i \tag{12.44}$$

where $\hat{p}_j$ is the proportion of the sample possessing the characteristic of interest. The "hat" over the $p$ indicates that it is an estimate of the population proportion. Thus, in the case of the ten log trucks, $\hat{p}$ is the proportion of the sample made up of diesels. The complete set of such sample proportions is a *sampling distribution of proportions*. Since each sample proportion is also the sample mean, a sampling distribution of proportion is simply a special case of, and possesses the characteristics of, a sampling distribution of means. Thus, regardless of the method of sampling, the expected value of the mean of the distribution is (see Equation 12.2),

$$E(\hat{p}) = \mu_{\hat{p}} = \mu_Y = p \tag{12.45}$$

If the sampling is with replacement (see Equation 12.3),

$$V(\hat{p}) = \sigma^2_{\hat{p}} = \frac{\sigma^2_Y}{n} = \frac{pq}{n} \tag{12.46}$$

whereas if the sampling is without replacement (see Equation 12.5),

$$V(\hat{p}) = \sigma^2_{\hat{p}} = \frac{\sigma^2_y}{n}\left(\frac{N-n}{N-1}\right) = \frac{pq}{n}\left(\frac{N-n}{N-1}\right) \tag{12.47}$$

Fundamentally, if the sampling is with replacement, the sampling distribution of proportions is distributed *binomially*. If the sampling is without replacement, the distribution is *hypergeometric*. For example, assume that the set of ten log trucks is sampled using sampling with replacement and samples of size $n = 3$. Let $Y$ equal the number of diesel trucks in the sample so that $Y$ can take on the values 0, 1, 2, and 3.

| $Y = y_1$ | $\hat{p}_i = \frac{y_i}{n}$ | $P(\hat{p}_i)$ | $\hat{p}_i - p$ | $(\hat{p}_i - p)^2$ |
|---|---|---|---|---|
| 0 | $\frac{0}{3} = 0.000$ | $\binom{3}{0} * 0.6^0 * 0.4^3 = 0.064$ | –0.600 | 0.360000 |
| 1 | $\frac{1}{3} = 0.333$ | $\binom{3}{1} * 0.6^1 * 0.4^2 = 0.288$ | –0.267 | 0.071111 |
| 2 | $\frac{2}{3} = 0.667$ | $\binom{3}{2} * 0.6^2 * 0.4^1 = 0.432$ | 0.067 | 0.004444 |
| 3 | $\frac{3}{3} = 1.000$ | $\binom{3}{3} * 0.6^3 * 0.4^0 = 0.216$ | 0.400 | 0.160000 |

The expected value or mean in this case is

$$\mu_{\hat{p}} = \sum_{i=1}^{n+1} [\hat{p}_i * P(\hat{p}_i)]$$

$$= 0(0.064) + \left(\frac{1}{3}\right)(0.288) + \left(\frac{2}{3}\right)(0.432) + 1(0.216)$$

$$= 0.600, \quad \text{which equals } p$$

The variance is

$$\sigma^2_{\hat{p}} = \sum_{i-1}^{n+1} [(\hat{p} - p)^2 * p(\hat{p}_i)]$$

$$= 0.360000(0.064) + 0.071111(0.288) + 0.004444(0.432)$$

$$+ 0.160000(0.216)$$

$$= 0.08 = \frac{pq}{n} = \frac{0.6(0.4)}{3}$$

If the sampling had been without replacement, the same sample proportions would have been encountered but the probabilities would be different (see Section 10.4):

| $Y = y_i$ | $\hat{p}_i = \frac{y_i}{n}$ | $\binom{N_s}{y_i}$ | $\binom{N - N_s}{n - y_i}$ | $P(y_i)$ |
|---|---|---|---|---|
| 0 | $\frac{0}{3}$ | $\frac{6!}{0!(6-0)!} = 1$ | $\frac{4!}{3!(4-3)!} = 4$ | $\frac{1(4)}{120} = 0.033$ |
| 1 | $\frac{1}{3}$ | $\frac{6!}{1!(6-1)!} = 6$ | $\frac{4!}{2!(4-2)!} = 6$ | $\frac{6(6)}{120} = 0.300$ |
| 2 | $\frac{2}{3}$ | $\frac{6!}{2!(6-2)!} = 15$ | $\frac{4!}{1!(4-1)!} = 4$ | $\frac{15(4)}{120} = 0.500$ |
| 3 | $\frac{3}{3}$ | $\frac{6!}{3!(6-3)!} = 20$ | $\frac{4!}{0!(4-0)!} = 1$ | $\frac{20(1)}{120} = 0.167$ |

and

$$\binom{N}{n} = \frac{10!}{3!(10-3)!} = 120$$

The expected value or mean is

$$\mu_{\hat{p}} = \sum_{i=1}^{n+1} [\hat{p}_i * P(\hat{p}_i)]$$

$$= 0(0.033) + \left(\frac{1}{3}\right)(0.300) + \left(\frac{2}{3}\right)(0.500) + 1(0.167)$$

$$= 0.600, \quad \text{which equals } p$$

The variance is

$$\sigma^2_{\hat{p}} = \sum_{i=1}^{n+1} [(\hat{p}_i - p)^2 * P(\hat{p}_i)]$$

$$= 0.360000(0.033) + 0.071111(0.300) + 0.004444(0.500)$$

$$+ 0.160000(0.167)$$

$$= 0.06215 = \frac{pq}{n}\left(\frac{N-n}{N-1}\right) = \frac{0.6(0.4)}{3}\left(\frac{10-3}{10-1}\right) = 0.06222$$

The difference is due to rounding errors.

If the sampling distribution of proportions is based on sampling with replacement or, if the sampling is without replacement, the range space or parent population is infinitely large, its distribution will be binomial. As $n$ increases, the distribution will tend toward the normal (see Section 10.3) and when $n \to \infty$, the sampling distribution will converge to the normal. Thus, while the central limit theorem is not applicable, since the random variable is discrete, the normal can be used as a model for the sampling distribution of proportions, provided $n$ is sufficiently large. As in the case of sampling distributions of means, this is usually assumed to be the case when $n > 30$.

When the sampling is without replacement, the picture is less clear. The sampling distributions arising from this type of sampling are hypergeometrically distributed. If $N$, the size of the parent population, is large and $n$ is small, the sampling distribution will tend toward a binomial distribution which is based on the same sample size. The approximation of the hypergeometric by the binomial is quite good if the sampling function, $n/N$, is 0.1 or less (see Section 10.4). Note that the tendency is toward the binomial — not the normal. However, according to Equation 13.68, if the sample size, $n$, and the magnitude of $p$ are such that

$$np(1-p)\left(\frac{N-n}{N}\right) > 9$$

the normal can be used as an approximation of the hypergeometric distribution. If the sampling fraction is 0.1 or less and if $n$ is greater than 30, it usually is assumed that the normal can be used as the model.

## 12.15 SAMPLING DISTRIBUTION OF COVARIANCES

The covariance was defined in Section 9.8 and further discussed in Sections 9.7 and 9.10. By definition the covariance is the mean or expected cross-product of the errors of the two primary random variables making up a two-dimensional random variable. Thus, from Equation 9.28, the covariance of $(X, Y)$ is

$$\text{COV}(XY) = \sigma_{XY} = E\{[X - E(X)][Y - E(Y)]\}$$
$$= E(\varepsilon_X * \varepsilon_Y)$$

As was brought out in Item 6 of Section 9.7, when every variate in population $X$ is free to pair with every variate in population $Y$, the covariance of $(X,Y)$ is equal to zero, but when the variates of the two primary random variables are linked together in some manner the covariance takes on a magnitude other than zero. Thus, covariance exists only when links exist between the individual variates of the primary random variables, and the following discussion is based on the assumption that linkages exist.

If a sample of size $n$ is drawn from the range space of $(X,Y)$, the mean cross-product of the sample, the *sample covariance*, is

$$s_{XY} = \sum_{i=1}^{n} \frac{(x_i - \bar{x})(y_i - \bar{y})}{n} \tag{12.48}$$

If all possible samples of size $n$ are drawn from the range space of $(X,Y)$, using either sampling with or without replacement, the set of sample covariances so generated constitutes a *sampling distribution of covariances*. Regardless of the method of sampling, the mean or expected value of this sampling distribution will not be equal to the population covariance, and consequently the sample covariances are biased.

$$s_{XY} = \sum_{i=1}^{n} \frac{(x_i - \bar{x})(y_i - \bar{y})}{n} \xrightarrow{\text{BIASED}} \sigma_{XY} \tag{12.49}$$

The source of this bias is the same as that causing the bias of sample variances. The sample covariances are based on residuals from sample means rather than errors from the population means. Because the sample means occupy the least squares positions within their respective sample sets, the residuals, on the average, are smaller than their equivalent errors. Consequently, sample covariances are biased downward.

Correction of this bias is the same as that for sample variances, dividing the sum of cross-products by the degrees of freedom, $n - 1$, rather than by the sample size. Thus, when the two-dimensional random variable is infinitely large or, if finite, the sampling is with replacement,

$$s_{XY} = \sum_{i=1}^{n} \frac{(x_i - \bar{x})(y_i - \bar{y})}{(n-1)} \xrightarrow{\text{UNBIASED}} \sigma_{XY} \tag{12.50}$$

The derivation of this correction follows that of the correction of the sample variances. If a random sample of size $n$ is drawn from $(X,Y)$, then

$$\varepsilon_{Xi} = x_i - \mu_X \qquad \bar{\varepsilon}_X = \bar{x} - \mu_X$$
$$\varepsilon_{Yi} = y_i - \mu_Y \qquad \bar{\varepsilon}_Y = \bar{y} - \mu_Y$$

$$\mu_X = x_i - \varepsilon_{Xi} \qquad \mu_X = \bar{x} - \bar{\varepsilon}_X$$
$$\mu_Y = y_i - \varepsilon_{Yi} \qquad \mu_Y = \bar{y} - \bar{\varepsilon}_Y$$

$$x_i - \varepsilon_{Xi} = \bar{x} - \bar{\varepsilon}_X \qquad y_i - \varepsilon_{Yi} = \bar{y} - \bar{\varepsilon}_X$$
$$x_i - \bar{x} = \varepsilon_{Xi} - \bar{\varepsilon}_X \qquad y_i - \bar{y} = \varepsilon_{Yi} - \bar{\varepsilon}_Y$$

$$s_{XY} = \sum_{i=1}^{n} \frac{(x_i - \bar{x})(y_i - \bar{y})}{n}$$

$$E\left[\sum_{i=1}^{n}(x_i - \bar{x})(y_i - \bar{y})\right] = E\left[\sum_{i=1}^{n}(x_i - \mu_X + \mu_X - \bar{x})(y_i - \mu_Y + \mu_Y - \bar{y})\right]$$

$$= E\left\{\sum_{i=1}^{n}[(x_i - \mu_X) - (\bar{x} - \mu_X)][(y_i - \mu_Y) - (\bar{y} - \mu_Y)]\right\}$$

$$= E\left\{\sum_{i=1}^{n}[(\varepsilon_{Xi} - \bar{\varepsilon}_X)(\varepsilon_{Yi} - \bar{\varepsilon}_Y)]\right\}$$

$$= E\left[\sum_{i=1}^{n}(\varepsilon_{Xi}\varepsilon_{Yi} - \varepsilon_{Xi}\bar{\varepsilon}_Y - \varepsilon_{Yi}\bar{\varepsilon}_X + \bar{\varepsilon}_X\bar{\varepsilon}_Y)\right]$$

$$= E\left(\sum_{i=1}^{n}\varepsilon_{Xi}\varepsilon_{Yi} - \sum_{i=1}^{n}\varepsilon_{Xi}\bar{\varepsilon}_Y - \sum_{i=1}^{n}\varepsilon_{Yi}\bar{\varepsilon}_X + \sum_{i=1}^{n}\bar{\varepsilon}_X\bar{\varepsilon}_Y\right)$$

$$= E\left[\sum_{i=1}^{n}\varepsilon_{Xi}\varepsilon_{Yi} - \bar{\varepsilon}_Y\sum_{i=1}^{n}(x_i - \mu_X) - \bar{\varepsilon}_X\sum_{i=1}^{n}(y_i - \mu_X) + n\bar{\varepsilon}_X\bar{\varepsilon}_Y\right]$$

$$= E\left[\sum_{i=1}^{n}\varepsilon_{Xi}\varepsilon_{Yi} - \bar{\varepsilon}_Y\sum_{i=1}^{n}x_i - n\mu_X - \bar{\varepsilon}_X\sum_{i=1}^{n}y_i - n\mu_Y + n\bar{\varepsilon}_X\bar{\varepsilon}_Y\right]$$

$$= E\left[\frac{n\sum_{i=1}^{n}\varepsilon_{Xi}\varepsilon_{Yi}}{n} - \bar{\varepsilon}_Y\left(\frac{n\sum_{i=1}^{n}x_i}{n} - n\mu_X\right) - \bar{\varepsilon}_X\left(\frac{n\sum_{i=1}^{n}y_i}{n} - n\mu_Y\right) + n\bar{\varepsilon}_X\bar{\varepsilon}_Y\right]$$

$$= E\left[n\sum_{i=1}^{n}\varepsilon_{Xi}\varepsilon_{Yi} - n\bar{\varepsilon}_Y(\bar{x} - \mu_X) - n\bar{\varepsilon}_X(\bar{y} - \mu_Y) + n\bar{\varepsilon}_X\bar{\varepsilon}_Y\right]$$

$$= E\left(\frac{n\sum_{i=1}^{n}\varepsilon_{Xi}\varepsilon_{Yi}}{n} - n\bar{\varepsilon}_X\bar{\varepsilon}_Y - n\bar{\varepsilon}_X\bar{\varepsilon}_Y + n\bar{\varepsilon}_X\bar{\varepsilon}_Y\right)$$

$$= E\left(\frac{n\sum_{i=1}^{n}\varepsilon_{Xi}\varepsilon_{Yi}}{n} - n\bar{\varepsilon}_X\bar{\varepsilon}_Y\right)$$

$$= nE\frac{\left(\sum_{i=1}^{n} n\varepsilon_{Xi}\varepsilon_{Yi}\right)}{n} - nE(\bar{\varepsilon}_X\bar{\varepsilon}_Y) \tag{12.51}$$

$$= n\text{COV}(XY) - n\text{COV}(\bar{x}\,\bar{y})$$

where $\text{COV}(\bar{x}\bar{y})$ is the *covariance of the sample means*:

$$\text{COV}(\bar{x}\,\bar{y}) = \sigma_{\bar{x}\bar{y}} = \frac{\sigma_{XY}}{n} \tag{12.52}$$

The derivation of Equation 12.52 is as follows:

$$\text{COV}(\bar{x}\,\bar{y}) = E\{[\bar{x} - E(\bar{x})][\bar{y} - E(\bar{y})]\}$$

$$= E[(\bar{x} - \mu_x)(\bar{y} - \mu_y)]$$

$$= E\left[\left(\frac{\sum_{i=1}^{n} x_i}{n} - \mu_X\right)\left(\frac{\sum_{i=1}^{n} y_i}{n} - \mu_Y\right)\right]$$

$$= E\left[\left(\frac{\sum_{i=1}^{n} x_i}{n} - \frac{n\mu_X}{n}\right)\left(\frac{\sum_{i=1}^{n} y_i}{n} - \frac{n\mu_Y}{n}\right)\right]$$

$$= \frac{1}{n^2}E\{[(x_i - \mu_X) + (x_2 - \mu_X) + \cdots + (x_n - \mu_X)]$$

$$*[(y_1 - \mu_Y) + (y_2 - \mu_Y) + \cdots + (y_n - \mu_Y)]\}$$

Since each of the individual pairings is the result of a single draw, the only cross-products to be considered are those where the subscripts of $x$ and $y$ are the same. Thus,

$$\text{COV}(\bar{x}\,\bar{y}) = \frac{1}{n^2}E[(x_1 - \mu_X)(y_1 - \mu_Y) + (x_2 - \mu_X)(y_2 - \mu_Y) + \cdots$$

$$+ (x_n - \mu_X)(y_n - \mu_Y)]$$

$$= \frac{1}{n^2}\{E[(x_1 - \mu_X)(y_1 - \mu_Y)] + E[(x_2 - \mu_X)(y_2 - \mu_Y)] + \cdots$$

$$+ E[(x_n - \mu_X)(y_n - \mu_Y)]\}$$

Since $E[(x - \mu_X)(y - \mu_Y)] = \sigma_{XY}$,

$$\text{COV}(\bar{x}\,\bar{y}) = \frac{1}{n^2}[(\sigma_{XY} + \sigma_{XY}) + \cdots + \sigma_{XY}]$$

$$= \frac{1}{n^2}(n\sigma_{XY})$$

$$= \frac{\sigma_{XY}}{n}$$

which is Equation 12.52. Then, returning to Equation 12.51,

$$E\left[\sum_{i=1}^{n}(x_i - \bar{x})(y_i - \bar{y})\right] = n\text{COV}(XY) - n\text{COV}(\bar{x}\,\bar{y})$$

$$= n\text{COV}(XY) - \frac{n\text{COV}(XY)}{n}$$

$$= \text{COV}(XY)(n-1)$$

Then

$$\text{COV}(XY) = \frac{E\left[\sum_{i=1}^{n}(x_i - \bar{x})(y_i - \bar{y})\right]}{(n-1)} \tag{12.53}$$

Thus,

$$s_{XY} = \sum_{i=1}^{n}\frac{(x_i - \bar{x})(y_i - \bar{y})}{(n-1)} \xrightarrow{\text{UNBIASED}} \sigma_{XY}$$

which is Equation 12.50.

When the two-dimensional random variable $(X,Y)$ is finite and the sampling is without replacement, it is necessary to include a finite population correction factor. Without proof, since the derivation is too lengthy for inclusion here, the covariance of the sample means is

$$\text{COV}(\bar{x}\,\bar{y}) = \sigma_{\bar{x}\bar{y}} = \frac{\sigma_{XY}}{n}\left(\frac{N-n}{N-1}\right) \tag{12.54}$$

Introducing this into Equation 12.51,

$$E\left[\sum_{i=1}^{n}(x_i - \bar{x})(y_i - \bar{y})\right] = n\text{COV}(XY) - n\text{COV}(\bar{x}\,\bar{y})$$

$$= n\sigma_{XY} - \frac{n\sigma_{XY}}{n}\left(\frac{N-n}{N-1}\right)$$

$$= \sigma_{XY}\left[n - \left(\frac{N-n}{N-1}\right)\right]$$

$$= \sigma_{XY}\left(\frac{nN - n - N + n}{N - 1}\right)$$

$$= \sigma_{XY}\left(\frac{nN - N}{N - 1}\right)$$

$$= \sigma_{XY}\left[N\left(\frac{n-1}{N-1}\right)\right]$$

$$= (n-1)\sigma_{XY}\left(\frac{N}{N-1}\right)$$

Then

$$\sigma_{XY}\left(\frac{N}{N-1}\right) = \frac{E\left[\sum_{i=1}^{n}(x_i - \bar{x})(y_i - \bar{y})\right]}{(n-1)}$$

and

$$s_{XY} = \frac{\sum_{i=1}^{n}(x_i - \bar{x})(y_i - \bar{y})}{(n-1)} \xrightarrow{\text{UNBIASED}} \sigma_{XY}\left(\frac{N}{N-1}\right)$$

and

$$s_{XY}\left(\frac{N-1}{N}\right) = \frac{\sum_{i=1}^{n}(x_i - \bar{x})(y_i - \bar{y})}{n-1}\left(\frac{N-1}{N}\right) \xrightarrow{\text{UNBIASED}} \sigma_{XY} \tag{12.55}$$

## 12.16 SAMPLING DISTRIBUTION OF DIFFERENCES

Assume two independent random variables, $X$ and $Y$, with the parameters ($\mu_X$, $\sigma_X^2$) and ($\mu_Y$, $\sigma_Y^2$). The distributions of the two random variables are not specified. Now assume that every variate in the range space of $X$ is paired with every variate in the range space of $Y$. The resulting set of pairs forms the range space of the two-dimensional random variable ($X$, $Y$). Let $D = X - Y$ be a function of ($X$, $Y$). Then, according to Item 3 in Section 9.7

$$\mu_D = \mu_X - \mu_Y \tag{12.56}$$

Remember that subtraction is the equivalent of adding a negative number. Furthermore, according to Item 4 in Section 9.9,

$$\sigma_D^2 = \sigma_X^2 + \sigma_Y^2 - 2\sigma_{XY} \tag{12.57}$$

As was brought out in Item 6 of Section 9.7, if all the variates of one of the primary variables of a two-dimensional random variable are free to pair with all the variates of the second primary variable, the primary variables are independent and the covariance is equal to zero. Thus, in the above case,

$$\sigma_D^2 = \sigma_X^2 + \sigma_Y^2 \tag{12.58}$$

If all the variates of the two primary variables are *not* free to pair but are linked so that the drawing of a variate from the range space of one automatically yields a corresponding variate of the other, the size of the range space of $D$ is reduced and its parameters are as shown in Equations 12.56 and 12.57.

Now, let all possible samples of size $n$ be drawn from $(X, Y)$. The mean difference for an individual sample is

$$\bar{d}_i = \frac{\sum_{i=1}^{n}(x-y)_{ij}}{n} = \frac{\sum_{i=1}^{n} d_{ij}}{n} = \left(\frac{\sum_{i=1}^{n} x_j}{n} - \frac{\sum_{i=1}^{n} y_j}{n}\right)_i = (\bar{x} - \bar{y})_i \tag{12.59}$$

The set of all such mean differences forms a *sampling distribution of mean differences*. This is actually a sampling distribution of means based on the random variable $D$. The mean or expected value of the sampling distribution is

$$\mu_{\bar{d}} = \mu_D = \mu_X - \mu_Y = \mu_{\bar{x}} - \mu_{\bar{y}} \tag{12.60}$$

The variance of the sampling distribution depends on how the sampling is carried out. If the sampling permits any variate of $X$ to be paired with any variate of $Y$ and if the sampling is with replacement or population $D$ is infinitely large, the variance is

$$\sigma_{\bar{d}}^2 = \frac{\sigma_D^2}{n} = \frac{\sigma_X^2 + \sigma_Y^2}{n} = \sigma_{\bar{x}}^2 + \sigma_{\bar{y}}^2 \tag{12.61}$$

Since the sampling procedure permits any variate of $X$ to be paired with any variate of $Y$, the covariance is equal to zero and, consequently, does not appear in Equation 12.61. Furthermore, since the sampling is with replacement or the population is infinitely large, there is no need for finite population correction factors.

If the sampling permits any variate of $X$ to be paired with any variate of $Y$, and if population $D$ is finite and the sampling is without replacement, the variance of the sampling distribution of mean differences is

$$\begin{aligned}\sigma_{\bar{d}}^2 &= \frac{\sigma_D^2}{n} = \left(\frac{N_D - n}{N_D - 1}\right) \\ &= \frac{\sigma_X^2}{n}\left(\frac{N_x - n}{N_X - 1}\right) + \frac{\sigma_Y^2}{n}\left(\frac{N_Y - n}{N_Y - 1}\right) \\ &= \sigma_{\bar{x}}^2 + \sigma_{\bar{y}}^2\end{aligned} \tag{12.62}$$

Again the covariance is equal to zero and does not appear.

If the variates of $X$ and $Y$ are linked through common sampling units, it becomes necessary to include covariance. Thus, if population $D$ is infinitely large or, if finite, the sampling is with replacement, the variance of the sampling distribution is

$$\sigma_{\bar{d}}^2 = \frac{\sigma_D^2}{n} = \frac{\sigma_X^2 + \sigma_Y^2 - 2\sigma_{XY}}{n} = \sigma_{\bar{x}}^2 + \sigma_{\bar{y}}^2 - 2\sigma_{\bar{x}\bar{y}} \tag{12.63}$$

On the other hand, if the distribution $D$ is finite and the sampling is without replacement, the variance is

$$\begin{aligned} \sigma_{\bar{d}}^2 &= \frac{\sigma_D^2}{n}\left(\frac{N-n}{N-1}\right) \\ &= \frac{\sigma_X^2}{n}\left(\frac{N-n}{N-1}\right) + \frac{\sigma_Y^2}{n}\left(\frac{N-n}{N-1}\right) - \frac{2\sigma_{XY}}{n}\left(\frac{N-n}{N-1}\right) \\ &= \sigma_{\bar{x}}^2 + \sigma_{\bar{y}}^2 - 2\sigma_{\bar{x}\bar{y}} \end{aligned} \tag{12.64}$$

To illustrate these situations, assume that populations $X$ and $Y$ are made up as follows:

| **X** | **Y** |
|---|---|
| 1 | 5 |
| 2 | 8 |
| 3 | 9 |

Their parameters are

$$\begin{aligned} N_X &= 3 & N_Y &= 3 \\ \mu_X &= 2.0000 & \mu_Y &= 7.3333 \\ \sigma_X^2 &= 0.6667 & \sigma_Y^2 &= 2.8889 \end{aligned}$$

Then, if $D = X - Y$, and the variates of $X$ and $Y$ are not linked in any way, population $D$ is

$$\begin{array}{lll} 1-5=-4 & 2-5=-3 & 3-5=-2 \\ 1-8=-7 & 2-8=-6 & 3-8=-5 \\ 1-9=-8 & 2-9=-7 & 3-9=-6 \end{array}$$

and its parameters are

$$N_D = N_X N_Y = 9$$

$$\mu_D = \mu_X - \mu_Y = 2.000 - 7.3333 = -5.3333$$

$$\sigma_D^2 = \sigma_X^2 + \sigma_Y^2 = 0.6667 + 2.8889 = 3.5556$$

These values can be confirmed using the basic formulae. Let all possible samples of size $n = 2$ be drawn from $D$, using sampling with replacement, and the mean computed for each sample. The resulting set of means is the sampling distribution of mean differences.

$$\frac{[(-4)+(-4)]}{2} = -4.0 \quad \frac{[(-7)+(-4)]}{2} = -5.5 \quad \frac{[(-8)+(-4)]}{2} = -6.0$$

$$\frac{[(-4)+(-7)]}{2} = -5.5 \quad \frac{[(-7)+(-7)]}{2} = -7.0 \quad \frac{[(-8)+(-7)]}{2} = -7.5$$

$$\frac{[(-4)+(-8)]}{2} = -6.0 \quad \frac{[(-7)+(-8)]}{2} = -7.5 \quad \frac{[(-8)+(-8)]}{2} = -8.0$$

$$\frac{[(-4)+(-3)]}{2} = -3.5 \quad \frac{[(-7)+(-3)]}{2} = -5.0 \quad \frac{[(-8)+(-3)]}{2} = -5.5$$

$$\frac{[(-4)+(-6)]}{2} = -5.0 \quad \frac{[(-7)+(-6)]}{2} = -6.5 \quad \frac{[(-8)+(-6)]}{2} = -7.0$$

$$\frac{[(-4)+(-7)]}{2} = -5.5 \quad \frac{[(-7)+(-7)]}{2} = -7.0 \quad \frac{[(-8)+(-7)]}{2} = -7.5$$

$$\frac{[(-4)+(-2)]}{2} = -3.0 \quad \frac{[(-7)+(-2)]}{2} = -4.5 \quad \frac{[(-8)+(-2)]}{2} = -5.0$$

$$\frac{[(-4)+(-5)]}{2} = -4.5 \quad \frac{[(-7)+(-5)]}{2} = -6.0 \quad \frac{[(-8)+(-5)]}{2} = -6.5$$

$$\frac{[(-4)+(-6)]}{2} = -5.0 \quad \frac{[(-7)+(-6)]}{2} = -6.5 \quad \frac{[(-8)+(-6)]}{2} = -7.0$$

$$\frac{[(-3)+(-4)]}{2} = -3.5 \quad \frac{[(-6)+(-4)]}{2} = -5.0 \quad \frac{[(-7)+(-4)]}{2} = -5.5$$

$$\frac{[(-3)+(-7)]}{2} = -5.0 \quad \frac{[(-6)+(-7)]}{2} = -6.5 \quad \frac{[(-7)+(-7)]}{2} = -7.0$$

$$\frac{[(-3)+(-8)]}{2} = -5.5 \quad \frac{[(-6)+(-8)]}{2} = -7.0 \quad \frac{[(-7)+(-8)]}{2} = -7.5$$

$$\frac{[(-3)+(-3)]}{2} = -3.0 \quad \frac{[(-6)+(-3)]}{2} = -4.5 \quad \frac{[(-7)+(-3)]}{2} = -5.0$$

$$\frac{[(-3)+(-6)]}{2} = -4.5 \quad \frac{[(-6)+(-6)]}{2} = -6.0 \quad \frac{[(-7)+(-6)]}{2} = -6.5$$

$$\frac{[(-3)+(-7)]}{2} = -5.0 \quad \frac{[(-6)+(-7)]}{2} = -6.5 \quad \frac{[(-7)+(-7)]}{2} = -7.0$$

$$\frac{[(-3)+(-2)]}{2}=-2.5 \quad \frac{[(-6)+(-2)]}{2}=-4.0 \quad \frac{[(-7)+(-2)]}{2}=-4.5$$

$$\frac{[(-3)+(-5)]}{2}=-4.0 \quad \frac{[(-6)+(-5)]}{2}=-5.5 \quad \frac{[(-7)+(-5)]}{2}=-6.0$$

$$\frac{[(-3)+(-6)]}{2}=-4.5 \quad \frac{[(-6)+(-6)]}{2}=-6.0 \quad \frac{[(-7)+(-6)]}{2}=-6.5$$

$$\frac{[(-2)+(-4)]}{2}=-3.0 \quad \frac{[(-5)+(-4)]}{2}=-4.5 \quad \frac{[(-6)+(-4)]}{2}=-5.0$$

$$\frac{[(-2)+(-7)]}{2}=-4.5 \quad \frac{[(-5)+(-7)]}{2}=-6.0 \quad \frac{[(-6)+(-7)]}{2}=-6.5$$

$$\frac{[(-2)+(-8)]}{2}=-5.0 \quad \frac{[(-5)+(-8)]}{2}=-6.5 \quad \frac{[(-6)+(-8)]}{2}=-7.0$$

$$\frac{[(-2)+(-3)]}{2}=-2.5 \quad \frac{[(-5)+(-3)]}{2}=-4.0 \quad \frac{[(-6)+(-3)]}{2}=-4.5$$

$$\frac{[(-2)+(-6)]}{2}=-4.0 \quad \frac{[(-5)+(-6)]}{2}=-5.5 \quad \frac{[(-6)+(-6)]}{2}=-6.0$$

$$\frac{[(-2)+(-7)]}{2}=-4.5 \quad \frac{[(-5)+(-7)]}{2}=-6.0 \quad \frac{[(-6)+(-7)]}{2}=-6.5$$

$$\frac{[(-2)+(-2)]}{2}=-2.0 \quad \frac{[(-5)+(-2)]}{2}=-3.5 \quad \frac{[(-6)+(-2)]}{2}=-4.0$$

$$\frac{[(-2)+(-5)]}{2}=-3.5 \quad \frac{[(-5)+(-5)]}{2}=-5.0 \quad \frac{[(-6)+(-5)]}{2}=-5.5$$

$$\frac{[(-2)+(-6)]}{2}=-4.0 \quad \frac{[(-5)+(-6)]}{2}=-5.5 \quad \frac{[(-6)+(-6)]}{2}=-6.0$$

The parameters of the sampling distributions are

$$N_{\bar{d}} = N_D^n = 9^2 = 81$$

$$\mu_{\bar{d}} = \mu_D = -5.3333$$

$$\sigma_{\bar{d}}^2 = \frac{\sigma_D^2}{n} = \frac{\sigma_X^2}{n} + \frac{\sigma_Y^2}{n} = \frac{3.5556}{2} = 1.7778$$

If the sampling had been without replacement, the sampling distribution of mean differences would be

$$\frac{[(-4)+(-7)]}{2} = -5.5$$

$$\frac{[(-4)+(-8)]}{2} = -6.0 \quad \frac{[(-7)+(-8)]}{2} = -7.5$$

$$\frac{[(-4)+(-3)]}{2} = -3.5 \quad \frac{[(-7)+(-3)]}{2} = -5.0 \quad \frac{[(-8)+(-3)]}{2} = -5.5$$

$$\frac{[(-4)+(-6)]}{2} = -5.0 \quad \frac{[(-7)+(-6)]}{2} = -6.5 \quad \frac{[(-8)+(-6)]}{2} = -7.0$$

$$\frac{[(-4)+(-7)]}{2} = -5.5 \quad \frac{[(-7)+(-7)]}{2} = -7.0 \quad \frac{[(-8)+(-7)]}{2} = -7.5$$

$$\frac{[(-4)+(-2)]}{2} = -3.0 \quad \frac{[(-7)+(-2)]}{2} = -4.5 \quad \frac{[(-8)+(-2)]}{2} = -5.0$$

$$\frac{[(-4)+(-5)]}{2} = -4.5 \quad \frac{[(-7)+(-5)]}{2} = -6.0 \quad \frac{[(-8)+(-5)]}{2} = -6.5$$

$$\frac{[(-4)+(-6)]}{2} = -5.0 \quad \frac{[(-7)+(-6)]}{2} = -6.5 \quad \frac{[(-8)+(-6)]}{2} = -7.0$$

$$\frac{[(-3)+(-6)]}{2} = -4.5$$

$$\frac{[(-3)+(-7)]}{2} = -5.0 \quad \frac{[(-6)+(-7)]}{2} = -6.5$$

$$\frac{[(-3)+(-2)]}{2} = -2.5 \quad \frac{[(-6)+(-2)]}{2} = -4.0 \quad \frac{[(-7)+(-2)]}{2} = -4.5$$

$$\frac{[(-3)+(-5)]}{2} = -4.0 \quad \frac{[(-6)+(-5)]}{2} = -5.5 \quad \frac{[(-7)+(-5)]}{2} = -6.0$$

$$\frac{[(-3)+(-6)]}{2} = -4.5 \quad \frac{[(-6)+(-6)]}{2} = -6.0 \quad \frac{[(-7)+(-6)]}{2} = -6.5$$

$$\frac{[(-2)+(-5)]}{2} = -3.5$$

$$\frac{[(-2)+(-6)]}{2} = -4.0 \quad \frac{[(-5)+(-6)]}{2} = -5.5$$

The parameters of this sampling distribution are

$$N_{\bar{d}} = \binom{N_D}{n} = \binom{9}{2} = 36$$

$$\mu_{\bar{d}} = \mu_D = -5.3333$$

$$\sigma_{\bar{d}}^2 = \frac{\sigma_D^2}{n}\left(\frac{N-n}{N-1}\right) = \left(\frac{\sigma_X^2}{n} + \frac{\sigma_Y^2}{n}\right)\left(\frac{N_D - n}{N_D - 1}\right) = \frac{3.5556}{2}\left(\frac{9-2}{9-1}\right) = 1.5556$$

If the values of $X$ and $Y$ are linked through common sampling units, as follows, the values of $D$ would be as shown.

| Sampling Unit | *X* | *Y* | *D* |
|---|---|---|---|
| 1 | 1 | 5 | –4 |
| 2 | 2 | 8 | –6 |
| 3 | 3 | 9 | –6 |

The parameters of $D$ are

$$N_D = N_X = N_Y = 3$$

$$\mu_D = \mu_X - \mu_Y = -5.3333$$

$$\begin{aligned}\sigma_D^2 &= \sigma_X^2 + \sigma_Y^2 - 2\sigma_{XY} \\ &= 0.6667 + 2.8889 - 2(1.3334) = 0.8888\end{aligned}$$

$$\sigma_{XY} = E(XY) - E(X)E(Y) = 16 - (2.0000)(7.3333) = 1.3334$$

Then if all possible samples of size $n - 2$ were drawn from $D$, using sampling with replacement, the sampling distribution of mean differences is

$$\frac{[(-4)+(-4)]}{2} = -4 \qquad \frac{[(-6)+(-4)]}{2} = -5 \qquad \frac{[(-6)+(-4)]}{2} = -5$$

$$\frac{[(-4)+(-6)]}{2} = -5 \qquad \frac{[(-6)+(-6)]}{2} = -6 \qquad \frac{[(-6)+(-6)]}{2} = -6$$

$$\frac{[(-4)+(-6)]}{2} = -5 \qquad \frac{[(-6)+(-6)]}{2} = -6 \qquad \frac{[(-6)+(-6)]}{2} = -6$$

The parameters are

$$N_{\bar{d}} = N_D^n = 3^2 = 9$$

$$\mu_{\bar{d}} = \mu_D = -5.3333$$

$$\sigma_{\bar{d}}^2 = \frac{\sigma_D^2}{n} = \frac{\sigma_X^2 + \sigma_Y^2 - 2\sigma_{XY}}{n} = \frac{0.8888}{2} = 0.4444$$

Finally, if the sampling is without replacement, the sampling distribution of mean differences is

$$\frac{[(-4)+(-6)]}{2} = -5$$

$$\frac{[(-4)+(-6)]}{2} = -5 \qquad \frac{[(-6)+(-6)]}{2} = -6$$

and its parameters are

$$N_{\bar{d}} = \binom{N_D}{n} = \binom{3}{2} = 3$$

$$\mu_{\bar{d}} = \mu_D = -5.3333$$

$$\sigma^2_{\bar{d}} = \frac{\sigma^2_D}{n}\left(\frac{N_D - n}{N_D - 1}\right) = \frac{0.8888}{2}\left(\frac{3-2}{3-1}\right) = 0.2222$$

Up to this point no references have been made to the distributional patterns of the random variables or of the sampling distributions of mean differences. If $X$ and $Y$ are normally distributed, population $D$ will also be normally distributed because of the reproductive property of the normal distribution (see Section 8.13). This can be shown as follows. The moment-generating function of $X$ (see Equation 11.15) is

$$M_X(t) = e^{t\mu_X + \sigma^2_X t^2/2}$$

while that of $Y$ is

$$M_Y(t) = e^{t\mu_Y + \sigma^2_Y t^2/2}$$

then according to Equation 8.46

$$M_D(t) = M_X(t) * M_Y(t)$$

$$= (e^{t\mu_X + \sigma^2_X t^2/2}) * (e^{t\mu_Y + \sigma^2_Y t^2/2})$$

$$= e^{t(\mu_X + \mu_Y) + t^2(\sigma^2_X + \sigma^2_Y)/2}$$

which is the mgf of a normal distribution having the parameters

$$\mu_D = \mu_X - \mu_Y \quad \text{and} \quad \sigma^2_D = \sigma^2_X + \sigma^2_Y$$

If $D$ is normally distributed, then, according to the central limit theorem, the sampling distribution of mean differences is also normally distributed.

If $X$ and $Y$ are continuous random variables but at least one is not normally distributed, the distribution of $D$ will be nonnormal. In such a case the sampling distribution of mean differences will, according to the central limit theorem, tend toward being normally distributed with the approximation becoming closer as the sample size increases. As was brought out in Section 12.10, when $n$ is 30 or greater, it usually is assumed that the approximation is sufficiently good that the sampling distribution can be modeled by the normal.

If $X$ and $Y$ are discrete, the central limit theorem is not applicable. However, as was brought out in Section 12.10, sampling distributions of means based on simple random sampling still tend to assume the shape of the normal, with the approximation becoming better as sample size increases. If the sampling is with replacement and $n$ exceeds 30, it usually is assumed that the normal can be used as the model. When sampling is without replacement, however, the situation is less clear-cut, especially when the parent populations are small and the sampling fractions are relatively large. Usually it is assumed that, if $n$ exceeds 30 and the sampling fraction is less than 0.1, it is safe to use the normal as the model. When these conditions cannot be met, no assumptions can be made with respect to the distributional pattern of the sampling distribution. If the distributions of the parent random variables are known, the distributions of $D$ and $\bar{d}$ can be determined mathematically but only very rarely is this feasible.

# 13 Estimation of Parameters

## 13.1 INTRODUCTION

It was brought out in Chapter 1 that the primary reason for sampling is to obtain estimates of population parameters that are of interest to a decision-maker. The decision-maker, naturally, would prefer knowledge of the parameters themselves but only rarely is it possible to obtain such knowledge at an acceptable cost in time or money. Consequently, the decision-maker usually must be satisfied with sample-based estimates of those parameters. In this chapter the estimation process will be explored and the foundation laid for later discussion of some of the sampling designs that may find application in land management operations.

## 13.2 ESTIMATORS AND ESTIMATES

Let us assume a 100-acre tract of land bearing a stand of timber which is to be sold. To set a price on the timber it is necessary to have an estimate of the volume of timber on the tract. To obtain such an estimate a sampling frame could be established consisting of a grid with a certain specified spacing. The grid intersections would be the potential centers of 1/5-acre plots. The volumes per acre on these plots are variates of the population of interest. What is needed is an estimate of the per acre volume on the tract, which could be used to obtain an estimate of the total volume. The estimate can be based either on a single randomly drawn plot, yielding a single variate from the population, or on the mean of several randomly drawn variates. In either case, the single variate or the mean of several variates, the value used for the estimate is a function of the sample data. In the context of sampling such a *function* is known as an *estimator*. Thus,

$$\bar{y} = \sum_{i=1}^{n} y_i / n$$

is an estimator since with it one can estimate the population mean.

Generalizing, let $Y$ be the random variable of interest which has a known probability distribution function but an unknown parameter $\theta$ which is to be estimated. Then, let $y_1, y_2, \ldots, y_n$ be a random sample drawn from population $Y$. The estimator, $t$, is then a function of the random sample:

$$t = t(y_1, y_2, \ldots, y_n) \tag{13.1}$$

e.g., $t = \bar{y} = (1/n)(y_1 + y_2 + \ldots + y_n)$. It is obvious that the unknown parameter $\theta$ cannot be a factor in the estimator. The elements of the sample are random variables in their own right since they vary in magnitude from sample to sample. Each has its own probability distribution function, which is identical with that of $Y$ if population $Y$ is infinitely large or, if finite, the sampling was with replacement. If, however, $Y$ is finite and the sampling is without replacement the probabilities associated with the elements will be influenced by the order of the draws.

When the data from a *specific* sample are entered into an estimator, the resulting value, $t$, is an *estimate* or a *statistic*.

## 13.3 POINT ESTIMATES

If the position of the estimate is plotted on a number line and it occupies only a single point, the estimate is known as a *point* estimate. Thus, in the case of the 100-acre tract in Section 13.2, both the single variate estimate and the sample mean are point estimates.

## 13.4 CRITERIA FOR JUDGING ESTIMATORS OF PARAMETERS

In some cases there may be more than one possible estimator of a given parameter. For example, the mean of a normally distributed population can be estimated either by a single randomly drawn variate or by the mean or the median of a random sample. When such a situation exists, one should use the most effective of the estimators. The choice should be based on the following criteria.

1. The estimator should be *unbiased*. In other words,

$$E(t) = \theta \tag{13.2}$$

   This was discussed in Chapter 12, where it was brought out that the sample mean is an unbiased estimater of the population mean but the sample variance, unless corrected, is a biased estimator of the population variance.

2. The estimator should have the *smallest variance* of any of alternative unbiased estimators. For example, as was mentioned earlier, the mean of a normally distributed population can be estimated by a single observation $y_i$, the sample mean, $\bar{y}$, or the sample median. The respective variances of these three possible estimators are

$$V(y) = \sigma_Y^2$$

$$V(\bar{y}) = (\sigma_Y^2 / n)$$

$$V(\text{median}) = 1.57(\sigma_Y^2 / n)$$

   Thus, the sample mean has the smallest variance and would be the preferred estimator. Another way of looking at this is to consider it a matter of *efficiency*. The estimator having the smallest variance is the most efficient of the estimators. Putting this into more concrete terms, consider the situation where a variance is specified and the sample size needed to obtain that variance is computed. If $V(\bar{y})$ and $V(\text{median})$ are both set equal to $k$, the needed sample sizes are

$$V(\bar{y}) = k = \sigma_Y^2 / n_y$$

$$n_y = \sigma_Y^2 / k$$

$$V(\text{median}) = k = 1.57(\sigma_Y^2 / n_{\text{median}})$$

$$n_{\text{median}} = 1.57(\sigma_Y^2 / k)$$

Obviously $n_{median} > n_{\bar{y}}$ and, thus, the sample mean is more efficient than the sample median when estimating the population mean.

3. The estimator should be *consistent*. An estimator is said to be consistent if, as $n$ approaches infinity, the variance of the estimator approaches zero and its expected value converges in probability to the parameter being estimated. In other words, the estimate should become better as the sample size increases.

   In Section 13.10 Chebyshev's inequality is discussed. The inequality states that (see Equation 13.31)

$$P(|y_i - \mu_Y| \geq r\sigma_Y) \leq 1/r^2; \quad r > 0$$

where $\mu_Y$ and $\sigma_Y^2$ are finite and $r$ is an arbitrarily chosen value. In words, the probability of a randomly drawn variate, $y_i$, deviating from the expected value, $E(Y) = \mu_Y$, by at least $r\sigma_Y$ is less than or equal to $1/r^2$. This is true regardless of the distributional pattern of the random variable. Generalizing, let $t = y_i$; $\theta = \mu_Y$; $\sigma_t = \sigma_Y$; and $k = r\sigma_Y = r\sigma_t$; then,

$$P(|t - \theta| \geq k) \leq \sigma_t^2 / k^2; \quad k > 0 \tag{13.3}$$

An estimator, $t$, is said to be consistent if

$$\lim_{n\to\infty} P(|t - \theta| > k) = 0; \quad k > 0 \tag{13.4}$$

thus, in the case of the sample mean, $t = \bar{y}$, $\theta = \mu_Y$, and $\sigma_t^2 = \sigma_{\bar{y}}^2 = \sigma_Y^2 / n$. Then,

$$\begin{aligned}\lim_{n\to\infty} P(|\bar{y} - \mu_y| > 1) &= \sigma_{\bar{y}}^2 / k^2 \\ &= \sigma_y^2 / nk^2 = 0\end{aligned}$$

As can be seen, the sample mean is consistent.

## 13.5 LAWS OF LARGE NUMBERS

In the previous section it was shown that the sample mean is consistent. As the sample size increases, the sample mean converges in probability to the population mean and the variance of the sample mean converges to zero. When this occurs the estimator obeys the *weak law of large numbers*. The weak law applies to sample-based relative frequencies or proportions, $p$, and the sample variances $s_y^2$, as well as to sample means. An example of how the law functions is to be found in Figure 7.4, where the empirical probability of heads was determined for a certain coin. As the sample size increased, the variability of $p$ decreased to the point where the relative frequency was considered "stable." At this point it was assumed that the relative frequency had converged to the true probability of heads.

Applying the weak law to sample variances, let $t = s_Y^2$; $\theta = \sigma_Y^2$; and $k = r\sigma_t = r\sigma_{s_Y^2}$. Then,

$$\lim_{n\to\infty} P(|s_Y^2 - \sigma_Y^2| \geq k) \leq \sigma_{s_Y^2}^2 / k^2$$

Substituting Equation 12.17 for $\sigma_{s_Y^2}^2$,

$$\lim_{n\to\infty} P(|s_Y^2 - \sigma_Y^2| \geq k) \leq \left[\frac{\mu_4}{n} - \frac{(\sigma_Y^2)^2}{n}\left(\frac{n-3}{n-1}\right)\right] \Big/ k^2$$

As $n \to \infty$, both the fourth moment about the mean, $\mu_4$, and the term $(\mu_Y^4 / n)[(n-3)/(n-1)]$ approach zero but cannot become negative. Thus,

$$\lim_{n\to\infty} P(|s_Y^2 - \sigma_Y^2| \geq k) = 0$$

indicating that the sample variance obeys the weak law, and therefore is consistent.

The *strong law of large numbers* is not a factor in evaluating the consistency of an estimator. It states that for any pair of arbitrarily chosen small numbers, $a$ and $b$, an $n$ can always be found so that

$$P(|t - \theta| \leq a) \leq 1 - b; \quad m \geq n \tag{13.5}$$

where $t$ = estimate of the parameter based on $m$ observations
$a = \sigma_t\sqrt{2c\ln(n)}$
$b = 1/[(c-1)n^{c-1}\sqrt{c\pi\ln(n)}]$
$c$ = a constant greater than 1

In other words, the probability of $|t - \theta|$ being less than $a$ is less than $1 - b$ if the sample size is equal to or less than $n$. Whenever $c > 1$, the inequalities can be satisfied by using a sufficiently large $n$. As $n$ increases, $b$ becomes less and in the limit, as $n \to \infty$, $b$ converges to zero. Thus, as the sample size increases the probability that $|t - \theta|$ is less than $a$ becomes greater and in the limit, as $n \to \infty$, it converges to 1 or 100%. This is true regardless of how small $a$ is made. Note, however, that $a$ must equal or exceed $\sigma_t\sqrt{2c\ln(n)}$. In the case of the sample mean: $t = \bar{y}$, based on $m$ observations; $\theta = \mu_{\bar{y}} = \mu_Y$; and $\sigma_t = \sigma_{\bar{y}} = \sigma_Y / \sqrt{n}$. Then, when $c > 1$, the expression becomes

$$\sigma_Y\sqrt{2c\ln(n)} / \sqrt{n}$$

and as $n$ increases the expression becomes smaller. In the limit, as $n \to \infty$, it converges to zero. Thus, regardless of how small $a$ is made, an $n$ can be found so that Equation 13.5 applies. Consequently, it can be said that the sample mean obeys the strong law of large numbers.

## 13.6 MAXIMUM LIKELIHOOD ESTIMATORS

Fisher devised a means of identifying efficient, but not always unbiased, estimators, which is known as the *method of maximum likelihood*. The procedure first requires the development of a *likelihood function*, $L$, for the random sample.

$$L(Y_1, Y_2, \ldots, y_n; \theta) = P(Y_1 = y_1; \theta) * P(Y_2 = y_2; \theta) \ldots P(Y_n = y_n; \theta) \tag{13.6}$$

where $y_i$ = $i$th variate in the sample of size $n$ drawn from population $Y$
$\theta$ = the unknown parameter of interest from population $Y$
$P(Y_i = y_i;\ \theta)$ = the probability of randomly drawing $Y_i = y_i$, when $\theta$ is equal to a specific value.

Then $L$ is the probability of a specific sample being drawn from population $Y$. It must be realized that each position in the sample can be filled by any of the variates in population $Y$. Consequently, each position is filled by a value of the random variable associated with that position. Position 1 is filled by a value of $Y_1$, position 2 by a value of $Y_2$, and so forth. The populations associated with these random variables are coincident with the parent population $Y$ and all possess the same parameters and distributional characteristics. $\theta$ is common to all these random variables.

The magnitude of $\theta$ is unknown and must be estimated. The maximum likelihood estimate of $\theta$ is obtained when the likelihood function, $L$, is maximized with respect to $\theta$. The maximization process can be carried out using either $L$ or $\ln L$ since both yield the same estimate of $\theta$. However, the process is usually simpler using $\ln L$. Thus, the estimate of $\theta$ is obtained when the *likelihood equation* is solved.

$$\frac{\partial \ln L}{\partial \theta} = \sum_{i=1}^{n} \left[ \frac{\partial \ln P(Y_i = y_i);\theta}{\partial \theta} \right] = 0 \tag{13.7}$$

For example, a certain proportion of a given tract of land is occupied by water features (ponds and streams). Thus, each of the infinitely large set of points making up the surface falls in either water or on land. The random variable is then

$$y_i = \begin{cases} 1, & \text{if the point falls on a water surface} \\ 0, & \text{if the point does not fall on a water surface} \end{cases}$$

The population $Y$ is then made up of variates possessing the values 0 or 1. The proportion of the population possessing the value 1, $p$, is then equal to the proportion of the universe of points associated with water surfaces. Let this be the parameter of interest

$$\theta = p$$

If a sample of $n$ randomly chosen points is obtained and each point is classified whether or not it falls on water,

$$P(Y_i = 1;\theta) = \theta$$

$$P(Y_i = 0;\theta) = 1 - \theta$$

Then,

$$\begin{aligned} L &= P(Y_1 = y_i;\theta) * P(Y_2 = y_2;\theta) * \cdots * P(Y_n = y_n;\theta) \\ &= \theta_1 * \theta_2 * \cdots * \theta_a * (1-\theta)_1 * (1-\theta)_2 * \cdots * (1-\theta)_{n-a} \\ &= \theta^a (1-\theta)^{n-a} \end{aligned}$$

where $a$ = number of points in the sample falling on water (or the number of variates in the sample that are equal to 1)

$n - a$ = number of points not falling on water (or the number of variates that are equal to 0)

Then, $\ln L = a \ln \theta + (n - a) \ln (1 - \theta)$ and

$$\frac{\partial \ln L}{\partial \theta} = a\frac{\partial \ln \theta}{\partial \theta} + (n-a)\frac{\partial \ln(1-\theta)}{\partial \theta}$$

$$= a\left(\frac{1}{\theta}\right)\frac{d\theta}{d\theta} + (n-a)\left(\frac{1}{1-\theta}\right)\frac{d(1-\theta)}{d\theta}$$

$$= \frac{a}{\theta} - \frac{(n-a)}{1-\theta}$$

Setting this equal to zero and solving for $\theta$,

$$\frac{a}{\theta} - \frac{(n-a)}{1-\theta} = 0$$

$$\frac{a(1-\theta) - \theta(n-a)}{\theta(1-\theta)} = 0$$

$$a - a\theta - n\theta + a\theta = 0[\theta(1-\theta)] = 0$$

$$a - n\theta = 0$$

$$\frac{a}{n} = \hat{P} \rightarrow \theta = 0$$

Thus, the sample proportion, $\hat{p}$, is the maximum likelihood estimator of the population proportion $p$. It is efficient since no other estimator will produce a greater $L$. In Section 12.13 it was shown that $E(\hat{p}) = p$. Thus, $\hat{p}$ is an unbiased estimator of $p$. It also was shown in Section 12.13 that $V(\hat{p}) = p(1-p)/n$ and that the distribution of $\hat{p}$ tends to take on the shape of the normal as $n$ becomes large.

If population $Y$ is normally distributed (see Equation 11.8), its pdf is

$$f(y;\ \mu;\ \sigma^2) = \frac{1}{\sigma\sqrt{2\pi}} e^{-1/2[y-u)/\sigma]^2}$$

If both $\mu$ and $\sigma^2$ must be estimated, let $\theta_1 = \mu$ and $\theta_2 = \sigma^2$. Then

$$L = (2\pi\theta_2)^{-n/2} e^{-1/(2\theta_2)\Sigma_{i=1}^{n}(y_i-\theta_1)^2}$$

and

$$\ln L = -\frac{n}{2}\ln(2\pi\theta_2) - \frac{1}{2\theta_2}\sum_{i=1}^{n}(y_i-\theta_1)^2$$

Then

$$\frac{\partial \ln L}{\partial \theta_1} = -\frac{n}{2}\frac{\partial(2\pi\theta_2)}{\partial \theta_1} - \frac{1}{2\theta_1} - \frac{\partial \sum_{i=1}(y_i-\theta_1)^2}{\partial \theta_1}$$

$$= \frac{1}{\theta_2}\sum_{i=1}^{n}(\theta_1 - y_i)$$

and

$$\frac{\partial \ln L}{\partial \theta_2} = -\frac{n}{2}\frac{\partial(2\pi\theta_2)}{\partial \theta_2} - \frac{\sum_{i=1}^{n}(y_i - \theta_1)^2}{2}\frac{\partial \theta_2^{-1}}{2\theta_2}$$

$$= -\frac{n}{2\theta_2} + \frac{\sum_{i=1}^{n}(y_i - \theta_1)^2}{2\theta_2^2}$$

Setting these partial derivatives equal to zero and solving for $\theta_1$ and $\theta_2$,

$$-\frac{1}{\theta_2}\sum_{i=1}^{n}(\theta_1 - y_i) = 0$$

$$\sum_{i=1}^{n}(\theta_1 - y_i) = 0$$

$$n\theta_1 - \sum_{i=1}^{n} y_i = 0$$

$$\theta_1 = \frac{\sum_{i=1}^{n} y_i}{n} = \bar{y} \rightarrow \mu$$

and

$$-\frac{n}{2\theta_2} + \frac{\sum_{i=1}^{n}(y_i - \theta_1)^2}{2\theta_2^2} = 0$$

$$\frac{\sum_{i=1}^{n}(y_i - \theta_1)^2}{2\theta_2^2} = \frac{n}{2\theta_2}$$

$$\theta_2 = \frac{\sum_{i=1}^{n}(y_i - \theta_1)^2}{n} = \frac{\sum_{i=1}^{n}(y_i - \bar{y})^2}{n} \rightarrow \sigma_2$$

Thus, the sample mean is the maximum likelihood estimator of the mean of a normally distributed population. It was shown in Section 12.10 (Equation 12.2) that $E(\bar{y}) = \mu_Y$ and consequently $\bar{y}$ is unbiased. However, the maximum likelihood estimator of the variance of a normally distributed population is biased, as was shown in Section 12.11 (Equation 12.7). As can be seen from this example, not all maximum likelihood estimators are unbiased.

In Section 5.2.4 it was shown that one of the characteristics of the arithmetic mean is that it occupies the *least squares* position within the data upon which it is based. The process of determining the least squares position or value involves the minimization of the sum of squared residuals (or errors) which corresponds to what is done to arrive at the maximum likelihood estimate of $\mu$. Thus, when population $Y$ is *normally* distributed, the maximum likelihood and least squares methods yield the same estimator of the population mean. Further reference will be made to their relationship in later segments of this text.

In general, maximum likelihood estimators possess the following properties:

1. They are not always unbiased.
2. They usually are consistent.
3. They are efficient when samples are large.
4. They are *sufficient*, meaning that no other estimator, based on the same sample data and independent of the maximum likelihood estimator, is able to yield more information about the parameter being estimated than does the maximum likelihood estimator. Stated mathematically, if $t$ is an estimate of $\theta$ and its sampling distribution has the pdf $f(t; \theta)$ and if the pdf of the sample can be written as

$$P(Y_1 = y_1; \theta) * P(Y_2 = y_2; \theta) * \ldots, P(Y_n = y_n; \theta) = f(t; \theta) * f(y_1, y_2, \ldots, y_n) \tag{13.8}$$

where $f(y_1, y_2, \ldots, y_n)$ does not contain $\theta$, the estimate $t$ is sufficient. For example, consider the sample mean, $\bar{y}$, as the estimator of the mean of the normally distributed population $Y$. The pdf of the sampling distribution of the mean is then

$$f(\bar{y};\ \mu_{\bar{y}} = \mu_Y = \mu,\ \sigma_{\bar{y}}^2 = \sigma_Y^2 / n) = \frac{1}{\sqrt{2\pi\sigma_Y^2 / n}} e^{-[(\bar{y}-\mu)^2/(\sigma_Y^2/n)]/2}$$

or $(2\pi\sigma_Y^2 / n)^{1/2} e^{-(n/2\sigma_Y^2)(\bar{y}-\mu)^2}$, while the pdf of the sample is the product of $n$ normal pdfs:

$$(2\pi\sigma_Y^2)^{n/2} e^{-(1/2\sigma_Y^2)\Sigma_{i=1}^n (y_i-\mu)^2}$$

Then,

$$f(y_1, y_2, \ldots, y_n) = \frac{(2\pi\sigma_Y^2)^{-n/2} e^{-(1/2\sigma_Y^2)\Sigma_{i=1}^n (y_i-\mu)^2}}{(2\pi\sigma_Y^2 / n)^{-1/2} e^{-(n/2\sigma_Y^2)(\bar{y}-\mu)^2}}$$

$$= \frac{(2\pi\sigma_Y^2) / n^{1/2} e^{-(n/2\sigma_Y^2)(\bar{y}-\mu)^2}}{(2\pi\sigma_Y^2)^{n/2} e^{(1/2\sigma_Y^2)\Sigma_{i=1}^n (y_i-\mu)^2}}$$

$$= (2\pi\sigma_Y^2)^{(1-n)/2} n^{-1/2} e^{[(n/2\sigma_Y^2)(\bar{y}-\mu)^2 - (1/2\sigma_Y^2)\Sigma_{i=1}^n (y_i-\mu)_2]}$$

$$= \frac{(2\pi\sigma_Y^2)}{\sqrt{n}^{(1-n)/2}} e^{(1/2\sigma_y^2)[n(\bar{y}-\mu)^2 - \Sigma_{i=1}^n (y_i-\mu)^2]}$$

Considering only the term within the brackets in the exponent of $e$

$$n(\bar{y} - \mu)^2 - \sum_{i=1}^{n} (y_i - \mu)^2$$

$$n(\bar{y}^2 - 2\mu\bar{y} + \mu^2) - \sum_{i=1}^{n} (y_i^2 - 2\mu y_i + \mu^2)$$

$$n\bar{y}^2 - 2n\mu\bar{y} + n\mu^2 - \sum_{i=1}^{n} (y_i^2 + 2\mu \sum_{i=1}^{n} y_i - n\mu^2)$$

$$n\bar{y}^2 - 2\mu \sum_{i=1}^{n} y_i - \sum_{i=1}^{n} y_i^2 + 2\mu \sum_{i=1}^{n} y_i$$

$$n\bar{y}^2 - \sum_{i=1}^{n} y_i^2$$

Thus,

$$f(y_i, y_2, \ldots, y_n) = \frac{(2\pi\sigma_Y^2)^{(1-n)/2}}{\sqrt{n}} e^{(1/2\sigma_Y^2)(n\bar{y}^2) - \sum_{i=1}^{n}(y_i^2)}$$

which does not contain $\mu$. Consequently, $\bar{y}$ is sufficient.

5. They are *invariant*, meaning that $f(t)$ is a maximum likelihood estimate of $f(\theta)$. For example, if $f(\theta) = \theta^2$, then $f(t) = t^2$ and $t^2$ will be a maximum likelihood estimator of $\theta^2$.
6. They are approximately normally distributed when samples are large. In other words, if $t$ is a maximum likelihood estimator of $\theta$ and $n$ is large, the random variable $t$ will be distributed approximately as the normal with the parameters

$$\mu_t = E(t) = \theta$$

$$\sigma_t^2 = V(t) \succeq \frac{1}{nE\left[\frac{\partial \ln f(Y;\ \theta)}{\partial \theta}\right]^2}$$

## 13.7 INTERVAL ESTIMATES

Return to the situation described in Section 13.2 where an estimate was needed of the volume of timber on a certain 100-acre tract. To obtain this estimate a sampling frame was formed by overlaying the area with a grid and using the grid intersections as possible plot centers. A random sample was obtained of these grid intersections and 1/5-acre plots were established and cruised at these points. As was mentioned earlier, each of the plots yielded an estimate of the actual per acre volume. If the smallest and largest of these estimates are identified, one could say that according to the available evidence the actual value probably lies *somewhere in the range between the minimum and maximum estimates*. If this range is plotted along a number line, there will be an interval within which the actual value might lie. An estimate of this type is known as an *interval estimate*.

Obviously, the likelihood of an interval including the actual value of the parameter is greater than the likelihood of a point estimate coinciding with the parameter. Furthermore, the larger the interval, the greater the likelihood of inclusion. Thus, one might say that one is more confident that an interval spanning two units includes the parameter than one would be with an interval spanning only one unit. If this confidence idea is inverted, one can say, if confidence is the same for two different sets of estimates but one set has a smaller interval than the other, that the set with the smaller interval is the more *precise* (see Chapter 2).

The discussion up to this point has not considered the idea of quantifying the level of confidence which might be applied to an interval estimate. Much of the remainder of this chapter will be concerned with this possibility.

## 13.8 CONFIDENCE INTERVAL ESTIMATES OF POPULATION PARAMETERS USING THE NORMAL DISTRIBUTION

In Section 11.2, it was shown that a random variable, $Y$, which is normally distributed with parameters $\mu_Y$ and $\sigma_Y^2$, can be converted to the normally distributed random variable, $Z$, through the use of the transformation equation (see Equation 11.10):

$$Z = (Y - \mu_Y)/\sigma_Y$$

The parameters of the Z population are $\mu_Z = 0$ and $\sigma_Z^2 = 1$. When this is done, it is said that the distribution of $Y$ has been put in standard form, which permits the use of Table A.4 in the Appendix to determine probabilities. This process is illustrated in Figures 11.1 and 11.2. By using these methods, it can be shown that the probability of drawing a value of $Z$ lying in the interval $-1 < z < +1$ is 68.27%. When the interval is $-2 < z < +2$, the probability increases to 95.45%, and when it is $-3 < z < +3$, the probability is 99.73%.

Now, assume that every item in the range space of $Z$ forms the center of a bracket that extends one unit of $Z$ (or one standard deviate) to each side of the value of the item (see Figure 13.1). Thus, the bracket extends from $(z - 1)$ to $(z + 1)$ or $z \pm 1$. If a random draw is to be made from the range space of $Z$, the probability is 68.27% that the value of the item drawn will be in the interval $\mu_Z = -1$ to $\mu_Z = +1$, or $0 \pm 1$. *If the item drawn does have a value in this interval, its associated bracket will enclose the value* $Z = 0 = \mu_Z$. Thus, the probability is 68.27% that a randomly drawn bracket or interval will include $\mu_Z$ or 0. This can be seen in Figure 13.1. If the bracket attached to each item in the range space extends two units to each side of the value of the item or $z \pm 2$, the probability of a randomly drawn bracket including $\mu_Z$ or 0 is 95.45%. This concept can be extended to any level of probability through the use of the appropriate number of standard deviates or units of $Z$. For example, if the level of probability is set at 90%, the number of standard deviates needed can be obtained from Table A.4 in the Appendix. Since the normal distribution is symmetrical, centered on $Z = 0$, the table provides areas under the curve only for one half of the distribution. Thus, 45% (half of the 90%) of the area under the curve lies between $Z = 0$ and $Z = 1.645$ (using straight-line interpolation between the tabular values 0.4495 and 0.4505, which correspond, respectively, to $Z$ values of 1.64 and 1.65). Consequently, if a draw is to be made from the range space or population of $Z$, the probability is 90% that the interval $z \pm 1.645$ will include $\mu_Z = 0$.

When the distribution of $Y$ is put in standard form, the mean $\mu_Y$ is made to correspond to $\mu_Z = 0$, and each unit on the $Z$ scale corresponds to $\sigma_Y$ units on the $Y$ scale. Consequently, *if it can be said that the probability is 90% that the randomly drawn bracket,* $z \pm 1.645$, *will include* $\mu_Z = 0$, *it also can be said that the probability is 90% that the bracket* $y \pm 1.645\sigma_Y$ *will include* $\mu_Y$. This can be stated in the following manner:

$$P(z - 1.645 < \mu_Z < z + 1.645) = 0.90 \text{ or } 90\%$$

or

$$P(y - 1.645\sigma_Y < \mu_Y > y + 1.645\sigma_Y) = 0.90 \text{ or } 90\%$$

The relationship between these expressions can be seen in Figure 13.1.

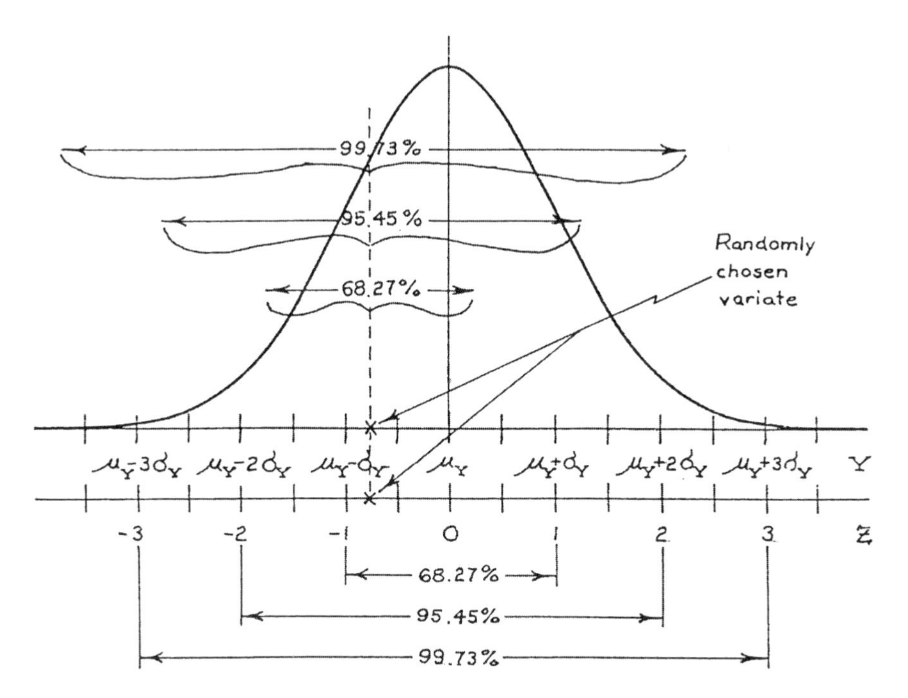


**FIGURE 13.1** The frequency distribution curve for a normally distributed random variable, *Y*, which has been standardized to show how a confidence interval can be used as an estimate of the population mean.

An interval of this type is known as a *confidence interval*. The probability level is referred to as the *probability level, confidence level*, or *significance level*. The end values of the interval are called *confidence limits*. The number of standard deviates (*Z* units) used to define the width of the bracket or interval is known as the *confidence coefficient* or *critical value*. The confidence coefficients for a selected set of confidence levels are shown in Table 13.1. Generalizing,

$$P[y - z_C\sigma_Y < \mu_Y < y + z_C\sigma_Y] = C \tag{13.9}$$

where $C$ = confidence level

$z_C$ = confidence coefficient for the $C$ level of confidence

**TABLE 13.1**
**Confidence Coefficients for a Selected Set of Confidence Levels**

| Confidence Level (C) (%) | Confidence Coefficient ($z_C$) |
|---|---|
| 99.73 | 3.00 |
| 99 | 2.58 |
| 98 | 2.33 |
| 96 | 2.05 |
| 95.45 | 2.00 |
| 95 | 1.96 |
| 90 | 1.645 |
| 80 | 1.28 |
| 68.27 | 1.00 |
| 50 | 0.6745 |

It should be noted at this time that the use of the term *probability* in a situation such as this is appropriate only *prior* to the draw. After the draw the position of the interval along the number line is fixed and it either includes or does not include the population mean. Probability is no longer a factor. The situation is comparable to one where a horse race has been run but the *results have not been made public*. In such a case it would be possible for persons to speculate on the results, indicate levels of confidence in the several horses, and even place bets on the outcome, even though the event has *already occurred*. In the same manner, after a draw has been made and the position of the confidence interval has been fixed along the number line, one can express an opinion whether the interval includes the population mean. Often this takes the form of a statement, such as, "The probability is 0.95 that the interval includes the population mean" (assuming that the confidence level was 0.95). When this is done, it is said that the expression is in terms of the *fiducial probability*. Insofar as the *existing* confidence interval is concerned, the fiducial probability is not at all a probability. It merely is a means of expressing a level of confidence. If, however, one states that if the trial were repeated indefinitely (actually until all possible samples were drawn), under the same circumstances as those prevailing when the draw was made leading to the existing confidence interval, a certain proportion (in the above case, 0.95) of the resulting confidence intervals would include the population mean. That proportion would be the fiducial probability and it would be equal in magnitude to the confidence level. Thus, fiducial probabilities refer to the likelihood that statistical statements are correct *in the long run*, not to a specific case.

## 13.9 CONFIDENCE INTERVAL ESTIMATES OF POPULATION MEANS USING STUDENT'S *t* DISTRIBUTION

The preceding development of the concept of the confidence interval was based on the assumption that the random variable, $Y$, was normally distributed and that the parameters of its distribution, $\mu_Y$ and $\sigma_Y^2$, were known. In real life, however, these conditions are rarely, if ever, encountered. Furthermore, in real life, the objective of the sampling operation is to obtain estimates of these unknown parameters, especially the mean. How then can valid interval estimates be made of these unknown parameters?

Fortunately, as was brought out in Section 12.10, sampling distributions of means tend to be normally distributed. If the random variable is continuous, the central limit theorem states that, if the parent population is normally distributed, the sampling distribution of means will also be normally distributed and even if the parent population is not normally distributed the sampling distribution of means will tend toward normality with the approximation becoming closer as the sample size increases. In the limit, as $n \to \infty$, the sampling distribution converges to the normal. This convergence to the normal is relatively rapid. Most workers in the field assume normality if $n$ exceeds 30. However, some set the point at which normality can be assumed at 60 and a few set it at 120. In many cases the sampling distribution, because the parent population has finite limits, cannot extend to infinity and the lack of coincidence with the normal is due to truncated tails. It usually is assumed that the approximation is acceptable in the interval $\pm 3\sigma_{\bar{y}}$ when $n$ is greater than 30. Even when the sample size is less than 30, the approximation is quite good in the central portion of the distribution.

When the random variable is discrete, the central limit theorem is not applicable but empirical evidence, such as that in Figure 12.1, indicates that sampling distributions of means based on discrete random variables tend to form histograms that follow the normal curve and that this tendency becomes stronger as the sample size increases. Consequently, most workers treat sampling distributions of means based on discrete random variables in the same manner as they do those based on continuous random variables.

This tendency of sampling distributions of means to approach the shape of the normal means that it is possible to make probability statements based on the sampling distributions of means and not be seriously in error even when the distributional patterns of the parent populations are not known. In this process the parameters of the *sampling distribution* are estimated and these estimates are then extended to the parameters of the parent population through the use of the known relationship between the parameters. Thus,

$$P(\bar{y} - z_C\sigma_{\bar{y}} < \mu_{\bar{y}} = \mu_Y < \bar{y} + z_C\sigma_{\bar{y}}) = C$$

or

$$\bar{y} \pm z_C\sigma_{\bar{y}} \rightarrow \mu_{\bar{y}} = \mu_Y \tag{13.10}$$

In Equation 13.10 the magnitude of the bracket or interval is a function of the true standard error of the mean, $\sigma_{\bar{y}}$, which is as unknown as the mean. Consequently, Equation 13.10 states the *concept* of using the sampling distribution of means to estimate the mean of the parent population, but it has little real-life significance. In real life the best that can be done is to use the sample variance, $s_Y^2$, as an estimate of the variance of the parent population, $\sigma_Y^2$ (see Equations 12.9 and 12.12). Thus, if the sampling is with replacement,

$$s_Y^2 = \frac{\sum_{i=1}^{n}(y_i - \bar{y})^2}{n-1} \xrightarrow{\text{UNBIASED}} \sigma_Y^2$$

and, if the sampling is without replacement,

$$s_Y^2 = \frac{\sum_{i=1}^{n}(y_i - \bar{y})^2}{n-1}\left(\frac{N-1}{N}\right) \xrightarrow{\text{UNBIASED}} \sigma_Y^2$$

Then, using the relationships discussed in Section 12.10, if the sampling is with replacement (see Equation 12.3),

$$\sigma_{\bar{y}}^2 = \sigma_Y^2 / n$$

Substituting $s_Y^2$ for $\sigma_Y^2$,

$$s_{\bar{y}}^2 = \frac{s_Y^2}{n} \rightarrow \sigma_{\bar{y}}^2 \tag{13.11}$$

If the sampling is without replacement (see Equation 12.5),

$$\sigma_{\bar{y}}^2 = \frac{\sigma_Y^2}{n}\left(\frac{N-n}{N-1}\right)$$

Substituting $s_Y^2(N - 1/N)$ for $\sigma_Y^2$,

$$s_{\bar{y}}^2 = \frac{s_Y^2}{n}\left(\frac{N-1}{N}\right)\left(\frac{N-n}{N-1}\right)$$
$$= \frac{s_Y^2}{n}\left(\frac{N-n}{N}\right) \to \sigma_{\bar{y}}^2 \quad (13.12)$$

When the sample standard error of the mean is substituted into Equation 13.10, it becomes

$$\bar{y} \pm z_C s_{\bar{y}} \to \mu_{\bar{y}} = \mu_Y \quad (13.13)$$

However, in this expression there are *two* random variables, $y$ and $s_{\bar{y}}$. Under these circumstances the assumption of normality and the use of $Z$ is inappropriate. Instead, it is necessary to use a value, corresponding to $Z$, which takes into account the variability of both $y$ and $s_{\bar{y}}$.

If $y$ is the mean of a random sample of size $n$ drawn from a population that has a *normal* distribution with the parameters $\mu_Y$ and $\sigma_Y^2$, then, according to the central limit theorem, it is a value from a normally distributed sampling distribution of means having the parameters $\mu_{\bar{y}}$ and $\sigma_Y^2$. Furthermore, $\mu_y = \mu_Y$ and $\sigma_{\bar{y}}^2 = \sigma_Y^2 / n$. Then,

$$t = (\bar{y} - \mu_y) / s_y \quad (13.14)$$

is a value of a random variable, $T$, having the pdf

$$f(t) = \frac{\Gamma[(k+1)/2]}{\sqrt{\pi k}\,\Gamma(k/2)}[(t^2/k)+1]^{-(k+2)/2}, \quad -\infty < t < \infty, \quad k = 1, 2, \ldots, \infty \quad (13.15)$$

where $k$ is the number of degrees of freedom. The derivation of $f(t)$ is as follows. Since $s_{\bar{y}} = s_Y / \sqrt{n}$,

$$t = (\bar{y} - \mu_{\bar{y}}) / (s_Y / \sqrt{n}) \quad (13.16)$$

Multiplying $s_Y / \sqrt{n}$ by $\sigma_Y/\sigma_Y$ yields

$$t = \frac{\bar{y} - \mu_{\bar{y}}}{\left(\frac{s_Y}{\sqrt{n}}\right)\left(\frac{\sigma_Y}{\sigma_Y}\right)} = \frac{\bar{y} - \mu_{\bar{y}}}{\left(\frac{s_Y}{\sigma_Y}\right)\left(\frac{\sigma_Y}{\sqrt{n}}\right)}$$

Since $\sigma_Y / \sqrt{n} = \sigma_{\bar{y}}^2$,

$$t = \frac{\bar{y} - \mu_{\bar{y}}}{\sigma_{\bar{y}}}\left|\frac{\sigma_Y}{s_Y}\right|$$

Let

$$z = \frac{\bar{y} - \mu_y}{\sigma_y} \quad (13.17)$$

which standardizes the error $\bar{y} - \mu_y$.

Since the sampling distribution of which $y$ is a value is normally distributed, $Z$ is also normally distributed with the parameters $\mu_Z = 0$ and $\sigma_Z^2 = 1$ (see Section 11.2). Thus, the pdf is

$$g(z) = \frac{1}{\sqrt{2\pi}} e^{-z^2/2}; \quad -\infty < Z < \infty$$

Then $t = z(\sigma_Y/s_Y) = \sigma_Y(z/s_Y)$.

The variance of the sample is

$$s_Y^2 = \frac{\sum_{i=1}^{n}(y_i - \bar{y})^2}{n-1}$$

As was brought out in Section 12.11, $s_Y^2$ is also equal to

$$\begin{aligned} s_Y^2 &= \frac{\sigma_Y^2}{n-1}\left[\left(\frac{y_1 - \bar{y}}{\sigma_Y}\right)^2 + \left(\frac{y_2 - \bar{y}}{\sigma_Y}\right)^2 + \cdots + \left(\frac{y_n - \bar{y}}{\sigma_Y}\right)^2\right] \\ &= \frac{\sigma_Y^2}{n-1}\sum_{i=1}^{n}\left(\frac{y_i - \bar{y}}{\sigma_Y}\right)^2 \end{aligned}$$

In this expression, each residual has been transformed into standard form. Let

$$v = \sum_{i=1}^{n}\left(\frac{y_i - \bar{y}}{\sigma_Y}\right) \tag{13.18}$$

and $k = n - 1$. Then, according to Section 12.11, $V$ is distributed as $\chi^2(k)$:

$$h(v) = \frac{1}{2^{k/2}\Gamma(k/2)} v^{(k/2)-1} e^{-v/2}, \quad 0 < V < \infty$$

Then $s_Y^2 = \sigma_Y^2(v/k)$ and $s_Y = \sigma_Y\sqrt{(v/k)}$.

Consequently, $t = \sigma_Y(z/\sigma_Y\sqrt{v/k}) = z/\sqrt{v/k}$ and

$$T = Z/\sqrt{V/k} \tag{13.19}$$

Let $U = V$. Then $u = v$, $t = z/\sqrt{u/k}$, $z = t\sqrt{u/k}$, and $v = u$.

$$\frac{\partial z}{\partial t} = \sqrt{u/k}, \quad \frac{\partial z}{\partial u} = \frac{t}{2\sqrt{ku}}, \quad \frac{\partial v}{\partial t} = 0; \quad \frac{\partial v}{\partial u} = 1$$

The Jacobian is then

$$J(t, u) = \begin{vmatrix} \sqrt{u/k} & t/2\sqrt{ku} \\ 0 & 1 \end{vmatrix} = \sqrt{u/k}$$

and the joint pdf of $T$ and $U$ is

$$\ln(t, u) = g(t\sqrt{u/k})h(u)|J|$$

$$= \left[\frac{1}{\sqrt{2\pi}} e^{-(t\sqrt{u/k})^2/2}\right] * \left[\frac{1}{2^{k/2}\Gamma(k/2)} u^{(k/2)-1} e^{-u/2}\right] * \sqrt{u/k}, \quad -\infty < t < \infty; \ 0 < u < \infty$$

Considering the exponents of $e$:

$$-(t\sqrt{u/k})^2/2 - u/2 = -\left(\frac{t^2 u}{2k} + \frac{u}{2}\right) = -\frac{u}{2}\left(\frac{t^2}{k} + 1\right)$$

and the product

$$u^{(k/2)-1} * (u/k)^{1/2} = u^{[(k+1)/2]-1} k^{-1/2}$$

Then

$$\ln(t, u) = \frac{1}{\sqrt{2\pi} 2^{k/2}\Gamma(k/2)} u^{[(k+1)/2]-1} e^{-(u/2)[(t^2/k)+1]}$$

$$= \frac{1}{\sqrt{2\pi k} 2^{k/2}\Gamma(k/2)} u^{[(k+1)/2]-1} e^{-(u/2)[(t^2/k)+1]}$$

The pdf of $T$ is then (see Section 9.6)

$$f(t) = \int_{-\infty}^{\infty} \ln(t, u) du$$

Since $t\sqrt{u/} = z$ and since $-\infty < Z < \infty$, $-\infty < t\sqrt{u/} < \infty$. Then

$$\begin{array}{lll} & -\infty < t\sqrt{u/k} \quad \text{and} & t\sqrt{u/k} < \infty \\ & -\infty / \sqrt{u/k} < t & t < \infty / \sqrt{u/k} \\ & -\infty < t & t < \infty \\ \text{and} & -\infty < t < \infty & \end{array}$$

Furthermore, since $u = v$, and since $0 < V < \infty$, $0 < U < \infty$. Then

$$f(t) = \frac{1}{\sqrt{2\pi k} 2^{k/2}\Gamma(k/2)} \int_{0}^{\infty} u^{[(k+1)/2]-1} e^{-(u/2)[(t^2/k)+1]} du$$

In this integral, let $w = (u/2)[(t^2/k) + 1]$. Then $u = 2w/[(t^2/k) + 1]$.

Converting the units of integration from $u$ to $w$:

$$\frac{dw}{du} = \frac{1}{2}[(t^2/k)+1] \quad \text{and} \quad du = \{2/[(t^2/k)+1]\}dw$$

Then

$$f(t) = \frac{1}{\sqrt{2\pi k}\, 2^{k/2}\Gamma(k/2)} \int_0^\infty \left[\frac{2w}{(t^2/k)+1}\right]^{[(k+1)/2]-1} e^{-w}\left[\frac{2}{(t^2/k)+1}\right]dw$$

$$= \left[\frac{1}{\sqrt{2\pi k}\, 2^{k/2}\Gamma(k/2)}\right] * \left[\frac{2}{(t^2/k)+1}\right] * \left[\frac{2}{(t^2/k)+1}\right]^{[(k+1)/2]-1} \int_0^\infty w^{[(k+1)/2]-1} e^{-w} dw$$

Since $\int_0^\infty w^{[(k+1)/2]-1} e^{-w} dw = \Gamma[(k+1)/2]$,

$$f(t) = \left\{\frac{2\Gamma[(k+1)/2]}{\sqrt{2\pi k}\, 2^{k/2}\Gamma(k/2)[(t^2/k)+1]}\right\} * \left[\frac{2}{(t^2/k)+1}\right]^{[(k+1)/2]-1}$$

Considering only the terms

$$2^{-k/2} * 2^{[(k+1)/2]-1} = 2^{-k/2+k/2+1/2-1} = 2^{-1/2}$$

Then the terms

$$\frac{2}{\sqrt{2\pi k}} * \frac{1}{\sqrt{2}} = \frac{2}{2\sqrt{\pi k}} = \frac{1}{\sqrt{\pi k}}$$

and the terms

$$[(t^2/k)+1]^{-1} * [(t^2/k)+1]^{-[(k+1)/2]-1} = [(t^2/k)+1]^{-(k+1)/2}$$

Thus,

$$f(t) = \frac{\Gamma[(k+1)/2][(t^2/k)+1]^{-(k+1)/2}}{\sqrt{\pi k}\,\Gamma(k/2)}, \quad -\infty < t < \infty, \; k = 1, 2, \ldots, \infty$$

which is Equation 13.15.

The distribution described by this pdf is known as Student's $t$ distribution.* As can be deduced from the pdf, the $t$ distribution is actually a set of distributions, each being defined in terms of a specific number of degrees of freedom, $k$. A representative set of these distributions is shown in Figure 13.2. The curves of the distributions are symmetrical, bell shaped, and asymptotic to the horizontal axis. All are platykurtic but, as the number of degrees of freedom increases, the degree of platykurtosis decreases until, in the limit as $k \to \infty$, the curves converge on that of the normal. This convergence is relatively rapid so that most workers assume normality when $k$ exceeds 30. In all cases,

* "Student" was the pseudonym for W. S. Gosset, who first described the $t$ distribution in 1908. The pseudonym was used since Gosset's employers forbade their employees to publish research results.

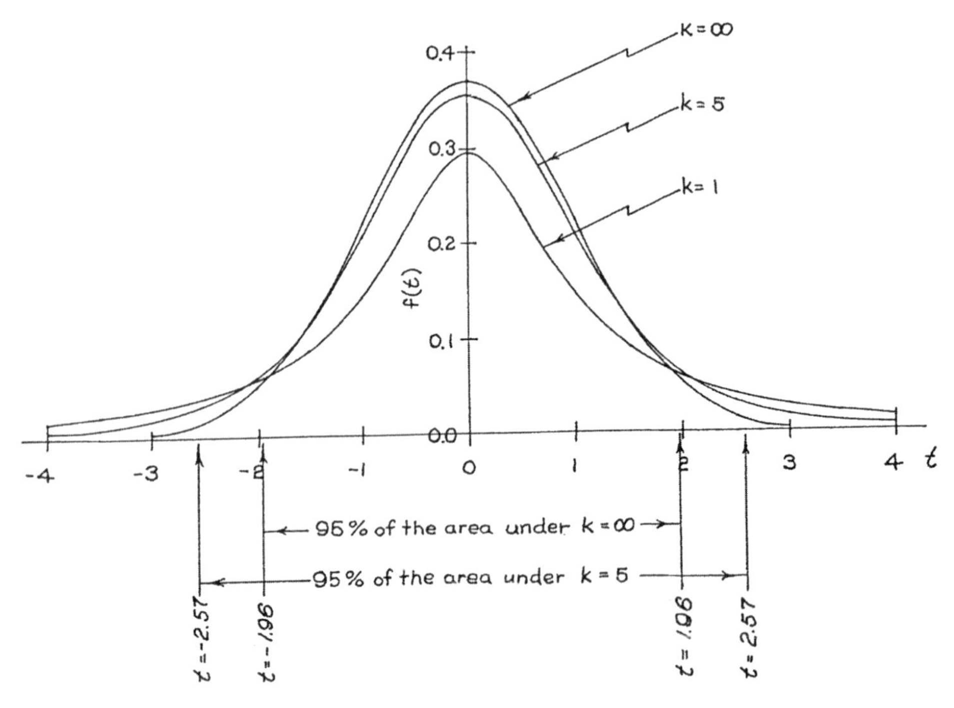


**FIGURE 13.2** Curves of the $t$ distribution when $k = \infty$, 5, and 1. Note the platykurtic nature of the distributions. As $k$ becomes smaller, the curves become flatter and the interval along the number line needed to include a specific area under the curves increases. When $k = 1$ the interval for 95% of the area extends from −12.71 to +12.71 on the $t$ scale.

$$\mu_t = 0 \tag{13.20}$$

and

$$\sigma_t^2 = k/(k-2), \quad k > 2 \tag{13.21}$$

As in the case of the normal and several other probability distributions described in this book, tables have been prepared that facilitate the use of the $t$ distributions. Table A.7 in the Appendix is such a table. Table A.7 is comparable with Table 13.1, which lists confidence or probability levels, but instead of doing this for only one distribution it does it for 31 different $t$ distributions. When $k \to \infty$, the $t$ distributions converge to the normal and, consequently, in the bottom row in Table A.7, which is for the $t$ distribution based on $k = \infty$, the confidence coefficients are the same as in Table 13.1.

It should be noted that the fractiles in $t$ tables are computed from the means of their respective distributions, $\mu_t = 0$, and that only positive values are shown. Because of the symmetry of the distributions about their means, the negative fractiles have the same absolute magnitudes as their positive counterparts and, consequently, do not need to be listed.

The confidence or probability levels can be based on either of two approaches. In the case of Table A.8 in the Appendix, the area under the curve corresponding to the confidence level is centered on $t = 0$ and, consequently, there are two fractiles associated with a given confidence level, $C$:

$$t_a = t_{(1-c)/2}(k), \quad \text{on the lower extreme} \tag{13.22}$$

where

$$P[t < t_{(1-C)/2}(k)] = (1 - C)/2 \tag{13.23}$$

and

$$t_b = t_{(C+1)/2}(k), \quad \text{on the upper extreme} \tag{13.24}$$

where

$$P([t < t_{(C+1)/2}(k)] = (C + 1)/2 \tag{13.25}$$

In other words,

$$P(t_a < t < t_b) = C \tag{13.26}$$

For example, if $C = 0.95$, $(1 - C)/2 = 0.025$, $(C + 1)/2 = 0.975$, and $P[t_{0.025}(k) < t < t_{0.975}(k)] = 0.95$.

Because of the symmetry of the $t$ distributions,

$$|t_a - 0| = |t_b - 0| \tag{13.27}$$

and, thus, only the $t_b$ values are shown in the table. Furthermore, these $t_b$ values are listed in the column under $C$, *not* under $(C + 1)/2$. If $C = 0.95$ the $t_b$ values are actually the fractiles for 0.975 but are listed under $C = 0.95$. Since the portion of the area under the curve that is associated with the probability of *failure* of a random value of $t$ to occur in the indicated interval is split into two parts, one in each tail of the distribution, this type of table is referred to as a *two-tailed* table.

The alternate approach to $t$ table construction is the conventional one in which the area under the curve corresponding to the confidence level extends from the lower extreme, $-\infty$, to a single fractile,

$$t_b = t_c(k), \quad \text{on the upper extreme} \tag{13.28}$$

where

$$P[-\infty < t < t_C(k)] = C \tag{13.29}$$

Tables based on this approach are known as *one-tailed* tables. If a one-tailed table is to be used when developing a confidence interval, it will be necessary to use the $t(k)$ value in the column headed $(C + 1)/2$ rather than the one headed $C$, in order to obtain the $t_a$ and $t_b$ value shown in Equation 13.26.

Returning to the development of a confidence interval, if the confidence level has been set at 0.90 and $k = 15$, the confidence coefficient is obtained from Table A.8 in the column headed 0.90 and the row labeled 15. Thus,

$$P[\bar{y} - t_{0.90}(15)s_{\bar{y}} < \mu_Y < \bar{y} + t_{0.90}(15)s_{\bar{y}}] = 0.90$$

or

$$\bar{y} \pm t_{0.90}(15)s_{\bar{y}} = \bar{y} \pm 1.753s_{\bar{y}} \rightarrow \mu_Y \tag{13.30}$$

In other words, when the curve appropriate for 15 degrees of freedom is used, 90% of the $t$ values occur in the interval from $-1.753$ to $+1.753$ and, consequently, 90% of the brackets of size $\pm 1.753$, attached to the $t$ values will enclose the mean, $\mu_t = 0$. Since, as in the case of $Z$, the mean of the $t$ distribution corresponds to the mean being estimated, $\mu_{\bar{y}}$, if the bracket ±1.753 encloses $\mu_t$, the mean bracket $\pm 1.753\, s_{\bar{y}}$ will enclose $\mu_{\bar{y}}$.

As defined, $t$ distributions should only be used when the parent population is normally distributed. However, statistical operations based on the use of the normal distribution are usually *robust*, meaning that they usually yield sufficiently satisfactory results for most purposes, even if the parent population is only approximately normal. Thus, the use of the $t$ distribution for constructing confidence intervals has become common practice, regardless of the nature and distributional pattern of the parent random variable.

## 13.10 CHEBYSHEV'S THEOREM

The concept of and the procedures associated with confidence intervals were explored in the preceding sections. At the heart of the process is the probability distribution of the random variable. If this is not known, the process is more or less crippled. However, it is possible to develop a form of confidence interval using Chebyshev's theorem.

The theorem states that if a random variable, say, $Y$, has a finite mean, $\mu_Y$, and a finite standard deviation, $\sigma_Y$, the probability that a variate, $y_i$, will deviate from $\mu_Y$ by more than $r\sigma_Y$, where $r$ is an arbitrary value, is equal to or less than $1/r^2$. In algebraic form,

$$P(|y_i - \mu_Y| \geq r\sigma_Y) \leq 1/r^2 \tag{13.31}$$

This inequality gives rise to the term *Chebyshev's inequality.*

The proof, in the case of a discrete distribution, is as follows. According to Equation 8.32,

$$\sigma_Y^2 = \sum_{i=1}^{N} (y_i - \mu_Y)^2 P(y_i)$$

Let the range space of $Y$ be divided into three subsets: $R_1$, where $y_i \leq \mu_Y - r\sigma_Y$; $R_2$, where $(\mu_Y - r\sigma_Y) < y_i < (\mu_Y + r\sigma_Y)$; and $R_3$, where $y_i \geq \mu_Y + r\sigma_Y$. Then

$$\sigma_Y^2 = \sum^{R_1} (y_i - \mu_Y)^2 P(y_i) + \sum^{R_2} (y_i - \mu_Y)^2 P(y_i) + \sum^{R_3} (y_i - \mu_Y)^2 P(y_i)$$

None of the terms in this expression can be negative since the $(y_i - \mu)$ terms are squared and probabilities are always positive. Then,

$$\sigma_Y^2 \geq \sum^{R_1} (y_i - \mu_Y)^2 P(y_i) + \sum^{R_3} (y_i - \mu_Y)^2 P(y_i) \tag{13.32}$$

Since in $R_1$, $(y_i - \mu_Y) \leq -r\sigma_Y$, and in $R_3$, $(y_i - \mu_Y) \geq r\sigma_Y$.

$$|\, y_i - \mu_Y \,| \geq r\sigma_Y \quad \text{and} \quad (y_i - \mu_Y)^2 \geq r^2\sigma_Y^2$$

Returning to Equation 13.32 and substituting $r^2\sigma_Y^2$ for $(y_i - \mu_Y)^2$,

$$\sigma_Y^2 \geq \sum^{R_1} r^2\sigma_Y^2 P(y_i) + \sum^{R_3} r^2\sigma_Y^2 P(y_i)$$

Dividing both sides by $r^2\sigma_Y^2$ and simplifying,

$$\frac{1}{r^2} \geq \sum^{R_1} P(y_i) + \sum^{R_3} P(y_i)$$

Consequently, the probability of $y_i$ occurring in the region $R_1 \cup R_3$ is no more than $1/r^2$.

$$P(|y_i - \mu_Y| \geq r\sigma_Y) \leq 1/r^2$$

Conversely, the probability of $y_i$ occurring in region $R_2$ is at least $1 - 1/r^2$,

$$P(|y_i - \mu_Y| < r\sigma_Y) > 1 - 1/r^2 \tag{13.33}$$

or

$$P[(\mu_Y - r\sigma_Y) < y_i < (\mu_Y + r\sigma_Y)] > 1 - 1/r_2 \tag{13.34}$$

The proof of the continuous case follows the same pattern.

Equation 13.31 is used to arrive at the probability of drawing a value of the random variable which differs from the mean by *at least* $r\sigma_Y$. For example, if $\mu_Y = 100$, $\sigma_Y = 10$, and $r$ is set at 3, the probability of drawing a value of $Y$ that differs from $\mu_Y = 100$ by at least $r\sigma_Y = 3(10) = 30$ is *at most* $1/(3)^2 = 1/9 = 0.1111$. Thus, Chebyshev's theorem provides a maximum bound for the probability rather than a specific probability. This bound can be misleading. This can be seen if one assumes that the distribution in the preceding example is normal. If this is the case, the probability of drawing a value of $Y$ less than or equal to $\mu_Y - r\sigma_Y = 100 - 3(10) = 70$ or more than or equal to $\mu + r\sigma_Y = 100 + 3(10) = 130$, using the methods described in Section 11.2, is 0.0026. This is much less than 0.1111, and if 0.1111 is used in place of 0.0026, it could cause a serious error.

Equation 13.34 is used to find the probability of occurrence of a value of the random variable that differs from the mean by *less than* $r\sigma_Y$. Using the same data as before, the probability of $y_i$ being greater than $\mu_Y - r\sigma_Y = 100 - 3(10) = 70$, but less than $\mu_Y + r\mu_Y = 100 + 3(10) = 130$, is less than $1 - 1/r^2 = 1 - 1/(3)^2 = 0.8889$. Again Chebyshev's theorem provides a bound on the probability rather than a specific probability. In this case it is a minimum bound and usually it is quite conservative. For example, if the distribution in the example is actually normally distributed, the actual probability is 0.9974.

In Equation 13.34,

$$P(\mu_Y - r\sigma_Y) < y_i < \mu_Y + r\sigma_Y)$$

Solving for $\mu_Y$ yields

$$P(y_i - r\sigma_Y) < \mu_Y < (y_i + r\sigma_Y)$$

and, thus,

$$P[(y_i - r\sigma_Y) < \mu_Y < (y_i + r\sigma_Y)] > 1 - 1/r^2 \tag{13.35}$$

Equation 13.35 is the equivalent of Equation 13.9 except that instead of a definite confidence level a minimum bound to the confidence level is used. In most cases, the actual probability of the confidence interval including the mean would be greater than $1 - 1/r^2$, but when the shape of the distribution of the random variable cannot be ascertained an interval estimate of this type may be the only one available.

If one wished to develop a confidence interval using a confidence level of 0.95 *as the minimum*, the corresponding value of $r$ is obtained as follows.

$$1 - 1/r_2 = 0.95$$

$$1 - 0.95 = 0.05 = 1/r^2$$

$$r = 4.4721$$

If $\sigma = 10$, the confidence interval would be

$$P[(y_i - 44.721) < \mu_Y < (y_i + 44.721)] > 0.95 \tag{13.36}$$

This can be compared with the confidence interval if the random variable is normally distributed:

$$P[(y_i - 1.96) < \mu_Y < (y_i + 1.96)] = 0.95 \tag{13.37}$$

The interval is much wider in Equation 13.36 than it is in Equation 13.37. If the distribution is normal and a confidence coefficient of 4.4721 is used, the actual probability is greater than 0.9999.

The *Camp–Meidell extension* of Chebyshev's theorem can be substituted for the Chebyshev inequality if the random variable is known to have a distribution that is unimodal. This extension states that, if the random variable has a unimodal distribution, the probability that a variate will deviate from the mean by more than $r\sigma_Y$ is equal to or less than $1/2.25r^2$. The algebraic form of this statement is

$$P(|y_i - \mu_Y| \geq r\sigma_Y) \leq 1/2.25r^2 \tag{13.38}$$

Obviously, as written, both the Chebyshev theorem and the Camp–Meidell extension require knowledge of the population standard deviation, $\sigma_Y$, which usually is as unknown as is the mean. Consequently, it must be estimated. The sample standard deviation is a random variable in its own right and its contribution to the variance is not known nor is any compensation for this included in the theorems.

## 13.11 CONFIDENCE INTERVAL ESTIMATES OF POPULATION PROPORTIONS

It was brought out in Section 12.13 that sample proportions are associated with values of a discrete random variable, say, $Y$, which is the number of successes in $n$ binomial trials. The distribution of the sample proportions follows that of $Y$ and may be binomial or Poisson, if the sampling is with replacement or the parent population is infinitely large, or hypergeometric, if the sampling is without replacement. In any of these cases the histogram of the distribution is made up of distinct bars and does not form a smooth curve. As a result, it often is impossible to develop an exact confidence interval at a specified confidence level. Furthermore, when developing or describing a confidence interval involving proportions, it is necessary to define precisely what is meant by the confidence interval. If the confidence limits are *included* in the interval, the definition is

$$P(\hat{p}_a \leq p \leq \hat{p}_b) = C_b - C_a = C \tag{13.39}$$

where $\hat{p}_a$ = lower confidence limit
$\hat{p}_b$ = upper confidence limit

$$C_a = P(Y / n < \hat{p}_a) \tag{13.40}$$

$$C_b = P(Y / n \leq \hat{p}_b) \tag{13.41}$$

$C$ = confidence level

On the other hand, if the limits are *outside* the interval, it is defined as

$$P(\hat{p}_a < p < \hat{p}_b) = C_b - C_a = C \tag{13.42}$$

where

$$C_a = P(Y / n \leq \hat{p}_a) \tag{13.43}$$

$$C_b = P(Y / n < \hat{p}_b) \tag{13.44}$$

The difference between the two definitions can be noticeable when sample sizes are small.

Confidence limits should be values that can actually occur. In other words, each limit should be associated with a definite possible sample proportion. A stated confidence limit that is not associated with a possible outcome is, strictly speaking, meaningless. In some of the approaches used to arrive at confidence interval estimates of population proportions, the gaps between the possible sample proportions are bridged using smooth curves and the confidence limits are obtained from the smooth curves. These usually do not coincide with possible outcomes and consequently are approximations of the actual limits. Furthermore, the confidence levels associated with confidence intervals obtained in this way are also only approximations.

### 13.11.1 True Confidence Intervals in the Case of the Binomial Distribution

The closest one can come to a true confidence interval in the case of proportions is to determine the actual proportions which can serve as confidence *limits*. Assuming that the sampling is with replacement, so that the sampling distribution is binomial, and that the confidence interval is defined as ($\hat{p}_a < p < \hat{p}_b$), the lower limit, $\hat{p}_a = y_a/n$, is that which is associated with the *largest* value of $Y$ that will satisfy the following equation when $p$ is fixed at some value and $C_a = (1 - C)/2$,

$$\sum_{y=0}^{y_a} \binom{n}{y} p^y (1-p)^{n-y} \leq C_a \tag{13.45}$$

To find $\hat{p}_b = y_b/n$, the upper limit, one determines the *smallest* value of $Y$ which will satisfy Equation 13.46 or 13.47, when $C_b = (C + 1)/2$,

$$\sum_{y=y_b}^{n} \binom{n}{y} p^y (1-p)^{n-y} \leq 1 - C_b \tag{13.46}$$

Isolating $C_b$ yields

$$C_b \leq 1 - \sum_{y=y_b}^{n} \binom{n}{y} p^y (1-p)^{n-y}$$

Since

$$\sum_{y=0}^{y_b-1} \binom{n}{y} p^y (1-p)^{n-y} + \sum_{y=y_b}^{n} \binom{n}{y} p^y (1-p)^{n-y} = 1$$

$$C_b \leq \sum_{y=0}^{y_b-1} \binom{n}{y} p^y (1-p)^{n-y} \tag{13.47}$$

For example, if $p = 0.4$, $n = 20$, and $C = 0.95$,

$$C_a = (1-C)/2 = 0.025$$

$$C_b = (C+1)/2 = 0.975$$

A series of trial runs will show that the largest value of $Y$ having a cumulative probability less than or equal to 0.025 is $y_a = 3$. Then,

$$\hat{p}_a = y_a / n = 3/20 = 0.15$$

since

$$\sum_{y=0}^{3} \binom{20}{y} 0.4^y (1-0.4)^{20-y} = 0.01597 < 0.025$$

and

$$\sum_{y=0}^{4} \binom{20}{y} 0.4^y (1-0.4)^{20-y} = 0.05096 > 0.025$$

Similarly, the smallest value of $Y$ having a cumulative probability greater than or equal to 0.975 is $y_b - 1 = 12$. Consequently, $y_b = 12 + 1 = 13$.

$$\hat{p}_b = y_b / n = 13/20 = 0.65$$

Since

$$\sum_{y=0}^{13-1} \binom{20}{y} 0.4^y (1-0.4)^{20-y} = 0.97898 > 0.975$$

and

$$\sum_{y=0}^{12-1} \binom{20}{y} 0.4^y (1-0.4)^{20-y} = 0.94348 < 0.975$$

The confidence interval would then be

$$P(0.15 < p < 0.65) \cong 0.95$$

As can be seen, this process can be applied to each of the $n + 1 = 21$ possible sample proportions. These $n + 1$ intervals are the only *true* confidence intervals that can be obtained. The actual fiducial probabilities associated with these intervals are only rarely equal to the target or desired confidence levels. For example, the actual fiducial probability in the case of the preceding example is (see Equation 13.42):

$$P(\hat{p}_a < p < \hat{p}_b) = C_b - C_a = C$$

According to Equation 13.43,

$$C_a = P(Y / n \le \hat{p}_a) = P(Y / n \le 0.15) = 0.01597$$

and according to Equation 13.44,

$$C_b = P(Y / n < \hat{p}) = P(Y / n < 0.65) = 0.97898$$

Then

$$P(\hat{p}_a < p < \hat{p}_b) = 0.97898 - 0.01597 = 0.96301$$

which is somewhat more than the desired level of 0.95.

If the confidence interval had been defined as in Equation 13.39, the interval and its associated fiducial probability would have been:

$$P(0.20 \le p \le 0.60) = 0.96301$$

since, according to Equation 13.40,

$$C_a = P(Y / n < \hat{p}_a) = P(Y / n < 0.20) = 0.01597$$

and, according to Equation 13.41,

$$C_b = P(Y / n \le \hat{p}_b) = P(Y / n \le 0.60) = 0.96301$$

which are the same probabilities obtained using the alternate definition and, thus, the actual fiducial probability is not affected by the definition.

However, one should be aware of the fact that alternative definitions exist and should be considered when interpreting a confidence interval.

### 13.11.2 True Confidence Intervals in the Case of the Hypergeometric Distribution

If the sampling is without replacement, the sampling distribution of proportions is hypergeometrically distributed. However, the process used to obtain the confidence limits is essentially the same as that described above when the sampling distribution was binomial. The lower limit, $\hat{p}_a = y_a/n$, is that associated with the *largest* value of $Y$ that will satisfy the following equation when $C_a = (1 - C)/2$:

$$\frac{1}{\binom{N}{n}} \sum_{y=0}^{y_a} \binom{N_s}{y} \binom{N - N_s}{n - y} \leq C_a \tag{13.48}$$

The upper limit, $\hat{p}_b = y_b/n$, is obtained using the *smallest* value of $Y$ that will satisfy the following equation when $C_b = (C + 1)/2$:

$$\frac{1}{\binom{N}{n}} \sum_{y=0}^{n} \binom{N_s}{y} \binom{N - N_s}{n - y} \leq 1 - C_b$$

Since

$$\begin{gathered}\frac{1}{\binom{N}{n}} \sum_{y=0}^{y_b-1} \binom{N_s}{y} \binom{N - N_s}{n - y} + \frac{1}{\binom{N}{n}} \sum_{y=y_b}^{n} \binom{N_s}{y} \binom{N - N_s}{n - y} = 1 \\ C_b \leq \frac{1}{\binom{N}{n}} \sum_{y=0}^{y_b-1} \binom{N_s}{y} \binom{N - N_s}{n - y}\end{gathered} \tag{13.49}$$

For example, if $N = 40$, $N_s = 16$, $n = 20$, and $C = 0.95$, then

$$p = N_s / N = 16.40 = 0.4$$

$$C_a = (1 - C)/2 = 0.025$$

$$\text{and} \quad C_b = (C + 1)/2 = 0.975$$

Following the same pattern used earlier, the largest value of $Y$ having a cumulative probability less than or equal to 0.025 is $y_a = 4$. Then,

$$\hat{p}_a = y_a / n = 4/20 = 0.20$$

Since

$$\frac{1}{\binom{40}{20}}\sum_{y=0}^{4}\binom{16}{y}\binom{40-16}{20-y} = 0.01124 < 0.025$$

and

$$\frac{1}{\binom{40}{20}}\sum_{y=0}^{5}\binom{16}{y}\binom{40-16}{20-y} = 0.05267 > 0.025$$

The smallest value of $Y$ having a cumulative probability greater than or equal to 0.975 is $y_b - 1 = 11$. As a result, $y_b = 11 + 1 = 12$.

$$\hat{p}_b = y_b / n = 12/20 = 0.60$$

Since

$$\frac{1}{\binom{40}{20}}\sum_{y=0}^{12-1}\binom{16}{y}\binom{40-16}{20-y} = 0.98877 > 0.975$$

and

$$\frac{1}{\binom{40}{20}}\sum_{y=0}^{11-1}\binom{16}{y}\binom{40-16}{20-y} = 0.94734 < 0.975$$

The confidence interval is then

$$P(0.20 < P < 0.60) \cong 0.95$$

It should be noted that when the sampling is without replacement, the number of possible outcomes is equal to $N_s + 1$. In this case $N_s = 16$ and the number of possible outcomes, or sample proportions, is equal to 17, ranging from 0 to 16 successes or proportions from 0 to 0.80. As in the case of the binomial, there is a set of confidence limits for each of the possible sample proportions and these form the only true confidence intervals.

It also should be noted that the confidence intervals associated with sample proportions obtained using sampling without replacement are somewhat smaller than their equivalents obtained using sampling with replacement. This follows the pattern in Section 13.9. Confidence intervals computed from samples obtained using sampling without replacement are never larger than those obtained using sampling with replacement and usually are smaller. The difference between the magnitudes of the confidence intervals is a function of the sampling fraction, $n/N$, becoming greater as the sampling fraction increases.

### 13.11.3 Confidence Intervals Based on the Normal Distribution

In Section 13.5, in the derivation of the $t$ distributions, use was made of the central limit theorem, which states that samples drawn from normally distributed populations have means which are normally distributed and even if the parent population is not normally distributed, but is continuous, the sampling distribution will tend toward the normal with the convergence increasing as the sample size increases. Sample proportions are special cases of sample means but, because they are discrete, one cannot use the central limit theorem to arrive at an indication of their distribution. However, it was shown in Section 11.2 that the normal can be derived from the binomial and is equivalent to the special case of the binomial when $n \to \infty$ and $p = 0.5$. Under these circumstances, the random variable $Y$ is distributed both binomially and normally. However, when $n$ is a finite quantity, the two distributions differ. The magnitude of the difference is a function of both $n$ and $p$, but $n$ is dominant since with increasing $n$ the binomial distribution converges toward the normal regardless of the magnitude of $p$. Thus, when $n$ is sufficiently large, $Y$ is approximately normally distributed with mean $np$ and variance $np(1 - p)$.

$$\sum_{y=0}^{y_o} \binom{n}{y} p^y (1-p)^{n-y} \cong \frac{1}{\sqrt{2\pi np(1-p)}} e^{-[(y_o - np)^2 / 2np(1-p)]} \tag{13.50}$$

where $y_o$ = observed value of $Y$.

Putting the distribution of $Y$ into standard form,

$$Z = \frac{Y - np}{\sqrt{np(1-p)}} \tag{13.51}$$

$Z$ is then approximately normally distributed with $\mu_Z = 0$ and $\sigma_Z^2 = 1$. Multiplying the right-hand term by $n/n$ yields

$$Z = \frac{Y / n - p}{\sqrt{p(1-p) / n}} \tag{13.52}$$

Now $Z$ is approximately normally distributed with $\mu_Z = 0$ and $\sigma_Z^2 = p(1 - p)/n$. Since individual values of $Y$ (e.g., $Y = y_o$) give rise to sample proportions (e.g., $\hat{p}_o = y_o/n$), Equation 13.52 refers to an approximately normal distribution of sample proportions.

Then, following the pattern used in Section 13.9, the confidence interval estimate of the population proportion is

$$\hat{p} \pm z_C \sigma_{\hat{p}} \to p \tag{13.53}$$

where

$$\sigma_{\hat{p}} = \sqrt{p(1-p) / n} \tag{13.54}$$

However, $p$ is not known and must be estimated by $\hat{p}_o$. Then, assuming sampling with replacement, $\sigma_{\hat{p}}$ is replaced by its estimator, $s_{\hat{p}}$:

$$s_{\hat{p}} = \sqrt{\frac{\hat{p}_o(1-p)_o}{n}} \tag{13.55}$$

But $\hat{p}_o(1 - p_o)/n$ is the equivalent of $\Sigma_{i=1}^{n}(y_i - \bar{y})^2/n$, which is biased (see Equation 12.7). In other words,

$$E[\hat{p}(1-p)] \neq p(1-p) \tag{13.56}$$

If $\Sigma_{i=1}^{n}(y_i - \bar{y})^2/n$ is multiplied by $n$, the resulting product is the sum of squared residuals. Thus, when $\hat{p}(1-\hat{p})/n$ is multiplied by $n$, the product is the equivalent of the sum of squares, $n\hat{p}_o(1-\hat{p}_o)$.

Following the pattern of Equation 12.9, dividing the sum of squares by the degrees of freedom, $n - 1$, yields an unbiased estimate of $p(1 - p)$:

$$E\left[\frac{n\hat{p}(1-\hat{p})}{n-1}\right] = p(1-p) \tag{13.57}$$

Then,

$$s_{\hat{p}} = \sqrt{\frac{n\hat{p}_o(1-\hat{p}_o)}{n-1}\left(\frac{1}{n}\right)} = \sqrt{\frac{\hat{p}_o(1-\hat{p}_o)}{n-1}} \tag{13.58}$$

Returning to Equation 13.54, if the sampling had been without replacement, the equivalent equation (see Equation 12.47) would be

$$\sigma_{\hat{p}} = \sqrt{\frac{p(1-p)}{n}\left(\frac{N-n}{N-1}\right)} \tag{13.59}$$

Again $p$ is not known and must be estimated by $\hat{p}_o$. Then

$$s_{\hat{p}} = \sqrt{\frac{\hat{p}_o(1-\hat{p}_o)}{n}\left(\frac{N-n}{N-1}\right)} \tag{13.60}$$

But, as before, $\hat{p}_o(1 - \hat{p}_o)/n$ is biased and the $n/(n - 1)$ correction must be applied. Furthermore, a finite population correction must be included to compensate for the changes in the sampling frame as the sampling without replacement proceeds (see Equation 12.12):

$$E\left[\frac{n\hat{p}(1-\hat{p})}{n-1}\left(\frac{N-1}{N}\right)\right] = p(1-p) \tag{13.61}$$

Substituting into Equation 13.59 yields

$$\begin{aligned} s_{\hat{p}} &= \sqrt{\frac{n\hat{p}_o(1-\hat{p}_o)}{n-1}\left(\frac{N-1}{N}\right)\left(\frac{N-n}{N-1}\right)\left(\frac{1}{n}\right)} \\ &= \sqrt{\frac{\hat{p}_o(1-\hat{p}_o)}{n-1}\left(\frac{N-n}{N}\right)} \end{aligned} \tag{13.62}$$

Then, when the sampling is with replacement,

$$\hat{p}_o \pm z_C\sqrt{\frac{\hat{p}_o(1-\hat{p}_o)}{n-1}} \rightarrow \mu_{\hat{p}} = p \tag{13.63}$$

and when the sampling is without replacement

$$\hat{p}_o \pm z_C\sqrt{\frac{\hat{P}_o(1-\hat{p}_o)}{n-1}\left(\frac{N-n}{N}\right)} \rightarrow \mu_{\hat{p}} = p \tag{13.64}$$

As an example of application of Equation 13.63 assume that $n = 100$, $y_o = 30$, $C = 0.95$.

$$\hat{p}_o = 30/100 = 0.3$$

$$1-\hat{p}_o = 0.7$$

$$\hat{p}_o(1-p_o) = 0.3(0.7) = 0.21$$

$$s_{\hat{p}} = \sqrt{\frac{0.21}{100-1}} = 0.0460566$$

Then

$$\hat{p}_o \pm z_{C=0.95}s_{\hat{p}} = 0.3 \pm 1.96(0.0460566) = 0.3 \pm 0.090271 \rightarrow \mu$$

The fiducial probability is 0.95 that $0.209729 \leq p \leq 0.390271$.

If the sampling had been without replacement and $N = 10{,}000$:

$$s_{\hat{p}} = \sqrt{\frac{0.21}{100-1}\left(\frac{10{,}000-100}{10{,}000}\right)} = 0.0458258$$

and

$$\hat{p}_o \pm z_{C=0.95}s_{\hat{p}} = 0.3 \pm 1.96(0.0458258) = 0.3 \pm 0.0898185 \rightarrow p$$

The fiducial probability is 0.95 that $0.2101815 \leq p \leq 0.3898185$.

It should be pointed out that many authors do not correct the sample variance for bias. In such a case, assuming sampling with replacement,

$$s_{\hat{p}} = \sqrt{\frac{\hat{p}_o(1-\hat{p})}{n}} = \sqrt{\frac{0.21}{100}} = 0.0458258$$

and

$$\hat{p}_o \pm z_{C=0.95}s_{\hat{p}} = 0.3 \pm 1.96(0.0458258) = 0.3 \pm 0.0898185 \rightarrow p$$

The agreement between this result and that obtained when the sampling was without replacement is by happenstance. However, as can be seen, the difference between the confidence intervals obtained when correcting for bias and when not correcting for bias is negligible, provided that the sample size is large.

Many authors are of the opinion that the correction for continuity (Section 11.2) should be applied whenever a continuous distribution is used as a model of a discrete distribution. Since there are $n + 1$ values of $Y$, there are $n + 1$ bars in the histogram. A half bar width must be added to the bracket size at each end of the interval to compensate for the lack of continuity. Thus, the correction for continuity is

$$\frac{1}{2(n+1)} = \frac{1}{2n+2} \tag{13.65}$$

Applying this to the development of a confidence interval, assuming the situation where the sampling was with replacement,

$$s_{\hat{p}} = \sqrt{\frac{\hat{p}_o(1-\hat{p}_o)}{n-1}} + \frac{1}{2n+2}$$

$$= \sqrt{\frac{0.21}{1-100}} + \frac{1}{2(100)+2}$$

$$= 0.510071$$

and

$$\hat{p}_o \pm z_{C=0.95} s_{\hat{p}} = 0.3 \pm 1.96(0.0510071) = 0.3 \pm 0.0999739 \rightarrow p$$

The fiducial probability is 0.95 that $0.2000261 \leq p \leq 0.3999739$. As can be seen, this interval is somewhat wider than that obtained where the correction for continuity was not used.

The use of the normal as a model of the binomial distribution is considered acceptable provided the sample size is sufficiently large. When $p \neq 0.5$, the binomial distribution is skewed. The skew increases as $p$ deviates farther from 0.5. However, as $n$ increases, for any given $p$, the distribution tends toward the shape of the normal. Consequently, if the normal is to be used as the model, the sample size must be sufficiently large so that the matching of the two distributions is acceptably close. It has been found (Hald, 1952) that this occurs when

$$np(1 - p) > 9 \tag{13.66}$$

Thus, the minimum sample size is a function of $p$:

$$n > 9/p(1 - p) \tag{13.67}$$

For example, if $p = 0.3$,

$$n > 9/0.3(1.0 - 0.3) = 43$$

Table 13.2 lists a representative set of minimum sample sizes derived using Equation 13.67. It should be understood that even with these sample sizes the match is not perfect and error is present.

**TABLE 13.2**
**Minimal Sample Sizes When the Normal Distribution is Being Used as the Model of the Binomial Distribution**

| $p$ | or | $1 - p$ | $n$ |
|---|---|---|---|
| 0.50 | | 0.50 | 36 |
| 0.40 | | 0.60 | 38 |
| 0.30 | | 0.70 | 43 |
| 0.20 | | 0.80 | 56 |
| 0.15 | | 0.85 | 71 |
| 0.10 | | 0.90 | 100 |
| 0.05 | | 0.95 | 189 |
| 0.02 | | 0.98 | 459 |
| 0.01 | | 0.99 | 909 |

Adapted from Hald, 1952, p. 680.

Furthermore, the relative error is not constant, being very small when $p = 0.5$ but increasing as $p$ deviates farther from 0.5. When skew is present, the relative error is not the same in both tails. Consequently, the use of the normal as a model of the binomial should be limited to the central portion of the distribution in the range $np \pm 3np(1 - p)$.

In this discussion of minimal sample sizes, $n$ was shown to be a function of $p$, the population proportion, which is unknown and is the parameter being estimated. How then can an appropriate $n$ be determined? In some cases the person making the estimate may have a subjective opinion about the magnitude of $p$, or data from previous estimates may provide the basis for deciding on a value of $p$, or it may be necessary to obtain a light preliminary sample (or *presample*) and base the sample size on the evidence from the presample.

It should be noted from Table 13.2 that all the minimal sample sizes exceed 30 and, consequently, the $t$ distributions do not enter into the confidence interval construction process. With sample sizes of the magnitudes shown, the appropriate $t$ distributions so closely approach the normal that normality is simply assumed.

If the population or sampling frame is finite and the sampling is without replacement, the sampling distribution of proportions is hypergeometric. When this is the case, the normal can still be used as the model, provided that (Hald, 1952)

$$np(1-p)\left(\frac{N-n}{N}\right) > 9 \tag{13.68}$$

It also is preferable that the sampling fraction, $n/N$, be less than 0.1.

### 13.11.4 Confidence Limits Based on the Use of the *F* Distribution

The relationship between the $F$ and binomial distributions is explored in Section 14.15 (see Equations 14.64 to 14.71). When this relationship is used for the development of confidence interval estimates of population proportions, the process is similar to that used in Section 13.11.1 in that the confidence limits are identified rather than determining the magnitude of an interval centered on the point estimate.

According to Equations 14.64 to 14.68 the relationship between the two distributions takes the following form:

$$P(Y \le y) = \sum_{w=0}^{y} \binom{y}{w} p^{w}(1-p)^{n-w}$$

$$= 1 - I_p(k_1/2,\ k_2/2) = 1 - I_p(y+1,\ n-y)$$

$$= 1 - P\left[F < \left(\frac{n-y}{y+1}\right)\left(\frac{p}{1-p}\right)\right]$$

Then

$$P\left[F < \left(\frac{n-y}{y+1}\right)\left(\frac{p}{1-p}\right)\right] = 1 - P(Y \le y)$$

Since

$$\left(\frac{n-y}{y+1}\right)\left(\frac{p}{1-p}\right) = F_{1-P(Y\le y)}[2(y+1),\ 2(n-y)]$$

Then

$$P\{F < F_{1-P(Y\le y)}[2(y+1),\ 2(n-y)]\} = 1 - P(Y \le y)$$

and

$$P(Y \le y) = 1 - P\{F < F_{1-P(Y\le y)}[2(y+1),\ 2(n-y)]\}$$

$$= P\{F < F_{1-P(Y\le y)}[2(y+1),\ 2(n-y)]\} \tag{13.69}$$

$$= \sum_{w=0}^{y} \binom{n}{w} p^{w}(1-p)^{n-w}$$

Thus, if $P(Y \le y)$, $p$, and $n$ are specified, the fractiles are the same for both distributions.

Defining the confidence interval as in Equation 13.39,

$$P(\hat{p}_a \le p \le \hat{p}_b) = C_b - C_a$$

Then $\hat{P}_b = y_b / n$, where $y_b$ is the largest value of $Y$ satisfying the equation

$$P(y/n) \le \hat{p}_o = y_o/n$$

assuming $p = \hat{p}_b = C_a = (1 - C)/2$.

Then

$$C_a = \sum_{y=0}^{y_o} \binom{n}{y} \hat{p}_b^{y}(1-\hat{p}_b)^{n-y} \tag{13.70}$$

$$= 1 - P\left[F < \left(\frac{n - y_o}{y_o + 1}\right)\left(\frac{\hat{p}_o}{1 - \hat{p}_o}\right)\right] \tag{13.71}$$

and $P\{F < F_{1-C_a}[2(y_o + 1),\ 2(n - y_o)]\} = 1 - C_o$

Let

$$F_{1-C_a}[2(y_o + 1),\ 2(n - y_o)] = \left(\frac{n - y_o}{y_o + 1}\right)\left(\frac{\hat{p}_o}{1 - \hat{p}_o}\right)$$

Solving for $\hat{p}$

$$\hat{p}_b = \frac{(y_o + 1)F_{1-C_a}[2(y_o + 1),\ 2(n - y_o)]}{(n - y_o) + (y_o + 1)F_{1-C_a}[2(y_o + 1),\ 2(n - y_o)]} \tag{13.72}$$

$\hat{p}_a = y_a/n$, where $y_a$ is the smallest value of $Y$ satisfying the equation

$$P(Y/n \geq \hat{p}_o = y_o/n, \text{ assuming } p = \hat{p}_a) = 1 - C_b = 1 - (C + 1)/2 = (1 - C)/2$$

Then

$$P[Y / n \leq (y_o - 1)] / n, \text{ assuming } p = \hat{p} = C_b = \sum_{y=0}^{y_o - 1} \binom{n}{y} \hat{p}_a^y (1 - \hat{p}_a)^{n-y} \tag{13.73}$$

$$k_1 = 2[(y_o - 1) + 1] = 2y_o \tag{13.74}$$

and

$$k_2 = 2[n - (y_o - 1)] = 2(n - y_o + 1) \tag{13.75}$$

Then,

$$C_b = 1 - P\left[F < \left(\frac{n - y_o + 1}{y_o}\right)\left(\frac{\hat{p}_a}{1 - \hat{p}_a}\right)\right] \tag{13.76}$$

$$P\left[F < \left(\frac{n - y_o + 1}{y_o}\right)\left(\frac{\hat{p}_a}{1 - \hat{p}_a}\right)\right] = 1 - C_b \tag{13.77}$$

and

$$\left(\frac{n - y_o + 1}{y_o}\right)\left(\frac{\hat{p}_a}{1 - \hat{p}_a}\right) = F_{1-C_b}[2y_o,\ 2(n - y_o + 1)] \tag{13.78}$$

According to Equation 14.51,

$$F_{C_b}[2(n-y_o+1),\ 2y_0] = \frac{1}{F_{1-C_b}[2y_o,\ 2(n-y_o+1)]}$$

$$= \left(\frac{y_o}{n-y_o+1}\right)\left(\frac{1-\hat{p}_a}{\hat{p}_a}\right)$$

Solving for $\hat{p}_a$,

$$\hat{p}_a = \frac{y_o}{y_o + (n-y_o+1)F_{C_b}[2(n-y_o+1),\ 2y_o]} \tag{13.79}$$

As an example, let $C = 0.95$, $C_a = (1 - C)/2 = 0.025$, $C_b = (C + 1)/2 = 0.975$, $n = 20$, and $y_o = 8$. Then,

$$\hat{p}_o = 0.4$$

$$\hat{p}_a = \frac{8}{8 + 13F_{0.975}(26,16)}$$

From Table A.9B in the Appendix, remembering that the probability levels are in terms of the complement of the cumulative probability, and using straight-line interpolation:

$$F_{0.975}\ (26,\ 16) = F_{\alpha=0.025}(26,\ 16) = 2.61$$

Then

$$\hat{p}_a = \frac{8}{8 + 13(2.61)} = 0.1908$$

Compare this with the true lower confidence limit, $\hat{p}_a = 0.20$.

The upper confidence limit,

$$\hat{p}_b = \frac{9F_{0.975}(18,24)}{12 + 9F_{0.975}(18,24)}$$

From Table A.9B,

$$F_{0.975}(18,\ 24) = 2.37$$

Then

$$\hat{p}_b = \frac{9(2.37)}{12 + 9(2.37)} = 0.6400$$

Compare this with the true upper confidence level, $\hat{p} = 0.60$.

It is interesting to note that when the $F$ distribution is used to obtain the confidence limits, the confidence interval may not be symmetrical about the point estimate. In the above example, the point estimate $\hat{p}_o = 0.4$. Then

$$\hat{p}_a - \hat{p}_b = 0.1908 - 0.4000 = -0.2092 \quad \text{and} \quad \hat{p}_b - \hat{p}_o = 0.6400 - 0.4000 = +0.2400$$

Such a lack of symmetry can occur whenever the process involves the determination of the confidence limits rather than the magnitude of the bracket centered on the point estimate.

### 13.11.5 Confidence Limits in the Case of the Poisson Distribution

In Section 10.5 it was shown how the Poisson distribution is a special case of the binomial distribution where $p \to 0$ and $n \to \infty$ so that $E(Y) = V(Y) = np = \lambda$. Thus, it is appropriate to use the cumulative distribution function of the binomial to express cumulative probabilities in the case of the Poisson. Then, following the pattern of development used in Section 13.11.4,

$$P(Y \leq y_o, \text{ assuming } p = \hat{p}_b) = C_a$$

$$C_a = 1 - P\left[F < \left(\frac{n - y_o}{y_o + 1}\right)\left(\frac{\hat{p}_b}{1 - \hat{p}_b}\right)\right]$$

$$P\left[F < \left(\frac{n - y_o}{y_o + 1}\right)\left(\frac{\hat{p}_b}{1 - \hat{p}_b}\right)\right] = 1 - C_a$$

Let

$$\hat{p}_b = \frac{\hat{\lambda}_b}{n} \quad \text{and} \quad n \to \infty$$

Then

$$\left(\frac{n - y_o}{y_o + 1}\right)\left(\frac{\hat{p}_b}{1 - \hat{p}_b}\right) = \frac{\hat{\lambda}_b}{y_o + 1} \tag{13.80}$$

and

$$P\left(F < \frac{\hat{\lambda}_b}{y_o + 1}\right) = 1 - C_a$$

Furthermore,

$$\frac{\hat{\lambda}_b}{y_o + 1} = F_{1-C}[2(y_o + 1), 2(n - y_o)] = F_{1-C}[2(y_o + 1), \infty]$$

According to Equation 14.52,

$$F(k_1, \infty) = \chi^2(k_1)/k_1$$

Then

$$\frac{\hat{\lambda}_b}{y_o + 1} = \chi^2_{1-C_a}[2(y_o + 1)]/2(y_o + 1)$$

Multiplying both sides of the equation by $(y_o + 1)$ yields

$$\hat{\lambda}_b = \chi^2_{1-C_a}[2(y_o + 1)]/2 \tag{13.81}$$

Then, $P(Y \leq y_o$, assuming $p = \hat{p}_a) = 1 - C_b = (1 - C)/2$ and $P(Y \leq y_o - 1$, assuming $p = \hat{p}_a) = C_b$.
According to Equations 13.74 and 13.75

$$k_1 = 2y_o \quad \text{and} \quad k_2 = 2(n - y_o + 1)$$

Then, according to Equation 13.77,

$$P\left[F < \left(\frac{n - y_o + 1}{y_o}\right)\left(\frac{\hat{p}_a}{1 - \hat{p}_a}\right)\right] = 1 - C_b$$

Letting

$$\hat{p}_a = \frac{\hat{\lambda}_a}{n} \quad \text{and} \quad n \rightarrow \left(\frac{n - y_o + 1}{y_o}\right)\left(\frac{\hat{p}_a}{1 - \hat{p}_a}\right) = \frac{\hat{\lambda}_a}{y_o}$$

and

$$P\left(F < \frac{\hat{\lambda}_a}{y_o}\right) = 1 - C_b$$

Now,

$$k_1 = 2_o \quad \text{and} \quad \frac{\hat{\lambda}_a}{y_o} = \chi^2_{1-C_b}(2y_o)/2y_o$$

Multiplying both sides of the equation by $y_o$ yields

$$\hat{\lambda}_a = \chi^2_{1-C_b}(2y_o)/2 \tag{13.82}$$

Thus,

$$P(\hat{\lambda}_a \leq \lambda \leq \hat{\lambda}_b) = P\left\{\frac{\chi^2_{1-C_b}(2y_o)}{2} \leq \lambda \leq \frac{\chi^2_{1-C_a}[2(y_o + 1)]}{2}\right\} \tag{13.83}$$

For example, assume that an estimate is desired of the proportion of the pine trees on a certain ownership in the Lower Coastal Plain of Alabama, above a minimum dbh of 10 in., which are spruce pines. To this end the tally sheets for a systematic cruise covering 10% of the area were examined. The total count of pine trees was 10,000, of which 10 were spruce pines. Then $y_o = 10$ and the confidence limits at the 0.95 level of confidence are determined as follows:

$$C = 0.95;\ C_a = (1 - C)/2 = 0.025;\ C_b = (C + 1)/2 = 0.975$$

Thus, $1 - C_a = 0.975$ and $1 - C_b = 0.025$. Then, using Table A.6 in the Appendix,

$$\hat{\lambda}_a = \frac{\chi^2_{0.025}(20)}{2} = 9.59 / 2 = 4.795$$

and

$$\hat{\lambda}_b = \frac{\chi^2_{0.975}(22)}{2} = 36.8 / 2 = 18.400$$

Consequently,

$$P\left(\frac{4.795}{10,000} \leq p \leq \frac{18,400}{10,000}\right) = 0.95$$

or $P(0.0004795 \leq p \leq 0.00184) = 0.95$.

When $k$ is in excess of 100, tables such as Table A.6 cannot supply the needed information and an alternate means of developing a confidence interval must be used. In Section 13.11.3 it was stated that the normal distribution can be used to model the binomial distribution provided that the sample size was such that $np(1 - p) > 9$. Since the Poisson is a special case of the binomial, the Poisson also can be modeled by the normal, provided that $\lambda$ is sufficiently large. In the case of the Poisson, the moment coefficient of skewness is $\alpha_3 = 1/\lambda$ (see Equation 10.29) while the moment coefficient of kurtosis is $\alpha_4 = 3 + 1/\lambda$ (see Equation 10.30). As can be seen, as $\lambda \to \infty$, $\alpha_3 \to 0$, and $\alpha_4 \to 3$, indicating that the Poisson distribution is converging on the normal. There are no firm guidelines regarding the minimum size of $\lambda$ that would permit the use of the normal as a model without incurring undue error.

Without proof, assuming that $\lambda$ is sufficiently large,

$$\hat{\lambda}_a \cong y_o - \frac{1}{2} + \frac{z_{C_b^2}}{2} - z_{C_a}\sqrt{y_o - \frac{1}{2} + \frac{z_{C_b^2}}{4}} \tag{13.84}$$

and

$$\hat{\lambda}_b \cong y_o - \frac{1}{2} + \frac{z_{C_b^2}}{2} - z_{C_a}\sqrt{y_o - \frac{1}{2} + \frac{z_{C_a^2}}{4}} \tag{13.85}$$

Using the data from the previous example, $n = 10,000$, $y_o = 10$, $\hat{p}_o = 0.001$, $\lambda_o = n\hat{p}_o = 10,000(0.001) = 10$, and $n\hat{p}_o(1 - \hat{p}_o) = 10,000(0.001)(0.999) = 9.99 > 9$. Consequently, the approximation using the normal should be acceptable. Since $C_b = (C + 1)/2 = 0.975$, the corresponding value of $Z$ is obtained by working backwards in Table A.4 in the Appendix from the percentage area value of 0.975 – 0.500 = 0.475.

$$z_{0.975} = 1.96$$

Then

$$\hat{\lambda}_a \cong 10 - 0.5 + \frac{(1.96)^2}{2} - 1.96\sqrt{10 - 0.5 + \frac{(1.96)^2}{4}}$$
$$\cong 5.082$$

The value of $z_{0.025}$ corresponds to a percentage area value of 0.025 – 0.500 = –0.475 and, thus,

$$z_{0.025} = -1.96$$

and

$$\hat{\lambda}_b \cong 10 - 0.5 + \frac{(-1.96)^2}{2} - (-1.96)\sqrt{10 - 0.5 + \frac{(-1.96)^2}{4}}$$
$$\cong 19.056$$

Then

$$P\left(\frac{5.082}{10,000} \leq p \leq \frac{19.056}{10,000}\right) \cong 0.95$$

or $P(0.0005082 \leq p \leq 0.0019056) \cong 0.95$.

### 13.11.6 Confidence Intervals from Published Curves and Tables

If confidence limits are obtained from every possible outcome, using any of the methods described, they can be plotted and the points connected by straight-line segments or by smooth curves. Figure 13.3 was developed under the definition ($\hat{p}_a \leq p \leq \hat{p}_b$) using the true confidence limits and straight-line segments. Such lines form the boundaries of confidence belts. From figures of this type one can obtain the confidence limits for any given proportion, even proportions that cannot be obtained. For example, one can use Figure 13.3 to obtain the confidence limits for $p = 0.63$, $\hat{p}_a = 0.43$, and $\hat{p}_b = 0.83$, even though there is no value of $Y$ that will yield a proportion of 0.63 when $n = 20$.

Pearson and Hartley (1954) published a set of confidence belts for proportions for $C = 0.95$ (see Table 9.10A in the Appendix) and another for $C = 0.99$ (see Table A.10B) which are based on the definition ($\hat{p}_a \leq p \leq \hat{p}_b$), sampling with replacement, and the use of the method of confidence limit determination making use of the $F$ distribution (Section 13.11.4). As an example of how these figures are used, assume that $C = 0.95$, $n = 20$, and $y_o = 8$. Then $\hat{p}_o = y_o/n = 8/20 = 0.4$. Enter Table A.10A with $y_o/n = 0.4$ along the horizontal axis. Note that at the bottom of the figure the values increase from left to right from 0 to 0.5 and then the progression continues at the top of the figure, increasing from right to left, from 0.5 to 1.0. The vertical scale on the left is used in conjunction with the horizontal scale along the bottom while that on the right is used with the horizontal scale along the top of the figure. The confidence limits are read from the appropriate vertical scale (the left in this case) at the points where the vertical line corresponding to $y_o/n = 0.4$ intersects the two

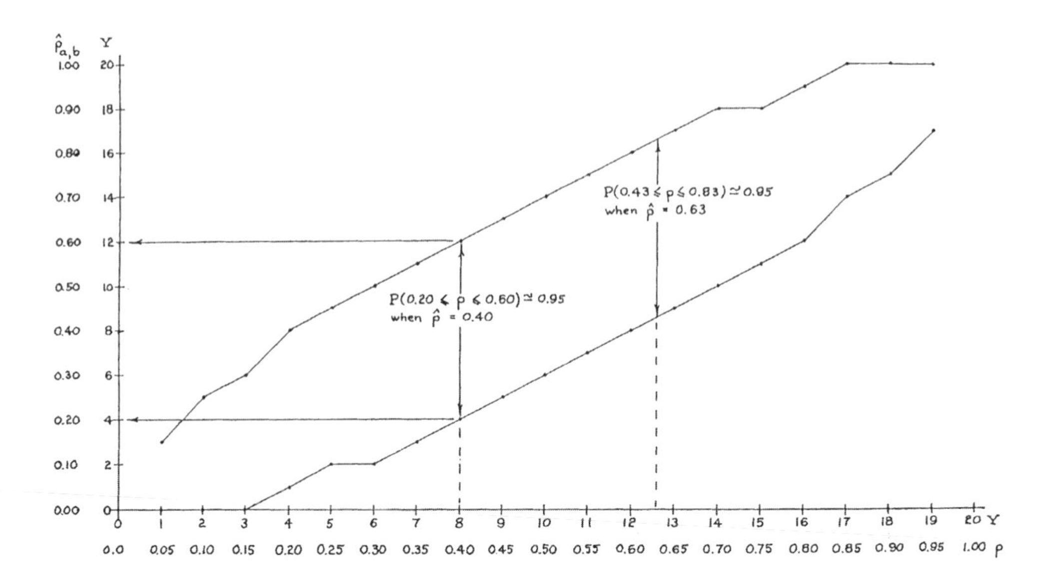


**FIGURE 13.3** True confidence limits for the proportions $(\hat{p}_a \leq p \leq \hat{p}_b)$ when $n = 20$, sampling is with replacement, and $C = 0.95$.

curves labeled 20. The confidence limits obtained this way are $\hat{p}_a = 0.19$ and $\hat{p}_b = 0.64$, which agrees with those obtained in the comparable example in Section 13.11.4.

Hald (1952b), in the separate appendix to his textbook (Hald, 1952a), has tables listing the 0.95 and 0.99 confidence limits upon which the curves in Tables A.9A and A.9B in the Appendix are based. The tables are somewhat easier to use, are more precise, and include only the limits associated with $Y$ values that can occur.

## 13.12 CONFIDENCE INTERVAL ESTIMATES OF POPULATION VARIANCES AND STANDARD DEVIATIONS

According to Equation 12.25, when the parent population is normally distributed with mean $\mu_Y$ and variance $\sigma_Y^2$ and when $k = n - 1$,

$$s_Y^2 = \sigma_Y^2 \chi^2(k) / k$$

This is true whether the sampling is with or without replacement because when the parent population is normal, $N = \infty$ and the finite population correction equals unity. As was shown in Equation 12.26:

$$\chi^2(k) = s_Y^2(k / \sigma_Y^2)$$

and consequently the fractiles of the chi-square distribution can be used to obtain cumulative probabilities of sample variances. Thus, the probability of obtaining a value lying in the range from $\chi_{C_a}^2(K)$ to $\chi_{C_b}^2(K)$, in a single random draw, is

$$P[\chi_{C_a}^2(k) < \chi^2 < \chi_{C_b}^2(k)] = C_b - C_a = C \tag{13.86}$$

Substituting $s_Y^2 k / \sigma_Y^2$ for $\chi^2$

$$P[\chi^2_{C_a}(k) < s_Y^2 k / \sigma_Y^2 < \chi^2_{C_b}(k)] = C \tag{13.87}$$

Solving for $\sigma_Y^2$

$$\begin{aligned} \chi^2_{C_a}(k) &< s_Y^2 k / \sigma_Y^2 \\ \sigma_Y^2 &< s_Y^2 k / \chi^2_{C_a}(k) \end{aligned} \tag{13.88}$$

and

$$\begin{aligned} s_Y^2 k / \sigma_Y^2 &< \chi^2_{C_b}(k) \\ s_Y^2 k / \chi^2_{C_b}(k) &< \sigma_Y^2 \end{aligned} \tag{13.89}$$

Thus, using Equations 13.88 and 13.89, the confidence interval for the population variance is

$$P[s_Y^2 k / \chi^2_{C_b}(k) < \sigma_Y^2 < s_Y^2 k / \chi^2_{C_a}(k)] = C \tag{13.90}$$

As an example, assume a random sample of size $n = 20$, with a variance $s_Y^2 = 100$. Then $k = n - 1 = 19$. If the confidence level is set at $C = 0.95$, $C_a = (1 - C)/2 = 0.025$, and $C_b = (C + 1)/2 = 0.975$. From Table A.6,

$$\chi^2_{0.025}(19) = 8.91 \quad \text{and} \quad \chi^2_{0.975}(19) = 32.9$$

Then,

$$s_Y^2 k / \chi^2_{C_b}(k) = 100(19)/32.9 = 57.75$$

$$s_Y^2 k / \chi^2_{C_a}(k) = 100(19)8.91 = 213.2$$

and

$$P(57.75 < \sigma_Y^2 < 213.24) = 0.95$$

Thus, the fiducial probability that the population variance, $\sigma_Y^2$, lies in the interval 57.75 to 213.24, is 0.95. Note that the confidence interval is asymmetric about the point estimate, $s_Y^2 = 100$. This arises from the fact that the chi-square distribution is skewed. Whenever the sampling distribution, or its model, is skewed the confidence interval will be asymmetric.

Confidence interval estimates of the population standard deviation follow directly from Equation 13.90:

$$P\left[\sqrt{s_Y^2 k / \chi^2_{C_b}(k)} < \sigma_Y < \sqrt{s_Y^2 k / \chi^2_{C_a}(k)}\right] = C \tag{13.91}$$

For example, further developing that used to illustrate the confidence interval estimate of the variance,

$$P(\sqrt{57.75} < \sigma_Y < \sqrt{213.24}) = 0.95$$

$$P(7.60 < \sigma_Y < 14.60) = 0.95$$

When the parent population is normally distributed and $k \rightarrow \infty$, the distribution of $\chi^2(k)$ converges to the normal, as can be seen from the moment coefficients of skew (Equation 11.48) and kurtosis (Equation 11.49).

$$\lim_{k\to\infty} \alpha_3 = \lim_{k\to\infty} 2 / \sqrt{k/2} = 0$$

and

$$\lim_{k\to\infty} \alpha_4 = \lim_{k\to\infty}(3 + 12/k) = 3$$

The rate of convergence is relatively rapid and it usually is assumed that the chi-square distribution can be modeled by the normal if $k > 30$. Thus, when $k > 30$, the sampling distributions of variances and standard deviations can be considered to be normal. The sampling distribution of variances would have the mean $\mu_{s^2} = E(s_Y^2) = \sigma_Y^2$ and the variance

$$\sigma^2_{s_Y^2} = V(s_Y^2) = \frac{\mu_4}{n} - \frac{(\sigma_Y^2)^2}{n}\left(\frac{n-3}{n-1}\right) \cong \frac{\mu_y - (\sigma_Y^2)^2}{n}$$

while that of the standard deviations would have the mean

$$\mu_{s_Y} = E(s_Y) = \sigma_Y \sqrt{\frac{2}{k}\left\{\frac{[(k-1)/2]!}{[(k-2)/2]!}\right\}} \cong \sigma_Y$$

and the variance

$$\sigma^2_{s_Y} = V(s_Y) = \sigma_Y^2 - 2\left(\frac{\sigma_Y^2}{k}\right)\left\{\frac{[(k-1)/2]!}{[(k-2)/2]!}\right\}^2 \cong \sigma_Y^2 / 2k$$

Thus, in the case of the sampling distribution of the variance, let (from Equation 12.6)

$$z_j = \frac{s_Y^2 - \sigma_Y^2}{\sigma^2_{s_Y}} \tag{13.92}$$

where, according to Equation 12.18,

$$\sigma_{s_Y^2} = \sqrt{V(s_Y^2)} \cong \sqrt{\frac{\mu_4 - (\sigma_Y^2)^2}{n}}$$

Since the sampling distribution is approximately normal, according to Equations 11.19 and 11.20,

$$\mu_4 \cong 3$$

Consequently,

$$z_j = \frac{s_Y^2 - \sigma_Y^2}{\sqrt{\dfrac{3 - (\sigma_Y^2)^2}{n}}}$$

and

$$P\left[-z_C < \frac{s_Y^2 - \sigma_Y^2}{\sqrt{\dfrac{3 - (\sigma_Y^2)^2}{n}}} < +z_C\right] = C \tag{13.93}$$

At this point the usual procedure is to solve for $\sigma_Y^2$ across both inequalities. However, $\sigma_Y^2$ cannot be algebraically isolated. As a result, the use of the normal distribution to develop confidence interval estimates of $\sigma_Y^2$ is not possible. It is possible, however, to use it to obtain confidence interval estimates of $\sigma_Y$. Let

$$z_j = \frac{s_Y - \sigma_Y}{\sigma_{s_Y}}$$

where, according to Equation 12.41,

$$\sigma_{s_Y} = \sqrt{V(s_Y)} \cong \sqrt{\sigma_Y^2 / 2k} \tag{13.94}$$

Then,

$$P\left[-z_C < \frac{s_Y - \sigma_Y}{\sqrt{\sigma_Y^2 / 2k}} < +z_C\right] = C \tag{13.95}$$

Solving for $\sigma_Y$ yields

$$P\left[\frac{s_Y}{1 + z_C / \sqrt{2k}} < \sigma_Y < \frac{s_Y}{1 - Z_C / \sqrt{2k}}\right] \tag{13.96}$$

For example, if $s_Y = 10$, $k = 50$, and $C = 0.95$,

$$P\left[\frac{10}{1 + 1.96 / \sqrt{100}} < \sigma_Y < \frac{10}{1 - 1.96 / \sqrt{100}}\right] = 0.95$$

$$P(4.5537 < \sigma_Y < 12.4378) = 0.95$$

When the parent population is not normally distributed, neither the chi-square nor the normal distribution can be used to develop confidence interval estimates of population variances or standard deviations. Of course, if the parent population is approximately normal, the use of the preceding procedures will produce estimates that are somewhat in error but may be usable. As the distribution of the parent population shifts farther and farther from the normal, the error increases. The point at which the error becomes excessive is dependent on many factors and cannot be formally identified.

When the distribution of the parent population is not known, one can assume normality and use the procedures described above to obtain confidence interval estimates of population variances and standard deviations. If such an assumption cannot be made, the only recourse is to use Chebyshev's theorem (see Section 13.10 or the Camp–Meidell extension of that theorem). According to Equation 13.35,

$$P[(\sigma_Y^2 - r\sigma_{s_Y^2}^2) < s_Y^2 < (\sigma_Y^2 + r\sigma_{s_Y^2})] > 1 - 1/r^2$$

where, according to Equation 12.18,

$$\sigma_{s_Y^2}^2 = V(s_Y^2) \cong \frac{\mu_4 - (\sigma_Y^2)^2}{n} = \frac{(\sigma_Y^2)^2}{n}\left[\frac{\mu_4}{(\sigma_Y^2)^2} - 1\right]$$

Then

$$\sigma_{s_Y^2}^2 = \sqrt{V(s_Y^2)} \cong \frac{(\sigma_Y^2)}{\sqrt{n}}\sqrt{\frac{\mu_4}{(\sigma_Y^4) - 1}} \tag{13.97}$$

and

$$P\left[\sigma_Y^2 - \frac{r\sigma_Y^2}{\sqrt{n}}\sqrt{\frac{\mu_4}{\sigma_Y^2} - 1} < s_Y^2 < \sigma_Y^2 + \frac{r\sigma_Y^2}{\sqrt{n}}\sqrt{\frac{\mu_4}{\sigma_Y^4} - 1}\right] > 1 - 1/r^2$$

Solving for $\sigma_Y^2$ and rearranging yields

$$P\left[\frac{s_Y^2}{1 + \frac{r}{\sqrt{n}}\sqrt{\frac{\mu_4}{\sigma_Y^2} - 1}} < \sigma_Y^2 < \frac{s_Y^2}{1 - \frac{r}{\sqrt{n}}\sqrt{\frac{\mu_4}{\sigma_Y^4} - 1}}\right] > 1 - 1/r^2 \tag{13.98}$$

To find $r$ for a specific confidence level, $C$, set

$$1 - 1/r^2 = C$$

Then

$$r = \sqrt{\frac{1}{1 - C}} \tag{13.99}$$

To use Equation 13.98 one must have values for $\mu_4$ and $\sigma_Y^4$, both of which are unknown. Consequently, they must be estimated using sample data. The sample must be large in order to obtain sound values.

$$m_4 = \sum_{i=1}^{n} (y_i - \bar{y})^4 / n \rightarrow \mu_4 \tag{13.100}$$

and

$$s_Y^4 = \left[ \sum_{i=1}^{n} (y_i - \bar{y})^2 / (n-1) \right]^2 \rightarrow \sigma_Y^4 \tag{13.101}$$

If the distribution of the parent population is known to be unimodal, the Camp–Meidell extension of Chebyshev's theorem can be used. In such a case

$$P\left[ \frac{s_Y^2}{1 + \frac{r}{\sqrt{n}}\sqrt{\frac{\mu_4}{\sigma_Y^4} - 1}} < \sigma_Y^2 < \frac{s_Y^2}{1 - \frac{r}{\sqrt{n}}\sqrt{\frac{\mu_4}{\sigma_Y^4} - 1}} \right] > 1 - 1/22.5r^2 \tag{13.102}$$

This, of course, is a much less conservative estimate but still is more conservative than what would be obtained if the estimate could be based on a known distribution of the parent population.

## 13.13 ONE-SIDED CONFIDENCE INTERVALS

In the preceding discussion, the confidence intervals all had two limits. For example, in the estimate

$$P[\bar{y} - t_C(k)s_{\bar{y}} < \mu_Y < \bar{y} + t_C(k)s_{\bar{y}}] = C$$

there are two limits: the lower limit, $\bar{y} - t_C(k)s_{\bar{y}}$, and $\bar{y} + t_C(k)s_{\bar{y}}$. Furthermore, this estimate is symmetric, extending along the number line the same distance in both directions from the point estimate, $y$. It has been shown that in some cases the intervals are not symmetric, as in the case of estimates of the population variance where the confidence interval is based on the chi-square distribution which is skewed. Asymmetry in confidence intervals can be extended to the point where one or the other of the confidence limits coincides with a limit of the distribution of the parent population. For example,

$$P[\mu < \bar{y} + t_C(k)s_{\bar{y}}] = C \tag{13.103}$$

states that the fiducial probability is $C$ that $\mu_Y$ lies in the interval from $-\infty$ to $\bar{y} + t_C(k)s_{\bar{y}}$. In the case of

$$P[\mu > \bar{y} - t_C(k)s_{\bar{y}}] = C \tag{13.104}$$

the fiducial probability is C that $\mu_Y$ lies in the interval from $\bar{y} - t_C(k)s_{\bar{y}}$ to $+\infty$. These are called *single limit* confidence intervals with Equation 13.103 representing an *upper limit* interval and Equation 13.104 a *lower limit* interval. These examples make use of the $t$ distributions. One should realize that the concept is not limited to situations involving the $t$ distributions. Single limit estimates can be made using any distributional model.

## 13.14 SAMPLING ERRORS, ALLOWABLE ERRORS, AND NEEDED SAMPLE SIZE

The *sampling error* of an estimate, SE, is defined as the confidence interval:

$$SE = \pm t_C(k)s_{EST} \tag{13.105}$$

In the case of an arithmetic mean, the sampling error is

$$SE_{\bar{y}} = \pm t_C(k)s_{\bar{y}} \tag{13.106}$$

while in the case of a proportion, provided that the sample size is sufficiently large so that the sampling distribution of proportions can be modeled by the normal,

$$SE_{\hat{p}} = \pm t_C(k)s_{\hat{p}}, \qquad k > 30 \tag{13.107}$$

Thus, every estimator has a sampling error. In many cases the symbol ± is eliminated and the sampling error then consists simply of the magnitude of the positive half of the confidence interval. In any case, the smaller the sampling error, at a given level of confidence, the more precise the estimate. It does not, however, provide any information regarding the presence and/or magnitude of any bias which may be associated with the point estimate.

The magnitude of the sampling error is a function of sample size. Consider Equation 13.106. It was brought out in Section 13.9 that the $t$ distribution approaches the normal as the number of degrees of freedom increase, and in the limit, as $k \to \infty$, the $t$ distribution coincides with the normal. It was also brought out that most workers assume normality when $k > 30$. Then, if k is kept greater than 30, $t_c(k)$ can be considered a constant and $SE_{\bar{y}}$ is a function only of the standard error, $s_{\bar{y}}$. According to Equation 13.11,

$$s_{\bar{y}} = \sqrt{\frac{s_Y^2}{n}}$$

Since $s_Y^2$ is an estimate of a constant, $\sigma_Y^2$, and is not a function of sample size, it can be considered a constant in the following discussion. Then, if $s_Y^2 = 1000^2$ and $n = 100$,

$$s_{\bar{y}} = \sqrt{\frac{1000^2}{100}} = 100$$

If $n = 400$,

$$s_{\bar{y}} = \sqrt{\frac{1000^2}{400}} = 50$$

and if $n = 1600$,

$$s_{\bar{y}} = \sqrt{\frac{1000^2}{1600}} = 25$$

As can be seen, the standard error is inversely proportional to the square of the sample size.

The relationship between the sampling error and sample size provides a mean for controlling the precision of an estimate. A maximum sampling error, at a given confidence level, known as an *allowable sampling error* or, simply, an *allowable error*, can be specified and the sample size needed to meet that allowable error can be computed. This, of course, presupposes a knowledge of the sample variance *prior* to the sampling. In some cases reference can be made to the records of earlier similar sampling oeprations carried out in conditions comparable with those existing in the universe that is to be sampled. In the absence of such records it will be necessary to obtain an estimate of $s_Y^2$ from a light *preliminary sample* or *presample*. Then, substituting the allowable error, AE, for the sampling error in Equation 13.106,

$$\begin{aligned} \text{AE}_{\bar{y}} &= \text{maximum SE}_{\bar{y}} = t_C(k)s_{\bar{y}} \\ &= t_C(k)s_Y / \sqrt{n} \end{aligned} \tag{13.108}$$

$$n = t_C^2(k)s_Y^2 / \text{AE}_{\bar{y}}^2 \tag{13.109}$$

Equation 13.109 is used to compute the needed sample size when the sampling is with replacement or when the sampling is without replacement and the sampling fraction, $n/N$, is negligible.

As an example, assume that a 1000-acre tract is to be cruised. The species of primary interest is pine. Consequently, the allowable error is in terms of pine. It was set at ±1000 bd. ft/acre at a confidence level of 0.80. The sampling frame is a grid overlying a U.S. Geological Survey quadrangle map on which the boundaries of the tract have been drawn. The grid spacing is such that there is one intersection per acre. At the intersections chosen in the sampling process $^1/_5$-acre plots will be established and cruised. The sampling will be random, with replacement. The plot volumes will be put on a per acre basis prior to any calculation of sample statistics. The presample data are $n = 36$ and $s_Y^2 = 25{,}000{,}000$. Then, from Table A.7 in the Appendix,

$$t_{0.80}(35) = 1.282$$

and

$$n_{\text{Needed}} = (1.282)^2(25{,}000{,}000)/)1000)^2 = 41.0881$$

In sample size calculations the computed value is always rounded *up* to be sure that the sample is sufficiently large. Consequently,

$$n_{\text{Needed}} = 42 \text{ plots}$$

Theoretically, the presample data should now be discarded and 42 new sampling points chosen from the sampling frame. However, because sampling costs are often high, as when cruising plots, the presample may simply be augmented. This practice introduces a factor that affects the confidence level. A discussion of this factor is beyond the scope of this book. However, workers should be aware that a problem is associated with the practice and it should be avoided if at all possible.

If the sampling had been without replacement, the use of Equation 13.108 would be inappropriate. When the sampling is without replacement (see Equation 13.12),

$$s_{\bar{y}} = \sqrt{\frac{s_Y^2}{n}\left(\frac{N-n}{N}\right)}$$

Substituting this in Equation 13.108,

$$
\begin{aligned}
\text{AE}_{\bar{y}} &= t_C(k)\sqrt{(s_Y^2 / n)[(N-n)/N]} \\
\text{AE}_{\bar{y}}^2 &= t_C^2(k)(s_Y^2 / n)[(N-n)/N] \\
\text{AE}_{\bar{y}}^2 N / t_C^2(k)s_Y^2 &= (N-n)/n = (N/n) - 1 \\
[\text{AE}_{\bar{y}}^2 N / t_C^2(k)s_Y^2] + 1 &= N/n \\
[\text{AE}_{\bar{y}}^2 N + t_C^2(k)s_Y^2] / t_C^2(k)s_Y^2 &= N/n \\
n &= t_C^2(k)s_Y^2 N / [\text{AE}_{\bar{y}}^2 N + t_C^2(k)s_Y^2]
\end{aligned}
\tag{13.111}
$$

Thus, Equation 13.111 would be appropriate if the sampling were without replacement. In the preceding example, $N = 1000$, because there was one grid intersection per acre. Then

$$
\begin{aligned}
n_{\text{Needed}} &= \frac{(1.282)^2(25,000,000)(1000)}{(1000)^2(1000) + (1.282)^2(25,000,000)} \\
&= 39.4665 \text{ or } 40 \text{ plots}
\end{aligned}
$$

Note that the needed sample size in this case is smaller than when the sampling was with replacement. This again illustrates that sampling without replacement is more efficient than sampling with replacement.

In the preceding discussion the sample sizes were all sufficiently large that normality could be assumed. In some cases, however, the sample sizes are too small for this assumption and the $t$ value becomes variable. For example, if in the case of the timber cruising problem, using sampling with replacement, the presample variance was 16,000,000 based on ten observations, the $t$ value would be

$$t_{0.80}(9) = 1.383$$

and

$$n_{\text{Needed}} = \frac{(1.383)^2 16,000,000}{(1000)^2} = 30.60 \text{ or } 31 \text{ plots}$$

However, the $t$ value used in this computation was appropriate for 9 degrees of freedom, not $31 - 1 = 30$ degrees of freedom. Something must be done to bring the $t$ value into agreement with the computed sample size. Since there are two unknowns and only one equation, one cannot solve for the unknowns directly. Instead one must iterate, using successive approximations. Thus, in the first iteration it was found that 31 plots would be needed. Then $k = 31 - 1 = 30$ and

$$t_{0.80}(30) = 1.310$$

Substituting this into Equation 13.109,

$$n_{\text{Needed}} = \frac{(1.310)^2 16,000,000}{(1000)^2} = 27.46 \text{ or } 28 \text{ plots}$$

Again the $t$ value is not appropriate and another iteration must be carried out. Now $k = 28 - 1 = 27$,

$$t_{0.80}(27) = 1.314$$

and

$$n_{\text{Needed}} = \frac{(1.314)^2 16,000,000}{(1000)^2} = 27.63 \text{ or } 28 \text{ plots}$$

At this point the $t$ value is appropriate and the iteration process is terminated.

In some cases one cannot obtain a clear-cut agreement. Instead the computed sample sizes flipflop back and forth between two values. If this occurs, one should use the larger of the two computed sample sizes.

In the case of a sampling operation involving the estimation of a proportion the problem of sample size determination becomes complex. First, the sample size must be of sufficient size to permit the use of the normal as the model of the sampling distribution of proportions. If the sampling is with replacement, this minimum sample size can be determined using Equation 13.66,

$$np(1 - p) > 9$$

or Equation 13.67,

$$n > 9/p(1 - p)$$

When the sampling is without replacement reference should be made to Equation 13.68,

$$np(1-p)\left(\frac{N-n}{N}\right) > 9$$

In addition, when the sampling is without replacement, the sample fraction, $n/N$, should be less than 0.1.

The determination of the sample size needed to meet the allowable error requires the use of Equation 13.112, when the sampling is with replacement, or Equation 13.113 when the sampling is without replacement. When the sampling is with replacement (see Equation 13.58),

$$s_{\hat{p}} = \sqrt{\frac{\hat{p}(1-\hat{p})}{n-1}}$$

Substituting this in Equation 13.107,

$$\text{SE}_{\hat{p}} = t_c(k)\sqrt{\frac{\hat{p}(1-\hat{p})}{n-1}}$$

Introducing the allowable error,

$$
\begin{aligned}
\text{AE}_{\hat{p}} &= \text{maximum SE}_{\hat{p}} = \pm t_C(k)\sqrt{[\hat{p}(1-\hat{p})/(n-1)]} \\
\text{AE}_{\hat{p}}^2 &= t_C^2(k)[\hat{p}(1-\hat{p})/(n-1)] \\
\text{AE}_{\hat{p}}^2 / t_C^2(k)[\hat{p}(1-\hat{p})] &= 1/(n-1) \\
n\{t_C^2(k)[\hat{p}(1-\hat{p})] &+ \text{AE}_{\hat{p}}^2\} / \text{AE}_{\hat{p}}^2
\end{aligned}
\tag{13.112}
$$

When the sampling is without replacement (see Equation 13.62),

$$
s_{\hat{p}} = \sqrt{[\hat{p}(1-\hat{p})/(n-1)][(N-n)/N]}
$$

and

$$
\begin{aligned}
\text{AE}_{\hat{p}} &= \text{maximum SE}_{\hat{p}} = \pm t_C(k)s_{\hat{p}} \\
\text{AE}_{\hat{p}}^2 &= t_C^2(k)\{[\hat{p}(1-\hat{p})/(n-1)][(N-n)/N]\} \\
\text{AE}_{\hat{p}}^2 N / t_C^2(k)\hat{p}(1-\hat{p}) &= (N-n)/(n-1) \\
(\text{AE}_{\hat{p}}^2 Nn - \text{AE}_{\hat{p}}^2 N)/[t_C^2(k)\hat{p}(1-\hat{p})] &= N-n \\
n &= N - \{\text{AE}_{\hat{p}}^2 Nn - \text{AE}_{\hat{p}}^2 N)/[t_C^2(k)\hat{p}(1-\hat{p})]\} \\
&= [Nt_C^2(k)\hat{p}(1-\hat{p}) + \text{AE}_{\hat{p}}^2 Nn + \text{AE}_{\hat{p}}^2 N] / t_C^2(k)\hat{p}(1-\hat{p}) \\
t_C^2(k)\hat{p}(1-\hat{p}) &= t_C^2(k)\hat{p}(1-\hat{p})N - \text{AE}_{\hat{p}}^2 Nn + \text{AE}_{\hat{p}}^2 N \\
t_C^2(k)\hat{p}(1-\hat{p}) + \text{AE}_{\hat{p}}^2 Nn &= t_C^2(k)\hat{p}(1-\hat{p})N + \text{AE}_{\hat{p}}^2 N \\
n[t_C^2(k)\hat{p}(1-\hat{p}) + \text{AE}_{\hat{p}}^2 N] &= t_C^2(k)\hat{p}(1-\hat{p})N + \text{AE}_{\hat{p}}^2 N \\
n &= [t_C^2(k)\hat{p}(1-\hat{p})N + \text{AE}_{\hat{p}}^2 N] / [t_C^2(k)\hat{p}(1-\hat{p}) + \text{AE}_{\hat{p}}^2 N]
\end{aligned}
\tag{13.113}
$$

As an example, assume a 10,000-acre tract of land. An estimate is to be made of the proportion of the tract that is occupied by bottomland hardwoods. The allowable error has been set at ±0.1 at a confidence level of 0.90. A stand map, prepared from aerial photographs, is available. The sampling frame is a set of points, defined by a square grid laid over the map. The grid interval is such that there is one intersection (or point) per acre. Consequently, $N$ = 10,000 points. The sample will be a randomly chosen subset of these points, each of which will be evaluated to determine whether or not it lies in an area classified as bottomland hardwoods. First consider the situation when the sampling is to be with replacement.

To obtain an estimate of the variance, a presample is drawn and evaluated. Assume that $n$ = 10 and that two points fall into bottomland hardwood stands. Then $\hat{p}$ = 2/10 = 0.2. According to our previous discussion, when $p = 0.2$, the minimum sample size needed to permit use of the normal is 57. Consequently, $k = 57 - 1 = 56$ and

$$
t_{0.90}(56) \cong t_{0.90}(\infty) = 1.65
$$

Then

$$n_{\text{Needed}} = [(1.65)^2(0.2)(0.8) + (0.1)^2]/(0.2)^2 = 44.56 \text{ or } 45 \text{ points}$$

However, the minimum required is 57. This overrides the computed sample size. As a result, the sample size to be used is 57 points.

If the sampling is to be without replacement, the minimum sample size must be determined using Equation 13.68. Assuming the same presample data, the minimum sample size is 57. Using it in Equation 13.68

$$57(0.2)(0.8)[(10{,}000 - 57)/10{,}000] = 9.02 > 9$$

This indicates that a sample size of 57 may be larger than necessary. To find the minimum needed requires the use of successive approximations. Thus, arbitrarily reducing 57 to 56 yields

$$56(0.2)(0.8)[(10{,}000 - 56)/10{,}000] = 8.91 < 9$$

Now the sample size to too small. Consequently, the minimum sample size satifying the normality requirement is 57. Now, $k = 57 - 1 = 56$ and

$$t_{0.90}(56) \cong t_{0.90} = 1.65$$

Substituting this in Equation 13.113 yields

$$n_{\text{Needed}} = \frac{(1.65)^2(0.2)(0.8)10{,}000 + (0.1)^2 10{,}000}{(1.65)^2(0.2)(0.8) + (0.1)^2 10{,}000} = 44.37 \text{ or } 45 \text{ points}$$

Again the sample size obtained by the formula based on the confidence interval is too small to ensure normality and one must make use of that obtained using Equation 13.68.

It is possible that when the computed sample size is used the final sampling error will be larger than the allowable error. This occurs when the initial estimate of the population variance is smaller than the estimate obtained in the sampling operation proper. Theoretically, in such a case the sample obtained in the operation should be considered a presample and the variance obtained from it should be used to recompute the needed sample size. The entire sampling operation should then be redone, discarding the previously obtained data. Whether or not this is actually done in any specific case depends largely on the costs involved. If the cost is slight, the work probably would be redone as theory dictates. If the cost is not slight, the difference in the new and old needed sample sizes could be determined, this number of new sampling units chosen and evaluated, and the estimate recomputed. This, of course, would introduce a source of bias since the old and new estimates are not independent. However, it may be the only feasible solution to the problem if the allowable error must be met. If the difference between the sampling error and allowable error is not great, it probably would be best simply to accept the sampling error. The decision of what path to take in cases such as this depends on the policies set forth by the decision makers using the estimates generated by the sampling operation.

# 14 Fundamentals of Hypothesis Testing

## 14.1 INTRODUCTION

In Section 1.3 it was pointed out that the subject matter of "statistics" is very broad and has been subdivided into the more or less distinct subdisciplines of "parameter estimation," "hypothesis testing," "regression analysis," and "quality control." The subject matter of this book is oriented primarily toward parameter estimation or "sampling." However, persons engaged in parameter estimation must have some background in the concepts underlying the procedures used in the testing of hypotheses, which forms the framework within which much research and development are done. This chapter is designed to provide this fundamental background. It certainly will not be complete. For more complete coverage, the student is referred to standard statistical texts such as those by Snedecor and Cochran (1967), Steel and Torrie (1960), Sokel and Rohlf (1969), Miller and Freund (1977), or Mendenhall and Shaeffer (1973).

## 14.2 STATISTICAL HYPOTHESES

In research or development, questions are asked: "Is method A a better method of controlling hardwoods than method B?" "Does the application of fertilizer, of a certain composition, to a particular forest type increase the yield?" To arrive at answers to questions of this type it is necessary to carry out tests and then to analyze the results of those tests. In this process one must make assumptions or guesses regarding the situation being evaluated. Such an assumption or guess is called a *statistical hypothesis*. After the hypothesis has been stated, it is tested to see if it is supported by the evidence.

The basic hypothesis, which is always used, is the *null* hypothesis or the hypothesis of *no difference from expected results*, which is symbolized by $H_0$. It states that there is no *real* difference between the *observed* and *expected* results and any difference that does occur is due to the variability of the variates making up the population. In other words, the difference is due to random processes. For example, in a trial of the effect of nitrogen fertilization on the yield of loblolly pine plantations, the null hypothesis might be:

$H_0$: The application of the fertilizer has no effect on the yield.

If a difference occurs between the yields of the treated and untreated *sample* plantations, it is assumed that it was due only to random processes. It should be noted that the null hypothesis does not necessarily mean that there are no differences at all between the population represented by the samples. It states an *expected* result. For example, in the case of the fertilizer trial, the null hypothesis could be

$H_0$: The application of the fertilizer will increase the pulpwood yield by *no more* than 5%.

In this case an increase in yield is expected, but it is not expected to exceed 5%. In the majority of cases, however, the null hypothesis is used in the context where no real difference between the sets of data is expected.

For every null hypothesis there are one or more *alternate* hypotheses, symbolized by $H_1$, $H_2$, etc. If, as the result of the test, the null hypothesis is rejected, the alternate hypothesis is accepted. If the alternate hypothesis is so accepted, it can lead to further tests, each with its own null and alternate hypotheses, depending on the design of the experiment. In the case of the fertilization experiment the initial hypotheses might be

$H_0$: The application of the fertilizer has *no effect* on the yield of the loblolly pine plantations.
$H_1$: The application of the fertilizer *has an effect* on the yield.

If $H_0$ is rejected and $H_1$ accepted, the next step might be to test the hypotheses:

$H_0$: The application of the fertilizer has a *negative* effect on the yield.
$H_2$: The application has a *positive* effect on the yield.

Assuming $H_2$ is accepted, the test could continue:

$H_0$: The application increases the yield by *no more* than 5%.
$H_3$: The application increases the yield by *more* than 5%.

In cases such as this, the testing process resembles a cascade, with subsequent steps taking place upon the acceptance of an alternate hypothesis. It should be noted that the null and alternate hypotheses that are to be tested should be formally stated *prior* to the designing of the experiment so that it can be carried out in a logical and efficient manner.

## 14.3 TYPE I AND TYPE II ERRORS

When a test has been made and a hypothesis is accepted, the conclusion reached may or may not be correct. Remember, the conclusion is based on sample data, which are incomplete and may be unrepresentative. For example, in the case of a certain coin the hypotheses might be

$H_0$: $P(\text{Heads}) = 0.5$.
$H_1$: $P(\text{Heads}) \neq 0.5$.

If 100 tosses are made and 60 heads appear, one might conclude that $H_0$ should be rejected and $H_1$ accepted. However, the test involved only 100 tosses. It is entirely possible that the results were due only to random chance and that if another test of 100 tosses were to be made 40 heads would appear, indicating that the coin again is not fair, but the bias is in the opposite direction! Finally, a third series of 100 tosses could be made and 50 heads could appear, which, on the basis of this set alone, would indicate that the coin is fair. If indeed the coin is fair, the conclusion that bias exists, resulting from the first and second tests, would be incorrect. On the other hand, if the coin is *really biased*, the conclusion reached on the basis of the *third* test would be incorrect.

As can be seen, whenever a hypothesis is subjected to a statistical test, it is possible to arrive at an incorrect conclusion in either one of two ways: one can *reject a correct hypothesis*, in which case a *Type I* error has been made; or one can *accept an incorrect hypothesis*, in which case a *Type II* error has been made. The probability of a Type I error occurring is usually symbolized by $\alpha$ while that of a Type II error occurring is symbolized by $\beta$.

## 14.4 STATISTICAL SIGNIFICANCE

When a hypothesis is tested the *maximum* probability of a Type I error that one is *willing to risk* is called the *level of significance*. As will be seen, the level of significance is comparable with the level of confidence or fiducial probability in the case of confidence intervals. Significance levels always should be stated *prior* to the drawing of the samples upon which the tests are to be based so that the sample data will not influence the decision of the level which is to be used. Commonly used levels are 0.05 and 0.01. Other levels may be used, depending on the discretion of the person responsible for the work. If, on the basis of the evidence of the sample data, the probability of a Type I error turns out to be less than the stated level of significance, it is said that the difference from the expected result is *significant*, meaning that it is assumed that the difference is *real* and not caused by random variation.

If the usual significance levels (0.05 and 0.01) are being used and the sample evidence indicates that the probability of a Type I error is in the range

$$0.01 < P(\text{Type I error}) \leq 0.05 \tag{14.1}$$

it is said that the difference from the expected result is *significant* and the test result is flagged with a single asterisk. If the evidence indicates that

$$P(\text{Type I error}) \leq 0.01 \tag{14.2}$$

it is said that the difference is *highly significant* and the test result is flagged with a double asterisk. On the other hand, if the evidence indicates

$$P(\text{Type I error}) > 0.05 \tag{14.3}$$

it is said that the difference is *not significant* and the test result bears the flag "*N.S.*"

## 14.5 HOMOGENEITY TESTS

The fundamental idea underlying many statistical tests is summed up in the hypothesis:

$H_0$: The variates associated with the sampling units in the sampling frame developed for the experiment *are* from the specified population.
$H_1$: The variates *are not* from the specified population.

Tests made to evaluate hypotheses of this type are known as *homogeneity* or *uniformity* tests. For example, assume an infinitely large universe and that a certain characteristic of the elements of this universe is of interest. Furthermore, assume that this characteristic of interest gives rise to the random variable $X$ which is known to be normally distributed with mean $\mu_X$ and variance $\sigma_X^2$. Furthermore, assume that the standard method of measuring individual elements to obtain values of $X$ is slow and cumbersome. However, a new procedure has been devised which is much easier and faster but may be biased. A test must be made to determine if a bias is present.

If the bias exists, the population made up of the measurements made using the new procedure will have a mean that is different from that of the true population. Furthermore, if only a bias is involved, the variance and distributional pattern of the "new" population will be the same as those of the original or true population. Refer to the population arising from the new procedure as population $Y$.

The first step in the testing process is to put population $X$ into standard form (see Section 11.2), so that each variate, $x$, is converted to its equivalent $z$ using Equation 11.10,

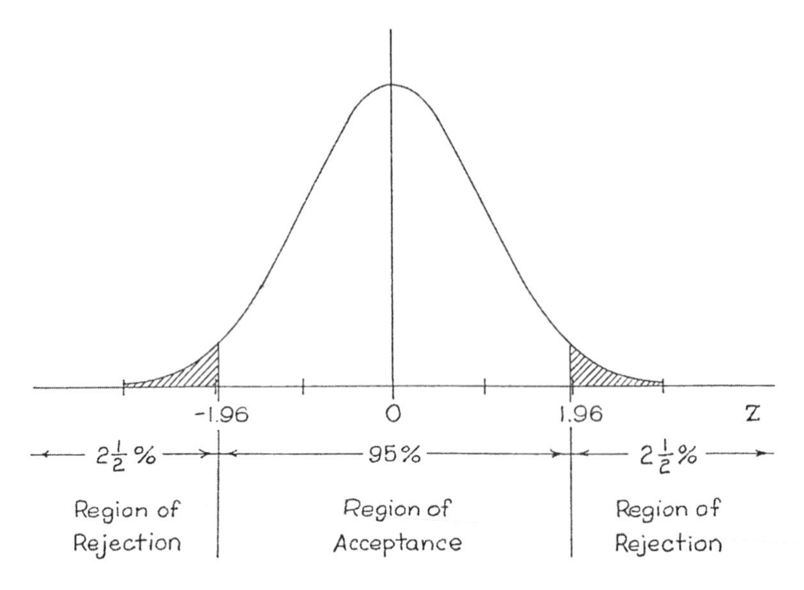


**FIGURE 14.1** The normal distribution showing the regions of acceptance and rejection of the null hypothesis in a two-tailed test when $\alpha = 0.05$.

$$z = (x - \mu_X)/\sigma_X$$

Then $\mu_z = 0$ and $\sigma_Z^2 = 1$. Assume that the sampling frame is coincident with the universe. The sampling consists of a single random draw from the frame. The sampling unit drawn is then evaluated using the new procedure. This experimental result is then converted to its equivalent $Z$ value. This test will be based on $Z$ and thus $Z = z$ is the *test statistic*. Prior to the draw, the probability of drawing a value of $Z$ in the interval $0 \pm 1.96$ is 0.95, while that of drawing a value of $Z$ outside of this interval is 0.05 (Figure 14.1). Consequently, if the sample yields a value that falls outside of $0 \pm 1.96$, it is concluded that the null hypothesis (that no bias exists) should be rejected and the alternate hypothesis (that a bias exists) accepted. This conclusion is reached because *there is only one chance in 20 that such a result would have been obtained if no bias existed.*

As an example of this process, assume that a certain population of *true* values is known to be normally distributed with mean $\mu_X = 100$ and variance $\sigma_X^2 = 2500$. Using the new method of evaluation, which may be biased, gives rise to the second population ($Y$), which also is normally distributed with the variance $\sigma_Y^2 = 2500$, but its mean is not known. It is hypothesized that no bias exists and thus $\mu_Y = 100$. This is to be tested so that the probability of making a Type I error is less than 0.05. To this end a random draw is made from the population generated by the new method of evaluation. Assume that the value drawn was $y = 90$. Then,

$$z = \frac{90 - 100}{50} = -0.2 \text{ N.S.}$$

which lies inside the interval $0 \pm 1.96$. Consequently, the null hypothesis, that $\mu_Y = 100$, is accepted and it is concluded that the new evaluation method does not yield biased results. The N.S. is a flag indicating that the difference between the sample and expected results is *not significant* (see Equation 14.3). On the other hand, if the value drawn had been $y = 220$, then

$$z = \frac{220 - 100}{50} = +2.4\,*$$

which lies *outside* the interval 0 ± 1.96. This would lead to the rejection of the null hypothesis and the conclusion reached would be that the new method of evaluation leads to biased results. The asterisk is a flag indicating that the difference between the sample and expected results is *significant*. In other words, the probability of a Type I error lies in the range from 0.05 to 0.01 (see Equation 14.1). Two asterisks would have indicated that the probability of a Type I error was less than 0.01, and it would be said that the difference between the sample and expected results is *highly significant* (see Equation 14.2).

It should be noted that 5% of the variates in population $X$ have equivalent $Z$ values that occur outside of the interval 0 ± 1.96 and it is possible that one of these was obtained in the sampling process. Thus, even if no bias exists, $H_0$ can be rejected. If this occurs, a Type I error has been committed. On the other hand, if indeed a bias exists, the randomly chosen $Z$ value could occur in the interval 0 ± 1.96. This would lead to acceptance of the null hypothesis and the commission of a Type II error.

This process can be put into the context of the confidence interval. In the sampling process a random draw is made from the frame and a variate, $y$, is obtained. Attached to $y$ is the bracket $\pm 1.96\,\sigma_Y$, assuming a confidence level of 0.95. Then, if the bracket does not include $\mu_X$, it is assumed that the variate is not from population $X$ and, consequently, that a bias exists. The logic is precisely the same as that used in uniformity trials — that the likelihood of membership in population $X$ is so small that it can be disregarded.

In the preceding example, the specified population, $X$, was normally distributed with a known mean and variance. It must be realized that the *concept* of the homogeneity test is not limited to normally distributed populations. It can be applied to situations involving virtually any probability distribution. The distributions most commonly encountered in tests of this type, in addition to the normal, are the $t$, $F$, and chi-square distributions.

## 14.6 ONE- AND TWO-TAILED TESTS

The interval within which a value of the test statistic can occur without causing rejection of the null hypothesis is known as the *region of acceptance* or *nonsignificance*. The region outside the region of acceptance is the *region of rejection* or *significance* (Figure 14.1). In the example in Section 14.5 there were two segments to the region of rejection, one in each tail of the distribution. When this is the case, it is said that the test is *two-tailed*. There are cases, however, where the region of rejection is concentrated in one tail. When this is the case, the test is *one-tailed*. Such one-tailed tests are related to single-limit confidence intervals in the same manner as two-tailed tests are related to two-limit confidence intervals. In many ways, homogeneity tests and confidence intervals are two sides of the same coin.

To illustrate a one-tailed test, the example in Section 14.5 will be modified. Assume that the new method of evaluation of the characteristic of interest is suspected of yielding results that are biased upwards. If this is true, $\mu_Y$ would be greater than $\mu_X$ but the variance and distributional pattern would not be affected. Consequently, the test is set up so that the null hypothesis would be rejected if $z > +1.645$. The region of rejection is concentrated in the right tail. If $z$ turned out to be negative, the null hypothesis would be accepted regardless of the absolute magnitude of $z$. Regions of rejections can be concentrated in the left tails as well as in the right. In such cases the bias would be assumed to be negative so that $\mu_Y < \mu_X$. The absolute magnitude of any region of rejection is a function of the specific level of significance, but its location depends on the hypotheses being tested.

## 14.7 MORE ON TYPE I AND TYPE II ERRORS

In the preceding discussion the probability of incurring a Type II error was not a factor. However, one must realize that the acceptance of a hypothesis is not free of the possibility of error. In the

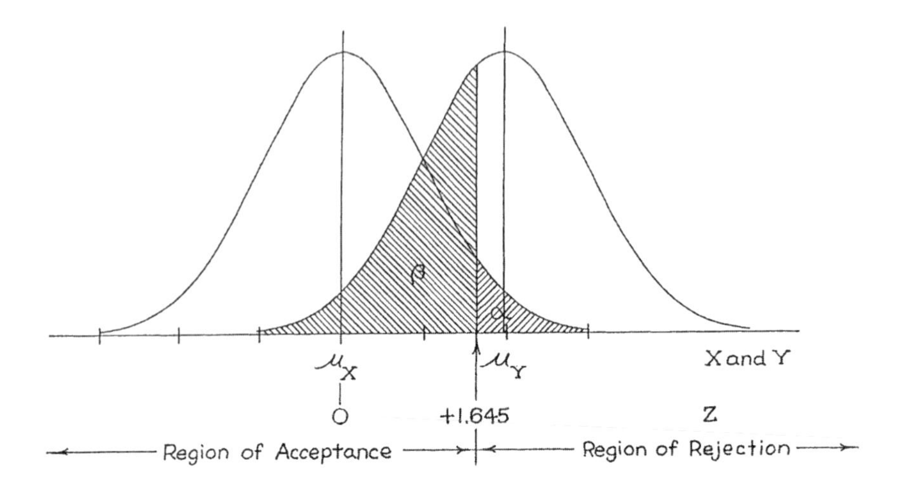


**FIGURE 14.2** Relationship between Type I and Type II errors when the test is one-tailed.

homogeneity test described in Section 14.5, the existence of a population other than that specified was implied. It was labeled "population $Y$." No assumption was made regarding the mean of this population except that it differed from that of population $X$. The variance and distributional pattern of $Y$ were assumed to be the same as those of $X$ since only a bias was suspected. In the context of the test in Section 14.4, if the null hypothesis was rejected, indicative that bias probably was present, it was assumed that $z$ belonged to population $Y$. However, $z$ might actually be a member of population $X$ in which case a Type II error would have been committed. *Only if the mean, variance, and distributional pattern of population Y are known can the probability of making a Type II error be determined.*

Figure 14.2 illustrates the situation when there are two populations, $X$ and $Y$, and the null hypothesis is that $Y$ does not exist (i.e., $\mu_Y = \mu_X$) while the alternate hypothesis is that $\mu_Y > \mu_X$. Both populations are normally distributed and $\sigma_Y^2 = \sigma_X^2$. The region of rejection is consequently concentrated in the right tail of population $X$. The relative area under the curve within this region is equal to $\alpha$, the level of significance. The position of the regional limit or *critical value* depends on $\alpha$. If $\alpha = 0.05$, the limit is located at $\mu_X + 1.645\sigma_X$ (if the test statistic is $y$) or at +1.645 (if the test statistic is $z$). *The relative area under the curve of population Y, lying to the left of the critical value, is equal to* $\beta$*, the probability of making a Type II error.* Thus, if the sample value $y$, or $z$, occurs to the *left* of the critical value, the null hypothesis would be accepted. However, $y$ (or $z$) could actually be a member of population $Y$ and, if this is the case, a Type II error would be made. If $\alpha$ is decreased, reducing the probability of a Type I error, the critical value is moved farther to the right. When this is done, $\beta$ increases. Conversely, if $\alpha$ is increased, $\beta$ is decreased. As can be seen, decreasing the probability of one type of error increases the probability of the other type of error. It must be emphasized, however, that decreasing $\alpha$ by one percentage point is not necessarily accompanied by a one percentage point increase in $\beta$. Only when $\mu_Y = \mu_X$ and $\sigma_Y^2 = \sigma_X^2$ will $\alpha + \beta = 1$.

It can be inferred from Figure 14.2 that, if $\alpha$ is held constant, $\beta$ will increase if the difference between $\mu_A$ and $\mu_B$ is decreased and will decrease if the difference between the two means is increased. If the test is one-tailed and the magnitude of $B$ is plotted against the standardized difference between the population mean,

$$\psi = (\mu_Y - \mu_X)/\sigma \tag{14.4}$$

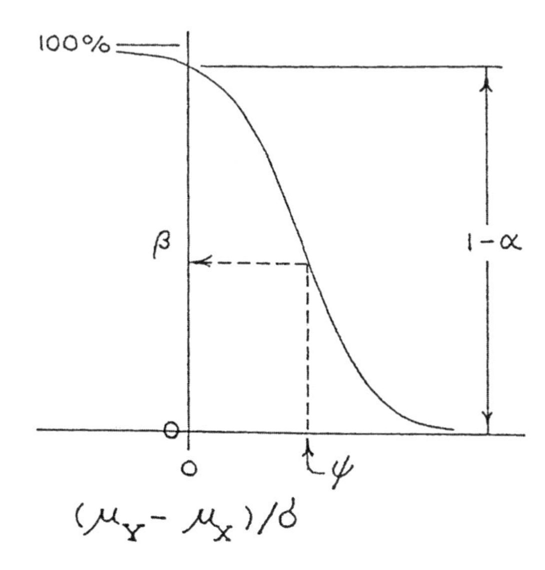


**FIGURE 14.3** Operating characteristic curve for a one-tailed test when populations $X$ and $Y$ are normally distributed and $\sigma_Y^2 = \sigma_X^2$. (From Miller, I. and Freund, J. E., *Probability and Statistics for Engineers,* 2nd ed., Prentice-Hall, Englewood Cliffs, NJ, 1977. With permission.)

a curve similar to that in Figure 14.3 will be generated. Such a curve is known as an *operating characteristic curve, or OC curve*. If the test is two-tailed, the OC curve will appear as in Figure 14.4. Curves of this type have been published by Bowyer and Lieberman (1972) for normal distributions with *homogeneous* variances (when $\sigma_Y^2 = \sigma_X^2$). OC curves are used to determine the magnitude of β when conditions match those upon which the curves are based. Obviously, a curve based on a normal distribution would be different from one based on a chi-square distribution. Furthermore, a curve based on normal distributions with homogeneous variances will be different from one based on normal distributions with *heterogeneous* variances (when $\sigma_Y^2 \neq \sigma_X^2$).

OC curves are used to assess the probabilities of Type II errors when α has been set at some magnitude. However, to use them it is necessary to have knowledge of the difference between the population means. Such knowledge is available in some contexts, notably in acceptance sampling where testing is carried out to see if manufactured products meet specified standards. A discussion of this use of OC curves can be found in Miller and Freund (1977).

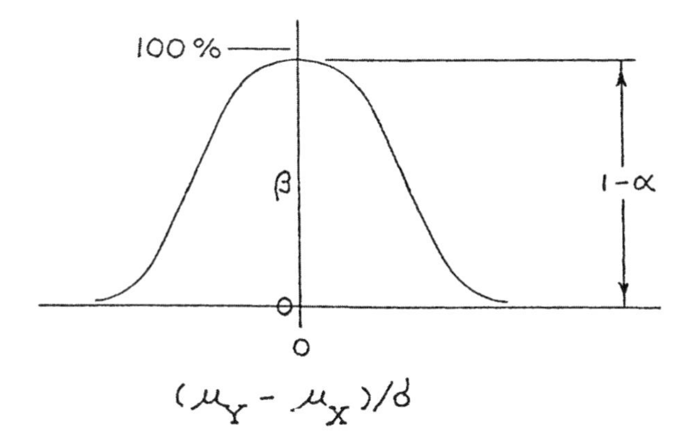


**FIGURE 14.4** Operating characteristic curve for a two-tailed test when populations $X$ and $Y$ are normally distributed and $\sigma_Y^2 = \sigma_X^2$. (From Miller, I. and Freund, J. E., *Probability and Statistics for Engineers,* 2nd ed., Prentice-Hall, Englewood Cliffs, NJ, 1977. With permission.)

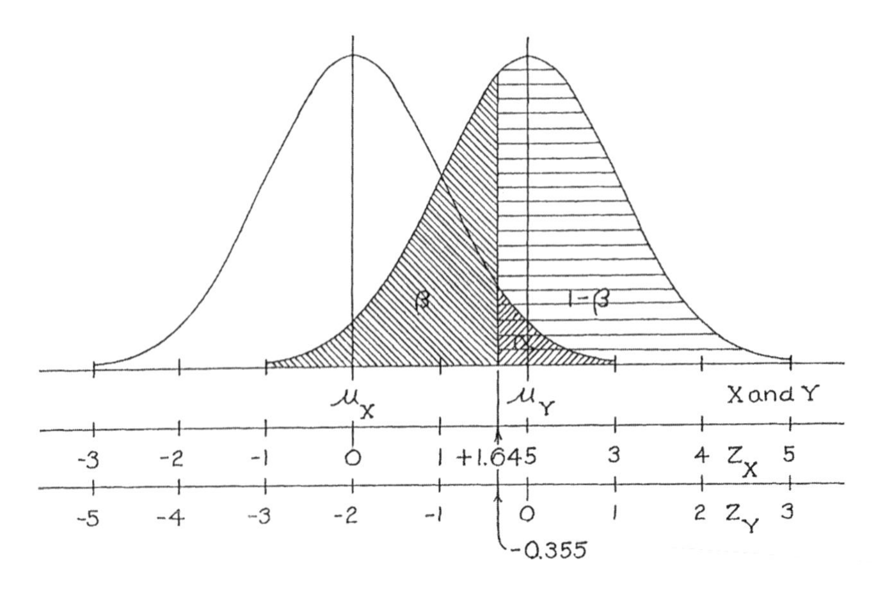


**FIGURE 14.5** Diagram showing the power of the test (horizontal crosshatch).

It should be noted that, even when the two population means in Figure 14.3 are equal, there is a probability of rejecting the null hypothesis, and that as the mean of population $Y$ becomes less than that of population $X$, the *actual* probability of a Type I error falls below $\alpha$, the specified probability limit. This does not occur when the test is two-tailed, as can be seen in Figure 14.4.

## 14.8 THE POWER OF THE TEST

Closely related to $\beta$, the probability of accepting a false hypothesis, is the *probability of accepting a true hypothesis* when rejecting the null hypothesis. This probability is known as the *power of the test*. As can be seen in Figure 14.5, the power of the test is the complement of $\beta$ and is equal to $1 - \beta$.

In the case of Figure 14.5, $X$ is the hypothesized population with mean $\mu_X = 100$ and standard deviation $\sigma_X = 50$. The hypotheses are

$H_0$: $\mu_Y = 100$.
$H_1$: $\mu_Y > 100$.

$\alpha$ is set at 0.05. In terms of $X$ the critical value for the one-tailed test is $z_X = +1.645$. The equivalent value in the case of $Y$ is $z_Y = (\mu_Y - \mu_X)/\sigma = -0.355$. The relative area under the curve for population $Y$ from $-0.355$ to $+\infty$, from Table A.4 in the Appendix, is 0.6387. Thus, the power of the test in this case is 0.6387 or 63.87%. This is the probability that the test will result in the acceptance of the true hypothesis that $\mu_Y > 100$.

Obviously, as in the case of $\beta$, the power of the test cannot be evaluated unless the difference between the two population means is known. However, it must be recognized that even if this difference is not known, if methods can be derived to reduce $\beta$, the power of the test will be increased, even though its magnitude remains unknown. Consequently, efforts are made, in the designing of tests, to minimize $\beta$ as much as possible within the constraints imposed on the test.

## 14.9 DECISION RULES

Prior to the testing of a hypothesis, it is necessary to define the rules that will govern the actions that are to be taken when the test is made. These are the *decision rules*. For example, if the test is to be based on the normal distributions, as was the example in Section 14.5, the test is to be two-tailed, and the level of significance, $\alpha$, has been set at 0.05, the decision rules are

1. *Reject* $H_0$ and accept $H_1$ if $|z| \geq 1.96$.
2. *Accept* $H_0$ if $|z| < 1.96$.

On the other hand, if the test is to be one-tailed and $\mu_Y$ may be greater than $\mu_A$, the decision rules are

1. *Reject* $H_0$ and accept $H_1$ if $z \geq 1.645$.
2. *Accept* $H_0$ if $z < 1.645$.

If the test is to be one-tailed and $\mu_Y$ may be less than $\mu_X$, the decision rules are

1. *Reject* $H_0$ if $z \leq -1.645$.
2. *Accept* $H_0$ if $z > -1.645$.

These rules provide the critical values defining the regions of acceptance and rejection and indicate the actions that are to be taken when the evidence from the sample becomes available. It must be emphasized that these rules are stated prior to the sampling stage so that the sample evidence will not influence the level of significance or even the hypotheses themselves.

## 14.10 TEST PROCEDURES

It is desirable to recognize that there is a definite set of steps that should be taken when setting up and executing a statistical test. They are

1. Define the universe of interest.
2. Define the variable of interest.
3. Define the sampling units.
4. Define the sampling frame.
5. State the null and alternative hypotheses.
6. State the type of test to be used (one- or two-tailed, $z$, $t$, $F$, $\chi^2$, etc.).
7. State the level of significance.
8. State the decision rules.
9. Carry out the sampling operation.
10. Compute the test statistic.
11. Apply the decision rules.

Putting the homogeneity test in Section 14.5 into this context:

1. Universe of interest: Assumed.
2. Variable of interest: Assumed.
3. Sampling units: Individual elements of the universe of interest.
4. Sampling frame: Assumed to be coincident with universe of interest.
5. Hypothesis:
   $H_0$: $\mu_Y = \mu_X$ (i.e., there is no bias and consequently population $Y$ is coincident with population $X$).
   $H_1$: $\mu_Y \neq \mu_X$ (i.e., bias exists and populations $Y$ and $X$ are not coincident).
6. Type of test: Two-tailed $z$ ($z$, because the test is based on the standard normal distribution, and two-tailed, because $\mu_Y$ may be either larger or smaller than $\mu_X$).
7. Level of significance: $\alpha = 0.05$.

8. Decision rules:
   *Reject* $H_0$ and *accept* $H_1$ if $|z| \geq 1.96$ (i.e., if $z$ is outside the interval $0 \pm 1.96$).
   *Accept* $H_0$ if $|z| < 1.96$.
9. Drawing of sample: Sample consists of a single random draw from the sampling frame.
10. Computations: $z = (y - \mu_X)/\sigma$.
11. Apply the decision rules.

## 14.11 HOMOGENEITY TEST USING A SAMPLE MEAN

In the example of a homogeneity test in Section 14.5, the test was based on a single randomly drawn variate $y$ (or its equivalent $z$). The parent population was known to be normally distributed and to have a specific variance. A test made using a single observation such as this is theoretically sound, but the power of the test is low. To improve the power of the test, it is necessary to replace the individual variate, $y$, as the estimator of $\mu_Y$, with the mean, $\bar{y}$, of a random sample of size $n$ drawn from the parent population. The sample mean is an individual variate from the sampling distribution of means based on samples of size $n$ drawn from the parent population $Y$, and is an unbiased estimator of $\mu_{\bar{y}}$ and, since $\mu_{\bar{y}} = \mu_Y$, it also is an unbiased estimator of $\mu_Y$. According to the central limit theorem, the sampling distribution of means will be normally distributed if the parent population is normally distributed. Consequently, the test described in Section 14.5 can be used in the context of the sample mean being a variate from the normally distributed sampling distribution. The gain in the power of the test is caused by the fact that the variance of a sampling distribution of means is much smaller than that of the parent population (see Equation 12.3).

The gain in the power of the test is demonstrated in Figure 14.6. In this case, population $X$ is known to be normally distributed with the mean $\mu_X = 100$ and the variance $\sigma_X^2 = 2500$. The population based on the alternative (and biased) measurement procedure, $Y$, is also normally distributed with the variance $\sigma_X^2 = 2500$ but its mean is $\mu_Y = 200$. The bias has no effect on the variance. Consequently, $\sigma_X^2 = \sigma_Y^2 = \sigma^2$. When this is the case, it is said that the variances are homogeneous, meaning that they are equal.* Then, using the procedure outlined in Sections 14.5 and 14.9, the power of the test in the case of the single variate, $y$, is obtained as follows.

The hypotheses:

$H_0$: $\mu_Y = \mu_X$ (= 100).
$H_1$: $\mu_Y > \mu_X$ (i.e., the bias is expected to be positive).
Type of test: One-tailed $z$ (one-tailed because the alternate hypothesis is that the bias is positive and thus $\mu_Y$ would be greater than $\mu_X$. The $z$ test is based on the standard normal distribution).
Significance level: $\alpha = 0.05$.
The decision rules:
   *Reject* $H_0$ and accept $H_1$ if $z \geq z = 1.645$.
   *Accept* $H_0$ if $z < 1.645$.

$$\text{The critical value for population } X = \mu_X + 1.645\sigma = 100 + 1.645(50) = 182.25$$

$$\text{The equivalent } z_Y = (182.25 - 200)/50 = -0.355$$

$$\text{The relative area from } -0.355 \text{ to } +\infty = 0.1387 + 0.5000 = 0.6387 \text{ or } 63.87\%$$

* A test for homogeneity of variance is described in Section 14.16.

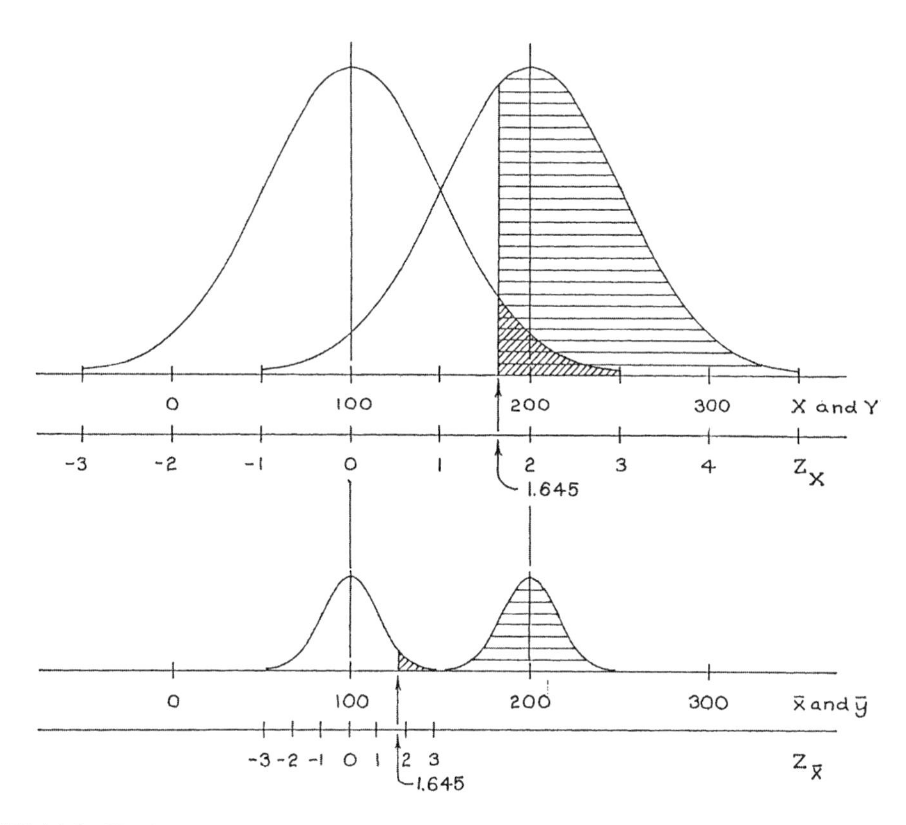


**FIGURE 14.6** The increase in the power of the test associated with the use of the sample mean instead of a single variate. The power of the test increases as $n$ increases. In this case, when $n$ = 10, the power of the test is practically 100%.

The power of the test is 63.87%. Compare this with the power of the test if a sample of size $n$ = 10 is used.

The hypotheses:

$H_0$: $\mu_{\bar{y}} = \mu_{\bar{x}}$ $(= \mu_X = 100)$.

$H_1$: $\mu_{\bar{y}} > \mu_{\bar{x}}$.

Type of test: One-tailed $z$.

Significance level: $\alpha = 0.05$.

The decision rules:

*Reject* $H_0$ and accept $H_1$ if $z \geq z_{0.05} = 1.645$.

*Accept* $H_0$ if $z < 1.645$.

The critical value for sampling distribution:

$$\bar{X} = \mu_{\bar{x}} + 1.645\sigma_{\bar{y}}$$

$$= 100 + 1.645\sqrt{\frac{2500}{10}}$$

$$= 126.01$$

The equivalent $z_{\bar{y}} = (126.01 - 200) / \sqrt{2500/10}$

$= -4.680$

The relative area from –4.680 to +∞ ≅ 1.000 or 100%.

The power of this test is practically 100%. The gain in power is obvious.

If the parent population is known to be normally distributed but its mean and variance are not known and one wishes to test the hypothesis that it has a certain mean, it becomes necessary to use the sample variance, $s_Y^2$, appropriately corrected for bias, as an estimate of $\sigma^2$. When this is done, it is necessary to use the appropriate $t$ distribution, rather than the normal, in the testing process (see Section 13.5). Thus, if in the previous example $\sigma^2$ was not known, the test of the hypothesis that $\mu_X = 100$ would take the following form, assuming a sample of size $n = 10$.

The hypotheses:

$H_0$: $\mu_{\bar{y}} = \mu_x$ $(= \mu_X = 100)$.

$H_1$: $\mu_{\bar{y}} > \mu_{\bar{x}}$.

Type of test: One-tailed $t$ test.

Significance level: $\alpha = 0.05$.

The decision rules:

*Reject* $H_0$ and *accept* $H_1$ if $t \geq t_{0.05}(k) = 1.833$; $k = n - 1 = 9$.

*Accept* $H_0$ if $t < 1.833$.

Then, if $n = 10$, $\bar{y} = 110$, and $s_Y^2 = 2600$,

$$s_{\bar{y}} = \sqrt{s_Y^2 / n} = \sqrt{2600/10} = 16.1245$$

$$t = (\bar{y} - \mu_{\bar{y}}) / s_{\bar{y}} \tag{14.5}$$

$$= (110 - 100) / 16.1245 = 0.620 \text{ N.S.}$$

Since 0.620 < 1.833, the null hypothesis is accepted and it is said that the difference between the test and expected results is not significant (see Section 14.4). The flag N.S. is attached to the $t$ value to indicate this fact.

If the sample mean had been 175,

$$t = (175 - 100)/16.1245 = 4.651^{**}$$

and the null hypothesis would have been rejected. Since $t_{0.01}(9) = 2.821$, the probability of a Type I error is less than 0.01 and it is said that the difference between test and expected results is highly significant. This fact is indicated by the two asterisks following the computed $t$ value.

In most situations the distributional pattern, as well as the mean and variance, of the parent population is not known. In such cases it becomes necessary to make use of the tendency of sampling distributions of means to assume distributional patterns which approximate the normal. This was discussed in Section 12.10. The central limit theorem refers to situations where the random variable is continuous but, as was seen in Section 12.10, the tendency is also present where the random variable is discrete. When the random variable is continuous or, if discrete, the population is infinitely large or, if finite, the sampling is with replacement, it usually is assumed that, if $n >$ 30, the sampling distribution is sufficiently close to being normal that it can be modeled by the normal. Since the sampling distribution is only approximately normal, there will be an unknown component of error in the testing process. However, the rate of closure to the normal, as sample size increases, is much more rapid in the middle portion of the distribution, in the range from $-3\sigma_{\bar{y}}$ to $+3\sigma_{\bar{y}}$, than it is in the tails. Consequently, if the level of significance, $\alpha$, is in the usual range, from 0.05 to 0.01, the error is usually of little importance and is ignored. Because of this, it is said that tests based on the assumption of normality of sampling distributions of means are *robust*, meaning that the tests will yield acceptable results even if the parent populations exhibit considerable divergence from the normal.

If the parent random variable is discrete, the population is finite, and the sampling is without replacement, the situation becomes much less clear-cut. For example, if $Y$ is distributed binomially, the distributions of sample proportions, which are means, are distributed binomially. As the sample size increases, the distribution of sample proportions approaches normality. However, even if $\mu_Y = p = 0.5$, the sample size must be greater than 35 before the binomial distribution can be modeled by the normal. As $\mu_Y$ shifts away from 0.5, the minimum sample size needed increases until, when $\mu_Y$ 0.01, $n$ must exceed 909. If, in the above case, the sampling had been without replacement, the distribution of sample proportions would have been hypergeometric. The minimum sample size would be somewhat smaller than when the sampling was with replacement (see Equation 13.68), but there would be an additional proviso that the sampling fraction, $n/N$, would have to be less than 0.1. These may be extreme cases, but they do indicate that a sample size which is simply greater than 30 is not adequate in all cases. Since finite populations and sampling without replacement are fairly common, this problem is not trivial. In most cases when the distribution of the parent population is not known investigators simply assume that, if the largest feasible sample is obtained, the robustness of the $t$ test will be sufficient to make the test meaningful. A further discussion of this matter is beyond the scope of this book.

When the distributional pattern of the parent population is not known, the testing process is exactly the same as was shown in the last example.

If, in the last example, the mean of $Y$ could have been either less than or more than 100, the test would have been two-tailed and taken the following form.

The hypotheses:

$H_0$: $\mu_{\bar{y}} = \mu_{\bar{x}}$ $(= \mu_X = 100)$.

$H_1$: $\mu_{\bar{y}} \neq \mu_{\bar{x}}$.

Type of test: Two-tailed $t$.

Significance level: $\alpha = 0.05$.

The decision rules:

*Reject* $H_0$ and *accept* $H_1$ if $|t| \geq t_{0.05}(9) = 2.262$.

*Accept* $H_0$ if $|t| < 2.262$.

Then, if $n = 10$, $\bar{y} = 85$, and $s_Y^2 = 2600$,

$$s_{\bar{y}} = \sqrt{s_Y^2 / n} = \sqrt{2600 / 10} = 16.1245$$

$$t = (\bar{y} - \mu_{\bar{y}}) / s_{\bar{y}} = (85 - 100) / 16.1245 = -0.930$$

$$|t| = 0.930 < 2.262 \text{ N.S.}$$

Since the absolute value of $t$ is less than 2.262, the null hypothesis is accepted.

## 14.12 TEST OF A DIFFERENCE (UNPAIRED)

Assume that the population associated with random variable $X$ is made up of the total heights of 10-year-old loblolly pine trees grown in plantations under a specified set of environmental conditions. Further assume that the population associated with random variable $Y$ is the set of total tree heights of similar loblolly pines to which nitrogen fertilizer was applied at age 5 years. Now assume that a test is to be made to determine whether the fertilizer increased height growth. The test can take on the form of the homogeneity tests previously described to see if the variate drawn from population $Y$, or from the sampling distribution of $\bar{y}$, comes from a population having a mean or expected value greater than $E(X)$. However, the test can also be made using the random variable $D = Y - X$ and the null hypothesis that $E(D) = 0$.

Assume that both parent populations are normally distributed. The assumption that tree heights are normally distributed is, of course, invalid, but, for the purpose of illustration, accept that assumption and proceed. The parameters of the two parent populations are $\mu_X$; $\sigma_X^2$; $\mu_Y$; and $\sigma_Y^2$. Assume that all but $\mu_Y$ are known. Population $D$ would then also be normally distributed and have the parameters $\mu_D = \mu_Y - \mu_X$ and $\sigma_D^2 = \sigma_X^2 + \sigma_Y^2$. $\mu_D$ is, of course, unknown. There is no covariance term in the expression for $\sigma_D^2$ since any variate in $Y$ is free to pair with any variate in $X$. The homogeneity test would be made to see if $\mu_D = 0$. Since one expects fertilizer to increase growth, the alternate hypothesis would be that $\mu_D$ is greater than zero. A *single* draw is made from $X$ and from $Y$. The resulting difference, $d$, is a variate from $D$.

The hypotheses:
- $H_0$: $\mu_D = 0$.
- $H_1$: $\mu_D > 0$.

Type of test: One-tailed $z$.
Significance level: $\alpha$.
The decision rules:
- *Reject* $H_0$ and *accept* $H_1$ if $z \geq z_\alpha$.
- *Accept* $H_0$ if $z < z_\alpha$.

$z = (d - 0)/\mu_D$

This is equivalent to the homogeneity test in Section 14.11 when only one variate was drawn from the parent population.

This procedure, although sound when the required conditions exist, is unrealistic. Except in very unusual situations the parameters and distributional patterns of the parent populations are not known. Furthermore, even if only $\mu_Y$ and $\mu_D$ were unknown and the test could be carried out, the power of the test would be low. To overcome these problems the test must be based on sampling distributions of means rather than the parent populations. Thus, if $X$ and $Y$ are the random variables: $\mu_{\bar{x}} = \mu_Y$; and $\sigma_{\bar{y}}^2 = \sigma_Y^2 / n_Y$. The parameters of the sampling distribution of differences between means would be

$$\mu_{(\bar{y}-\bar{x})} = \mu_{\bar{y}} - \mu_{\bar{x}} = \mu_Y - \mu_X \tag{14.6}$$

Its variance depends on whether $\sigma_X^2 = \sigma_X^2$ and whether $n_X = n_Y$. If the variances are homogeneous, which can be ascertained using the $F$ test described in Section 14.16, $\sigma_X^2 = \sigma_Y^2 = \sigma^2$, and if the sample sizes are equal, $n_X = n_Y = n$,

$$\sigma_{(\bar{y}-\bar{x})}^2 = \sigma_{\bar{x}}^2 + \sigma_{\bar{y}}^2 = \frac{\sigma_X^2}{n} + \frac{\sigma_Y^2}{n} = \frac{2\sigma^2}{n} \tag{14.7}$$

Since the pairings of $\bar{x}$ and $\bar{y}$ are random, there is no covariance term. If the sample sizes are not equal,

$$\sigma^2_{(\bar{y}-\bar{x})} = \sigma^2_{\bar{x}} + \sigma^2_{\bar{y}} = \frac{\sigma^2_X}{n_X} + \frac{\sigma^2_Y}{n_Y} = \sigma^2\left(\frac{n_X + n_Y}{n_X n_Y}\right) \tag{14.8}$$

If the variances are heterogeneous and the sample sizes are equal,

$$\sigma^2_{(\bar{y}-\bar{x})} = \sigma^2_{\bar{x}} + \sigma^2_{\bar{y}} = \frac{\sigma^2_X}{n_X} + \frac{\sigma^2_Y}{n_Y} = \frac{(\sigma^2_X + \sigma^2_Y)}{n} \tag{14.9}$$

Finally, if the variances are heterogeneous and the sample sizes are not equal,

$$\sigma^2_{(\bar{y}-\bar{x})} = \sigma^2_{\bar{x}} + \sigma^2_{\bar{y}} = \frac{\sigma^2_X}{n_X} + \frac{\sigma^2_Y}{n_Y} \tag{14.10}$$

Obviously, Equation 14.10 is general and could be used in all cases.

If $X$ and $Y$ are normally or nearly normally distributed, $\bar{x}$ and $\bar{y}$ would also be normally or nearly normally distributed. In turn, the sampling distribution of differences between means would also be normally or nearly normally distributed. Under these circumstances, the test would take the following form:

The hypotheses:

$H_0$: $\mu_{(\bar{y}-\bar{x})} = 0$.

$H_1$: $\mu_{(\bar{y}-\bar{x})} > 0$.

Type of test: One-tailed $t$.

Significance level: $\alpha$.

Decision rules:

*Reject* $H_0$ and *accept* $H_1$ if $t \geq t_\alpha(k)$ or $t'_\alpha$.

*Accept* $H_0$ if $t < t'_\alpha$.

The critical value, $t_\alpha(k)$ or $t'_\alpha$, depends on the situation. If the variances are homogeneous and the sample sizes are equal, $k = 2(n - 1)$. If the variances are homogeneous but the sample sizes are not equal, $k = n_X + n_Y - 2$. If the variances are heterogeneous, regardless of whether or not sample sizes are equal, it is impossible to determine exactly which $t$ distribution is appropriate. Welch (1949) suggested the use of a $k$ computed as follows:

$$k = 1\bigg/\left[\frac{u^2}{n_X - 1} + \frac{(1-u)^2}{n_Y - 1}\right] \tag{14.11}$$

where

$$u = \frac{s^2_X}{n_X}\bigg/\left(\frac{s^2_X}{n_X} + \frac{s^2_Y}{n_Y}\right) \tag{14.12}$$

Cochran and Cox (1957) proposed the following as an approximation of the appropriate critical value:

$$t'_{\alpha} \cong [s_{\bar{x}}^2 t_{\alpha}(k_X) + s_{\bar{y}}^2 t_{\alpha}(k_Y)] / (s_{\bar{x}}^2 + s_{\bar{y}}^2) \tag{Eq.14.13}$$

The latter approach is slightly more conservative than that suggested by Welch, in that it yields a critical value that is somewhat larger, making the rejection of the null hypothesis somewhat more difficult.

Then, in the sample being discussed, if $A$ is set at 0.05, the variances of $X$ and $Y$ are homogeneous, the sample sizes are equal, and the sample data are $n = 10$, $\bar{x} = 27$ ft, $s_X^2 = 9$, $\bar{y} = 32$ ft, and $s_Y^2 = 12$,

$$k = 2(n-1) = 2(10-1) = 18$$

$$t_{\alpha}(k) = t_{0.05}(18) = 1.734$$

$$\bar{d} = \bar{y} - \bar{x} = 32 - 27 = 5 \text{ ft}$$

Since $\sigma_X^2 = \sigma_Y^2 = \sigma^2$, both $s_X^2$ and $s_Y^2$ are estimates of $\sigma^2$. Consequently, the sums of squared residuals and the degrees of freedom from both samples can be combined or *pooled* to arrive at a single estimate, $s^2$, which is known as the *pooled* variance:

$$s^2 = \frac{\sum_{i=1}^{n}(x_i - \bar{x})^2 + \sum_{i=1}^{n}(y_i - \bar{y})^2}{(n_X - 1) + (n_Y - 1)} \tag{14.14}$$

Thus, in the case of the example

$$\sum_{i=1}^{n}(x_i - \bar{x}) = (n_X - 1)s_X^2 = 9(9) = 81$$

$$\sum_{i=1}^{n}(y_i - \bar{y}) = (n_Y - 1)s_Y^2 = 9(12) = 108$$

$$s^2 = (81 + 108)/2(9) = 189/18 = 10.5$$

Using $s^2$ to arrive at $s_{\bar{d}}^2$,

$$\begin{aligned} s_{\bar{d}}^2 &= s^2 / n \\ &= 10.5/10 = 1.05 \\ s_{\bar{d}} &= \sqrt{1.05} = 1.0247 \text{ ft} \\ t &= (\bar{d} - 0)/s_{\bar{d}} = 5/1.0247 = 4.879^{**} \end{aligned} \tag{14.15}$$

Then $t = 4.879 > t_{0.05}(18) = 1.734$. Consequently, the null hypothesis is rejected and it is concluded that nitrogen fertilizer has a positive effect on the height growth of 5-year-old loblolly pines. Since $t = 4.879$ is also greater than $t_{0.01}(18) = 2.552$, it would be said that the difference is highly significant.

If the variances are homogeneous but the sample sizes are different, the test proceeds according to the following pattern. Let the sample data remain the same except that $n_X = 10$ and $n_Y = 12$. Then,

$$k = n_X + n_Y - 2 = 10 + 12 - 2 = 20$$

$$t_{0.05}(20) = 1.725$$

$$\sum_{i=1}^{n_X} (x_i - \bar{x})^2 = (n_X - 1)s_X^2 = 9(9) = 81$$

$$\sum_{i=1}^{n_Y} (y_i - \bar{y})^2 = (n_Y - 1)s_Y^2 = 11(12) = 132$$

$$s^2 = (81 + 132) / [(10 - 1) + (12 - 1)] = 10.65$$

Using Equation 14.8,

$$s_{\bar{d}}^2 = s^2\left(\frac{n_X + n_Y}{n_X n_Y}\right) = 10.65\left[\frac{10 + 12}{10(12)}\right] = 1.9525$$

$$s_{\bar{d}} = \sqrt{1.9525} = 1.397 \text{ ft}$$

$$t = 5 / 1.397 = 3.578^{**}$$

Again the null hypothesis is rejected and the difference is said to be highly significant.

If the variances of the parent populations are heterogeneous, the test would be as follows. Let the sample data be $n_X = 10$, $\bar{x} = 27$ ft, $s_X^2 = 9$, $n_Y = 12$, $\bar{y} = 32$ ft, and $s_Y^2 = 30$. Then,

$$\sum_{i=1}^{n_X} (x_i - \bar{x})^2 = (n_X - 1)s_X^2 = 9(9) = 81$$

$$\sum_{i=1}^{n_Y} (y_i - \bar{y})^2 = (n_Y - 1)s_Y^2 = 11(30) = 330$$

$$s^2 = (81 + 330) / [(10 - 1) + (12 - 1)] = 20.55$$

$$s_{\bar{d}}^2 = 20.55\left[\frac{10 + 12}{10(12)}\right] = 3.7675 = 20.55(10 + 12) / 120 = 3.7675$$

$$s_{\bar{d}} = \sqrt{3.7675} = 1.941$$

$$t = 5 - 0 / (1.941) = 2.576^{**}$$

Using the Welch (1949) method of determining the approximate number of degrees of freedom (see Equation 14.11) to obtain a critical $t$ value,

$$u = \frac{s_X^2}{n_X} \bigg/ \left( \frac{s_X^2}{n_X} + \frac{s_Y^2}{n_Y} \right) = \frac{9}{10} \bigg/ \left( \frac{9}{10} + \frac{30}{12} \right) = 0.2647059$$

$$(1 - u) = 1 - 0.2647059 = 0.7352941$$

$$k = 1 \bigg/ \left[ \frac{(0.2647059)^2}{9} + \frac{(0.7352941)^2}{11} \right] = 17.563 \text{ or } 18$$

Then $t_{0.05}(18) = 1.734$ and $t = 2.576 > t_{0.05}(18) = 1.734$ and the null hypothesis is rejected. Since $t = 2.576 > t_{0.01}(18) = 2.552$, it is said that the difference is highly significant.

If the Cochran and Cox (1957) method of arriving at the critical $t$ value is used,

$$k_X = 10 - 1 = 9$$

$$k_Y = 12 - 1 = 11$$

$$t_{0.05}(9) = 1.833$$

$$t_{0.05}(11) = 1.796$$

$$T'_{0.05} \cong \left[ \left( \frac{9}{10} \right) 1.833 + \left( \frac{30}{12} \right) 1.796 \right] \bigg/ \left[ \left( \frac{9}{10} \right) + \left( \frac{30}{12} \right) \right] = 1.806$$

Now, $t = 2.576 > t'_{0.05} = 1.806$, and again the null hypothesis is rejected. If $t'_{0.01}$ is computed, it will be found to be 2.745, which is greater than $t = 2.576$. Consequently, when the Cochran and Cox method is used, the difference is said to be significant but not highly significant. This bears out what was said earlier that the Cochran and Cox procedure is more conservative than Welch's, making it more difficult to reject the null hypothesis.

If the parent populations, $X$ and $Y$, are not normally or nearly normally distributed or if their distributional patterns are not known, it becomes necessary, as in Section 14.11, to fall back on the robustness of the $t$ test. Sample sizes should be as large as is feasible and should, as a minimum, exceed 30.

## 14.13 TEST OF A DIFFERENCE (PAIRED)

In the above example, the individual tree heights making up $X$ come from a different set of trees than those whose heights make up population $Y$. Thus, the set of variates making up population $D$ arises from the pairing of every variate of $X$ with every variate of $Y$ and, when a single draw is made from each of the two parent populations, the result is a random pairing. The same situation exists when the sampling distribution of differences between means is generated. Each mean in sampling distribution $\bar{X}$ is free to pair with any mean in the set making up sampling distribution $Y$, and then each $d$ is the result of a random pairing. When the pairing of variates is random, it is said that the variates are *not paired* and that the test is an *unpaired* $z$ or $t$ test. This is unfortunate terminology because pairing certainly takes place. The implication, however, is that the pairing is random.

In some cases the pairing is not random. In other words, a linkage exists between the two primary random variables. In some manner, the drawing of a given sampling unit gives rise to a variate from each of the two parent populations. For example, the universe (or sampling frame) could consist of 1/5-acre sample plots within a given tract of land. These plots could be cruised conventionally on the ground, to obtain a per acre volume estimate. The set of such values would form population $X$. The plots could also be cruised using aerial photographic evidence. The set of values obtained in this manner would form population *Y*. Then, the drawing of a given sample plot would give rise to a variate *X and* a variate of *Y*. These two variates would be linked through their common sample plot. When a test is made using variates linked together in this manner, it is said that the test is a *paired z* or *t* test. The implication is that the pairing is *not* random. When the sampling is of this type, the testing procedure must be modified.

As before, $D = Y - X$ and $\mu_D = \mu_Y - \mu_X$. However, from Section 9.9, Item 4,

$$\sigma_D^2 = \sigma_X^2 + \sigma_Y^2 - 2\sigma_{XY}$$

where $\sigma_{XY}$ is the covariance of *X* and *Y* (see Section 9.8 and Equation 9.30). Covariance is a factor since the pairings of the variates of *X* and *Y* are not random and independence between the parent random variables cannot be assumed.

The distributional patterns and variances of the parent populations are rarely known and, consequently, they are not known for the population of differences. Again it becomes necessary to use sampling distributions of means to obtain a basis upon which to construct a test. In the context of the problem involving ground cruising and cruising from aerial photographs, assume that a sample of size $n > 1$ is drawn from the universe of sample plots. The plots were cruised in two ways and data of the following type obtained.

| ***Plot*** | $y_i$ **(bd. ft)** | $x_i$ **(bd. ft)** | $d_i$ **(bd. ft)** |
|---|---|---|---|
| 1 | 1540 | 1660 | –120 |
| 2 | 1350 | 1440 | –90 |
| 3 | 1260 | 1320 | –60 |
| . | . | . | . |
| . | . | . | . |
| . | . | . | . |
| *n* | 310 | 360 | –50 |

The mean difference is then

$$\bar{d} = \sum_{i=1}^{n} d_i \Big/ n \tag{14.16}$$

If this is done for all possible samples of size *n* from the stated universe, the resulting set of mean differences would form a sampling distribution. Since it is made up of means, it would tend to have a distributional pattern which more or less approximates the normal, with the approximation improving as the sample size increases. The parameters of the sampling distribution are as follows.

$$\mu_{\bar{d}} = \mu_{\bar{y}} - \mu_{\bar{x}} = \mu_Y - \mu_X \tag{14.17}$$

If the parent random variables are continuous or if they are discrete and their populations are infinitely large or, if finite, the sampling is with replacement,

$$\begin{aligned} \sigma_{\bar{d}}^2 &= \sigma_D^2 / n = (\sigma_X^2 + \sigma_Y^2 - 2\sigma_{XY}) / n \\ &= \sigma_{\bar{x}}^2 + \sigma_{\bar{y}}^2 - 2\sigma_{\bar{x}\bar{y}} \end{aligned} \tag{14.18}$$

On the other hand, if the parent random variables are discrete, their populations are finite, and the sampling is without replacement.

$$\sigma_{\bar{d}}^2 = \frac{\sigma_D^2}{n}\left(\frac{N-n}{N-1}\right) = \frac{1}{n}(\sigma_X^2 + \sigma_Y^2 - 2\sigma_{XY})\left(\frac{N-n}{N-1}\right) \quad (14.19)$$

The derivation of Equation 14.19 takes the following pattern. From Equation 9.28, the correlation coefficient is

$$\rho = \sigma_{XY} / \sigma_X\sigma_Y \quad (14.20)$$

Then

$$\sigma_{XY} = \rho\sigma_X\sigma_Y \quad (14.21)$$

Also, from Equation 9.28

$$\rho = \sigma_{\bar{x}\bar{y}} \Big/ \sqrt{\sigma_{\bar{x}}^2\sigma_{\bar{y}}^2}$$

$$\sigma_{\bar{x}\bar{y}} = \rho\sqrt{\sigma_{\bar{x}}^2\sigma_{\bar{y}}^2} = \rho\sqrt{\frac{\sigma_X^2}{n}\left(\frac{N-n}{N-1}\right)\frac{\sigma_Y^2}{n}\left(\frac{N-n}{N-1}\right)}$$

$$\sigma_{\bar{x}\bar{y}}^2 = \rho^2\left(\frac{\sigma_X^2}{n}\right)\left(\frac{N-n}{N-1}\right)\left(\frac{\sigma_Y^2}{n}\right)\left(\frac{N-n}{N-1}\right) = \rho^2\sigma_X^2\sigma_Y^2\left(\frac{1}{n^2}\right)\left(\frac{N-n}{N-1}\right)^2$$

Then

$$\sigma_{\bar{x}\bar{y}} = \rho\left(\frac{\sigma_X\sigma_Y}{n}\right)\left(\frac{N-n}{N-1}\right) \quad (14.22)$$

Returning to Equation 14.18 and substituting,

$$\sigma_{\bar{d}}^2 = \sigma_{\bar{x}}^2 + \sigma_{\bar{y}}^2 - 2\sigma_{\bar{x}\bar{y}} = \frac{\sigma_X^2}{n}\left(\frac{N-n}{N-1}\right) + \frac{\sigma_Y^2}{n}\left(\frac{N-n}{N-1}\right) - 2\rho\left(\frac{\sigma_X\sigma_Y}{n}\right)\left(\frac{N-n}{N-1}\right) \quad (14.23)$$

Substituting for $\rho$ using Equation 14.20,

$$\sigma_{\bar{d}}^2 = \frac{\sigma_X^2}{n}\left(\frac{N-n}{N-1}\right) + \frac{\sigma_Y^2}{n}\left(\frac{N-n}{N-1}\right) - 2\left(\frac{\sigma_{XY}}{\sigma_X\sigma_Y}\right)\left(\frac{\sigma_X\sigma_Y}{n}\right)\left(\frac{N-n}{N-1}\right)$$

$$= \frac{1}{n}(\sigma_X^2 + \sigma_Y^2 - 2\sigma_{XY})\left(\frac{N-n}{N-1}\right)$$

which is Equation 14.19. This expression is valid only when the relationship between $X$ and $Y$ takes the form of a simple linear function (see Section 9.10). If the relationship is not linear, the situation becomes obscure and generalizations cannot be made.

It should be pointed out that finite population correction factors are rarely used. It usually is assumed that their effects are negligible, which is true if the sampling fraction is small. To have a meaningful test, however, the sample size should be large enough so that the sampling distribution can be assumed to be normal. Most workers assume that $n$ should be greater than 30. There is no real consensus with respect to the maximum sampling fraction, but it appears that it never should exceed 0.1.

The test is a conventional $t$ test with the decision rules depending on the situation. In some cases there is no reason to believe $\mu_Y$ to be either larger or smaller than $\mu_X$ and the test would simply be to determine whether a difference exists. On the other hand, as in the case of the two different approaches to cruising, there may be a reason for expecting a difference in a particular direction. In the former situation a two-tailed test is indicated, while in the latter a one-tailed test would be appropriate.

As an example, consider the two methods of cruising. Assume that $Y$ and $X$ are both normally distributed. Since it is anticipated that $\mu_Y$ will be less than $\mu_X$,

The hypotheses:

$H_0$: $\mu_{\bar{d}} \geq 0$.

$H_1$: $\mu_{\bar{d}} < 0$.

Type of test: One-tailed $t$.

Significance level: $\alpha = 0.05$.

Decision rules (assuming a sample of size $n = 10$):

*Reject* $H_0$ and *accept* $H_1$ if $t \leq -t_{0.05}(9) = -1.833$.

*Accept* $H_0$ if $t > -1.833$.

Assume that $X$ and $Y$ are normally distributed and that the sample yielded the following data:

| Plot | $y_i$ (bd. ft) | $x_i$ (bd. ft) | $d_i$ (bd. ft) |
|---|---|---|---|
| 1 | 1540 | 1660 | –120 |
| 2 | 1350 | 1440 | –90 |
| 3 | 1260 | 1320 | –60 |
| 4 | 1080 | 1180 | –100 |
| 5 | 640 | 700 | –60 |
| 6 | 690 | 740 | –50 |
| 7 | 1080 | 1180 | –100 |
| 8 | 1000 | 1100 | –100 |
| 9 | 1200 | 1300 | –100 |
| 10 | 310 | 360 | –50 |

Then,

$$\sum_{i=1}^{n} d_i = -830; \quad \sum_{i=1}^{n} d_i^2 = 74,700$$

$$d = -830 / 10 = -83.0 \text{ bd. ft / plot}$$

$$s_D^2 = [74,700 - (830)^2 / 10] / (10 - 1) = 645.56$$

$$s_{\bar{d}}^2 = 645.56 / 10 = 64,556 \text{ (see Equation 14.18)}$$

$$s_{\bar{d}} = \sqrt{64.556} = 8.03 \text{ bd. ft / plot}$$

$$t = (-83.0 - 0) / 8.03 = -10.336^{**}$$

Since –10.336 < –1.833, the null hypothesis is rejected and the alternate accepted. Since –10.336 < $-t_{0.01}(9) = -2.821$, it is said that the difference is highly significant.

It should be noted that in the preceding procedure covariance was hidden. Using the following alternate computations reveals its presence.

$$\sum_{i=1}^{n}(x_i - \bar{x})^2 = 1,234,050$$

$$\sum_{i=1}^{n}(y_i - \bar{y})^2 = 1,367,560$$

$$\sum_{i=1}^{n}(x_i - \bar{x})(y_i - \bar{y}) = 1,297,900$$

$$s_Y^2 = 1,234,050 / (10 - 1) = 137,116.67$$

$$s_X^2 = 1,367,560 / (10 - 1) = 151,951.11$$

$$s_{XY} = 1,297,900 / (10 - 1) = 144,211.11$$

$$s_D^2 = 137,116.67 + 151,951.11 - 2(144,211.11)$$

$$= 645.56$$

The inclusion of the covariance term reduces the magnitude of $S_{\bar{D}}^2$ and, consequently, increases the power of the test. This, of course, only occurs when the covariance is positive (when one variable increases, the other also increases). Consequently, when the situation permits linked pairings and the correlation between the parent random variable is positive, paired $t$ tests are more efficient than unpaired tests and are to be preferred.

The increased sensitivity of the paired $t$ test over the unpaired can be seen if one assumes that the sample plot data used in the preceding example had resulted from observing the photo-based volumes ($Y$) on one set of randomly chosen plots and the ground-based volumes ($X$) on a separate set of randomly chosen plots. In such a case the plot pairings are random and the covariance term is equal to zero. Then, from Equation 14.14,

$$s^2 = \frac{1,234,050 + 1,347,500}{9 + 9} = 144,533.89$$

and from Equation 14.15,

$$s_{\bar{d}}^2 = 144,533.89 / 10 = 14,453.389$$

$$s_{\bar{d}} = \sqrt{14,453.389} = 120.22 \text{ bd. ft / plot}$$

and

$$t = (-83.0 - 0)/120.22 = -0.6903 \text{ N.S.}$$

since the critical value $t_{0.05}(18) = -1.729$.

As can be seen, if the data had not been paired, the null hypothesis would be accepted. With pairing it was rejected.

It should also be noted that, in order to bring $s^2_{\bar{d}}$ down to the level encountered when the observations were paired, the sample size would have to be increased from 10 to 2,239:

$$n_{\text{needed}} = s^2 / s^2_{\bar{d}}$$

$$= 144,533.89 / 64.556 = 2238.9 \text{ or } 2239$$

The gain in efficiency is obvious.

If the parent random variables are the number of trees per plot instead of board foot volume per plot, the random variables are discrete instead of continuous. As with the volume per plot values, one would expect the tree counts per plot obtained from the photographs to be less than those obtained on the ground because small trees simply cannot be distinguished on the photographs. Thus, if a test is to be made of the difference between the results of the two methods, it would be one-tailed with $\mu_y < \mu_x$.

The hypotheses:

$H_0$: $\mu_{\bar{d}} = 0$.

$H_1$: $\mu_{\bar{d}} < 0$.

Type of test: One-tailed $t$.

Significance level: $\alpha = 0.05$.

Decision rules:

*Reject* $H_0$ and *accept* $H_1$ if $t \leq -t_{0.05}(49) = 2.010$.

*Accept* $H_0$ if $t > 2.010$.

If the universe (or sample frame) contained 1000 plots, the sample size was $n = 50$, the sampling was without replacement, and the sample data were $\bar{y} = 19.7$ trees per plot, $s^2_{\bar{Y}} = 53.344444$, $\bar{x} = 25.6$ trees per plot, $s^2_{\bar{x}} = 87.822222$; and $s_{XY} = 7.565432$,

$$\bar{d} = \bar{y} - \bar{x} = 2.56 - 19.7 = 5.9 \text{ trees per plot}$$

$$s^2_{\bar{d}} = \frac{1}{n}(s^2_{\bar{X}} + s^2_{\bar{Y}} - 2s_{XY})\left(\frac{N-n}{N}\right)$$

$$= \frac{1}{50}[87.822222 + 53.344444 - 2(7.565432)]\left(\frac{1000-50}{1000}\right) \quad (14.24)$$

$$= 2.394680$$

and

$$s_{\bar{d}} = \sqrt{2.394680} = 1.547475 \text{ trees per plot}$$

$$t = (\bar{d} - 0) / s_{\bar{d}} = 5.9 / 1.547475 = 3.8127^{**}$$

The null hypothesis is rejected and the difference is said to be highly significant.

In this case the sample size was well above 30 and the sampling fraction was only $n/N = 50/1000 = 0.05$, well below the usual maximum. Consequently, the test should be meaningful.

## 14.14 THE LINEAR ADDITIVE MODEL

In Section 5.3.5 the concept of the *error* was introduced. Reference is made to Equation 5.6:

$$\varepsilon_i = y_i - \mu_Y$$

which can be rewritten as

$$y_i = \mu_Y + \varepsilon_i \tag{14.25}$$

This is the simplest form of the *linear additive model,* which describes the composition of a given variate from a specific population or range space. It shows that $y_i$ is made up of the mean of the population plus a component, $\varepsilon_i$, which varies in a random manner from variate to variate. It is these errors that give rise to the population variance (see Equations 8.32 and 8.33):

$$\sigma_Y^2 = E(y - \mu_Y)^2 = E(\varepsilon^2) = V(Y) \tag{14.26}$$

A somewhat more fundamental form of the model recognizes that populations differ and that the population mean itself consists of two components:

$$y_{ij} = \mu + \tau_i + \varepsilon_{ij} \tag{14.27}$$

where $\mu$ is the *general* mean, across all the populations involved. For example, assuming $M$ finite populations

$$\mu = \left( \sum_{j=1}^{N_1} y_{1j} + \sum_{j=1}^{N_2} y_{2j} + \cdots + \sum_{j=1}^{N_m} y_{mj} \right) \Bigg/ (N_1 + N_2 + \cdots + N_m) \tag{14.28}$$

$\tau_i$ is the component that distinguishes population $i$ from the remaining populations.

$$\mu_i = \mu + \tau_i \tag{14.29}$$

In experimental work these populations are associated with different treatments, hence the use of the symbol $\tau$.

If the test is to be made between specific treatments, which would be the same if the test were repeated time after time, $\tau_i$ would be a fixed quantity and the model would be referred to as a *fixed effects model* (or Model I). On the other hand, if the test is to be made between randomly selected treatments out of a population of possible treatments and if the test were repeated (the treatments involved might be different from those in the first test), $\tau_i$ would be a randomly chosen quantity and the model would be a *random effects model* (or Model II). The analyses involved would depend on the model. In this book only fixed effects models will be considered since the discussion of hypothesis testing in it is of very limited scope. For more information relating to this subject, reference is made to standard statistical texts such as Steel and Torrie (1960) or Snedecor and Cochran (1967).

It usually is assumed, in the case of fixed effects models, that $\Sigma_{i=1}^{m}\tau_i = 0$, where $m$ is the number of populations involved. For example, with two populations

$$\tau_1 + \tau_2 = 0 \tag{14.30}$$

and

$$\tau_1 = -\tau_2 \quad \text{or} \quad \tau_2 = -\tau_1 \tag{14.31}$$

For example, let population 1 be made up of the variates 4, 6, 6, 7, 9, and 10, and population 2 be made up of the variates 5, 6, 6, 9, 10, and 12. Then $\mu_1 = 7$ and $\mu_2 = 8$. The general or overall mean $\mu = (\mu_1 + \mu_2)/2 = (7 + 8)/2 = 7.5$. Then,

$$\tau_1 = \mu_1 - \mu = 7 - 7.5 = -0.5$$

and

$$\tau_2 = \mu_2 - \mu = 8 - 7.5 = 0.5$$

As can be seen, $\tau_1 + \tau_2 = 0$ and $\tau_1 = -\tau_2$.

The $\tau$ terms are considered to be deviations from the general mean. The difference between the two population means is

$$\begin{aligned} \mu_1 - \mu_2 &= (\mu + \tau_1) - (\mu + \tau_2) = \tau_1 - \tau_2 \\ &= \tau_1 - \tau_2 \end{aligned} \tag{14.32}$$

which can be written

$$\mu_1 - \mu_2 = 2\tau_1 \tag{14.33}$$

In the set of variates forming population $Y_i$, $\mu$ and $\tau_i$ are constants, but $\varepsilon_{ij}$ varies from variate to variate, giving rise to the variance of $Y_i$. For example, assume $Y_i$ to be discrete with a finite population of size $N$. Then,

$$\begin{aligned} y_{i1} &= \mu + \tau_i + \varepsilon_{i1} \\ y_{i2} &= \mu + \tau_i + \varepsilon_{i2} \\ &\vdots \\ y_{iN} &= \mu + \tau_i + \varepsilon_{iN} \end{aligned}$$

$$\sum_{j=1}^{N} y_{ij} = N\mu + N\tau_i + \sum_{j=1}^{N} \varepsilon_{ij}$$

and

$$\mu_i = \mu + \tau_i + \bar{\varepsilon}_i$$

Then,

$$y_{i1} - \mu_i = \mu + \tau_i + \varepsilon_{i1} - \mu - \tau_i - \bar{\varepsilon}_i = \varepsilon_{i1} - \bar{\varepsilon}_i$$

$$y_{i2} - \mu_i = \mu + \tau_i + \varepsilon_{i2} - \mu - \tau_i - \bar{\varepsilon}_i - \varepsilon_{i2} - \bar{\varepsilon}_i$$

$$\vdots$$

$$y_{iN} - \mu_i = \mu + \tau_i + \varepsilon_{iN} - \mu - \tau_i - \bar{\varepsilon}_i = E_{iN} - \bar{\varepsilon}_i$$

Since all $(y_{ij} - \mu_i)$ equal their equivalent $(\varepsilon_{ij} - \bar{\varepsilon}_i)$,

$$\sigma^2_{y_i} = \sum_{j=1}^{N} (y_{ij} - \mu)^2 \bigg/ N = \sum_{j=1}^{N} (\varepsilon_{ij} - \bar{\varepsilon}_i)^2 \bigg/ N$$

As can be seen, the variance is caused by the $\varepsilon$ terms alone, and the distribution of the random variable $Y_i$ is that of its associated $\varepsilon$.

The ideas represented by the linear additive model are utilized in conceptualizing the process used in statistical testing. For example, in the case of an unpaired $t$ test of the difference between two sample means where the parent populations are infinitely large, the variances are homogeneous, and the sample sizes are equal, as in Section 14.12.

| **Sample from *Y*** | **Sample from *X*** |
|---|---|
| $y_1 = \mu + \tau_Y + \varepsilon_{Y1}$ | $x_1 = \mu + \tau_X + \varepsilon_{X1}$ |
| $y_2 = \mu + \tau_Y + \varepsilon_{Y2}$ | $x_2 = \mu + \tau_X + \varepsilon_{X2}$ |
| . | . |
| . | . |
| . | . |
| $y_n = \mu + \tau_Y + \varepsilon_{Yn}$ | $x_n = \mu + \tau_X + \varepsilon_{Xn}$ |
| $\sum_{i=1}^{n} y_i = n\mu + n\tau_Y + \sum_{i=1}^{n} \varepsilon_{Yi}$ | $\sum_{i=1}^{n} x_i = n\mu + n\tau_X + \sum_{i=1}^{n} \varepsilon X_{Yi}$ |

Then

$$\bar{y} = \mu + \tau_Y + \bar{\varepsilon}_Y \quad \text{and} \quad \bar{x} = \mu + \tau_X + \bar{\varepsilon}_X$$

and

$$\begin{aligned} (\bar{y} - \bar{x}) &= (\mu + \tau_Y + \bar{\varepsilon}_X) - (\mu + \tau_X + \bar{\varepsilon}_X) \\ &= (\tau_Y - \tau_X) + (\bar{\varepsilon}_Y - \bar{\varepsilon}_X) \end{aligned} \tag{14.34}$$

If the null hypothesis is correct, $\tau_y = \tau_x = 0$ and the difference between $\bar{y}$ and $\bar{x}$ is caused only by the random variation of the $\varepsilon$ terms. Since $E(\varepsilon_Y) = E(\varepsilon_X) = 0$, $E(\bar{\varepsilon}_Y) = E(\bar{\varepsilon}_X) = 0$. Furthermore, since $V(\varepsilon_Y) = V(\varepsilon_X) = \sigma^2$, $V(\bar{\varepsilon}_Y) = V(\bar{\varepsilon}_X) = \sigma^2/n$, and $V(\bar{y} - \bar{x}) = (\sigma^2/n) + (\sigma^2/n) = 2\sigma^2/n$. The latter is relatively small. Consequently, one would expect a small difference between the two sample means even though $\mu_Y = \mu_X$.

On the other hand, if the alternate hypothesis is correct, $\tau_Y \neq - \tau_X$ and the difference between $\bar{y}$ and $\bar{x}$ is made up of the two components shown in Equation 14.34. The magnitude of the difference between $\tau_Y$ and $\tau_X$ would have an effect on the ease by which the null hypothesis would be rejected. If it is great and $V(\bar{y} - \bar{x})$ is small, the value of $t$ would be very likely to exceed the critical value. If the difference is small, $\sigma^2$ is large, and $n$ is relatively small, $V(\bar{y} - \bar{x})$ would be relatively large and the likelihood of making a Type II error would also be relatively great.

In the ease of a paired comparison, there is an additional component in the linear additive model:

$$y_{ij} = \mu + \tau_i + \beta_j + \varepsilon_{ij} \tag{14.35}$$

The new component $\beta_j$ is associated with the linkage between the two observations making up a pair. In other words, when pairing is carried out, the two observations are obtained from the same sampling unit. Sampling units differ from one another even if no treatments are involved. For example, consider the sample plots used in the example illustrating the paired $t$ test in Section 14.13. The actual per acre timber volume on these plots varies regardless of the method used to estimate that volume. It is this type of basic variation that is represented by $\beta_j$.

The effect of $\beta_j$ is removed during the computation of the individual differences:

$$\begin{aligned} d_j &= y_j - x_j \\ &= (\mu + \tau_Y + \beta_j + \varepsilon_{Yj}) - (\mu + \tau_X + \beta_j + \varepsilon_{Xj}) \\ &= (\tau_Y - \tau_X) + (\varepsilon_{Yj} - \varepsilon_{Xj}) \end{aligned} \tag{14.36}$$

Since $\tau_Y$ and $\tau_X$ are constants, it is the $\varepsilon$ terms alone that give rise to the variance. Since $\beta_j$ has been eliminated, it is not a factor influencing the variance. If, however, one considers only the $Y$ values (or only the $X$ values), the $\beta_j$ effect is not canceled out and it contributes to the variance of that variable.

## 14.15 THE *F* DISTRIBUTION

In Section 13.11, confidence limits for proportions were determined using the $F$ statistic.* Reference at that time was made to this section in which the statistic and its distribution are described. The $F$ statistic plays a very small role in parameter estimation, which is the primary thrust of this book, but, in conjunction with the analysis of variance, it is a major tool for statisticians engaged in hypothesis testing.

Let $X_1$ and $X_2$ be independent random variables that are normally distributed with the parameters $(\mu_{X_1}, \sigma^2_{X_1})$ and $(\mu_{X_2}, \sigma^2_{X_2})$. Furthermore, let $\sigma^2_{X_1} = \sigma^2_{X_2} = \sigma^2_X$. Then, let random samples of size $n_1$ and $n_2$, respectively, be drawn from the range spaces of the two random variables. $s_1^2$ and $s_2^2$ are the variances of the two samples. They are individual values of the random variables:

$$Y_1 = \sum_{j=1}^{n_1} (X_{1_j} - \bar{X}_1)^2 / (n_1 - 1)$$

and

$$Y_2 = \sum_{j=1}^{n_2} (X_{2_j} - \bar{X}_2)^2 / (n_2 - 1)$$

* The $F$ statistic and its distribution were first described by R. A. Fisher in the early 1920s. The label "$F$" was given the statistic by G. W. Snedecor in honor of Fisher, who was a dominant figure among the pioneer statisticians (Snedecor, 1934).

$F$ is then defined as

$$F = Y_1/Y_2 \tag{14.37}$$

As can be seen, $F$ is a ratio of sample variances and is often referred to as the *variance ratio*. It, like $t$, is a statistic and possesses a characteristic probability distribution. Its pdf is

$$f(F) = \frac{\Gamma(k/2)k_1^{k_1/2}k_2^{k_2/2}F^{(k_1/2)-1}}{\Gamma(k_1/2)\Gamma(k_2/2)(k_1F + k_2)^{k/2}}; \quad 0 \leq F \leq \infty \quad k_1 = 1, 2, \ldots, \infty \text{ and } k_2 = 1, 2, \ldots, \infty \tag{14.38}$$

where $k_1 = n_1 - 1$, the degrees of freedom associated with the sample from population $X_1$; $k_2 = n_2 - 1$, the degrees of freedom associated with the sample from population $X_2$; and $k = k_1 + k_2$. Note that, unlike most random variables, the capital $F$ is used to symbolize individual values of $F$ as well as the random variable in general.

According to Section 12.11,

$$s_Y^2 = \frac{\sigma_Y^2}{n-1}\left[\left(\frac{Y_1 - \bar{Y}}{\sigma_Y}\right)^2 + \left(\frac{Y_2 - \bar{Y}}{\sigma_Y}\right)^2 + \cdots + \left(\frac{Y_n - \bar{Y}}{\sigma_Y}\right)^2\right]$$

As was done in Section 12.11, let

$$W_i = \left(\frac{Y_i - \bar{Y}}{\sigma_Y}\right)^2$$

Then,

$$s_Y^2 = \frac{\sigma_Y^2}{n-1}\sum_{i=1}^{n} W_i$$

By definition (Equation 11.47 and Section 12.11),

$$\chi^2(n-1) = \sum_{i=1}^{n} /W_i$$

Consequently,

$$s_Y^2 = \frac{\sigma_Y^2}{n-1}\chi^2(n-1)$$

and, returning to the two samples,

$$s_1^2 - \frac{\sigma_Y^2}{k_1}\chi^2(k_1) \quad \text{and} \quad s_2^2 = \frac{\sigma_Y^2}{k_2}\chi^2(k_2)$$

Then,

$$
\begin{aligned}
F &= y_i / y_2 = s_1^2 / s_2^2 \\
&= \frac{\sigma_Y^2}{k_1}\chi^2(k_1) \bigg/ \frac{\sigma_Y^2}{k_2}\chi^2(k_2) \\
&= \frac{\chi^2(k_1)}{k_1} \bigg/ \frac{\chi^2(k_2)}{k_2}
\end{aligned}
\tag{14.39}
$$

and

$$y_1 = \chi^2(k_1)/k_1 \quad \text{and} \quad y_2 = \chi^2(k_2)/k_2$$

$Y_1$ is now a function of $\chi^2(k_1)$ and $Y_2$ is a function of $\chi^2(k_2)$. The pdf's of $Y_1$ and $Y_2$ are, consequently, obtained from the pdf of $\chi^2$. Thus, using Equation 8.26,

$$g(y_1) = f[\chi^2(k_1)]\left|\frac{d}{dy_1}[\chi^2(k_1)]\right|$$

Since $y_1 = \chi^2(k_1)/k_1$, $k_1 y_1 = \chi^2(k_1)$,

$$\frac{d[\chi^2(k_1)]}{dy_1} = \frac{dk_1 y_1}{dy_1} = k_1$$

Substituting into the pdf of $\chi^2$ (Equation 11.44),

$$
\begin{aligned}
g(y_1) &= \frac{1}{2^{k_1/2}\Gamma(k_1/2)}(k_1 y_1)^{(k_1/2)-1} e^{-k_1 y_1/2} k_1 \\
&= \frac{k^{k_1/2}}{2^{k_1/2}\Gamma(k_1/2)} y_1^{(k_1/2)-1} e^{k_1 y_1/2}; \quad 0 < Y_1 < \infty
\end{aligned}
$$

Since $Y_1$ is a variance and, thus, must be positive, its range of values is from zero to infinity. Similarly,

$$h(y_2) = \frac{k_2^{k_2/2}}{2^{k_2/2}\Gamma(k_2/2)} y_2^{(k_2/2)-1} e^{-k_2 y_2/2}; \quad 0 < Y_2 < \infty$$

Following the procedure described in Section 9.6.3:

$$z = y_2, \quad y_2 = z, \quad F = y_1/y_2 = y_1/z, \quad y_1 = Fz, \quad \text{and} \quad J = z$$

$$f(F) = \int_0^\infty g(Fz)h(z)zdz$$

$$= \int_0^\infty \left[\frac{k_1^{k_1/2}}{2^{k_1/2}\Gamma(k_1/2)}\right](Fz)^{(k_1/2)-1}e^{-k_1Fz/2}$$

$$* \left[\frac{k_2^{k_2/2}}{2^{k_2/2}\Gamma(k_2/2)}\right]z^{(k_2/2)-1}e^{-k_2Fz/2}zdz$$

$$= \frac{k_1^{k_1/2}k_2^{k_2/2}}{2^{(k_1+k_2)/2}\Gamma(k_1/2)\Gamma(k_2/2)}\int_0^\infty (Fz)^{(k_1/2)-1}$$

$$* e^{-(k_1Fz+k_2z)/2}z^{k_2/2}dz$$

Let $k = k_1 + k_2$. Then,

$$f(F) = \frac{k_1^{k_1/2}k_2^{k_2/2}F^{(k_1/2)-1}}{2^{k/2}\Gamma(k_1/2)\Gamma(k_2/2)}\int_0^\infty z^{(k_1/2)-1} * e^{-(k_1F+k_2)/2}dz$$

Let $m = (k_1F + k_2)z$. Then $z = m/(k_1F + k_2)$, and, using Equation 8.26,

$$\frac{dz}{dm} = (k_1F + k_2)^{-1} \quad \text{and} \quad dz = (k_1F + k_2)^{-1}dm$$

$$f(F) = \frac{k_1^{k_1/2}k_2^{k_2/2}F^{(k_1/2)-1}}{2^{k/2}\Gamma(k_1/2)\Gamma(k_2/2)}\int_0^\infty \left[\frac{m}{(k_1F + k_2)}\right]^{(k/2)-1}$$

$$* e^{-m/2}(k_1F + k_2)^{-1}dm$$

$$= \left[\frac{k_1^{k_1/2}k_2^{k_2/2}F^{(k_1/2)-1}}{\Gamma(k_1/2)\Gamma(k_2/2)(k_1/F + k_2)^{k/2}}\right]\frac{1}{2^{k/2}}$$

$$* \int_0^\infty m^{(k/2)-1}e^{-m/2}dm$$

According to Equation 11.45, the cdf of $\chi^2$,

$$\frac{1}{2^{k/2}\Gamma(k/2)}\int_0^\infty m^{(k/2)-1}e^{-m/2}dm = 1$$

Then,

$$\frac{1}{2^{k/2}}\int_0^\infty m^{(k/2)-1}e^{-m/2}dm = \Gamma(k/2)$$

and

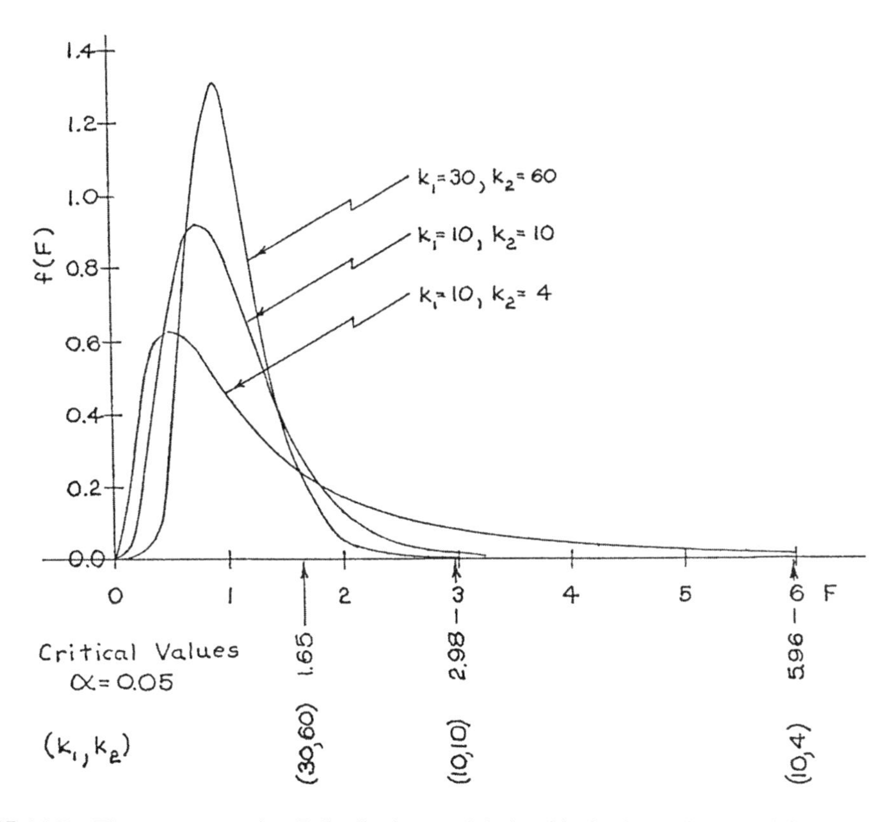


**FIGURE 14.7** Three representative $F$ distributions and their critical values when $\alpha = 0.05$.

$$f(F) = \frac{k_1^{k_1} k_2^{k_2} F^{(k_1/2)-1} \Gamma(k/2)}{\Gamma(k_1/2)\Gamma(k_2/2)(k_1F + k_2)}$$

or

$$f(F) = \frac{\Gamma(k/2)}{\Gamma(k_1/2)\Gamma(k_2/2)} k_1^{k_1/2} k_2^{k_2/2} \frac{F^{(k_1/2)-1}}{(k_1F + k_2)^{k/2}}$$

which is Equation 14.38.

As can be seen from Equation 14.38, the $F$ distribution is actually an infinitely large set of distributions, each being defined by $k_1$ and $k_2$, where $k_1$ is associated with the sample variance forming the numerator of the ratio, and $k_2$ with the sample variance forming the denominator. Unlike the $t$ distributions, the $F$ distributions are asymmetrical, ranging in shape from a reversed J, when $k_1 = 1$, to mounds that are skewed to the right. The curves of a representative set of $F$ distributions are shown in Figure 14.7.

Without proofs, which are complex and tedious, the expected value and the variance of a given $F$ distribution are

$$E[F(k_1, k_2)] = k_2/(k_2 - 2);\ k_2 > 2 \tag{14.40}$$

and

$$V[F(k_1,\ k_2)] = \frac{2k_2^2(k_1 + k_2 - 2)}{k_1(k_2 - 2)^2(k_2 - 4)};\quad k_2 > 4 \tag{14.41}$$

It can be shown, although the process is too lengthy for inclusion here, that Equation 14.38 can be transformed, using the function

$$U = k_1F/(k_2 + k_1F) \tag{14.42}$$

and the procedure described in Section 8.8, into

$$f(u) = \frac{\Gamma(k/2)}{\Gamma(k_1/2)\Gamma(k_2/2)} u^{(k_1/2)-1}(1-u)^{(k_2/2)-1};\quad 0 < U < 1 \tag{14.43}$$

which is the pdf of the beta distribution (Equation 11.55), with $k_1/2 = r_1$ and $k_2/2 = r_2$. Then, as is shown in Section 11.6, $\beta(k_1/2, k_2/2)$ represents the total area under the curve of the distribution and the incomplete beta function, $B_{y_b}(k_1/2,\ k_1/2)$ represents the area under the curve from $U = 0$ to $U = u$.

The ratio

$$I_u(k_1/2, k_2/2) = \beta_u(k_1/2, k_2/2)/\beta(k_1/2, k_2/2)$$

is the cdf of $U$,

$$F(U < u) = F(u) = I_u(k_1/2, k_2/2) \tag{14.44}$$

If Equation 14.42 is rearranged so that

$$F = (k_2/k_1)[u/(1-u)] \tag{14.45}$$

one can find the probabilities of fractiles of $F$ using the cumulated probabilities of the corresponding fractiles of $U$. Since there is an infinite number of $F$ distributions, it is obvious that tables of fractiles, comparable to that for the standard normal (see Table A.4 in the Appendix), cannot be developed. Furthermore, because of the way the $F$ statistic is usually used, tables of $F$ list the critical values, $F_c$, for specific levels of $\alpha$, usually 0.05 and 0.01, for a selected subset of the possible $F$ distributions defined by $k_1$, $k_2$ (see Table A.9 in the Appendix). Figure 14.7 shows the curves of three typical $F$ distributions and that the critical values, $F_c$, are for one-tailed homogeneity tests. It should be noted that $F_c$ is the fractile for $c$ and

$$c = P\ (F < F_c) = 1 - \alpha \tag{14.46}$$

Thus, using Table A.9A, if $k_1 = 10$ and $k_2 = 20$, the fractile associated with $c = 0.95$ and $\alpha = 0.05$ is 2.35, while that associated with $c = 0.99$ and $\alpha = 0.01$ is 3.37. The levels of $\alpha$ that are commonly used are 0.05 and 0.01 and tables limited to these two levels are standard. Some tables, however, also show the fractiles for additional levels of $\alpha$ such as 0.10, 0.025, and 0.001.

Tables A.7 and A.9A can be used to demonstrate how the fractiles of the $F$ distribution can be obtained from those of the beta distribution. However, since the fractiles of the beta distribution are in terms of $c$ and those of the $F$ distribution are in terms of $\alpha = 1 - c$, they will have to be put on a common basis. To do this it will be necessary to determine the relationship between the fractiles of the $F$ distribution obtained using $c$ and those obtained using $\alpha$. When $F$ is defined as in Equation 14.37

$$F = s_1^2 / s_2^2$$

then

$$P[(s_1^2 / s_2^2) < F_c(k_1,\ k_2)] = c \tag{14.47}$$

where $F_c(k_1, k_2)$ is the fractile at which the cumulative probability is equal to $c$ in the $F$ distribution defined by $k_1$ and $k_2$. Taking the reciprocal on each side of the inequality yields

$$P[(s_2^2 / s_1^2) > 1 / F_c(k_1,\ k_2)] = c \tag{14.48}$$

which means that

$$P[(s_2^2 / s_1^2) < 1 / F_c(k_1,\ k_2)] = 1 - c = \alpha \tag{14.49}$$

If $F$ is defined as $s_2^2 / s_1^2$, rather than $s_1^2 / s_2^2$,

$$P[(s_2^2 / s_1^2) < F_\alpha(k_1,\ k_2)] = \alpha \tag{14.50}$$

where $F_\alpha(k_2, k_1)$ is the fractile at which the cumulative probability is equal to $\alpha$ in the $F$ distribution defined by $k_2$ and $k_1$. From Equations 14.49 and 14.50

$$F_c(k_1, k_2) = 1/F_\alpha(k_2, k_1) \tag{14.51}$$

Thus, if $k_1 = 10$, $k_2 = 20$, and $\alpha = 0.05$,

$$P[\mathrm{F} < \mathrm{F}_{\alpha=0.05}(20,10)] = I_{u_b}(10,5) = 0.05$$

From Table A.7, the fractile corresponding to $I_{u_b}(10,5) = 0.05$ is

$$u_b = 0.45999$$

The equivalent $F$ fractile, from Equation 14.45, is

$$F_{\alpha=0.05}(20,10) = (10/20)(0.45999/0.54001) = 0.42591$$

Then, using Equation 14.51,

$$F_{\alpha=0.05}(10,20) = F_{c=0.95}(10,20) = 1/[F_{\alpha=0.05}(20,10)] = 2.34792$$

From Table A.9

$$F_{\alpha=0.05}(10,20) = 2.35$$

which agrees with the value obtained using the beta distribution.

As was brought out in the previous discussion, the $F$ and beta distributions are related. The $F$ is also related to the $\chi^2$, Student's $t$, the normal, and the binomial distributions, and these relationships are utilized by statistical workers to arrive at solutions to problems that otherwise would be very difficult to solve. In the case of chi-square, the development of the relationship is as follows. Equation 14.39 states that

$$F(k_1, k_2) = [\chi^2(k_1)/k_1]/[\chi^2(k_2)/k_2]$$

As $n_2 \to \infty$, $k_2 \to \infty$, and $\chi^2(k_2) \to \infty$, since (see Section 12.11),

$$\chi^2(k_2) = \sum_{i=1}^{\infty} W_i = \sum_{i=1}^{\infty} [(Y_i - \mu_Y)/\sigma_Y]^2$$

Then

$$F(k_1, \infty) = [\chi^2(k_1)/k_1]/(\infty/\infty) = \chi^2(k_1)/k_1 \tag{14.52}$$

and

$$P[F < F_c(k_1, \infty)] = P[\chi^2 < \chi_C^2(k_1)/k_1] = c \tag{14.53}$$

If, on the other hand, $n_1 \to \infty$, $k_1 \to \infty$, and $\chi^2(k_1) \to \infty$, then,

$$F(\infty, k_2) = (\infty/\infty)/[\chi^2(k_2)/k_2] = k_2/\chi^2(k_2) \tag{14.54}$$

and

$$\begin{aligned} P[F < F_c(\infty, k_2)] &= P\{[k_2/\chi^2(k_2)] < F_c(\infty, k_2)\} \\ &= P\{[\chi^2(k_2)/k_2] > 1/[F_c(\infty, k_2)]\} \\ &= c \end{aligned}$$

Consequently,

$$P\{[\chi^2(k_2)/k_2] < 1/[F_c(\infty, k_2)]\} = 1 - c = \alpha$$

and, thus,

$$F_c(\infty, k_2) = k_2/\chi_\alpha^2(k_2) \tag{14.55}$$

As can be seen from this development, $\chi^2$ can be considered a special case of $F$. The distribution of $\chi_c^2(k_1)/k_1$ matches that of $F_c(k_1, \infty)$ while, the distribution of $k_2/\chi_\alpha^2(k_2)$ matches that of $F_c(\infty, k_2)$. The relationship can be demonstrated using Tables A.6 and A.9 in the Appendix. For example, let $k_1 = 20$, $k_2 \to \infty$, and $C = 0.95$. Then from Table A.9A,

$$F_{\alpha=0.05}(20, \infty) = F_{c=0.95}(20, \infty) = 1.57$$

and from Table A.6,

$$\chi_{c=0.95}(20) = 31.4$$

Then, following up on Equation 14.52

$$\chi^2(k_1)/k_1 = [\chi^2_{c=0.95}(20)]/20 = 31.4/20 = 1.57$$

Thus, $P(\text{F} < 1.57) = p\{[\chi^2_{c=0.95}(20)]/20 < 1.57\} = 0.95$.

If $k_1 \rightarrow \infty$, $k_2 = 20$, $c = 0.95$, and $\alpha = 0.05$, from Table A.9A,

$$F_{\alpha=0.05}(\infty, 20) = F_{c=0.95}(\infty, 20) = 1.84$$

and from Table A.6,

$$\chi^2_{\alpha=0.05}(20) = 10.9$$

From Equation 14.54

$$k_2/\chi^2(k_2) = 20/[\chi^2_{\alpha=0.05}(20)] = 20/10.9 = 1.83$$

Then $P(F < 1.83) = P\{20/[\chi^2(20)] < 1.83\} = 0.95$.

The relationship between the $F$ and $t$ distributions arises when one or the other of the two $k$ values is held constant at 1. Returning to Equation 14.39,

$$F(k_1, k_2) = [\chi^2(k_1)/k_1]/[\chi^2(k_2)/k_2]$$

If $k_1 = 1$, the expression becomes

$$F(1, k_2) = \chi^2(1)/[\chi^2(k_2)/k_2] \tag{14.56}$$

According to Sections 11.3 and 11.5, if a random variable is normally distributed with $\mu = 0$ and $\sigma^2 = 1$, the square of the variable is distributed as chi-square based on $k = 1$. Thus, using the symbology in Equation 13.17,

$$z^2[(\bar{y} - \mu_{\bar{y}}/\sigma_{\bar{y}})]^2 = \chi^2(1) \tag{14.57}$$

By using Equation 13.18 and setting $k_2 = n_2 - 1$,

$$v = \sum_{i=1}^{n_2} [(y_i - \mu_Y)/\sigma_Y]^2 = \chi^2(k_2) \tag{14.58}$$

Then, according to Equation 13.19,

$$t = t(k_2) = z/\sqrt{v/k_2} = \pm\sqrt{F(1,\ k_2)} \tag{14.59}$$

and

$$P[-\sqrt{F_c(1,\ k_2)} < t(k_2) < +\sqrt{F_c(1,\ k_2)} = c \tag{14.60}$$

which corresponds to Equation 13.26:

$$P(t_a < t < t_b) = c$$

According to Equation 13.22,

$$t_a = t_{\alpha/2}(k_2) = -\sqrt{F_c(1,\ k_2)}$$

and according to Equation 13.24,

$$t_b = t_{(C+1)/2}(k_2) = +\sqrt{F_c(1,\ k_2)}$$

Since $t$ distributions are symmetrical about $t = 0$ (see Equation 13.27),

$$(t_a)^2 = (t_b)^2 = F_c(1,\ k_2)$$

$$[t_{(c+1)/2}(k_2)]^2 = F_c(1,\ k_2)$$

Demonstrating this, let $k_2 = 20$ and $c = 0.95$. Then,

$$P[F_{0.95}(1,20)] = P[t_{0.975}(20)]^2 = 0.95$$

From Table A.9A,

$$F_{c=0.95}(1,20) = F_{\alpha=0.05}(1,20) = 4.35$$

while, from Table A.8, remembering that this is a one-tailed problem,

$$t_{0.975}(20) = 2.086$$

Then $(2.086)^2 = 4.35$.

Referring back to Equations 14.39, 14.57, 14.58, and 14.59 and setting $k_2 = 1$:

$$F(k_1,\ 1) = [\chi^2(k_1)/k_1][1/\chi^2(1)]$$

$$= (v/k_1)(1/z^2)$$

$$= 1/[t(k_1)]^2$$

According to Equation 14.40,

$$F_c(k_1, k_2) = 1/F_\alpha(k_2, k_1)$$

and since $[(c-1)+1]/2 = c/2$

$$F_c(k_1, 1) = 1/F_\alpha(1, k_1) = 1/[t_{c/2}(k_1)]^2 \tag{14.61}$$

Consequently, the $t$ distribution can be considered a special case of the $F$ distribution. The distribution of $[t_{(c+1)/2}(k_2)]^2$ matches that of $F_c(1, k_2)$, while the distribution of $[t_{c/2}(k_1)]^2$ matches that of $1/F_c(k_1, 1)$. In other words, when $s_1^2$ is based on only two observations, the square of $t$ is distributed as $F$.

Using the same arguments, when $k_1 = 1$ and $k_2 \to \infty$,

$$F(1, \infty) = \chi^2(1)/[\chi^2(\infty)/\infty] = z^2$$

showing that a relationship between the $F$ and normal distributions exists. Thus,

$$F_c(1, \infty) = z^2_{(c+1)/2} \tag{14.62}$$

If $k_1 \to \infty$ and $k_2 = 1$,

$$F_c(\infty, 1) = \frac{1}{F_\alpha(1, \infty)} = \frac{1}{z^2_{c/2}} \tag{14.63}$$

Finally, a relationship exists between the binomial and $F$ distributions. The cdf of the binomial (Equation 10.12) is

$$P(0 \preceq Y \preceq y) = \sum_{2=0}^{y} \binom{n}{w} p^w (1-p)^{n-w}$$

It can be shown that

$$n\binom{n-1}{y} \int_p^1 x^y (1-x)^{n-y-1}\,dx = \sum_{w=0}^{y} \binom{n}{w} p^w (1-p)^{n-w} \tag{14.64}$$

The proof is too lengthy for inclusion here since the integration by parts results in a recursion yielding an integral in each cycle which differs from that in the previous cycle only in that the exponent of $x$ has been reduced by 1. The cycling continues until the exponent equals zero. It also can be shown that

$$n\binom{n-1}{y} = \frac{\Gamma(y+1+n-y)}{\Gamma(y+1)\Gamma(n-y)} = \frac{1}{\beta(y+1,\ n-y)} \tag{14.65}$$

where $\beta(y+1, n-y)$ is the beta function (Equation 11.55) with $r_1 = y + 1$ and $r_2 = n - y$. Furthermore, it can be shown that

$$\begin{aligned} \int_p^1 x^y (1-x)^{n-y-1}\,dx &= \int_0^1 x^y (1-x)^{n-y-1}\,dx - \int_0^p x^y (1-x)^{n-y-1}\,dx \\ &= \beta(y+1,\ n-y) - \beta_p(y+1,\ n-y) \end{aligned} \tag{14.66}$$

where $\beta_p(y + 1, n - y)$ is the incomplete beta function (Equation 11.56) where $X = p$. Substituting into Equation 14.64,

$$n\binom{n-1}{y}\int_p^1 x^y(1-x)^{n-y-1}\,dx = \frac{\beta(y+1,\ n-1)-\beta_p(y+1,\ n-1)}{\beta(y+1,\ n-y)}$$

$$= 1 - I_p(y+1,\ n-y) \tag{14.67}$$

$$= P(0 \preceq Y \preceq y)$$

It was shown earlier (see Equations 14.42 and 14.44) that the cdf of

$$X = k_1\ F/(k_2 + k_1F)$$

is

$$P(X < x) = I_x(k_1/2,\ k_2/2)$$

If $k_1 = 2(y + 1)$, $k_2 = 2(n - y)$, and $p$ is a specific value of $X$,

$$P(X < p) = I_p(y + 1,\ n - y)$$

and from Equation 14.67

$$P(0 \leq Y \leq y) = 1 - I_p(y + 1,\ n - y) = 1 - p(X < p)$$

Putting $X$ in terms of $F$,

$$P(0 \leq Y \leq y) = 1 - P\{[k_1/F(k_2 + k_1F)] < p\}$$

which, after some algebraic operations and remembering that $k_1 = 2(y + 1)$ and $k_2 = 2(n - y)$, becomes

$$P(0 \leq Y \leq y) = 1 - P\{F < [(n - y)/(y + 1)][p/(1 - p)]\} \tag{14.68}$$

or

$$P\{F < [(n - y)/(y + 1)][p/(1 - p)]\} = 1 - P(0 \leq Y \leq y)$$

Setting $P(0 \leq Y \leq y) = c$, then

$$P\{F < [(n - y)/(y + 1)][p/(1 - p)]\} = 1 - c \tag{14.69}$$

According to Equation 14.50,

$$P[F < F_{1-c}(k_2,\ k_1)] = 1 - c$$

Consequently,

$$[(n - y_c)/(y_c + 1)][p/(1-p)] = F_{1-c}(k_2, k_1) = F_{1-c}[2(n - y_c), 2(y_c + 1)] \tag{14.70}$$

Solving for $y_c$ yields

$$y_c = \frac{np - (1-p)F_\alpha(k_2, k_1)}{p + (1-p)F_\alpha(k_2, k_1)}; \quad k_1 = 2(y_c + 1), \; k_2 = 2(n - y_c) \tag{14.71}$$

Thus, each value of $Y$ has a corresponding value of $F$ and, consequently, the fractiles of the binomial distribution can be obtained using the $F$ distribution. For example, if $n = 20$, $y = 16$, and $p = 0.6$,

$$\begin{aligned}
k_1 &= 2(y+1) = 2(16+1) &&= 34 \\
k_2 &= 2(n-y) = 2(20-16) &&= 8 \\
[(n-y)/(y+1)][p/(1-p)] & &&= [(20-16)/(16+1)][0.6/(1.0-0.6)] \\
& &&= 0.3529 = F_\alpha(k_2, k_1)
\end{aligned}$$

Then $P(0 \leq Y \leq 16) = 1 - P(F < 0.3529)$.

The magnitude of $P(F < 0.3529)$ must be determined using numerical integration of the incomplete beta distribution (see Johnson and Kotz, 1970, pp. 48–57). If the process is reversed, as when developing confidence intervals, as in Section 13.6.4, tables of $F$ can be used since then the probability is fixed and the value of $y_c$ is that being determined.

## 14.16 TESTING FOR HOMOGENEITY OF VARIANCE USING F

Referring back to Table 6.1, which shows the diameter distributions of two stands, assume that the data in Table 6.1 are from random samples drawn from the two stands. Furthermore, assume that the diameter distributions of the two stands are approximately normal. Then, the sample statistics are

$$\bar{y}_A = 15.0 \text{ in.}; \quad s_A^2 = 12.73; \quad k_A = 30$$

and

$$\bar{y}_B = 15.0 \text{ in.}; \quad s_B^2 = 1.60; \quad k_B = 30$$

The sample variances are obviously different; but are the actual variances, $\sigma_A^2$ and $\sigma_B^2$, different? After all, $s_A^2$ and $s_B^2$ are only estimates and are subject to variation from sample to sample. Might not the difference be due only to random variation? Thus, one can set up a homogeneity test as follows:

The hypotheses:

$H_0$: $\sigma_A^2 = \sigma_B^2$.

$H_1$: $\sigma_A^2 \neq \sigma_B^2$.

Type of test: Two-tailed $F$.

The test will be based on the $F$ distribution since variances are involved. Since the alternate hypothesis simply states that $\sigma_A^2$ is different than $\sigma_B^2$ and does not specify which is larger or smaller, the test must be two-tailed.

Significance level: $\alpha = 0.05$.
The decision rules:
*Reject* $H_0$ and *accept* $H_1$ if $F \geq F_{0.025}(30, 30) = F_c = 2.07$.
*Accept* $H_0$ if $F < 2.07$.

Note that the critical value $F_c$ is based on a significance level of $\alpha/2 = 0.025$ rather than the specified $\alpha = 0.05$. The reason for this is that the test statistic is defined as

$$F = \text{larger } s^2/\text{smaller } s^2 \tag{14.72}$$

Consequently, the half of the $F$ distribution that involves variance ratios where the smaller $s^2$ is the numerator rather than the denominator is not a factor. Thus, if a critical value, $F_c$, forms the bound for $100\alpha\%$ of the area under the entire curve, it forms the bound for $2(100\alpha\%)$ of the area under the segment of the curve where the variance ratios equal or exceed unity. As a result, if one wishes to operate at a significance level of $\alpha$, the critical value used will have to be the tabular value associated with $\alpha/2$.

The test statistic in this case is

$$F = s_A^2/s_B^2 = 12.73/1.60 = 7.95625^{**}$$

Since $7.95625 > 2.07$, the null hypothesis is rejected and it is assumed that the two stands indeed have different variances.

In the above example, $k_A = k_B$. If they differed, as, for example, $k_A = 20$ and $k_B = 15$, the test would be the same but the critical value, $F_c$, would be

$$F_c = F_{\alpha/2}(k_A, k_B) = F_{0.025}(20, 15) = 2.76$$

## 14.17 BARTLETT'S TEST FOR HOMOGENEITY OF VARIANCE

The $F$ test for homogeneity of variance is not appropriate when more than two sample variances are involved. Instead, use may be made of several alternative procedures of which one, suggested by M. S. Bartlett (1937a and b), is probably the most widely used. The test is only approximate and is not free of ambiguity. Consequently, one should approach its use with caution.

Assume a situation when samples are drawn from different populations which are normally distributed. Normality is a key factor in Bartlett's test. As the populations deviate more and more from normality, the more unreliable becomes the test. The test of the homogeneity of the variances of these populations is based on the test statistic:

$$M = 2.3026\left[\log \bar{s}^2 \sum_{i=1}^{m} k_i - \sum_{i=1}^{m} (k_i \log s_i^2)\right] \tag{14.73}$$

where

$$\bar{s}^2 = \sum_{i=1}^{m} k_i s_i^2 \Bigg/ \sum_{i=1}^{m} k_i$$

If natural or Naperian logarithms are used, the constant 2.3026 is eliminated and Equation 14.73 becomes

$$M = \ln \bar{s} \sum_{i=1}^{m} k_i - \sum_{i=1}^{m} (k_i \ln s_i^2) \tag{14.74}$$

The sampling distribution of $M$ is distributed approximately as $\chi^2(m - 1)$. The introduction of the correction term, $C$, improves the approximation.

$$C = \left\{ 1 + [1/3(m-1)] \left[ \sum_{i=1}^{m} (l/k_i) - 1 \sum_{i=1}^{m} k_i \right] \right\}^{-1} \tag{14.75}$$

Thus, if $MC$ exceeds the critical value $\chi^2_\alpha(m - 1)$, it is assumed that at least one of the populations has a variance that is different from that of the remaining populations.

As an example of the use of Bartlett's test, assume these hypothetical forest condition classes: $A$ = loblolly pine sawtimber with stand basal area exceeding 90 ft²/acre; $B$ = loblolly pine sawtimber with from 50 to 90 ft²/acre of stand basal area; and $C$ = loblolly pine sawtimber with less than 50 ft²/acre of stand basal area. Further, assume that ten sample plots were randomly located and cruised in each of the three condition classes. The plot volumes are shown below (in terms of hundreds of board feet per acre in order to reduce the computation complexity):

| | Classes | | |
|---|---|---|---|
| | **$A(i = 1)$ (100 bd. ft/acre)** | **$B(i = 2)$ (100 bd. ft/acre)** | **$C(i = 3)$ (100 bd. ft/acre)** |
| | 10.7 | 8.5 | 4.5 |
| | 10.9 | 3.3 | 1.2 |
| | 9.7 | 4.1 | 1.2 |
| | 9.9 | 2.1 | 3.1 |
| | 7.0 | 5.9 | 5.3 |
| | 10.9 | 4.5 | 1.5 |
| | 13.5 | 7.3 | 3.6 |
| | 10.9 | 6.9 | 2.0 |
| | 10.4 | 6.9 | 3.1 |
| | 11.1 | 4.2 | 2.1 |
| $\sum_{i=1}^{n_i} y_{ij}$ | 105.0 | 53.7 | 27.6 |
| $\bar{y}_i$ | 10.5 | 5.37 | 2.76 |
| $\sum_{i=1}^{n_i} (y_{ij} - \bar{y}_i)^2$ | 23.14 | 37.20 | 17.88 |
| $s_i^2$ | 2.5711111 | 4.1334444 | 1.9871111 |

A test is to be made of the sample variances to see if they are all estimates of the same population variance.

The hypotheses:

$H_0$: $\sigma_A^2 = \sigma_B^2 = \sigma_C^2 = \sigma^2$.

$H_1$: At least one of the variances is not equal to $\sigma^2$.

Significance level: $\alpha = 0.05$.

Decision rules:

*Reject* $H_0$ and *accept* $H_1$ if $MC \geq \chi^2_{0.05}(3-1) = 5.99$.

*Accept* $H_0$ if $MC < 5.99$.

$$k_A = k_B = k_C = 10 - 1 = 9$$

$$\bar{s}^2 = [9(2.5711111) + 9(4.1334444) + 9(1.9871111)] / 3(9)$$

$$= 2.897037$$

$$M = 2.3026[27 \ \log\ 2.897037 - (9 \ \log\ 2.5711111 + 9 \ \log\ 4.1334444 + 9 \ \log\ 1.9871111)]$$

$$= 1.2684178$$

$$C = \{1 + [1/3(3-1)][3(1/9) - 1/3(9)]\}^{-1}$$

$$= 0.9529412$$

$$MC = 1.2684178(0.9529412) = 1.2087276 < 5.99 \text{ N.S.}$$

The null hypothesis is accepted and it is assumed that all three condition classes possess the same basic variance.

It should be noted that $C$ can never be greater than 1 and, consequently, $MC$ can never be greater than $M$. For this reason, if the uncorrected $M$ is less than the critical value, there is no need to apply the correction. If, however, $M$ is greater than the critical value, the correction should be applied to reduce the likelihood of incurring a Type I error.

It also should be noted that Bartlett's test is only an approximate test and under some circumstances will yield ambiguous or misleading results. Consequently, one should approach its use with caution. The test is based on the assumption that the parent populations are normally distributed and it is quite sensitive to nonnormality of the data, especially when positive kurtosis is present. As a result, if the test indicates heterogeneity among the class variances, it may actually be indicating a lack of normality rather than a lack of homogeneity. This ambiguity in the interpretation of the results of the test has caused many statisticians to choose alternate methods of testing for homogeneity of variance. Unfortunately, the remaining options are limited. The most promising involves the transformation of the data, through a mathematical conversion process, which normalizes it. This is discussed briefly in Section 14.21.

It must be remembered that Bartlett's test simply tests whether or not the several classes possess a common variance. It does not isolate or identify the class or classes with divergent variances. There is no way to sort out such classes in an objective manner. The use of repetitive $F$ tests is not appropriate because the tests would not be independent and consequently the probabilities involved would be undefined.

## 14.18 TESTING FOR GOODNESS OF FIT USING $\chi^2$

In many cases in this and the previous chapter, references have been made to the distributional pattern of the populations from which samples are drawn for the purpose of making estimates or population parameters or for testing to see if a sample-based statistic is an element of a specified population. In most real-life situations, the distributional pattern of the parent population is not known and reliance must be placed on the central limit theorem or other known relationships. In many cases, however, it is necessary to determine how the values of the random variable actually are distributed. To do this, one must hypothesize a distribution and then see if the sample data support the hypothesis. There are several ways by which this can be done. The following is a method that is essentially a homogeneity test based on chi-square.

Let us assume a situation where the investigator wishes to test the hypothesis that the diameters breast high in the forest condition class labeled "all-aged" (see Table 14.1) are distributed as an exponential distribution where the parameter $a = 0.2$.

The hypotheses:

$H_0$: The dbhs of the trees in the all-aged condition class have an exponential distribution where $a = 0$.

$H_1$: The dbhs do not have such a distribution.

Type of test: $\chi^2$.

Significance level: $\alpha = 0.05$.

Decision rules:

*Reject* $H_0$ and *accept* $H_1$ if $\chi^2 \geq \chi^2_\alpha(m-1) = \chi^2_{0.05}(12) = 21.0$.

*Accept* $H_0$ if $\mu^2 < 21.0$

This test is based on the idea that the theoretical distribution (in this case the exponential distribution with parameter $a = 0.2$) will be divided into classes corresponding to the classes used with the empirical data (in this case the sample-based stand table), the theoretical frequency computed for each such class, and then the actual frequencies compared with the theoretical frequencies. It is assumed that the difference between the actual and theoretical frequencies in a given class is a variate from a normally distributed population of such differences. When the theoretical frequency is equal to or greater than 5, it usually is assumed that normality is present. If the theoretical frequency is less than 5, the class is combined with an adjacent class or classes until the theoretical frequency equals or exceeds 5 (see dbh classes 15, 16, and 17 in. in Table 14.1). As was brought out in Section 11.5, if a variable is normally distributed, then its square is distributed as chi-square with one degree of freedom. Then, if each class difference is squared and these squares summed across all the classes, the sum becomes a variate in a chi-square distribution with $m - 1$ degrees of freedom ($m$ = number of classes recognized in test), *provided that the null hypothesis is correct*. Thus, if this sum is less than the critical value of $\chi^2_\alpha(m-1)$, it is assumed that it is from the specified population and the null hypothesis is accepted. On the other hand, if it equals or exceeds $\chi^2_\alpha(m-1)$, it is assumed that the sum is not from the specified population and the null hypothesis is rejected. Again, the basic idea behind the test is that of the homogeneity test.

Returning to the example, if the stand table is made up of 1-in. dbh classes, the theoretical proportion of the population falling into each of the dbh classes, $\Theta_i$, is

$$\Theta_i = P(Y \leq y_i + 0.5) - P(Y \leq y_i - 0.5) \tag{14.76}$$

where $Y$ is the dbh, $y_i$ is the midpoint of the $i$th class, and the class limits are $y_i \pm 0.5$ in. Then, using the cdf of the exponential distribution (see Equation 11.40)

$$\begin{aligned}\Theta_i &= \int_0^{y_i+0.5} ae^{-ay}dy - \int_0^{y_i-0.05} ae^{-ay}dy \\ &= [1 - e^{-a(y_i+0.05)}] - [1 - e^{-a(y_i-0.05)}] \\ &= e^{-a(y_i-0.05)} - e^{-a(y_i-0.5)}; \quad i = 1, 2, \ldots, m\end{aligned}$$

**TABLE 14.1**
**Stand Table (on a per acre basis) for a Hypothetical All-Aged Stand and the Computations Needed to Determine the Goodness of Fit of These Data to an Exponential Distribution with $a$ = 0.02**

| i | dbh $y_i$ (in.) | $\gamma_i$ | $\rho_1$ | $\phi_i$ | | $f_i$ | | $(fi - \phi_i)$ | $z_i$ |
|---|---|---|---|---|---|---|---|---|---|
| 1 | 3 | 0.1099454 | 0.1907670 | 48.84 | | 57 | | 8.16 | 1.3633415 |
| 2 | 4 | 0.0900156 | 0.1561867 | 39.98 | | 47 | | 7.02 | 1.2326263 |
| 3 | 5 | 0.0736986 | 0.1278750 | 37.74 | | 31 | | −1.74 | 0.0924740 |
| 4 | 6 | 0.0603393 | 0.1046951 | 26.80 | | 29 | | 2.20 | 0.1805970 |
| 5 | 7 | 0.0494016 | 0.0857171 | 21.94 | | 22 | | 0.06 | 0.0001641 |
| 6 | 8 | 0.0404467 | 0.0701794 | 17.97 | | 7 | | −10.97 | 6.6967668 |
| 7 | 9 | 0.0331149 | 0.0574579 | 14.71 | | 18 | | 3.29 | 0.7358328 |
| 8 | 10 | 0.0271122 | 0.0470426 | 12.04 | | 9 | | −3.04 | 0.7675748 |
| 9 | 11 | 0.0221976 | 0.0385152 | 9.86 | | 10 | | 0.14 | 0.0019878 |
| 10 | 12 | 0.0181738 | 0.0315335 | 8.07 | | 5 | | −3.07 | 1.1678934 |
| 11 | 13 | 0.0148795 | 0.0258175 | 6.61 | | 7 | | 0.39 | 0.0230106 |
| 12 | 14 | 0.0121823 | 0.0211376 | 5.41 | | 5 | | 0.41 | 0.0310721 |
| 13 | 15 | 0.0099740 | 0.0173060 | 4.43 | | 4 | | | |
| 14 | 16 | 0.0081660 | 0.0141689 | 3.63 | 11.03 | 4 | 11 | −0.03 | 0.0000816 |
| 15 | 17 | 0.0066858 | 0.0116006 | 2.97 | | 3 | | | |
| | | $v$ = 0.5763333 | | 256.00 | | 256 | | | $\chi^2$ = 12.293423 |

The values of $\Theta_i$ for the dbh classes included in the stand table are shown in column 3 of Table 14.1. These values are then summed:

$$v = \sum_{i=1}^{m} \Theta_i = 0.5763333 \tag{14.77}$$

The ratio of $\Theta_i$ to $v$ yields the proportion of the total count *in the stand table* that is in dbh class $i$:

$$\rho_i = \Theta_i / v \tag{14.78}$$

The total count in the stand table, $n$, is the sum of the dbh class frequencies, $f_i$:

$$n = \sum_{i=1}^{m} f_i = 256 \tag{14.79}$$

The theoretical frequency in class $i$, $\phi_i$, is then

$$\phi_i = n\rho_i = 256\rho_i \tag{14.80}$$

Since the theoretical frequencies in dbh classes 15, 16, and 17 in. are all less than 5, they are all combined into a single class. As a result, from this point on, $m$, the number of classes, is equal to 13 instead of 15. The difference between the actual, $f_i$, and theoretical frequencies in dbh class $i$ is

$$d_i = f_i - \phi_i \tag{14.81}$$

Then the difference is squared and standardized:

$$z_i = (f_i - \phi_i)^2/\phi \tag{14.82}$$

The sum of these $z$ values forms the test statistic:

$$\chi^2 = \sum_{i=1}^{m} z_i = 12.293423 \text{ N.S.} \tag{14.83}$$

Since $\chi^2 < \chi^2_{0.05}(12) - 21.0$, the null hypothesis is accepted and it is assumed that indeed the diameter distribution in the all-aged stand has an exponential distribution with $a = 0.2$.

The degree of freedom used to determine the critical value is equal to the number of classes, $m$, less 1:

$$k = m - 1 \tag{14.84}$$

In this case, $k = 13 - 1 = 12$.

The utility of a goodness of fit test lies in the area of modeling. In some cases it is known from theoretical reasoning that a certain set of data comes from a population having a known distribution, e.g., sample means tend to be normally distributed. However, in many other cases the distributional pattern of the population must be estimated. The example used to describe the test procedure fits into this category. Since the test indicates that the fit is good enough for the null hypothesis to be accepted, it is assumed that diameter distribution of all-aged stands, comparable with those from which the sample data were obtained, can be modeled by the exponential distribution with $a = 0.2$. The test procedure would be the same regardless of the theoretical distribution. For example, one might wish to obtain a model of the diameter distribution in pine plantations. Since such diameter distributions are known to be mound shaped and skewed it would be logical to use one of the $\beta$ or Weibull distributions as the model. In such a case the theoretical frequencies for the classes would be determined using $\beta$ or the Weibull, but the remainder of the test process would be as described.

It should be noted that the chi-square goodness of fit test is highly sensitive to deviations from the theoretical form and, consequently, the null hypothesis is often rejected when in actuality the fit is quite good. There are no rules governing this situation and investigators must make use of their best judgment in cases where this occurs.

## 14.19 TESTING FOR NORMALITY

An underlying assumption in much of the hypothesis testing discussed in this chapter is that the populations involved are normally distributed. If they are not normally distributed, an element of bias is introduced into the testing process which has the potential of causing misinterpretations of the test results. As the population distributions deviate more and more from the normal, the likelihood of misinterpretation increases. Fortunately, most tests of hypotheses based on the assumption of normality are reasonably *robust*, meaning the likelihood of misinterpretation of their results is quite small even when the populations diverge to some extent from the normal. There are times, however, when the existence or nonexistence of normality must or should be established. A number of procedures exist for making such a determination.

First, the chi-square test for goodness of fit, described in Section 14.18, can be used, provided sufficient data are available. This test is based on the assumption that the sampling distribution of differences between the theoretical and actual class counts for each class is normally distributed.

That this is the case is supported by the central limit theorem, but the sample size must be large (see Par 12.15).

The second, and least satisfactory, of normality testing procedures makes use of *normal probability paper*, which also is known as *arithmetic probability paper* or, simply, *probability paper*. Such paper consists of printed coordinate sheets in which the graduations along one axis have constant spacing, while those along the other axis have been anamorphosed so that if a cumulative normal distribution (ogive) is plotted on the paper, it forms a straight line.

As an example of the use of normal probability paper, use the data from the three loblolly pine stand density classes used as an example in Sections 14.17 and 14.19. In Section 14.17 it was determined that the three class variances were homogeneous. Consequently, the within-class residuals can be considered as arising from a common source. Then,

$$(y_{ij} - \bar{y}_i)$$

| $j$ | $A(i = 1)$ | $B(i = 2)$ | $C(i = 3)$ |
|---|---|---|---|
| 1 | 10.7 – 10.5 = 0.20 | 8.5 – 5.37 = 3.13 | 4.5 – 2.76 = 1.74 |
| 2 | 10.9 – 10.5 = 0.40 | 3.3 – 5.37 = –2.07 | 1.2 – 2.76 = –1.56 |
| 3 | 9.7 – 10.5 = – 0.80 | 4.1 – 5.37 = –1.27 | 1.2 – 2.76 = –1.56 |
| 4 | 9.9 – 10.5 = – 0.60 | 2.1 – 5.37 = –3.27 | 3.1 – 2.76 = 0.34 |
| 5 | 7.0 – 10.5 = – 3.50 | 5.9 – 5.37 = 0.53 | 5.3 – 2.76 = 2.54 |
| 6 | 10.9 – 10.5 = 0.40 | 4.5 – 5.37 = –0.87 | 1.5 – 2.76 = –1.26 |
| 7 | 13.5 – 10.5 = 3.00 | 7.3 – 5.37 = 1.93 | 3.6 – 2.76 = 0.84 |
| 8 | 10.9 – 10.5 = 0.40 | 6.9 – 5.37 = 1.53 | 2.0 – 2.76 = 0.76 |
| 9 | 10.4 – 10.5 = – 0.10 | 6.9 – 5.37 = 1.53 | 3.1 – 2.76 = 0.34 |
| 10 | 11.1 – 10.5 = 0.60 | 4.2 – 5.37 = –1.17 | 2.1 – 2.76 = –0.66 |

Arranging the residuals into an array in ascending order,

| | | |
|---|---|---|
| –3.50 | –0.76 | 0.53 |
| –3.27 | –0.66 | 0.60 |
| –2.07 | –0.60 | 0.84 |
| –1.56 | –0.10 | 1.53 |
| –1.56 | 0.20 | 1.53 |
| –1.27 | 0.34 | 1.74 |
| –1.26 | 0.34 | 1.93 |
| –1.17 | 0.40 | 2.54 |
| –0.87 | 0.40 | 2.54 |
| –0.80 | 0.40 | 3.13 |

the array is then partitioned into ten classes, each spanning 10% of the total range of magnitude of the residuals.

$$R = 3.13 - (-3.50) = 6.63$$

The lower boundaries of the classes will begin with –3.50 and increase in multiples of 6.63/10 = 0.663. The "less-than" cumulative distribution of the residuals, by counts and percents, is shown below.

| Less Than | Count | Percent | Less Than | Count | Percent |
|---|---|---|---|---|---|
| –3.500 | 0 | 0 | 0.183 | 14 | 47 |
| –2.763 | 2 | 7 | 0.920 | 23 | 77 |
| –2.027 | 3 | 10 | 1.657 | 25 | 83 |
| –1.290 | 5 | 17 | 2.393 | 27 | 90 |
| –0.553 | 13 | 43 | 3.130 | 30 | 100 |

The class boundaries are then entered along the equal interval scale on the normal probability paper and the cumulative percentages are plotted against them as shown in Figure 14.8. If the set of plotted points form a straight line, it is assumed that the distribution is normal. If the points do not fall on a straight line, as in this case, the interpretation is not clear. One would expect sample data to vary from sample to sample and to form different plots. However, no method exists that would indicate when a plot differed sufficiently from a straight line so that a hypothesis of normality could be tested in an objective manner. Consequently, in the case of this example, there is uncertainty whether or not the distribution is normal.

The third method of testing for normality makes use of the moment coefficients of skew and kurtosis. In the case of the normal distribution, the moment coefficient of skew, $\alpha_3$, is equal to zero and the moment coefficient of kurtosis, $\alpha_4$, is equal to 3. The sample-based estimates of these parameters, as in the case of the sample variance (see Section 12.11), have distributional patterns that depend on the distributional pattern of the parent population. However, they, like the variance, are arithmetic means and, according to the central limit theorem, tend to be normally distributed with the approximation improving as the sample size increases. When the sample size is sufficiently large, which is usually assumed to be greater than 30, the approximation is such that normality can be assumed. Consequently, if the sample size is sufficiently large, a two-tailed homogeneity test, based on the normal distribution, can be made of each of the estimators.

The sample-based estimator of the moment coefficient of skew is

$$a_3 = \sum_{i=1}^{n} (y_i - \bar{y})^3 / (n-1)s^3 \rightarrow \alpha_3 \tag{14.85}$$

where $s$ is the standard deviation of the sample. The estimator of the moment coefficient of kurtosis is

$$a_4 = \sum_{i=1}^{n} (y_i - \bar{y})^4 / (n-1)s^4 \rightarrow \alpha_4 \tag{14.86}$$

For large samples, the variance of the estimated moment coefficient of skew is

$$s_{a_3}^2 = 6 / n \tag{14.87}$$

while that of the moment coefficient of kurtosis is

$$s_{a_4}^2 = 24 / n \tag{14.88}$$

Then, following the pattern of homogeneity tests

$$z_3 = (a_3 - \alpha_3) / s_{a_3} = (a_3 - 0) / s_{a_3} \tag{14.89}$$

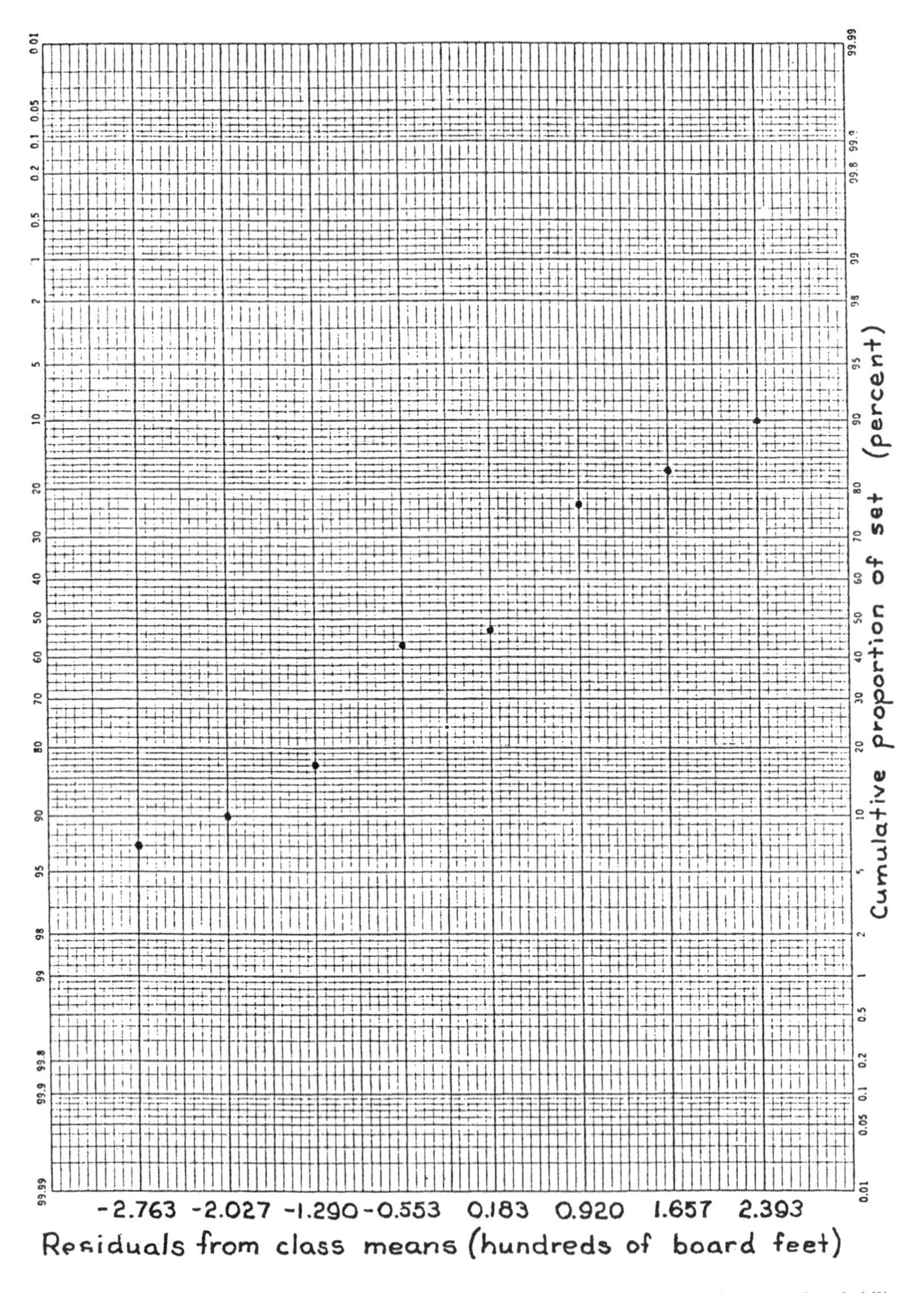


**FIGURE 14.8** The evaluation of the normality of the distribution of plot volumes using normal probability paper.

and

$$z_4 = (a_4 - \alpha_4) / s_{a_4} = (a_4 - 3) / s_{a_4} \tag{14.90}$$

The absolute values of these computed $z$ values are then compared with the critical values of $z$ at the specified level of significance. The absolute values of $z$ are used in this comparison since the test is two-tailed.

As an example, assume that the normality of the population of per acre volumes, based on 1/5-acre plots, for a certain loblolly pine condition class, is to be tested. Then, for the moment coefficient of skew,

The hypotheses:
$H_0$: $\alpha_3 = 0$.
$H_1$: $\alpha_3 \neq 0$.
Significance level: 0.05.
Decision rules:
*Reject* $H_0$ and *accept* $H_1$ if $|z| \geq z_{0.05} = 1.96$.
*Accept* $H_0$ if $|z| < 1.96$.

To save space, the individual data items will be omitted and only the appropriate sums shown. The data have been coded to reduce computation problems. They are in terms of hundreds of board feet per acre.

$$\sum_{i=1}^{n} (y_i - \bar{y})^2 = 78.225$$

$$s^2 = 78.225 / 29 = 2.6974138$$

$$s = 1.6423805$$

$$\sum_{i=1}^{n} (y_i - \bar{y})^3 = -7.159$$

$$a_3 = -7.159 / (30 - 1)(1.6423805)^3 = -0.0557228$$

$$s_{\alpha_3} = \sqrt{6 / 30} = 0.4472136$$

$$z = (-0.0557228 - 0) / 0.4472136 = -0.1246$$

Since $|z| = 0.1246 < F_{0.05} = 1.96$, the difference is considered nonsignificant and it is assumed that the population is symmetrically distributed. Then, for the moment coefficient of kurtosis,

The hypotheses:
$H_0$: $\alpha_4 = 3$.
$H_1$: $\alpha_4 \neq 3$.
Significance level: 0.05.
Decision rules:
*Reject* $H_0$ and *accept* $H_1$ if $|z| \geq z_{0.05} = 1.96$.
*Accept* $H_0$ if $|z| < 1.96$.

$$\sum_{i=1}^{n} (y_i - \bar{y})^4 = 556.652$$

$$a_4 = 556.652 / (30 - 1)(1.6423805)^4 = 2.6380964$$

$$s_{\alpha_4} = \sqrt{24 / 30} = 0.8944272$$

$$z = (2.6380964 - 3) / 0.8944272 = -0.4046206$$

The null hypothesis is accepted and it is assumed that the population is distributed normally.

## 14.20 THE ANALYSIS OF VARIANCE

$t$ tests are appropriate mechanisms for testing to see if the means of samples drawn from two different sampling frames are or are not estimates of the same population mean. However, $t$ tests cannot be made when more than two sample means are involved. The standard procedure for testing the differences between three or more sample means involves a process known as the *analysis of variance* followed by the tests proper, which make use of the $F$ distribution.

As an example of this process, assume that the loblolly pine stands on a certain industrial forest ownership have been classified into three volume classes, $A$, $B$, and $C$, using aerial photographic evidence. A test is to be made to see if the classification procedure is effective and that the mean volumes per acre for the three classes actually differ. Let the subscript $i$ represent the following:

$i = 1$, volume class $A$
$i = 2$, volume class $B$
$i = 3$, volume class $C$

Furthermore, assume that the variable of interest is the board foot volume per acre on 1/5-acre circular plots centered on the intersections of a 105.32 × 105.32 ft grid which permits the adjacent plot circles to touch each other. It also is assumed that the plot volumes within each of the classes are normally distributed with the following parameters:

$\mu_1$ and $\sigma_1^2$, for class $A$
$\mu_2$ and $\sigma_2^2$, for class $B$
$\mu_3$ and $\sigma_3^2$, for class $C$

In addition, it is assumed that the class variances are homogeneous:

$$\sigma_1^2 = \sigma_2^2 = \sigma_3^2 = \sigma^2$$

The test is to be made to see if differences exist between the three class means. Thus, the hypotheses are

$H_0$: $\mu_1 = \mu_2 = \mu_3 = \mu$.
$H_1$: All the class means, $\mu_i$, are not equal.

Assume that $n_i$ plots were randomly chosen from each of the volume classes and then cruised. Note that the sample size can be either constant or variable across the several classes. However, it usually is considered desirable to have equal sample sizes in the classes. Then,

$m$ = the number of classes

$n_i$ = the sample size in the $i$th classes

$y_{ij}$ = volume per acre on the $j$th plot in the $i$th class

$$\bar{y}_o = \sum_{i=1}^{m}\sum_{j=1}^{n_i} y_{ij} \Bigg/ \sum_{i=1}^{m} n_i = \text{the overall sample mean} \tag{14.91}$$

which is an unbiased estimator of the overall mean, μ.

$$\bar{y}_i = \sum_{j=1}^{n_i} y_{ij} / n_i = \text{the sample mean from the } i\text{th class} \tag{14.92}$$

which is an unbiased estimator of the mean of the $i$th class, $\mu_i$

$$\begin{aligned} y_{ij} - \bar{y}_o &= y_{ij} - \bar{y}_i + \bar{y}_i - \bar{y}_o \\ &= (y_{ij} - \bar{y}) - (\bar{y}_i - \bar{y}_o) \end{aligned}$$

Squaring,

$$\begin{aligned} (y_{ij} - \bar{y}_o)^2 &= [(y_{ij} - \bar{y}_i) - (\bar{y}_i - \bar{y}_o)]^2 \\ &= (y_{ij} - \bar{y}_i)^2 + 2(y_{ij} - \bar{y}_i)(\bar{y}_i - \bar{y}_o) + (\bar{y}_i - \bar{y}_0)^2 \end{aligned}$$

Summing,

$$\sum_{i=1}^{m}\sum_{j=1}^{n_i}(y_{ij} - \bar{y}_o)^2 = \sum_{i=1}^{m}\sum_{j=1}^{n_i}(y_{ij} - \bar{y}_i)^2 + 2\sum_{i=1}^{m}\sum_{j=1}^{n_i}(y_{ij} - \bar{y}_i)(\bar{y}_i - \bar{y}_o) + \sum_{i=1}^{m} n_i(\bar{y}_i - \bar{y}_o)^2$$

Since within each class $(\bar{y}_i - \bar{y}_o)^2$ is a constant and $\Sigma_{j=1}^{n_i}(\bar{y}_{ij} - \bar{y}_i) = 0$

$$\sum_{i=1}^{m}\sum_{j=1}^{n_i}(\bar{y}_{ij} - \bar{y}_i)(\bar{y}_i - \bar{y}_o) = 0$$

and

$$\sum_{i=1}^{m}\sum_{j=1}^{n_i}(\bar{y}_{ij} - \bar{y}_o)^2 = \sum_{i=1}^{m}\sum_{j=1}^{n_i}(\bar{y}_{ij} - \bar{y}_o)^2 + \sum_{i=1}^{m} n_i(\bar{y}_i - \bar{y}_o)^2 \tag{14.93}$$

Within this expression

$$\sum_{i=1}^{m}\sum_{j=1}^{n_i}(y_{ij} - \bar{y}_0)^2 = \text{the } \textit{total sum of squares} \tag{14.94}$$

$$\sum_{i=1}^{m}\sum_{j=1}^{n_i}(y_{ij}-\bar{y}_i)^2 = \text{the } \textit{within class} \text{ or } \textit{error sum of squares} \tag{14.95}$$

and

$$\sum_{i=1}^{m} n_i(\bar{y}_i-\bar{y}_o)^2 = \text{the } \textit{between classes sum of squares} \tag{14.96}$$

The degrees of freedom associated with each of these sums of square are

$$k_T = \sum_{i=1}^{m} n_i - 1, \text{ for the total sum of squares}$$

$$k_w = \sum_{i=1}^{m}(n_i - 1), \text{ for the within class or error sum of squares}$$

$$k_B = m - 1, \text{ for the between class sum of squares}$$

Furthermore,

$$k_T = k_w + k_B = (n_1-1)+(n_2-1)+\cdots+(n_m-1)+(m-1) = \sum_{i=1}^{m} n_i - 1 \tag{14.97}$$

Then,

$$s_w^2 = \sum_{i=1}^{m}\sum_{j=1}^{n_i}(y_{ij}-\bar{y}_i)^2 / k_w = \text{within class or error mean square} \tag{14.98}$$

This is a pooled variance since the class variances are homogeneous (see Equation 14.14). Consequently, $s_w^2$ is an unbiased estimator of $\sigma^2$.

Continuing,

$$s_B^2 = \sum_{i=1}^{m} n_i(\bar{y}_i-\bar{y}_o)^2 / k_B = \textit{between class mean square} \tag{14.99}$$

If the $n_i$ are constant across all classes so that $n_1 = n_2 = n_3 = n$, the above expression becomes

$$s_B^2 = n\sum_{i=1}^{m}(\bar{y}_i-\bar{y}_o)^2 / k_B \tag{14.100}$$

In such a case $\Sigma_{i=1}^{m}(\bar{y}_i-\bar{y}_o)^2 / k_B$ is an unbiased estimator of $\sigma_{\bar{y}}^2 = \sigma^2 / n$ *provided* that the class means, $\mu_i$, are all equal to the overall mean, $\mu$, so that all the sample means, $\bar{y}_i$, are estimators of the same quantity. If this is the case,

$$s_B^2 = n\sum_{i=1}^{m}(\bar{y}_i - \bar{y}_o)^2 / k_B \xrightarrow{\text{UNBIASED}} n\sigma^2 / n = \sigma^2$$

It can be shown that when the class sample sizes $n_i$ are not equal

$$s_B^2 = \sum_{i=1}^{m} n_i(\bar{y}_i - \bar{y}_o)^2 / k_B$$

is still an unbiased estimator of $\sigma^2$.

This development shows how the variance based on the total sum of squares can be partitioned into two components, each of which is an unbiased estimator of $\sigma^2$, *provided* that $\mu_1 = \mu_2 = \mu_3 = \mu$. Table 14.2 shows the above analysis of variance (ANOVA) in standard tabular form.

It should be recalled that it is assumed that the variates within the classes are normally distributed and the class variances are homogeneous. Proceeding under these assumptions, if one can consider only one of the classes (say, class $i$), the sample consists of $y_{i1}, y_{i2}, \ldots, y_{in}$. Then,

$$\varepsilon_{ij} = y_{ij} - \mu_i$$

Dividing $\varepsilon_{ij}$ by $\sigma$ converts it to standard form:

$$z_{ij} = \varepsilon_{ij}/\sigma \tag{14.101}$$

The random variable $Z_i$ then is normally distributed with the parameters $\mu_Z = 0$ and $\sigma_Z^2 = 1$:

$$\begin{aligned} z_{ij} &= (y_{ij} - \mu_i)/\sigma = (y_{ij} - \bar{y}_i + \bar{y}_i - \mu_i)/\sigma \\ &= [(y_{ij} - \bar{y}_i)/\sigma] + [(\bar{y}_i - \mu_i)/\sigma] \end{aligned} \tag{14.102}$$

Squaring,

$$z_{ij}^2 = [(y_{ij} - \bar{y}_i)/\sigma]^2 + 2[(y_{ij} - \bar{y}_i)/\sigma][(\bar{y}_i - \mu_i)/\sigma] + [(\bar{y}_i - \mu_i)/\sigma]^2 \tag{14.103}$$

Summing,

$$\sum_{j=1}^{n_i} z_{ij}^2 = \sum_{j=1}^{n_i}[(y_{ij} - \bar{y}_i)/\sigma]^2 + (2/\sigma)[(y_{ij} - \bar{y}_i)/\sigma]\sum_{j=1}^{n_i}(y_{ij} - \bar{y}_i)^2 + n_i[(\bar{y}_i - \mu_i)/\sigma]^2$$

Since $\Sigma_{j=1}^{n_i}(y_{ij} - \bar{y}_i) = 0$, the cross-product term equals zero. Then,

$$\sum_{j=1}^{n_i} z_{ij}^2 = \sum_{j=1}^{n_i}[(y_{ij} - \bar{y}_i)/\sigma]^2 + n_i[(\bar{y}_i - \mu_i)/\sigma]^2 \tag{14.104}$$

$$\bar{z}_i = \sum_{j=1}^{n_i} z_i / n_i$$

$$= \frac{1}{n_i} \sum_{j=1}^{n_i} \{[(y_{ij} - \bar{y}_i)/\sigma] + [(\bar{y}_i - \mu_i)/\sigma]\} \tag{14.105}$$

$$= \frac{1}{n_i\sigma} \sum_{j=1}^{n_i} (y_{ij} - \bar{y}_i) + \frac{n_i}{n_i}[(\bar{y}_i - \mu_i)/\sigma]$$

$$= (\bar{y}_i - \mu_i)/\sigma$$

$\bar{z}_i$ is a sample mean and, according to the central limit theorem, is normally distributed with the parameters:

$$\mu_{\bar{Z}_i} = \mu_{Z_i} = 0 \quad \text{and} \quad \sigma^2_{\bar{Z}_i} = \sigma^2_{Z_i} / n_i = \sigma^2 / n_i = 1 / n_i$$

Then,

$$z_{ij} - \bar{z}_i = [(y_{ij} - \mu_i)/\sigma] - [(\bar{y}_i - \mu_i)/\sigma] = (y_{ij} - \bar{y}_i)/\sigma \tag{14.106}$$

Since both $z_{ij}$ and $\bar{z}_i$ are normally distributed, the distribution of $z_{ij} - \bar{z}_i$ is also normal because of the reproductive property of the normal (see Section 8.13).

Let

$$\ell_{ij} = z_{ij} - \bar{z}_i \tag{14.107}$$

then $\ell_{i1}, \ell_{i2}, \ldots, \ell_{in_i}$ are linear functions of $z_{i1}, z_{i2}, \ldots, z_{in_i}$:

$$\ell_{ij} = -(z_{i1}/n_i) - (z_{i2}/n_i) - \cdots - (z_{i(j-1)}/n_i) + [1 - (1/n_i)z_{ij}] - \cdots - (z_{1n_i}/n_i) \tag{14.108}$$

There is a linear relationship between the $\ell$ terms since they must sum to zero. This means that the $\ell$ terms are linearly dependent and the degrees of freedom associated with the sum of the $\ell$ terms, or of their squares, are equal to the number of $\ell$ terms minus the number of constraints. Thus, $\Sigma_{j=1}^{n_i}(z_{ij} - \bar{z}_i)$ and $\Sigma_{j=1}^{n_i}(z_{ij} - \bar{z}_i)^2$ have associated with them $(n_i - 1)$ degrees of freedom.

In Section 12.11 (see Equations 12.20 through 12.24) it is brought out that the squared variates of normally distributed random variables are distributed as $\chi^2(1)$ and the sums of $n$ such squared variates are distributed as $\chi^2(n)$. Consequently, $z_{ij}^2$ is distributed as $\chi^2(1)$ and $\Sigma_{j=1}^{n_i} z_{ij}^2$ is distributed as $\chi^2(n_i)$. The sum of squared $z$ terms can be partitioned (see Equation 14.104):

$$\sum_{j=1}^{n_i} z_{ij}^2 = \sum_{i=1}^{n_i} (z_{ij} - \bar{z}_i)^2 + n_i \bar{z}_i^2$$

Since both $(z_{ij} - \bar{z}_i)$ and $z_i$ are normally distributed, $\Sigma_{j=1}^{n_i}(z_{ij} - \bar{z}_i)^2$ is distributed as $\chi^2(n_i - 1)$ and $z_i^2$ is distributed as $\chi^2(1)$. Then, according to Equation 14.106,

$$\sum_{j=1}^{n_i}(z_{ij}-\bar{z}_i)^2=\sum_{j=1}^{n_i}[(y_{ij}-\bar{y}_i)/\sigma]^2=\frac{1}{\sigma^2}\sum_{j=1}^{n_i}(y_{ij}-\bar{y}_i)^2$$

and

$$\sum_{j=1}^{n_i}(y_{ij}-\bar{y}_i)^2=\sigma^2\chi^2(n_i-1) \tag{14.109}$$

Thus, $\Sigma_{j=1}^{n_i}(y_{ij}-\bar{y}_i)^2$ is distributed as $\sigma^2\chi^2(n_i-1)$.

According to Equation 14.105,

$$n_i\bar{z}_i^2=n_i[(\bar{y}_i-\mu_i)/\sigma]^2$$

and

$$n_i(\bar{y}_i-\mu_i)^2=\sigma^2\chi^2(1) \tag{14.110}$$

Consequently, $n_i(\bar{y}_i-\mu_i)^2$ is distributed as $\sigma^2\chi^2(1)$.

Now, if one considers all the classes instead of just one,

$$\sum_{i=1}^{m}\sum_{j=1}^{n_i}(y_{ij}-\bar{y}_i)^2=\sigma^2\chi^2(kw) \tag{14.111}$$

and

$$\sum_{i=1}^{m}n_i(\bar{y}_i-\mu_i)^2=\sigma^2\chi^2(m) \tag{14.112}$$

Equation 14.111 indicates that the within or error sum of squares in the analysis of variance, Equation 14.95, is distributed as $\sigma^2\chi^2(kw)$. The between sum of squares, Equation 14.96, is analogous to Equation 14.112 but differs in that $\bar{y}_o$ appears in place of $\mu$. The use of $\bar{y}_o$ requires a modification of the associated degrees of freedom. When $\mu$ is used, there is no requirement that $\Sigma_{i=1}^{m}n_i(\bar{y}_i-\mu_i)$ be equal to zero, since $\mu_i$ is not the mean of the set of $\bar{y}_i$, and, thus, there is no loss of degrees of freedom. However, when $\bar{y}_o$, the weighted mean of the $\bar{y}_i$, is used in place of $\mu_i$, $\Sigma_{i=1}^{m}n_i(\bar{y}_i-\bar{y}_o)$ must equal zero, making the class means, $\bar{y}_i$, linearly dependent and forming a constraint. Consequently, the number of degrees of freedom is equal to $m-1$ and

$$\sum_{i=1}^{m}n_i(\bar{y}_i-\bar{y}_o)^2=\sigma^2\chi^2(m-1)=\sigma^2\chi^2(k_B) \tag{14.113}$$

and thus the between sum of squares is distributed as $\sigma^2\chi^2(k_B)$.

Consider the within or error mean square, $s_w^2$ (see Equation 14.98). It is based on the variability of the individual variates about their class means and is independent of $\mu$. Since the class variances are homogeneous, all equal to $\sigma^2$, the within class or error sums of squares and degrees of freedom can be pooled to produce an estimate of $\sigma^2$ which is stronger than any one of the individual class estimates.

The between mean square, $s_w^2$ (see Equation 14.99), on the other hand, is based on the variability of the class means about the overall mean, $\bar{y}_o$, which is an estimate of the true overall mean, $\mu$. If the true class means, $\mu_i$, are all equal to $\mu$, each of the class means, $\bar{y}_i$, and the overall mean, $\bar{y}_o$, is an unbiased estimate of $\mu$. Furthermore, if the sample sizes are the same in all classes so that $n_i = n$, the variance of the sampling distribution of means would be

$$\sigma_{\bar{y}}^2 = \sigma^2 / n \tag{14.114}$$

and

$$n\sigma_{\bar{y}}^2 = \sigma^2 \tag{14.115}$$

In such a case, the between mean square would be

$$s_B^2 = \sum_{i=1}^{m} n(\bar{y}_i - \bar{y}_o)^2 / k_B = n\sum_{i=1}^{m}(\bar{y}_i - \bar{y}_o)^2 / k_B = ns_{\bar{y}}^2 \tag{14.116}$$

and the between mean square would be an unbiased estimator of $\sigma^2$. It should be noted that when the sample sizes vary from class to class, the between mean square remains an unbiased estimator of $\sigma^2$.

If the true class means, $\mu_i$, are not equal to $\mu$, each $\bar{y}_i$ is an unbiased estimate of its class mean but is not an estimate of $\mu$. The effect of the inequality of the class means with $\mu$ can be seen in the following development.

First, consider the situation in class $i$ if $\mu_i = \mu$.

$$s_{Bi}^2 = n_i(\bar{y}_i - \mu)^2$$

$$E(s_{Bi}^2) = n_i E(\bar{y}_i - \mu)^2$$

Since $\bar{y}_i$ is an estimate of $\mu_i = \mu$,

$$E(s_{Bi}^2) - n_i\sigma_{\bar{y}}^2 = n_i\sigma^2 / n_i = \sigma^2 \tag{14.117}$$

Thus, $s_{Bi}^2$ is an unbiased estimator of $\sigma^2$. However, if $\mu_i \neq \mu$,

$$\begin{aligned} s_{Bi}^2 &= n_i[(\bar{y}_i - \mu)^2 = n_i(\bar{y}_i - \mu_i + \mu_i - \mu)]^2 \\ &= n_i[(\bar{y}_i - \mu_i) + (\mu_i - \mu)]^2 \\ &= n_i[(\bar{y}_i - \mu_i)^2 + 2(\bar{y}_i - \mu_i)(\mu_i - \mu) + (\mu_i - \mu)^2] \end{aligned}$$

Then,

$$E(s_{Bi}^2) = n_i[E(\bar{y}_i - \mu_i)^2 + 2(\mu_i - \mu)E(\bar{y}_i - \mu_i) + (\mu_i - \mu)^2]$$

Since the class variances are homogeneous,

$$E(\bar{y}_i - \mu_i)^2 = V(\bar{y}_i) = \sigma^2_{\bar{y}_i} = \sigma^2 / n_i$$

and since $E(\bar{y}_i - \mu_i) = 0$,

$$E(s^2_{Bi}) = n_i[\sigma^2 / n_i + (\mu_i - \mu)^2] = \sigma^2 + n_i(\mu_i - \mu)^2 \tag{14.118}$$

As can be seen, when there is a difference between a class mean and the overall mean, the between mean square as an estimator of $\sigma^2$, will be biased upward. Extending this to all the classes yields,

$$E(s^2_B) = \sum_{i=1}^{m}[\sigma^2 + n_i(\mu_i - \mu)^2] / m = \sigma^2 + \sum_{i=1}^{m} n_i(\mu_i - \mu)^2 / m \tag{14.119}$$

Then, according to Equation 14.111,

$$\sum_{i=1}^{m}\sum_{j=1}^{n_i}(y_{ij} - \bar{y}_i)^2 = \sigma^2\chi^2(k_w)$$

and

$$s^2_w = \sum_{i=1}^{m}\sum_{j=1}^{n_i}[(y_{ij} - \bar{y}_i)^2] / k_w = \sigma^2\chi^2(k_w) / k_w \tag{14.120}$$

Furthermore, according to Equation 14.112,

$$\sum_{i=1}^{m} n_i(\bar{y}_i - \bar{y}_o)^2 = \sigma^2\chi^2(k_B)$$

and

$$s^2_B = \sum_{i=1}^{m} n_i(\bar{y}_i - \bar{y}_o)^2 / k_B = \sigma^2\chi^2(k_B) / k_B \tag{14.121}$$

Referring to Equation 14.39,

$$F(k_b,\ k_w) = s^2_B / s^2_w$$

If all the true class means, $\mu_i$, are equal to the true overall mean, $\mu$, both the between and within mean squares are unbiased estimators of $\sigma^2$. In such a case, from a *theoretical* point of view,

$$F = \sigma^2/\sigma^2 = 1 \tag{14.122}$$

However, if the class means are not all equal to the overall mean, the $F$ ratio would be greater than 1 because of the positive bias of the between mean square.

$$F = \left[\sigma^2 + \sum_{i=1}^{m} n_i(\mu_i - \mu)^2 / m\right] \Big/ \sigma^2 > 1 \tag{14.123}$$

When the $F$ ratio is based on sample statistics, rather than the population parameters, it will be subject to random variation, as was described in Section 14.15. Consequently, one cannot assume that any sample-based $F$ ratio that exceeds 1 indicates that the class means differ. The computed $F$ ratio must exceed the critical value $F_c$, based on the specified level of significance, before the decision can be made that the class means differ. Note that the test is one-tailed. Any difference between a class mean and the overall mean will cause an upward bias in the between class mean square, causing the $F$ ratio to be excessively large. The reverse is not possible. Any sample-based $F$ ratio that is less than 1 is simply caused by random variation.

As an example of this process, return to the hypothetical loblolly pine stand density classes used in Section 14.17 to illustrate the use of Bartlett's test of the homogeneity of the class variances. In the present case,

The hypotheses:
$H_0$: $\mu_A = \mu_B = \mu_C = \mu$.
$H_1$: At least one of the class means is not equal to $\mu$.
Level of significance: $\alpha = 0.05$.
Decision rules:
*Reject* $H_0$ and *accept* $H_1$ if $F \geq F_{0.05}(2,27) = 3.35$
*Accept* $H_0$ if $F < 3.35$.

The critical value is based on $m - 1 = 3 - 1 = 2$ and $(10 - 1) + (10 - 1) = 27$ degrees of freedom.
Then,

| | Class | | |
|---|---|---|---|
| | A ($i$ = 1) | B ($i$ = 2) | C ($i$ = 3) |
| $\sum_{j=1}^{n_i} y_{ij}$ | 105.0 | 53.7 | 27.6 |
| $\bar{y}_i$ | 10.50 | 5.37 | 2.76 |
| $\sum_{j=1}^{n_i} (y_{ij} - \bar{y}_i)^2$ | 23.14 | 37.20 | 17.88 |

$$\sum_{i=1}^{m}\sum_{j=1}^{n_i} (y_{ij} - \bar{y}_i)^2 = 78.22, \text{ the within classes sum of squares}$$

$$s_w^2 = \sum_{i=1}^{m}\sum_{j=1}^{n_i} (y_{ij} - \bar{y}_i)^2 \Big/ \sum_{i=1}^{m} (n_i - 1) = 78.22/27 = 2.897, \text{ the within classes mean square}$$

$$\sum_{i=1}^{m}\sum_{j=1}^{n_i} y_{ij} = 186.3$$

$$\bar{y}_o = \sum_{i=1}^{m}\sum_{j=1}^{n_i} y_{ij} \Big/ \sum_{i=1}^{m} (n_i - 1) = 186.3/30 = 6.21$$

$$\sum_{i=1}^{m} n_i(\bar{y}_i - \bar{y}_o)^2 = 310.12, \text{ the between classes sum of squares}$$

$$s_B^2 = \sum_{i=1}^{m} n_i(\bar{y}_i - \bar{y}_o)^2 \Big/ (m-1) = 310.12/2 = 155.061, \text{ the between classes mean square}$$

$$\sum_{i=1}^{m}\sum_{j=1}^{n_i} (y_{ij} - \bar{y}_o)^2 = 388.34, \text{ the total sum of squares}$$

**ANOVA**

| Source | df | SS | MS | F |
|---|---|---|---|---|
| Between classes | 2 | 310.12 | 155.061 | 53.52** |
| Within classes | 27 | 78.22 | 2.897 | |
| Total | 29 | 388.34 | | |

$F$ obviously exceeds the critical value, $F_{0.05}(2,27) = 3.35$, and thus the null hypothesis is rejected. It also exceeds the critical value when $\alpha = 0.01$, $F_{0.01}(2,27) = 5.46$. Consequently, the differences between the class means are considered highly significant.

The preceding discussion of the analysis of variance and associated $F$ test is appropriate when the class variances are homogeneous, the variates within the classes are normally distributed, and the treatment effects are fixed. The experimental design is known as a "completely randomized design." It is the simplest of a large number of designs that make use of the analysis of variance. As the designs become more complex, the analyses of variance become more complex. For more information regarding experimental designs and their analyses, reference is made to such standard statistical tests as Snedecor and Cochran (1967), Steel and Torrie (1960), or Cochran and Cox (1957).

## 14.21 TRANSFORMATIONS

Heteroscedasticity can take two basically different forms. It can be *regular* or *irregular*. In the case of regular heteroscedasticity there is some form of relationship between the class variances and the class means. For example, small class means are apt to be associated with small class variances and large class means with large class variances. This occurs when the variates within the classes have Poisson distributions where the class means and variances are equal. If the distributional patterns of the class populations are known, it may be possible to convert or *transform* those populations to normally distributed populations and in the process homogenize the class variances. Such a conversion is known as a *transformation*. There are a number of more or less standard transformations that can be used under specified conditions. A discussion of them is beyond the scope of this book and reference is made to standard texts such as Steel and Torrie (1960).

When the distributional patterns of the class populations cannot be deduced, which often is the case, the standard transformations can be tried on a trial-and-error basis. There is no guarantee, however, that a suitable transformation can be found.

In the case of irregular heteroscedasticity, transformations are of no value. In such cases it may be necessary to exclude the errant class or classes from analyses dependent on homogeneity of variance. Methods do exist, however, that permit analysis of experimental data when heteroscedasticity is present. Descriptions of such methods are, as with transformations, beyond the scope of this book and reference is made to such sources as Steel and Torrie (1960).

## 14.22 MULTIPLE COMPARISONS

The analysis of variance and associated $F$ tests are used to test the hypothesis that the set of sample-based class or treatment means are *all* estimates of the hypothesized overall population mean. If the null hypothesis is rejected, it is concluded that at least one of the classes or treatments has a population mean that differs from the hypothesized mean. However, the test provides no information that can be used to identify the aberrant mean or means. To identify these means, one must use one or another of several procedures, which collectively are referred to as *multiple comparisons*.

If, at the time the experiment is being planned, the investigator decides that the means of certain of the classes or treatments will be compared, regardless of the outcome of the overall $F$ test, such comparisons become part of the experiment plan and are known as *planned comparisons*. In contrast to these are comparisons which become interesting to the investigator only after the data have been collected and summarized. Such are known as *unplanned comparisons*. Testing procedures appropriate for planned comparisons are usually not appropriate for unplanned comparisons. In any case, unplanned comparisons should never be made if the overall $F$ test leads to an acceptance of the null hypothesis. If the null hypothesis is accepted, it is assumed that the class or treatment means are all estimates of the same population mean and the differences are due only to random processes.

As was previously stated, if an $F$ test is made and the computed $F$ is greater than the critical value, it is assumed that at least one of the classes or treatments has a population mean that differs from the hypothesized mean. This may or may not be true. By random chance the sample data may be such that leads to the rejection of the null hypothesis when, in fact, the null hypothesis is correct. If the level of significance, $\alpha$, is 0.05, and the *experiment* is repeated an indefinitely large number of times, one would expect 5 such errors per 100 repeats of the experiment. Thus, the expected *per experiment* error rate is 5 out of 100.

If, instead of using the analysis of variance and $F$ test, one attempts to test the null hypothesis by pairing every class or treatment mean with every other class or treatment mean and testing the difference between the members of the pairs using the $t$ test, the error rate would be on a *per comparison* basis. Each comparison would be made with an expected error rate of $\alpha$. If $m$ classes or treatments are involved, $\binom{m}{2}$ comparisons would be made. As can be seen, if the overall null hypothesis is to be rejected if any one of the individual comparisons indicates significance, the likelihood of rejection will be much higher when individual comparisons are made than when one overall test is made. The difference in the two error rates cannot be disregarded.

### 14.22.1 The Least Significant Difference

According to Equation 14.5,

$$t = (\bar{y} - \mu_{\bar{y}}) / s_{\bar{y}}$$

In Section 14.12 this was modified to

$$t = (\bar{d} - \mu_d) / s_{\bar{d}}$$

Then, since $s_{\bar{d}}$ is a square root and can be either positive or negative,

$$\begin{aligned} \pm ts_{\bar{d}} &= \bar{d} - s_{\bar{d}} \\ \mu_d &= \bar{d} \pm ts_{\bar{d}} \end{aligned} \tag{14.124}$$

If $t$ is replaced by the critical value of $t$ at a given significance level, the expression becomes a confidence interval:

$$\mu_{\bar{d}} = \bar{d} \pm t_\alpha(k)s_{\bar{d}} \tag{14.125}$$

or

$$P[\bar{d} - t_\alpha(k)s_{\bar{d}} \leq \mu_{\bar{d}} \leq \bar{d} + t_\alpha(k)s_{\bar{d}}] = C = 1 - \alpha \tag{14.126}$$

If $\mu_{\bar{d}}$ is equal to zero and $\alpha$ is set equal to 0.05, the probability of such a confidence interval *not* including zero is 0.05. It also can be said that if $\bar{d} > t_\alpha(k)s_{\bar{d}}$, the confidence interval will not include zero. Thus, if $\bar{d}$ is the difference between two class or treatment means and it is hypothesized that no difference exists between the two corresponding population means, if $d$ exceeds $t_\alpha(k)s_{\bar{d}}$, the hypothesis would be rejected.

The term $t_\alpha(k)s_{\bar{d}}$ is known as the *least significant difference* or *lsd*. It is a constant when a series of multiple comparisons are to be made. If it is to be used, the class or treatment variances must be homogeneous so that they can be pooled. If the class variances are heterogeneous, one must take this into account and modify what is done using the procedures discussed in Section 14.12. If the sample sizes in the classes being compared are equal, the process takes the form used in the following example.

Assume a situation when the forest cover on a certain ownership has been subdivided into eight volume classes: *A, B, C, D, E, F, G,* and *H*. This subdivision was made using aerial photographic evidence and there is reason to believe that the subdivision is into too many classes and that some of the classes should be lumped together since there are no real differences in their mean volumes per acre. Consequently, it is proposed that a test be made of the differences between the class means *A* and *B*, *C* and *D*, *E* and *F*, and *G* and *H*. To this end, 20 plots were randomly located and cruised in each of the eight classes. The resulting class statistics and computations are

| Class | $\bar{y}_i$ (cords/acre) | $s_i$ | $k_i = n_i - 1$ | $k_i s_i^2$ |
|---|---|---|---|---|
| A | 46.2 | 16.6569 | 19 | 316.4811 |
| B | 44.1 | 15.5281 | 19 | 295.0339 |
| C | 40.3 | 16.7225 | 19 | 317.7275 |
| D | 36.7 | 15.9683 | 19 | 303.3977 |
| E | 29.2 | 16.5921 | 19 | 315.2499 |
| F | 28.6 | 16.4649 | 19 | 312.8331 |
| G | 22.5 | 15.8244 | 19 | 300.6636 |
| H | 20.4 | 16.5600 | 19 | 314.6400 |
| | | | 152 | 2476.0268 |

When tested using Bartlett's test the class variances were found to be homogeneous. Consequently, they can be pooled.

$$k_w = \sum_{i=1}^{m} k_i = 152$$

$$s_w^2 = \sum_{i=1}^{m} k_i s_i^2 \Big/ k_w = 2476.0268 / 152 = 16.28965$$

$$s_{\bar{d}}^2 = 2s_w^2 / n = 2(16.28965) / 20 = 1.628965$$

$$s_{\bar{d}} = \sqrt{1.628965} = 1.2763091 \text{ cords / acre}$$

If $\alpha = 0.05$,

$$\text{lsd} = t_{0.05}(152)s_{\bar{d}} = \sqrt{1.628965} = 1.96(1.2763091) = 2.502 \text{ cords / acre}$$

$$\bar{y}_A - \bar{y}_B = 46.2 - 44.1 = 2.1 < 2.502 \text{ cords / acre}$$

$$\bar{y}_C - \bar{y}_D = 40.3 - 36.7 = 3.6 > 2.502$$

$$\bar{y}_E - \bar{y}_F = 29.2 - 28.6 = 0.6 < 2.502$$

$$\bar{y}_G - \bar{y}_H = 22.5 - 20.4 = 2.1 < 2.502$$

As a consequence of this test, it would be logical to lump $A$ and $B$, $E$ and $F$, and $G$ and $H$. $C$ and $D$, however, appear to have different population means and should be kept separate.

It should be noted that these tests were set up prior to data collection and consequently are *planned* comparisons. Furthermore, no analysis of variance or $F$ test was involved and no attempt was made to test an overall null hypothesis. Finally, the error rate was on a per comparison basis.

It might be asked, "Why weren't the differences between $E$ and $G$ or $F$ and $H$ also tested?" The reason is that in order for the results to be meaningful the several must be *independent* of one another. This means that no class mean can appear in more than one comparison. In other words, $\bar{y}_A$ can be compared with $\bar{y}_B$ but then neither $\bar{y}_A$ nor $\bar{y}_B$ can be compared with any other mean if the tests are to be valid.

Some investigators use the lsd approach to test *all* the possible pairs of class means as a means of testing the overall null hypothesis. This approach is flawed because of the lack of independence between the individual tests, which affects the probabilities, and also because the error rate would be on a per comparison rather than a per experiment basis.

Other investigators have used the lsd approach to test the overall null hypothesis by comparing the largest and smallest of the sample-based class means. This is flawed because the test is based on the assumption that *all* the possible differences, not just those between the largest and smallest means, are included in the sampling distribution of differences. In this procedure, only the maximum differences are considered. Consequently, the probability of a difference exceeding the lsd is much greater than the stated significance level. Cochran and Cox (1957) state that when 3 classes or treatments are involved the $\text{lsd}_{0.05}$ will be exceeded about 13% of the time; when 6 classes are involved the $\text{lsd}_{0.05}$ will be exceeded about 40% of the time; with 10 classes, 60%; and with 20 classes, 90%. Obviously, this procedure for testing the null hypothesis should not be used.

The lsd is easy to compute and use. Because of this it is tempting to use it *after* the data have been gathered and summarized to "hunt" for differences suggested by the data rather than by reasoning prior to data collection. When such hunting is carried out, the probabilities are distorted and the results of the tests unreliable. Investigators should be aware of this problem and should govern their actions accordingly.

### 14.22.2 The Studentized Range

Assume a random sample of size $n$ drawn from population $Y$: $y_1, y_2, \ldots, y_n$. Arranging these values into an ascending array, $y_{(1)}, y_{(2)}, \ldots, y_{(n)}$; where the subscripts in parentheses indicate the position of the value in the array. Then,

$$w_i = y_{(n)} - y_{(i)}; \quad i = (n-1), \ldots, (2), (1) \tag{14.127}$$

$w_i$ is the range across a certain number of values $p$. For example, if $(n) = 6$ and $i = (2)$, $p = (n) - i + 1 = 6 - 2 + 1 = 5$. In other words, there are five values in the range in question: $y_{(6)}$, $y_{(5)}$, $y_{(4)}$, $y_{(3)}$, and $y_{(2)}$. As can be seen, $w_i$ cannot be negative. If all possible samples of size $n$ are drawn from population $Y$, the set of ranges, $W_p$, associated with a specific $p$, forms a sampling distribution with its own distributional pattern, which is dependent on $p$, $n$, and the pdf of $Y$. If each $w_i$ is divided by $\sigma_Y$, the resulting variable

$$r_i = [y_{(n)} - y_{(i)}]/\sigma_Y = w_i/\sigma_Y; \quad i = (n - 1), \ldots, (2), (1) \tag{14.128}$$

is referred to as the *standardized range*, which, again, forms a sampling distribution with its own parameters and distributional pattern. If $\sigma_Y$ is estimated by $s_Y$, from a sample which is independent of that supplying the $w$ values,

$$q_i = [y_{(n)} - y_{(i)}]/s_Y = w_i/s_Y; \quad i = (n - 1), \ldots, (2), (1) \tag{14.129}$$

the resulting variable, $q$, is known as the *studentized range*. Like $w$ and $r$, it has a distributional pattern that is dependent on $p$ and the pdf of $Y$, but instead of $n$, it is dependent on $k$, the number of degrees of freedom associated with $s_Y$. Reference is made to Beyer (1968) for these pdfs.

In the case of an experiment involving several classes or treatments, the investigator is interested in the differences between class or treatment means rather than between individual variates. If the class or treatment means are from samples which, in fact, are drawn from the same parent population, $Y$, the means are variates from a sampling distribution of means which, according to the central limit theorem, is normally or nearly normally distributed. The set of class means constitutes a sample of size $m$ from that sampling distribution. Then,

$$q_i = [\bar{y}_{(m)} - \bar{y}_{(i)}]s_{\bar{y}}; \quad i = (m-1), \ \ldots, \ (2), \ (1) \tag{14.130}$$

If the experiment is repeated an indefinitely large number of times and the $q$ values are computed and then segregated according to $p$, each of these segregated sets forms a sampling distribution of studentized ranges. Table A.11 in the Appendix contains the critical values of $q$ from $\alpha$ levels 0.05 and 0.01, by $p$ and $k = \Sigma_{i=1}^{m}(n_i - 1)$. As usual, if $\alpha$ equals 0.05, one would expect 5% of the studentized ranges to exceed this critical value, even when all the class means are drawn from the same parent population.

Tukey (1953) suggested the use of the studentized range when testing the overall null hypothesis using multiple comparisons. If it is so used, the expected error rate is comparable with that incurred when the $F$ test is used. In other words, the error rate is on a per experiment, rather than a per comparison, basis. In Tukey's procedure all the possible differences are considered. The class means are ranked in ascending order, $\bar{y}_{(1)}$, $\bar{y}_{(2)}$, ..., $\bar{y}_{(m)}$, and then paired in the following manner:

$$\begin{gathered}
\bar{y}_{(m)} - \bar{y}_{(1)} \\
\bar{y}_{(m)} - \bar{y}_{(2)} \\
\vdots \\
\bar{y}_{(m)} - \bar{y}_{(m-1)} \\
\bar{y}_{(m-1)} - \bar{y}_{(1)} \\
\bar{y}_{(m-1)} - \bar{y}_{(2)} \\
\vdots \\
\bar{y}_{(2)} - \bar{y}_{(1)}
\end{gathered}$$

The differences between the paired values are determined, and each difference is compared with the *honestly significant difference*, or hsd. According to Equation 14.129

$$q = [\bar{y}_{(i)} - \bar{y}_{(j)}] / s_{\bar{y}}$$

Then

$$qs_{\bar{y}} = \bar{y}_{(i)} - \bar{y}_{(j)}$$

Setting $q$ equal to $q_\alpha(p, k)$, the critical value,

$$q_\alpha(p,\ k)s_{\bar{y}} = \bar{y}_{(i)} - \bar{y}_{(j)} \tag{14.131}$$

The term $W = q_\alpha(p,\ k_w)s$ is the hsd. If $\bar{y}_{(i)} - \bar{y}_{(j)} > W$, it is assumed that the range is too great for all the class means within it to be variates from the same sampling distribution of means. On the other hand, if $\bar{y}_{(i)} - \bar{y}_{(j)} < W$, it is assumed that all the class means are from the same sampling distribution and, consequently, are estimates of the same population mean.

As an example, return to the eight volume classes. Arranging the class means in ascending order: 20.4, 22.5, 28.6, 29.2, 36.7, 40.3, 44.1, and 46.2 cords/acre. Pairing them:

| | | | | | | |
|---|---|---|---|---|---|---|
| A,B | B,C | C,D | D,E | E,F | F,G | G,H |
| A,C | B,D | C,E | D,F | E,G | F,H | |
| A,D | B,E | C,F | D,G | E,H | | |
| A,E | B,F | C,G | D,H | | | |
| A,F | B,G | C,H | | | | |
| A,G | B,H | | | | | |
| A,H | | | | | | |

Then

$$s_{\bar{y}} = \sqrt{s_w^2 / m} = \frac{\sqrt{16.28965 / 8}}{m} = 1.426957 \text{ cords / acre}$$

From Table A.11, assuming $\alpha = 0.05$,

$$q_\alpha(p, k_w) = q_{0.05}(8152) = 4.29$$

Then $W = q_\alpha(p,\ k_w)s_{\bar{y}} = 4.29(1.426957) = 6.12$ cords/acre.

All the ranges exceeding 6.12 cords/acre are too large for the class means within those ranges to be estimates of the same population mean.

| **Classes** | **Range** |
|---|---|
| *A,B* | 46.2 – 44.1 = 2.1 < 6.12 |
| *A,C* | 46.2 – 40.3 = 5.9 < 6.12 |
| *A,D* | 46.2 – 36.7 = 9.5 > 6.12 |
| *A,E* | 46.2 – 29.2 = 17.0 > 6.12 |
| *A,F* | 46.2 – 28.6 = 17.6 > 6.12 |

| Classes | Range |
|---|---|
| *A,G* | 46.2 – 22.5 = 23.7 > 6.12 |
| *A,H* | 46.2 – 20.4 = 25.8 > 6.12 |
| *B,C* | 44.1 – 40.3 = 3.8 < 6.12 |
| *B,D* | 44.1 – 36.7 = 7.4 > 6.12 |
| *B,E* | 44.1 – 29.2 = 14.9 > 6.12 |
| *B,F* | 44.1 – 28.6 = 15.5 > 6.12 |
| *B,G* | 44.1 – 22.5 = 21.6 > 6.12 |
| *B,H* | 44.1 – 20.4 = 23.7 > 6.12 |
| *C,D* | 40.3 – 36.7 = 3.6 < 6.12 |
| *C,E* | 40.3 – 29.2 = 11.1 > 6.12 |
| *C,F* | 40.3 – 28.6 = 11.7 > 6.12 |
| *C,G* | 40.3 – 22.5 = 17.8 > 6.12 |
| *C,H* | 40.3 – 20.4 = 19.9 > 6.12 |
| *D,E* | 36.7 – 29.2 = 7.5 > 6.12 |
| *D,F* | 36.7 – 28.6 = 8.1 > 6.12 |
| *D,G* | 36.7 – 22.5 = 14.2 > 6.12 |
| *D,H* | 36.7 – 20.4 = 16.3 > 6.12 |
| *E,F* | 29.2 – 28.6 = 0.6 < 6.12 |
| *E,G* | 29.2 – 22.5 = 6.7 > 6.12 |
| *E,H* | 29.2 – 20.4 = 8.8 > 6.12 |
| *F,G* | 28.6 – 22.5 = 6.1 < 6.12 |
| *F,H* | 28.6 – 20.4 = 8.2 > 6.12 |
| *G,H* | 22.5 – 20.4 = 2.1 < 6.12 |

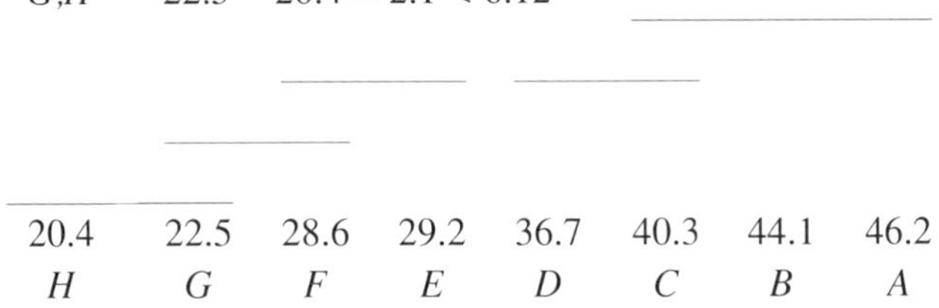

| 20.4 | 22.5 | 28.6 | 29.2 | 36.7 | 40.3 | 44.1 | 46.2 |
|---|---|---|---|---|---|---|---|
| *H* | *G* | *F* | *E* | *D* | *C* | *B* | *A* |

The horizontal lines above the means indicate the ranges within which the means are estimates of the same population mean. Thus, in this case, the overall null hypothesis would be rejected at the 0.05 level of significance because all the means are not included in a single range. Furthermore, the evidence indicates that *G* and *H, E* and *F, C* and *D*, and *A* and *B* could be lumped to form volume classes rather than the original eight. Where the lines overlap, the lumping is done arbitrarily. The class means under a given line segment are estimates of the same population mean. Populations overlap, however, and consequently a given mean in an overlap area could be an estimate of either of the population means.

As has been mentioned Tukey's method can be used to test the overall null hypothesis. If all the means are not included in a single range, the null hypothesis is rejected. Another way of looking at this is to use the hsd to test the difference between the largest and smallest of the means. If the difference exceeds the hsd, the null hypothesis is rejected. The significance level used to obtain the critical value of $q$ would be appropriate in such a case. If the ranges being tested do not span the entire set of class means, the actual significance level is somewhat less than *A*. In other words the test is conservative, meaning that the likelihood of rejecting the null hypothesis is less than is indicated.

Tukey's procedure can be viewed from another perspective. The hsd is, in a sense, comparable with the sampling error, $\pm t_\alpha(k)s_{\bar{d}}$, of a difference. Consequently, it can be used to develop a confidence interval:

$$\mu_i - \mu_j = (\bar{y}_i - \bar{y}_j) \pm q_\alpha(p,\ k_w)s_{\bar{y}} \tag{14.132}$$

Then, if it is hypothesized that $\mu_i = \mu_j$ or $\mu_i - \mu_j = 0$,

$$0 = (\bar{y}_i - \bar{y}_j) \pm q_\alpha(p,\ k^w)s_{\bar{y}}$$

In such a case, if $(\bar{y}_i - \bar{y}_j) > q_\alpha(p,\ k_w)s_{\bar{y}}$, the confidence interval cannot include zero and the null hypothesis would be rejected.

The *Student–Newman–Keul* test, or, simply, *Keul's* test (Newman, 1939; Keuls, 1952), is similar to Tukey's test in that it makes use of the studentized range but differs from Tukey's in that a separate critical value of $q$ is used for each level of $p$, the number of class or treatment means included in the range. For example, using the eight volume classes:

$$W_2 = q_{0.05}(2,152)s_{\bar{y}} = 2.77(1.426957) = 3.95$$

$$W_3 = q_{0.05}(3,152)s_{\bar{y}} = 3.31(1.426957) = 4.72$$

$$W_4 = q_{0.05}(4,152)s_{\bar{y}} = 3.63(1.426957) = 5.18$$

$$W_5 = q_{0.05}(5,152)s_{\bar{y}} = 3.86(1.426957) = 5.51$$

$$W_6 = q_{0.05}(6,152)s_{\bar{y}} = 4.03(1.426957) = 5.75$$

$$W_7 = q_{0.05}(7,152)s_{\bar{y}} = 4.17(1.426957) = 5.95$$

$$W_8 = q_{0.05}(8,152)s_{\bar{y}} = 4.29(1.426957) = 6.12$$

Then, when $p = 2$, the ranges are

| *A,B* | *B,C* | *C,D* | *D,E* | *E,F* | *F,G* | *G,H* |
|---|---|---|---|---|---|---|
| 2.1 | 3.8 | 3.6 | 7.5 | 0.6 | 6.1 | 2.1 |
| $<W_2$ | $<W_2$ | $<W_2$ | $>W_2$ | $<W_2$ | $>W_2$ | $<W_2$ |

and when $p = 3$,

| *A,C* | *B,D* | *C,E* | *D,F* | *E,G* | *F,H* |
|---|---|---|---|---|---|
| 5.9 | 7.4 | 11.1 | 8.1 | 6.7 | 8.2 |
| $>W_3$ | $>W_3$ | $>W_3$ | $>W_3$ | $>W_3$ | $>W_3$ |

Since all the ranges involving three class means are greater than $W_3$, the ranges involving more than three means would also be significant. This can be confirmed by continuing the process until all the ranges have been tested. In graphical form the results of the Student–Newman–Keul test are

| 20.4 | 22.5 | 28.6 | 29.2 | 36.7 | 40.3 | 44.1 | 46.2 |
|---|---|---|---|---|---|---|---|
| *H* | *G* | *F* | *E* | *D* | *C* | *B* | *A* |

As can be seen, $W_8$ is the same as $W$ in Tukey's test. Since it is the largest of the critical values in the Student–Newman–Keul test, it is to be expected that the latter test is more sensitive than in Tukey's and is likely to indicate more ranges differing significantly from zero. In the case of the eight volume classes the difference between the several ranges are made clearer. However, the final interpretation would be much the same. Classes $A$ and $B$, $C$ and $D$, $E$ and $F$, and $G$ and $H$ could be lumped.

Since the critical values differ according to $p$, the error rates indicated by the significance level apply only to the levels of $p$. Neither the idea of a per comparison nor that of a per experiment error rate is applicable in the case of the Student–Newman–Keul test.

Another test that makes use of a separate critical value for each level of $p$ and which involves studentized ranges is Duncan's *new multiple-range test* (Duncan, 1955). It is a further development of an earlier test (Duncan, 1951). This is the reason for the adjective "new" in the name of the test. Unlike the Student–Newman–Keul test, Duncan's new multiple-range test does not use conventional significance levels for determining the critical studentized range. Instead, he uses what he labels *special protection levels*. The rationale behind the special protection levels is beyond the scope of this book. However, it can be said that it stems from the problem of independence. Because of this, the standard tables of critical values of the studentized range, such as Table A.11, cannot be used. Instead, use is made of tables prepared by Duncan. Table A.12 is such a table. It provides values of the *significant studentized range* by levels of $p$ and residual degrees of freedom for protection levels of 0.05 and 0.01.

The procedure for using Duncan's new multiple range test is similar to that of the Student–Newman–Keul test. The critical value is the *least significant range*, $R_p$:

$$R_p = \text{significant studentized range} * s_{\bar{y}} \tag{14.133}$$

Thus, in the case of the eight volume classes, using a protection level of 0.05 and 152 degrees of freedom:

$$\begin{aligned}
R_2 &= 2.77(1.426957) = 3.95 \\
R_3 &= 2.92(1.426957) = 4.17 \\
R_4 &= 3.02(1.426957) = 4.31 \\
R_5 &= 3.09(1.426957) = 4.41 \\
R_6 &= 3.15(1.426957) = 4.49 \\
R_7 &= 3.19(1.426957) = 4.55 \\
R_8 &= 3.23(1.426957) = 4.61
\end{aligned}$$

When $p = 2$,

| $A,B$ | $B,C$ | $C,D$ | $D,E$ | $E,F$ | $F,G$ | $G,H$ |
|---|---|---|---|---|---|---|
| 2.1 | 3.8 | 3.6 | 7.5 | 0.6 | 6.1 | 2.1 |
| $<R_2$ | $<R_2$ | $<R_2$ | $>R_2$ | $<R_2$ | $>R_2$ | $<R_2$ |

When $p = 3$,

| $A,C$ | $B,D$ | $C,E$ | $D,F$ | $E,G$ | $F,H$ |
|---|---|---|---|---|---|
| 5.9 | 7.4 | 11.1 | 8.1 | 6.7 | 8.2 |
| $>R_3$ | $>R_3$ | $>R_3$ | $>R_3$ | $>R_3$ | $>R_3$ |

Since all the ranges involving three means exceed $R_3$, it is assumed that all three of the means in each range do not estimate the same population mean. Increasing the range will show the same results at each level, making further comparisons unnecessary. In this example, Duncan's new multiple-range test yielded results that were exactly the same as those obtained with the Student–Newman–Keul test. However, because the $R_p$ values never exceed the $W_p$ values and usually are smaller, Duncan's new multiple-range test is more sensitive than the Student–Newman–Keul test and, consequently, often shows more significant differences. This does not exhaust the subject of multiple comparisons. For further information, reference is made to Harter (1957) and to standard statistical texts such as Steel and Torrie (1960), Snedecor and Cochran (1967), and Sokal and Rohlf (1969).

# 15 Regression and Correlation

## 15.1 INTRODUCTION

Regression and correlation are concepts that are associated with the relationships between two or more random variables. Both were discussed briefly in Chapter 9, correlation in Section 9.8 and regression in Section 9.10. In this chapter the coverage will be expanded and made more complete. The subject matter associated with regression and correlation is, however, far too extensive for complete coverage here. For more information, reference is made to such standard texts as Draper and Smith (1981).

*Regression* is concerned with the development of graphs, or mathematical expressions corresponding to graphs, showing the relationship between a variable of interest and one or more other variables. The utility of such graphs or mathematical expressions arises from the fact that they make it possible to

1. *Visualize* the relationship (e.g., between stand volume and stand age, stand density, and site index),
2. *Make indirect estimates* of variables of interest which are difficult to measure in a direct manner (e.g., merchantable volume of a tree using dbh and merchantable or total tree height), and
3. *Make predictions* regarding future developments (e.g., future dbh based on past dbh as revealed by increment borer cores).

As was brought out in Section 9.8, *correlation* is concerned with the strength of the relationship between the variables. A relationship between two or more variables may exist and take the form indicated by a regression analysis, but if it is to be used to make indirect estimates or predictions, it must be strong enough so that the estimates or predictions are meaningful. Therefore, a measure of the degree of correlation existing between the variables is usually included in a statement based on the regression.

## 15.2 THEORY OF REGRESSION

As in the case of univariate sampling, as discussed in Chapter 13, one must realize that what is of interest to a decision maker is the situation at the *population* level. When this involves the description of the relationship between a variable of interest and one or more other variables, the decision maker would like to have information about the actual situation. This would involve complete enumerations, which may or may not be possible to obtain and certainly would be expensive in terms of both money and time. Consequently, the decision maker must make use of estimates of the situation that are based on sample data. To comprehend what is being done in the estimation process, one must understand what it is that is being estimated. To that end, this chapter will consider the situation at the population level before proceeding to the estimation procedures. By necessity, this discussion will involve some material already presented in Section 9.10. However, the coverage in Section 9.10 is minimal and should be considered a subset of that discussed in this chapter.

### 15.2.1 The Variables

Fundamental to a discussion of regression is the assumption that the variables are *continuous*. Where regression methods are used with discrete variables, the situation parallels that where the normal distribution is used as a substitute for the binomial. The results may be useful but in actuality are only approximations.

As was brought out earlier, regression methods are concerned with the relationship between a variable of interest and one or more other variables. In this context of regression, the variable of interest is known as the *dependent*, *predicted*, or *response* variable and is usually symbolized by $Y$. The remaining variable or variables are known as *independent* or *predicting* variables and are usually symbolized by $X$ or $X_i$. The terms *dependent* and *independent* do *not* imply that a *causal* relationship exists between the variables. They simply indicate which variable(s) is being predicted and which variable(s) is doing the predicting.

### 15.2.2 Coordinate Systems

Graphical representations of relationships between variables must be made within the context of a specific coordinate system. The most commonly encountered system in regression operations is the familiar *rectangular* system. In some cases it may be appropriate to use a *polar* coordinate system as, for example, when the relationship between site index and aspect is being described. However, through the case of appropriate transformations, such as the use of the sine or cosine of the azimuth of an aspect, most of these can be shown in the context of the rectangular system.

It should be pointed out that *anamorphosed* scales, as on semilog, log–log, or arithmetic probability papers, are actually rectangular coordinate systems in which the scale graduations are varied in such a manner as to show nonlinear (curved) relationships as straight lines. These types of coordinate systems have great practical value, but since they are still rectangular systems, they are of significance only in the display phase and do not require special consideration in the following discussion.

Within the rectangular coordinate system, the response variable, $Y$, is assigned to the *vertical* axis while the predicting variable(*s*), $X$ or $X_i$, is assigned to the *horizontal* axis or axes, which follows the convention used in algebra and analytical geometry.

### 15.2.3 Two-Variable Regression

Assume that all the elements of a given universe have been evaluated with respect to two variable characteristics, one of which is designated the response variable and is labeled $Y$. The second variable is then the predicting variable $X$. Let the data be sorted by values of $X$. Theoretically, there is an infinite number of such $X$ values, with adjacent values differing only by an infinitesimal. There will be a number of $Y$ values, $N_i$, also infinite in number, associated with each individual value of $X$, $x_i$. Consider each of these subsets of $Y$ as a separate population with its own mean,

$$\mu_{Y \cdot X = x_i} = E(Y \mid X = x_i) \tag{15.1}$$

variance,

$$\sigma^2_{Y \cdot X = x} = V(Y \mid X = x_i) \tag{15.2}$$

and distributional pattern (e.g., normal, uniform, etc.).

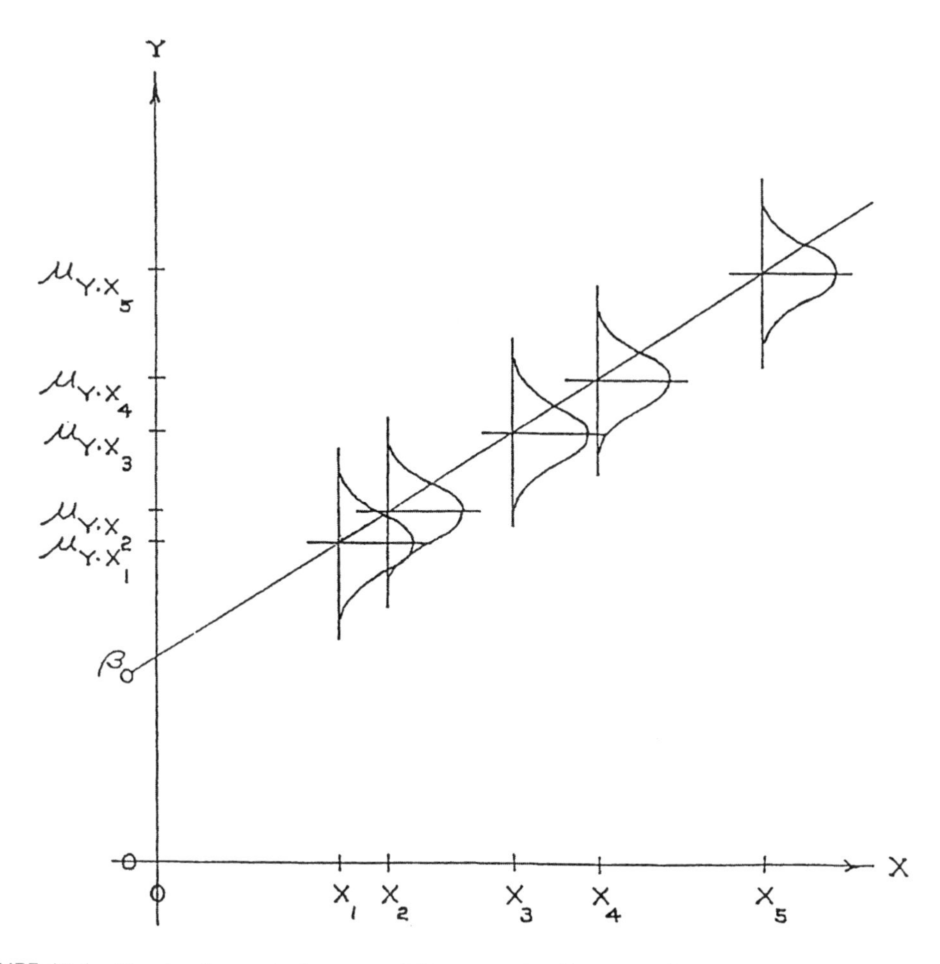


**FIGURE 15.1** The development of a two-variable regression line or moving mean.

Assume that the subpopulation means differ from one another and that their magnitudes are in some way associated with the magnitudes of their corresponding values of $X$. If the magnitudes of these subpopulation or *conditional* means increase as the $X$ values increase, they would show, if plotted within a rectangular coordinate system, a pattern as in Figure 15.1. It must be remembered that both $X$ and $Y$ are continuous variables and, consequently, Figure 15.1 shows only a representative few of the infinite number of points which should be shown. The line formed by the infinitely large number of subpopulation or conditional means is known as the *regression line*. It is the trace of the arithmetic mean of $Y$ which changes magnitude as $X$ changes. Because of this the line is sometimes referred to as the *moving mean*.

The mathematical expression describing the regression line is known as the *regression equation* or *regression function*. If, as in Figure 15.1, the line is *straight*, the regression equation would be

$$\mu_{Y \cdot X} = E(Y \mid X) = \beta_0 + \beta_1 X; \quad x_A < X < x_B \tag{15.3}$$

where $\beta_0$ is the expected value of $Y$ when $X = 0$, $E(Y|X = 0)$, referred to as the $Y$-intercept, and $\beta_1$ indicates the change in the value of $E(Y|X)$ *per unit change in* $X$, which is known as the *regression coefficient*, and $x_A$ and $x_B$ are, respectively, the smallest and largest possible values of $X$. Equation 15.3 is identical to Equation 9.32.

The relationship between $Y$ and $X$ may or may not form a straight line. If it does form a straight line, it is referred to as a *linear* relationship. If not, the relationship is *curvilinear*. If curvilinear, the relationship may take on the form of a parabola (J shaped or reversed J), a hyperbola, an S (sigmoid), or any one of an innumerable number of possible curve forms. Thus, to generalize,

$$\mu_{Y \cdot X} = E(Y \mid X) = g(X); \quad x_A < X < x_B \tag{15.4}$$

where $g(X)$ is a continuous function of $X$ in the region $x_A$ to $x_B$.

Consider the linear model of an individual variate of $Y$ when $X = x_i$:

$$y_{ij} = E(Y \mid X = x_i) + \varepsilon_{ij} = g(X = x_i) + \varepsilon_{ij} \tag{15.5}$$

Rearranging,

$$\varepsilon_{ij} = y_{ij} - g(X = x_i)$$

The conditional variance of subpopulation $Y_i$ is then

$$\begin{aligned} V(X = x_i) &= \int_{Y_{iA}}^{Y_{iB}} \varepsilon^2 f(y)_i \, dy \\ &= \int_{y_{iA}}^{Y_{iB}} [y_i - g(X = x_i)] f(y)_i \, dy \end{aligned} \tag{15.6}$$

where $y_{iA}$ and $y_{iB}$ are, respectively, the smallest and largest values of $Y$ when $x = x_i$, and $f(y)_i$ is the pdf of $Y$ when $X = x_i$. If the subpopulation or conditional variances are equal for all values of $X$, it is said that they are *homogeneous* and that *homoscedasticity* is present. On the other hand, if the conditional variances are heterogeneous, it is said that *heteroscedasticity* exists. When the conditional variances are homogeneous, the overall *variance about the regression line is*

$$V(Y \mid X) = \sigma^2_{Y \cdot X} = \int_{x_A}^{x_B} \int_{y_A}^{y_B} [y + g(x)]^2 f(y) dy dx \tag{15.7}$$

### 15.2.4 Multiple Variable Regression

When more than one independent variable is involved,

$$\mu_{Y \cdot X_i \cdot X_2 \ldots X_M} = E(Y \mid X_1 X_2 \ldots X_M) = g(X_1, \ X_2, \ \ldots, \ X_M) \tag{15.8}$$

Where the relationship involves two predicting variables, as when total height of the dominant stand is being related to tree age and site index, the regression would generate a three-dimensional surface rather than a two-dimensional line. Conceptually, both the surface and the line represent the same thing. They both are moving means. A surface generated in this manner is referred to as a *response surface*.

Displaying a three-dimensional surface in the context of a two-dimensional graph requires the plotting of a series of cross sections of the surface, each associated with a specific value or level of the second predicting variable. The intervals between the levels of the second predicting variable should be constant. The result is a set of curves such as those in Figure 15.2. If the spacing of the

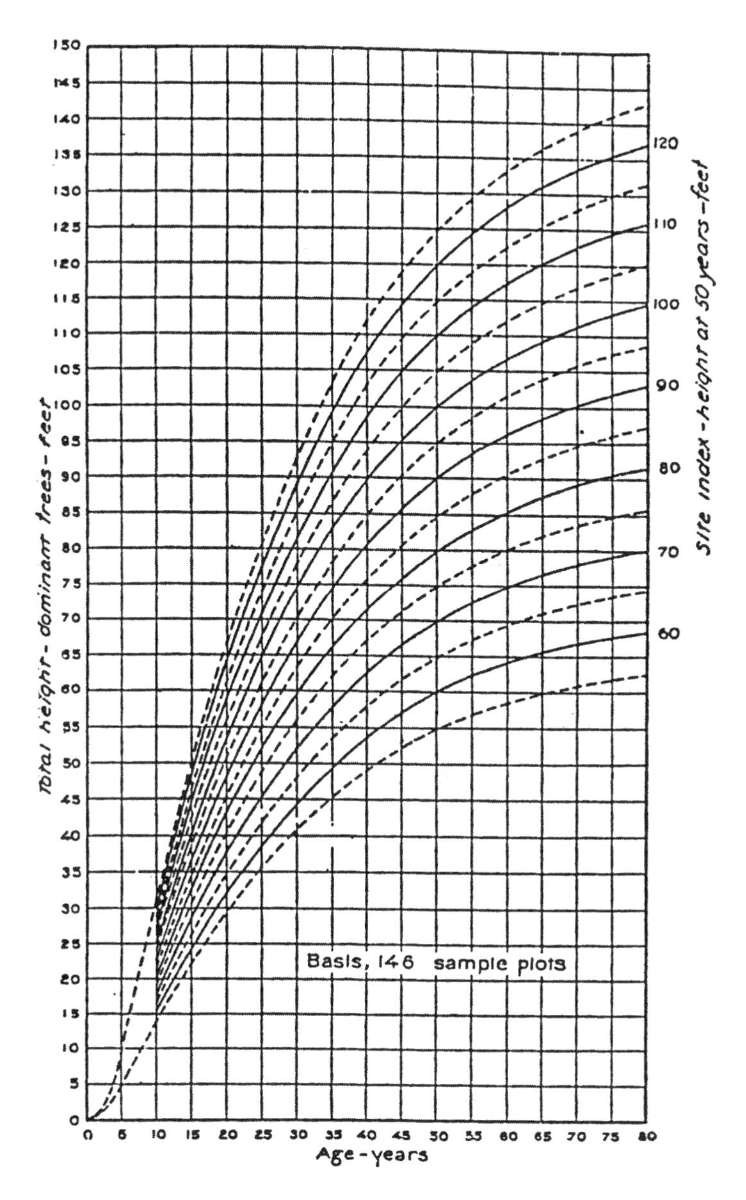


**FIGURE 15.2** (A) Site index curves for longleaf pine. This relationship is anamorphic since the ratio of the height of any of the curves to the height of any other of the curves is constant for all values of $X_1$ (age). (From USDA Forest Service, 1928.)

curves is such that the height of any one of the curves remains a constant proportion of the height of any one of the remaining curves, across the entire range of values of the first predicting variable, the relationship is said to be anamorphic (see Figure 15.2A). If this condition is not met, the relationship is *polymorphic* (see Figure 15.2B). Furthermore, in the latter case, if the curves do not cross one another in the range $x_{1A}$ to $x_{1B}$, the relationship is *polymorphic–disjoint*. If they cross the relationship is *polymorphic–nondisjoint*.

In the case of a multiple regression the relationship between the response variable, $Y$, and one of the predicting variables, say, $X_1$, might be a simple linear as shown in Figure 15.3A. The regression equation would be

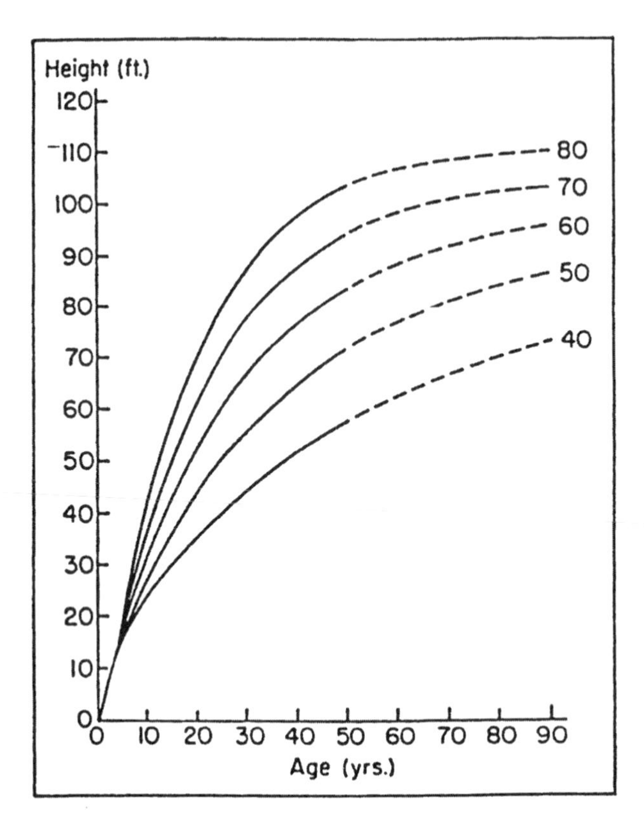


**FIGURE 15.2** *Continued.* (B) Site index curves for sweetgum. This relationship is polymorphic–disjoint. The ratios of line heights are not constant for all values of $X_1$ (age). It is disjoint because the lines do not cross one another. (From Lyle, E. A. et al., Alabama Agricultural Experimental Station Circ. 220, 1975.)

$$\mu_{Y \cdot X_1} = \beta_0 + \beta_1 X_1 \tag{15.9}$$

When a second predicting variable is introduced, say, $X_2$, and a graphical plot is made of the relationship, using a set interval between the levels of $X_2$, the graph might appear as in Figure 15.3B. Note that the cross sections of the response surface form a set of parallel lines. All these lines have the same slope but intersect the $Y$-axis at different points. In other words, the $Y$-intercept is a function of $X_2$.

$$\beta_0 = h(X_2) \tag{15.10}$$

If the interval between the levels of $X_2$ is constant, the spacing of the lines will be constant, and the relationship between $\beta_0$ and $X_2$ will be linear:

$$\beta_0 = \beta_2 + \beta_3 X_1 \tag{15.11}$$

Substituting Equation 15.11 for $\beta_0$ in Equation 15.9,

$$\mu_{Y \cdot X_1 \cdot X_2} = \beta_2 + \beta_3 X_2 + \beta_1 X_1 \tag{15.12}$$

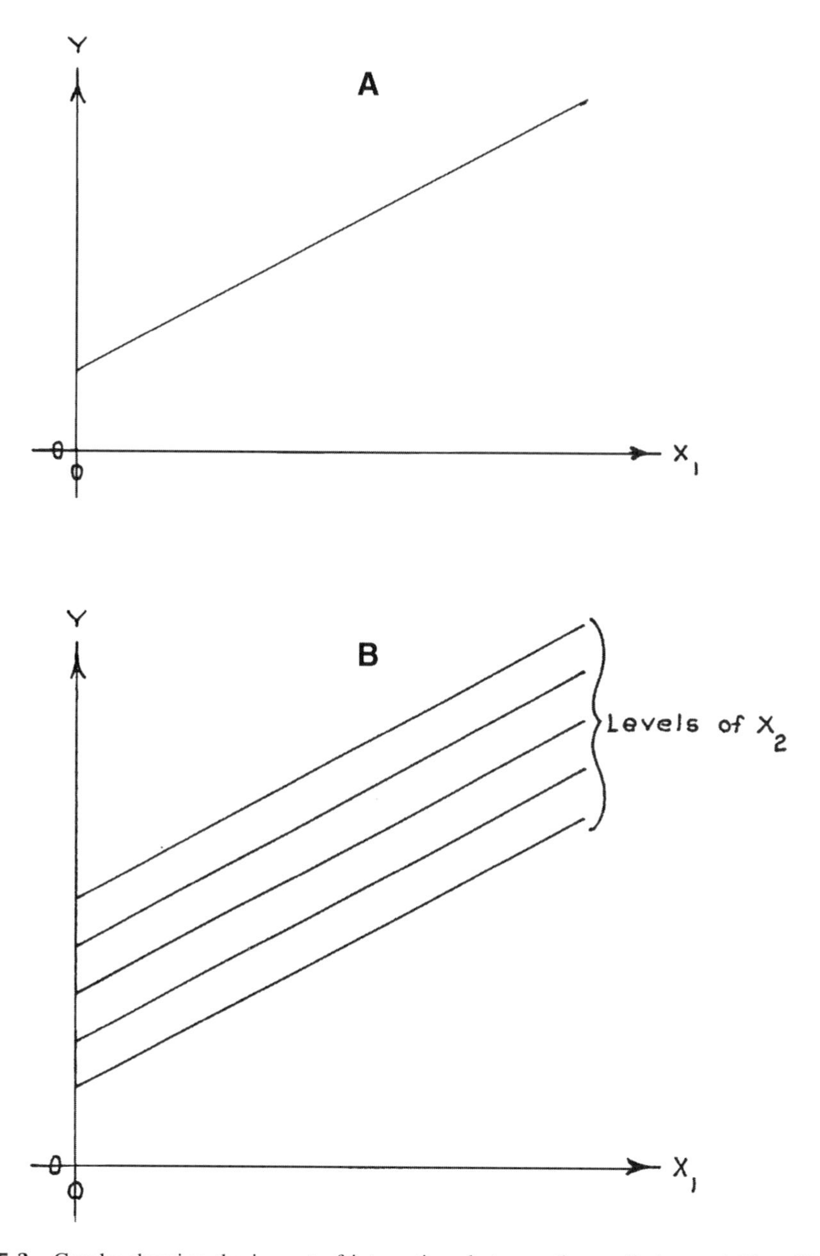


**FIGURE 15.3** Graphs showing the impact of interactions between the predicting variables. (From Freese, USDA Forest Service Res. Pap. FPL 17, 1964.) (A) The simple relationship between $Y$ and $X_1$. (B) The $Y$-intercept is a function of $X_2$. There are no interactions between $X_1$ and $X_2$. (C) Slope of the lines is a function of $X_2$. In this case there is an interaction between $X_1$ and $X_2$. (D) Both the $Y$-intercept and the slopes of the lines are functions of $X_2$, and an interaction between $X_1$ and $X_2$ is present.

In such a case the effects of $X_1$ and $X_2$ are additive and, furthermore, the presence of $X_2$ has no effect on the coefficient of $X_1$ nor does $X_1$ have an effect on the coefficient of $X_2$. Consequently, it is said that $X_1$ and $X_2$ are *independent of one another* and *are not correlated.*

Returning to Equation 15.9, if, when $X_2$ is introduced into the graph, the lines representing the levels of $X_2$ radiate from a common $Y$-intercept, as in Figure 15.3C, the slopes of the lines change as $X_2$ changes. Consequently, the coefficient $\beta_1$ in Equation 15.9 is a function of $X_2$:

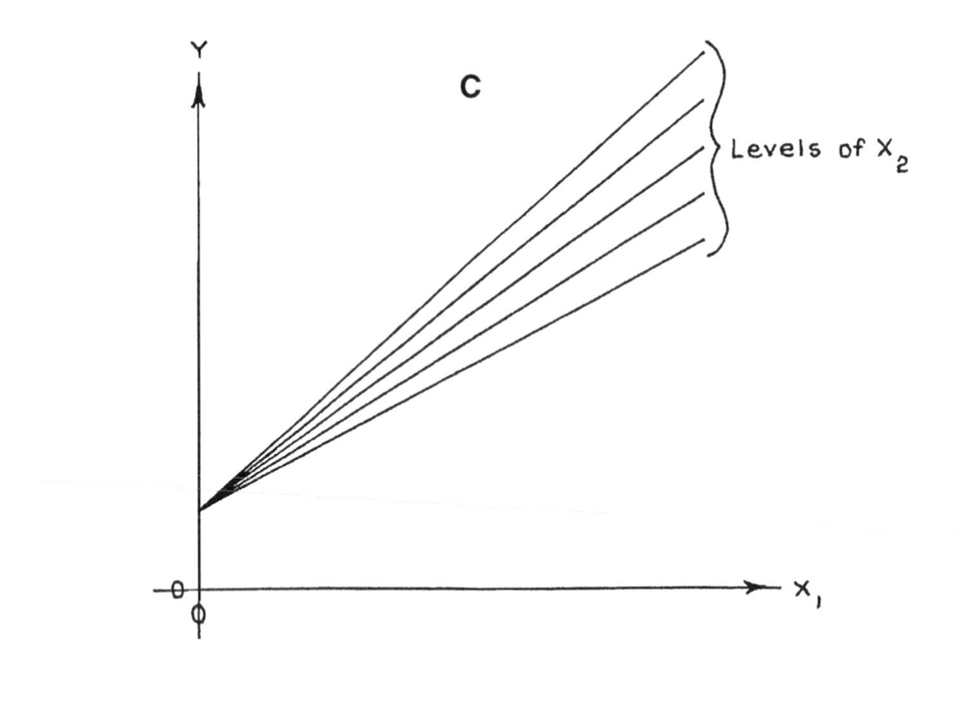


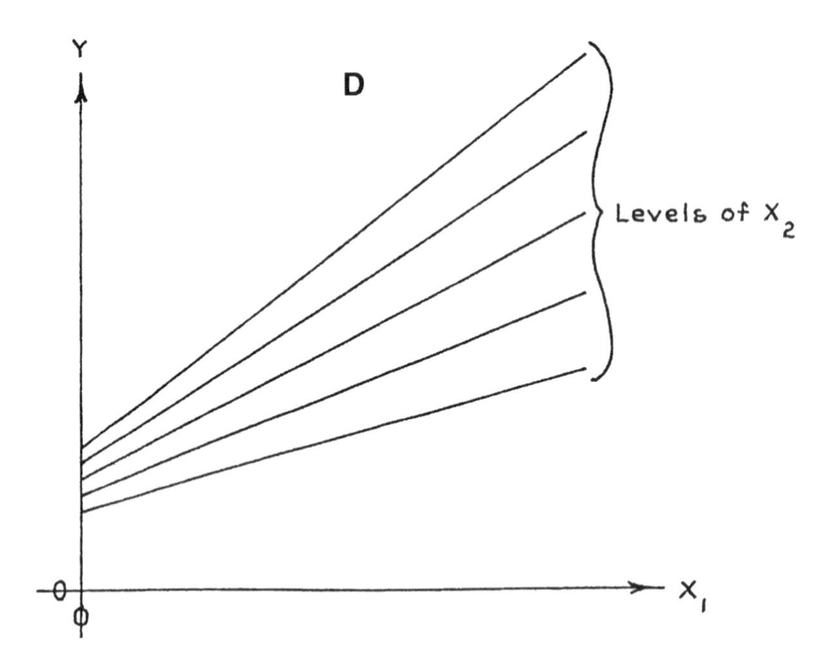


**FIGURE 15.3** *Continued.*

$$\beta_1 = h(X_2) \tag{15.13}$$

If the intervals between the levels of $X_2$ are constant and the spacing of the lines is constant at a given value of $X_1$, the relationship between $\beta_1$ and $X_2$ is linear:

$$\beta_1 = \beta_4 + \beta_5 X_2 \tag{15.14}$$

Substituting Equation 15.14 for $\beta_1$ in Equation 15.9,

$$\mu_{Y \cdot X_1 \cdot X_2} = \beta_0 + (\beta_4 + \beta_5 X_2)X_1 = \beta_0 + \beta_4 X_1 + \beta_5 X_1 X_2 \quad (15.15)$$

In this case the effect of $X_2$ is not additive. Instead, $X_2$ forms a product with $X_1$, $X_1X_2$. Such a product indicates that an *interaction* exists between the predicting variables and that they are *not independent* of one another. The term $\beta_5 X_1 X_2$ in Equation 15.15 is known as an *interaction term*.

In the majority of cases *both* $\beta_0$ and $\beta_1$ in Equation 15.9 are functions of $X_2$ and, if the relationships are as previously described, the graph of $Y$ or $X_1$ and $X_2$ would take on the form shown in Figure 15.3D. The regression equation would be

$$\mu_{Y \cdot x_1 x_2} = \beta_2 + \beta_3 X_2 + (\beta_4 + \beta_5 X_2)X_1 = \beta_2 + \beta_3 X_2 + \beta_4 X_1 + \beta_5 X_1 X_2 \quad (15.16)$$

Again, an interaction term appears indicating that a relationship exists between the two predicting variables.

The conditional variance about a multiple regression is analogous to that about a regression line:

$$\begin{aligned} V(Y \mid X_1 X_2 \ldots X_M) &= \sigma^2_{Y \cdot X} \\ &= \int_{x_{MA}}^{x_{MB}} \cdots \int_{x_{2A}}^{x_{2B}} \int_{x_{1A}}^{x_{1B}} [y - g(x)]^2 f(y) dy dx_1 \ldots dx_M \end{aligned} \quad (15.17)$$

As in the case of the variance about a regression line, either homoscedasticity or heteroscedasticity may be present.

It should be noted that $f(y)$ in Equations 15.7 and 15.17 is the pdf of the response variable. In these equations it is assumed that it is the same for all values of the predicting variable or variables. If this assumption is not valid, Equations 15.7 and 15.17 are also not valid. However, taking variability of $f(y)$ into account would introduce problems which probably are insurmountable and, consequently, in regression operations the possibility of a variable pdf is not taken into consideration. As a matter of fact, in most cases it is assumed that the distributional patterns of the subpopulations of $Y$ are constant and that in each case that distribution is normal. A further discussion of this point is beyond the scope of this book.

## 15.3 FITTING A REGRESSION

In Chapter 13 the concept of parameter estimation using sample statistics was introduced. That discussion involved only single variables. The basic idea of parameter estimation using sample data can, however, be extended to situations where the relationships between two or more variables are concerned. The process leading to a graphical or mathematical estimate of such a relationship is known as *fitting a curve* or *fitting a regression*.

### 15.3.1 The Controlling Criteria

When a regression line or surface is to be fitted to sample data, the fitting must be done in such a manner that the following criteria are satisfied.

1. The criterion of *appropriateness*. This involves the *shape* of the line or surface. The shape is appropriate if it makes sense when the general laws governing the situation are considered. For example, the gross cubic foot volume of a tree increases as its dbh and/or

total height increases. A regression depicting this relationship that indicated a reduction in volume as size increased would be obviously erroneous. It is essential that the regression makes sense.

2. The *least squares* criterion. The *position* of the most nearly representative line or surface is usually defined in terms of least squares (see Section 5.3.6), where the sum of squared residuals from the regression is at a minimum. This is in recognition of the fact that the regression line or surface is a form of arithmetic mean. The arithmetic mean occupies the least squares position along a number line (a one-dimensional space). In the same manner, the regression should pass through the scatter of data points so as to occupy the least squares position in the multidimensional space.

   When the conditional variances about the regression are homogeneous and the subpopulations are normally distributed about their respective means, the least squares–fitting process yields a maximum likelihood estimate (see Section 13.6) of the population parameters.

   As will be seen, in some cases the least squares criterion, as stated, cannot be met. In such cases the criterion is eased to the extent that the sum of squared residuals does not need to be an absolute minimum. Instead, the sum should be as small as is practicable within the limits, usually cost, imposed.

### 15.3.2 Manual or Freehand Fitting

Prior to the development of digital computers the mathematical fitting of a regression was not a task to be taken lightly. It was a lengthy, tedious operation subject to human error. Consequently, much use was made of manual or freehand curve fitting in which the curve is drawn directly onto the coordinate system through the scatter of the data in such a manner as to have a logical shape and so positioned that the sum of the residuals from the curve is equal to zero. The procedure is seldom used at the present time, but it does have some potential when computing facilities are limited. It also can be used as a preliminary step when one is seeking an appropriate model which is to be fitted mathematically.

Manual or freehand fitting is a trial-and-error procedure that is practical only when two variables are involved, but methods exist through which the process can be extended to more variables (Ezekiel, 1941). Since trial and error is involved, each fit is an individual estimate of the situation. It cannot be exactly reproduced, even by the same person using the same data.

The process is as follows:

1. The coordinate system is laid out on suitable cross-sectioned paper with the independent variable, $X$, assigned to the horizontal axis and the response variable, $Y$, assigned to the vertical axis, as is shown in Figure 15.4.
2. The data points are plotted within the coordinate system. Any points representing more than one observation are identified by subscripts indicating their weights, as is shown at point A in Figure 15.4. The resulting set of plotted points form a *scatter diagram* or *scattergram*.
3. A straight line is drawn, parallel to the vertical axis, through the scatter diagram to divide it into two more or less equal parts, as at point B in Figure 15.4.
4. Using a flexible curve, a ship curve, or a french curve, a line is passed through the scatter diagram so as to occupy a position that is logical and is apparently close to where the regression line should be (see point C in Figure 15.4). This is a trial line and should be very lightly drawn. It should curve smoothly, without any abrupt changes in direction. Furthermore, it should not wander up and down in order to pass close to the individual plotted points.
5. Determine the residual, $d$, from the trial line for each of the plotted points,

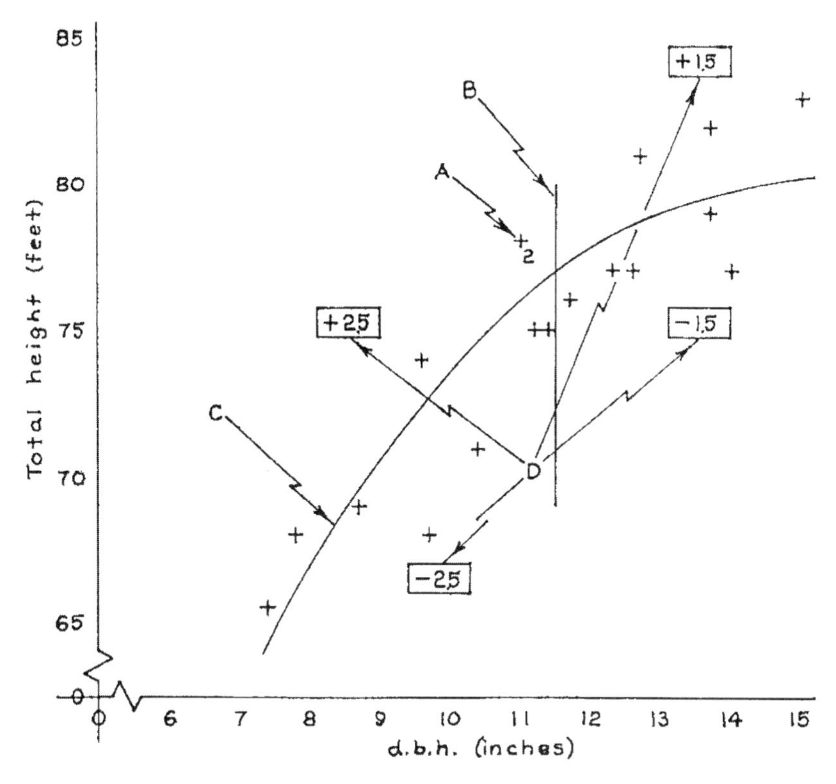


**FIGURE 15.4** Regression of total height on diameter breast height developed using manual fitting procedures. A = Point represents two data items. B = Vertical dividing line. C = Balanced regression line. D = Balances for each half of curve.

$$d_i = y_i - \hat{y}_i \tag{15.18}$$

where $d_i$ = residual for observation $i$
$y_i$ = $Y$ coordinate of observation $i$
$\hat{y}_i$ = the curve value at $i$

The points lying above the regression line have positive residuals while the points lying below the regression line have negative residuals. If a point represents more than one observation, as at point A in Figure 15.4, its residual must be accordingly weighted.

6. A separate algebraic sum of the residuals is obtained for the data points lying on either side of the vertical line B. These sums are obtained by adding the positive and negative sums, such as those shown at D in Figure 15.4. If these two algebraic sums are equal to zero, it is assumed that the regression is correctly positioned and the fitting is completed.
7. If, however, the algebraic sums are not equal to zero, it becomes necessary to move the regression line in such a manner that the algebraic sums do equal zero. If a sum is positive, it means that the line is too low on that side of the vertical line. If it is negative, the line is too high. This process may require several trial lines before the algebraic sums go to zero. It must be emphasized that the shifting must be done to meet the criterion of appropriateness. The curve cannot show a break or change in direction as it crosses the vertical line B.
8. The final line is made permanent and is ready for use.

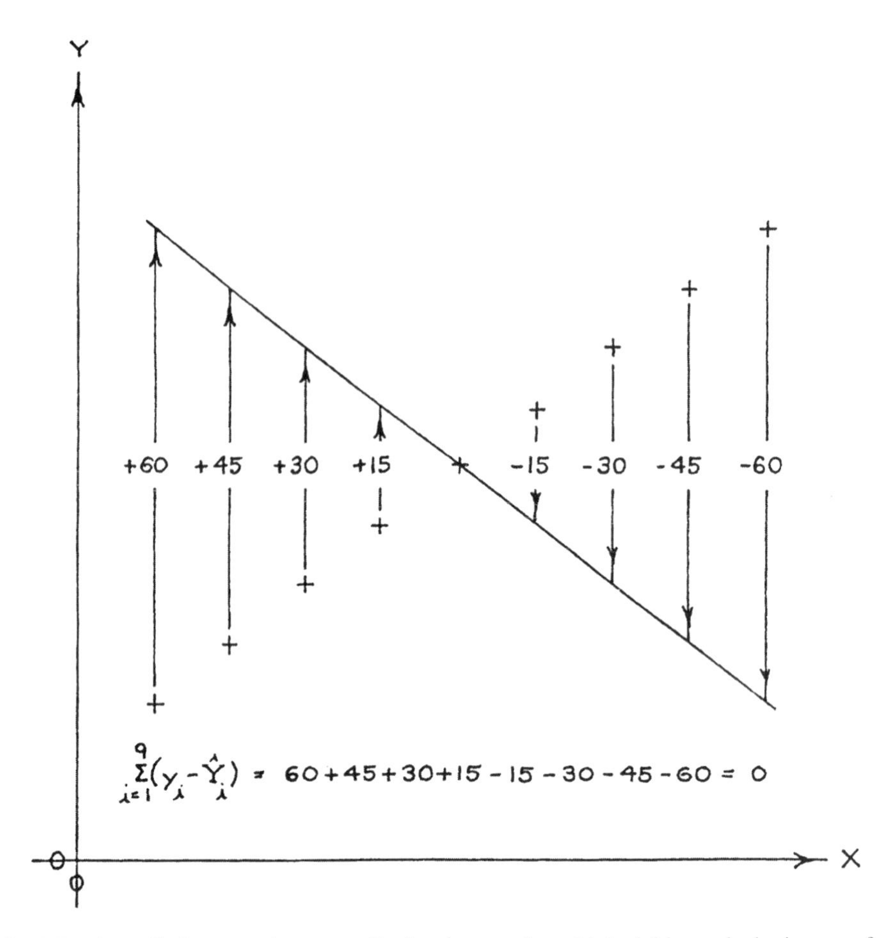


**FIGURE 15.5** A totally inappropriate manually fitted regression which yields an algebraic sum of residuals that is equal to zero.

Obviously the freehand procedure does not make use of the sum of squared residuals and one cannot make any statement to the effect that the least squares criterion has been met. All that the procedure requires is that the algebraic sum of residuals be equal to zero. This requirement could be met with a curve that is totally inappropriate, as can be seen in Figure 15.5. To forestall this type of outcome, the vertical line B in Figure 15.4 is inserted into the scatter diagram and the algebraic sums of the residuals on both sides of the vertical line must be equal to zero.

Manual or freehand curve fitting is a trial-and-error procedure that is practical only when two variables are involved, but methods have been devised by which the process can be extended to more variables (Ezekiel, 1941). Since trial and error are involved, each fit is an individual estimate of the situation. It cannot be exactly reproduced, even by the same person using the same data. For this reason freehand fitted curves cannot be used in scientific papers or statements. However, where the curves are used internally or for preliminary studies, this flaw is not significant and the curves can be quite valuable. Ironically, because of the flexibility inherent in the process it often is possible to develop a freehand curve which has a smaller sum of squared residuals than does any reasonable mathematical fit.

### 15.3.3 Modeling

A *model*, in the context of regression, is an algebraic expression that the investigator believes describes the relationship between an individual observation of the response variable, $y_i$, and the independent variable or variables. This expression consists of two parts: the algebraic function describing the regression line or surface, known as the *regression function* or *regression equation*; and the error of the individual observation. Thus, the regression model is

$$y_{ij} = f(X_1, X_2, \ldots)_i + \varepsilon_{ij} \quad (15.19)$$

where $f(X_1, X_2, \ldots)$ = the regression function
$\varepsilon_{ij}$ = the error of the $j$th observation in the $i$th $X_1, X_2, \ldots,$ class

*Modeling* is the process through which the investigator arrives at the general form of the regression function which is to be used in the fitting operation. In this process all the information that can be gathered which bears on the problem should be considered so that the choice is logical and the resulting regression will accurately describe the relationship between the chosen variables. The choice is critical. It is through this choice that the criterion of appropriateness is satisfied.

If the situation being represented by the regression is poorly understood, it may not be possible to choose a regression function using logic. When this occurs, the investigator is forced into building a model that is *based on the data*. Since sample data are not necessarily representative of the populations from which they are drawn, it is possible that regressions generated in this manner may be misleading. Consequently, every effort should be made to begin the fitting process with a logical model that can be defended if challenged. Data-based models are built making use of so-called *stepwise* procedures, which are described in Section 15.9.

There are two categories of regression models, *linear* and *nonlinear*. Linear models contain linear regression functions such as

$$f(X) = \beta_0 + \beta_1 X \quad (15.20)$$

$$f(X) = \beta_0 + \beta_1 X + \beta_2 X^2 \quad (15.21)$$

$$f(X) = \beta_0 + \beta_1(1/X) \quad (15.22)$$

$$f(X) = \beta_0 + \beta_1(\log X) \quad (15.23)$$

In these expressions the coefficients are nonzero constants, the exponents (if present) are constants, and the terms are additive.

It should be noted that linear regression functions can produce curved lines, as can be seen in Figure 15.6. Consider Equation 15.23 as an example. Let $U = \log X$. Then

$$f(U) = \beta_0 + \beta_1 U \quad (15.24)$$

The independent variable is now $U$ and the horizontal axis of the coordinate system must be graduated in terms of $U$ rather than $X$. When this is done, as is shown in Figure 15.7, the regression line will be a straight line. Thus, the term *linear* is not a misnomer. This transformation leading to a straight regression line is known as *anamorphosis*.

The $X_A$ scale in Figure 15.7 is comparable to one cycle on semilogarithmic cross-section paper.

The least squares–fitting process is based on the assumption that the models being fitted are linear. Consequently, the fact that a given algebraic expression being proposed as a regression function is or is not linear is not trivial.

Some examples of nonlinear functions are

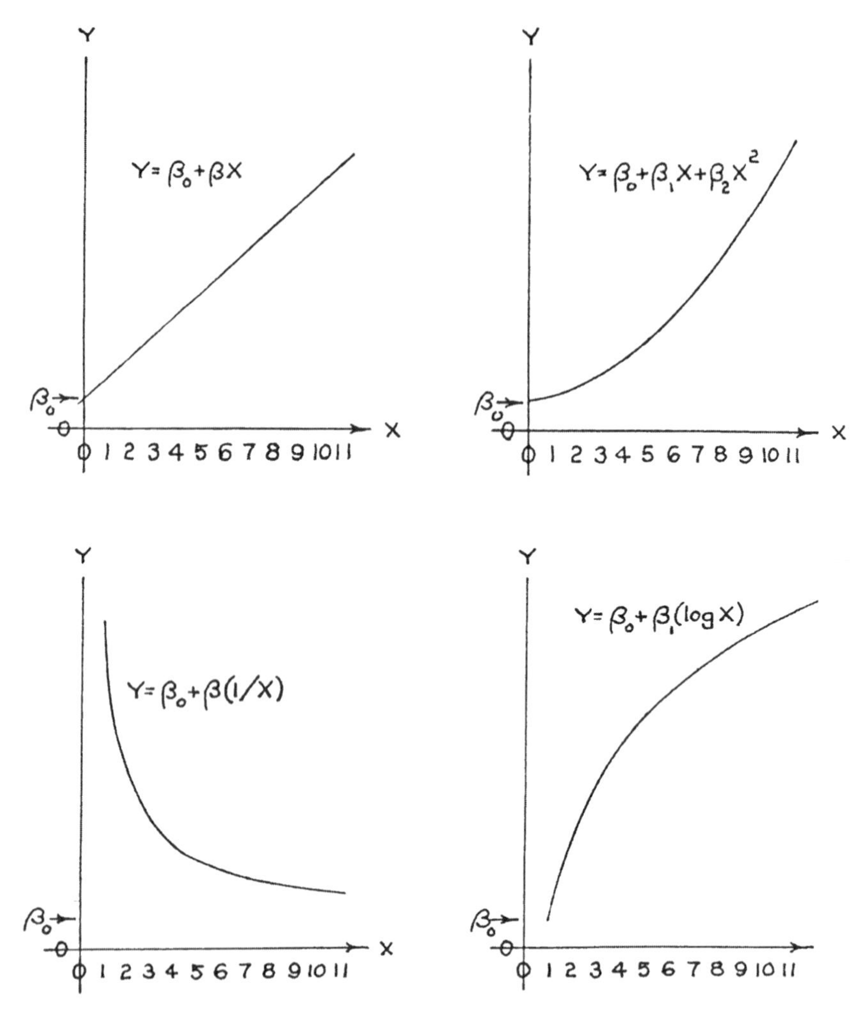


**FIGURE 15.6** A representative set of linear regression functions. Note that some functions can produce curved regression lines.

$$f(X) = \beta_0 + \beta_1^X \tag{15.25}$$

$$f(X) = \beta_0 X^{\beta_1} \tag{15.26}$$

$$f(X) = e^{-\beta X} \tag{15.27}$$

and

$$f(X) = K / [1 + e^{-(\beta_0 + \beta_1 X)}] \tag{15.28}$$

The highly flexible frequency functions or probability density functions for the gamma, beta, and Weibull probability distributions (see Sections 11.3, 11.6, and 11.8) are often used as regression functions and they are nonlinear functions. As can be seen, each of these expressions, in one way or another, fails to meet the criteria for a linear function.

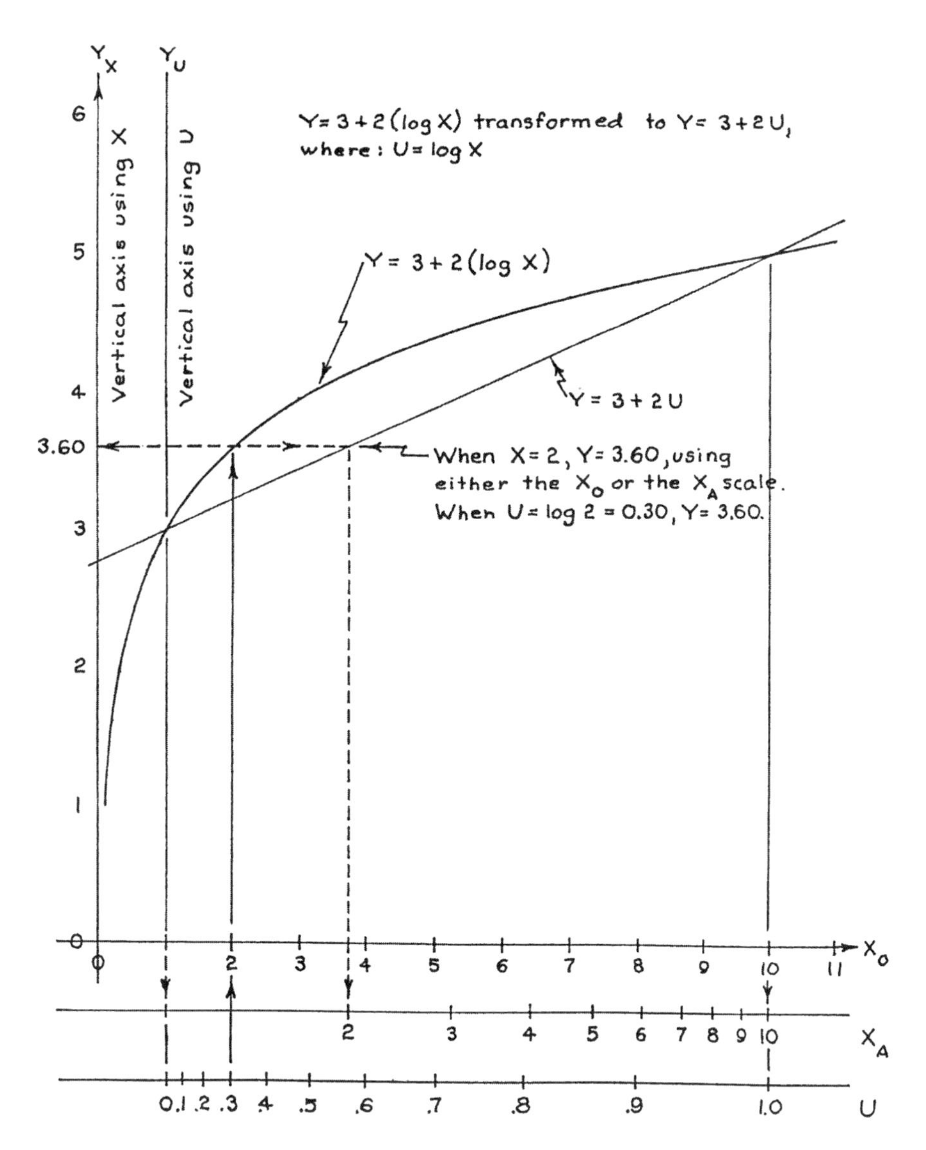


**FIGURE 15.7** An example of anamorphosis in which a curved regression is transformed into a straight line by regraduation of the horizontal axis. $X_A$ is the regraduated axis when the original $X$ units are used. The $U$ axis is in terms of $U = \log X$. If the *figure* is to be used to obtain values of $Y$, one can enter either with the original units of $X$, using the $X_A$ scale, or with the $U$ units.

In some cases nonlinear functions can be transformed into expressions that are linear. For example, consider Equation 15.26

$$f(X) = \beta_0 X^{\beta_1}$$

If one disregards the error term,

$$Y = \beta_0 X^{\beta_1} \tag{15.29}$$

In logarithmic form, this expression becomes

$$\log Y = \log \beta_0 + \beta_1 \log X \tag{15.30}$$

which is linear. However, Equation 15.25

$$Y = \beta_0 + \beta_1^X$$

cannot be so transformed.

In some instances it is possible to restrict the study to a small part of a nonlinear relationship and to use a linear function to describe the relationship within that narrow range. For example, the relationship between the size of a population of plants or animals and its age is sigmoid, as is shown in Figure 15.8, but the segment between $x_A$ and $x_B$ is almost straight and can be approximated using the simple linear function, $Y = \beta_0 + \beta_1 X$. One must realize, however, that any statements made regarding the relationship expressed by $Y = \beta_0 + \beta_1 X$ should be confined to the region $x_A$ to $x_B$. Extrapolating in either direction beyond these limits can result in serious errors. This cannot be too strongly emphasized. Many prediction operations, such as stand table projection, estimation of future market trends, etc., are based on extrapolation of regressions. If these regressions are improperly modeled, the predictions will be in error. Consequently, if one intends to extrapolate the regression being developed, the model used should portray the situation accurately. Predictions based on short segments of the values of the independent variable are not to be trusted.

If a suitable linear regression function cannot be found that will describe the relationship with suitable accuracy, it will be necessary to select an appropriate nonlinear function. However, when this is done, the least squares–fitting process cannot be used. It will be necessary to use much more complicated and costly iterative methods which close in on the least squares fit through a series of approximations. There is no way of knowing what the true minimum sum of squared residuals from the regression should be, so the interactive approximations are continued until the sum of squared residuals is "satisfactorily" small. This maximum-allowable sum of squares is set arbitrarily by the investigator. Computer algorithms are available for such nonlinear fits but it must be realized that they will not yield true least squares fits. A detailed discussion of the fitting of nonlinear models is beyond the scope of this book. For a discussion, one should refer to a text such as that by Draper and Smith (1981).

### 15.3.4 Developing a Model

To do an effective job of model development one must be able to visualize the curve forms associated with algebraic expressions. Figure 15.9 is intended to provide some insight into this matter. Additional curve forms can be found in Figures 11.5, 11.6, 11.7, 11.8A, and 11.8B, which illustrate the gamma, exponential, beta, and Weibull probability distributions. An excellent source of information is Jensen's (1964) paper describing algebraic forms in space. Still others may be found in standard statistical texts such as Steel and Torrie (1960), Snedecor and Cochran (1967), or Draper and Smith (1981).

It is convenient to follow a formalized series of steps when developing a model. For example,

1. Specify the response variable and the independent variable(*s*).
2. Set up an appropriate coordinate system and label the axes.
3. Indicate the *Y*-intercept if the nature of the problem requires a specific intercept.
4. Sketch the general curve on the coordinate system basing the curve shape on what is known about the relationship. In the case of a two-independent-variable problem this might require sketches of how cross sections of the surface should look. The curve or surface should be sketched so that it passes through the specified *y*-intercept.

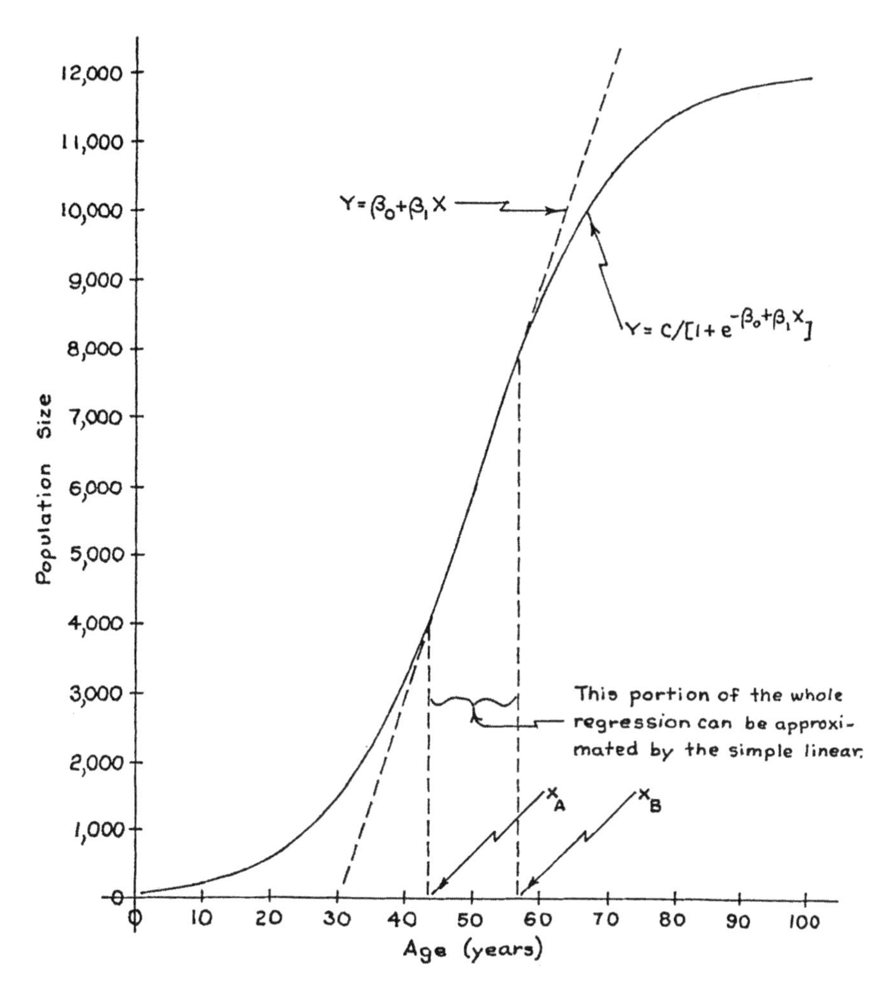


**FIGURE 15.8** Using a straight line to model a short segment of a curvilinear relationship.

5. Select one *or more* algebraic expressions that could yield curves similar to those sketched. These should be ranked according to how their inherent characteristics relate to the problem requirements and constraints. *The algebraic expression chosen is the regression function in the model.*

As an example of this process, consider the relationship between inside and outside bark diameter at breast height. The relationship is used in stand table projection procedures and elsewhere and thus this is not a trivial exercise. In stand table projection the growth prediction is based on radial growth *of the wood* as seen in increment cores. In essence, in this process the bark is disregarded until after the growth projection has been made. Then the bark thickness has to be added to the predicted inside bark diameter. Consequently, the variable being predicted, $Y$, is *outside bark* dbh and the predicting variable $X$, is *inside bark* dbh.

The relationship can best be shown in the context of a rectangular coordinate system with $Y$ on the vertical axis and $X$ on the horizontal axis.

When the inside bark diameter is equal to zero, the outside bark diameter is, practically speaking, equal to zero.

If the bark has no thickness, $Y$ would equal $X$ and the regression of dbhob on dbhib would have a slope of 1 and would pass through the origin of the coordinate system. This is shown on

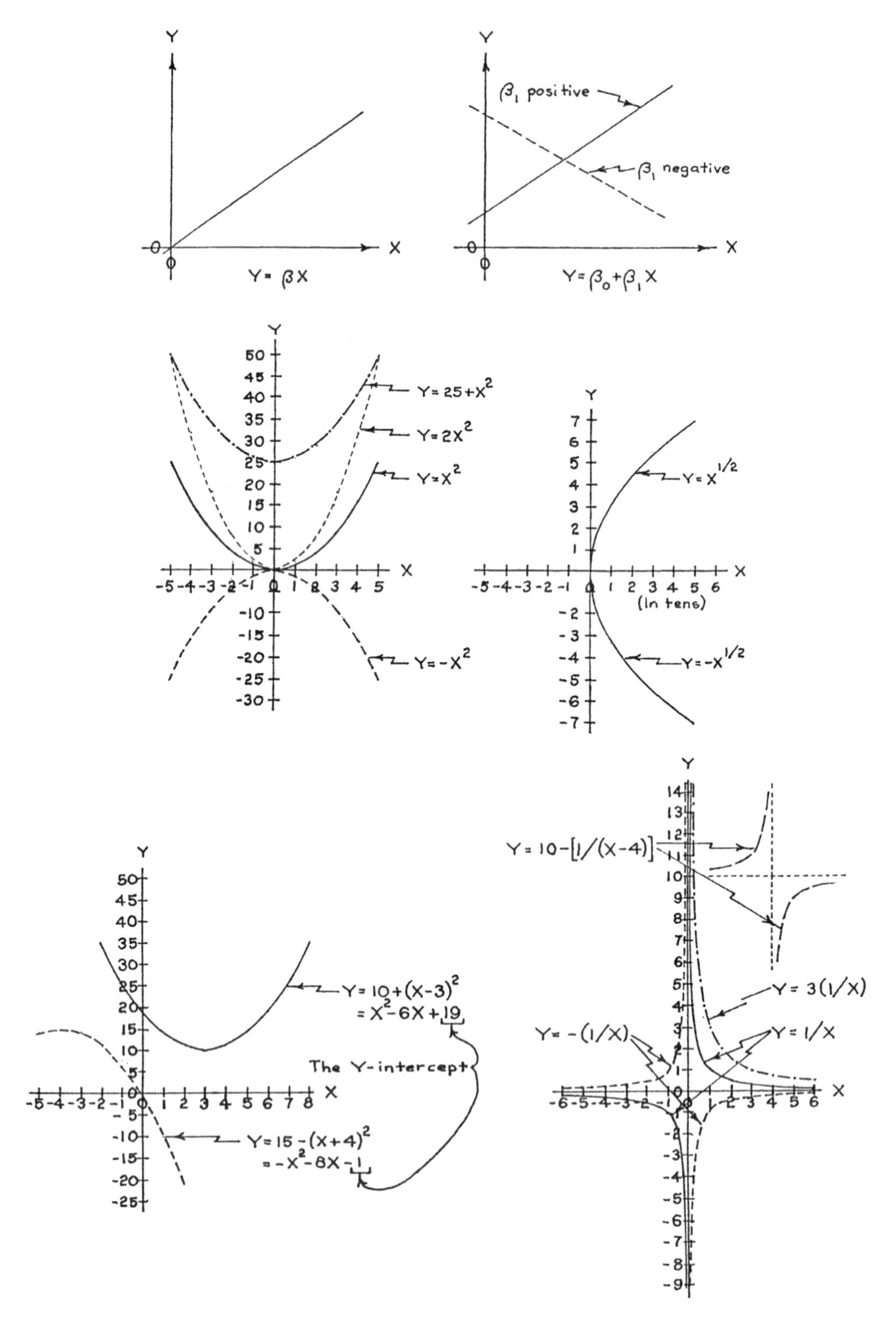


FIGURE 15.9 Examples of curve forms.

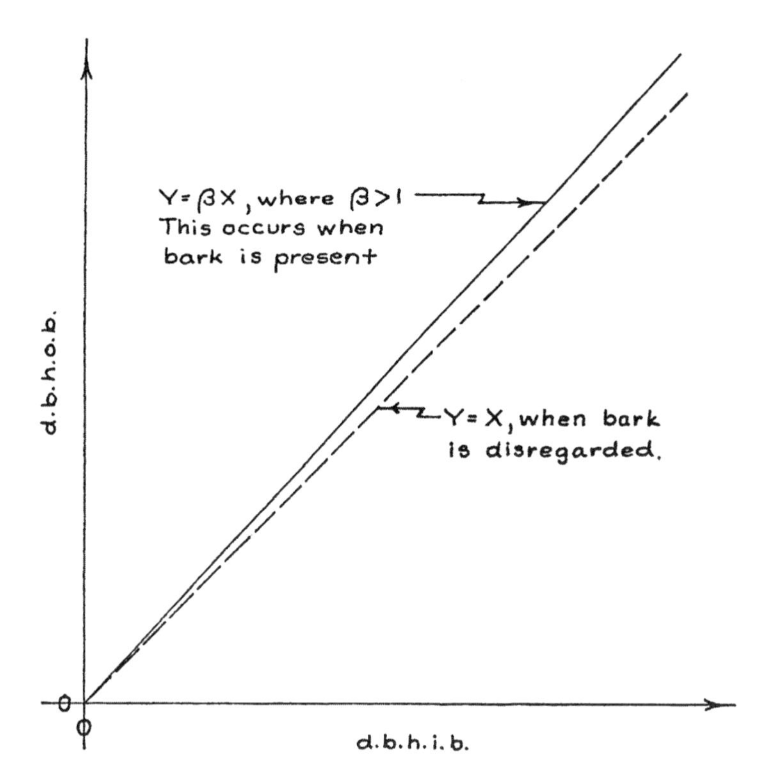


**FIGURE 15.10** Sketch of relationship between dbhob and dbhib.

Figure 15.10. When bark is added to the dbhib, the regression is forced to lie above the $Y = X$ line but still must pass through the origin. The regression function would then be

$$Y = \beta X; \quad \beta > 1 \tag{15.31}$$

The above model does not consider the sloughing of the bark as the tree increases in size. If sloughing is to be included as a factor, it would require that the slope of the line be somewhat reduced. If this loss of diameter is considered to be a constant proportion of the diameter, the sloughing would reduce the magnitude of β but the regression line would remain straight. If, however, it is felt that the rate of sloughing increases with tree size, the portion of bark thickness to inside bark diameter would no longer be constant and an extra term would have to be introduced into the model to take this into account. A possible solution might be

$$Y = \beta_1 X + \beta_2 X^{\beta_3}; \quad 0 < \beta_3 < 1 \tag{15.32}$$

This, of course, is a nonlinear function.

Another example of modeling might involve the relationship of site index to aspect while holding slope within certain limits (e.g., 20 to 30%). This problem could arise when developing procedures for evaluating site quality using aerial photographic evidence. The response variable is site index, while the independent or predicting variable is aspect, $A$.

Again, a rectangular coordinate system can be used. However, it also would be possible to use a polar coordinate system. This section considers only the case where the rectangular system is used.

There is no reason to specify a $Y$-intercept. Consequently, its value will have to be determined from the data.

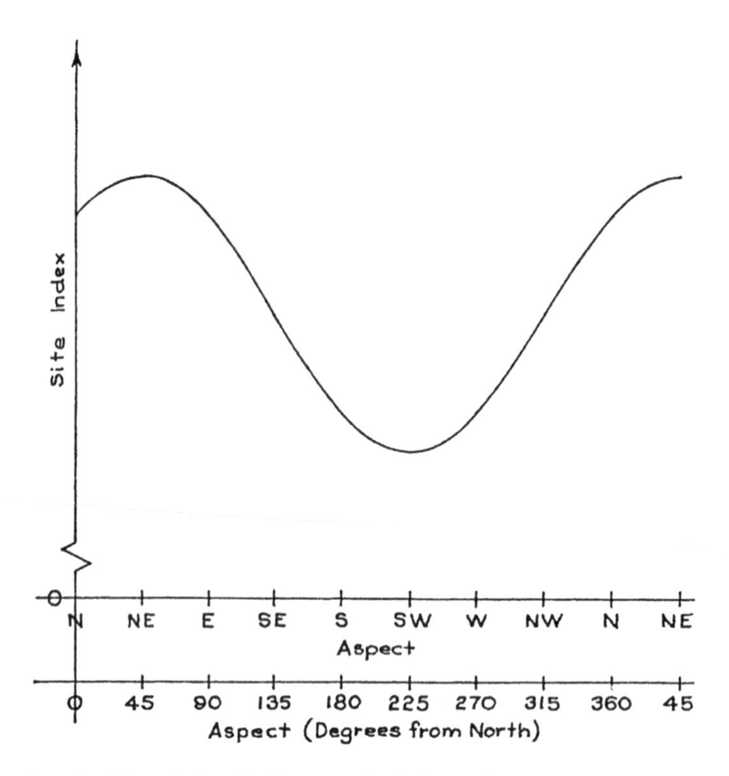


**FIGURE 15.11** Sketch of the relationship between site index and aspect.

It is known that, in the Northern Hemisphere, the coolest and dampest sites are to be found on northeastern-facing slopes and the warmest and driest on southwestern-facing slopes. The better sites are associated with coolness and dampness and the poorer sites with heat and dryness. The model chosen will have to recognize these truths. The regression, therefore, should take on a shape as shown in Figure 15.11. The sketch has the appearance of a sine or cosine wave. The peak of the curve should be when the azimuth of the slope is equal to 45° and the lowest point when the azimuth equals 225°. If one makes use of the cosine wave, the regression function takes on the form:

$$Y = \beta_0 + \beta_1 \cos(A - 45) \tag{15.33}$$

If the sine wave is used, the function is

$$Y = \beta_0 + \beta_1 \sin(A + 45) \tag{15.34}$$

As can be seen, the regression is not a straight line, but the model is that of a simple linear function where $U = \cos(A - 45)$ or $\sin(A + 45)$.

If one wishes to introduce slope into the model, it can be assumed that the effect would be additive. From much evidence it is known that site quality decreases as slope increases. The rate of falloff increases as the slope steepens. This follows the pattern of a downward-opening parabola (see Figure 15.9) so that

$$Y = \beta_0 + \beta_1 S + \beta_2 S^2 \tag{15.35}$$

where $S$ = slope in percent (or degrees).

The fitting process will assign the appropriate signs to the regression coefficients.

When a second-order polynomial is used in this manner, it often produces a result, for a certain range of values of the independent variable, which is not appropriate. In this case, it would show site index equaling zero at some degree of slope. To eliminate this problem, it should be stated that the regression is to be used *only within a specified range of slopes.*

Incorporating Equation 15.35 into the model represented by Equation 15.33 yields the function

$$Y = \beta_0 + \beta_1 \cos(A - 45) + \beta_2 S + \beta_2 S^2; \quad s_A \leq S \leq s_B \tag{15.36}$$

where $s_A$ and $s_B$ = appropriate slope limits.

Since there is now more than one independent variable in the expression, it is appropriate to include the interaction terms $S[\cos(A - 45)]$ and $S^2[\cos(A - 45)]$, making the function

$$\begin{aligned} Y &= \beta_0 + \beta_1 \cos(A-45) + \beta_2 S + \beta_3 S^2 + \beta_4 S[\cos(A-45)] \\ &\quad + \beta_5 S^2[\cos(A-45)]; \quad s_A \leq S \leq s_B \end{aligned} \tag{15.37}$$

This is a linear function which can be fitted using the standard least squares procedure.

Finally, consider the development of a tree volume equation. Furthermore, for simplicity, this discussion will be restricted to total bole volume above stump height. From basic mensurational principles it can be deduced that the volume of a tree is a function of its cross-sectional area, height, and form. The basic cubic volume function is then

$$Y = BHF \tag{15.38}$$

where $Y$ = volume of tree in cubic units
$B$ = cross-sectional area of tree; conventionally this is at breast height so $B$ represents basal area
$H$ = total height of tree above some reference; in this case it is above stump height
$F$ = breast-high form factor (the ratio of the actual volume of a right cylinder with an end area equal to the basal area, $B$, and a length equal to the height of the tree, $H$)

This, of course, is not a linear function. However, form, as defined here, cannot be evaluated unless the volume is known. Consequently, it must be deleted from the function and its effect assigned to a regression coefficient, $\beta$, whose magnitude will be determined empirically from the data. In essence, this regression coefficient will be the equivalent of the average breast-high form factor for the set of trees included in the sample. Usually these data are for a given species or species group growing in a specific region and, consequently, the final volume equation will be appropriate only for that species or species group in that region. The volume function then becomes

$$Y = \beta BH \tag{15.39}$$

The basal area, $B$, is a function of the dbh, $D$, of the tree:

$$B = \pi(D/2)^2 = (\pi/4)D^2 \tag{15.40}$$

The term $(\pi/4)$ is a constant and can be merged with the regression coefficient $\beta$ to form a new regression coefficient, $\beta_1$:

$$\beta_1 = (\pi/4)\beta \tag{15.41}$$

The volume function then becomes

$$Y = \beta_1 D^2 H \tag{15.42}$$

When $H$ is less than breast height, there is some cubic volume even though $D$ is equal to zero. Consequently, there should be a positive $Y$-intercept. Its magnitude cannot be deduced. The function volume would then be

$$Y = \beta_0 + \beta_1 D^2 H \tag{15.43}$$

The magnitude of $\beta_0$ is minuscule and probably would be ignored, leaving the regression function in the form of Equation 15.42.

It must be remembered that there is only one independent variable in these models, $D^2H$. The horizontal axis of a graph of the relationship would be *graduated in terms of* $D^2H$. Thus, two very different-appearing trees, one tall and slender and the other short and stout, could have the same volume.

When merchantable, rather than total, volume is the variable of interest, the problem becomes more complex. The usual approach is to continue to use total height in the independent variable, $D^2H$, and to use Equation 15.43, containing the $Y$-intercept, $\beta_0$, as the regression function. When there is no merchantable volume in a tree until $D \geq d$, where $d$ is the minimum acceptable dbh, consequently, the regression line crosses the horizontal axis some distance to the right of the origin and the vertical axis below the origin (see Figure 15.12A).

The problem with this model is that when the dbh of a tree is an infinitesimal below the minimum acceptable dbh, there is *no* merchantable volume but *at* the minimum dbh there is, relatively speaking, a substantial amount of merchantable volume. This sudden jump in volume is not incorporated into the model. This problem can be alleviated by incorporating the merchantability constraint into the model.

Since there is no merchantable volume when $D < d$, $D^2H$ must equal zero when $D < d$. Then, using Equation 15.42,

$$Y = \beta_1 (D - d)^2 H \tag{15.44}$$

When $D = d$, the above function indicates that no merchantable volume is present. However, some *is* present. This should be included in the model in the form of the $Y$-intercept, $\beta_0$. Then the function becomes

$$Y = \beta_0 + \beta_1 (D - d)^2 H \tag{15.45}$$

A complicating factor is that the volume $\beta_0$ is not constant because $H$ is free to vary. In Equation 15.44, when $D = d$, the effect of $H$ on $\beta_0$ is linear:

$$\beta_0 = \beta_2 + \beta_3 H \tag{15.46}$$

Incorporating this into the regression function yields (see Figure 15.12B)

$$Y = \beta_2 + \beta_3 H + \beta_1 (D - d)^2 H \tag{15.47}$$

If a minimum total height is specified, it is possible for a tree to be classed as not merchantable, even if $D > d$, because it is too short. In such a case the regression function would have to be modified to

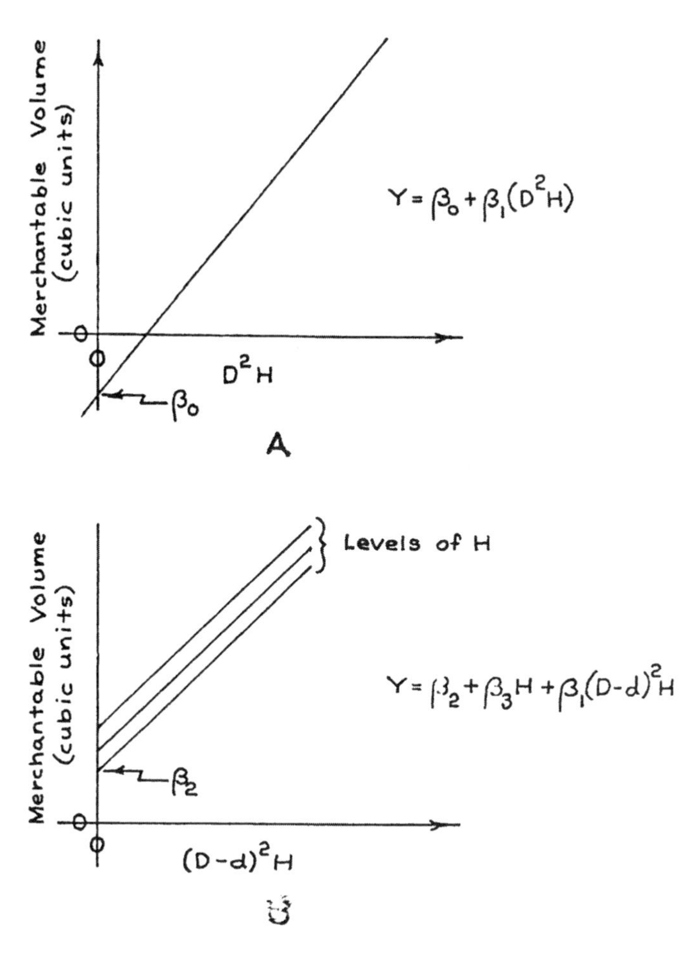


**FIGURE 15.12** Graphical representations of two possible models for a merchantable volume equation.

$$Y = \beta_2 + \beta_3 H + \beta_1 (D - d)^2 (H - h) \tag{15.48}$$

where $h$ = minimum acceptable height.

The problem would be further complicated if merchantable height, rather than total height, is to be used because merchantable height, as usually expressed in terms of logs or bolts, is not a continuous variable. One can consider developing a separate volume equation for each log or bolt length or the discrete nature of merchantable height can be ignored and Equation 15.48 used as the regression function.

The terms $D^2H$, $(D - d)^2H$, and $(D - d)^2(H - h)$ are interaction terms which themselves are random variables. They can be defined as

$$Z = D^2H, \quad (D - d)^2\text{H}, \quad \text{or} \quad (D - d)^2(H - h)$$

Equations 15.42 and 15.44 become

$$Y = \beta Z$$

Equations 15.43 and 15.45 become

$$Y = \beta_0 + \beta_1 Z$$

and Equations 15.47 and 15.48 become

$$Y = \beta_0 + \beta_1 H + \beta_2 Z$$

These functions are all linear and can be fitted using standard least squares procedures. Equations 15.42 and 15.43 are in common use. $Y = \beta D^2 H$ is referred to as a *constant form factor volume equation* and $Y = \beta_0 + \beta_1 D^2 H$ as a *combined variable volume equation*.

### 15.3.5 Logarithmic Transformations of Nonlinear Models

The basic tree volume function, Equation 15.39, used as an example in Section 15.3.4, is nonlinear:

$$Y = \beta BH$$

If it can be assumed that the volume does not vary exactly as the square of the dbh, $D$, and directly as the height, Equation 15.42 can be rewritten:

$$Y = \beta_0 D^{\beta_1} H^{\beta_2} \tag{15.49}$$

This also is a nonlinear function. However, as is described in Section 15.3.3, it can be transformed into a linear function using logarithms:

$$\log Y = \log \beta_0 + \beta_1 \log D + \beta_2 \log H \tag{15.50}$$

This function can be fitted using least squares.

Situations where logarithmic transformations of this type can be made are not uncommon. However, one must realize that when a logarithmic transformation is made and the transformed model is fitted to the data, the fit is of the transformed model and the statistics generated apply to the transformed model. If one were to convert the fitted model back to nonlogarithmic form, the statistics would not be applicable. On the average values of $\hat{Y}$ would be low; i.e., they would be negatively biased. The bias arises from the fact that regression lines or surfaces fitted using least squares pass through the point in the coordinate system defined by the arithmetic means of the variables. In the case of logarithmic transformations, the variables are logarithms of the original data values. When these arithmetic means are converted to their equivalent antilogs, these antilogs are geometric means of the original data. Section 5.6 pointed out that geometric means are always less than their corresponding arithmetic means. Thus, the antilogs are negatively biased. In turn, this bias causes the variance about the regression, based on the antilogs, to be positively biased. Corrections for this transformation bias have been devised, but a discussion of them is beyond the scope of this book. Reference is made to Baskerville (1972), Beauchamp and Olson (1973), Flewelling and Pienaar (1981), Meyer (1941), and Zar (1968). Meyer's correction was designed specifically for the tree volume function, Equation 15.50. It assumes that common logarithms were used in the transformation and that the logarithmic data are normally distributed about the regression. The correction term, which is to be applied to the antilogs, is then

$$10^{1.51 s^2} \tag{15.51}$$

where $s^2$ is the variance about the logarithmic regression.

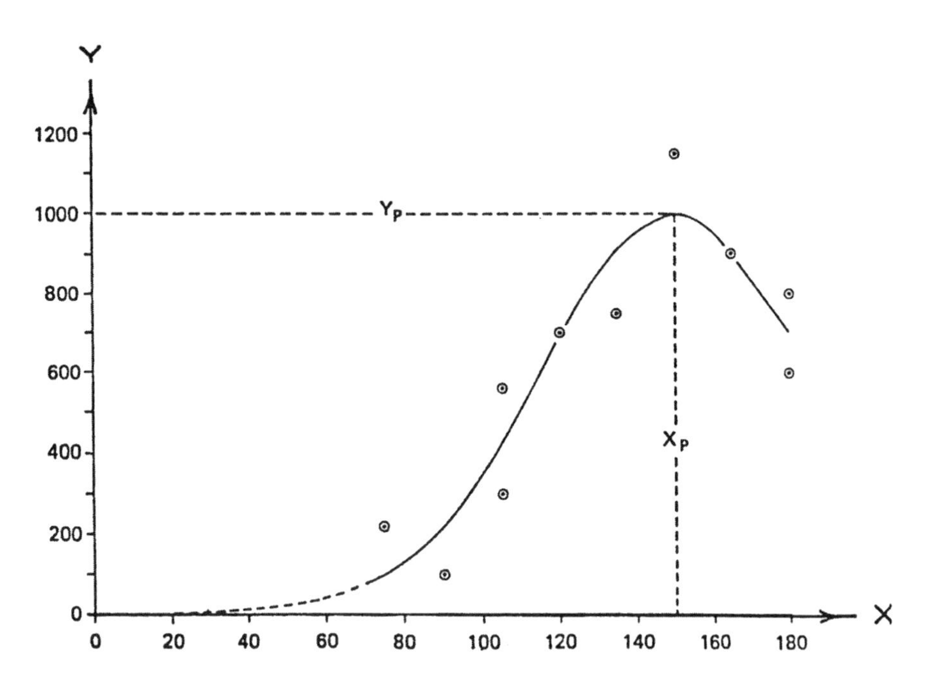


**FIGURE 15.13** The initial scatter diagram and the manually fitted curve. Note that $(X_p,Y_p)$ locates the extreme end of the sigmoid portion of the curve. (From Jensen, C. E. and Homeyer, J. W., USDA Forest Service Intermountain Forest and Range Experiment Station, 1970.)

The corrected prediction expression is then

$$\hat{Y} = 10^{\hat{Y}} * 10^{1.1513s^2} = 10^{b_0+b_1 \log D+b_2 \log H} \tag{15.52}$$

### 15.3.6 The MATCHACURVE Procedure

Certain nonlinear functions produce curve forms which are very useful to investigators making use of regression analysis. Among these are exponential functions that produce sigmoid and bell-shaped curves or hyperbolic J and reversed-J curves. Since these curves are nonlinear, they cannot be fitted using conventional least squares procedures. However, Jensen and Homeyer (1970; 1971) and Jensen (1973; 1976; 1979) have developed a procedure, known as MATCHACURVE, which makes these functions accessible.

In Section 15.3.3 it was brought out that, through the process of anamorphosis, curved regression lines can be made straight by transforming or modifying the graduations along the $X$-axis. In this process the variable $X$ was transformed into a new independent variable, $U$, and the horizontal axis was graduated in terms of $U$ producing a straight regression line (see Figure 15.6). This concept lies at the heart of the MATCHACURVE process. The MATCHACURVE process identifies the appropriate function which then is fitted to the data using the method of least squares.

This discussion follows the example used by Jensen and Homeyer (1970) in their initial paper. In this paper the transforms described sigmoid and bell-shaped curves.

1. A scatter diagram of the original two-dimensional sample data is constructed and a manually balanced regression line is drawn through the scatter, as in Figure 15.13.
2. The values of $X$ and $Y$ at the peak of the curve are determined from the plot. These are $X_p$ and $Y_p$.

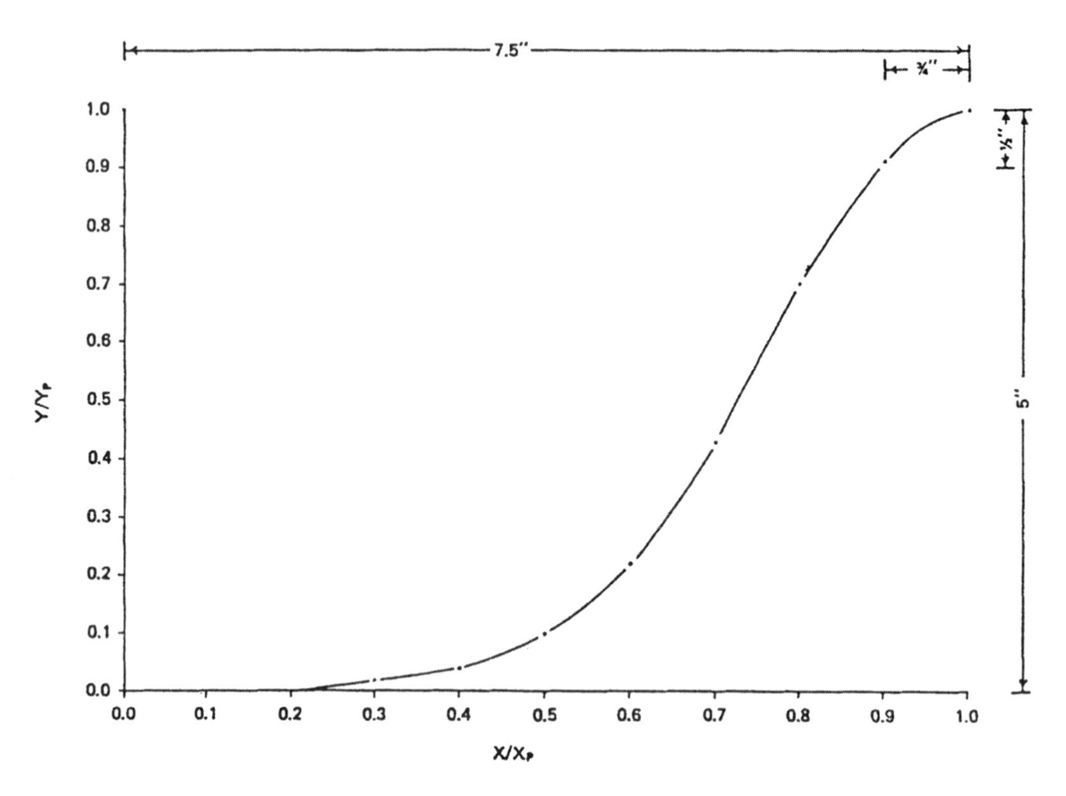


**FIGURE 15.14** The manually fitted curve has been put into standard form. (From Jensen, C. E. and Homeyer, J. W., USDA Forest Service Intermountain Forest and Range Experiment Station, 1970.)

3. The scale along the $X$-axis from 0 to $X_p$ is divided into ten equal parts and the $X$ value at each of division points is determined.
4. At each of these 11 $X$ values, the corresponding $Y$ value is read from the curve.
5. $X/X_p$ and $Y/Y_p$ are then computed using the values obtained in steps 3 and 4. This converts the raw values to proportions of their respective maxima.
6. Using these proportional values a plot as in Figure 15.14 is constructed. The dimensions of the coordinate system should be 7½ in. in width and 5 in. in height. The graduations along the horizontal axis must be ¾ in. while those along the vertical axis must be ½ in. The reason for these specific dimensions will become clear in step 8.
7. Draw a smooth curve through the plotted points. This is *not* a curve fitting process. The curve should cross each of the plotted points.
8. Lay this curve over the set of prepared (standard) curves (e.g., Figure 15.15) to identify which of the standard curves most closely matches the plotted curve. Jensen and Homeyer (1970) provided ten sets of standard curves, each consisting of nine curves. These curves were based on the function

$$Y / Y_p = \frac{e^{-T} - e^{-T_0}}{1 - e^{-T_0}} \tag{15.53}$$

where $e$ = base of natural logarithms

$$T = \left| \frac{(X / X_p) - 1}{(X_I / X_p) - 1} \right|^n$$

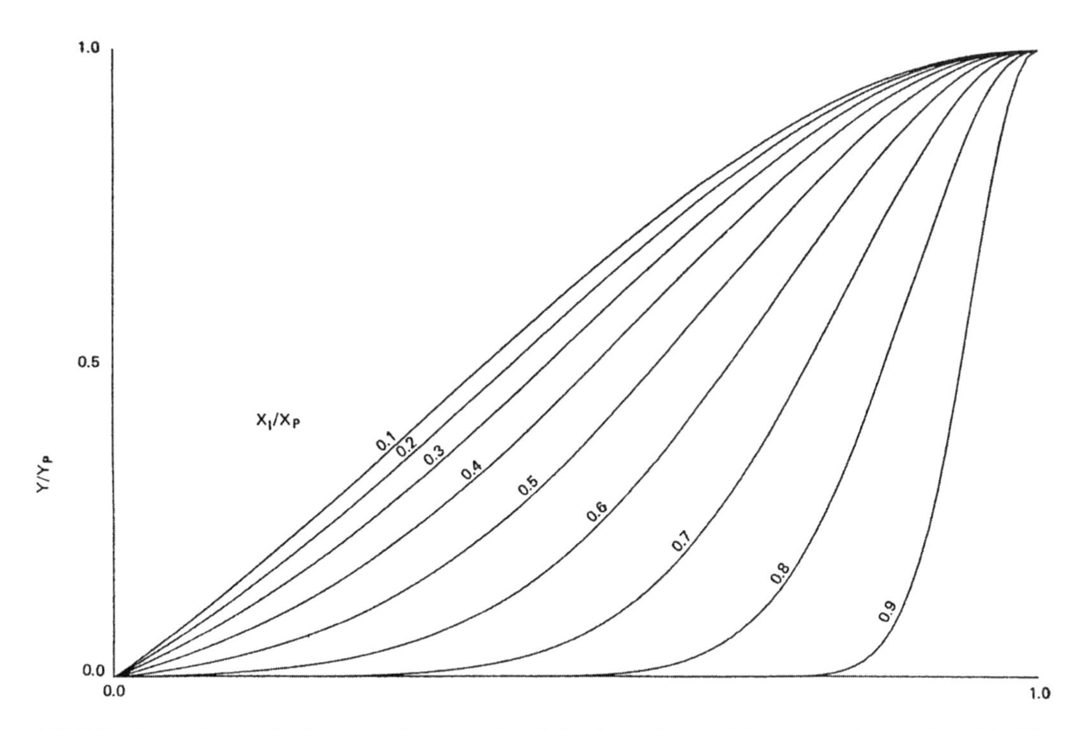


**FIGURE 15.15** Standards for $n = 2.0$ and scaled inflection points $X_1/X_p$, ranging from 1.0 to 0.9. The dimensions in both $X/X_p$ and $Y/Y_p$ in Figure 15.13 must match those in this figure. (From Jensen, C. E. and Homeyer, J. W., USDA Forest Service Intermountain Forest and Range Experiment Station, 1970.)

$T_0 = T$, evaluated at $X = 0$; then $(X/X_p) = 0$
$X/X_p$ = scaled $X$ values, 0.0 to 1.0
$X_I/X_p$ = scaled inflection point when $0 < X_I/X_p < 1$

The parameters giving rise to the several curves are the exponent $n$ (with $n$ assuming the values 1.5, 2.0, 3.0, 4.0, 5.0, 6.0, 7.0, 8.0, 9.0, and 10.0) and the relative point of inflection $X_I/X_p$ (ranging from 0.0 to 0.9 in $^1/_{10}$ increments). Figure 15.15 shows the set when $n = 2.0$

9. The plotted curve falls between the $X_I/X_p = 0.6$ and 0.7 curves as shown in Figure 15.16. Using straight line interpolation between the curves, $X_I/X_p$ is found to be 0.67. Thus, in this manner the parameters defining the curve shape have been determined.
10. Then,

$$T = \left|\frac{(X/150)-1}{(0.67)-1}\right|^{2.0}$$

and

$$Y/Y_p = \frac{e^{-\left|\frac{(X/150)-1}{0.33}\right|^{2.0}} - e^{-\left|\frac{0-1}{0.33}\right|^{2.0}}}{1-e^{-\left|\frac{0-1}{0.33}\right|^{2.0}}}$$

Since $e^{-9.18} < 0.0002$, the expression can be simplified to

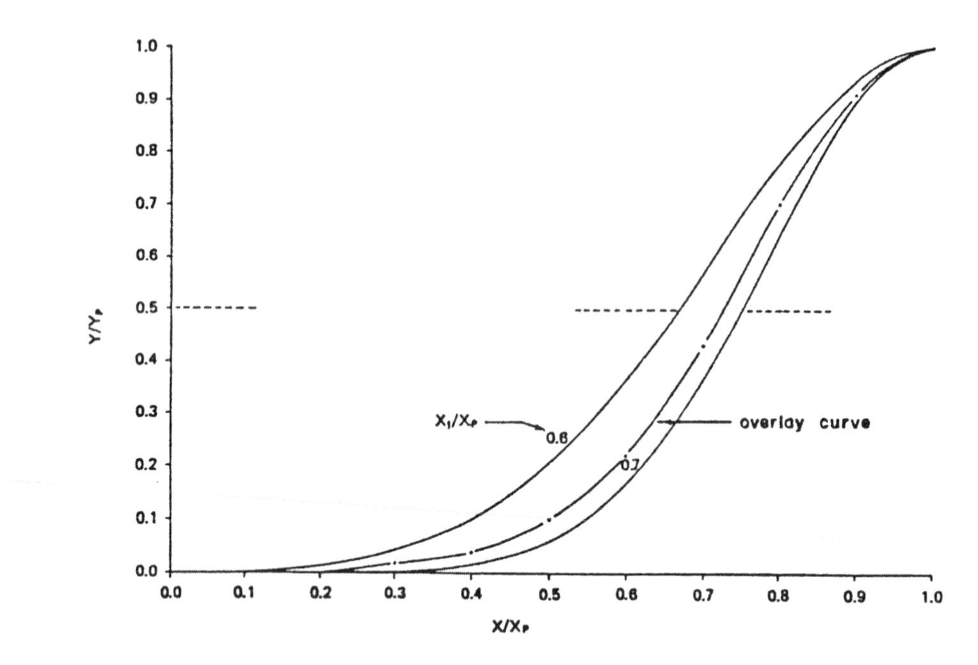


**FIGURE 15.16** The standards best matched to the original curve bracket the overlay curve. (From Jensen, C. E. and Homeyer, J. W., USDA Forest Service Intermountain Forest and Range Experiment Station, 1970.)

$$Y/Y_p = e^{-|(X/150)-1|2.0/0.1089}$$

Since $Y/Y_p$ is only a scaling change for $Y$, $Y$ can be substituted for $Y/Y_p$ producing the model that is to be fitted using least squares,

$$Y = \beta_0 + \beta_1 U$$

where $U = e^{-|(X/150)-1|^{2.0}/0.1089}$.

In this case the final regression equation is

$$\hat{Y} = -2.05 + 998.7e^{-|(X/150)-1|^{2.0}/0.1089}$$

The plot of this expression is shown in Figure 15.17.

The process described in Jensen and Homeyer (1970; 1971) is limited to two-dimensional models. Later Jensen (1973; 1976; 1979) expanded the concept to include multidimensional and asymmetric models. In Jensen (1979) the improved sensitivity to multiple variable interactions when using MATCHACURVE, as compared with that obtained using conventional least squares fitting, was brought out. This echoes the statement in Section 15.3.2 that subtleties can be brought out in manual fitting that are difficult to detect using conventional mathematical fitting procedures. If one has access to Jensen and Homeyer's publications, their procedures can open up the use of a number of models which probably would otherwise be ignored because of the difficulties of nonlinear curve fitting.

### 15.3.7 Least Squares Fitting

The concept of least squares was explored in Section 5.3.4 where it was found that the arithmetic mean occupies the least squares position within the data set where the data are plotted along a

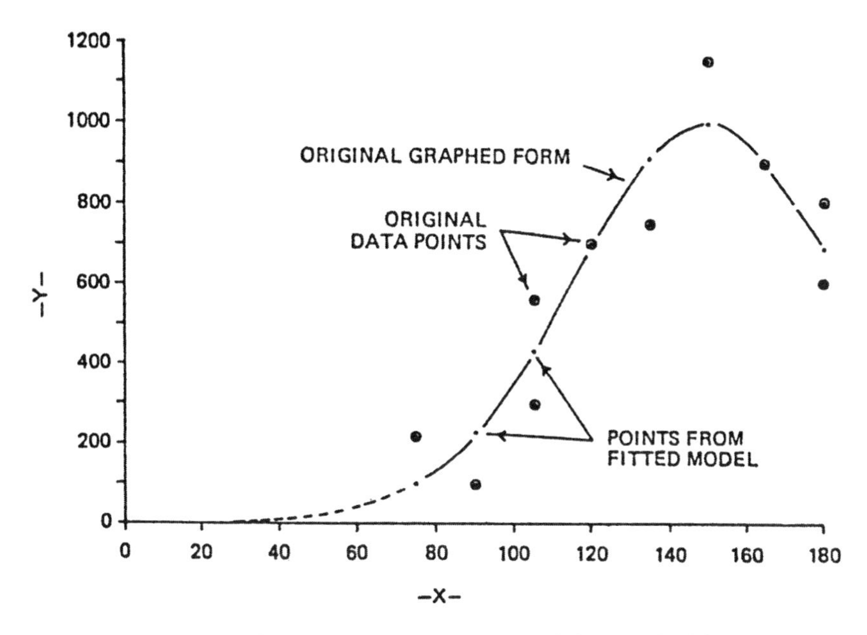


**FIGURE 15.17** Points from the fitted model plotted over the original graphed form. (From Jensen, C. E. and Homeyer, J. W., USDA Forest Service Intermountain Forest and Range Experiment Station, 1970.)

number line (i.e., within a one-dimensional space). It also was brought out in Section 13.6 that, if the sample had been drawn from a normally distributed population, the arithmetic mean of the sample would be a maximum likelihood estimate of the population mean. Extending the concept to regressions, one finds that when a linear function has been fitted into the scatter, in multidimensional space, of the data from the sample, using the least squares process, the resulting regression line or surface occupies the least squares position *for the chosen linear function.* However, as an estimate of the true relationship between the variables, it is ambiguous. If the linear function being fitted differs appreciably from the true regression function, the results of the fit are meaningless insofar as estimation of the true relationship is concerned. It must be realized at the outset that *any* linear function can be fitted, using the least squares process, to any set of data regardless of the true population relationships. If the function chosen to be the model is appropriate, the difference between the estimated and true relationship will be due only to random chance and, if the sampling were to be continued indefinitely, the means of the sampling distributions of the coefficients would approach, in the limit, the true coefficients.

Statistical tests can be made of the validity of the regression and its component coefficients provided that the following conditions exist.

1. The sample was actually drawn from the population of interest. This implies that the variables to be considered have been clearly defined as has the sampling frame from which the sampling units were drawn.
2. The measurements of the variables are free of error. Measurement errors may lead to biases and also contribute unknown and possibly important components to the variance of the response variable about the regression line or surface. Insofar as the response variable is concerned, errors of measurement can be tolerated if their expected values are zero for all values of the independent variables.

   The requirement that the measurements be without error leads to a fundamental assumption about regression analyses. It is assumed that the values of the *independent* variables appearing in the sample are *fixed* and are not random variables. In other words, if the sampling were repeated so as to obtain a second, third, or other set of data, the only values of the independent variables appearing in these samples would be those occurring in the initial sample. As a matter of fact, those values of the independent variables can be specified. If the values of the independent variables are fixed, they contribute nothing to the variance of the response variable about the regression. Valid test results can be obtained only if this assumption is correct.

3. The values of the response variable appearing in the sample are independent of one another. In other words the sampling is truly random so that the drawing of a given value of the response variable has no effect on the probability of drawing any other value in a subsequent draw.
4. The conditional variances are homogeneous. If meaningful tests are to be made, the variance of the response variable must be the same for all values of the independent variables. If such homogeneity is not present, it may be possible to make use of weighted least squares (Section 15.3.10) to arrive at a testable estimate of the population relationships.
5. The subpopulations of values of the response variable are normally distributed. The central limit theorem is not applicable in the case of regression unless the observations are made up of means and, consequently, the population of values of the response variable itself must be normally distributed if conventional testing is to be used.
6. The sample size, $n$, is greater than $M$, the number of independent variables. This requirement arises from the fact that a degree of freedom is lost for each independent variable in the evaluation process.

### 15.3.8 Fitting a Two-Dimensional Model

The concept of a residual, within the context of a two-dimensional regression, is illustrated in Figure 15.18. Algebraically,

$$d_i = y_i - \hat{y}_i \tag{15.54}$$

Thus, in the case of the simple two-variable linear model,

$$\hat{Y} = b_0 + b_1 X$$

and

$$d_i = y_i - b_0 = b_1 x_i \tag{15.55}$$

The sum of squared residuals is

$$\sum_{i=1}^{n} d_i^2 = \sum_{i=1}^{n} (y_i - b_0 - b_1 x_i)^2 \tag{15.56}$$

It is this sum that must be minimized. The partial derivative with respect to $b_0$ is

$$\frac{\partial \sum_{i=1}^{n} (y_i - b_0 - b_1 x_i)^2}{\partial b_0} = 2 \sum_{i=1}^{n} (b_0 + b_1 x_i - y_i)$$

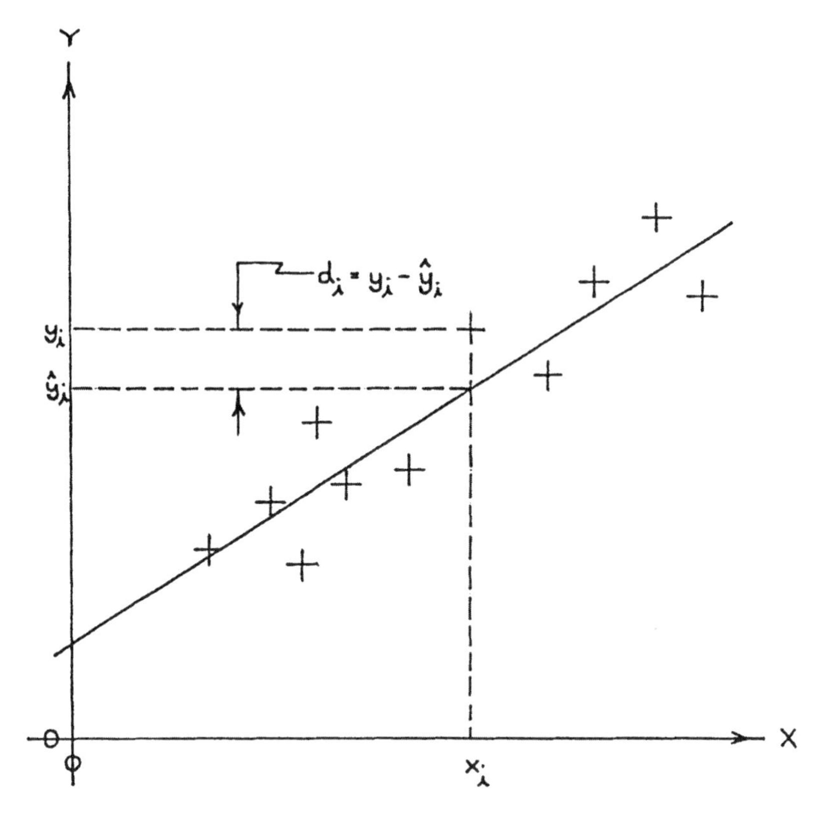


**FIGURE 15.18** Graphic representative of the residual $d_i$ within the context of a two-variable linear regression.

and with respect to $b_1$

$$\frac{\partial \sum_{i=1}^{n}(y_i - b_0 - b_1 x_i)^2}{\partial b_1} = 2\sum_{i=1}^{n}(b_0 x_i + b_1 x_i^2 - x_i y_i)$$

Setting these partial derivatives equal to zero

$$\left[\begin{array}{l} 2\sum_{i=1}^{n}(b_0 + b_1 x_i - y_i) = 0 \\ 2\sum_{i=1}^{n}(b_0 x_i + b_1 x_i^2 - x_i y_i) = 0 \end{array}\right.$$

yields

$$\left[ nb_0 + b_1 \sum_{i=1}^{n} x_i = \sum_{i=1}^{n} y_i \right. \tag{15.57}$$

$$\left[ b_0 \sum_{i=1}^{n} x_i + b_1 \sum_{i=1}^{n} x_i^2 = \sum_{i=1}^{n} x_i y_i \right. \tag{15.58}$$

Equations 15.57 and 15.58 are *normal* equations. The equations must be solved simultaneously to obtain the numerical values of the $b$ coefficients. In the case of two-dimensional models, this can be done algebraically. Thus, solving for $b_0$, using the first normal equation,

$$nb_0 + b_1 \sum_{i=1}^{n} x_i = \sum_{i=1}^{n} y_i$$

$$nb_0 = \sum_{i=1}^{n} y_i - b_1 \sum_{i=1}^{n} x_i \tag{15.59}$$

$$b_0 = \frac{\sum_{i=1}^{n} y_i}{n} - b_1 \frac{\sum_{i=1}^{n} x_i}{n} = \bar{y} - b_1 x$$

Solving for $b_1$, using the second normal equation,

$$b_0 \sum_{i=1}^{n} x_i + b_1 \sum_{i=1}^{n} x_i^2 = \sum_{i=1}^{n} x_i y_i$$

$$b_1 \sum_{i=1}^{n} x_i^2 = \sum_{i=1}^{n} x_i y_i - b_0 \sum_{i=1}^{n} x_i$$

Substituting for $b_0$ using Equation 15.59

$$b_1 \sum_{i=1}^{n} x_i^2 = \sum_{i=1}^{n} x_i y_i - \left( \frac{\sum_{i=1}^{n} y_i}{n} - b_1 \frac{\sum_{i=1}^{n} x_i}{n} \right) \sum_{i=1}^{n} x_i$$

$$= \sum_{i=1}^{n} x_i y_i - \frac{\sum_{i=1}^{n} x_i \sum_{i=1}^{n} y_i}{n} + b_1 \frac{\left( \sum_{i=1}^{n} x_i \right)^2}{n}$$

$$b_1 \sum_{i=1}^{n} x_i^2 - b_1 \frac{\left( \sum_{i=1}^{n} x_i \right)^2}{n} = \sum_{i=1}^{n} x_i y_i - \frac{\sum_{i=1}^{n} x_i \sum_{i=1}^{n} y_i}{n}$$

According to Equation 6.10,

$$\sum_{i=1}^{n} x_i^2 - \frac{\left(\sum_{i=1}^{n} x_i\right)^2}{n} = \sum_{i=1}^{n} (x_i - \bar{x})^2$$

In the same manner,

$$\sum_{i=1}^{n} x_i y_i - \frac{\sum_{i=1}^{n} x_i \sum_{i=1}^{n} y_i}{n} = \sum_{i=1}^{n} (x_i - \bar{x})(y_i - \bar{y}) \tag{15.60}$$

Thus,

$$b_1 \sum_{i=1}^{n} (x_i - \bar{x})^2 = \sum_{i=1}^{n} (x_i - \bar{x})(y_i - \bar{y})$$

and

$$b_1 = \sum_{i=1}^{n} (x_i - \bar{x})(y_i - \bar{y}) \Bigg/ \sum_{i=1}^{n} (x_i - \bar{x})^2 \tag{15.61}$$

Equation 15.61 can take on other forms. Consider that

$$\begin{aligned}(x_i - \bar{x})(y_i - \bar{y}) &= x_i y_i - y_i \bar{x} - x_i \bar{y} + \overline{xy} \\ &= x_i (y_i - \bar{y}) - \bar{x}(y_i - \bar{y})\end{aligned}$$

and

$$b_1 = \sum_{i=1}^{n} [x_i (y_i - \bar{y}) - \bar{x}(y_i - \bar{y})] \Bigg/ \sum_{i=1}^{n} (x_i - \bar{x})^2$$

However,

$$\sum_{i=1}^{n} (y_i - \bar{y}) = 0$$

Therefore,

$$\bar{x} \sum_{i=1}^{n} (y_i - \bar{y}) = 0$$

and

$$b_1 = \sum_{i=1}^{n} x_i(y_i - \bar{y}) \Big/ \sum_{i=1}^{n} (x_i - \bar{x})^2 \tag{15.62}$$

In the same manner,

$$b_1 = \sum_{i=1}^{n} y_i(x_i - \bar{x}) \Big/ \sum_{i=1}^{n} (x_i - \bar{x})^2 \tag{15.63}$$

As an example of this process, assume the following data:

| X | 3 | 5 | 9 | 11 | 15 | 18 | 21 |
|---|---|---|---|---|---|---|---|
| Y | 4 | 7 | 10 | 9 | 12 | 19 | 18 |

Then,

$$n = 7, \quad \sum_{i=1}^{7} x_i = 82, \quad \sum_{i=1}^{7} y_i = 79, \quad \sum_{i=1}^{7} x_i^2 = 1226, \quad \sum_{i=1}^{7} y_i^2 = 1075, \quad \text{and} \quad \sum_{i=1}^{7} x_i y_i = 1136$$

The normal equations are

$$\begin{bmatrix} 7b_0 + 82b_1 = 79 \\ 82b_0 + 1226b_1 = 1136 \end{bmatrix}$$

and

$$\bar{x} = 11.714286, \quad \bar{y} = 11.282514, \quad \sum_{i=1}^{n} (x_i - \bar{x})^2 = 265.42857$$

and

$$\sum_{i=1}^{n} (x_i - \bar{x})(y_i - \bar{y}) = 210.57143$$

Solving for the $b$ coefficients,

$$b_1 = 210.57143 / 265.42857 = 0.7933262$$

$$b_0 = 11.285714 - 0.7933262(11.714286) = 1.9924644$$

The regression equation is then

$$\hat{Y} = 1.9925 + 0.7933X$$

The plot of the data, the *scatter diagram*, and the regression line through those data are shown in Figure 15.19.

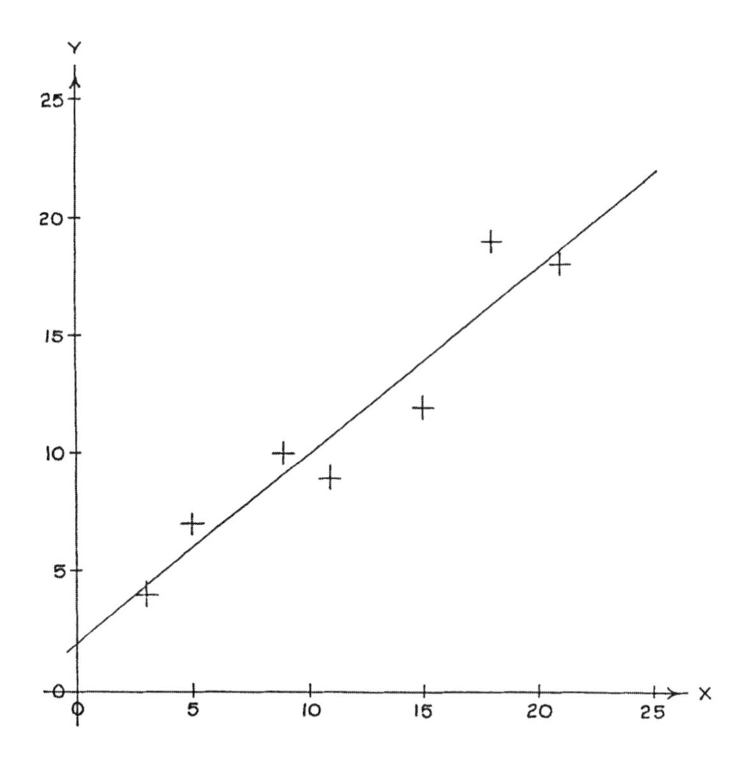


**FIGURE 15.19** The scatter diagram and the associated least squares regression line for the example in the text.

The determination of the numerical values of the $b$ coefficients can also be done using matrix algebra.* Let

$$E = \begin{bmatrix} e_{00} & e_{01} \\ e_{10} & e_{11} \end{bmatrix} = \begin{bmatrix} 7 & 82 \\ 82 & 1226 \end{bmatrix} \tag{15.64}$$

$E$ is then the square matrix of the sums making up the left-hand terms of the normal equations. Let

$$B = \begin{bmatrix} b_0 \\ b_1 \end{bmatrix} \tag{15.65}$$

$B$ is the column vector of the unknown $b$ coefficients. Finally, let

$$R = \begin{bmatrix} r_0 \\ r_1 \end{bmatrix} = \begin{bmatrix} 79 \\ 1136 \end{bmatrix} \tag{15.66}$$

is the column vector of the right-hand terms in the normal equations. Then

$$E * B = R \tag{15.67}$$

* Sufficient background to follow the discussion involving matrix algebra can be obtained from any college review text such as Ayres (1962).

Multiplying both sides of this equation by the inverse matrix of $E$:

$$E^{-1} * E * B = E^{-1}R \tag{15.68}$$

But $E^{-1} * E = I$, the identity matrix. Thus,

$$I * B = B = E^{-1}R$$

$$\begin{bmatrix} b_0 \\ b_1 \end{bmatrix} = \begin{bmatrix} c_{00} & C_{01} \\ c_{10} & C_{11} \end{bmatrix} * \begin{bmatrix} r_0 \\ r_1 \end{bmatrix} \tag{15.69}$$

Then,

$$b_0 = c_{00}r_0 + c_{01}r_1 \tag{15.70}$$

and

$$b_1 = c_{10}r_0 + c_{11}r_1 \tag{15.71}$$

The key to this process is $E^{-1}$, the inverse of $E$. The inversion can be carried out in more than one way. Perhaps the simplest to use when $E$ is a $2 \times 2$ matrix is as follows. The determinant of $E$ is

$$|E| = \begin{vmatrix} e_{00} & e_{01} \\ e_{10} & e_{11} \end{vmatrix} = e_{00}e_{11} - e_{10}e_{01} \tag{15.72}$$

and

$$E^{-1} = \begin{bmatrix} e_{11}/|E| & -e_{01}/|E| \\ -e_{10}/|E| & e_{00}/|E| \end{bmatrix} \tag{15.73}$$

In the case of our example,

$$|E| = \begin{vmatrix} 7 & 82 \\ 82 & 1226 \end{vmatrix} = 7(1226) - 82(82) = 1858$$

and

$$E^{-1} \begin{bmatrix} 1226/1858 & -82/1858 \\ -82/1858 & 7/1858 \end{bmatrix} = \begin{bmatrix} 0.6599493 & -0.0441335 \\ -0.0441335 & 0.0337675 \end{bmatrix}$$

Then,

$$b_0 = 0.6598493(79) + (-0.0441335)(1136) = 1.9924$$

and

$$b_1 = -0.0441335(79) + (0.0037675)(1336) = 0.7933$$

The differences between these values of the $b$ coefficients and those calculated using Equations 15.59 and 15.61 are due to rounding errors.

The elements of the inverse matrix, $E^{-1}$, are known as *c-multipliers*. Their utility extends beyond the computation of the $b$ coefficients, as will be brought out later.

It also should be pointed out that a matrix cannot be inverted if its determinant is equal to zero. When this is the case, the matrix is said to be *singular*. When this condition exists, it is also impossible to obtain a unique solution to the normal equations using the alternate methods of calculation. Thus, when this condition exists, it is impossible to arrive at a unique regression equation that meets the least squares criterion.

As can be seen from this discussion, least squares curve fitting is reasonably straightforward. Because of this, least squares procedures dominate regression fitting operations. It must be admitted, however, that the computations are tedious and time-consuming. Consequently, every effort should be made to enlist the aid of computers. Many regression fitting programs are available for use on either personal or mainframe computers. As a matter of fact, most scientific handheld calculators have the capability of fitting two-dimensional models.

### 15.3.9 Fitting a Multidimensional Model

Multidimensional models contain two or more independent or predicting variables. Thus, a regression function such as

$$Y = \beta_0 + \beta_1 X_1 + \beta_2 X_2 + \beta_3 X_3$$

is a multidimensional linear function with three independent variables. These variables may be unrelated as when $X_1$ = altitude above mean sea level, $X_2$ = azimuth from true north, and $X_3$ = degree of slope. On the other hand, they may be powers, reciprocals, logarithms, or other transformations of a single independent variable as when $X_1$ is some variable, $X_2 = X_1^2$, or $X_3 = X_1^3$, or $X_2 = \log X_1$ and $X_3 = 1/\log X_1$. Insofar as the fitting process is concerned, the nature of the variable is of no consequence.

The process of fitting a multidimensional model is simply an expansion of that used with a two-variable model. The function to be fitted is

$$Y = \beta_0 + \beta_1 X_1 + \beta_2 X_2 + \cdots + \beta_m X_m \tag{15.74}$$

A residual is

$$d_i = y_i - b_0 - b_1 x_{1i} - b_2 x_{2i} - \cdots - b_m x_{mi} \tag{15.75}$$

and the sum to be minimized is

$$\sum_{i=1}^{n} d_i^2 = \sum_{i=1}^{n} (y_i - b_0 - b_1 x_{1i} - b_2 x_{2i} - \cdots - b_m x_{mi})^2 \tag{15.76}$$

The partial derivatives of this expression with respect to the regression coefficient are

$$
\begin{aligned}
\frac{\partial \sum_{i=1}^{n} d_i^2}{\partial b_0} &= 2\sum_{i=1}^{n} (b_0 + b_1 x_{1i} + b_2 x_{2i} + \cdots + b_m x_{mi} - y_i) \\
\frac{\partial \sum_{i=1}^{n} d_i^2}{\partial b_1} &= 2\sum_{i=1}^{n} (b_0 x_{1i} + b_1 x_{1i}^2 + b_2 x_{1i} x_{2i} + \cdots + b_m x_{1i} x_{mi} - x_{1i} y_i) \\
\frac{\partial \sum_{i=1}^{n} d_i^2}{\partial b_2} &= 2\sum_{i=1}^{n} (b_0 x_{2i} + b_1 x_{1i} x_{2i} + b_2 x_{2i}^2 + \cdots + b_m x_{2i} x_{mi} - x_{2i} y_i) \qquad (15.77) \\
&\vdots \\
\frac{\partial \sum_{i=1}^{n} d_i^2}{\partial b_m} &= 2\sum_{i=1}^{n} (b_0 x_{mi} + b_1 x_{1i} x_{mi} + b_2 x_{2i} x_{mi} + \cdots + b_m x_{mi}^2 - x_{mi} y_i)
\end{aligned}
$$

Setting the partial derivatives equal to zero and simplifying yields the normal equations:

$$
\left[
\begin{array}{l}
nb_0 + b_1 \sum_{i=1}^{n} x_{1i} + b_2 \sum_{i=1}^{n} x_{2i} + \cdots + b_m \sum_{i=1}^{n} x_{mi} = \sum_{i=1}^{n} y_i \\
b_0 \sum_{i=1}^{n} x_{1i} + b_1 \sum_{i=1}^{n} x_{1i}^2 + b_2 \sum_{i=1}^{n} x_{1i} x_{2i} + \cdots + b_m \sum_{i=1}^{n} x_{1i} x_{mi} = \sum_{i=1}^{n} x_{1i} y_i \\
b_0 \sum_{i=1}^{n} x_{2i} + b_1 \sum_{i=1}^{n} x_{1i} x_{2i} + b_2 \sum_{i=1}^{n} x_{2i}^2 + \cdots + b_m \sum_{i=1}^{n} x_{2i} x_{mi} = \sum_{i=1}^{n} x_{2i} y_i \\
\vdots \\
b_m \sum_{i=1}^{n} x_{mi} + b_1 \sum_{i=1}^{n} x_{1i} x_{mi} + b_2 \sum_{i=1}^{n} x_{2i} x_{mi} + \cdots + \sum_{i=1}^{n} x_{mi}^2 = \sum_{i=1}^{n} x_{mi} y_i
\end{array}
\right. \qquad (15.78)
$$

Note the pattern of sums of squares and cross-products. This pattern is constant. Consequently, the normal equations can be generated without having to go through the partial derivative stage.

The normal equations must be solved simultaneously. Algebraic expressions for the $b$ coefficients cannot be generated as was done with the two variable regressions. Many ways of solving for the $b$ coefficients have been devised. All are laborious and time-consuming if a computer is not available. The sheer amount of time required for the computations did much to discourage the use of multidimensional regressions prior to the advent of the computer. Now the work can be done on computers of all sizes with readily available software. These software packages are all based on the use of matrix algebra. Consequently, the following is in terms of matrix algebra.

Returning to the normal equations. Substituting the symbols $e_{00}$, $e_{01}$, etc., for the sums of squares and cross-products in the left-hand members and $r_0$, $r_1$, etc., for the right-hand members yields:

$$\begin{bmatrix} b_0e_{00} + b_1e_{01} + b_2e_{02} + \cdots + b_me_{0m} = r_0 \\ b_0e_{10} + b_1e_{11} + b_2e_{12} + \cdots + b_me_{1m} = r_1 \\ b_0e_{20} + b_1e_{21} + b_2e_{22} + \cdots + b_me_{2m} + r_2 \\ \vdots \\ b_0e_{m1} + b_1e_{m2} + b_2e_{m3} + \cdots + b_me_{mm} = r_m \end{bmatrix} \tag{15.79}$$

The subscripting of the $e$ and $r$ elements reflects the subscripting of the unknown $b$ coefficients. These equations can be written in matrix form as:

$$\begin{bmatrix} e_{00} & e_{01} & e_{02} \cdots e_{0m} \\ e_{10} & e_{11} & e_{12} \cdots e_{1m} \\ e_{20} & e_{21} & e_{22} \cdots e_{2m} \\ \vdots & & \\ e_{m0} & e_{m2} & e_{m3} \cdots e_{mm} \end{bmatrix} * \begin{bmatrix} b_0 \\ b_1 \\ b_2 \\ \vdots \\ b_m \end{bmatrix} = \begin{bmatrix} r_0 \\ r_1 \\ r_2 \\ \vdots \\ r_m \end{bmatrix} \tag{15.80}$$

or

$$E * B = R \tag{15.81}$$

The inverse of $E = E^{-1}$. Then,

$$E^{-1} * E * B = E^{-1}R \tag{15.82}$$

But $E^{-1} * E = I$, the identity matrix. Thus,

$$I * B = E^{-1} * R \tag{15.83}$$

Since $I * B = B$

$$B = E^{-1} * R \tag{15.84}$$

or

$$\begin{bmatrix} b_0 \\ b_1 \\ b_2 \\ \vdots \\ b_m \end{bmatrix} = \begin{bmatrix} c_{00} & c_{01} & c_{02} \cdots c_{0m} \\ c_{10} & c_{11} & c_{12} \cdots c_{1m} \\ c_{20} & c_{21} & c_{22} \cdots c_{2m} \\ \vdots & & \\ c_{m1} & c_{m2} & c_{m3} \cdots c_{mm} \end{bmatrix} * \begin{bmatrix} r_0 \\ r_1 \\ r_2 \\ \vdots \\ r_m \end{bmatrix} \tag{15.85}$$

$E^{-1}$ is the inverse matrix of $E$. Then, as was done in the case of the two-dimensional regression,

$$\begin{aligned} b_0 &= c_{00}r_0 + c_{01}r_1 + c_{02}r_2 + \cdots + c_{0m}r_m \\ b_1 &= c_{01}r_0 + c_{11}r_1 + c_{12}r_2 + \cdots + c_{1m}r_m \\ b_2 &= c_{20}r_0 + r_{21}r_1 + c_{22}r_2 + \cdots + c_{2m}r_m \\ &\vdots \\ b_m &= c_{m0}r_0 + c_{m1}r_1 + c_{m2}r_2 + \cdots + c_{mm}r_m \end{aligned} \tag{15.86}$$

As an example of this process, assume the following data (which is in whole numbers to simplify calculations):

| $X_1$ | $X_2$ | $Y$ | $X_1$ | $X_2$ | $Y$ |
|---|---|---|---|---|---|
| 4 | 3 | 5 | 2 | 2 | 3 |
| 6 | 4 | 8 | 4 | 5 | 9 |
| 3 | 5 | 9 | 3 | 2 | 3 |
| 4 | 2 | 3 | 7 | 2 | 4 |
| 7 | 5 | 11 | 7 | 4 | 9 |
| 6 | 3 | 6 | 5 | 2 | 3 |
| 2 | 4 | 6 | 2 | 5 | 8 |
| 5 | 3 | 5 | 5 | 4 | 7 |
| 2 | 2 | 2 | 3 | 4 | 7 |
| 6 | 5 | 11 | 4 | 4 | 7 |
| 7 | 3 | 6 | 6 | 5 | 10 |
| 5 | 5 | 10 | 2 | 3 | 4 |
| 3 | 3 | 4 | 6 | 2 | 4 |

$$n = 26, \quad \sum_{i=1}^{26} x_{1i} = 116, \quad \sum_{i=1}^{26} x_{2i} = 91, \quad \sum_{i=1}^{26} y_i = 164$$

$$\sum_{i=1}^{26} x_{1i}^2 = 596, \quad \sum_{i=1}^{26} x_{2i}^2 = 353, \quad \sum_{i=1}^{26} y_i^2 = 1222$$

$$\sum_{i=1}^{26} (x_1 x_2)_i = 412, \quad \sum_{i=1}^{26} (x_1 x_i) = 780, \quad \sum_{i=1}^{26} (x_2 x_i) = 650$$

The normal equations are

$$\begin{bmatrix} b_0 26 + b_1 116 + b_2 91 = 164 \\ b_0 116 + b_1 596 + b_2 412 = 780 \\ b_0 91 + b_1 412 + b_2 353 = 650 \end{bmatrix}$$

Then

$$E = \begin{bmatrix} 26 & 116 & 91 \\ 116 & 596 & 412 \\ 91 & 412 & 353 \end{bmatrix}$$

$$E^{-1} = \begin{bmatrix} 0.5852773 & -0.0497667 & -0.0927942 \\ -0.0497667 & 0.0129169 & -0.0022464 \\ -0.0927942 & -0.0022464 & 0.0293762 \end{bmatrix}$$

The regression coefficients are

$$b_0 = 0.5852773(164) - 0.0497667(780) - 0.0927942(650) = -3.1488$$

$$b_1 = -0.0497667(164) + 0.0129169(780) - 0.0022464(650) = 0.4533$$

$$b_2 = -0.0927942(164) - 0.0022464(780) + 0.0293762(650) = 2.1241$$

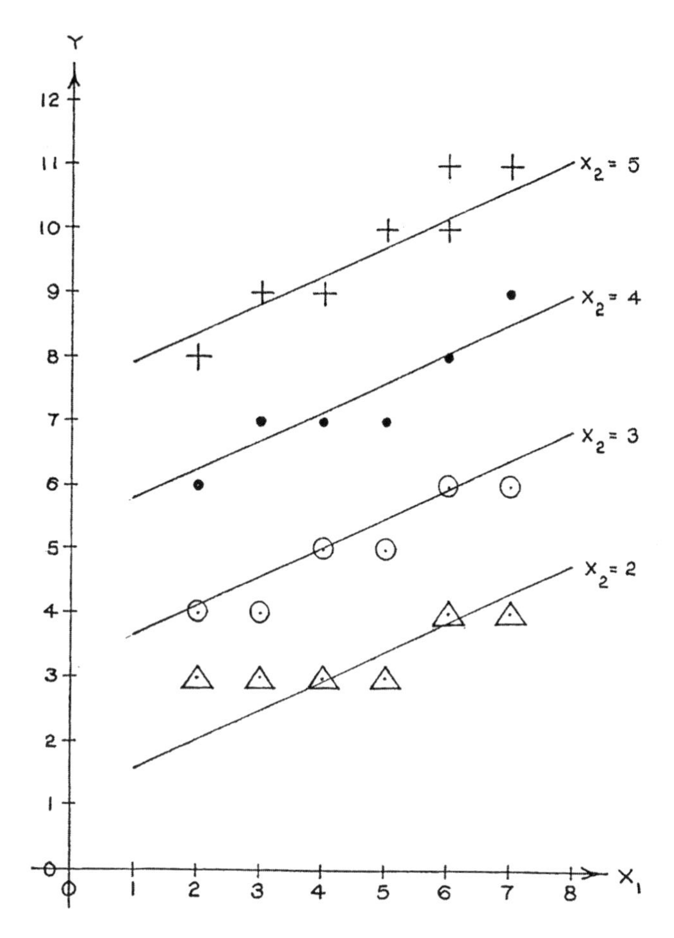


**FIGURE 15.20** The scatter diagram and associated least square regression for the multiple regression example in the text.

and the regression equation is

$$\hat{Y} = 0.4533X_1 + 2.1241X_2 - 3.1488$$

The plot of this regression is shown in Figure 15.20.

The key to this process is the inversion of $E$. The inversion procedure is too lengthy for inclusion here. There are several alternatives that can be used, which are described in standard texts dealing with matrix algebra. One must realize that, because of the sheer amount of computation that is involved, the only feasible way to carry out the fitting of a multidimensional model is to use the regression analysis algorithm from one of the existing statistical computer packages. Such algorithms are invariably based on the use of matrix methods and, in essence, follow the pattern described above.

### 15.3.10 Weighted Regression

In the discussion of freehand curve fitting (Section 15.3.2) a situation arose where two data items occupied the same position in the scatter diagram (see A in Figure 15.4). When the residuals were summed, that for the two data item point was included in the summation twice. In other words, it was given a *weight* of 2, compared with all the remaining data points. Obviously, such weighting would be used whenever data points represented multiple observations, with the weights being equal to the number of data items involved. Similarly, if each data point were an arithmetic mean of several observations and the sample size varied from mean to mean, it would be necessary to weight the data points according to the sample size.

If the variance about the regression is not homogeneous, the position of the regression is more likely to be correct where the scatter of the data is small than where it is large. Consequently, one could say that the data are more reliable or precise where the scatter is small and therefore should be given more weight than the data with a greater scatter. Furthermore, if confidence statements are to be made regarding the regression or any of its components, the variance must be homogeneous. By appropriate weighting of the residuals from the regression, the problem of heterogeneous variances can be brought under control.

In general, the weighting process is as follows. Equation 15.55 defined a residual, in the case of a two-variable linear regression as

$$d_i = y_i - b_0 - b_1 x_i$$

and the sum of squared residuals which is to be minimized as (see Equation 15.56):

$$\sum_{i=1}^{n} d_i^2 = \sum_{i=1}^{n} (y_i - b_0 - b_1 x_i)^2$$

When weights are introduced, each *squared residual* is appropriately weighted so that

$$\sum_{i=1}^{n} w_i d_i^2 = \sum_{i=1}^{n} w_i (y_i - b_0 - b_1 x_i)^2 \tag{15.87}$$

The partial derivatives with respect to $b_0$ and $b_1$ are

$$\frac{\partial \sum_{i=1}^{n} w_i d_i^2}{\partial b_0} = 2 \sum_{i=1}^{n} w_i (b_0 + b_1 x_i - y_i)$$

and

$$\frac{\partial \sum_{i=1}^{n} w_i d_i^2}{\partial b_1} = 2 \sum_{i=1}^{n} w_i (b_0 x_i + b_1 x_i^2 - x_i y_i)$$

The normal equations are

$$\begin{cases} b_0 = \sum_{i=1}^{n} w_i + b_1 \sum_{i=1}^{n} w_i x_i = \sum_{i=1}^{n} w_i y_i \\ b_0 = \sum_{i=1}^{n} w_i x_i + b_1 \sum_{i=1}^{n} w_i x_i^2 = \sum_{i=1}^{n} w_i x_i y_i \end{cases} \tag{15.88}$$

Solving for the $b$ coefficients yields

$$b_0 = \frac{\sum_{i=1}^{n} w_i y_i}{\sum_{i=1}^{n} w_i} - \frac{b_1 \sum_{i=1}^{n} w_i x_i}{\sum_{i=1}^{n} w_i} = \bar{y}_w - b_1 \bar{x}_w \tag{15.89}$$

$$b_1 = \left[ \sum_{i=1}^{n} w_i x_i y_i - \frac{\left(\sum_{i=1}^{n} w_i x_i\right)\left(\sum_{i=1}^{n} w_i y_i\right)}{\sum_{i=1}^{n} w_i} \right] \Bigg/ \left[ \sum_{i=1}^{n} w_i x_i^2 - \frac{\left(\sum_{i=1}^{n} w_i x_i\right)^2}{\sum_{i=1}^{n} w_i} \right] \tag{15.90}$$

$$= \sum_{i=1}^{n} w_i (x_i - \bar{x}_w)(y_i - \bar{y}_w) \Bigg/ \sum_{i=1}^{n} w_i (x_i - \bar{x}_w)^2$$

where

$$\bar{x}_w = \sum_{i=1}^{n} w_i x_i \Bigg/ \sum_{i=1}^{n} w_i$$

$$\bar{y}_w = \sum_{i=1}^{n} w_i y_i \Bigg/ \sum_{i=1}^{n} w_i$$

In the case of data made up of sample means based on differing sample sizes the weights are the sample sizes:

$$w_i = n_i \tag{15.91}$$

As an example, assume that an experiment has been conducted where the response variable was evaluated at six levels of the independent variable, as is shown in the following computations and in Figure 15.21.

| *x* | *y* | *n* | *wx* | *wx*² | *wy* | *wy*² | *wxy* |
|---|---|---|---|---|---|---|---|
| 5 | 33 | 5 | 25 | 125 | 165 | 5445 | 825 |
| 8 | 36 | 10 | 80 | 640 | 360 | 12,960 | 2880 |
| 11 | 46 | 12 | 132 | 1452 | 552 | 25,392 | 6072 |
| 14 | 63 | 7 | 98 | 1372 | 441 | 27,783 | 6174 |
| 17 | 60 | 9 | 153 | 2601 | 540 | 32,400 | 9180 |
| 20 | 73 | 4 | 80 | 1600 | 292 | 21,316 | 5840 |

$$\sum_{i=1}^{n} w_i = 47, \quad \sum_{i=1}^{n} w_i x_i = 568, \quad \sum_{i=1}^{n} w_i x_i^2 = 7790, \quad \sum_{i=1}^{n} w_i y_i = 2350$$

$$\sum_{i=1}^{n} w_i y_i^2 = 125{,}296, \quad \sum_{i=1}^{n} w_i x_i y_i = 30{,}971$$

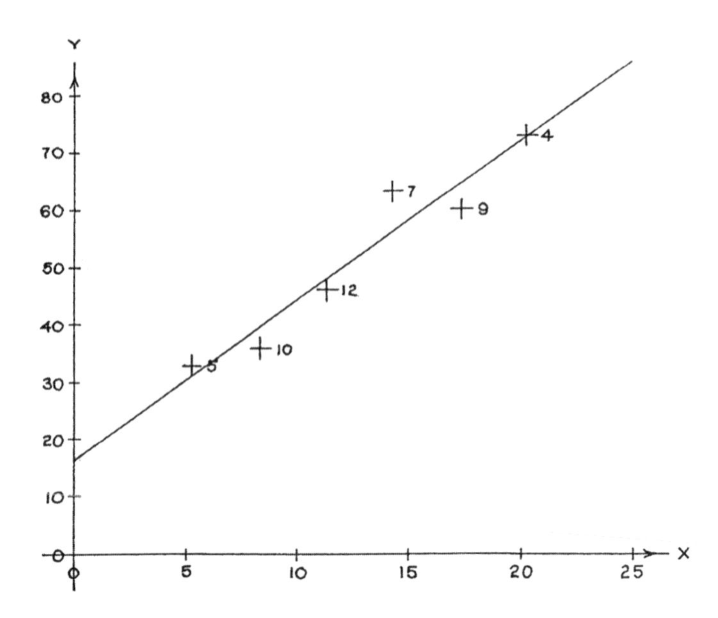


**FIGURE 15.21** Weighted regression where the data points are means of samples with varying sample sizes.

Then,

$$\bar{x}_w = 568/47 = 12.0851, \quad \bar{y}_w = 2350/47 = 50.0000$$

$$\sum_{i=1}^{n} w_i(x_i - \bar{x}_w)^2 = 7790 - (568)^2/47 = 925.6596$$

$$\sum_{i=1}^{n} w_i(x_i - \bar{x}_w)(y_i - \bar{y}_w) = 30{,}971 - (568)(2350)/47 = 2{,}571.0000$$

$$b_1 = 2571.0000/925.6596 = 2.7775$$

$$b_0 = 50.0000 - 2.7775(12.0851) = 16.4339$$

and

$$\hat{Y} = 16.4339 + 2.7775X$$

The trace of this regression equation is shown in Figure 15.21.

If the variance about the regression is not homogeneous and the data points are means based on a constant sample size, the appropriate weights would be the reciprocals of the variances:

$$w_i = 1/s_i^2 \tag{15.92}$$

For example, assume the following data (see Figure 15.22)

| x | y | x | y | x | y | x | y | x | y | x | y |
|---|---|---|---|---|---|---|---|---|---|---|---|
| 5 | 16 | 8 | 18 | 11 | 16 | 14 | 17 | 17 | 39 | 20 | 27 |
| 5 | 8 | 8 | 21 | 11 | 28 | 14 | 28 | 17 | 49 | 20 | 37 |

| | $x$ | $y$ | $x$ | $y$ | $x$ | $y$ | $x$ | $y$ | $x$ | $y$ | $x$ | $y$ |
|---|---|---|---|---|---|---|---|---|---|---|---|---|
| | 5 | 11 | 8 | 14 | 11 | 23 | 14 | 37 | 17 | 17 | 20 | 56 |
| | 5 | 13 | 8 | 25 | 11 | 34 | 14 | 41 | 17 | 20 | 20 | 19 |
| | 5 | 18 | 8 | 10 | 11 | 15 | 14 | 21 | 17 | 32 | 20 | 46 |
| $n =$ | 5 | | 5 | | 5 | | 5 | | 5 | | 5 | |
| $s_Y^2 =$ | 15.7 | | 34.3 | | 64.7 | | 104.2 | | 176.3 | | 216.5 | |
| $w =$ | 0.0636943 | | 0.0291545 | | 0.015456 | | 0.0095969 | | 0.0056721 | | 0.0046189 | |

Note that for each value of $X$, there are five observations of $Y$. The reciprocals of the variances of these sets of five observations are the weights. Following the pattern of computation used previously,

$$\sum_{i=1}^{n} w_i = 0.6409635, \quad \sum_{i=1}^{n} w_i x_i = 5.2245, \quad \sum_{i=1}^{n} w_i x_i^2 = 53.4815$$

$$\sum_{i=1}^{n} w_i y_i = 11.6892, \quad \sum_{i=1}^{n} w_i x_i y_i = 112.8419$$

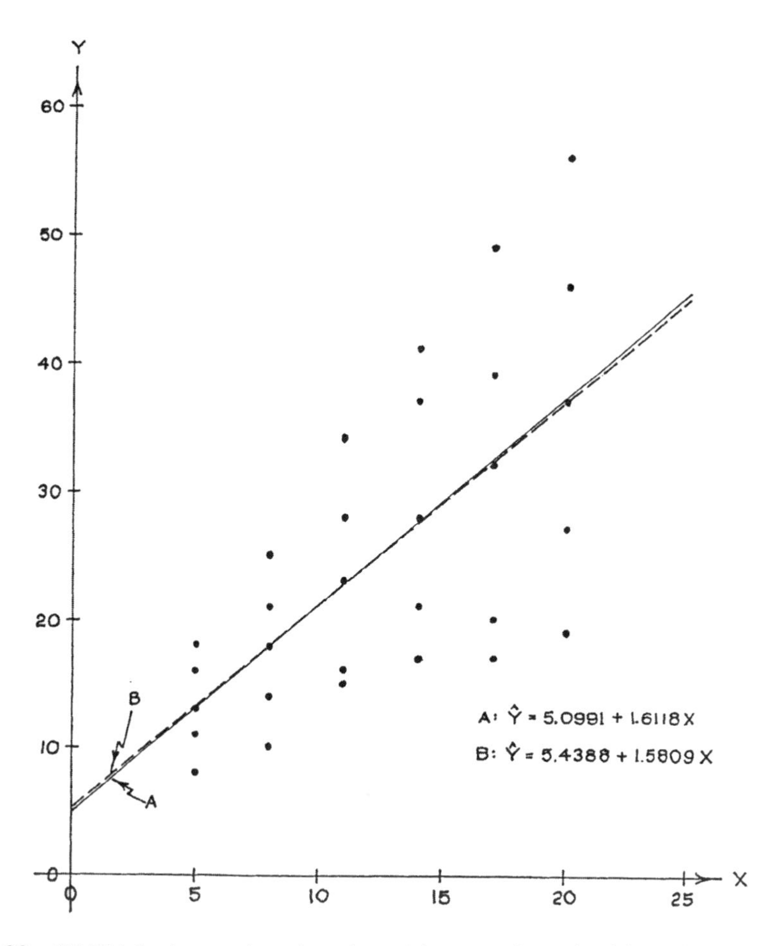


**FIGURE 15.22** (A) Weighted regression where the weights are reciprocals of the variances. (B) Unweighted regression through the same data points.

Then,

$$\bar{x}_w = 5.2245 / 0.6409635 = 8.1510$$

$$\bar{y}_w = 11.6892 / 0.6409635 = 18.2369$$

$$\sum_{i=1}^{n} w_i(x_i - \bar{x}_w)^2 = 53.4815 - (5.2245)^2 / 0.6409635 = 10.8965$$

$$\sum_{i=1}^{n} w_i(x_i - \bar{x}_w)(y_i - \bar{y}_w) = 112.8419 - (5.2245)(11.6892) / 0.6409635 = 17.5631$$

$$b_1 = 17.5631 / 10.8965 = 1.6118$$

$$b_0 = 18.2369 - 1.6118(8.1510) = 5.0991$$

and

$$\hat{Y} = 5.0991 + 1.6118X$$

This regression is labeled A in Figure 15.22. Regression B is the trace of

$$\hat{Y} = 5.4388 + 1.5809X$$

which was computed in the conventional manner without weighting. The regressions are different but both are least squares fits. However, A was fitted through weighted points and B was not. Regression A is more nearly correct because the widely scattered points at the upper end of the scatter diagram have less impact on the final result than do the closely grouped points at the lower end.

In the case of regression A the objective was not only to obtain better estimates of the regression coefficients but to homogenize the variance. The following examines how well this was done.

$$d = \sum_{i=1}^{n} w_i d_i^2/(n-1)$$

| $x$ | $y$ | $\hat{Y}$ | $Y - \hat{Y}$ | $d^2$ | $w$ | $wd^2$ | $s^2$ |
|---|---|---|---|---|---|---|---|
| 5 | 16 | 13.1581 | 2.8419 | 8.0764 | 0.0636943 | 0.5144 | |
| 5 | 8 | 13.1581 | –5.1581 | 26.6060 | 0.0636943 | 1.6947 | |
| 5 | 11 | 13.1581 | –2.1581 | 4.6574 | 0.0636943 | 0.2966 | 1.0001 |
| 5 | 13 | 13.1581 | –0.1581 | 0.0250 | 0.0636943 | 0.0016 | |
| 5 | 18 | 13.1581 | 4.8419 | 23.4440 | 0.0636943 | 1.4932 | |
| 8 | 18 | 17.9935 | 0.0065 | 0.0000 | 0.0291545 | 0.0000 | |
| 8 | 21 | 17.9935 | 3.0065 | 9.0390 | 0.0291545 | 0.2635 | |
| 8 | 14 | 17.9935 | –3.9935 | 15.9480 | 0.0291545 | 0.4650 | 1.0057 |
| 8 | 25 | 17.9935 | 7.0065 | 49.0910 | 0.0291545 | 1.4312 | |
| 8 | 10 | 17.9935 | –7.9935 | 63.8960 | 0.0291545 | 1.8629 | |
| 11 | 16 | 22.8289 | –6.8289 | 46.6339 | 0.015456 | 0.7208 | |
| 11 | 28 | 22.8289 | 5.1711 | 26.7403 | 0.015456 | 0.4133 | |

$$d = \sum_{i=1}^{n} w_i d_i^2/(n-1)$$

| x | y | $\hat{Y}$ | $Y - \hat{Y}$ | $d^2$ | w | $wd^2$ | $s^2$ |
|---|---|---|---|---|---|---|---|
| 11 | 23 | 22.8289 | 0.1711 | 0.0293 | 0.015456 | 0.0005 | 1.0027 |
| 11 | 34 | 22.8289 | 11.1711 | 124.7935 | 0.015456 | 1.9288 | |
| 11 | 15 | 22.8289 | −7.8289 | 61.2917 | 0.015456 | 0.9473 | |
| 14 | 17 | 27.6643 | −10.6643 | 113.7273 | 0.0095969 | 1.0914 | |
| 14 | 28 | 27.6643 | 0.3357 | 0.1127 | 0.0095969 | 0.0011 | |
| 14 | 37 | 27.6643 | 9.3357 | 87.1553 | 0.0095969 | 0.8364 | 1.0155 |
| 14 | 41 | 27.6643 | 13.3357 | 177.8409 | 0.0095969 | 1.7067 | |
| 14 | 21 | 27.6643 | −6.6643 | 44.4129 | 0.0095969 | 0.4262 | |
| 17 | 39 | 32.4997 | 6.5003 | 42.2539 | 0.0056721 | 0.2397 | |
| 17 | 49 | 32.4997 | 16.5003 | 272.2599 | 0.0056721 | 1.5443 | |
| 17 | 17 | 32.4997 | −15.4997 | 240.2407 | 0.0056721 | 1.3627 | 1.0086 |
| 17 | 20 | 32.4997 | −12.4997 | 156.2425 | 0.0056721 | 0.8862 | |
| 17 | 32 | 32.4997 | −0.4997 | 0.2497 | 0.0056721 | 0.0014 | |
| 20 | 27 | 37.3351 | −10.3351 | 106.8143 | 0.0046189 | 0.4934 | |
| 20 | 37 | 37.3351 | −0.3351 | 0.1123 | 0.0046189 | 0.0005 | |
| 20 | 56 | 37.3351 | 18.6649 | 348.3785 | 0.0046189 | 1.6091 | 1.0007 |
| 20 | 19 | 37.3351 | −18.3351 | 336.1759 | 0.0046189 | 1.5528 | |
| 20 | 46 | 37.3351 | 8.6649 | 75.0805 | 0.0046189 | 0.3468 | |

As can be seen from the column headed $s^2$, the variances are essentially equal. The differences are caused by rounding errors.

If the data points are means based on different-sized samples and the variances are not homogeneous, the weights are

$$w_i = n_i / s_i^2 \tag{15.93}$$

When the data points are not means but the variance is not homogeneous, as shown in Figure 15.23, the problem of assigning a weight is more complex. One solution is to group the data into several classes and then to determine the mean of $X$ and the variance of $Y$ in each class. The variances are then plotted against their corresponding mean $X$ values, as shown in Figure 15.24, and a freehand curve balanced through the scatter diagram. The variances of $Y$ for the individual values of $X$ are read from the curve. The weights used are the reciprocals of the curved variance values. As an example, assume the following data:

| x | y | x | y | x | y | x | y |
|---|---|---|---|---|---|---|---|
| 5 | 28 | 9 | 33 | 13 | 42 | 17 | 39 |
| 6 | 28 | 9 | 38 | 13 | 49 | 17 | 58 |
| 6 | 29 | 10 | 46 | 14 | 35 | 18 | 35 |
| 7 | 27 | 11 | 48 | 14 | 52 | 18 | 70 |
| 7 | 34 | 11 | 31 | 15 | 60 | 19 | 58 |

Dividing the data into four classes, as shown in Figure 15.23, yields the following:

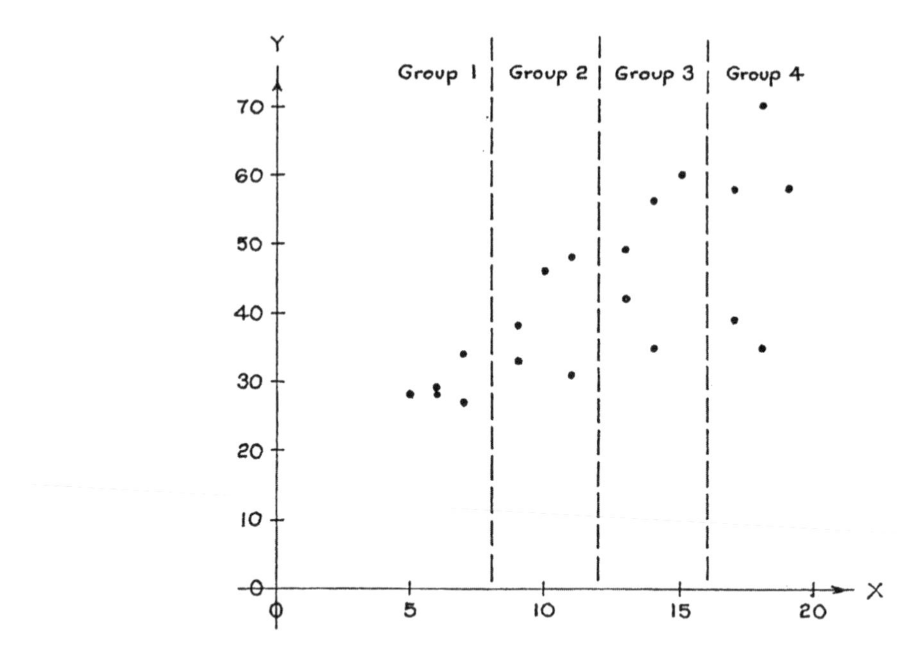


FIGURE 15.23 Scatter diagrams are divided into four classes.

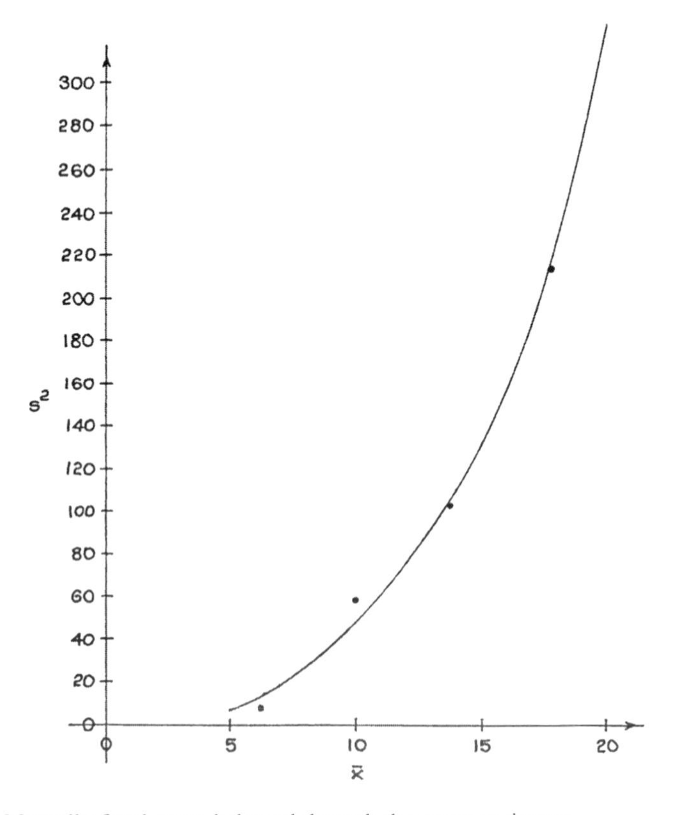


FIGURE 15.24 Manually fitted curve balanced through the group variances.

$$\bar{x}_1 = 6.2 \qquad \bar{x}_2 = 10.0 \qquad \bar{x}_3 = 13.8 \qquad \bar{x}_4 = 17.8$$

$$s_{Y1}^2 = 7.7 \qquad s_{Y2}^2 = 57.7 \qquad s_{Y3}^2 = 103.3 \qquad s_{Y4}^2 = 213.5$$

Figure 15.24 shows the plot of these values and the freehand curve balanced through the points. The variances and weights are then as shown below (only the $X$ values that occur are shown):

| x | $s_Y^2$ | w | x | $s_Y^2$ | w |
|---|---|---|---|---|---|
| 5 | 7 | 0.1429 | 13 | 91 | 0.0110 |
| 6 | 13 | 0.0769 | 14 | 111 | 0.0090 |
| 7 | 19 | 0.0526 | 15 | 132 | 0.0076 |
| 9 | 37 | 0.0270 | 17 | 187 | 0.0053 |
| 10 | 48 | 0.0208 | 18 | 235 | 0.0043 |
| 11 | 60 | 0.0167 | 19 | 274 | 0.0036 |

Using the weights, the resulting regression equation is

$$\hat{Y} = 16.1653 + 2.1941X$$

### 15.3.11 Conditioned Regression

In some cases the nature of the problem requires one or more of the $b$ coefficients to assume specific values or other constraints may exist. This occurred, for example, in Section 15.3.3 when the model for dbhob on dbhib was developed. If a total tree height on dbh regression is generated, the line would have to have a $Y$-intercept of 4.5 ft since a tree has no dbh until it reaches a height of 4.5 ft. When constraints such as these are imposed on the model, it is said that the regression is *conditioned*.

When a regression is conditioned, it requires some changes in the fitting process but basically it remains much the same. For a two-variable regression when the $Y$-intercept must be equal to $k$, the model is

$$Y = k + \beta X \tag{15.94}$$

and the $i$th residual is

$$d_i = y_i - k - bx_i \tag{15.95}$$

The sum of squared residuals is

$$\sum_{i=1}^{n} d_i^2 = \sum_{i=1}^{n} (y_i - k - bx_i)^2 \tag{15.96}$$

Now, since there is only one unknown, $b$, there is only one derivative,

$$\frac{d\sum_{i=1}^{n} d_i^2}{db_i} = 2\sum_{i=1}^{n} (kx_i + bx_i^2 - x_i y_i) \tag{15.97}$$

Setting this equal to zero and solving for $b_1$ yields:

$$k\sum_{i=1}^{n} x_i + b\sum_{i=1}^{n} x_i^2 = \sum_{i=1}^{n} x_i y_i \tag{15.98}$$

$$b = \left(\sum_{i=1}^{n} x_i y_i - k\sum_{i=1}^{n} x_i\right) \Bigg/ \sum_{i=1}^{n} x_i^2 \tag{15.99}$$

Thus, if $k = 0$,

$$b = \sum_{i=1}^{n} x_i y_i \Bigg/ \sum_{i=1}^{n} x_i^2 \tag{15.100}$$

In the case of a multidimensional regression with a fixed $y$-intercept, the regression function is

$$Y = k + \beta_1 X_1 + \beta_2 X_2 + \cdots + \beta_m X_m \tag{15.101}$$

The sum of squared residuals that is to be minimized is

$$\sum_{i=1}^{n} (y_i - k - b_1 x_{1i} - b_2 x_{2i} - \cdots - b_m x_{mi})^2 \tag{15.102}$$

The $m - 1$ partial deviatives are

$$\begin{aligned}
\frac{\partial \sum_{i=1}^{n} d_i^2}{\partial b_1} &= 2\sum_{i=1}^{n} (kx_{1i} + b_1 x_{1i}^2 + b_2 x_{1i} x_{2i} + \cdots + b_m x_{1i} x_{mi} - x_{1i} y_i) \\
\frac{\partial \sum_{i=1}^{n} d_i^2}{\partial b_2} &= 2\sum_{i=1}^{n} (kx_{2i} + b_1 x_{1i} x_{2i} + b_2 x_{2i}^2 + \cdots + b_m x_{2i} x_{mi} - x_{2i} y_i) \\
&\vdots \\
\frac{\partial \sum_{i=1}^{n} d_i^2}{\partial b_m} &= 2\sum_{i=1}^{n} (kx_{mi} + b_1 x_{1i} x_{mi} + b_2 x_{2i} x_{mi} + \cdots + b_m x_{mi} x_{mi}^2 - x_{mi} y_i)
\end{aligned} \tag{15.103}$$

and the normal equations are

$$
\begin{aligned}
&b_1\sum_{i=1}^{n}x_{1i}^2 + b_2\sum_{i=1}^{n}x_{1i}x_{2i} + \cdots + b_m\sum_{i=1}^{n}x_{1i}x_m = \sum_{i=1}^{n}x_{1i}y_i - k\sum_{i=1}^{n}x_{1i}\\
&b_1\sum_{i=1}^{n}x_{1i}x_{2i} + b_2\sum_{i=1}^{n}x_{2i}^2 + \cdots + b_m\sum_{i=1}^{n}x_{2i}x_m = \sum_{i=1}^{n}x_{2i}y_i - k\sum_{i=1}^{n}x_{2i}\\
&\vdots\\
&b_1\sum_{i=1}^{n}x_{1i}x_{mi} + b_2\sum_{i=1}^{n}x_{2i}x_{mi} + \cdots + b_m\sum_{i=1}^{n}x_{mi}^2 = \sum_{i=1}^{n}x_{mi}y_i - k\sum_{i=1}^{n}x_{mi}
\end{aligned}
\qquad (15.104)
$$

In terms of matrix algebra, now each of the elements of the $R$ vector is made up of two, instead of one, terms. Since these values are known, coming from the sample data, they can be combined into single numerical values and the matrix inversion and solution of the $b$ coefficients can proceed as before.

These conditioned regressions are least squares fits *with the* k *constraint*. Because the constraint exists, the regression may or may not pass through the point in the coordinate system defined by the means of the several variables. Consequently, the algebraic sum of residuals may not be equal to zero.

## 15.4 DUMMY VARIABLES

In the previous discussion the variables, both predicting and response, were defined as, or assumed to be, continuous and measurable. However, situations arise when it becomes necessary to incorporate into a regression a recognition of some attribute of the sampling units that is not measurable. For example, one might wish to introduce species, sex, type of vehicle, or geologic formation into the regression. To do this, each attribute is treated as a variable. Such a variable has but two numerical values: 1, if the sampling unit possesses the attribute; and 0, if it does not possess the attribute. Such a variable is a *dummy variable*. In essence this permits the incorporation into a single regression the relationships between a response variable and a given set of predicting variables for two different populations, one with the attribute and one without it. Insofar as computational procedures are concerned, dummy variables are no different from conventional predicting variables.

If the problem requires the recognition of characteristics of the attribute type but involves more than the presence or absence of one trait, a separate dummy variable must be assigned to each of the traits. For example, consider the situation where the volume–weight relationship is to be described in a simple regression and both longleaf and loblolly pine in the Alabama Coastal Plain and the Alabama Piedmont are involved. In such a case four dummy variables would have to be recognized: $X_A$, for longleaf pine in the Coastal Plain; $X_B$, for longleaf pine in the Piedmont; $X_C$, for loblolly pine in the Coastal Plain; and $X_D$, for loblolly pine in the Piedmont. Thus, for sampling unit $i$, which contains loblolly pine from the Coastal Plain,

$$X_A = x_{Ai} = 0, \quad X_B = x_{Bi} = 0, \quad X_C = x_{Ci} = 1, \quad \text{and} \quad X_D = x_{Di} = 0$$

In practice one usually would drop out the equivalent of $X_D$ because it actually acts as a catchall variable becoming operational only when a data item fails to fit into any of the previous categories. Thus, in general one fewer dummy variable is established than there are traits or categories. The rationale for this can be seen when one examines the following distribution of numerical values by species–region categories:

| Category | $X_A$ | $X_B$ | $X_C$ |
|---|---|---|---|
| Longleaf, Coastal Plain | 1 | 0 | 0 |
| Longleaf, Piedmont | 0 | 1 | 0 |
| Loblolly, Coastal Plain | 0 | 0 | 1 |
| Loblolly, Piedmont | 0 | 0 | 0 |

If a regression equation containing such dummy variables is used to predict the volume in a given load of logs, the term representing the applicable dummy variable ($X_C$ in the above example) would take on the value of its regression coefficient. All the other dummy variables would become equal to zero. Consequently, the effect of the effective dummy variable is equal to its regression coefficient and is additive to the value derived from the conventional variables. This, in essence, is an effect on the $y$-intercept. The general regression line or surface simply is moved up or down from its average position. In some cases, however, an interaction exists between the dummy variable and one or more of the conventional variables. When this is the case, the dummy variable affects the slope of the regression line or surface as well as its intercept.

## 15.5 VARIANCE ABOUT THE REGRESSION

The process used to obtain the least squares estimates of the $b$ coefficients is purely mathematical and is not constrained in any way except that the model being fitted be linear. The resulting equation, however, may or may not provide an appropriate "picture" of the situation. A nonrepresentative sample, especially one containing one or more extreme values, can influence the result in much the same way as an extreme value in a sample can distort the arithmetic mean. If the fitting process is repeated indefinitely, each time with data from a new random sample, it is expected that, on the average, the regression would match the actual regression, *provided that the model being used is appropriate.*

In the case of the variance, however, one must observe certain constraints if meaningful results are to be obtained:

1. The set of values of the *independent* variable, or variables, must be assumed to be fixed. In other words, it must be assumed that, if the sampling process is to be repeated an indefinite number of times, in every case the same values *and only the same values* of the independent variable would appear. This constraint is imposed in order to prevent the appearance of a component of variability caused by the variation of the values of the independent variable.
2. The variances of the populations of values of the dependent variable associated with the specific values of the independent variables must be homogeneous.
3. If confidence intervals, bands, or surfaces are to be computed or if statistical tests are to be made of the regression or any of the $b$ coefficients, the populations of values of the dependent variables must be normally distributed.

It is under the assumption that these conditions exist that the following discussion is generated.

The variance of the sample values of the dependent variable, $Y$, about the *fixed mean*, $y$, disregarding the independent variables, is referred to as the *total* or *overall* variance, and is an unbiased estimate of the population variance $\sigma_Y^2$,

$$s_0^2 = \sum_{i=1}^{n} (y_i - \bar{y})^2 / (n-1) \rightarrow \sigma_Y^2 \tag{15.105}$$

When the presence of the independent variables is recognized, the variance of the sample values of $Y$ is *about the regression,* $\hat{Y}$, is referred to as the *residual* variance, and is an unbiased estimate of the population variance $\sigma^2_{Y \cdot X}$:

$$s_R^2 = s_{Y \cdot X}^2 = \sum_{i=1}^{n} (y_i - \hat{y}_i)^2 / (n - m - 1) \rightarrow \sigma^2_{Y \cdot X} \tag{15.106}$$

The residual variance is the estimate of the population variance about the regression and is the basic variance upon which all the subsequent operations are based. The relationship between the two variances can be seen through the following development. Returning to Equation 15.54,

$$d_i = y_i - \hat{y}_i$$

In the case of a simple linear regression,

$$\hat{y}_i = b_0 + b_1 x_i$$

Then,

$$d_i = y_i - b_0 - b_1 x_i \tag{15.107}$$

and

$$\sum_{i=1}^{n} d_i^2 = \sum_{i=1}^{n} (y_i - b_0 - b_1 x_i)^2 \tag{15.108}$$

But from the first normal equation (Equation 15.59),

$$b_0 = \bar{y} - b_1 x$$

Then,

$$\begin{aligned}
\sum_{i=1}^{n} d_i^2 &= \sum_{i=1}^{n} [y_i - (\bar{y} - b_1 \bar{x}) - b_1 x_i]^2 \\
&= \sum_{i=1}^{n} [(y_i - \bar{y})^2 - 2b_1 (x_i - \bar{x})(y_i - \bar{y}) + b_1^2 (x_i - \bar{x})^2] \qquad (15.109) \\
&= \sum_{i=1}^{n} (y_i - \bar{y})^2 - 2b_1 \sum_{i=1}^{n} (x_i - \bar{x})(y_i - \bar{y}) + b_1^2 \sum_{i=1}^{n} (x_i - \bar{x})^2
\end{aligned}$$

But, from the second normal equation (Equation 15.61),

$$b_1 = \sum_{i=1}^{n} (x_i - \bar{x})(y_i - \bar{y}) \bigg/ \sum_{i=1}^{n} (x_i - \bar{x})^2$$

Then,

$$b_1 \sum_{i=1}^{n} (x_i - \bar{x})^2 = \sum_{i=1}^{n} (x_i - \bar{x})(y_i - \bar{y}) \tag{15.110}$$

Multiplying both sides by $b_1$,

$$b_1^2 \sum_{i=1}^{n} (x_i - \bar{x})^2 = b_1 \sum_{i=1}^{n} (x_i - \bar{x})(y_i - \bar{y}) \tag{15.111}$$

Substituting $b_1 \Sigma_{i=1}^{n} (x_i - \bar{x})(y_i - \bar{y})$ for $b_1^2 \Sigma_{i=1}^{n} (x_i - \bar{x})$ in Equation 15.109 yields

$$\sum_{i=1}^{n} d_i^2 = \sum_{i=1}^{n} (y_i - \bar{y})^2 - b_1 \sum_{i=1}^{n} (x_i - \bar{x})(y_i - \bar{y}) \tag{15.112}$$

or, substituting the other way,

$$\sum_{i=1}^{n} d_i^2 = \sum_{i=1}^{n} (y_i - \bar{y})^2 - b_1^2 \sum_{i=1}^{n} (x_i - \bar{x})^2 \tag{15.113}$$

In Equations 15.112 and 15.113, $\Sigma d_i^2$ represents the sum of squared residuals from the *moving mean* and is the *error* or *residual* sum of squares. $\Sigma(y_i - \bar{y})^2$ is the sum of squared residuals from the *overall* fixed mean and is the *total* sum of squares. $b_1 \Sigma_{i=1}^{n} (x_i - \bar{x})(y_i - \bar{y})$, or $b_1^2 \Sigma_{i=1}^{n} (x_i - \bar{x})^2$, represents the portion of the total sum of squares which is attributable to, or explained by, the regression. Thus,

$$\text{Total SS} = \text{Error SS} + \text{Explained SS} \tag{15.114}$$

This can be expressed as an analysis of variance.

**ANOVA**

| **Source** | **df** | **SS** | **MS** |
|---|---|---|---|
| Total | $n - 1$ | $\sum_{i=1}^{n} (y_i - \bar{y})^2$ | $s_Y^2$ |
| Regression (explained by) | 1 | $b_1 \sum_{i=1}^{n} (x_i - \bar{x})(y_i - \bar{y})$ or $b_1^2 \sum_{i=1}^{n} (x_i - \bar{x})^2$ | |
| Error (or residual) | $n - 2$ | $\sum_{i=1}^{n} d_i^2$ | $s_{Y \cdot X}^2$ |

It should be remembered that a degree of freedom is lost whenever a sample mean is calculated. Thus, in the case of the "Total," $\bar{y}$ was calculated and one degree of freedom was lost. In the case of "Regression," $\bar{x}$ was calculated and another degree of freedom was lost. Both $\bar{y}$ and $\bar{x}$ are needed to compute the "Error" sum of squares and thus two degrees of freedom are lost.

Applying this development to the earlier sample where

$$\hat{Y} = 1.9925 + 0.7933X$$

$$\sum_{i=1}^{7}(y_i - \bar{y})^2 = 1075 - (79)^2 / 7 = 183.4286$$

$$\sum_{i=1}^{7}(x_i - \bar{x})^2 = 1226 - (82)^2 / 7 = 265.4286$$

Then, using Equation 15.113,

$$\sum_{i=1}^{7} d_i^2 = 183.4286 - (0.7933)^2(265.4286) = 16.3878$$

$$s_Y^2 = 183.4286 / (7 - 1) = 30.5714$$

$$s_{Y \cdot X}^2 = 16.3878 / (7 - 1) = 3.2776$$

The analysis of variance is

**ANOVA**

| Source | df | SS | MS |
|---|---|---|---|
| Total | 6 | 183.4286 | 30.5714 |
| Regression | 1 | 167.0408 | 167.0408 |
| Residual | 5 | 16.3878 | 3.2776 |

In the case of a multidimensional regression, an expression for the residual sum of squares, comparable to Equation 15.112, can be developed:

$$\sum_{i=1}^{n} d_i^2 = \sum_{i=1}^{n}(y_i - \bar{y})^2 - b_1\sum_{i=1}^{n}(x_{1i} - \bar{x}_1)(y_i - \bar{y}) - b_2\sum_{i=1}^{n}(x_{2i} - \bar{x}_2)(y_i - \bar{y}) - \cdots - b_m\sum_{i=1}^{n}(x_{mi} - \bar{x}_m)(y_i - \bar{y}) \quad (15.115)$$

It must be emphasized that the $b$ coefficients in Equation 15.115 are those that arise from the fitting of the multidimensional model. They are not those that would be obtained if each independent variable were used in a separate fit of a simple linear model. As can be seen in Equation 15.115, each independent variable is represented. Presumably each independent variable "explains" a portion of the scatter of the data about $\bar{y}$. Consequently, the sum of squared residuals from the regression, $\Sigma_{i=1}^{n} d_i^2$ should become smaller as more independent variables are added to the model. This is an important concept that plays a key role in the statistical testing of regressions.

When there is more than one independent variable in the regression, the analysis of variance takes the following form.

**ANOVA**

| Source | df | SS | MS |
|---|---|---|---|
| Total | $n-1$ | $\sum_{i=1}^{n}(y_i-\bar{y})^2$ | $s_Y^2$ |
| Regression | $m$ | $b_1\sum_{i=1}^{n}(x_{1i}-\bar{x}_1)(y_i-\bar{y})+b_2\sum_{i=1}^{n}(x_{2i}-\bar{x}_2)(y_i-\bar{y}) +\cdots+b_m\sum_{i=1}^{n}(x_{mi}-\bar{x}_m)(y_i-\bar{y})$ | |
| Error | $n-m-1$ | $\sum_{i=1}^{n}d_i^2$ | $s_{Y\cdot X}^2$ |

Use, as an example, the multiple regression developed in Section 15.3.9, where

$$\hat{Y} = 0.4533X_i + 2.1241X_2 - 3.1488$$

Then

$$\sum_{i=1}^{26} y_i^2 = 1222$$

$$\sum_{i=1}^{26}(y_i-\bar{y})^2 = 1222-(164)^2/26 = 187.5385$$

$$\sum_{i=1}^{26}(x_{1i}-\bar{x}_1)^2 = 596-(116)^2/26 = 78.4615$$

$$\sum_{i=1}^{26}(x_{2i}-\bar{x}_2)^2 = 353-(91)^2/26 = 34.5000$$

$$\sum_{i=1}^{26}(x_{1i}-\bar{x}_1)(y_i-\bar{y}) = 780-(116)(164)/26 = 48.3077$$

$$\sum_{i=1}^{26}(x_{2i}-\bar{x}_2)(y_i-\bar{y}) = 650-(91)(164)/26 = 76.0000$$

The residual sum of squares, using Equation 15.115, is

$$\sum_{i=1}^{26} d_i^2 = 187.5385 - (0.4533)(48.3077) - (2.1241)(76.0000)$$

$$= 4.2090$$

The variance of the estimate is

$$s_{Y \cdot X}^2 = 4.2090(26 - 3) = 0.1830009$$

and the standard error of the estimate is

$$s_{Y \cdot X} = \sqrt{0.183009} = \pm 0.4278$$

The analysis of variance is

**ANOVA**

| Source | df | SS | MS |
|---|---|---|---|
| Total | 25 | 187.5385 | 7.5015 |
| Regression | 2 | 183.3295 | |
| Residual | 23 | 4.2090 | 0.1830 |

When the regression has been conditioned so as to have a specific $y$-intercept (e.g., $b_0 = 0$, $b_0 = 4.5$ ft, or $b_0 = k$), Equations 15.112, 15.113, or 15.115 are no longer appropriate. When a two-dimensional linear model is involved and the specific $y$-intercept is $k$,

$$\hat{Y} = k + bX$$

$$d_i = y_i - \hat{y}_i = y_i - k - bx_i$$

$$d_i^2 = y_i^2 + k^2 + b^2 x_i^2 - 2ky_i - 2bx_i y_i + 2kbx_i \tag{15.116}$$

$$\sum_{i=1}^{n} d_i^2 = \sum_{i=1}^{n} y_i^2 + nk^2 + b^2 \sum_{i=1}^{n} x_i^2 - 2k \sum_{i=1}^{n} y_i - 2b \sum_{i=1}^{n} x_i y_i + 2kb \sum_{i=1}^{n} x_i$$

$$= \sum_{i=1}^{n} y_i^2 + b\left(b \sum_{i=1}^{n} x^2 - \sum_{i=1}^{n} x_i y_i\right) - b \sum_{i=1}^{n} x_i y_i + nk^2 - 2k \sum_{i=1}^{n} y_i + 2kb \sum_{i=1}^{n} x_i$$

According to Equation 15.98

$$k \sum_{i=1}^{n} x_i + b \sum_{i=1}^{n} x_i^2 = \sum_{i=1}^{n} x_i y_i$$

Then

$$b \sum_{i=1}^{n} x_i^2 - \sum_{i=1}^{n} x_i y_i = -k \sum_{i=1}^{n} x_i \tag{15.117}$$

Substituting Equation 15.117 in 15.116 yields

$$\begin{aligned}\sum_{i=1}^{n} d_i^2 &= \sum_{i=1}^{n} y_i^2 + b\left(-k\sum_{i=1}^{n} x_i\right) - b\sum_{i=1}^{n} x_i y_i + nk^2 - 2k\sum_{i=1}^{n} y_i + 2kb\sum_{i=1}^{n} x_i \\ &= \sum_{i=1}^{n} y_i^2 - b\sum_{i=1}^{n} x_i y_i + nk^2 - 2k\sum_{i=1}^{n} y_i + kb\sum_{i=1}^{n} x_i\end{aligned} \tag{15.118}$$

Equation 15.118 would be appropriate when a two-dimensional regression contains a specified $y$-intercept. Note that where $k = 0$

$$\sum_{i=1}^{n} d_i^2 + = \sum_{i=1}^{n} y_i^2 - b\sum_{i=1}^{n} x_i y_i \tag{15.119}$$

In the case of a multiple regression, the equivalent equation can be shown to be

$$\begin{aligned}\sum_{i=1}^{n} d_i^2 &= \sum_{i=1}^{n} y_i^2 - b_1\sum_{i=1}^{n} x_{1i} y_i - b_2^2\sum_{i=1}^{n} x_{2i} y_i - \cdots - b_m\sum_{i=1}^{n} x_{mi} y_i \\ &\quad + nk^2 - 2k\sum_{i=1}^{n} y_i + kb_1\sum_{i=1}^{n} x_{1i} + kb_2\sum_{i=1}^{n} x_{2i} + \cdots + kb_m\sum_{i=1}^{n} x_{mi}\end{aligned} \tag{15.120}$$

When $k = 0$,

$$\sum_{i=1}^{n} d_i^2 = \sum_{i=1}^{n} y_i^2 - b_1\sum_{i=1}^{n} x_{1i} y_i - b_2\sum_{i=1}^{n} x_{2i} y_i - \cdots - b_m\sum_{i=1}^{n} x_{mi} y_i \tag{15.121}$$

## 15.6 CONFIDENCE INTERVALS FOR REGRESSION ESTIMATES

A sample-based regression is an estimate of the true population regression. If the model is appropriate, each $b$ coefficient is an estimate of its corresponding $\beta$ coefficient. If the sampling is continued indefinitely, each time with the same sample size and the same set of $X$ values, a sampling distribution will be generated for each of the $\beta$ coefficients and for the regression as a whole. Estimates of the standard errors of these sampling distributions can be used in conjunction with Student's $t$ to arrive at confidence intervals.

First consider confidence intervals in the context of the simple linear regression. According to Item 3 in Section 9.9, the variance of the sum of two random variables is

$$V(X_1 + X_2) = V(X_1) + V(X_2) + 2\ \mathrm{COV}(X_1X_2)$$

Since the $b$ coefficients are being considered as elements in their respective sampling distributions, each becomes a random variable. Consequently, one can consider the regression equation as the sum of two random variables. The variance of the regression estimates is then

$$\begin{aligned}V(\hat{Y}) &= V(b_0 + b_1X) \\ &= V(b_0 + X^2V(b_1) + 2X\,\mathrm{COV}(b_0b_1)\end{aligned} \tag{15.122}$$

$V(b_0)$, $V(b_1)$, and $\mathrm{COV}(b_0b_1)$ can best be determined using matrix algebra. According to Equation 15.73,

$$E^{-1} = \begin{bmatrix} c_{00} = e_{11}/|E| & c_{01} = -e_{01}/|E| \\ c_{10} = -e_{10}/|E| & c_{11} = e_{00}/|E| \end{bmatrix} \tag{15.123}$$

Multiplying $E^{-1}$ by $\sigma^2_{Y \cdot X}$ produces the *variance–covariance* matrix.

$$\sigma^2_{Y \cdot X} E^{-1} = \begin{bmatrix} \sigma^2_{Y \cdot X} c_{00} = V(b_0) & \sigma^2_{Y \cdot X} c_{01} = \mathrm{COV}(b_0 b_1) \\ \sigma^2_{Y \cdot X} c_{10} = \mathrm{COV}(b_1 b_0) & \sigma^2_{Y \cdot X} c_{11} = V(b_1) \end{bmatrix} \tag{15.124}$$

Consequently, Equation 15.122 can be written

$$V(\hat{Y}) = \sigma^2_{Y \cdot X}(c_{00} + X^2 c_{11}^2 + 2Xc_{01}) \tag{15.125}$$

In the case of a sample, $e_{00} = n$, $e_{01} = e_{10} = \Sigma_{i=1}^{n} x_i$, $e_{11} = \Sigma_{i=1}^{n} x_i^2$, and

$$\begin{aligned} |E| &= e_{00}e_{11} - e_{01}e_{10} \\ &= n\sum_{i=1}^{n} x_i^2 - \left(\sum_{i=1}^{n} x_i\right)^2 \\ &= n\sum_{i=1}^{n} (x_i - \bar{x})^2 \end{aligned}$$

Then,

$$E^{-1} = \begin{bmatrix} c_{00} = \sum_{i=1}^{n} x_i^2 \Big/ n\sum_{i=1}^{n}(x_i - \bar{x})^2 & c_{01} = -\sum_{i=1}^{n} x_i \Big/ n\sum_{i=1}^{n}(x_i - \bar{x})^2 \\ c_{10} = -\sum_{i=1}^{n} x_i \Big/ n\sum_{i=1}^{n}(x_i - \bar{x})^2 & c_{11} = 1\Big/\sum_{i=1}^{n}(x_i - \bar{x})^2 \end{bmatrix} \tag{15.126}$$

The variance–covariance matrix is

$$s^2_{Y \cdot X} E^{-1} = \begin{bmatrix} s^2_{Y \cdot X} c_{00} = s^2_{b_0} & s^2_{Y \cdot X} c_{01} = s_{b_0 b_1} \\ s^2_{Y \cdot X} c_{10} = s_{b_1 b_0} & s^2_{Y \cdot X} c_{11} = s^2_{b_1} \end{bmatrix} \tag{15.127}$$

Then the variance and standard error of $b_0$ are

$$s^2_{b_0} = s^2_{Y \cdot X}\left[\sum_{i=1}^{n} x_i^2 \Big/ n\sum_{i=1}^{n}(x_i - \bar{x})^2\right] \tag{15.128}$$

$$s_{b_0} = \sqrt{s_{b_0}^2} = s_{Y \cdot X} \sqrt{\sum_{i=1}^{n} x_i^2 \Big/ n \sum_{i=1}^{n} (x_i - \bar{x})^2} \tag{15.129}$$

and the confidence interval associated with $b_0$ is

$$b_0 \pm t_c(k) s_{b_0} \rightarrow \beta_0 \tag{15.130}$$

where $k = n - m = n - 2$.

The variance and standard error of $b_1$ are

$$s_{b_1}^2 = s_{Y \cdot X}^2 \Big/ \sum_{i=1}^{n} (x_i - \bar{x})^2 \tag{15.131}$$

$$s_{b_1} = \sqrt{s_{b_1}^2} = s_{Y \cdot X} \sqrt{1 \Big/ \sum_{i=1}^{n} (x_i - \bar{x})^2} \tag{15.132}$$

and the confidence interval associated with $b_1$ is

$$b_1 \pm t_c(k) s_{b_1} \rightarrow \beta_1 \tag{15.133}$$

where $k = n - m = n - z$.

The covariance of $b_0$ and $b_1$ is

$$s_{b_0 b_1} = s_{b_1 b_0} = s_{Y \cdot X}^2 \left[ -\sum_{i=1}^{n} x_i \Big/ n \sum_{i=1}^{n} (x_i - \bar{x})^2 \right] \tag{15.134}$$

The variance and standard error of the regression estimate are

$$s_{\hat{y}i}^2 = s_{Y \cdot X}^2 (c_{00} + x_i^2 c_{11} + 2 x_i c_{01}) \quad \text{or}$$

$$= s_{Y \cdot X}^2 \left\{ \left[ \sum_{i=1}^{n} x_i^2 \Big/ n \sum_{i=1}^{n} (x_i - \bar{x})^2 \right] + \left[ x_i^2 \Big/ \sum_{i=1}^{n} (x_i - \bar{x})^2 \right] \right. \tag{15.135}$$

$$\left. + 2 \left[ x_i \left( -\sum_{i=1}^{n} x_i \right) \Big/ n \sum_{i=1}^{n} (x_i - \bar{x})^2 \right] \right\} \tag{15.136}$$

and

$$s_{\hat{y}i} = \sqrt{s_{\hat{y}i}^2} \tag{15.137}$$

It should be noted that the variance of $\hat{y}_i$ varies according to the magnitude of $X$. Consequently, the confidence interval about the regression line, $\mathrm{CI}_{X=x_i}$, is not constant for all values of $X$.

$$
\begin{aligned}
CI_{X=x_i} &= \hat{y}_i \pm t_c(k)s_{\hat{y}i} \\
&= \hat{y}_i \pm t_c(k)s_{Y\cdot X}\sqrt{c_{00} + x_i^2 c_{11} + 2x_i c_{01}} \quad or \qquad (15.138) \\
&= \hat{y} \pm t(k)s_{Y\cdot X}\sqrt{\frac{\sum_{i=1}^{n} x_i^2 + nx_i^2 - 2x_i \sum_{i=1}^{n} x_i}{n\sum_{i=1}^{n}(x_i - \bar{x})^2}}
\end{aligned}
$$

where $k = n - m - 1 = n - 1 - 1 = n - 2$.

When values of $X$ are substituted into this expression, it is found that the confidence limits form lines, called *confidence bands*, which bend away from the regression line as shown in Figure 15.25. The minimum interval is associated with the minimum value of $s_{\hat{y}}^2$.

$$
\frac{ds_{\hat{y}}^2}{dx} = \left[\frac{s_{Y\cdot X}^2}{n\sum_{i=1}^{n}(x_i - \bar{y})^2}\right]\frac{d}{dx}\left(\sum_{i=1}^{n} x_i^2 + nx_i^2 - 2x_i \sum_{i=1}^{n} x_i\right)
$$

In this context $s_{Y\cdot X}^2$, $\Sigma_{l=1}^{n} x_i$, $\Sigma_{i=1}^{n} x_i^2$, and $n\Sigma_{i=1}^{n}(x_i - \bar{x})^2$ are constants. Then,

$$
\frac{d}{dx}\left(\sum_{i=1}^{n} x_i^2 + nx_i^2 - 2x_i \sum_{i=1}^{n} x_i\right) = 2nx - 2\sum_{i=1}^{n} x_i
$$

Setting the derivative equal to zero

$$
2nx - 2\sum_{i=1}^{n} x_i = 0
$$

$$
2nx = 2\sum_{i=1}^{n} x_i
$$

$$
x = \sum_{i=1}^{n} x_i / n = x
$$

Thus, the variance of $\hat{y}$ is at a minimum when $X = \bar{x}$. At that point the variance of $\hat{y}$ is

$$s_{\hat{y}_{\bar{x}}}^2 = \left[\frac{s_{Y \cdot X}^2}{n\sum_{i=1}^{n}(x_i - \bar{y})^2}\right]\left(\sum_{i=1}^{n} x_i^2 + nx_i^2 - 2\bar{x}\sum_{i=1}^{n} x_i\right)$$

$$= \left[\frac{s_{Y \cdot X}^2}{n\sum_{i=1}^{n}(x_i - \bar{x})^2}\right]\left[\sum_{i=1}^{n} x_i^2 + n\left(\frac{\sum_{i=1}^{n} x_i}{n}\right)^2 - 2\left(\frac{\sum_{i=1}^{n} x_i}{n}\right)\sum_{i=1}^{n} x_i\right]$$

$$= \left[\frac{s_{Y \cdot X}^2}{n\sum_{i=1}^{n}(x_i - \bar{x})^2}\right]\left[\sum_{i=1}^{n} x_i^2 - \frac{\sum_{i=1}^{n} x_i}{n}\right]^2 \tag{15.139}$$

$$= \left[\frac{s_{Y \cdot X}^2}{n\sum_{i=1}^{n}(x_i - \bar{x})^2}\right]\left[\sum_{i=1}^{n}(x_i - \bar{x})^2\right]$$

$$= s_{Y \cdot X}^2 / n$$

and the confidence interval is

$$\text{CI}_{X=\bar{x}} = \hat{y}_{\bar{x}} \pm t_c(k) s_{X \cdot Y} / \sqrt{n} \tag{15.140}$$

where $k = n - m = n - 2$.

As $|X - \bar{x}|$ increases, the magnitude of the confidence interval increases, as is shown in Figure 15.25.

An examination of why the confidence bands are not straight follows. Returning to the regression equation,

$$\hat{Y} = b_0 + b_1 X$$

Subtracting $\hat{y}$ from both sides

$$\hat{Y} - \bar{y} = b_0 + b_1 X - y$$

But, according to Equation 15.59, $b_0 = \bar{y} - b_1\bar{x}$. Consequently,

$$\hat{Y} - \bar{y} = \bar{y} - b_1\bar{x} + b_1 X - \bar{y}$$

When $X = \bar{x}$, $(X - \bar{x}) = 0$ and $\hat{Y} - \bar{y} = 0$, then, $\hat{Y} = \hat{y}_{\bar{x}}$ and

$$\hat{y}_{\bar{x}} = \bar{y} \tag{15.141}$$

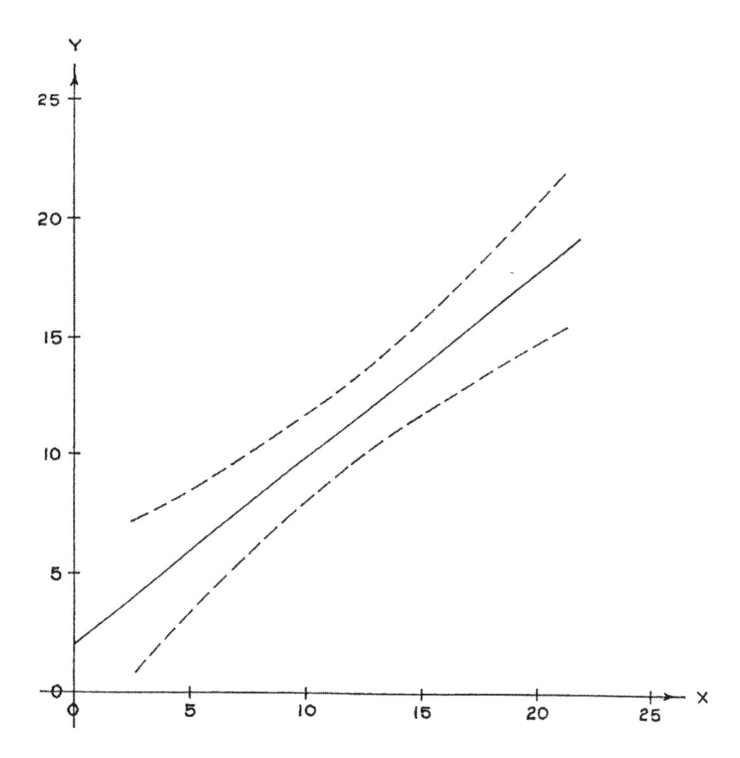


**FIGURE 15.25** A least squares regression line and its associated 0.95 confidence bands.

Furthermore, if $\bar{\hat{y}}$ is the arithmetic mean of the regression values corresponding to the set of $X$ values used in the previous development, then

$$\begin{aligned}
\bar{\hat{y}} &= \sum_{i=1}^{n} \hat{y}_i \Big/ n \\
&= \sum_{i=1}^{n} (b_0 + b_1 x_i) \Big/ n \\
&= \frac{\sum_{i=1}^{n} b_0}{n} + \frac{b_1 \sum_{i=1}^{n} x_i}{n} \\
&= b_0 + b_1 x \\
&= \bar{y} - b_1 \bar{x} + b_1 x \\
\bar{\hat{y}} &= y
\end{aligned} \tag{15.142}$$

Thus, the mean estimated $Y$ is equal to $\bar{y}$. This implies that each regression resulting from a sample of size $n$ drawn from the specified population will pass through the coordinate position $(\bar{x}, \bar{y})$. Since the set of $X$ values is assumed to be fixed, $\bar{x}$ is a constant, $\bar{y}$, however, will vary randomly from sample to sample. This is illustrated in Figure 15.26. The set of $\bar{y}$ values from all possible samples of size $n$ drawn from the specified population constitutes a sampling distribution of regression line elevations above (or below) the horizontal axis. The parent population from which

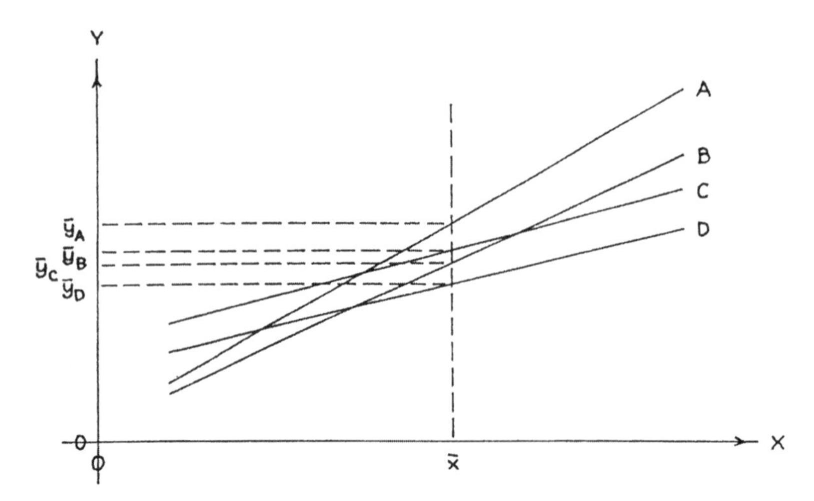


**FIGURE 15.26** A set of representative regression lines. Each passes through the position corresponding to the coordinates $(\bar{x}, \bar{y})$.

this sampling distribution is derived can be equated to the subpopulation of $Y$ values which is associated with $X = \bar{x}$. Since homoscedasticity is assumed, the variance of the subpopulation is equal to $\sigma^2_{Y \cdot X}$. Consequently,

$$V(\bar{y}) = \sigma^2_{Y \cdot X}/n \tag{15.143}$$

Then, since $\bar{y} = \hat{y}_{\bar{x}} = \bar{\hat{y}}$,

$$s^2_{\bar{y}} = s^2_{\hat{y}_{\bar{x}}} = s^2_{\bar{\hat{y}}} = s^2_{Y \cdot X} / n \rightarrow V(\bar{y}) \tag{15.144}$$

As can be seen from this development, when $X = \bar{x}$, the slope of the line, $b_1$, has no effect on the magnitude of the confidence interval.

If all possible samples of size $n$ were drawn from the specified population and the regression equations developed, each of the regressions would have its own slope or $b_1$ coefficient. If each of these $b_1$ values were used to plot a line passing through the coordinate position $[\bar{x}, E(\bar{y})]$, the lines would form a fanlike array pivoting on $[\bar{x}, E(\bar{y})]$. This is illustrated in Figure 15.27. Because of this fanlike distribution, the confidence bands about a given regression line must bow outward, away from the regression line, to take into account the fact that the slope of the line, as well as its elevation about the horizontal axis, is subject to random variation.

Using the data from the example in Section 15.3.8 and a confidence level of 0.95, $k = n - m = 7 - 2 = 5$, and

$$t_{0.95}(5) = 2.571$$

Furthermore,

$$E^{-1} = \begin{bmatrix} 0.6598493 & -0.0441335 \\ -0.0441335 & 0.0037675 \end{bmatrix}$$

Then,

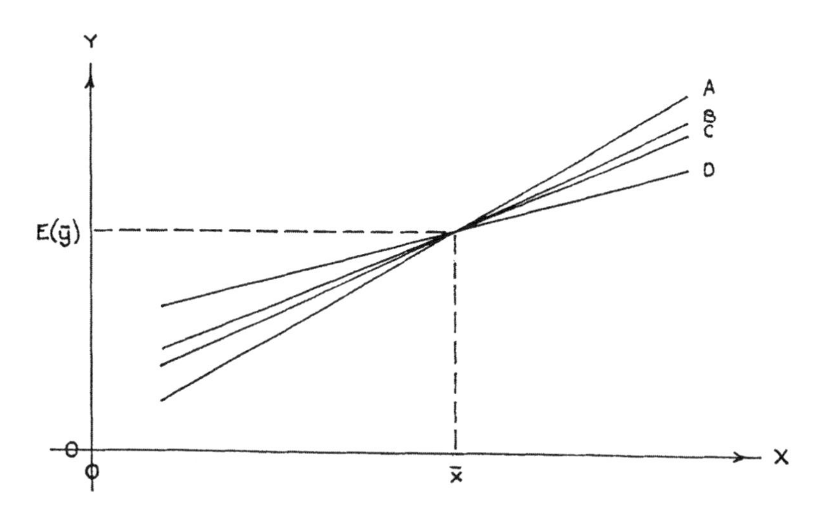


**FIGURE 15.27** A set of representative regression lines showing the variability of estimated slopes.

$$s_{b_0} = \sqrt{s_{Y \cdot X}^2 c_{00}}$$

$$= \sqrt{3.2776(0.6598493)} = 1.4706196$$

$$b_0 \pm t_{0.95}(5) s_{b_0} = 1.9925 \pm 2.571(1.4706196)$$

$$= 1.9925 \pm 3.7810 \rightarrow \beta$$

$$s_{b_1} = \sqrt{s_{Y \cdot X}^2 c_{11}} = \sqrt{3.2776(0.0037675)}$$

$$= 0.1111232$$

and

$$b_0 \pm t_{0.95}(5) s_{b_1} = 0.7933 \pm 2.571(0.1111232) = 0.7933 \pm 0.2857 \rightarrow \beta_1$$

The computations for the confidence bands are as follows:

| $X$ | CI |
|---|---|
| 3 | $\pm 2.571\sqrt{3.2776[0.6598493 + (3)^2(0.0037675 + 2(3)(-0.0441335)]} = \pm 3.0485$ |
| 5 | $\pm 2.571\sqrt{3.2776[0.6598493 + (5)^2(0.0037675 + 2(5)(-0.0441335)]} = \pm 2.6028$ |
| 9 | $\pm 2.571\sqrt{3.2776[0.6598493 + (9)^2(0.0037675 + 2(9)(-0.0441335)]} = \pm 1.9226$ |
| 11 | $\pm 2.571\sqrt{3.2776[0.6598493 + (11)^2(0.0037675 + 2(11)(-0.0441335)]} = \pm 1.7711$ |
| $\bar{x}$ = 11.714286 | $\pm 2.571\sqrt{3.2776[0.6598493 + \bar{x}^2(0.0037675 + 2\bar{x}(-0.0441335)]} = \pm 1.7593$ |
| 15 | $\pm 2.571\sqrt{3.2776[0.6598493 + (15)^2(0.0037675 + 2(15)(-0.0441335)]} = \pm 1.9940$ |

| $X$ | CI |
|---|---|
| 18 | $\pm 2.571\sqrt{3.2776[0.6598493 + (18)^2(0.0037675 + 2(18)(-0.0441335)]} = \pm 2.5140$ |
| 20 | $\pm 2.571\sqrt{3.2776[0.6598493 + (20)^2(0.0037675 + 2(20)(-0.0441335)]} = \pm 2.9494$ |
| 21 | $\pm 2.571\sqrt{3.2776[0.6598493 + (21)^2(0.0037675 + 2(21)(-0.0441335)]} = \pm 3.1832$ |

Then, when $X =$

| $X$ | $\hat{y}_i \pm \text{CI}$ |
|---|---|
| 3 | $= 4.3724 \pm 3.0485$ |
| 5 | $= 5.9590 \pm 2.6028$ |
| 9 | $= 9.1322 \pm 1.9226$ |
| 11 | $= 10.7188 \pm 1.7711$ |
| 11.714286 | $= 11.2854 \pm 1.7593$ |
| 15 | $= 13.8920 \pm 1.9940$ |
| 18 | $= 16.2719 \pm 2.5140$ |
| 20 | $= 17.8585 \pm 2.9494$ |
| 21 | $= 18.6518 \pm 3.1832$ |

These values were used to plot the confidence bands on Figure 15.25. The fiducial probability is 0.95 that the true regression line lies within the plotted bands.

The pattern of computation of confidence intervals described above for simple linear regression is generally applicable in the case of multiple regressions. The computations are, of course, more complex. The key lies in the variance–covariance matrix:

$$s_{Y \cdot X}^2 * E_{-1} = \begin{bmatrix} s_{Y \cdot X}^2 c_{00} = s_{b_0}^2 & s_{Y \cdot X}^2 c_{01} = s_{b_0 b_1} & \cdots & s_{Y \cdot X}^2 c_{0m} = s_{b_0 b_m} \\ s_{Y \cdot X}^2 c_{10} = s_{b_1 b_0} & s_{Y \cdot X}^2 c_{11} = s_{b_1}^2 & \cdots & s_{Y \cdot X}^2 c_{1m} = s_{b_1 b_m} \\ \vdots & \vdots & & \vdots \\ s_{Y \cdot X}^2 c_{m0} = s_{b_m b_0} & s_{Y \cdot X}^2 c_{m1} = s_{b_m b_1} & \cdots & s_{Y \cdot X} c_{mm} = s_{b_m}^2 \end{bmatrix} \quad (15.145)$$

where, according to Equations 15.106 and 15.115,

$$s_{Y \cdot X}^2 = \sum_{i=1}^{n} d_i^2 \Big/ k$$

$$= \left[\sum_{i=1}^{n} (y_i - \bar{y})^2 - b_1 \sum_{i=1}^{n} (x_{1i} - \bar{x}_1)(y_i - \bar{y}) - b_2 \sum_{i=1}^{n} (x_{2i} - \bar{x}_2)(y_i - \bar{y}) - \cdots \right.$$

$$\left. - b_m \sum_{i=1}^{n} (x_{mi} - \bar{x}_m)(y_i - \bar{y}) \right] \Big/ k$$

and $k = n - m - 1$.

The confidence intervals for the regression coefficients are then

$$b_0 \pm t_c(k) s_{b_0} = b_0 \pm t_c(k) \sqrt{s_{Y \cdot X}^2 c_{00}} \rightarrow \beta_0 \quad (15.146)$$

$$b_1 \pm t_c(k) s_{b_1} = b_1 \pm t_c(k) \sqrt{s_{Y \cdot X}^2 c_{11}} \rightarrow \beta_1 \quad (15.147)$$

$$b_m \pm t_c(k)s_{b_m} = b_m \pm t_c(k)\sqrt{s^2_{Y \cdot X} c_{mm}} \rightarrow \beta_0 \tag{15.148}$$

Confidence bands cannot be satisfactorily plotted about a multiple regression. However, the confidence limits can be determined for specific sets of values of the independent variables using Equation 15.138,

$$\text{CI} = \hat{y}_i \pm t_c(k)s_{\hat{y}_i}$$

where, expanding Equation 15.135 to recognize the additional independent variables,

$$s_{\hat{y}_i} = s_{Y \cdot X}\sqrt{c_{00} + x^2_{1i}c_{11} + x^2_{2i}c_{22} + \cdots + x^2_{mi}c_{mm} + 2[x_{1i}c_{01} + x_{2i}c_{01} + \cdots + x_{mi}c_{0m} + x_{1i}x_{2i}c_{12} + x_{1i}x_{3i}c_{13} + \cdots + x_{1i}x_{mi}c_{1m} + \cdots + x_{(m-1)i}x_{mi}c_{(m-1)m}} \tag{15.149}$$

Applying these formulations to the data for the example in Section 15.3.9,

$$\hat{y} = 0.4533X_1 + 2.1241X_2 - 3.1488$$

If the confidence level is set at 0.95 and the confidence limits when $X_1 = 5$ and $X_2 = 3$ are to be determined,

$$\hat{y} = 0.4533(5) + 2.1241(3) - 3.1488 = 5.4900$$

$$k = 26 - 3 = 23$$

$$t_{0.95}(23) = 2.069$$

Then,

$$s^2_{\hat{y}} = s^2_{Y \cdot X}[c_{00} + x^2_{1i}c_{11} + x^2_{2i}c_{22} + 2(x_1c_{01} + x_2c_{02} + x_1x_2c_{12})]$$

$$= 0.1830009\{0.5852773 + (5)^2(0.0129169) + (3)^2(0.0293762)$$

$$+ 2[5(-0.0497667) + 3(-0.0927942) + 5(3)(-0.0022464)]\}$$

$$= 0.0092894$$

$$s_{\hat{y}} = 0.0963814$$

and

$$\hat{y} \pm t_c(k)s_{\hat{y}} = 5.4900 \pm 2.069(0.0963814)$$

$$= 5.4900 \pm 0.1994$$

or

$$p(5.6894 < \mu_{Y|X_1=5,X_{3=3}} < 29.0587) = 0.95$$

Occasions arise when one possesses a regression that describes the relationship between a certain response variable (e.g., site index) and one or more independent variables (e.g., relative slope position, slope steepness, aspect, etc., as seen on vertical aerial photographs). If observations are made of these independent variables at a given location and these observations are then entered into the regression equation, the result will be an estimate of the *mean* value of the response variable under those conditions. The actual value of the response variable at that specific locality is apt to be different from the regression estimate. If one wishes to make an interval estimate of this actual value rather than the mean value, it is necessary to modify Equations 15.132 and 15.143 by adding a 1 under the radical. For example, in the case of Equation 15.132,

$$\mathrm{CI}_{X=x_i} = \hat{y}_i \pm t_c(k)s_{Y\cdot X}\sqrt{1 + c_{00} + x_i^2 c_{11} + 2x_i c_{01}} \tag{15.150}$$

This indicates that the fiducial probability is $c$ that the true value of the response variable at that specific locality lies in the interval defined by Equation 15.150. The added component recognizes the variability of the individual data points as well as the variability of the regression as a whole.

## 15.7 CORRELATION

As was brought out in the introduction to this chapter and earlier in Section 9.8, *correlation* is concerned with the *strength* of relationships between variables. In other words, when the magnitudes of two or more variables are determined from each sampling unit and these are arranged in vectors, how well do the values keep in step with one another from vector to vector? For example, assume that observations were made of two variables on each sampling unit and the data took the following form:

$$[X_1, X_2] = [2,3];\ [4,5];\ [7,8];\ \ldots;\ [30,31]$$

It is obvious that as $X_1$ increases, $X_2$ also increases or keeps in step with $X_1$. Correlation is concerned with the degree of such relationships.

When correlation is under consideration, the *scatter diagram,* rather than the moving mean, provides the relevant picture of the relationship. When the scatter is circular, as in Figure 15.28A, there is no evidence of a trend and it is said that no correlation exists between the variables. When the scatter forms a broad ellipse, as in Figure 15.28B, a trend is evident but is poorly defined and it is said that the correlation is poor. If the scatter forms a narrow ellipse, as in Figure 15.28C, the trend is well defined and the correlation between the variables is good. If the scatter forms a single line (a degenerate ellipse) as in Figure 15.28D, the correlation is perfect. The remainder of this discussion is devoted to measures quantifying images of this type.

Before a valid measure can be developed, certain conditions must exist:

1. No variable is to be considered as the response or dependent variable. Consequently, no variable is identified by $Y$. Instead, all are identified by a subscripted $X$. Thus, if two variables are involved they are designated $X_1$ and $X_2$.
2. All the values for *all* the variables must be randomly drawn. This is in contrast to regression, where values of the independent or predicting variables can be deliberately chosen.
3. The variances of the variables must be homogeneous, or nearly so, across the ranges of values of the companion variable(s).

In Section 15.3.8 a set of data was assumed which involved two random variables. One of these was designated the response variable, $Y$. Thus, the data took the form:

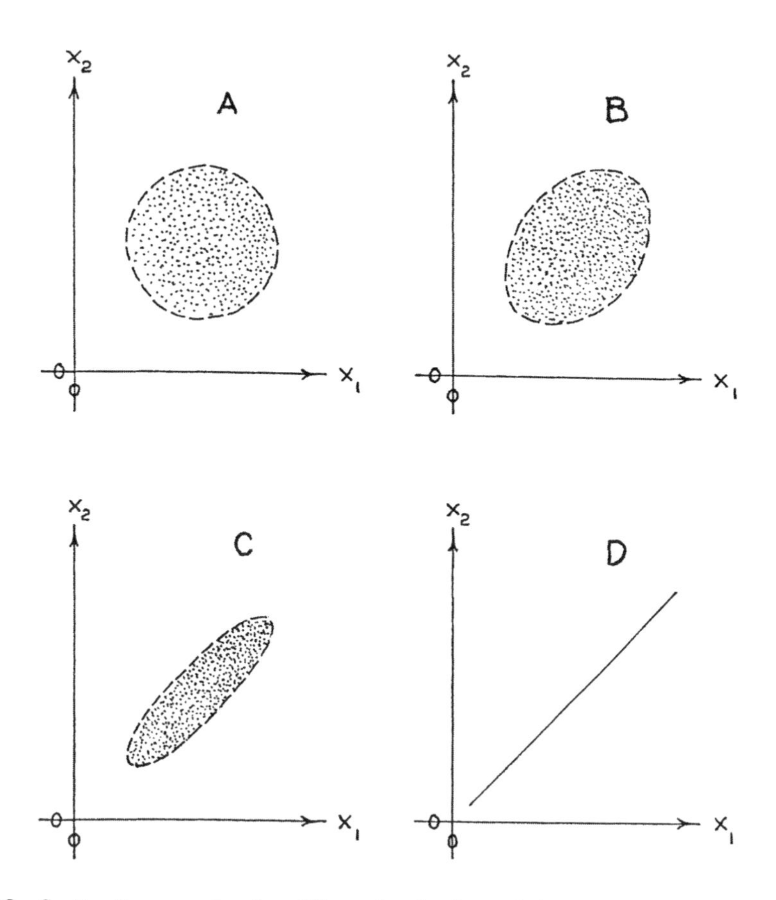


**FIGURE 15.28** Scatter diagrams showing different levels of correlation between two variables.

| $X$ | 3 | 5 | 9 | 11 | 15 | 18 | 21 |
|---|---|---|---|---|---|---|---|
| $Y$ | 4 | 7 | 10 | 9 | 12 | 19 | 18 |

In correlation operations, there is no response variable. However, the mathematical operations carried out in Section 15.3.8 can still be carried out if the labeling of the variables is changed to

| $X_1$ | 3 | 5 | 9 | 11 | 15 | 18 | 21 |
|---|---|---|---|---|---|---|---|
| $Y_2$ | 4 | 7 | 10 | 9 | 12 | 19 | 18 |

When this is done, Equation 15.112, which yields the sum of squared residuals about the regression line, becomes

$$\sum_{i=1}^{n} d_i^2 = \sum_{i=1}^{n} (x_{2i} - \bar{x}_2)^2 - b_1 \sum_{i=1}^{n} (x_{1i} - \bar{x}_1)(x_{2i} - \bar{x}_2) \tag{15.151}$$

or

$$\text{Error SS} = \text{Total SS} - \text{Explained SS} \tag{15.152}$$

Then, the ratio

$$r^2 = \frac{\text{Explained SS}}{\text{Total SS}} \tag{15.153}$$

is referred to as the *coefficient of determination*. It indicates the *proportion of the total variation which can be attributed to the relationship between the two variables.*

The ratio can be written

$$r^2 = \frac{\sum_{i=1}^{n}(x_{2i} - \bar{x}_2)^2 - \sum_{i=1}^{n} d_i^2}{\sum_{i=1}^{n}(x_{2i} - \bar{x}_2)^2} \tag{15.154}$$

But,

$$\sum_{i=1}^{n} d_i^2 = \sum_{i=1}^{n}(x_{2i} - \bar{x}_2)^2 - b_1 \sum_{i=1}^{n}(x_{1i} - \bar{x}_1)(x_{2i} - \bar{x}_2)$$

Then,

$$r^2 = \frac{\sum_{i=1}^{n}(x_{2i} - \bar{x}_2)^2 - \sum_{i=1}^{n}(x_{2i} - \bar{x}_2)^2 + b_1 \sum_{i=1}^{n}(x_{1i} - \bar{x}_1)(x_{2i} - \bar{x}_2)}{\sum_{i=1}^{n}(x_{2i} - \bar{x}_2)^2}$$

$$= \frac{b_1 \sum_{i=1}^{n}(x_{1i} - \bar{x}_1)(x_{2i} - \bar{x}_2)}{\sum_{i=1}^{n}(x_{2i} - \bar{x}_2)^2}$$

But,

$$b_1 = \frac{\sum_{i=1}^{n}(x_{1i} - \bar{x}_1)(x_{2i} - \bar{x}_2)}{\sum_{i=1}^{n}(x_{1i} - \bar{x}_1)^2}$$

Then,

$$r^2 = \frac{\left[\sum_{i=1}^{n}(x_{1i} - \bar{x}_1)(x_{2i} - \bar{x}_2)\right]^2}{\sum_{i=1}^{n}(x_{1i} - \bar{x}_1)^2 \sum_{i=1}^{n}(x_{2i} - \bar{x}_2)^2} \tag{15.155}$$

If the right-hand member is multiplied by

$$\frac{|n-2|}{|n-2|}$$

then

$$r^2 = \frac{(s_{X_1X_2})^2}{s_{X_1}^2 s_{X_2}^2} \tag{15.156}$$

$$r = \sqrt{r^2} = \frac{s_{X_1X_2}}{\sqrt{s_{X_1}^2 s_{X_2}^2}} \rightarrow \rho(X_1X_2) = \frac{\text{COV}(X_1X_2)}{\sqrt{V(X_1)V(X_2)}} \tag{15.157}$$

is the sample-based *correlation coefficient,* which is an estimator of ρ, the population correlation coefficient (see Equation 9.32).

If $X_1$ had been used as the "response variable" in the computations instead of $X_2$, the regression coefficients would have changed but the coefficient of determination and the correlation coefficient would remain unchanged. Thus, it can be said that there is no dependent variable insofar as correlation operations are concerned.

In the case where more than two variables are involved, the basic formula, Equation 15.153, remains unchanged except for the label for the ratio:

$$R^2 = \frac{\text{Explained SS}}{\text{Total SS}} \tag{15.158}$$

$R^2$ is known as the *coefficient of multiple determination* and $R$, the square root of $R^2$, is called the *coefficient of multiple correlation*. When more than two variables are involved, one must arrive at the explained and total sums of squares using the conventional formulae:

$$\text{Explained SS} = b_1\sum_{i=1}^{n}(x_{1i}-\bar{x}_1)(x_{2i}-\bar{x}_2) + b_2\sum_{i=1}^{n}(x_{1i}-\bar{x}_1)(x_{3i}-\bar{x}_3) + b_3\sum_{i=1}^{n}(x_{1i}-\bar{x}_1)(x_{3i}-\bar{x}_3) + \cdots$$

and

$$\text{Total SS} = \sum_{i=1}^{n}(x_{i1}-\bar{x}_1)^2$$

The coefficients of determination, $r^2$, and multiple determination, $R^2$, are proportions and range in magnitude from 0 to 1. In other words, the proportion of the total sum of squares that can be attributed to the relationship between the variables can only range from 0 to 1, or from 0 to 100%. Obviously, neither can be negative. On the other hand, $r$ and $R$ are square roots and may be either positive or negative. Only the positive root of $R^2$ is used because no meaning can be assigned to the negative root. In the case of $r$, however, the signs may be either positive or negative, indicating the direction of slope of the associated regression. These signs appear when $r$ is computed using the "product–moment" formula (which is the square root of Equation 15.155):

$$r = \frac{\sum_{i=1}^{n}(x_{1i} - \bar{x}_1)(x_{2i} - \bar{x}_2)}{\sqrt{\sum_{i=1}^{n}(x_{1i} - \bar{x}_1)^2 \sum_{i=1}^{n}(x_{2i} - \bar{x}_2)^2}} \tag{15.159}$$

It also must be emphasized that since $r$ and $R$ are square roots of proportions *they cannot be expressed as percents.*

As an example, one can determine the coefficient of determination and the correlation coefficient in the case of the data set used earlier in this section. Reference is made to Section 15.3.8 for the sums of squares and cross-products. When $X_2$ acts as the "response" variable,

$$\sum_{i=1}^{n}(x_{1i} - \bar{x}_1)^2 = 265.42857$$

$$\sum_{i=1}^{n}(x_{2i} - \bar{x}_2)^2 = 183.42857$$

$$\sum_{i=1}^{n}(x_{1i} - \bar{x}_1)(x_{2i} - \bar{x}_2) = 210.57143$$

and

$$b_1 = 0.7933262$$

From these one can compute

$$\text{Explained SS} = b_1 \sum_{i=1}^{n}(x_{1i} - \bar{x}_1)(x_{2i} - \bar{x}_2)$$

$$= 0.7933262(210.57143) = 167.05183$$

$$r^2 = 167.05183 / 183.42857 = 0.9107 \quad \text{or} \quad 91.07\%$$

and

$$r = \sqrt{0.9107} = 0.9543$$

If $X_1$ had been chosen as the "response" variable,

$$b_1 = \sum_{i=1}^{n}(x_{1i} - \bar{x}_1)(x_{2i} - \bar{x}_2) \Big/ \sum_{i=1}^{n}(x_{2i} - \bar{x}_2)^2$$

$$= 210.57143 / 183.42857 = 1.1479751$$

Then

$$\text{Explained SS} = 1.1479751(210.57143) = 241.73076$$

$$r^2 = 241.73076/265.42857 = 0.9107 \text{ or } 91.07\%$$

and

$$r = \sqrt{0.9107} = 0.9543$$

As can be seen, the $r^2$ and $r$ values are the same regardless of which variable was chosen as the "response" variable. $r^2$ and $r$ values as high as these indicate a strong relationship between the variables. If they had equaled 1, the relationship would be perfect, while it would be nonexistent if they had equaled zero. The terms *poor* and *good* are not well defined in terms of any of the measures of correlation. Consequently, each person must evaluate the magnitudes of $r^2$ and $r$ in the context of the problem under investigation.

As an example of the computation of the coefficients of multiple determination and multiple correlation, one can use the data from the multiple regression example in Sections 15.3.9 and 15.4. The relevant data are

$$\sum_{i=1}^{n} (x_{3i} - \bar{x}_3)^2 = \sum_{i=1}^{n} (y_i - \bar{y})^2 = 187.5385$$

and

$$\sum_{i=1}^{n} d_i^2 = 4.2090$$

Then

$$\text{Explained SS} = \sum_{i=1}^{n} (x_{3i} - \bar{x}_3)^2 - \sum_{i=1}^{n} d_i^2$$

$$= 187.5385 - 4.2090 = 183.3295$$

$$R^2 = 183.3295 / 187.5385 = 0.9776$$

and

$$R = \sqrt{0.9776} = 0.9887$$

The magnitude of the correlation coefficient indicates how close a match would be obtained if a regression of $X_1$ on $X_2$ were superimposed on a regression of $X_2$ on $X_1$. If $r = 1$, the two lines would coincide. If $0 < r < 1$, the two lines would form an angle $\alpha$ such that $0° < \alpha < 90°$. Finally, if $r = 0$, the two lines would form an angle of 90°. This points up why an independent variable cannot be predicted from a dependent variable by working backwards through a regression. For example, if a regression has been fitted where total tree height is the response variable and dbh is

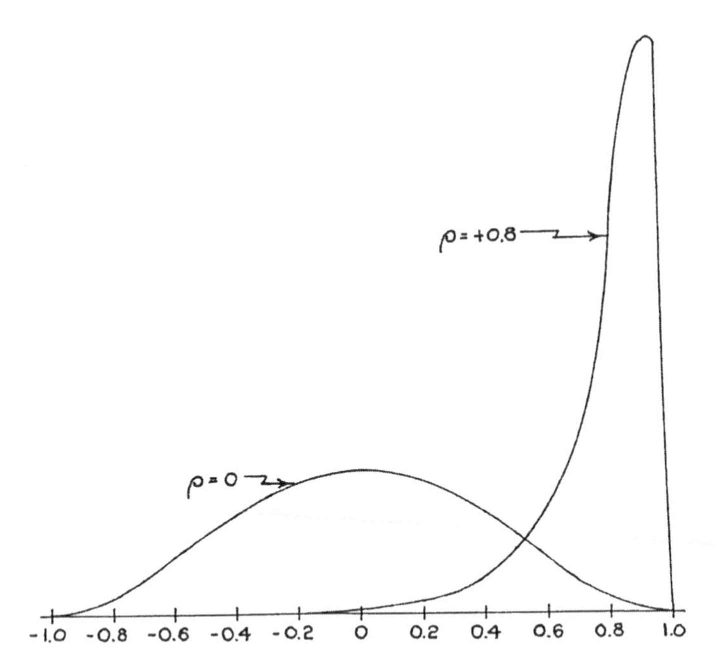


**FIGURE 15.29** Representative sampling distributions of the correlation coefficient, *r*. (From Snedecor, 1946.)

the predicting or independent variable, one cannot work backwards through that regression to predict dbh from total tree height *unless the two are perfectly correlated.*

If it becomes necessary to predict the initial predicting variable using the initial response variable, it is necessary to fit a new regression with the roles of the two variables reversed from those they played when the original regression was fitted.

The pdfs of the sampling distributions of none of the measures of correlation have been satisfactorily determined. In the case of the correlation coefficient, $r$, the sampling distribution is mound shaped but is symmetrical only when $\mu_r = \rho = 0$. Furthermore, the range of possible values of $r$ is only from –1 to +1, which means that the tails of the distribution terminate at these points. When $\rho \neq 0$ the distribution is skewed with the skew becoming extreme as ρ approaches the limits –1 and +1, as can be seen in Figure 15.29. Because of this, $r$ cannot be modeled by the normal. Fisher (1921), however, has suggested that $r$ could be transformed into a variable that would be almost normally distributed using the transformation

$$z = \frac{1}{2}[\ln(1+r) - \ln(1-r)] = \frac{1}{2}\ln\left(\frac{1+r}{1-r}\right) \tag{15.160}$$

or

$$z = 1.1513\log\left(\frac{1+r}{1-r}\right) \tag{15.161}$$

The variance of $z$ is

$$V(z) = 1/(n-3) \tag{15.162}$$

A confidence interval can then be determined for $z$:

$$\text{CI} = z \pm t_c(k = \infty)s_z \tag{15.163}$$

where

$$s_z = \sqrt{\frac{1}{n-3}} \tag{15.164}$$

The limits of this interval can then be converted into their equivalent $r$ values:

$$r = (e^{2z} - 1)/(1 + e^{2z}) \tag{15.165}$$

Continuing the previous example where $r = 0.9543$ and $n = 7$,

$$z = \frac{1}{2}\ln\left(\frac{1+0.9543}{1-0.9543}\right) = 1.8778$$

$$s_z = \sqrt{\frac{1}{7-3}} = 0.5$$

$$t_{0.95}(\infty) = 1.96$$

Then, the confidence interval estimate for $Z$ is

$$1.8778 \pm 1.96(0.5) = 1.8778 \pm 0.98 \rightarrow Z$$

or

$$P(0.898 < Z < 2.858) \cong 0.95$$

and the confidence limits for the estimate of $\rho$ are

$$\begin{aligned} r_a &= [e^{2(0.898)} - 1]/[1 + e^{2(0.898)}] \\ &= (6.0255 - 1)/(1 + 6.0255) = 0.7153 \\ r_b &= [e^{2(2.858)} - 1]/[1 + e^{2(2.858)} \\ &= (303.6877 - 1)/(1 + 303.6877) = 0.9934 \end{aligned}$$

and

$$P(0.7153 < \rho < 0.9934) \cong 0.95$$

Notice that the confidence interval is asymmetric, which is to be expected when the sampling distribution of $r$ is skewed. It also should be noted that the size of the confidence interval is dependent only on sample size. The greater the sample, the smaller the confidence interval. Finally, it must be emphasized that the procedure for obtaining confidence interval estimates of $\rho$ is applicable only in the case where two variables are involved. It cannot be used with the coefficient of multiple correlation, $R$.

## 15.8 HYPOTHESIS TESTING IN REGRESSION OPERATIONS

Sample-based estimates of the regression coefficients are random variables and tests can be made of hypotheses concerning their magnitudes using methods similar to those described in Chapter 14. For example, it can be hypothesized that $b_1$ in a simple linear regression is equal to zero. This is another way of hypothesizing that the independent variable really has no effect on the response variable and that $b_1$ has a magnitude other than zero only because of random chance. To demonstrate this, consider a situation where a regression is to be developed where the response variable is pine volume per acre obtained on randomly located sample plots using conventional procedures and the independent variable is the volume on the same plots obtained by a certain procedure making use of vertical aerial photographs. Since the aerial photographic procedure is less time-consuming and less expensive than the ground procedure, it would be a desirable alternative if proven reliable. To this end ten plots were randomly chosen (in real life this would be too light a sample) and the timber volumes determined using both methods. The data are as follows:

| X<br>Photo-Based Volume<br>(bd. ft/acre) | Y<br>Ground-Based Volume<br>(bd. ft/acre) |
|---|---|
| 1710 | 1820 |
| 2200 | 2380 |
| 2160 | 2360 |
| 2500 | 2640 |
| 2400 | 2560 |
| 1340 | 1300 |
| 2500 | 2350 |
| 1900 | 1980 |
| 2600 | 2860 |
| 1300 | 1340 |

$$n = 10, \quad \sum_{i=1}^{10} x_i = 20,610, \quad \sum_{i=1}^{10} y_i = 21,590, \quad \sum_{i=1}^{10} x_i^2 = 44,545,300$$

$$\sum_{i=1}^{10} y_i^2 = 49,177,700, \quad \sum_{i=1}^{10} x_i y_i = 46,746,800$$

$$\bar{x} = 2,061 \text{ bd. ft / acre}, \quad \bar{y} = 2,159 \text{ bd. ft / acre}$$

$$\sum_{i=1}^{10} (x_i - \bar{x})^2 = 2,068,090, \quad \sum_{i=1}^{10} (y_i - \bar{y})^2 = 2,564,890$$

$$\sum_{i=1}^{n} (x_i - \bar{x})(y_i - \bar{y}) = 2,249,810$$

$$b_0 = -83.097012, \quad b_1 = 1.0878685$$

$$\hat{Y} = 1.088X - 83.1 \text{ bd. ft / acre}$$

The inverse matrix with the $c$-multipliers:

$$E^{-1} = \begin{bmatrix} 2.1539343 & -0.0009966 \\ -0.0009966 & 0.0000004835 \end{bmatrix}$$

It should be noted that if one were to compute $b_0$ and $b_1$ using matrix algebra the results would be different from those shown because of rounding. When very large numbers are multiplied by very small numbers, rounding back can cause the results to drift substantially. Continuing,

$$\sum_{i=1}^{10} d_i^2 = 2,564,890 - 1.0878685(2,249,810) = 117,392.57$$

$$s_{Y\cdot X}^2 = 117,392.57 / (10.2) = 14,674.071$$

$$s_{b_1}^2 = 14,674.071(0.0000004835) = 0.007095$$

$$s_{b_1} = \sqrt{0.007095} = 0.08423$$

$$s_{b_0}^2 = 14,674.071(2.1539343) = 31,606.985$$

$$s_{b_0} = \sqrt{31,606.985} = 177.78354$$

The analysis of variance is as follows:

**ANOVA**

| Source | df | SS | MS |
|---|---|---|---|
| Total | 10 – 1 = 9 | 2,564,890 | |
| Model | 1 | 2,447,497 | 2,447,497 |
| Error | 10 – 2 = 8 | 117,393 | 14,674.125 |

Then,

The hypotheses:

$H_0$: $\beta_1 = 0$.

$H_1$: $\beta_1 \neq 0$.

Level of significance: 0.05.

Type of test: $t$, two-tailed, because $\beta_1$ may be less than or greater than zero.

Decision rule:

*Reject* $H_0$ and *accept* $H_1$ if $|t| \geq t_{0.05}(8) = 2.306$.

*Accept* $H_0$ if $|t| < 2.306$.

Computations:

$$t = (b_1 - \beta_1) / s_{b_1} = (1.0878685 - 0) / 0.08423$$

$$= 12.915452 > 2.306^{**}$$

The null hypothesis is rejected and it is assumed that a real relationship exists between the two variables.

Another possible hypothesis that could be tested might be that $\beta_1 = 1$, indicating that, as the aerial photographic estimate increases 1 bd. ft/acre, the ground cruise estimate also increases 1 bd. ft/acre. Then,

The hypotheses:

$H_0$: $\beta_1 = 0$.

$H_1$: $\beta_1 \neq 0$.

Level of Significance: 0.05.

Type of test: $t$, two-tailed, since $\beta_1$ may be either less than or more than 1.

Decision rules:

*Reject* $H_0$ and *accept* $H_1$ if $|t| \geq t_{0.05}(8) = 2.306$.

*Accept* $H_0$ if $0.5\ |t| < 2.306$.

Computations:

$$t = (b_1 - \beta_1)/s_{b_1} = (1.0878685 - 1)/0.08423$$

$$= 1.0431972 < 2.306 \text{ N.S.}$$

The null hypothesis is accepted and it is assumed that the ground and aerial photographic estimates increase at the same rate.

One can continue this problem and test for the presence of a bias. It would be best to do this using a paired $t$ test as was done with similar data in Section 14.13. However, one can test for bias using $t$ tests involving the $y$-intercept, $b_0$. If, prior to the test, one cannot logically assign a sign to the suspected bias, one would use a two-tailed $t$ test to test the hypotheses: $H_0$: $\beta_0 = 0$ and $H_1$: $\beta_0 \neq 0$. However, if there is reason to believe that a bias exists and it possesses a specific sign, the test would have to be one-tailed. In the case of the present example, assume that the aerial photographic method makes use of an individual tree volume table which requires the measurement of the crown width and total height of each tree on each plot. When this is done, it usually is impossible to evaluate many of the trees because they are hidden in the canopy or are merged into tree clumps. Consequently, one would expect the volumes obtained from the ground cruise to be somewhat higher than those obtained from the aerial photographic cruise. Then,

The hypotheses:

$H_0$: $\beta_1 \leq 0$ (i.e., there is no bias or, if one exists, its sign is negative).

$H_1$: $\beta_0 > 0$.

Level of significance: 0.05.

Type of test: $t$, one-tailed.

Decision rules:

*Reject* $H_0$ and *accept* $H_1$ if $t \geq t_{0.05}(8) = 1.860$.

*Accept* $H_0$ if $t < 1.860$.

Computations:

$$t = (b_0 - \beta_0)/|s_{b_0}| = (-83.097012 - 0)/117.78354$$

$$= -0.7055062 < 1.860 \text{ N.S.}$$

The null hypothesis is accepted and it is assumed that no bias exists. It should be noted that if these data are used in a paired $t$ test the null hypothesis would be rejected and a positive bias would be assumed. The difference between the two test results arises from the greater sensitivity of the paired $t$ test.

The process is similar in the case of a multiple regression. As a basis for discussion, consider a situation where cubic foot volume per acre of loblolly pine in fully stocked stands, $Y$, is being predicted using stand age, $X_1$, and site index, $X_2$. The regression model is

$$Y = \beta_0 + \beta_1 X_1 + \beta_2 X_2$$

The data are from 15 1/5-acre plots located so as to obtain a good spread of values of the two independent variables. The values of $Y$ are rounded to simplify calculations.

| $X_1$ (years) | $X_2$ | $Y$ (ft³/acre) | $X_1$ | $X_2$ | $Y$ | $X_1$ | $X_2$ | $Y$ |
|---|---|---|---|---|---|---|---|---|
| 10 | 74 | 150 | 30 | 85 | 300 | 50 | 69 | 480 |
| 19 | 85 | 240 | 34 | 85 | 310 | 56 | 67 | 470 |
| 25 | 87 | 400 | 40 | 86 | 540 | 58 | 73 | 800 |
| 28 | 79 | 300 | 44 | 69 | 300 | 83 | 78 | 630 |
| 29 | 77 | 280 | 45 | 84 | 540 | 98 | 80 | 700 |

$$\sum_{i=1}^{15} x_{1i} = 655, \quad \sum_{i=1}^{15} x_{2i} = 1178, \quad \sum_{i=1}^{15} y_i = 6240$$

$$\sum_{i=1}^{15} x_{1i}^2 = 35{,}977, \quad \sum_{i=1}^{15} x_{2i}^2 = 93{,}166, \quad \sum_{i=1}^{15} y_i^2 = 2{,}966{,}000$$

$$\sum_{i=1}^{15} (x_1 x_2)_i = 50{,}865, \quad \sum_{i=1}^{15} (x_1 y)_i = 318{,}130, \quad \sum_{i=1}^{15} (x_2,\ y)_i = 489{,}460$$

$$\bar{x}_1 = 43.7, \quad \bar{x}_2 = 78.5, \quad \bar{y} = 416.0$$

In matrix form, the normal equations are

$$\begin{bmatrix} 15 & 655 & 1178 \\ 655 & 35{,}977 & 50{,}865 \\ 1178 & 50{,}865 & 93{,}166 \end{bmatrix} * \begin{bmatrix} b_0 \\ b_1 \\ b_2 \end{bmatrix} = \begin{bmatrix} 6240 \\ 318{,}130 \\ 489{,}460 \end{bmatrix}$$

The inverted matrix, $E^{-1}$, is then

$$\begin{bmatrix} 11.348260 & -0.016397 & -0.134536 \\ -0.016297 & 0.000146 & 0.000128 \\ -0.134536 & 0.000128 & 0.001642 \end{bmatrix}$$

and the regression coefficients are

$$b_0 = 11.348260(6240) - 0.016397(318{,}130) - 0.134536(489{,}460) = -253.2$$

$$b_1 = -0.016397(6240) + 0.000146(318{,}130) + 0.000128(489{,}460) = 6.78058$$

$$b_2 = -0.134536(6240) + 0.000128(318{,}130) + 0.001642(489{,}460) = 4.90932$$

Consequently, the regression equation is

$$\hat{Y} = 6.78058X_1 + 4.90932X_2 - 253.2 \text{ ft}^3\text{/acre}$$

The analysis of variance is

**ANOVA**

| Source | df | SS | MS |
|---|---|---|---|
| Total | 15 – 1 = 14 | 370,160 | |
| Model | 2 | 306,647 | 153,323.5 |
| Error | 12 | 63,513 | 5292.8 |

$$\text{Total SS} = \sum_{i=1}^{15} (y - \bar{y})_i^2 = 2,966,000 - (6240)^2 / 15 = 370,160$$

$$\text{Model SS} = b_1 \sum_{i=1}^{15} [(x_1 - \bar{x})(y - \bar{y})]_i + b_2 \sum_{i=1}^{15} [(x_2 - \bar{x}_2)(y - \bar{y})]_i$$

$$= 6.78058(45,650) + 4.90932(-588) = 306,647$$

$$\text{Error SS} = \text{Total SS} - \text{Model SS}$$

$$= 370,160 - 306,647 = 63,513$$

One of the objectives of this investigation is to determine if site index, $X_2$, is an effective predicting variable in this context. The test takes the following form:

The hypotheses:
- $H_0$: $\beta_2 = 0$.
- $H_1$: $\beta_2 \neq 0$.

Level of significance: 0.05.
Type of test: $t$, two-tailed, because $\beta_2$ could have either a positive or negative effect.
Decision rules:
- *Reject* $H_0$ and *accept* $H_1$ if $|t| \geq t_{0.05}(12) = 2.179$.
- *Accept* $H_0$ if $|t| < 2.179$.

Computations:

$$s_{Y \cdot X}^2 = \sum_{i=1}^{15} d_i^2 / 12 = 63,513 / 12 = 5292.8$$

$$s_{b_2} = \sqrt{s_{Y \cdot X}^2 c_{33}} = \sqrt{5292.8(0.001642)} = 2.9480125$$

$$t = (b_2 - \beta_2) / s_{b_2} = (4.90932 - 0.00000) / 2.9480125 = 1.665 \text{ N.S.}$$

The null hypothesis is accepted. It is assumed that site index has no effect on the cubic foot volume per acre of loblolly pine. In such a case it would be logical to delete site index as a predicting variable and to recompute $b_0$ and $b_1$, thus simplifying the regression equation. Incidentally, one should not use this example as an argument that site index has no effect on stand volume. The sample is very light and probably not representative of the actual situation.

Another hypothesis that can be tested is that an increase of 1 year of stand age will increase the stand volume by a minimum of some number of cubic feet per acre. One can set that magnitude at 7 ft³/acre. Furthermore, one can continue to use site index as a predicting variable. Then,

The hypotheses:

$H_0$: $\beta_1 \geq 7.0$.

$H_1$: $\beta_1 < 7.0$.

Level of significance: 0.05.

Type of test: $t$, one-tailed.

Decision rules:

*Reject* $H_0$ and *accept* $H_1$ if $t \leq -t_{0.05}(12) = -1.782$.

*Accept* $H_0$ if $t > -1.782$.

Computations:

$$s_{b_1} = \sqrt{s^2_{Y \cdot X} c_{22}} = \sqrt{5292.8(0.000146)} = 0.8790613$$

$$t = (b_1 - \beta_1 / | s_{b_1} | = (6.78058 - 7.00000)/ | 0.8790613 |$$

$$= 0.2496072 \text{ N.S.}$$

The null hypothesis, that as stand age is increased by 1 year the stand volume increases by a minimum of 7 ft³/acre, is accepted.

As can be seen from these examples, $t$ tests are used to test *components* of regressions. They cannot, however, be used when one wishes to evaluate an entire regression *model* or when alternative models are being compared.

First consider the evaluation of a certain model. Reference is made to Equation 15.115 and its associated analysis of variance. These refer to multiple regressions but the concepts that will be based on them are equally applicable to simple regressions. As can be seen, the "Total" sum of squares is made up of two components: the "Model" and the "Error" sums of squares. Dividing these component sums of squares by their associated degrees of freedom gives rise to their respective mean squares which, in this context, can be considered variances. Theoretically, if the regression model is inappropriate, the "Model" component would be equal to zero and the "Total" and "Error" sums of squares would be equal. However, since the fitted regression is based on incomplete data, which can and does vary from sample to sample, the fact of an actual zero effect can be masked. The $F$ test is used to determine whether a model effect is real or spurious.

In a case such as this, the hypotheses would be

$$H_0\colon\ b_1 \sum_{i=1}^{n} [(x_1 - \bar{x}_1)(y - \bar{y})]_i + \cdots + b_m \sum_{i=1}^{n} [(x_m - \bar{x}_m)(y - \bar{y})]_i \leq 0$$

$$H_1\colon\ b_1 \sum_{i=1}^{n} [(x_1 - \bar{x}_1)(y - \bar{y})]_i + \cdots + b_m \sum_{i=1}^{n} [(x_m - \bar{x}_m)(y - \bar{y})]_i > 0$$

The test would be a conventional one-tailed $F$ test because the model mean square cannot be negative. Then,

$$F = \text{Model MS/Error MS} \tag{15.166}$$

and the decision rules would be to

*Reject* $H_0$ and *accept* $H_1$ if $F \geq F_b\,(k_{\text{Model}}, k_{\text{Error}})$.
*Accept* $H_0$ if $F < F_b(k_{\text{Model}}, k_{\text{Error}})$.

As an example, one can use the regression of cubic foot volume per acre of loblolly pine on stand age and site index that was used earlier. Reference is made to the analysis of variance developed in that discussion. Set $\alpha$ at 0.05. Then,

$$F_{0.05}(2, 12) = 3.88 \quad \text{and} \quad F_{0.01}(2, 12) = 6.93$$

and

$$F = 153{,}323.5/5292.8 = 28.97^{**}$$

The null hypothesis is rejected and it is assumed that the chosen model describes the relationship adequately.

Obviously, as the effect of the model increases, the magnitude of the computed $F$ will also increase. The computed $F$ is then a measure of the strength of the relationship, but it should not be used in such a role. The coefficient of determination is to be preferred because it is a proportion of the total variability, which is, within the context of a specific data set, fixed. $F$, on the other hand, is a proportion of the error mean square, which is variable from model to model even when the data set remains constant.

The second way in which the $F$ test can be used to test regression models occurs when two different models can be used to produce a predicting equation for a certain response variable and a decision must be made regarding which one should be used. If both perform equivalently, it would be logical to use the simplest of the two when the fitted regression is used in an operational situation. This situation is not unusual. Often an investigator will use a model with a large number of independent variables, many of which are transformations of basic variables (e.g.; $X_1$, $X_2 = X_1^2$, $X_3 = \log X_1$, etc.). Some of these variables may appear to be superfluous. As an alternative a second model is developed which has been stripped of these possibly superfluous variables. A test must then be made to see if, indeed, those variables are unnecessary. This involves the testing of the models as wholes.

The test hinges on the relative magnitudes of the two error sums of squares when both models are fitted to the same data set. If the two sums are essentially equal, the two models perform essentially equally well and the simpler should be chosen because it would be easier to use. If they differ, the model with the smaller error sum of squares should be chosen because it does a better job of prediction.

In the following discussion, the model with the possibly superfluous variables will be referred to as the maximum model and will be labeled Max. The alternative model will be labeled Alt.

To carry out this test, it is necessary to fit both models to the same data set and then to compute their error sums of squares using Equation 15.115.

$$\text{Error SS} = \sum_{i=1}^{n} d_i^2$$

The absolute difference between the two sums is the *difference sum of squares* or the *reduction due to the added variable*

$$\text{Diff SS} = |\text{Error SS}_{\text{Max}} - \text{Error SS}_{\text{Alt}}| \tag{15.167}$$

Dividing the difference sum of squares by the difference between the numbers of degrees of freedom associated with the two error sums of squares gives rise to the *difference mean square*

$$\text{Diff MS} = \text{Diff SS}/|k_{\text{Max}} - k_{\text{Alt}}| \tag{15.168}$$

Theoretically, if the two models are equally good at predicting the response variable, the difference mean square would be equal to zero. However, since the test is based on sample data, the difference mean square is subject to random variation and must be appropriately tested. The hypotheses to be tested are

$H_0$: Diff MS = 0.
$H_1$: Diff MS > 0.

Then,

$$F = \text{Diff MS}/\text{Error MS}_{\text{Max}} \tag{15.169}$$

and the decision rules are to

*Reject* $H_0$ and *accept* $H_1$ if $F \geq F_b(1, k_{\text{Max}})$.
*Accept* $H_0$ if $F < F_b(1, k_{\text{Max}})$.

As an example of this process, one can again make use of the data for the cubic foot volume per acre of loblolly pine on stand age and site index as the maximum model. The alternative model will use only stand age as the predicting variable. The level of significance will be set at 0.05. Fitting both models to the original data set yields the two regression equations:

$$\hat{Y}_{\text{Max}} = 6.78058X_1 + 4.90932X_2 - 253.2 \text{ ft}^3/\text{acre}$$

and

$$\hat{Y}_{\text{Alt}} = 6.18955X_1 + 145.5 \text{ ft}^3/\text{acre}$$

where $X_1$ is stand age and $X_2$ is site index. The computations for the alternative model are as follows.

$$\sum_{i=1}^{n}(x_i - \bar{x})^2 = 35{,}977 - (655)^2/15 = 7375.3333$$

$$\sum_{i=1}^{n}(x_i - \bar{x})(y_i - \bar{y}) = 318{,}130 - (655)(6240)/15 = 45{,}650$$

$$b_1 = 45{,}650/375.3333 = 6.18955$$

$$b_0 = 416.0 - 6.18955(43.7) = 145.5$$

$$\text{Error SS}_{\text{Alt}} = 370{,}160 - 6.18955(45{,}650) = 87{,}607$$

Then,

$$\text{Diff SS} = |63{,}513 - 87{,}607| = 24{,}094$$

and

$$\text{Diff MS} = 24{,}094/|12 - 13| = 24{,}094$$

The decision rules are

*Reject* $H_0$ and *accept* $H_1$ if $F \geq F_{0.05}$ (1, 12) = 4.75.
*Accept* $H_0$ if $F < 4.75$.

Then,

$$F = 24{,}094/5292.8 = 4.55 \text{ N.S.}$$

The null hypothesis is accepted and it is assumed that both models yield essentially equivalent predictions. Because of its greater simplicity the alternative model would probably be used if predictions were to be made.

It should be noted that this example parallels the $t$ test made of the hypothesis that $\beta_2 = 0$, and the results were the same. In both cases it was concluded that site index was superfluous and could be eliminated from the predicting equation. These results can be verified graphically. Figure 15.30A shows the scattergram and regression of volume per acre on stand age. Note that a definite trend is evident. Compare this with Figure 15.30B, which shows volume per acre on site index. The regression line is essentially horizontal, indicating that volume per acre is about the same regardless of site index. Furthermore, the scattergram shows no evidence of a trend.

## 15.9 STEPWISE VARIABLE SELECTION

There are times when an investigator cannot logically develop a model for a predicting equation because the controlling conditions are not understood and the investigator is really feeling his or her way through a maze. In such a case the model must be generated *from the data*. These data should include values from as many of the potential independent variables as is possible. Additional potential independent variables are synthesized using functions of the primary variables (e.g., powers, roots, logarithms, etc.) and interactions among the variables. The list of candidate predicting variables can become quite large, often far too large for a feasible predicting equation. It then becomes necessary to screen these candidate predicting variables and include in the final regression equation only those needed to meet some arbitrarily chosen criterion, such as a given $R^2$. Even when a logical model can be developed, it may be necessary to reduce the number of predicting variables to arrive at a feasible operational version of the equation. This again would involve screening the predicting variables to determine which are of minor importance and can be deleted. There are several approaches that can be used to sort through the candidate variables to arrive at the final predicting equation. Collectively, they are referred to as *stepwise variable selection* methods. Unfortunately, there is a terminology problem with the term *stepwise* because it is used in a generic sense, as above, and also as the name of one of the fitting procedures. Failure to realize that this situation with two meanings exists can lead to confusion. In the following development stepwise variable selection in its generic sense will be discussed first, and then specific fitting procedures, including the stepwise method, will be described.

The "explained" portion of the total sum of squares, from Equation 15.115, is equal to

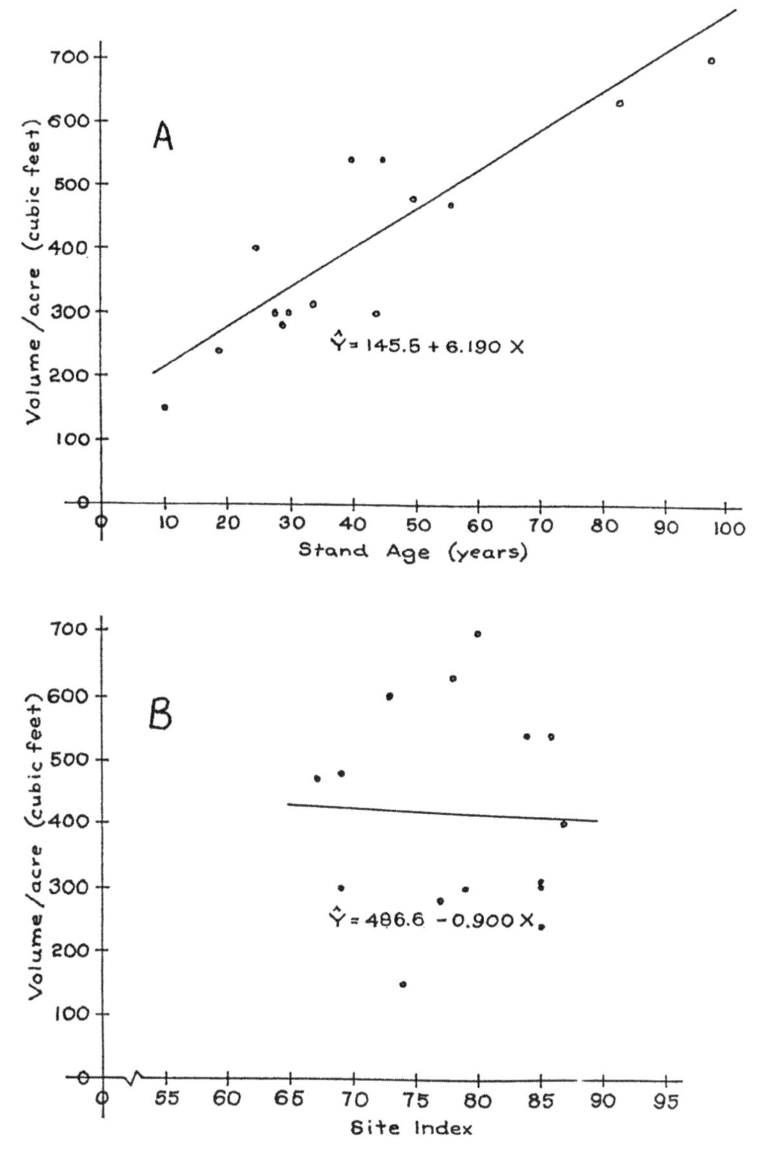


**FIGURE 15.30** Simple linear regressions of volume per acre of loblolly pine on A, stand age and B, site index. Note that the regression B is essentially horizontal or flat, indicating that volume per acre is the same for all site indices.

$$b_1 \sum_{i=1}^{n} (x_i - \bar{x})_1 (y_i - \bar{y}) + \cdots + b_m \sum_{i=1}^{n} (x_i - \bar{x})_m (y_i - \bar{y})$$

Each of the terms in this expression is associated with one of the independent variables and represents the *reduction* in the error sum of squares attributable to that variable. As was made evident in Section 15.7, when cubic foot volume per acre of loblolly pine was being predicted by stand age and site index, some predicting variables, such as stand age, produce large reductions, while others, such as site index, have little or no effect on the error sum of squares. The *potency* of a variable is its ability to reduce the error sum of squares. Stepwise variable selection involves the use of the potency of the variables to identify which variables should be included in the final predicting equation and which should be discarded. The process of accepting or deleting predicting variables is cyclic and continues until the investigator's objective has been reached. This objective may be to include the fewest predicting variables that will produce an error sum of squares equal to or less than some arbitrary magnitude or to produce a coefficient of multiple determination ($R^2$) equal to or more than some arbitrary limit. The objective also might be to include only those variables in the predicting equation that produce statistically significant reductions in the error sum of squares. The objective could also involve some combination of these criteria.

In *additive* or *step-up* modeling, also known as the *forward solution*, one first computes the simple linear regression linking each of the candidate predicting variables with the response variable. The error sum of squares is computed for each of these regressions and these sums are then ranked. The predicting variable associated with the *smallest* sum is considered to be the *most potent* individual variable. In other words, it is associated with the greatest reduction in the sum of squared residuals. One can refer to this most potent variable as $X_A$. Usually, a test is made of this variable to see if its associated reduction in the sum of squares is statistically significant (i.e., is real). This usually involves an $F$ test of the null hypothesis:

$$H_0\text{: } \beta_A \sum_{i=1}^{n} (x_i - \mu_X)_A (y_i - \mu_Y) = 0$$

but a $t$ test of the hypothesis:

$$H_0\text{: } \beta_A = 0$$

can also be used. In either case, the test procedure is as described in Section 15.8.

If $X_A$ is significant, each of the remaining candidate variables is *paired with* $X_A$ in a multiple regression. Again the error sums of squares are computed and ranked. The pair of variables associated with the smallest sum is judged to be the *most potent pair* ($X_A$ and $X_B$). As at the end of the first cycle, a test is made to see if the reduction in the sum of squares associated with $X_B$ is statistically significant. This usually takes the form of an $F$ test of the null hypothesis:

$$H_0\text{: } \beta'_A \sum_{i=1}^{n} (x_i - \mu_X)_A (y_i - \mu_Y) + \beta_B \sum_{i=1}^{n} (x_i - \mu_X)_B (y_i - \mu_Y)$$

$$- \; \beta_A (x_i - \mu_X)_A (y_i - \mu_Y) = \text{Diff SS} \; = \; 0$$

It should be noted that $\beta'_A$ is not the same as $\beta_A$. This reflects the fact that the effect of $X_A$ usually is different when it is paired with another predicting variable than it is when $X_A$ is acting above. The difference occurs when the predicting variables are correlated. When this is the case, it is said that *multicollinearity* is present.

If the effect of the pair is significant and/or the target magnitude of the error sum of squares has not been reached, each of the remaining candidate variables are teamed with the most potent pair in multiple regressions. The error sums of squares of these regressions are ranked and the *most potent trio*, ($X_A$, $X_B$, and $X_C$), is identified and the reduction in the sum of squares associated with $X_C$ is tested for significance.

This process is continued until the target error sum of squares is reached or until the candidate variables cease having significant effects, depending on how the investigator defined the objective.

As an example of the additive process, assume five independent variables: $X_1$, $X_2$, $X_3$, $X_4$, and $X_5$. Further assume a maximum allowable error sum of squares of 2000, a sample size of 30, a level of significance of 0.05, and a sum of squared residuals from $\bar{y}$ of 3,000,000. The simple linear regressions and their associated error sums of squares are

| | $\Sigma d^2$ |
|---|---|
| $\hat{Y} = b_0 + b_1X_1$ | 3800 |
| $\hat{Y} = b_0 + b_1X_2$ | 3500 |
| $\hat{Y} = b_0 + b_1X_3$ | 3000 ← |
| $\hat{Y} = b_0 + b_1X_4$ | 3200 |
| $\hat{Y} = b_0 + b_1X_5$ | 3300 |

$X_3$ has the smallest error sum of squares and is the most potent predicting variable. Testing it for significance:

**ANOVA**

| Source | df | SS | MS |
|---|---|---|---|
| Total (Alt) | $k_{Alt} = 20 - 1 = 29$ | 3,000,000 | |
| $\text{Error}_{Max}$ | $k_{Max} = 30 - 2 = 28$ | 3000 | 107.14286 |
| Difference (Max and Alt) | = 1 | 2,997,000 | 2,997,000 |

$$H_0\text{: Diff MS} = 0; \quad H_1\text{: Diff MS} > 0; \quad F_{0.05}(1 - 28) = 4.20$$

$$F = 2{,}997{,}000/107.14286 = 27{,}972^{**}$$

$X_3$ is associated with a highly significant reduction in the error sum of squares. Pairing $X_3$ with each of the remaining candidate variables and computing *new* regression coefficients and error sums of squares yields

| | $\Sigma d^2$ |
|---|---|
| $\hat{Y} = b_0' + b_1'X_3 + b_2'X_1$ | 2800 |
| $\hat{Y} = b_0' + b_1'X_3 + b_2'X_2$ | 2400 ← |
| $\hat{Y} = b_0' + b_1'X_3 + b_2'X_4$ | 2500 |
| $\hat{Y} = b_0' + b_1'X_3 + b_2'X_5$ | 2700 |

$X_3$ and $X_2$ form the most potent pair of variables. Testing them for significance:

**ANOVA**

| Source | df | SS | MS |
|---|---|---|---|
| $X_3$ alone (Alt) | $k_{Alt} = 28$ | 3000 | |
| $\text{Error}_{Max}$ | $k_{Max} = 30 - 3 = 27$ | 2400 | 88.888887 |
| Difference (Max and Alt) | = 1 | 600 | 600 |

$$H_0\text{: Diff MS} = 0, \quad H_1\text{: Diff MS} > 0, \quad F_{0.05}(1, 27) = 4.21$$

$$F = 600/88.888889 = 6.75^*$$

The reduction in the error sum of squares associated with $X_2$ is significant. Since the target error sum of squares has not yet been reached, the process is continued. Teaming $X_3$ and $X_2$ with each of the remaining candidate variables yields

| | $\Sigma d^2$ |
|---|---|
| $\hat{Y} = b_{0+}'' + b_1''X_3 + b_2''X_2 + b_3''X_1$ | 2400 |
| $\hat{Y} = b_0'' + b_1''X_3 + b_2''X_2 + b_3''X_4$ | 2100 ← |
| $\hat{Y} = b_0'' + b_1''X_3 + b_2''X_2 + b_3''X_5$ | 2500 |

$X_3$, $X_2$, and $X_4$ form the most potent trio. Testing for significance:

**ANOVA**

| Source | df | SS | MS |
|---|---|---|---|
| $X_3$ and $X_2$ (Alt) | $k_{Alt} = 27$ | 2400 | |
| $Error_{Max}$ | $k_{Max} = 30 - 4 = 26$ | 2100 | 80.769231 |
| Difference (Max and Alt) | $= 1$ | 300 | 300 |

$$H_0\text{: Diff MS} = 0, \quad H_1\text{: Diff MS} > 0, \quad F_{0.05}(1, 26) = 4.22$$

$$F = 300/80.769231 = 3.71 \text{ N.S.}$$

While the use of $X_4$ in the regression equation reduced the error sum of squares to 2100, which is greater than the target value, the reduction was not significant. On the basis of the $F$ test one would conclude that the reduction was due to random variation and was not real. In such a case the investigator would have to conclude that the target error sum of squares is unattainable with the available candidate predicting variables and that the investigator will have to settle for the regression equation making use of $X_3$ and $X_2$.

In *deletion* or *step-down* fitting, also known as the *backward solution*, the preceding procedure is essentially reversed. First a regression is computed making use of *all* the candidate variables. Then, if there are $M$ candidate variables, $M$ multiple regressions and their associated error sums of squares are computed. Each of these regressions contains *all but one* of the candidate variables. This makes it possible to ascertain the power of each of these variables, in the presence of all the others, to reduce the error sum of squares. If the variable associated with the smallest reduction is found to have a significant effect, it is assumed that all the candidate variables are needed and the initial regression is retained in unmodified form. On the other hand, if the weakest variable has a nonsignificant effect, it is deleted and the process is repeated with the shortened model, cycle after cycle, until the remaining weakest variable has a significant effect. If the investigator has set a target magnitude for the error sum of squares, the process terminates when that target is reached, provided all the remaining candidate variables have significant effects. It may, of course, be impossible to reach the target using only variables with significant effects.

One can carry out a deletion procedure using the assumed data used when describing the additive process. One can further use the same maximum allowable error sum of squares (2000) and the same level of significance ($\alpha = 0.05$). The regression equation and error sum of squares when all five predicting variables are included are

$$\hat{Y} = b_0 + b_1X_1 + b_2X_2 + b_3X_3 + b_4X_4 + b_5X_5, \quad \Sigma d^2 = 1900$$

The set of regressions in which one variable is deleted are

| | $\Sigma d^2$ |
|---|---|
| $\hat{Y} = b_0' + b_1'X_1 + b_2'X_2 + b_3'X_3 + b_4'X_4$ | 2200 |
| $\hat{Y} = b_0' + b_1'X_1 + b_2'X_2 + b_3'X_3 + b_4'X_5$ | 2300 |
| $\hat{Y} = b_0' + b_1'X_1 + b_2'X_2 + b_3'X_4 + b_4'X_5$ | 2600 |
| $\hat{Y} = b_0' + b_1'X_1 + b_2'X_3 + b_3'X_4 + b_4'X_5$ | 2500 |
| $\hat{Y} = b_0' + b_1'X_2 + b_2'X_3 + b_3'X_4 + b_4'X_5$ | 2000 ← |

$X_1$ is the weakest because in its absence the error sum of squares is least. If not significant, it should be deleted. Testing for significance,

**ANOVA**

| Source | df | SS | MS |
|---|---|---|---|
| $X_2$, $X_3$, $X_4$, and $X_5$(Alt) | $k_{\text{Alt}} = 30 - 5 = 25$ | 2000 | |
| $X_1$, $X_2$, $X_3$, $X_4$, and $X_5$(Max) | $k_{\text{Max}} = 30 - 6 = 24$ | 1900 | 79.166667 |
| Difference (Max and Alt) | 1 | 100 | 100 |

$$H_0\text{: Diff MS} = 0, \quad H_1\text{: Diff MS} > 0, \quad F_{0.05}(1, 24) = 4.26$$

$$F = 100/79.166667 = 1.26 \text{ N.S.}$$

The reduction in the error sum of squares associated with $X_1$ is not significant, confirming the conclusion that it should be deleted. At this point the error sum of squares equals the target value and the deletion process could be terminated. However, there may be additional nonsignificant variables to be identified so it is logical to continue. Then the set of regressions are:

| | $\Sigma d^2$ |
|---|---|
| $\hat{Y} = b_0'' + b_1''X_2 + b_2''X_3 + b_3''X_4$ | 2100 ← |
| $\hat{Y} = b_0'' + b_1''X_2 + b_2''X_3 + b_3''X_5$ | 2400 |
| $\hat{Y} = b_0'' + b_1''X_2 + b_2''X_4 + b_3''X_5$ | 2700 |
| $\hat{Y} = b_0'' + b_1''X_3 + b_2''X_4 + b_3''X_5$ | 2600 |

$X_5$ is the weakest and, if not significant, should be deleted. Testing for significance,

ANOVA

| Source | df | SS | MS |
|---|---|---|---|
| $X_2$, $X_3$, and $X_4$ (Alt) | $k_{\text{Alt}} = 30 - 4 = 26$ | 2100 | |
| $X_2$, $X_3$, $X_4$, and $X_5$(Max) | $k_{\text{Max}} = 30 - 5 = 25$ | 2000 | 80.00000 |
| Difference (Max and Alt) | 1 | 100 | 100 |

$$H_0\text{: Diff MS} = 0, \quad H_1\text{: Diff MS} > 0, \quad F_{0.05}(1, 25) = 4.24$$

$$F = 100/80.00000 = 1.25 \text{ N.S.}$$

The reduction associated with $X_5$ is not significant and it should be deleted. Continuing,

| | $\Sigma d^2$ |
|---|---|
| $\hat{Y} = b_0''' + b_1'X_2 + b_2'''X_3$ | 2400 ← |
| $\hat{Y} = b_0''' + b_1'X_2 + b_2'''X_4$ | 2800 |
| $\hat{Y} = b_0''' + b_1''X_3 + b_2'''X_4$ | 2700 |

$X_4$ is the weakest. Testing for significance,

ANOVA

| Source | df | SS | MS |
|---|---|---|---|
| $X_2$ and $X_3$ (Alt) | $k_{\text{Alt}} = 30 - 3 = 27$ | 2400 | |
| $X_2$, $X_3$, and $X_4$ (Max) | $k_{\text{Max}} = 30 - 4 = 26$ | 2100 | 80.769231 |
| Difference (Max and Alt) | 1 | 300 | 300 |

$$H_0\text{: Diff MS} = 0, \quad H_1\text{: Diff MS} > 0, \quad F_{0.05}(1, 26) = 4.22$$

$$F = 300/80.769231 = 3.71 \text{ N.S.}$$

Again, the reduction is not significant, confirming that $X_4$ should be deleted. Continuing,

| | $\Sigma d^2$ |
|---|---|
| $\hat{Y} b_0'''' + b_1''''X_2$ | 3200 |
| $\hat{Y} b_0'''' + b_1''''X_3$ | 3000 ← |

$X_2$ is the weakest. Testing for significance,

ANOVA

| Source | df | SS | MS |
|---|---|---|---|
| $X_3$ (Alt) | $k_{\text{Alt}} = 30 - 2 = 28$ | 3000 | |
| $X_2$ and $X_3$ (Max) | $k_{\text{Max}} = 30 - 3 = 27$ | 2400 | 88.888889 |
| Difference (Max and Alt) | 1 | 600 | 600 |

$$H_0\text{: Diff MS} = 0, \quad H_1\text{: Diff MS} > 0, \quad F_{0.05}(1, 27) = 4.21$$

$$F = 600/88.888889 = 6.75^*$$

The reduction in the error sum of squares associated with $X_2$ is significant, and therefore $X_2$ should be retained in the final equation,

$$\hat{Y} = b_0''' + b_1''X_2 + b_2''X_3$$

Thus, the deletion process led to the same conclusion as was reached using the additive process. This type of agreement rarely occurs.

The *stepwise* procedure ("stepwise" now being used in its specific sense) is a further development of the additive method which provides some protection against the effects of multicollinearity. It was brought out earlier that, as predicting variables move into or out of a regression, these respective regression coefficients are apt to change. These changes occur because the predicting variables are to some extent correlated. If two of these variables are highly correlated, each would have about the same power to reduce the error sum of squares. If both are used in the regression at the same time, their effects are shared and the total effect on the sum of squares is not much different from what it would be if only one of them were involved. Thus, in the additive process, the stronger of the two would enter the regression first. The other would probably never be able to enter the model because its effect, in the presence of the first, would be too small. However, when used with a specific set of variables, *not including the "stronger" variable*, its effect might be greater than that of the "stronger" variable when the latter is used with that set of variables. In both the additive and deletion methods, this result would never have an opportunity to appear. The so-called "weaker" variable would not have a second chance to show its strength. The stepwise procedure provides this chance.

To prevent such oversights the stepwise method requires the testing for significance of *all* the predicting variables appearing in the regression *in each cycle*, not only the variable just added. This testing process is equivalent to that used in the deletion procedure. If in the course of this testing any variable is found to have a nonsignificant effect on the error sum of squares, it is deleted and cannot appear in any subsequent cycles. Thus, if the investigator begins with $M$ candidate variables, some of which are correlated (e.g., $X_B$ and $X_D$), the process would proceed in the following manner. In the first cycle it is found that $X_A$ is the most potent and, when tested, is found to possess a significant effect on the error sum of squares. In the second cycle the pair $X_A$ and $X_B$ are found to be the most potent and when tested individually both are found have significant effects. In the third cycle the trio $X_A$, $X_B$, and $X_C$ are found to be the most potent with all three possessing significant effects. In the fourth cycle $X_A$, $X_B$, $X_C$, and $X_D$ are the most potent. However, when their respective individual effects on the error sum of squares are tested, it is found that $X_B$ is not significant. $X_B$ is then dropped from further consideration because, in the presence of $X_A$ and $X_C$, $X_D$ proved to be dominant over $X_B$. Consequently, in the fifth cycle the in-place predicting variables would be $X_A$, $X_C$, and $X_D$. This process continues until all the available candidate variables are found to be either significant, and in the predicting equation, or not significant, and excluded from the final equation. Along the way more of the previously selected variables may be replaced as was $X_B$. It must be understood, however, this procedure does not guarantee an absolutely most potent predicting equation because once a variable is found to be not significant it is permanently rejected. It is possible that, in some combinations, such a rejected variable might be a stronger candidate than one that is in the final equation.

Evaluating the three methods of variable selection, one finds that in the case of the additive method the computation time per cycle, and, thus, the cost of computation is usually less than it is with the other method. In addition, since the matrices are smaller, rounding errors are less of a problem with the additive and stepwise methods than they are in the deletion method. On the other side of the coin, highly correlated variables are removed early and variables with small, but independent, effects are less likely to be overlooked in the deletion model; the maximum model is generated and evaluated and this does not ordinarily occur when the additive or stepwise methods are used. The stepwise method is probably the most commonly used of these three procedures but questions have been raised about the benefits gained from the added computations. All three of these procedures are usually available in statistical analysis computer packages and the investigator can choose that which appears to best suit the needs of the problem.

It must be understood that none of these variable selection methods can guarantee the generation of the absolutely most potent model or the most efficient model in terms of easily measured predicting variables. It is likely that, given the same set of candidate variables, the three methods would generate three different models. Multicollinearity plays a big role in these effects and so does the level of significance that is chosen. The only sure way to obtain the desired model is to compute the regressions, with their associated statistics, for all the possible combinations of the candidate variables and then to sort through these results. As the number of candidate variables increases, it becomes less and less feasible to do this. The number of regressions that would have to be computed and analyzed would be equal to

$$\binom{m}{m} + \binom{m}{m-1} + \binom{m}{m-2} + \cdots + \binom{m}{1}$$

where $m$ is the number of candidate variables. Thus, if $m = 5$, 31 regressions would have to be computed. Furnival (1964) graphically describes the magnitude of the task if $m = 40$. He stated,

> More than a million possible regressions that can be found from these same 40 variables. The time required to compute all these regressions would be astronomical on any computer. The printed output with 60 regressions on a page would make a pile of paper more than a thousand miles high.

This problem has been attacked by many workers, among them Grosenbaugh (1958) and Furnival (1961; 1964). In some cases these efforts have resulted in programs that will indeed generate all the possible regressions but with severe constraints on the number of candidate variables. In other cases filters have been built into the programs to remove the combinations less likely to be of value. A discussion of these is beyond the scope of this book. Most investigators make use of proprietary statistical computing packages and these all have regression analysis capabilities. An inspection of the instructions for the package being used will reveal the available options and the constraints that apply.

# 16 Stratified Random Sampling

## 16.1 INTRODUCTION

In many cases the universe being sampled can be divided into subdivisions and the sampling carried out in such a manner that each of these subdivisions is represented in the sample. When this is done, it is said that *stratified* sampling is being used. The subdivisions of the universe or sampling frame are known as *strata*. An individual subdivision is a *stratum*. Examples of such stratification are:

1. The divisions of a student body into two strata, *male* and *female*;
2. The divisions of pine stands on a certain tract into three volume classes or strata, *high*, *medium*, and *low*; and
3. The divisions of a forest ownership into its component administrative units, such as *compartments*.

There are two reasons for stratification. It ensures that the sample will be well distributed across the sampling frame, thus increasing the probability of obtaining a representative sample. For example, if the variable of interest is student weight and the two strata mentioned above are recognized, *male* and *female,* the sample weights will include some weights from males and also from some females. If the sampling is unrestricted, all the weights could possibly be from males or all from females, which would lead to a distorted picture of the student body.

The second reason for stratification is that it will reduce the magnitude of the standard error of the mean. Since some of the sample must come from each of the strata, the likelihood of a sample containing all high values or all low values is greatly reduced and consequently repeated samples will have more nearly alike means than would be the case if the sampling was unrestricted. Thus, the variance and standard error of the mean would be less in the case of the stratified sample than it would be if the sampling is unrestricted.

## 16.2 THE STRATIFICATION PROCESS

Stratification involves the subdivision of a *specified* sampling frame, which must be in existence before the stratification can proceed. The way by which the frame is divided depends on the objectives of the sampling operation. If an objective is to obtain an estimate with a high degree of precision (i.e., with a small standard error of the mean) the stratification should be done so that the differences between the stratum means are maximized and that the values of the variable of interest within each stratum are as nearly alike (i.e., are homogeneous) as is feasible. This, of course, requires knowledge of the population. In the case of a student population, the student directory might serve as the sampling frame but it would have to be supplemented by the school records before it could be subdivided into the two strata: *male* and *female*. In the case of a timber inventory of a forest tract, existing stand maps, such as that in Figure 12.2, or aerial photographs,

| 2.55 | 3.00 | 1.60 | 1.80 | 2.65 | 2.10 | 1.90 | 3.40 | 3.65 | 4.00 | 4.00 | 1.75 | 2.60 | 2.85 | 2.00 | 2.10 |
|---|---|---|---|---|---|---|---|---|---|---|---|---|---|---|---|
| 3.15 | 2.95 | 2.40 | 2.25 | 3.20 | 2.80 | 2.40 | 4.55 | 4.50 | 3.25 | 4.00 | 1.50 | 1.80 | 2.70 | 1.70 | 2.50 |
| 3.05 | 2.10 | 2.35 | 2.45 | 2.90 | 1.10 | 4.30 | 4.00 | 4.10 | 2.85 | 3.40 | 1.85 | 2.00 | 2.30 | 2.20 | 2.50 |
| 1.55 | 1.55 | 2.55 | 2.15 | 3.10 | 2.75 | 4.50 | 4.60 | 3.30 | 4.30 | 3.75 | 2.30 | 1.80 | 2.00 | 2.10 | 1.80 |
| 2.55 | 0.90 | 2.25 | 2.35 | 2.45 | 2.90 | 4.65 | 1.90 | 3.35 | 4.60 | 2.60 | 2.35 | 2.60 | 2.75 | 2.70 | 3.25 |
| 2.70 | 2.50 | 1.90 | 2.30 | 2.95 | 3.15 | 3.00 | 4.55 | 2.65 | 4.40 | 4.20 | 2.20 | 1.90 | 1.70 | 2.30 | 1.70 |
| 2.50 | 1.50 | 2.00 | 2.55 | 3.15 | 3.30 | 4.05 | 4.50 | 4.35 | 3.80 | 4.50 | 2.00 | 3.65 | 1.35 | 3.15 | 1.60 |
| 2.35 | 2.90 | 2.05 | 2.10 | 3.05 | 3.35 | 4.15 | 4.60 | 4.00 | 3.45 | 1.50 | 1.75 | 2.20 | 2.85 | 3.45 | 2.40 |
| 2.85 | 1.80 | 1.50 | 2.05 | 2.25 | 4.80 | 4.40 | 3.15 | 3.00 | 3.65 | 2.30 | 1.35 | 1.90 | 2.40 | 2.20 | 0.90 |
| 3.15 | 1.60 | 2.40 | 2.60 | 3.00 | 1.90 | 2.55 | 3.55 | 4.00 | 2.75 | 2.65 | 1.90 | 2.60 | 2.70 | 1.50 | 2.65 |
| 1.75 | 0.25 | 1.05 | 0.95 | 3.00 | 2.90 | 3.75 | 3.15 | 3.35 | 2.30 | 3.25 | 2.30 | 2.30 | 1.90 | 1.60 | 0.00 |
| 0.00 | 0.00 | 0.00 | 0.45 | 1.20 | 1.90 | 2.90 | 4.10 | 4.50 | 3.00 | 2.65 | 2.50 | 1.00 | 1.40 | 0.00 | 0.05 |
| 0.60 | 0.00 | 0.50 | 0.55 | 2.05 | 3.10 | 4.65 | 2.30 | 3.00 | 2.20 | 1.90 | 1.30 | 0.55 | 0.75 | 0.15 | 0.20 |
| 0.30 | 0.00 | 0.30 | 0.40 | 1.50 | 1.15 | 2.35 | 1.00 | 2.55 | 2.45 | 2.25 | 0.85 | 0.40 | 0.15 | 1.05 | 0.10 |
| 0.00 | 0.05 | 0.00 | 0.40 | 1.40 | 2.20 | 2.40 | 2.00 | 1.40 | 1.50 | 1.25 | 0.00 | 0.10 | 1.35 | 1.45 | 0.60 |
| 0.10 | 0.00 | 0.00 | 1.00 | 1.95 | 2.40 | 3.60 | 1.60 | 1.90 | 2.40 | 1.00 | 1.15 | 0.40 | 0.20 | 0.05 | 0.00 |
| 0.55 | 0.15 | 0.10 | 0.85 | 2.05 | 1.60 | 1.90 | 2.75 | 1.75 | 2.00 | 0.55 | 0.00 | 0.30 | 0.00 | 1.60 | 0.30 |
| 0.60 | 0.50 | 0.00 | 0.65 | 1.80 | 1.65 | 2.35 | 2.45 | 1.90 | 1.15 | 0.00 | 1.50 | 0.85 | 0.05 | 0.50 | 0.00 |
| 0.60 | 1.00 | 0.15 | 0.00 | 1.70 | 2.50 | 1.25 | 2.40 | 0.80 | 0.50 | 0.10 | 0.75 | 0.30 | 0.35 | 0.20 | 0.10 |
| 0.55 | 0.20 | 0.50 | 0.30 | 1.50 | 2.05 | 1.80 | 1.40 | 1.05 | 0.10 | 0.00 | 0.45 | 0.50 | 0.40 | 0.75 | 0.40 |

**FIGURE 16.1** Hypothetical 16-ha tract divided into 320 $^1/_{20}$-ha square plots. The numbers refer to the volume of timber in terms of cubic meters per hectare.

could be used to divide the sampling frame into a series of volume classes, which would be used as strata. If knowledge, such as in these examples, is available, the stratification can be made in a logical manner, which leads to higher levels of precision than can be achieved in the absence of such knowledge.

If knowledge of the population is scant or nonexistent, it may be possible to divide the frame in some arbitrary manner and obtain some benefits beyond that possible using unrestricted random sampling. This is what is done when a survey is to be made, for example, of the characteristics of small nonindustrial forestland owners in a certain state. The sampling frame is made up of the individual landowners in the state. A possible stratification would be to recognize the counties in the state as strata. Such a division is usually not correlated with the variable of interest, but it does ensure that the sampling will be widespread and, presumably, representative. The effect on the precision of the estimate, if numerical, will be less than if the stratification were correlated with the variable of interest, but it will be better than that which would be obtained using simple random sampling.

As an example of the stratification process, consider the hypothetical tract shown in Figure 16.1. This tract covers 16 ha. The variable of interest is the volume of pine timber in terms of cubic meters per ha. The sampling units are 0.05-ha square plots. The tract is divided into these sampling units and the volume on each, in terms of cubic meters, is shown. These values are unknown to the sampler but are present on the ground. The parameters of the population of per plot volumes

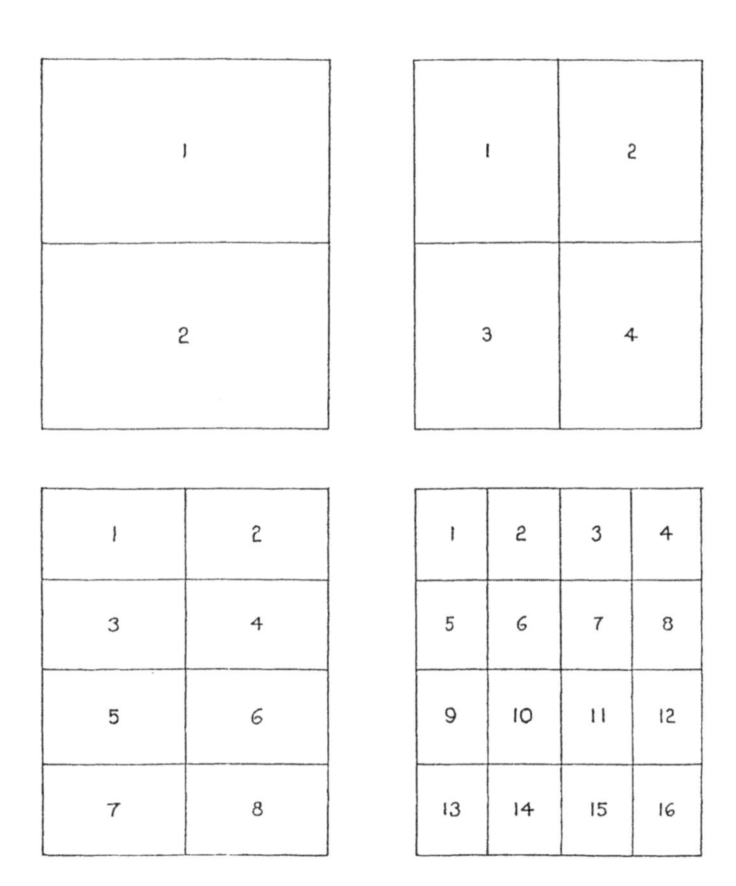


**FIGURE 16.2** Arbitrary division of the 16-ha tract into 2, 4, 8, and 16 strata.

are $\mu_Y = 1.970156$ m³/plot and $\sigma_Y^2 = 1.595367$. The total volume on the tract is 630.45 m³. The sampling frame size, $N$, is 320.

This frame could be divided arbitrarily into $M = 2$, 4, 8, or 16 strata as shown in Figure 16.2. It also could be divided into volume classes, provided an existing stand map is present or aerial photographs are available from which a stand map can be constructed. If this is the case, the stratification could look like Figure 16.3. Here four volume classes are recognized:

*A*, 76–100 m³/ha (3.8–5.0 m³/plot)
*B*, 51 – 75 m³/ha (2.55–3.75 m³/plot)
*C*, 26 – 50 m³/ha (1.3–2.5 m³/plot)
*D*, 0 – 25 m³/ha (0–1.25 m³/plot)

Minimum stand size is 0.5 ha, which means that at least ten plots within a class must touch for the group to be recognized as a stand. The shaded plots in Figure 16.3 are plots that either contain too much or too little volume to be properly assigned to the strata in which they fall but are included in those strata because an insufficient number are in contact with others of their condition to form a separate stand. Such anomalies have little effect on the stratum means, but they do increase the stratum variances. However, it is not feasible in most cases, such as this, to recognize individual plots as stands.

It should be noted that all members of a given stratum do not need to be concentrated in one group. Stratum *A* is made up of only one stand, as is stratum *B*, but both stratum *C* and stratum *D* contain two stands. This is typically the case when forest cover is involved.

| | | | | | | | | | | | | | | | |
|---|---|---|---|---|---|---|---|---|---|---|---|---|---|---|---|
| 2.55 | 3.00 | 1.60 | 1.80 | 2.65 | 2.10 | 1.90 | 3.40 | 3.65 | 4.00 | 4.00 | 1.75 | 2.60 | 2.85 | 2.00 | 2.10 |
| 3.15 | 2.95 | 2.40 | 2.25 | 3.20 | 2.80 | 2.40 | 4.55 | 4.50 | 3.25 | 4.00 | 1.50 | 1.80 | 2.70 | 1.70 | 2.50 |
| 3.05 | 2.10 | 2.35 | 2.45 | 2.90 | 1.10 | 4.30 | 4.00 | 4.10 | 2.85 | 3.40 | 1.85 | 2.00 | 2.30 | 2.20 | 2.50 |
| 1.55 | 1.55 | 2.55 | 2.15 | 3.10 | 2.75 | 4.50 | 4.60 | 3.30 | 4.30 | 3.75 | 2.30 | 1.80 | 2.00 | 2.10 | 1.80 |
| 2.55 | 0.90 | 2.25 | 2.35 | 2.45 | 2.90 | 4.65 | 1.90 | 3.35 | 4.60 | 2.60 | 2.35 | 2.60 | 2.75 | 2.70 | 3.25 |
| 2.70 | 2.50 | 1.90 | 2.30 | 2.95 | 3.15 | 3.00 | 4.55 | 2.65 | 4.40 | 4.20 | 2.20 | 1.90 | 1.70 | 2.30 | 1.70 |
| 2.50 | 1.50 | 2.00 | 2.55 | 3.15 | 3.30 | 4.05 | 4.50 | 4.35 | 3.80 | 4.50 | 2.00 | 3.65 | 1.35 | 3.15 | 1.60 |
| 2.35 | 2.90 | 2.05 | 2.10 | 3.05 | 3.35 | 4.15 | 4.60 | 4.00 | 3.45 | 1.50 | 1.75 | 2.20 | 2.85 | 3.45 | 2.40 |
| 2.85 | 1.80 | 1.50 | 2.05 | 2.25 | 4.80 | 4.40 | 3.15 | 3.00 | 3.65 | 2.30 | 1.35 | 1.90 | 2.40 | 2.20 | 0.90 |
| 3.15 | 1.60 | 2.40 | 2.60 | 3.00 | 1.90 | 2.55 | 3.55 | 4.00 | 2.75 | 2.65 | 1.90 | 2.60 | 2.70 | 1.50 | 2.65 |
| 1.75 | 0.25 | 1.05 | 0.95 | 3.00 | 2.90 | 3.75 | 3.15 | 3.35 | 2.30 | 3.25 | 2.30 | 2.30 | 1.90 | 1.60 | 0.00 |
| 0.00 | 0.00 | 0.00 | 0.45 | 1.20 | 1.90 | 2.90 | 4.10 | 4.50 | 3.00 | 2.65 | 2.50 | 1.00 | 1.40 | 0.00 | 0.05 |
| 0.60 | 0.00 | 0.50 | 0.55 | 2.05 | 3.10 | 4.65 | 2.30 | 3.00 | 2.20 | 1.90 | 1.30 | 0.55 | 0.75 | 0.15 | 0.20 |
| 0.30 | 0.00 | 0.30 | 0.40 | 1.50 | 1.15 | 2.35 | 1.00 | 2.55 | 2.45 | 2.25 | 0.85 | 0.40 | 0.15 | 1.05 | 0.10 |
| 0.00 | 0.05 | 0.00 | 0.40 | 1.40 | 2.20 | 2.40 | 2.00 | 1.40 | 1.50 | 1.25 | 0.00 | 0.10 | 1.35 | 1.45 | 0.60 |
| 0.10 | 0.00 | 0.00 | 1.00 | 1.95 | 2.40 | 3.60 | 1.60 | 1.90 | 2.40 | 1.00 | 1.15 | 0.40 | 0.20 | 0.05 | 0.00 |
| 0.55 | 0.15 | 0.10 | 0.85 | 2.05 | 1.60 | 1.90 | 2.75 | 1.75 | 2.00 | 0.55 | 0.00 | 0.30 | 0.00 | 1.60 | 0.30 |
| 0.60 | 0.50 | 0.00 | 0.65 | 1.80 | 1.65 | 2.35 | 2.45 | 1.90 | 1.15 | 0.00 | 1.50 | 0.85 | 0.05 | 0.50 | 0.00 |
| 0.60 | 1.00 | 0.15 | 0.00 | 1.70 | 2.50 | 1.25 | 2.40 | 0.80 | 0.50 | 0.10 | 0.75 | 0.30 | 0.35 | 0.20 | 0.10 |
| 0.55 | 0.20 | 0.50 | 0.30 | 1.50 | 2.05 | 1.80 | 1.40 | 1.05 | 0.10 | 0.00 | 0.45 | 0.50 | 0.40 | 0.75 | 0.40 |

**FIGURE 16.3** Division of the 16-ha tract into four timber volume classes: A, 76–100 $m^3$/ha; B, 51–75 $m^3$/ha; C, 26–50 $m^3$/ha; and D, 0–25 $m^3$/ha. The shaded areas are anomalies that cover insufficient contiguous area to be mapped as separate stands.

There must be at least two strata. From a theoretical point of view there is no upper limit on the number of strata that could be established. However, as the number of strata increases, the magnitude of the variance of the mean decreases and the cost of the sampling operation increases. Consequently, the choice of the number of strata to use is not a trivial exercise. Cochran (1963) discusses this matter but the formulae he presents are only of limited application since, if the stratification is made in terms of the variance of interest, they require knowledge of the frequency distribution and variance of that variable or, if the stratification is made in terms of a variable other than the variable of interest, the relationship between the two variables must be linear and the correlation coefficient known. These conditions rarely exist in real life. However, through the use of these formulae, using a number of different distributions, Cochran reached the conclusion that, in the majority of cases, the reduction in the variance of the mean becomes negligible when more than six strata are used.

The impact of number of strata on the cost of the sampling operations can be visualized through the use of a cost function suggested by Dalenius (1957):

$$c = Mc_{\text{st}} = nc_n \tag{16.1}$$

where $c$ = total cost of the sampling operation
$c_{st}$ = cost of establishing a stratum
$c_n$ = cost of locating and evaluating a sampling unit (assumed to be constant across all strata)

then,

$$M = (c - nc_n)/c_{st} \tag{16.2}$$

and

$$n = (c - Mc_{st})/c_n \tag{16.3}$$

Obviously, if $c$ is fixed, increasing $M$ will reduce the sample size, $n$. Cochran (1963) discusses this relationship and states that, if increasing $M$ beyond 6 reduces $n$ substantially, the increase in precision of the inventory brought about by the increases in $M$ will rarely be sufficient to be of any significance. Compounding the problem is the fact that sample size also has an effect on the precision of the inventory, with the variance of the mean decreasing in magnitude as $n$ increases. At some point increasing $M$ may reduce $n$ to the point that the net effect is an increase in the size of the variance of the mean. For this reason, if the total cost of the inventory cannot exceed a specified amount, one should keep the number of strata at or below six.

The preceding discussion of the number of strata to use was based on the assumption that only overall statistics were to be generated. If estimates are needed of the stratum parameters, as well as the overall parameters, the sample size must be adequate to obtain satisfactory stratum estimates. In such a case the strata have significance beyond the stratified sampling operation and their number is controlled by considerations other than those associated with the sampling operation. The total cost of such a project would be much higher than that of an inventory where only overall statistics are obtained. This type of situation is not unusual in forestry. Forest managers usually classify the forest cover into a series of condition classes using a system more or less similar to the following:

| Species | Size | Density |
|---|---|---|
| Pine | Reproduction | Poor |
| Pine–hardwoods | Saplings | Medium |
| Upland hardwoods | Poles | Good |
| Bottomland hardwoods | Sawtimbe | |

In this case there are 4 * 4 * 3 = 48 condition classes which could be used for stratification purposes. The forest manager would need information of various kinds about each in order to do a proper job of management. As can be seen, an adequate inventory involving all these classes would be an expensive operation. In some cases, all 48 classes would have to be used in the inventory process but in others it might be possible to reduce the number of strata and, consequently, the cost of the operation. If, for example, volume per unit area of *pine* sawtimber is the variable of interest, the number of strata could be reduced without damaging the overall estimate by eliminating from consideration those condition classes where pine sawtimber does not exist or is so rare that it can be considered to be nonexistant. Thus, one could disregard all reproducing and sapling classes and also all bottomland hardwood classes. This would cut the number of strata to 3 * 2 * 3 = 18, a considerable reduction. While this does not reduce $M$ to 6, it is much more economically feasible than using all 48 classes. In essence, this process of elimination reduces the size of the sampling frame so that it includes only classes likely to contain pine sawtimber.

The basis for the stratification should be as closely related to the variable of interest as is possible. In the case of the preceding example, the variable of interest was pine sawtimber. The final 16 strata represented different levels of sawtimber volume and, thus, followed the above rule. In many real-life cases, however, foresters stratify on forest *types*, not *condition classes* (as was done above). Forest types refer to species composition alone and have some relationship to a variable such as pine sawtimber, but it is a loose relationship since tree size and stand density are not considered. Thus, the use of forest types results in great variability in per unit area volumes within the strata and a low precision of estimate.

If the stratification is not going to be arbitrary, the sampler must have sufficiently good knowledge of the situation to permit an accurate delineation of the several strata. If the sampler's knowledge is faulty, or if the stratification is not carefully made, serious errors may appear in the final estimates. These errors arise from the use of incorrect stratum size in the computation process. Thus, in forestry it is bad practice to use old stand maps or old aerial photographs that do not portray the actual situation at the time of the estimate. Furthermore, even if the maps and/or photographs are current, if they are distorted (and virtually all vertical aerial photographs are distorted), the results of the sampling process can be seriously flawed. Whenever aerial photographs are used as the basis of a stratified sampling scheme, they should be converted to planimetric maps in which the distortions have been removed. Making such maps may not be possible if the camera were handheld or if nonstabilized camera mounts were used. Tilt distortions are usually severe in such cases. This subject is discussed in greater detail in Section 16.8.

## 16.3 THE STRATIFIED POPULATION

One can examine stratification at the population level so that one can understand what is being estimated when stratified sampling in being used.

The sampling frame is divided into $M$ strata. The total frame size is $N$ and the individual stratum size is $N_i$. Then, within stratum $i$, the expected value or mean is

$$E(y)_i = \mu_i = \sum_j^{N_i} y_{ij} \Big/ N_i \tag{16.4}$$

the total is

$$\tau_i = \sum_j^{N_i} y_{ij} = N_i E(y)_i = N_i \mu_i \tag{16.5}$$

and the variance is

$$V(y)_i = \sigma_i^2 = \sum_j^{N_i} (y_{ij} - \mu_i)^2 \Big/ N_i \tag{16.6}$$

If a sample of size $n_i$ is drawn from stratum $i$, and the sampling is with replacement, according to Equation 12.3,

$$V(y)_i = \sigma_{\bar{y}_i}^2 = \sigma_i^2 / n_i \tag{16.7}$$

and if the sampling is without replacement, according to Equation 12.5,

$$V(y)_i = \sigma^2_{\bar{y}_i} = \frac{\sigma^2_i}{n_i}\left(\frac{N_i - n_i}{N_i - 1}\right) \tag{16.8}$$

The overall population mean is

$$E(y) = \mu_{yst} = \sum_{h}^{N} y_h \Big/ N \tag{16.9}$$

$$= \sum_{i}^{M}\sum_{j}^{N_i} y_{ij} \Big/ N \tag{16.10}$$

$$\begin{aligned} &= \sum_{i}^{M} N_i E(y)_i \Big/ N \\ &= \sum_{i}^{M} (N_i / N)\mu_i \end{aligned} \tag{16.11}$$

where $(N_i/N)$ is the *relative weight* of stratum $i$. As defined in Equation 16.11, $\mu_Y$ is the weighted mean of the stratum means. It should be noted that the overall mean is the same regardless of the presence or absence of stratification.

The overall total is

$$\tau = \sum_{i}^{M}\sum_{j}^{N_i} y_{ij} = \sum_{i}^{M} N_i E(y)_i = \sum_{i}^{M} \tau_i = N\mu \tag{16.12}$$

In the *absence* of stratification, the variance of the population is

$$V(y) = \sigma^2_Y = \frac{1}{N}\sum_{h}^{N}(y_h - \mu_Y)^2 \tag{16.13}$$

When stratification is being used, the variance of the population is

$$V(y)_{St} = \sigma^2_{St} = \frac{1}{N}\sum_{i}^{M}\sum_{j}^{N_i}(y_{ij} - \mu_i)^2 \tag{16.14}$$

Multiplying each squared error by $(N_i/N_i)$,

$$V(y)_{St} = \sigma^2_{St} = \frac{1}{N}\sum_i^M \sum_j^{N_i} [(y_{ij} - \mu_i)^2 (N_i / N_i)]$$

$$= \frac{1}{N}\sum_i^M N_i \left[ \sum_j^{N_i} (y_{ij} - \mu_i)^2 \middle/ N_i \right] \tag{16.15}$$

$$= \frac{1}{N}\sum_i^M N_i \sigma_i^2 = \sum_i^M (N_i / N)\sigma_i^2$$

Note that the errors in this case are from the *stratum* means, not the population mean (compare Equations 16.13 and 16.14). According to the theory of least squares, the sum of squared errors will be at the minimum when those errors are measured from the mean of that specific set of variates. Within a stratum the mean of the set is the stratum mean, $\mu_i$, not the population mean, $\mu_Y$. Only rarely do $\mu_i$ and $\mu_y$ agree. Consequently, $\sigma^2_{st}$ will usually be smaller than $\sigma^2_Y$ and never will it exceed $\sigma^2_Y$.

Equation 16.13 can be written

$$\sigma^2_Y = \frac{1}{N}\sum_i^M \sum_j^{N_i} (y_{ij} - \mu_Y)^2 \tag{16.16}$$

Then,

$$N\sigma^2_Y = \sum_i^M \sum_j^{N_i} (y_{ij} - \mu_Y)^2 \tag{16.17}$$

where the right-hand member is the sum of squared errors from the overall mean.

$$N\sigma^2_y = \sum_i^M \sum_j^{N_i} [(y_{ij} - \mu_Y) + (\mu_i - \mu_i)]^2$$

$$= \sum_i^M \sum_j^{N_i} [(y_{ij} - \mu_i) + (\mu_i - \mu_Y)]^2$$

$$= \sum_i^M \sum_j^{N_i} [(y_{ij} - \mu_i)^2 + 2(y_{ij} - \mu_i)(\mu_i - \mu_Y) + (\mu_i - \mu_Y)^2]$$

$$= \sum_i^M \left[ \sum_j^{N_i} (y_{ij} - \mu_i)^2 + 2(\mu_i - \mu_Y)\sum_j^{N_i} (y_{ij} - \mu_i) + \sum_j^{N_i} (\mu_i - \mu_Y)^2 \right]$$

Since $\Sigma_j^{N_i} (y_{ij} - \mu_i) = 0$ and since $(\mu_i - \mu_Y)^2$ is a constant within stratum $i$,

$$N\sigma^2_Y = \sum_i^M \sum_j^{N_i} (y_{ij} - \mu_i)^2 + \sum_i^M N_i(\mu_i - M_Y)^2 \tag{16.18}$$

As can be seen, the sum of squared errors from the population mean ($N\sigma_Y^2$), which is referred to as the *total sum of squares*, is made up of two components: the *within sum of squares*,

$$\sum_i^M \sum_j^{N_i} (y_{ij} - \mu_i)^2 \tag{16.19}$$

which reflects the variability of the data within the strata, and the *between sum of squares*,

$$\sum_i^M N_i(\mu_i - \mu_Y)^2 \tag{16.20}$$

which reflects the variability of the stratum means. *The more the stratum means differ, the greater the difference between the population variance, $\sigma_y^2$, and the variance of stratified population, $\sigma_{st}^2$.*

Dividing both sides of Equation 16.18 by $N$ yields

$$\begin{aligned} \sigma_y^2 &= \frac{1}{N}\sum_i^M \sum_j^{N_i} (y_{ij} - \mu_i)^2 + \frac{1}{N}\sum_i^M N_i(\mu_i - \mu_Y)^2 \\ &= \sigma_w^2 + \sigma_B^2 \end{aligned} \tag{16.21}$$

where $\sigma_w^2$ = *within* stratum variance (which is the variance of the stratified population)
$\sigma_B^2$ = *between* stratum variance.

Rearranging Equation 16.19,

$$\sigma_w^2 = \sigma_y^2 - \sigma_B^2 \tag{16.22}$$

Thus, the within or stratified population variance, $\sigma_w^2$, is comparable with the variance about a regression line or surface (see Sections 15.2.3 and 15.2.4) or to the error mean square in the case of a completely randomized statistical design (see Section 14.20).

When samples of size $n_i$ are drawn from the several strata and the sampling is with replacement, the stratum variances of the mean must be weighted by the squares of the relative stratum sizes and then summed:

$$V(\bar{y}) = \sigma_{\bar{y}}^2 = \sum_i^M [\sigma_{\bar{y}}^2 (N_i / N)^2] \tag{16.23}$$

$$= \sum_i^M \left[\frac{\sigma_i^2}{n_i}\left(\frac{N_i}{N}\right)^2\right] \tag{16.24}$$

If the sampling is without replacement,

$$V(\bar{y}) = \sigma_{\bar{y}}^2 = \sum_i^M \left[\frac{\sigma_i^2}{n_i}\left(\frac{N_i - n_i}{N_i - 1}\right)\left(\frac{N_i}{N}\right)^2\right] \tag{16.25}$$

The variance of the total is the product of the appropriate variance of the mean and the square of the sampling frame size:

$$V(T) = \sigma_T^2 = N^2\sigma_{\bar{y}}^2 \tag{16.26}$$

## 16.4 ESTIMATION OF THE PARAMETERS

The sampling within the strata must be random but the sampling may be with or without replacement. However, sampling with replacement is rarely used.

The estimates of the stratum mean and total are

$$\bar{y}_i = \sum_j^{n_i} y_{ij} \Big/ n_i \to E(y)_i = \mu_i \tag{16.27}$$

and

$$T_i = N_i\bar{y}_i \to \tau_i \tag{16.28}$$

When the sampling is with replacement, the estimate of the stratum variance is

$$s_i^2 = \sum_j^{n_i} (y_{ij} - \bar{y}_i)^2 \Big/ (n_i - 1) \to V(y)_i = \sigma_i^2 \tag{16.29}$$

and when the sampling is without replacement,

$$s_i^2 = \sum_j^{n_i} (y_{ij} - \bar{y}_i)^2 \Big/ (n_i - 1) \to \sigma_i^2[N_i / (N_i - 1)] \tag{16.30}$$

Then,

$$s_i^2[(N_i - 1) / N_i] \to V(y)_i = \sigma_i^2 \tag{16.31}$$

As can be seen by an examination of Equations 16.29 and 16.30 *$n_i$ must be greater than 1 if a variance is to be computed.* If $n_i = 1$, $s_i^2 = 0/0$, which is indeterminate and meaningless. However, when $n_i = 1$, the mean, $\bar{y}_i$, is valid and can be used as an estimator, but an estimator without an accompanying expression of precision such as the standard error of the mean, $s_{\bar{y}_i}$, is often viewed with skepticism.

The estimate of the stratum variance of the mean, when the sampling is with replacement is

$$s_{\bar{y}_i}^2 = s_i^2 / n_i \to V(\bar{y})_i = \sigma_{\bar{y}_i}^2 \tag{16.32}$$

and when the sampling is without replacement:

$$s_{\bar{y}_i}^2 = \frac{s_i^2}{n_i}\left(\frac{N_i - 1}{N_i}\right)\left(\frac{N_i - n_i}{N_i - 1}\right)$$
$$= \frac{s_i^2}{n_i}\left(\frac{N_i - n_i}{N_i}\right) \rightarrow V(\bar{y})_I = \sigma_{\bar{y}_i}^2 \tag{16.33}$$

Completing the stratum statistics is the estimate of the variance of the stratum total:

$$s_{T_i}^2 = N_i^2 s_{\bar{y}_i}^2 \tag{16.34}$$

The estimate of the overall or combined mean and total are then

$$\bar{y} = \sum_i^M (N_i / N)\bar{y}_i = \frac{1}{N}\sum_i^M N_i\bar{y}_i \rightarrow E(\bar{y}) = \mu \tag{16.35}$$

$$T = \sum_i^M T_i = \sum_i^M N_i\bar{y}_i \rightarrow E(T) = \tau \tag{16.36}$$

The estimate of the within variance, or the variance of the stratified sample, across all strata, where the sampling is with replacement, is

$$s_w^2 = \sum_i^M \sum_j^{n_i} (y_{ij} - \bar{y}_i)^2 \Bigg/ \left(\sum_i^M n_i - M\right) \rightarrow V(y)_w = \sigma_w^2 \tag{16.37}$$

When the sampling is without replacement,

$$s_w^2 = \sum_i^M \sum_j^{n_i} (y_{ij} - \bar{y}_i)^2 \Bigg/ \left(\sum_i^M n_i - M\right) \rightarrow \sigma_w^2\left(\frac{N}{N-1}\right) \tag{16.38}$$

and thus,

$$s_W^2\left(\frac{N}{N-1}\right) \rightarrow V(y)_W = \sigma_w^2 \tag{16.39}$$

The estimate of the variance of the mean, when the sampling is with replacement, is obtained when $s_i^2$ is substituted for $\sigma_i^2$ in Equation 16.24:

$$s_{\bar{y}}^2 = \sum_i^M \left[\frac{s_i^2}{n_i}\left(\frac{N_i}{N}\right)^2\right] \rightarrow V(\bar{y}) = \sigma_{\bar{y}}^2 \tag{16.40}$$

and when the sampling is without replacement, $s_i^2[(N_i - 1)/N_i]$ is substituted for $\sigma_i^2$ in Equation 16.25:

$$
\begin{aligned}
s_{\bar{y}}^2 &= \sum_i^M \left[ \frac{s_i^2}{n_i} \left( \frac{N_i - 1}{N_i} \right) \left( \frac{N_i - n_i}{N_i - 1} \right) \left( \frac{N_i}{N} \right)^2 \right] \\
&= \sum_i^M \left[ \frac{s_i^2}{n_i} \left( \frac{N_i - n_i}{N_i} \right) \left( \frac{N_i}{N} \right)^2 \right] \rightarrow V(\bar{y}) = \sigma_{\bar{y}}^2
\end{aligned}
\tag{16.41}
$$

Equation 16.41 is the general-case expression for the estimated variance of the mean. In the special case where the sampling fractions ($n_i/N_i$) are constant across all strata and stratum variances are homogeneous, it can be simplified. Equation 16.41 can be written

$$
s_{\bar{y}}^2 = \sum_i^M \left[ \frac{s_i^2}{n_i} \left( \frac{N_i - 1}{N_i} \right) \left( \frac{N_i}{N} \right) \left( \frac{N_i}{N} \right) \right]
$$

Since the sampling fraction is constant

$$
n_i/N_i = n/n \quad \text{and} \quad 1/n_i = N/nN_i
$$

Substituting this expression in Equation 16.41 and doing some canceling

$$
\begin{aligned}
s_{\bar{y}}^2 &= \sum_i^M \left[ s_i^2 \left( \frac{N}{nN_i} \right) \left( \frac{N_i - n_i}{N_i} \right) \left( \frac{N_i}{N} \right) \left( \frac{N_i}{N} \right) \right] \\
&= \sum_i^M \left[ \frac{s_i^2}{n} \left( \frac{N_i - n_i}{N_i} \right) \right]
\end{aligned}
$$

Since the stratum variances are homogeneous, one can substitute $s_w^2$ for the $s_i^2$ values.

$$
\begin{aligned}
s_{\bar{y}}^2 &= \frac{s_w^2}{n} \left( \frac{1}{N} \right) \sum_i^M (N_i - n_i) \\
&= \frac{s_W^2}{n} \left( \frac{1}{N} \right) \left[ \sum_i^M N_i - \sum_i^M n_i \right]
\end{aligned}
$$

Since $\Sigma_i^M N_i = N$ and $\Sigma_i^M n_i = n$, this expression becomes

$$
s_{\bar{y}}^2 = \frac{s_w^2}{n} \left( \frac{N - n}{N} \right) \tag{16.42}
$$

If the sampling is with replacement, Equation 16.42 can be further reduced to

$$
s_{\bar{y}}^2 = s_w^2 / n \tag{16.43}
$$

When these conditions, homogeneity of stratum variances and constant sampling intensity across all strata, are met, it becomes possible to develop a valid sampling error.

$$SE_c = \pm t_c(k)s_{\bar{y}} \tag{16.44}$$

where

$$k = \sum_i^M n_i - M \tag{16.45}$$

When the stratum variances are heterogeneous and/or the sampling intensity varies from stratum to stratum, it is impossible to determine the appropriate $t$ distribution to use. If the stratum populations are *all normally* distributed, one can use a procedure proposed by Satterthwaite (1946) to arrive at an approximation of the proper or effective number of degrees of freedom.

$$k_e = \left(\sum_i^M f_i s_i^2\right) \Bigg/ \sum_i^M [f_i^2 s_i^4 / (n_i - 1)] \tag{16.46}$$

where $k_e$ = effective degrees of freedom

$$f_i = N_i(N_i - n_i)/n_i$$

If the stratum populations are in fact normally distributed, $k_e$ will lie in the range between the smallest $(n_i - 1)$ and the sum $\Sigma_i^M(n_i - 1)$. The procedure recognizes that the stratum variances may differ, but it is quite intolerant of deviations from normality.

Since normality cannot be assumed in most real-life situations, one probably should simply use the standard error of the mean, $s_{\bar{y}}$, as the measure of precision and not attempt to compute a sampling error. If a sampling error is required, however, one should use a sample that is large enough to ensure normality of the sampling distribution of means. There are no rules to guide one to that sample size. However, it should be sufficiently large so that $n_i - M > 30$ or, preferably, 60. Even then, the situation would be sufficiently obscure so that the use of confidence coefficients from the $t$ table would be inappropriate. Instead the usual practice is to set the confidence coefficient simply at 1, when the desired confidence or level is 0.66, and at 2, when the level is 0.95.

To complete the listing and descriptions of the sample statistics, the estimate of the variance of the total is

$$s_T^2 = N^2 s_{\bar{y}}^2 \rightarrow V(T) = \sigma_t^2 \tag{16.47}$$

where $s_{\bar{y}}^2$ depends on whether the sampling was with replacement (see Equation 16.40) or without replacement (see Equation 16.41).

One can now carry out a stratified sampling operation on the 16-ha tract shown in Figure 16.1. The sampling frame is the set of 0.05-ha square plots making up the tract. The total frame size, $N$, equals $(1/0.05)16 = 320$. Assume that the available knowledge of the tract is insufficient to divide the frame logically into strata and consequently the division must be done arbitrarily. This would at least assure a broad distribution of the sample. Then set $M = 4$ and divide the sampling frame into four equal-sized strata as shown in Figure 16.4. Then, $N_1 = N_2 = N_3 = N_4 = 80$. Assume that a 10% cruise is desired. Then, the total sample size is $n = 0.1(320) = 32$ plots. Dividing these equally among the four strata results in a sample size of 8 in each of the strata and $n_1 = n_2 = n_3 = n_4 = 8$. Using standard random sampling procedures, a sample is obtained from each of the strata as shown in Figure 16.4. The data (cubic meters per plot) obtained are

| | Stratum | | | |
|---|---|---|---|---|
| | 1 | 2 | 3 | 4 |
| | 3.35 | 2.85 | 0.40 | 0.35 |
| | 1.90 | 1.75 | 1.75 | 0.40 |
| | 2.50 | 1.50 | 0.60 | 1.50 |
| | 2.55 | 1.90 | 2.35 | 0.40 |
| | 2.55 | 2.20 | 2.90 | 2.30 |
| | 2.90 | 1.70 | 1.00 | 1.35 |
| | 3.15 | 3.65 | 3.60 | 0.55 |
| | 4.65 | 2.70 | 1.65 | 0.00 |
| $\bar{y}_i$ | = 2.94375 | 2.28125 | 1.78125 | 0.85625 |
| $s_i^2$ | = 0.673884 | 0.536384 | 1.259241 | 0.606027 |

$$\bar{y} = \sum_i^4 (N_i / N)\bar{y}_i$$

$$= (80/320)(2.94375 + 2.28125 + 1.78125 + 0.85625)$$

$$= 1.965625 \text{ m}^3/\text{plot}$$

$$s_{\bar{y}}^2 = \sum_i^4 \left[ \frac{s_i^2}{n_i} \left( \frac{N_i - n_i}{N_i} \right) \left( \frac{N_i}{N} \right)^2 \right]$$

$$= \left( \frac{1}{8} \right) \left( \frac{80-8}{80} \right) \left( \frac{80}{320} \right)^2 (0.673884 + 0.536384 + 1.259244 + 0.606027)$$

$$= 0.0216249$$

$$s_{\bar{y}} = \pm 0.147054 \text{ m}^3/\text{plot}$$

The sample evidence suggests that the stratum variances are heterogeneous. This hypothesis can be tested by using Bartlett's test (see Section 14.17). If the test indicates heterogeneity, the statistical analysis ends with the computations of the standard error of the mean. The conditions will not permit the computation of a *valid* sampling error.

If the stratum populations could have been considered to be normally distributed (which they are not, being finite and irregular), one could have used Satterthwaite's procedure to obtain an approximate value for the effective degrees of freedom, and using this value, arrive at an approximate sampling error. Make the assumption of normality. Then, $N_1 = N_2 = N_3 = N_4 = 80$, and $n_1 = n_2 = n_3 = n_4 = 8$. Consequently, $f_1 = f_2 = f_3 = f_4 = 80(80 - 8)/8 = 720$, and $(n_i - 1) = 7$. According to Equation 16.46,

$$k_e = \left( \sum_i^M f_i s_i^2 \right)^2 \bigg/ \sum_i^M [f_i^2 s_i^4 / (n_i - 1)$$

$$= 720[0.673884 + 720(0.536384) + 720(1.259241) + 720(0.606027)]^2$$

$$\left[ \frac{(720)^2(0.673884)^2}{7} + \frac{(720)^2(0.536384)^2}{7} + \frac{(720)^2(1.259241)^2}{y} + \frac{(720)^2(0.606027)^2}{7} \right]$$

$$= 24.z57 \text{ or } 25 \text{ degrees of freedom}$$

| | | | | | | | | | | | | | | | |
|---|---|---|---|---|---|---|---|---|---|---|---|---|---|---|---|
| 2.55 | 3.00 | 1.60 | 1.80 | 2.65 | 2.10 | 1.90 | 3.40 | 3.65 | 4.00 | 4.00 | 1.75 | 2.60 | 2.85 | 2.00 | 2.10 |
| 3.15 | 2.95 | 2.40 | 2.25 | 3.20 | 2.80 | 2.40 | 4.55 | 4.50 | 3.25 | 4.00 | 1.50 | 1.80 | 2.70 | 1.70 | 2.50 |
| 3.05 | 2.10 | 2.35 | 2.45 | 2.90 | 1.10 | 4.30 | 4.00 | 4.10 | 2.85 | 3.40 | 1.85 | 2.00 | 2.30 | 2.20 | 2.50 |
| 1.55 | 1.55 | 2.55 | 2.15 | 3.10 | 2.75 | 4.50 | 4.60 | 3.30 | 4.30 | 3.75 | 2.30 | 1.80 | 2.00 | 2.10 | 1.80 |
| 2.55 | 0.90 | 2.25 | 2.35 | 2.45 | 2.90 | 4.65 | 1.90 | 3.35 | 4.60 | 2.60 | 2.35 | 2.60 | 2.75 | 2.70 | 3.25 |
| 2.70 | 2.50 | 1.90 | 2.30 | 2.95 | 3.15 | 3.00 | 4.55 | 2.65 | 4.40 | 4.20 | 2.20 | 1.90 | 1.70 | 2.30 | 1.70 |
| 2.50 | 1.50 | 2.00 | 2.55 | 3.15 | 3.30 | 4.05 | 4.50 | 4.35 | 3.80 | 4.50 | 2.00 | 3.65 | 1.35 | 3.15 | 1.60 |
| 2.35 | 2.90 | 2.05 | 2.10 | 3.05 | 3.35 | 4.15 | 4.60 | 4.00 | 3.45 | 1.50 | 1.75 | 2.20 | 2.85 | 3.45 | 2.40 |
| 2.85 | 1.80 | 1.50 | 2.05 | 2.25 | 4.80 | 4.40 | 3.15 | 3.00 | 3.65 | 2.30 | 1.35 | 1.90 | 2.40 | 2.20 | 0.90 |
| 3.15 | 1.60 | 2.40 | 2.60 | 3.00 | 1.90 | 2.55 | 3.55 | 4.00 | 2.75 | 2.65 | 1.90 | 2.60 | 2.70 | 1.50 | 2.65 |
| 1.75 | 0.25 | 1.05 | 0.95 | 3.00 | 2.90 | 3.75 | 3.15 | 3.35 | 2.30 | 3.25 | 2.30 | 2.30 | 1.90 | 1.60 | 0.00 |
| 0.00 | 0.00 | 0.00 | 0.45 | 1.20 | 1.90 | 2.90 | 4.10 | 4.50 | 3.00 | 2.65 | 2.50 | 1.00 | 1.40 | 0.00 | 0.05 |
| 0.60 | 0.00 | 0.50 | 0.55 | 2.05 | 3.10 | 4.65 | 2.30 | 3.00 | 2.20 | 1.90 | 1.30 | 0.55 | 0.75 | 0.15 | 0.20 |
| 0.30 | 0.00 | 0.30 | 0.40 | 1.50 | 1.15 | 2.35 | 1.00 | 2.55 | 2.45 | 2.25 | 0.85 | 0.40 | 0.15 | 1.05 | 0.10 |
| 0.00 | 0.05 | 0.00 | 0.40 | 1.40 | 2.20 | 2.40 | 2.00 | 1.40 | 1.50 | 1.25 | 0.00 | 0.10 | 1.35 | 1.45 | 0.60 |
| 0.10 | 0.00 | 0.00 | 1.00 | 1.95 | 2.40 | 3.60 | 1.60 | 1.90 | 2.40 | 1.00 | 1.15 | 0.40 | 0.20 | 0.05 | 0.00 |
| 0.55 | 0.15 | 0.10 | 0.85 | 2.05 | 1.60 | 1.90 | 2.75 | 1.75 | 2.00 | 0.55 | 0.00 | 0.30 | 0.00 | 1.60 | 0.30 |
| 0.60 | 0.50 | 0.00 | 0.65 | 1.80 | 1.65 | 2.35 | 2.45 | 1.90 | 1.15 | 0.00 | 1.50 | 0.85 | 0.05 | 0.50 | 0.00 |
| 0.60 | 1.00 | 0.15 | 0.00 | 1.70 | 2.50 | 1.25 | 2.40 | 0.80 | 0.50 | 0.10 | 0.75 | 0.30 | 0.35 | 0.20 | 0.10 |
| 0.55 | 0.20 | 0.50 | 0.30 | 1.50 | 2.05 | 1.80 | 1.40 | 1.05 | 0.10 | 0.00 | 0.45 | 0.50 | 0.40 | 0.75 | 0.40 |

**FIGURE 16.4** Arbitrary division of the 16-ha tract into four equal-sized strata. The plots chosen in the stratified random sample are shaded in.

The sampling error, at the 95% level of confidence, is computed using $t_{0.95}(k = 25) = 2.060$. Then,

$$SE = \pm 2.060(0.147054) = \pm 0.302931 \text{ m}^3/\text{plot}$$

It must be emphasized that this sampling error is *not valid* because the stratum populations are not normally distributed. The computations are valid only up to and including the determination of the standard error of the mean.

One can compare the results of this stratified sampling operation with those of an unrestricted random sample of the same total size ($n = 32$). The sample is shown in Figure 16.5 and the statistics are as follows.

$$\bar{y} = 19.6 \text{ m}^3 / \text{plot}$$

$$s^2 = 1.448889$$

$$s_{\bar{y}}^2 = \frac{1.448889}{32}\left(\frac{320 - 32}{320}\right) = 0.04075$$

and

| 2.55 | 3.00 | 1.60 | 1.80 | 2.65 | 2.10 | 1.90 | 3.40 | 3.65 | 4.00 | 4.00 | 1.75 | 2.60 | 2.85 | 2.00 | 2.10 |
|---|---|---|---|---|---|---|---|---|---|---|---|---|---|---|---|
| 3.15 | 2.95 | 2.40 | 2.25 | 3.20 | 2.80 | 2.40 | 4.55 | 4.50 | 3.25 | 4.00 | 1.50 | 1.80 | 2.70 | 1.70 | 2.50 |
| 3.05 | 2.10 | 2.35 | 2.45 | 2.90 | 1.10 | 4.30 | 4.00 | 4.10 | 2.85 | 3.40 | 1.85 | 2.00 | 2.30 | 2.20 | 2.50 |
| 1.55 | 1.55 | 2.55 | 2.15 | 3.10 | 2.75 | 4.50 | 4.60 | 3.30 | 4.30 | 3.75 | 2.30 | 1.80 | 2.00 | 2.10 | 1.80 |
| 2.55 | 0.90 | 2.25 | 2.35 | 2.45 | 2.90 | 4.65 | 1.90 | 3.35 | 4.60 | 2.60 | 2.35 | 2.60 | 2.75 | 2.70 | 3.25 |
| 2.70 | 2.50 | 1.90 | 2.30 | 2.95 | 3.15 | 3.00 | 4.55 | 2.65 | 4.40 | 4.20 | 2.20 | 1.90 | 1.70 | 2.30 | 1.70 |
| 2.50 | 1.50 | 2.00 | 2.55 | 3.15 | 3.30 | 4.05 | 4.50 | 4.35 | 3.80 | 4.50 | 2.00 | 3.65 | 1.35 | 3.15 | 1.60 |
| 2.35 | 2.90 | 2.05 | 2.10 | 3.05 | 3.35 | 4.15 | 4.60 | 4.00 | 3.45 | 1.50 | 1.75 | 2.20 | 2.85 | 3.45 | 2.40 |
| 2.85 | 1.80 | 1.50 | 2.05 | 2.25 | 4.80 | 4.40 | 3.15 | 3.00 | 3.65 | 2.30 | 1.35 | 1.90 | 2.40 | 2.20 | 0.90 |
| 3.15 | 1.60 | 2.40 | 2.60 | 3.00 | 1.90 | 2.55 | 3.55 | 4.00 | 2.75 | 2.65 | 1.90 | 2.60 | 2.70 | 1.50 | 2.65 |
| 1.75 | 0.25 | 1.05 | 0.95 | 3.00 | 2.90 | 3.75 | 3.15 | 3.35 | 2.30 | 3.25 | 2.30 | 2.30 | 1.90 | 1.60 | 0.00 |
| 0.00 | 0.00 | 0.00 | 0.45 | 1.20 | 1.90 | 2.90 | 4.10 | 4.50 | 3.00 | 2.65 | 2.50 | 1.00 | 1.40 | 0.00 | 0.05 |
| 0.60 | 0.00 | 0.50 | 0.55 | 2.05 | 3.10 | 4.65 | 2.30 | 3.00 | 2.20 | 1.90 | 1.30 | 0.55 | 0.75 | 0.15 | 0.20 |
| 0.30 | 0.00 | 0.30 | 0.40 | 1.50 | 1.15 | 2.35 | 1.00 | 2.55 | 2.45 | 2.25 | 0.85 | 0.40 | 0.15 | 1.05 | 0.10 |
| 0.00 | 0.05 | 0.00 | 0.40 | 1.40 | 2.20 | 2.40 | 2.00 | 1.40 | 1.50 | 1.25 | 0.00 | 0.10 | 1.35 | 1.45 | 0.60 |
| 0.10 | 0.00 | 0.00 | 1.00 | 1.95 | 2.40 | 3.60 | 1.60 | 1.90 | 2.40 | 1.00 | 1.15 | 0.40 | 0.20 | 0.05 | 0.00 |
| 0.55 | 0.15 | 0.10 | 0.85 | 2.05 | 1.60 | 1.90 | 2.75 | 1.75 | 2.00 | 0.55 | 0.00 | 0.30 | 0.00 | 1.60 | 0.30 |
| 0.60 | 0.50 | 0.00 | 0.65 | 1.80 | 1.65 | 2.35 | 2.45 | 1.90 | 1.15 | 0.00 | 1.50 | 0.85 | 0.05 | 0.50 | 0.00 |
| 0.60 | 1.00 | 0.15 | 0.00 | 1.70 | 2.50 | 1.25 | 2.40 | 0.80 | 0.50 | 0.10 | 0.75 | 0.30 | 0.35 | 0.20 | 0.10 |
| 0.55 | 0.20 | 0.50 | 0.30 | 1.50 | 2.05 | 1.80 | 1.40 | 1.05 | 0.10 | 0.00 | 0.45 | 0.50 | 0.40 | 0.75 | 0.40 |

**FIGURE 16.5** Unrestricted random sample of size 32 drawn from the 16-ha tract.

$$s_{\bar{y}} = \pm 0.201866 \text{ m}^3 / \text{plot}$$

When this standard error is compared with that of the stratified sample, both based on the same sample size, the greater efficiency of stratified over unrestricted random sampling becomes evident:

$$s_{\bar{y}_{\text{St}}} = \pm 0.147054 < s_{\bar{y}_{\text{Ran}}} = \pm 0.210866 \text{ m}^3 / \text{plot}$$

Now consider the situation when the stratification is carried out with prior knowledge of the structure of the forest cover as might be revealed on vertical aerial photography. This is essentially what was done in Figure 16.3. Four timber volume classes or strata were recognized:

Stratum *A*: 76–100 $m^3$/ha (3.8–5.0 $m^3$/plot)
Stratum *B*: 51–75 $m^3$/ha (2.55–3.75 $m^3$/plot)
Stratum *C*: 26–50 $m^3$/ha (1.30–2.50 $m^3$/plot)
Stratum *D*: 0–25 $m^3$/ha (0–1.25 $m^3$/plot)

Thus, $M = 4$. Using the same sampling intensity as in the previous examples, the total sample size, $n = 32$. Dividing these equally among the strata:

$$n_i = n/M = 32/4 = 8 \text{ plots/stratum}$$

These were randomly chosen from the sampling frame as shown in Figure 16.6. The data (in cubic meters per plot) and computations are shown below.

| | Stratum | | | |
|---|---|---|---|---|
| | *A* | *B* | *C* | *D* |
| | 4.30 | 2.75 | 2.00 | 0.15 |
| | 4.55 | 3.25 | 1.50 | 0.60 |
| | 4.20 | 3.10 | 2.40 | 0.50 |
| | 2.65 | 3.00 | 1.50 | 0.50 |
| | 4.00 | 3.75 | 2.40 | 0.60 |
| | 4.00 | 3.45 | 2.45 | 1.35 |
| | 4.50 | 4.65 | 2.90 | 1.05 |
| | 4.80 | 2.65 | 1.70 | 0.40 |
| $\bar{y}_i$ | = 4.125 | 3.325 | 2.10625 | 0.64375 |
| $s_i^2$ | = 0.431429 | 0.414286 | 0.261741 | 0.144598 |
| $N_i$ | = 36 | 44 | 145 | 95 |

$$\bar{y} = \sum_i^4 (N_i / N)\bar{y} = [36(4.125) + 44(3.325) + 145(2.10625) + 95(0.63275)] / 320$$

$$= 2.066758 \text{ m}^3 / \text{plot}$$

$$s_{\bar{y}}^2 = \sum_i^M \left[ \frac{s_i^2}{n_i} \left( \frac{N_i - n_i}{N_i} \right) \left( \frac{N_i}{N} \right)^2 \right]$$

$$= \frac{0.431429}{8} \left( \frac{36-8}{36} \right) \left( \frac{36}{320} \right)^2 + \frac{0.414286}{8} \left( \frac{44-8}{44} \right) \left( \frac{44}{320} \right)^2$$

$$+ \frac{0.261741}{8} \left( \frac{145-8}{145} \right) \left( \frac{145}{320} \right)^2 + \frac{0.144598}{8} \left( \frac{95-8}{95} \right) \left( \frac{95}{320} \right)^2$$

$$= 0.009138$$

$$s_{\bar{y}} = \pm 0.095592 \text{ m}^3/\text{plot}$$

Compare this standard error with that obtained when the stratification was arbitrary.

$$s_{\bar{y}_{\text{Logical}}} = \pm 0.096 < s_{\bar{y}_{\text{Arbitrary}}} = \pm 0.147054 \text{ m}^3 / \text{plot}$$

In this case the gain in precision using logical rather than arbitrary stratification is substantial. The reason for this gain in precision is that when logical stratification is used the plot volumes within each stratum are more nearly uniform than when the stratification is arbitrary and thus the stratum variances are smaller.

Since the stratum variances are not homogeneous and the sampling intensity differs from stratum to stratum, it is not possible to compute a valid sampling error.

## 16.5 ALLOCATION OF THE SAMPLE

*Allocation* of the sample refers to the procedures used to determine the stratum sample sizes, $n_i$. There are four recognized ways by which this can be done. They usually result in different sample

| 2.55 | 3.00 | 1.60 | 1.80 | 2.65 | 2.10 | 1.90 | 3.40 | 3.65 | 4.00 | 4.00 | 1.75 | 2.60 | 2.85 | 2.00 | 2.10 |
|---|---|---|---|---|---|---|---|---|---|---|---|---|---|---|---|
| 3.15 | 2.95 | 2.40 | 2.25 | 3.20 | 2.80 | 2.40 | 4.55 | 4.50 | 3.25 | 4.00 | 1.50 | 1.80 | 2.70 | 1.70 | 2.50 |
| 3.05 | 2.10 | 2.35 | 2.45 | 2.90 | 1.10 | 4.30 | 4.00 | 4.10 | 2.85 | 3.40 | 1.85 | 2.00 | 2.30 | 2.20 | 2.50 |
| 1.55 | 1.55 | 2.55 | 2.15 | 3.10 | 2.75 | 4.50 | 4.60 | 3.30 | 4.30 | 3.75 | 2.30 | 1.80 | 2.00 | 2.10 | 1.80 |
| 2.55 | 0.90 | 2.25 | 2.35 | 2.45 | 2.90 | 4.65 | 1.90 | 3.35 | 4.60 | 2.60 | 2.35 | 2.60 | 2.75 | 2.70 | 3.25 |
| 2.70 | 2.30 | 1.90 | 2.30 | 2.95 | 3.15 | 3.00 | 4.55 | 2.65 | 4.40 | 4.20 | 2.20 | 1.90 | 1.70 | 2.30 | 1.70 |
| 2.50 | 1.50 | 2.00 | 2.55 | 3.15 | 3.30 | 4.05 | 4.50 | 4.35 | 3.80 | 4.50 | 2.00 | 3.65 | 1.35 | 3.15 | 1.60 |
| 2.35 | 2.90 | 2.05 | 2.10 | 3.05 | 3.35 | 4.15 | 4.60 | 4.00 | 3.45 | 1.50 | 1.75 | 2.20 | 2.85 | 3.45 | 2.40 |
| 2.85 | 1.80 | 1.50 | 2.05 | 2.25 | 4.80 | 4.40 | 3.15 | 3.00 | 3.65 | 2.30 | 1.35 | 1.90 | 2.40 | 2.20 | 0.90 |
| 3.15 | 1.60 | 2.40 | 2.60 | 3.00 | 1.90 | 2.55 | 3.55 | 4.00 | 2.75 | 2.65 | 1.90 | 2.60 | 2.70 | 1.50 | 2.65 |
| 1.75 | 0.25 | 1.05 | 0.95 | 3.00 | 2.90 | 3.75 | 3.15 | 3.35 | 2.30 | 3.25 | 2.30 | 2.30 | 1.90 | 1.60 | 0.00 |
| 0.00 | 0.00 | 0.00 | 0.45 | 1.20 | 1.90 | 2.90 | 4.10 | 4.50 | 3.00 | 2.65 | 2.50 | 1.00 | 1.40 | 0.00 | 0.05 |
| 0.60 | 0.00 | 0.50 | 0.55 | 2.05 | 3.10 | 4.65 | 2.30 | 3.00 | 2.20 | 1.90 | 1.30 | 0.55 | 0.75 | 0.15 | 0.20 |
| 0.30 | 0.00 | 0.30 | 0.40 | 1.50 | 1.15 | 2.35 | 1.00 | 2.55 | 2.45 | 2.25 | 0.85 | 0.40 | 0.15 | 1.05 | 0.10 |
| 0.00 | 0.05 | 0.00 | 0.40 | 1.40 | 2.20 | 2.40 | 2.00 | 1.40 | 1.50 | 1.25 | 0.00 | 0.10 | 1.35 | 1.45 | 0.60 |
| 0.10 | 0.00 | 0.00 | 1.00 | 1.95 | 2.40 | 3.60 | 1.60 | 1.90 | 2.40 | 1.00 | 1.15 | 0.40 | 0.20 | 0.05 | 0.00 |
| 0.55 | 0.15 | 0.10 | 0.85 | 2.05 | 1.60 | 1.90 | 2.75 | 1.75 | 2.00 | 0.55 | 0.00 | 0.30 | 0.00 | 1.60 | 0.30 |
| 0.60 | 0.50 | 0.00 | 0.65 | 1.80 | 1.65 | 2.35 | 2.45 | 1.90 | 1.15 | 0.00 | 1.50 | 0.85 | 0.05 | 0.50 | 0.00 |
| 0.60 | 1.00 | 0.15 | 0.00 | 1.70 | 2.50 | 1.25 | 2.40 | 0.80 | 0.50 | 0.10 | 0.75 | 0.30 | 0.35 | 0.20 | 0.10 |
| 0.55 | 0.20 | 0.50 | 0.30 | 1.50 | 2.05 | 1.80 | 1.40 | 1.05 | 0.10 | 0.00 | 0.45 | 0.50 | 0.40 | 0.75 | 0.40 |

**FIGURE 16.6** An equally allocated stratified random sample drawn from the 16-ha tract.

sizes and, consequently, in different standard errors of the mean. There is a pattern in these differences and the sampler should be able to pick the method of allocation that best meets the conditions surrounding a given project.

### 16.5.1 Total Sample Size

Before the stratum sample sizes can be computed, the total sample size must be determined. There are three options available.

First, a sampling intensity may be specified:

$$n = \phi N \tag{16.48}$$

where $\phi$ = specified sampling intensity. This might occur when a paper company specifies that timber cruises include 10% of tracts up to 50 ha or 100 acres, and 5% of larger tracts. This procedure was used in the examples shown in Section 16.4.

$$n = 0.1(320) = 32 \text{ plots}$$

The second method of arriving at the total sample size involves the use of a *cost function* which models the distribution of costs within a project. The basic cost function for a sampling operations is

$$c = c_o + c_F = c_o + \sum_i^M n_i c_i \tag{16.49}$$

where $c$ = total cost of project
$c_o$ = overhead costs (such things as project design, computer costs, photointerpretation, report preparation, and other office expenses assigned to the project)
$c_i$ = cost of evaluating an individual sampling unit in stratum $i$
$n_i$ = sample size in stratum $i$
$c_F$ = total field cost = $c - c_o$

In simple random sampling, $M = 1$ and the cost of evaluating a sampling unit, $c_i$, is a constant. The cost function then becomes

$$c = c_o + nc_i \tag{16.50}$$

This form of the cost function is also applicable to stratified random sampling where the cost of evaluating a sampling unit is constant across all strata. This sample size can be deduced from Equation 16.50,

$$n = (c - c_o)/c_i = c_F/c_i \tag{16.51}$$

As an example, assume that \$400 has been made available for a certain project. It is estimated that the overhead charges will total \$80. From previous experience with similar projects, it is known that the cost of a sampling unit will be approximately \$10; then, using Equation 16.51,

$$n = (400 - 80)/10 = 32 \text{ sampling units}$$

This budget and sample size fits the examples shown in Section 16.4. The actual monetary values used are low but they illustrate the procedure.

When the cost of evaluation of a sampling unit differs from stratum to stratum, the problem of determining the total sample size becomes more difficult. Only in the case of optimum allocation (Section 16.5.4) can differing sampling unit costs be incorporated into the computation of the total sample size.

Finally, one can compute the sample size needed to meet a specified maximum allowable standard error of the mean or, if conditions permit, a specified maximum allowable sampling error. These possibilities will be considered in the context of the four different methods of allocation.

### 16.5.2 Equal Allocation

In the case of *equal* allocation, the sample size is the same in all strata:

$$n_i = n/M \tag{16.52}$$

This form of allocation was used in both examples shown in Section 16.4. The variance of the mean, when the allocation is equal and the sampling is with replacement or the population is infinitely large, is obtained when the allocation expression, $n/M$, is substituted for $n_i$ in Equation 16.24.

$$V(\bar{y})_{St_{(Eq)}} = \sum_i^M \sigma_i^2 \left(\frac{M}{n}\right)\left(\frac{N_i}{N}\right)^2 = \frac{M}{N^2 n} \sum_i^M \sigma_i^2 N_i^2 \tag{16.53}$$

When the sampling is without replacement, from Equation 16.25

$$V(\bar{y})_{\text{St(Eq)}} = \sum_i^M \sigma_i^2 \left(\frac{M}{n}\right)\left(\frac{1}{N_i - 1}\right)\left(N_i - \frac{n}{M}\right)\left(\frac{N_i}{N}\right)^2$$
$$= \frac{M}{N^2 n}\sum_i^M \left(\frac{\sigma_i^2 N_i^3}{N_i - 1}\right) - \frac{1}{N^2}\sum_i^M \left(\frac{\sigma_i^2 N_i^2}{N_i - 1}\right) \tag{16.54}$$

The sample-based estimates of these variances of the mean are obtained when $s_i^2$ is substituted for $\sigma_i^2$ in Equation 16.53 and $s_i^2[(N_i - 1)/N_i)]$ for $\sigma_i^2$ (see Equation 16.31) in Equation 16.54. Thus, when the sampling is with replacement,

$$s^2_{\bar{y}_{\text{St(Eq)}}} = \frac{M}{N^2 n}\sum_i^M s_i^2 N_i^2 \rightarrow V(\bar{y})_{\text{St(Eq)}} \tag{16.55}$$

and when the sampling is without replacement

$$s^2_{\bar{y}_{\text{St(Eq)}}} = \frac{M}{N^2 n}\sum_i^M s_i^2 N_i^2 - \frac{1}{N^2}\sum_i^M s_i^2 N_i \rightarrow V(\bar{y})_{\text{St(Eq)}} \tag{16.56}$$

The total sample size, as can be seen from the examples in Section 16.4, can be determined from either a specified sampling intensity or from the modified cost function, Equation 16.51. In addition, it can be determined when a maximum allowable standard error is specified. To do this, the allowable standard error, $A$, is substituted for $s\bar{y}$ in Equation 16.55 or 16.56 and the equation is then solved for $n$. This follows the pattern established in Section 13.14 for determining the sample size. When the sampling is with replacement, using Equation 16.55,

$$A^2 = \frac{M}{N^2 n}\sum_i^M s_i^2 N_i$$
$$n = \frac{M}{N^2 A^2}\sum_i^M s_i^2 N_i \tag{16.57}$$

and when the sampling is without replacement, using Equation 16.56,

$$A^2 = \frac{M}{N^2 n}\sum_i^M s_i^2 N_i - \frac{1}{N^2}\sum_i^M s_i^2 N_i$$
$$= \frac{1}{N^2 n}\left[M\sum_i^M s_i^2 N_i^2 - n\sum_i^M s_i^2 N_i\right]$$
$$N^2 A^2 n = M\sum_i^M s_i^2 N_i^2 - n\sum_i^M s_i^2 N_i$$

$$n\left(N^2A^2 + \sum_i^M s_i^2 N_i\right) = M\sum_i^M s_i^2 N_i^2$$

$$n = M\sum_i^M s_i^2 N_i^2 \Big/ \left(N^2A^2 + \sum_i^M s_i^2 N_i\right) \qquad (16.58)$$

Before Equations 16.57 and 16.58 can be used, one must have knowledge of the stratum variances. In some cases this can be obtained from records of previous sampling operations in similar conditions as, for example, old timber cruises. In most cases, however, a light preliminary sample, or *presample*, will be used. This presample probably should be equally allocated so that equally strong estimates of the stratum variances are obtained from all strata.

As an example of the use of Equation 16.57, one can apply it to the problem of the 16-ha tract when it has been stratified into the four volume classes. Consider the data from the previous example as coming from a presample. If the maximum allowable standard error is set at ±2.00 m³/ha, $A$ is equal to ±0.1 m³/plot and the total sample size needed is

$$n = \frac{4[(0.431429)(36)^2 + (0.414286)(44)^2 + (0.261741)(145)^2 + (0.144598)(95)^2]}{(320)^2(0.1)^2 + [(0.431429)36 + (0.414286)44 + (0.261741)145 + (0.144598)95]}$$

$$= 29.45 \text{ or } 30 \text{ plots}$$

Note that the rounding was up to 30, rather than down to 29, in spite of the fact that $0.45 < 0.5$. Rounding up is always done when an allowable standard error is specified because the specification must be met. In this case, if 29 plots were used, all else remaining the same, the specifications would not be met.

Equally allocating these 30 plots:

$$n_A = n_B = n_C = n_D = 30/4 = 7.5$$

Fractional sampling units are usually not acceptable. Consequently, 7.5 must be rounded off. If the allocation is to be equal and the allowable standard error specification is to be met, the rounding will have to be *upward* to 8. Thus,

$$n_A = n_B = n_C = n_D = 8$$

and the total sample size becomes 32.

Rounding off stratum sample sizes can cause problems. In the proceeding discussion, where the total sample size was controlled by a specified sampling intensity and by the available funds, $n = 32$ and $M = 4$. Then, with equal allocation, $n_i = 8$. If, however, there were three instead of four strata, then $M$ would equal 3, and

$$n_A = n_B = n_C = n_D = 32/3 = 10.67$$

Rounding up to 11 yields a total of 33, one more than the 32 determined at the beginning of the operation. If funds are sufficient to permit the evaluation of the additional sampling unit, the extra unit should not present a problem. Presumably, this would be the case when a minimum sampling intensity is specified. If funds are limited, however, it may not be possible to pay for the extra unit. Sometimes the costs of evaluation are heavy and this problem may not be trivial. In such a situation the rounding should be *downward*, so that $n_i = 10$ and the total sample size is 30, which fits into the budget.

When the rounding off results in a total sample size that is smaller than that originally computed, the options described above are essentially reversed. Assume that $n = 32$ and $M = 5$. Then $n_i = 32/5 = 6.4$, which would be rounded off to $n_i = 6$. The total sample size would then be 30. If the availability of funds controls the sample size the results of this rounding off would be acceptable. If, however, a minimum sampling intensity had been specified, the total sample would be too small. It would be necessary to round *upward*, so that $n_i = 7$ and the total sample size would be 35.

### 16.5.3 Proportional Allocation

In *proportional* allocation the sampling intensity is constant across all strata.

$$n_i = n(N_i/N) \tag{16.59}$$

The variance of the mean is obtained in the usual manner. When the sampling is with replacement, $nN_i/N$ is substituted for $n_i$ in Equation 16.24.

$$V(\bar{y})_{\text{St(Prop)}} = \sigma^2_{\bar{y}_{\text{St(Prop)}}} = \sum_i^M \sigma_i^2 \left(\frac{n}{nN_i}\right)\frac{N_i^2}{N^2} = \frac{1}{Nn}\sum_i^M \sigma_i^2 N_i \tag{16.60}$$

When the sampling is without replacement, using Equation 16.25,

$$\begin{aligned} V(\bar{y})_{\text{St(Prop)}} = \sigma^2_{\bar{y}_{\text{St(Prop)}}} &= \sum_i^M \sigma_i^2 \left(\frac{N}{nN_i}\right)\left(\frac{1}{N_i - 1}\right)\left(N_i - \frac{nN_i}{N}\right)\left(\frac{N_I}{N}\right)^2 \\ &= \frac{1}{Nn}\left(\frac{N-n}{N}\right)\sum_i^M \sigma_i^2 \left(\frac{N_i}{N_i - 1}\right) N_i \end{aligned} \tag{16.61}$$

which can be written

$$V(\bar{y})_{\text{St(Prop)}} = \frac{1}{Nn}\sum_i^M \left[\sigma_i^2 \left(\frac{N_i}{N_i - 1}\right) N_i\right] - \frac{1}{N^2}\sum_i^M \left[\sigma_i^2 \left(\frac{N_i}{N_i - 1}\right) N_i\right] \tag{16.62}$$

The estimates of these variances are generated when $s_i^2$ is substituted for $\sigma_i^2$ in Equation 16.60 and $s_i^2[(N_i - 1)/N_i]$ is substituted for $\sigma_i^2$ in Equations 16.61 and 16.52. Thus, where the sampling is with replacement,

$$s^2_{\bar{y}_{\text{St(Prop)}}} = \frac{1}{Nn}\sum_i^M s_i^2 N_i \rightarrow V(\bar{y})_{\text{St(Prop)}} \tag{16.63}$$

and when the sampling is without replacement, from Equation 16.61,

$$s^2_{\bar{y}_{\text{St(Prop)}}} = \frac{1}{Nn}\left(\frac{N-n}{N}\right)\sum_i^M s_i^2 N_i \rightarrow V(\bar{y})_{\text{St(Prop)}} \tag{16.64}$$

or from Equation 16.62,

$$s^2_{\bar{y}_{\mathrm{St(Prop)}}} = \frac{1}{Nn}\sum_i^M s_i^2 N_i - \frac{1}{N^2}\sum_i^M s_i^2 N_i \rightarrow V(\bar{y})_{\mathrm{St(Prop)}} \tag{16.65}$$

The total sample size can be determined using a specified sampling intensity, by the availability of funds, or by a specified maximum allowable standard error of the mean. In the latter case, when the sampling is with replacement, the allowable error, $A$, is substituted for $s_{\bar{y}}$ in Equation 16.63 and the equation is then solved for $n$.

$$A^2 = \frac{1}{Nn}\sum_i^M s_i^2 N_i$$

and

$$n = \frac{1}{NA^2}\sum_i^M s_i^2 N_i \tag{16.66}$$

When the sampling is without replacement, using Equation 16.65,

$$A^2 = \frac{1}{Nn}\sum_i^M s_i^2 N_i - \frac{1}{N^2}\sum_i^M s_i^2 N_i$$

Multiplying the first term on the right side of the equation by $N/N$ and the second term by $n/n$,

$$A^2 = \frac{N}{N^2 n}\sum_i^M s_i^2 N_i - \frac{n}{N^2 n}\sum_i^M s_i^2 N_i$$

Then,

$$N^2 n A^2 = N\sum_i^M s_i^2 N_i - n\sum_i^M s_i^2 N_i$$

$$n\left(N^2 A^2 + \sum_i^M s_i^2 N_i\right) = N\sum_i^M s_i^2 N_i$$

and

$$n = N\sum_i^M s_i^2 N_i \bigg/ \left(N^2 A^2 + \sum_i^M s_i^2 N_i\right) \tag{16.67}$$

It was brought out in Section 16.4 (Equations 16.44 through 16.47) that when the stratum variances are homogeneous and the allocation is proportional, one can compute a valid sampling error. Let AE represent the allowable sampling error at the $C$ level of confidence. From Equation 13.108,

$$\text{AE} = t_c(k)s_{\bar{y}_{\text{st(Prop)}}} \quad \text{or simply } tA$$

where $k = n - M$
$A = s_{\bar{y}_{\text{st(Prop)}}}$

Then $A = \text{AE}/t$.

Substitute AE/$t$ for $A$ and $s_w^2$ for $s_i^2$ in Equation 16.66,

$$n = \frac{1}{N(\text{AE}/t)^2}\sum_i^M s_w^2 N_i$$

Since $s_w^2$ is a constant, it can be factored out of the summations leaving $\Sigma_i^M N_i$, which is equal to $N$. Thus,

$$n = \frac{s_w^2 N}{N(\text{AE}/t)^2} = t^2 s_w^2 / \text{AE}^2 \tag{16.68}$$

When the sampling is without replacement, AE/$t$ is substituted for $A$ and $s_w^2$ for $s_i^2$ in Equation 16.67.

$$n = N\sum_i^M s_w^2 N_i \Bigg/ \left[N^2(\text{AE}/t)^2 + \sum_i^M s_w^2 N_i\right] = N^2 s_w^2 / [N^2(\text{AE}/t)^2 + N s_w^2] \tag{16.69}$$

Equations 16.66 through 16.69 require a knowledge of the stratum variances. As was brought out earlier, this information could come from old records but, preferably, it would come from a presample. This presample probably should be equally allocated so as to obtain equally strong estimates of the variances from all the strata.

As an example of proportional allocation, one can again use the 16-ha tract which has been stratified into four volume classes. Then, if $n = 32$, and the sampling is without replacement,

$$n_A = 32(36/360) = 3.6 \text{ or } 4 \text{ plots}$$

$$n_B = 32(44/360) = 4.4 \text{ or } 4 \text{ plots}$$

$$n_C = 32(145/360) = 14.5 \text{ or } 15 \text{ plots}$$

$$n_D = 32(95/360) = 9.5 \text{ or } 9 \text{ plots}$$

Note the means of rounding in the case of strata $C$ and $D$. $n_C$ was rounded up and $n_D$ down. This was done to keep the total sample at 32 and $n_D$ was discriminated against because it was the least heavily timbered (hence, least valuable) stratum.

The sample yielded the following data (in cubic meters per plot):

| | Stratum | | | |
|---|---|---|---|---|
| | *A* | *B* | *C* | *D* |
| | 3.40 | 2.90 | 1.75 | 1.60 |
| | 3.25 | 3.00 | 1.60 | 0.50 |
| | 4.50 | 2.90 | 2.20 | 0.60 |
| | 4.55 | 3.55 | 1.60 | 0.40 |
| | | | 2.05 | 1.00 |
| | | | 2.85 | 0.20 |
| | | | 2.85 | 0.40 |
| | | | 2.60 | 0.00 |
| | | | 1.25 | 1.15 |
| | | | 1.30 | |
| | | | 2.60 | |
| | | | 1.60 | |
| | | | 2.40 | |
| | | | 2.70 | |
| | | | 2.30 | |
| $\bar{y}_i$ | = 3.925 | 3.0875 | 2.11 | 0.65 |
| $s_i^2$ | = 0.484167 | 0.097292 | 0.311500 | 0.256250 |

$$\bar{y} = \sum_i^M (N_i / N)\bar{y}_i$$

$$= (36/320)3.925 + (44/320)3.0875 + (145/320)2.11 + (95/320)0.65$$

$$= 2.015156 \text{ m}^3 / \text{plot}$$

$$s_{\bar{y}}^2 = \sum_i^M \left[ \frac{s_i^2}{n_i} \left( \frac{N_i - n_i}{N_i} \right) \left( \frac{N_i}{N} \right)^2 \right]$$

$$= \frac{0.484167}{4} \left( \frac{36-4}{36} \right) \left( \frac{36}{320} \right)^2 + \frac{0.09792}{4} \left( \frac{44-4}{44} \right) \left( \frac{44}{320} \right)^2$$

$$+ \frac{0.3115}{15} \left( \frac{145-15}{145} \right) \left( \frac{145}{320} \right)^2 + \frac{0.25625}{9} \left( \frac{95-9}{95} \right) \left( \frac{95}{320} \right)^2$$

$$= 0.007874$$

$$s_{\bar{y}} = \pm 0.088737 \text{ m}^3/\text{plot}$$

If a maximum allowable standard error of the mean of ±2.00 m³/ha (±0.1 m³/plot) is specified and the sampling is without replacement, the total sample size would be determined using Equation 16.64. As a presample one can again use the data from the equal allocation example: $s_A^2 = 0.431429$; $s_B^2 = 0.414286$; $s_C^2 = 0.261741$; and $s_D^2 = 0.144598$. Then,

$$n = \frac{320[(0.431429)36 + (0.414286)44 + (0.261741)145 + (0.144598)95]}{(320)^2(2)^2 + [(0.431429)36 + (0.414286)44 + (0.261741)145 + (0.144598)95]}$$

$$= 24.64 \text{ or } 25 \text{ plots}$$

If the stratum variances are homogeneous, which can be ascertained using Bartlett's test, an allowable sampling error can be used to control the sample size. In this case, assume homogeneity and an allowable sampling error, AE, of ±3.5 m³/ha (±0.175 m³/plot) at a confidence level of 0.95. Then using Equation 16.66, $t_{0.95}(k = \infty) = 1.96$, and substituting AE/$t$ for A,

$$n = \frac{320[(0.431429)36 + (0.414286)44 + (0.261741)145 + (0.144598)95]}{[(320)^2(0.175)^2 / (1.96)^2] + (0.431429)36 + (0.414286)44 + (0.261741)145 + (0.144598)95}$$

$$= 30.32 \text{ or } 31 \text{ plots}$$

Since the degrees of freedom $k = n - M = 31 - 4 = 27$, it will be necessary to iterate until the $t$ value used is appropriate for the computed sample size. Using $t_{0.95}(27) = 2.052$,

$$n = 32.94 \text{ or } 33 \text{ plots}$$

Iterating again with $t_{0.95}(29) = 2.045$,

$$n = 32.73 \text{ or } 33 \text{ plots}$$

The value of $n$ has stabilized at 33 plots. In this case, of course, the stratum variances are not homogeneous and a sampling error computed using proportional allocation of the 33 plots would not be valid.

### 16.5.4 Value Allocation

A variant of proportional allocation has been used on occasion by foresters in timber inventory operations. In it the per unit *value*, in terms of volume or dollars, is incorporated into the allocation process, giving rise to the label *value* allocation. In essence this is proportional allocation with modified stratum weights. Conventionally, the stratum weight is the ratio of the size of the stratum to the total size. In value allocation it is the ratio of stratum value to the total value. Thus, if $N_i$ is the size of stratum $i$ and $W_i$ is its value on a per sampling unit basis, the value of stratum $i$ would be $Q_i = N_iW_i$. Then the allocation expression becomes

$$n_i = n\,(Q_i/Q) \tag{16.72}$$

where

$$Q = \sum_{i}^{M} Q_i$$

The variance of the mean, when the sampling is with replacement, is derived from Equation 16.24,

$$V(\bar{y})_{\text{st(Va)}} = \frac{Q}{N^2 n} \sum_{i}^{M} (\sigma_i^2 N_i^2 / Q_i) \tag{16.73}$$

and when the sampling is without replacement, Equation 16.25 becomes

$$V(\bar{y})_{\text{st(Va)}} = \frac{1}{N^2}\sum_{i}^{M}\left[\frac{\sigma_i^2}{n}\left(\frac{Q}{Q_i}\right)\left(\frac{N_i^3}{N_i - 1}\right) - \frac{\sigma_i^2 N_i^2}{(N_i - 1)}\right] \tag{16.74}$$

The estimates of these variances of the mean are then, when the sampling is with replacement,

$$s^2_{\bar{y}_{\text{st(Va)}}} = \frac{Q}{N^2 n}\sum_{i}^{M}(s_i^2 N_i^2 / Q_i) \rightarrow V(\bar{y})_{\text{st(Va)}} \tag{16.75}$$

and when the sampling is without replacement,

$$s^2_{\bar{y}_{\text{st(Va)}}} = \frac{1}{N^2}\sum_{i}^{M}\left[\frac{s_i^2}{n}\left(\frac{Q}{q_i}\right)N_i^2 - s_i^2 N_i\right] \rightarrow V(\bar{y})_{\text{st(Va)}} \tag{16.76}$$

As with any proportional allocation method, the total sample size can be determined using a specified sampling intensity, by the cost function, Equation 16.51, by a maximum allowable standard error, or, if the stratum variances are homogeneous, by a maximum allowable sampling error. The usual formulae used with proportional allocation (Equations 16.66 through 16.69) are applicable.

As an example of this method of allocation, one can apply it to the problem of the timber inventory on the 16-ha tract. Assume that the results of the equal allocation example form a presample. From this presample it was found that the mean stratum volumes were

$$\bar{y}_A = 4.125 \text{ m}^3\text{/plot}$$

$$\bar{y}_B = 3.325 \text{ m}^3\text{/plot}$$

$$\bar{y}_C = 2.10625 \text{ m}^3\text{/plot}$$

$$\bar{y}_D = 0.64375 \text{ m}^3\text{/plot}$$

For convenience in calculations these per plot volumes have been rounded to

$$W_A = 4 \text{ m}^3\text{/plot} \qquad W_C = 2 \text{ m}^3\text{/plot}$$

$$W_C = 3 \text{ m}^3\text{/plot} \qquad W_D = 1 \text{ m}^3\text{/plot}$$

Then,

$$Q_A = W_A N_A = 4(36) = 144$$

$$Q_B = W_B N_B = 3(44) = 132$$

$$Q_C = W_C N_C = 2(145) = 290$$

$$Q_D = W_D N_D = 1(95) = \underline{95}$$

$$Q = 661$$

If the total sample size is 32, the allocation is

$$n_A = 32(144/661) = 6.97 \text{ or } 7 \text{ plots}$$

$$n_B = 32(132/661) = 6.39 \text{ or } 6 \text{ plots}$$

$$n_C = 32(290/661) = 14.93 \text{ or } 14 \text{ plots}$$

$$n_D = 32(95/661) = 4.60 \text{ or } \underline{5} \text{ plots}$$

$$n = 32 \text{ plots}$$

Note that there is a definite difference between this allocation and that obtained using conventional proportional allocation. The value allocation provides a heavier sample in the valuable strata and a lighter sample in the low-value strata and, thus, counters one of the major flaws foresters associate with proportional allocation — oversampling large, low-value strata and undersampling small, high-value strata.

### 16.5.5 Optimum Allocation

This form of allocation is known as *optimum* because it produces, on the average, for a given sample size and a specific set of sampling unit evaluation costs, the smallest possible standard error of the mean. Furthermore, when a maximum allowable standard error is specified, the use of optimum allocation will ensure the meeting of the specification at the lowest possible cost. This efficiency is achieved by including the stratum standard deviations and costs, as well as the stratum sizes, in the weighting process.

To simplify the mathematical development, let

$$S_i^2 = \sigma_i^2\left(\frac{N_i}{N_i - 1}\right)$$

and

$$w_i = N_i/N$$

Equation 16.25 then becomes

$$\begin{aligned} V(\bar{y})_{st} &= \sum_i^M \left[\frac{S_i^2}{n_i}\left(\frac{N_i - n_i}{N_i}\right)w_i^2\right] \\ &= \sum_i^M S_i^2 w_i^2 n_i^{-1} - \sum_i^M S_i^2 w_i^2 N_i^{-1} \end{aligned} \tag{16.77}$$

According to Equation 16.49, the cost function is

$$\begin{aligned} c &= c_o + \sum_i^M n_i c_i \\ c - c_o &= c_F = \sum_i^M n_i c_i \end{aligned} \tag{16.78}$$

where $c_F$ = cost of field operations.

Combining the variance of the mean and cost functions, using the Lagrangian multiplier (see, e.g., Chiang, 1974),

$$V = V(\bar{y}_o)_{st} + \lambda\left(\sum_i^M h_i c_i - c_f\right)$$
$$= \sum_i^M S_i^2 w_i^2 n_i^{-1} - \sum_i^M S_i^2 w_i^2 N_i^{-1} + \lambda\left(\sum_i^M n_i c_i - c_F\right) \quad (16.79)$$

Then

$$\frac{dV}{dn_i} = -S_1^2 w_1^2 n_1^{-2} + \lambda c_1 = 0$$

$$\frac{dV}{dn^2} = -S_2^2 w_2^2 n_2^{-2} + \lambda c_2 = 0$$

$$\vdots$$

$$\frac{dV}{dn_M} = -S_M^2 w_M^2 n_M^{-2} + \lambda c_M = 0$$

Generalizing,

$$\frac{dV}{dn_i} = -S_i^2 w_i^2 n_i^{-2} + \lambda c_i = 0 \quad (16.80)$$

Then, for stratum $i$

$$\lambda c_i = S_i^2 w_i^2 n_i^{-2}$$

$$\lambda n_i^2 = S_i^2 w_i^2 / c_i$$

$$n_i \sqrt{\lambda} = S_i w_i / \sqrt{c_i}$$

Since

$$n\sqrt{\lambda} = \sqrt{\lambda} \sum_i^M n_i$$

$$\frac{n_i}{n} = \frac{n_i\sqrt{\lambda}}{n\sqrt{\lambda}} = (S_i w_i / \sqrt{c_i}) \Big/ \sum_i^M (S_i w_i / \sqrt{c_i})$$

The optimum allocation is then

$$n_i = (nS_iw_i / \sqrt{c_i}) \Big/ \sum_i^M (S_iw_i / \sqrt{c_i}) \tag{16.81}$$

which, since $w_i = N_i/N$, can be written, after simplication,

$$n_i = (nS_iN_i / \sqrt{c_i}) \Big/ \sum_i^M (S_iN_i / \sqrt{c_i}) \tag{16.82}$$

In the derivation of Equations 16.81 and 16.82, the second term of Equation 16.77, which is the finite population correction, drops out. Consequently, Equations 16.81 and 16.82 are applicable whether the sampling is with or without replacement.

To obtain $n_i$ it is necessary to determine the magnitude of the total sample size, $n$. When the cost of field operations, $c_F$, is fixed by a budget, $n$ is obtained by substituting the right side of Equation 16.82 in the cost function (Equation 16.78) and solving for $n$.

$$c_F = \sum_i^M \left[ (nS_iN_i / \sqrt{c_i}) \Big/ \sum_i^M (S_iN_i / \sqrt{c_i}) \right] c_i$$

Since $c_i / \sqrt{c_i} = \sqrt{c_i}$,

$$c_F = n \sum_i^M S_iN_i\sqrt{c_i} \Big/ \sum_i^M (S_iN_i / \sqrt{c_i})$$

Then,

$$n = c_F \sum_i^M (S_iN_i\sqrt{c_i}) \Big/ \sum_i^M (S_iN_i / \sqrt{c_i}) \tag{16.83}$$

The use of this expression will ensure that the variance and/or the standard error of the mean will be at the minimum achievable with the available funds. Furthermore, it is applicable whether the sampling is with or without replacement.

In practice, the actual stratum standard deviations, $S_i$, are not known and must be estimated by $s_i$.

When a maximum allowable standard error of the mean, $A$, has been specified, and the sampling is with replacement, $n$ is obtained by substituting $A^2$ for $V(\bar{y})$ and the right side of Equation 16.82 for $n_i$ in Equation 16.24. If the sampling is without replacement, these substitutions are made in Equation 16.25. These modified equations are then solved for $n$.

When the sampling is with replacement, $N = \infty$, $[N/(N - 1)] = 1$, and $S_i^2 = \sigma_i^2$. Equation 16.24 then becomes

$$V(\bar{y}_o)_{st} = \sum_i^M \left[ \frac{S_i^2}{n_i} \left( \frac{N_i}{N} \right)^2 \right] \tag{16.84}$$

Then,

$$A^2 = \sum_i^M S_i^2 \frac{N_i^2}{N_2}\left[\frac{\sum_i^M (S_i N_i / \sqrt{c_i})}{n S_i N_i / \sqrt{c_i}}\right]$$

$$= \frac{1}{N^2 n}\sum_i^M S_i N_i \sqrt{c_i} \sum_i^M (S_i N_i / \sqrt{c_i})$$

and

$$n = \frac{1}{N^2 A^2}\sum_i^M S_i N_i \sqrt{c_i} \sum_i^M (S_i N_i / \sqrt{c_i}) \tag{16.85}$$

When the sampling is without replacement,

$$A^2 = \sum_i^M S_i^2 w_i^2 \left[\frac{\sum_i^M (S_i w_i / \sqrt{c_i})}{n S_i w_i / \sqrt{c_i}}\right] - \sum_i^M \left(\frac{S_i^2 w_i^2}{N_i}\right)$$

Since $w_i^2/N_i = (N_i^2/N^2)/N_i = w_i(N_i/N)/N_i = w_i/N$,

$$A^2 = \frac{1}{n}\sum_i^M S_i w_i \sqrt{c_i} \sum_i^M (S_i w_i / \sqrt{c_i}) - \frac{1}{N}\sum_i^M S_i^2 w_i$$

$$A^2 + \frac{1}{n}\sum_i^M S_i^2 w_i = \frac{1}{n}\sum_i^M S_i w_i \sqrt{c_i} \sum_i^M (S_i w_i / \sqrt{c_i})$$

and

$$\begin{aligned} n &= N\sum_i^M S_i w_i \sqrt{c_i} \sum_i^M (S_i w_i / \sqrt{c_i}) \Bigg/ \left(A^2 N + \sum_i^M S_i^2 w_i\right) \\ &= \sum_i^M S_i w_i \sqrt{c_i} \sum_i^M (S_i w_i / \sqrt{c_i}) \Bigg/ \left[A^2 + (1/N)\sum_i^M S_i^2 w_i\right] \end{aligned} \tag{16.86}$$

which, since $w_i = N_i/N$, can be written

$$n = \sum_i^M S_i N_i \sqrt{c_i} \sum_i^M (S_i N_i / \sqrt{c_i}) \Bigg/ \left(N^2 A^2 + \sum_i^M S_i^2 N_i\right) \tag{16.87}$$

It should be noted that the sample-based stratum variance, $s_i^2$, is

$$s_i^2 = \frac{1}{n_i - 1}\sum_j^{n_i}(y_{ij} - \bar{y}_i)^2 \to \sigma_i^2\left(\frac{N_i}{N_i - 1}\right) = S_i^2$$

and, consequently, $s_i$ can be substituted without any correction for $S_i$ in Equations 16.81 through 16.83 and 16.85 through 16.87 to obtain estimates of $n_i$ and $n$.

As can be seen, $n_i$ is positively proportional to the stratum size, $N_i$, and the stratum standard deviation, $S_i$. It is, however, negatively proportional to the square root of the cost of evaluating a sampling unit, $c_i$. This means that large strata and/or strata with large variances will be more heavily sampled than small strata and/or strata that are relatively homogeneous. Furthermore, strata that are expensive to sample will be assigned relatively smaller samples than strata that are inexpensive to sample.

The variance of the mean is obtained when the sampling is with replacement, by substituting the right side of Equation 16.82 for $n_i$ in Equation 16.84, and when the sampling is without replacement, by substituting the right side of Equation 16.82 for $n_i$ in Equation 16.77, and then solving for $n$. Thus, when the sampling is with replacement,

$$V(\bar{y}_o)_{\text{St(Op)}} = \frac{1}{N^2 n}\sum_i^M S_i N_i \sqrt{c_i}\sum_i^M (S_i N_i / \sqrt{c_i}) \tag{16.88}$$

and when the sampling is without replacement,

$$V(\bar{y}_o)_{\text{St(Op)}} = \frac{1}{n}\sum_i^M S_i w_i \sqrt{c_i}\sum_i^M (S_i w_i / \sqrt{c_i}) - \frac{1}{N}\sum_i^M S_i^2 w_i \tag{16.89}$$

or

$$V(\bar{y}_o)_{\text{St(Op)}} = \frac{1}{N^2 n}\sum_i^M S_i N_i \sqrt{c_i}\sum_i^M (S_i N_i / \sqrt{c_i}) - \frac{1}{N^2}\sum_i^M S_i^2 N_i \tag{16.90}$$

The estimated variance of the mean is then, when the sampling is with replacement,

$$s^2_{\bar{y}_{\text{st(Op)}}} = \frac{1}{N^2 n}\sum_i^M s_i N_i \sqrt{c_i}\sum_i^M (s_i N_i / \sqrt{c_i}) \to V(\bar{y}_o)_{\text{st(Op)}} \tag{16.91}$$

and when the sampling is without replacement,

$$s^2_{\bar{y}_{\text{st(Op)}}} = \frac{1}{N^2 n}\sum_i^M s_i N_i \sqrt{c_i}\sum_i^M (s_i N_i / \sqrt{c_i}) - \frac{1}{N^2}\sum_i^M s_i^2 N_i \to V(\bar{y}_o)_{\text{st(Op)}} \tag{16.92}$$

As an example of optimum allocation, one can return to the timber inventory of the 16-ha tract. If \$200 has been made available for the project, the overhead costs are estimated to be \$40, and the costs of plot evaluation are $c_A$ = \$8.50, $c_B$ = \$6.50, $c_C$ = \$5.00, and $c_D$ = \$3.00, the total sample size, assuming sampling without replacement, is computed using equation 16.83. The stratum variances are from the equal allocation example, which is being used as a presample.

| Stratum | $N_i$ | $s_i$ | $\sqrt{c_i}$ | $s_iN_i$ | $s_i^2N_i$ | $s_iN_i\sqrt{c_i}$ | $s_iN_i/\sqrt{c_i}$ |
|---|---|---|---|---|---|---|---|
| *A* | 36 | 0.656832 | 2.915476 | 23.645952 | 15.53143 | 68.939206 | 8.110495 |
| *B* | 44 | 0.643650 | 2.549510 | 28.320600 | 18.22857 | 72.203653 | 11.108252 |
| *C* | 145 | 0.511606 | 2.236068 | 74.182870 | 37.95246 | 165.877940 | 33.175588 |
| *D* | 95 | 0.380261 | 1.732051 | 36.124795 | 13.73683 | 62.569987 | 20.856658 |
| | | | | 162.274217 | 85.44929 | 369.590786 | 73.250993 |

$$n = (200 - 40)73.250993/369.590786$$
$$= 31.71 \text{ or } 32 \text{ plots}$$
$$n_A = 32\ (8.110495/73.250993) = 3.55 \text{ or } 4$$
$$n_B = 32(11.108252/73.250993) = 4.85 \text{ or } 5$$
$$n_C = 32(33.175588/73.250993) = 14.49 \text{ or } 14$$
$$n_D = 32(20.856658/73.250993) = 9.11 \text{ or } \underline{9}$$
$$32$$

These 32 plots were randomly chosen. The cruise data (in cubic meters per plot) and computations are as follows.

| | Stratum | | | |
|---|---|---|---|---|
| | *A* | *B* | *C* | *D* |
| | 2.65 | 3.05 | 2.75 | 1.05 |
| | 4.80 | 2.30 | 1.40 | 0.50 |
| | 3.40 | 3.10 | 1.55 | 0.60 |
| | 4.35 | 3.35 | 2.40 | 0.15 |
| | | 3.00 | 2.55 | 0.00 |
| | | | 1.40 | 0.00 |
| | | | 2.20 | 1.20 |
| | | | 1.90 | 0.00 |
| | | | 2.50 | 0.60 |
| | | | 2.30 | |
| | | | 3.15 | |
| | | | 2.10 | |
| | | | 2.25 | |
| | | | 2.30 | |
| $\bar{y}_i =$ | 3.8 | 2.96 | 2.196429 | 0.455556 |
| $s_i^2 =$ | 0.928333 | 0.154250 | 0.252486 | 0.208403 |

$$\bar{y} = \sum_{i}^{4} (N_i / N)\bar{y}_i$$
$$= (36/320)3.8 + (44/320)2.96 + (145/320)2.196429 + (95/320)0.455556$$
$$= 1.965 \text{ m}^3/\text{plot}$$

| | Stratum | | |
|---|---|---|---|
| *A* | *B* | *C* | *D* |

$$s_{\bar{y}}^2 = \sum_i^M \left[ \frac{s_i^2}{n_i} \left( \frac{N_i - n_i}{N_i} \right) \left( \frac{N_i}{N} \right)^2 \right]$$

$$= \frac{0.928333}{4}\left(\frac{36-4}{36}\right)\left(\frac{36}{320}\right)^2 + \frac{0.154250}{5}\left(\frac{44-5}{44}\right)\left(\frac{44}{320}\right)^2$$

$$+ \frac{0.252486}{14}\left(\frac{145-14}{145}\right)\left(\frac{145}{320}\right)^2 + \frac{0.208403}{9}\left(\frac{95-9}{95}\right)\left(\frac{95}{320}\right)^2$$

$$= 0.008321$$

$$s_{\bar{y}} = \pm 0.091220 \text{ m}^3\text{/plot}$$

If, instead of a budgetary limit, the project is being controlled by the specification that the maximum allowable standard error is ±2.00 m$^3$/ha (±0.1 m$^3$/plot), the total sample size, using equation 16.91 is,

$$n = 369.590786(73.250993)/[(320)^2(0.1)^2 + 85.44929]$$
$$= 24.40 \text{ or } 25 \text{ plots}$$

In many cases the cost of calculating a sampling unit will vary from stratum to stratum. This certainly occurs in many forest inventory operations. Cruising timber in open mature stands on flat terrain is much easier and less expensive than cruising dense pole-sized stands on steep slopes or in deep swamps. The problem in these cruising situations is not simply that of evaluation of the sampling units themselves; it also involves the problem of traveling from plot to plot. This traveling time must be considered a factor when arriving at values for $c_i$. These costs can usually be estimated using past experiences in similar situations. If such experiences are not available, it will be necessary to carry out time and cost studies while the presample is being obtained. In the absence of hard data and costs, one may be able to make estimates of the relative costs and use these estimates in the computation process. For example, in the case of the four volume classes on the 16-ha tract, one might estimate that it costs twice as much to evaluate a plot in stratum *C* as it does in stratum *D*, three times as much in stratum *B*, and four times as much in stratum *A*. The stratum costs, as used in the computations, would then be $c_A = 4$, $c_B = 3$, $c_C = 2$, $c_D = 1$.

### 16.5.6 Neyman Allocation

If the costs of sampling unit evaluations are the same, or nearly the same, across all the strata, the computations associated with optimum allocation can be simplified. In the case of Equation 16.82, when the $c_i$ are constant, $\sqrt{c_i}$ can be factored out of the summation and then

$$n_i = \frac{nS_iN_i}{\sqrt{c_i}} \left( \frac{\sqrt{c_i}}{\sum_i^M S_iN_i} \right) = nS_iN_i \Big/ \sum_i^M S_iN_i \tag{16.93}$$

This form of optimum allocation is known as *Neyman* allocation* or *optimum allocation with constant costs*.

* After J. Neyman, who described it in 1934 (Neyman, 1934).

Because the stratum costs are constant, the expressions for the variance of the mean under optimum allocation can be simplified. Thus, Equation 16.88 becomes

$$V(\bar{y}_o)_{\text{st(Ney)}} = \frac{1}{N^2 n}\sqrt{c_i}\sum_i^M S_i N_i \left(\frac{1}{\sqrt{c_i}}\right)\sum_i^M S_i N_i$$
$$= \frac{1}{N^2 n}\left(\sum_i^M S_i N_i\right)^2 \tag{16.94}$$

When the sampling is without replacement, Equation 16.89 becomes

$$V(\bar{y}_o)_{\text{st(Ney)}} = \frac{1}{N^2 n}\sqrt{c_i}\sum_i^M S_i N_i \left(\frac{1}{\sqrt{c_i}}\right)\sum_i^M S_i N_i - \frac{1}{N}\sum_i^M S_i^2 N_i$$
$$= \frac{1}{N^2 n}\left(\sum_i^M S_i N_i\right)^2 - \frac{1}{N}\sum_i^M S_i^2 N_i \tag{16.95}$$

The estimate of the variance of the mean, when the sampling is with replacement, from Equation 16.94, is

$$s^2_{\bar{y}_{\text{st(Ney)}}} = \frac{1}{N^2 n}\left(\sum_i^M s_i N_i\right)^2 \tag{16.96}$$

and when the sampling is without replacement, from Equation 16.95,

$$s^2_{\bar{y}_{\text{st(Ney)}}} = \frac{1}{N^2 n}\left(\sum_i^M s_i N_i\right)^2 - \frac{1}{N}\sum_i^M s_i^2 N_i \tag{16.97}$$

When the total sample size is constrained by the available funds and the stratum costs are constant, Equation 16.83 reduces to Equation 16.51.

$$n = c_F\left(\frac{1}{\sqrt{c_i}}\right)\sum_i^M S_i N_i \left(\frac{1}{\sqrt{c_i}}\right) \Big/ \sum_i^M S_i N_i$$

The summations cancel, leaving

$$n = c_F/c_i$$

which is Equation 16.51.

When a maximum allowable standard error is specified and the sampling is with replacement, the total sample size is obtained when the allowable error, $A$, is substituted for $S_{\bar{y}_{\text{st(Ney)}}}$ in Equation 16.96 and the equation is solved for $n$.

$$A^2 = \frac{1}{N^2 n}\left(\sum_i^M s_i N_i\right)^2$$

and

$$n = \frac{1}{N^2 A^2}\left(\sum_i^M s_i N_i\right)^2 \tag{16.98}$$

When the sampling is without replacement, using Equation 16.97,

$$A^2 = \frac{1}{N^2 n}\left(\sum_i^M s_i N_i\right)^2 - \frac{1}{N}\sum_i^M s_i^2 N_i$$

and

$$n = \left(\sum_i^M s_i N_i\right)^2 \Bigg/ \left(N^2 A^2 + \sum_i^M s_i^2 N_i\right) \tag{16.99}$$

When the stratum variances are homogeneous, the Neyman allocation reduces to the proportional. Substituting $S_w$ for $S_i$ in Equation 16.93, the Neyman allocation formula becomes

$$n_i = nS_w N_i \Big/ \sum_i^M S_w N_i$$

Since $S_w$ is a constant and $\Sigma_i^M N_i = N$,

$$\begin{aligned} n_i &= nS_w N_i / S_w N \\ &= n(S_w / S_w)(N_i / N) \\ &= n(N_i / N) \end{aligned}$$

which is Equation 16.59, the proportional allocation expression. Similarly, the expression for the total sample size, when a maximum allowable standard error has been specified, Equation 16.99, becomes

$$n = N^2 S_w^2 / (N^2 A^2 + NS_w^2) \tag{16.100}$$

As in the case of the optimum allocation, these expressions for $n_i$ and $n$ make use of the $S$ rather than the $\sigma$ symbology. Consequently, sample variances can be substituted for the $S$ values, without corrections, to obtain estimates of $n_i$ and $n$. The sample variances would be obtained from old records or a presample.

One can now apply the Neyman allocation to the timber inventory of the 16-ha tract. Assume a total sample size of 32, based on available funds, and that the data from the equal allocation example are from a presample.

| Stratum | $N_i$ | $s_i$ | $N_is_i$ | $N_is_i/\Sigma_i^M N_is_i$ | $n_i$ |
|---|---|---|---|---|---|
| *A* | 36 | 0.656832 | 23.645952 | 0.145716 | 32(0.145716) = 4.66 or 5 |
| *B* | 44 | 0.643650 | 28.320600 | 0.174523 | 32(0.174523) = 5.58 or 6 |
| *C* | 145 | 0.511606 | 74.182870 | 0.457145 | 32(0.457145) = 14.63 or 15 |
| *D* | 95 | 0.380261 | 36.124795 | 0.222616 | 32(0.222616) = 7.12 or 7 |

$$\sum_i^M N_i s_i = 162.274217.4854 \qquad \sum_i^M n_i = 33$$

Since the total sample size is limited by the availability of funds to 32, one sampling unit must be deleted from one of the strata. The decision of which stratum sample is to be reduced is arbitrary. In this case, reduce $n_c$ from 15 to 14. The rationale behind this decision is that the computed sample size for stratum $C$, 14.63, is only slightly more than 14.5, below which the $n_c$ would be 14. The final allocation is then $n_A = 5$, $n_B = 6$, $n_C = 14$, and $n_D = 7$. Samples of these magnitudes were randomly drawn from the four strata and found and evaluated in the field. The resulting data (in cubic meters per plot) and computations are as follows.

| | Stratum | | | |
|---|---|---|---|---|
| | *A* | *B* | *C* | *D* |
| | 4.65 | 2.65 | 2.20 | 0.15 |
| | 4.60 | 3.00 | 2.10 | 0.40 |
| | 4.60 | 2.65 | 2.85 | 0.80 |
| | 4.20 | 3.65 | 2.00 | 1.25 |
| | 4.30 | 1.90 | 1.90 | 0.60 |
| | | 3.10 | 2.05 | 0.00 |
| | | | 1.90 | 0.40 |
| | | | 2.45 | |
| | | | 1.40 | |
| | | | 2.30 | |
| | | | 2.25 | |
| | | | 2.00 | |
| | | | 2.60 | |
| | | | 2.40 | |
| $\bar{y}_i =$ | 4.47 | 2.825 | 2.171429 | 0.514286 |
| $s_i^2 =$ | 0.042000 | 0.340750 | 0.124506 | 0.175595 |

$$\bar{y} = \sum_i^M (N_i / N)\bar{y}_i$$

$$= (36/320)4.47 + (44/320)2.825 + (145/320)2.171429 + (95/320)0.514286$$

$$= 2.027920 \text{ m}^3/\text{plot}$$

$$s_{\bar{y}}^2 = \sum_i^4 \left[ \frac{s_i^2}{n_i} \left( \frac{N_i - n_i}{N_i} \right) \left( \frac{N_i}{N} \right)^2 \right]$$

$$= \frac{0.042000}{5} \left( \frac{36-5}{36} \right) \left( \frac{36}{320} \right)^2 + \frac{0.340750}{6} \left( \frac{44-6}{44} \right) \left( \frac{44}{320} \right)^2$$

$$+ \frac{0.124506}{14} \left( \frac{145-14}{145} \right) \left( \frac{145}{320} \right)^2 + \frac{0.175595}{7} \left( \frac{95-7}{95} \right) \left( \frac{95}{320} \right)^2$$

$$= 0.004719$$

$$s_{\bar{y}} = \pm 0.068695 \text{ m}^3/\text{plot}$$

If the maximum allowable standard error of the mean is set at ±2.00 m³/ha (±0.01 m³/plot), the total sample size needed would be computed using Equation 16.99 and the data from the equal allocation example.

$$n = \frac{(162.274217)^2}{(320)^2(0.01)^2 + (0.431429)36 + (0.414286)44 + (0.261741)145 + (0.144598)95}$$

$$= 23.74 \text{ or } 24 \text{ plots}$$

## 16.6 COMPARISON OF THE METHODS OF ALLOCATION

There are two measures by which one can judge the relative efficiencies of the different methods of allocation. One is to use the relative magnitudes of the variances or standard errors of the mean and the other is the sample sizes needed to meet a specified maximum allowable standard error. These data, along with those of a simple random sample, have been generated for the 16-ha tract when it was divided into the four volume classes. They are as follows:

| **Method of Sampling** | $s_{\bar{y}}$ **(m³/plot)** | ***n*(Needed) (plots)** |
|---|---|---|
| Simple random | ±0.201866 | 100 |
| Equal allocation | ±0.095592 | 30 |
| Proportional allocation | ±0.088737 | 25 |
| Neyman allocation | ±0.068695 | 24 |

Comparable standard errors and sample sizes cannot be generated for the value or optimum allocation methods because of, in the case of value allocation, the varying stratum values and, in the case of optimum allocation, the varying stratum costs. Notice that the efficiencies improve as one goes from simple random to equal allocation to proportional allocation and, finally, to the Neyman allocation. The greatest gain is obviously associated with stratification itself. Gains comparable with these are not uncommon, provided that the stratification is appropriately done.

It is also obvious that the methods of allocation have differing effects on the efficiency of the sampling operation. To explore the reasons for these differences in efficiency, this section will base the explanation on the effects of the methods of allocation on the variance of the mean.

According to Equation 12.5, the variance of the mean of an unrestricted random sample when the sampling is without replacement is

$$V(\bar{y})_{\text{Ran}} = \sigma^2_{\bar{y}_{\text{Ran}}} = \frac{\sigma^2_Y}{n}\left(\frac{N-n}{N-1}\right)$$

If the sampling is with replacement or $N$ is much larger than $n$, the finite population correction can be assumed to be equal to 1. Then, according to Equation 12.3,

$$V(\bar{y})_{\text{Ran}} = \sigma^2_{\bar{y}_{\text{Ran}}} = \sigma^2_Y / n$$

According to Equation 16.21,

$$\sigma^2_Y = \sigma^2_W + \sigma^2_B$$

Substituting this in Equation 12.3 yields the expression:

$$V(\bar{y})_{\text{Ran}} = \frac{1}{n}(\sigma^2_W + \sigma^2_B) \tag{16.101}$$

Thus, the variance of the mean of a simple random sample is made up of two components, the *within-strata* and the *between-strata* variances of the mean.

For simplicity in the following discussion, it will be assumed that the sampling is with replacement or that $N$ is sufficiently large, relative to $n$, that the finite population corrections are negligible and $S$ can be considered the equivalent of $\sigma$. Then, according to Equation 16.60,

$$V(\bar{y})_{\text{st(Prop)}} = \frac{1}{Nn}\sum_i^M \sigma^2_i N_i$$

Since $\sigma^2_i = (1/N)\Sigma_j^{N_i}(y_{ij} - \mu_i)^2$,

$$\sigma^2_i N_i = \sum_j^M (y_{ij} - \mu_i)^2$$

Substituting this into Equation 16.60,

$$V(\bar{y})_{\text{st(Prop)}} = \frac{1}{Nn}\sum_i^M \sum_j^{N_i} (y_{ij} - \mu_i)^2$$

According to Equation 16.18,

$$\sigma^2_W = \frac{1}{N}\sum_i^M \sum_j^{N_i} (y_{ij} - \mu_i)^2$$

Consequently,

$$V(\bar{y})_{\text{st(Prop)}} = \frac{1}{N}\sigma_W^2 \tag{16.102}$$

Substituting this into Equation 16.100,

$$V(\bar{y})_{\text{Ran}} = V(\bar{y})_{\text{st(Prop)}} + \frac{1}{N}\sigma_B^2 \tag{16.103}$$

Thus, the variance of the mean of the proportionally allocated stratified sample is one component of the variance of the mean of the unrestricted random sample. Consequently, $V(\bar{y})_{\text{st(Prop)}} \leq V(\bar{y})_{\text{Ran}}$. The more homogeneous the values of the random variance within the strata, the smaller the $V(\bar{y})_{\text{st(Prop)}}$ and the greater the difference in efficiency between the two methods of sampling. Only if all the strata have the same mean would $V(\bar{y})_{\text{Ran}} = V(\bar{y})_{\text{st(Prop)}}$. This supports the argument that the stratification should be carefully done. It should be done in such a manner that the differences between the stratum means are maximized.

In the case of the 16-ha tract the between strata variance of the mean, when using proportional allocation, is

$$\frac{1}{n}s_B^2 = s_{\bar{y}_{\text{Ran}}} - s^2_{\bar{y}_{\text{st}}(\text{prop})} = 0.040750 - 0.007874 = 0.032876$$

indicating that the stratification was quite successful. Almost 81% of the variance of the means of the unrestricted random sample was removed by the stratification.

When the Neyman allocation is used, the variance and standard error of the mean are even smaller than when proportional allocation is used. According to Equation 16.94, the variance of the means of a Neyman allocated stratified sample, when the sampling is with replacement, is

$$V(\bar{y})_{\text{st(Ney)}} = \frac{1}{N^2 n}\left(\sum_i^M \sigma_i N_i\right)^2$$

The corresponding variance of the means of a proportionally allocated sample is given by Equation 16.60:

$$V(\bar{y})_{\text{st(Prop)}} = \frac{1}{Nn}\sum_i^M \sigma_i^2 N_i$$

Then,

$$V(\bar{y})_{\text{st(Prop)}} = \frac{1}{Nn}\sum_i^M \sigma_i^2 N_i + \frac{1}{N^2 n}\left(\sum_i^M \sigma_i N_i\right)^2 - \frac{1}{N^2 n}\left(\sum_i^M \sigma_i N_i\right)$$

Rearranging,

$$V(\bar{y})_{\text{st(Prop)}} = \frac{1}{N^2 n}\left(\sum_i^M \sigma_i N_i\right)^2 + \left[\frac{1}{Nn}\sum_i^M \sigma_i^2 N_i - \frac{1}{N^2 n}\left(\sum_i^M \sigma_i N_i\right)^2\right]$$

$$= V(\bar{y})_{\text{st(Ney)}} + \left[\frac{1}{Nn}\sum_i^M \sigma_i^2 N_i - \frac{1}{N^2 n}\left(\sum_i^M \sigma_i N_i\right)^2\right] \tag{16.104}$$

Considering now only the segment within the brackets, let

$$D = \frac{1}{Nn}\left[\sum_i^M (\sigma_i^2 N_i) - \frac{1}{N}\left(\sum_i^M \sigma_i N_i\right)^2 + \frac{1}{N}\left(\sum_i^M \sigma_i N_i\right)^2 - \frac{1}{N}\left(\sum_i^M \sigma_i N_i\right)^2\right]$$

$$= \frac{1}{Nn}\left[\sum_i^M (\sigma_i^2 N_i) - \frac{2}{N}\left(\sum_i^M \sigma_i N_i\right)^2 + \frac{1}{N}\left(\sum_i^M \sigma_i N_i\right)^2\right]$$

Multiplying the last term within the brackets by $\Sigma_i^M N_i/N$, which equals 1,

$$D = \frac{1}{Nn}\left[\sum_i^M (\sigma_i^2 N_i) - \frac{2}{N}\left(\sum_i^M \sigma_i N_i\right)^2 + \frac{1}{N^2}\sum_i^M N_i\left(\sum_i^M \sigma_i N_i\right)^2\right]$$

Since

$$\left(\sum_i^M \sigma_i N_i\right)^2 = \sum_i^M\left(\sigma_i N_i \sum_i^M \sigma_i N_i\right)$$

then

$$D = \frac{1}{Nn}\left[\sum_i^M (\sigma_i^2 N_i) - \frac{2}{N}\sum_i^M\left(\sigma_i N_i \sum_i^M \sigma_i N_i\right) + \frac{1}{N^2}\sum_i^M N_i\left(\sum_i^M \sigma_i N_i\right)^2\right]$$

Factoring out $\Sigma_i^M N_i$

$$D = \frac{1}{Nn}\sum_i^M N_i\left[\sigma_i^2 - \frac{2}{N}\sigma_i \sum_i^M \sigma_i N_i + \frac{1}{N^2}\left(\sum_i^M \sigma_i N_i\right)^2\right]$$

Since

$$\left(\sigma_i - \frac{1}{N}\sum_i^M \sigma_i N_i\right)^2 = \left[\sigma_i^2 - \frac{2}{N}\sigma_i \sum_i^M \sigma_i N_i + \frac{1}{N^2}\left(\sum_i^M \sigma_i N_i\right)^2\right]$$

then

$$D = \frac{1}{Nn}\sum_{i}^{M} N_i\left(\sigma_i - \frac{1}{N}\sum_{i}^{M}\sigma_i N_i\right)^2$$

Since

$$\frac{1}{N}\sum_{i}^{M}\sigma_i N_i = \bar{\sigma} \tag{16.105}$$

which is the weighted mean stratum standard deviation,

$$D = \frac{1}{Nn}\sum_{i}^{M} N_i(\sigma_i - \bar{\sigma})^2 \tag{16.106}$$

Then, substituting the above expression into Equation 16.103 yields

$$V(\bar{y})_{\text{st(Prop)}} = V(\bar{y})_{\text{st(Ney)}} + \frac{1}{Nn}\sum_{i}^{M} N_i(\sigma_i - \bar{\sigma})^2 \tag{16.107}$$

As can be seen, the variance of the mean of the proportionally allocated sample consists of two components, one of which is the variance of the mean of the Neyman allocated sample. Consequently,

$$V(\bar{y})_{\text{st(Ney)}} \leq V(\bar{y})_{\text{st(Prop)}}$$

The two are equal only when $\Sigma_i^M N_i(\sigma_i - \bar{\sigma})^2$ equals zero, which occurs only when all the stratum variances are homogeneous, so that all $\sigma_i = \bar{\sigma}$.

One can now compare the variance of the mean of an equally allocated stratified sample to that of a proportionally allocated sample. According to Equation 16.53, the variance of the mean of an equally allocated sample is

$$V(\bar{y})_{\text{st(Eq)}} = \frac{M}{N^2 n}\sum_{i}^{M}\sigma_i^2 N_i^2 = \frac{1}{N^2 n}\sum_{i}^{M}\sigma_i^2 N_i^2 M$$

Then, remembering that $V(\bar{y})_{\text{st(Prop)}} = (1/Nn)\Sigma_i^M \sigma_i^2 N_i$,

$$V(\bar{y})_{\text{st(Eq)}} = \frac{1}{N^2 n}\sum_{i}^{M}(\sigma_i^2 N_i M) + \frac{1}{Nn}\sum_{i}^{M}(\sigma_i^2 N_i^2) - \frac{1}{Nn}\sum_{i}^{M}(\sigma_i^2 N_i)$$

Rearranging,

$$V(\bar{y})_{\text{st(Eq)}} = \frac{1}{Nn}\sum_{i}^{M}(\sigma_i^2 N_i) + \frac{1}{N^2 n}\sum_{i}^{M}(\sigma_i^2 N_i^2 M) - \frac{1}{Nn}\sum_{i}^{M}(\sigma_i^2 N_i)$$

$$= V(\bar{y})_{\text{st(Prop)}} + \frac{1}{N^2 n}\left[\sum_{i}^{M}(\sigma_i^2 N_i M) - N\sum_{i}^{M}(\sigma_i^2 N_i^2)\right]$$

**TABLE 16.1**
**Values of $N_i[(N_iM - N)/N]$ for a Selected Series of Values of $N_i$ When $M = 4$ and $N = 320$**

| $N_i$ | $N_i\left(\frac{N_iM - N}{N}\right)$ | $N_i$ | $N_i\left(\frac{N_iM - N}{N}\right)$ |
|---|---|---|---|
| 10 | –8.75 | 130 | 81.25 |
| 20 | –15.00 | 140 | 105.00 |
| 30 | –18.75 | 150 | 131.25 |
| 40 | –20.00 | 160 | 160.00 |
| 50 | –18.75 | 170 | 191.25 |
| 60 | –15.00 | ↓ | ↓ |
| 70 | –8.75 | 200 | 300.00 |
| 80 | 0.00 | ↓ | ↓ |
| 90 | 11.25 | 260 | 585.00 |
| 100 | 25.00 | ↓ | ↓ |
| 110 | 41.25 | 290 | 761.25 |
| 120 | 60.00 | | |

Factoring out $\Sigma_i^M(\sigma_i^2 N_i)$,

$$V(\bar{y})_{\text{st(Eq)}} = V(\bar{y})_{\text{st(Prop)}} + \frac{1}{Nn}\sum_i^M (\sigma_i^2 N_i)\left(\frac{N_iM - N}{N}\right) \tag{16.108}$$

As can be seen, $V(\bar{y})_{\text{st(Eq)}}$ consists of two parts, $V(\bar{y})_{\text{st(Prop)}}$ and an additional component. In this component the sum of squared errors of each stratum is modified by the multiplier $[(N_iM - N)/N]$ whose magnitude is dependent on the stratum size. When $N_i < N/M$, the multiplier is negative, when $N_i > N/M$ it is positive, and when $N_i = N/M$ it is equal to zero. Thus, when all the strata are of the same size, all $N_i = N/M$ and all the multipliers are equal to zero. The second term in Equation 16.108 then equals zero and $V(\bar{y})_{\text{st(Eq)}} = V(\bar{y})_{\text{st(Prop)}}$. In all other cases the second term will be positive, making $V(\bar{y})_{\text{st(Eq)}} > V(\bar{y})_{\text{st(Prop)}}$. The rationale behind this statement is best explained through the use of an example. Assume in the case of the 16-ha tract that, regardless of stratum size, the stratum variances are homogeneous, thus removing a variable which does not play a major role in this development. The second term of Equation 16.108 then becomes

$$\frac{1}{Nn}\sum_i^M (\sigma_i^2 N_i)\left(\frac{N_iM - N}{N}\right) = \frac{1}{Nn}\sigma_W^2 \sum_i^M N_i\left(\frac{N_iM - N}{N}\right) \tag{16.109}$$

Set $M = 4$. $N = 320$. The numerical value of $N_i[(N_iM - N)/N]$ for a series of stratum sizes is shown in Table 16.1 Note that the negative values occur when $N_i < N/M = 320/4 = 80$. Furthermore, note that the absolute magnitudes of these negative values are at a peak at $N_i = N/2M = 40$ and diminish to zeros at $N_i = 0$ and $N_i = N/M = 80$. On the other hand, the magnitudes of the positive values rise monotonically as $N_i$ increases beyond $N/M = 80$. If one applies the values from Table 16.1 to different combinations of stratum sizes, remembering that they must sum to $N$, one can see that Equation 16.109 will always be positive, except when all the strata are the same size. For example, if three of the strata have the same size, $N_i = 40$, the remaining stratum would have to have a size of $320 - 3(40) = 200$. Then, using data from Table 16.1,

$$\sum_{i}^{M} N_i \left( \frac{N_i M - N}{N} \right) = 3(-20.00) + 300.00 = +240.00$$

Another example might be when $N_i = 10$, $N_2 = 20$, $N_3 = 30$, and $N_4 = 260$. In such a case the sum would be +735. As you can see, the sum is positive in each case. If $N_1 = N_2 = N_3 = N_4 = 80$, the sum would be zero. There is no way by which the sum can be negative. Consequently, it can be said that $V(\bar{y})_{St(Eq)}$ is always larger than $V(\bar{y})_{St(Prop)}$ except when the stratum sizes are all equal and when that occurs, $V(\bar{y})_{St(Eq)} = V(\bar{y})_{St(Prop)}$.

## 16.7 STRATIFICATION AND THE ESTIMATION OF PROPORTIONS

When the variable of interest is an attribute, such as the presence of a disease in a tree, and one desires an estimate of the proportion of the universe possessing that attribute, stratification may improve the efficiency of the sampling process. The random variable in such a case can take only two values (Section 12.13):

$$Y = \begin{cases} 1, & \text{if the attribute is present} \\ 0, & \text{if it is absent} \end{cases}$$

The population mean of these values is the proportion of the universe possessing the attribute. Then, according to Equation 12.42,

$$E(y) = \mu_y = p \quad \text{and} \quad q = 1 - p$$

According to Equation 12.43,

$$V(y) = \sigma_y^2 = pq$$

When the universe is stratified, within stratum $i$,

$$E(y_i) = \mu_i = p_i \tag{16.110}$$

$$q_i = 1 - p_i \tag{16.111}$$

and

$$V(y_i) = \sigma_i^2 = p_i q_i \tag{16.112}$$

Since a proportion is simply a special case of the arithmetic mean, the parameters of the stratified populations are obtained when $p$ is substituted for $\mu_y$, $pq$ for $\sigma_y^2$, $p_i$ for $\mu_i$, and $p_i q_i$ for $\sigma_i^2$ in the appropriate equations in Section 16.3. Thus, from Equation 16.11,

$$E(y) = \mu_y = \sum_{i}^{M} (N_i / N) p_i = p \tag{16.113}$$

and from Equation 16.15,

$$V(y)_{\text{St}} = \sigma^2_{\bar{y}_{\text{st}}} = \sum_{i}^{M} (N_i / N) p_i q_i \tag{16.114}$$

When the stratified population is sampled, $\hat{p}_i$ is the equivalent of $\bar{y}_i$. Then, according to Equation 12.45,

$$E(\hat{p}_i) = \mu_{\hat{p}_i} = p_i \tag{16.115}$$

and

$$E(1 - \hat{p}_i) = E(\hat{q}_i) = q_i \tag{16.116}$$

In Section 12.11 it is shown (see Equation 12.7) that one cannot simply substitute $n$ for $N$ and $\bar{y}$ for $\mu_y$ in Equation 6.7 to obtain an unbiased estimate of the variance, $\sigma^2_{\bar{y}}$. To overcome the bias associated with the substitution of $\bar{y}$ for $\mu_y$, the sum of squared residuals must be divided by the degrees of freedom, $n - 1$, rather than by the sample size, $n$. Furthermore, if the sampling is without replacement and the population is finite, the correction $(N - 1)/N$ must be applied (see Equation 12.12). When $\hat{p}_i\hat{q}_i$ is used as an estimate of $p_i q_i$, it is the equivalent of the sum of squared residuals divided by $n$ and is, thereby, biased. To overcome this bias one must multiply $\hat{p}_i\hat{q}_i$ by $n_i$ to obtain the equivalent of the sum of squared residuals and then this sum must be divided by $n - 1$. Thus, when the sampling is with replacement or the population is infinitely large,

$$s_i^2 = \hat{p}_i\hat{q}_i\left(\frac{n_i}{n_i - 1}\right) \rightarrow p_i q_i = V(y_i) = \sigma_i^2 \tag{16.117}$$

When the sampling is without replacement, the finite population correction is also included:

$$s_i^2 = \hat{p}_i\hat{q}_i\left(\frac{n_i}{n_i - 1}\right)\left(\frac{N_i - 1}{N_i}\right) \rightarrow p_i q_i = V(y_i) = \sigma_i^2 \tag{16.118}$$

As an example, consider the population in stratum $i$ ($i$ will be left out of the subscript to simplify the symbology): $y_1 = 0$, $y_2 = 1$, $y_3 = 1$, $y_4 = 0$, and $y_5 = 1$. Then $p_i = 0.6$, $q_i = 0.4$, and $p_i q_i = 0.24$. If $n = 2$ and the sampling is with replacement,

| Sample | $\hat{p}$ | $\hat{q}$ | $\hat{p}\hat{q}$ | $\hat{p}\hat{q}\left(\frac{n}{n-1}\right)$ |
|---|---|---|---|---|
| $(y_1,y_1) = (0,0)$ | 0.0 | 1.0 | 0.00 | 0.00 |
| $(y_1,y_2) = (0,1)$ | 0.5 | 0.5 | 0.25 | 0.50 |
| $(y_1,y_3) = (0,1)$ | 0.5 | 0.5 | 0.25 | 0.50 |
| $(y_1,y_4) = (0,0)$ | 0.0 | 1.0 | 0.00 | 0.00 |
| $(y_1,y_5) = (0,1)$ | 0.5 | 0.5 | 0.25 | 0.50 |
| $(y_2,y_1) = (1,0)$ | 0.5 | 0.5 | 0.25 | 0.50 |
| $(y_2,y_2) = (1,1)$ | 1.0 | 0.0 | 0.00 | 0.00 |
| $(y_2,y_3) = (1,1)$ | 1.0 | 0.0 | 0.00 | 0.00 |

| Sample | $\hat{p}$ | $\hat{q}$ | $\hat{p}\hat{q}$ | $\hat{p}\hat{q}\left(\frac{n}{n-1}\right)$ |
|---|---|---|---|---|
| $(y_2,y_4) = (1,0)$ | 0.5 | 0.5 | 0.25 | 0.50 |
| $(y_2,y_5) = (1,1)$ | 1.0 | 0.0 | 0.00 | 0.00 |
| $(y_3,y_1) = (1,0)$ | 0.5 | 0.5 | 0.25 | 0.50 |
| $(y_3,y_2) = (1,1)$ | 1.0 | 0.0 | 0.00 | 0.00 |
| $(y_3,y_3) = (1,1)$ | 1.0 | 0.0 | 0.00 | 0.00 |
| $(y_3,y_4) = (1,0)$ | 0.5 | 0.5 | 0.25 | 0.50 |
| $(y_3,y_5) = (1,1)$ | 1.0 | 0.0 | 0.00 | 0.00 |
| $(y_4,y_1) = (0,0)$ | 0.0 | 1.0 | 0.00 | 0.00 |
| $(y_4,y_2) = (0,1)$ | 0.5 | 0.5 | 0.25 | 0.50 |
| $(y_4,y_3) = (0,1)$ | 0.5 | 0.5 | 0.25 | 0.50 |
| $(y_4,y_4) = (0,0)$ | 0.0 | 1.0 | 0.00 | 0.00 |
| $(y_4,y_5) = (0,1)$ | 0.5 | 0.5 | 0.25 | 0.50 |
| $(y_5,y_1) = (1,0)$ | 0.5 | 0.5 | 0.25 | 0.50 |
| $(y_5,y_2) = (1,1)$ | 1.0 | 0.0 | 0.00 | 0.00 |
| $(y_5,y_3) = (1,1)$ | 1.0 | 0.0 | 0.00 | 0.00 |
| $(y_5,y_4) = (1,0)$ | 0.5 | 0.5 | 0.25 | 0.50 |
| $(y_5,y_5) = (1,1)$ | 1.0 | 0.0 | 0.00 | 0.00 |
| | 15.0 | | 3.00 | 6.00 |

$$E(\hat{p}) = \sum_j^{25} \hat{p}_j / 25 = 15.0 / 25 = 0.6 = p$$

$$E(\hat{p}\hat{q}) = \sum_j^{25} \hat{p}_j\hat{q}_j / 25 = 3.00 / 25 = 0.12 \neq pq = 0.24$$

but

$$E\left[\hat{p}\hat{q}\left(\frac{n}{n-1}\right)\right] = \sum_j^{25} \hat{p}_j\hat{q}_j\left(\frac{2}{2-1}\right) \Big/ 25 = 6.00 / 25 = 0.24 = pqV(y)$$

If the sampling is without replacement,

| Sample | $\hat{p}$ | $\hat{q}$ | $\hat{p}\hat{q}$ | $\hat{p}\hat{q}\left(\frac{n}{n-1}\right)$ | $\hat{p}\hat{q}\left(\frac{n}{n-1}\right)\left(\frac{N-1}{N}\right)$ |
|---|---|---|---|---|---|
| $(y_1,y_2) = (0,1)$ | 0.5 | 0.5 | 0.25 | 0.50 | 0.40 |
| $(y_1,y_3) = (0,1)$ | 0.5 | 0.5 | 0.25 | 0.50 | 0.40 |
| $(y_1,y_4) = (0,0)$ | 0.0 | 1.0 | 0.00 | 0.00 | 0.00 |
| $(y_1,y_5) = (0,1)$ | 0.5 | 0.5 | 0.25 | 0.50 | 0.40 |
| $(y_2,y_3) = (1,1)$ | 1.0 | 0.0 | 0.00 | 0.00 | 0.00 |
| $(y_2,y_4) = (1,0)$ | 0.5 | 0.5 | 0.25 | 0.50 | 0.40 |

| Sample | $\hat{p}$ | $\hat{q}$ | $\hat{p}\hat{q}$ | $\hat{p}\hat{q}\left(\frac{n}{n-1}\right)$ | $\hat{p}\hat{q}\left(\frac{n}{n-1}\right)\left(\frac{N-1}{N}\right)$ |
|---|---|---|---|---|---|
| $(y_2,y_5) = (1,1)$ | 1.0 | 0.0 | 0.00 | 0.00 | 0.00 |
| $(y_3,y_4) = (1,0)$ | 0.5 | 0.5 | 0.25 | 0.50 | 0.40 |
| $(y_3,y_5) = (1,1)$ | 1.0 | 0.0 | 0.00 | 0.00 | 0.00 |
| $(y_4,y_5) = (0,1)$ | 0.5 | 0.5 | 0.25 | 0.50 | 0.40 |
| | 6.0 | | 1.50 | 3.00 | 2.40 |

$$E(\hat{p}) = \sum_{h}^{10} \hat{p}_h / 10 = 6.0 / 10 = 0.6 = p$$

$$E(\hat{p}\hat{q}) = \sum_{j}^{10} (\hat{p}_j\hat{q}_j) / 10 = 1.50 / 10 = 0.15 \neq pq = 0.24$$

$$E\left[\hat{p}\hat{q}\left(\frac{n}{n-1}\right)\right] = \sum_{j}^{10}\left[\hat{p}\hat{q}\left(\frac{2}{2-1}\right)\right] \Big/ 10 = 3.00 / 10 = 0.30 \neq pq = 0.24$$

$$E\left[\hat{p}\hat{q}\left(\frac{n}{n-1}\right)\left(\frac{N-1}{N}\right)\right] = \sum_{j}^{10}\left[\hat{p}\hat{q}\left(\frac{2}{2-1}\right)\left(\frac{5-1}{5}\right)\right] \Big/ 1$$

$$= 2.40 / 10 = 0.24 = pq = V(y)$$

Within stratum $i$, when the sampling is with replacement, the variance of the stratum proportion, from Equation 16.7 is

$$V(\hat{p}_i) = \sigma^2_{\hat{p}_i} = \sigma^2_i / n_i = p_i q_i / n_i \tag{16.119}$$

and when the sampling is without replacement, from Equation 16.8, it is

$$V(\hat{p}_i) = \sigma^2_{\hat{p}_i} = \frac{\sigma^2_i}{n_i}\left(\frac{N_i - n_i}{N_i - 1}\right) = \frac{p_i q_i}{n_i}\left(\frac{N_i - n_i}{N_i - 1}\right) \tag{16.120}$$

The sample-based estimate of $V(\hat{p}_i)$, when the sampling is with replacement, is obtained by substituting $\hat{p}_i\hat{q}_i[n_i / (n_i - 1)]$ for $p_i q_i$ in Equation 16.120:

$$s^2_{\hat{p}_i} = \frac{\hat{p}_i\hat{q}_i}{n_i}\left(\frac{n_i}{n_i - 1}\right) = \frac{\hat{p}_i\hat{q}_i}{n_i - 1} \rightarrow V(\hat{p}_i) \tag{16.121}$$

When the sampling is without replacement,

$$\hat{p}_i\hat{q}_i\left(\frac{n_i}{n_i - 1}\right)\left(\frac{N_i - 1}{N_i}\right)$$

is substituted for $p_i q_i$ in Equation 16.120,

$$s_{\hat{p}_i}^2 = \frac{\hat{p}_i\hat{q}_i}{n_i}\left(\frac{n_i}{n_i - 1}\right)\left(\frac{N_i - 1}{N_i}\right)\left(\frac{N_i - n_i}{N_i - 1}\right)$$
$$= \frac{\hat{p}_i\hat{q}_i}{n_i - 1}\left(\frac{N_i - n_i}{N_i}\right) \rightarrow V(\hat{p}_i) \tag{16.122}$$

Both of these estimates are unbiased. The proofs are too lengthy for inclusion here.

The estimate of the proportions of the entire stratified population possessing the attribute of interest is obtained by substituting $\hat{p}_i$ for $\bar{y}_i$ in Equation 16.32.

$$\hat{p} = \sum_i^M \hat{p}_i(N_i / N) \rightarrow E(\hat{p}) = p \tag{16.123}$$

and

$$\hat{q} = 1 - p$$

The estimate of the variance of the proportion, when sampling with replacement, is obtained when $\hat{p}_i\hat{q}_i/(n_i - 1)$ is substituted for $s_i^2/n_i$ in Equation 16.40.

$$s_{\hat{p}}^2 = \sum_i^M \left[\frac{\hat{p}_i\hat{q}_i}{n_i - 1}\left(\frac{N_i}{N}\right)^2\right] \rightarrow V(\hat{p}) = \sigma_{\hat{p}}^2 \tag{16.124}$$

When the sampling is without replacement, $\hat{p}_i\hat{q}_i/(n_i - 1)$ is substituted for $s_i^2/n_i$ in Equation 16.41.

$$s_{\hat{p}}^2 = \sum_i^M \left[\frac{\hat{p}_i\hat{q}_i}{n_i - 1}\left(\frac{N_i - n_i}{N_i}\right)\left(\frac{N_i}{N}\right)^2\right] \rightarrow V(\hat{p}) = \sigma_{\hat{p}}^2 \tag{16.125}$$

It was shown in Section 16.3 that Equation 16.41 could be simplified to Equation 16.43,

$$s_{\bar{y}}^2 = s_w^2 / n$$

and that a valid sampling error could be computed if the stratum variances were homogeneous and the allocation was proportional. Theoretically, the same simplification could be made when the sampling is for proportions rather than means. However, the likelihood of meeting the homogeneity of variance constraint is so small that the possibility can be ignored. In stratified sampling for proportions, the stratum proportions must differ if the stratification is to be meaningful. If the stratum proportions do not differ, neither do the stratum variances. Only in the special case when two strata are used (one can label them $A$ and $B$) and when $p_B = 1 - p_A$, can the stratum means be different and the stratum variances equal. This is the reason the probability of computing a valid sampling error is so small.

It should be noted that Equations 16.124 and 16.125 are general equations that can be used regardless of the method of allocation. When one wishes to express algebraically the variance of the proportion under a specific method of allocation, as, for example, a preliminary step in the derivation of a formula for total sample size, the allocation expression must be substituted for $n_i$

in the appropriate equations, whether 16.124 or 16.125. In both of these equations, $\hat{p}_i\hat{q}_i$ is multiplied by $1/(n_i - 1)$. Thus, when equal allocation is involved, $n/M$ is substituted for $n_i$ and

$$1/(n_i - 1) = M/(n - M) \tag{16.126}$$

In the case of proportional allocation,

$$1/(n_i - 1) = N/(nN_i - N) \tag{16.127}$$

When this is attempted with the expression for the Neyman allocation, a problem develops. According to Equation 16.93,

$$n_i = ns_iN_i \Big/ \sum_i^M s_iN_i$$

Substituting $\sqrt{\hat{p}_i\hat{q}_i[n_i/(n_i - 1)]}$ for $s_i$ yields

$$n_i = nN_i\sqrt{\hat{p}_i\hat{q}_i[n_i/(n_i - 1)]} \Big/ \sum_i^M N_i\sqrt{\hat{p}_i\hat{q}_i[n_i/(n_i - 1)]}$$

but $n_i$ appears on both sides of the equation and cannot be isolated. The same problem occurs when optimum allocation is involved. Consequently, algebraic expressions for the estimates of the variance of the proportion cannot be developed for either the Neyman or the optimum allocations. This does not mean, however, that *numerical* values for the variances of the proportions cannot be computed using Equations 16.124 or 16.125.

When the allocation is equal and the sampling is with replacement, $M/(n - M)$ is substituted for $1/(n_i - 1)$ in Equation 16.124 yielding the estimate

$$s^2_{\hat{p}_{\text{st(Eq)}}} = \sum_i^M \left[\hat{p}_i\hat{q}_i\left(\frac{M}{n-M}\right)\left(\frac{N_i}{N}\right)^2\right] = \frac{M}{(n-M)N^2}\sum_i^M \hat{p}_i\hat{q}_iN_i^2 \tag{16.128}$$

When the sampling is without replacement $M/(n - M)$ is substituted for $1/(n_i - 1)$ in Equation 16.125.

$$s^2_{\hat{p}_{\text{st(Eq)}}} = \sum_i^M \left[\hat{p}_i\hat{q}_i\left(\frac{M}{n-M}\right)\left(\frac{n_i - (n/M)}{N_i}\right)\left(\frac{N_i}{N}\right)^2\right]$$

Since

$$[N_i - (n/M)]/N_i = (N_iM - n)/N_iM \tag{16.129}$$

then

$$\begin{aligned} s^2_{\hat{p}_{\text{st(Eq)}}} &= \sum_i^M \left[\hat{p}_i\hat{q}_i\left(\frac{M}{n-M}\right)\left(\frac{N_iM - n}{N_iM}\right)\left(\frac{N_i}{N}\right)^2\right] \\ &= \frac{1}{(n-M)N^2}\sum_i^M [\hat{p}_i\hat{q}_i(N_iM - n)N_i] \end{aligned} \tag{16.130}$$

In the case of proportional allocation, when the sampling is with replacement, $nN_i/N$ is substituted for $1/(n_i - 1)$ in Equation 16.124.

$$s^2_{\hat{p}_{st(Prop)}} = \sum_i^M \left[\hat{p}_i\hat{q}_i\left(\frac{N}{nN_i - N}\right)\left(\frac{N_i}{N}\right)^2\right] = \frac{1}{N}\sum_i^M \left(\frac{\hat{p}_i\hat{q}_iN_i^2}{nN_i - N}\right) \tag{16.131}$$

When the sampling is without replacement, $nN_i/N$ is substituted for $n_i$ and $N/(nN_i - N)$ for $1/(n_i - 1)$ in Equation 16.125.

$$s^2_{\hat{p}_{st(Prop)}} = \sum_i^M \left[\hat{p}_i\hat{q}_i\left(\frac{N}{nN_i - N}\right)\left[\frac{N_i - (nN_i / N)}{N_i}\right]\left(\frac{N_i}{N}\right)^2\right]$$

Since

$$\frac{N_i - (nN_i / N)}{N_i} = \frac{N_iN - nN_i}{N_iN} = \frac{N - n}{N}$$

then

$$s^2_{\hat{p}_{st(Prop)}} = \left(\frac{N - n}{N^2}\right)\sum_i^M \left(\frac{\hat{p}_i\hat{q}_iN_i^2}{nN_i - N}\right) \tag{16.132}$$

When a maximum allowable standard error of the proportion, $A$, is specified, the total sample size needed to meet that specification is obtained when $A^2$ is substituted for $s^2_{\hat{p}}$ in the appropriate equation and the equation is then solved for $n$. Thus, when the allocation is equal and the sampling is with replacement, using Equation 16.128,

$$n = M + \frac{M}{N^2A^2}\sum_i^M \hat{p}_i\hat{q}_iN_i^2 \tag{16.133}$$

When the sampling is without replacement, $A^2$ is substituted for $s^2_{\hat{p}}$ in Equation 16.130,

$$n = \frac{M\left(\sum_i^M \hat{p}_i\hat{q}_iN_i^2 + N^2A^2\right)}{N^2A^2 + \sum_I^M \hat{p}_i\hat{q}_iN_i} \tag{16.134}$$

Comparable expressions cannot be developed for the remaining methods of allocation because $n$ cannot be isolated in the case of Equations 16.131 and 16.132, which apply to proportional allocation, and because expressions for $s^2_{\hat{p}}$ cannot be developed for the Neyman and optimum allocations.

One can often bypass the problems just discussed by assuming that the bias correction, $n_i/(n_i - 1)$, is equal to 1. In such a case Equation 16.117 becomes

$$s_i^2 \doteq \hat{p}_i\hat{q}_i \tag{16.135}$$

Since, from Equation 16.117,

$$\hat{p}_i\hat{q}_i[n_i/(n_i-1)] \rightarrow p_iq_i$$

then

$$\hat{p}_i\hat{q}_i \rightarrow p_iq_i[(n_i-1)/n_i] < p_iq_i \tag{16.136}$$

The magnitude of the bias is dependent on the stratum sample size. When $n_i = 10$, the estimate of $p_iq_i$ is 10.00% low; when $n_i = 20$, it is 5.00%; when $n_i = 30$, it is 3.33% low; and when $n_i = 100$, it is 1.00% low. Thus, when $\hat{p}_i\hat{q}_i$ is used in place of $\hat{p}_i\hat{q}_i[n_i/(n_i - 1)]$ to replace $s_i^2$ in Equation 16.40,

$$s_{\hat{p}}^2 \doteq \sum_i^M \left[\frac{\hat{p}_i\hat{q}_i}{n_i}\left(\frac{N_i}{N}\right)^2\right] \tag{16.137}$$

or in Equation 16.41,

$$s_{\hat{p}}^2 \doteq \sum_i^M \left[\frac{\hat{p}_i\hat{q}_i}{n_i}\left(\frac{N_i-n_i}{N_i}\right)\left(\frac{N_i}{N}\right)^2\right] \tag{16.138}$$

the estimate will be too low because $\hat{p}_i\hat{q}_i$ is being divided by $n_i$, not by $n_i - 1$. Furthermore, when it is used to arrive at the total sample size needed to meet an allowable standard error specification, that sample size will be too small.

It was brought out earlier that allowable expressions for the Neyman and optimum allocation methods cannot be derived when $\hat{p}_i\hat{q}_i[n_i/(n_i - 1)]$ is used as the substituting for $s_i^2$. However, when the bias correction is assumed to be equal to 1, the problem is resolved. Thus, the Neyman allocation, Equation 16.93, when $\sqrt{\hat{p}_i\hat{q}_i}$ is substituted for $s_i$ becomes

$$n_i = n\sqrt{\hat{p}_i\hat{q}_i}N_i \Big/ \sum_i^M \sqrt{\hat{p}_i\hat{q}_i}N_i \tag{16.139}$$

and the optimum allocation, from Equation 16.82, becomes

$$n_i = (n\sqrt{\hat{p}_i\hat{q}_i}N_i/\sqrt{c_i} \Big/ \sum_i^M (\sqrt{\hat{p}_i\hat{q}_i}N_i/\sqrt{c_i}) \tag{16.140}$$

When the sampling is with replacement, the approximate total sample size needed to meet an allowable standard error specification becomes:

For proportional allocation, from Equation 16.66,

$$n_{\text{st(Prop)}} \doteq \frac{1}{NA^2}\sum_i^M \hat{p}_i\hat{q}_iN_i \tag{16.141}$$

For the Neyman allocation, from Equation 16.98,

$$n_{\text{st(Ney)}} \doteq \frac{1}{N^2A^2}\left(\sum_i^M \sqrt{\hat{p}_i\hat{q}_i}N_i\right)^2 \tag{16.142}$$

For optimum allocation, from Equation 16.85,

$$n_{\text{st(Op)}} \doteq \frac{1}{N^2A^2}\sum_i^M \sqrt{\hat{p}_i\hat{q}_i}N_i\sqrt{c_i}\sum_i^M (\sqrt{\hat{p}_i\hat{q}_i}N_i / \sqrt{c_i}) \tag{16.143}$$

When the sampling is without replacement:

For proportional allocation, from Equation 16.67,

$$n_{\text{st(Prop)}} \doteq N\sum_i^M \hat{p}_i\hat{q}_iN_i \Bigg/ \left(N^2A^2 + \sum_i^M \hat{p}_i\hat{q}_iN_i\right) \tag{16.144}$$

For the Neyman allocation, from Equation 16.99,

$$n_{\text{st(Ney)}} \doteq \left(\sum_i^M \sqrt{\hat{p}_i\hat{q}_i}N_i\right)^2 \Bigg/ \left(N^2A^2 + \sum_i^M \hat{p}_i\hat{q}_iN_i\right) \tag{16.145}$$

For optimum allocation, from Equation 16.87,

$$n_{st(Op)} \doteq \sum_i^M \sqrt{\hat{p}_i\hat{q}_i}N_i\sqrt{c_i}\sum_i^M (\sqrt{\hat{p}_i\hat{q}_i}N_i / \sqrt{c_i}) \Bigg/ \left(N^2A^2 + \prod_i^M \hat{p}_i\hat{q}_iN_i\right) \tag{16.146}$$

These sample sizes are too low. Their use will result in estimates of the variance and standard error of the proportion that are too high. Consider Equation 16.124:

$$s_{\hat{p}}^2 = \sum_i^M \left[\frac{\hat{p}_i\hat{q}_i}{n_i - 1}\left(\frac{N_i}{N}\right)^2\right]$$

Notice that, if the $n_i$ are too low, the estimates of the stratum variances of the proportion will be too high, causing the overall estimate to be too high. It is possible to make corrections to the stratum sample sizes that would bring them up to approximately their proper magnitudes. To do this the approximate total sample size is computed using the appropriate expression and data from a presample. The allocation is then made. The sample size in each stratum is then multiplied by the bias correction, $n_i/(n_i - 1)$, to arrive at the corrected stratum sample sizes. As an example of this process, assume a situation when proportional allocation is to be used, the sampling is to be with replacement, the allowable standard error $A = 0.05$, and the following stratum sizes are presample data.

| Stratum | $\hat{p}_i$ | $\hat{q}_i$ | $\hat{p}_i\hat{q}_i$ | | $N_i$ | $\hat{p}_i\hat{q}_iN_i$ | $N_i/N$ |
|---|---|---|---|---|---|---|---|
| 1 | 0.2 | 0.8 | 0.16 | | 250 | 40.0 | 0.25 |
| 2 | 0.3 | 0.7 | 0.21 | | 350 | 73.5 | 0.35 |
| 3 | 0.5 | 0.5 | 0.25 | | 400 | 100.0 | 0.40 |
| | | | | $N =$ | 1000 | | |

$$\sum_i^M \hat{p}_i\hat{q}_iN_i = 213.5$$

$$N^2A^2 = (1006)^2(0.05)^2 = 2500$$

$$n = 1000(213.5)/2500 = 85.4 \text{ or } 86$$

The allocation is then

| Stratum | $n_i = n(N_i/N)$ |
|---|---|
| 1 | 86(0.25) = 21.5 or 22 |
| 2 | 86(0.35) = 30.1 or 30 |
| 3 | 86(0.40) = 34.4 or 34 |
| | $n = 86$ |

Since an effort is being made to meet an allowable standard error and, presumably, cost is not a controlling factor, all the rounding should have been upward. However, to do so in this case would tend to mask the effect being demonstrated and consequently the rounding was to the nearest whole number so that the total sample size would be as computed. If one now assumes that the sample was obtained and that the statistics from that sample showed $\hat{p}_i$ and $\hat{q}_i$ to be the same as they were in the presample, the estimated standard error of the proportion, using Equation 16.123, would be computed in the following manner.

| Stratum | $\hat{p}_i\hat{q}_i/(n_i - 1)$ | $(N_i/N)^2$ | $[\hat{p}_i\hat{q}_i/(n_i - 1)](N_i/N)^2$ |
|---|---|---|---|
| 1 | 0.16/(22 – 1) = 0.007619 | 0.0625 | 0.0004762 |
| 2 | 0.21/(30 – 1) = 0.0072414 | 0.1225 | 0.0008871 |
| 3 | 0.25/(34 – 1) = 0.0075758 | 0.1600 | 0.0012121 |
| | | | $s_{\hat{p}}^2 = 0.0025754$ |

and $s_{\hat{p}} = \pm\sqrt{0.0025754} = \pm 0.0507486$, which is marginally larger than the specified maximum allowable error.

Now, applying the bias correction to each of the stratum sizes and recomputating the standard error,

| Stratum | $n_i[n_i/(n_i - 1)]$ | $[\hat{p}_i\hat{q}_i/(n_i - 1)](N_i/N)^2$ |
|---|---|---|
| 1 | 22(22/21) = 23.04 or 23 | (0.16/22)(0.0625) = 0.0004545 |
| 2 | 30(30/29) = 31.03 or 31 | (0.21/30)(0.1225) = 0.0008575 |
| 3 | 34(34/33) = 35.03 or 35 | (0.25/34)(0.1600) = 0.0011765 |
| | $n = 89$ | $s_{\hat{p}}^2 = 0.0024885$ |

and $s_{\hat{p}} = \pm\sqrt{0.0024885} = \pm 0.0498846$, which meets the specification.

## 16.8 EFFECTIVENESS OF STRATIFICATION

Stratification is of value only if the strata actually differ. Thus, stratification can be considered effective if

1. The stratum means differ and the stratum variances are homogeneous.
2. The stratum means differ and the stratum variances are heterogeneous.
3. The stratum means do not differ but the stratum variances are heterogeneous.

Obviously, if the stratum means do not differ and the stratum variances are homogeneous, the stratification has accomplished nothing.

The first two of the above situations have been explored in the earlier sections of this chapter. The third rarely occurs but should not be ignored. As an example of this third situation, in a forestry context, one can have a stratum that is made up of stands of trees which are more or less uniformly spaced, as in a plantation, while a second stratum, *with the same mean volume per unit acre*, is made up of stands within which the trees form clumps, as might be found in a natural stand. The variance of the first stratum would be relatively small because the sample plot volumes would exhibit little variation. In contrast, the variance of the second stratum would be relatively large because some sample plots would fall in clumps and contain heavy per unit acre volumes while other plots would fall into openings or thinly stocked areas and would contain little or no volume. It is obvious that the two strata are different in spite of the fact that their means do not differ. When this type of situation exists and one wishes to reduce the magnitude of the standard error of the means, when compared with that which would be obtained if unrestricted random sampling is used, one must use Neyman allocation. According to Equation 16.103,

$$V(\bar{y})_{\text{Ran}} = V(\bar{y})_{\text{st(Prop)}} + \frac{1}{n}\sigma_B^2$$

where $\sigma_B^2$ is the variance of the stratum means. Since they do not differ, $\sigma_B^2 = 0$ and

$$V(\bar{y})_{\text{Ran}} = V(\bar{y})_{\text{st(Prop)}} \tag{16.147}$$

Thus, if one uses proportional allocation, there is no gain in precision. Consider now Equation 16.107,

$$V(\bar{y})_{\text{st(Prop)}} = V(\bar{y})_{\text{st(Ney)}} + \frac{1}{Nn}\sum_{i}^{M} N_i(\sigma_i - \bar{\sigma})^2$$

which can be written, in this situation,

$$V(\bar{y})_{\text{Ran}} = V(\bar{y})_{\text{st(Ney)}} + \frac{1}{Nn}\sum_{i}^{M} N_i(\sigma_i - \bar{\sigma})^2 \tag{16.148}$$

As can be seen $V(\bar{y})_{\text{Ran}}$ is greater than $V(\bar{y})_{\text{st(Ney)}}$ by the magnitude of the second term in Equation 16.148. This term involves only the stratum variances and increases in magnitude as the variability of the stratum variances increases. The fact that the stratum means do not differ is not a factor. Thus, using the Neyman allocation would measure the precision of the estimate.

When the allocation is equal, the variance (and standard error) of the mean will be greater than or equal to that obtained when proportional allocation is used. This was brought out in Section

16.6. Consequently, the use of equal allocation when the stratum means do not differ and the stratum variances are homogeneous would produce an estimate with a precision poorer than that which would be obtained from an unrestricted random sample.

## 16.9 STRATUM SIZES

In most of the formulae used to compute the statistics for stratified samples, the sizes of the strata appear as weighting factors (e.g., $N_i/N$). If these sizes are not accurately known, at least in a relative sense, the computed statistics can be severely distorted even if the sampling unit evaluations have been made without error. As an example of this, assume two strata $A$ and $B$. The *true* means of these strata are $\mu_A = 5$ and $\mu_B = 10$. The *true* sizes of the strata are $N_A = 20$ and $N_B = 10$. The *true* total for the two strata is then

$$T = N_A\mu_A = N_B\mu_B = 20(5) = 10(10) = 200$$

Now assume that the stratum sizes are imperfectly known and are thought to be $N'_A = 25$ and $N'_B = 5$. Then

$$T' = 25(5) + 5(10) = 175$$

which is 25 units less than the true total, even though the true means were used in the calculations.

These results illustrate a serious source of error in timber cruises in which stratified sampling is used. Timber cruisers usually do a very good job of dbh, merchantable height, and defect estimation and observe the boundaries of their sampling units with sufficient care so that one can consider the sampling unit values to be accurate within quite narrow limits. However, the determination of stratum sizes is often treated in a much more casual manner. The outlining of the stands on aerial photographs is often crudely done and, in addition, the distortion inherent in the photographs (and their enlargements) are often ignored. Consequently, the measured stratum sizes are in error and this carries over into the computations for the mean and total volumes. The importance of using accurate stratum sizes is these computations cannot be overemphasized.

Mathematically, the bias, $B$, resulting from the use of incorrect stratum sizes is

$$\begin{aligned} B &= T' - T \\ &= \sum_i^M N'_i\mu_i - \sum_i^M N_i\mu_i \\ &= \sum_i^M (N'_i - N_i)\mu_i \end{aligned} \tag{16.149}$$

Using relative weights, $N'_i/N$ and $N_i/N$, rather than the stratum sizes themselves, puts the bias on a per unit of stratum size basis:

$$\begin{aligned} B_{\text{Rel}} &= \sum_i^M (N'_i/N)\mu_i - \sum_i^M (N_i/N)\mu_i \\ &= \frac{1}{N}\sum_i^M (N'_i - N_i)\mu_i \end{aligned} \tag{16.150}$$

Thus, for the example being used,

$$B_{\text{Rel}} = \frac{1}{30}[(25-20)5+(5-10)10] = -\frac{5}{6}$$

per unit of stratum size.

It has been shown by Stephan (1941) that the mean square error of the mean of the stratified sample, when the stratum sizes are in error, is

$$MSE(\bar{y}_{st}) = \sum_i^M \left[\frac{\sigma_i^2}{n_i}\left(\frac{N_i'-n_i}{N_i-1}\right)\left(\frac{N_i'}{N}\right)^2\right] + \left[\frac{1}{N}\sum_i^M (N_i'-N_i)\mu_i\right]^2 \quad (16.151)$$

Obviously, the use of incorrect stratum sizes decreases the precision of the estimate as well as introducing a bias. As the differences from the two stratum sizes increase, both the bias and the mean square error of the mean increase. The net effect can be seriously flawed estimates of the population parameters, which can lead, in a forestry context, to financial problems.

## 16.10 MISCELLANY

**1.** Stratified sampling can be equated with a completely randomized statistical design (see Section 14.20) where the strata act as treatments. If the stratum variances are homogeneous and the stratum populations are normally distributed, one can evaluate the effectiveness of the stratification in the same manner as one can test for treatment effects. One would evaluate the differences in the sample estimates of the stratum variances using Bartlett's test (Section 14.17) and, if found to be homogeneous, the differences between the sample estimates of the stratum means would be evaluated using an analysis of variance, $F$ test, and Duncan's new multiple-range test (Section 14.22.2). If the stratum variances are tested and found to be heterogeneous it still may be possible to test the stratum means because the indicated tests are quite robust.

**2.** When a population has been stratified and a sample drawn, neither the boundaries of strata nor the assignment of sampling units should be changed because of the data obtained in the course of sampling unit evaluations. If a sampling unit appears to be abnormal, in the context of the stratum to which it is assigned, it should not be deleted from the sample nor reassigned to another stratum. Strata are not absolutely uniform and often contain sampling units that appear to be out of place. This was brought out in the discussion of stratification in Section 16.2 and is illustrated in Figure 16.3, where a minimum mapping unit size was specified. This caused "abnormal" sampling units to be included within the boundaries of the strata. Deleting the "abnormal" sampling units from the sample would result in the lack of representation in the sample of some of the conditions existing in each of the strata and, consequently, would result in biased estimates of the stratum and overall parameters.

**3.** *Post-stratification* is the stratification of the population after the sample has been obtained or while the sampling process is in progress. This might be done when sufficient information about the population is not available to permit stratification in the normal manner. It is an acceptable procedure if the sampling is random and the stratum sizes are known within relatively narrow limits. The allocation cannot be controlled, but if the sample size is such that at least 20 sampling units have been evaluated in each of the strata, the results would be about as precise as those that would be obtained using proportional allocation of a sample of the same total size. When post-stratification is carried out and the sampling is random and the stratum sizes are known, the estimation of the population parameters follows the standard pattern described in Section 16.4. In forestry, post-stratification has been used in the context of line-plot cruising where stand maps are made in the course of the cruise. The stratification is then carried out using the stand map. This procedure is flawed in that the line–plot procedure is a form of systematic sampling and thus the requirement of a random sample is violated. Consequently, as is described in Chapter 18, valid variability statistics cannot be calculated and the estimates of the stratum and overall means may be biased.

**4.** When, prior to the actual sampling, information on stratum variances is questionable it is best to use proportional allocation since stratum variances are not used in the allocation computation.

**5.** When the stratum variances are homogeneous, one may be tempted to use proportional or Neyman allocation because then a valid sampling error can be computed. However, if the cost of sampling unit evaluation varies from stratum to stratum it may be preferable to use optimum allocation to keep the cost of the operation at a minimum.

**6.** Proportional allocation is not as efficient as Neyman or optimum allocation but is often used. It does not require a presample since stratum variances are not used in the computations. It is self-weighting. However, in forestry it may be considered undesirable because strata with large areas but with small per acre volumes or values will be assigned large sample sizes while strata with small area but large per unit area volume or values are assigned small sample sizes.

**7.** Whenever a total sample size computation indicates a sample larger than the population, the surplus above 100% of the population is ignored and the entire population is included in the evaluation process.

# 17 Cluster Sampling

## 17.1 INTRODUCTION

Up to this point, when sampling has been discussed, the sampling units have been made up of single elements of the universe of interest. In some cases, however, it is advantageous to use sampling units that contain one or more elements or may even be empty. Such sampling units are known as *clusters,* and when they are used the sampling process is known as *cluster sampling.*

Cluster sampling becomes an attractive procedure when the universe of interest is large and/or its elements are widely dispersed and/or it is difficult or impossible to identify or locate individual elements. When these conditions exist, it usually is impossible to construct a sampling frame using single element sampling units. Cluster sampling also becomes attractive when the time and costs associated with travel between and location of single element sampling units becomes excessive. One can use a forest as an example of such a situation. Consider the trees of a certain species, possessing a dbh equal to or greater than some minimum, occurring on a 1000-ha tract, as the elements of the universe of interest. Such a universe would be large. Its elements could be widely and irregularly dispersed. The trees do not possess recognizable identities, such as numbers or names, and their locations within the tract are not known. It would be impossible to construct a sampling frame using these individual trees as sampling units. Furthermore, even if such a frame could be constructed, the job of physically finding the sample trees would be difficult, time-consuming, and expensive. Because of this, timber volume estimation is rarely based on individual tree sampling units. The main exception to this rule is when a complete enumeration, a 100% cruise, is made. In such a case *all* the trees fitting the description of an element of the universe of interest are evaluated and there is no search made for specified trees. To overcome the problems associated with the use of single tree sampling units, timber cruises are usually carried out using plots (patches of ground) as sampling units with the volume of timber on the plots as the variable of interest. In such a case, the trees are no longer the elements of the universe of interest. Their volumes are simply aggregated within the plots to form the variate. This is what was done in the case of the 16-ha tract shown in Figure 16.1. One can, however, consider the trees on a plot as forming a cluster. Some forms of forest sampling are based on such an assumption.

Even if the universe is defined as being made up of plots, one can still encounter the problem of traveling to and locating the plots chosen for the sample. If the tract lies in one of the deep coastal swamps along the eastern seaboard of the United States or in the essentially trackless boreal forest of eastern Canada, the travel from point to point can be extremely difficult. Under such circumstances it may be desirable to evaluate several plots in close proximity to one another about each sampling point rather than to have the same number of plots scattered as individuals. The plots associated with a sampling point constitute a cluster. If this approach is adopted the travel time between sampling units can be materially reduced. Even if helicopters are used to leapfrog crews from sampling point to sampling point, the ratio of productive time to travel time will favor cluster sampling over simple random sampling.

## 17.2 TYPES OF CLUSTER SAMPLING

When the universe of interest is divided into subsets that are to be used as sampling units, those subsets are *primary* clusters. Each of these primary clusters can be considered a miniuniverse which can be made up either of individual elements or groups of individual elements. Those *groups* are *secondary* clusters. Secondary clusters can also be considered miniuniverses containing either individual elements or groups of elements. Such groups are *tertiary* clusters. This process can be continued until there are not enough elements within a cluster to form groups.

When the sampling process is limited to drawing primary clusters, it is referred to as *simple*, *one-stage*, or *single-stage* cluster sampling. If subsamples are drawn from randomly drawn primary clusters, either individual elements or secondary clusters, *two-stage* sampling or *cluster sampling with subsampling* is being used. If subsamples are drawn from randomly chosen secondary clusters, either individual elements or tertiary clusters, *three-stage* sampling is being used. There is no limit to the number of stages that can be used except for the availability of elements to form clusters. The general term for sampling designs involving such repeated subsampling is *multistage* sampling.

## 17.3 CLUSTER CONFIGURATION

There are no standard cluster configurations. They can take on a multitude of different forms. The configuration used in any specific situation is controlled largely by the nature of the sampling problem and the universe of interest. However, even with these constraints, there usually are several possible configurations which could be used. The choice made should, however, be based on some general rules.

Since statistical efficiency is measured by the magnitude of the standard error of the mean or total, an effort should be made to keep this measure as small as possible. In general, this is accomplished by using sampling units that include within themselves as much of the heterogeneity of the population as is possible so as to minimize the differences between the sampling units. In the context of conventional timber cruising, this can be accomplished by using either large sample plots or, if plot size is to be held constant, by using rectangular rather than circular plots. Large plots include more conditions (e.g., differing stand densities) than do small plots simply because they cover larger areas. Rectangular plots include more conditions than similarly sized circular plots because they are less compact and extend laterally across greater distances. Since these sampling unit configurations reduce the standard error, they are more efficient and fewer sampling units are needed to meet an allowable standard error requirement. This same idea should be applied to cluster sampling design. Include as much as possible of the population heterogeneity *within* the clusters to keep heterogeneity *between* the clusters at a minimum. This is the reverse of the proper procedure when stratifying a population.

The second general rule is that the cluster configuration should be such that field operations are kept as simple and convenient as is possible. This means that the elements making up a cluster should be in relatively close proximity to one another. This requirement tends to conflict with the first since universes are rarely randomly distributed and elements that are close to one another are more likely to be similar than elements that are remote from one another. Consequently, when designing cluster configurations, trade-offs must be made. Whether statistical efficiency or sampling convenience dominates the trade-off depends on the situation.

Within a given sampling operation, the clusters may or may not contain the same number of elements. As will be brought out later, in some cases equal-sized clusters provide advantages but equality of cluster size is not essential.

The following are a number of possible cluster configurations that might be used. Except under item 4, in each case the cluster is identified by underlining.

1. Clusters consisting of elements falling inside some physical or geographic boundaries:
   a. Trees in plots,
   b. Small private forest landowners in G.L.O. sections,
   c. People in city blocks,
   d. Test pieces cut from the same sheet of plywood.
2. Clusters consisting of elements within some organizational structure:
   a. Foresters within national forests,
   b. Students within classrooms.
3. Clusters consisting of events occurring with specified time spans:
   a. Logs passing through a sawmill during 4-hour time periods,
   b. Deer killed during days.
4. Clusters of sample plots within some geometric configuration, centered on randomly selected points (see Figure 17.1):
   a. Four plots in contact with each other.
   b. Five plots spread out forming a crosslike constellation. The plots could be circular as well as square.
   c. Set of plots at some arbitrary spacing. Again, the plots could be circular. The number of plots could be extended indefinitely.

This certainly does not exhaust the possibilities. Each sampler must devise the configuration that will be appropriate for the problem being addressed.

It should be noted that all of the above examples can be used in the context of simple or two-stage cluster sampling. They may also be used within a multistage sampling scheme but would form only a part of that scheme.

## 17.4 POPULATION PARAMETERS

A characteristic of multistage sampling is that the formulae for the population parameters in the case of a $Z$-stage sampling scheme can be used when less than $Z$ stages are involved but not when there are more than $Z$ stages. The following development is appropriate for two-stage sampling. Thus the formulae can be used with simple cluster sampling but not when more than two stages are involved.

In the following development, the total population size is equal to $N$ and the sampling frame contains $M$ clusters or primary sampling units. The number of elements or secondary sampling units in cluster $i$ is equal to $N_i$, with $i - 1, 2, \ldots, M$. Cluster size may or may not be constant across all $i$.

### 17.4.1 Cluster Parameters

Within cluster $i$, the cluster mean is

$$\mu_i = E(y_i) = \frac{1}{N} \sum_{j}^{N_i} y_{ij} \tag{17.1}$$

the cluster total is

$$\tau_i = \sum_{j}^{N_i} y_{ij} \tag{17.2}$$

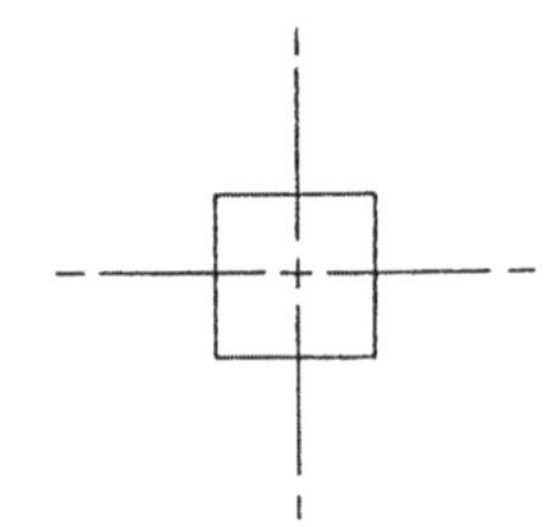

a. Four square plots in contact

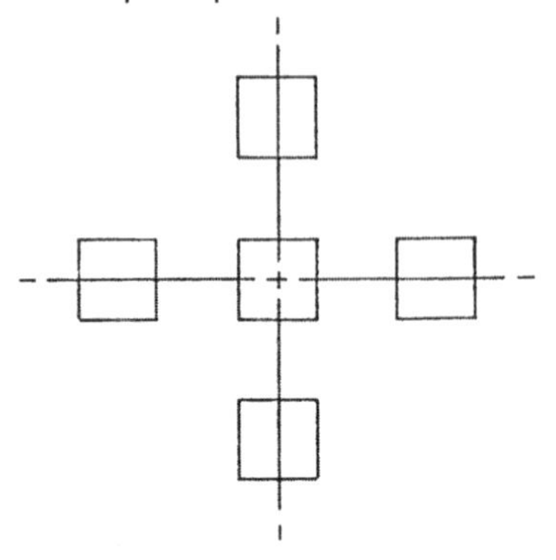

b. Five plots forming a cross

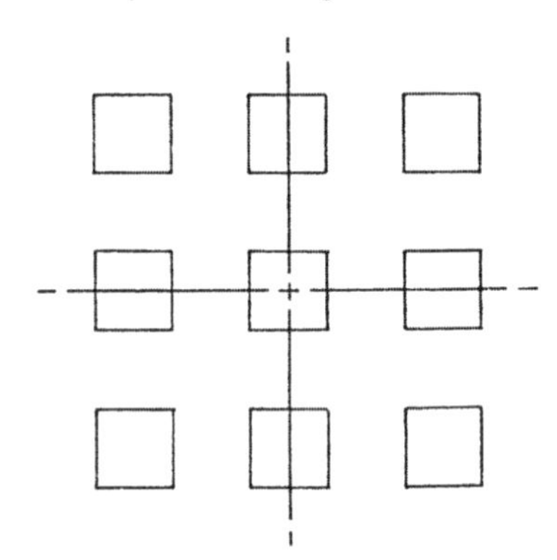

**FIGURE 17.1** Examples of possible cluster configurations when the elements are plots. Each cluster is centered on a randomly selected sampling point. (a) Four square plots in contact. (b) Five plots forming a cross. (c) Set of plots at some arbitrary spacing.

or, multiplying by $N_i/N_i$,

$$\tau_i \frac{N_i}{N_i} \sum_{j}^{N_i} y_{ij} = N_i \mu_i \tag{17.3}$$

and the variance within the cluster is

$$V(y_i) = \sigma_i^2 = E(y_{ij} - \mu_i)^2 = \frac{1}{N_i} \sum_{j}^{N_i} (y_{ij} - \mu_i)^2 \tag{17.4}$$

### 17.4.2 Overall Parameters

Across all $M$ clusters, the overall or population mean is

$$\mu_o = E(y) = \frac{1}{N}\sum_i^M \sum_j^{N_i} y_{ij} \tag{17.5}$$

the population total is

$$\tau_o = \sum_i^M \sum_j^{N_i} y_{ij} = \sum_i^M N_i \mu_i = \sum_i^M \tau_i \tag{17.6}$$

and the mean cluster total is

$$\mu_\tau = \bar{\tau} = E(\tau_i) = \frac{1}{M}\sum_i^M \tau_i \tag{17.7}$$

Since $\Sigma_i^M \tau_i = \tau_o$,

$$\bar{\tau} = \frac{1}{M}\tau_o \tag{17.8}$$

The variance of the cluster totals is

$$V(\tau) = \sigma_\tau^2 = E(\tau_i - \bar{\tau})^2 = \frac{1}{M}\sum_i^M (\tau_i - \bar{\tau})^2 \tag{17.9}$$

According to Equation 17.8, $\bar{\tau} = \tau_o / M$. Then,

$$\begin{aligned} V(\tau) &= \frac{1}{M}\sum_i^M (\tau_i - \tau_o / M)^2 \\ &= \frac{1}{M}\sum_i^M (\tau_i^2 - 2\tau_i\tau_o / M + \tau_o^2 / M^2) \\ &= \frac{1}{M}\sum_i^M \tau_i^2 - \frac{2\tau_o}{M^2}\sum_i^M \tau_i + \tau_o^2 / M^2 \end{aligned}$$

Since $\Sigma_i^M \tau_i = \tau_o$,

$$V(\tau) = \frac{1}{M}\sum_{i}^{M}\tau_i^2 - 2\tau_o^2/M^2 + \tau_o^2/M^2$$

$$= \frac{1}{M}\sum_{i}^{M}\tau_i^2 - \tau_o^2/M$$

$$= \frac{1}{M}\sum_{i}^{M}\tau_i^2 - \bar{\tau}^2 \tag{17.10}$$

## 17.5 ESTIMATION OF THE PARAMETERS

In two-stage cluster sampling, $m$ clusters are randomly chosen from the $M$ clusters in the *primary* or *cluster* sampling frame. Each cluster is made up of $N_i$ elements from the *basic* sampling frame, which consists of all the elements of the universe of interest that are available for sampling. The total count of the elements in the basic frame is then

$$N = \sum_{i}^{M} N_i \tag{17.11}$$

In the second stage of the sampling operation, a random sample is drawn from each of the $m$ chosen primary clusters. The number of units drawn from cluster $i$, $n_i$, may vary from cluster to cluster. The sampling at both stages is assumed to be without replacement.

### 17.5.1 Cluster Statistics

Since simple random sampling is used within the clusters, the cluster statistics, according to the arguments in Chapter 12, are unbiased estimators of their corresponding parameters.

The estimated mean of cluster $i$ is then

$$\bar{y}_i = \frac{1}{n_i}\sum_{j}^{n_i} y_{ij} \rightarrow \mu_i = E(y_i) \tag{17.12}$$

The estimated total in cluster $i$ is

$$T_i = N_i\bar{y}_i \tag{17.13}$$

and its expected value is

$$E(T_i) = E(N_i\bar{y}_i) = N_iE(\bar{y}_i) = N_i\mu_i = \tau_i \tag{17.14}$$

Therefore, $T_i$ is an unbiased estimator of $\tau_i$. The estimated variance of cluster $i$ is

$$s_i^2 = \frac{1}{n_i - 1}\sum_{j}^{n_i}(y_{ij} - \bar{y}_i)^2$$

or

$$s_i^2 = \frac{1}{n_i - 1}\left[\sum_j^{n_i} y_{ij}^2 - \left(\sum_j^{n_i} y_{ij}\right)^2 \Big/ n_i\right] \rightarrow \sigma_i^2\left(\frac{N_i}{N_i - 1}\right) \tag{17.15}$$

Then,

$$s_i^2\left(\frac{N_i - 1}{N_i}\right) \rightarrow \sigma_i^2 = V(y_i) \tag{17.16}$$

The variance of the mean in cluster $i$ is

$$V(\bar{y}_i) = \sigma_{\bar{y}_i}^2 = \frac{\sigma_i^2}{n_i}\left(\frac{N_i - n_i}{N_i - 1}\right) \tag{17.17}$$

Its estimate is

$$s_{\bar{y}_i}^2 = \frac{s_i^2}{n_i}\left(\frac{N_i - 1}{N_i}\right)\left(\frac{N_i - n_i}{N_i - 1}\right) = \frac{s_i^2}{n_i}\left(\frac{N_i - n_i}{N_i}\right) \rightarrow \sigma_{\bar{y}_i}^2 \tag{17.18}$$

The variance of the total in cluster $i$ is

$$\begin{aligned} V(T_i) = \sigma_{T_i}^2 &= V(N_i\bar{y}_i) \\ &= E(N_i\bar{y}_i - N_i\mu_i)^2 \\ &= N_i^2 E(\bar{y}_i - \mu_i)^2 \end{aligned}$$

By definition, $E(\bar{y}_i - \mu_i)^2 = V(\bar{y}_i) = \sigma_{\bar{y}_i}^2$. Then,

$$V(T_i) = N_i^2\sigma_{\bar{y}_i}^2 = N_i^2\,\frac{\sigma_i^2}{n_i}\left(\frac{N_i - n_i}{N_i - 1}\right) \tag{17.19}$$

The estimate of the variance of the total in cluster $i$ is then

$$s_{T_i}^2 = N_i^2 s_{\bar{y}_i}^2 \tag{17.20}$$

or

$$s_{T_i}^2 = N_i^2\,\frac{s_i^2}{n_i}\left(\frac{N_i - n_i}{N_i}\right) \tag{17.21}$$

Its expected value is

$$E(s_{T_i}^2) = E(N_i^2 s_{\bar{y}_i}^2)$$
$$= N_i^2 E(s_{\bar{y}_i}^2) = N_i^2 \sigma_{\bar{y}_i}^2 = V(T_i) \tag{17.22}$$

and $s_{T_i}^2$ is an unbiased estimator of $\sigma_{\tau_i}^2 = V(T_i)$.

### 17.5.2 Overall Statistics

The estimate of the mean cluster total is

$$\bar{T} = \frac{1}{m}\sum_{I}^{m} T_i = \frac{1}{m}\sum_{i}^{m} N_i \bar{y}_i \tag{17.23}$$

and its expected value is

$$E(\bar{T}) = E\left(\frac{1}{m}\sum_{I}^{m} N_i \bar{y}_i\right)$$
$$= \frac{1}{m}\sum_{i}^{m} E(N_i \bar{y}_i) = \frac{1}{m}(m)\frac{1}{M}\sum_{i}^{M} N_i E(\bar{y}_i) \tag{17.24}$$
$$= \frac{1}{M}\sum_{i}^{M} N_i \mu_i = \frac{1}{M}\sum_{i}^{M} \tau_i = \bar{\tau}$$

Consequently, $\bar{T}$ is an unbiased estimator of $\bar{\tau}$. The estimate of the overall total is

$$T_o = M\bar{T} = \frac{M}{m}\sum_{i}^{m} T_i = \frac{M}{m}\sum_{i}^{m} N_i \bar{y}_i \tag{17.25}$$

Its expected value is

$$E(T_o) = E(M\bar{T}) = ME(\bar{T}) = M\bar{\tau} = \tau_o \tag{17.26}$$

Thus, $T_o$ is an unbiased estimator of $\tau_o$. The estimate of the overall or population mean is

$$\bar{y}_o = T_o / N \tag{17.27}$$

$$= \frac{M}{Nm}\sum_{i}^{m} T_i$$
$$= \frac{M}{Nm}\sum_{i}^{m} N_i \bar{y}_i \tag{17.28}$$
$$= \frac{M}{Nm}\sum_{i}^{m} \frac{N_i}{n_i}\sum_{j}^{n_i} y_{ij}$$

Its expected value is

$$E(\bar{y}_o) = E(T_o / N) = E(T_o) / N = \tau_o / N = \mu_o \tag{17.29}$$

indicating that $\bar{y}_o$ is an unbiased estimator of $\mu_o$.

If cluster and subsample sizes are held constant so that $N_i = A$ and $n_i = a$, Equation 17.28 becomes

$$\bar{y}_o = \frac{M}{Nm} \sum_i^m \frac{A}{a} \sum_j^a y_{ij}$$

Since $A$ and $a$ are constants that can be factored out,

$$\bar{y}_o = \frac{MA}{Nma} \sum_i^m \sum_j^a y_{ij}$$

and since $MA = N$,

$$\bar{y}_o = \frac{1}{ma} \sum_i^m \sum_j^a y_{ij} \tag{17.30}$$

To obtain the expected value of $\bar{y}_o$ under these circumstances, one must recognize that two stages are involved with $i$ referring to the first stage and $j$ to the second. Thus,

$$\begin{aligned} E(\bar{y}_o) &= E_i E_j \left( \frac{1}{ma} \sum_i^m \sum_j^a y_{ij} \right) \\ &= E_i \left[ \frac{1}{ma} \sum_i^m a E_j(y_{ij}) \right] \\ &= E_i \left[ \frac{1}{m} \sum_i^m E_j(y_{ij}) \right] \end{aligned}$$

When $i$ is held constant,

$$E_j(y_{ij}) = \frac{1}{A} \sum_j^A y_{ij} = \mu_i$$

Then,

$$E(\bar{y}_o) = E_i \left[ \frac{1}{m} (m) \mu_i \right] = E_i(\mu_i)$$

Since all clusters are the same size, their weights are equal and

$$E_i(\mu_i) = \frac{1}{M}\sum_i^M \mu_i = \mu_o$$

Then $E(\bar{y}_o) = \mu_o$ and

$$\bar{y}_o = \frac{1}{ma}\sum_i^m \sum_j^a y_{ij}$$

is an unbiased estimator of $\mu_o$. The significance of Equation 17.30 is that it permits the computation of an estimate of the overall mean when the total size of the population is not known. It must be emphasized that its use is appropriate only when cluster and subsample sizes are constant.

The estimate of the variance of the cluster totals is

$$s_T^2 = \frac{1}{(m-1)}\sum_i^m (T_i - \bar{T})^2 \to \sigma_\tau^2\left(\frac{M}{M-1}\right) \tag{17.31}$$

and

$$s_T^2\left(\frac{M-1}{M}\right) \to \sigma_\tau^2 = V(\tau) \tag{17.32}$$

### 17.5.3 The Variance of the Overall Total

By definition, the variance of the overall total is

$$V(T_o)_{2s} = \sigma_{T_{o2s}}^2 = E(T_o - \tau_o)^2 \tag{17.33}$$

The subscript $2s$ identifies this as the variance of the overall total of a two-stage cluster sampling operation. According to Equation 17.25,

$$T_o = \frac{M}{m}\sum_i^m T_i$$

Thus,

$$V(T_o)_{2s} = E\left(\frac{M}{m}\sum_i^m T_i - \tau_o\right)^2$$

Here again two stages are involved, which must be recognized when deriving this expected value. Thus,

$$E\left(\frac{M}{m}\sum_i^m T_i - \tau_o\right)^2 = E_iE_j\left(\frac{M}{m}\sum_i^m T_i - \tau_o\right)^2$$

$$= E_iE_j\left[\left(\frac{M}{m}\sum_i^m T_i - \frac{M}{m}\sum_i^m \tau_i\right) + \left(\frac{M}{m}\sum_i^m \tau_i - \tau_o\right)\right]$$

$(M/m)\Sigma_i^m \tau_i$ is a random variable since only $m$ of the $M$ clusters in the sampling frame are included in the sample. Expanding the binomial:

$$V(T_o)_{2s} = E_iE_j\left\{\left(\frac{M}{m}\right)^2\left[\sum_i^m (T_i - \tau_i)\right]^2 + 2\left(\frac{M}{m}\right)\left[\sum_i^m (T_i - \tau_i)\right]\left(\frac{M}{m}\sum_i^m \tau_i - \tau_o\right) + \left(\frac{M}{m}\sum_i^m \tau_i - \tau_o\right)^2\right\} \tag{17.34}$$

Since

$$\left[\sum_i^m (T_i - \tau_i)\right]^2 = \sum_i^m (T_i - \tau_i)^2 + \sum_{i\neq h}^m (T_i - \tau_i)(T_h - \tau_h)$$

Equation 17.34 can be written:

$$V(T_o)_{2s} = E_i\left\{\left(\frac{M}{m}\right)^2\left[\sum_i^m E_j(T_i - \tau_i)^2\right] + \left(\frac{M}{m}\right)^2\left[\sum_{i\neq h}^m E_j(T_i - \tau_i)(T_h - \tau_h)\right] + 2\left(\frac{M}{m}\right)\left[\sum_i^m E_j(T_i - \tau_i)\right]\left[\frac{M}{m}\sum_i^m E_j(\tau_i - \tau_o)\right] + E_j\left(\frac{M}{m}\sum_i^m \tau_i - \tau_o\right)^2\right\} \tag{17.35}$$

Then, in the first term, by definition,

$$E_j(T_i - \tau_i)^2 = V(T_i) = \sigma_{T_i}^2$$

which is the variance of the total in cluster $i$. In the second term

$$E_j(T_i - \tau_i)(T_h - \tau_h) = E_j(T_iT_h - T_i\tau_h - \tau_iT_h + \tau_i\tau_h)$$

$$= E_j(T_i)E_j(T_h) - E_j(T_i)E_j(\tau_h) - E_j(\tau_i)E_j(T_h) + E_j(\tau_i)E_j(\tau_h)$$

Since $E_j(T_i) = \tau_i$ and $E_j(T_h) = \tau_h$,

$$E_j(T_i - \tau_i)(T_h - \tau_h) = 0$$

and the second term vanishes. In the third term

$$E_j(T_i - \tau_i) = E_j(T_i) - E_j(\tau_i) = \tau_i - \tau_i = 0$$

and the third term also vanishes. In the fourth term, since $i$ is being held constant and since $\tau_i$ is a parameter and not an estimate,

$$E_j\left(\frac{M}{m}\sum_i^m \tau_i - \tau_o\right)^2 = \left(\frac{M}{m}\sum_i^m \tau_i - \tau_o\right)^2$$

Consequently,

$$V(T_o)_{2s} = E_i\left[\left(\frac{M}{m}\right)^2 \sum_i^m \sigma_{T_i}^2 + \left(\frac{M}{m}\sum_i^m \tau_i - \tau_o\right)^2\right] \tag{17.36}$$

Since the sample includes only $m$ clusters,

$$\frac{1}{m}\sum_i^m \tau_i = \bar{T}$$

which is the estimated mean cluster total and is a random variable. Then, since $M\bar{T} = T_o$, the estimated overall total,

$$\left(\frac{M}{m}\sum_i^m \tau_i - \tau_o\right)^2 = (M\bar{T} - \tau_o)^2$$

and

$$V(T_o)_{2s} = \left(\frac{M}{m}\right)^2 \sum_i^m E_i(\sigma_{T_i}^2) + E_i(M\bar{T} - \tau_o)^2 \tag{17.37}$$

Consider the first term in Equation 17.37.

$$E_i(\sigma_{T_i}^2) = \frac{1}{M}\sum_i^M \sigma_{T_i}^2$$

According to Equation 17.19,

$$\sigma_{T_i}^2 = \frac{\sigma_i^2}{n_i}\left(\frac{N_i - n_i}{N_i - 1}\right)N_i^2$$

which is the variance of the total in cluster $i$. Thus,

$$E_i(\sigma^2_{T_i}) = \frac{1}{M}\sum_i^M \left[\frac{\sigma_i^2}{n_i}\left(\frac{N_i - n_i}{N_i - 1}\right)N_i^2\right]$$

and

$$\left(\frac{M}{m}\right)^2 \sum_i^m E_i(\sigma^2_{T_i}) = \frac{M}{m}\sum_i^M \left[\frac{\sigma_i^2}{n_i}\left(\frac{N_i - n_i}{N_i - 1}\right)N_i^2\right] \tag{17.38}$$

The component of variance shown in Equation 17.38 is due to the variability of the variates *within* the clusters. Now consider the second term in Equation 17.37. By definition,

$$E_i(M\bar{T} - \tau_o)^2 = E_i(T_o - \tau_o)^2 = \sigma^2_{T_o} = M^2\sigma^2_{\bar{T}} = M^2\frac{\sigma^2_\tau}{m}\left(\frac{M - m}{M - 1}\right) \tag{17.39}$$

where $\sigma^2_{T_o}$ = variance of the overall total

$\sigma^2_{\bar{T}}$ = variance of the mean of the cluster totals

$\sigma^2_\tau = \frac{1}{M}\sum_i^M (\tau_i - \tau_o)^2$ = variance of the cluster totals

The component of variance shown in Equation 17.39 is due to the variability *between* the cluster totals. Then,

$$V(T_o)_{2s} = \frac{M}{m}\sum_i^m \left[\frac{\sigma_i^2}{n_i}\left(\frac{N_i - n_i}{N_i - 1}\right)N_i^2\right] + M^2\frac{\sigma^2_\tau}{m}\left(\frac{M - m}{M - 1}\right) \tag{17.40}$$

The estimate of $V(T_o)_{2s}$ is obtained by substituting $s_i^2[(N - 1)/N_i]$ for $\sigma_i^2$ and $s_T^2[(M - 1)/M]$ for $\sigma_\tau^2$ in Equation 17.40.

$$s^2_{T_o} = \frac{M}{m}\sum_i^m \left[\frac{s_i^2}{n_i}\left(\frac{N_i - n_i}{N_i}\right)N_i^2\right] + M^2\frac{s_T^2}{m}\left(\frac{M - m}{M}\right) \tag{17.41}$$

The determination of the expected value of $s^2_{T_o}$ requires recognition of the fact that the sampling involved two stages. Thus,

$$E(s^2_{T_o}) = E_iE_j(s^2_{T_o}) = E_i\left\{\frac{M}{m}\sum_i^m \left[\frac{N_i^2}{n_i}\left(\frac{N_i - n_i}{N_i}\right)E_j(s_i^2)\right]\right\} + \frac{M^2}{m}\left(\frac{M - m}{M}\right)E(s_T^2) \tag{17.42}$$

Consider the first term in Equation 17.42. Its expected value, in terms of clusters, is the mean across all $M$ clusters. The expected value of $s_i^2$, according to Equation 12.12, is $\sigma_i^2[N_i/(N_i - 1)]$. Thus, the first term becomes

$$\frac{1}{M}\sum_{i}^{M}\left\{\frac{M}{m}m\left[\frac{\sigma_i^2}{n_i}\left(\frac{N_i - n_i}{N_i - 1}\right)N_i^2\right]\right\} = \sum_{i}^{M}\left[\frac{\sigma_i^2}{n_i}\left(\frac{N_i - n_i}{N_i - 1}\right)N_i^2\right] \tag{17.43}$$

Now consider the second term in Equation 17.42. Since the subsampling has no effect on this term, the expectation is only across $i$. According to Equation 17.31,

$$s_T^2 = \frac{1}{(m-1)}\sum_{i}^{m}(T_i - \bar{T})^2, \quad \text{where} \quad \bar{T} = \frac{1}{m}\sum_{i}^{m}T_i$$

$$= \frac{1}{(m-1)}\sum_{i}^{m}(T_i^2 - 2T_i\bar{T} + \bar{T}^2) = \frac{1}{(m-1)}\left(\sum_{i}^{m}T_i^2 - 2\bar{T}\sum_{i}^{m}T_i + \sum_{i}^{m}\bar{T}^2\right)$$

Multiplying the right side of the equation by $m/m$ and recognizing that $(1/m)\Sigma_i^M T_i = \bar{T}$ yields

$$s_T^2\left(\frac{m-1}{m}\right) = \frac{1}{m}\sum_{i}^{m}T_i^2 - \bar{T}^2$$

$$E\left[s_T^2\left(\frac{m-1}{m}\right)\right] = E_i\left[E_j\left(\frac{1}{m}\sum_{i}^{m}T_i^2 - \bar{T}^2\right)\right] = E_iE_j(T_i^2) - E(\bar{T}^2) \tag{17.44}$$

Since $\bar{T} = T_o/M$,

$$E\left[s_T^2\left(\frac{m-1}{m}\right)\right] = \left(\frac{m-1}{m}\right)E(s_T^2) = E_iE_j(T_i^2) - \frac{1}{M^2}E(T_o^2) \tag{17.45}$$

Let

$$T_i = T_i - \tau_i + \tau_i = (T_i - \tau_i) + \tau_i$$

Then

$$T_i^2 = (T_i - \tau_i)^2 + 2(T_i - \tau_i)\tau_i + \tau_i^2$$

and

$$E_j(T_i^2) = E_j(T_i - \tau_i)^2 + 2E_j(T_i - \tau_i)E_j(\tau_i) + E_j(\tau_i^2)$$

Since $E_j(T_i - \tau_i)^2 = V_j(T_i)$, $E_j(T_i - \tau_i) = 0$, $E_j(\tau_i) = \tau_i$, and $E_j(\tau_i^2) = \tau_i^2$,

$$E_j(T_i^2) = V_j(T_i) + \tau_i^2$$

According to Equation 17.19, where $i$ is held constant,

$$V_j(T_i) = \frac{\sigma_i^2}{n_i}\left(\frac{N_i - n_i}{N_i - 1}\right)N_i^2$$

Consequently,

$$E_iE_j(T_i^2) = \frac{1}{M}\sum_i^M\left[\frac{\sigma_i^2}{n_i}\left(\frac{N_i - n_i}{N_i - 1}\right)N_i^2\right] + \frac{1}{M}\sum_i^M \tau_i^2 \tag{17.46}$$

Now, returning to Equation 17.45, let

$$T_o = T_o - \tau_o + \tau_o = (T_o - \tau_o) + \tau_o$$

Then

$$T_o^2 = (T_o - \tau_o)^2 + 2(T_o - \tau_o)\tau_o + \tau_o^2$$

and

$$E(T_o^2) = E(T_o - \tau_o)^2 + 2E(T_o - \tau_o)E(\tau_o) + E(\tau_o^2) = V(T_o) + \tau_o^2 \tag{17.47}$$

Then, substitute Equations 17.46 and 17.47 into Equation 17.45 after substituting Equation 17.40 for $V(T_o)$ in Equation 17.47.

$$E\left[s_T^2\left(\frac{m-1}{m}\right)\right] = \left(\frac{1}{M} - \frac{1}{Mm}\right)\sum_i^M\left[\frac{\sigma_i^2}{n_i}\left(\frac{N_i - n_i}{N_i - 1}\right)N_i^2\right] + \left(\frac{1}{M}\sum_i^M \tau_i^2 - \frac{\tau_o^2}{M^2}\right) - \frac{\sigma_\tau^2}{m}\left(\frac{M-m}{M-1}\right)$$

According to Equation 17.10,

$$\frac{1}{M}\sum_i^M \tau_i^2 - \tau_o^2 / M^2 = \sigma_\tau^2$$

Furthermore,

$$\left(\frac{1}{M} - \frac{1}{Mm}\right) = \frac{1}{M}\left(\frac{m-1}{m}\right)$$

Consequently,

$$E\left[s_T^2\left(\frac{m-1}{m}\right)\right] = \frac{1}{M}\left(\frac{m-1}{m}\right)\sum_i^M\left[\frac{\sigma_i^2}{n_i}\left(\frac{N_i - n_i}{N_i - 1}\right)N_i^2\right] + \sigma_\tau^2 - \frac{\sigma_\tau^2}{m}\left(\frac{M-m}{M-1}\right)$$

Since

$$\sigma_\tau^2 - \frac{\sigma_\tau^2}{m}\left(\frac{M-m}{M-1}\right) = \sigma_\tau^2\left(\frac{M}{M-1}\right)\left(\frac{m-1}{m}\right)$$

then

$$E\left[s_T^2\left(\frac{m-1}{m}\right)\right] = \frac{1}{M}\left(\frac{m-1}{m}\right)\sum_i^M\left[\frac{\sigma_i^2}{n_i}\left(\frac{N_i-n_i}{N_i-1}\right)N_i^2\right] + \sigma_\tau^2\left(\frac{M}{M-1}\right)\left(\frac{m-1}{m}\right)$$

and

$$E(s_T^2) = \frac{1}{M}\sum_i^M\left[\frac{\sigma_i^2}{n_i}\left(\frac{N_i-n_i}{N_i-1}\right)N_i^2\right] + \sigma_\tau^2\left(\frac{M}{M-1}\right) \tag{17.48}$$

The second term in Equation 17.42 is then

$$\frac{M^2}{m}\left(\frac{M-m}{M}\right)E(s_T^2) = \left(\frac{M-m}{m}\right)\sum_i^M\left[\frac{\sigma_i^2}{n_i}\left(\frac{N_i-n_i}{N_i-1}\right)N_i^2\right] + M^2\frac{\sigma_\tau^2}{m}\left(\frac{M-m}{M-1}\right) \tag{17.49}$$

Substituting Equations 17.43 and 17.49 for their respective terms in Equation 17.42 yields

$$E(s_{T_o}^2) = \sum_i^M\left[\frac{\sigma_i^2}{n_i}\left(\frac{N_i-n_i}{N_i-1}\right)N_i^2\right] + \left(\frac{M-m}{m}\right)\sum_i^M\left[\frac{\sigma_i^2}{n_i}\left(\frac{N_i-n_i}{N_i-1}\right)N_i^2\right] + M^2\frac{\sigma_\tau^2}{m}\left(\frac{M-m}{M-1}\right)$$

Multiplying the first terms by $m/m$ and simplifying:

$$E(s_{T_o}^2) = \frac{M}{m}\sum_i^M\left[\frac{\sigma_i^2}{n_i}\left(\frac{N_i-n_i}{N_i-1}\right)N_i^2\right] + M^2\frac{\sigma_\tau^2}{m}\left(\frac{M-m}{M-1}\right) \tag{17.50}$$

which agrees with Equation 17.40, indicating that $s_{T_o}^2$ is an unbiased estimator of $V(T_o)$.

### 17.5.4 The Variance of the Overall Mean

By definition, the variance of the estimated population mean is

$$V(\bar{y}_o)_{2s} = \sigma_{\bar{y}_o}^2 = E(\bar{y}_o - \mu_o)^2$$

According to Equation 17.27, $\bar{y}_o = T_o/N$. Furthermore, according to Equation 17.5,

$$\mu_o = \frac{1}{N}\sum_i^M\sum_j^{N_i} y_{ij}$$

and, according to Equation 17.6,

$$\tau_o = \sum_i^M \sum_j^{N_i} y_{ij}$$

Then $\mu_o = \tau_o/N$. Consequently,

$$V(\bar{y}_o)_{2s} = E[(T_o / N) - (\tau_o / N)]^2 = \frac{1}{N^2} E(T_o - \tau_o)^2 = \frac{1}{N^2} V(T_o)$$
$$-\frac{1}{N^2} \left\{ \frac{M}{m} \sum_i^M \left[ \frac{\sigma_i^2}{n_i} \left( \frac{N_i - n_i}{N_i - 1} \right) N_i^2 \right] + M^2 \frac{\sigma_\tau^2}{m} \left( \frac{M-m}{M-1} \right) \right\} \tag{17.51}$$

The estimate of $V(\bar{y}_o)$ is obtained by substituting $s_{\tau_o}^2$ for $V(T_o)_{2s}$ in Equation 17.51. Thus,

$$s_{\bar{y}_o}^2 = s_{T_o}^2 / N^2$$
$$= \frac{1}{N^2} \left\{ \frac{M}{m} \sum_i^M \left[ \frac{s_i^2}{n_i} \left( \frac{N_i - n_i}{N_i} \right) N_i^2 \right] + M^2 \frac{s_T^2}{m} \left( \frac{M-m}{M} \right) \right\} \tag{17.52}$$

The expected value of $s_{\bar{y}_o}^2$ is

$$E(s_{\bar{y}_o})^2 = E(s_{T_o}^2 / N^2) = \frac{1}{N^2} E(s_{T_o}^2) = \frac{1}{N^2} V(T_o) = V(\bar{y}_o)$$

indicating that $s_{\bar{y}_o}^2$ is an unbiased estimator of $V(\bar{y}_o)$.

When cluster and subsample sizes are held constant so that $N_i = N/M = A$ and $n_i = n/m = a$, Equation 17.51 can be written

$$V(\bar{y}_o)_{2s} = \frac{M}{N^2 m} \sum_i^M \left[ \frac{\sigma_i^2}{a} \left( \frac{A-a}{A-1} \right) A^2 \right] + \frac{M^2}{N^2} \frac{\sigma_\tau^2}{m} \left( \frac{M-m}{M-1} \right) \tag{17.53}$$

Consider now only the first term in Equation 17.53. Substituting $N^2/M^2$ for $A^2$ and simplifying yields

$$\frac{1}{ma} \left( \frac{A-a}{A-1} \right) \left( \frac{1}{M} \right) \sum_i^M \sigma_i^2$$

Since

$$\frac{1}{M} \sum_i^M \sigma_i^2 = \frac{1}{M} \sum_i^M \left( \frac{1}{A} \right) \sum_j^A (y_{ij} - \mu_i)^2 = \sigma_w^2 \tag{17.54}$$

which is the *pooled within cluster variance*, the first term in Equation 17.53 can be written

$$\frac{1}{ma}\left(\frac{A-a}{A-1}\right)\sigma_w^2 \tag{17.55}$$

Now consider the second term in Equation 17.53. It can be written

$$\frac{1}{A^2}\frac{\sigma_\tau^2}{m}\left(\frac{M-m}{M-1}\right) \tag{17.56}$$

Since

$$\sigma_\tau^2 = \frac{1}{M}\sum_i^M (\tau_i - \bar{\tau})^2$$

then

$$\frac{\sigma_\tau^2}{A^2} = \frac{1}{M}\sum_i^M \left(\frac{\tau_i}{A} - \frac{\bar{\tau}}{A}\right)^2 = \frac{1}{M}\sum_i^M (\mu_i - \mu_o)^2 = \sigma_B^2 \tag{17.57}$$

which is the *between cluster variance*. Substituting $\sigma_B^2$ for $\sigma_\tau^2/A^2$ in Equation 17.56 yields

$$\frac{\sigma_B^2}{m}\left(\frac{M-m}{M-1}\right) \tag{17.58}$$

Thus, substituting Equations 17.55 and 17.58 for their respective terms in equation 17.53 yields

$$V(\bar{y}_o)_{2s} = \frac{\sigma_w^2}{ma}\left(\frac{A-a}{A-1}\right) + \frac{\sigma_B^2}{m}\left(\frac{M-m}{M-1}\right) \tag{17.59}$$

which clearly shows the two components that make up the variance of the overall mean.

Equation 17.59 can be written

$$V(\bar{y}_o)_{2s} = \frac{1}{M}\sum_i^M \left[\frac{\sigma_i^2}{ma}\left(\frac{A-a}{A-1}\right)\right] + \frac{\sigma_B^2}{m}\left(\frac{M-m}{M-1}\right) \tag{17.60}$$

The estimator of $V(\bar{y}_o)$ is obtained by substituting $s_i^2[(A-1)/A]$ for $\sigma_i^2$, $(s_T^2/A^2)[(M-1)/M]$ for $\sigma_B^2$, and summing over $m$ instead of $M$ in Equation 17.60.

$$s_{\bar{y}_{o2s}}^2 = \frac{1}{M}\sum_i^m \left[\frac{s_i^2}{ma}\left(\frac{A-a}{A}\right)\right] + \frac{1}{m}\frac{s_T^2}{A^2}\left(\frac{M-m}{M}\right) \tag{17.61}$$

Since $(1/m)\Sigma_i^m s_i^2 = s_w^2$ and $s_T^2/A^2 = s_B^2$,

$$s_{\bar{y}_{o2s}}^2 = \frac{s_w^2}{Ma}\left(\frac{A-a}{A}\right) + \frac{s_B^2}{m}\left(\frac{M-m}{M}\right) \tag{17.62}$$

The expected value of $s^2_{\bar{y}_o}$ is based on Equation 17.61.

$$E(s^2_{\bar{y}_o}) = E_i E_j \left\{ \frac{1}{Mma}\left(\frac{A-a}{A}\right)\sum_i^m s_i^2 + \frac{1}{m}\left(\frac{M-m}{M}\right)\frac{s_T^2}{A^2} \right\} \tag{17.63}$$

Considering only the first term in Equation 17.63,

$$\begin{aligned} E_i\left[\frac{1}{Mma}\left(\frac{A-a}{A}\right)\sum_i^m E_j(s_i^2)\right] &= E_i\left[\frac{1}{Mma}\left(\frac{A-a}{A}\right)\sum_i^m \sigma_i^2\left(\frac{A}{A-1}\right)\right] \\ &= \frac{1}{Mma}\left(\frac{A-a}{A-1}\right) mE_i(\sigma_i^2) \\ &= \frac{1}{Ma}\left(\frac{A-a}{A-1}\right)\frac{1}{M}\sum_i^M \sigma_i^2 \\ &= \frac{1}{M^2}\sum_i^M\left[\frac{\sigma_i^2}{a}\left(\frac{A-a}{A-1}\right)\right] \end{aligned} \tag{17.64}$$

Now consider the second term in Equation 17.63. Since no subsampling is involved, the expectation involves only the clusters. Thus,

$$\begin{aligned} E_i E_j\left[\frac{1}{m}\left(\frac{M-m}{M}\right)\frac{s_T^2}{A^2}\right] &= E\left[\frac{1}{m}\left(\frac{M-m}{M}\right)\frac{s_T^2}{A^2}\right] \\ &= \frac{1}{mA^2}\left(\frac{M-m}{M}\right)E(s_T^2) \end{aligned}$$

Substituting Equation 17.48 for $E(s_T^2)$ yields

$$E\left[\frac{1}{m}\left(\frac{M-m}{M}\right)\frac{s_T^2}{A^2}\right] = \frac{1}{mA^2}\left(\frac{M-m}{M}\right)\left\{\frac{1}{M}\sum_i^M\left[\frac{\sigma_i^2}{a}\left(\frac{A-a}{A-1}\right)A^2\right] + \sigma_\tau^2\left(\frac{M}{M-1}\right)\right\} \tag{17.65}$$

Substituting Equations 17.64 and 17.65 for their corresponding terms in Equation 17.63 and simplifying,

$$\begin{aligned} E(s^2_{\bar{y}_o}) &= \left[\frac{1}{M} + \frac{1}{m}\left(\frac{M-m}{M}\right)\right]\left\{\frac{1}{M^2}\sum_i^M\left[\frac{\sigma_i^2}{a}\left(\frac{A-a}{A-1}\right)\right]\right\} + \frac{1}{m}\frac{\sigma_\tau^2}{A^2}\left(\frac{M-m}{M-1}\right) \\ &= \frac{1}{M}\sum_i^M\left[\frac{\sigma_i^2}{ma}\left(\frac{A-a}{A-1}\right)\right] + \frac{\sigma_B^2}{m}\left(\frac{M-m}{M-1}\right) \end{aligned} \tag{17.66}$$

which agrees with Equation 17.60, indicating that $s^2_{\bar{y}_o}$, as defined in Equations 17.61 and 17.62, is an unbiased estimator of $V(\bar{y}_o)_{2S}$.

When the total size of the population is not known, we can obtain an estimate of the population mean by keeping the clusters and subsample sizes constant and using Equation 17.30. However, even if the cluster and subsample sizes are held constant, an estimate of the variance of the mean cannot be determined if the population size is not known, because $MA = N$ and both $M$ and A appear in Equations 17.61 and 17.62.

## 17.6 THE INTRACLASS CORRELATION COEFFICIENT

The *intraclass*, *intracluster*, or *spatial correlation coefficient*, ρ, is a measure of the degree of homogeneity of the variates within the clusters making up a population.

$$\rho = E(y_{ij} - \mu_o)(y_{ik} - \mu_o) / E(y_{ij} - \mu_o)^2 \tag{17.67}$$

Since the total number of combinations of $y_{ij}$ and $y_{ik}$ is equal to

$$\binom{A}{2} = A(A-1)/2$$

then

$$E(y_{ij} - \mu_o)(y_{ik} - \mu_o) = \frac{2}{MA(A-1)} \sum_{i}^{M} \sum_{j<k}^{A(A-1)/2} (y_{ij} - \mu_o)(y_{ik} - \mu_o) \tag{17.68}$$

and

$$E(y_{ij} - \mu_o)^2 = \frac{1}{MA} \sum_{i}^{M} \sum_{j}^{A} (y_{ij} - \mu_o)^2 = V(y) = \sigma_o^2 \tag{17.69}$$

Then,

$$\rho = \frac{2}{MA(A-1)\sigma_o^2} \sum_{i}^{M} \sum_{j<k}^{A(A-1)/2} (y_{ij} - \mu_o)(y_{ik} - \mu_o) \tag{17.70}$$

In the case of a single cluster,

$$\rho_i = 2 \sum_{j<k}^{A(A-1)/2} (y_{ij} - \mu_o)(y_{ik} - \mu_o) \Bigg/ (A-1) \sum_{j}^{A} (y_{ij} - \mu_o)^2 \tag{17.71}$$

For example, given the following population which is divided into two clusters ($\mu_o = 5.5$ and $\sigma_o^2 = 8.25$):

| Cluster | Variates | $y_{ij}, y_{ik}$ | $(y_{ij} - \mu_o)(y_{ik} - \mu_o)$ |
|---|---|---|---|
| 1 | $y_{11} = 1$ | 1,2 | (1 – 5.5)(2 – 5.5) = 15.75 |
| | $y_{12} = 2$ | 1,3 | (1 – 5.5)(3 – 5.5) = 11.25 |

| Cluster | Variates | $y_{ij}, y_{ik}$ | $(y_{ij} - \mu_o)(y_{ik} - \mu_o)$ |
|---|---|---|---|
| 1 | $y_{13} = 3$ | 1,4 | $(1 - 5.5)(4 - 5.5) = 6.75$ |
| | $y_{14} = 4$ | 1,5 | $(1 - 5.5)(5 - 5.5) = 2.25$ |
| | $y_{15} = 5$ | 2,3 | $(2 - 5.5)(3 - 5.5) = 8.75$ |
| | | 2,4 | $(2 - 5.5)(4 - 5.5) = 5.25$ |
| | $\mu_1 = 3.0$ | 2,5 | $(2 - 5.5)(5 - 5.5) = 1.75$ |
| | | 3,4 | $(3 - 5.5)(4 - 5.5) = 3.75$ |
| | | 3,5 | $(3 - 5.5)(5 - 5.5) = 1.25$ |
| | | 4,5 | $(4 - 5.5)(5 - 5.5) = 0.75$ |
| | | | $\sum_{j<k}^{A(A-1)/2} (y_{ij} - \mu_o)(y_{ik} - \mu_o) = 57.50$ |
| 2 | $y_{21} = 6$ | 6,7 | $(6 - 5.5)(7 - 5.5) = 0.75$ |
| | $y_{22} = 7$ | 6,8 | $(6 - 5.5)(8 - 5.5) = 1.25$ |
| | $y_{23} = 8$ | 6,9 | $(6 - 5.5)(9 - 5.5) = 1.75$ |
| | $y_{24} = 9$ | 6,10 | $(6 - 5.5)(10 - 5.5) = 2.25$ |
| | $y_{25} = 10$ | 7,8 | $(7 - 5.5)(8 - 5.5) = 3.75$ |
| | | 7,9 | $(7 - 5.5)(9 - 5.5) = 5.25$ |
| | $\mu_2 = 8.0$ | 7,10 | $(7 - 5.5)(10 - 5.5) = 6.75$ |
| | | 8,9 | $(8 - 5.5)(9 - 5.5) = 8.75$ |
| | | 8,10 | $(8 - 5.5)(10 - 5.5) = 11.25$ |
| | | 9,10 | $(9 - 5.5)(10 - 5.5) = 15.75$ |
| | | | $\sum_{j<k}^{A(A-1)/2} (y_{ij} - \mu_o)(y_{ik} - \mu_o) = 57.50$ |

$$E(y_{ij} - \mu_o)(y_{ik} - \mu_o) = \frac{2}{2(5)(4)}(57.70 + 57.50) = 5.75$$

$$\rho = E(y_{ij} - \mu_o)(y_{ik} - \mu_o)/\sigma_o^2 = 5.75/8.25 = 0.697$$

Consider the pairs of variates making up the possible combinations. When $y_{ik}$ is plotted against $y_{ij}$ and the trend indicates that $y_{ik}$ increases as $y_{ij}$ increases, the intraclass correlation coefficient will be positive. This is the case in the above example. The cluster regression coefficients are $\beta_1 = \beta_2 = 0.5$.

When all the variates within each of the clusters are equal, $y_{ik} = y_{ij}$, the trend lines are horizontal ($\beta_1$ and $\beta_2 = 0$), $E(y_{ij} - \mu_o)(y_{ik} - \mu_o) = E(y_{ij} - \mu_o)^2$, and $\rho = 1$. For example, ($\mu_o = 40$; $\sigma_o^2 = 1.00$):

| Cluster | Variates | $y_{ij}, y_{ik}$ | $(y_{ij} - \mu_o)(y_{ik} - \mu_o)$ |
|---|---|---|---|
| 1 | $y_{11} = 3$ | 3,3 | $(3 - 4)(3 - 4) = 1$ |
| | $y_{12} = 3$ | 3,3 | $(3 - 4)(3 - 4) = 1$ |
| | $y_{13} = 3$ | 3,3 | $(3 - 4)(3 - 4) = 1$ |
| | $y_{14} = 3$ | 3,3 | $(3 - 4)(3 - 4) = 1$ |
| | $y_{15} = 3$ | 3,3 | $(3 - 4)(3 - 4) = 1$ |
| | | 3,3 | $(3 - 4)(3 - 4) = 1$ |
| | | 3,3 | $(3 - 4)(3 - 4) = 1$ |
| | | 3,3 | $(3 - 4)(3 - 4) = 1$ |
| | | 3,3 | $(3 - 4)(3 - 4) = 1$ |

| Cluster | Variates | $y_{ij}, y_{ik}$ | $(y_{ij} - \mu_o)(y_{ik} - \mu_o)$ |
|---|---|---|---|
| | | 3,3 | $(3 - 4)(3 - 4) = 1$ |
| | | | $\sum_{j<k}^{A(A-1)/2}(y_{ij} - \mu_o)(y_{ik} - \mu_o) = 10$ |
| 2 | $y_{21} = 5$ | 5,5 | $(5 - 4)(5 - 4) = 1$ |
| | $y_{22} = 5$ | 5,5 | $(5 - 4)(5 - 4) = 1$ |
| | $y_{23} = 5$ | 5,5 | $(5 - 4)(5 - 4) = 1$ |
| | $y_{24} = 5$ | 5,5 | $(5 - 4)(5 - 4) = 1$ |
| | $y_{25} = 5$ | 5,5 | $(5 - 4)(5 - 4) = 1$ |
| | | 5,5 | $(5 - 4)(5 - 4) = 1$ |
| | | 5,5 | $(5 - 4)(5 - 4) = 1$ |
| | | 5,5 | $(5 - 4)(5 - 4) = 1$ |
| | | 5,5 | $(5 - 4)(5 - 4) = 1$ |
| | | 5,5 | $(5 - 4)(5 - 4) = 1$ |
| | | | $\sum_{j<k}^{A(A-1)/2}(y_{ij} - \mu_o)(y_{ik} - \mu_o) = 10.00$ |

$$E(y_{ij} - \mu_o)(y_{ik} - \mu_o) = \frac{2}{2(5)(4)}(10 + 10) = 1.00$$

$$\rho = E(y_{ij} - \mu_o)(y_{ik} - \mu_o)/\sigma_o^2 = 1.00/1.00 = 1.00$$

Thus, $\rho = 1$ is associated with perfect homogeneity *within each of the clusters*. Note that the cluster means can differ. The entire population may or may not be homogeneous.

When the trend is such that $y_{ik}$ decreases as $y_{ij}$ increases, the intraclass correlation coefficient will be negative. For example, consider the following population containing ten variates which are divided into two clusters.

Cluster 1: $y_{11} = 2$, $y_{12} = -6$, $y_{13} = -3$, $y_{14} = 5$, and $y_{15} = 1$
Cluster 2: $y_{21} = 5$, $y_{22} = -10$, $y_{23} = -7$, $y_{24} = 2$, and $y_{25} = 3$

The population mean and variance are $\mu_0 = -0.7$ and $\sigma_o^2 = 26.21$. The slopes of the cluster trends are $\beta_1 = -0.189$ and $\beta_2 = 0.181$. The intraclass correlation coefficient $\rho = -0.2262$.

Finally, when the scattergrams of $y_{ik}$ vs. $y_{ij}$ show no trend, the variates within the clusters are randomly distributed (i.e., they are completely heterogeneous). Then, $E(y_{ij} - \mu_o)(y_{ik} - \mu_o) = 0$, and $\sigma_o^2 > 0$, and $\rho = 0$. Thus, $-1 \leq \rho \leq 1$.

In most natural universes, adjacent elements tend to be similar but the similarity decreases if an interval occurs between the elements being compared. Thus, the intraclass correlation will be less in clusters having an open or spread-out configuration than in compact clusters containing the same number of elements.

In two-stage cluster sampling, where cluster and subsample sizes are constant, according to Equation 17.59:

$$V(\bar{y}_o)_{2s} = \frac{\sigma_w^2}{ma}\left(\frac{A-a}{A-1}\right) + \frac{\sigma_B^2}{m}\left(\frac{M-m}{M-1}\right)$$

Consider now only $\sigma_B^2$. From Equation 17.57,

$$\sigma_B^2 = \frac{1}{M}\sum_i^M(\mu_i - \mu_o)^2 = \frac{1}{M}\sum_i^M\left(\frac{\tau_i}{A} - \frac{\bar{\tau}}{A}\right)^2$$
$$= \frac{1}{MA^2}\sum_i^M = (\tau_i - \bar{\tau})^2 \qquad (17.72)$$

Since

$$\tau_i = \sum_j^A y_{ij} \quad \text{and} \quad \bar{\tau} = \frac{\tau_o}{M} = \frac{1}{M}\sum_i^M\sum_j^A y_{ij} = A\mu_o$$

then

$$(\tau_i - \bar{\tau}) = \sum_j^A y_{ij} - A\mu_o = \sum_j^A (y_{ij} - \mu_o)$$
$$= (y_{i1} - \mu_o) + (y_{i2} - \mu_o) + \cdots + (y_{iA} - \mu_o)$$

Then

$$\sigma_B^2 = \frac{1}{MA^2}\sum_i^M[(y_{i1} - \mu_o) + (y_{i2} - \mu_o) + \cdots + (y_{iA} - \mu_o)]^2$$
$$= \frac{1}{MA^2}\left[\sum_i^M\sum_j^A(y_{ij} - \mu_o)^2 + 2\sum_i^M\sum_{j<k}^{A(A-1)/2}(y_{ij} - \mu_o)(y_{ik} - \mu_o)\right] \qquad (17.73)$$

The first term within the brackets can be written

$$\sum_i^M\sum_j^A(y_{ij} - \mu_o)^2 = MA\sigma_o^2$$

and the second term

$$2\sum_i^M\sum_{j<k}^{A(A-1)/2}(y_{ij} - \mu_o)(y_{ik} - \mu_o) = \rho MA(A-1)\sigma_o^2$$

Then Equation 17.73 becomes

$$\sigma_B^2 = \frac{1}{MA^2}[(MA\sigma_o^2 + \rho MA(A-1)\sigma_o^2]$$
$$= \frac{\sigma_o^2}{A}[1 + \rho(A-1)] \qquad (17.74)$$

Now consider $\sigma_w^2$. The total sum of squared errors is partitioned as in Equation 16.18.

$$\sum_i^M \sum_j^A (y_{ij} - \mu_o)^2 = A \sum_i^M (\mu_i - \mu_o)^2 + \sum_i^M \sum_j^A (y_{ij} - \mu_i)^2 \tag{17.75}$$

Since $\Sigma_i^M \Sigma_j^A (y_{ij} - \mu_o)^2 = MA\sigma_o^2$, $A\Sigma_i^M (\mu_i - \mu_o)^2 = MA\sigma_B^2$, and $\Sigma_i^M \Sigma_j^A (y_{ij} - \mu_i)^2 = MA\sigma_w^2$, Equation 17.75 can be written

$$MA\sigma_o^2 = MA\sigma_B^2 + MA\sigma_w^2$$

or

$$\sigma_o^2 = \sigma_B^2 + \sigma_w^2 \tag{17.76}$$

Then,

$$\sigma_w^2 = \sigma_o^2 - \sigma_B^2$$

Substituting the right side of Equation 17.74 for $\sigma_B^2$,

$$\sigma_w^2 = \sigma_o^2 - \left\{ \frac{\sigma_o^2}{A} [1 + \rho(A-1)] \right\}$$

Then

$$\sigma_w^2 = \sigma_o^2 \left( \frac{A-1}{A} \right) (1 - \rho) \tag{17.77}$$

Returning to Equation 17.59 and substituting the right side of Equation 17.77 for $\sigma_w^2$ and the right side of Equation 17.74 for $\sigma_B^2$ and then simplifying,

$$V(\bar{y}_o)_{2s} = \frac{\sigma_w^2}{ma} \left( \frac{A-a}{A} \right) (1 - \rho) + \frac{\sigma_o^2}{mA} [1 + \rho(A-1)] \left( \frac{M-m}{M-1} \right) \tag{17.78}$$

which expresses the variance of the overall mean in terms of the intraclass correlation coefficient.

## 17.7 OVERLAPPING AND INTERLOCKING OF CLUSTERS

To prevent bias from creeping into estimates of population parameters, every element in the universe of interest must have some chance of being included in a sample. If some elements are more likely to be drawn, this fact must be recognized by suitable weighting of their associated variates. In the discussion of sampling to this point it has been assumed that the probabilities for all variates are the same. The computational patterns reflect that assumption. Thus, the population total is

$$\tau_Y = \sum_i^N y_i$$

the population mean is

$$\mu_Y = \frac{1}{N}\sum_i^N y_i$$

and the population variance is

$$\sigma_Y^2 = \frac{1}{N}\sum_i^N (y_i - \mu_Y)^2$$

These expressions are actually simplifications of more general ones in which the probabilities of occurrence of the variates are taken into account. This recognition of the differences in probability takes the form of weights which are the reciprocals of the probabilities. Thus, the weight for variate $i$ is

$$w_i = 1/P(y_i) \tag{17.79}$$

Then the population mean is

$$\mu_Y = \sum_i^N w_i y_i \Bigg/ \sum_i^N w_i \tag{17.80}$$

the population total is

$$\tau_Y = N\mu_Y \tag{17.81}$$

and the population variance is

$$\sigma_Y^2 = \sum_i^N w_i (y_i - \mu_Y)^2 \Bigg/ \sum_i^N w_i \tag{17.82}$$

If the probabilities are constant, the weights become constant, $w_i = k$. Then Equation 17.80 becomes

$$\mu_Y = \sum_i^N k y_i \Bigg/ \sum_i^N k = k\sum_i^N y_i \Bigg/ Nk = \frac{1}{N}\sum_i^N y_i$$

and Equation 17.82 becomes

$$\sigma_Y^2 = \sum_i^N k(y_i - \mu_Y)^2 \Bigg/ \sum_i^N k = k\sum_i^N (y_i - \mu_Y)^2 \Bigg/ Nk = \frac{1}{N}\sum_i^N (y_i - \mu_Y)^2$$

which are the expressions usually used.

It should be noted that when a variate is multiplied by the reciprocal of its probability of occurrences, the product is an estimate of the population total.

$$T_i = w_i y_i \tag{17.83}$$

The reciprocal of the probability is then, in essence, an expansion term, similar in some respects to those used to bring plot totals to a per acre or per hectare basis.

One can put this discussion into a numerical context. Assume the following universe, made up of 0.05-ha plots, and its associated population, made up of the timber volumes on the plots, measured in cubic meters.

| Plot | 1 | 2 | 3 | 4 | 5 | 6 | 7 | 8 | 9 | 10 | 11 | 12 |
|---|---|---|---|---|---|---|---|---|---|---|---|---|
| $y$ | 2.55 | 3.15 | 3.05 | 1.55 | 2.55 | 2.70 | 2.50 | 2.25 | 2.85 | 3.15 | 1.75 | 0.00 |

The total volume on the tract is 28.05 m³, the mean volume per plot is 2.3375 m³/plot, the mean volume per hectare is 46.75 m³/ha, and the variance is 0.7313021.

In this case, the probabilities are constant at $^1/_{12}$. Then,uj

| **Plot** | $y_i$ **(m³)** | $w_i$ | $T_i$ **(m³)** | $w_i(y_i - \mu_Y)^2$ |
|---|---|---|---|---|
| 1 | 2.55 | 12 | 30.6 | 0.541875 |
| 2 | 3.15 | 12 | 37.8 | 7.921875 |
| 3 | 3.05 | 12 | 36.6 | 6.091875 |
| 4 | 1.55 | 12 | 18.6 | 7.441875 |
| 5 | 2.55 | 12 | 30.6 | 0.541875 |
| 6 | 2.70 | 12 | 32.4 | 1.576875 |
| 7 | 2.50 | 12 | 30.0 | 0.316875 |
| 8 | 2.25 | 12 | 27.0 | 0.091875 |
| 9 | 2.85 | 12 | 34.2 | 3.151875 |
| 10 | 3.15 | 12 | 37.8 | 7.921875 |
| 11 | 1.75 | 12 | 21.0 | 4.141875 |
| 12 | 0.00 | 12 | 0.0 | 65.566875 |

$$\sum_i^N w_i = 144$$

$$\sum_i^N w_i y_i = 336.6$$

$$\sum_i^N w_i (y_i - \mu_Y)^2 = 105.3075$$

$$\mu_Y = 336.6 / 144 = 2.3375 \text{ m}^3 / \text{plot}$$

$$\tau_Y = 12(2.3375) = 28.05 \text{ m}^3 \text{ on tract}$$

$$\sigma_Y^2 = 105.3075 / 144 = 0.7313021$$

If cluster sampling is being used and the clusters are compact, each cluster, as a whole, serves as a sampling unit, its total is the variate of interest and the reciprocal of its probability of being chosen is its weight. If no subsampling is involved, the elements making up the cluster are certain to be evaluated and consequently the variates associated with these elements require no weighting. Let us use compact clusters of size $A = 2$ and compute the parameters of the 12 plots.

$$M = 12/2 = 6$$

$$P\ (\text{cluster}_i) = {}^1/_6$$

$$w_i = 6$$

Then,

| Cluster$_i$ | Plot | $y_{ij}$ (m³) | $z_i$ = Cluster Total (m³) | $w_i$ | $w_i z_i = T_i$ (m³) | $w_i(z_i - \bar{\tau})^2$ |
|---|---|---|---|---|---|---|
| 1 | 1 | 2.55 | 5.70 | 6 | 34.2 | 6.30375 |
| | 2 | 3.15 | | | | |
| 2 | 3 | 3.05 | 4.60 | 6 | 27.6 | 0.03375 |
| | 4 | 1.55 | | | | |
| 3 | 5 | 2.55 | 5.25 | 6 | 31.5 | 1.98375 |
| | 6 | 2.70 | | | | |
| 4 | 7 | 2.50 | 4.75 | 6 | 28.5 | 0.03375 |
| | 8 | 2.25 | | | | |
| 5 | 9 | 2.85 | 6.00 | 6 | 36.0 | 10.53375 |
| | 10 | 3.15 | | | | |
| 6 | 11 | 1.75 | 1.75 | 6 | 10.5 | 57.33375 |
| | 12 | 0.00 | | | | |

$$\sum_i^M w_i = 36$$

$$\sum_i^M w_i z_i = 168.3$$

$$\sum_i^M w_i (z_i - \bar{\tau})^2 = 70.2225$$

$$\bar{\tau} = 168.3 / 36 = 4.675 \text{ m}^3 / \text{cluster}$$

$$\tau_Y = 6(4.675) = 28.05 \text{ m}^3 \text{ on tract}$$

$$\mu_Y = 28.05 / 12 = 2.3375 \text{ m}^3 / \text{plot}$$

$$\sigma_T^2 = 70.2225 / 36 = 1.950625$$

The between cluster variance is then

$$\sigma_B^2 = \sigma_T^2 / A^2 = 1.950625/4 = 0.4876562$$

and the within cluster variance is

$$\sigma_w^2 = \sum_i^M \sum_j^A (y_{ij} - \mu_i)^2 / N = 0.2436458$$

Then $\sigma_o^2 = \sigma_B^2 + \sigma_w^2 = 0.4876562 + 0.2436458 = 0.7313021$. As can be seen, no bias is present.

Now use clusters of size $A = 2$, which are spread out so there is an interval of two plots between the chosen plots.

| Cluster$_i$ | Plot | $y_{ij}$ (m$^3$) | $z_i$ = Cluster Total (m$^3$) | $w_i$ | $w_iz_i = T_i$ (m$^3$) | $w_i(z_i - \bar{\tau})^2$ |
|---|---|---|---|---|---|---|
| 1 | 1 | 2.55 | 4.10 | 6 | 24.6 | 1.98375 |
| | 4 | 1.55 | | | | |
| 2 | 2 | 3.15 | 5.70 | 6 | 34.2 | 6.30375 |
| | 5 | 2.55 | | | | |
| 3 | 3 | 3.05 | 5.75 | 6 | 34.5 | 6.93375 |
| | 6 | 2.70 | | | | |
| 4 | 7 | 2.50 | 5.65 | 6 | 33.9 | 5.70375 |
| | 10 | 3.15 | | | | |
| 5 | 8 | 2.25 | 4.00 | 6 | 24.0 | 2.73375 |
| | 11 | 1.75 | | | | |
| 6 | 9 | 2.85 | 2.85 | 6 | 17.1 | 19.98375 |
| | 12 | 0.00 | | | | |

$$\sum_i^M w_i = 36$$

$$\sum_i^M w_i z_i = 168.3$$

$$\sum_i^M w_i (z_i - \bar{\tau})^2 = 43.6425$$

$$\bar{\tau} = 168.3 / 36 = 4.675 \text{ m}^3 / \text{cluster}$$

$$\tau_Y = 6(4.675) = 28.05 \text{ m}^3 \text{ on tract}$$

$$\mu_Y = 28.05 / 12 = 2.3375 \text{ m}^3 / \text{plot}$$

$$\sigma_T^2 = 43.6425 / 36 = 1.2122917$$

The between cluster variance is

$$\sigma_B^2 = \sigma_T^2 / A^2 = 1.2122917 / 4 = 0.3030729$$

the within cluster variance is

$$\sigma_w^2 = \sum_i^M \sum_j^A (y_{ij} - \mu_i)^2 / N = 0.4282292$$

and the population variance is

$$\sigma_o^2 = \sigma_B^2 + \sigma_w^2 = 0.3030729 + 0.4282292 = 0.7313021$$

Again, no bias is present. Note that in this case the clusters *interlock* but *no plot appears in more than one cluster*. Assembling the plots into clusters of the same size but a different configuration affected only the magnitudes of the within and between cluster variances.

Now permit the clusters to *overlap*, again using clusters of size $A = 2$, with an interval of two plots between the chosen plots.

| Cluster$_i$ | Plot | $y_{ij}$ (m³) | $z_i$ = Cluster Total (m³) | $w_i$ | $w_iz_i = T_i$ (m³) |
|---|---|---|---|---|---|
| 1 | 1 | 2.55 | 4.10 | 9 | 36.90 |
| | 4 | 1.55 | | | |
| 2 | 2 | 3.15 | 5.70 | 9 | 51.30 |
| | 5 | 2.55 | | | |
| 3 | 3 | 3.05 | 5.75 | 9 | 51.75 |
| | 6 | 2.70 | | | |
| 4 | 4 | 1.55 | 4.05 | 9 | 36.45 |
| | 7 | 2.50 | | | |
| 5 | 5 | 2.55 | 4.80 | 9 | 43.20 |
| | 8 | 2.25 | | | |
| 6 | 6 | 2.70 | 5.55 | 9 | 49.95 |
| | 9 | 2.85 | | | |
| 7 | 7 | 2.50 | 5.65 | 9 | 50.85 |
| | 10 | 3.15 | | | |
| 8 | 8 | 2.25 | 4.00 | 9 | 36.00 |
| | 11 | 1.75 | | | |
| 9 | 9 | 2.85 | 2.85 | 9 | 25.65 |
| | 12 | 0.00 | | | |

$$\sum_i^M w_i = 81$$

$$\sum_i^M w_i z_i = 382.05$$

$$\bar{\tau} = 382.05 / 81 = 4.7166667 \text{ m}^3 \text{ / cluster}$$

$$\tau_Y = 9(4.7166667) = 42.45 \text{ m}^3 \text{ on tract}$$

$$\mu_Y = 42.45 / 12 = 3.5375 \text{ m}^3 \text{ / plot}$$

Obviously, these results are wrong. The overall total, 42.45 m³, is 51.34% greater than the actual total. What caused this error? All the variates have been included in the computations and the pattern of computations is the same as that used in the preceding example. The problem arises because some of the variates were included in more than one cluster. This influenced the results in two ways. First, because of the repeated use of some variates, the number of clusters increased from six to nine. The total, $\tau_Y$, is the total of the cluster totals. In this case there were three extra clusters, which forced the magnitude of $\tau_Y$ upward. If one modifies the computational patterns to obtain the mean volume per cluster and from that the mean volume per hectare one can compare the result with the actual per hectare volume.

$$42.45/9 = 4.7166667 \text{ m}^3/\text{cluster}$$

$$4.7166667\ (10) = 47.16667 \text{ m}^3/\text{ha}$$

where the expansion factor is the reciprocal of the total plot area in the cluster

$$F = 1/(0.05 \text{ ha} + 0.05 \text{ ha}) = 10$$

When the computed volume per hectare is compared with the true volume per hectare it is found to be 0.8913% high. Thus, even with this approach, some bias still exists. To overcome the problem one must take into account the probabilities of occurrence of the *variates*, not the clusters.

The rates and probabilities of occurrence of the 12 variates, when the clusters were permitted to overlap in the manner shown, are

| $y_{ij}$ | 2.55 | 3.15 | 3.05 | 1.55 | 2.55 | 2.70 | 2.50 | 2.25 | 2.85 | 3.15 | 1.75 | 0.00 |
|---|---|---|---|---|---|---|---|---|---|---|---|---|
| Occurred | 1 | 1 | 1 | 2 | 2 | 2 | 2 | 2 | 2 | 1 | 1 | 1 |
| $P(y_{ij})$ | 1/18 | 1/18 | 1/18 | 1/9 | 1/9 | 1/9 | 1/9 | 1/9 | 1/9 | 1/18 | 1/18 | 1/18 |

The sum of the occurrences is equal to 18. Consequently, the probability of occurrence of a given variate is $^1/_{18}$ when the variate occurs once and $^2/_{18}$ or $^2/_9$ when the variate occurs twice. The proper computation process making use of these probabilities is as follows.

| Cluster | Plot | $y_{ij}$ (m³) | $w_{ij}$ | $T_{ij}$ (m³) | $w_{ij}(y_{ij} - \mu_Y)^2$ |
|---|---|---|---|---|---|
| 1 | 1 | 2.55 | 18 | 45.90 | 0.8128125 |
| | 4 | 1.55 | 9 | 13.95 | 5.5814063 |
| 2 | 2 | 3.15 | 18 | 56.70 | 11.8828130 |
| | 5 | 2.55 | 9 | 22.95 | 0.4064063 |
| 3 | 3 | 3.05 | 18 | 54.90 | 9.1378125 |
| | 6 | 2.70 | 9 | 24.30 | 1.1826563 |
| 4 | 4 | 1.55 | 9 | 13.95 | 5.5814063 |
| | 7 | 2.50 | 9 | 22.50 | 0.2376563 |
| 5 | 5 | 2.55 | 9 | 22.95 | 0.4064063 |
| | 8 | 2.25 | 9 | 20.25 | 0.0689063 |
| 6 | 6 | 2.70 | 9 | 24.30 | 1.1826563 |
| | 9 | 2.85 | 9 | 25.65 | 2.3639063 |
| 7 | 7 | 2.50 | 9 | 22.50 | 0.2376563 |
| | 10 | 3.15 | 18 | 56.70 | 11.8828130 |
| 8 | 8 | 2.25 | 9 | 20.25 | 0.0689063 |
| | 11 | 1.75 | 18 | 31.50 | 6.2128125 |

| Cluster | Plot | $y_{ij}$ ($m^3$) | $w_{ij}$ | $T_{ij}$ ($m^3$) | $w_{ij}(y_{ij} - \mu_Y)^2$ |
|---|---|---|---|---|---|
| 9 | 9 | 2.85 | 9 | 25.65 | 2.3639063 |
| | 12 | 0.00 | 18 | 0.00 | 98.3503130 |

$$\sum_i^M \sum_j^A w_{ij} = 216$$

$$\sum_i^M \sum_j^A T_{ij} = 504.90$$

$$\sum_i^M \sum_j^A w_{ij}(y_{ij} - \mu_Y)^2 = 157.96125$$

$$\mu_Y = 504.90/216 = 2.3375 \text{m}^3/\text{plot}$$

$$\tau_Y = 2.3375\ (12) = 28.05\ \text{m}^3 \text{ on tract}$$

$$\sigma_Y^2 = 157.96125/216 = 0.7313021$$

Using this procedure, which properly weights the variates, yields unbiased results.

As can be seen from this example, the variates that occurred only once had twice the weight of those occurring twice. This fact can be used to simplify, to some extent, the computation process by using weights of 2 and 1 instead of 18 and 9. Thus,

| Cluster | Plot | $y_{ij}$ ($m^3$) | $w_{ij}$ | $w_{ij}y_{ij}$ | $w_{ij}(y_{ij} - \mu_o)^2$ |
|---|---|---|---|---|---|
| 1 | 1 | 2.55 | 2 | 5.10 | 0.0903125 |
| | 4 | 1.55 | 1 | 1.55 | 0.6201563 |
| 2 | 2 | 3.15 | 2 | 6.30 | 1.3203125 |
| | 5 | 2.55 | 1 | 2.55 | 0.0451563 |
| 3 | 3 | 3.05 | 2 | 6.10 | 1.0153125 |
| | 6 | 2.70 | 1 | 2.70 | 0.1314063 |
| 4 | 4 | 1.55 | 1 | 1.55 | 0.6201563 |
| | 7 | 2.50 | 1 | 2.50 | 0.0264063 |
| 5 | 5 | 2.55 | 1 | 2.55 | 0.0451563 |
| | 8 | 2.25 | 1 | 2.25 | 0.0076563 |
| 6 | 6 | 2.70 | 1 | 2.70 | 0.1314063 |
| | 9 | 2.85 | 1 | 2.85 | 0.2626563 |
| 7 | 7 | 2.50 | 1 | 2.50 | 0.0264063 |
| | 10 | 3.15 | 2 | 6.30 | 1.3203125 |
| 8 | 8 | 2.25 | 1 | 2.25 | 0.0076563 |
| | 11 | 1.75 | 2 | 3.50 | 0.6903125 |
| 9 | 9 | 2.85 | 1 | 2.85 | 0.2626563 |
| | 12 | 0.00 | 2 | 0.00 | 10.9278130 |

$$\sum_i^M \sum_j^A w_{ij} = 24$$

$$\sum_{i}^{M}\sum_{j}^{A} w_{ij}y_{ij} = 56.10$$

$$\sum_{i}^{M}\sum_{j}^{A} w_{ij}(y_{ij} - \mu_o)^2 = 17.551251$$

$$\mu_o = 56.10/24 = 2.3375 \text{ m}^3\text{/plot}$$

$$\sigma_o^2 = 17.551251/24 = 0.7313021$$

This latter procedure brings out the fact that, from a *computational* point of view, the clusters vary in size, even though, from an *operational* point of view, they have a constant size. This means that Equation 17.76,

$$\sigma_o^2 = \sigma_B^2 + \sigma_w^2$$

and its related expressions are not applicable when the clusters overlap. Generalizing, when clusters overlap,

$N$ becomes $\Sigma_i^M \Sigma_j^{N_i} w_{ij}$,

$N_i$ becomes $\Sigma_j^{N_i} w_{ij}$, and

$n_i$ becomes $\Sigma_j^{n_i} w_{ij}$.

Then,

$$\mu_o = \sum_{i}^{M}\sum_{j}^{N_i} w_{ij}y_{ij} \Big/ \sum_{i}^{M}\sum_{j}^{N_i} w_{ij} \tag{17.84}$$

$$\bar{y}_o = \sum_{i}^{m}\sum_{j}^{n_i} w_{ij}y_{ij} \Big/ \sum_{i}^{m}\sum_{j}^{n_i} w_{ij} \tag{17.85}$$

$$\mu_i = \sum_{j}^{N_i} w_{ij}y_{ij} \Big/ \sum_{j}^{N_i} w_{ij} \tag{17.86}$$

$$\bar{y}_i = \sum_{j}^{n_i} w_{ij}y_{ij} \Big/ \sum_{j}^{n_i} w_{ij} \tag{17.87}$$

$$\sigma_o^2 = \sum_{i}^{M}\sum_{j}^{N_i} w_{ij}(y_{ij} - \mu_o)^2 \Big/ \sum_{i}^{M}\sum_{j}^{N_i} w_{ij} \tag{17.88}$$

$$s_o^2 = \sum_i^m \sum_j^{n_i} w_{ij}(y_{ij} - \bar{y}_o)^2 \Bigg/ \left( \sum_i^m \sum_j^{n_i} w_{ij} - 1 \right) \tag{17.89}$$

$$\sigma_i^2 = \sum_j^{N_i} w_{ij}(y_{ij} - \mu_i)^2 \Bigg/ \sum_j^{N_i} w_{ij} \tag{17.90}$$

$$s_i^2 = \sum_j^{n_i} w_{ij}(y_{ij} - \bar{y}_i)^2 \Bigg/ \left( \sum_j^{n_i} w_{ij} - 1 \right) \tag{17.91}$$

$$V(T_o)_{2s} = \frac{M}{m} \sum_i^M \left[ \frac{\sigma_i^2}{\sum_j^{n_i} w_{ij}} \left( \frac{\sum_j^{N_i} w_{ij} - \sum_j^{n_i} w_{ij}}{\sum_j^{N_i} w_{ij} - 1} \right) \left( \sum_j^{N_i} w_{ij} \right)^2 \right] - M^2 \frac{\sigma_\tau^2}{m} \left( \frac{M - m}{M - 1} \right) \tag{17.92}$$

$$s_{T_o}^2 = \frac{M}{m} \sum_i^m \left[ \frac{s_i^2}{\sum_j^{n_i} w_{ij}} \left( \frac{\sum_j^{N_i} w_{ij} - \sum_j^{n_i} w_{ij}}{\sum_j^{N_i} w_{ij}} \right) \left( \sum_j^{N_i} w_{ij} \right)^2 \right] - M^2 \frac{s_\tau^2}{m} \left( \frac{M - m}{M} \right) \tag{17.93}$$

$$V(\bar{y}_o)_{2s} = V(T_o)_{2s} \Bigg/ \left( \sum_j^M \sum_j^{N_i} w_{ij} \right)^2 \tag{17.94}$$

and

$$s_{\bar{y}_o}^2 = s_{T_o}^2 \Bigg/ \left( \sum_j^M \sum_j^{N_i} w_{ij} \right)^2 \tag{17.95}$$

Cluster overlap can occur when the clusters are compact but, since the problem can be easily avoided, it rarely occurs. If it does occur, unbiased estimates of the population parameters can be obtained by using the preceding approach which makes use of variate probabilities.

As can be seen, when clusters overlap, it is possible to obtain unbiased estimates of the population parameters but the process is cumbersome, especially when the universe of interest is multidimensional, as is the case when it is the set of plots making up a tract of forestland, especially when the tract boundaries are irregular. Because of this, many involved with sampling simply establish a policy which states that sampling units, including clusters, should not overlap. This is a sound policy when clusters are not involved, but when cluster sampling is to be used and the clusters are not compact, it will lead to biased estimates. For example, again use the 12-plot universe and clusters of size $A = 2$, with an interval of two plots between the chosen plots. If interlocking is permitted there would be no problem, but if it is not allowed the results would be as shown below.

| Cluster | Plot | $y_{ij}$ (m³) | $z_i$ = Cluster Total (m³) |
|---|---|---|---|
| 1 | 1 | 2.55 | 4.10 |
| | 4 | 1.55 | |
| 2 | 5 | 2.55 | 4.80 |
| | 8 | 2.25 | |
| 3 | 9 | 2.85 | 2.85 |
| | 12 | 0.00 | |

$$\bar{T}_Y = \sum_i^M z_i = 11.75 \text{ m}^3 \neq 28.05 \text{ m}^3 \text{ on tract} = \tau_Y$$

$$\bar{y} = 11.75/3(2) = 1.9583333 \text{ m}^3 \neq 2.3375 \text{ m}^3/\text{plot} = \mu_Y$$

The bias, in percent, is

$$[(1.9583333 - 2.3375)/2.3375]\ 100 = -16.2210\%$$

Such a result is inevitable since a large proportion of the elements in the universe of interest have no chance at all of being included in a sample. The policy prohibiting overlapping clusters automatically results in an abbreviated sampling frame, which ensures bias.

Another "practical" method of avoiding the labor involved with determining the probabilities of element occurrence is to assume that no differences in probabilities exist (i.e., that the probabilities are constant) and simply take the data as it is. For example,

| Cluster | Plot | $y_{ij}$ (m³) | $z_i$ = Cluster Total (m³) |
|---|---|---|---|
| 1 | 1 | 2.55 | 4.10 |
| | 4 | 1.55 | |
| 2 | 2 | 3.15 | 5.70 |
| | 5 | 2.55 | |
| 3 | 3 | 3.05 | 5.75 |
| | 6 | 2.70 | |
| 4 | 4 | 1.55 | 4.05 |
| | 7 | 2.50 | |
| 5 | 5 | 2.55 | 4.80 |
| | 8 | 2.25 | |
| 6 | 6 | 2.70 | 5.55 |
| | 9 | 2.85 | |
| 7 | 7 | 2.50 | 5.65 |
| | 10 | 3.15 | |
| 8 | 8 | 2.25 | 4.00 |
| | 11 | 1.75 | |
| 9 | 9 | 2.85 | 2.85 |
| | 12 | 0.00 | |
| | | | $\sum_i^M z_i = 42.45$ |

$$\bar{y} = 42.45/9(2) = 2.3583333 \text{ m}^3 \neq 2.3375 \text{ m}^3/\text{plot} = \mu_Y$$

The bias is

$$[(2.3583333 - 2.3375)/2.3375]\ 100 = +0.8913\%$$

As can be seen, if the differences in element probability are ignored, the resulting estimates will be biased but much less so than when overlapping is prohibited. When a constant probability is assumed, the amount of bias decreases as the size of the universe increases since an increasing proportion of the elements will have the same probability of being chosen. For example, in the case of the 12-plot universe, when cluster size $A = 2$, with an interval of two plots between chosen plots, six, or 50%, of the plots will have a probability of 1/18 and six, or 50%, will have a probability of 1/9. If the same cluster configuration is used with a universe consisting of 1200 plots, six of the plots will occur once while the remaining 1194 will occur twice. The total number of occurrences will then be 6 + 2(1194) = 2394. Thus, $P$ (occurring once) = 1/2394 and $P$ (occurring twice) = 1/1197. Consequently, all but six of the plots, or 99.5%, will have a probability of 1/1197. The fact that six plots have a probability of 1/2394 will cause a bias, but it will be negligible.

When overlapping and interlocking are prohibited, the degree of bias is a function of cluster configuration instead of universe size. In the example shown, two out of every four elements, or 50%, were excluded from possible sampling. If the universe included 1200 elements and the same cluster configuration is used, 50% would still be blocked out. For these reasons it is preferable to permit overlapping and the ignoring of differences in element probabilities rather than to use the policy prohibiting overlapping and interlocking of clusters.

## 17.8 EXAMPLES

All the following examples make use of the 16-ha tract shown in Figure 16.1. As shown in that figure the tract has been divided into $N = 320$ $^1/_{20}$-ha square plots. These plots form the elements of the universe of interest. The universe has been divided into $M = 20$ clusters, or primary sampling units, each containing $N_i = A = 16$ plots or elements. This division is shown in Figure 17.2. The clusters are numbered as shown in Figure 17.3a, which serves as the primary sampling frame. The 16 plots within each cluster are numbered as shown in Figure 17.3b, which is the secondary sampling frame.

### 17.8.1 Two-Stage Sampling When $1 < m < M$, $1 < n_i < N_i$, and $N_i$ is Constant

Using random numbers and sampling without replacement , three of the clusters were selected for sampling. Thus, $M = 20$ and $m = 3$. The clusters chosen were 10, 3, and 18. In the following discussion, let

Cluster 10 be designated $i = 1$,
Cluster 3 be designated $i = 2$,
Cluster 18 be designated $i = 3$.

Then, using random numbers and the sampling frame shown in Figure 17.3b, five plots were chosen from each of the three clusters, using sampling without replacement. Thus, $n_1 = n_2 = n_3 = a = 5$. The plots chosen in each of the clusters are shown in Figure 17.4 and listed below:

| | | | | | | | | | | | | | | | |
|---|---|---|---|---|---|---|---|---|---|---|---|---|---|---|---|
| 2.55 | 3.00 | 1.60 | 1.80 | 2.65 | 2.10 | 1.90 | 3.40 | 3.65 | 4.00 | 4.00 | 1.75 | 2.60 | 2.85 | 2.00 | 2.10 |
| 3.15 | 2.95 | 2.40 | 2.25 | 3.20 | 2.80 | 2.40 | 4.55 | 4.50 | 3.25 | 4.00 | 1.50 | 1.80 | 2.70 | 1.70 | 2.50 |
| 3.05 | 2.10 | 2.35 | 2.45 | 2.90 | 1.10 | 4.30 | 4.00 | 4.10 | 2.85 | 3.40 | 1.85 | 2.00 | 2.30 | 2.20 | 2.50 |
| 1.55 | 1.55 | 2.55 | 2.15 | 3.10 | 2.75 | 4.50 | 4.60 | 3.30 | 4.30 | 3.75 | 2.30 | 1.80 | 2.00 | 2.10 | 1.80 |
| 2.55 | 0.90 | 2.25 | 2.35 | 2.45 | 2.90 | 4.65 | 1.90 | 3.35 | 4.60 | 2.60 | 2.35 | 2.60 | 2.75 | 2.70 | 3.25 |
| 2.70 | 2.50 | 1.90 | 2.30 | 2.95 | 3.15 | 3.00 | 4.55 | 2.65 | 4.40 | 4.20 | 2.20 | 1.90 | 1.70 | 2.30 | 1.70 |
| 2.50 | 1.50 | 2.00 | 2.55 | 3.15 | 3.30 | 4.05 | 4.50 | 4.35 | 3.80 | 4.50 | 2.00 | 3.65 | 1.35 | 3.15 | 1.60 |
| 2.35 | 2.90 | 2.05 | 2.10 | 3.05 | 3.35 | 4.15 | 4.60 | 4.00 | 3.45 | 1.50 | 1.75 | 2.20 | 2.85 | 3.45 | 2.40 |
| 2.85 | 1.80 | 1.50 | 2.05 | 2.25 | 4.80 | 4.40 | 3.15 | 3.00 | 3.65 | 2.30 | 1.35 | 1.90 | 2.40 | 2.20 | 0.90 |
| 3.15 | 1.60 | 2.40 | 2.60 | 3.00 | 1.90 | 2.55 | 3.55 | 4.00 | 2.75 | 2.65 | 1.90 | 2.60 | 2.70 | 1.50 | 2.65 |
| 1.75 | 0.25 | 1.05 | 0.95 | 3.00 | 2.90 | 3.75 | 3.15 | 3.35 | 2.30 | 3.25 | 2.30 | 2.30 | 1.90 | 1.60 | 0.00 |
| 0.00 | 0.00 | 0.00 | 0.45 | 1.20 | 1.90 | 2.90 | 4.10 | 4.50 | 3.00 | 2.65 | 2.50 | 1.00 | 1.40 | 0.00 | 0.05 |
| 0.60 | 0.00 | 0.50 | 0.55 | 2.05 | 3.10 | 4.65 | 2.30 | 3.00 | 2.20 | 1.90 | 1.30 | 0.55 | 0.75 | 0.15 | 0.20 |
| 0.30 | 0.00 | 0.30 | 0.40 | 1.50 | 1.15 | 2.35 | 1.00 | 2.55 | 2.45 | 2.25 | 0.85 | 0.40 | 0.15 | 1.05 | 0.10 |
| 0.00 | 0.05 | 0.00 | 0.40 | 1.40 | 2.20 | 2.40 | 2.00 | 1.40 | 1.50 | 1.25 | 0.00 | 0.10 | 1.35 | 1.45 | 0.60 |
| 0.10 | 0.00 | 0.00 | 1.00 | 1.95 | 2.40 | 3.60 | 1.60 | 1.90 | 2.40 | 1.00 | 1.15 | 0.40 | 0.20 | 0.05 | 0.00 |
| 0.55 | 0.15 | 0.10 | 0.85 | 2.05 | 1.60 | 1.90 | 2.75 | 1.75 | 2.00 | 0.55 | 0.00 | 0.30 | 0.00 | 1.60 | 0.30 |
| 0.60 | 0.50 | 0.00 | 0.65 | 1.80 | 1.65 | 2.35 | 2.45 | 1.90 | 1.15 | 0.00 | 1.50 | 0.85 | 0.05 | 0.50 | 0.00 |
| 0.60 | 1.00 | 0.15 | 0.00 | 1.70 | 2.50 | 1.25 | 2.40 | 0.80 | 0.50 | 0.10 | 0.75 | 0.30 | 0.35 | 0.20 | 0.10 |
| 0.55 | 0.20 | 0.50 | 0.30 | 1.50 | 2.05 | 1.80 | 1.40 | 1.05 | 0.10 | 0.00 | 0.45 | 0.50 | 0.40 | 0.75 | 0.40 |

**FIGURE 17.2** The 16-ha tract divided into 20 clusters.

| | *i* | | |
|---|---|---|---|
| *j* | 1 | 2 | 3 |
| 1 | 09 | 09 | 10 |
| 2 | 02 | 05 | 13 |
| 3 | 15 | 10 | 03 |
| 4 | 03 | 04 | 02 |
| 5 | 12 | 16 | 08 |

Each plot was located in the field and cruised. Because of the clustering, the field work was concentrated in three small areas and travel time (unproductive time) between plots was greatly reduced from that which would have been needed if the sampling were unrestricted. The data from the cruise and the computations are as follows.

$$
\begin{array}{lll}
y_{11} = 3.00 & y_{21} = 4.10 & y_{31} = 2.50 \\
y_{12} = 4.80 & y_{22} = 4.50 & y_{32} = 1.50 \\
y_{13} = 2.90 & y_{23} = 2.85 & y_{33} = 1.90 \\
y_{14} = 4.40 & y_{24} = 1.75 & y_{34} = 1.60
\end{array}
$$

| | | | |
|---|---|---|---|
| | $y_{15} =$ <u>3.15</u> | $y_{25} =$ <u>2.30</u> | $y_{35} =$ <u>2.45</u> |
| $\sum_{j}^{5} y_{ij} =$ | 18.25 | 15.50 | 9.95 |
| $\sum_{j}^{5} y_{ij}^2 =$ | 69.7325 | 53.535 | 20.6725 |
| $\bar{y}_i =$ | 3.65 | 3.10 | 1.99 $m^3$/plot |
| $T_i = N_i \bar{y}_i =$ | 58.4 | 49.6 | 31.84 |

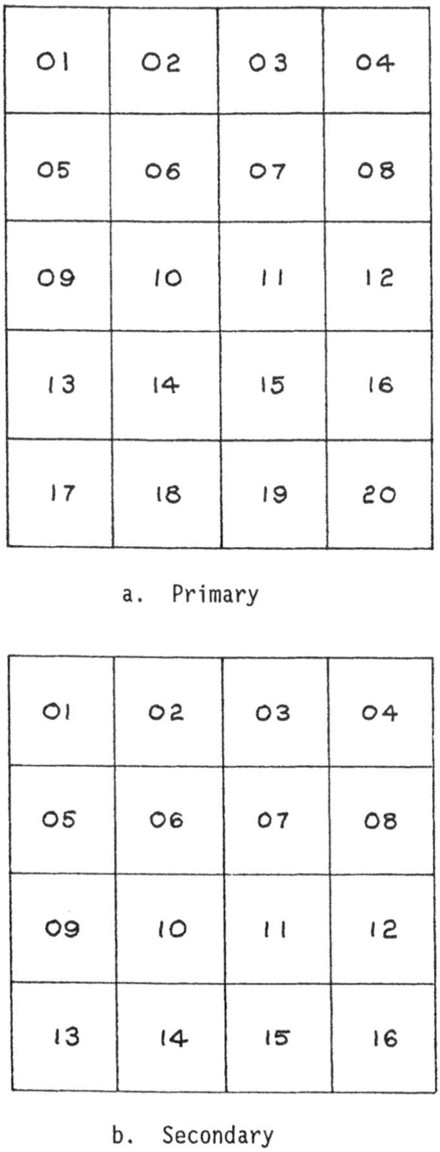

| | | | |
|---|---|---|---|
| 01 | 02 | 03 | 04 |
| 05 | 06 | 07 | 08 |
| 09 | 10 | 11 | 12 |
| 13 | 14 | 15 | 16 |
| 17 | 18 | 19 | 20 |

a. Primary

| | | | |
|---|---|---|---|
| 01 | 02 | 03 | 04 |
| 05 | 06 | 07 | 08 |
| 09 | 10 | 11 | 12 |
| 13 | 14 | 15 | 16 |

b. Secondary

**FIGURE 17.3** The primary (a) and secondary (b) sampling frames.

| | | | | | | | | | | | | | | | |
|---|---|---|---|---|---|---|---|---|---|---|---|---|---|---|---|
| 2.55 | 3.00 | 1.60 | 1.80 | 2.65 | 2.10 | 1.90 | 3.40 | 3.65 | 4.00 | 4.00 | 1.75 | 2.60 | 2.85 | 2.00 | 2.10 |
| 3.15 | 2.95 | 2.40 | 2.25 | 3.20 | 2.80 | 2.40 | 4.55 | 4.50 | 3.25 | 4.00 | 1.50 | 1.80 | 2.70 | 1.70 | 2.50 |
| 3.05 | 2.10 | 2.35 | 2.45 | 2.90 | 1.10 | 4.30 | 4.00 | 4.10 | 2.85 | 3.40 | 1.85 | 2.00 | 2.30 | 2.20 | 2.50 |
| 1.55 | 1.55 | 2.55 | 2.15 | 3.10 | 2.75 | 4.50 | 4.60 | 3.30 | 4.30 | 3.75 | 2.30 | 1.80 | 2.00 | 2.10 | 1.80 |
| 2.55 | 0.90 | 2.25 | 2.35 | 2.45 | 2.90 | 4.65 | 1.90 | 3.35 | 4.60 | 2.60 | 2.35 | 2.60 | 2.75 | 2.70 | 3.25 |
| 2.70 | 2.50 | 1.90 | 2.30 | 2.95 | 3.15 | 3.00 | 4.55 | 2.65 | 4.40 | 4.20 | 2.20 | 1.90 | 1.70 | 2.30 | 1.70 |
| 2.50 | 1.50 | 2.00 | 2.55 | 3.15 | 3.30 | 4.05 | 4.50 | 4.35 | 3.80 | 4.50 | 2.00 | 3.65 | 1.35 | 3.15 | 1.60 |
| 2.35 | 2.90 | 2.05 | 2.10 | 3.05 | 3.35 | 4.15 | 4.60 | 4.00 | 3.45 | 1.50 | 1.75 | 2.20 | 2.85 | 3.45 | 2.40 |
| 2.85 | 1.80 | 1.50 | 2.05 | 2.25 | 4.80 | 4.40 | 3.15 | 3.00 | 3.65 | 2.30 | 1.35 | 1.90 | 2.40 | 2.20 | 0.90 |
| 3.15 | 1.60 | 2.40 | 2.60 | 3.00 | 1.90 | 2.55 | 3.55 | 4.00 | 2.75 | 2.65 | 1.90 | 2.60 | 2.70 | 1.50 | 2.65 |
| 1.75 | 0.25 | 1.05 | 0.95 | 3.00 | 2.90 | 3.75 | 3.15 | 3.35 | 2.30 | 3.25 | 2.30 | 2.30 | 1.90 | 1.60 | 0.00 |
| 0.00 | 0.00 | 0.00 | 0.45 | 1.20 | 1.90 | 2.90 | 4.10 | 4.50 | 3.00 | 2.65 | 2.50 | 1.00 | 1.40 | 0.00 | 0.05 |
| 0.60 | 0.00 | 0.50 | 0.55 | 2.05 | 3.10 | 4.65 | 2.30 | 3.00 | 2.20 | 1.90 | 1.30 | 0.55 | 0.75 | 0.15 | 0.20 |
| 0.30 | 0.00 | 0.30 | 0.40 | 1.50 | 1.15 | 2.35 | 1.00 | 2.55 | 2.45 | 2.25 | 0.85 | 0.40 | 0.15 | 1.05 | 0.10 |
| 0.00 | 0.05 | 0.00 | 0.40 | 1.40 | 2.20 | 2.40 | 2.00 | 1.40 | 1.50 | 1.25 | 0.00 | 0.10 | 1.35 | 1.45 | 0.60 |
| 0.10 | 0.00 | 0.00 | 1.00 | 1.95 | 2.40 | 3.60 | 1.60 | 1.90 | 2.40 | 1.00 | 1.15 | 0.40 | 0.20 | 0.05 | 0.00 |
| 0.55 | 0.15 | 0.10 | 0.85 | 2.05 | 1.60 | 1.90 | 2.75 | 1.75 | 2.00 | 0.55 | 0.00 | 0.30 | 0.00 | 1.60 | 0.30 |
| 0.60 | 0.50 | 0.00 | 0.65 | 1.80 | 1.65 | 2.35 | 2.45 | 1.90 | 1.15 | 0.00 | 1.50 | 0.85 | 0.05 | 0.50 | 0.00 |
| 0.60 | 1.00 | 0.15 | 0.00 | 1.70 | 2.50 | 1.25 | 2.40 | 0.80 | 0.50 | 0.10 | 0.75 | 0.30 | 0.35 | 0.20 | 0.10 |
| 0.55 | 0.20 | 0.50 | 0.30 | 1.50 | 2.05 | 1.80 | 1.40 | 1.05 | 0.10 | 0.00 | 0.45 | 0.50 | 0.40 | 0.75 | 0.40 |

**FIGURE 17.4** The sample chosen for the two-stage sampling example ($1 < m < M$ and $1 < n_i = a < N_i = A$).

Using Equation 17.28,

$$\bar{y}_o = \frac{M}{Nm}\sum_i^m N_i \bar{y}_i = \frac{20}{320(3)}(58.4 + 49.6 + 31.84)$$

$$= 2.913333 \text{ m}^3/\text{plot}$$

or, since each plot is $^1/_{20}$ ha,

$$\bar{y}_o = 20(2.913333) = 58.266667 \text{ or } 58.27 \text{ m}^3/\text{ha}$$

Since cluster and subsample sizes are constant, one can also use Equation 17.30 to obtain the estimate of the overall mean. Thus,

$$\bar{y}_o = \frac{1}{ma}\sum_i^m \sum_j^a y_{ij} = \frac{1}{3(5)}(18.25 + 15.50 + 9.95)$$

$$= 2.913333 \text{ m}^3/\text{plot}$$

which agrees with that obtained using Equation 17.28. Continuing,

$$\bar{T} = \frac{1}{m}\sum_{i}^{m} T_i = \frac{1}{3}(58.4 + 49.6 + 31.84) = 46.613333 \text{ m}^3\text{/cluster}$$

$$T_o = N\bar{y}_o = 320(2.913333) = 932.26667 \text{ or } 932 \text{ m}^3 \text{ on tract}$$

or

$$T_o = M\bar{T} = 20(46.613333) = 932.26667 \text{ or } 932 \text{ m}^3 \text{ on tract}$$

$$s_1^2 = \frac{1}{(5-1)}[69.7325 - (18.25)^2/5] = 3.12/4 = 0.78$$

$$s_2^2 = \frac{1}{(5-1)}[53.535 - (15.5)^2/5] = 5.485/4 = 1.37125$$

$$s_3^2 = \frac{1}{(5-1)}[20.6725 - (9.95)^2/5] = 0.872/4 = 0.218$$

$$s_T^2 = \frac{1}{(m-1)}\sum_{i}^{m}(T_i - \bar{T})^2$$

$$= \frac{1}{(3-1)}[(58.4 - 46.613333)^3 + (49.6 - 46.613333)^2 = (31.84 - 46.613333)^2]$$

$$= \frac{1}{2}(366.09707) = 183.04853$$

Using Equation 17.52,

$$s_{\bar{y}_o}^2 = \frac{1}{N^2}\left\{\frac{M}{m}\sum_{i}^{m}\left[\frac{s_i^2}{n_i}\left(\frac{N_i - n_i}{N_i}\right)N_i^2\right] + M^2\frac{s_T^2}{m}\left(\frac{M-m}{M}\right)\right\}$$

For cluster $i = 1$, $\frac{s_i^2}{n_i}\left(\frac{N_i - n_i}{N_i}\right)N_i^2 = \frac{0.78}{5}\left(\frac{16-5}{16}\right)(16)^2 = 27.456$

For cluster $i = 2$, $= \frac{1.37125}{5}\left(\frac{16-5}{16}\right)(16)^2 = 48.268$

For cluster $i = 3$, $= \frac{0.218}{5}\left(\frac{16-5}{16}\right)(16)^2 = 7.6736$

$$\sum_{i}^{m}\left[\frac{s_i^2}{n_i}\left(\frac{N_i - n_i}{N_i}\right)N_i^2\right] = 83.3976$$

Then

$$s_{\bar{y}_o}^2 = \frac{1}{(320)^2}\left[\frac{20}{3}(83.3976) + (20)^2\left(\frac{183.0528}{3}\right)\left(\frac{20-3}{20}\right)\right]$$

$$= 0.208022$$

$$s_{\bar{y}_o} = \pm\sqrt{0.208022} = \pm 0.456094 \text{ cm}^3\text{/plot}$$

or, since each plot is 1/20 ha,

$$s_{\bar{y}_o} = 20(\pm 0.456094) = \pm 9.12 \text{ m}^3\text{/ha}$$

and

$$s_{T_o} = Ns_{\bar{y}_o} = 320(\pm 0.45094) = \pm 145.9503 \text{ or } 2919.0 \text{ m}^3 \text{ on tract}$$

Since the cluster and subsample sizes are constant, it is possible, as in the case of the overall mean, to use a simplified expression to compute the variance of the overall mean. Thus, using Equations 17.62,

$$s_{\bar{y}_o}^2 = \frac{s_w^2}{Ma}\left(\frac{A-a}{A}\right) + \frac{s_B^2}{m}\left(\frac{M-m}{M}\right)$$

Since

$$s_w^2 = \frac{1}{3}(0.78 + 1.37125 + 0.218) = 0.78975$$

and

$$s_B^2 = \frac{1}{3-1}[(3.65 - 2.913333)^2 + (3.1 - 2.913333)^2 + (1.99 - 2.913333)^2]$$

$$= 0.715033$$

$$s_{\bar{y}_o}^2 = \frac{0.78975}{20(5)}\left(\frac{16-5}{16}\right) + \frac{0.715033}{3}\left(\frac{20-3}{3}\right) = 0.208022$$

which agrees with the value obtained using Equations 17.52.

A valid sampling error cannot be computed because the cluster variances are not homogeneous (see the data in Section 17.8.2). As in stratified sampling, homogeneity of variance and a constant sampling intensity are required before a valid sampling error can be computed. In this example, the sampling intensity was constant, which may or may not be true in other situations, but the cluster variances are heterogeneous. The conditions required for computing valid sampling errors rarely occur in cluster sampling and, consequently, the usual measure of the precision of an estimate is the standard error of either the mean or the total. If the degrees of freedom, $k = \Sigma_i^m n_i - m$, are sufficiently numerous, which is usually assumed to be 30 or more, an approximate $t$ may be used to obtain an approximate sampling error. In this case $k = 15 - 3 = 12$, which is less than the minimum required. Consequently, it would be inappropriate to compute even an approximate sampling error.

| 2.55 | 3.00 | 1.60 | 1.80 | 2.65 | 2.10 | 1.90 | 3.40 | 3.65 | 4.00 | 4.00 | 1.75 | 2.60 | 2.85 | 2.00 | 2.10 |
|---|---|---|---|---|---|---|---|---|---|---|---|---|---|---|---|
| 3.15 | 2.95 | 2.40 | 2.25 | 3.20 | 2.80 | 2.40 | 4.55 | 4.50 | 3.25 | 4.00 | 1.50 | 1.80 | 2.70 | 1.70 | 2.50 |
| 3.05 | 2.10 | 2.35 | 2.45 | 2.90 | 1.10 | 4.30 | 4.00 | 4.10 | 2.85 | 3.40 | 1.85 | 2.00 | 2.30 | 2.20 | 2.50 |
| 1.55 | 1.55 | 2.55 | 2.15 | 3.10 | 2.75 | 4.50 | 4.60 | 3.30 | 4.30 | 3.75 | 2.30 | 1.80 | 2.00 | 2.10 | 1.80 |
| 2.55 | 0.90 | 2.25 | 2.35 | 2.45 | 2.90 | 4.65 | 1.90 | 3.35 | 4.60 | 2.60 | 2.35 | 2.60 | 2.75 | 2.70 | 3.25 |
| 2.70 | 2.50 | 1.90 | 2.30 | 2.95 | 3.15 | 3.00 | 4.55 | 2.65 | 4.40 | 4.20 | 2.20 | 1.90 | 1.70 | 2.30 | 1.70 |
| 2.50 | 1.50 | 2.00 | 2.55 | 3.15 | 3.30 | 4.05 | 4.50 | 4.35 | 3.80 | 4.50 | 2.00 | 3.65 | 1.35 | 3.15 | 1.60 |
| 2.35 | 2.90 | 2.05 | 2.10 | 3.05 | 3.35 | 4.15 | 4.60 | 4.00 | 3.45 | 1.50 | 1.75 | 2.20 | 2.85 | 3.45 | 2.40 |
| 2.85 | 1.80 | 1.50 | 2.05 | 2.25 | 4.80 | 4.40 | 3.15 | 3.00 | 3.65 | 2.30 | 1.35 | 1.90 | 2.40 | 2.20 | 0.90 |
| 3.15 | 1.60 | 2.40 | 2.60 | 3.00 | 1.90 | 2.55 | 3.55 | 4.00 | 2.75 | 2.65 | 1.90 | 2.60 | 2.70 | 1.50 | 2.65 |
| 1.75 | 0.25 | 1.05 | 0.95 | 3.00 | 2.90 | 3.75 | 3.15 | 3.35 | 2.30 | 3.25 | 2.30 | 2.30 | 1.90 | 1.60 | 0.00 |
| 0.00 | 0.00 | 0.00 | 0.45 | 1.20 | 1.90 | 2.90 | 4.10 | 4.50 | 3.00 | 2.65 | 2.50 | 1.00 | 1.40 | 0.00 | 0.05 |
| 0.60 | 0.00 | 0.50 | 0.55 | 2.05 | 3.10 | 4.65 | 2.30 | 3.00 | 2.20 | 1.90 | 1.30 | 0.55 | 0.75 | 0.15 | 0.20 |
| 0.30 | 0.00 | 0.30 | 0.40 | 1.50 | 1.15 | 2.35 | 1.00 | 2.55 | 2.45 | 2.25 | 0.85 | 0.40 | 0.15 | 1.05 | 0.10 |
| 0.00 | 0.05 | 0.00 | 0.40 | 1.40 | 2.20 | 2.40 | 2.00 | 1.40 | 1.50 | 1.25 | 0.00 | 0.10 | 1.35 | 1.45 | 0.60 |
| 0.10 | 0.00 | 0.00 | 1.00 | 1.95 | 2.40 | 3.60 | 1.60 | 1.90 | 2.40 | 1.00 | 1.15 | 0.40 | 0.20 | 0.05 | 0.00 |
| 0.55 | 0.15 | 0.10 | 0.85 | 2.05 | 1.60 | 1.90 | 2.75 | 1.75 | 2.00 | 0.55 | 0.00 | 0.30 | 0.00 | 1.60 | 0.30 |
| 0.60 | 0.50 | 0.00 | 0.65 | 1.80 | 1.65 | 2.35 | 2.45 | 1.90 | 1.15 | 0.00 | 1.50 | 0.85 | 0.05 | 0.50 | 0.00 |
| 0.60 | 1.00 | 0.15 | 0.00 | 1.70 | 2.50 | 1.25 | 2.40 | 0.80 | 0.50 | 0.10 | 0.75 | 0.30 | 0.35 | 0.20 | 0.10 |
| 0.55 | 0.20 | 0.50 | 0.30 | 1.50 | 2.05 | 1.80 | 1.40 | 1.05 | 0.10 | 0.00 | 0.45 | 0.50 | 0.40 | 0.75 | 0.40 |

**FIGURE 17.5** The sample chosen for the simple cluster sampling example ($1 < m < M$ and $n_i = a = N_i = A$).

### 17.8.2 Simple Cluster Sampling When $1 < m < M$ (Example I)

Using random sampling without replacement, three of the clusters in the sampling frame were selected for evaluation (the same three as in the previous example), as shown in Figure 17.5. There was no subsampling (i.e., all the plots within each of the clusters were cruised). Thus, $M = 20$; $m = 3$; and $n_i = a = N_i = A = 16$, $i = 1, 2, 3$; with the chosen plots identified as in the previous example. Since all the plots within the clusters were cruised, the cluster values are parameters and not estimates. The data and computations are as follows.

| | $i$ = 1 | 2 | 3 |
|---|---|---|---|
| $\sum_{j}^{16} y_{ij} =$ | 48.5 | 52.5 | 31.15 |
| $\sum_{j}^{16} y_{ij}^2 =$ | 160.765 | 186.095 | 63.5375 |
| $\mu_i =$ | 3.03125 | 3.28125 | 1.946875 m³/plot |

| | *i* | | |
|---|---|---|---|
| | **1** | **2** | **3** |
| $t_i = N_i\mu_i =$ | 48.5 | 52.5 | 31.15 m³ |

$$\bar{y}_o = \frac{1}{N}\left(\frac{M}{m}\right)\sum_i^m N_i\bar{y}_i = \frac{1}{N}\left(\frac{M}{m}\right)\sum_i^m T_i$$

but in this case the $\bar{y}_i$ and $T_i$ values are parameters. However, since only 3 of the 20 clusters were sampled, $\bar{y}_o$ is still an estimate. Thus,

$$\bar{y}_o = \frac{1}{N}\left(\frac{M}{m}\right)\sum_i^m N_i\mu_i = \frac{1}{N}\left(\frac{M}{m}\right)\sum_i^m \tau_i$$

$$= \frac{1}{320}\left(\frac{20}{3}\right)(48.5 + 52.5 + 31.15) = 2.753125 \text{ m}^3\text{/plot}$$

or 20(2.753125) = 55.0625 m³/ha.

$$\bar{T} = \frac{1}{m}\sum_i^m \tau_i = \frac{1}{3}(48.5 + 52.5 + 31.15) = 44.05 \text{ m}^3\text{/cluster}$$

$$T_o = M\bar{T} = 20(44.05) = 881 \text{ m}^3\text{/on tract}$$

or

$$T_o = N\bar{y}_o = 320(2.753125) = 881 \text{ m}^3 \text{ on tract}$$

Since no subsampling was done within the clusters the cluster variances are parameters.

$$\sigma_1^2 = \frac{1}{16}[160.765 - (48.5)^2/16] = 0.859336$$

$$\sigma_2^2 = \frac{1}{16}[186.095 - (52.5)^2/16] = 0.864336$$

$$\sigma_3^2 = \frac{1}{16}[63.5375 - (31.15)^2/16] = 0.180772$$

The variance of the total is, however, an estimate since only three of the clusters were in the sample.

$$s_T^2 = \frac{1}{(m-1)}\sum_i^m (\tau_i - \bar{T})^2$$

$$= \frac{1}{(3-1)}[(48.5 - 44.05)^2 + (52.5 - 44.05)^2 + (31.15 - 44.05)^2]$$

$$= \frac{1}{2}(257.615) = 128.8075$$

According to Equation 17.52,

$$s_{\bar{y}_o}^2 = \frac{1}{N^2}\left\{\frac{M}{m}\sum_i^m\left[\frac{s_i^2}{n_i}\left(\frac{N_i - n_i}{N_i}\right)N_i^2\right] + M^2\frac{s_T^2}{m}\left(\frac{M-m}{M}\right)\right\}$$

However, since the cluster variances are parameters, the first term within the braces becomes

$$\frac{M}{m}\sum_i^M\left[\frac{\sigma_i^2}{n_i}\left(\frac{N_i - n_i}{N_i - 1}\right)N_i^2\right]$$

Then

For cluster $i = 1$, $\left[\frac{\sigma_i^2}{n_i}\left(\frac{N_i - n_i}{N_i - 1}\right)N_i^2\right]$

$$= \frac{0.859336}{16}\left(\frac{16-16}{16-1}\right)16^2$$

$$= 0.859336(0)(16/15)$$

$$= 0$$

For cluster $i = 2$,

$$= \frac{0.864336}{16}\left(\frac{16-16}{16-1}\right)16^2$$

$$= 0.864336(0)(16/15)$$

$$= 0$$

For cluster $i = 3$,

$$= \frac{0.180772}{16}\left(\frac{16-16}{16-1}\right)16^2$$

$$= 0.180772(0)(16/15)$$

$$= 0$$

Consequently, the sum is equal to zero. Then,

$$s_{\bar{y}_o}^2 = \frac{1}{(320)^2}\left[\frac{20}{3}(0) + 20^2\left(\frac{128.8075}{3}\right)\left(\frac{20-3}{20}\right)\right] = 0.142560$$

$$s_{\bar{y}_o} = \pm\sqrt{0.142560} = \pm 0.377571 \text{ m}^3\text{/plot}$$

or

$$20(\pm 0.377571) = \pm 7.55 \text{ m}^3\text{/ha}$$

and

$$s_{T_o} = Ns_{\bar{y}_o} = 320(\pm(0.377571) = \pm 120.82272 \text{ or } \pm 120.8 \text{ m}^3 \text{ / on tract}$$

As can be seen, when all the elements within a cluster are included in the sample, the within cluster component of the variance of the overall mean equals zero and only the variability of the cluster totals affects the variance of the overall mean. In essence, the cluster means or totals, as the case may be, become the variable of interest and the computations appropriate to simple random sampling can be used. In this case, the data and computations are as follows.

$$\tau_1 = 48.5 \text{ m}^3$$

$$\tau_2 = 52.5 \text{ m}^3$$

$$\tau_3 = 31.15 \text{ m}^3$$

$$\bar{T} = (1/3)(48.5 + 52.5 + 31.15) = 44.05 \text{ m}^3 / \text{cluster}$$

$$s_T^2 = \frac{1}{(3-1)}[(48.5 - 44.05)^2 + (52.5 - 44.05)^2 + (31.15 - 44.05)^2] = 128.8075$$

$$s_{\bar{T}}^2 = \frac{128.8075}{3}\left(\frac{20-3}{20}\right) = 36.495458$$

$$s_{\bar{T}} = \pm\sqrt{36.495458} = \pm 6.041147 \text{ or } \pm 6.04 \text{ m}^3 / \text{cluster}$$

or

$$\pm 6.041147/16 = \pm 0.377572 \text{ m}^3/\text{plot}$$

or

$$20(0.377572) = \pm 7.55144 \text{ or } \pm 7.55 \text{ m}^3/\text{ha}$$

which agrees with the previous results.

Since this is a simple random sampling operation, a valid sampling can be computed. At the 90% level of confidence,

$$\text{SE}_{0.90} = \pm t_{0.90}(k = 2)s_{\bar{y}} = \pm 2.920(6.041147)$$

$$= \pm 17.640149 \text{ or } \pm 17.64 \text{ m}^3/\text{cluster}$$

or

$$\pm 17.640149/16 = \pm 1.1025097 \text{ m}^3/\text{plot}$$

or

$$20\ (\pm 1.102509) = \pm 22.05 \text{ m}^3/\text{ha}$$

### 17.8.3 Two-Stage Sampling When $m = M$ and $1 < n_i < N_i$

In this example all 20 of the clusters or primary sampling units were included in the sample and a subsample of 4 plots was shown from each cluster using random sampling without replacement. Consequently, $M = 20$; $m = 20$; $N_i = A = 16$, $i = 1, 2, \ldots 20$; and $n_i = a = 4$, $i = 1, 2, \ldots 20$. The plots chosen in each cluster are listed below and shown in Figure 17.6.

| | | | | | | | | | | | | | | | |
|---|---|---|---|---|---|---|---|---|---|---|---|---|---|---|---|
| 2.55 | 3.00 | 1.60 | 1.80 | 2.65 | 2.10 | 1.90 | 3.40 | 3.65 | 4.00 | 4.00 | 1.75 | 2.60 | 2.85 | 2.00 | 2.10 |
| 3.15 | 2.95 | 2.40 | 2.25 | 3.20 | 2.80 | 2.40 | 4.55 | 4.50 | 3.25 | 4.00 | 1.50 | 1.80 | 2.70 | 1.70 | 2.50 |
| 3.05 | 2.10 | 2.35 | 2.45 | 2.90 | 1.10 | 4.30 | 4.00 | 4.10 | 2.85 | 3.40 | 1.85 | 2.00 | 2.30 | 2.20 | 2.50 |
| 1.55 | 1.55 | 2.55 | 2.15 | 3.10 | 2.75 | 4.50 | 4.60 | 3.30 | 4.30 | 3.75 | 2.30 | 1.80 | 2.00 | 2.10 | 1.80 |
| 2.55 | 0.90 | 2.25 | 2.35 | 2.45 | 2.90 | 4.65 | 1.90 | 3.35 | 4.60 | 2.60 | 2.35 | 2.60 | 2.75 | 2.70 | 3.25 |
| 2.70 | 2.50 | 1.90 | 2.30 | 2.95 | 3.15 | 3.00 | 4.55 | 2.65 | 4.40 | 4.20 | 2.20 | 1.90 | 1.70 | 2.30 | 1.70 |
| 2.50 | 1.50 | 2.00 | 2.55 | 3.15 | 3.30 | 4.05 | 4.50 | 4.35 | 3.80 | 4.50 | 2.00 | 3.65 | 1.35 | 3.15 | 1.60 |
| 2.35 | 2.90 | 2.05 | 2.10 | 3.05 | 3.35 | 4.15 | 4.60 | 4.00 | 3.45 | 1.50 | 1.75 | 2.20 | 2.85 | 3.45 | 2.40 |
| 2.85 | 1.80 | 1.50 | 2.05 | 2.25 | 4.80 | 4.40 | 3.15 | 3.00 | 3.65 | 2.30 | 1.35 | 1.90 | 2.40 | 2.20 | 0.90 |
| 3.15 | 1.60 | 2.40 | 2.60 | 3.00 | 1.90 | 2.55 | 3.55 | 4.00 | 2.75 | 2.65 | 1.90 | 2.60 | 2.70 | 1.50 | 2.65 |
| 1.75 | 0.25 | 1.05 | 0.95 | 3.00 | 2.90 | 3.75 | 3.15 | 3.35 | 2.30 | 3.25 | 2.30 | 2.30 | 1.90 | 1.60 | 0.00 |
| 0.00 | 0.00 | 0.00 | 0.45 | 1.20 | 1.90 | 2.90 | 4.10 | 4.50 | 3.00 | 2.65 | 2.50 | 1.00 | 1.40 | 0.00 | 0.05 |
| 0.60 | 0.00 | 0.50 | 0.55 | 2.05 | 3.10 | 4.65 | 2.30 | 3.00 | 2.20 | 1.90 | 1.30 | 0.55 | 0.75 | 0.15 | 0.20 |
| 0.30 | 0.00 | 0.30 | 0.40 | 1.50 | 1.15 | 2.35 | 1.00 | 2.55 | 2.45 | 2.25 | 0.85 | 0.40 | 0.15 | 1.05 | 0.10 |
| 0.00 | 0.05 | 0.00 | 0.40 | 1.40 | 2.20 | 2.40 | 2.00 | 1.40 | 1.50 | 1.25 | 0.00 | 0.10 | 1.35 | 1.45 | 0.60 |
| 0.10 | 0.00 | 0.00 | 1.00 | 1.95 | 2.40 | 3.60 | 1.60 | 1.90 | 2.40 | 1.00 | 1.15 | 0.40 | 0.20 | 0.05 | 0.00 |
| 0.55 | 0.15 | 0.10 | 0.85 | 2.05 | 1.60 | 1.90 | 2.75 | 1.75 | 2.00 | 0.55 | 0.00 | 0.30 | 0.00 | 1.60 | 0.30 |
| 0.60 | 0.50 | 0.00 | 0.65 | 1.80 | 1.65 | 2.35 | 2.45 | 1.90 | 1.15 | 0.00 | 1.50 | 0.85 | 0.05 | 0.50 | 0.00 |
| 0.60 | 1.00 | 0.15 | 0.00 | 1.70 | 2.50 | 1.25 | 2.40 | 0.80 | 0.50 | 0.10 | 0.75 | 0.30 | 0.35 | 0.20 | 0.10 |
| 0.55 | 0.20 | 0.50 | 0.30 | 1.50 | 2.05 | 1.80 | 1.40 | 1.05 | 0.10 | 0.00 | 0.45 | 0.50 | 0.40 | 0.75 | 0.40 |

**FIGURE 17.6** The sample chosen when subsampling was used in all the clusters ($m = M$ and $1 < n_i = a < N_i = A$).

| | *i* | | | | | | | | | |
|---|---|---|---|---|---|---|---|---|---|---|
| *j* | **1** | **2** | **3** | **4** | **5** | **6** | **7** | **8** | **9** | **10** |
| 1 | 10 | 04 | 15 | 09 | 14 | 05 | 15 | 11 | 15 | 10 |
| 2 | 09 | 05 | 07 | 06 | 08 | 15 | 13 | 16 | 05 | 14 |
| 3 | 11 | 02 | 13 | 16 | 15 | 08 | 05 | 01 | 10 | 08 |
| 4 | 12 | 09 | 16 | 08 | 04 | 07 | 10 | 12 | 03 | 03 |

| | *i* | | | | | | | | | |
|---|---|---|---|---|---|---|---|---|---|---|
| | **11** | **12** | **13** | **14** | **15** | **16** | **17** | **18** | **19** | **20** |
| 1 | 15 | 09 | 12 | 06 | 16 | 10 | 07 | 05 | 16 | 03 |
| 2 | 11 | 05 | 06 | 08 | 15 | 08 | 01 | 13 | 15 | 07 |
| 3 | 09 | 11 | 02 | 15 | 01 | 13 | 12 | 01 | 02 | 14 |
| 4 | 10 | 13 | 10 | 14 | 03 | 14 | 02 | 08 | 06 | 05 |

Each plot was located and cruised. The data (in cubic meters per plot) and computations are as shown below.

| $j$ \ $i$ | 1 | 2 | 3 | 4 | 5 | 6 | 7 | 8 | 9 | 10 |
|---|---|---|---|---|---|---|---|---|---|---|
| 1 | 2.10 | 3.40 | 3.75 | 2.00 | 2.90 | 2.95 | 1.50 | 3.15 | 0.00 | 2.90 |
| 2 | 3.05 | 3.20 | 4.00 | 2.70 | 2.30 | 4.15 | 4.00 | 2.40 | 3.15 | 1.90 |
| 3 | 2.35 | 2.10 | 3.30 | 1.80 | 2.05 | 4.55 | 2.65 | 2.60 | 0.25 | 3.55 |
| 4 | 2.45 | 2.90 | 2.30 | 2.50 | 2.35 | 3.00 | 3.80 | 1.60 | 1.50 | 4.40 |
| $\sum_{j}^{4} y_{ij} =$ | 9.95 | 11.6 | 13.35 | 9.00 | 9.60 | 14.65 | 11.95 | 9.75 | 4.90 | 12.75 |
| $\sum_{j}^{4} y_{ij}^2 =$ | 25.2375 | 34.6200 | 46.2425 | 20.7800 | 23.4250 | 55.6275 | 39.7175 | 25.0025 | 12.2350 | 43.9825 |
| $\bar{y}_i =$ | 2.4875 | 2.9000 | 3.3375 | 2.2500 | 2.4000 | 3.6625 | 2.9875 | 2.4375 | 1.2250 | 3.1875 |
| $T_i =$ | 39.8 | 46.4 | 53.4 | 36.0 | 38.4 | 58.6 | 47.8 | 39.0 | 19.6 | 51.0 |
| $s_i^2 =$ | 0.162292 | 0.326667 | 0.562292 | 0.17667 | 0.128333 | 0.657292 | 0.1337292 | 0.412292 | 2.077500 | 1.113958 |

| $j$ \ $i$ | 11 | 12 | 13 | 14 | 15 | 16 | 17 | 18 | 19 | 20 |
|---|---|---|---|---|---|---|---|---|---|---|
| 1 | 2.65 | 2.30 | 0.40 | 1.15 | 1.15 | 1.35 | 0.00 | 1.80 | 0.45 | 1.60 |
| 2 | 3.25 | 2.60 | 0.00 | 1.00 | 1.00 | 0.10 | 0.55 | 1.50 | 0.00 | 0.50 |
| 3 | 3.35 | 1.60 | 0.00 | 3.60 | 3.00 | 0.40 | 0.00 | 2.05 | 2.00 | 0.40 |
| 4 | 2.30 | 1.00 | 0.05 | 2.40 | 1.90 | 0.20 | 0.15 | 2.45 | 1.15 | 0.85 |
| $\sum_{j}^{4} y_{ij} =$ | 11.55 | 7.50 | 0.45 | 8.15 | 7.05 | 2.05 | 0.70 | 7.80 | 3.60 | 3.35 |
| $\sum_{j}^{4} y_{ij}^2 =$ | 34.0975 | 15.6100 | 0.1625 | 21.0425 | 14.9325 | 2.0325 | 0.3250 | 15.6950 | 5.5250 | 3.6925 |
| $\bar{y}_i =$ | 2.8875 | 1.8750 | 0.1125 | 2.0375 | 1.7625 | 0.5125 | 0.1750 | 1.9500 | 0.90000 | 0.8375 |
| $T_i =$ | 46.2 | 30.0 | 1.8 | 32.6 | 28.2 | 8.2 | 2.8 | 31.20 | 14.4 | 13.4 |
| $s_i^2 =$ | 0.248958 | 0.515833 | 0.037292 | 1.478958 | 0.835625 | 0.327292 | 0.067500 | 0.161667 | 0.761667 | 0.295625 |

$$\bar{y}_o = \frac{1}{N}\left(\frac{M}{m}\right)\sum_{i}^{m} N_i \bar{y}_i = \frac{1}{320}\left(\frac{20}{20}\right)(39.8 + 46.4 + \cdots + 13.4)$$

$$= \frac{1}{320}(1)(638.8) = 1.99625 \text{ m}^3 \text{ / plot}$$

or 20(1.99625) = 39.925 $m^3$/ha.

$$\bar{T} = \frac{1}{m}\sum_{i}^{m} T_i = \frac{1}{20}(39.8 + 46.4 + \cdots + 13.4)$$

$$= \frac{1}{20}(638.8) = 37.94 \text{ m}^3\text{/cluster}$$

$$T_o = N\bar{y}_o = 320(1.99625) = 638.8 \text{ m}^3 \text{ on tract}$$

or

$$T_o = M\bar{T} = 20(31.94) = 638.8 \text{ m}^3 \text{ on tract}$$

$$s_T^2 = \frac{1}{(m-1)}\sum_{i}^{m}(T_i - \bar{T})^2$$

$$= \frac{1}{(20-1)}[(39.80 - 31.94)^2 + (46.40 - 31.94)^2 + \cdots + (13.40 - 31.94)^2]$$

$$= 290.10148$$

$$s_{\bar{y}_o}^2 = \frac{1}{N^2}\left\{\frac{M}{m}\sum_{i}^{m}\left[\frac{s_i^2}{n_i}\left(\frac{N_i - n_i}{N_i}\right)N_i^2\right] + M^2\frac{s_T^2}{m}\left(\frac{M-m}{M}\right)\right\}$$

$$\frac{s_i^2}{n_i}\left(\frac{N_i - n_i}{N_i}\right)N_i^2 =$$

| ***i*** | | | | | | | | | |
|---|---|---|---|---|---|---|---|---|---|
| **1** | **2** | **3** | **4** | **5** | **6** | **7** | **8** | **9** | **10** |
| 7.78992 | 15.68016 | 26.98992 | 8.48016 | 6.15984 | 31.54992 | 64.18992 | 19.78992 | 99.7200 | 53.47008 |

| ***i*** | | | | | | | | | |
|---|---|---|---|---|---|---|---|---|---|
| **11** | **12** | **13** | **14** | **15** | **16** | **17** | **18** | **19** | **20** |
| 11.95008 | 24.75984 | 1.78992 | 70.99008 | 40.11024 | 15.70992 | 3.24000 | 7.76016 | 36.56016 | 14.19024 |

Then,

$$\sum_{i}^{m}\left[\frac{s_i^2}{n_i}\left(\frac{N_i - n_i}{N_i}\right)N_i^2\right] = 7.78992 + 15.68016 + \cdots + 14.19024 = 560.88048$$

and

$$s_{\bar{y}_o}^2 = \frac{1}{(320)^2}\left[\frac{20}{20}(560.88048) + 20^2\left(\frac{290.10148}{20}\right)\left(\frac{20-20}{20}\right)\right]$$

$$= \frac{1}{102,400}(560.88048 + 0) = 0.005477$$

$$s_{\bar{y}_o} = \sqrt{0.005477} = \pm 0.074007 \text{ m}^3\text{/plot}$$

or

$$20\ (\pm 0.074007) = \pm 1.48014 \text{ or } \pm 1.48 \text{ m}^3\text{/ha}$$

and

$$s_{T_o} = Ns_{\bar{y}_o} = 320(\pm 0.074009) = \pm 23.68288 \text{ or } \pm 23.7 \text{ m}^3 \text{ on tract}$$

In this example the between cluster component of the variance of the overall mean was reduced to zero and dropped out of the equation leaving only the within cluster component.

$$s^2_{\bar{y}_o} = \frac{1}{N^2}\left\{\frac{M}{m}\sum_i^m\left[\frac{s_i^2}{n_i}\left(\frac{N_i - n_i}{N_i}\right)N_i^2\right]\right\}$$

Since all the clusters in the frame were included in the sample, $m = M$, and $M/m = 1$. Then,

$$s^2_{\bar{y}_o} = \frac{1}{N^2}\sum_i^m\left[\frac{s_i^2}{n_i}\left(\frac{N_i - n_i}{N_i}\right)N_i^2\right]$$

$$= \sum_i^m\left[\frac{s_i^3}{n_i}\left(\frac{N_i - n_i}{N_i}\right)\left(\frac{N_i^2}{N}\right)^2\right]$$

which is Equation 16.41, the variance of the mean in the case of a stratified random sample. Thus, stratified random sampling is a special case of two-stage cluster sampling where strata correspond to clusters, all the strata (clusters) are included in the sample, and subsampling takes place within the strata (clusters). In this example the stratification was arbitrary but the mathematics would be unchanged if the strata (clusters) were constructed logically, as done in Figure 16.3.

Since this example is actually an exercise in stratified sampling, the statements made in Chapter 16 regarding the conditions under which a valid sampling error can be computed apply. The sampling intensity was constant in all the strata (clusters) but, as can be seen from the evidence in the example in Section 17.8.2, the stratum (clusters) variances are not homogeneous. Consequently, a proper sampling error cannot be computed. However, the degrees of freedom ($k = \Sigma_i^m n_i - m = 80 - 20 = 60$) are sufficiently numerous to justify the use of an approximate $t$ value. Thus, for an approximate confidence level of 0.95, $t = 2$, and the sampling error is

$$\text{SE}_{0.95} = \pm 2s_{\bar{y}_o} = \pm 2(0.074007) = \pm 0.148014 \text{ m}^3\text{/plot}$$

or

$$20(\pm 0.148014) = \pm 2.96028 \text{ or } \pm 2.96 \text{ m}^3\text{/ha}$$

### 17.8.4 Two-Stage Sampling When $m = M$ and $n_i = 1$

As in the previous example, all 20 clusters were included in the sample but only one element or secondary sampling unit was sampled in each cluster. Then $M = 20$; $N_i = A = 16$, $i = 1, 2, \ldots, 20$; and $n_i = a = 1$, $i = 1, 2, \ldots, 20$. The plots chosen are listed below and shown in Figure 17.7.

| $i$ | $j$ | $i$ | $j$ | $i$ | $j$ | $i$ | $j$ |
|---|---|---|---|---|---|---|---|
| 1 | 07 | 6 | 11 | 11 | 07 | 16 | 02 |
| 2 | 05 | 7 | 13 | 12 | 08 | 17 | 07 |
| 3 | 12 | 8 | 15 | 13 | 12 | 18 | 03 |
| 4 | 09 | 9 | 10 | 14 | 03 | 19 | 10 |
| 5 | 04 | 10 | 03 | 15 | 16 | 20 | 03 |

The plot data and computations follow:

| $i$ | 1 | 2 | 3 | 4 | 5 | 6 | 7 | 8 | 9 | 10 |
|---|---|---|---|---|---|---|---|---|---|---|
| $y_{ij} =$ | 2.40 | 3.20 | 1.85 | 2.00 | 2.35 | 4.05 | 4.00 | 3.45 | 0.25 | 4.40 |
| $\sum_j^1 y_{ij} =$ | 2.40 | 3.20 | 1.85 | 2.00 | 2.35 | 4.05 | 4.00 | 3.45 | 0.25 | 4.40 |
| $\sum_j^1 y_{ij}^2 =$ | 5.7600 | 10.2400 | 3.4225 | 4.0000 | 5.5225 | 16.4025 | 16.0000 | 11.9025 | 0.0625 | 19.3600 |
| $\bar{y}_i =$ | 2.40 | 3.20 | 1.85 | 2.00 | 2.35 | 4.05 | 4.00 | 3.45 | 0.25 | 4.40 |
| $T_i =$ | 38.4 | 51.2 | 29.6 | 32.0 | 37.6 | 64.8 | 64.0 | 55.2 | 4.40 | 70.4 |
| $s_i^2 =$ | $\frac{0}{0}$ | $\frac{0}{0}$ | $\frac{0}{0}$ | $\frac{0}{0}$ | $\frac{0}{0}$ | $\frac{0}{0}$ | $\frac{0}{0}$ | $\frac{0}{0}$ | $\frac{0}{0}$ | $\frac{0}{0}$ |

| $i$ | 11 | 12 | 13 | 14 | 15 | 16 | 17 | 18 | 19 | 20 |
|---|---|---|---|---|---|---|---|---|---|---|
| $y_{ij} =$ | 2.65 | 2.65 | 0.40 | 4.65 | 1.15 | 0.75 | 0.00 | 1.90 | 0.50 | 1.60 |
| $\sum_j^1 y_{ij} =$ | 2.65 | 2.65 | 0.40 | 4.65 | 1.15 | 0.75 | 0.00 | 1.90 | 0.50 | 1.60 |
| $\sum_j^1 y_{ij}^2 =$ | 7.0225 | 7.0225 | 0.1600 | 21.6225 | 1.3225 | 0.5625 | 0.0000 | 3.6100 | 0.2500 | 2.5600 |
| $\bar{y}_i =$ | 2.65 | 2.65 | 0.40 | 4.65 | 1.15 | 0.75 | 0.00 | 1.90 | 0.50 | 1.60 |
| $T_i =$ | 42.4 | 42.4 | 6.4 | 74.4 | 18.4 | 12.0 | 0.00 | 30.4 | 8.0 | 25.6 |
| $s_i^2 =$ | $\frac{0}{0}$ | $\frac{0}{0}$ | $\frac{0}{0}$ | $\frac{0}{0}$ | $\frac{0}{0}$ | $\frac{0}{0}$ | $\frac{0}{0}$ | $\frac{0}{0}$ | $\frac{0}{0}$ | $\frac{0}{0}$ |

| | | | | | | | | | | | | | | | |
|---|---|---|---|---|---|---|---|---|---|---|---|---|---|---|---|
| 2.55 | 3.00 | 1.60 | 1.80 | 2.65 | 2.10 | 1.90 | 3.40 | 3.65 | 4.00 | 4.00 | 1.75 | 2.60 | 2.85 | 2.00 | 2.10 |
| 3.15 | 2.95 | 2.40 | 2.25 | 3.20 | 2.80 | 2.40 | 4.55 | 4.50 | 3.25 | 4.00 | 1.50 | 1.80 | 2.70 | 1.70 | 2.50 |
| 3.05 | 2.10 | 2.35 | 2.45 | 2.90 | 1.10 | 4.30 | 4.00 | 4.10 | 2.85 | 3.40 | 1.85 | 2.00 | 2.30 | 2.20 | 2.50 |
| 1.55 | 1.55 | 2.55 | 2.15 | 3.10 | 2.75 | 4.50 | 4.60 | 3.30 | 4.30 | 3.75 | 2.30 | 1.80 | 2.00 | 2.10 | 1.80 |
| 2.55 | 0.90 | 2.25 | 2.35 | 2.45 | 2.90 | 4.65 | 1.90 | 3.35 | 4.60 | 2.60 | 2.35 | 2.60 | 2.75 | 2.70 | 3.25 |
| 2.70 | 2.50 | 1.90 | 2.30 | 2.95 | 3.15 | 3.00 | 4.55 | 2.65 | 4.40 | 4.20 | 2.20 | 1.90 | 1.70 | 2.30 | 1.70 |
| 2.50 | 1.50 | 2.00 | 2.55 | 3.15 | 3.30 | 4.05 | 4.50 | 4.35 | 3.80 | 4.50 | 2.00 | 3.65 | 1.35 | 3.15 | 1.60 |
| 2.35 | 2.90 | 2.05 | 2.10 | 3.05 | 3.35 | 4.15 | 4.60 | 4.00 | 3.45 | 1.50 | 1.75 | 2.20 | 2.85 | 3.45 | 2.40 |
| 2.85 | 1.80 | 1.50 | 2.05 | 2.25 | 4.80 | 4.40 | 3.15 | 3.00 | 3.65 | 2.30 | 1.35 | 1.90 | 2.40 | 2.20 | 0.90 |
| 3.15 | 1.60 | 2.40 | 2.60 | 3.00 | 1.90 | 2.55 | 3.55 | 4.00 | 2.75 | 2.65 | 1.90 | 2.60 | 2.70 | 1.50 | 2.65 |
| 1.75 | 0.25 | 1.05 | 0.95 | 3.00 | 2.90 | 3.75 | 3.15 | 3.35 | 2.30 | 3.25 | 2.30 | 2.30 | 1.90 | 1.60 | 0.00 |
| 0.00 | 0.00 | 0.00 | 0.45 | 1.20 | 1.90 | 2.90 | 4.10 | 4.50 | 3.00 | 2.65 | 2.50 | 1.00 | 1.40 | 0.00 | 0.05 |
| 0.60 | 0.00 | 0.50 | 0.55 | 2.05 | 3.10 | 4.65 | 2.30 | 3.00 | 2.20 | 1.90 | 1.30 | 0.55 | 0.75 | 0.15 | 0.20 |
| 0.30 | 0.00 | 0.30 | 0.40 | 1.50 | 1.15 | 2.35 | 1.00 | 2.55 | 2.45 | 2.25 | 0.85 | 0.40 | 0.15 | 1.05 | 0.10 |
| 0.00 | 0.05 | 0.00 | 0.40 | 1.40 | 2.20 | 2.40 | 2.00 | 1.40 | 1.50 | 1.25 | 0.00 | 0.10 | 1.35 | 1.45 | 0.60 |
| 0.10 | 0.00 | 0.00 | 1.00 | 1.95 | 2.40 | 3.60 | 1.60 | 1.90 | 2.40 | 1.00 | 1.15 | 0.40 | 0.20 | 0.05 | 0.00 |
| 0.55 | 0.15 | 0.10 | 0.85 | 2.05 | 1.60 | 1.90 | 2.75 | 1.75 | 2.00 | 0.55 | 0.00 | 0.30 | 0.00 | 1.60 | 0.30 |
| 0.60 | 0.50 | 0.00 | 0.65 | 1.80 | 1.65 | 2.35 | 2.45 | 1.90 | 1.15 | 0.00 | 1.50 | 0.85 | 0.05 | 0.50 | 0.00 |
| 0.60 | 1.00 | 0.15 | 0.00 | 1.70 | 2.50 | 1.25 | 2.40 | 0.80 | 0.50 | 0.10 | 0.75 | 0.30 | 0.35 | 0.20 | 0.10 |
| 0.55 | 0.20 | 0.50 | 0.30 | 1.50 | 2.05 | 1.80 | 1.40 | 1.05 | 0.10 | 0.00 | 0.45 | 0.50 | 0.40 | 0.75 | 0.40 |

**FIGURE 17.7** The sample chosen when only one element was chosen from each cluster ($m = M$, $N_i = A = 16$, and $n_i = a = 1$).

$$\bar{y}_o = \frac{1}{N}\left(\frac{M}{m}\right)\sum_i^m N_i \bar{y}_i = \frac{1}{320}\left(\frac{20}{20}\right)(38.4 + 51.2 + \cdots + 25.6)$$

$$= \frac{1}{320}(1)(707.2) = 2.21 \text{ m}^3\text{/plot}$$

$$20(2.21) = 44.2 \text{ m}^3\text{/ha}$$

$$\bar{T} = \frac{1}{m}\sum_i^m T_i = \frac{1}{20}(38.4 + 51.2 + \cdots + 25.6)$$

$$= \frac{1}{20}(707.2) = 35.36 \text{ m}^3\text{/cluster}$$

$$T_o = N\bar{y}_o = 320(2.21) = 707.2 \text{ m}^3, \quad \text{or}$$

$$T_o = M\bar{T} = 20(35.36) = 707.2 \text{ m}^3 \text{ on tract}$$

$$\frac{s_i^2}{n_i}\left(\frac{N_i - n_i}{N_i}\right)N_i^2 = \frac{0}{0} \text{ in all clusters}$$

$$s_T^2 = \frac{1}{(m-1)}\sum_i^m (T_i - \bar{T})^2 = \frac{1}{(20-1)}[(38.4 - 35.36)^2 + (51.2 - 35.36)^2 + \cdots + (25.6 - 36.36)^2]$$

$$s_{\bar{y}_o}^2 = \frac{1}{N^2}\left\{\frac{M}{m}\sum_i^m\left[\frac{s_i^2}{n_i}\left(\frac{N_i - n_i}{N_i}\right)N_i^2\right] + M^2\frac{s_T^2}{m}\left(\frac{M-m}{M}\right)\right\}$$

$$= \frac{1}{(320)^2}\left[\frac{20}{20}\left(\frac{0}{0}\right) + 20^2\left(\frac{527.13095}{20}\right)\left(\frac{20-20}{20}\right)\right]$$

$$= \frac{1}{(320)^2}\left(\frac{0}{0} + 0\right) = \frac{0}{0}, \text{ which is indeterminate}$$

In this case the sample was well distributed over the sampling frame and the estimate of the overall mean and/or total should be excellent. However, the variance of the overall mean and/or total is indeterminate. Consequently, there is no measure of precision available to evaluate the results of the inventory. This points up the fact that in cluster sampling (as in stratified random sampling) if subsampling is used the subsample must contain at least two observations if a within cluster (or stratum) component of variance is to be obtained.

### 17.8.5 Simple Cluster Sampling When $m = 1$

In this example only one cluster was randomly chosen from the sampling frame and all the plots within the cluster were cruised. Consequently, $M = 20$; $m = 1$; and $n_i = a = N_i = A = 16$, $i = 1$. The cluster chosen was number 10 and was designated $i = 1$ (see Figure 17.8). The data and computations are as follows.

$$\sum_j^{16} y_{1j} = 48.5$$

$$\sum_j^{16} y_{1j}^2 = 160.765$$

Since there was no subsampling the cluster values are parameters, not estimates. Thus,

$$\mu_1 = 3.03125 \text{ m}^3\text{/plot}$$

$$\tau_1 = 16(3.03125) = 48.5 \text{ m}^3$$

According to Equation 17.47,

$$\bar{y}_o = \frac{1}{N}\left(\frac{M}{m}\right)\sum_i^m N_i\bar{y}_i = \frac{1}{N}\left(\frac{M}{m}\right)\sum_i^m T_i$$

| | | | | | | | | | | | | | | | |
|---|---|---|---|---|---|---|---|---|---|---|---|---|---|---|---|
| 2.55 | 3.00 | 1.60 | 1.80 | 2.65 | 2.10 | 1.90 | 3.40 | 3.65 | 4.00 | 4.00 | 1.75 | 2.60 | 2.85 | 2.00 | 2.10 |
| 3.15 | 2.95 | 2.40 | 2.25 | 3.20 | 2.80 | 2.40 | 4.55 | 4.50 | 3.25 | 4.00 | 1.50 | 1.80 | 2.70 | 1.70 | 2.50 |
| 3.05 | 2.10 | 2.35 | 2.45 | 2.90 | 1.10 | 4.30 | 4.00 | 4.10 | 2.85 | 3.40 | 1.85 | 2.00 | 2.30 | 2.20 | 2.50 |
| 1.55 | 1.55 | 2.55 | 2.15 | 3.10 | 2.75 | 4.50 | 4.60 | 3.30 | 4.30 | 3.75 | 2.30 | 1.80 | 2.00 | 2.10 | 1.80 |
| 2.55 | 0.90 | 2.25 | 2.35 | 2.45 | 2.90 | 4.65 | 1.90 | 3.35 | 4.60 | 2.60 | 2.35 | 2.60 | 2.75 | 2.70 | 3.25 |
| 2.70 | 2.50 | 1.90 | 2.30 | 2.95 | 3.15 | 3.00 | 4.55 | 2.65 | 4.40 | 4.20 | 2.20 | 1.90 | 1.70 | 2.30 | 1.70 |
| 2.50 | 1.50 | 2.00 | 2.55 | 3.15 | 3.30 | 4.05 | 4.50 | 4.35 | 3.80 | 4.50 | 2.00 | 3.65 | 1.35 | 3.15 | 1.60 |
| 2.35 | 2.90 | 2.05 | 2.10 | 3.05 | 3.35 | 4.15 | 4.60 | 4.00 | 3.45 | 1.50 | 1.75 | 2.20 | 2.85 | 3.45 | 2.40 |
| 2.85 | 1.80 | 1.50 | 2.05 | 2.25 | 4.80 | 4.40 | 3.15 | 3.00 | 3.65 | 2.30 | 1.35 | 1.90 | 2.40 | 2.20 | 0.90 |
| 3.15 | 1.60 | 2.40 | 2.60 | 3.00 | 1.90 | 2.55 | 3.55 | 4.00 | 2.75 | 2.65 | 1.90 | 2.60 | 2.70 | 1.50 | 2.65 |
| 1.75 | 0.25 | 1.05 | 0.95 | 3.00 | 2.90 | 3.75 | 3.15 | 3.35 | 2.30 | 3.25 | 2.30 | 2.30 | 1.90 | 1.60 | 0.00 |
| 0.00 | 0.00 | 0.00 | 0.45 | 1.20 | 1.90 | 2.90 | 4.10 | 4.50 | 3.00 | 2.65 | 2.50 | 1.00 | 1.40 | 0.00 | 0.05 |
| 0.60 | 0.00 | 0.50 | 0.55 | 2.05 | 3.10 | 4.65 | 2.30 | 3.00 | 2.20 | 1.90 | 1.30 | 0.55 | 0.75 | 0.15 | 0.20 |
| 0.30 | 0.00 | 0.30 | 0.40 | 1.50 | 1.15 | 2.35 | 1.00 | 2.55 | 2.45 | 2.25 | 0.85 | 0.40 | 0.15 | 1.05 | 0.10 |
| 0.00 | 0.05 | 0.00 | 0.40 | 1.40 | 2.20 | 2.40 | 2.00 | 1.40 | 1.50 | 1.25 | 0.00 | 0.10 | 1.35 | 1.45 | 0.60 |
| 0.10 | 0.00 | 0.00 | 1.00 | 1.95 | 2.40 | 3.60 | 1.60 | 1.90 | 2.40 | 1.00 | 1.15 | 0.40 | 0.20 | 0.05 | 0.00 |
| 0.55 | 0.15 | 0.10 | 0.85 | 2.05 | 1.60 | 1.90 | 2.75 | 1.75 | 2.00 | 0.55 | 0.00 | 0.30 | 0.00 | 1.60 | 0.30 |
| 0.60 | 0.50 | 0.00 | 0.65 | 1.80 | 1.65 | 2.35 | 2.45 | 1.90 | 1.15 | 0.00 | 1.50 | 0.85 | 0.05 | 0.50 | 0.00 |
| 0.60 | 1.00 | 0.15 | 0.00 | 1.70 | 2.50 | 1.25 | 2.40 | 0.80 | 0.50 | 0.10 | 0.75 | 0.30 | 0.35 | 0.20 | 0.10 |
| 0.55 | 0.20 | 0.50 | 0.30 | 1.50 | 2.05 | 1.80 | 1.40 | 1.05 | 0.10 | 0.00 | 0.45 | 0.50 | 0.40 | 0.75 | 0.40 |

**FIGURE 17.8** The sample chosen when only one cluster was included in the sample and no subsampling took place ($M = 20$; $m = 1$; and $n_i = a = N_i = A = 16$).

but in this case $\bar{y}_i$ is $\mu_i$ and $T_i$ is $\tau_i$ so that the expression becomes

$$\bar{y}_o = \frac{1}{N}\left(\frac{M}{m}\right)\sum_i^m N_i\mu_i = \frac{1}{N}\left(\frac{M}{m}\right)\sum_i^m \tau_i$$

Then,

$$\bar{y}_o = \frac{1}{320}\left(\frac{20}{1}\right)(16)(3.03125) = 3.03125 \text{ m}^3\text{/plot}$$

or

$$20(3.03125) = 60.625 \text{ m}^3\text{/ha}$$

In this case

$$\bar{T} = \frac{1}{m}\sum_{i}^{m} T_i = \frac{1}{m}\sum_{i}^{m} \tau_i = \frac{1}{1}(48.5) = 4.85 \text{ m}^3\text{/cluster}$$

$$T_o = N\bar{y}_o = 320(3.03125) = 970 \text{ m}^3$$

or

$$T_o = M\bar{T} = 20(48.5) = 970 \text{ m}^3 \text{ on tract}$$

$$\sigma_1^2 = \frac{1}{N_1}\sum_{i}^{N_1}(y_{ij} - \mu_1)^2$$

$$= \frac{1}{16}[160.765 - (48.5)^2 / 16)] = 0.859336$$

In this case

$$s_T^2 = \frac{1}{(m-1)}\sum_{i}^{m}(T_i - \bar{T})^2 = \frac{1}{(m-1)}\sum_{i}^{m}(\tau_1 - \bar{T})^2$$

$$= \frac{1}{(1-1)}(48.5 - 48.5)^2 = \frac{0}{0}$$

which is indeterminate. Substituting

$$\frac{\sigma_i^2}{n_i}\left(\frac{N_i - n_i}{N_i - 1}\right)N_i^2$$

for

$$\frac{s_i^2}{n_i}\left(\frac{N_i - n_i}{N_i}\right)N_i^2$$

in Equation 17.68 yields

$$s_{\bar{y}_o}^2 = \frac{1}{N^2}\left\{\frac{M}{m}\sum_{i}^{m}\left[\frac{\sigma_i^2}{n_i}\left(\frac{N_i - n_i}{N_i - 1}\right)N_i^2\right] + M^2\frac{s_T^2}{m}\left(\frac{M-m}{M}\right)\right\}$$

$$= \frac{1}{(320)^2}\left\{\frac{20}{1}\left[\frac{0.859336}{16}\left(\frac{16-16}{16-1}\right)(16)^2\right] + 20^2\left(\frac{0}{0}\right)\left(\frac{1}{1}\right)\left(\frac{20-1}{20}\right)\right\}$$

$$= \frac{1}{(320)^2}\left(0 + \frac{0}{0}\right)$$

$$= \frac{0}{0}$$

which is indeterminate.

As can be seen, under the conditions of this example, with only one cluster and no subsampling, it is impossible to obtain a variance or standard error of the overall mean or total. Essentially what the sampler has, in a case such as this, is a single observation and one cannot obtain an estimate of variability from a single variate.

In this example the elements of the cluster were in contact with one another making it quite obvious that only one observation was being made. However, what was demonstrated here applies regardless of the cluster configuration. Clusters could have been set up to take the forms shown in Figure 17.1b or c. In such cases, if only one cluster is chosen, the elements making up the cluster would no longer be in contact with one another and superficially the situation would appear to be different from that in this example. However, from a mathematical point of view, there is no difference. A standard error of the mean or of the total could not be obtained. If the clusters making up the sampling frame do not overlap, the mean of the chosen cluster would be an unbiased estimate of the population mean, as is any single variate drawn from the population of interest.

### 17.8.6 Simple Cluster Sampling When $1 < m < M$ (Example II)

In this example the cluster size will remain at $N_i = A = 16$ but instead of being compact, the clusters will be spread out with a three-element gap in the east–west direction and a four-element gap in the north–south direction. This cluster configuration is being used to avoid the problems associated with cluster overlap. The primary sampling frame is the $4 \times 5$ element rectangle in the upper left corner of the tract, as shown in Figure 17.9. The elements within this frame are numbered as is shown in Figure 17.3a. The random drawing of one of these elements identifies the upper left corner of the 16-element cluster. Since there are 20 positions in the frame, $M = 20$. Figure 17.9 shows three such clusters which have been randomly chosen. The positions chosen from the sampling frame were 11, 04, and 17. In the following discussion,

$i = 1$, refers to cluster 11,
$i = 2$, refers to cluster 04, and
$i = 3$, refers to cluster 17.

From this point the computational procedure is exactly the same as that in Example I:

| | $i$ = 1 | 2 | 3 |
|---|---|---|---|
| $\sum_j^{16} y_{ij} =$ | 33.45 | 29.95 | 30.2 |
| $\sum_j^{16} y_{ij}^2 =$ | 107.2675 | 79.9925 | 80.0050 |
| $\mu_i =$ | 2.090625 | 1.871875 | 1.887500 m³/plot |
| $\tau_i =$ | 33.45 | 29.95 | 30.20 m³ |

$$\bar{y}_o = 1.95 \text{ m}^3/\text{plot}, \quad \text{or}$$

$$20(1.95) = 39.0 \text{ m}^3/\text{ha}$$

$$\bar{T} = \frac{1}{3}(33.45 + 29.95 + 30.20) = 31.2 \text{ m}^3/\text{cluster}$$

| 2.55 | 3.00 | 1.60 | 1.80 | 2.65 | 2.10 | 1.90 | 3.40 | 3.65 | 4.00 | 4.00 | 1.75 | 2.60 | 2.85 | 2.00 | 2.10 |
|---|---|---|---|---|---|---|---|---|---|---|---|---|---|---|---|
| 3.15 | 2.95 | 2.40 | 2.25 | 3.20 | 2.80 | 2.40 | 4.55 | 4.50 | 3.25 | 4.00 | 1.50 | 1.80 | 2.70 | 1.70 | 2.50 |
| 3.05 | 2.10 | 2.35 | 2.45 | 2.90 | 1.10 | 4.30 | 4.00 | 4.10 | 2.85 | 3.40 | 1.85 | 2.00 | 2.30 | 2.20 | 2.50 |
| 1.55 | 1.55 | 2.55 | 2.15 | 3.10 | 2.75 | 4.50 | 4.60 | 3.30 | 4.30 | 3.75 | 2.30 | 1.80 | 2.00 | 2.10 | 1.80 |
| 2.55 | 0.90 | 2.25 | 2.35 | 2.45 | 2.90 | 4.65 | 1.90 | 3.35 | 4.60 | 2.60 | 2.35 | 2.60 | 2.75 | 2.70 | 3.25 |
| 2.70 | 2.50 | 1.90 | 2.30 | 2.95 | 3.15 | 3.00 | 4.55 | 2.65 | 4.40 | 4.20 | 2.20 | 1.90 | 1.70 | 2.30 | 1.70 |
| 2.50 | 1.50 | 2.00 | 2.55 | 3.15 | 3.30 | 4.05 | 4.50 | 4.35 | 3.80 | 4.50 | 2.00 | 3.65 | 1.35 | 3.15 | 1.60 |
| 2.35 | 2.90 | 2.05 | 2.10 | 3.05 | 3.35 | 4.15 | 4.60 | 4.00 | 3.45 | 1.50 | 1.75 | 2.20 | 2.85 | 3.45 | 2.40 |
| 2.85 | 1.80 | 1.50 | 2.05 | 2.25 | 4.80 | 4.40 | 3.15 | 3.00 | 3.65 | 2.30 | 1.35 | 1.90 | 2.40 | 2.20 | 0.90 |
| 3.15 | 1.60 | 2.40 | 2.60 | 3.00 | 1.90 | 2.55 | 3.55 | 4.00 | 2.75 | 2.65 | 1.90 | 2.60 | 2.70 | 1.50 | 2.65 |
| 1.75 | 0.25 | 1.05 | 0.95 | 3.00 | 2.90 | 3.75 | 3.15 | 3.35 | 2.30 | 3.25 | 2.30 | 2.30 | 1.90 | 1.60 | 0.00 |
| 0.00 | 0.00 | 0.00 | 0.45 | 1.20 | 1.90 | 2.90 | 4.10 | 4.50 | 3.00 | 2.65 | 2.50 | 1.00 | 1.40 | 0.00 | 0.05 |
| 0.60 | 0.00 | 0.50 | 0.55 | 2.05 | 3.10 | 4.65 | 2.30 | 3.00 | 2.20 | 1.90 | 1.30 | 0.55 | 0.75 | 0.15 | 0.20 |
| 0.30 | 0.00 | 0.30 | 0.40 | 1.50 | 1.15 | 2.35 | 1.00 | 2.55 | 2.45 | 2.25 | 0.85 | 0.40 | 0.15 | 1.05 | 0.10 |
| 0.00 | 0.05 | 0.00 | 0.40 | 1.40 | 2.20 | 2.40 | 2.00 | 1.40 | 1.50 | 1.25 | 0.00 | 0.10 | 1.35 | 1.45 | 0.60 |
| 0.10 | 0.00 | 0.00 | 1.00 | 1.95 | 2.40 | 3.60 | 1.60 | 1.90 | 2.40 | 1.00 | 1.15 | 0.40 | 0.20 | 0.05 | 0.00 |
| 0.55 | 0.15 | 0.10 | 0.85 | 2.05 | 1.60 | 1.90 | 2.75 | 1.75 | 2.00 | 0.55 | 0.00 | 0.30 | 0.00 | 1.60 | 0.30 |
| 0.60 | 0.50 | 0.00 | 0.65 | 1.80 | 1.65 | 2.35 | 2.45 | 1.90 | 1.15 | 0.00 | 1.50 | 0.85 | 0.05 | 0.50 | 0.00 |
| 0.60 | 1.00 | 0.15 | 0.00 | 1.70 | 2.50 | 1.25 | 2.40 | 0.80 | 0.50 | 0.10 | 0.75 | 0.30 | 0.35 | 0.20 | 0.10 |
| 0.55 | 0.20 | 0.50 | 0.30 | 1.50 | 2.05 | 1.80 | 1.40 | 1.05 | 0.10 | 0.00 | 0.45 | 0.50 | 0.40 | 0.75 | 0.40 |

**FIGURE 17.9** The three-cluster sample drawn when the cluster configuration was an open rectangle. The sampling frame is outlined in the upper left corner. (a) Cluster $i$ = 1, originating in position 11. (b) Cluster $i$ = 2, originating in position 04. (c) Cluster $i$ = 3, originating in position 17.

$$T_o = 20(31.2) = 624 \text{ m}^3 \text{ on tract} \quad \text{or}$$

$$T_o = 320(1.95) = 624 \text{ m}^3 \text{ on tract}$$

$$s_T^2 = \frac{1}{(3-1)}[(33.45-31.20)^2 + (29.95-31.20)^2 + (30.20-31.20)^2] = 3.8125$$

$$s_{\bar{y}_o}^2 = \frac{1}{(320)^2}\left[\frac{20}{3}(0) + 20^2\left(\frac{3.8125}{3}\right)\left(\frac{20-3}{20}\right)\right] = 0.004220$$

$$s_{\bar{y}_o} = \pm\sqrt{0.004220} = \pm 0.064962 \text{ m}^3\text{/plot}$$

or

$$s_{\bar{y}_o} = 20(\pm 0.064962) = \pm 1.299231 \text{ or } 1.30 \text{ m}^3\text{/ha}$$

and

| 2.55 | 3.00 | 1.60 | 1.80 | 2.65 | 2.10 | 1.90 | 3.40 | 3.65 | 4.00 | 4.00 | 1.75 | 2.60 | 2.85 | 2.00 | 2.10 |
|---|---|---|---|---|---|---|---|---|---|---|---|---|---|---|---|
| 3.15 | 2.95 | 2.40 | 2.25 | 3.20 | 2.80 | 2.40 | 4.55 | 4.50 | 3.25 | 4.00 | 1.50 | 1.80 | 2.70 | 1.70 | 2.50 |
| 3.05 | 2.10 | 2.35 | 2.45 | 2.90 | 1.10 | 4.30 | 4.00 | 4.10 | 2.85 | 3.40 | 1.85 | 2.00 | 2.30 | 2.20 | 2.50 |
| 1.55 | 1.55 | 2.55 | 2.15 | 3.10 | 2.75 | 4.50 | 4.60 | 3.30 | 4.30 | 3.75 | 2.30 | 1.80 | 2.00 | 2.10 | 1.80 |
| 2.55 | 0.90 | 2.25 | 2.35 | 2.45 | 2.90 | 4.65 | 1.90 | 3.35 | 4.60 | 2.60 | 2.35 | 2.60 | 2.75 | 2.70 | 3.25 |
| 2.70 | 2.50 | 1.90 | 2.30 | 2.95 | 3.15 | 3.00 | 4.55 | 2.65 | 4.40 | 4.20 | 2.20 | 1.90 | 1.70 | 2.30 | 1.70 |
| 2.50 | 1.50 | 2.00 | 2.55 | 3.15 | 3.30 | 4.05 | 4.50 | 4.35 | 3.80 | 4.50 | 2.00 | 3.65 | 1.35 | 3.15 | 1.60 |
| 2.35 | 2.90 | 2.05 | 2.10 | 3.05 | 3.35 | 4.15 | 4.60 | 4.00 | 3.45 | 1.50 | 1.75 | 2.20 | 2.85 | 3.45 | 2.40 |
| 2.85 | 1.80 | 1.50 | 2.05 | 2.25 | 4.80 | 4.40 | 3.15 | 3.00 | 3.65 | 2.30 | 1.35 | 1.90 | 2.40 | 2.20 | 0.90 |
| 3.15 | 1.60 | 2.40 | 2.60 | 3.00 | 1.90 | 2.55 | 3.55 | 4.00 | 2.75 | 2.65 | 1.90 | 2.60 | 2.70 | 1.50 | 2.65 |
| 1.75 | 0.25 | 1.05 | 0.95 | 3.00 | 2.90 | 3.75 | 3.15 | 3.35 | 2.30 | 3.25 | 2.30 | 2.30 | 1.90 | 1.60 | 0.00 |
| 0.00 | 0.00 | 0.00 | 0.45 | 1.20 | 1.90 | 2.90 | 4.10 | 4.50 | 3.00 | 2.65 | 2.50 | 1.00 | 1.40 | 0.00 | 0.05 |
| 0.60 | 0.00 | 0.50 | 0.55 | 2.05 | 3.10 | 4.65 | 2.30 | 3.00 | 2.20 | 1.90 | 1.30 | 0.55 | 0.75 | 0.15 | 0.20 |
| 0.30 | 0.00 | 0.30 | 0.40 | 1.50 | 1.15 | 2.35 | 1.00 | 2.55 | 2.45 | 2.25 | 0.85 | 0.40 | 0.15 | 1.05 | 0.10 |
| 0.00 | 0.05 | 0.00 | 0.40 | 1.40 | 2.20 | 2.40 | 2.00 | 1.40 | 1.50 | 1.25 | 0.00 | 0.10 | 1.35 | 1.45 | 0.60 |
| 0.10 | 0.00 | 0.00 | 1.00 | 1.95 | 2.40 | 3.60 | 1.60 | 1.90 | 2.40 | 1.00 | 1.15 | 0.40 | 0.20 | 0.05 | 0.00 |
| 0.55 | 0.15 | 0.10 | 0.85 | 2.05 | 1.60 | 1.90 | 2.75 | 1.75 | 2.00 | 0.55 | 0.00 | 0.30 | 0.00 | 1.60 | 0.30 |
| 0.60 | 0.50 | 0.00 | 0.65 | 1.80 | 1.65 | 2.35 | 2.45 | 1.90 | 1.15 | 0.00 | 1.50 | 0.85 | 0.05 | 0.50 | 0.00 |
| 0.60 | 1.00 | 0.15 | 0.00 | 1.70 | 2.50 | 1.25 | 2.40 | 0.80 | 0.50 | 0.10 | 0.75 | 0.30 | 0.35 | 0.20 | 0.10 |
| 0.55 | 0.20 | 0.50 | 0.30 | 1.50 | 2.05 | 1.80 | 1.40 | 1.05 | 0.10 | 0.00 | 0.45 | 0.50 | 0.40 | 0.75 | 0.40 |

**FIGURE 17.9** *Continued.*

$$s_{\bar{T}}^2 = \frac{3.8125}{3}\left(\frac{20-3}{20}\right) = 1.080208$$

$$s_{\bar{T}} = \pm\sqrt{1.080208} = \pm 1.039331 \text{ or } \pm 1.04 \text{ m}^3\text{/cluster}$$

As in Example I, a valid sampling error can be computed. At the 90% level of confidence,

$$\text{SE}_{0.90} = \pm t_{0.90}(k=2)s_{\bar{T}} = \pm 2.920(1.039331)$$

$$= \pm 3.034847 \text{ or } \pm 3.035 \text{ m}^3\text{/cluster}$$

or

$$\pm 3.034847/16 = \pm 0.189678 \text{ m}^3\text{/plot}$$

or

$$20(\pm 0.189678) = \pm 3.794 \text{ m}^3\text{/ha}$$

| 2.55 | 3.00 | 1.60 | 1.80 | 2.65 | 2.10 | 1.90 | 3.40 | 3.65 | 4.00 | 4.00 | 1.75 | 2.60 | 2.85 | 2.00 | 2.10 |
|---|---|---|---|---|---|---|---|---|---|---|---|---|---|---|---|
| 3.15 | 2.95 | 2.40 | 2.25 | 3.20 | 2.80 | 2.40 | 4.55 | 4.50 | 3.25 | 4.00 | 1.50 | 1.80 | 2.70 | 1.70 | 2.50 |
| 3.05 | 2.10 | 2.35 | 2.45 | 2.90 | 1.10 | 4.30 | 4.00 | 4.10 | 2.85 | 3.40 | 1.85 | 2.00 | 2.30 | 2.20 | 2.50 |
| 1.55 | 1.55 | 2.55 | 2.15 | 3.10 | 2.75 | 4.50 | 4.60 | 3.30 | 4.30 | 3.75 | 2.30 | 1.80 | 2.00 | 2.10 | 1.80 |
| 2.55 | 0.90 | 2.25 | 2.35 | 2.45 | 2.90 | 4.65 | 1.90 | 3.35 | 4.60 | 2.60 | 2.35 | 2.60 | 2.75 | 2.70 | 3.25 |
| 2.70 | 2.50 | 1.90 | 2.30 | 2.95 | 3.15 | 3.00 | 4.55 | 2.65 | 4.40 | 4.20 | 2.20 | 1.90 | 1.70 | 2.30 | 1.70 |
| 2.50 | 1.50 | 2.00 | 2.55 | 3.15 | 3.30 | 4.05 | 4.50 | 4.35 | 3.80 | 4.50 | 2.00 | 3.65 | 1.35 | 3.15 | 1.60 |
| 2.35 | 2.90 | 2.05 | 2.10 | 3.05 | 3.35 | 4.15 | 4.60 | 4.00 | 3.45 | 1.50 | 1.75 | 2.20 | 2.85 | 3.45 | 2.40 |
| 2.85 | 1.80 | 1.50 | 2.05 | 2.25 | 4.80 | 4.40 | 3.15 | 3.00 | 3.65 | 2.30 | 1.35 | 1.90 | 2.40 | 2.20 | 0.90 |
| 3.15 | 1.60 | 2.40 | 2.60 | 3.00 | 1.90 | 2.55 | 3.55 | 4.00 | 2.75 | 2.65 | 1.90 | 2.60 | 2.70 | 1.50 | 2.65 |
| 1.75 | 0.25 | 1.05 | 0.95 | 3.00 | 2.90 | 3.75 | 3.15 | 3.35 | 2.30 | 3.25 | 2.30 | 2.30 | 1.90 | 1.60 | 0.00 |
| 0.00 | 0.00 | 0.00 | 0.45 | 1.20 | 1.90 | 2.90 | 4.10 | 4.50 | 3.00 | 2.65 | 2.50 | 1.00 | 1.40 | 0.00 | 0.05 |
| 0.60 | 0.00 | 0.50 | 0.55 | 2.05 | 3.10 | 4.65 | 2.30 | 3.00 | 2.20 | 1.90 | 1.30 | 0.55 | 0.75 | 0.15 | 0.20 |
| 0.30 | 0.00 | 0.30 | 0.40 | 1.50 | 1.15 | 2.35 | 1.00 | 2.55 | 2.45 | 2.25 | 0.85 | 0.40 | 0.15 | 1.05 | 0.10 |
| 0.00 | 0.05 | 0.00 | 0.40 | 1.40 | 2.20 | 2.40 | 2.00 | 1.40 | 1.50 | 1.25 | 0.00 | 0.10 | 1.35 | 1.45 | 0.60 |
| 0.10 | 0.00 | 0.00 | 1.00 | 1.95 | 2.40 | 3.60 | 1.60 | 1.90 | 2.40 | 1.00 | 1.15 | 0.40 | 0.20 | 0.05 | 0.00 |
| 0.55 | 0.15 | 0.10 | 0.85 | 2.05 | 1.60 | 1.90 | 2.75 | 1.75 | 2.00 | 0.55 | 0.00 | 0.30 | 0.00 | 1.60 | 0.30 |
| 0.60 | 0.50 | 0.00 | 0.65 | 1.80 | 1.65 | 2.35 | 2.45 | 1.90 | 1.15 | 0.00 | 1.50 | 0.85 | 0.05 | 0.50 | 0.00 |
| 0.60 | 1.00 | 0.15 | 0.00 | 1.70 | 2.50 | 1.25 | 2.40 | 0.80 | 0.50 | 0.10 | 0.75 | 0.30 | 0.35 | 0.20 | 0.10 |
| 0.55 | 0.20 | 0.50 | 0.30 | 1.50 | 2.05 | 1.80 | 1.40 | 1.05 | 0.10 | 0.00 | 0.45 | 0.50 | 0.40 | 0.75 | 0.40 |

**FIGURE 17.9** *Continued.*

These sampling errors are much smaller than those obtained in Example I, reflecting the fact that with the spread-out cluster configuration the sample was more widely distributed over the population and more variability was contained within the clusters than was the case with the compact clusters. This can be further demonstrated by comparing the within cluster variances:

| | $\sigma_w^2$ | |
|---|---|---|
| **Cluster $i$** | **Compact Clusters** | **Spread Clusters** |
| 1 | 0.859336 | 2.333506 |
| 2 | 0.864336 | 1.308115 |
| 3 | 0.180772 | 1.437656 |

### 17.8.7 Two-Stage Sampling When $m = M = 1$ and $1 < n_i < N_i$

In this example there is only one cluster which includes all 320 plots in the tract. Consequently, $M = 1$ and $N_i = A = N = 320$. Since there is only one cluster in the sampling frame, when the sampling is carried out, $m = M = 1$ and $1 < n_i < N_i$. Obviously, what one has here is simple random sampling with the sampling frame acting as a cluster. This is the situation with all simple random sampling operations. The sampling frame is, in essence, a cluster that may or may not include the entire universe. Whenever the universe is infinite, the sampling frame is a subset and is, in actuality, a cluster. In such a case an infinitely large set of possible sampling frames exists, but only one is

| 2.55 | 3.00 | 1.60 | 1.80 | 2.65 | 2.10 | 1.90 | 3.40 | 3.65 | 4.00 | 4.00 | 1.75 | 2.60 | 2.85 | 2.00 | 2.10 |
|---|---|---|---|---|---|---|---|---|---|---|---|---|---|---|---|
| 3.15 | 2.95 | 2.40 | 2.25 | 3.20 | 2.80 | 2.40 | 4.55 | 4.50 | 3.25 | 4.00 | 1.50 | 1.80 | 2.70 | 1.70 | 2.50 |
| 3.05 | 2.10 | 2.35 | 2.45 | 2.90 | 1.10 | 4.30 | 4.00 | 4.10 | 2.85 | 3.40 | 1.85 | 2.00 | 2.30 | 2.20 | 2.50 |
| 1.55 | 1.55 | 2.55 | 2.15 | 3.10 | 2.75 | 4.50 | 4.60 | 3.30 | 4.30 | 3.75 | 2.30 | 1.80 | 2.00 | 2.10 | 1.80 |
| 2.55 | 0.90 | 2.25 | 2.35 | 2.45 | 2.90 | 4.65 | 1.90 | 3.35 | 4.60 | 2.60 | 2.35 | 2.60 | 2.75 | 2.70 | 3.25 |
| 2.70 | 2.50 | 1.90 | 2.30 | 2.95 | 3.15 | 3.00 | 4.55 | 2.65 | 4.40 | 4.20 | 2.20 | 1.90 | 1.70 | 2.30 | 1.70 |
| 2.50 | 1.50 | 2.00 | 2.55 | 3.15 | 3.30 | 4.05 | 4.50 | 4.35 | 3.80 | 4.50 | 2.00 | 3.65 | 1.35 | 3.15 | 1.60 |
| 2.35 | 2.90 | 2.05 | 2.10 | 3.05 | 3.35 | 4.15 | 4.60 | 4.00 | 3.45 | 1.50 | 1.75 | 2.20 | 2.85 | 3.45 | 2.40 |
| 2.85 | 1.80 | 1.50 | 2.05 | 2.25 | 4.80 | 4.40 | 3.15 | 3.00 | 3.65 | 2.30 | 1.35 | 1.90 | 2.40 | 2.20 | 0.90 |
| 3.15 | 1.60 | 2.40 | 2.60 | 3.00 | 1.90 | 2.55 | 3.55 | 4.00 | 2.75 | 2.65 | 1.90 | 2.60 | 2.70 | 1.50 | 2.65 |
| 1.75 | 0.25 | 1.05 | 0.95 | 3.00 | 2.90 | 3.75 | 3.15 | 3.35 | 2.30 | 3.25 | 2.30 | 2.30 | 1.90 | 1.60 | 0.00 |
| 0.00 | 0.00 | 0.00 | 0.45 | 1.20 | 1.90 | 2.90 | 4.10 | 4.50 | 3.00 | 2.65 | 2.50 | 1.00 | 1.40 | 0.00 | 0.05 |
| 0.60 | 0.00 | 0.50 | 0.55 | 2.05 | 3.10 | 4.65 | 2.30 | 3.00 | 2.20 | 1.90 | 1.30 | 0.55 | 0.75 | 0.15 | 0.20 |
| 0.30 | 0.00 | 0.30 | 0.40 | 1.50 | 1.15 | 2.35 | 1.00 | 2.55 | 2.45 | 2.25 | 0.85 | 0.40 | 0.15 | 1.05 | 0.10 |
| 0.00 | 0.05 | 0.00 | 0.40 | 1.40 | 2.20 | 2.40 | 2.00 | 1.40 | 1.50 | 1.25 | 0.00 | 0.10 | 1.35 | 1.45 | 0.60 |
| 0.10 | 0.00 | 0.00 | 1.00 | 1.95 | 2.40 | 3.60 | 1.60 | 1.90 | 2.40 | 1.00 | 1.15 | 0.40 | 0.20 | 0.05 | 0.00 |
| 0.55 | 0.15 | 0.10 | 0.85 | 2.05 | 1.60 | 1.90 | 2.75 | 1.75 | 2.00 | 0.55 | 0.00 | 0.30 | 0.00 | 1.60 | 0.30 |
| 0.60 | 0.50 | 0.00 | 0.65 | 1.80 | 1.65 | 2.35 | 2.45 | 1.90 | 1.15 | 0.00 | 1.50 | 0.85 | 0.05 | 0.50 | 0.00 |
| 0.60 | 1.00 | 0.15 | 0.00 | 1.70 | 2.50 | 1.25 | 2.40 | 0.80 | 0.50 | 0.10 | 0.75 | 0.30 | 0.35 | 0.20 | 0.10 |
| 0.55 | 0.20 | 0.50 | 0.30 | 1.50 | 2.05 | 1.80 | 1.40 | 1.05 | 0.10 | 0.00 | 0.45 | 0.50 | 0.40 | 0.75 | 0.40 |

**FIGURE 17.10** The sampling frame when the clusters differ in size, showing the subsamples drawn from each of the chosen clusters.

chosen. As an example, consider the infinity of points making up a surface of a map such as that in Figure 12.2. In that figure the only points considered were those at the intersections of the grid lines. A random sample of those points was chosen to determine the proportion of the tract included in each of the condition classes. If the grid was shifted slightly, the frame (cluster) would be different. Thus, one has the situation described in this example.

### 17.8.8 Two-Stage Sampling When $1 < m < M$, $1 < n_i < N_i$, and $N_i$ Is Not Constant

In this example, the 16-ha tract has been divided into $M = 28$ clusters which vary in size, as shown in Figure 17.10. The example is not realistic within the context of a timber cruise, but it shows the process that would be appropriate if a universe were divided into clusters of differing sizes.

The clusters were numbered from 01 to 28, with the numbers increasing from left to right. Four clusters were randomly chosen using sampling without replacement. Thus, $m = 4$. Using a constant sampling intensity, $n_i/N_i = 0.25$, subsamples were randomly drawn, again using sampling without replacement, from the chosen clusters, as is shown in Figure 17.10. It must be pointed out that a constant sampling intensity is *not* a requirement. It is essential, however, that the subsample sizes equal or exceed 2, so that the within cluster variance can be estimated.

The chosen plots were located in the field and cruised. The data (in cubic meters per plot) obtained and the resulting computations are as shown below.

| $j$ | $i$ = 1 | 2 | 3 | 4 |
|---|---|---|---|---|
| 1 | 4.55 | 1.90 | 4.05 | 0.30 |
| 2 | 4.00 | 3.65 | 4.50 | 0.20 |
| 3 | 3.30 | | 4.00 | 0.05 |
| 4 | | | | 0.00 |
| $N_i =$ | 12 | 8 | 12 | 16 |
| $n_i =$ | 3 | 2 | 3 | 4 |
| $\sum_j^{n_i} y_{ij} =$ | 11.85 | 5.55 | 12.55 | 0.55 |
| $\sum_j^{n_i} y_{ij}^2 =$ | 47.5925 | 6.9325 | 52.6525 | 0.1325 |
| $\bar{y}_i =$ | 3.9500 | 2.7750 | 4.1830 | 0.1375 |
| $T_i = N_i\bar{y}_i =$ | 47.4 | 22.2 | 50.2 | 2.2 |
| $s_i^2 =$ | 0.392500 | 1.531250 | 0.075833 | 0.018958 |
| $\frac{s_i^2}{n_i}\left(\frac{N_i - n_i}{N_i}\right)N_i^2 =$ | 14.13 | 36.75 | 2.73 | 0.91 |

Then,

$$\bar{T} = \frac{1}{m}\sum_i^m T_i = 122.0/4 = 30.5 \text{ m}^3\text{/cluster}$$

$$T_o = M\bar{T} = 28(30.5) = 854 \text{ m}^3 \text{ on tract}$$

$$\bar{y}_o = T_o / N = 854/320 = 2.66875 \text{ m}^3\text{/plot}$$

or, since each plot is $^1/_{20}$ ha,

$$\bar{y}_o = 20(2.66875) = 53.375 \text{ m}^3\text{/ha}$$

$$\sum_i^m \left[\frac{s_i^2}{n_i}\left(\frac{N_i - n_i}{n_i}\right)^2 N_i^2\right] = 54.52$$

$$s_T^2 = \frac{1}{m-1}\sum_i^m (T_i - \bar{T})^2 = 514.493333$$

Using Equation 17.58,

$$s_{T_o}^2 = \frac{28}{4}(54.52) + \frac{28^2}{4}(514.493333) = 101,222.33$$

Then, from Equation 17.78,

$$s^2_{\bar{y}_o} = 101,222.33/(320)^2 = 0.988499$$

and

$$s_{\bar{y}_o} = \pm\sqrt{0.988499} = \pm 0.994233 \text{ m}^3/\text{plot}$$

or

$$s_{\bar{y}_o} = \pm 20(0.994233) = \pm 19.88466 \text{ or } \pm 19.885 \text{ m}^3/\text{ha}$$

Again, a valid sampling error cannot be computed.

## 17.9 VARIANCE FUNCTIONS

An algebraic expression that relates variance to cluster or sampling unit size is known as a *variance function*. Under some circumstances such functions can be used to estimate variances of the mean or the total, between cluster variance, within cluster variance, or the variance of cluster or sampling unit totals.

To simplify the following mathematical development, expanded use will be made of the symbology used in Section 16.5.5 where $S^2$ was substituted for $\sigma^2[N/(N-1)]$. According to Equation 17.75, when cluster size is constant, the total sum of squared errors associated with two-stage cluster sampling is partitioned in the following manner:

$$\sum_i^M \sum_j^A (y_{ij} - \mu_o)^2 = A\sum_i^M (\mu_i - \mu_o)^2 + \sum_i^M \sum_j^A (y_{ij} - \mu_i)^2$$

Putting this into the form of an analysis of variance:

**ANOVA**

| Source | df | SS | MS |
|---|---|---|---|
| Total | $MA - 1$ | $\sum_i^M \sum_j^A (y_{ij} - \mu_o)^2$ | $S_o^2 = \sigma_o^2\left(\frac{MA}{MA-1}\right)$ |
| Between clusters | $M - 1$ | $A\sum_i^M (\mu_i - \mu_o)^2$ | $AS_B^2 = A\sigma_B^2\left(\frac{M}{M-1}\right)$ |
| Within clusters | $M(A - 1)$ | $\sum_i^M \sum_j^A (y_{ij} - \mu_i)^2$ | $S_w^2 = \sigma_w^2\left(\frac{A}{A-1}\right)$ |

The use of the $S$ symbols, representing, except in the case of the between cluster symbol, *mean squares*, permits the application of the analysis of variance to the population. Furthermore, a sample-based variance is actually an estimate of the quantity represented by its associated $S^2$, e.g.,

$$s_Y^2 = \frac{1}{(n-1)}\sum_i^n (y_i - \bar{y})^2 \rightarrow \sigma_Y^2\left(\frac{N}{N-1}\right) = S_Y^2$$

thus simplifying the estimation process. The $S$ symbology is widely used by statisticians. It should be noted, however, that whenever $S$ symbols are substituted for their associated $\sigma$ symbols, the appropriate corrections must be made.

To set the stage for a discussion of variance functions, consider Equation 17.59:

$$V(\bar{y}_o)_{2s} = \frac{\sigma_w^2}{ma}\left(\frac{A-a}{A-1}\right) + \frac{\sigma_B^2}{m}\left(\frac{M-m}{M-1}\right)$$

In single-stage cluster sampling there is no subsampling and thus $a = A$. Consequently, $A - a = 0$ and the first term in the expression vanishes.

$$V(\bar{y}_o)_{1s} = \frac{\sigma_B^2}{m}\left(\frac{M-m}{M-1}\right)$$

which can be written

$$V(\bar{y}_o)_{1s} = \frac{S_B^2}{m}\left(\frac{M-m}{M}\right) = \frac{S_B^2}{m} - \frac{S_B^2}{M} \tag{17.96}$$

Since $1/M = A/N$,

$$V(\bar{y}_o)_{1s} = \frac{S_B^2}{m} - S_B^2\left(\frac{A}{N}\right) \tag{17.97}$$

Thus, the key to variance of the mean in single-stage cluster sampling is the between cluster variance.

Using an argument that has its roots in a paper by Smith (1938), if a population is *randomly* distributed and the variance of the mean is being estimated, a cluster of size $A$ is the equivalent of a random sample of size $n = A$. Consequently, one can substitute $A$ for $n$ in Equation 12.5:

$$\sigma_{\bar{y}}^2 = \frac{\sigma_o^2}{A}\left(\frac{N-A}{N-1}\right)$$

which can be written

$$V(\bar{y}_o) = \frac{S_o^2}{A}\left(\frac{N-A}{N}\right) \tag{17.98}$$

Then, according to Equations 17.96 and 17.98,

$$\frac{S_B^2}{m}\left(\frac{M-m}{M}\right) = \frac{S_o^2}{A}\left(\frac{N-A}{N}\right) \tag{17.99}$$

Since only one cluster in involved, $m = 1$. Furthermore, $M = N/A$. Then,

$$\left(\frac{M-m}{M}\right) = \left(\frac{M-1}{M}\right) = \frac{(N/A)-(A/A)}{(N/A)} = \left(\frac{N-A}{N}\right)$$

Equation 17.99 can then be written

$$S_B^2\left(\frac{N-A}{N}\right) = \frac{S_o^2}{A}\left(\frac{N-A}{N}\right)$$

and

$$S_B^2 = S_o^2 / A \tag{17.100}$$

Since most real populations are not randomly distributed and adjacent elements tend to be similar, the variance within a cluster of size $A$ from a real population will be, on the average, less than that from the same population which has been randomly distributed. According to Equation 17.76,

$$\sigma_o^2 = \sigma_B^2 + \sigma_w^2$$

Since the overall population variance, $\sigma_o^2$, is independent of the distributional pattern of the population, if the within cluster variance, $\sigma_w^2$, is reduced, the between cluster variance, $\sigma_B^2$, must increase. In such a case, the denominator in Equation 17.100 must be decreased. Consequently,

$$S_B^2 = S_o^2/A^\beta \tag{17.101}$$

Substituting the right side of Equation 17.101 for $S_B^2$ in Equation 17.97, and remembering that only one cluster is involved, yields

$$V(\bar{y}_o) = \frac{S_o^2}{A^\beta} - \frac{S_o^2}{A^\beta}\left(\frac{A}{N}\right) = \frac{S_o^2}{A^\beta}\left(\frac{N-A}{N}\right) \tag{17.102}$$

Usually it is assumed that $N$ is large relative to $A$ and that the finite population correction is negligible. Then,

$$V(\bar{y}_o) \doteq S_o^2 / A^\beta \tag{17.103}$$

Equations 17.101 through 17.103 are variance equations.

The exponent $\beta$ is obviously a function of the intraclass correlation coefficient. According to Equation 17.74,

$$\sigma_B^2 = \frac{\sigma_o^2}{A}[1 + \rho(A-1)]$$

which, since $S_B^2[(M-1)/M] = \sigma_B^2$ and $S_o^2[(N-1)/N] = \sigma_o^2$, can be written

$$S_B^2\left(\frac{M-1}{M}\right) = \frac{S_o^2}{A}\left(\frac{N-1}{N}\right)[1+\rho(A-1)]$$

$$S_B^2 = \frac{S_o^2}{A}\left(\frac{N-1}{N}\right)\left(\frac{M}{M-1}\right)[1+\rho(A-1)]$$

According to Equation 17.101,

$$S_B^2 = S_o^2 / A^\beta$$

Then,

$$\frac{S_o^2}{A^\beta} = \frac{S_o^2}{A}\left(\frac{N-1}{N}\right)\left(\frac{M}{M-1}\right)[1+\rho(A-1)]$$

Dividing both sides by $S_o^2$,

$$\frac{1}{A^\beta} = \frac{1}{A}\left(\frac{N-1}{N}\right)\left(\frac{M}{M-1}\right)[1+\rho(A-1)]$$

Since $N = MA$, $M/(M - 1) = N/(N - A)$, and

$$\frac{1}{A^\beta} = \frac{1}{A}\left(\frac{N-1}{N-A}\right)[1+\rho(A-1)]$$

Then,

$$A^\beta = A\left(\frac{N-A}{N-1}\right)\frac{1}{[1+\rho(A-1)]}$$

The logarithmic equivalent of the expression is

$$\beta \log A = \log A + \log\left(\frac{N-A}{N-1}\right) - \log[1+\rho(A-1)]$$

Then

$$\beta = 1 + \log\left(\frac{N-A}{N-1}\right)\left(\frac{1}{\log A}\right) - \frac{\log[1+\rho(A-1)]}{\log A} \tag{17.104}$$

The term $[(N - A)/(N - 1)]$ is always less than 1 but approaches zero as $N$ approaches infinity. Consequently, $\log[(N - A)/(N - 1)] \leq 0$. Since $A$ is always equal to or greater than 1, its logarithm will be positive. Consequently, the second term in Equation 17.104 will always be equal to or less than zero.

When $\rho = 0$, the third term in Equation 17.104 becomes $\log 1/\log A$, which is equal to zero. Thus, $\beta$ will be equal to the sum of the first two terms. If $N = \infty$, $\beta$ will be equal to 1, but if $N$ is finite β will be somewhat less than 1. For example, if $N = 1000$ and $A = 20$, β would be equal to 0.9936 and

$$S_B^2 = S_o^2 / A^{0.9936}$$

When $\rho = 1$, $\rho(A - 1) = A - 1$ and $[1 + \rho(A - 1)] = A$. The third term in Equation 17.104 is then equal to 1 and β would be equal to the second term. For example, if $N = \infty$ and $A = 20$,

$$\beta = 1 + \frac{\log 1}{\log 20} - \frac{\log 20}{\log 20} = 1 + 0 - 1 = 0$$

and

$$S_B^2 = S_o^2 / A^o = S_o^2$$

If $N$ is finite, $\log[(N - A)/(N - 1)]$ and, thus, β, will be negative.

When β is negative, it indicates that $S_B^2$ is greater than $S_o^2$, which is not possible. Consequently, there is a range of ρ values for which valid β values cannot be determined. Setting $\beta = 0$ in Equation 17.104,

$$0 = 1 + \log\left(\frac{N-A}{N-1}\right)\frac{1}{\log A} - \frac{\log[1+\rho(A-1)]}{\log A}$$

$$-1 = \log\left(\frac{N-A}{N-1}\right)\frac{1}{\log A} - \frac{\log[1+\rho(A-1)]}{\log A}$$

$$-\log A = \log\left(\frac{N-A}{N-1}\right) - \log[1+\rho(A-1)]$$

Then

$$\log[1+\rho(A-1)] = \log\left(\frac{N-A}{N-1}\right) + \log A$$

$$\text{antilog}\left[\log\left(\frac{N-A}{N-1}\right) + \log A\right] = 1 + \rho(A-1)$$

$$\frac{\text{antilog}\left[\log\left(\frac{N-A}{N-1}\right) + \log A\right] - 1}{A-1} = \rho$$

Thus, if

$$\rho > \frac{\text{antilog}\left[\log\left(\frac{N-A}{N-1}\right) + \log A\right] - 1}{A-1} \tag{17.105}$$

a valid β cannot be computed. For example, if $N = 1000$ and $A = 20$, the critical value is

$$\frac{\text{antilog}\left[\log\left(\frac{980}{999}\right) + \log 20\right] - 1}{19} = 0.97998$$

When ρ is negative, β becomes greater than 1, indicating that $S_B^2 < V(\bar{y}_o)_{\text{Ran}}$, when $A = n$. In other words, the cluster means vary relatively little. For example, if ρ = –0.05; $N$ = 1000; and $A$ = 20:

$$\beta = 1 + \log\left(\frac{980}{999}\right)\frac{1}{\log 20} - \frac{\log[+(-0.05)(A-1)]}{\log 20}$$

$$= 1.9935902$$

and

$$S_B^2 = S_o^2 / A^{1.9935902}$$

Obviously, if β keeps increasing $S_B^2$ will decrease. In the limit, as $\beta \to \infty$, $S_B^2 \to 0$.

When ρ is negative, the term $[1 + \rho(A - 1)]$ is equal to zero or is negative if $\rho \leq -1/(A - 1)$. Since logarithms for zero and negative numbers do not exist, a magnitude for the third term in Equation 17.104 cannot be obtained. In turn this means that a value for β cannot be computed. For example, if $A$ = 20, ρ cannot be less than or equal to –1/(20 – 1) = –0.0526316. If ρ = –0.06,

$$[1 + (-0.06)(20 - 1)] = 1 - 1.14 = -0.14$$

There is no logarithm for –0.14. The maximum allowable cluster size is

$$A = (1 + |\rho|)/|\rho|$$

Thus, if ρ = –0.06, the maximum cluster size would be

$$A = (1 + 0.06)/0.06 = 17.7 \text{ or } 17$$

The rounding off must be downward so that $[1 + \rho(A - 1)]$ will be positive.

Since ρ decays as the interval between the elements within the cluster increases, the magnitude of β increases as the cluster configuration shifts from a compact to an open or extended format. In other words, as the elements making up the cluster become more dispersed, the variates tend to become more varied. This causes the within cluster variance to increase and the between cluster variance to decrease, which, in turn, causes the variance of the mean to decrease. Cluster configuration can have a considerable impact on the magnitude of the between cluster variance and the variance of the mean, but there is no way to quantify it so that it can be included as a factor in a variance function. The only way to handle this problem is to recognize a series of configuration classes and to determine the appropriate magnitude of β for each class. Some possible configuration classes are (1) compact square; (2) compact strip; (3) open square, with a specified interval, in both directions, between the elements making up the square; and (4) open strip, with a specified interval between the elements making up the strip. It must be remembered that the clusters can interlock but not overlap. Overlapping causes differences in variate weights which violates the assumption of constant weight upon which the variance function concept is based.

In some cases the variance of the total, rather than that of the mean, is to be estimated. According to Equation 17.57

$$\sigma_\tau^2 / A^2 = \sigma_B^2$$

which, since the correction terms cancel out, can be written

$$S_\tau^2 / A^2 = S_B^2$$

Then

$$S_\tau^2 = A^2 S_B^2 = A^2 S_o^2 / A^\beta = s_o^2 A^{2-\beta} \tag{17.106}$$

According to Equation 17.40, when $N_i = n_i = A$,

$$V(T_o)_{1S} = M^2 \frac{\sigma_\tau^2}{m}\left(\frac{M-m}{M-1}\right)$$

which can be written

$$V(T_o)_{1S} = M^2 \frac{S_\tau^2}{m}\left(\frac{M-m}{M}\right)$$

In the context of this discussion, $m = 1$. Thus,

$$V(T_o)_{1S} = M^2 S_\tau^2 \left(\frac{M-1}{M}\right)$$

or

$$= M^2 S_o^2 A^{2-\beta}\left(\frac{M-1}{M}\right)$$

Let $\alpha = 2 = \beta$ and $\gamma = M^2 S_o^2$. Then,

$$V(T_o)_{1S} = \gamma A^\alpha \left(\frac{M-1}{M}\right), \quad 2 \geq \alpha \geq 1 \tag{17.107}$$

Often the correction term is essentially equal to unity. Then,

$$V(T_o)_{1S} \doteq \gamma A^\alpha \tag{17.108}$$

Mahalanobis (1940), working independently, arrived at essentially the same model as did Smith, but stated it in terms of proportions:

$$V(\hat{p}_o) = pq / A^{\beta} \quad (17.109)$$

The logarithmic transformation of Smith's model (Equation 17.103) is

$$\log V(\bar{y}_o) = \log S_B^2 = \log S_o^2 + \beta \log A \quad (17.110)$$

while that of Mahalanobis's model (Equation 17.109) is

$$\log V(\hat{p}_o) = \log(pq) + \beta \log A \quad (17.111)$$

These are linear functions that can be fitted to actual data using conventional least squares procedures. Mahalanobis (1940) tested his model and Jessen (1942) tested Smith's model, using data from agricultural surveys, and found they predicted $S_B^2$ with acceptable precision. However, Hendricks (1944) and Asthana (1950) found that Smith's model underestimated the actual $S_B^2$ value when cluster size became large.

Jessen (1942) approached the problem of developing a variance function from a direction different from that used by Smith and Mahalanobis. He suggested that

$$\log S_w^2 = \log \alpha + \beta \log A \quad (17.112)$$

where $\alpha$ and $\beta$ are constants to be determined by fitting the model to sample data. In antilog form the function is

$$S_w^2 = \alpha A^{\beta} \quad (17.113)$$

According to Equation 17.76,

$$\sigma_o^2 = \sigma_B^2 + \sigma_w^2$$

Then,

$$\sigma_B^2 = \sigma_o^2 - \sigma_w^2$$

which can be written

$$S_B^2\left(\frac{M-1}{M}\right) = S_o^2\left(\frac{MA-1}{MA}\right) - S_w^2\left(\frac{A}{A-1}\right)$$

$$S_B^2 = \left(\frac{M}{M-1}\right)\left[S_o^2\left(\frac{MA-1}{MA}\right) - S_w^2\left(\frac{A}{A-1}\right)\right]$$

Substituting the right side of Equation 17.113 for $S_w^2$,

$$S_B^2 = \left(\frac{M}{M-1}\right)\left[S_o^2\left(\frac{MA-1}{MA}\right) - \alpha A^{\beta}\left(\frac{A-1}{A}\right)\right]$$

Since $MA = N$ and $M = N/A$,

$$S_B^2 = S_o^2\left(\frac{N-1}{N-A}\right) - \alpha A^{\beta}\left(\frac{N}{A}\right)\left(\frac{A-1}{N-A}\right) \tag{17.114}$$

According to Equation 17.96,

$$V(\bar{y}_o) = \frac{S_B^2}{m} - \frac{S_B^2}{M} = S_B^2\left(\frac{1}{m} - \frac{1}{M}\right)$$

In this case $m = 1$ and $1/M = A/n$. Then,

$$V(\bar{y}_o) = S_B^2\left(1 - \frac{A}{N}\right) = S_B^2\left(\frac{N-A}{N}\right) \tag{17.115}$$

Substituting the right side of Equation 17.114 for $S_B^2$

$$\begin{aligned} V(\bar{y}_o) &= \left[S_o^2\left(\frac{N-1}{N-A}\right) - \alpha A^{\beta}\left(\frac{N}{A}\right)\left(\frac{A-1}{N-A}\right)\right]\left(\frac{N-A}{N}\right) \\ &= S_o^2\left(\frac{N-1}{N}\right) - \alpha A^{\beta}\left(\frac{A-1}{A}\right) \end{aligned} \tag{17.116}$$

If $N$ is large, $[(N - 1)/N]$ can be considered to be equal to unity. Then

$$V(\bar{y}_o) \doteq S_o^2 - \alpha A^{\beta}\left(\frac{A-1}{A}\right) \tag{17.117}$$

Again, the intracluster correlation controls the magnitude of the exponent $\beta$, so that it increases as $\rho$ increases. Again, cluster configuration is a factor controlling $V(\bar{y}_o)$ which can be incorporated into the variance functions only through the use of configuration classes.

Hansen et al. (1953) described a variance function which made use of the relationship between the intraclass correlation coefficient and cluster size:

$$\log \rho = \log \alpha + \beta \log A \tag{17.118}$$

or

$$\rho = \alpha A^{\beta} \tag{17.119}$$

where the constants $\alpha$ and $\beta$ are determined, as in the case of Jessen's model, by fitting the logarithmic model to field data. According to Equation 17.74,

$$\sigma_B^2 = \frac{\sigma_o^2}{A}[1 + \rho(A-1)]$$

or

$$S_B^2\left(\frac{M-1}{M}\right)=\frac{S_o^2}{A}\left(\frac{N-1}{N}\right)[1+\rho(A-1)]$$

Since $M = N/A$, $[(M - 1)/M] = [(N - A)/N]$. Then,

$$S_B^2=\frac{S_o^2}{A}\left(\frac{N-1}{N-A}\right)[1+\rho(A-1)]$$

Substituting the right side of Equation 17.119 for ρ,

$$S_B^2=\frac{S_o^2}{A}\left(\frac{N-1}{N-A}\right)[1+\alpha A^{\beta}(A-1)] \tag{17.120}$$

Substituting the right side of Equation 17.120 for $S_B^2$ in Equation 17.115:

$$\begin{aligned} V(\bar{y}_o)&=\frac{S_o^2}{A}\left(\frac{N-1}{N-A}\right)[1+\alpha A^{\beta}(A-1)]\left(\frac{N-A}{N}\right)\\ &=\frac{S_o^2}{A}\left(\frac{N-1}{N}\right)[1+\alpha A^{\beta}(A-1)] \end{aligned} \tag{17.121}$$

When $N$ is large, $[(N - 1)/N] \to 1$, and

$$V(\bar{y}_o)\doteq\frac{S_o^2}{A}[1+\alpha A^{\beta}(A-1)]$$

Unlike the Smith, Mahalanobis, and Jessen models, this one by Hansen, Hurwitz, and Madow does take cluster configuration into account. The intraclass correlation coefficient is based on the similarities of the variates within a cluster. As cluster configuration changes from compact to open, the variates become less alike and the intraclass correlation decays, which is recognized by the model.

To construct one of these variance functions one preferably should have a complete enumeration, *by position within the frame*, of a population which is typical of the populations which are going to be subsequently sampled. An example is the 16-ha tract shown in Figure 16.1. Such data are rarely available to persons involved with forest inventory. They can be found, however, for certain fields, greenhouses, and growth chambers used in agricultural research. The data, in such cases, are obtained as the result of *uniformity trials*, which are carried out to determine if the environmental conditions throughout the test area are sufficiently uniform that they will not compromise the results of experiments carried out in those areas.

As an example of variance function construction, one can make use of Smith's model. According to Equation 17.110,

$$\log V(\bar{y}_o)=\log S_B^2=\log S_o^2+\beta\log A$$

To obtain the most nearly correct value of β, one must obtain an $S_B^2$ value for each possible cluster size within a specified cluster configuration class. The logarithm of each of these $S_B^2$ values must be appropriately weighted. A regression of log $S_B^2$ (weighted) on log $A$, conditioned to pass through the point (log $A = 0$, log $S_B^2$ = log $S_o^2$), is then fitted and β determined. The question of appropriate weights is complex, and a meaningful discussion of it would be too lengthy for inclusion here. For more information see Smith (1938), Koch and Rigney (1951), and Hatheway and Williams (1958).

Because complete enumerations are rarely available, those interested in variance functions usually make use of sample data to obtain estimates of the needed constants. In the case of Smith's model one only needs an estimate of the exponent β. To do this, $m$ clusters of a specified size and configuration are randomly chosen from the population. Preferably the chosen cluster size should be at or near the maximum which would be considered feasible to use. Then,

$$s_o^2 = \frac{1}{(mA-1)} \sum_i^m \sum_j^A (y_{ij} - \bar{y}_o)^2 \rightarrow S_o^2$$
$$s_B^2 = \frac{1}{(m-1)} \sum_i^m (\bar{y}_i - \bar{y}_o)^2 \rightarrow S_B^2 \tag{17.122}$$

The estimate of β is then obtained using the method of moments:

$$b = \frac{\log s_o^2 - \log s_B^2}{\log 1 - \log A} = \frac{\log s_o^2 - \log S_B^2}{-\log A} \rightarrow \beta \tag{17.123}$$

As can be seen, $b$ is negative, as indicated by the model. Then,

$$s_{\bar{y}_o}^2 = s_o^2 A^{-b} = s_o^2 / A^b \rightarrow V(\bar{y}_o) \tag{17.124}$$

As an example of this process, one can use the 16-ha tract and the compact square clusters of size $A = 16$ shown in Figure 17.2. Then, from Section 17.8.2

| Cluster | $\mu_i$ (m³/plot) | $\sum_j^{16} y_{ij}$ | $\sum_j^{16} y_{ij}^2$ |
|---|---|---|---|
| 1 | 3.03125 | 48.50 | 160.7650 |
| 2 | 3.28125 | 52.50 | 186.0950 |
| 3 | 1.94688 | 31.15 | 63.5375 |

$$\bar{y}_o = 2.753125 \text{ m}^3 / \text{plot}$$

$$s_B^2 = \frac{1}{(3-1)}[(3.03125 - 2.753125)^2 + (3.28125 - 2.753125)^2 + (1.94688 - 2.753125)^2]$$
$$= 0.503150$$

$$s_o^2 = \frac{1}{3(16)-1}[(160.1650 + 186.0950 + 63.5375) - (48.50 + 52.50 + 31.15)^2 / 48]$$
$$= 0.990894$$

$$b = \frac{\log 0.990894 - \log 0.503150}{-\log 16}$$
$$= \frac{-0.0039728 - (-0.298303)}{-1.20412} = -0.2444359$$

and

$$s_{\bar{y}_o}^2 = 0.990894A^{-0.2444359}$$

Then, when $A = 1$, $s_{\bar{y}_c}^2 = 0.990894$, as it should; when $A = 4$, $s_{\bar{y}_c}^2 = 0.706093$; and when $A = 25$, $s_{\bar{y}_o}^2 = 0.451149$.

To illustrate the effect of cluster configuration, the following repeats the above procedure using the data from Section 17.8.6, where the cluster size $A = 16$ and the sample size $m = 3$, as before, but the clusters are spread out (see Figure 17.9) instead of being compact. Then, from Section 17.8.6:

| Cluster | $\mu_i$ | $\sum_j^{16} y_{ij}$ | $\sum_j^{16} y_{ij}^2$ |
|---|---|---|---|
| 1 | 2.090625 | 33.45 | 107.2675 |
| 2 | 1.871875 | 29.95 | 76.9925 |
| 3 | 1.887500 | 30.20 | 80.0050 |

$$\bar{y}_o = 1.95 \text{ m}^3 / \text{plot}$$

$$s_B^2 = \frac{1}{(3-1)}[(2.090625-1.95)^2 + (1.871875-1.95)^2 + (1.887500-1.95)^2]$$

$$= 0.014893$$

$$s_o^2 = \frac{1}{3(16)-1}[(107.2675+76.9925+80.0050) - (33.45+29.95+30.20)^2 / 48]$$

$$= 1.739255$$

$$b = \frac{\log 1.739255 - \log 0.014893}{-\log 16}$$

$$= \frac{0.2403633 - (-1.8270178)}{-1.20412} = -1.7169228$$

$$s_{\bar{y}_o}^2 = 1.739255A^{-1.7169228}$$

When $A = 1$, $s_{\bar{y}_o}^2 = 1.739255$; when $A = 4$, $s_{\bar{y}_o}^2 = 0.1609432$; and when $A = 25$, $s_{\bar{y}_o}^2 = 0.006922$.

As can be seen, the magnitudes of the two $b$ are dramatically different, even though the cluster and sample sizes used in the two examples were the same. As the clusters are spread out, the intraclass correlation weakens and the $\beta$ value increases. These results demonstrate why cluster configuration classes should be recognized where developing variance functions.

Now consider Jessen's model. It will be necessary to compute values for, or estimates of, $\alpha$ and $\beta$. This can be done by fitting a regression through the set of $S_w^2$ values associated with all, or as many as is feasible, possible cluster sizes within a specified cluster configuration class, plotted against the cluster sizes. To linearize the model it is transformed into its logarithmic form (see Equation 17.112). A simpler approach is to use only two cluster sizes, which should be as widely different as is feasible, to develop two equations that can be solved simultaneously for the two constants.

As an example of this latter procedure one can use, instead of samples, *all* the possible clusters of sizes $A = 4$ and $A = 80$, from the 16-ha tract. Where $A = 4$, the clusters are in the form of $2 \times 2$ element compact squares and when $A = 80$ they are in the form of $8 \times 10$ element compact rectangles. These 80-element clusters are sufficiently similar to squares that they can be considered to be in the same configuration class. In the case of the 4-element clusters the analysis of variance is

ANOVA

| Source | df | SS | MS |
|---|---|---|---|
| Total | 320 – 1 = 319 | 510.51751 | $1.600368 = S_o^2$ |
| Within clusters | 80(4 – 1) = 240 | 101.17313 | $0.421555 = S_w^2$ |
| Between clusters | 80 – 1 = 79 | 409.34438 | $5.181574 = AS_B^2$ |

The analysis of variance for the 80-element clusters is:

ANOVA

| Source | df | SS | MS |
|---|---|---|---|
| Total | 320 – 1 = 319 | 510.51751 | $1.600368 = S_o^2$ |
| Within clusters | 4(80 – 1) = 316 | 315.00303 | $0.996845 = S_w^2$ |
| Between clusters | 4 – 1 = 3 | 195.51448 | $65.171492 = AS_B^2$ |

The two equations are then

$$\begin{bmatrix} \log 0.421555 = \log a + b \log 4 \\ \log 0.996845 = \log a + b \log 80 \end{bmatrix}$$

$$\begin{bmatrix} -0.3750994 = \log a + 0.60206b \\ -0.001372 = \log a + 1.90309b \end{bmatrix}$$

Multiplying the first equation by 1.90309 and the second by –0.60206 and then adding the resulting equation yields:

$$0.713111 = 1.30103 \log a$$

$$\log a = -0.713111/1.30103 = -0.548113$$

and

$$a = 0.283066$$

Substituting –0.548113 for log $a$ in the first original equation and solving for $b$:

$$b = 0.287292$$

Then $s_w^2 = 0.283066A^{0.287292}$

Substituting these values of $a$ and $b$ into Equation 17.96 yields

$$V(\bar{y}_o) = 1.600368\left(\frac{319}{320}\right) - 0.283066A^{0.287292}$$

When $A = 1$, $V(\bar{y}_o) = 1.312301$; when $A = 4$, $V(\bar{y}_o) = 1.173812$; and when $A = 25$, $V(\bar{y}_o) = 0.881685$.

## 17.10 COST FUNCTIONS

The costs associated with simple cluster sampling can be broken down into the following categories:

$c_T$ = total cost of the inventory
$c_o$ = overhead or nonfield costs
$c_F$ = field costs, which should include the cost of travel from a base of operations to and from the field

The fundamental cost function is then

$$c_T = c_o + c_F \tag{17.125}$$

The costs that are of interest to the person designing a sampling operation are those associated with the field operations. The basic field cost function, assuming constant cluster size, is

$$c_F = c_1 mA + c_2(m - 1) + c_3 + c_4 \tag{17.126}$$

where $c_1$ = cost of locating and evaluating one element within a cluster
$c_2$ = average cost of travel between two clusters
$c_3$ = cost of travel from vehicle to first cluster and from last cluster to vehicle
$c_4$ = cost of travel to and from field from base of operations

Note that the equation recognizes that there are $(m - 1)$ intervals between clusters. Equation 17.126 can be made more detailed:

$$\begin{aligned} c_F = {} & mA[d_w c_w + t_w c_L + c_L(d_w / s_w)] + (m-1)[d_B c_B + c_L(d_B / s_B)] \\ & + c_L(d_D / s_D) + [c_L(d_T / s_T) + c_T d_T] \end{aligned} \tag{17.127}$$

where $d_w$ = average distance between element centers *within* a cluster, expressed in feet, meters, chains, etc. When a cluster has a fixed configuration, this can be obtained from a plan of the cluster.
$c_w$ = cost per unit distance traveled within the cluster. *This does not refer to labor costs*. It is appropriate only if a vehicle would be needed to move from element to element. If the travel is on foot, the time needed would be included in $t_w$.
$t_w$ = time spent evaluating an element within a cluster, expressed in units such as seconds, minutes, or hours.
$c_L$ = labor costs per unit time, including *per diem* expenses.
$s_w$ = speed of travel within a cluster, in distance units per time unit such as feet, meters, or chains per minute.
$d_B$ = average distance between clusters, expressed in feet, meters, chains, etc. This is probably best obtained by plotting the cluster positions on a map of the tract, laying out what appears to be a logical path connecting the clusters, measuring the distance to be traveled between the clusters, and determining the mean of those distances. In a

general sense, $d_B$ is a function of tract size and shape and mode of travel (on foot, by car or truck, by boat, or by helicopter). The mode of travel determines whether the path to be traveled will be beeline or crooked, while tract size and shape control the distances between the clusters. According to Jessen (1942) if a tract is rectangular, with a length of $x$ units and breadth of $y$ units, the mean distance between pairs of randomly located points is

$$d = \sqrt{\frac{x^2 + y^2}{6}} \tag{17.128}$$

If the tract is square, with sides $x$ units in length,

$$d = \sqrt{\frac{x^2}{3}} \tag{17.129}$$

In the case of a tract that is not a rectangle or square, $d$ can only be estimated by measuring, on a map, the distance between a set of randomly located points.

$c_B$ = cost of crew per unit traveled between cluster centers. The term refers *only* to *nonfoot* travel. It does not include labor costs for those being transported.

$s_B$ = speed of travel between cluster centers, in distance units per time units.

$d_D$ = distance traveled from parked vehicle to the first cluster and from the last cluster back to the vehicle.

$d_T$ = distance from tract to base of operations.

$s_T$ = speed of travel to and from the field.

$c_T$ = cost of operating vehicle to and from the field.

As an example, one can estimate the cost of the inventory of the 16-ha tract described in Section 17.8.6. In this inventory the sample consists of three 16-element clusters. The clusters have a spread configuration, being made up of four rows of four 0.05-ha plots (elements) forming a rectangle, as can be seen in Figure 17.9. The within-row interval between plots is 3 elements and the between-row interval is 4 elements.

Assume that the tract is 10 miles from the cruiser's office, that one person will carry out the cruise, and that his or her vehicle will be brought to the northwestern corner of the tract. Further assume that the cruiser will begin working through each cluster at the plot forming the northwestern corner of the cluster. The cruiser's path will be eastward along the first row of plots, then south to the next row, then westward, etc., forming a zigzag pattern which ends at the southwestern corner of the cluster. After completing the cluster, the cruiser will proceed to the northeastern plot of the next cluster. More efficient between-cluster travel patterns can be developed, but this assumption keeps the problem simple. All the travel within and between the clusters will be on foot.

Each plot is a 0.05-ha or 500-m$^2$ square. Each side has a length of 22.4 m. Since within a cluster there are 12 intervals of 4(22.4) = 896 meters to be traveled between plots in an east–west direction and 3 intervals of 5(22.4) = 112.0 m in a southerly direction, the total distance traveled within a cluster is 12(89.6) + 3(112.0)\1411.2 or 1411 m. The average distance between plots is then

$$d_w = 1411/15 = 94.1 \text{ m}$$

Since all travel will be on foot,

$$c_w = 0$$

The time needed to cruise a plot will be assumed to be

$$t_w = 10 \text{ min}$$

It will be assumed that the cruiser can travel at an average speed of 2 mph. Since 2 miles = 2(5280) = 10,560 ft or 3218.7 m, 2 mph = 3218.7 m/hour or

$$s_w = 53.645 \text{ m/min}$$

As was mentioned earlier, it is assumed that the cruiser will complete the evaluation of each cluster at its southwestern corner plot and then will proceed to the northwestern corner plot of the next cluster. These distances will be obtained from measurements on the map:

SW corner of cluster 1 to NW corner of cluster 2 = 395 m
SW corner of cluster 1 to NW corner of cluster 3 = 262 m

Then, $d_B = (395+262)/2 = 328.5$ m.

Since the travel between clusters will be on foot,

$$c_B = 0$$

Assume that speed of travel between clusters will be the same as that within the clusters. Then,

$$s_B = 53.645 \text{ m/min}$$

It was stated earlier that the cruiser would be able to drive a vehicle to the northwestern corner of the tract. Then the distance from the vehicle to the northwestern plot in cluster 1 is, from map measurement, 81 m. The distance from the southwestern plot in cluster 3 to the vehicle is 450 m. Then,

$$d_D = 81 + 450 = 531 \text{ m}$$

Since the cruiser will be walking out,

$$s_D = 53.645 \text{ m/min}$$

The distance from the home office to the tract is 10 miles. Thus, the roundtrip will cover

$$d_T = 2(10) = 20 \text{ miles}$$

It will be assumed that the average speed of the vehicle in this trip will be

$$s_T = 50 \text{ miles/hour or } 0.8333 \text{ miles/min}$$

The cost of operating the vehicle is

$$c_T = \$0.25\text{/mile}$$

The cruiser will be paid \$15/hour or

$$c_L = \$0.25\text{/min}$$

The cost of evaluating one element within a cluster is then

$$\begin{aligned} c_1 &= [d_w c_w + t_w c_L + c_L(d_w / s_w)] \\ &= [94.1(0) + 10(\$0.25) + \$0.25(94.1/53.645)] \\ &= \$2.67\text{/plot cruised} \end{aligned}$$

The cost of cruising one cluster:

$$Ac_1 = 16(\$2.67) = \$42.72$$

and the cost of cruising the three clusters is

$$mAc_1 = 3(\$42.72) = \$128.16$$

Since all travel will be on foot there will be no between-cluster transportation costs:

$$d_B c_B = 328.5(0) = 0$$

The expected average travel time between clusters will be

$$d_B/s_B = 328.5/53.645 = 6.12 \text{ min}$$

The average cost of such travel is

$$c_2 = \$0.25(6.12) = \$1.53$$

and

$$(m - 1)c_2 = (3 - 1)(\$1.53) = \$3.06$$

Travel into and out of the woods will require

$$d_D/s_D = 531/53.645 = 9.9 \text{ min}$$

and its cost will be

$$c_3 = c_L(d_D/s_D) = \$0.25(9.9) = \$2.48$$

Travel time to and from the job will be

$$d_T/s_T = 20/0.8333 = 24 \text{ min}$$

Labor costs while traveling to and from the woods are

$$c_L(d_T/s_T) = \$0.25(24) = \$6.00$$

Vehicle operation cost will be

$$c_4 = c_T d_T = \$0.25(20) = \$5.00$$

The total cost of the field operations in the cruise described in Section 17.8.6 is then estimated to be

$$c_F = \$128.16 + \$3.06 + \$2.48 + \$6.00 + \$5.00 = \$144.70$$

## 17.11 EFFICIENCY OF CLUSTER SAMPLING

The relative efficiencies of two different sampling designs can be ascertained by comparing their respective variances of the mean while holding sample size, in terms of elements of the universe of interest, constant. In this discussion of the efficiency of cluster sampling, it will be compared with that of simple random sampling.

### 17.11.1 Two-Stage Cluster Sampling

It will be assumed that $N_i = A = N/M$ and $n_i = a = n/m$. According to Equation 17.78,

$$V(\bar{y}_o)_{2s} = \frac{\sigma_o^2}{ma}\left(\frac{A-a}{A}\right)(1-\rho) + \frac{\sigma_o^2}{mA}[1+\rho(A-1)]\left(\frac{M-m}{M-1}\right)$$

Multiplying the second term by $(a/a)$,

$$V(\bar{y}_o)_{2s} = \frac{\sigma_o^2}{ma}\left(\frac{A-a}{A}\right)(1-\rho) + \frac{\sigma_o^2}{mA}\left(\frac{M-m}{M-1}\right)\left(\frac{a}{A}\right)[1+\rho(A-1)] \tag{17.130}$$

The first term on the right is the within-cluster component and the second term is the between-cluster component.

When simple random sampling without replacement is used

$$V(\bar{y}_o)_{\text{Ran}} = \frac{\sigma_o^2}{n}\left(\frac{N-n}{N-1}\right) \tag{17.131}$$

Under the conditions of this development $MA = N$ and $ma = n$. Substituting these in Equation 17.131,

$$V(\bar{y}_o)_{\text{Ran}} = \frac{\sigma_o^2}{n}\left(\frac{MA-ma}{MA-1}\right) \tag{17.132}$$

Then

$$\frac{\sigma_o^2}{ma} = V(\bar{y}_o)_{\text{Ran}}\left(\frac{MA-1}{MA-ma}\right) \tag{17.133}$$

Substituting the right side of Equation 17.133 for $\sigma_o^2$/ma in Equation 17.130 and then factoring out $V(\bar{y}_o)_{\text{Ran}}$ yields

$$V(\bar{y}_o)_{2S} = V(\bar{y}_o)_{\text{Ran}} \left\{ \left( \frac{MA-1}{MA-ma} \right) \left( \frac{A-a}{A} \right) (1-\rho) + \left( \frac{MA-1}{MA-ma} \right) \left( \frac{M-m}{M-1} \right) \left( \frac{a}{A} \right) [1+\rho(A-1)] \right\} \quad (17.134)$$

The efficiency of two-stage cluster sampling compared with that of simple random sampling is then measured by the reciprocal of the ratio $V(\bar{y}_o)_{2S} / V(\bar{y}_o)_{\text{Ran}}$. Thus, when $V(\bar{y}_o)_{2S} > V(\bar{y}_o)_{\text{Ran}}$ the relative efficiency of two-stage cluster sampling will be less than that of simple random sampling. Only when the term within the braces in Equation 17.134 equals unity will $V(\bar{y}_o)_{2S} = V(\bar{y}_o)_{\text{Ran}}$. When $M$, $A$, $m$, and $a$ are held constant and the term within the braces is set equal to 1, the intraclass correlation coefficient is

$$\rho_{cR} = \frac{(MA-ma)(M-1)A - (MA-1)(A-a)(M-1) - (MA-1)(M-m)a}{(MA-1)(M-m)(A-1)a - (MA-1)(A-a)(M-1)} \quad (17.135)$$

This computed coefficient can be referred to as the *critical* value of $\rho$ (hence the subscript $cR$). When the actual value of $\rho = \rho_{cR}$, $V(\bar{y}_o)_{2S} = V(\bar{y}_o)_{\text{Ran}}$ and the efficiency of two-stage cluster sampling will be equal to that of a simple random sampling operation based on the same sample size.

The critical value of $\rho$ is not a constant. It is actually a function of population size. Regardless of the individual magnitudes of $M$, $A$, $m$, and $a$, as long as $MA$ remains constant, $\rho_{cR}$ will remain constant. For example, if $MA = 15$, $\rho_{cR} = -0.071429$; if $MA = 1000$, $\rho_{cR} = -0.001001$; and if $MA = \infty$, $\rho_{cR} = 0$.

It should be noted that when $\rho$ is negative, in most cases the efficiency of two-stage cluster sampling will be greater than that of simple random sampling. This is demonstrated in Table 17.1, which shows the percent increase or decrease in efficiency of two-stage cluster sampling, relative to simple random sampling, as $M$, $A$, $m$, $a$, and $\rho$ change magnitudes. For example, when $MA = 100$, $M = 20$, $A = 5$, $m = 10$, $a = 2$, and $\rho = 0.1$, two-stage cluster sampling is 3.30% less efficient than simple random sampling. Usually the efficiency of two-stage cluster sampling is poorer than that of simple random sampling, with the difference increasing as $\rho$ increases. This trend is accentuated as cluster size, $A$, increases.

The evidence provided by Equation 17.134 and Table 17.1 leads to the following conclusions:

1. Two-stage cluster sampling is usually less efficient than simple random sampling.
2. As $\rho$ increases (i.e., as the variates within the cluster become more nearly alike), the relative efficiency of two-stage cluster sampling decreases.
3. When $\rho$ is positive, regardless of magnitude, the relative efficiency of two-stage cluster sampling decreases as $a$ increases and $m$ decreases.
4. When $\rho$ is negative the relative efficiency of two-stage cluster sampling is greater than that of simple random sampling.
5. When $\rho = 0$, the difference between $V(\bar{y}_o)_{2S}$ and $V(\bar{y}_o)_{\text{Ran}}$ is small and it decreases as the population size increases, becoming equal to zero when $N = \infty$.
6. Sample size, per se, has little effect on the relative efficiency of two-stage cluster sampling.
7. When sample size is held constant, the relative efficiency of two-stage cluster sampling declines as subsample size, $a$, increases. It is poorest when $a = A$.
8. When $MA$, $m$, and $a$ are held constant, the relative efficiency of two-stage cluster sampling increases as cluster size, $A$, increases.

**TABLE 17.1**
**A Demonstration of the Effects of $M$, $A$, $m$, $a$, and $\rho$ on the Efficiency of Two-Stage Cluster Sampling Compared with Simple Random Sampling When Sample Size is Held Constant**

| | | $\rho$ | | | | |
|---|---|---|---|---|---|---|
| $m$ | $a$ | −0.1 | 0.0 | 0.1 | 0.2 | 1.0 |
| | | $MA = 100$; $M = 20$; $A = 5$; $ma = 5$ | | | | |
| 5 | 1 | −1.58 | +0.18 | +1.93 | +3.69 | +17.73 |
| 1 | 5 | +37.47 | −4.21 | −45.90 | −87.58 | −421.10 |
| | | $MA = 100$; $M = 20$; $A = 5$; $ma = 10$ | | | | |
| 10 | 1 | −3.75 | +0.42 | +4.59 | +8.76 | +42.11 |
| 5 | 2 | +6.56 | −0.74 | −8.03 | −15.33 | −73.67 |
| 2 | 5 | +37.47 | −4.21 | −45.90 | −87.58 | −421.10 |
| | | $MA = 100$; $M = 20$; $A = 5$; $ma = 20$ | | | | |
| 20 | 1 | −8.90 | +1.00 | +10.90 | +20.80 | — |
| 10 | 2 | +2.69 | −0.30 | −3.30 | −6.29 | −30.26 |
| 5 | 4 | +25.88 | −2.91 | −31.70 | −60.49 | −290.78 |
| 4 | 5 | +37.47 | −4.21 | −45.90 | −87.58 | −421.10 |
| | | $MA = 100$; $M = 10$; $A = 10$; $ma = 10$ | | | | |
| 10 | 1 | −8.90 | +1.00 | +10.90 | +20.80 | — |
| 5 | 2 | +1.98 | −0.22 | −2.42 | −4.62 | −22.22 |
| 2 | 5 | +34.61 | −3.89 | −42.39 | −80.90 | −389.00 |
| 1 | 10 | +89.00 | −10.00 | −108.99 | −207.98 | −1000.11 |
| | | $MA = 100$; $M = 10$; $A = 10$; $ma = 20$ | | | | |
| 10 | 2 | −8.90 | +1.00 | +10.90 | +20.80 | — |
| 5 | 4 | +15.57 | −1.75 | −19.08 | −36.41 | −175.03 |
| 4 | 5 | +27.81 | −3.72 | −34.07 | −64.99 | −312.54 |
| 2 | 10 | +89.00 | −10.00 | −108.99 | −207.98 | −1000.11 |
| | | $MA = 1000$; $M = 50$; $A = 20$; $ma = 20$ | | | | |
| 20 | 1 | −3.72 | +0.04 | +3.79 | +7.55 | +37.59 |
| 10 | 2 | +6.57 | −0.07 | −6.70 | −13.34 | −66.42 |
| 5 | 4 | +27.15 | −0.27 | −27.70 | −55.11 | −274.53 |
| 4 | 5 | +37.43 | −0.38 | −38.20 | −75.99 | −378.47 |
| 2 | 10 | +89.81 | −0.90 | −90.66 | −180.43 | −899.00 |
| 1 | 20 | — | −1.94 | −195.60 | −389.24 | −1940.82 |

*Note:* The tabular values are $[1 - V(\bar{y}_o)_{1s} / V(\bar{y}_o)_{\text{Ran}}]100$, which show the percent increase or decrease in efficiency when two-stage cluster sampling is used.

Another factor which must be recognized is that adjacent elements in most real-life universes are more likely to be similar than are nonadjacent elements. Consequently, compact clusters, where the elements are close together, as in Figure 17.5, are apt to have smaller within-cluster variances than similar-sized clusters in which the elements are spread out, as in Figure 17.9. Furthermore, large clusters (when $A$ is large) are apt to have larger within-cluster variances than small clusters since they are more likely to include a set of variates exhibiting a wider range of magnitudes.

Thus, cluster size and configuration will have significant impacts on the efficiency of two-stage cluster sampling.

The designer of a two-stage cluster sampling operation must recognize the influence cluster size and configuration have on the efficiency of the operation. Since cluster sampling is usually chosen because it can result in less time being spent in nonproductive travel than would be the case with simple random sampling, one would probably choose to develop a sampling frame made up of relatively large compact clusters. They would be large so as to reduce the between-cluster variance as much as possible and compact so as to reduce travel time from element to element within the clusters. $M$ would then be relatively small and $A$ relatively large. Sample size, $n = ma$, is usually controlled by the availability of funds. The evidence indicates that the efficiency of two-stage cluster sampling is greatest, regardless of $\rho$, when $m$ is as large as possible and $a$ as small as possible, remembering that $a$ must be 2 or greater if variability statistics are to be generated. However, if $\rho$ is low, $V(\bar{y}_o)_{2S}$ is not much different from $V(\bar{y}_o)_{\text{Ran}}$ and $m$ and $a$ can be manipulated to meet best the needs of the sampler. Thus, when $\rho$ is low, $m$ can be made relatively small and $a$ relatively large. This would have a beneficial effect on between-cluster travel time since fewer clusters would be involved. When $\rho$ approaches 1, however, the disparity between $V(\bar{y}_o)_{2S}$ and $V(\bar{y}_o)_{\text{Ran}}$ is great and, if the sampling fraction, $ma/MA$, is being held at some constant, it would be preferable to use a relatively large $m$ and a relatively small $a$, to keep $V(\bar{y}_o)_{2S}$ as small as possible. Needless to say, balancing statistical precision and travel time costs in this context is not a simple process. It is aggravated, of course, by the fact that $\rho$ is not known and would have to be estimated using historical or presample data. Since presamples are expensive, especially when a range of cluster sizes and configurations are involved, they are rarely obtained and the design process usually proceeds as though $\rho = 0$ (i.e., as though the variates within the clusters are completely heterogeneous). Consequently, the designer of such a sampling operation has no idea of its efficiency.

### 17.11.2 Simple Cluster Sampling

In simple cluster sampling there is no subsampling and $a = A$. Consequently, the first term within the braces in Equation 17.134 vanishes and the equation becomes

$$V(\bar{y}_o)_{1S} = V(\bar{y}_o)_{\text{Ran}}\left(\frac{MA-1}{MA-mA}\right)\left(\frac{M-m}{M-1}\right)[1+\rho(A-1)] \tag{17.136}$$

Then

$$V(\bar{y}_o)_{1S} / V(\bar{y}_o)_{\text{Ran}} = \left(\frac{MA-1}{MA-mA}\right)\left(\frac{M-m}{M-1}\right)[1+\rho(A-1)] \tag{17.137}$$

Only when the right side of Equation 17.137 equals 1 will $V(\bar{y}_o)_{1S} = V(\bar{y}_o)_{\text{Ran}}$, which will only occur when $\rho = \rho_{cR}$ or when $a = A = 1$.

The percent increase or decrease in the efficiency of simple cluster sampling, compared with that of simple random sampling, is shown in Table 17.2 for a selected set of magnitudes of MA, $M$, $A$, $m$, and $\rho$. As in the case of two-stage cluster sampling, the efficiency of simple cluster sampling is usually poorer than that of simple random sampling and the difference widens as $\rho$ and/or $a$ increases. It also can be seen that as long as $M$, $A$, and $\rho$ remain constant, regardless of sample size, the difference between $V(\bar{y}_o)_{1S}$ and $V(\bar{y}_o)_{\text{Ran}}$ remain constant. If $A$ and $m$ are held constant, the difference between the two decreases very slowly as the population size, $N = MA$, increases. In the limit, as $N \rightarrow \infty$, the difference stabilizes at $[1 + \rho(A - 1)]$.

In general, the recommendations made regarding the design of two-stage cluster sampling operations in Section 17.11.1 are applicable when simple cluster sampling is to be used.

**TABLE 17.2**
**A Demonstration of the Effects of *M*, *A*, *m*, and ρ on the Efficiency of Simple Cluster Sampling Compared with Simple Random Sampling When Sample Size is the Same**

| | | | ρ | | | | |
|---|---|---|---|---|---|---|---|
| *M* | *A* | *m* | −0.1 | 0.0 | 0.1 | 0.2 | 1.0 |
| | | | *MA* = 100; *ma* = *n* = 5 | | | | |
| 100 | 1 | 5 | 0.00 | 0.00 | 0.00 | 0.00 | 0.00 |
| | | | *MA* = 100; *ma* = *n* = 10 | | | | |
| 100 | 1 | 10 | 0.00 | 0.00 | 0.00 | 0.00 | 0.00 |
| 50 | 2 | 5 | +9.08 | −1.02 | −11.12 | −21.22 | −102.04 |
| 20 | 5 | 2 | +37.47 | −4.21 | −45.89 | −87.58 | −421.05 |
| | | | MA = 100; *ma* = *n* = 20 | | | | |
| 100 | 1 | 20 | 0.00 | 0.00 | 0.00 | 0.00 | 0.00 |
| 50 | 2 | 10 | +9.08 | −1.02 | −11.12 | −21.22 | −102.04 |
| 25 | 4 | 5 | +27.81 | −3.13 | −34.06 | −65.00 | −312.50 |
| 20 | 5 | 4 | +37.47 | −4.21 | −45.89 | −87.58 | −421.05 |
| 10 | 10 | 2 | +89.00 | −10.00 | −109.00 | −208.00 | −1,000.00 |
| 5 | 20 | 1 | — | −23.75 | −258.88 | −494.00 | −2,375.00 |
| | | | *MA* = 1000; *ma* = *n* = 20 | | | | |
| 1000 | 1 | 20 | 0.00 | 0.00 | 0.00 | 0.00 | 0.00 |
| 500 | 2 | 10 | +9.91 | −0.10 | −10.11 | −20.12 | −100.20 |
| 250 | 4 | 5 | +29.79 | −0.30 | −30.39 | −60.48 | −301.20 |
| 200 | 5 | 4 | +39.76 | −0.40 | −40.56 | −80.72 | −402.01 |
| 100 | 10 | 2 | +89.91 | −0.91 | −91.73 | −182.55 | −909.09 |
| 50 | 20 | 1 | — | −1.94 | −195.62 | −389.31 | −1,938.78 |

*Note*: The tabular values are $[1 - V(\bar{y}_o)_{1S} / V(\bar{y}_o)_{\text{Ran}}]100$, which show the percent increase or decrease in efficiency when simple cluster sampling is used.

It should be obvious from the evidence presented that, in general, cluster sampling is less efficient, from a statistical point of view, than is simple random sampling. The desirability of cluster sampling depends on nonstatistical factors, such as deadhead costs and the difficulties associated with the construction of an appropriate sampling frame.

## 17.12 SAMPLE SIZE DETERMINATION

Cluster sampling is often the procedure of choice because it provides a means of reducing the cost of nonproductive travel between sampling units. Cost of field operations then becomes a factor when determining the size and allocation of the sample. The objective of the sample size computation operation is to obtain the smallest possible variance (or standard error) of the mean, given a specified field operation budget, the cost of locating a chosen cluster in the field, and the cost of locating and evaluating an element within a chosen cluster. Thus, the process to be described yields an optimum allocation of the sample. It should be noted, however, that the procedure is appropriate only when the cluster size, $N_i$, and, in the case of two-stage sampling, the subsample size, $n_i$, are constant. If they are not constant, the procedure will, in most cases, yield usable approximations of the optimum sample sizes if the average size of the clusters, $\bar{N} = (1/M)\Sigma_i^M N_i$, is used in place of $N_i = A$, and the average subsample size, $\bar{n} = (1/M)\Sigma_i^M n_i$, is used in place of $n_i = a$.

### 17.12.1 Two-Stage Cluster Sampling

According to Equation 17.126, the cost function for field operations is

$$c_F = c_1 mA + c_2(m - 1) + c_3 + c_4$$

Since the costs for travel into and out of the tract and from the base of operations to and from the tract can be considered overhead when determining sample size, they can be eliminated from the cost function, which then becomes

$$c'_F = c_1 mA + c_2(m-1) \tag{17.138}$$

This function is not appropriate when subsampling is involved since it indicates that all $A$ elements in a cluster will be evaluated. The proper expression for two-stage cluster sampling is

$$c'_F = c_1 ma + c_2(m-1) \tag{17.139}$$

According to Equation 17.59, when $N_i = A$ and $n_i = a$,

$$V(\bar{y}_o)_{2S} = \frac{\sigma_w^2}{ma}\left(\frac{A-a}{A-1}\right) + \frac{\sigma_B^2}{m}\left(\frac{M-m}{M-1}\right)$$

which can be written

$$\begin{aligned} V(\bar{y}_o)_{2S} &= \frac{S_w^2}{ma}\left(\frac{A-a}{A}\right) + \frac{S_B^2}{m}\left(\frac{M-m}{M}\right) \\ &= \frac{S_w^2}{ma} - \frac{S_w^2}{mA} + \frac{S_B^2}{m} - \frac{S_B^2}{M} \end{aligned} \tag{17.140}$$

Combining the variance and cost function, using the Lagrangian multiplier, $\lambda$ (see, e.g., Chiang, 1974),

$$\begin{aligned} V &= V(\bar{y}_o)_{2S} + \lambda[c_1 ma + c_2(m-1) = c'_F] \\ &= \frac{S_w^2}{ma} - \frac{S_w^2}{mA} + \frac{S_B^2}{m} - \frac{S_B^2}{M} + \lambda c_1 ma + \lambda c_2 m - \lambda c_2 - \lambda c'_F \end{aligned} \tag{17.141}$$

Then,

$$\frac{\partial V}{\partial m} = \frac{-S_w^2}{m^2 a} + \frac{S_w^2}{m^2 A} - \frac{S_B^2}{m^2} - \lambda c_1 a - \lambda c_2 = 0 \tag{17.142}$$

$$\frac{\partial V}{\partial a} = \frac{-S_w^2}{ma^2} - \lambda c_1 m = 0 \tag{17.143}$$

Multiplying Equation 17.142 by $-(1/a)$, Equation 17.143 by $(1/m)$, and then adding the two resulting equations yields

$$\frac{(S_B^2 - S_w^2 / A)}{m^2 a} + \frac{\lambda c_2}{a} = 0 \tag{17.144}$$

Multiplying Equation 17.143 by $-[(S_B^2 - S_w^2/A)/m]$, Equation 17.144 by $-(S_w^2/a)$, and then adding the resulting equations:

$$\lambda c_1 (S_B^2 - S_w^2 / A) - \lambda c_2 S_w^2 / a^2 = 0 \tag{17.145}$$

which can be written

$$\lambda c_1 (S_B^2 - S_w^2 / A) = \frac{\lambda c_2 S_w^2}{a^2}$$

Then,

$$a^2 = \frac{\lambda c_2 S_w^2}{\lambda c_1 (S_B^2 - S_w^2 / A)} = \frac{c_2 S_w^2}{c_1 (S_B^2 - S_w^2 / A)}$$

and

$$a = \sqrt{\frac{c_2 S_w^2}{c_1 (S_B^2 - S_w^2 / A)}} \tag{17.146}$$

It can also be shown that the magnitude of $a$ can be obtained using the intraclass correlation coefficient:

$$a = \sqrt{\frac{c_1 (1 - \rho)}{c_2 \rho}} \tag{17.147}$$

However, since $\rho$ is rarely, if ever, known in real-life situations, the utility of Equation 17.147 is limited to showing the relationship of $a$ to $\rho$.

The computed magnitude of $a$ will rarely be a whole number and will have to be rounded back to a whole number. This rounding off must be done in such a manner that the following criterion is met:

$$a(a+1) \geq \frac{c_2 S_w^2}{c_1 (S_B^2 - S_w / A)} \geq a(a-1) \tag{17.148}$$

Returning to Equation 17.139 and solving for $m$:

$$c_F' = c_1 ma + c_2 (m - 1) = c_1 ma + c_2 m - c_2 = m(c_1 a + c_2) - c_2$$

Then

$$m = \frac{c'_F + c_2}{c_1 a + c_2} \tag{17.149}$$

From the evidence of Equation 17.146, it can be seen that $a$ is proportional to $\sqrt{c_2 / c_1}$ and will increase as $c_1$ increases relative to $c_2$. According to Equation 17.149, as $a$ increases, $m$ decreases. Furthermore, from Equation 17.147, it can be seen that $a$ will decrease as $\rho$ increases (i.e., as the variates within the clusters become more homogeneous).

Estimates of $a$ and $m$ can be made using presample data. To obtain the estimate of $a$, substitute $s_B^2$ for $S_B^2$ and $s_w^2$ for $S_w^2$ in Equation 17.146. Since these sample values are unbiased estimators of the mean squares, no corrections are needed. $m$ is then estimated using Equation 17.149.

As an example of the sample size computation process, consider the data obtained in the sample in Section 17.8.1 as coming from a presample. Further assume that the amount budgeted for the fieldwork is \$200; the cost of locating a cluster in the field is \$20; and the cost of locating and cruising an individual 1/20-ha plot in a cluster is \$10. Then, $c'_F$ = \$200, $c_2$ = \$20, and $c_1$ = \$10. Then,

$$s_T^2 = 183.04853$$

$$s_B^2 = s_T^2 / A^2 = 183.04853/(16)^2 = 0.715033$$

$$s_1^2 = 0.78000; \quad s_2^2 = 1.37125; \quad s_3^2 = 0.21800$$

$$s_w^2 = (0.78000 + 1.37125 + 0.21800)/3 = 0.78975$$

$$a = \sqrt{\frac{20}{10}\left(\frac{0.78975}{0.715033 - 0.78975/16}\right)} = 1.540384$$

Using the rounding criterion, let

$$\begin{aligned} Q &= s_w^2 c_2 / (s_B^2 - s_w^2 / A) c_1 \\ &= 0.78975(20)/(0.715033 - 0.78975/16)10 = 2.372784 \end{aligned}$$

Rounding up from 1.540384 makes $a$ = 2. Then,

$$a(a + 1) = 2(2 + 1) = 6 > Q$$

and

$$a(a - 1) = 2(2 - 1) = 2 < Q$$

Consequently, $a$ = 2 plots/cluster is appropriate. Then,

$$m = \frac{200 + 20}{10(2) + 20} = 5.5 \text{ or } 5$$

$m$ is rounded back to 5 because the available funds will not permit the inclusion of the extra cluster.

Thus, the design would call for two randomly chosen plots from each of the five clusters.

### 17.12.2 Simple Cluster Sampling

Since simple cluster sampling does not involve subsampling, sample size is determined using the methods of simple random sampling. Thus, if a budgetary constraint controls the sample size, one would use the following cost function (see Equation 17.138) to arrive at the sample size:

$$c'_F = c_2(m-1) \tag{17.150}$$

Then,

$$m = \frac{c'_F - c_2}{c_2} \tag{17.151}$$

For example, if the amount allotted to field operations, exclusive of travel from and to the vehicle used to bring the crew to the inventory area, is \$200 and the cost of locating and evaluating one cluster is, on the average, \$20, then the sample size would be

$$m = (200 - 20)/20 = 9 \text{ clusters}$$

If a sampling intensity is specified, the sample size would be obtained using Equation 16.48:

$$m = \psi M$$

Thus, if the sampling intensity, $\psi$, is to be 10% and the sampling frame contains 100 clusters, the sample size would be

$$m = 0.1(100) = 10 \text{ clusters}$$

If a maximum allowable sampling error is specified and the sampling is without replacement, it would be appropriate to use Equation 13.111:

$$m = t_c^2(k)s_\tau^2 M / [\text{AE}_\tau^2 M + \tau_c^2(k)s_\tau^2$$

Assume that the data in the example in Section 17.8.2 is from a presample and that the allowable sampling error has been set at ±3 m³/ha at 0.90. Since the sampling unit is the cluster, the allowable sampling error must be restated to recognize that the cluster covers only 16/20 of 1 ha. Thus,

$$\text{AE} = \pm(16/20)(3) = \pm 2.4 \text{ m}^3/\text{cluster at } 0.90$$

The variates are then the cluster totals: $\tau_1 = 48.5$; $\tau_2 = 52.5$; and $\tau_3 = 31.15$ m³. The sample mean is

$$\bar{T} = (48.50 + 52.50 + 31.15)/3 = 44.05 \text{ m}^3/\text{cluster}$$

and the sample variance is

$$s_T^2 = [(48.50 - 44.05)^2 + (52.50 - 44.05)^2 + (31.15 - 44.05)^2]/(3 - 1) = 128.8075$$

$$t_{0.90}(2) = 2.920$$

$$m = (2.920)^2(128.8075)(20)/[(2.4)^2(20) + (2.920)^2(128.8075)]$$

$$= 18.10 \text{ or } 19 \text{ clusters}$$

Iterating, $t_{0.90}(18) = 1.734$, and

$$m = (1.734)^2(128.8075)(20)/[(2.4)^2(20) + (1.734)^2(128.8075)]$$
$$= 15.41 \text{ or } 16 \text{ clusters}$$

Iterating, $t_{0.90}(15) = 1.753$, and

$$m = (1.753)^2(128.8075)(20)/[(2.4)^2(20) + (1.753)^2(128.8075)]$$
$$= 15.49 \text{ or } 16 \text{ clusters}$$

Thus, the sample size stabilized at 16 clusters.

This procedure would remain the same regardless of the cluster configuration. For example, if the presample came from the clusters in the example in Section 17.8.6, the sample size would be computed as follows. The variates are now: $\tau_1 = 33.45$; $\tau_2 = 29.95$; and $\tau_3 = 30.20$ m³. The sample mean is

$$\bar{T} = (33.45 + 29.95 + 30.20)/3 = 31.20 \text{ m}^3\text{/cluster}$$

and the sample variance is

$$s_{\bar{T}}^2 = [(33.45 - 31.20)^2 + (29.95 - 31.20)^2 + (30.20 - 31.20)^2]/(3 - 1) = 3.8125$$
$$t_{0.90}(2) = 2.920$$
$$m = (2.920)^2(3.8125)(20)/[(2.4)^2(20) + (2.920)^2(3.8125)]$$
$$= 4.40 \text{ or } 5 \text{ clusters}$$

Iterating, $t_{0.90}(4) = 2.132$

$$m = (2.132)^2(3.8125)(20)/[(2.4)^2(20) + (2.132)^2(3.8125)]$$
$$= 2.62 \text{ or } 3 \text{ clusters}$$

The magnitude of $m$ has not stabilized, but if the iteration process is continued it would cycle between $m = 3$ and $m = 5$. The specification would be met when $m = 5$ but not when $m$ is less than 5. Consequently, $m$ is set at five clusters. It should be noted that the spread configuration used in this example is statistically more efficient than the compact configuration (since fewer clusters are needed to meet the allowable error). However, the cost of locating and evaluating the plots in the spread configuration would be greater than it would be with the compact configuration. As a consequence, the efficiency in terms of field costs could possibly be greater in the case of the compact configuration.

If a maximum allowable sampling error is specified and the sampling frame is infinitely large or, if finite, is very large, it would be appropriate to use Equation 13.109:

$$m = t_c(k)s_{\bar{T}}^2 / \text{AE}_{\bar{T}}^2 \quad \text{or} \quad t_c(k)s_o^2 / \text{AE}_{\bar{y}}^2$$

As an example, assume that the supervisor of a certain national forest wished to determine the area of blown-down timber following a hurricane. This information was needed so that planning for rehabilitation could proceed. To this end aerial photographs were taken of the forest and a map prepared showing the blowdowns. Since the blowdown occurred in patches it was felt that it would be inefficient to use a planimeter to measure these numerous bits and pieces. Instead, the area information would be obtained using a dot grid. The grid was dropped onto the map so as to obtain a random location and orientation. *The set of dots of the grid constituted a cluster.* No subsampling was done.

The allowable sampling error was set at ±10 acres at 0.90. The map was at a scale of 1:7920 (1 in. = 10 chains). The grid had four dots per square inch. As a presample the grid was dropped three times and the dots in the blowdown counted. It should be realized that the sampling frame in this case is infinitely large since there is no limit to the number of grid drops that could be made. Furthermore, the sampling is with replacement because it is possible for a random drop to coincide with another random drop.

An inch on the map is equivalent to 10 chains on the ground. Thus, a square inch is equivalent to 10 * 10 = 100 square chains or 10 acres. Each dot on the grid then represents 10/4 = 2.5 acres. The data obtained from the presample and the associated computations are as follows.

| Grid Drop = Cluster | Number of Dots in Blowdown | $Y$ = Acres in Blowdown |
|---|---|---|
| 1 | 220 | 2.5 (220) = 550 |
| 2 | 236 | 2.5 (236) = 590 |
| 3 | 232 | 2.5 (232) = 580 |

$$\bar{y} = (550 + 590 + 580)/3 = 573.3 \text{ acres of blowdown}$$

$$s_o^2 = [(550 - 573.3)^2 + (590 - 573.3)^2 + (580 - 573.3)^2]/(3 - 1) = 433.335$$

$$t_{0.90}(2) = 2.920$$

$$m = (2.920)^2(433.335)/(10)^2$$

$$= 36.95 \text{ or } 37 \text{ drops}$$

Iterating,

$$t_{0.90}(36) = 1.645$$

$$m = (1.645)^2(433.335)/(10)^2$$

$$= 11.73 \text{ or } 12 \text{ drops}$$

Iterating,

$$t_{0.90}(11) = 1.796$$

$$m = (1.796)^2(433.335)/(10)^2$$

$$= 13.98 \text{ or } 14 \text{ drops}$$

Iterating,

$$t_{0.90}(13) = 1.771$$

$$m = (1.771)^2(433.335)/(10)^2$$

$$= 13.59 \text{ or } 14 \text{ drops}$$

Thus, 14 drops of the dot grid would be needed to meet the allowable specification. It should be realized that, if the areas had been obtained by polar planimeter, there still would be a need for repeated estimates if a sampling error is to be computed.

## 17.13 OPTIMUM CLUSTER SIZE

The discussion in Sections 17.12.1 and 17.12.2 was concerned with optimum sample sizes. The following will explore the possibility of determining the optimum cluster size. In this process it will be assumed that cluster size is constant so that $N_i = A$ and $MA = N$.

The fundamental problem is to arrive at magnitudes for $m$ and A such that $V(\bar{y}_o)_{1S}$ is minimized for a specific cost. Jessen (1942) did this empirically using data from a survey of agriculture in Iowa. These data had been gathered in such a manner that he could aggregate them into different cluster sizes. He found that the efficiency of the sampling process improved as the cluster size increased up to a maximum and then began to decline. He interpreted this to mean that an optimum set of $m$ and $A$ magnitudes existed. He was not able to pinpoint it using his empirical approach but was able to arrive at useful approximations.

One can now see if an algebraic solution for the problem can be found. Again Jessen (1942) provides guidance.

According to Equation 17.97,

$$V(\bar{y}_o)_{1S} = \frac{S_B^2}{m} - S_B^2\left(\frac{A}{N}\right)$$

When $A$ is small compared to $N$, it can be assumed that $A/N \to 0$ and that

$$V(\bar{y}_o)_{1S} \doteq \frac{S_B^2}{m} \tag{17.152}$$

Since the second term in Equation 17.97 causes algebraic difficulties, the following development makes use of Equation 17.152.

Turning now to the cost function. As was done in Section 17.12, one can eliminate the third and fourth terms in Equation 17.126 because the travel costs into and out of the tract and from the base of operations to and from the tract are not functions of the size and number of clusters. Consequently, Equation 17.126 is modified to

$$c_F' = c_1 mA + c_2(m-1)$$

According to Equation 17.127,

$$c_2 = d_B c_B + c_L(d_B / s_B)$$

Note that $c_B$, $c_L$, and $s_B$ are constants but $d_B$ is dependent on the number of clusters. Since $m$ is being sought, a substitute for $d_B$ must be found. Let $D$ be the expected minimum total length of route connecting $m$ clusters. Depending on the situation, the route may be a series of beelines or it can be quite crooked, following roads, trails, or even streams. An estimate of the $D$ associated with a specific number of clusters is obtained by randomly locating $m$ points on a map of the tract in question and measuring the distances between the points following the most logical path. This is done for a range of $m$ values. Thus, one has one or more $D$ magnitudes for each $m$ value. These are related through a regression of $D$ on $\sqrt{m}$. Then

$$D = b\sqrt{m} \tag{17.153}$$

where $b$ is the regression coefficient. The cost function can then be written

$$c'_F = c_1 mA + D(c_B + c_L / s_L)$$
$$= c_1 mA + b(c_B + c_L / S_B)\sqrt{m}$$

Now $b$, $c_B$, $c_L$, and $s_B$ are all constants and one can define

$$c'_2 = b(c_B + c_L / s_B)$$

and

$$c'_F = c_1 mA + c'_2\sqrt{m} \tag{17.154}$$

or

$$c_1 mA + c'_2\sqrt{m} - c'_F = 0 \tag{17.155}$$

Solving for $\sqrt{m}$, using the quadratic equation

$$\sqrt{m} = \frac{-c_2 + \sqrt{c_2^2 + 4c_1 c'_F A}}{2c_1 A} \tag{17.156}$$

and

$$m = (\sqrt{m})^2 \tag{17.157}$$

Combining Equations 17.152 and 17.155, using the Lagrangian multiplier:

$$\begin{aligned} V &= V(\bar{y}_o) + \lambda(c_1 mA + c'_2\sqrt{m} - c'_F) \\ &= \frac{S_B^2}{m} + \lambda c_1 mA + \lambda c'_2\sqrt{m} - \lambda c'_F \end{aligned} \tag{17.158}$$

At this point a variance function is substituted for $S_B^2$. Using Smith's model, Equation 17.103, then

$$V = \frac{S_o^2 A^{-\beta}}{m} + \lambda c_1 mA + \lambda c'_2\sqrt{m} - \lambda c'_F \tag{17.159}$$

and

$$\frac{\partial V}{\partial m} = \frac{-S_o^2 A^{-\beta}}{m^2} + \lambda c_1 A + \frac{\lambda c'_2}{2m} = 0 \tag{17.160}$$

$$\frac{\partial V}{\partial A} = \frac{-\beta S_o^2 A^{-\beta-1}}{m} + \lambda c_1 m = 0 \tag{17.161}$$

Multiplying Equation 17.160 by $-m^2$, Equation 17.161 by $mA/\beta$, and adding the resulting equations, yields

$$\frac{\lambda c_1 A m^2}{\beta} - \lambda c_1 A m^2 - \frac{\lambda c_2' m^2}{2\sqrt{m}} = 0$$

Dividing both sides by $m^2$ and rearranging

$$\frac{\lambda c_1 A}{\beta} = \lambda c_1 A + \frac{\lambda c_2'}{2\sqrt{m}}$$

Dividing both sides by $\lambda$ and rearranging,

$$A(c_1 - c_1\beta) = \frac{\beta c_2'}{2\sqrt{m}}$$

Then,

$$A = \frac{\beta c_2'}{2\sqrt{m}(c_1 - c_2\beta)} = \frac{c_2'}{c_1}\left[\frac{\beta}{2(1-\beta)}\right]\frac{1}{\sqrt{m}} \tag{17.162}$$

Substituting the right side of Equation 17.156 for $\sqrt{m}$,

$$A = \frac{c_2'}{c_1}\left[\frac{\beta}{2(1-\beta)}\right]\frac{2c_1 A}{(-c_2' + \sqrt{(c_2')^2 + 4c_1 c_F' A})}$$

Dividing both sides by $A$,

$$1 = \frac{c_2'\beta}{(1-\beta)[-c_2' + \sqrt{(c_2')^2 + 4c_1 c_F' A}}$$

$$-c_2' + \sqrt{(c_2')^2 + 4c_1 c_F' A} = c_2'\beta/(1-\beta)$$

$$\sqrt{(c_2')^2 + 4c_1 c_F' A} = \frac{c_2'\beta + c_2'(1-\beta)}{(1-\beta)} = \frac{c_2'}{(1-\beta)}$$

Squaring both sides,

$$(c_2')^2 + 4c_1 c_F' A = \frac{(c_2')^2}{(1-\beta)^2}$$

$$4c_1 c_F' A = \frac{(c_2')^2}{(1-\beta)^2} - (c_2')^2 = \frac{(c_2')^2[1-(1-\beta)^2]}{(1-\beta)^2}$$

and

$$A = \frac{(c_2')^2[1-(1-\beta)^2]}{c_1 4c_F(1-\beta)^2} \tag{17.163}$$

It should be noted that this development was carried out in the context of Smith's model of a variance function. If Jessen's or another model is used, the development process would be similar but the expression for $A$ will be different. It also should be noted that if the second term in Equation 17.97 is retained in the expression for $V(\bar{y}_o)_{1S}$, it becomes impossible to eliminate the ρ multiplier and the result becomes indeterminate.

## 17.14 CLUSTER SAMPLING FOR PROPORTIONS

When the variable of interest is an attribute, such as sex, species, or presence of a certain disease, and the parameter of interest is the proportion of the specified universe possessing that attribute, the evaluation phase of the sampling operation usually is much less costly in time and money than is the location in the field of the sampling units that are to be evaluated. Consequently, cluster sampling is seen as a means by which the cost of the sampling operation can be reduced because, if appropriately designed, the clusters concentrate the elements to be evaluated into relatively fewer locations than would be needed with simple random sampling.

For example, consider the problem of estimating the proportion of a certain tract that is stocked with lodgepole pine seedlings 2 years after a forest fire. A classic approach is to assume full stocking if every square milacre (0.001 acre) plot making up the surface of the tract contains at least one live seedling. The degree of stocking is then expressed as the proportion of the milacre plots containing live seedlings:

$$p_o = \tau_o/N$$

where $p_o$ = proportion stocked (which is usually expressed as a percent)
$\tau_o$ = number of plots containing live seedlings
$N$ = total number of plots in tract

To estimate $p_o$, one would have to estimate the relative frequency of milacre plots containing seedlings. The most straightforward approach would be to construct the sampling frame by overlaying a map of the tract with a square grid, the squares of which would be correctly scaled to represent 0.001 acre on the ground. Frame size would then be equal to $N$ = 1000 (area of tract, in acres), assuming no partial squares along the tract boundaries. Simple random sampling, either with or without replacement, could be used to select the plots for evaluation. Each plot selected would have to be located in the field and examined for the presence of live lodgepole pine seedlings. Obviously, the cost in time and money for the location phase would be much greater than the cost of the evaluation phase. It would be preferable to limit the number of positions that would have to be located and to evaluate a number of plots at each position. This would be cluster sampling.

Regardless of the method of sampling to be used, the random variable, $Y$, can take on only two values:

$$y = \begin{cases} 0, & \text{if the attribute of interest is absent} \\ 1, & \text{if the attribute of interest is present} \end{cases}$$

When the random variable is of this type, the population mean is a proportion (see Equation 12.42).

$$\mu_Y = \frac{1}{N}\sum_i^N y_i = p_o \tag{17.165}$$

and the variance, according to Equation 12.43, is

$$\sigma_Y^2 = p_o(1 - p_o) \tag{17.166}$$

The objective of the sampling operation is to obtain an estimate of $p_o$.

### 17.14.1 Using Two-Stage Cluster Sampling, When $1 < m < M$ and $1 < n_i < N_i$

The design of the sampling frame would depend on the situation. In any case, the frame contains $M$ clusters. Cluster size, $N_i$, is not constant. The parameters for cluster $i$ are then

$$p_i = \frac{1}{N_i}\sum_j^{N_i} y_{ij}; \quad i = 1,\ 2,\ \ldots,\ M; \quad j = 1,\ 2,\ \ldots,\ N_i \tag{17.167}$$

$$\tau_i = \sum_j^{N_i} y_{ij} = N_i p_i \tag{17.168}$$

and

$$V(y)_i = \sigma_i^2 = \frac{1}{N_i}\sum_j^{N_i}(y_{ij} - p_i)^2 = p_i(1 - p_i) \tag{17.169}$$

The overall parameters are then

$$N = \sum_i^M N_i \tag{17.170}$$

$$p_o = \frac{1}{N_i}\sum_i^M \sum_j^{N_i} y_{ij} = \frac{1}{N}\sum_i^M \tau_i = \frac{1}{N}\sum_i^M N_i p_i \tag{17.171}$$

which is a weighted mean;

$$\bar{N} = N / M \tag{17.172}$$

$$\tau_o = \sum_i^M \sum_j^{N_i} y_{ij} = \sum_i^M \tau_i = \sum_i^M N_i p_i = N p_o \tag{17.173}$$

$$\bar{\tau}_o = \frac{1}{M}\sum_i^M \tau_i = \tau_o / M = \bar{N}p_o \tag{17.174}$$

$$V(y) = \sigma_Y^2 = \frac{1}{N}\sum_i^M \sum_j^{N_i} (y_{ij} - p_o)^2 = p_o(1 - p_o) \tag{17.175}$$

and

$$V(\tau) = \sigma_\tau^2 = \frac{1}{M}\sum_i^M (\tau_i - \bar{\tau})^2 = \frac{1}{M}\sum_i^M \tau_i - \bar{\tau}^2 \tag{17.176}$$

which can be written

$$V(\tau) = \sigma_\tau^2 = \frac{1}{M}\sum_i^M (N_i p_i - \bar{N}p_o)^2 \tag{17.177}$$

If $m$ clusters, when $1 < m < M$, are randomly drawn from the sampling frame, using sampling without replacement, and a subsample of size $n$, when $1 < n_i < N_i$, which is not constant, is drawn from each selected cluster, again using sampling without replacement, the variance of the estimate of the overall proportion is

$$V(\hat{p}) = \sigma_{\hat{p}}^2 = \frac{1}{N^2}\left\{M^2 \frac{\sigma_\tau^2}{m}\left(\frac{M-m}{M-1}\right) + \frac{M}{m}\sum_i^m \left[\frac{\sigma_i^2}{n_i}\left(\frac{N_i - n_i}{N_i - 1}\right)N_i^2\right]\right\} \tag{17.178}$$

The statistics for cluster $i$ would be

$$\hat{p}_i = \frac{1}{n_i}\sum_j^{n_i} y_{ij} \xrightarrow{\text{unbiased}} p_i \tag{17.179}$$

$$T_i = N_i \hat{p}_i \xrightarrow{\text{unbiased}} \tau_i \tag{17.180}$$

and

$$s_i^2 = \frac{1}{n_i - 1}\sum_j^{n_i} (y_{ij} - \hat{p}_i)^2 = \frac{\hat{p}_i(1 - \hat{p}_i)n_i}{n_i - 1} \xrightarrow{\text{unbiased}} \sigma_i^2\left(\frac{N_i}{N_i - 1}\right) \tag{17.181}$$

The estimates of the overall parameters are

$$\bar{T} = \frac{1}{m}\sum_i^m T_i \xrightarrow{\text{unbiased}} \bar{\tau} \tag{17.182}$$

$$T_o = M\overline{T} \xrightarrow{\text{unbiased}} \tau_o \tag{17.183}$$

$$\hat{p}_o = \frac{M}{Nm}\sum_i^m \frac{N}{n_i}\sum_j^{n_i} y_{ij} = \frac{M}{Nm}\sum_i^m N_i\hat{p}_i = \frac{M}{Nm}\sum_i^m T_i = \frac{M}{N}\overline{T} = \frac{T_o}{N} \xrightarrow{\text{unbiased}} p_o \tag{17.184}$$

and

$$s_T^2 = \frac{1}{(m-1)}\sum_i^m (T_i - \overline{T})^2 = \frac{1}{(m-1)}\sum_i^m (N_i\hat{p}_i - \overline{N}\hat{p}_o)^2 \xrightarrow{\text{unbiased}} \sigma_{\overline{T}}^2\left(\frac{M}{M-1}\right) \tag{17.185}$$

Then, according to Equation 17.58, the estimate of the variance of the total is

$$s_{T_o}^2 = \frac{M}{m}\sum_i^m\left[\frac{s_i^2}{n_i}\left(\frac{N_i - n_i}{N_i}\right)N_i^2\right] + M^2\frac{s_T^2}{m}\left(\frac{M-m}{M}\right) \xrightarrow{\text{unbiased}} \sigma_{T_o}^2 \tag{17.186}$$

and, according to Equation 17.69, the estimate of the variance of the overall proportion is

$$s_{\hat{p}_o}^2 = s_{T_o}^2 / N^2 \xrightarrow{\text{unbiased}} \sigma_{\hat{p}_o}^2 \tag{17.187}$$

The standard error of the proportion is then

$$s_{\hat{p}_o} = \pm\sqrt{s_{T_o}^2 / N^2} + \frac{1}{2(n+1)} \xrightarrow{\text{unbiased}} \pm\sqrt{V(\hat{p}_o)} \tag{17.188}$$

where $\frac{1}{2(n+1)}$ = correction for continuity (see Equation 13.65)

In most cases, however, the correction for continuity is ignored.

As an example, one can use the universe shown in Figure 17.11. The presence or absence of the attribute of interest is indicated, respectively, by the numerals 1 and 0. The total size of the universe $N$ = 320 elements; 187 of these elements possess the attribute of interest and thus $\tau_o$ = 187. Then, the proportion of the universe possessing the attribute is

$$p_o = 187/320 = 0.584375$$

and the proportion not possessing the attribute is

$$q_o = 1 - p_o = 1 - 0.584375 = 0.415625$$

The variance of the population is then

$$\sigma_Y^2 = p_o q_o = 0.584375(0.415625) = 0.2428809$$

| | | | | | | | | | | | | | | | |
|---|---|---|---|---|---|---|---|---|---|---|---|---|---|---|---|
| 0 | 1 | 1 | 1 | 0 | 1 | 1 | 0 | 1 | 1 | 0 | 0 | 0 | 1 | 1 | 1 |
| 1 | 1 | 0 | 0 | 1 | 1 | 0 | 1 | 1 | 1 | 1 | 0 | 1 | 0 | 0 | 1 |
| 1 | 0 | 1 | 1 | 0 | 1 | 1 | 0 | 0 | 1 | 1 | 0 | 0 | 1 | 1 | 0 |
| 0 | 1 | 1 | 1 | 1 | 0 | 1 | 0 | 0 | 1 | 1 | 1 | 1 | 1 | 1 | 1 |
| 0 | 0 | 1 | 1 | 0 | 1 | 1 | 1 | 0 | 1 | 1 | 0 | 1 | 1 | 1 | 0 |
| 0 | 0 | 0 | 1 | 0 | 0 | 0 | 0 | 1 | 1 | 0 | 0 | 0 | 1 | 0 | 1 |
| 0 | 0 | 0 | 1 | 1 | 0 | 1 | 1 | 0 | 1 | 1 | 1 | 0 | 0 | 1 | 1 |
| 1 | 1 | 1 | 1 | 0 | 1 | 1 | 0 | 0 | 1 | 0 | 1 | 1 | 1 | 1 | 1 |
| 1 | 1 | 0 | 1 | 1 | 0 | 1 | 0 | 1 | 0 | 0 | 1 | 1 | 1 | 1 | 0 |
| 0 | 1 | 1 | 0 | 0 | 1 | 1 | 1 | 0 | 1 | 0 | 1 | 0 | 1 | 0 | 0 |
| 0 | 0 | 0 | 0 | 0 | 1 | 1 | 1 | 0 | 1 | 1 | 0 | 0 | 1 | 0 | 1 |
| 0 | 0 | 0 | 0 | 1 | 0 | 0 | 1 | 1 | 0 | 0 | 1 | 1 | 1 | 1 | 1 |
| 0 | 0 | 0 | 1 | 1 | 1 | 1 | 1 | 0 | 1 | 0 | 1 | 1 | 1 | 1 | 1 |
| 0 | 1 | 1 | 0 | 0 | 1 | 0 | 1 | 1 | 1 | 1 | 0 | 1 | 1 | 1 | 1 |
| 0 | 0 | 0 | 0 | 1 | 0 | 1 | 0 | 1 | 1 | 0 | 1 | 0 | 1 | 1 | 0 |
| 0 | 1 | 0 | 1 | 1 | 0 | 0 | 1 | 0 | 1 | 0 | 1 | 1 | 1 | 1 | 1 |
| 1 | 0 | 1 | 1 | 1 | 1 | 1 | 0 | 0 | 1 | 0 | 0 | 0 | 1 | 1 | 0 |
| 0 | 1 | 1 | 0 | 1 | 1 | 0 | 1 | 1 | 1 | 1 | 0 | 1 | 0 | 0 | 1 |
| 0 | 0 | 0 | 1 | 1 | 0 | 0 | 1 | 1 | 1 | 1 | 1 | 1 | 1 | 1 | 0 |
| 1 | 1 | 1 | 0 | 1 | 1 | 1 | 0 | 0 | 1 | 0 | 0 | 1 | 0 | 0 | 1 |

**FIGURE 17.11** Hypothetical universe. The numeral 1 indicates the presence of the attribute of interest, while the numeral 0 indicates its absence.

An estimate is to be made of $p_o$, using cluster sampling. Figure 17.12 shows the division of the universe into $M = 53$ clusters and serves as the primary sampling frame. As can be seen, the clusters vary in size. Figure 17.13 shows the numbering pattern within each of the four cluster configurations present in the primary sampling frame and serves as the secondary sampling frame. The mean cluster size is then

$$\bar{N} = N / M = 320 / 53 = 6.0377358 \text{ elements}$$

The primary sample consists of $m = 6$ clusters that were randomly selected using sampling without replacement. These clusters are 14, 11, 32, 25, 52, and 2. Using a constant sampling intensity of 0.5, the following subsamples were drawn from the primary sampling units using random sampling without replacement, as shown in Figure 17.14:

| Cluster | $N_i$ | $n_i$ | Elements Chosen | $y_{ij}$ | $\Sigma y_{ij}$ | $p_i$ | $T_i$ | $s_i^2$ |
|---|---|---|---|---|---|---|---|---|
| 2 | 4 | 2 | 2,4 | 0,1 | 1 | 0.500 | 2 | 0.500 |
| 11 | 4 | 2 | 3,4 | 1,1 | 2 | 1.000 | 4 | 0.000 |
| 14 | 8 | 4 | 1,8,7,3 | 1,1,1,0 | 3 | 0.750 | 6 | 0.250 |
| 25 | 8 | 4 | 4,7,8,1 | 0,1,1,1 | 3 | 0.750 | 6 | 0.250 |

| Cluster | $N_i$ | $n_i$ | Elements Chosen | $y_{ij}$ | $\Sigma y_{ij}$ | $p_i$ | $T_i$ | $s_i^2$ |
|---|---|---|---|---|---|---|---|---|
| 32 | 12 | 6 | 4,12,2,10,3,7 | 1,1,1,1,1,0 | 5 | 0.833 | 10 | 0.167 |
| 52 | 4 | 2 | 4,3 | 0,0 | 0 | 0.000 | 0 | 0.000 |
| | | | | | | | 28 | |

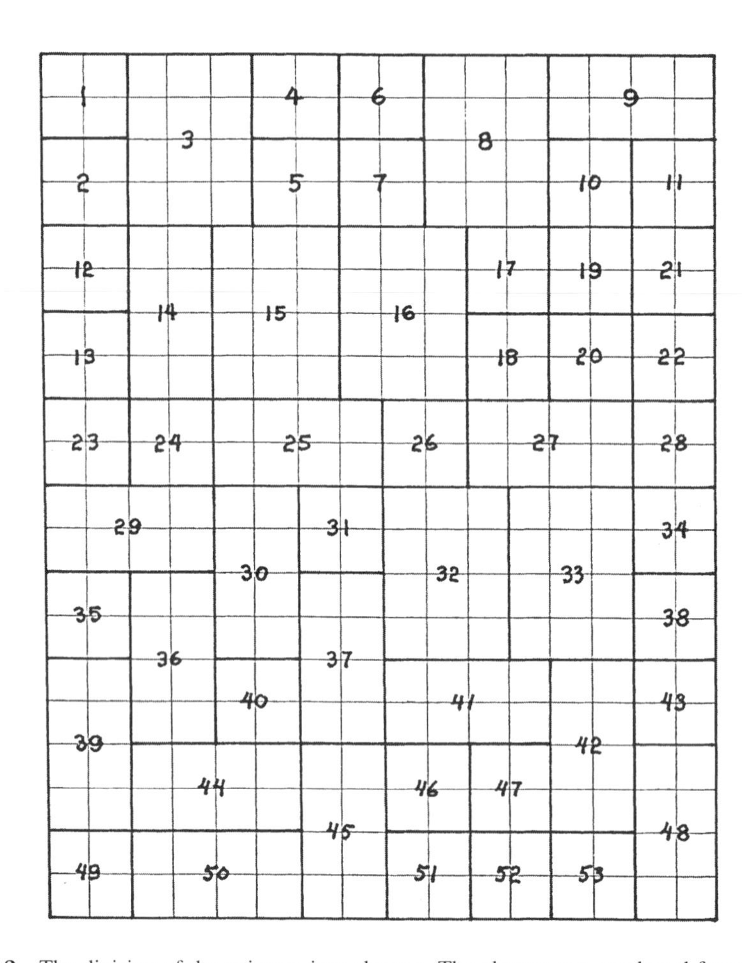


**FIGURE 17.12** The division of the universe into clusters. The clusters are numbered from 1 to 53. Consequently, this figure serves as the sampling frame.

| 1 | 2 |
|---|---|
| 3 | 4 |

| 1 | 2 | 3 | 4 |
|---|---|---|---|
| 5 | 6 | 7 | 8 |

| 1 | 2 |
|---|---|
| 3 | 4 |
| 5 | 6 |
| 7 | 8 |

| 01 | 02 | 03 |
|---|---|---|
| 04 | 05 | 06 |
| 07 | 08 | 09 |
| 10 | 11 | 12 |

**FIGURE 17.13** The four cluster configurations found in the sampling frame. The element numbering provides the secondary sampling frame.

The mean total is

| | | | | | | | | | | | | | | | |
|---|---|---|---|---|---|---|---|---|---|---|---|---|---|---|---|
| 0 | 1 | 1 | 1 | 0 | 1 | 1 | 0 | 1 | 1 | 0 | 0 | 0 | 1 | 1 | 1 |
| 1 | 1 | 0 | 0 | 1 | 1 | 0 | 1 | 1 | 1 | 1 | 0 | 1 | 0 | 0 | 1 |
| 1 | 0 | 1 | 1 | 0 | 1 | 1 | 0 | 0 | 1 | 1 | 0 | 0 | 1 | 1 | 0 |
| 0 | 1 | 1 | 1 | 1 | 0 | 1 | 0 | 0 | 1 | 1 | 1 | 1 | 1 | 1 | 1 |
| 0 | 0 | 1 | 1 | 0 | 1 | 1 | 1 | 0 | 1 | 1 | 0 | 1 | 1 | 1 | 0 |
| 0 | 0 | 0 | 1 | 0 | 0 | 0 | 0 | 1 | 1 | 0 | 0 | 0 | 1 | 0 | 1 |
| 0 | 0 | 0 | 1 | 1 | 0 | 1 | 1 | 0 | 1 | 1 | 1 | 0 | 0 | 1 | 1 |
| 1 | 1 | 1 | 1 | 0 | 1 | 1 | 0 | 0 | 1 | 0 | 1 | 1 | 1 | 1 | 1 |
| 1 | 1 | 0 | 1 | 1 | 0 | 1 | 0 | 1 | 0 | 0 | 1 | 1 | 1 | 1 | 0 |
| 0 | 1 | 1 | 0 | 0 | 1 | 1 | 1 | 0 | 1 | 0 | 1 | 0 | 1 | 0 | 0 |
| 0 | 0 | 0 | 0 | 0 | 1 | 1 | 1 | 0 | 1 | 1 | 0 | 0 | 1 | 0 | 1 |
| 0 | 0 | 0 | 0 | 1 | 0 | 0 | 1 | 1 | 0 | 0 | 1 | 1 | 1 | 1 | 1 |
| 0 | 0 | 0 | 1 | 1 | 1 | 1 | 1 | 0 | 1 | 0 | 1 | 1 | 1 | 1 | 1 |
| 0 | 1 | 1 | 0 | 0 | 1 | 0 | 1 | 1 | 1 | 1 | 0 | 1 | 1 | 1 | 1 |
| 0 | 0 | 0 | 0 | 1 | 0 | 1 | 0 | 1 | 1 | 0 | 1 | 0 | 1 | 1 | 0 |
| 0 | 1 | 0 | 1 | 1 | 0 | 0 | 1 | 0 | 1 | 0 | 1 | 1 | 1 | 1 | 1 |
| 1 | 0 | 1 | 1 | 1 | 1 | 1 | 0 | 0 | 1 | 0 | 0 | 0 | 1 | 1 | 0 |
| 0 | 1 | 1 | 0 | 1 | 1 | 0 | 1 | 1 | 1 | 1 | 0 | 1 | 0 | 0 | 1 |
| 0 | 0 | 0 | 1 | 1 | 0 | 0 | 1 | 1 | 1 | 1 | 1 | 1 | 1 | 1 | 0 |
| 1 | 1 | 1 | 0 | 1 | 1 | 1 | 0 | 0 | 1 | 0 | 0 | 1 | 0 | 0 | 1 |

**FIGURE 17.14** The elements chosen in the subsampling phase.

$$\bar{T} = 28/6 = 4.666667$$

the estimate of the overall total (the total number of elements in the universe possessing the attribute of interest) is

$$T_o = 53(4.666667) = 247.3333$$

and the estimate of the overall proportion (the proportion of the universe possessing the attribute of interest) is

$$\hat{p}_o = 247.3333/320 = 0.7729167$$

The estimate of the variance of the cluster totals is

$$s_T^2 = \frac{1}{(6-1)}[(2-4.666667)^2 + (4-4.666667)^2 + (6-4.666667)^2$$

$$+(6-4.66667)^2 + (10-4.666667)^2 + (0-4.666667)^2]$$

$$= 13.3666667$$

the estimate of the variance of the overall total is

$$s_{T_o}^2 = \frac{53}{6}\left[\frac{0.5}{2}\left(\frac{4-2}{4}\right)4^2 + \frac{0.0}{4}\left(\frac{4-2}{4}\right)4^2 + \frac{0.25}{4}\left(\frac{8-4}{8}\right)8^2 + \frac{0.25}{4}\left(\frac{8-4}{8}\right)8^2\right.$$

$$\left. + \frac{0.167}{6}\left(\frac{12-6}{12}\right)12^2 + \frac{0.00}{2}\left(\frac{4-2}{4}\right)42\right]$$

$$+ 53^2\left(\frac{12.2666667}{6}\right)\left(\frac{53-6}{53}\right) + 5163.3775$$

the estimate of the overall proportion is

$$s_{\hat{p}_o}^2 = 5163.3775/320^2 = 0.0504236$$

and the standard error of the estimate is

$$s_{\hat{p}_o} = \sqrt{0.0504236} + \frac{1}{2(6+1)} = \pm(0.224552 + 0.071429)$$

$$= \pm 0.295981$$

### 17.14.2 Using Simple Cluster Sampling, When $1 < m < M$ and $n_i = a = N_i = A$

When cluster size is constant and no subsampling is involved, the sampling process parallels that as was described in Section 17.8.2. In essence, the cluster proportion, $p_i$, becomes the variable of interest. A key point is that the cluster proportions are all based on the same number of observations and, consequently, have the same weight.

The size of the sampling frame, $M$, is

$$M = N/A \tag{17.189}$$

The proportion of the elements in cluster $i$ that possess the attribute of interest is

$$p_i = \mu_i = \frac{1}{A}\sum_j^A y_{ij}, \quad i = 1, 2, \ldots, M, \quad j = 1, 2, \ldots, A \tag{17.190}$$

the cluster total is

$$\tau_i = \sum_j^A y_{ij} = Ap_i \tag{17.191}$$

and the cluster variance is

$$V(y_i) = \sigma_i^2 = \frac{1}{A}\sum_j^A (y_{ij} - p_i)^2 = p_i(1 - p_i) \tag{17.192}$$

Since the clusters are all of the same size, the overall proportion is simply the mean of the cluster proportions:

$$p_o = \mu_o = \frac{1}{N}\sum_i^M \sum_j^A y_{ij} = \frac{1}{MA}\sum_i^M \sum_j^A y_{ij} = \frac{1}{M}\sum_i^M p_i \tag{17.193}$$

The population total is then

$$\tau_o = \sum_i^M \sum_j^A y_{ij} = \sum_i^M Ap_i = A\sum_i^M p_i \tag{17.194}$$

The overall variance of the random variable, $Y$, is

$$V(y) = \sigma_Y^2 = \frac{1}{N}\sum_i^M \sum_j^A (y_{ij} - p_o)^2 = \frac{1}{N}\sum_i^M \sum_j^A y_{ij}^2 - p_o^2 = p_o(1 - p_o) \tag{17.195}$$

The variance of the cluster totals is then

$$V(\tau) = \sigma_\tau^2 = \frac{1}{M}\sum_i^M (\tau_i - \tau_o / M)^2 = \frac{1}{M}\sum_i^M (\tau_i - \bar{\tau})^2 \tag{17.196}$$

and the variance of the cluster proportions, which, according to Equation 17.84, is also the between cluster variance, is

$$V(p) = \sigma_p^2 = \sigma_B^2 = \frac{1}{M}\sum_i^M (p_i - p_o)^2 = \sigma_\tau^2 / A \tag{17.197}$$

If $m$ clusters are randomly drawn from the sampling frame, using sampling without replacement, and no subsampling is done, the cluster proportions, totals, and variances are parameters. The estimate of the population proportion is then

$$\hat{p}_o = \frac{1}{mA}\sum_i^m \sum_j^A y_{ij} = \frac{1}{m}\sum_i^m p_i \xrightarrow{\text{unbiased}} p_o \tag{17.198}$$

The variance of the estimate of the overall proportion is obtained using Equation 17.178. However, since no subsampling is involved, the within-cluster component vanishes and the expression becomes

$$V(\hat{p}) = \sigma_{\hat{p}_o}^2 = \frac{M}{N^2}\frac{\sigma_\tau^2}{m}\left(\frac{M - m}{M - 1}\right) \tag{17.199}$$

Since the between-cluster variance, $\sigma_B^2 = \sigma_\tau^2 / A^2$, $\sigma_\tau^2 = A^2\sigma_B^2$. Substituting $A^2\sigma_B^2$ for $\sigma_\tau^2$ in Equation 17.199 yields

$$V(\hat{p}_o) = \sigma_{\hat{p}_o}^2 = \frac{M^2 A^2}{N^2} \frac{\sigma_B^2}{m}\left(\frac{M-m}{M-1}\right) = \frac{\sigma_B^2}{m}\left(\frac{M-m}{M-1}\right) = \frac{\sigma_p^2}{m}\left(\frac{M-m}{M-1}\right) \tag{17.200}$$

It should be noted that this expression is not the equivalent of Equation 12.5. In other words,

$$V(\hat{p}_o) \neq \sigma_y^2 = \frac{\sigma_Y^2}{n}\left(\frac{N-n}{N-1}\right) \tag{17.201}$$

The estimate of the between cluster variance is

$$s_B^2 = \frac{1}{m-1}\sum_i^m (p_i - \hat{p}_o)^2 \xrightarrow{\text{unbiased}} \sigma_B^2\left(\frac{M}{M-1}\right) \tag{17.202}$$

Substituting $s_B^2[(M-1)/M]$ for $\sigma_B^2$ in Equation 17.200 yields the estimate of the variance of the estimate of the overall proportion.

$$s_{\hat{p}_o}^2 = \frac{s_B^2}{m}\left(\frac{M-1}{M}\right)\left(\frac{M-m}{M-1}\right) = \frac{s_B^2}{m}\left(\frac{M-m}{M}\right) \xrightarrow{\text{unbiased}} \sigma_{\hat{p}_o}^2 \tag{17.203}$$

A confidence interval can be computed for the estimate of the overall proportion, $p_o$, using the standard procedure based on the use of Student's $t$ distribution as described in Section 13.09. Since no subsampling is involved, the basic sampling unit is the cluster and the variable of interest is the proportion of the elements making up the cluster that possesses the attribute of interest. Note the shift in the variable of interest. The initial variable of interest yielded values within each cluster that were used to generate the values of the second variable of interest. The universe of interest is then the set of possible clusters. As can be seen in Equation 17.193, the overall mean of the cluster proportions is equal to the overall proportion computed as though no clusters existed. Thus, when one computes a confidence interval estimate of the overall mean of the cluster proportions, that confidence interval also applies to the overall proportion.

Since the population of cluster proportions is finite and its variates can only take on value from zero to one, it is not normally distributed and the sample size must be large enough so that the sampling distribution of mean proportions can be assumed to be normally distributed. Usually this requires a sample in excess of 30 clusters. Some workers prefer sample sizes in excess of 60 clusters.

If a maximum allowable sampling error has been specified, the sample size needed to meet this specification can be computed using the procedure described in Section 13.14. However, if this procedure indicates a sample size that is less than the 30 (or 60) required for the assumption of normality, the standard minima (30 or 60) should be given precedence over the computed needed sample size.

As an example, assume that the universe shown in Figure 17.11 represents a tract of land. Each of the squares represents one milacre quadrat (0.001 acre). A survey was made of the tract to determine the percent stocking of lodgepole pine seedlings. If a milacre quadrat contained at least one seedling, it was considered stocked. Cluster sampling was used to reduce travel time between sampling units. The clusters were made up of four milacre quadrats arranged to form a compact square. Figure 17.15 shows the sampling frame and the $m = 8$ clusters that were randomly drawn, using sampling without replacement. The data and standard computations are as follows:

| | | | | | | | | | | | | | | | |
|---|---|---|---|---|---|---|---|---|---|---|---|---|---|---|---|
| 0 | 1 | 1 | 1 | 0 | 1 | 1 | 0 | 1 | 1 | 0 | 0 | 0 | 1 | 1 | 1 |
| 1 | 1 | 0 | 0 | 1 | 1 | 0 | 1 | 1 | 1 | 1 | 0 | 1 | 0 | 0 | 1 |
| 1 | 0 | 1 | 1 | 0 | 1 | 1 | 0 | 0 | 1 | 1 | 0 | 0 | 1 | 1 | 0 |
| 0 | 1 | 1 | 1 | 1 | 0 | 1 | 0 | 0 | 1 | 1 | 1 | 1 | 1 | 1 | 1 |
| 0 | 0 | 1 | 1 | 0 | 1 | 1 | 1 | 0 | 1 | 1 | 0 | 1 | 1 | 1 | 0 |
| 0 | 0 | 0 | 1 | 0 | 0 | 0 | 0 | 1 | 1 | 0 | 0 | 0 | 1 | 0 | 1 |
| 0 | 0 | 0 | 1 | 1 | 0 | 1 | 1 | 0 | 1 | 1 | 1 | 0 | 0 | 1 | 1 |
| 1 | 1 | 1 | 1 | 0 | 1 | 1 | 0 | 0 | 1 | 0 | 1 | 1 | 1 | 1 | 1 |
| 1 | 1 | 0 | 1 | 1 | 0 | 1 | 0 | 1 | 0 | 0 | 1 | 1 | 1 | 1 | 0 |
| 0 | 1 | 1 | 0 | 0 | 1 | 1 | 1 | 0 | 1 | 0 | 1 | 0 | 1 | 0 | 0 |
| 0 | 0 | 0 | 0 | 0 | 1 | 1 | 1 | 0 | 1 | 1 | 0 | 0 | 1 | 0 | 1 |
| 0 | 0 | 0 | 0 | 1 | 0 | 0 | 1 | 1 | 0 | 0 | 1 | 1 | 1 | 1 | 1 |
| 0 | 0 | 0 | 1 | 1 | 1 | 1 | 1 | 0 | 1 | 0 | 1 | 1 | 1 | 1 | 1 |
| 0 | 1 | 1 | 0 | 0 | 1 | 0 | 1 | 1 | 1 | 1 | 0 | 1 | 1 | 1 | 1 |
| 0 | 0 | 0 | 0 | 1 | 0 | 1 | 0 | 1 | 1 | 0 | 1 | 0 | 1 | 1 | 0 |
| 0 | 1 | 0 | 1 | 1 | 0 | 0 | 1 | 0 | 1 | 0 | 1 | 1 | 1 | 1 | 1 |
| 1 | 0 | 1 | 1 | 1 | 1 | 1 | 0 | 0 | 1 | 0 | 0 | 0 | 1 | 1 | 0 |
| 0 | 1 | 1 | 0 | 1 | 1 | 0 | 1 | 1 | 1 | 1 | 0 | 1 | 0 | 0 | 1 |
| 0 | 0 | 0 | 1 | 1 | 0 | 0 | 1 | 1 | 1 | 1 | 1 | 1 | 1 | 1 | 0 |
| 1 | 1 | 1 | 0 | 1 | 1 | 1 | 0 | 0 | 1 | 0 | 0 | 1 | 0 | 0 | 1 |

**FIGURE 17.15** The sampling frame, showing the chosen clusters.

| | $i$ | | | | | | | |
|---|---|---|---|---|---|---|---|---|
| | **1** | **2** | **3** | **4** | **5** | **6** | **7** | **8** |
| Row | 4 | 8 | 8 | 4 | 2 | 6 | 5 | 6 |
| Column | 3 | 2 | 6 | 7 | 2 | 4 | 5 | 6 |
| $\tau_i$ | 2 | 1 | 2 | 2 | 4 | 3 | 2 | 2 |
| $p_i$ | 0.50 | 0.25 | 0.50 | 0.50 | 1.00 | 0.75 | 0.50 | 0.50 |

$$\sum_i^8 \tau_i = 18$$

$\bar{T} = 18/8 = 2.25$ stocked quadrats/cluster

$T_o = 80\,(2.25) = 180$, the estimate of the total number of stocked quadrats in the tract

$\hat{p}_o = 180/320 = 0.5625$, the estimate of the proportion of the population of the tract stocked with lodgepole pine seedlings

$$s_B^2 = \frac{1}{(8-1)}[(0.50-0.5625)^2 + (0.25-0.5625)^2 + (0.50-0.5625)^2 + (0.50-0.5625)^2 + (1.00-0.5625)^2 + (0.75-0.5625)^2 + (0.50-0.5625)^2 + (0.50-0.5625)^2] = 0.0491071$$

$$s_{\hat{p}_o}^2 = \frac{0.0491071}{8}\left(\frac{80-8}{80}\right) = 0.0055246$$

$$s_{\hat{p}_o} = \pm\sqrt{0.0055246} = 0.0743273$$

If the confidence level is set at 0.95, $t_{0.95}(k = 8 - 1) = 2.365$, and the sampling error is

$$\text{SE} = \pm 2.365(0.0743273) = \pm 0.1757842$$

and

$$P(0.5675 \pm 0.1758) = 0.95$$

If an allowable error of ±0.15, at a confidence level of 0.95, has been specified, the needed sample size can be computed using the eight cluster samples as a presample. According to Equation 13.111,

$$n_{\text{Needed}} = t_C^2(k)s_B^2 M / [\text{AE}_{\hat{p}_o}^2 M + t_C^2(k)s_B^2]$$

Then, using $t_{0.95}(7) = 2.365$,

$$n_{\text{Needed}} = (2.365)^2(0.0491071)80/[(0.15)^2\, 80 + (2.365)^2 0.0491071]$$
$$= 10.59 \text{ or } 11 \text{ clusters}$$

Iterating and using $t_{0.95}(10) = 2.228$ yields

$$n_{\text{Needed}} = 9.54 \text{ or } 10 \text{ clusters}$$

Iterating and using $t_{0.95}(9) = 2.262$ yields

$$n_{\text{Needed}} = 9.80 \text{ or } 10 \text{ clusters}$$

The procedure stabilizes at a needed sample size of 10 clusters.

## 17.15 THREE-STAGE SAMPLING

As was mentioned in Section 17.2, it is possible to extend the subsampling process to three, four, or more stages. As an example, consider the extension to three stages. This might arise in the case of an inventory of the timber on a large tract of land such as a major industrial holding or a governmental entity such as a county. The primary sampling frame could be made up of the General Land Office townships in the tract, the secondary sampling frame could be the sections within the townships, and the tertiary sampling frame could be a set of points arranged in a square pattern within each section. Each of these points would serve as the center of a sample plot. A random sample would be drawn from the set of townships, then a random sample of sections would be drawn from each of the chosen townships, and finally a random sample of points would be drawn from each of the chosen sections. Such a process would tend to concentrate the sampling operations into a set of clumps, which, in turn, would reduce the travel time between plots when compared with that needed in the case of a simple random sample.

### 17.15.1 The Parameters

The pattern begun in Section 17.4.1 is simply expanded. Let

$$L = \text{number of primary clusters}$$

$$M_h = \text{number of secondary clusters in primary cluster } h$$

$$N_{hi} = \text{number of ultimate sampling units in secondary cluster } hi \tag{17.204}$$

$$N = \sum_h^L \sum_i^{M_h} N_{hi} = \text{population size}$$

Then, without proof:

$$\mu_{hi} = \frac{1}{N_{hi}} \sum_j^{N_{hi}} y_{hij} = \text{mean of the variates in secondary cluster } hi \tag{17.205}$$

$$\sigma_{hi}^2 = \frac{1}{N_{hi}} \sum_j^{N_{hi}} (y_{hij} - \mu_{hi})^2 = \text{variance of the variates in secondary cluster } hi \tag{17.206}$$

$$\tau_{hi} = \sum_j^{N_{hi}} y_{hij} = N_{hi}\mu_{hi} = \text{total in secondary cluster } hi \tag{17.207}$$

$$\bar{\tau}_h = \frac{1}{M_h} \sum_i^{M_h} \tau_{hi} = \text{mean of the secondary cluster totals in primary cluster } h \tag{17.208}$$

$$\sigma_{\tau_h}^2 = \frac{1}{M_h} \sum_i^{M_h} (\tau_{hi} - \bar{\tau}_h)^2 = \text{variance of the secondary cluster totals in primary cluster } h \tag{17.209}$$

$$t_h = \sum_i^{M_h} \tau_{hi} = \sum_i^{M_h} \sum_j^{N_{hi}} y_{hij} = \text{total in primary cluster } h \tag{17.210}$$

$$\bar{\tau} = \frac{1}{L} \sum_h^L \tau_h = \frac{1}{L} \sum_h^L \sum_i^{M_h} \sum_j^{N_{hi}} y_{hij} = \text{mean of the primary cluster totals} \tag{17.211}$$

$$\sigma_\tau^2 = \frac{1}{L} \sum_h^L (\tau_h - \bar{\tau})^2 = \text{variance of the primary cluster totals} \tag{17.212}$$

$$\tau_{O_{3s}} = \sum_h^L \tau_h = \sum_h^L \sum_i^{M_h} \sum_j^{N_{hi}} y_{hij} = \text{population total} \tag{17.213}$$

$$\mu_{o_{3S}} = \frac{\tau_o}{N} = \text{population mean} \tag{17.214}$$

### 17.15.2 Estimation of the Parameters

Let

$L'$ = number of primary clusters chosen

$m_h$ = number of secondary clusters chosen in chosen primary cluster $h$

$n_{hi}$ = number of ultimate sampling units chosen in chosen secondary cluster $hi$

Then, without proof:

$$\bar{y}_{hi} = \frac{1}{n_{hi}} \sum_{j}^{n_{hi}} y_{hij} \xrightarrow{\text{unbiased}} \mu_{hi} \tag{17.215}$$

$$s_{hi}^2 = \frac{1}{n_{hi} - 1} \sum_{j}^{n_{hi}} (y_{hij} - \bar{y}_{hi})^2 \xrightarrow{\text{unbiased}} \sigma_{hi}^2 \left( \frac{N_{hi}}{N_{hi} - 1} \right) \tag{17.216}$$

$$T_{hi} = N_{hi}\bar{y}_{hi} \xrightarrow{\text{unbiased}} \tau_{hi} \tag{17.217}$$

$$\bar{T}_h = \frac{1}{m_h} \sum_{i}^{m_h} T_{hi} \xrightarrow{\text{unbiased}} \bar{\tau}_h \tag{17.218}$$

$$s_{\bar{T}_h}^2 = \frac{1}{m_h - 1} \sum_{i}^{m_h} (T_{hi} - \bar{T}_h)^2 \xrightarrow{\text{unbiased}} \sigma_{\tau_h}^2 \left( \frac{M_h}{M_h - 1} \right) \tag{17.219}$$

$$T_h = M_h \bar{T}_h \xrightarrow{\text{unbiased}} \tau_h \tag{17.220}$$

$$\bar{T} = \frac{1}{L'} \sum_{h}^{L'} T_h \xrightarrow{\text{unbiased}} \bar{\tau} \tag{17.221}$$

$$s_{\bar{T}}^2 = \frac{1}{L' - 1} \sum_{h}^{L'} (T_h - \bar{T})^2 \xrightarrow{\text{unbiased}} \sigma_{\tau}^2 \left( \frac{L}{L - 1} \right) \tag{17.222}$$

$$T_{o_{3S}} = L\bar{T} \xrightarrow{\text{unbiased}} \tau_o \tag{17.223}$$

$$\bar{y}_{o_{3S}} = \frac{T_o}{N} = \frac{L}{NL'} \sum_{h}^{L'} \frac{M_h}{m_h} \sum_{i}^{m_h} \frac{N_{hi}}{n_{hi}} \sum_{j}^{n_{hi}} y_{hij} \xrightarrow{\text{unbiased}} \tau_o \tag{17.224}$$

### 17.15.3 Variance of the Overall Mean

Whenever a stage is added to a multistage sampling design, it becomes necessary to add a component to the formula for the variance of the overall mean. Thus, in the case of three-stage sampling,

$$V(\bar{y}_o)_{3S} = \frac{1}{N^2}\left[\frac{L^2}{L'}\left(\frac{L-L'}{L-1}\right)\sigma_\tau^2 + \frac{L}{L'}\sum_h^L \frac{M_h^2}{m_h}\left(\frac{M_h-m_h}{M_h-1}\right)\sigma_{\tau_h}^2 + \frac{L}{L'}\sum_h^L \frac{M_h}{m_h}\sum_i^M \frac{\sigma_{hi}^2}{n_{hi}}\left(\frac{N_{hi}-n_{hi}}{N_{hi}-1}\right)N_{hi}^2\right] \tag{17.225}$$

The derivation of Equation 17.225 follows the pattern described in Sections 17.5.3 and 17.5.4.

If the primary and secondary clusters are constant in size, $M_h = M$, $m_h = m$, $N_{hi} = A$, and $n_{hi} = a$. Furthermore, $N = LMA$ and $N^2 = L^2M^2A^2$. Then,

$$V(\bar{y}_o)_{3S} = \frac{1}{L'}\left(\frac{L-L'}{L-1}\right)\frac{\sigma_\tau^2}{M^2A^2} + \frac{1}{LL'm}\left(\frac{M-m}{M-1}\right)\sum_h^L \frac{\sigma_{\tau_h}^2}{A^2} + \frac{1}{LML'ma}\left(\frac{A-a}{A-1}\right)\sum_h^L\sum_i^M \sigma_h^2 \tag{17.226}$$

Since

$$\frac{\sigma_\tau^2}{M^2A^2} = \frac{1}{L}\sum_h^L\left(\frac{\tau_h}{MA} - \frac{\bar{\tau}}{MA}\right)^2 = \frac{1}{L}\sum_h^L(\mu_h - \mu_o)^2 = \sigma_p^2 \tag{17.227}$$

the between primary cluster variance, and

$$\frac{\sigma_{\tau_h}^2}{A^2} = \frac{1}{M}\sum_i^M\left(\frac{\tau_{hi}}{A} - \frac{\bar{\tau}_h}{A}\right)^2 = \frac{1}{M}\sum_i^M(\mu_{hi} - \mu_h)^2 = \sigma_{B_h}^2 \tag{17.228}$$

the between secondary cluster variance in primary cluster $h$,

$$V(\bar{y}_o)_{3S} = \frac{1}{L'}\left(\frac{L-L'}{L-1}\right)\sigma_p^2 + \frac{1}{LL'm}\left(\frac{M-m}{M-1}\right)\sum_h^L \sigma_{B_h}^2 + \frac{1}{LML'ma}\left(\frac{A-a}{A-1}\right)\sum_h^L\sum_i^M \sigma_{hi}^2 \tag{17.229}$$

Then, since

$$\frac{1}{L}\sum_h^L \sigma_{B_h}^2 = \sigma_B^2 \tag{17.230}$$

the between secondary cluster variance across all primary clusters and

$$\frac{1}{LM}\sum_h^L\sum_i^M \sigma_{hi}^2 = \frac{1}{L}\sum_h^L \sigma_{w_h}^2 = \sigma_w^2 \tag{17.231}$$

the pooled within secondary cluster variance,

$$V(\bar{y}_o)_{3S} = \frac{1}{L'}\left(\frac{L-L'}{L-1}\right)\sigma_p^2 + \frac{1}{L'm}\left(\frac{M-m}{M-1}\right)\sigma_B^2 + \frac{1}{L'ma}\left(\frac{A-a}{A-1}\right)\sigma_w^2 \tag{17.232}$$

Without proof, the unbiased estimator of the variance of the overall mean, where the primary and secondary clusters are *not* constant in size, is obtained by substituting $s_T^2[(L-1)/L]$ for $\sigma_\tau^2$, $s_{T_h}^2[(M-1)/M]$ for $\sigma_{\tau_h}^2$, and $s_{hi}^2[(A-1)/A]$ for $\sigma_{hi}^2$, and summing over $L'$ instead of $L$ in Equation 17.225.

$$s_{\bar{y}_{o3S}}^2 = \frac{1}{N^2}\left[\frac{L^2}{L'}\left(\frac{L-L'}{L}\right)s_T^2 + \frac{L}{L'}\sum_h^{L'}\frac{M_h^2}{m_h}\left(\frac{M_h-m_h}{M_h}\right)s_{T_h}^2 + \frac{L}{L'}\sum_h^{L'}\frac{M_h}{m_h}\sum_i^{m_h}\frac{s_{hi}^2}{n_{hi}}\left(\frac{N_{hi}-n_{hi}}{N_{hi}}\right)N_{hi}^2\right] \tag{17.233}$$

When the primary and secondary cluster are constant,

$$\bar{y}_h = \frac{1}{m}\sum_i^M \bar{y}_{hi} \xrightarrow{\text{unbiased}} \mu_h \tag{17.234}$$

$$\bar{y}_o = \frac{1}{L'}\sum_h^{L'} \bar{y}_h \xrightarrow{\text{unbiased}} \mu_o \tag{17.235}$$

$$s_T^2\left(\frac{L-1}{L}\right) = \left(\frac{L-1}{L}\right)\frac{1}{(L'-1)}\sum_h^{L'}(T_h-\bar{T})^2 \xrightarrow{\text{unbiased}} \sigma_\tau^2 \tag{17.236}$$

$$s_p^2\left(\frac{L-1}{L}\right) = \left(\frac{L-1}{L}\right)\frac{s_T^2}{M^2A^2} = \left(\frac{L-1}{L}\right)\frac{1}{(L'-1)}\sum_h^{L'}\left(\frac{T_h}{A}-\frac{\bar{T}}{A}\right)^2 \xrightarrow{\text{unbiased}} \sigma_p^2 \tag{17.237}$$

$$s_{T_h}^2\left(\frac{M-1}{M}\right) = \left(\frac{M-1}{M}\right)\frac{1}{(m-1)}\sum_i^m(T_{hi}-\bar{T}_h) \xrightarrow{\text{unbiased}} \sigma_{\tau_h}^2 \tag{17.238}$$

$$s_{B_h}^2\left(\frac{M-1}{M}\right) = \left(\frac{M-1}{M}\right)\frac{s_{T_h}^2}{A^2} = \left(\frac{M-1}{M}\right)\frac{1}{(m-1)}\sum_i^m\left(\frac{T_{hi}}{A}-\frac{\bar{T}_h}{A}\right)^2 \xrightarrow{\text{unbiased}} \sigma_{B_h}^2 \tag{17.239}$$

$$s_B^2\left(\frac{M-1}{M}\right) = \left(\frac{M-1}{M}\right)\frac{1}{L'}\sum_h^{L'} s_{B_h}^2 \xrightarrow{\text{unbiased}} \sigma_B^2 \tag{17.240}$$

$$s_{w_h}^2\left(\frac{A-1}{A}\right) = \left(\frac{A-1}{A}\right)\frac{1}{m}\sum_i^m s_{hi}^2 \xrightarrow{\text{unbiased}} \sigma_{w_h}^2 \tag{17.241}$$

$$s_w^2\left(\frac{A-1}{A}\right)=\left(\frac{A-1}{A}\right)\frac{1}{L'}\sum_h^{L'} s_{w_h}^2 \xrightarrow{\text{unbiased}} \sigma_w^2 \qquad (17.242)$$

Substituting these sample statistics for their corresponding parameters in Equation 17.233 and summing over $L'$ and $m$ instead of $L$ and $M$:

$$\begin{aligned} s_{\bar{y}_{o3S}}^2 &= \frac{1}{L'}\left(\frac{L-L'}{L}\right)s_p^2 + \frac{1}{LL'm}\left(\frac{M-m}{M}\right)\sum_h^{L'} s_{B_h}^2 + \frac{1}{LML'ma}\left(\frac{A-a}{A}\right)\sum_h^{L'}\sum_i^{m} s_{hi}^2 \\ &= \frac{1}{L'}\left(\frac{L-L'}{L}\right)s_p^2 + \frac{1}{Lm}\left(\frac{M-m}{M}\right)s_B^2 + \frac{1}{LMa}\left(\frac{A-a}{A}\right)s_w^2 \end{aligned} \qquad (17.243)$$

### 17.15.4 Example of Three-Stage Sampling

One can again make use of the 16-ha tract shown in Figure 16.1. Figure 17.16 shows the primary, secondary, and tertiary sampling frames and the numbering system used in the random sampling process. As configured, $L = 4$, $M_h = 20$, and $N_{hi} = 4$. Setting the sample sizes at $L' = 2$, $m_h = 8$, and $n_{hi} = 2$, will produce a total sample of size $n = 32$ and a sampling intensity of 10%. The sample drawn is shown in Figure 17.17.

First analyze the sample ignoring the fact that $M_h$, $m_h$, $N_{hi}$, and $n_{hi}$ are constants, which will provide an example of the computational proceeding which would be used if they were not constant. The data from the sample and the computation are as follows.

| Primary Cluster | $h$ | Secondary Cluster | $i$ | Element | $j$ | $y_{hij}$ | $y_{hi}$ | $N_{hi}\bar{y}_{hi} = T_{hi}$ | $s^2_{hi}$ |
|---|---|---|---|---|---|---|---|---|---|
| 1 | 1 | 06 | 1 | 2 | 1 | 2.45 | | | |
| | | | | 1 | 2 | 2.35 | 2.400 | 9.6 | 0.00500 |
| | | 15 | 2 | 2 | 1 | 3.30 | | | |
| | | | | 3 | 2 | 3.05 | 3.175 | 12.7 | 0.03125 |
| | | 09 | 3 | 1 | 1 | 2.55 | | | |
| | | | | 2 | 2 | 0.90 | 1.725 | 6.9 | 1.36125 |
| | | 04 | 4 | 2 | 1 | 3.40 | | | |
| | | | | 3 | 2 | 2.40 | 2.900 | 11.6 | 0.50000 |
| | | 05 | 5 | 3 | 1 | 1.55 | | | |
| | | | | 1 | 2 | 3.05 | 2.300 | 9.2 | 1.12500 |
| | | 07 | 6 | 2 | 1 | 1.10 | | | |
| | | | | 3 | 2 | 3.10 | 2.100 | 8.4 | 2.00000 |
| | | 16 | 7 | 3 | 1 | 2.00 | | | |
| | | | | 2 | 2 | 4.15 | 3.075 | 12.3 | 2.31125 |
| | | 17 | 8 | 1 | 1 | 2.85 | | | |
| | | | | 2 | 2 | 1.80 | 2.325 | 9.3 | 0.55125 |
| 4 | 2 | 14 | 1 | 3 | 1 | 0.00 | | | |
| | | | | 1 | 2 | 0.55 | 0.275 | 1.1 | 0.15125 |
| | | 16 | 2 | 2 | 1 | 0.30 | | | |
| | | | | 1 | 2 | 1.60 | 0.950 | 3.8 | 0.84500 |

| Primary Cluster | $h$ | Secondary Cluster | $i$ | Element | $j$ | $y^{hij}$ | $y^{hi}$ | $N_{hi}\bar{y}_{hi} = T_{hi}$ | $s^{2hi}$ |
|---|---|---|---|---|---|---|---|---|---|
| | | 07 | 3 | 4 | 1 | 0.15 | | | |
| | | | | 3 | 2 | 0.40 | 0.275 | 1.1 | 0.03125 |
| | | 20 | 4 | 1 | 1 | 0.20 | | | |
| | | | | 4 | 2 | 0.40 | 0.300 | 1.2 | 0.02000 |
| | | 19 | 5 | 2 | 1 | 0.35 | | | |
| | | | | 4 | 2 | 0.40 | 0.375 | 1.5 | 0.00125 |
| | | 02 | 6 | 4 | 1 | 2.50 | | | |
| | | | | 1 | 2 | 3.25 | 2.875 | 11.5 | 0.28125 |
| | | 15 | 7 | 1 | 1 | 0.30 | | | |
| | | | | 4 | 2 | 0.05 | 0.175 | 0.7 | 0.03125 |
| | | 03 | 8 | 1 | 1 | 1.90 | | | |
| | | | | 2 | 2 | 2.30 | 2.100 | 8.4 | 0.08000 |

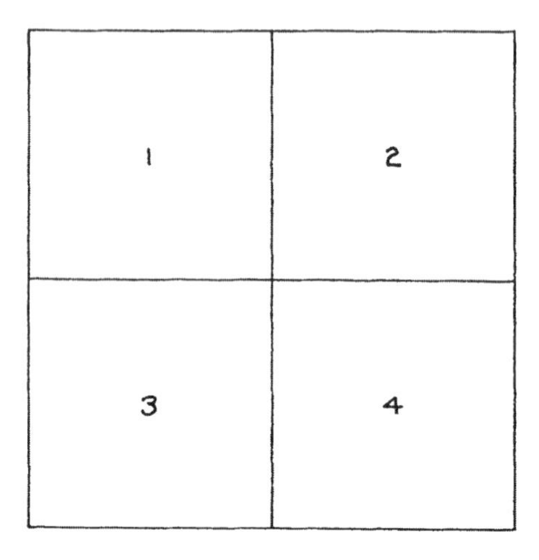


a. Primary

| | | | |
|---|---|---|---|
| 01 | 02 | 03 | 04 |
| 05 | 06 | 07 | 08 |
| 09 | 10 | 11 | 12 |
| 13 | 14 | 15 | 16 |
| 17 | 18 | 19 | 20 |

b. Secondary

| | |
|---|---|
| 1 | 2 |
| 3 | 4 |

c. Tertiary

**FIGURE 17.16** The primary (a), secondary (b), and tertiary (c) sampling frames.

| 2.55 | 3.00 | 1.60 | 1.80 | 2.65 | 2.10 | 1.90 | 3.40 | 3.65 | 4.00 | 4.00 | 1.75 | 2.60 | 2.85 | 2.00 | 2.10 |
|---|---|---|---|---|---|---|---|---|---|---|---|---|---|---|---|
| 3.15 | 2.95 | 2.40 | 2.25 | 3.20 | 2.80 | 2.40 | 4.55 | 4.50 | 3.25 | 4.00 | 1.50 | 1.80 | 2.70 | 1.70 | 2.50 |
| 3.05 | 2.10 | 2.35 | 2.45 | 2.90 | 1.10 | 4.30 | 4.00 | 4.10 | 2.85 | 3.40 | 1.85 | 2.00 | 2.30 | 2.20 | 2.50 |
| 1.55 | 1.55 | 2.55 | 2.15 | 3.10 | 2.75 | 4.50 | 4.60 | 3.30 | 4.30 | 3.75 | 2.30 | 1.80 | 2.00 | 2.10 | 1.80 |
| 2.55 | 0.90 | 2.25 | 2.35 | 2.45 | 2.90 | 4.65 | 1.90 | 3.35 | 4.60 | 2.60 | 2.35 | 2.60 | 2.75 | 2.70 | 3.25 |
| 2.70 | 2.50 | 1.90 | 2.30 | 2.95 | 3.15 | 3.00 | 4.55 | 2.65 | 4.40 | 4.20 | 2.20 | 1.90 | 1.70 | 2.30 | 1.70 |
| 2.50 | 1.50 | 2.00 | 2.55 | 3.15 | 3.30 | 4.05 | 4.50 | 4.35 | 3.80 | 4.50 | 2.00 | 3.65 | 1.35 | 3.15 | 1.60 |
| 2.35 | 2.90 | 2.05 | 2.10 | 3.05 | 3.35 | 4.15 | 4.60 | 4.00 | 3.45 | 1.50 | 1.75 | 2.20 | 2.85 | 3.45 | 2.40 |
| 2.85 | 1.80 | 1.50 | 2.05 | 2.25 | 4.80 | 4.40 | 3.15 | 3.00 | 3.65 | 2.30 | 1.35 | 1.90 | 2.40 | 2.20 | 0.90 |
| 3.15 | 1.60 | 2.40 | 2.60 | 3.00 | 1.90 | 2.55 | 3.55 | 4.00 | 2.75 | 2.65 | 1.90 | 2.60 | 2.70 | 1.50 | 2.65 |
| 1.75 | 0.25 | 1.05 | 0.95 | 3.00 | 2.90 | 3.75 | 3.15 | 3.35 | 2.30 | 3.25 | 2.30 | 2.30 | 1.90 | 1.60 | 0.00 |
| 0.00 | 0.00 | 0.00 | 0.45 | 1.20 | 1.90 | 2.90 | 4.10 | 4.50 | 3.00 | 2.65 | 2.50 | 1.00 | 1.40 | 0.00 | 0.05 |
| 0.60 | 0.00 | 0.50 | 0.55 | 2.05 | 3.10 | 4.65 | 2.30 | 3.00 | 2.20 | 1.90 | 1.30 | 0.55 | 0.75 | 0.15 | 0.20 |
| 0.30 | 0.00 | 0.30 | 0.40 | 1.50 | 1.15 | 2.35 | 1.00 | 2.55 | 2.45 | 2.25 | 0.85 | 0.40 | 0.15 | 1.05 | 0.10 |
| 0.00 | 0.05 | 0.00 | 0.40 | 1.40 | 2.20 | 2.40 | 2.00 | 1.40 | 1.50 | 1.25 | 0.00 | 0.10 | 1.35 | 1.45 | 0.60 |
| 0.10 | 0.00 | 0.00 | 1.00 | 1.95 | 2.40 | 3.60 | 1.60 | 1.90 | 2.40 | 1.00 | 1.15 | 0.40 | 0.20 | 0.05 | 0.00 |
| 0.55 | 0.15 | 0.10 | 0.85 | 2.05 | 1.60 | 1.90 | 2.75 | 1.75 | 2.00 | 0.55 | 0.00 | 0.30 | 0.00 | 1.60 | 0.30 |
| 0.60 | 0.50 | 0.00 | 0.65 | 1.80 | 1.65 | 2.35 | 2.45 | 1.90 | 1.15 | 0.00 | 1.50 | 0.85 | 0.05 | 0.50 | 0.00 |
| 0.60 | 1.00 | 0.15 | 0.00 | 1.70 | 2.50 | 1.25 | 2.40 | 0.80 | 0.50 | 0.10 | 0.75 | 0.30 | 0.35 | 0.20 | 0.10 |
| 0.55 | 0.20 | 0.50 | 0.30 | 1.50 | 2.05 | 1.80 | 1.40 | 1.05 | 0.10 | 0.00 | 0.45 | 0.50 | 0.40 | 0.75 | 0.40 |

**FIGURE 17.17** The three-stage sample.

$$\bar{T}_1 = (1/8)(9.6 + 12.7 + 6.9 + 11.6 + 9.2 + 8.4 + 12.3 + 9.3) = 10.0 \text{ m}^3$$

$$\bar{T}_2 = (1/8)(1.1 + 3.8 + 1.1 + 1.2 + 1.5 + 11.5 + 0.7 + 8.4) = 3.6625 \text{ m}^3$$

$$T_1 = 20(10.0) = 200.00 \text{ m}^3 \text{ in primary cluster 1 } (h = 1)$$

$$T_2 = 20(3.6625) = 73.25 \text{ m}^3 \text{ in primary cluster 4 } (h = 2)$$

$$\bar{T} = (1/2)(200.00 + 73.25) = 136.625 \text{ m}^3 \text{ / primary cluster}$$

$$T_{o_{3S}} = 4(136.625) = 546.5 \text{ or } 547 \text{ m}^3 \text{ on the tract}$$

$$\bar{y}_{o_{3S}} = 546.5/320 = 1.707813 \text{ m}^3 \text{ / plot, or}$$

$$\bar{y}_{o_{3S}} = 546.5/16 = 34.15625 \text{ or } 34.16 \text{ m}^3 \text{ / ha}$$

| Primary Cluster | Secondary Cluster | $\frac{s_{hi}^2}{n_{hi}}\left(\frac{N_{hi}-n_{hi}}{N_{hi}}\right)N_{hi}^2$ | $\frac{M_h}{m_h}\sum_i^{m_h}\left[\frac{s_{hi}^2}{n_{hi}}\left(\frac{N_{hi}-n_{hi}}{N_{hi}}\right)N_{hi}^2\right]$ |
|---|---|---|---|
| $h = 1$ | $i = 1$ | $\frac{0.00500}{2}\left(\frac{4-2}{4}\right)4^2 = 0.020$ | |
| | 2 | $\frac{0.03125}{2}\left(\frac{4-2}{4}\right)4^2 = 0.125$ | |
| | 3 | $\frac{1.36125}{2}\left(\frac{4-2}{4}\right)4^2 = 5.445$ | |
| | 4 | $\frac{0.50000}{2}\left(\frac{4-2}{4}\right)4^2 = 2.000$ | |
| | 5 | $\frac{1.12500}{2}\left(\frac{4-2}{4}\right)4^2 = 4.500$ | |
| | 6 | $\frac{2.00000}{2}\left(\frac{4-2}{4}\right)4^2 = 8.000$ | |
| | 7 | $\frac{2.31125}{2}\left(\frac{4-2}{4}\right)4^2 = 9.245$ | |
| | 8 | $\frac{0.55125}{2}\left(\frac{4-2}{4}\right)4^2 = 2.205$ | $\frac{20}{8}(31.54) = 78.8500$ |
| $h = 2$ | $i = 1$ | $\frac{0.15125}{2}\left(\frac{4-2}{4}\right)4^2 = 0.605$ | |
| | 2 | $\frac{0.84500}{2}\left(\frac{4-2}{4}\right)4^2 = 3.380$ | |
| | 3 | $\frac{0.03125}{2}\left(\frac{4-2}{4}\right)4^2 = 0.125$ | |
| | 4 | $\frac{0.02000}{2}\left(\frac{4-2}{4}\right)4^2 = 0.080$ | |
| | 5 | $\frac{0.00125}{2}\left(\frac{4-2}{4}\right)4^2 = 0.005$ | |
| | 6 | $\frac{0.28125}{2}\left(\frac{4-2}{4}\right)4^2 = 1.125$ | |
| | 7 | $\frac{0.03125}{2}\left(\frac{4-2}{2}\right)4^2 = 0.125$ | |
| | 8 | $\frac{0.08000}{2}\left(\frac{4-2}{2}\right)4^2 = 0.320$ | $\frac{20}{8}(5.765) = 14.4125$ |

$$\frac{L}{L'}\sum_h^{L'}\left\{\frac{M_h}{m_h}\sum_i^{m_h}\left[\frac{s_{hi}^2}{n_{hi}}\left(\frac{N_{hi}-n_{hi}}{N_{hi}}\right)N_{hi}^2\right]\right\} = \frac{4}{2}(78.8500 + 14.4125) = 186.525$$

$$s_{T_1}^2 = \frac{1}{(8-1)}[(9.6-10.0)^2 + (12.7-10.0)^2 + (6.9-10.0)^2 + (11.6-10.0)^2$$
$$+(9.2-10.0)^2 + (8.4-10.0)^2 + (12.3-10.0)^2 + (9.3-10.0)^2]$$
$$= 4.085714$$

$$s_{T_2}^2 = \frac{1}{(8-1)}[(1.1-3.6625)^2 + (3.8-3.6625)^2 + (1.1-3.6625)^2 + (1.2-3.6625)^2$$
$$+(1.5-3.6625)^2 + (11.5-3.6625)^2 + (0.7-3.6625)^2 + (8.4-3.6625)^2]$$
$$= 16.648393$$

| **Primary Cluster** | $\frac{M_h^2}{m_h}\left(\frac{M_h - m_h}{M_h}\right)s_{T_h}^2$ |
|---|---|
| $h = 1$ | $\frac{20^2}{8}\left(\frac{20-8}{20}\right)4.085714 = 122.57142$ |
| 2 | $\frac{20^2}{8}\left(\frac{20-8}{20}\right)16.648393 = 499.45179$ |

$$\frac{L}{L'}\sum_h^{L'}\left[\frac{M_h^2}{m_h}s_{T_h}^2\left(\frac{M_h - m_h}{M_h}\right)\right] = \frac{4}{2}(122.57142 + 499.45179) = 1244.0464$$

$$s_T^2 = \frac{1}{2-1}[(200.00-136.625)^2 + (73.25-136.625)^2] = 8032.7812$$

$$s_{\bar{y}_{o3S}}^2 = \frac{1}{(320)^2}\left\{\left[\frac{4^2}{2}\left(\frac{4-2}{4}\right)8,032.7812\right] + 1244.0464 + 186.525\right\} = 0.327751$$

$$s_{\bar{y}_{o3S}} = \pm\sqrt{0.3696622} = \pm 0.6079985 \text{ m}^3 \text{ / plot, or}$$

$$s_{\bar{y}_{o3S}} = \pm 20(0.6079985) = \pm 12.159971 \text{ or } \pm 12.16 \text{ m}^3 \text{ / ha, and}$$

$$s_{T_{o3S}} = \pm 320(0.6079985) = \pm 194.5592 \text{ or } \pm 194.5 \text{ m}^3 \text{ on the tract}$$

Now one can analyze the data recognizing that $M_h = M$, $m_h = m$, $N_{hi} = A$, and $n_{hi} = a$.

$$\bar{y}_i = \frac{1}{8}(2.400 + 3.175 + 1.725 + 2.900 + 2.300 + 2.100 + 3.075 + 2.325)$$
$$= 2.500 \text{ m}^3 \text{ / plot in primary cluster 1}$$

$$\bar{y}_2 = \frac{1}{8}(0.275 + 0.950 + 0.275 + 0.300 + 0.375 + 2.875 + 0.175 + 2.100)$$
$$= 0.915625 \text{ m}^3 \text{ / plot in primary cluster 2}$$

$$\bar{y}_0 = \frac{1}{2}(2.5 + 0.915625) = 1.707813 \text{ m}^3 \text{ / plot}$$

$$T_o = 320(1.707813) = 546.5 \text{ or } 547 \text{ m}^3 \text{ on the tract}$$

$$s_p^2 = \frac{2}{(2-1)}[(2.5 - 1.707813)^2 + (0.915625 - 1.707813)^2]$$

$$= 1.255122$$

$$s_{B_1}^2 = \frac{1}{(8-1)}[(2.400 - 2.500)^2 + (3.175 - 2.500)^2 + \cdots + (2.325 - 2.500)^2]$$

$$= 0.255357$$

$$s_{B_2}^2 = \frac{1}{(8-1)}[(0.275 - 0.915625)^2 + (0.950 - 0.915625)^2 + \cdots + (2.100 - 0.915625)^2]$$

$$= 1.040525$$

$$s_B^2 = \frac{1}{2}(0.255357 + 1.040525) = 0.647941$$

$$s_{w_1}^2 = \frac{1}{8}(0.005 + 0.03125 + 1.36125 + 0.5 + 1.125 + 2.0 + 2.31125 + 0.55125)$$

$$= 0.985625$$

$$s_{w_2}^2 = \frac{1}{8}(0.15125 + 0.845 + 0.03125 + 0.02 + 0.00125 + 0.28125 + 0.03125 + 0.08)$$

$$= 0.180156$$

$$s_w^2 = \frac{1}{2}(0.985625 + 0.180156) = 0.582891$$

$$s_{\bar{y}_{o3S}}^2 = \frac{1}{2}\left(\frac{4-2}{4}\right)1.255122 + \frac{1}{4(8)}\left(\frac{20-8}{20}\right)0.647941 + \frac{1}{4(20)(2)}\left(\frac{4-2}{4}\right)0.582891$$

$$= 0.327751$$

which agrees with the previous result.

### 17.15.5 Sample Sizes

The procedure used to obtain the optimum sample sizes for the three stages is an extension of the procedure described in Section 17.12.1 for two-stage sampling. As in that case, the sample sizes obtained are appropriate only when the primary, secondary, and tertiary cluster sizes are constant but can be used with differing cluster sizes if mean cluster sizes are used in the computation. In such cases the computed sample sizes would be means and can be considered to be equivalent to expected sample sizes.

According to Equation 17.232,

$$V(\bar{y}_o)_{3S} = \frac{1}{L'}\left(\frac{L-L'}{L-1}\right)\sigma_p^2 + \frac{1}{L'm}\left(\frac{M-m}{M-1}\right)\sigma_B^2 + \frac{1}{L'ma}\left(\frac{A-a}{A-1}\right)$$

Let

$$S_p^2 = \sigma_p^2\left(\frac{L}{L-1}\right), \quad \sigma_p^2 = S_p^2\left(\frac{L-1}{L}\right)$$

$$S_B^2 = \sigma_B^2\left(\frac{M}{M-1}\right), \quad \sigma_B^2 = S_B^2\left(\frac{M-1}{M}\right)$$

$$S_w^2 = \sigma_w^2\left(\frac{A}{A-1}\right), \quad \sigma_w^2 = S_w^2\left(\frac{A-1}{A}\right)$$

Then

$$V(\bar{y}_o)_{3S} = \frac{1}{L'}\left(\frac{L-L'}{L}\right)S_p^2 + \frac{1}{L'm}\left(\frac{M-m}{M}\right)S_B^2 + \frac{1}{L'ma}\left(\frac{A-a}{A}\right) \tag{17.244}$$

which can be written

$$V(\bar{y}_o)_{3S} = \frac{S_p^2}{L'} - \frac{S_p^2}{L} + \frac{S_B^2}{L'm} - \frac{S_B^2}{L'M} + \frac{S_w^2}{L'ma} - \frac{S_w^2}{L'mA} \tag{17.245}$$

In Section 17.10 it was brought out that, in most cases, a cruiser upon finishing the evaluation of the last unit within a cluster would proceed directly to the next cluster. Consequently, if there are $m$ clusters, the cruiser would cross $(m-1)$ between-cluster intervals. Only if the cruiser returned to the initial unit within the cluster would $m$ intervals be traversed. In the case of three-stage cluster sampling there are $L'$ primary clusters and the cruiser would have $(L'-1)$ intervals to cross. Within each primary cluster there are $m$ secondary clusters. To keep the following development simple, it will be assumed that the cruiser will return to the initial unit within each secondary cluster and consequently will cross $m$ intervals. The cost function is then

$$\begin{aligned} c_F' &= c_1L'ma + c_2L'm + c_3(L'-1) \\ &= c_1L'ma + c_2L'm + c_3L' - c_3 \end{aligned} \tag{17.246}$$

where $c_1$ = average cost of evaluating one tertiary unit
$c_2$ = average cost of travel between two consecutive secondary clusters
$c_3$ = average cost of travel between two consecutive primary clusters

Combining Equations 17.245 and 17.246, using the Lagrangian multiplier

$$V = V(\bar{y}_o)_{3S} + \lambda(c_1L'ma + c_2L'm + c_3L' - c_3 - c_F') \tag{17.247}$$

Then

$$\frac{\partial V}{\partial L'} = -\frac{S_p^2}{L'^2} - \frac{S_B^2}{L'^2m} + \frac{S_B^2}{L'^2M} - \frac{S_w^2}{L'^2ma} + \frac{S_w^2}{L'^2mA} + \lambda c_1ma + \lambda c_2m + \lambda c_3 = 0 \tag{17.248}$$

$$\frac{\partial V}{\partial m} = -\frac{S_B^2}{L'm^2} - \frac{S_w^2}{L'm^2a} + \frac{S_w^2}{L'm^2A} + \lambda c_1L'a + \lambda c_2L' = 0 \tag{17.249}$$

$$\frac{\partial V}{\partial a} = -\frac{S_w^2}{L'ma^2} + \lambda c_1 L'm = 0 \tag{17.250}$$

Multiplying Equation 17.248 by $-1/m$, Equation 17.249 by $1/L'$, and adding the resulting equations yields

$$\frac{Sp^2}{L'^2m} - \frac{S_B^2}{L'^2Mm} - \frac{\lambda c_3}{m} = \frac{(S_p^2 - S_B^2 / M)}{L'^2m} - \frac{\lambda c_3 L'^2}{L'^2m} \tag{17.251}$$

Multiplying Equation 17.249 by $-1/a$, Equation 17.250 by $1/m$, and adding the resulting equations:

$$\frac{S_B^2}{L'm^2a} - \frac{S_w^2}{L'm^2Aa} - \frac{\lambda c_2 L'}{a} = \frac{(S_B^2 - S_w^2 / A)}{L'm^2a} - \frac{\lambda c_2 L'^2m^2}{L'm^2a} = 0 \tag{17.252}$$

Multiplying Equation 17.251 by $(S_B^2 - S_w^2 / A)/ma$, Equation 17.252 by $-(S_p^2 - S_B^2 / M)/L'$, and adding the resulting equations,

$$\frac{\lambda c_2 L'^2m^2(S_p^2 - S_B^2 / M)}{L'^2m^3a} - \frac{\lambda c_3 L'^2(S_B^2 - S_w^2 / A)}{L'^2m^2a} = 0$$

or

$$\lambda c_2 L'^2m^2(S_p^2 - S_B^2 / M) = \lambda c_3 L'^2(S_B^2 - S_w^2 / A)$$

Solving for $m^2$ and simplifying:

$$m^2 = \frac{c_3}{c_2}\left(\frac{S_B^2 - S_w^2 / A}{S_p^2 - S_B^2 / M}\right)$$

Then,

$$m = \sqrt{\frac{c_3}{c_2}\left(\frac{S_B^2 - S_W^2 / A}{S_p^2 - S_B^2 / M}\right)} \tag{17.253}$$

Multiplying Equation 17.252 by $S_w^2/a$, Equation 17.250 by $(S_B^2 - S_w^2 / A)/m$, and adding the resulting equations:

$$\lambda c_1 L'(S_B^2 - S_w^2 / A) = \frac{\lambda c_2 L'S_w^2}{a^2}$$

Solving for $a^2$ and simplifying:

$$a^2 = \frac{c^2 S_w^2}{c_1(S_B^2 - S_w^2 / A)}$$

Then

$$a = \sqrt{\frac{c_2 s_w^2}{c_1(s_B^2 - s_w^2 / A)}} \tag{17.254}$$

Returning to the cost function, into which the values of $m$ and $a$ can be substituted, and solving for $L'$:

$$L' = \frac{c_F' + c_3}{c_1 ma + c_2 m + c_3} \tag{17.255}$$

According to Equations 17.237, 17.240, and 17.242,

$$s_p^2\left(\frac{L-1}{L}\right) \to \sigma_p^2 = S_p^2\left(\frac{L-1}{L}\right), \quad \text{thus } s_p^2 \to S_p^2$$

$$s_B^2\left(\frac{M-1}{M}\right) \to \sigma_B^2 = S_B^2\left(\frac{M-1}{M}\right), \quad \text{thus } s_B^2 \to S_B^2$$

and

$$s_w^2\left(\frac{A-1}{A}\right) \to \sigma_w^2 = S_w^2\left(\frac{A-1}{A}\right), \quad \text{thus } s_w^2 \to S_w^2$$

Consequently, the optimum sample sizes $m$ and $a$ can be estimated by substituting the above sample-based variances into Equations 17.253 and 17.254:

$$a = \sqrt{\frac{c_2 s_w^2}{c_1(s_B^2 - s_w^2 / A)}} \tag{17.256}$$

and

$$a = \sqrt{\frac{c_2 s_w^2}{c_1(s_B^2 - s_w^2 / A)}} \tag{17.257}$$

These magnitudes of $L'$, $m$, and $a$ will rarely be whole numbers and they must be rounded off. The rounding must be done in such a manner that the $c_F$ constraint is not exceeded. Each of the computed sample sizes can be rounded either up or down. Let a + associated with a sample size indicate rounding up and – indicate rounding down. Then the costs associated with each combination of rounded sample sizes would be

$$c_F = c_1 L'^+ + c_2 L'^+ m^+ + c_3 L'^+ m^+ a^+$$

$$c_F = c_1 L'^+ + c_2 L'^+ m^+ + c_3 L'^+ m^+ a^-$$

$$c_F = c_1 L'^+ + c_2 L'^+ m^- + c_3 L'^+ m^- a^+$$

$$c_F = c_1L'^{+} + c_2L'^{+}m^{-} + c_3L'^{+}m^{-}a^{-}$$

$$c_F = c_1L'^{-} + c_2L'^{-}m^{+} + c_3L'^{-}m^{+}a^{+}$$

$$c_F = c_1L'^{-} + c_2L'^{-}m^{+} + c_3L'^{-}m^{+}a^{-}$$

$$c_F = c_1L'^{-} + c_2L'^{-}m^{-} + c_3L'^{-}m^{-}a^{+}$$

$$c_F = c_1L'^{-} + c_2L'^{-}m^{-} + c_3L'^{-}m^{-}a^{-}$$

Some of these $c_F$ magnitudes will be less than or equal to the specified $c_F$. Their associated expression would indicate the sample size combination that would be acceptable. The choice between these candidate combinations is arbitrary. Ideally, each of the candidate combinations should be used in Equation 17.232 to determine which would yield the smallest variance of the mean. Unfortunately, $\sigma_p^2$, $\sigma_B^2$, and $\sigma_w^2$ are not known. Sample-based estimates could be used but since such estimates are subject to random variations, one would never be certain that a choice of sample sizes based on such information would yield the most efficient sampling operation.

If the sample sizes are so small relative to their respective frame sizes that their associated finite population correction factors are essentially equal to unity, the expression for $m$ and $a$ simplify to

$$m = \sqrt{\frac{c_3 s_B^2}{c_2 s_p^2}} \tag{17.258}$$

and

$$a = \sqrt{\frac{c_2 s_w^2}{c_1 s_B^2}} \tag{17.259}$$

If the secondary and tertiary sampling units and sample size are not constant, their respective mean sizes can be considered to be their expected sizes and be substituted for the corresponding constant sizes used in the generation of the sample size formulae:

$$\frac{1}{L}\sum_h^L M_h = \overline{M} \rightarrow M$$

$$\frac{1}{LM}\sum_h^L \sum_i^M N_{hi} = \overline{N} \rightarrow A$$

$$\frac{1}{L}\sum_h^L m_h = \overline{m} \rightarrow m$$

and

$$\frac{1}{LM}\sum_h^L \sum_i^M n_{hi} = \overline{n} \rightarrow a$$

where $\rightarrow$ indicates substitution. Those substitutions will not change the derivation process. Consequently, under these circumstances

$$m = \sqrt{\frac{c_3(s_B^2 - s_w^2 / \bar{N})}{c_2(s_p^2 - s_B^2 / \bar{M})}} \tag{17.260}$$

$$a = \sqrt{\frac{c_2 s_w^2}{c_1(s_B^2 - s_w^2 / \bar{N})}} \tag{17.261}$$

and

$$L' = \frac{c_F' + c_3}{c_1 \bar{m}\bar{n} + c_2 \bar{m} + c_3} \tag{17.262}$$

As an example of the sample size determination process, assume the following: $s_p^2 = 35$; $s_B^2 = 330$; $s_w^2 = 1000$; $L = 40$; $M = 20$; $A = 4$; $c_1 = \$3$; $c_2 = \$1$; $c_3 = \$2$; and $c_F' = \$200$. Then,

$$m = \sqrt{\frac{2(330 - 1000/4)}{1(35 - 330/20)}} = 2.941$$

$$a = \sqrt{\frac{1(1000)}{3(330 - 1000/4)}} = 2.041$$

and

$$L' = \frac{200 + 2}{3(2.941)(2.041) + 1(2.941) + 2} = 11.217$$

Then,

$$L'^{+} = 12, \quad L'^{-} = 11, \quad m^{+} = 3, \quad m^{-} = 2, \quad a^{+} = 3, \quad \text{and} \quad a^{-} = 2$$

and

$$c_F = 3(12)(3)(3) + 1(12)(3) + 2(12 - 1) = 382 > 200$$

$$c_F = 3(12)(3)(2) + 1(12)(3) + 2(12 - 1) = 274 > 200$$

$$c_F = 3(12)(2)(3) + 1(12)(2) + 2(12 - 1) = 262 > 200$$

$$c_F = 3(12)(2)(2) + 1(12)(2) + 2(12 - 1) = 190 < 200^*$$

$$c_F = 3(11)(3)(3) + 1(11)(3) + 2(11 - 1) = 350 > 200$$

$$c_F = 3(11)(3)(2) + 1(11)(3) + 2(11 - 1) = 251 > 200$$

$$c_F = 3(11)(2)(3) + 1(11)(2) + 2(11 - 1) = 240 > 200$$

$$c_F = 3(11)(2)(2) + 1(11)(2) + 2(11 - 1) = 174 < 200^*$$

Only the two combinations identified by asterisks are viable options. If the computed sample sizes and the presample variances are substituted into Equation 17.243, the resulting estimates of the variance of the mean can be used to make the final decision. Thus, using $L' = 12$, $m = 2$, and $a = 2$,

$$s_{\bar{y}_o}^2 = \frac{1}{12}\left(\frac{40-12}{40}\right)35 + \frac{1}{40(2)}\left(\frac{20-2}{20}\right)330 + \frac{1}{40(20)(2)}\left(\frac{4-2}{4}\right)1000 = 6.0667$$

and when $L' = 11$, $m = 2$, and $a = 2$,

$$s_{\bar{y}_o}^2 = \frac{1}{11}\left(\frac{40-11}{40}\right)35 + \frac{1}{40(2)}\left(\frac{20-2}{20}\right)330 + \frac{1}{40(20)(2)}\left(\frac{4-2}{4}\right)1000 = 6.3318$$

The available evidence indicates that the most efficient combination of sample sizes would be $L' = 12$, $m = 2$, and $a = 2$.

## 17.16 STRATIFIED CLUSTER SAMPLING

When the universe of interest is stratified, either logically or arbitrarily, and a sample involving clusters is randomly drawn from each of the strata, the sampling design is known as *stratified cluster sampling*. The clusters may or may not be subsampled and there is no limit to the number of stages that can be used. The design is appealing because it is efficient in both a statistical and an economic sense. The probability is high that a representative sample will be obtained, especially when the stratification is logical, because every stratum (condition) is represented in the sample. Thus, repeated sampling will yield similar results and the variance of the mean or the total will be small, relative to what would be obtained from repeated simple random samples of the same size. The use of clusters instead of simple element sampling units tends to concentrate the data collection operations into relatively few locations, which reduces travel time between sampling units, which, in turn, reduces costs. Consequently, stratified cluster sampling is often the design of choice.

### 17.16.1 THE PARAMETERS AND THEIR ESTIMATION

In a mathematical sense stratified cluster sampling is simply a form of multistage sampling in which the strata serve as the primary sampling units and in which each of these primary sampling units is represented in the sample. The secondary sampling units are clusters, which may have a compact or some form of open configuration. Subsampling may be used, through as many stages as are deemed desirable. Consequently, one can use the mathematical development in Sections 17.12.1 and 17.12.2 as being appropriate through the situation when subsampling is done from the secondary sampling units. As was pointed out in those paragraphs, the approaches used therein can be extended to any member of stages.

Whether all or only some of the primary sampling units (the strata) are included in a sample is of no significance insofar as the population parameters are concerned. Thus, Equations 17.204 through 17.214 are appropriate. However, the appropriateness of the estimators in Equations 17.215 through 17.224 depends on how the sampling is carried out.

If subsampling from the secondary sampling units is assumed, Equations 17.215 through 17.220 remain unchanged when $L' = L$. However, Equation 17.221 becomes

$$\bar{T} = \frac{1}{L}\sum_{h}^{L} T_h \tag{17.263}$$

Equation 17.224 becomes

$$\bar{y}_o = \frac{1}{N}\sum_{h}^{L}\frac{M_h}{m_h}\sum_{i}^{m_h}\frac{N_{hi}}{n_{hi}}\sum_{j}^{n_{hi}} y_{hij} \tag{17.264}$$

Equation 17.225 becomes

$$V(\bar{y}_o)_{SC} = \frac{1}{N^2}\left[\sum_h^L \frac{M_h^2}{m_h}\left(\frac{M_h - m_h}{M_h - 1}\right)\sigma_{\tau_h}^2 + \sum_h^L \frac{M_h}{m_h}\sum_i^{M_h}\frac{\sigma_{hi}^2}{n_{hi}}\left(\frac{N_{hi} - n_{hi}}{N_{hi} - 1}\right)N_{hi}^2\right] \tag{17.265}$$

and Equation 17.233 becomes

$$s_{\bar{y}_{oSC}}^2 = \frac{1}{N^2}\left[\sum_h^L \frac{M_h^2}{m_h}\left(\frac{M_h - m_h}{M_h}\right)s_{T_h}^2 + \sum_h^L \frac{M_h}{m_h}\sum_i^{M_h}\frac{s_{hi}^2}{n_{hi}}\left(\frac{N_{hi} - n_{hi}}{N_{hi}}\right)N_{hi}^2\right] \tag{17.266}$$

Since stratum sizes are rarely constant in size, especially when the stratification is logical, the number of clusters in a stratum, $M_h$, will not be constant from stratum to stratum. The number of clusters included in a sample, $m_h$, will also vary from stratum to stratum. Cluster size, $N_{hi}$, may or may not be constant. It depends on the nature of the universe of interest, the cluster configuration, and how the stratification is carried out. Ideally, the stratification should be done in such a manner that all the clusters within a stratum are of the same size. This often is not possible when logical stratification is used as, for example, when the stratification is on the basis of forest condition classes.

If no subsamples are drawn from the secondary sampling units, $n_{hi} = N_{hi}$ and the data collection operations yield complete numerations. Consequently, under those circumstances,

$$\bar{y}_{hi} = \frac{1}{n_{hi}}\sum_j^{n_{ij}} y_{hij} = \frac{1}{N_{hi}}\sum_j^{N_{hi}} y_{hij} = \mu_{hi} \tag{17.267}$$

$$T_{hi} = \sum_j^{n_{hi}} y_{hij} = \sum_j^{N_{hi}} y_{hij} = \tau_{hi} \tag{17.268}$$

$$s_{hi}^2 = \frac{1}{(n_{hi} - 1)}\sum_j^{n_{hi}}(y_{hij} - \bar{y}_{hi})^2 = \frac{1}{(N_{hi} - 1)}\sum_j^{N_{hi}}(y_{hij} - \mu_{hi})^2 = \sigma_{hi}^2\left(\frac{N_{hi}}{N_{hi} - 1}\right) \tag{17.269}$$

and, since $(N_{hi} - n_{hi}) = (N_{hi} - N_{hi}) = 0$, the estimate of the variance of the overall mean becomes

$$s_{\bar{y}_{oSC}}^2 = \frac{1}{N^2}\left[\sum_h^L \frac{M_h^2}{m_h}(M_h - m_h)s_{T_h}^2\right] \tag{17.270}$$

### 17.16.2 Example of Stratified Cluster Sampling Using Subsampling

This section again uses the 16-ha tract shown in Figure 16.1 and again assumes that a sampling intensity of 10% has been specified. The stratification has been into the four timber volume classes recognized in Figure 16.3: $A$ ($h = 1$), 76 to 100 $m^3$/ha (3.8 to 5.0 $m^3$/plot); $B$ ($h = 2$), 51 to 75 $m^3$/ha (2.55 to 3.75 $m^3$/plot); $C$ ($h = 3$), 26 to 50 $m^3$/ha (1.3 to 2.5 $m^3$/plot); and $D$ ($h = 4$), 0 to 25 $m^3$/ha (0 to $^1/_{25}$ $m^3$/plot). The stratification is shown in Figure 17.18. It should be noted that the stratum

| | | | | | | | | | | | | | | | |
|---|---|---|---|---|---|---|---|---|---|---|---|---|---|---|---|
| 2.55 | 3.00 | 1.60 | 1.80 | 2.65 | 2.10 | 1.90 | 3.40 | 3.65 | 4.05 | 4.00 | 1.75 | 2.60 | 2.85 | 2.00 | 2.10 |
| 3.15 | 2.95 | 2.40 | 2.25 | 3.20 | 2.80 | 2.40 | 4.55 | 4.50 | 3.25 | 4.00 | 1.50 | 1.80 | 2.70 | 1.70 | 2.50 |
| 3.05 | 2.10 | 2.35 | 2.45 | 2.90 | 1.10 | 4.30 | 4.00 | 4.10 | 2.85 | 3.40 | 1.85 | 2.00 | 2.30 | 2.20 | 2.50 |
| 1.55 | 1.55 | 2.55 | 2.15 | 3.10 | 2.75 | 4.50 | 4.60 | 3.30 | 4.30 | 3.75 | 2.30 | 1.80 | 2.00 | 2.10 | 1.80 |
| 2.55 | 0.90 | 2.25 | 2.35 | 2.45 | 2.90 | 4.65 | 1.90 | 3.35 | 4.60 | 2.60 | 2.35 | 2.60 | 2.75 | 2.70 | 3.25 |
| 2.70 | 2.50 | 1.90 | 2.30 | 2.95 | 3.15 | 3.00 | 4.55 | 2.65 | 4.40 | 4.20 | 2.20 | 1.90 | 1.70 | 2.30 | 1.70 |
| 2.50 | 1.50 | 2.00 | 2.55 | 3.15 | 3.30 | 4.05 | 4.50 | 4.35 | 3.80 | 4.50 | 2.00 | 3.65 | 1.35 | 3.15 | 1.60 |
| 2.35 | 2.90 | 2.05 | 2.10 | 3.05 | 3.35 | 4.15 | 4.60 | 4.00 | 3.45 | 1.50 | 1.75 | 2.20 | 2.85 | 3.45 | 2.40 |
| 2.85 | 1.80 | 1.50 | 2.05 | 2.25 | 4.80 | 4.40 | 3.15 | 3.00 | 3.65 | 2.30 | 1.35 | 1.90 | 2.40 | 2.20 | 0.90 |
| 3.15 | 1.60 | 2.40 | 2.60 | 3.00 | 1.90 | 2.55 | 3.55 | 4.00 | 2.75 | 2.65 | 1.90 | 2.60 | 2.70 | 1.50 | 2.65 |
| 1.75 | 0.25 | 1.05 | 0.95 | 3.00 | 2.90 | 3.75 | 3.15 | 3.35 | 2.30 | 3.25 | 2.30 | 2.30 | 1.90 | 1.60 | 0.00 |
| 0.00 | 0.00 | 0.00 | 0.45 | 1.20 | 1.90 | 2.90 | 4.10 | 4.50 | 3.00 | 2.65 | 2.50 | 1.00 | 1.40 | 0.00 | 0.05 |
| 0.60 | 0.00 | 0.50 | 0.55 | 2.05 | 3.10 | 4.65 | 2.30 | 3.00 | 2.20 | 1.90 | 1.30 | 0.55 | 0.75 | 0.15 | 0.20 |
| 0.30 | 0.00 | 0.30 | 0.40 | 1.50 | 1.15 | 2.35 | 1.00 | 2.55 | 2.45 | 2.25 | 0.85 | 0.40 | 0.15 | 1.05 | 0.10 |
| 0.00 | 0.05 | 0.00 | 0.40 | 1.40 | 2.20 | 2.40 | 2.00 | 1.40 | 1.50 | 1.25 | 0.00 | 0.10 | 1.35 | 1.45 | 0.60 |
| 0.10 | 0.00 | 0.00 | 1.00 | 1.95 | 2.40 | 3.60 | 1.60 | 1.90 | 2.40 | 1.00 | 1.15 | 0.40 | 0.20 | 0.05 | 0.00 |
| 0.55 | 0.15 | 0.10 | 0.85 | 2.05 | 1.60 | 1.90 | 2.75 | 1.75 | 2.00 | 0.55 | 0.00 | 0.30 | 0.00 | 1.60 | 0.30 |
| 0.60 | 0.50 | 0.00 | 0.65 | 1.80 | 1.65 | 2.35 | 2.45 | 1.90 | 1.15 | 0.00 | 1.50 | 0.85 | 0.05 | 0.50 | 0.00 |
| 0.60 | 1.00 | 0.15 | 0.00 | 1.70 | 2.50 | 1.25 | 2.40 | 0.80 | 0.50 | 0.10 | 0.75 | 0.30 | 0.35 | 0.20 | 0.10 |
| 0.55 | 0.20 | 0.50 | 0.30 | 1.50 | 2.05 | 1.80 | 1.40 | 1.05 | 0.10 | 0.00 | 0.45 | 0.50 | 0.40 | 0.75 | 0.40 |

**FIGURE 17.18** The 16-ha tract logically stratified, conditioned by the constraint that four-element clusters will be used.

boundaries in Figure 17.18 differ somewhat from those shown in Figure 16.3. These differences arose because of a specification that the cluster size be kept constant. The clusters were made up of four of the 1/20-ha plots arranged to form a 2 × 2 square. The clusters were numbered, within each stratum, as shown in Figure 17.18. The plots within the cluster were numbered as were those shown in Figure 17.16. Since the 16 ha contain 320 plots, there were 80 clusters divided as follows: $M_1 = 5$; $M_2 = 21$; $M_3 = 31$; and $M_4 = 23$. The total sample size was $n = 0.1(320) = 32$ plots. A subsample of size $n_{hi} = 2$ was drawn from each of the clusters chosen in the sampling operation. Consequently, $n/n_{hi} = 32/2 = 16$ clusters were sampled. If the allocations had been proportional, the number of clusters chosen in each of the strata would have been

$$m_1 = (5/80)16 = 1.0 \text{ or } 1$$

$$m_2 = (21/80)16 = 4.2 \text{ or } 4$$

$$m_3 = (31/80)16 = 6.2 \text{ or } 6$$

$$m_4 = (23/80)16 = 4.6 \text{ or } 5$$

However, since at least two clusters are needed per stratum if stratum variances are to be estimated, the above allocation must be modified so that $m_1 \geq 2$. The extra cluster was taken from stratum *D*

| | | | | | | | | | | | | | | | |
|---|---|---|---|---|---|---|---|---|---|---|---|---|---|---|---|
| 2.55 | 3.00 | 1.60 | 1.80 | 2.65 | 2.10 | 1.90 | 3.40 | 3.65 | 4.00 | 4.00 | 1.75 | 2.60 | 2.85 | 2.00 | 2.10 |
| 3.15 | 2.95 | 2.40 | 2.25 | 3.20 | 2.80 | 2.40 | 4.55 | 4.50 | 3.25 | 4.00 | 1.50 | 1.80 | 2.70 | 1.70 | 2.50 |
| 3.05 | 2.10 | 2.35 | 2.45 | 2.90 | 1.10 | 4.30 | 4.00 | 4.10 | 2.85 | 3.40 | 1.85 | 2.00 | 2.30 | 2.20 | 2.50 |
| 1.55 | 1.55 | 2.55 | 2.15 | 3.10 | 2.75 | 4.50 | 4.60 | 3.30 | 4.30 | 3.75 | 2.30 | 1.80 | 2.00 | 2.10 | 1.80 |
| 2.55 | 0.90 | 2.25 | 2.35 | 2.45 | 2.90 | 4.65 | 1.90 | 3.35 | 4.60 | 2.60 | 2.35 | 2.60 | 2.75 | 2.70 | 3.25 |
| 2.70 | 2.50 | 1.90 | 2.30 | 2.95 | 3.15 | 3.00 | 4.55 | 2.65 | 4.40 | 4.20 | 2.20 | 1.90 | 1.70 | 2.30 | 1.70 |
| 2.50 | 1.50 | 2.00 | 2.55 | 3.15 | 3.30 | 4.05 | 4.50 | 4.35 | 3.80 | 4.50 | 2.00 | 3.65 | 1.35 | 3.15 | 1.60 |
| 2.35 | 2.90 | 2.05 | 2.10 | 3.05 | 3.35 | 4.15 | 4.60 | 4.00 | 3.45 | 1.50 | 1.75 | 2.20 | 2.85 | 3.45 | 2.40 |
| 2.85 | 1.80 | 1.50 | 2.05 | 2.25 | 4.80 | 4.40 | 3.15 | 3.00 | 3.65 | 2.30 | 1.35 | 1.90 | 2.40 | 2.20 | 0.90 |
| 3.15 | 1.60 | 2.40 | 2.60 | 3.00 | 1.90 | 2.55 | 3.55 | 4.00 | 2.75 | 2.65 | 1.90 | 2.60 | 2.70 | 1.50 | 2.65 |
| 1.75 | 0.25 | 1.05 | 0.95 | 3.00 | 2.90 | 3.75 | 3.15 | 3.35 | 2.30 | 3.25 | 2.30 | 2.30 | 1.90 | 1.60 | 0.00 |
| 0.00 | 0.00 | 0.00 | 0.45 | 1.20 | 1.90 | 2.90 | 4.10 | 4.50 | 3.00 | 2.65 | 2.50 | 1.00 | 1.40 | 0.00 | 0.05 |
| 0.60 | 0.00 | 0.50 | 0.55 | 2.05 | 3.10 | 4.65 | 2.30 | 3.00 | 2.20 | 1.90 | 1.30 | 0.55 | 0.75 | 0.15 | 0.20 |
| 0.30 | 0.00 | 0.30 | 0.40 | 1.50 | 1.15 | 2.35 | 1.00 | 2.55 | 2.45 | 2.25 | 0.85 | 0.40 | 0.15 | 1.05 | 0.10 |
| 0.00 | 0.05 | 0.00 | 0.40 | 1.40 | 2.20 | 2.40 | 2.00 | 1.40 | 1.50 | 1.25 | 0.00 | 0.10 | 1.35 | 1.45 | 0.60 |
| 0.10 | 0.00 | 0.00 | 1.00 | 1.95 | 2.40 | 3.60 | 1.60 | 1.90 | 2.40 | 1.00 | 1.15 | 0.40 | 0.20 | 0.05 | 0.00 |
| 0.55 | 0.15 | 0.10 | 0.85 | 2.05 | 1.60 | 1.90 | 2.75 | 1.75 | 2.00 | 0.55 | 0.00 | 0.30 | 0.00 | 1.60 | 0.30 |
| 0.60 | 0.50 | 0.00 | 0.65 | 1.80 | 1.65 | 2.35 | 2.45 | 1.90 | 1.15 | 0.00 | 1.50 | 0.85 | 0.05 | 0.50 | 0.00 |
| 0.60 | 1.00 | 0.15 | 0.00 | 1.70 | 2.50 | 1.25 | 2.40 | 0.80 | 0.50 | 0.10 | 0.75 | 0.30 | 0.35 | 0.20 | 0.10 |
| 0.55 | 0.20 | 0.50 | 0.30 | 1.50 | 2.05 | 1.80 | 1.40 | 1.05 | 0.10 | 0.00 | 0.45 | 0.50 | 0.40 | 0.75 | 0.40 |

**FIGURE 17.19** The stratified cluster sample.

($h = 4$) because it exhibits the least volume per hectare and therefore is the least valuable. The final allocation was then $m_1 = 2$; $m_2 = 4$; $m_3 = 6$; and $m_4 = 4$. These were randomly drawn from their respective strata and a subsample of size 2 was randomly drawn from each of the chosen clusters. These are shown in Figure 17.19. The data and computations are as follows:

| Stratum | $h$ | Cluster | $i$ | Element | $j$ | $y_{hij}$ | $\bar{y}_i$ | $N_{hi}\ \bar{y}_i = T_{hi}$ | $s^2_{hi}$ |
|---|---|---|---|---|---|---|---|---|---|
| $A$ | 1 | 03 | 1 | 1 | 1 | 4.65 | | | |
| | | | | 2 | 2 | 1.90 | 3.275 | 13.1 | 3.78125 |
| | | 05 | 2 | 2 | 1 | 3.80 | | | |
| | | | | 3 | 2 | 4.00 | 3.900 | 15.6 | 0.02000 |
| $B$ | 2 | 09 | 1 | 2 | 1 | 3.65 | | | |
| | | | | 3 | 2 | 4.00 | 3.825 | 15.3 | 0.06125 |
| | | 11 | 2 | 1 | 1 | 3.35 | | | |
| | | | | 3 | 2 | 4.50 | 3.925 | 15.7 | 0.66125 |
| | | 19 | 3 | 3 | 1 | 4.00 | | | |
| | | | | 4 | 2 | 1.50 | 2.750 | 11.0 | 3.12500 |
| | | 20 | 4 | 3 | 1 | 1.80 | | | |
| | | | | 4 | 2 | 2.70 | 2.250 | 9.0 | 0.40500 |

| Stratum | $h$ | Cluster | $i$ | Element | $j$ | $y_{hij}$ | $\bar{y}_i$ | $N_{hi}\ \bar{y}_i = T_{hi}$ | $s^2_{hi}$ |
|---|---|---|---|---|---|---|---|---|---|
| $C$ | 3 | 03 | 1 | 1 | 1 | 2.35 | | | |
| | | | | 3 | 2 | 2.55 | 2.450 | 9.8 | 0.02000 |
| | | 07 | 2 | 1 | 1 | 2.00 | | | |
| | | | | 2 | 2 | 2.55 | 2.275 | 9.1 | 0.15125 |
| | | 09 | 3 | 2 | 1 | 2.05 | | | |
| | | | | 4 | 2 | 2.60 | 2.325 | 9.3 | 0.15125 |
| | | 14 | 4 | 2 | 1 | 1.50 | | | |
| | | | | 3 | 2 | 1.65 | 1.575 | 6.3 | 0.01125 |
| | | 19 | 5 | 1 | 1 | 1.25 | | | |
| | | | | 3 | 2 | 1.80 | 1.525 | 6.1 | 0.15125 |
| | | 26 | 6 | 1 | 1 | 3.65 | | | |
| | | | | 4 | 2 | 2.85 | 3.250 | 13.0 | 0.32000 |
| $D$ | 4 | 03 | 1 | 2 | 1 | 0.00 | | | |
| | | | | 4 | 2 | 0.00 | 0.000 | 0.0 | 0.00000 |
| | | 08 | 2 | 1 | 1 | 0.10 | | | |
| | | | | 3 | 2 | 0.00 | 0.050 | 0.2 | 0.00500 |
| | | 11 | 3 | 1 | 1 | 0.80 | | | |
| | | | | 4 | 2 | 0.10 | 0.450 | 1.8 | 0.24500 |
| | | 18 | 4 | 2 | 1 | 0.00 | | | |
| | | | | 4 | 2 | 1.15 | 0.575 | 2.3 | 0.66125 |

$$\bar{T}_1 = (1/2)(13.1 + 15.6) = 14.35$$

$$\bar{T}_2 = (1/4)(15.3 + 15.7 + 11.0 + 9.0) = 12.75$$

$$\bar{T}_3 = (1/6)(9.8 + 9.1 + 9.3 + 6.3 + 6.1 + 13.0) = 8.93$$

$$\bar{T}_4 = (1/4)(0.0 + 0.2 + 1.8 + 2.3) = 1.075$$

$$T_1 = 5(14.35) = 71.75 \text{ m}^3 \text{ in stratum } A$$

$$T_2 = 21(12.75) = 267.75 \text{ m}^3 \text{ in stratum B}$$

$$T_3 = 31(8.93) = 276.83 \text{ m}^3 \text{ in stratum } C$$

$$T_4 = 23(1.075) = 24.725 \text{ m}^3 \text{ in stratum } D$$

$$\bar{T} = (1/4)(71.75 + 267.75 + 276.83 + 24.725) = 160.26375$$

$$T_{o_{SC}} = 4(160.26.375) = 641.055 \text{ or } 641 \text{ m}^3 \text{ on tract}$$

$$\bar{y}_{o_{SC}} = 641.055 / 320 = 2.00330 \text{ m}^3 / \text{plot} \quad \text{or}$$

$$\bar{y}_{o_{SC}} = 641.055 / 16 = 40.06594 \text{ or } 40.1 \text{ m}^3 / \text{ha}$$

Then,

| Stratum | Cluster | $\frac{s_{hi}^2}{n_{hi}}\left(\frac{N_{hi}-n_{hi}}{N_{hi}}\right)N_{hi}^2$ | $\frac{M_h}{m_h}\sum_i^{m_h}\left[\frac{s_{hi}^2}{n_{hi}}\left(\frac{N_{hi}-n_{hi}}{N_{hi}}\right)N_{hi}^2\right]$ |
|---|---|---|---|
| 1 | 1 | $\frac{3.78125}{2}\left(\frac{4-2}{4}\right)4^2 = 15.125$ | |
| | 2 | $\frac{0.02}{2}\left(\frac{4-2}{4}\right)4^2 = 0.08$ | $\frac{5}{2}(15.125+0.08) = 38.0125$ |
| 2 | 1 | $\frac{0.06125}{2}\left(\frac{4-2}{4}\right)4^2 = 0.245$ | |
| | 2 | $\frac{0.66125}{2}\left(\frac{4-2}{4}\right)4^2 = 2.645$ | |
| | 3 | $\frac{3.125}{2}\left(\frac{4-2}{4}\right)4^2 = 12.5$ | |
| | 4 | $\frac{0.405}{2}\left(\frac{4-2}{4}\right)4^2 = 1.62$ | $\frac{21}{4}(0.245+2.645+12.5+1.62) = 89.3025$ |
| 3 | 1 | $\frac{0.02}{2}\left(\frac{4-2}{4}\right)4^2 = 0.08$ | |
| | 2 | $\frac{0.15125}{2}\left(\frac{4-2}{4}\right)4^2 = 0.605$ | |
| | 3 | $\frac{0.15125}{2}\left(\frac{4-2}{4}\right)4^2 = 0.605$ | |
| | 4 | $\frac{0.01125}{2}\left(\frac{4-2}{4}\right)4^2 = 0.045$ | |
| | 5 | $\frac{0.15125}{2}\left(\frac{4-2}{4}\right)4^2 = 0.605$ | |
| | 6 | $\frac{0.32}{2}\left(\frac{4-2}{4}\right)4^2 = 1.28$ | $\frac{31}{6}(0.08+0.605+0.605+0.045+0.605+1.28) = 16.63667$ |
| 4 | 1 | $\frac{0.0}{2}\left(\frac{4-2}{4}\right)4^2 = 0.0$ | |
| | 2 | $\frac{0.005}{2}\left(\frac{4-2}{4}\right)4^2 = 0.02$ | |
| | 3 | $\frac{0.245}{2}\left(\frac{4-2}{4}\right)4^2 = 0.98$ | |
| | 4 | $\frac{0.66125}{2}\left(\frac{4-2}{4}\right)4^2 = 2.645$ | $\frac{23}{4}(0.0+0.02+0.98+2.645) = 20.95875$ |

$$\sum_h^L\left\{\frac{M_h}{m_h}\sum_i^{m_h}\left[\frac{s_{hi}^2}{n_{hi}}\left(\frac{N_{hi}-n_{hi}}{N_{hi}}\right)N_{hi}^2\right]\right\} = 38.0125+89.3025+16.63667+20.95875$$

$$= 164.91042$$

$$s_{T_1}^2 = \frac{1}{(2-1)}[(13.1-14.35)^2 + (15.6-14.35)^2] = 3.125$$

$$s_{T_2}^2 = \frac{1}{(4-1)}[(15.3-12.75)^2 + (15.7 = 12.75)^2 + (11.0-12.75)^2 + (9.0-12.75)^2]$$

$$= 10.77667$$

$$s_{T_3}^2 = \frac{1}{(6-1)}[(9.8-8.93333)^2 + (9.1-8.93333)^2 + (9.3-8.93333)^3$$

$$+ (6.3-8.93333)^2 + (6.1-8.93333)^2 + (13.0-8.93333)^2]$$

$$= 6.48267$$

$$s_{T_4}^2 = \frac{1}{(4-1)}[(0-1.075)^2 + (0.2-1.075)^2 + (1.8-1.075)^2 + (2.3-1.075)^2]$$

$$= 1.31583$$

| **Stratum** | $\frac{M_h^2}{m_h} s_{T_h}^2 \left(\frac{M_h - m_h}{M_h}\right)$ |
|---|---|
| 1 | $\frac{5^2}{2}(3.125)\left(\frac{5-2}{5}\right) = 23.4375$ |
| 2 | $\frac{21^2}{4}(10.77667)\left(\frac{21-4}{21}\right) = 961.8178$ |
| 3 | $\frac{31^2}{6}(6.48267)\left(\frac{31-6}{31}\right) = 837.34487$ |
| 4 | $\frac{23^2}{4}(1.31583)\left(\frac{23-4}{23}\right) = 143.75443$ |

$$\sum_h^L \left[\frac{M_h^2}{m_h} s_{T_h}^2 \left(\frac{M_h - m_h}{M_h}\right)\right] = 23.4375 + 961.8178 + 837.34487 + 143.75443$$

$$= 1966.3546$$

$$s_{\bar{y}_{oSC}}^2 = \frac{1}{(320)^2}(164.91042 + 1966.3546) = 0.0208131$$

$$s_{\bar{y}_{oSC}} = \pm\sqrt{0.0208131} = \pm 0.1442676 \text{ m}^3/\text{plot} \quad \text{or}$$

$$s_{\bar{y}_{oSC}} = \pm 20(0.1442676) = \pm 2.8853516 \text{ or } \pm 2.89 \text{ m}^3/\text{ha} \quad \text{and}$$

$$s_{T_{oSC}} = \pm 320(0.1442676) = \pm 46.165626 \text{ or } \pm 46.2 \text{ m}^3 \text{ on tract}$$

Thus, according to this inventory there is a total of 641 ± 46.2 m³ (or 40.1 ± 2.89 m³/ha) of timber on the tract.

| | | | | | | | | | | | | | | | |
|---|---|---|---|---|---|---|---|---|---|---|---|---|---|---|---|
| 2.55 | 3.00 | 1.60 | 1.80 | 2.65 | 2.10 | 1.90 | 3.40 | 3.65 | 4.00 | 4.00 | 1.75 | 2.60 | 2.85 | 2.00 | 2.10 |
| 3.15 | 2.95 | 2.40 | 2.25 | 3.20 | 2.80 | 2.40 | 4.55 | 4.50 | 3.25 | 4.00 | 1.50 | 1.80 | 2.70 | 1.70 | 2.50 |
| 3.05 | 2.10 | 2.35 | 2.45 | 2.90 | 1.10 | 4.30 | 4.00 | 4.10 | 2.85 | 3.40 | 1.85 | 2.00 | 2.30 | 2.20 | 2.50 |
| 1.55 | 1.55 | 2.55 | 2.15 | 3.10 | 2.75 | 4.50 | 4.60 | 3.30 | 4.30 | 3.75 | 2.30 | 1.80 | 2.00 | 2.10 | 1.80 |
| 2.55 | 0.90 | 2.25 | 2.35 | 2.45 | 2.90 | 4.65 | 1.90 | 3.35 | 4.60 | 2.60 | 2.35 | 2.60 | 2.75 | 2.70 | 3.25 |
| 2.70 | 2.50 | 1.90 | 2.30 | 2.95 | 3.15 | 3.00 | 4.55 | 2.65 | 4.40 | 4.20 | 2.20 | 1.90 | 1.70 | 2.30 | 1.70 |
| 2.50 | 1.50 | 2.00 | 2.55 | 3.15 | 3.30 | 4.05 | 4.50 | 4.35 | 3.80 | 4.50 | 2.00 | 3.65 | 1.35 | 3.15 | 1.60 |
| 2.35 | 2.90 | 2.05 | 2.10 | 3.05 | 3.35 | 4.15 | 4.60 | 4.00 | 3.45 | 1.50 | 1.75 | 2.20 | 2.85 | 3.45 | 2.40 |
| 2.85 | 1.80 | 1.50 | 2.05 | 2.25 | 4.80 | 4.40 | 3.15 | 3.00 | 3.65 | 2.30 | 1.35 | 1.90 | 2.40 | 2.20 | 0.90 |
| 3.15 | 1.60 | 2.40 | 2.60 | 3.00 | 1.90 | 2.55 | 3.55 | 4.00 | 2.75 | 2.65 | 1.90 | 2.60 | 2.70 | 1.50 | 2.65 |
| 1.75 | 0.25 | 1.05 | 0.95 | 3.00 | 2.90 | 3.75 | 3.15 | 3.35 | 2.30 | 3.25 | 2.30 | 2.30 | 1.90 | 1.60 | 0.00 |
| 0.00 | 0.00 | 0.00 | 0.45 | 1.20 | 1.90 | 2.90 | 4.10 | 4.50 | 3.00 | 2.65 | 2.50 | 1.00 | 1.40 | 0.00 | 0.05 |
| 0.60 | 0.00 | 0.50 | 0.55 | 2.05 | 3.10 | 4.65 | 2.30 | 3.00 | 2.20 | 1.90 | 1.30 | 0.55 | 0.75 | 0.15 | 0.20 |
| 0.30 | 0.00 | 0.30 | 0.40 | 1.50 | 1.15 | 2.35 | 1.00 | 2.55 | 2.45 | 2.25 | 0.85 | 0.40 | 0.15 | 1.05 | 0.10 |
| 0.00 | 0.05 | 0.00 | 0.40 | 1.40 | 2.20 | 2.40 | 2.00 | 1.40 | 1.50 | 1.25 | 0.00 | 0.10 | 1.35 | 1.45 | 0.60 |
| 0.10 | 0.00 | 0.00 | 1.00 | 1.95 | 2.40 | 3.60 | 1.60 | 1.90 | 2.40 | 1.00 | 1.15 | 0.40 | 0.20 | 0.05 | 0.00 |
| 0.55 | 0.15 | 0.10 | 0.85 | 2.05 | 1.60 | 1.90 | 2.75 | 1.75 | 2.00 | 0.55 | 0.00 | 0.30 | 0.00 | 1.60 | 0.30 |
| 0.60 | 0.50 | 0.00 | 0.65 | 1.80 | 1.65 | 2.35 | 2.45 | 1.90 | 1.15 | 0.00 | 1.50 | 0.85 | 0.05 | 0.50 | 0.00 |
| 0.60 | 1.00 | 0.15 | 0.00 | 1.70 | 2.50 | 1.25 | 2.40 | 0.80 | 0.50 | 0.10 | 0.75 | 0.30 | 0.35 | 0.20 | 0.10 |
| 0.55 | 0.20 | 0.50 | 0.30 | 1.50 | 2.05 | 1.80 | 1.40 | 1.05 | 0.10 | 0.00 | 0.45 | 0.50 | 0.40 | 0.75 | 0.40 |

**FIGURE 17.20** The sampling frame showing the strata and the chosen four-element compact clusters.

### 17.16.3 Example of Stratified Cluster Sampling without Subsampling

This example uses the 16-ha tract, stratified as in Figure 17.18. Cluster size and configuration are also the same as in the previous example. Consequently, the number of clusters remain at 320/4 = 80. With a sampling intensity of 0.1, the total sample size is 0.1(80) = 8 clusters. Since no subsampling is involved, all four plots within each chosen cluster will be cruised. Equal allocation will be used to ensure a minimum of two cluster observations per stratum, permitting the computation of an estimated stratum variance. Consequently, $m_1 = m_2 = m_3 = m_4 = 2$. The clusters randomly chosen from the four strata are shown in Figure 17.20. The data and computations are as follows.

| Stratum | $h$ | Cluster | $i$ | $y_{hij}$ | $T_{hi} = \tau_{hi}$ | $\bar{y}_{hi} = \mu_{hi}$ | $s_{hi}^2 = \sigma_{hi}^2\left(\frac{N_{hi}}{N_{hi}-1}\right)$ |
|---|---|---|---|---|---|---|---|
| $A$ | 1 | 2 | 1 | 4.30 | | | |
| | | | | 4.00 | | | |
| | | | | 4.50 | | | |
| | | | | 4.60 | 17.40 | 4.3500 | 0.05250 |
| | | 5 | 2 | 4.35 | | | |
| | | | | 3.80 | | | |

| Stratum | $h$ | Cluster | $i$ | $y_{hij}$ | $T_{hi} = \tau_{hi}$ | $\bar{y}_{hi} = \mu_{hi}$ | $s^2_{hi} = \sigma^2_{hi}\left(\frac{N_{hi}}{N_{hi}-1}\right)$ |
|---|---|---|---|---|---|---|---|
| | | | | 4.00 | | | |
| | | | | 3.45 | 15.60 | 3.900 | 0.10625 |
| $B$ | 2 | 12 | 1 | 3.25 | | | |
| | | | | 2.30 | | | |
| | | | | 2.65 | | | |
| | | | | 2.50 | 10.70 | 2.6750 | 0.12563 |
| | | 17 | 2 | 4.10 | | | |
| | | | | 2.85 | | | |
| | | | | 3.30 | | | |
| | | | | 4.30 | 14.55 | 3.6375 | 0.34672 |
| $C$ | 3 | 9 | 1 | 1.50 | | | |
| | | | | 2.05 | | | |
| | | | | 2.40 | | | |
| | | | | 2.60 | 8.55 | 2.1375 | 0.17422 |
| | | 16 | 2 | 1.90 | | | |
| | | | | 2.75 | | | |
| | | | | 2.35 | | | |
| | | | | 2.45 | 9.45 | 2.3625 | 0.09297 |
| $D$ | 4 | 15 | 1 | 0.55 | | | |
| | | | | 0.00 | | | |
| | | | | 0.00 | | | |
| | | | | 1.50 | 2.05 | 0.5125 | 0.37547 |
| | | 19 | 2 | 0.10 | | | |
| | | | | 1.35 | | | |
| | | | | 0.40 | | | |
| | | | | 0.20 | 2.05 | 0.5125 | 0.24547 |

$\bar{T}_1 = \frac{1}{2}(17.40 + 15.60) = 16.500$ m³/cluster in stratum $A$ ($h = 1$)

$\bar{T}_2 = \frac{1}{2}(10.70 + 14.55) = 12.625$ m³/cluster in stratum $B$ ($h = 2$)

$\bar{T}_3 = \frac{1}{2}(8.55 + 9.45) = 9.000$ m³/cluster in stratum $C$ ($h = 3$)

$\bar{T}_4 = \frac{1}{2}(2.05 + 2.05) = 2.050$ m³/cluster in stratum $D$ ($h = 4$)

$T_1 = 5(16.500) = 82.5$ m³ in stratum $A$

$T_2 = 21(12.625) = 265.1$ m³ in stratum $B$

$T_3 = 31(9.000) = 279.0$ m³ in stratum $C$

$T_4 = 23(2.050) = 47.2$ m³ in stratum $D$

$T_{o_{SC}} = 82.5 + 265.1 + 279.0 + 47.2 = 674$ m³ on tract

$\bar{y}_{o_{SC}} = 674/320 = 2.10625$ m³/plot or

$\bar{y}_{o_{SC}} = 674/16 = 42.125$ or 42.1 m³/ha.

| Stratum | $\frac{M_h^2}{m_h}\left(\frac{M_h - m_h}{M_h}\right)s_{T_h}^2$ |
|---|---|
| 1 | $\frac{5^2}{2}\left(\frac{5-2}{5}\right)1.62 = 12.15$ |
| 2 | $\frac{21^2}{2}\left(\frac{21-2}{21}\right)7.41125 = 1478.5444$ |
| 3 | $\frac{31^2}{2}\left(\frac{31-2}{31}\right)0.405 = 182.0475$ |
| 4 | $\frac{23^2}{2}\left(\frac{23-2}{23}\right)0.0 = 0.00$ |

$$s_{\bar{y}_{oSC}}^2 = \frac{1}{(320)^2}(12.15 + 1,478.5444 + 182.0475 + 0.00)$$

$$= 0.0163354$$

$$s_{\bar{y}_{oSC}} = \pm\sqrt{0.0163354} = \pm 0.1278099 \text{ m}^3 \text{ / plot} \quad \text{or}$$

$$s_{\bar{y}_{oSC}} = \pm 20(0.1278099) = \pm 2.556198 \text{ or } \pm 2.56 \text{ m}^3 \text{ / ha} \quad \text{and}$$

$$s_{T_{oSC}} = \pm 320(0.1278099) = \pm 40.899167 \text{ or } \pm 40.9 \text{ m}^3 \text{ on tract}$$

According to this inventory, there is a total of 674 ± 40.9 m³ of timber on the tract or 42.1 ± 2.56 m³/ha.

### 17.16.4 Sample Size

In the case of stratified cluster sampling all $L$ of the first stage sampling units (strata) are represented in the sample so that $L' = L$. Then, if $N_{hi} = A$ and $n_{hi} = a$, or if $N_{hi}$ and $n_{hi}$ are not constants, their respective average values, $\bar{N}$ and $\bar{n}$, are used in place of $A$ and $a$, the derivation of the sample size formulae will parallel that associated with two-stage cluster sampling, thus, from Equation 17.146

$$a = \sqrt{\frac{c_1 s_w^2}{C_2(s_B^2 - s_w^2 / A)}} \tag{17.271}$$

and from Equation 17.149,

$$m = \frac{c_F' + c_2}{c_1 a + c_2} \tag{17.272}$$

where $c_F^1$ refers to the field costs after the stratification is complete, $c_1$ to the secondary units, and $c_2$ to the tertiary units.

The rounding-off rules described in Section 17.12.1 are also applicable in this case.

# 18 Systematic Sampling

## 18.1 INTRODUCTION

Systematic sampling is a special case of cluster sampling where the elements of the sample are drawn from the frame *at some specified interval*. Examples of systematic sampling drawn from forestry are as follows:

1. When one evaluates every tenth tree marked in a timber marking operation to obtain an estimate of the volume of timber that will be removed from the tract.
2. When the "line–plot" method of timber cruising is used, the plots are equidistantly spaced along equidistantly spaced cruise lines.
3. When the "strip" method of timber cruising is used, the strips are laid out to be parallel to each other and the spacing between them is constant.
4. When the area of a certain condition class is estimated using a dot grid laid over a stand map of the tract, the points making up the grid are equidistantly spaced.

Systematic sampling is commonly used in forestry. It is a mechanical method that is easily taught to field personnel, is relatively easy to field-check, and, in general, is easy to use. Furthermore, since the sample is spread across the entire sampling frame, it is likely to be more representative of the population than would be a random sample of the same size. However, because the selected sampling units are not independent of each other, a systematic sample is fundamentally different from simple random sampling.

## 18.2 APPROACHES TO SYSTEMATIC SAMPLING

While simple to apply in the field, systematic sampling requires thought when interpreting the results. It can be considered to be either (1) a form of simple cluster sampling (hereafter referred to as "Approach 1" or (2) a form of two-stage cluster sampling (hereafter referred to as "Approach 2"). Before one can construct a sampling frame or interpret the results of a systematic sampling operation, one must decide which of these two approaches is to be used.

## 18.3 ELEMENT SPACING AND SAMPLING INTENSITY

As in conventional cluster sampling there are two sampling frames. The first is the set of elements of the universe of interest which are available for sampling. This is the *basic* or *secondary* sampling frame. The second is made up of the available clusters. It is referred to as the *cluster* or *primary* frame. The choice of interpretation of the nature of systematic sampling will determine the character of the cluster frame.

When the basic sampling frame is in the form of a list (e.g., a listing of a set of landowners) or consists of the elements of the universe of interest that form a stream passing a certain point during a specified period of time (e.g., logs passing into a sawmill during a specified 8-hour working day), the sample is generated by choosing elements from the frame that are separated by a constant interval, $L$. In such a case, it is said that a "1 in $L$" systematic sample has been obtained. This implies a sampling intensity, $n/N$, which is equal to $1/L$. However, this is true only when the population size, $N$, is an integral (i.e., whole number) multiple of $L$. When this occurs, all the possible samples are the same size so that $n$ is constant. For example, if the basic sampling frame contains 24 elements,

$$(1,2,3,4,5,6,7,8,9,10,11,12,13,14,15,16,17,18,19,20,21,22,23,24)$$

where each number refers to the position of an element within the list, and if $L = 4$, the following subsets will be generated: (1,5,9,13,17,21), (2,6,10,14,18,22), (3,7,11,15,19,23), and (4,8,12,16,20,24). *In the sampling operation one of these subsets will be chosen.* Thus, *a subset is a sample*. All contain $n = 6$ elements and the sampling intensity in each case is $n/N = 6/24 = 1/4 = 1/L$.

If $N$ is not an integral multiple of $L$, the subsets will vary in size and, consequently, the sampling intensity will also vary. For example, if in the previous example, the universe consisted of $N = 22$ elements, the possible subsets would be (1,5,9,13,17,21), (2,6,10,14,18,22), (3,7,11,15,19), and (4,8,12,16,20). Then $n_1 = n_2 = 6$ and $n_3 = n_4 = 5$. The sampling intensity is 6/22 for the first two subsets and 5/22 for the remaining two subsets.

The problem of varying sample sizes and sampling intensities associated with the times when $N$ is not an integral multiple of $L$ can be resolved, provided that $N$ and $L$ are relatively prime (i.e., they are divisible only by themselves and by 1), by using what is known as *circular* systematic sampling (Sukhatmi et al., 1984). In this procedure the basic sampling frame is one dimensional (i.e., a list). It is, in essence, looped back on itself to form a circle, hence the name "circular" systematic sampling. A position within the circular frame is designated as position number 1. Every $L$th position from that point on is included in the cluster until the desired sample size is obtained. Then a second cluster is generated starting at position 2. This is continued until $N$ different clusters have been generated, The set of clusters forms the cluster or primary sampling frame. The probability of a given cluster being randomly drawn from this frame is $1/N$ and the sampling intensity for each cluster is $n/N$.

As an example of this process, assume $N = 11$, $L = 3$, and $n = 4$. The cluster frame would then contain the following clusters. The numbers within the cluster refer to the positions within the basic frame.

| Cluster | Positions Included $n = 4$ | Positions Included $n = 6$ |
|---|---|---|
| 1 | 1, 4, 7, 10 | (1, 4, 7, 10, 2, 5) |
| 2 | 2, 5, 8, 11 | (2, 5, 8, 11, 3, 6) |
| 3 | 3, 6, 9, 1 | (3, 6, 9, 1, 4, 7) |
| 4 | 4, 7, 10, 2 | (4, 7, 10, 2, 5, 8) |
| 5 | 5, 8, 11, 3 | (5, 8, 11, 3, 6, 9) |
| 6 | 6, 9, 1, 4 | (6, 9, 1, 4, 7, 10) |
| 7 | 7, 10, 2, 5 | (7, 10, 2, 5, 8, 11) |
| 8 | 8, 11, 3, 6 | (8, 11, 3, 6, 9, 1) |
| 9 | 9, 1, 4, 7 | (9, 1, 4, 7, 10, 2) |
| 10 | 10, 2, 5, 8 | (10, 2, 5, 8, 11, 3) |
| 11 | 11, 3, 6, 9 | (11, 3, 6, 9, 12, 4) |

| | B | | | | | | | | | |
|---|---|---|---|---|---|---|---|---|---|---|
| L | 1 | 2 | 3 | 4 | 5 | 6 | 7 | 8 | 9 | 10 |
| 1 | · | · | · | · | · | · | · | · | · | · |
| 2 | · | · | · | · | · | · | · | · | · | · |
| 3 | · | · | · | · | · | · | · | · | · | · |
| 4 | · | · | · | · | · | · | · | · | · | · |
| 5 | · | · | · | · | · | · | · | · | · | · |
| 6 | · | · | · | · | · | · | · | · | · | · |
| 7 | · | · | · | · | · | · | · | · | · | · |
| 8 | · | · | · | · | · | · | · | · | · | · |
| 9 | · | · | · | · | · | · | · | · | · | · |
| 10 | · | · | · | · | · | · | · | · | · | · |

**FIGURE 18.1** A two-dimensional basic sampling frame.

Note that all the clusters are the same size and that all are different. If a sample size of 6 had been desired, the clusters in the third column would be generated. Note that all these clusters would also be different.

When the basic sampling frame is in the form of a map, the universe is two dimensional, "North to South" (N to S) and "West to East" (W to E), as in the case of the 16-ha tract shown in Figure 16.1. Consequently, there are two spacing constants, $L$ for the (N to S) spacing and $B$ for the (W to E) spacing. $L$ and $B$ may or may not be equal. The sampling design is then a "1 in ($L * B$) systematic sample." For example, if $L = 5$ and $B = 4$, it would be a 1 in (5 * 4) or 1 in 20 systematic sample. When the (N to S) extent is an integral multiple of $L$ and the (W to E) extent is an integral multiple of $B$, all the possible subsets (samples) are the same size and the sampling intensity, for all subsets (samples), is equal to the reciprocal of the product of $L$ and $B$. For example, if the basic sampling frame is in the form of a square made up of 10 rows and 10 columns, as shown in Figure 18.1, and $L = 5$ and $B = 5$, (N to S)/$L$ = 10/5 = 2 and (W to E)/$B$ = 10/5 = 2, both of which are whole numbers. Consequently, the subset (sample) sizes are constant with $n_i = 4$, as shown below:

(1,1;1,6;6,1;6,6)
(2,1;2,6;7,1;7,6)
(3,1;3,6;8,1;8,6)
(4,1;4,6;9,1;9,6)
(5,1;5,6;10,1;10,6)
(1,2;1,7;6,2;6,7)
(2,2;2,7;7,2;7,7)
(3,2;3,7;8,2;8,7)
(4,2;4,7;9,2;9,7)
(5,2;5,7,10,2;10,7)
⋮
(1,5;1,10;6,5;6,10)

(2,5;2,10;7,5;7,10)
(3,5;3,10;8,5;8,10)
(4,5;4,10;9,5;9,10)
(5,5;5,10;10,5;10,10)

The sampling intensity in all cases is 4/100 = 1/25 = 1/($L * B$).

If the (N to S) and/or the (W to E) extents are not integral multipliers of $L$ and $B$, respectively, the subset (sample) sizes will vary and the sampling intensity will vary from subset to subset. For example, if in the previous example, the (W to E) extent had been equal to 12, instead of 10, it would not have been an integral multiple of $B$. Consequently, the subset sizes would have varied as shown below:

(1,1;1,6;6,1;6,6;6,11) $N_1 = 6$
(2,1;2,6;7,1;7,6;7,11) $N_2 = 6$
(3,1;3,6;8,1;8,6;8,11) $N_3 = 6$
(4,1;4,6;9,1;9,6;9,11) $N_4 = 6$
(5,1;5,6;10,1;10,6;10,11) $N_5 = 6$
(1,2;1,7;6,2;6,7;6,12) $N_6 = 6$
(2,2;2,7;7,2;7,7;7,12) $N_7 = 6$
(3,2;3,7;8,2;8,7;8,12) $N_8 = 6$
(4,2;4,7;9,2;9,7;9,12) $N_9 = 6$
(5,2;5,7,10,2;10,7;10,12) $N_{10} = 6$
(1,3;1,8;6,3;6,8) $N_{11} = 6$
(2,3;2,8;7,3;7,8) $N_{12} = 6$
(3,3;3,8;8,3;8,8) $N_{13} = 6$
(4,3;4,8;9,3;9,8) $N_{14} = 6$
(5,3;5,8;10,3;10,8) $N_{15} = 6$
⋮
(1,5;1,10;6,5;6,10) $N_{21} = 4$
(2,5;2,10;7,5;7,10) $N_{22} = 4$
(3,5;3,10;8,5;8,10) $N_{23} = 4$
(4,5;4,10;9,5;9,10) $N_{24} = 4$
(5,5;5,10;10,5;10,10) $N_{25} = 4$

The sampling intensity for subsets 1 to 10 would be 6/120 = 1/20, while that for the remaining 15 subsets would be 4/120 = 1/30.

The discrepancy between 1/($L * B$) and the actual sampling intensity becomes greater and more varied if the basic sampling frame is not a square or rectangle. For example, if it is as shown in Figure 18.2, $N$ would be equal to 65. If $L$ = 3 and $B$ = 3 the possible subsets and their sizes and associated sampling intensities would be

| | | $n_i$ | $n_i/N$ |
|---|---|---|---|
| 1 | (- - -;1,4;- - -;4,1;4,4;- - -;7,1;7,4;7,7;10,1;10,4;10,7) | 9 | 9/65 |
| 2 | (- - -;2,4;- - -;5,1;5,4;5,7;8,1;8,4;8,7;- - -;- - -;11,7) | 8 | 8/65 |
| 3 | (3,1;3,4;- - -;6,1;6,4;6,7;9,1;9,4;9,7;- - -;- - -;12,7) | 9 | 9/65 |
| 4 | (1,2;- - -;- - -;4,2;4,5;4,8;7,2;7,5;7,8;- - -;10,5;- - -) | 8 | 8/65 |
| 5 | (2,2;- - -;- - -;5,2;5,5;5,8;8,2;8,5;8,8;- - -;- - -;- - -) | 7 | 7/65 |
| 6 | (3,2;- - -;- - -;6,2;6,5;6,8;9,2;9,5;9,8;- - -;- - -;- - -) | 7 | 7/65 |
| 7 | (- - -;- - -;- - -;4,3;4,6;- - -;7,3;7,6;- - -;10,3;10,6;- - -) | 6 | 6/65 |
| 8 | (- - -;- - -;- - -;5,3;5,6;- - -;8,3;8,6;- - -;11,3;11,6;- - -) | 6 | 6/65 |
| 9 | (3,3;- - -;- - -;6,3;6,6;- - -;9,3;9,6;- - -;- - -;- - -;- - -) | 5 | 5/65 |

| L | B 1 | 2 | 3 | 4 | 5 | 6 | 7 | 8 |
|---|---|---|---|---|---|---|---|---|
| 1 | | • | | • | | | | |
| 2 | | • | | • | | | | |
| 3 | • | • | • | • | | | | |
| 4 | • | • | • | • | • | • | | • |
| 5 | • | • | • | • | • | • | • | • |
| 6 | • | • | • | • | • | • | • | • |
| 7 | • | • | • | • | • | • | • | • |
| 8 | • | • | • | • | • | • | • | • |
| 9 | • | • | • | • | • | • | • | • |
| 10 | • | | • | • | • | • | • | |
| 11 | | | • | | | • | • | |
| 12 | | | | | | | • | |

**FIGURE 18.2** A two-dimensional basic sampling frame with irregular boundaries.

Thus, it can be seen that the statement that a "1 in $L$" or "1 in $(L * B)$" systematic sample has been drawn does not necessarily mean that the sampling intensity is equal to the indicated ratio. A recognition of this fact is important because the intervals are specified by the designer of the inventory and those intervals are usually based on the designer's opinion of what sampling intensity would be appropriate. In most cases the actual sampling intensity will differ to some extent from that implied by the interval specification. The significance of the difference depends on the situation. For example, if a timber cruise is a 1/10 systematic sample, the implication is that 10% of the tract will be evaluated and the total volume on the tract is ten times the cruise total. If the tract size and/or boundaries are such that the possible sample sizes vary, this assumption is incorrect and conceivably could lead to erroneous conclusions. For this reason, the actual sampling intensity should be used in any analysis of the cruise data rather than the intensity implied by the interval specification.

## 18.4 CLUSTER SAMPLING FRAMES

In the previous discussion the possible subsets that could be obtained from a basic sampling frame, given a spacing interval, were identified as *samples*. The nature of these samples depends on the perception of the nature of systematic sampling. This is manifested in the design of the cluster sampling frame.

### 18.4.1 Under Approach 1

In this case *each subset is considered to be a cluster* and *only one such cluster is chosen in the sampling operation*. For example, consider the following basic and cluster sampling frames. If $L = 4$,

| Cluster and Sample | Position in Basic Frame | | | | | | | | | | | | | | | | | | | | | | | |
|---|---|---|---|---|---|---|---|---|---|---|---|---|---|---|---|---|---|---|---|---|---|---|---|---|
| | 1, | 2, | 3, | 4, | 5, | 6, | 7, | 8, | 9, | 10, | 11, | 12, | 13, | 14, | 15, | 16, | 17, | 18, | 19, | 20, | 21, | 22, | 23, | 24, |
| 1 | 1 | | | | 5 | | | | 9 | | | | 13 | | | | 17 | | | | 21 | | | |
| 2 | | 2 | | | | 6 | | | | 10 | | | | 14 | | | | 18 | | | | 22 | | |
| 3 | | | 3 | | | | 7 | | | | 11 | | | | 15 | | | | 19 | | | | 23 | |
| 4 | | | | 4 | | | | 8 | | | | 12 | | | | 16 | | | | 20 | | | | 24 |

Thus, there are $M = 4$ clusters, each containing $N_i = A = 6$ elements. Notice that the clusters are *interlocking*, *Not compact*, and extend across the entire basic sampling frame. In a sampling operation, *one* of these clusters would be chosen. No subsampling is involved.

If the basic sampling frame is two dimensional, the clusters are made up of elements separated by the constant intervals $L$ and $B$, in the same manner as those in Figure 17.9, except that the clusters extend all the way to the boundaries of the basic sampling frame. Again, only one cluster would be chosen and there would be no subsampling.

### 18.4.2 Under Approach 2

The situation is quite different under this assumption. In this case, the process follows that of two-stage cluster sampling. The basic sampling frame is divided into $M$ *compact* clusters, each containing $N_i = L = A$ elements (provided that $N$ is an integral multiple of $L$). *A subsample of size* $n_i = a = 1$ *is drawn from each of the clusters*. For example, consider the following basic and cluster sampling frames, if $L = 4$:

| | Clusters | | | | | | | | | | | | | | | | | | | | | | | |
|---|---|---|---|---|---|---|---|---|---|---|---|---|---|---|---|---|---|---|---|---|---|---|---|---|
| | 1 | | | | 2 | | | | 3 | | | | 4 | | | | 5 | | | | 6 | | | |
| | Element No. | | | | | | | | | | | | | | | | | | | | | | | |
| Samples | 1, | 2, | 3, | 4 | 5, | 6, | 7, | 8 | 9, | 10, | 11, | 12 | 13, | 14, | 15, | 16 | 17, | 18, | 19, | 20 | 21, | 22, | 23, | 24 |
| 1 | 1 | | | | 5 | | | | 9 | | | | 13 | | | | 17 | | | | 21 | | | |
| 2 | | 2 | | | | 6 | | | | 10 | | | | 14 | | | | 18 | | | | 22 | | |
| 3 | | | 3 | | | | 7 | | | | 11 | | | | 15 | | | | 19 | | | | 23 | |
| 4 | | | | 4 | | | | 8 | | | | 12 | | | | 16 | | | | 20 | | | | 24 |

As can be seen, there are now $M = 6$ *compact* clusters, each containing $N_i = A = 4$ elements, and a subsample of size $n_i = a = 1$ has been drawn from each cluster to form the sample. The sample size is $n = 6$. Due to the subsampling, the sample is spread across the entire basic sampling frame in spite of the fact that the clusters are compact.

When the basic sampling frame is two dimensional the cluster frame is like that shown in Figure 17.7. The clusters are compact and only one element is chosen from each cluster. The difference is that the subsamples are systematically, not randomly, drawn.

The possible samples under this assumption are identical to those obtained under the assumption that systematic sampling is a form of simple cluster sampling. However, the analyses and interpretations will differ.

## 18.5 RANDOM AND ARBITRARY STARTS

Prior to the construction of the cluster sampling frame, one must have decided on which assumption (as to the nature of systematic sampling) is to be used and on the spacing interval so that the configuration of the clusters can be determined. In addition, a decision is needed on how the sample is to be identified. Usually this involves the selection of an element from which the rest of the sample can be constructed.

One can again use the basic 24-element basic sampling frame used in the previous examples. Number the elements consecutively from 01 to 24. Using random numbers, choose one of these elements to serve as the initial unit within the sample. Assume that element 07 is chosen. If $L = 4$, the sample would include elements 03. 07, 11, 15, 19, and 23; see cluster (and sample) 3 in Section 18.4.1 and sample 3 in Section 18.4.2. This sample has a *random start*. Note that the sample is generated in *both* directions from the chosen element. It should be further noted that, if $N$ is an integral multiple of $L$, the $M$ possible samples will all be of the same size. If $N$ is not an integral multiple of $L$ the sample size will vary.

Another way of making a random start is to number the *initial L elements* in the basic frame from 1 to $L$ and then randomly choose one of these as the initial element. For example, again using the same 24-element basic sampling frame, and setting $L = 4$, the first four elements would be numbered from 1 to 4. Then, if element 4 is randomly drawn, the sample would consist of elements 4, 8, 12, 16, 20, and 24. See cluster (and sample) 4 in Section 18.4.1 and sample 4 in Section 18.4.2. Again, if $N$ is an integral multiple of $L$ the sample sizes will all be the same, but if $N$ is not an integral multiple of $L$, the sample sizes will differ. This procedure was used in a two-dimensional sense in the example in Section 17.6.7.

This second method would be appropriate if one were breaking into an ongoing stream of elements as, for example, the logs passing through a sawmill that is operating on a nonstop basis. Under these circumstances one would have to specify the time period during which the sample would be obtained. This would have to be done arbitrarily since the stream of logs is continuous and there are no natural beginning or end points that could be used to define the sampling period. Obviously, in this situation the basic sampling frame and the universe will not be coincident. Prior to the sampling, a number is randomly drawn from the set 01, 02, …, $L$. The log passing the selection point at the time the operation begins is given the number 1. When the randomly chosen log passes the selection point, it is evaluated as will every $L$th log from that point on. It is obvious that in a situation such as this the only way to identify the initial element, and its associated sample members, is to choose it from among the first $L$ elements passing the selection point.

If the designer of the inventory specifies that the initial element shall be that occupying a *specific* position, it is said that the start is *arbitrary*. Usually this specification refers to a position within the first $L$ elements in the basic frame. Furthermore, in most cases the specified position is the midpoint of that initial set. In such a case the start is a *central* start. If a central start is to be used, $L$ must be an odd number. Consequently, one cannot use samples from the previously generated cluster sampling frames to demonstrate a central start sample. One can, however, use the same basic frame and set $L = 5$. The central position in the interval 1 to 5 is 3. Then the sample generated would consist of elements 3, 8, 13, 18, and 23. Obviously, *if an arbitrary start is used, there is no possibility of any of the alternative samples being chosen.*

Further examples of arbitrary starts can be found in common forest inventory practices. Typically, in line–plot timber cruising the cruise lines are drawn parallel to a tract boundary, a cardinal direction, or so as to cross, at right angles, the major streams and ridges within the tract. The first of these cruise lines is located at $^1/_2$ the cruise line spacing from the left-most point on the tract boundary, which is usually the westernmost point. The first plot on that line is centered on the point that is $^1/_2$ plot interval from either the upper (usually the northern) or the lower (usually the southern) tract boundary. This process freezes the position of the set of plots.

Similarly, if a dot grid is used to obtain an estimate of the proportion or area of a tract included in a given condition class and the grid is laid down over the stand map of the tract so that two edges of the grid coincide with two boundaries of the tract, the position of the grid is fixed and repeated trials will yield precisely the same results, barring operator error. In contrast, if the dot grid is simply dropped down onto the map without regard to position or orientation, the cluster would be one of an infinitely large number of possible clusters and the start would be random.

## 18.6 REPRESENTATIVENESS OF THE SAMPLE

A sample is *representative* of a population if it provides a relatively undistorted picture of the population. To be representative the sample should include data from all the conditions present in the population and these data should reflect the relative weights of these conditions within the population. This distinguishes representativeness from unbiasedness. Bias is present when, *on the average*, across all possible samples of the same type and size, the estimates of a population parameter are not equal to the parameter. Thus, simple random sampling yields unbiased estimates of the population mean but an individual random sample can be badly distorted because its component sampling units happen to come from only a portion of the several conditions present in the population.

The complexity of the population being sampled has an effect on the representativeness of the samples and, since complexity usually increases as population size increases, population size also has an effect on the representativeness of the samples. Consequently, a systematic sample, since it is spread widely across the population, usually will be more representative of the population than would be a simple random sample of the same size. As population complexity increases, the relative representativeness of systematic sampling increases relative to that of simple random sampling.

## 18.7 POPULATION TYPES

The representativeness of a sample, be it simple random, stratified random, or systematic with either a random or arbitrary start, depends on the way the variates occur relative to their positions in the basic sampling frame. When the variates are plotted against their positions in the frame, the resulting scatter diagram and general trend line can be used to place the population into a *population type*. Knowledge of population type is of great value to the sampler since it provides information regarding what sampling design would be most appropriate to use.

### 18.7.1 Random Populations

A population has a *random* distribution if any one of its constituent variates has the same probability of occurrence at a *given position* as does any of the remaining variates. A plot of the variates against their positions will show a horizontal trend (regression line or surface) through the scatter diagram, indicating that no correlation exists between variate magnitude and position. Furthermore, the degree of scatter of the variates about the trend should be homogeneous.

If the population is randomly distributed, both random and systematic samples will be representative of that population. Furthermore, this is true whether the systematic samples have random or arbitrary starts.

If the population is not randomly distributed, a systematic sample will usually be more representative of the population than will a simple random sample of the same size because the systematic sample is designed to be broadly spread throughout the population. Random samples may be as broadly spread, but in many cases they are not and fall in such a manner that some conditions are overrepresented while others are underrepresented or not represented at all.

Randomly distributed populations do exist as, for example, if the universe of interest is the set of undergraduate students at a certain university and the variable of interest is the student's entering SAT score. If the student directory is used as the basic sampling frame, and if there is no correlation between name and SAT score, the population of scores can be considered to be randomly distributed.

### 18.7.2 Ordered Populations

A population is *ordered* if its variates are arranged in such a manner that when they are plotted against their positions in the frame the trend takes a form other than a horizontal line or surface.

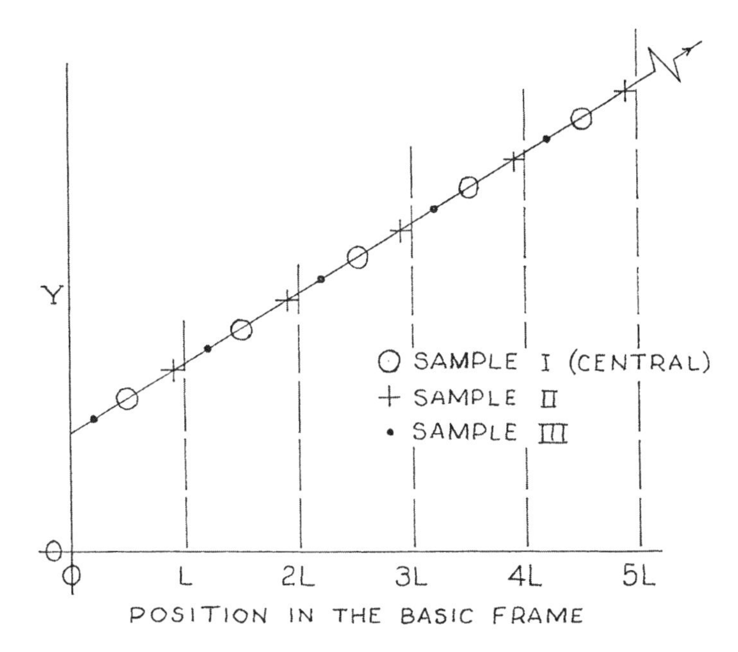


**FIGURE 18.3** A population with a linear trend. Sample I has a central start, whereas samples II and III have random starts.

The trend may be linear or curvilinear. For example, if the variates are arranged to form an ascending array, as in Figure 18.3, the trend line will rise monotonically as the position number increases. When a trend of this type is present, the overall population variance will be composed of two components: (1) that due to the trend and (2) that due to residual random variation about the trend line or surface, as was discussed in Chapter 15.

When the trend is linear, as in Figure 18.3, a systematic sample with a central start, such as Sample I, would probably be the most representative, because it would consist of variates occurring at, or close to, the median of each $L$ item subset. Sample II would be less representative because it would consist of variates lying *consistently* below the medians of the subsets and would yield a sample mean that would be less than the population mean. The reverse is true of Sample III. The means of all these systematic samples are unbiased estimates of the population mean provided that the sample starts are randomly chosen. If the start is arbitrary, only Sample I would yield an unbiased estimate of the population mean and that only if no residual random component is present.

### 18.7.3 Periodic Populations

A population is *periodic* if its trend undulates with a more or less constant wavelength, as in the case of a sine or cosine wave. Figure 18.4 provides a picture of such a trend. If the element spacing, $L$, of a systematic sample, or some integral multiple or fractional multiple of $L$, is such that the sample falls into step with the wave, the variates of the sample will occupy the same relative positions within the respective waves and consequently will have the same or nearly the same magnitude. This is shown in Figure 18.4. In Figure 18.4A the sample interval is equal to the wavelength of the trend. The three samples have different starting positions and consequently fall at different positions within the wave. Sample I lies above the line of the population mean. The reverse is true for Sample II. Sample III, however, occurs along the line of the population mean and its mean will be equal to the population mean. Insofar as the population mean is concerned, only Sample III would be representative of the populations. All are unbiased if the starts are random but only Sample III will yield an unbiased mean if the starts are arbitrary. Insofar as variance is

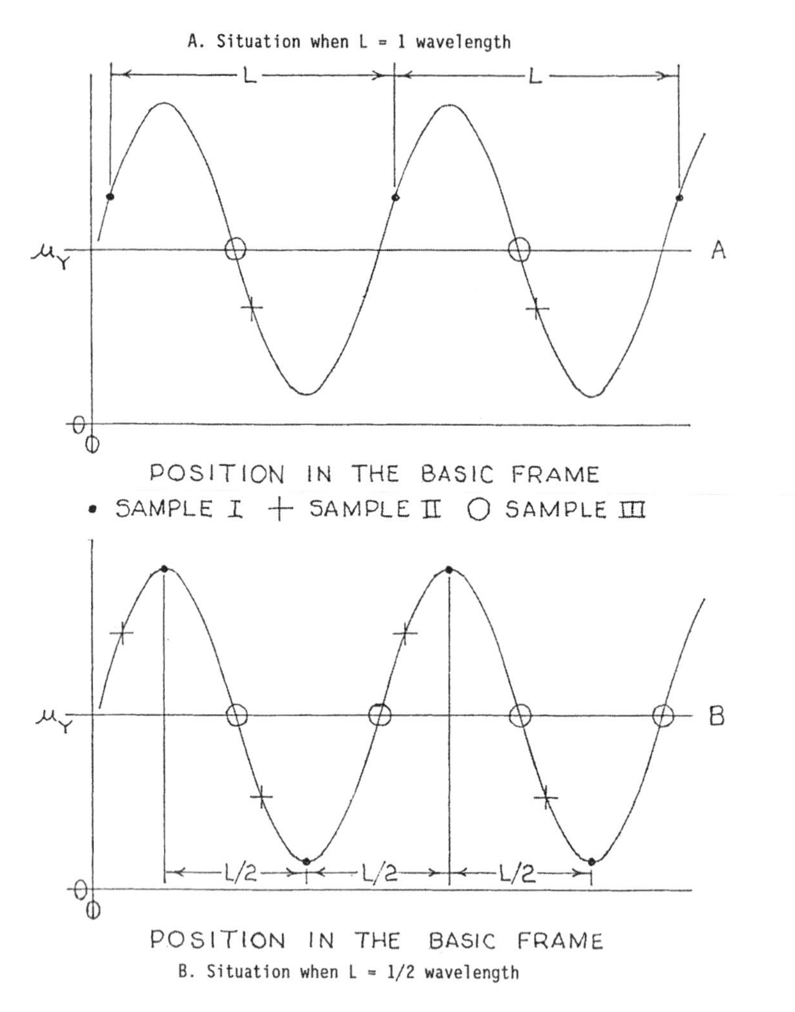


**FIGURE 18.4** A periodic population. The systematic samples are in step with its period. A. Situation when $L = 1$ wavelength. B. Situation when $L = {}^1/_2$ wavelength.

concerned, all would be biased. As a matter of fact, if the sample interval matches the wavelength of the trend, and there is no residual variation, the sample variances would all equal zero.

When the sample interval is some integral multiple or fractional multiple of the wavelength of the trend, as in Figure 18.4B, the situation becomes more complicated. In the case of Figure 18.4B the sample interval, $L$, is equal to one half the trend wavelength. As can be seen, the variates in a given sample lie alternately above and below the line of the population mean, as in the cases of Samples I and II, or they lie on the line of the population mean, as does Sample III. All the samples would be representative of the population insofar as the population mean is concerned, but their variances would be distorted, varying from zero, when the sample lies along the line of the population mean (Sample III), to a maximum when the sample occupies the extreme peaks and valleys of the trend (Sample I).

Periodicity does occur in populations of interest to foresters, especially where the spacing is the result of human activities, as in plantations, nursery seedbeds, and transplant beds, and in areas where in the past logging roads or railroads have been laid out in some regular pattern and have subsequently been overgrown. Periodicity is also present in natural populations where topography and/or altitude controls species distribution and development. As an illustration, consider a line–plot cruise of a forest plantation in which the trees are reasonably regularly spaced. This is shown in

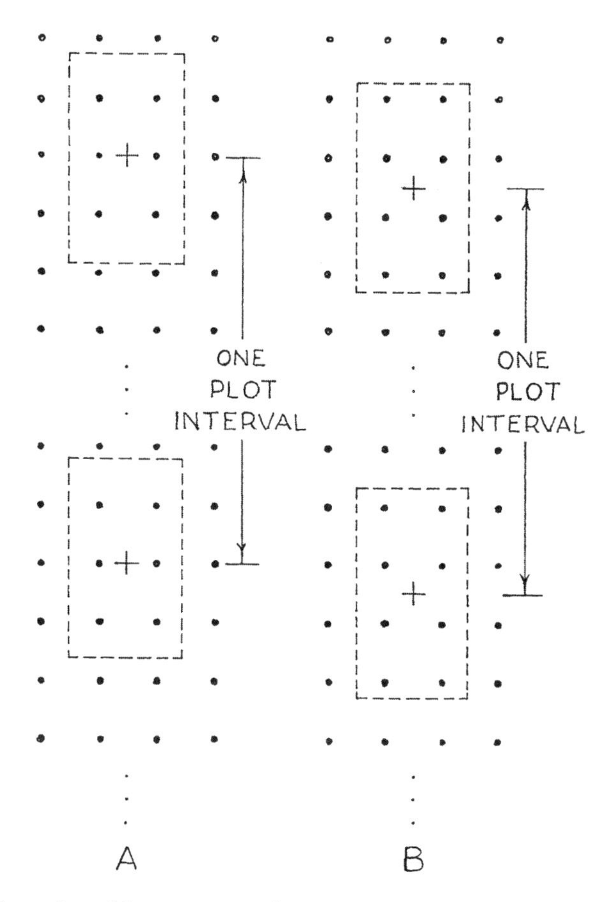


**FIGURE 18.5** An illustration of how a systematic sample can lead to an erroneous conclusion by falling into step with the periodicity in a forest plantation. Note that in example A each plot includes six trees while in example B each plot includes eight trees.

Figure 18.5. To keep things simple, assume that plots are rectangular, rather than circular, but are centered along the cruise lines in the usual manner. Figure 18.5 shows portions of two cruise lines. The starts of the two lines were slightly different. The plot intervals, however, are identical. The interval is such that the plot centers are in step with the tree spacing. As a result, assuming no mortality, each plot in cruise A will contain six trees while each plot in cruise B will contain eight trees. This is a 25% difference in tree count! Of course, most forest plantations are not this precisely spaced but, even if irregularities exist, the likelihood of the sample not being representative is sufficiently great that it should not be ignored.

In general, periodicity is most dangerous when the sampler is unaware of its presence. Compounding the problem is that only rarely will it be possible to detect its presence from the evidence of a sample, be it random or systematic. Consequently, the sampler must be willing to accept the risk of a nonrepresentative sample if he or she elects to use systematic sampling when evaluating a population whose distribution is not known.

### 18.7.4 Uniform Populations

A population is *uniform* if all of its constituent variates possess the same magnitude. Its overall mean will be equal to that magnitude and its overall variance will be equal to zero. All samples, regardless of type, drawn from such a population would be representative and would yield unbiased estimates of the population parameters.

### 18.7.5 Stratified Populations

Populations that possess subsets of contiguous variates which are internally reasonably homogeneous and whose means and/or variances differ are known as *stratified* populations. The population of plot timber volumes shown in Figure 16.3 is of this type. Such populations do not show clear-cut trends. Instead, their trend lines or surfaces show a series of plateaus, each representing a subset. With populations of this type, if the strata can be identified prior to the sampling (e.g., aerial photographic evidence can be used to stratify forest cover into condition classes), stratified random sampling would be more efficient than simple random sampling. A systematic sample would have many of the characteristics of a stratified sample designed to use proportional allocation. Within each stratum the systematic sample would probably yield a more representative sample than would random sampling, but whether systematic sampling would be more or less efficient than stratified random sampling using optimum allocation is open to question.

The distribution of forest cover is correlated with the availability of water. Consequently, it is correlated with topography, aspect, and altitude. Stands along stream bottoms are different in species composition and silvical characteristics from stands on slopes and these, in turn, are different from those along the ridges. Because of this, stands often form bands paralleling the streams and ridge lines. Systematic samples made along transects (cruise lines) that parallel the stream/ridge trend can miss important conditions and overemphasize others. Such samples would be highly unrepresentative of the forest population. On the other hand, if the cruise lines are run so as to cross the stream/ridge trend at right angles, the sample would be much more representative. Representativeness would probably be enhanced if the between-cruise line distance is increased and the between-plot/within-line distance correspondingly decreased. This is precisely what is done in most timber cruises. On the other hand, if the land is flat and there is no clear-cut stream/ridge trend, the between-line and between-plot spacing should be equalized to produce a square pattern. This would minimize the danger of missing some of the conditions within the population.

### 18.7.6 Autocorrelated Populations

When variates of a population tend to be more nearly like their near neighbors than they are like more distant members of the population, with the degree of likeness decreasing as the distance between their positions increase, it is said that the variates form an *autoregressive sequence* and that the population is *autocorrelated* or *serially correlated*. This implies that there is a functional relationship between the variates of the population:

$$y_i = g(y_{i-j}) + \varepsilon_i \tag{18.1}$$

where $i$ = position in the population
$j$ = the specified interval or "lag"
$\varepsilon$ = the error or random component as the $i$th variate

In other words, when a population is autocorrelated one can compute the magnitude of a given variate, except for its random component, using the magnitude of a variate appearing earlier in the sequence. For example, consider a population that exhibits a linear trend so that (ignoring the random component):

$$y = 10 + 0.5x$$

where $x$ = position in the sequence. If the interval $j$ is set at 1 so that the $i$th variate is being predicted by its immediate predecessor, $y_{i-1}$, then

$$y_i = 0.5 + y_{i-1}$$

As can be seen, in this case the variate $y_i$ can be determined using the variate $y_{i-1}$, indicating that the population is autocorrelated.

The concept of correlation was explored in Sections 9.8 and 15.7. The correlation coefficient is the most commonly used measure of the degree of relationship between two random variables. In its product–moment form the correlation coefficient is (from Equation 9.32):

$$\rho = \sigma_{12} / \sqrt{\sigma_1^2 \sigma_2^2}$$

$$= \sum_{i=1}^{N} (y_{1i} - \mu_1)(y_{2i} - \mu_2) \Bigg/ \left( \sqrt{\sum_{i=1}^{N} (y_{1i} - \mu_1)^2} \sqrt{\sum_{i=1}^{N} (y_{2i} - \mu_2)^2} \right)$$

Serial or autocorrelation is similar to conventional correlation, differing only in the nature of the two random variables. In the case of conventional correlation, the variables are distinctly different (e.g., one might be crown diameter while the other is bole diameter at breast height). In contrast, serial correlation is concerned with the relationship between members of the same population that are separated by some specified interval. Thus, the $j$th order serial correlation coefficient is

$$\rho_j = \frac{\sum_{i=1}^{N-j} (y_i - \mu_1)(y_{i-j} - \mu_2)}{\sqrt{\sum_{i=1}^{N-j} (y_i - \mu_1)^2} \sqrt{\sum_{i=j+1}^{N} (y_i - \mu_2)^2}} \tag{18.2}$$

where $Y_i = y_1, y_2, \ldots, y_{N-j}$
$Y_2 = y_{1+j}, y_{2+j}, \ldots, y_N$

$$\mu_1 = \sum_{i=1}^{N-j} y_i / (N - j) \tag{18.3}$$

$$\mu_2 = \sum_{i=1+j}^{N} y_i / (N - j) \tag{18.4}$$

As an example, assume the following population:

| $i =$ | 1 | 2 | 3 | 4 | 5 | 6 | 7 | 8 | 9 | 10 |
|---|---|---|---|---|---|---|---|---|---|---|
| $y_i =$ | 8 | 10 | 13 | 13 | 15 | 16 | 19 | 21 | 23 | 25 |

If $j = 1$, the pairings are $(y_1, y_2)$, $(y_2, y_3)$, $(y_3, y_4)$, $(y_4, y_5)$, $(y_5, y_6)$, $(y_6, y_7)$, $(y_7, y_8)$, $(y_8, y_9)$, and $(y_9, y_{10})$. Then,

| $Y_{i-1} =$ | 8 | 10 | 13 | 13 | 15 | 16 | 19 | 21 | 23 |
|---|---|---|---|---|---|---|---|---|---|
| $y_i =$ | 10 | 13 | 13 | 15 | 16 | 19 | 21 | 23 | 25 |

and $\rho_1 = 0.9831$. If $j = 2$, the pairings are: $(y_1, y_3)$, $(y_2, y_4)$, $(y_3, y_5)$, $(y_4, y_6)$, $(y_5, y_7)$, $(y_6, y_8)$, $(y_7, y_9)$, and $(y_8, y_{10})$ Then,

| $y_{i-2} =$ | 8 | 10 | 13 | 13 | 15 | 16 | 19 | 21 |
|---|---|---|---|---|---|---|---|---|
| $y_i =$ | 13 | 13 | 15 | 16 | 19 | 21 | 23 | 25 |

and $\rho_2 = 0.9745$. When $j = 3$, $\rho_3 = 0.9663$. As can be seen, the serial correlation decreases or decays as the interval widens. In the case of this particular population the serial correlation pattern becomes erratic beyond this point. Serial correlation can play a part in forest inventories when line–plot sampling is used. Since travel time is unproductive, cruise designers tend to widen the spacing between cruise line and shorten the distance between plots within the cruise lines. In some cases the plots touch or almost touch. This can introduce serial correlation into the sample. However, its effect on the inventory results is probably minimal.

## 18.8 POPULATION PARAMETERS

Since systematic sampling, under both approaches, is a form of cluster sampling, the discussion of population parameters in Sections 17.4, 17.4.1, and 17.4.2 is applicable here and requires no modification.

## 18.9 VARIANCE OF THE OVERALL TOTAL AND MEAN

It is at this point that the difference between the two approaches to systematic sampling become clearly evident.

### 18.9.1 Under Approach 1

According to Equation 17.40, the variance of the overall total of a cluster sample is

$$V(T_o) = \frac{M}{m}\sum_i^m \left[\frac{\sigma_i^2}{n_i}\left(\frac{N_i - n_i}{N_i - 1}\right)N_i^2\right] + M^2\frac{\sigma_\tau^2}{m}\left(\frac{M-m}{M-1}\right)$$

Under Approach 1, $m = 1$ and $n_i = N_i$. This results in $(N_i - n_i)/(N_i - 1) = 0$ and $(M - m)/(M - 1) = 1$. Consequently, the variance of the overall total is

$$V(T_0)_{\text{Sys1}} = M^2\sigma_\tau^2 = M\sum_i^M (\tau_i - \bar{\tau})^2 \tag{18.5}$$

The subscript "Sys 1" identifies this variance of the overall total as being appropriate under Approach 1.

According to Equation 17.51, the variance of the overall mean is

$$V(\bar{y}_o) = \frac{1}{N^2}V(T_o)$$

Then,

$$V(\bar{y}_o)_{\text{Sys1}} = \frac{M^2}{N^2}\sigma_\tau^2 \tag{18.6}$$

To simplify the mathematics, assume that cluster size is constant so that $N_i = A$. Then $N = MA$. As a result,

$$V(\bar{y}_o)_{\text{Sys1}} = \frac{M^2}{M^2 A^2}\sigma_\tau^2 = \frac{1}{A^2}\sigma \tag{18.7}$$

which, according to Equation 17.84, is

$$V(\bar{y}_o)_{\text{Sys1}} = \sigma_B^2 = \frac{1}{M}\sum_{i}^{M}(\mu_i - \mu_o)^2 \tag{18.8}$$

Thus, when systematic sampling is viewed as being a form of simple cluster sampling and the cluster size is constant, the variance of the overall mean is equal to the *between-cluster* variance.

As an example, consider the following population, which contains $N = 24$ variates, has a total $\tau_o = 13$, an overall mean $\mu_o = 0.54165$, and an overall variance $\sigma_o^2 = 0.248254$. Setting $L = 4$ results in $M = 4$ clusters, each containing $N_i = A = 6$ variates.

| Cluster and Sample | Population: 0, 1, 1, 0, 1, 1, 1, 0, 0, 1, 1, 1, 1, 0, 0, 1, 0, 1, 1, 0, 0, 0, 1, 0 |
|---|---|
| 1 | 0 1 0 1 0 0 |
| 2 | 1 1 1 0 1 0 |
| 3 | 1 1 1 0 1 1 |
| 4 | 0 0 1 1 0 0 |

The cluster parameters are

| Cluster | $\tau_i$ | $\mu_i$ | $\sigma_i^2$ |
|---|---|---|---|
| 1 | 2 | 0.3333 | 0.2222 |
| 2 | 4 | 0.6667 | 0.2222 |
| 3 | 5 | 0.8333 | 0.1389 |
| 4 | 2 | 0.3333 | 0.2222 |

The overall parameters from the cluster data are

$$\mu_o = {}^1/_4(0.3333 + 0.6667 + 0.8333 + 0.3333) = 0.54165$$

$$\sigma_w^2 = {}^1/_4(0.2222 + 0.2222 + 0.1389 + 0.2222) = 0.201375$$

(note that these are unweighted means, which is permissible since the cluster size is constant) and

$$\sigma_B^2 = \frac{1}{4}[(0.3333 - 0.5417)^2 + (0.6666 - 0.5417)^2 + (0.8333 - 0.5417)^2$$
$$+ (0.3333 - 0.5417)^2] = 0.046879$$

Thus,

$$V(\bar{y}_o)_{\text{Sys1}} = 0.046879$$

One can now relate the variance of the overall mean of the systematic sample, under Approach 1, to that of a simple random sample of the same size. To do this one must first determine the conditions under which the two variances are equal.

$$V(\bar{y}_o)_{\text{Sys1}} = \sigma_B^2 = V(\bar{y}_o)_{\text{Ran}} = \frac{\sigma_o^2}{n}\left(\frac{N-n}{N-1}\right) \tag{18.9}$$

Under Approach 1, $n = A$. Then,

$$\sigma_B^2 = \frac{\sigma_o^2}{A}\left(\frac{N-A}{N-1}\right) \tag{18. 10}$$

$$\frac{\sigma_B^2}{\sigma_o^2} = \frac{1}{A}\left(\frac{N-A}{N-1}\right) = \frac{1}{A}\left(\frac{MA-A}{N-1}\right) = \frac{A}{A}\left(\frac{M-1}{N-1}\right) = \left(\frac{M-1}{N-1}\right) \tag{18. 11}$$

Thus, when the ratio $\sigma_B^2/\sigma_0^2$ is equal to $(M - 1)/(N - 1)$, $V(\bar{y}_o)_{\text{Sys1}} = V(\bar{y}_o)_{\text{Ran}}$, indicating that the precision of the systematic sample, under Approach 1, will be equal to that of a simple random sample of the same size. Similarly, when the ratio is greater than $(M - 1)/(N - 1)$, $V(\bar{y}_o)_{\text{Sys1}} > V(\bar{y}_o)_{\text{Ran}}$ and the precision of the systematic sample will be poorer than that of a random sample. When the ratio is less than $(M - 1)/(N - 1)$, $V(\bar{y}_o)_{\text{Sys1}} < V(\bar{y}_o)_{\text{Ran}}$ and the systematic sample will be more precise than the random sample.

From this development it can be seen that the difference in precision between systematic sampling, under Approach 1, and simple random sampling, with samples of the same size, is a function of the nature of the population. If the within-cluster variance is large, the between-cluster variance will be correspondingly low and, thus, the ratio $\sigma_B^2/\sigma_0^2$ will be small and the precision of the estimate of the variance of the overall mean will be relatively great. The reverse is true if the within-cluster variance is small and the between-cluster variance large.

In the case of this example, $N = 24$; $M = 4$; $\sigma_o^2 = 0.248254$; $\sigma_B^2 = 0.046879$; $V(\bar{y}_o)_{\text{Sys1}} = 0.046879$; and $V(\bar{y}_o)_{\text{Ran}} = 0.032381$.

$$\frac{\sigma_B^2}{\sigma_0^2} = \frac{0.046879}{0.248254} = 0.188835 > \left(\frac{M-1}{N-1}\right) = \left(\frac{4-1}{24-1}\right) = 0.13043$$

Therefore $V(\bar{y}_o)_{\text{Sys1}}$ should be greater than $V(\bar{y}_o)_{\text{Ran}}$. Since $V(\bar{y}_o)_{\text{Sys1}} = 0.046879 > 0.032381 = V(\bar{y}_o)_{\text{Ran}}$, the prediction is correct.

The degree of variability of the variates within the clusters can be quantified through the use of the intraclass correlation coefficient, ρ (see Section 17.6). According to Equation 17.78,

$$V(\bar{y}_o)_{\text{cl}} = \frac{\sigma_o^2}{ma}\left(\frac{A-a}{A}\right)(1-\rho) + \frac{\sigma_o^2}{mA}\left(\frac{M-m}{M-1}\right)\left(\frac{a}{A}\right)[1+\rho(A-1)]$$

Since

$$V(\bar{y}_o)_{\text{Ran}} = \frac{\sigma_o^2}{n}\left(\frac{N-n}{N-1}\right)$$

and, since no subsampling is involved, $ma = mA = n$,

$$\frac{\sigma_o^2}{n} = \frac{\sigma_o^2}{ma} = V(\bar{y}_o)_{\text{Ran}}\left(\frac{N-1}{N-ma}\right)$$

Substituting this into Equation 17.78 yields

$$V(\bar{y}_o)_{\text{c1}} = V(\bar{y}_o)_{Ran}\left(\frac{N-1}{N-ma}\right)\left(\frac{A-a}{A}\right)(1-\rho) + V(\bar{y}_o)_{\text{Ran}}\left(\frac{N-1}{N-ma}\right)\left(\frac{M-m}{M-1}\right)\left(\frac{a}{A}\right)[1+\rho(A-1)] \tag{18.12}$$

Under Approach 1, $m = 1$ and $a = A$. Then

$$\left(\frac{A-A}{A}\right) = 0, \qquad \left(\frac{M-1}{M-1}\right) = 1, \qquad \left(\frac{A}{A}\right) = 1$$

and, consequently,

$$V(\bar{y}_o)_{\text{Sys1}} = V(\bar{y}_o)_{\text{Ran}}\left(\frac{N-1}{N-A}\right)[1+\rho(A-1)] \tag{18.13}$$

The precision of systematic sampling under Approach 1, relative to that of simple random sampling when the sample sizes are the same, is

$$RP_{\text{Sys1}} = V(\bar{y}_o)_{\text{Ran}} / V(\bar{y}_o)_{\text{Sys1}} \tag{18.14a}$$

which can be written

$$RP_{\text{Sys1}} = \frac{1}{\left(\frac{N-1}{N-A}\right)[1+\rho(A-1)]} \tag{18.14b}$$

Figure 18.6 shows how the relative precision changes as the intraclass correlation coefficient changes. As can be seen, the relationship is hyperbolic with a discontinuity at the value of $\rho$ at which the term $[1 + \rho(A - 1)] = 0$. At this point,

$$RP_{\text{Sys1}} = 1/0 \rightarrow \pm\infty$$

The value of $\rho$ at which this occurs is

$$\rho_{\text{dis}} = -1/(A-1)$$

The precision of systematic sampling under Approach 1 is equal to that of simple random sampling only when $\rho$ is equal to its critical value (see Section 17.11.1).

$$\rho_{\text{crit}} = \frac{(MA-ma)(M-1)A-(MA-1)(A-a)(M-1)-(MA-1)(M-m)a}{(MA-1)(M-m)(A-1)a-(MA-1)(A-a)(M-1)}$$

Under Approach 1, $m = 1$ and $a = A$. Making these substitutions in Equation 17.135 and simplifying,

$$\rho_{crit} = \frac{-(A-1)}{(MA-1)(A-1)}$$

Since $MA = N$,

$$\rho_{crit} = \frac{1}{-(N-1)} \tag{18.15}$$

Thus, when $\rho = \rho_{crit}$,

$$\begin{aligned}\left(\frac{N-1}{N-A}\right)[1+\rho(A-1)] &= \left(\frac{N-1}{N-A}\right)\left[1+\left(-\frac{1}{N-1}\right)(A-1)\right] \\ &= \left(\frac{N-1}{N-A}\right)\left[1-\left(\frac{A-1}{N-1}\right)\right] \\ &= \left(\frac{N-1}{N-A}\right)\left(\frac{N-1-A+1}{N-1}\right) \\ &= \left(\frac{N-1}{N-A}\right)\left(\frac{N-A}{N-1}\right) = 1\end{aligned}$$

and $V(\bar{y}_o)_{Sys1} = V(\bar{y}_o)_{Ran}$.

Note that the critical value of $\rho$ is dependent only on the population size and ranges from $-1$ when $N = 2$ to zero when $N = \infty$.

As can be seen from Figure 18.6 and Table 18.1, systematic sampling under Approach 1 is relatively more precise than simple random sampling only when $\rho_{dis} < \rho < \rho_{crit}$, and is equally as precise only when $\rho = \rho_{crit}$. The ranges of $\rho$ values within which $RP_{sys1} \geq 1$ are narrow, becoming narrower as $N$ and $A$ increase. In the case of $N = 1000$ and $A = 100$, the range is only from $\rho_{crit} = -0.001001$ to $\rho_{dis} = -0.010101$. Since most real-life populations being sampled are large, one can assume that systematic sampling under Approach 1 will yield a variance of the mean that will be larger than what would have been obtained using simple random sampling with a sample of the same size because the probability of having an intraclass correlation coefficient that lies within the $\rho_{dis}$ to $\rho_{crit}$ range will be very small.

Insofar as this example is concerned: $N = 24$, $M = 4$; $A = 6$; $\sigma_o^2 = 0.2482639$; $\Sigma_i^M \Sigma_{j<k}^{A(A-1)/2} (y_{ij} - \mu_o)(y_{ik} - \mu_o) = 0.3953$; and $V(\bar{y}_o)_{Ran} = 0.032381$. Then, according to Equation 17.70,

$$\begin{aligned}\rho &= \frac{2}{MA(A-1)\sigma_o^2} \sum_i^M \sum_{j<k}^{A(A-1)/2} (y_{ij} - \mu_o)(y_{ik} - \mu_o) \\ &= \frac{2(0.3953)}{4(6)(6-1)0.2482639} = +0.0265376\end{aligned}$$

The critical value of $\rho$ is

$$\rho_{crit} = -\frac{1}{(24-1)} = -0.0434783$$

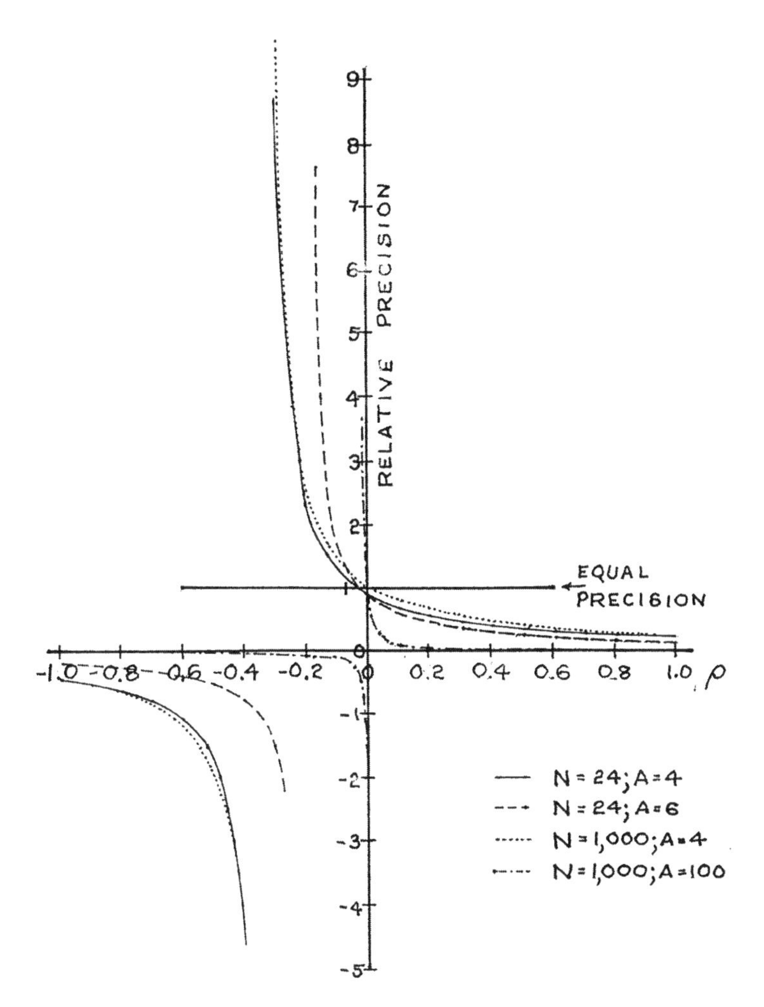


**FIGURE 18.6** The precision of systematic sampling, under Approach 1, relative to that of simple random sampling.

and the discontinuity lies at

$$\rho_{dis} = -\frac{1}{(6-1)} = -0.2$$

Then, $\rho_{Sys1} > \rho_{crit}$, indicating that $V(\bar{y}_o)_{Sys1}$ should be greater than $V(\bar{y}_o)_{Ran}$. According to Equation 18.13,

$$V(\bar{y}_o)_{Sys1} = V(\bar{y}_o)_{Ran}\left(\frac{N-1}{N-A}\right)[1+\rho_{csys}(A-1)]$$

$$= 0.032381\left(\frac{24-1}{24-6}\right)[1+0.0265376(6-1)]$$

$$= 0.046866$$

The difference between this result and that obtained earlier is due to rounding errors. In this case $V(\bar{y}_o)_{Sys1} = 0.046866 > V(\bar{y}_o)_{Ran} = 0.032381$ as was predicted.

**TABLE 18.1**
**Relative Precision of Systematic Sampling under Approach 1**

| | $N = 24$ | | $N = 1000$ | | |
|---|---|---|---|---|---|
| ρ | $A = 4$ | $A = 6$ | $A = 4$ | $A = 30$ | $A = 100$ |
| 1.0 | 0.217 | 0.130 | 0.249 | 0.032 | 0.009 |
| 0.9 | 0.235 | 0.142 | 0.269 | 0.036 | 0.010 |
| 0.8 | 0.256 | 0.157 | 0.293 | 0.040 | 0.011 |
| 0.7 | 0.281 | 0.174 | 0.322 | 0.046 | 0.013 |
| 0.6 | 0.311 | 0.196 | 0.356 | 0.053 | 0.015 |
| 0.5 | 0.348 | 0.224 | 0.399 | 0.063 | 0.018 |
| 0.4 | 0.395 | 0.261 | 0.453 | 0.077 | 0.022 |
| 0.3 | 0.458 | 0.313 | 0.524 | 0.100 | 0.029 |
| 0.2 | 0.543 | 0.391 | 0.623 | 0.143 | 0.043 |
| 0.1 | 0.669 | 0.522 | 0.767 | 0.249 | 0.083 |
| 0.0 | 0.870 | 0.783 | 0.997 | 0.972 | 0.901 |
| $\rho_{crit(1000)} =$ –0.001001 | | | 1.000 | 1.000 | 1.000 |
| $\rho_{dis(100)} =$ –0.010101 | | | | | ± ∞ |
| $\rho_{dis(30)} =$ –0.034483 | | | | ± ∞ | |
| $\rho_{crit(24)} =$ –0.043478 | 1.000 | 1.000 | | | |
| –0.1 | 1.242 | 1.565 | 1.424 | –0.512 | –0.101 |
| $\rho_{dis(6)} =$ –0.2 | 2.174 | ± ∞ | 2.492 | –0.202 | –0.048 |
| –0.3 | 8.696 | –1.567 | 9.970 | –0.126 | –0.031 |
| $\rho_{dis(4)} =$ 0.0333333 | ± ∞ | | ± ∞ | | |
| –0.4 | –4.378 | –0.783 | –4.985 | –0.092 | –0.023 |
| –0.5 | –1.739 | –0.522 | –1.994 | –0.072 | –0.019 |
| –0.6 | –1.087 | –0.391 | –1.246 | –0.059 | –0.015 |
| –0.7 | –0.791 | –0.313 | –0.906 | –0.050 | –0.013 |
| –0.8 | –0.621 | –0.261 | –0.712 | –0.044 | –0.012 |
| –0.9 | –0.512 | –0.224 | –0.586 | –0.039 | –0.010 |
| –1.0 | –0.435 | –0.196 | –0.498 | –0.035 | –0.009 |

One must realize that $\sigma_o^2$, $\sigma_w^2$, $\sigma_B^2$, or ρ are never known in real-life situations and that the precision of estimate of an overall mean from a real-life systematic sample relative to that from a comparable simple random sample cannot be determined using the procedure discussed here. What has been done here is to show the relationship when the needed population parameters are known.

### 18.9.2 Under Approach 2

Returning to Equation 17.40, the variance of the overall total of a cluster sample is

$$V(T_o) = \frac{M}{m}\sum_{i}^{M}\left[\frac{\sigma_i^2}{n_i}\left(\frac{N_i - n_i}{N_i - 1}\right)N_i^2\right] + M^2\frac{\sigma_\tau^2}{m}\left(\frac{M - m}{M - 1}\right)$$

Under Approach 2, $m = M$ and $n_i = 1$. Then $M/m = 1$, $(N_i - n_i)/(N_i - 1) = 1$, and $(M - m)/(M - 1) = 0$. Consequently,

$$V(T_o)_{\text{Sys2}} = \sum_{i}^{M} \sigma_i^2 N_i^2 \tag{18.16}$$

where the subscript "Sys 2" indicates that Approach 2 is being used. Note that in this case the second term of Equation 17.40 drops out because *all* the clusters were included in the sample and then there is no between-cluster component in the variance of the overall mean. From this point, to simplify the mathematics, it will be assumed, as was done before, that cluster size is constant so that $N_i = A$. Then,

$$V(T_o)_{\text{Sys2}} = A^2 \sum_{i}^{M} \sigma_i^2 \tag{18.17}$$

According to Equation 17.51,

$$V(y_o)_{\text{Sys2}} = \frac{1}{N^2} V(T_o)_{\text{Sys2}} = \frac{A^2}{N^2} \sum_{i}^{M} \sigma^2$$

Since $N = MA$,

$$V(\bar{y}_o)_{\text{Sys2}} = \frac{A^2}{M^2 A^2} \sum_{i}^{M} \sigma_i^2 = \frac{1}{M^2} \sum_{i}^{M} \sigma^2 \tag{18.18}$$

According to Equation 17.54,

$$\frac{1}{M} \sum_{i}^{M} \sigma_i^2 = \sigma_w^2$$

Then

$$V(\bar{y}_o)_{\text{Sys2}} = \frac{1}{M} \sigma_w^2 \tag{18.19}$$

which is not the same as when systematic sampling under Approach 1 is used (see Equation 18.7).

As an example, one can use the same population as in the example of Approach 1. The spacing interval, $L$, remains at 4 but now there are $M = 6$ clusters, each containing $N_i = A = 4$ variates.

| | **Clusters** | | | | | |
|---|---|---|---|---|---|---|
| | **1** | **2** | **3** | **4** | **5** | **6** |
| **Samples** | **0, 1, 1, 0** | **1, 1, 1, 0** | **0, 1, 1, 1** | **1, 0, 0, 1** | **0, 1, 1, 0** | **0, 0, 1, 0** |
| 1 | 0 | 1 | 0 | 1 | 0 | 0 |
| 2 | 1 | 1 | 1 | 0 | 1 | 0 |
| 3 | 1 | 1 | 1 | 0 | 1 | 1 |
| 4 | 0 | 0 | 1 | 1 | 0 | 0 |

The cluster parameters are

| Cluster | $\tau_i$ | $\mu_i$ | $\sigma_i^2$ |
|---|---|---|---|
| 1 | 2 | 0.50 | 0.2500 |
| 2 | 3 | 0.75 | 0.1875 |
| 3 | 3 | 0.75 | 0.1875 |
| 4 | 2 | 0.50 | 0.2500 |
| 5 | 2 | 0.50 | 0.2500 |
| 6 | 1 | 0.25 | 0.1875 |

The overall parameters, from the cluster data, are

$$\mu_o = \frac{1}{6}(0.50 + 0.75 + 0.75 + 0.50 + 0.50 + 0.25) = 0.54167$$

$$\sigma_w^2 = \frac{1}{6}(0.2500 + 0.1875 + 0.1875 + 0.2500 + 0.2500 + 0.1875) = 0.218750$$

and

$$\begin{aligned}\sigma_B^2 &= \frac{1}{6}[(0.50 - 0.5417)^2 + (0.75 - 0.5417)^2 + (0.75 - 0.5417)^2 \\ &+ (0.50 - 0.5417)^2 + (0.50 - 0.5417)^2 + (0.25 - 0.5417)^2] \\ &= 0.029514\end{aligned}$$

Then

$$V(\bar{y}_o)_{\text{Sys2}} = \frac{1}{6}(0.218750) = 0.036458$$

Note that $V(\bar{y}_o)_{\text{Sys2}} = 0.036458 \neq 0.046879 = V(\bar{y}_o)_{\text{Sys1}}$.

One can now relate the variance of the overall mean of the systematic sample, under Approach 2, to that of a random sample of the same size. As before, the conditions under which the two variances will be equal will be determined first.

In this case there is one observation per cluster, and thus $n = M$. Then,

$$V(\bar{y}_o)_{\text{Sys2}} = \frac{\sigma_w^2}{M} = V(\bar{y}_o)_{\text{Ran}} = \frac{\sigma_o^2}{M}\left(\frac{N - M}{N - 1}\right)$$

Multiplying both sides by $M$,

$$\sigma_w^2 = \sigma_o^2\left(\frac{N - M}{N - 1}\right)$$

Then,

$$\frac{\sigma_w^2}{\sigma_o^2} = \left(\frac{N - M}{N - 1}\right) \tag{18.20}$$

Thus, when the ratio $\sigma_w^2 / \sigma_o^2$ is equal to $(N - M)/(N - 1)$, $V(\bar{y}_o)_{\text{Sys2}} = V(\bar{y}_o)_{\text{Ran}}$ and the precision of the systematic sample is equal to that of the random sample. In like manner, when the ratio is greater than $(N - M)/(N - 1)$, $V(\bar{y}_o)_{\text{Sys2}} > V(\bar{y}_o)_{\text{Ran}}$ and the precision of the systematic sample is poorer than that of the random sample. Finally, when the ratio is less than $(N - M)/(N - 1)$, $V(\bar{y}_o)_{\text{Sys2}} < V(\bar{y}_o)_{\text{Ran}}$ and the precision of the systematic sample is greater than that of the random sample.

Extending this reasoning to the example, where $N = 24$, $M = 6$, $\sigma_o^2 = 0.248254$, $\sigma_w^2 = 0.218750$, $V(\bar{y}_o)_{\text{Sys2}} = 0.036458$, and $V(\bar{y}_o)_{\text{Ran}} = 0.032381$,

$$\frac{\sigma_w^2}{\sigma_o^2} = \frac{0.218750}{0.248254} = 0.881154 > \left(\frac{N-M}{N-1}\right) = \frac{24-6}{24-1} = 0.78360$$

This indicates that $V(\bar{y}_o)_{\text{Sys2}}$ should be greater than $V(\bar{y}_o)_{\text{Ran}}$. Since $V(\bar{y}_o)_{\text{Sys2}} = 0.036458 > 0.032381 = V(\bar{y}_o)_{\text{Ran}}$ the prediction is correct.

Again, as under Approach 1, the precision of the estimate of the variance of the overall mean is a function of the nature of the population. If the variability of the variates within the clusters is great compared with the variability of the cluster means, $\sigma_w^2$ will be large relative to $\sigma_o^2$ and the precision of the systematic sample will be relatively low. The reverse will be true if the within-cluster variabilities are small and the between-cluster variability is large. *It should be noted that this pattern is the opposite of that associated with Approach 1.*

Now consider the effect of the intraclass correlation coefficient, ρ, on the precision of the estimate of the overall mean. According to Equation 18.12,

$$\begin{aligned} V(\bar{y}_o)_{\text{cl}} &= V(\bar{y}_o)_{\text{Ran}}\left(\frac{N-1}{N-ma}\right)\left(\frac{A-a}{A}\right)(1-\rho) \\ &+ V(\bar{y}_o)_{\text{Ran}}\left(\frac{N-1}{N-ma}\right)\left(\frac{M-m}{M-1}\right)\left(\frac{a}{A}\right)[1+\rho(A+1)] \end{aligned}$$

Under Approach 2, $m = M$ and $a = 1$. Then, since $(M - M)/(M - 1) = 0$,

$$V(\bar{y}_o)_{\text{Sys2}} = V(\bar{y}_o)_{\text{Ran}}\left(\frac{N-1}{N-M}\right)\left(\frac{A-1}{A}\right)(1-\rho)$$

Since $N = MA$, $N - M = MA - M = M(A - 1)$. Then,

$$\begin{aligned} V(\bar{y}_o)_{\text{Sys2}} &= V(\bar{y}_o)_{\text{Ran}}\left(\frac{N-1}{MA}\right)(1-\rho) \\ &= V(\bar{y}_o)_{\text{Ran}}\left(\frac{N-1}{N}\right)(1-\rho) \end{aligned}$$

Then

$$\frac{V(\bar{y}_o)_{\text{Sys2}}}{V(\bar{y}_o)_{\text{Ran}}} = \left(\frac{N-1}{N}\right)(1-\rho)$$

and

$$RP_{\text{Sys2}} = \frac{V(\bar{y}_o)_{\text{Ran}}}{V(\bar{y}_o)_{\text{Sys2}}} = \frac{1}{\left(\frac{N-1}{N}\right)(1-\rho)} \tag{18.21}$$

Consequently, the relative precision of systematic sampling under Approach 2 is a function of only population size and the intraclass correlation coefficient. Cluster size and number of clusters have no effect, which can be seen in Table 18.2. Furthermore, under Approach 2, the critical value of $\rho$ is obtained when $m = M$ and $a = 1$ are substituted into Equation 17.135. After simplification

$$\rho_{\text{crit}} = \frac{1}{-(N-1)} \tag{18.22}$$

which is the same as that under Approach 1.

Thus, under Approach 2, when $\rho = \rho_{\text{crit}} = -[1/(MA - 1)]$

$$\left(\frac{N-1}{N-M}\right)\left(\frac{A-1}{A}\right)(1-\rho) = \left(\frac{MA-1}{MA-M}\right)\left(\frac{A-1}{A}\right)\left[1-\left(-\frac{1}{MA-1}\right)\right]$$

$$= \left(\frac{MA-1}{MA-M}\right)\left(\frac{A-1}{A}\right)\left(\frac{MA-1+1}{MA-1}\right)$$

$$= \left(\frac{A-1}{M(A-1)}\right)\left(\frac{MA}{A}\right) = 1$$

and $V(\bar{y}_o)_{\text{Sys2}} = V(\bar{y}_o)_{\text{Ran}}$.

The effect of the intraclass correlation on the relative precision of the overall mean can be seen in Figure 18.7. As in the case of Approach 1, the relationship is hyperbolic, but only one limb is involved since the discontinuity occurs at $\rho = +1.0$. Within the range $-1.0 \leq \rho \leq 1.0$, the curve rises monotonically, crossing the line of equal precision at $\rho_{\text{crit}}$. Consequently, when $\rho = \rho_{\text{crit}}$, systematic sampling under Approach 2 yields more precise estimates of the overall mean than does simple random sampling when sample size is held constant. Furthermore, the relative precision of the systematic sampling increases at an increasing rate as $\rho$ increases, becoming infinitely better as $\rho \rightarrow 1.0$. This latter characteristic is a reflection of the nature of $\rho$. When $\rho = 1$, all the variates within each of the clusters are equal but the cluster means can be different, as is shown below, where $N = 24$, $A = 4$, and $M = 6$.

| | Clusters | | | | | |
|---|---|---|---|---|---|---|
| | **1** | **2** | **3** | **4** | **5** | **6** |
| | 0,0,0,0 | 1,1,1,1 | 1,1,1,1 | 0,0,0,0 | 1,1,1,1 | 0,0,0,0 |
| $\mu_i$ | 0.0 | 1.0 | 1.0 | 0.0 | 1.0 | 0.0 |
| $\sigma_i^2$ | 0 | 0 | 0 | 0 | 0 | 0 |
| $\mu_0$ | 0.5 | | | | | |
| $\sigma_0^2$ | 0.25 | | | | | |

According to Equation 18.18,

$$V(\bar{y}_0)_{\text{Sys2}} = \frac{1}{M^2}\sum_i^M \sigma_i^2 = \frac{1}{6^2}(0+0+0+0+0+0) = 0$$

**TABLE 18.2**
**Relative Precision of Systematic Sampling under Approach 2**

| | $N = 24$ | | $N = 1000$ | | |
|---|---|---|---|---|---|
| $\rho$ | $A = 4$ $M = 6$ | $A = 8$ $M = 4$ | $A = 4$ $M = 250$ | $A = 20$ $M = 50$ | $A = 100$ $M = 10$ |
| 1.0 | $\infty$ | $\infty$ | $\infty$ | $\infty$ | $\infty$ |
| 0.9 | 10.435 | 10.435 | 10.010 | 10.010 | 10.010 |
| 0.8 | 5.217 | 5.217 | 5.005 | 5.005 | 5.005 |
| 0.7 | 3.478 | 3.478 | 3.337 | 3.337 | 3.337 |
| 0.6 | 2.609 | 2.609 | 2.503 | 2.503 | 2.503 |
| 0.5 | 2.087 | 2.087 | 2.002 | 2.002 | 2.002 |
| 0.4 | 1.739 | 1.739 | 1.668 | 1.668 | 1.668 |
| 0.3 | 1.491 | 1.491 | 1.430 | 1.430 | 1.430 |
| 0.2 | 1.304 | 1.304 | 1.251 | 1.251 | 1.251 |
| 0.1 | 1.159 | 1.159 | 1.112 | 1.112 | 1.112 |
| 0.0 | 1.043 | 1.043 | 1.001 | 1.001 | 1.001 |
| $\rho_{crit(1,000)} =$ −0.001001 | | | 1.000 | 1.000 | 1.000 |
| $\rho_{crit(24)} =$ −0.043478 | 1.000 | 1.000 | | | |
| −0.1 | 0.949 | 0.949 | 0.910 | 0.910 | 0.910 |
| −0.2 | 0.870 | 0.870 | 0.834 | 0.834 | 0.834 |
| −0.3 | 0.803 | 0.803 | 0.770 | 0.770 | 0.770 |
| −0.4 | 0.745 | 0.745 | 0.715 | 0.715 | 0.715 |
| −0.5 | 0.696 | 0.696 | 0.667 | 0.667 | 0.667 |
| −0.6 | 0.652 | 0.652 | 0.626 | 0.626 | 0.626 |
| −0.7 | 0.614 | 0.614 | 0.589 | 0.589 | 0.589 |
| −0.8 | 0.580 | 0.580 | 0.556 | 0.556 | 0.556 |
| −0.9 | 0.549 | 0.549 | 0.527 | 0.527 | 0.527 |
| −1.0 | 0.522 | 0.522 | 0.501 | 0.501 | 0.501 |

In the case of simple random sampling there are

$$\binom{24}{6} = 134,596$$

possible samples that can be drawn, without replacement, from the population. These samples are constructed without regard to the clustering.

$$V(\bar{y}_0)_{\text{Ran}} = \frac{1}{n}\sigma_0^2 = \frac{0.25}{6} = 0.041$$

and the relative precision of the systematic sample under Approach 2 is

$$V(\bar{y}_0)_{\text{Ran}} / V(\bar{y}_0)_{\text{Sys2}} = 0.0417/0 \to \infty$$

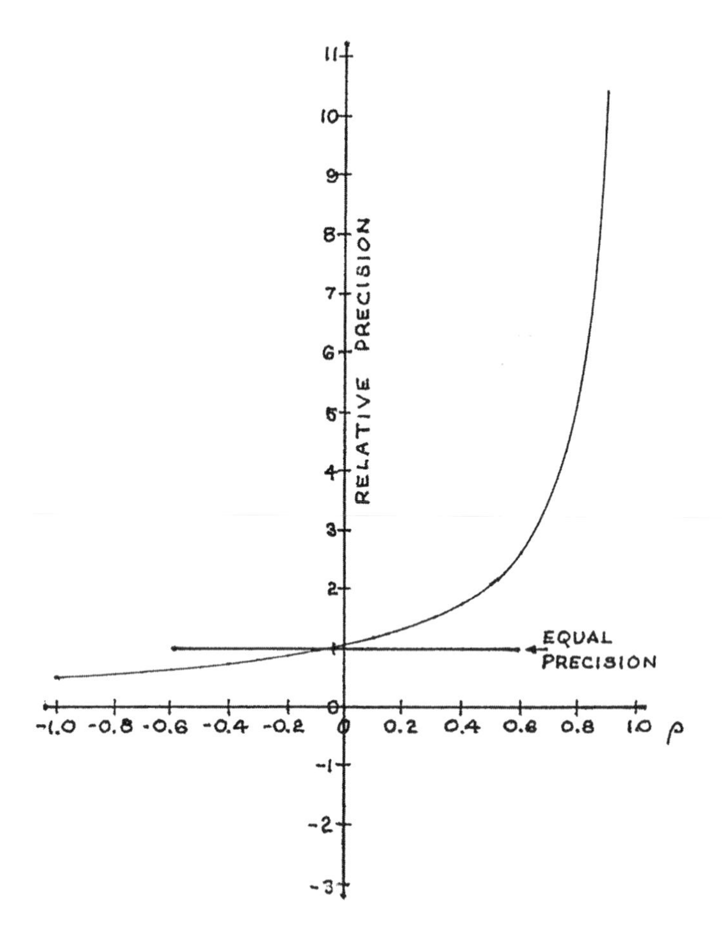


**FIGURE 18.7** The precision of systematic sampling, under Approach 2, relative to that of simple random sampling. The curve shown is for $N = 24$. That for $N = 1000$ is so similar that it cannot be distinguished at this scale.

The evidence of Figures 18.6 and 18.7 indicates that systematic sampling is more reliable under Approach 2 than it is under Approach 1.

### 18.9.3 Effect of a Linear Trend

As was brought out in Section 18.7.2, some populations exhibit linear trends when their variates are plotted against their respective positions within the population. This trend can be described using the expression

$$Y = \beta_o + \beta_1 X + \varepsilon \tag{18.23}$$

where $X$ refers to the position in the population and $\varepsilon$ represents the random component. In the following discussion the random component will be considered nonexistent.

If the coordinate axes are translated so as to eliminate the intercept, $\beta_o$, Equation 18.23 can be simplified to

$$Y = \beta_1 X$$

In terms of a specific observation, this expression becomes

$$y_i = \beta_1 x_i$$

Since $x_i = i$, the position within the population,

$$y_i = \beta_1 i$$

Since $\beta_1$ is a constant, $y_i$ can be represented simply by $i$. It can be shown, using methods involving fractional polynomials, that

$$\sum_i^N i = N(N+1)/2 \tag{18.24}$$

and

$$\sum_i^N i^2 = N(N+1)(2N+1)/6 \tag{18.25}$$

Then

$$\sigma_Y^2 = \frac{1}{N}\sum_i^N (y_i - \mu_Y)^2 = \frac{1}{N}\left[\sum_i^N y_i^2 - \left(\sum_i^N y_i\right)^2 \Big/ N\right] = \frac{1}{N}\left[\beta_1^2 \sum_i^N i^2 - \left(\beta_1 \sum_i^N i\right)^2 \Big/ N\right] \tag{18.26}$$

which can be written

$$\begin{aligned} \sigma_Y^2 &= \frac{\beta_1^2}{N}\left\{\frac{N(N+1)(2N+1)}{6} - \frac{[N(N+1)]^2}{4N}\right\} \\ &= \beta_1^2 \frac{(N^2-1)}{12} \end{aligned} \tag{18.27}$$

Then

$$V(\bar{y}_o)_{\text{Ran}} = \beta_1^2 \frac{(N^2-1)}{12n}\left(\frac{N-n}{N-1}\right) \tag{18.28}$$

Since $\beta_1$ is constant and the random component is assumed to be equal to zero, the cluster variances are homogeneous. Thus, under Approach 1, $n$ = A, the interval between elements of the cluster is equal to $L$, $L = M$,

$$\sum_i^A i = A(A+1)/2 \tag{18.29}$$

and

$$\sum_{i}^{A} i^2 = A(A+1)(2A+1)/6 \tag{18.30}$$

Then,

$$\sigma_w^2 = \frac{\beta_1^2 M^2 (A^2 - 1)}{12} \tag{18.31}$$

and

$$\begin{aligned}\sigma_B^2 = \sigma_Y^2 - \sigma_w^2 &= \frac{\beta_1^2 (N^2 - 1)}{12} - \frac{\beta_1^2 M^2 (A^2 - 1)}{12} \\ &= \beta_1 [M^2 A^2 - 1 - M^2 A^2 + M^2]/12 \\ &= \beta_1^2 (M^2 - 1)/12\end{aligned} \tag{18.32}$$

Then

$$V(\bar{y}_o)_{\text{Sys1}} = \sigma_B^2 = \beta_1^2 (M^2 - 1)/12 \tag{18.33}$$

According to the discussion associated with Equation 18.11, the ratio $\sigma_B^2 / \sigma_Y^2$ can be used to determine whether $V(\bar{y}_o)_{\text{Sys1}}$ is greater than, equal to, or less than $V(\bar{y}_o)_{\text{Ran}}$ by comparing it to the ratio $(M - 1)/(N - 1)$. Substituting Equation 18.31 for $\sigma_B^2$ and Equation 18.26 for $\sigma_Y^2$ yields

$$\frac{\sigma_B^2}{\sigma_Y^2} = \left[\frac{\beta_1^2 (M^2 - 1)}{12}\right]\left[\frac{12}{\beta_1 (N^2 - 1)}\right] = \frac{M^2 - 1}{N^2 - 1} \tag{18.34}$$

Since $(M^2 - 1)/(N^2 - 1) < (M - 1)/(N - 1)$, $V(\bar{y}_o)_{\text{Sys1}} < V(\bar{y}_o)_{\text{Ran}}$. As can be seen, the magnitude of $\beta_1$ does not affect this conclusion. Thus, whenever a linear trend is present in the population being sampled, systematic sampling, under Approach 1, will be more efficient than simple random sampling.

When Approach 2 is used the magnitudes of $A$ and $M$ are reversed from what they are under Approach 1. Consequently, $n = M$. Cluster size changes but $A$ remains as the symbol for cluster size. Then, as before,

$$\sum_{i}^{A} i = A(A+1)/2 \tag{18.35}$$

and

$$\sum_{i}^{A} i^2 = A(A+1)(2A+1)/6 \tag{18.36}$$

However, under Approach 2, since the interval between elements of the cluster is equal to unity,

$$\sigma_w^2 = \beta_1^2 (A^2 - 1)/12 \tag{18.37}$$

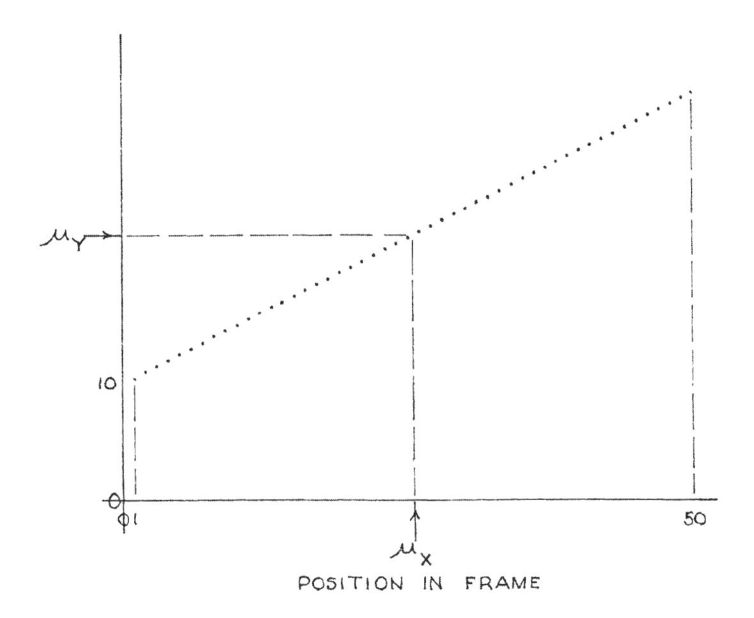


**FIGURE 18.8** A population distributed with the trend $Y = 10 + 0.5X$.

and

$$V(\bar{y}_o)_{\text{Sys2}} = \frac{\sigma_w^2}{M} = \frac{\beta_1^2(A^2 - 1)}{12M} \quad \text{or} \quad \frac{\beta_1^2(A^2 - 1)}{12n} \tag{18.38}$$

Then, since $M$ under Approach 1 is equal to $A$ under Approach 2,

$$V(\bar{y}_o)_{\text{Sys2}} = \frac{1}{n} V(\bar{y}_o)_{\text{Sys1}} \tag{18.39}$$

Thus, when a linear trend is present and there is no random component, the efficiency of systematic sampling under Approach 2 is $n$ times greater than that under Approach 1.

As an example, consider the population shown in Figure 18.8 where the trend is described by

$$Y = 10 + 0.5X$$

and the population size $N = 50$. The population parameters are $\mu_0 = 22.75$ and $\sigma_0^2 = 52.0625$. When $L = 5$ and Approach 1 is used, $M = 5$ and $A = 10$. When Approach 2 is used, $M = 10$ and $A = 5$.

**Positions Included in**

| Approach 1 Cluster | Approach 2 — Clusters | | | | | | | | | |
|---|---|---|---|---|---|---|---|---|---|---|
| | **1** | **2** | **3** | **4** | **5** | **6** | **7** | **8** | **9** | **10** |
| 1 | 1 | 6 | 11 | 16 | 21 | 26 | 31 | 36 | 41 | 46 |
| 2 | 2 | 7 | 12 | 17 | 22 | 27 | 32 | 37 | 42 | 47 |
| 3 | 3 | 8 | 13 | 18 | 23 | 28 | 33 | 38 | 43 | 48 |
| 4 | 4 | 9 | 14 | 19 | 24 | 29 | 34 | 39 | 44 | 49 |
| 5 | 5 | 10 | 15 | 20 | 25 | 30 | 35 | 40 | 45 | 50 |

| | Y values | | | | | | | | | | Approach 1 | |
|---|---|---|---|---|---|---|---|---|---|---|---|---|
| | | | | | | | | | | | $\mu_i$ | $\sigma_i^2$ |
| 1 | 10.5 | 13.0 | 15.5 | 18.0 | 20.5 | 23.0 | 25.5 | 28.0 | 30.5 | 33.0 | 21.75 | 51.5625 |
| 2 | 11.0 | 13.5 | 16.0 | 18.5 | 21.0 | 23.5 | 26.0 | 28.5 | 31.0 | 33.5 | 22.25 | 51.5625 |
| 3 | 11.5 | 14.0 | 16.5 | 19.0 | 21.5 | 24.0 | 26.5 | 29.0 | 31.5 | 34.0 | 22.75 | 51.5625 |
| 4 | 12.0 | 14.5 | 17.0 | 19.5 | 22.0 | 24.5 | 27.0 | 29.5 | 32.0 | 34.5 | 23.25 | 51.5625 |
| 5 | 12.5 | 15.0 | 17.5 | 20.0 | 22.5 | 25.0 | 27.5 | 30.0 | 32.5 | 35.0 | 23.75 | 51.5625 |

**Approach 2**

| | | | | | | | | | | |
|---|---|---|---|---|---|---|---|---|---|---|
| $\mu_i$ | 11.5 | 14.0 | 16.5 | 19.0 | 21.5 | 24.0 | 26.5 | 29.0 | 31.5 | 34.0 |
| $\sigma_i^2$ | 0.5 | 0.5 | 0.5 | 0.5 | 0.5 | 0.5 | 0.5 | 0.5 | 0.5 | 0.5 |

| Approach | $\mu_o$ | $\sigma_o^2$ | $\sigma_w^2$ | $\sigma_B^2$ | $V(\bar{y}_o)$ |
|---|---|---|---|---|---|
| 1 | 22.75 | | 51.5625 | 0.5 | $0.5 = V(\bar{y}_o)_{\text{Sys1}}$ |
| 2 | 22.75 | | 0.5 | 51.5625 | $0.5 = V(\bar{y}_o)_{\text{Sys2}}$ |
| Random | 22.75 | 52.0625 | | | $4.25 = V(\bar{y}_o)_{\text{Ran}}$ |

As can be seen, the variances of the mean follow the theoretical pattern. One should also note that the $\sigma_w^2$ and $\sigma_B^2$ values reverse magnitudes when going from one approach to the other.

One can now consider a population that possesses a linear trend and a random component. Assume the population shown in Figure 18.9, which is listed below. $N$ again is equal to 50 and $L = 5$. As before under Approach 1, $M = 5$ and $A = 10$, while under Approach 2, $M = 10$ and $A = 5$.

**Y Values**

| Approach 1 | Approach 2 — Clusters | | | | | | | | | | Approach 1 | |
|---|---|---|---|---|---|---|---|---|---|---|---|---|
| Cluster | 1 | 2 | 3 | 4 | 5 | 6 | 7 | 8 | 9 | 10 | $\mu_i$ | $\sigma_i^2$ |
| 1 | 12.5 | 12.5 | 13.5 | 17.5 | 22.0 | 23.5 | 25.0 | 26.5 | 29.5 | 35.0 | 21.75 | 52.9125 |
| 2 | 11.0 | 15.0 | 16.5 | 20.0 | 21.0 | 23.5 | 27.5 | 29.0 | 31.5 | 33.5 | 22.85 | 50.2025 |
| 3 | 10.0 | 13.5 | 16.5 | 19.5 | 21.0 | 24.5 | 26.5 | 28.5 | 31.5 | 33.5 | 22.50 | 54.2500 |
| 4 | 13.5 | 14.0 | 18.0 | 19.0 | 21.0 | 23.5 | 26.5 | 28.5 | 31.5 | 35.5 | 23.10 | 48.7400 |
| 5 | 12.5 | 15.0 | 17.5 | 19.5 | 22.0 | 25.0 | 27.5 | 31.0 | 31.5 | 35.5 | 23.70 | 52.9600 |

**Approach 2**

| | | | | | | | | | | |
|---|---|---|---|---|---|---|---|---|---|---|
| $\sigma_i$ | 11.9 | 14.0 | 16.4 | 19.1 | 21.4 | 24.0 | 26.6 | 28.7 | 31.1 | 34.6 |
| $\sigma_i^2$ | 1.54 | 0.90 | 2.44 | 0.74 | 0.24 | 0.40 | 0.84 | 2.06 | 0.64 | 0.84 |

| Approach | $\mu_o$ | $\sigma_o^2$ | $\sigma_w^2$ | $\sigma_B^2$ | $V(\bar{y}_o)$ |
|---|---|---|---|---|---|
| 1 | 22.78 | | 51.8130 | 0.4186 | $0.4186 = V(\bar{y}_o)_{\text{Sys1}}$ |
| 2 | 22.78 | | 1.0640 | 51.1676 | $0.1064 = V(\bar{y}_o)_{\text{Sys2}}$ |
| Random | 22.78 | 52.2316 | | | $4.2638 = V(\bar{y}_o)_{\text{Ran}}$ |

Again, $V(\bar{y}_o)_{\text{Sys2}} < V(\bar{y}_o)_{\text{Sys1}} < V(\bar{y}_o)_{\text{Ran}}$.

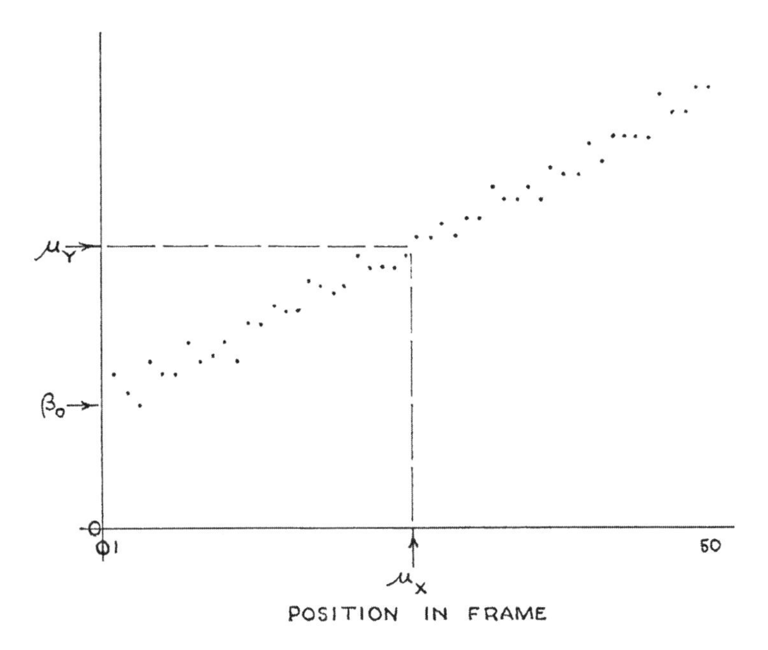


**FIGURE 18.9** A population exhibiting a linear trend ($Y = 10.11 + 0.497X$) and also a random component.

As can be seen from this discussion, the presence of a linear trend will have an effect on the magnitude of the variances of the overall mean. All increase as $\beta_1$ increases. Furthermore, when there is no random component $V(\bar{y}_o)_{\text{Sys2}} < V(\bar{y}_o)_{\text{Sys1}} < V(\bar{y}_o)_{\text{Ran}}$. The presence of a random component will influence the relative magnitudes of these variances, but the general picture will remain the same.

### 18.9.4 Effect of a Curvilinear Trend

Due to the great number of possible curvilinear trends that could be present in a population, it is impossible to arrive at generalizations regarding the effects of such trends. It has been shown that a periodic trend, which is curvilinear, has a definite impact on the variance of the mean. Other curve forms would have different effects. They would have to be determined on an individual case basis, using hypothetical populations representing specific curve forms (e.g., parabolic, hyperbolic, etc.).

### 18.9.5 Removing Trend Effects Using Regression

If the regression equation is known, whether linear or curvilinear, the trend can be removed and the variances of the mean computed using the trend-free data. In the case of the population shown in Figure 18.9, the trend is removed using the expression:

$$y'_{ij} = y_{ij} - [\beta_1(x_{ij} - \mu_X)] = y_{ij} - 0.4969028(x_{ij} - 25.5)$$

Applying this yields the following data set:

**$Y'$ Values**

| Approach 1 Cluster | Approach 2 — Clusters 1 | 2 | 3 | 4 | 5 | 6 | 7 | 8 | 9 | 10 | Approach 1 $\mu_i$ | $\sigma_i^2$ |
|---|---|---|---|---|---|---|---|---|---|---|---|---|
| 1 | 24.67 | 22.19 | 20.71 | 22.22 | 24.24 | 23.25 | 22.27 | 21.28 | 21.80 | 24.81 | 22.74 | 1.8450 |
| 2 | 22.68 | 24.19 | 23.21 | 24.22 | 22.74 | 22.75 | 24.27 | 23.29 | 23.30 | 22.82 | 23.35 | 0.3796 |
| 3 | 21.18 | 22.20 | 22.71 | 23.23 | 22.24 | 23.26 | 22.77 | 22.29 | 22.80 | 22.32 | 22.50 | 0.3294 |
| 4 | 24.18 | 22.20 | 23.71 | 22.23 | 21.75 | 21.76 | 22.28 | 21.79 | 22.31 | 22.82 | 22.60 | 0.7784 |
| 5 | 22.69 | 22.70 | 22.72 | 22.23 | 22.25 | 22.76 | 22.78 | 23.79 | 21.81 | 23.33 | 22.71 | 0.2811 |
| **Approach 2** | | | | | | | | | | | | |
| $\mu_i$ | 23.08 | 22.70 | 22.61 | 22.83 | 22.64 | 22.76 | 22.87 | 22.49 | 22.40 | 23.42 | | |
| $\sigma_i^2$ | 1.5320 | 0.5960 | 1.0404 | 0.6368 | 0.7348 | 0.2980 | 0.5372 | 0.8648 | 0.3372 | 0.7340 | | |

| Approach | $\mu_o$ | $\sigma_o^2 = \sigma_{Y \cdot X}^2$ | $\sigma_w^2$ | $\sigma_B^2$ | $V(\bar{y}_o)$ |
|---|---|---|---|---|---|
| 1 | 22.78 | | 0.7227 | 0.0884 | $0.0884 = V(\bar{y}_o)_{Sys1}$ |
| 2 | 22.78 | | 0.7312 | 0.0794 | $0.0731 = V(\bar{y}_o)_{Sys2}$ |
| Random | 22.78 | 0.8103 | | | $0.0661 = V(\bar{y}_o)_{Ran}$ |

Removing the trends, in essence, creates new populations, as was shown in this example. The new populations possess the same means as the originals but their variances are reduced to that of their random components. In other words, $\sigma_o^2$ becomes $\sigma_{Y \cdot X}^2$. This can be seen if one considers the sum of squares:

$$\text{Total SS} = \text{SS due to trend} + \text{Residual SS}$$

In the case of this example, the overall variance was reduced from 52.2316 to 0.8103 and all three of the variances of the overall mean were also reduced. It should be noted that removing the trend reduces the magnitudes of the three variances of the mean and can change their positions relative to level of precision. For example, considering only Approach 1, prior to removing the trend, using Equation 18.11:

$$\frac{\sigma_B^2}{\sigma_o^2} = \frac{0.4186}{52.2316} = 0.008014 < \left(\frac{M-1}{N-1}\right) = \left(\frac{5-1}{50-1}\right) = 0.08163$$

After removing the trend

$$\frac{\sigma_B^2}{\sigma_{Y \cdot X}^2} = \frac{0.0884}{0.8103} = 0.109095 > \left(\frac{M-1}{N-1}\right) = \left(\frac{5-1}{50-1}\right) = 0.08163$$

In this case, removing the trend made simple random sampling the most efficient sampling procedure.

Population trend is often suspected, or it may even be known to exist, but only rarely will the equation describing that trend be known. Consequently, it must be estimated using data from a sample. This can be done using conventional least squares, nonlinear curve fitting methods, or even graphical procedures (where the resulting graph serves as the describing equation). In any case, the observed values of the variable of interest must be paired with their positions within the *population*, not the sample. Use of sample positions will steepen or otherwise distort the resulting regression line or surface. Then,

$$s_{Y \cdot X}^2 = \frac{1}{k} \sum_i^n (y_i - \hat{y}_i)^2 \rightarrow \sigma_{Y \cdot X}^2 \left( \frac{N}{N-1} \right) \quad (18.40)$$

where $\hat{y}_i = g(x_i)$; $g(x_i)$ is the fitted regression equation; and $k$ = degrees of freedom (which would be indeterminate in the case of a graphical solution). For example, one can use the second sample (cluster 2 under Approach 1) from the population shown in Figure 18.9:

| $y$ | 11.0 | 15.0 | 16.5 | 20.0 | 21.0 | 23.5 | 27.5 | 29.0 | 31.5 | 33.5 |
|---|---|---|---|---|---|---|---|---|---|---|
| $x$ | 2 | 7 | 12 | 17 | 22 | 27 | 32 | 37 | 42 | 47 |

Then, $\hat{Y} = 10.807879 + 0.4915152X$ and $s_{Y \cdot X}^2 = 0.468935$.

When a regression is to be fitted using least squares, the values of the independent or predicting variable are assumed to be fixed and can, in fact, be deliberately chosen as long as the choice is not based on the magnitudes of the associated values of the dependent variable. It is desirable that the chosen values of the independent variable be distributed evenly across its entire range. Systematic sampling accomplishes this admirably.

The major problem in this fitting process is the choice of the regression model. Simple linear relationships rarely occur in forestry situations. However, Hasel (1942) did effectively use such a model in the case of timber volumes on strips run across a tract of virgin ponderosa pine in California. In most cases, the relationships are such that models generating curved regression are needed.

DeLury (1950) suggested the use of high-order polynomials, which might be flexible enough to follow the undulating trends commonly assumed to exist in forest populations, in other ecological complexes, and in geologic formations. Thus, for one-dimensional populations (e.g., timber volumes on parallel strips run across a tract of timber) the model would take the form:

$$\hat{Y} = \beta_0 + \beta_1 X + \beta_2 X^2 + \beta_3 X^3 + \cdots + \beta_p X^p + \varepsilon \quad (18.41)$$

while in the case of two-dimensional populations (e.g., a population of timber volumes or plots, as in Figure 16.1) the model would be

$$\hat{Y} = \beta_{00} + \beta_{10}\mu + \beta_{01}v + \beta_{20}\mu^2 + \beta_{11}uv + \beta_{02}v^2 + \cdots + \varepsilon \quad (18.42)$$

where $\mu$ refers to positions in the first dimension and $v$ to positions in the second dimension. DeLury suggested that the regression be fitted stepwise, using orthogonal polynomials,* but, with the aid of a computer, the fitting can be carried out using conventional polynomials. Terms should be added until reductions in the residual sum of squares cease being significant. The residual sum of squares remaining after the last significant term is added to the model is used to obtain the estimate of $\sigma_{Y \cdot X}^2$, as is shown in Equation 18.40. The degrees of freedom, $k = n - h - 1$, where $h$ is equal to the number of terms in the model which include $X$ or one of its powers.

DeLury assumed that the systematic sample would have a central start and that each observation of $Y$ would lie at the center of an interval $L$ units wide. Consequently, $L$ would have to be an odd number. He suggested that an estimate of the population total be obtained by integrating the regression equation. Thus, in the case of a one-dimensional population:

* A discussion of orthogonal polynomials is beyond the scope of this book. Reference is made to Fisher and Yates (1957), Wishart and Metakides (1953), or most standard statistical texts.

$$T = \int_{1/2}^{nL+1/2} g(x)dx \tag{18.43}$$

where $g(x)$ refers to the regression equation. The estimate of the population mean follows:

$$\bar{y} = T / N = T / nL \tag{18.44}$$

DeLury's proposal has drawn considerable interest. Finney (1952) criticized it sharply, doubting that it would ever lead to trustworthy results. Wheeler (1956) described how it was being used in the portion of the Forest Survey of the United States under the supervision of the Southern Forest Experiment Station. Sternitzke (1960; 1962) mentions its use in the Arkansas and Tennessee surveys. Swindel and Bamping (Swindel, 1961; Swindal and Bamping, 1963) tested it on a set of data from Georgia and obtained mixed results. The most exhaustive test of the procedure was carried out by Sayn-Wittgenstein (1966). The results of his work, like those of Swindel and Bamping, are mixed. His opinion was that it would not be generally satisfactory for forest surveys. He explained that its failures were probably due to forests not occurring in such a manner so that they can be described by smooth curves. Transitions from one condition to another are often too abrupt to permit a curve to follow the changes. He found, as did Swindel and Bamping, that in many cases the residual sum of squares increased, instead of decreasing, as the number of polynomial terms increased. He expected that a given polynomial would better describe a set of observations over a short rather than a long distance. Test results did not support that hypothesis. The only success be experienced in his series of tests of DeLury's proposal was in a large, undisturbed, homogeneous forest of overmature white spruce growing on a floodplain in northern Alberta. He attributed this result to slow changes in soil and microclimate as the distance from the stream edge increased. He states,

> Probably trends exist in all forests, but they are of such short duration that they cannot be detected by sampling observations at the intervals usual in forest surveys.

Many workers involved with forest inventories have found evidence of trends and have made use of regressions to remove such trends. Much of the early work was carried out by Scandinavian foresters such as Lindeberg (1923), Lappi-Seppälä (1924), Langsaeter (1926; 1927; 1932), Östlind (1932), and Näslund (1930; 1939). Matérn (1947; 1960) reviewed these efforts and suggested newer approaches to the problem.

The acquisition of an estimate of $\sigma^2_{Y \cdot X}$ does not solve the problems of computing a meaningful estimate of the variance of the mean, which are discussed in Sections 18.10, 18.10.1, and 18.10.2. Those who have used DeLury's and other regression procedures to remove the effects of trend have apparently simply fallen back on the assumption that the population of residuals from the regression are randomly distributed and that $s^2_{Y \cdot X}$ can be used in the expression.

$$s^2_{\bar{y}} = \frac{s^2_{y \cdot x}}{n}\left(\frac{N-n}{N}\right) \tag{18.45}$$

### 18.9.6 Removing Trend Effects without Using Regression

Several procedures have been suggested to remove the effect of linear trend without resorting to the development of regression equations. The simplest of these was proposed by Madow (1953). It is known as the *centered systematic sampling* method. In essence, Madow uses a central start under Approach 2. If $N = MA$ and $L$ is an odd number, each chosen variate is the median of its cluster. If no random component is present, the median will be equal to the cluster mean and the mean of the cluster means will be equal to the population mean. If, however, $L$ is an even number, there will be no median position or median and the chosen variates will not be equal to the means of their respective clusters. The chosen variates will be either all less than or all greater than their cluster means. Since the start is arbitrary, only one sample is possible. Consequently, the sample mean will be biased. In any case, there will be no way by which a variance of the mean can be determined.

Yates (1948) suggested the use of *end corrections* to eliminate linear trend effects. In this procedure a random start is made in the range 1 to $L$. The starting position is labeled $i$. All the observations, with the exception of the first and the last, are given a weight of 1. The weight of the first is given as

$$w_1 = 1 + \left[ \frac{A(2i - M - 1)}{2(A-1)M} \right] \tag{18.46}$$

while that for the last observation in the cluster is

$$w_{\text{Last}} = 1 - \left[ \frac{A(2i - M - 1)}{2(A-1)M} \right] \tag{18.47}$$

If no residual variance about the trend line is present, this weighting process results in sample means that are equal to the population mean and thus the $V(\bar{y}_o)$ will be equal to zero.

One can apply this to the 50-item population shown in Figure 18.8. In this population the variance is caused by trend alone. Then $N = 50$, $L = 5$, and, under Approach 1, $M = 5$ and $A = 10$. If $i = 1$,

$$w_1 = \frac{1 + 10[2(1) - 5 - 1]}{2(10-1)5} = 0.555556$$

and

$$w_{46} = \frac{1 - 10[2(1) - 5 - 1]}{2(10-1)5} = 1.444444$$

Then

$$w_1 y_1 = 0.555556(10.5) = 5.833333$$

$$w_{46} y_{46} = 1.444444(33.0) = 47.666665$$

$$\sum_j^A w_j y_j = 227.5$$

and

$$\bar{y}_i = \sum_j^A w_j y_j \Big/ \sum_j^A w_j = 227.5/10 = 22.75 = \mu_o$$

This same result would be obtained from each of the five possible samples, causing $V(\bar{y}_o)$ to be equal to zero.

When end corrections are used with a population with residual variance, the sample means will not equal the population mean but the variance of the mean would be considerably reduced from that obtained when end corrections are not used. For example, using the 50-item population with residual variance, when $M = 5$ and $A = 10$:

| | | First Observations | | | Last Observations | | | |
|---|---|---|---|---|---|---|---|---|
| Cluster | $i$ | $w$ | $y_{i1}$ | $wy_{i1}$ | $w$ | $y_{i10}$ | $wy_{i10}$ | $\bar{y}'_i$ |
| 1 | 1 | 0.555556 | 12.5 | 6.944445 | 1.444444 | 35.0 | 50.555554 | 22.75 |
| 2 | 2 | 0.777778 | 11.0 | 8.555556 | 1.222222 | 33.5 | 40.944444 | 23.35 |
| 3 | 3 | 1.000000 | 10.0 | 10.000000 | 1.000000 | 33.5 | 33.500000 | 22.50 |
| 4 | 4 | 1.222222 | 13.5 | 16.500000 | 0.777778 | 35.5 | 27.611112 | 22.61 |
| 5 | 5 | 1.444444 | 12.5 | 18.055555 | 0.555556 | 35.5 | 19.722224 | 22.68 |

The set of weighted variables would then be

**Y′ Values**

| Approach 1 Clusters | Approach 2 — Cluster 1 | 2 | 3 | 4 | 5 | 6 | 7 | 8 | 9 | 10 | Approach 1 $\mu_i$ | $\sigma_i^2$ |
|---|---|---|---|---|---|---|---|---|---|---|---|---|
| 1 | 6.9 | 12.5 | 13.5 | 17.5 | 22.0 | 23.5 | 25.0 | 26.5 | 29.5 | 50.6 | 22.75 | 131.08 |
| 2 | 8.6 | 15.0 | 16.5 | 20.0 | 21.0 | 23.5 | 27.5 | 29.0 | 31.5 | 40.9 | 23.35 | 77.45 |
| 3 | 10.0 | 13.5 | 16.5 | 19.5 | 21.0 | 24.5 | 26.5 | 28.5 | 31.5 | 33.5 | 22.50 | 54.25 |
| 4 | 16.5 | 14.0 | 18.0 | 19.0 | 21.0 | 23.5 | 26.5 | 28.5 | 31.5 | 27.6 | 22.61 | 30.29 |
| 5 | 18.1 | 15.0 | 17.5 | 19.5 | 22.0 | 25.0 | 27.5 | 31.0 | 31.5 | 19.7 | 22.68 | 30.19 |
| **Approach 2** | | | | | | | | | | | | |
| $\mu_i$ | 12.02 | 14.00 | 16.40 | 19.10 | 21.40 | 24.00 | 26.60 | 28.70 | 31.10 | 34.46 | | |
| $\sigma_i^2$ | 19.81 | 0.90 | 2.44 | 0.74 | 0.24 | 0.40 | 0.84 | 2.06 | 0.64 | 113.56 | | |

| Approach | $\mu'_o$ | $\sigma_w^{2'}$ | $\sigma_B^{2'}$ | $V(\bar{y}_o)'$ |
|---|---|---|---|---|
| 1 | 22.78 | 64.6520 | 0.0886 | $0.0886 = V(\bar{y}_o)'_{\text{Sys1}}$ |
| 2 | 22.78 | 14.1630 | 50.5789 | $1.4163 = V(\bar{y}_o)'_{\text{Sys2}}$ |

Compare $\mu'_o = 22.78$ with $\mu_o = 22.78$. There is no difference. Compare $V(\bar{y}_o)'_{\text{Sys1}} = 0.0886$ with $V(\bar{y}_o)_{\text{Sys1}} = 0.4186$. The improvement is substantial. However, when $V(\bar{y}_o)'_{\text{Sys2}} = 1.4163$ is compared to $V(\bar{y}_o)_{\text{Sys2}} = 0.1064$, it becomes apparent that the use of end corrections will have an adverse effect on $V(\bar{y}_o)'_{\text{Sys2}}$.

*Balanced systematic sampling* is a procedure developed by Sethi (1965) to overcome the problem of linear trends in populations being sampled. If the sample size, $n$, is an even number, the procedure yields sample means that are equal to the population mean and a variance of the mean equal to zero. If the sample size is odd, the results will not be as neat.

In Sethi's procedure the initial step is to divide the sampling frame into $n/2$ groups of $2L$ consecutive units each. For example, if $N = 50$, $L = 5$, and $n = 10$, the frame would be divided into $n/2 = 10/2 = 5$ groups, each containing 10 elements. Group 1 would consist of the elements in positions 1 through 10; group 2, positions 11 through 20; group 3, positions 21 through 30; group 4, positions 31 through 40; and group 5, positions 41 through 50. Two elements are chosen from each group. These elements occupy corresponding positions from each end of their respective groups. The positions are determined by a random draw from the range 1 to $L$. This draw fixes the position of the initial element ($i$). In the case of the 50-item population shown in Figure 18.8, the possible samples are as follows:

**Sample Positions**

| $i$ | Group | | | | |
|---|---|---|---|---|---|
| **Cluster** | 1 | 2 | 3 | 4 | 5 |
| 1 | 1,10 | 11,20 | 21,30 | 31,40 | 41,50 |
| 2 | 2, 9 | 12,19 | 22,29 | 32,39 | 42,49 |
| 3 | 3, 8 | 13,18 | 23,28 | 33,38 | 43,48 |
| 4 | 4, 7 | 14,17 | 24,27 | 34,37 | 44,47 |
| 5 | 5, 6 | 15,16 | 25,26 | 35,36 | 45,46 |

Applying this to the population that possesses no random component:

| Cluster | *Y* Values | | | | | | | | | | $\mu_i$ |
|---|---|---|---|---|---|---|---|---|---|---|---|
| 1 | 10.5 | 15.0 | 15.5 | 20.0 | 20.5 | 25.0 | 25.5 | 30.0 | 30.5 | 35.5 | 22.75 |
| 2 | 11.0 | 14.5 | 16.0 | 19.5 | 21.0 | 24.5 | 26.0 | 29.5 | 31.0 | 34.5 | 22.75 |
| 3 | 11.5 | 14.0 | 16.5 | 19.0 | 21.5 | 24.0 | 26.5 | 29.0 | 31.5 | 34.0 | 22.75 |
| 4 | 12.0 | 13.5 | 17.0 | 18.5 | 22.0 | 23.5 | 27.0 | 28.5 | 32.0 | 33.5 | 27.75 |
| 5 | 12.5 | 13.0 | 17.5 | 18.0 | 22.5 | 23.0 | 27.5 | 28.0 | 32.5 | 33.0 | 27.75 |

As can be seen, the mean of the cluster means is equal to the overall population mean, indicating that the estimates are unbiased. Furthermore, since all the means have the same magnitude, $\sigma_B^2 = 0.00$ and $V(\bar{y}_o)_{\text{Sethi}} = 0.00$.

Singh et al. (1968) developed a procedure, similar to Sethi's, which is known as *modified systematic sampling*. The results of this procedure are similar to those obtained using Sethi's method.

The procedures developed by Madow, Yates, Sethi, and Singh et al. are valuable when a linear trend is present in the population of interest but its parameters, $\beta_0$ and $\beta_1$, are not known.

### 18.9.7 Effect of a Naturally Stratified Population

When a population is naturally stratified and a systematic sample is appropriately obtained (e.g., cruise lines are run at right angles to the stream–ridge trend, etc.), the sample is naturally stratified with an approximate proportional allocation. However, it is questionable if its estimate would be as precise as one obtained using stratified random sampling when the stratification followed the natural divisions within the population.

As an example, consider, for simplicity's sake, that the plots in column 10 of the population in Figure 16.3 form a population. This population contains four natural strata.

| Plot | 1 | 2 | 3 | 4 | 5 | 6 | 7 | 8 | 9 | 10 | 11 | 12 | 13 | 14 | 15 | 16 | 17 | 18 | 19 | 20 |
|---|---|---|---|---|---|---|---|---|---|---|---|---|---|---|---|---|---|---|---|---|
| Stratum | D | D | D | C | C | C | C | C | B | B | B | B | B | A | A | A | A | A | A | A |
| *Y* | 2 | 10 | 23 | 40 | 48 | 30 | 49 | 44 | 60 | 46 | 55 | 73 | 69 | 76 | 88 | 92 | 86 | 57 | 65 | 80 |

Then, if $L = 5$, $M = 4$, and $A = 5$:

| Approach 1 Clusters | Approach 2 — Clusters | | | | Y Values Approach 1 | | Proportion in | | | |
|---|---|---|---|---|---|---|---|---|---|---|
| | 1 | 2 | 3 | 4 | $\mu_i$ | $\sigma_i^2$ | A | B | C | D |
| 1 | 2 | 30 | 55 | 92 | 44.75 | 1,095.6875 | 0.25 | 0.25 | 0.25 | 0.25 |
| 2 | 10 | 49 | 73 | 86 | 54.50 | 836.2500 | 0.25 | 0.25 | 0.25 | 0.25 |
| 3 | 23 | 44 | 69 | 57 | 48.25 | 290.6875 | 0.25 | 0.25 | 0.25 | 0.25 |
| 4 | 40 | 60 | 76 | 65 | 60.25 | 170.1875 | 0.50 | 0.25 | 0.25 | 0.00 |
| 5 | 48 | 46 | 88 | 80 | 65.50 | 350.7500 | 0.50 | 0.25 | 0.25 | 0.00 |

**Approach 2**

| | 1 | 2 | 3 | 4 |
|---|---|---|---|---|
| $\mu_i$ | 24.60 | 45.80 | 72.20 | 76.00 |
| $\sigma_i^2$ | 302.2400 | 92.9600 | 114.1600 | 170.8000 |

| Approach | $\mu_i$ | $\sigma_o^2$ | $\sigma_w^2$ | $\sigma_B^2$ | $V(\bar{y}_o)$ |
|---|---|---|---|---|---|
| 1 | 54.65 | | 548.7125 | 57.6150 | $57.6150 = V(\bar{y}_o)_{\text{Sys1}}$ |
| 2 | 54.65 | | 170.0400 | 436.2875 | $42.5100 = V(\bar{y}_o)_{\text{Sys2}}$ |
| Random | 54.65 | 606.3275 | | | $127.6479 = V(\bar{y})_{\text{Ran}}$ |
| Natural stratification (see Equation 16.25) | | | | | $27.6150 = V(\bar{y})_{\text{Strat.Nat.}}$ |
| Arbitrary stratification (see Equation 16.25) | | | | | $42.5100 = V(\bar{y})_{\text{Strat.Arb.}}$ |

The actual proportions of the population in the strata are *A*, 0.35; *B*, 0.25; *C*, 0.25; and *D*, 0.15. As can be seen, the five possible samples have proportions that are similar to these population values. If the population and samples were larger, the correspondence would be better.

As can be seen, when the population contains natural strata, systematic sampling, under both approaches, yields estimates of the variance of the overall mean that are more precise than that which would be obtained using simple random sampling. However, the precision of a naturally stratified random sample will be greater than that of the systematic sample under either approach. If the stratification had been arbitrary, using strata of size $N_i = 5$, the variance of the mean would be precisely what was obtained under Approach 2. The gain of efficiency associated with natural stratification is caused by the fact that the stratum variances are, on the average, smaller because the variates within the natural strata are more homogeneous than those within the arbitrary strata.

### 18.9.8 Effects of Periodicity

When the distribution of variates within the population is such that periodicity is present, as described in Section 18.7.3, the efficiency of systematic sampling, compared with that of simple random sampling, is dependent on the sampling interval, *L*. If *L* is equal to the period of the population, as is shown in Figure 18.4A, the sample will be made up of variates occurring at the same relative position in each cycle and consequently possessing the same magnitude. As a result, the sample variance will be equal to zero. Since the sample variance is the within-cluster variance in the case of Approach 1, $\sigma_w^2 = 0$ and $\sigma_B^2 = \sigma_o^2$. This makes sense because across all possible systematic samples with the same *L*, each sample mean will correspond to a given variate. Thus, the mean and variance of the sample means will be the same as the mean and variance of the variates. Then,

$$V(\bar{y}_o)_{\text{Sys1}} = \sigma_B^2 = \sigma_o^2 > V(\bar{y}_o)_{\text{Ran}} = \frac{\sigma_o^2}{n}\left(\frac{N-n}{N-1}\right) \tag{18.48}$$

Under Approach 2, the sample variance is an estimate of the between-cluster variance. Since each cluster contains the same mix of variates, the cluster means are all equal. Consequently, $\sigma_B^2 = 0$ and $\sigma_W^2 = \sigma_o^2$. Then, since $n = M$

$$V(\bar{y}_o)_{\text{Sys2}} = \frac{\sigma_w^2}{M} = \frac{\sigma_o^2}{n} > V(\bar{y}_o)_{\text{Ran}} = \frac{\sigma_o^2}{n}\left(\frac{N-n}{N-1}\right) \tag{18.49}$$

As can be seen, simple random sampling is more efficient than systematic sampling, regardless of approach, when the sampling interval is equal to the population period.

When $L$ is an integral multiple of the population period, regardless of the magnitude of the multiple, the effects will be the same as they are when $L$ is equal to the population period.

When $L$ is equal to one half the population period, as in Figure 18.4B and when $N$ is an integral multiple of $L$, there will be two observations per period, one greater and one less than the population mean. Their differences from the population mean will be equal in magnitude but opposite in sign so they cancel each other, leading the sample mean to be equal to the population mean. Then, under Approach 1, $\sigma_B^2 = 0$ and

$$V(\bar{y}_o)_{\text{Sys1}} = \sigma_B^2 = 0 < V(\bar{y}_o)_{\text{Ran}} = \frac{\sigma_o^2}{n}\left(\frac{N-n}{N-1}\right) \tag{18.50a}$$

Under Approach 2, $\sigma_w^2 = 0$ and

$$V(\bar{y}_o)_{\text{Sys2}} = \frac{\sigma_w^2}{M} = 0 < V(\bar{y}_o)_{\text{Ran}} = \frac{\sigma_o^2}{n}\left(\frac{N-n}{N-1}\right) \tag{18.50b}$$

When $L$ is equal to one third the population period, there are three observations per period. The mean of these observations is equal to the population mean. Thus, as before, under Approach 1, $\sigma_B^2 = 0$, and under Approach 2, $\sigma_w^2 = 0$, and Equations 18.50a and b are applicable. Furthermore, they are applicable whenever $L$ is an integral submultiple of the population period.

In most cases the sampling interval is such that the sample is not in step with the population period. When this situation occurs, one can generalize only to the extent that the efficiency of systematic sampling declines as the closeness of the in-step relationship increases. However, because of the possibility of obtaining a nonrepresentative sample and the uncertainty regarding its efficiency, if there is any reason to believe that periodicity is present in the population being sampled and its period is not known, it would be preferable to use random, rather than systematic, sampling.

### 18.9.9 Effect of Autocorrelation

As was discussed in Section 18.7.6, autocorrelation may be present in the population being sampled. Evidence of serial correlation in natural population has been found. For example, Osborne (1942), using the transect method to determine the areas of forest types, found that the length of line within a given type changed from transect to transect and that the serial correlation between such line lengths decreased exponentially as the spacing of the transects was increased. Similar results were obtained by Matérn (1947; 1960), working with data from the Swedish National Forest Survey. This pattern can be visualized if one considers the latitudinal zonation of forest types on a mountain ridge. If the transects are run parallel to the trend of the ridge, transects that are close together are going to be more nearly alike than transects that are separated by a wide interval. Sayn-Wittgenstein (1966), using timber volume data from sets of plots in Alberta and Oregon, found that autocorrelation was present between plot volumes but it decayed so rapidly that the effects were negligible beyond an interval of two plots. Distribution patterns in forests that give rise to autocorrelation are the results of many factors. Among these are site conditions, common plant parentage, and the activities of humans. The presence of such patterns is probably universal but the degree of serial correlation is usually quite low.

If, in the case of a specific population, the serial correlations are computed as $j$ is increased and then these serial correlations are plotted against $j$, the resulting scattergram is known as a *correlogram*. It has been shown (Sukhatmi et al., 1984) that, on the average, the variance of an arbitrarily stratified random sample will be less than or equal to that of a simple random sample when the trend of the correlogram decreases as $j$ increases. A similar general statement cannot be made with respect to the difference between the variances of systematic and other types of samples. However, Cochran (1946), using theoretical populations, proved that when the correlogram is concave upward systematic sampling yields smaller variances than arbitrarily stratified random samples. In practice, of course, the shape of the correlogram is not known and, consequently, little can be said relative to the efficiencies of the different sampling procedures in the presence of serial correlation. What can be said is that systematic samples tend to be more representative of the populations being sampled than are simple or stratified random samples because, for a given sample size, the distance between the systematically chosen elements is kept as great as possible, thus keeping the effects of autocorrelation at a minimum.

## 18.10 ESTIMATION OF THE PARAMETERS

Here, again, recognition must be made of the differences between the two approaches to systematic sampling.

### 18.10.1 UNDER APPROACH 1

Under this approach the sample consists of *one* of the $M$ available clusters and no subsampling is done. Thus, $m = 1$ and $n = N_i$. Since no subsampling is involved, the sample total, mean, and variance are based on *all* the variates within the chosen cluster. Consequently, they are parameters, not estimates and are, respectively, $\tau_i$, $\mu_i$, and $\sigma_i^2$. Then

$$\bar{T} = \frac{1}{m}\sum_i^m T_i = \frac{1}{1}\sum_i^1 \tau_i = \tau_i \tag{18.51}$$

If the start is *random*, $\bar{T}$ is an unbiased estimator of $\bar{\tau}$:

$$E(\bar{T}) = \frac{1}{M}\sum_i^M \tau_i = \bar{\tau} \tag{18.52}$$

However, if the start is *arbitrary*, $\bar{T}$ is biased because its expected value is equal to $\tau_i$, not to $\bar{\tau}$:

$$E(\bar{T}) = \frac{1}{1}\sum_i^M \tau_i = \tau_i \tag{18.53}$$

When the start is arbitrary, $M = 1$ because the cluster sampling frame contains only the clusters *available for sampling* and when an arbitrary start is used only one cluster is available. By specifying the position of the start, all the alternative clusters are excluded from consideration and, in essence, are removed from the frame.

In the same manner, the estimators of the overall total and mean will be *unbiased if the start is random* but *biased if the start is arbitrary*. According to Equation 17.30, the estimate of the overall total is

$$T_o = MT$$

In this case, since $\overline{T} = \tau_i$

$$T_o = M\tau_i \tag{18.54}$$

The estimate of the overall mean, according to Equation 17.34, is

$$\bar{y}_o = T_o / N$$

which becomes

$$\bar{y}_o = (M / N)\tau_i \tag{18.55}$$

Note that $\bar{y}_o$ is a weighted mean, reflecting the fact that the cluster sizes may differ. If the cluster size is constant across all the clusters in the sampling frame, the weights associated with the cluster means are constant and the estimate of the overall mean can be computed using

$$\bar{y}_o = \frac{1}{m}\sum_{i}^{m}\mu_i = \frac{1}{1}\sum_{i}^{1}\mu_i = \mu_i \tag{18.56}$$

while the estimate of the overall total is

$$T_o = N\bar{y}_o = N\mu_i \tag{18.57}$$

According to Equation 17.58, the variance of the overall total is

$$s_{T_o}^2 = \frac{M}{m}\sum_{i}^{m}\left[\frac{s_i^2}{n_i}\left(\frac{N_i - n_i}{N_i}\right)N_i^2\right] + M^2\frac{s_T^2}{m}\left(\frac{M-m}{M}\right)$$

Since $m = 1$ and $n_i = N_i$,

$$s_{T_o}^2 = M^2 s_T^2\left(\frac{M-1}{M}\right) = Ms_T^2(M-1) \tag{18.58}$$

According to Equation 17.38,

$$s_T^2 - \frac{1}{m-1}\sum_{i}^{m}(T_i - \overline{T})^2$$

and since $m = 1$, $T_i = \tau_i$, and $\overline{T} = \tau_i$,

$$s_T^2 = \frac{1}{1-1}\sum_i^1 (\tau_i - \tau_i)^2 = \frac{0}{0} \tag{18.59}$$

Substituting 0/0 for $s_T^2$ in Equation 18.58 yields

$$s_{T_o}^2 = M\left(\frac{0}{0}\right)(M-1) = \frac{0}{0}$$

which is indeterminate. As can be seen, an estimate cannot be made of the variance of the overall total. In the same manner, an estimate cannot be made of the variance of the overall mean. According to Equation 17.78,

$$s_{\bar{y}_o}^2 = s_{T_o}^2 / N^2 = \frac{0}{0} \Big/ N^2 = \frac{0}{0} \tag{18.60}$$

which again is indeterminate. Obviously, one cannot compute variances of the overall total or mean using the expressions for Approach 1.

### 18.10.2 Under Approach 2

When Approach 2 is being used, the sample is made up of *one element* from *each* of the *M clusters*. Consequently, $m = M$ and $n_i = 1$. The sample size $n = M$. In the following discussion the subscript $j$ refers to the same position within each of the clusters.

According to Equation 17.15, the estimated mean of cluster $i$ is

$$\bar{y}_i = \frac{1}{n_i}\sum_j^{n_i} y_{ij} = \frac{1}{1}\sum_j^1 y_{ij} = y_i \tag{18.61}$$

This estimate of $\bar{y}_i$ is unbiased if the start is random:

$$E(\bar{y}_i) = \frac{1}{N_i}\sum_j^{N_i} y_{ij} = \mu_i \tag{18.62}$$

If the start is arbitrary, the estimate of the cluster mean is biased since only one position within each cluster can be used. Then, its expected value is equal to $y_{ij}$, not $\mu_i$:

$$E(\bar{y}_i) = \frac{1}{1}\sum_j^1 y_{ij} = y_{ij} \tag{18.63}$$

The estimate of the cluster total, according to Equation 17.16, is

$$T_i = N_i\bar{y}_i = N_i y_{ij} \tag{18.64}$$

As in the case of the cluster mean, the cluster total will be unbiased if the start is random but biased if the start is arbitrary.

The estimate of the cluster variance, according to Equation 17.18, is

$$s_i^2 = \frac{1}{(n_i - 1)} \sum_j^{n_i} (y_{ij} - \bar{y}_i)^2 = \frac{1}{1-1} \sum_j^{1} (y_{ij} - y_{ij})^2 = \frac{0}{0} \tag{18.65}$$

which is indeterminate. Consequently, the estimator of the variances of the cluster mean and the cluster total are also indeterminate.

According to Equation 17.28, the estimate of the mean cluster total is

$$\bar{T} = \frac{1}{m} \sum_i^m T_i = \frac{1}{M} \sum_i^M T_i = \frac{1}{M} \sum_i^M N_i y_{ij} \tag{18.66}$$

and, according to Equation 17.30, the estimate of the overall total is

$$T_o = M\bar{T} = \sum_i^M N_i y_{ij} \tag{18.67}$$

Then, according to Equation 17.35, the estimate of the overall mean is

$$\bar{y}_o = \frac{M}{Nm} \sum_i^m \frac{N_i}{n_i} \sum_j^{n_i} y_{ij} = \frac{1}{N} \sum_i^M N_i y_{ij} \tag{18.68}$$

It should be clear that $\bar{T}$, $T_o$, and $\bar{y}_o$ will be unbiased if the start is random but biased if the start is arbitrary. Limiting the sample to the elements occupying a specific position within each of the clusters means that the remaining elements can never be considered, effectively reducing the clusters to one-element sets.

The estimate of the variance of the cluster totals, according to Equation 17.38, is

$$s_T^2 = \frac{1}{m-1} \sum_i^m (T_i - \bar{T})^2 = \frac{1}{M-1} \sum_i^M (T_i - \bar{T})^2 \tag{18.69}$$

According to Equation 17.58 the estimate of the variance of the overall total is

$$s_{T_o}^2 = \frac{M}{m} \sum_i^m \left[ \frac{s_i^2}{n_i} \left( \frac{N_i - n_i}{N_i} \right) N_i^2 \right] + M^2 \frac{s_T^2}{m} \left( \frac{M - m}{M} \right)$$

Since $m = M$, $(M - M)/M = 0$, and

$$\begin{aligned} s_{T_o}^2 &= \sum_i^M \left[ s_i^2 \left( \frac{N_i - 1}{N_i} \right) N_i^2 \right] \\ &= \sum_i^M \left[ \frac{0}{0} \left( \frac{N_i - 1}{N_i} \right) N_i^2 \right] = \frac{0}{0} \end{aligned} \tag{18.70}$$

which is indeterminate. The estimate of the variance of the overall mean is then, according to Equation 17.78,

$$s_{\bar{y}_o}^2 = s_{T_o}^2 / N^2 = \frac{0}{0} \tag{18.71}$$

As can be seen, valid estimates of the variances of neither the overall total nor overall mean cannot be obtained when systematic sampling is used, regardless of the approach.

## 18.11 APPROXIMATIONS OF THE VARIANCE OF THE OVERALL MEAN

A number of attempts have been made to circumvent this problem of indeterminate estimates of the variance of the mean when systematic sampling is used. Prior to the selection of one of these approximation procedures, one should have an idea of the type of population being sampled because some of the procedures are based on assumptions regarding population type.

### 18.11.1 USE OF MULTIPLE RANDOM STARTS

The only way to obtain a valid estimate of the variance of the mean when using systematic sampling is to use two or more samples using random starts. This is simply using cluster sampling with more than one cluster. Each of the systematic samples is a cluster, as in Approach 1. The clusters are separate from one another but interlock. Consequently, they are referred to as *interpenetrating* samples. As an example, consider the 24-item population in Section 18.4.1. When $L = 4$ there are four possible clusters or samples, each containing six variates. Say that Samples 2 and 4 are randomly chosen. Then the first cluster (Sample 2) contains items 2, 6, 10, 14, 18, and 22 while the second cluster (Sample 4) contains items 4, 8, 12, 16, 20, and 24. The computations leading to the estimate of the variance of the overall mean follow the standard cluster-sampling procedure.

According to Gautschi (1957), the use of multiple random starts was probably first suggested by J. W. Tukey. It is now the solution most often suggested by authorities in the field (e.g., Gautschi, 1957; Raj, 1972; Som, 1973; and Sukhatmi et al., 1984). It was first suggested in a forestry context by Shiue (1960).

While statistically valid, the use of multiple random starts has drawbacks. First, two or more independent samples must be obtained, which involves two or more passes through the universe of interest (i.e., if a timber inventory is being made, two or more separate timber cruises would have to be carried out). Since a large portion of the cost of any sample is that associated with travel time between observations, the cost of an inventory involving multiple random starts would be considerably greater than a single sample involving the same overall sampling intensity. Second, there may be very little physical separation of the two or more clusters, which could lead to essentially the same results and a misleadingly small between cluster variance. Finally, since there is only one degree of freedom available per cluster, the number of degrees of freedom available for computation of the variance of the sample are minimal, which would make the estimate of the variance of the mean inefficient.

### 18.11.2 USING RANDOM SAMPLING FORMULAE

The most commonly used of the approximation procedures is based on the assumption that the population is randomly distributed and that systematic samples can therefore be considered the equivalent of random samples. This leads to the assumption that the sample variance, $s_Y^2$, can be substituted into Equation 12.12 to obtain the needed estimate:

$$s_{\bar{y}_o}^2 = \frac{s_Y^2}{n}\left(\frac{N-n}{N}\right) \to \frac{\sigma_o^2}{n}\left(\frac{N-n}{N-1}\right) \tag{18.72}$$

If the systematic sampling is being carried out under Approach 1, the sample includes *all* the elements in *one* of the clusters in the frame. Thus,

$$\bar{y}_i = \frac{1}{N_i}\sum_j^{N_i} y_{ij} \tag{18.73}$$

and

$$s^2 = \frac{1}{(N_i - 1)}\sum_j^{N_i}(y_{ij} - \bar{y}_i)^2$$

However, since *all* the elements in the chosen cluster are included in the sample,

$$\bar{y} = \mu_i \tag{18.74}$$

and

$$s^2 = \sigma_i^2\left(\frac{N_i}{N_i - 1}\right) \tag{18.75}$$

Furthermore, since $\bar{y} = \mu_i$, there is no loss of a degree of freedom, and thus

$$s^2 = \sigma_i^2 = \frac{1}{N_i}\sum_j^{N_i}(y_{ij} - \mu_i)^2 \tag{18.76}$$

To keep the mathematics as simple as possible it will be assumed, from this point on, that $N = MA$ and $N_i = A$. Then, the within cluster variance is

$$\sigma_w^2 = \frac{1}{M}\sum_i^M \sigma_i^2 \tag{18.77}$$

The sample variance, $s^2 = \sigma_i^2$, is an estimate of $\sigma_w^2$, *not* $\sigma_o^2$. Since $\sigma_w^2$ is a component of $\sigma_o{}^2$, it can never exceed $\sigma_o^2$. Consequently, if $\sigma_w^2$ is substituted for $\sigma_o^2$ in the expression

$$V(\bar{y}_o)_{\text{Ran}} = \frac{\sigma_o^2}{n}\left(\frac{N-n}{N-1}\right) \tag{18.78}$$

the resulting estimate will be biased *downward*, except when $\sigma_w^2 = \sigma_o^2$. This can be seen in the following example which is based on the population of size $N = 24$ used in Section 18.9.1, with a variance $\sigma_o^2 = 0.248254$. With a sample size, $n = 6$, $V(\bar{y}_o)_{\text{Ran}} = 0.032381$. According to Equation 17.105

$$\sigma_w^2 = \sigma_o^2\left(\frac{A-1}{A}\right)(1-\rho)$$

Then, assuming three different intraclass correlation coefficients,

| | $\rho = -0.2$ | $\rho = \rho_{crit} = -0.0434783$ | $\rho = 0.2$ |
|---|---|---|---|
| $\sigma_w^2 =$ | 0.248254 | 0.215873 | 0.165503 |
| $\frac{\sigma_w^2}{6}\left(\frac{24-6}{24-1}\right) =$ | 0.032381 | 0.028157 | 0.021587 |
| $V(\bar{y}_o)_{\text{Ran}} =$ | 0.032381 | 0.032381 | 0.032381 |
| Bias | 0.000000 | −0.004224 | −0.010794 |

It should be noted that in this case, $\sigma_w^2 = \sigma_o^2$ when $\rho = -0.2$. Furthermore, when $\sigma_w^2 = \sigma_o^2$, there is no bias but in all other cases there is negative bias.

The sample variance can be used to obtain an unbiased estimate if Equation 18.78 is suitably modified. If

$$\sigma_w^2 = \sigma_o^2\left(\frac{A-1}{A}\right)(1-\rho)$$

then

$$\sigma_o^2 = \sigma_w^2\left(\frac{A}{A-1}\right)\left(\frac{1}{1-\rho}\right) \tag{18.79}$$

and since $n = A$,

$$V(\bar{y}_o)_{\text{Ran}} = \frac{\sigma_w^2}{A-1}\left(\frac{1}{1-\rho}\right)\left(\frac{N-A}{N-1}\right) \tag{18.80}$$

Then, since $s^2 = \sigma_i^2$, which is an estimate of $\sigma_w^2$,

$$s_{\bar{y}_{o\text{Ran}}}^2 = \frac{\sigma_i^2}{A-1}\left(\frac{1}{1-\rho}\right)\left(\frac{N-A}{N-1}\right) \xrightarrow{\text{unbiased}} V(\bar{y}_o)_{\text{Ran}} \tag{18.81}$$

This can be demonstrated using the data from the population introduced in Section 18.9.1 in which: $N = 24$; $\sigma_o^2 = 0.248254$; $\rho = 0.026538$; and $V(\bar{y}_o)_{\text{Ran}} = 0.032381$. Under Approach 1, $M = 4$ and $A = 6$. Then,

| Cluster | $\sigma_i^2$ | $s_{\bar{y}_{o\text{Ran}}}^2$ |
|---|---|---|
| 1 | 0.2222 | 0.0357273 |
| 2 | 0.2222 | 0.0357273 |
| 3 | 0.1389 | 0.0223336 |
| 4 | 0.2222 | 0.0357273 |
| | | 0.1295153 |

and 0.1295153/4 = 0.032379. The difference between this outcome and $V(\bar{y}_o)_{\text{Ran}} = 0.032381$ is caused by rounding errors. The big problem with using Equation 18.81 is that it requires a knowledge of ρ and that is rarely, if ever, available.

The proper expression for the variance of the overall mean when systematic sampling under Approach 1 is being used is

$$V(\bar{y}_o)_{\text{Sys1}} = \sigma_B^2$$

which, according to Equation 17.103, can be written

$$V(\bar{y}_o)_{\text{Sys1}} = \frac{\sigma_o^2}{A}[1 + \rho(A-1)]$$

According to Equation 18.74,

$$\sigma_o^2 = \sigma_w^2\left(\frac{A}{A-1}\right)\left(\frac{1}{1-\rho}\right)$$

Substituting this into Equation 17.103 yields

$$V(\bar{y}_o)_{\text{Sys1}} = \frac{\sigma_w^2}{A-1}\left[\frac{1+\rho(A-1)}{1-\rho}\right] \tag{18.82}$$

Since $s^2 = \sigma_i^2$ and $\sigma_i^2$ is an unbiased estimate of $\sigma_w^2$,

$$s^2_{\bar{y}_o\text{Sys1}} = \frac{\sigma_i^2}{A-1}\left[\frac{1+\rho(A-1)}{1-\rho}\right] \xrightarrow{\text{unbiased}} V(\bar{y}_o)_{\text{Sys1}} \tag{18.83}$$

Again, demonstrating using the population in Section 18.9.1, where $V(\bar{y}_o)_{\text{Sys1}} = \sigma_B^2 = 0.046879$, then

| Cluster | $\sigma_i^2$ | $s^2_{\bar{y}_o\text{Sys1}}$ |
|---|---|---|
| 1 | 0.2222 | 0.051709 |
| 2 | 0.2222 | 0.051709 |
| 3 | 0.1389 | 0.032324 |
| 4 | 0.2222 | 0.051709 |
| | | 4)0.187451 |
| | | 0.046863 |

The difference between $V(\bar{y}_o)_{\text{Sys1}} = 0.046879$ and 0.046863 is due to rounding errors. Again, an unbiased estimate of the variance of the overall mean can be made from a systematic sample, but it requires knowledge of the intraclass correlation coefficient.

If

$$\frac{\sigma_i^2}{A}\left(\frac{N-A}{N-1}\right)$$

is used to estimate $V(\bar{y}_o)_{\text{Sys1}}$ the direction and magnitude of the bias would again be dependent on the intraclass correlation coefficient, ρ. For example, using the 24-item population with $\sigma_o^2 =$ 0.248254 when $M = 4$ and $A = 6$,

| | $\rho = -0.2$ | $\rho = \rho_{crit} = -0.0434783$ | $\rho = 0.2$ |
|---|---|---|---|
| $\sigma_w^2 =$ | 0.248254 | 0.215873 | 0.165503 |
| $\frac{\sigma_w^2}{6}\left(\frac{24-6}{24-1}\right) =$ | 0.032381 | 0.028157 | 0.021587 |
| $V(\bar{y}_o)_{\text{Sys1}} = \sigma_{\bar{B}}^2 =$ | 0.000000 | 0.032381 | 0.082751 |
| Bias | +0.032381 | −0.004224 | −0.061164 |

Under the circumstances of this example, when $\rho = -0.0615385$ the bias is equal to zero. When $\rho < -0.0615385$ the bias is positive and when $\rho > -0.0615385$ the bias is negative.

As can be seen, in the absence of knowledge of the magnitude and sign of ρ, one cannot predict the magnitude and sign of the bias associated with the assumption that one can use the formulae appropriate for random sampling when the sampling is actually systematic. The situation when Approach 2 is used has not been explored. It is sufficient to say that the results would again be dependent on σ, which is not known.

### 18.11.3 Grouping Pairs of Clusters

Another procedure that has been suggested (Yates, 1949, 1981; Jessen, 1978) involves the grouping together of pairs of adjacent clusters and considering the groups as strata in a stratified sampling scheme. There would be two observations per stratum, permitting the computation of a within-stratum variance.

As an example one again uses the 24-item population described in Section 18.9.1. The basic frame, the cluster frame, the strata formed by the pairing of clusters, and the samples are shown below:

| Position: | 1 | 2 | 3 | 4 | 5 | 6 | 7 | 8 | 9 | 10 | 11 | 12 | 13 | 14 | 15 | 16 | 17 | 18 | 19 | 20 | 21 | 22 | 23 | 24 |
|---|---|---|---|---|---|---|---|---|---|---|---|---|---|---|---|---|---|---|---|---|---|---|---|---|
| y: | 0 | 1 | 1 | 0 | 1 | 1 | 1 | 0 | 0 | 1 | 1 | 1 | 1 | 0 | 0 | 1 | 0 | 1 | 1 | 0 | 0 | 0 | 1 | 0 |
| Cluster: | | | 1 | | | | 2 | | | | 3 | | | | 4 | | | | 5 | | | | 6 | |
| Stratum: | | | | | 1 | | | | | | | | 2 | | | | | | | | 3 | | | |
| Sample | | | | | | | | | | | | | | | | | | | | | | | | |
| 1 | 0 | | | | 1 | | | | 0 | | | | 1 | | | | 0 | | | | 0 | | | |
| 2 | | 1 | | | | 1 | | | | 1 | | | | 0 | | | | 1 | | | | 0 | | |
| 3 | | | 1 | | | | 1 | | | | 1 | | | | 0 | | | | 1 | | | | 1 | |
| 4 | | | | 0 | | | | 0 | | | | 1 | | | | 1 | | | | 0 | | | | 0 |

Thus, there are three strata, each with two observations.

$$s_{\bar{y}_i}^2 = \frac{1}{24^2}\sum_j^3\left[\frac{s_j^2}{2}\left(\frac{8-2}{8}\right)8^2\right]$$

| Sample | Stratum $j = 1$ | $j = 2$ | $j = 3$ | $s_{\bar{y}_i}^2$ |
|---|---|---|---|---|
| $i = 1$ | $\bar{y}_{11} = 0.5$ | $\bar{y}_{12} = 0.5$ | $\bar{y}_{13} = 0.0$ | |
| | $s_{11}^2 = 0.5$ | $s_{12}^2 = 0.5$ | $s_{13}^2 = 0.0$ | 0.041667 |
| $i = 2$ | $\bar{y}_{21} = 1.0$ | $\bar{y}_{22} = 0.5$ | $\bar{y}_{23} = 0.5$ | |
| | $s_{21}^2 = 0.0$ | $s_{22}^2 = 0.5$ | $s_{23}^2 = 0.5$ | 0.041667 |

| | Stratum | | | $s_{\bar{y}_i}^2 = \frac{1}{24^2}\sum_j^3 \left[\frac{s_j^2}{2}\left(\frac{8-2}{8}\right)8^2\right]$ |
|---|---|---|---|---|
| Sample | $j = 1$ | $j = 2$ | $j = 3$ | |
| $i = 3$ | $\bar{y}_{31} = 1.0$ | $\bar{y}_{32} = 0.5$ | $\bar{y}_{33} = 1.0$ | |
| | $s_{31}^2 = 0.0$ | $s_{32}^2 = 0.5$ | $s_{33}^2 = 0.0$ | 0.020833 |
| $i = 4$ | $\bar{y}_{41} = 0.0$ | $\bar{y}_{42} = 1.0$ | $\bar{y}_{43} = 0.0$ | |
| | $s_{41}^2 = 0.0$ | $s_{42}^2 = 0.0$ | $s_{43}^2 = 0.0$ | 0.000000 |
| $u_j =$ | 0.625 | 0.625 | 0.375 | |
| $\sigma_j^2 =$ | 0.234375 | 0.234375 | 0.234375 | |

$$V(\bar{y}_o)_{st} = \frac{1}{24^2}\sum_j^3 \left[\frac{\sigma_j^2}{2}\left(\frac{8-2}{8-1}\right)8^2\right] = 0.047619\sum_j^3 \sigma_j^2 = 0.033482$$

$$E(s_{\bar{y}}^2)\sum_i^4 s_{\bar{y}_i}^2 = 0.026042$$

As can be seen, the mean $s_{\bar{y}}^2$ is not equal to $V(\bar{y}_o)_{st}$. This result stems from the fact that the sampling was systematic, rather than random, within the strata. If all 28 possible samples of size 2 had been drawn from each stratum and an $s_{\bar{y}}^2$ computed for each of the $28^3 = 21,952$ possible stratified samples, the mean $s_{\bar{y}}^2$ would equal $V(\bar{y}_o)_{st}$. Thus, this procedure yields biased results but may be a useful way to arrive at an estimate of the variance of the overall mean. It should be pointed out that the bias can be either positive or negative.

In the case of two-dimensional samples using a square or rectangular pattern, Yates suggests that the sample be broken up into $2 \times 2$ unit blocks, each of which would be considered a stratum. This would take into account variability in both directions.

### 18.11.4 Method of Successive Differences

Von Neumann et al. (1941) suggested the use of the sum of squared differences between adjacent variates in the sample rather than the sum of squared residuals when computing the variances of the mean of stratified populations with homogeneous variances. The expression that was developed from this suggestion is

$$s\bar{y}_o^2 = \frac{1}{2n(n-1)}\left(\frac{N-n}{N}\right)\sum_i^{n-1}(y_{i+1} - y_i)^2 \tag{18.84}$$

Note that no degrees of freedom have been lost. This is because no means have been calculated.

As an example, use the 24-item population and the sampling interval $L = 4$ so that four samples, each containing six variates, are obtained. The estimate based on the first sample is

$$s_{\bar{y}_{o1}}^2 = \frac{1}{2(6)(6-1)}\left(\frac{24-6}{24}\right)[(0-1)^2 + (1-0)^2 + (0-1)^2 + (1-0)^2 + (0-0)^2] = 0.05$$

Similarly, for the remaining samples: $s_{\bar{y}_{o2}}^2 = 0.0375$; $s_{\bar{y}_{o3}}^2 = 0.025$; $s_{\bar{y}_{o4}}^2 = 0.025$. The expected value of the variance of the mean, using this procedure, is equal to the mean of the four estimates, 0.034375, which is greater than $V(\bar{y}_o)_{\text{Ran}} = 0.032381$, which indicates that positive bias is present. This agrees with general experience with the method.

Loetsch and Haller (1964) extended the idea to line–plot timber cruises where the cruise line interval and the between-plot interval are equal. In this procedure the differences are obtained within rows or within columns. After squaring and summing, the row (or column) totals are summed to obtain the sum of squared differences:

$$s_{\bar{y}_o}^2 = \frac{1}{2n}\left(\frac{N-n}{N}\right)\sum_i^m \frac{1}{(n_i - 1)} \sum_j^{n_i - 1} (y_{i(j+1)} - y_i)^2 \tag{18.85}$$

As an example, use the 16-ha tract shown in Figure 16.1. Assume $L = 4$ so that the cluster takes on a $4 \times 4$ pattern. If a random draw indicates that the starting position is row 3 and column 3, the sample is as follows:

| | Column | | | |
|---|---|---|---|---|
| **Row** | **3** | **7** | **11** | **15** |
| 3 | 47 | 86 | 68 | 44 |
| 7 | 40 | 81 | 90 | 63 |
| 11 | 21 | 75 | 65 | 32 |
| 15 | 0 | 48 | 25 | 29 |
| 19 | 3 | 25 | 2 | 4 |

There are $m = 4$ columns. Then,

$$\begin{aligned}
(47-40)^2 + (40-21)^2 + (21-0)^2 + (0-3)^2 &= 860 \\
(86-81)^2 + (81-75)^2 + (75-48)^2 + (48-25)^2 &= 1319 \\
(68-90)^2 + (90-65)^2 + (65-25)^2 + (25-2)^2 &= 3238 \\
(44-63)^2 + (63-32)^2 + (32-29)^2 + (29-4)^2 &= \underline{1956} \\
&\phantom{=} 7373
\end{aligned}$$

$$s_{\bar{y}_o}^2 = \frac{1}{2(20)}\left(\frac{320-20}{320}\right)\frac{7373}{(5-1)} = 43.20117$$

$$s_{\bar{y}_o}^2 = \pm\sqrt{43.201172} = \pm 6.572760 \text{ m}^3/\text{plot} \quad \text{or}$$

$$\pm 131.455 \text{ m}^3/\text{ha}$$

The actual population variance, $\sigma_o^2 = 638.1469$. Consequently,

$$V(\bar{y}_o)_{\text{Ran}} = \frac{638.1469}{20}\left(\frac{320-20}{320-1}\right) = 30.006908$$

Again, the method of successive differences yielded an estimate that was high, which is typical. Spurr (1952) states that experience in the United States and in Scandinavia indicates that the magnitude of the positive bias increases as the spacing interval increases.

Further information about the method of successive differences can be found in Yates (1946; 1948), Morse and Grubbs (1947), Quenouille (1953), Moore (1955), Meyer (1956), Cochran (1963), and Husch et al. (1972).

### 18.11.5 Method of Balanced Differences

Continuing his argument for the use of end corrections, Yates (1948; 1949; 1981) has suggested the use of *balanced differences* for the estimation of the variance of the overall mean. This procedure was originally proposed for situations where the population possesses a definite but unknown trend. However, it apparently is useful in all cases except where periodicity is present.

In this procedure the sample is divided into subsets known as *partial systematic samples*, each of which contributes one degree of freedom. The size of these subsets, $c$, is arbitrary. When the subsets are small, the number of degrees of freedom will be correspondingly high but the differences between subsets will tend to be large because they contain too few observations for compensation to take place. Yates suggested, as a compromise, that subsets of size 9 be used when the population is one dimensional. To increase the degrees of freedom the subsets may be overlapped, but overlapping is not a requirement. When the sample is small, the degrees of freedom can be maximized by beginning the first subset with $y_1$ (where the subscript refers to the position in the *sample*), the second subset with $y_2$, the third with $y_3$, etc., until as many subsets of the specified size as possible have been generated. When the sample is large, the amount of overlap can be reduced or even eliminated. A common approach is to overlap just the end element of adjacent subsets so that the last element in one subset becomes the first element in the subsequent subset.

For each subset a balanced difference is obtained.

$$d_i = \left(\frac{1}{2}\right)y_{i1} - (1)y_{i2} + (1)y_{i3} - \cdots - (1)y_{ic-1} + \left(\frac{1}{2}\right)y_{ic} \tag{18.86}$$

The values in parentheses are weights, $w_{ij}$. Note that the first and last observations have weights of $^1/_2$ while the remainder have weights of 1. Furthermore, note that the balanced difference is obtained by alternately subtracting and adding the constituent terms. The estimate of $V(\bar{y}_o)$ is then

$$s^2_{\bar{y}_o} = \frac{1}{nt\sum_j^c w_{ij}^2}\sum_i^t d_i^2\left(\frac{N-n}{N}\right) \tag{18.87}$$

where $t$ is the number of balanced differences formed.

As an example, consider the 50-item population shown in Figure 18.7. This population has a definite trend with no residual variance about that trend. Then, $N = 50$, $L = 5$, $n = 10$, subsample size set at 7. Assume that position 2 was chosen as the start in a random draw. The sample is then

11.0, 13.5, 16.0, 18.5, 21.0, 23.5, 26.0, 28.5, 31.0, 33.5

$$d_1 = {}^1/_2(11.0) - 13.5 + 16.0 - 18.5 + 21.0 - 23.5 + {}^1/_2(26.0) = 0$$

$$d_2 = {}^1/_2(13.5) - 16.0 + 18.5 - 21.0 + 23.5 - 26.0 + {}^1/_2(28.5) = 0$$

$$d_3 = {}^1/_2(16.0) - 18.5 + 21.0 - 23.5 + 26.0 - 28.5 + {}^1/_2(31.0) = 0$$

$$d_4 = {}^1/_2(18.5) - 21.0 + 23.5 - 26.0 + 28.5 - 31.0 + {}^1/_2(33.5) = 0$$

and

$$s_{\bar{y}_o}^2 = \frac{1}{10(4)(5.5)}(0^2 + 0^2 + 0^2 + 0^2)\left(\frac{50-10}{50}\right) = 0$$

As can be seen, the effect of the trend has been completely eliminated.

Now consider the population shown in Figure 18.8, which possesses a trend but there is residual variance about the trend line. Then, $N = 50$, $L = 5$, $n = 10$; subsample size set at 9. In this case, consider all the possible samples. Thus, when the start is in position 1, the sample is

12.5, 12.5, 13.5, 17.5, 22.0, 23.5, 25.0, 26.5, 29.5, 35.0

$$d_1 = {}^1/_2(12.5) - 12.5 + 13.5 - 17.5 + 22.0 - 23.5 + 25.0 - 26.5 + {}^1/_2(29.5) = 1.5$$

$$d_2 = {}^1/_2(12.5) - 13.5 + 17.5 - 22.0 + 23.5 - 25.0 + 26.5 - 29.5 + {}^1/_2(35.0) = 1.25$$

$$s_{\bar{y}_{o1}}^2 = \frac{1}{10(2)(7.5)}[(1.5)^2 + (1.25)^2]\left(\frac{50-10}{50}\right) = 0.0203333$$

When the start is in position 2, the sample is

11.0, 15.0, 16.5, 20.0, 21.0, 23.5, 27.5, 29.0, 31.5, 33.5

$$d_1 = {}^1/_2(11.0) - 15.0 + 16.5 - 20.0 + 21.0 - 23.5 + 27.5 - 29.0 + {}^1/_2(31.5) = -1.25$$

$$d_2 = {}^1/_2(15.0) - 16.5 + 20.0 - 21.0 + 23.5 - 27.5 + 29.0 - 31.5 + {}^1/_2(33.5) = 0.25$$

$$s_{\bar{y}_{o2}}^2 = \frac{1}{10(2)(7.5)}[(-1.25)^2 + (0.25)^2]\left(\frac{50-10}{50}\right) = 0.0086667$$

When the start is in position 3, the sample is

10.0, 13.5, 16.5, 19.5, 21.0, 24.5, 26.5, 28.5, 31.5, 33.5

$$d_1 = {}^1/_2(10.0) - 13.5 + 16.5 - 19.5 + 21.0 - 24.5 + 26.5 - 28.5 + {}^1/_2(31.5) = -1.25$$

$$d_2 = {}^1/_2(13.5) - 16.5 + 19.5 - 21.0 + 24.5 - 26.5 + 28.5 - 31.5 + {}^1/_2(33.5) = 0.50$$

$$s_{\bar{y}_{o3}}^2 = \frac{1}{10(2)(7.5)}[(-1.25)^2 + (0.5)^2]\left(\frac{50-10}{50}\right) = 0.0096667$$

When the start is in position 4, the sample is

13.5, 14.0, 18.0, 19.0, 21.0, 23.5, 26.5, 28.5, 31.5, 35.5

$$d_1 = {}^1/_2(13.5) - 14.0 + 18.0 - 19.0 + 21.0 - 23.5 + 26.5 - 28.5 + {}^1/_2(31.5) = 3$$

$$d_2 = {}^1/_2(14.0) - 18.0 + 19.0 - 21.0 + 23.5 - 26.5 + 28.5 - 31.5 + {}^1/_2(35.5) = -1.25$$

$$s^2_{\bar{y}_{o4}} = \frac{1}{10(2)(7.5)}[(3)^2 + (-1.25)^2]\left(\frac{50-10}{50}\right) = 0.0563333$$

When the start is in position 5, the sample is

12.5, 15.0, 17.5, 19.5, 22.0, 25.0, 27.5, 31.0, 31.5, 35.5

$$d_1 = {}^1\!/_2(12.5) - 15.0 + 17.5 - 19.5 + 22.0 - 25.0 + 27.5 - 31.0 + {}^1\!/_2(31.5) = -1.5$$

$$d_2 = {}^1\!/_2(15.0) - 17.5 + 19.5 - 22.0 + 25.0 - 27.5 + 31.0 - 31.5 + {}^1\!/_2(35.5) = 2.25$$

$$s^2_{\bar{y}_{o5}} = \frac{1}{10(2)(7.5)}[(-1.5)^2 + (2.25)^2]\left(\frac{50-10}{50}\right) = 0.0390000$$

$$E(sy_o)^2 = \frac{1}{5}(0.0203333 + 0.008667 + 0.0096667 + 0.0563333 + 0.0390000)$$

$$= 0.026800$$

Compare this with $V(\bar{y}_o)_{\text{Ran}} = 4.2628$ (trend not removed); $V(\bar{y}_o)'_{\text{Sys1}} = 0.1004$; and $V(\bar{y}_o)'_{\text{Sys2}} = 0.0795$, where trend was removed using regression (Section 18.9.5). The estimate using balanced differences is smaller than any of the above expressions for the variance of the overall mean. Thus, in this case the apparent precision is greater than the actual precision. The reverse could occur.

When the population is two dimensional and the systematic sample takes the form of a square grid (e.g., a line–plot cruise), Yates (1949; 1981) recommends the use of square 16-element (4 × 4) subsets. The corner observations are assigned a weight of ¹/₄, the side observations a weight of ¹/₂, and the rest a weight of 1. The sum of the squared weights would then be 6.25. As in the case of one-dimensional population, the subsets can be overlapped to obtain more degrees of freedom.

As an example, use the 16-ha tract shown in Figure 16.1. Use a spacing interval $L = 4$ and a random start in row 3 and column 3, as was done in Section 18.11.3. The first subset is then, with the weight in parentheses,

| | | | |
|---|---|---|---|
| 47(¹/₄) | 86(¹/₂) | 68(¹/₂) | 44(¹/₄) |
| 40(¹/₂) | 81(1) | 90(1) | 63(¹/₂) |
| 21(¹/₂) | 75(1) | 65(1) | 32(¹/₂) |
| 0(¹/₄) | 48(¹/₂) | 25(¹/₂) | 29(¹/₄) |

or

| | | | |
|---|---|---|---|
| 11.75 | 43.00 | 34.00 | 11.00 |
| 20.00 | 81.00 | 90.00 | 31.50 |
| 10.50 | 75.00 | 65.00 | 16.00 |
| 0.00 | 24.00 | 12.50 | 7.25 |

$$d_1 = 11.75 - 43.00 + 34.00 - 11.00 + 20.00 - 81.00 + 90.00 - 31.50$$
$$+ 10.50 - 75.00 + 65.00 - 16.00 + 0.00 - 24.00 + 12.50 - 7.25 = -45$$

The second subset is

| | | | |
|---|---|---|---|
| 40(1/4) | 81(1/2) | 90(1/2) | 63(1/4) |
| 21(1/2) | 75(1) | 65(1) | 32(1/2) |
| 0(1/2) | 48(1) | 25(1) | 29(1/2) |
| 3(1/4) | 25(1/2) | 2(1/2) | 4(1/4) |

or

| | | | |
|---|---|---|---|
| 10.00 | 40.50 | 45.00 | 15.75 |
| 10.50 | 75.00 | 65.00 | 16.00 |
| 0.00 | 48.00 | 25.00 | 14.50 |
| 0.75 | 12.50 | 1.00 | 1.00 |

$$d_2 = 10.00 - 40.50 + 45.00 - 15.75 + 10.50 - 75.00 + 65.00 - 16.00$$
$$+ 0.00 - 48.00 + 25.00 - 14.50 + 0.75 - 12.50 + 1.00 - 1.00 = -66$$

Then

$$s^2_{\bar{y}_o} = \frac{1}{20(2)(6.25)}[(-45)^2 + (-66)^2]\left(\frac{320 - 20}{320}\right) = 23.92875$$

$$s_{\bar{y}_o} = \pm 4.8917 \text{ m}^3 / \text{plot or } \pm 97.834 \text{ m}^3 / \text{ha}$$

Using the method of successive differences (Section 18.11.3), $s_{\bar{y}_o} = \pm 131.455$ m³/ha. As Yates predicted, the method of balanced differences yielded a more precise estimate than did the method of successive differences.

It should be pointed out that when autocorrelation is present and the serial correlations are strong, the method of balanced differences may yield substantially smaller estimates of the variances of the mean than will the method of successive differences.

# 19 Ratio Estimation

## 19.1 INTRODUCTION

When two random variables are correlated and the relationship between them can be described on a straight line, the regression equation is

$$\mu_{Y \cdot X} = Y = \beta_o + \beta_1 X$$

According to Equation 15.59,

$$\beta_o = \mu_Y - \beta_1 \mu_X$$

Thus, if the regression line passes through the origin, $\beta_o = 0$ and

$$\beta_1 = \mu_Y / \mu_X$$

As can be seen, $\beta_1$ is a ratio of the means of the two random variables. Then,

$$R = \beta_1 = \mu_Y / \mu_X \tag{19.1}$$

$$= \sum_i^N y_i \Big/ \sum_i^N x_i \quad \text{or} \tag{19.2}$$

$$= \tau_Y / \tau_X \tag{19.3}$$

Ratios of this sort are not uncommon in forestry. In some cases the ratios themselves are of interest while in others the ratios are used to obtain estimates of parameters that otherwise would be excessively difficult and/or expensive to obtain directly. Some examples are ratios of wood volume to wood weight, which are used to estimate wood volume on trucks; ratios of present volumes of timber per unit land area to past volumes, which are used to update timber cruises; ratios of dbh to crown diameter, which are used to estimate dbh from crown images on aerial photographs; and ratios of basal area per unit land area to percent crown closure, which are used to estimate stand density, in terms of basal area per unit land area, from aerial photographic evidence.

In addition to these forms and uses of ratios, there are times when they can be used to improve substantially the precision of estimates of population parameters.

It should be noted from the outset of this discussion that $R$ is not a population proportion (see Section 12.15). $R$ is a ratio of two random variables. $p$, the population proportion, is the rate of

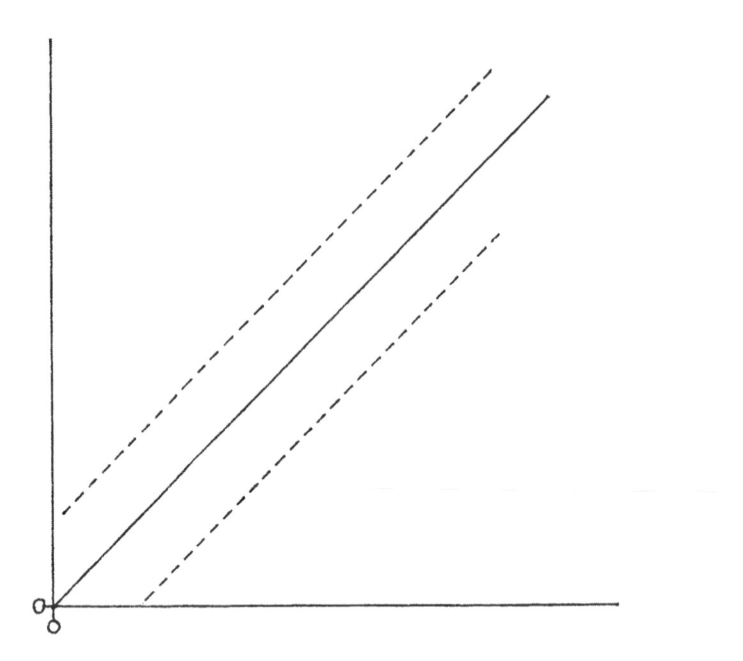


**FIGURE 19.1** Schematic scatter diagram illustrating $V(y_{ij}|x_i) = c$.

occurrence of a certain characteristic within a population and is obtained by dividing the count of occurrences of the characteristic of interest by the size of the population. Population size is a total count. It is not the mean of a random variable. Since $R$ is not the equivalent of $p$, one cannot use the statistical approaches and procedures associated with $p$, which are described in Sections 12.15 and 13.11, to evaluate estimates of $R$.

## 19.2 ESTIMATING *R*

The procedure used to arrive at valid sample-based estimates of $R$ involves weighted regression with the weights being dependent on how the data points are distributed about the regression line. According to Equations 15.90 and 15.100, *when the regression forms a straight line passing through the origin*,

$$b = r = \sum_i^n w_i y_i x_i \bigg/ \sum_i^n w_i x_i^2 \tag{19.4}$$

where, according to Equation 15.92,

$$w_i = 1/V(y_{ij} \mid x_i) \tag{19.5}$$

When the variance of the dependent variable about the regression line is homogeneous (see Figure 19.1),

$$V(y_{ij} \mid x_i) = c \tag{19.6}$$

$$w_i = 1/c \tag{19.7}$$

and

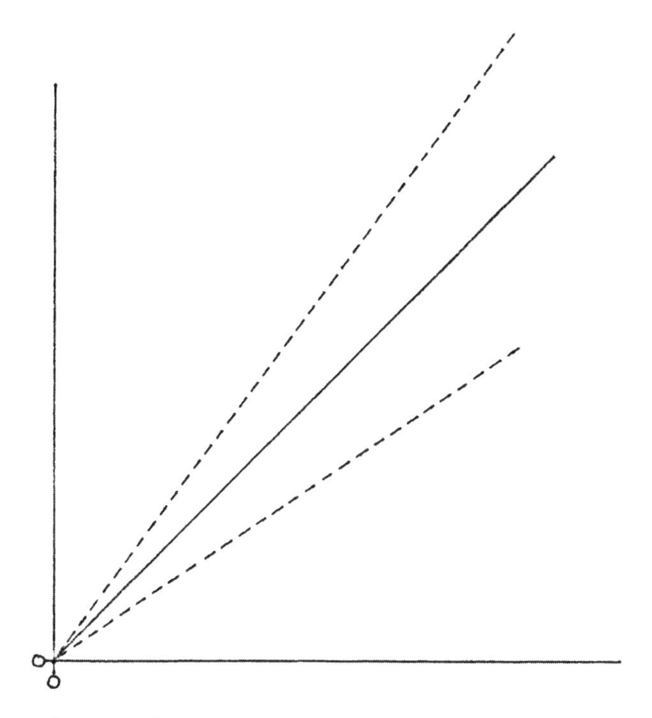

**FIGURE 19.2** Schematic scatter diagram illustrating $V(y_{ij}|x_i) = cx_i$.

$$r = \sum_i^n (1/c)y_i x_i \Big/ \sum_i^n (1/c)x_i^2$$

$$= \sum_i^n y_i \Big/ \sum_i^n x_i \tag{19.8}$$

$$r = r_R = \bar{y}/\bar{x}$$

Under these conditions, $r_R$ is referred to as the *ratio of means*.

When the variance about the regression is proportional to the independent variable, as shown in Figure 19.2,

$$V(y_{ij}|x_i) = cx_i \tag{19.9}$$

$$w_i = 1/cx_i \tag{19.10}$$

and

$$r = \sum_i^n (1/cx_i)y_i x_i \Big/ \sum_i^n (1/cx_i)x_i^2$$

$$= \sum_i^n y_i \Big/ \sum_i^n x_i \tag{19.11}$$

$$r = r_R = \bar{y}/\bar{x}$$

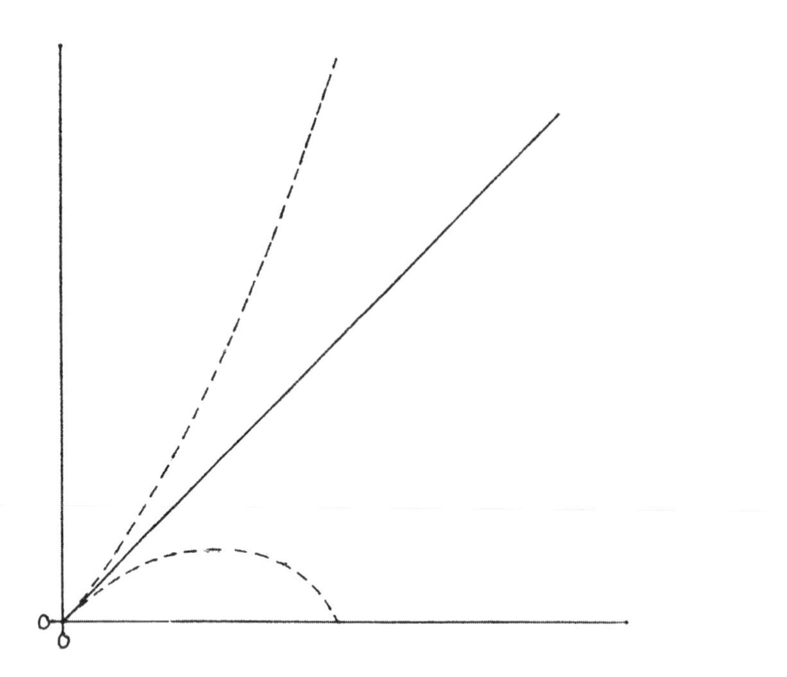


**FIGURE 19.3** Schematic scatter diagram illustrating $V(y_{ij}|x_i) = cx_i^2$.

Again, $r$ is a ratio of means.

When, however, the variance about the regression is a function of the *square* of the independent variable (see Figure 19.3),

$$V(y_{ij} \mid x_i) = cx_i^2 \tag{19.12}$$

$$w_i = 1/cx_i^2 \tag{19.13}$$

and

$$\begin{aligned} r &= \sum_i^n (1/cx_i^2)y_i x_i \Big/ \sum_i^n (1/cx_i^2)x_i^2 \\ &= \sum_i^n (1/cx_i)y_i \Big/ \sum_i^n (1/c) \\ &= \frac{1}{n}\sum_i^n \left(\frac{y_i}{x_i}\right) \Big/ \frac{1}{n} n \\ r = r_M &= \frac{1}{n}\sum_i^n \left(\frac{y_i}{x_i}\right) = \frac{1}{n}\sum_i^n r_i \end{aligned} \tag{19.14}$$

In this case, each draw yields a ratio, $r_i = y_i/x_i$. Consequently, $r_M$ is referred to as the *mean of ratios*.

As can be seen, the relationship between the variance of $Y$ and the independent variable $X$ controls the procedure that should be used to obtain an estimate of $R$. The distributional pattern of data points can be revealed if a scatter diagram of presample or historical data is constructed, as is discussed in Section 15.3.10 and shown in Figure 15.22 and 15.23.

These ways of estimating $R$ do not exhaust the possibilities. However, the ratio of means, $r_R$, is the most commonly encountered estimate and forms the standard against which the others are judged.

## 19.3 BIAS OF *r*

In the preceding discussion it was assumed that the *sample* regression line passed through the origin. When this assumption is correct, the regression also passes through the point $(\bar{x},\bar{y})$ and $r = b_1$. However, even if the population regression line passes through the origin, most sample regressions would not do so, in which case $r \neq b_1$. If the sample regression is conditioned to pass through the origin, it will not pass through $(\bar{x},\bar{y})$ and $r \neq b_1$. In addition to these conclusions, when one is concerned with ratios, the values of the independent variable, $X$, are not fixed as is required in standard regression theory, but change from sample to sample. Consequently, the shape of the sampling distribution of $r$ is difficult to define. It is, however, known to be positively skewed and that it tends to take on the shape of the normal distribution when the sample size becomes large. As a result of the skewness, $E(r)$ is usually not equal to $R$, making $r$ a biased estimator of $R$. The bias, however, usually becomes negligible when the sample size is large and it disappears when $n = N$. Most workers in the field consider the bias to be acceptably small if the sample size is large enough so that the distribution of the estimated ratios approximates the normal and the ratios $s_{\bar{y}} / \bar{y}$ and $s_{\bar{x}} / \bar{x}$ are both less than 0.1 (Cochran, 1963). A sample larger than 30 is usually considered adequate.

### 19.3.1 Ratio of Means

In the case of the ratio of means, the exact magnitude of the bias can be determined as follows. According to Equation 9.29,

$$\text{COV}(r_R, \bar{x}) = E(r_R\bar{x}) - E(r_R)E(\bar{x})$$

and according to Equations 19.8 and 19.11, the ratio of means is

$$r_R = \bar{y} / \bar{x}$$

Then

$$r_R\bar{x} = \bar{y}$$

$$\text{COV}(r_R, \bar{x}) = E(\bar{y}) - E(r_R)E(\bar{x}) = \mu_Y - \mu_{r_R}\mu_X$$

$$\mu_{r_R} = \frac{\mu_Y}{\mu_X} - \frac{\text{COV}(r_R, \bar{x})}{\mu_X} = R - \frac{\text{COV}(r_R, \bar{x})}{\mu_X} \tag{19.15}$$

and

$$\text{BIAS}_{r_R} = \mu_{r_R} - R = -\text{COV}(r_R, \bar{x}) / \mu_X \tag{19.16}$$

Since, according to Equation 9.32, $COV(r_R, \bar{x}) = \rho_{r_R} - \sigma_{r_R}\sigma_{\bar{x}}$, Equation 19.16 can be written

$$BIAS_{r_R} = \mu_{r_R} - R = -\rho_{r_R\bar{x}}(-\sigma_{r_R}\sigma_{\bar{x}}) / \mu_X \tag{19.17}$$

As can be seen, if there is no correlation between $r_R$ and $\bar{x}$, there will be no bias.

In the preceding development one should notice that $\mu_{r_R}$ is the expected value of the sample ratios and is not the equivalent of $R$. It also should be noted that the population correlation coefficient, $\rho_{r_R\bar{x}}$, refers to the correlation between $r_R$ and $\bar{x}$, not to that between $Y$ and $X$.

As an example, consider the two-dimensional population:

| $i$ | 1 | 2 | 3 | 4 | 5 |
|---|---|---|---|---|---|
| $X$ | 0 | 10 | 20 | 30 | 40 |
| $Y$ | 0.0 | 3.5 | 4.5 | 10.5 | 11.5 |

$$N = 5;\quad \mu_X = 20.0;\quad \sigma_X^2 = 200.0;\quad \sigma_X = 14.142136$$
$$\mu_Y = 6.0;\quad \sigma_Y^2 = 19.0;\quad \sigma_Y = 4.35890;\quad \sigma_{XY} = 60.0$$
$$\beta_o = 0;\quad \beta_1 = 0.3;\quad \rho_{XY} = 0.97333;\quad R = 0.3$$

Let $n$ = 3. The possible samples are then:

| Sample | $x$ | $y$ | $\bar{x}$ | $\bar{y}$ | $r_R$ |
|---|---|---|---|---|---|
| 1 | 0,10.20 | 0, 3.5, 4.5 | 10.00000 | 2.66667 | 0.26667 |
| 2 | 0,10,30 | 0, 3.5,10.5 | 13.33333 | 4.66667 | 0.35000 |
| 3 | 0,10,40 | 0, 3.5,11.5 | 16.66667 | 5.00000 | 0.30000 |
| 4 | 0,20,30 | 0, 4.5,10.5 | 16.66667 | 5.00000 | 0.30000 |
| 5 | 0,20,40 | 0, 4.5,11.5 | 20.00000 | 5.33333 | 0.26667 |
| 6 | 0,30,40 | 0, 10.5,11.5 | 23.33333 | 7.33333 | 0.31429 |
| 7 | 10,20,30 | 3.5, 4.5,10.5 | 20.00000 | 6.16667 | 0.30833 |
| 8 | 10,20,40 | 3.5, 4.5,11.5 | 23.33333 | 6.50000 | 0.27857 |
| 9 | 10,30,40 | 3.5,10.5,11.5 | 26.66667 | 8.50000 | 0.31875 |
| 10 | 20,30,40 | 4.5,10.5,11.5 | 30.00000 | 8.83333 | 0.29444 |

$$\mu_{r_R} = 0.299772;\quad \sigma_{r_R}^2 = 0.0005837;\quad \sigma_{r_R} = 0.024159$$

$$COV(r_R, \bar{x}) = \frac{1}{M}\sum_i^M [(r_R - \mu_{r_R})(\bar{x} - \mu_X)] = 0.0455667$$

$$\rho_{r_R\bar{x}} = COV(r_R, \bar{x}) / \sigma_{r_R}\sigma_{\bar{x}}$$
$$= 0.0455667 / (0.024159)(5.7735027) = 0.0326684$$

Then, according to Equation 19.16,

$$BIAS_{r_R} = -(0.024159)(5.7735027)(0.0326684) / 20 = -0.000228$$

which agrees with

$$\text{BIAS}_{r_R} = \mu_{r_R} - R = 0.299772 - 0.3 = -0.000228$$

Equations 19.16 and 19.17 yield the exact bias of $r$ but are not satisfactory for assessing the factors controlling that bias. To make such assessments it will be necessary to generate alternative expressions for the bias. Let

$$\bar{y} = \mu_Y + (\bar{y} - \mu_Y) = \mu_Y[1 + (\bar{y} - \mu_Y)/\mu_Y]$$

and

$$\bar{x} = \mu_X + (\bar{x} - \mu_X) = \mu_X[1 + (\bar{x} - \mu_X)/\mu_X]$$

Then,

$$r_R = \frac{\mu_Y}{\mu_X}\left[1 + \frac{(\bar{y} - \mu_Y)}{\mu_Y}\right]\left[1 + \frac{(\bar{x} - \mu_X)}{\mu_X}\right]^{-1}$$

Expanding

$$\left[1 + \frac{(\bar{x} - \mu_X)}{\mu_X}\right]^{-1}$$

using Taylor's formula,

$$\begin{aligned}
r_R &= R\left[1 + \frac{(\bar{y} - \mu_Y)}{\mu_Y}\right]\left[1 - \frac{(\bar{x} - \mu_X)}{\mu_X} + \frac{(\bar{x} - \mu_X)^2}{\mu_x^2} - \cdots\right] \\
&= R\left[1 + \frac{(\bar{y} - \mu_Y)}{\mu_Y} - \frac{(\bar{x} - \mu_X)}{\mu_X} + \frac{(\bar{x} - \mu_X)^2}{\mu_x^2} - \frac{(\bar{x} - \mu_X)^3}{\mu_X^3} + \cdots\right. \\
&\quad \left. - \frac{(\bar{y} - \mu_Y)(\bar{x} - \mu_X)}{\mu_Y\mu_X} + \frac{(\bar{y} - \mu_Y)(\bar{x} - \mu_X)^2}{\mu_Y\mu_X^2} - \frac{(\bar{y} - \mu_Y)(\bar{x} - \mu_X)^3}{\mu_Y\mu_x^3} + \cdots\right]
\end{aligned} \tag{19.18}$$

The expected value of $r_R$ is

$$\begin{aligned}
E(r_R) = R&\left[1 + \frac{E(\bar{y} - \mu_Y)}{\mu_Y} - \frac{E(\bar{x} - \mu_X)}{\mu_X} + \frac{E(\bar{x} - \mu_X)^2}{\mu_x^2} - \frac{E(\bar{x} - \mu_X)^3}{\mu_X^3} + \cdots\right. \\
&\left. - \frac{E(\bar{y} - \mu_Y)(\bar{x} - \mu_X)}{\mu_Y\mu_X} + \frac{E(\bar{y} - \mu_Y)(\bar{x} - \mu_X)^2}{\mu_Y\mu_X^2} - \frac{E(\bar{y} - \mu_Y)(\bar{x} - \mu_X)^3}{\mu_Y\mu_x^3} + \cdots\right]
\end{aligned} \tag{19.19}$$

Since

$$E(\bar{y}-\mu_Y) = \frac{1}{M}\sum_i^M(\bar{y}_i - \mu_Y) = 0$$

$$E(\bar{x}-\mu_X) = \frac{1}{M}\sum_i^M(\bar{x}_i - \mu_X) = 0 \tag{19.20}$$

$$E(\bar{x}-\mu_X)^2 = \frac{1}{M}\sum_i^M(\bar{x}_i - \mu_X)^2 = \sigma_{\bar{x}}^2 = \frac{\sigma_X^2}{n}\left(\frac{N-n}{N-1}\right)$$

and

$$E(\bar{y}-\mu_Y)(\bar{x}-\mu_X) = \frac{1}{M}\sum_i^M(\bar{y}-\mu_Y)(\bar{x}-\mu_X) = \text{COV}(\overline{xy})$$

$$E(r_R) = R\left[1 + \frac{\sigma_X^2}{\mu_X n}\left(\frac{N-n}{N-1}\right) - \cdots - \frac{\text{COV}(\bar{x}\,\bar{y})}{\mu_Y\mu_X} + \cdots\right] \tag{19.21}$$

It can be shown that

$$\text{COV}(\bar{x}\,\bar{y}) = \frac{\text{COV}(XY)}{n}\left(\frac{N-n}{N-1}\right) = \frac{\sigma_{XY}}{n}\left(\frac{N-n}{N-1}\right) \tag{19.22}$$

Then

$$E(r_R) = R\left[1 + \frac{\sigma_X^2}{\mu_X n}\left(\frac{N-n}{N-1}\right) - \cdots - \frac{\sigma_{XY}}{\mu_Y\mu_X n} + \cdots\right] \tag{19.23}$$

According to Equation 9.32,

$$\text{COV}(XY)/\sigma_X\sigma_Y = \rho_{XY}$$

where $\rho_{XY}$ is the population correlation coefficient. Then, if Equation 19.21 is multiplied by $\sigma_X\sigma_Y/\sigma_X\sigma_Y$, the term

$$\left[\frac{\text{COV}(\bar{x}\,\bar{y})}{\mu_Y\mu_X}\right]$$

becomes

$$\frac{\sigma_X\sigma_Y}{\sigma_X\sigma_Y}\left[\frac{\text{COV}(\bar{x}\,\bar{y})}{\mu_Y\mu_Y}\right] = \frac{\sigma_X\sigma_Y\text{COV}(XY)}{\sigma_X\sigma_Y\mu_Y\mu_X n}\left(\frac{N-n}{N-1}\right) = \frac{\sigma_X\sigma_Y\rho_{XY}}{\mu_Y\mu_X n}\left(\frac{N-n}{N-1}\right) \tag{19.24}$$

Consequently, Equation 19.21 can be written

$$E(r_R) = R\left[1 + \frac{\sigma_X^2}{\mu_X n}\left(\frac{N-n}{N-1}\right) - \cdots - \frac{\sigma_X \sigma_Y}{\mu_X \mu_Y n}\left(\frac{N-n}{N-1}\right)\rho_{XY} + \cdots\right] \quad (19.25)$$

The bias is then

$$\text{BIAS}_{r_R} = E(r_R - R) = E(r_R) - R \quad (19.26)$$

$$= \left(\frac{N-n}{N-1}\right)\left[\frac{R\sigma_X^2}{\mu_X^2 n} - \cdots - \frac{R\sigma_{XY}}{\mu_X \mu_Y n} + \cdots\right] \text{ or} \quad (19.27)$$

$$= \left(\frac{N-n}{N-1}\right)\left[\frac{R\sigma_X^2}{\mu_X^2 n} - \cdots - \frac{R\rho_{XY}\sigma_X\sigma_Y}{\mu_X \mu_Y n} + \cdots\right] \quad (19.28)$$

As can be seen from Equations 19.27 and 19.28, bias is a function of sample size. The only value in the expression that is subject to manipulation is $n$. Consider the finite population correction. As $n$ increases, the correction and the bias decrease, approaching zero as $n$ approaches $N$. Now consider the portion of the equation lying between the brackets. Sample size appears in the denominator of each term, causing the terms to decrease in absolute magnitude as $n$ increases. This leads to the general statement that the bias of the ratio of means varies approximately as $1/n$.

Sample size, however, is not the only determinant of the degree of bias. Where the parent populations are perfectly correlated, so that $\rho_{XY} = 1$, $r_R$ will be an unbiased estimator of $R$ regardless of the sample size. This is difficult to prove because of the nature of Equation 19.28. However, a hint of how it might be done is as follows. One can rewrite Equation 19.28

$$BIAS_{r_R} = \frac{R\sigma_X^2}{\mu_X^2 n}\left(\frac{N-n}{N-1}\right) - \frac{R\sigma_X\sigma_Y}{\mu_X\mu_Y n}\left(\frac{N-n}{N-1}\right) + \cdots$$

Since the coefficient of variation, CV, is equal to $(\sigma/\mu)$, let $CV_X = \sigma_X/\mu_X$ and $CV_Y = \sigma_Y/\mu_Y$. Then,

$$\text{BIAS}_{r_R} = \frac{R\text{CV}_X\text{CV}_X}{n}\left(\frac{N-n}{N-1}\right) - \frac{R\text{CV}_X\text{CV}_Y}{n}\left(\frac{N-n}{N-1}\right) + \cdots$$

It can be shown that when $\rho_{XY} = 1$, $CV_X = CV_Y$. Consequently,

$$\frac{R\text{CV}_X\text{CV}_X}{n}\left(\frac{N-n}{N-1}\right) = \frac{R\text{CV}_X\text{CV}_Y}{n}\left(\frac{N-n}{N-1}\right)$$

and they cancel each other in Equation 19.28. Such canceling occurs throughout the rest of the equation leaving $\text{BIAS}_{r_R} = 0$. This is demonstrated in the following example. Consider the two-dimensional population:

| $i$ | 1 | 2 | 3 | 4 | 5 |
|---|---|---|---|---|---|
| $X$ | 0 | 10 | 20 | 30 | 40 |
| $Y$ | 0 | 3 | 6 | 9 | 12 |

$$N = 5; \quad \mu_X = 20.0; \quad \sigma_X^2 = 200.0; \quad \sigma_X = 14.142136$$
$$\mu_Y = 6.0; \quad \sigma_Y^2 = 18.0; \quad \sigma_Y = 4.242641; \quad \sigma_{XY} = 60.0$$
$$\beta_0 = 0; \quad \beta_1 = 0.3; \quad \rho_{XY} = 1; \quad R = 0.3$$

Using sample size $n = 3$, the number of possible samples, when sampling without replacement, is

$$M = \binom{5}{3} = 10$$

These are

| Sample | $x$ | $y$ | $\bar{x}$ | $\bar{y}$ | $r_R$ |
|---|---|---|---|---|---|
| 1 | 0,10,20 | 0,3, 6 | 10.00000 | 3.0 | 0.3 |
| 2 | 0,10,30 | 0,3, 9 | 13.33333 | 4.0 | 0.3 |
| 3 | 0,10,40 | 0,3, 12 | 16.66667 | 5.0 | 0.3 |
| 4 | 0,20,30 | 0,6, 9 | 16.66667 | 5.0 | 0.3 |
| 5 | 0,20,40 | 0,6, 12 | 20.00000 | 6.0 | 0.3 |
| 6 | 0,30,40 | 0,9, 12 | 23.33333 | 7.0 | 0.3 |
| 7 | 10,20,30 | 3,6, 9 | 20.00000 | 6.0 | 0.3 |
| 8 | 10,20,40 | 3,6, 12 | 23.33333 | 7.0 | 0.3 |
| 9 | 10,30,40 | 3,9, 12 | 26.66667 | 8.0 | 0.3 |
| 10 | 20,30,40 | 6,9, 12 | 30.00000 | 9.0 | 0.3 |

$\mu_{r_R} = 0.3 = R$. Consequently, in this case, $r_R$ is an unbiased estimator of $R$. If a sample size of $n = 2$ is used, $\mu_{r_R}$ again equals $R$ and $r_R$ is an unbiased estimator of $R$. Whenever $\rho_{XY} = 1$, $r_R$ will be unbiased, regardless of sample size.

As an example of what occurs when $\rho_{XY}$ is not equal to 1, one can use the data from the earlier example where $\rho_{XY} = 0.97333$:

$$\mu_{r_R} = 0.29977 \quad \text{and}$$

$$\text{BIAS}_{r_R} = \mu_{r_R} - R = 0.29977 - 0.3 = -0.00023$$

### 19.3.2 Mean of Ratios

The bias in the case of the mean of ratios is determined using an approach that closely parallels that used with the ratio of means. An individual ratio is

$$r_i = y_i / x_i; \quad E(r_i) = \frac{1}{N} \sum_i^N r_i = \mu_r$$

Then,

$$r_i x_i = y_i; \quad E(r_i x_i) = E(y_i) = \frac{1}{N}\sum_i^N y_i = \mu_Y$$

$$r_M = \frac{1}{n}\sum_i^n r_i; \quad E(r_M) = \frac{1}{n}\sum_i^n [E(r_i)] = E(r_i) = \mu_{r_M}$$

$$\begin{aligned} \mathrm{COV}(r_i x_i) &= E(r_i x_i) - E(r_i)E(x_i) = E(y_i) - E(r_i)E(x_i) \\ &= \mu_Y - \mu_{r_M}\mu_X \end{aligned} \tag{19.29}$$

Then,

$$\mu_{r_M} = \frac{\mu_Y}{\mu_X} - \frac{\mathrm{COV}(r_i x_i)}{\mu_X} = R - \frac{\mathrm{COV}(r_i x_i)}{\mu_X} \tag{19.30}$$

and

$$\mathrm{BIAS}_{r_M} = \mu_{r_M} - R = -\mathrm{COV}(r_i x_i)/\mu_X \quad \text{or} \tag{19.31}$$

$$= -(\mu_Y - \mu_{r_M}\mu_X)/\mu_X \quad \text{or} \tag{19.32}$$

$$= -\rho_{r_i x_i}\sigma_{r_i}\sigma_{x_i}/\mu_X \tag{19.33}$$

As an example, use the data from the earlier example where $\rho_{XY} = 0.97333$.

| *i* | 1 | 2 | 3 | 4 | 5 | μ |
|---|---|---|---|---|---|---|
| $x_i$ | 0 | 10.000 | 20.000 | 30.000 | 40.0000 | 20.0 |
| $y_i$ | 0 | 3.500 | 4.500 | 10.500 | 11.5000 | 6.0 |
| $r_i$ | 0 | 0.350 | 0.255 | 0.350 | 0.2875 | 0.3 |

$$\mathrm{COV}(r_i x_i) = 6 - 0.2425(20) = 1.15$$

$$\mu_{r_M} = R - \frac{\mathrm{COV}(r_i x_i)}{\mu_X} = 0.3 - \frac{1.15}{20} = 0.2425$$

$$\mathrm{BIAS}_{r_M} = -\frac{\mathrm{COV}(r_i x_i)}{\mu_X} = -\frac{1.15}{20} = -0.0575$$

which agrees with

$$\mathrm{BIAS}_{r_M} = \mu_{r_M} - R = 0.2425 - 0.3 = -0.0575$$

As was the case with the ratio of means, these formulas for the bias do not permit an assessment of the factors controlling the bias. Consequently, it will be necessary to generate alternative expressions as was done in Section 19.3.1. The derivation of these expressions closely follows the pattern used in the case of the ratio of means.

$$r_i = y_i / x_i$$

$$r_i - R = \frac{y_i}{x_i} - R = \frac{y_i - Rx_i}{x_i}$$

$$y_i = \mu_Y + (y_i - \mu_Y) = \mu_Y[1 + (y_i - \mu_Y)/\mu_Y]$$

$$x_i = \mu_X + (x_i - \mu_X) = \mu_X[1 + (x_i - \mu_X)/\mu_X]$$

$$r_i = \frac{y_i}{x_i} = \frac{\mu_Y[1 + (y_i - \mu_Y)/\mu_Y]}{\mu_X[1 + (x_i - \mu_X)/\mu_X]} = R\left[\frac{1 + (y_i - \mu_Y)/\mu_Y}{1 + (x_i - \mu_X)/\mu_X}\right]$$

Expanding the denominator using Taylor's formula,

$$\begin{aligned} r_i &= R[1 + (y_i - \mu_Y)/\mu_Y][1 - (x_i - \mu_X)/\mu_X + (x_i - \mu_X)^2/\mu_X^2 - \cdots] \\ &= R\{1 + [(y_i - \mu_Y)/\mu_Y] - [(x_i - \mu_X)/\mu_X] + [(x_i - \mu_X)^2/\mu_X^2] - [(x_i - \mu_X)^3/\mu_X^3] \\ &\quad + \cdots - [(y_i - \mu_Y)(x_i - \mu_X)/\mu_Y\mu_X] + [(y_i - \mu_Y)(x_i - \mu_X)^2/\mu_Y\mu_X^2] \\ &\quad - [(y_i - \mu_Y)(x_i - \mu_X)^3/\mu_Y\mu_X^3] + \cdots\} \end{aligned} \tag{19.34}$$

$$\begin{aligned} E(r_i) = R\{&1 + E[(y_i - \mu_Y)/\mu_Y] - E[(x_i - \mu_X)/\mu_X] + E[(x_i - \mu_X)^2/\mu_X^2] \\ &- E[(x_i - \mu_X)^3/\mu_X^3] + \cdots - E[(y_i - \mu_Y)(x_i - \mu_X)/\mu_Y\mu_X] \\ &+ E[(y_i - \mu_Y)(x_i - \mu_X)^2/\mu_Y\mu_X^2] \\ &- E[(y_i - \mu_Y)(x_i - \mu_X)^3/\mu_Y\mu_X^3] + \cdots\} \end{aligned} \tag{19.35}$$

However,

$$E(y_i - \mu_Y) = \frac{1}{N}\sum_i^N (y_i - \mu_Y) = 0$$

$$E(x_i - \mu_X) = \frac{1}{N}\sum_i^N (x_i - \mu_X) = 0$$

$$E(x_i - \mu_X)^2 = \frac{1}{N}\sum_i^N (x_i - \mu_X)^2 = \sigma_X^2$$

and

$$E(y_i - \mu_Y)(x_i - \mu_X) = \frac{1}{N}\sum_i^N (y_i - \mu_Y)(x_i - \mu_X) = \sigma_{XY} = \text{COV}(XY)$$

Then

$$E(r_i) = R[1 + \sigma_X^2 / \mu_X^2 - \cdots - \sigma_{XY} / \mu_Y \mu_X + \cdots]$$

and

$$\text{BIAS}_{r_M} = E(r_M - R) = E(r_M) - R = E(r_i) - R$$

$$= R + R\sigma_X^2 / \mu_X^2 - \cdots - R\sigma_{XY} / \mu_Y \mu_X + \cdots - R \tag{19.36}$$

$$= R\sigma_X^2 / \mu_X^2 - \cdots - R\sigma_{XY} / \mu_Y \mu_X + \cdots, \quad \text{or}$$

$$\text{BIAS}_{r_M} = R\sigma_X^2 / \mu_X^2 - \cdots - R\rho_{XY}\sigma_X\sigma_Y / \mu_Y \mu_X + \cdots \tag{19.37}$$

In contrast to the bias of the ratio of means, the magnitude of the bias of the mean of ratios is not a function of sample size. Increasing sample size has no effect on the bias. This means that the estimate, $r_M$, will not converge on the parameter, $R$, as the sample size approaches the size of the population. Consequently, the mean of ratios is not a consistent estimator.

### 19.3.3 Unbiased Ratio-Type Estimators

If the bias of an estimator can be identified and quantified, it is logical to consider correcting the estimator by adding to it, or subtracting from it, the bias. This leads to a class of estimators referred to as *unbiased* or *almost unbiased*. For example, Hartley and Ross (1954) proposed such a correction for the mean of ratios. The uncorrected estimator is

$$r_M = \frac{1}{n}\sum_i^n \frac{y_i}{x_i} = \frac{1}{n}\sum_i^n r_i$$

According to Equation 19.31

$$\text{BIAS}_{r_M} = \mu_{r_M} - R = -\text{COV}(r_i x_i) / \mu_X = -\sigma_{r_i x_i} / \mu_X$$

$$s_{r_i x_i} = \frac{1}{(n-1)}\sum_i^n (r_i - r_M)(x_i - \bar{x})$$

Then, since $r_i x_i = y_i$,

$$s_{r_i x_i} = \frac{1}{(n-1)}\left(\sum_i^n y_i - r_M \sum_i^n x_i - \bar{x}\sum_i^n r_i + n r_M \bar{x}\right)$$

$$= \frac{n}{(n-1)}(\bar{y} - r_M \bar{x}) \to \sigma_{r_i x_i}\left(\frac{N}{N-1}\right)$$

$$\frac{n(N-1)}{N(n-1)}(\bar{y}_i - r_M \bar{x}) \to \sigma_{r_i x_i}$$

and the unbiased estimator is

$$r_{HR} = r_M + \frac{n(N-1)(\bar{y} - r_M \bar{x})}{N(n-1)\mu_X} \to R \tag{19.38}$$

Other unbiased estimators have been proposed, e.g., Lahiri (1951), Mickey (1959), Quenouille (1956), and Tin (1965), which have somewhat different characteristics and efficiencies. As an example of these, examine the estimator developed by Mickey, using Ek (1971), because of its simplicity, as a guide.

A random sample, without replacement, of size $n$, is drawn from the sampling frame or population. This sample is divided into $g$ subsets, each containing $m$ elements. The estimator of $R$ is then

$$r_{\text{Mick}} = \bar{r}_M + \frac{(N - n + m)g(\bar{y} - \bar{r}_g \bar{x})}{N\mu_X} \rightarrow R \tag{19.39}$$

where $\bar{r}_g = \frac{1}{g}\sum_i^g r_i$

$r_i = \frac{n\bar{y} - m\bar{y}_i}{n\bar{x} - m\bar{x}_i}$, the ratio of means after the $i$th subset has been excluded

$\bar{y}_i$ = mean of $Y$ in the $i$th subset

$\bar{x}_i$ = mean of $X$ in the $i$th subset

As an example, assume a population of size $N = 100$, with $\mu_X = 3.5$, from which a sample of size $n = 6$ has been drawn. The sample has been divided into $g = 3$ subsets, each containing $m = 2$ elements. The data are

| | Subset 1 | | Subset 2 | | Subset 3 | | | |
|---|---|---|---|---|---|---|---|---|
| $x =$ | 4 | 1 | 3 | 5 | 6 | 2 | $\bar{x} = 3.5$ | $n\bar{x} = 21.0$ |
| $Y =$ | 8 | 2 | 6 | 10 | 12 | 4 | $\bar{y} = 7.0$ | $n\bar{y} = 42.0$ |
| | $\bar{x}_1 = 2.5$ | | $\bar{x}_2 = 4.0$ | | $\bar{x}_3 = 4.0$ | | | |
| | $\bar{y}_1 = 5.0$ | | $\bar{y}_2 = 8.0$ | | $\bar{y}_3 = 8.0$ | | | |
| | $m\bar{x}_1 = 5.0$ | | $m\bar{x}_2 = 8.0$ | | $m\bar{x}_3 = 8.0$ | | | |
| | $m\bar{y}_1 = 10.0$ | | $m\bar{y}_2 = 16.0$ | | $m\bar{y}_3 = 16.0$ | | | |

$$r_1 = \frac{42 - 10}{21 - 5} = 2$$

$$r_2 = \frac{42 - 16}{21 - 8} = 2$$

$$r_3 = \frac{42 - 16}{21 - 8} = 2$$

$$\bar{r}_g = \frac{1}{3}(2 + 2 + 2) = 2$$

$$r_{\text{Mick}} = 2 + \frac{(100 - 6 - 2)3[7 - 2(3.5)]}{100(3.5)} = 2 + 0$$

$$= 2$$

The sample data indicate perfect correlation between the two variables. In such a case the bias should be equal to zero, as it is. For further discussions of the Mickey estimator see Ek (1971), and Sukhatme et al. (1984).

## 19.4 APPROXIMATIONS OF THE RATIO ESTIMATORS

The expressions developed in Section 19.3 which permit the identification of the factors controlling the magnitudes of the biases of the ratio estimators can also be used to generate approximations that may be easier to use than are the exact expressions.

### 19.4.1 Ratio of Means

When the sample size can be considered large enough that $\bar{x}$ can be assumed to be equal to $\mu_X$, Equation 19.18 reduces to

$$r_R \doteq R\left[1 + \left(\frac{(\bar{y} - \mu_Y)}{\mu_Y}\right)\right] = R + R\left(\frac{(\bar{y} - \mu_Y)}{\mu_Y}\right) \qquad (19.40)$$

because the terms containing $(\bar{x} - \mu_X)$ are considered equal to zero. Then,

$$E(r_R) = R + R\left[E\frac{(\bar{y} - \mu_Y)}{\mu_Y}\right] = R + R\left[\frac{1}{L}\sum_h^L (\bar{y}_h - \mu_Y)\right]$$

where

$$L = \binom{N}{n}$$

the total number of possible samples of size $n$. Since

$$\sum_h^L (\bar{y}_h - \mu_Y) = 0$$

then

$$E(r_R) \doteq R$$

and $r_R$ can be considered to be free of bias.

An examination of Equation 19.18 will reveal that the magnitudes of the individual terms decrease rapidly from the term $(\bar{x} - \mu_x)/\mu_x$. If the sample size is sufficiently large so that the difference between $\bar{x}$ and $\mu_X$ can be considered to be of no consequence, the terms quickly become so small that they can be ignored. For example, if $\bar{y} = 51$, $\mu_Y = 50$, $\bar{x} = 101$, and $\mu_X = 100$, the expression becomes

$$r_R = R\left[1 + \frac{1}{50} - \frac{1}{100} + \frac{1}{10{,}000} - \frac{1}{1{,}000{,}000} + \cdots \right.$$
$$\left. - \frac{1}{5000} + \frac{1}{500{,}000} - \frac{1}{50{,}000{,}000} \cdots \right]$$

Thus, if the terms beyond $(\bar{x} - \mu_X)/\mu_X$ can be ignored, Equation 19.18 can be written

$$
\begin{aligned}
r_R &\doteq R\left[1 + \frac{(\bar{y} - \mu_Y)}{\mu_Y} - \frac{(\bar{x} - \mu_X)}{\mu_X}\right] \\
&\doteq R + R\left(\frac{\bar{y} - \mu_Y}{\mu_Y}\right) - R\left(\frac{\bar{x} - \mu_X}{\mu_X}\right)
\end{aligned} \tag{19.41}
$$

Then

$$
\begin{aligned}
E(r_R) &= R + R\left[\frac{E(\bar{y} - \mu_Y)}{\mu_Y}\right] - R\left[\frac{E(\bar{x} - \mu_X)}{\mu_X}\right] \\
&= R + 0 - 0 \\
&= R
\end{aligned}
$$

and $r_R$ can be considered to be unbiased.

When the magnitudes of the terms in Equation 19.18 are such that the equation can be reduced to

$$
\begin{aligned}
r_R &\doteq R\left[1 + \frac{(\bar{y} - \mu_Y)}{\mu_Y} - \frac{(\bar{x} - \mu_X)}{\mu_X} + \frac{(\bar{x} - \mu_X)^2}{\mu_X^2} - \frac{(\bar{y} - \mu_Y)(\bar{x} - \mu_X)}{\mu_Y \mu_X}\right] \\
&\doteq R + R\left(\frac{\bar{y} - \mu_Y}{\mu_Y}\right) - R\left(\frac{\bar{x} - \mu_X}{\mu_X}\right) + R\left(\frac{\bar{x} - \mu_X}{\mu_X}\right)^2 - R\left[\frac{(\bar{y} - \mu_Y)(\bar{x} - \mu_X)}{\mu_Y \mu_X}\right]
\end{aligned} \tag{19.42}
$$

Then

$$
\begin{aligned}
E(r_R) &= R + R\left[\frac{E(\bar{y} - \mu_Y)}{\mu_Y}\right] - R\left[\frac{E(\bar{x} - \mu_X)}{\mu_X}\right] + R\left[\frac{E(\bar{x} - \mu_X)^2}{\mu_X^2}\right] - R\left[\frac{E(\bar{y} - \mu_Y)(\bar{x} - \mu_X)}{\mu_Y \mu_X}\right] \\
&= R + 0 - 0 + \frac{R}{M\mu_X^2}\sum_i^M (\bar{x}_i - \mu_X)^2 - \frac{R}{M\mu_Y\mu_X}\sum_i^M (\bar{y}_i - \mu_Y)(\bar{x}_i - \mu_X) \\
&= R + \frac{R\sigma_{\bar{x}}^2}{\mu_X^2} - \frac{R\,\mathrm{COV}(\bar{x}\,\bar{y})}{\mu_Y \mu_X} \\
&= R + \frac{R\sigma_X^2}{\mu_X^2 n}\left(\frac{N-n}{N-1}\right) - \frac{R\,\mathrm{COV}(XY)}{\mu_Y \mu_X n}\left(\frac{N-n}{N-1}\right)
\end{aligned} \tag{19.43}
$$

Since $R = \mu_Y/\mu_X$,

$$
\begin{aligned}
\mathrm{BIAS} = E(r_R) - R &\doteq \frac{R\sigma_X^2}{\mu_X^2 n}\left(\frac{N-n}{N-1}\right) - \frac{\mu_Y}{\mu_X}\left(\frac{1}{\mu_Y \mu_X}\right)\frac{\sigma_{XY}}{n}\left(\frac{N-n}{N-1}\right) \\
&\doteq \frac{1}{\mu_X^2 n}\left(\frac{N-n}{N-1}\right)[R\sigma_X^2 - \sigma_{XY}]
\end{aligned} \tag{19.44}
$$

which can be written

$$\text{BIAS}_{r_R} \doteq \frac{1}{\mu_X^2 n}\left(\frac{N-n}{N-1}\right)(R\sigma_X^2 - \rho_{XY}\sigma_X\sigma_Y) \tag{19.45}$$

When $R\sigma_X^2 = \rho_{XY}\sigma_X\sigma_Y$, one can consider $r_R$ to be free of bias. Note that if one ignores the finite population correction, the bias is inversely proportional to the sample size, so that if, for example, the sample size is doubled, the bias will be cut in half.

As can be seen from this discussion, $r_R$ is usually a biased estimator of $R$ but the magnitude of the bias is, to a large extent, a function of sample size. It usually is assumed that the bias will be acceptably low if the sample size exceeds 30.

### 19.4.2 Mean of Ratios

Where $n$ is sufficiently large that it can be assumed that $\bar{x} = \mu_X$, Equation 19.35 can be reduced to

$$E(r_i) = E(r_M) \doteq R\{1 + E[(y_i - \mu_Y)/\mu_Y]\}$$
$$\doteq R + 0 = R$$

Consequently,

$$\text{BIAS}_{r_M} \doteq E(r_M) - R = R - R = 0$$

indicating that the mean of ratios is essentially unbiased if $\bar{x}$ can be assumed to be equal to $\mu_X$.

If it can be assumed that the falloff in magnitude of the terms in Equation 19.34, as was demonstrated in Section 19.4.1, is such that Equation 19.35 can be shortened to

$$E(r_i) = E(r_M) \doteq R\{1 + E[(y_i - \mu_Y)/\mu_Y] - E[(x_i - \mu_X)/\mu_X] + E[(x_i - \mu_X)^2/\mu_X^2]$$
$$-E[(y_i - \mu_Y)(x_i - \mu_X)/\mu_Y\mu_X]\}$$
$$\doteq R + 0 - 0 + \frac{R}{N\mu_X^2}\sum_i^N (x_i - \mu_X)^2 - \frac{R}{N\mu_X\mu_Y}\sum_i^N (y_i - \mu_Y)(x_i - \mu_X)$$
$$\doteq R + \frac{R\mu_X^2}{\mu_X^2} - \frac{R\sigma_{XY}}{\mu_Y\mu_X}$$

and

$$\text{BIAS}_{r_M} \doteq E(r_M) - R$$
$$\doteq R + \frac{R\sigma_X^2}{\mu_X^2} - \frac{R\sigma_{XY}}{\mu_Y\mu_X} - R$$
$$\doteq \frac{R\sigma_X^2}{\mu_X^2} - \frac{R\sigma_{XY}}{\mu_Y\mu_X} \tag{19.46}$$
$$\doteq \frac{R\sigma_X^2}{\mu_X^2} - \frac{\mu_Y}{\mu_X}\frac{\sigma_{XY}}{\mu_Y\mu_X}$$
$$\doteq \frac{1}{\mu_X^2}(R\sigma_X^2 - \sigma_{XY}) \quad \text{or}$$

$$\doteq \frac{1}{\mu_X^2}(R\sigma_X^2 - \rho_{XY}\sigma_X\sigma_Y) \tag{19.47}$$

This development indicates that, in contrast to the ratio of means, the estimate will be bias free *only* when $\bar{x}$ *actually* is equal to $\mu_X$. The degree of proximity of $\bar{x}$ to $\mu_X$ is not a factor. If the two do not agree, the full bias will be present. Consequently, increasing sample size will not reduce the bias except when $n = N$.

## 19.5 USING A RATIO TO ESTIMATE A POPULATION TOTAL AND MEAN

Earlier it was stated that, under certain circumstances, one can take advantage of the correlation between the variable of interest and another, more easily evaluated, variable to obtain a better estimate of the population total than could be obtained using simple random sampling with a sample of the same size. This possibility is explored in this section.

### 19.5.1 RATIO OF MEANS

According to Equations 19.1 and 19.2,

$$R = \mu_Y/\mu_X = \tau_Y/\tau_X$$

Then,

$$\tau_Y = R\tau_X \tag{19.48}$$

According to Equation 19.8,

$$r_R = \bar{y}/\bar{x}$$

Then

$$r_R = N\bar{y}/N\bar{x} = T_Y/T_X$$

and

$$\begin{aligned} T_Y &= r_R T_X \\ &= (\bar{y}/\bar{x})N\bar{x} \end{aligned} \tag{19.49}$$

$$= N\bar{y} \to \tau_Y \tag{19.50}$$

which is the expression used to estimate $\tau_Y$ when using simple random sampling. Consequently, if this procedure is followed, the use of the ratio will not yield an estimate that is any better than one obtained using simple random sampling.

Situations arise, however, where the population total of the independent variable, $X$, can be obtained without difficulty. This rarely occurs in the context of forestry but the possibility exists. When it does occur, Equation 19.49 can be written

$$T_{Y_R} = r_R \tau_X \to \tau_Y \tag{19.51}$$

The expected value of $T_{Y_R}$ is then

$$E(T_{Y_R}) = E(r_R)\tau_X = \mu_{r_R}\tau_X$$

According to Equation 19.15,

$$\mu_{r_R} = \frac{\mu_Y}{\mu_X} - \frac{\text{COV}(r_R\bar{x})}{\mu_X}$$

Then,

$$\begin{aligned} E(T_{Y_R}) &= \left[\frac{\mu_Y}{\mu_X} - \frac{\text{COV}(r_R\bar{x})}{\mu_X}\right]N\mu_X \\ &= N\mu_Y - N\,\text{COV}(r_R\bar{x}) \qquad (19.52) \\ &= \tau_Y - N\,\text{COV}(r_R\bar{x}) \end{aligned}$$

which can be written

$$E(T_{Y_R}) = \tau_Y - N\rho_{r_R\bar{x}}\sigma_{r_R}\sigma_{\bar{x}} \tag{19.53}$$

As can be seen, only when the correlation between $r_R$ and $\bar{x}$ is equal to zero will $r_R\tau_X$ be an unbiased estimator of $\tau_Y$. However, if one uses the arguments from Section 19.4, one can see that, given the proper circumstances, the bias will be minimal. When the sample is large enough so that $\bar{x}$ can be considered equal to $\mu_X$, $E(r_R)$ can be considered equal to $R$ and $r_R\tau_X$ will be an essentially bias-free estimator of $\tau_Y$. In like manner, if the difference between $\bar{x}$ and $\mu_X$ is so small that the terms beyond $(\bar{x} - \mu_X)/\mu_X$ in Equation 19.18 can be considered to be negligible, $E(r_R)$ can again be considered to be equal to $R$ and $r_R\tau_X$ will be unbiased.

$$E(T_{Y_R}) = E(r_R)\tau_X$$

Then, from Equation 19.43,

$$\begin{aligned} E(T_{Y_R}) &= \tau_X\left[R + \frac{R\sigma_X^2}{\mu_X^2 n}\left(\frac{N-n}{N-1}\right) - \frac{R\,\text{COV}(XY)}{\mu_Y\mu_X n}\left(\frac{N-n}{N-1}\right)\right] \\ &= \tau_X R\left[1 + \frac{\sigma_X^2}{\mu_X^2 n}\left(\frac{N-n}{N-1}\right) - \frac{\text{COV}(XY)}{\mu_Y\mu_X n}\left(\frac{N-n}{N-1}\right)\right] \end{aligned}$$

Thus, when

$$\frac{\sigma_X^2}{\mu_X^2 n}\left(\frac{N-n}{N-1}\right) - \frac{\text{COV}(XY)}{\mu_Y\mu_X n}\left(\frac{N-n}{N-1}\right)$$

$r_R\tau_X$ will be an essentially bias-free estimator of $\tau_Y$.

Since $\tau_Y = N\mu_Y$, if one divides the estimator of $\tau_Y$ by $N$, the resulting quotient will be an estimator of the population mean, $\mu_Y$. According to Equation 19.51,

$$T_{Y_R} = r_R \tau_X$$

$$\frac{T_{Y_R}}{N} = \frac{r_R \tau_X}{N} \tag{19.54}$$

$$\hat{\bar{y}}_R = r_R(\tau_X / N) = r_R \mu_X \rightarrow \mu_Y$$

### 19.5.2 Mean of Ratios

The mean of ratios, $r_M$, is an estimate of $R$ and can be used to make estimates of $\tau_Y$ and $\mu_Y$ but the details are somewhat different. Substituting $r_M$ for $r_R$ in Equation 19.51,

$$T_{Y_M} = r_M \tau_X \rightarrow \tau_Y \tag{19.55}$$

Then,

$$E(T_{Y_M}) = E(r_M)\tau_X = E(r_M)N\mu_X$$

$$= \mu_{r_M} N\mu_X$$

Making use of Equation 19.30,

$$E(T_{Y_M}) = N\mu_X \left[ R - \frac{\text{COV}(r_i x_i)}{\mu_X} \right]$$

$$= N[R\mu_X - \text{COV}(r_i x_i)] \tag{19.56}$$

$$= N[\mu_Y - \text{COV}(r_i x_i)]$$

$$= \tau_Y - N\,\text{COV}(r_i x_i) \quad \text{or}$$

$$= \tau_Y - N\rho_{r_i X}\sigma_{r_i}\sigma_x \tag{19.57}$$

Again the estimator is biased. In the case of the mean of ratios, however, the magnitude of the bias cannot be reduced by increasing sample size. One has here an "either or" situation. If by chance $\bar{x}$ is equal to $\mu_X$, there will be no bias. If the two are not equal, the full-sized bias will be present regardless of how small the difference is between $\bar{x}$ and $\mu_X$. This means that if $V(y_{ij}|x_i) = cx_i^2$, as shown in Equation 19.13, and Figure 19.3, the appropriate estimator is the mean of ratios, but it will be almost impossible to escape its bias.

As in the case of the ratio of means, one can estimate the population mean of $Y$ using the mean of ratios.

$$\hat{\bar{y}}_M = r_M \mu_X \rightarrow \mu_Y \tag{19.58}$$

### 19.5.3 Unbiased Ratio-Type Estimators

By definition, these estimators are free of bias and can be used to obtain estimators of both the population total and the mean of the variable of interest. Thus,

$$T_{Y_{MC}} = r_{MC}\tau_X \rightarrow \tau_Y \tag{19.59}$$

and

$$\hat{\bar{y}}_{MC} = r_{MC}\mu_X \rightarrow \mu_Y \tag{19.60}$$

## 19.6 VARIANCE OF THE RATIO ESTIMATES

There are two obstacles blocking a clear-cut statement regarding these variances. The first is the nature of Equations 19.18 and 19.34 which contain an indefinite number of terms, forcing the use of approximations. Second, the values of the independent variable, $X$, are not fixed, as is required in standard regression analysis. Consequently, it will be assumed that the sample size is large enough that $\bar{x}$ is essentially equal to $\mu_X$. If this condition is not met, the formulae presented will not be valid. Furthermore, it is assumed that the variance about the regression line is homogeneous.

### 19.6.1 Ratio of Means

Fundamentally, the variance of the ratio of means is

$$V(r_R) = E(r_R - R)^2 \tag{19.61}$$

where $r_R = \bar{y}/\bar{x}$. Then,

$$r_R - R = \frac{\bar{y}}{\bar{x}} - R = \frac{\bar{y} - R\bar{x}}{\bar{x}} = \frac{1}{\bar{x}n}\sum_i^n (y_i - Rx_i) \tag{19.62}$$

According to Equation 19.40, when it can be assumed that $\bar{x} = \mu_X$, and since $\mu_y = R\mu_x$,

$$r_R - R \doteq \frac{\bar{y} - R\mu_X}{\mu_X}$$

Since it is assumed that $\bar{x} = \mu_X$,

$$r_R - R \doteq \frac{\bar{y} - R\bar{x}}{\mu_X}$$

According to Equation 19.62,

$$\bar{y} - R\bar{x} = \frac{1}{n}\sum_i^n (y_i - Rx_i)$$

Let $d_i = y_i - Rx_i$. Then,

$$r_R - R \doteq \frac{1}{\mu_X}\left(\frac{1}{n}\sum_i^n d_i\right) = \frac{\bar{d}}{\mu_X}$$
$$\mu_d = \frac{1}{N}\sum_i^N d_i = \frac{1}{N}\sum_i^N (y_i - Rx_i) = \mu_Y - R\mu_X \tag{19.63}$$

Since $R = \mu_Y/\mu_X$, $\mu_Y = R\mu_X$, and $\mu_d = \mu_Y - R\mu_X = 0$, Equation 19.63 can be written

$$r_R - R \doteq (\bar{d} - \mu_d)/\mu_X$$

Then,

$$E(r_R - R)^2 \doteq E[(\bar{d} - \mu_d)/\mu_X]^2$$
$$\doteq \frac{1}{\mu^2} E(\bar{d} - \mu_d)^2$$

$$E(\bar{d} - \mu_d)^2 - V(\bar{d}) = \frac{\sigma_d^2}{n}\left(\frac{N-n}{N-1}\right)$$

where

$$\sigma_d^2 = \frac{1}{N}\sum_i^N (d_i - \mu_d)^2 = \frac{1}{N}\sum_i^N d_i^2$$

Then,

$$E(r_R - R)^2 = V(r_R) = \sigma_{r_R}^2 \doteq V(\bar{d})/\mu_X^2$$
$$\doteq \frac{1}{\mu_X^2 n}\left(\frac{N-n}{N-1}\right)\left(\frac{1}{N}\right)\sum_i^N d_i^2 \tag{19.64}$$
$$\doteq \frac{1}{\mu_X^2 n}\left(\frac{N-n}{N-1}\right)\left(\frac{1}{N}\right)\sum_i^N (y_i - Rx_i)^2$$

It must be emphasized that Equation 19.64 is an approximation that is valid only if the sample size is large enough so that one can be reasonably certain that $\bar{x}$ is equal to, or almost equal to, $\mu_X$. The approximate nature of the equation can be demonstrated if it is used with the data from the example where $\rho_{XY} = 0.97333$

$$V(r_R) \doteq \frac{1}{(20)^2 3}\left(\frac{5-3}{5-1}\right)\left(\frac{5.00}{5}\right) = 0.0004167$$

When $V(r_R)$ is computed directly using the data from the $r$ column,

$$V(r_R) = \frac{1}{L}\sum_{h}^{L}(r_{R_h} - \mu_R)^2 = 0.0005837$$

The difference between the two values of $V(r_R)$ is caused by the general lack of agreement between the sample and population means, which violated the assumption upon which Equation 19.64 is based. Contributing to the problem is the extremely small size of the population which made it impossible to obtain samples large enough to provide stable $\bar{x}$ values.

Consider the term $\Sigma_i^N (y_i - Rx_i)^2$ in Equation 19.64. It is actually the sum of squared errors from the regression line.

$$\sum_{i}^{N}(y_R - Rx_i)^2 = \sum_{i}^{N}[(y_i - \mu_Y) - (Rx_i - \mu_Y)]^2$$

$$= \sum_{i}^{N}[(y_i - \mu_Y) - (Rx_i - R\mu_X)]^2$$

$$= \sum_{i}^{N}[(y_i - \mu_Y) - R(x_i - \mu_X)]^2 \quad (19.65)$$

$$= \sum_{i}^{N}(y_i - \mu_Y)^2 + R^2\sum_{i}^{N}(x_i - \mu_X)^2 - 2R\sum_{i}^{N}(y_i - \mu_Y)(x_i - \mu_X)$$

$$= N\sigma_Y^2 + R^2 N\sigma_X^2 - 2RN\,\mathrm{COV}(XY)$$

$$= N(\sigma_y^2 + R^2\sigma_X^2 - 2R\sigma_{XY}) \quad \text{or}$$

$$= N(\sigma_Y^2 + R^2\sigma_X^2 - 2R\rho_{XY}\sigma_X\sigma_Y) \quad (19.66)$$

Then Equation 19.64 can be written

$$V(r_R) = \sigma_{r_R}^2 \doteq \frac{1}{\mu_X^2 n}\left(\frac{N-n}{N-1}\right)(\sigma_Y^2 + R^2\sigma_X^2 - 2R\sigma_{XY}) \quad \text{or} \quad (19.67)$$

$$\doteq \frac{1}{\mu_X^2 n}\left(\frac{N-n}{N-1}\right)(\sigma_Y^2 + R^2\sigma_X^2 - 2R\rho_{XY}\sigma_X\sigma_Y) \quad (19.68)$$

It should be noted that $V(r_R)$ will equal zero when $\rho_{XY} = 1$.

$$R = \beta_1 = \sigma_{XY}/\sigma_X^2 = (\sigma_{XY}/\sigma_X)(1/\sigma_X)$$

$$\rho_{XY} = \sigma_{XY}/\sigma_X\sigma_Y = (\sigma_{XY}/\sigma_X)(1/\sigma_Y)$$

Then, when $\rho_{XY} = 1$,

$$1 = \left(\frac{\sigma_{XY}}{\sigma_X}\right)\frac{1}{\sigma_Y}$$

$$\sigma_Y = \frac{\sigma_{XY}}{\sigma_X}$$

$$R = \left(\frac{\sigma_{XY}}{\sigma_X}\right)\frac{1}{\sigma_X} = \frac{\sigma_Y}{\sigma_X}$$

Then,

$$\begin{aligned}&\sigma_Y^2 + R^2\sigma_X^2 - 2R\rho_{XY}\sigma_X\sigma_Y\\ &= \sigma_Y^2 + \frac{\sigma_Y^2}{\sigma_X^2}\sigma_X^2 - 2\frac{\sigma_Y}{\sigma_X}\sigma_x\sigma_Y\\ &= \sigma_Y^2 + \sigma_Y^2 - 2\sigma_Y^2 = 0\end{aligned}$$

Substituting the zero into Equation 19.68 will make $V(r_R) = 0$. As $\rho_{XY}$ diminishes, $V(r_R)$ will increase until $\rho_{XY} = 0$, at which point

$$V(r_R) \doteq \frac{1}{\mu_X^2 n}\left(\frac{N-n}{N-1}\right)(d_Y^2 + R^2\sigma_X^2) \tag{19.69}$$

Using the data from the example where $\rho_{XY} = 0.97333$ in Equation 19.68,

$$\begin{aligned}V(r_R) &\doteq \frac{1}{(20)^2 3}\left(\frac{5-3}{5-1}\right)[19.0 + (0.3)^2 200.0 - 2(0.3)(0.9733)(14.142136)(4.35890)]\\ &\doteq 0.0004167\end{aligned}$$

which agrees with the previous computations.

In the discussion of the bias of $r_R$, it was brought out that the bias is inversely proportional to the sample size. As can be seen from Equations 19.64, 19.67, and 19.68, the variance of $r_R$ is also essentially inversely proportional to the sample size. Furthermore, if $V(r_R)$ is computed under the assumption that Equation 19.43, rather than Equation 19.40, is appropriate, it will be seen that again it is inversely proportional to sample size. Then, the ratio of the bias to the standard error of $r_R$, $\sqrt{V(r_R)}$, known as the *relative bias*, is $(1/n)/(1/\sqrt{n}) = 1/\sqrt{n}$, which becomes negligible as $n$ increases.

To complete this discussion, the variance of the estimate of $\mu_Y$ is

$$V(\bar{y}_R) = \sigma_{\bar{y}_R}^2 \doteq V(r_R\mu_X) = \mu_X^2 V(r_R) \tag{19.70}$$

and the variance of the estimate of $\tau_Y$ is

$$V(T_{Y_R}) = \sigma_{T_{Y_R}}^2 \doteq V(r_R\tau_X) = \tau_X^2 V(r_R) = N^2\mu_X^2 V(r_R) \tag{19.71}$$

The sample-based estimate of $V(r_R)$ is derived from Equation 19.64, which is based on the assumption that $\bar{x}$ is equal to $\mu_X$ and that $\mu_X$ is known.

$$s_{r_R}^2 = \frac{1}{\mu_X^2 n}\left(\frac{N-n}{N}\right)\left(\frac{1}{n-1}\right)\sum_i^n (y_i - r_R x_i)^2 \rightarrow V(r_R) \quad (19.72)$$

or from Equation 19.67

$$s_{r_R}^2 = \frac{1}{\mu_X^2 n}\left(\frac{N-n}{N}\right)(s_Y^2 + r_R^2 s_X^2 - 2r_R s_{XY}) \quad (19.73)$$

Then, from Equation 19.70, the estimated variance of $\bar{y}_R$ is

$$s_{\hat{\bar{y}}_R}^2 = \mu_X^2 s_{r_R}^2 = \left(\frac{N-n}{Nn}\right)(s_Y^2 + r_R^2 s_X^2 - 2r_R s_{XY}) \rightarrow V(\bar{y}_R) \quad (19.74)$$

and from Equation 19.71, the estimated variance of $T_{Y_R}$ is

$$s_{T_{Y_R}}^2 = N^2 \mu_X^2 s_{r_R}^2 = N\left(\frac{N-n}{n}\right)(s_Y^2 + r_R^2 s_X^2 - 2r_R s_{XY}) \rightarrow V(T_{Y_R}) \quad (19.75)$$

### 19.6.2 Mean of Ratios

According to Equation 19.34, when $\bar{x}$ can be considered equal to $\mu_X$

$$\begin{aligned} r_i &= R + R\left(\frac{y_i - \mu_Y}{\mu_Y}\right) \\ &= R\left[1 + \left(\frac{y_i - \mu_Y}{\mu_Y}\right)\right] \\ &= R\left(\frac{\mu_Y + y_i - \mu_Y}{\mu_Y}\right) \\ &= \frac{Ry_i}{\mu_Y} = \left(\frac{\mu_Y}{\mu_X}\right)\frac{y_i}{\mu_Y} = \frac{y_i}{\mu_X} \end{aligned}$$

Then

$$\begin{aligned} E(r_i) &= \frac{1}{N}\sum_i^N r_i = \frac{1}{N}\sum_i^N\left(\frac{y_i}{\mu_X}\right) = \frac{\mu_Y}{\mu_X} = R \\ r_M &= \frac{1}{n}\sum_i^n r_i = \frac{1}{n}\sum_i^n\left(\frac{y_i}{\mu_X}\right) = \frac{\bar{y}}{\mu_X} \end{aligned} \quad (19.76)$$

Let

$$L = \binom{N}{n}$$

the total number of possible samples of size $n$ when the sampling is without replacement. Then,

$$E(r_M) \doteq \frac{1}{L}\sum_h^L r_{M_h} = \frac{1}{L}\sum_h^L \left(\frac{\bar{y}}{\mu_X}\right)_h = \frac{\mu_Y}{\mu_X} = R \tag{19.77}$$

$$E(r_i - R)^2 = V(r_i) = \sigma^2_{r_i} \doteq \frac{1}{N}\sum_i^N (r_i - R)^2 \tag{19.78}$$

and

$$E(r_M - R)^2 = V(r_M) = \sigma^2_{r_M} \doteq \frac{1}{L}\sum_h^L (r_{M_h} - R)^2 \tag{19.79}$$

which can be written, when the sampling is without replacement,

$$E(r_M - R)^2 = V(r_M) \doteq \frac{\sigma^2_{r_i}}{n}\left(\frac{N-n}{N-1}\right) \tag{19.80}$$

As an example, one can again use the data from the population where $\rho_{XY} = 0.97333$.

**$r_i$**

0/0 = 0
3.5/10 = 0.35
4.5/20 = 0.225
10.5/30 = 0.35
11.5/40 = 0.2875

$$\mu_{r_i} = 0.2425; \quad \sigma^2_{r_i} = 0.01685$$

| Sample | Ratios | $r_M$ |
|---|---|---|
| 1 | 0;0.35;0.225 | 0.191667 |
| 2 | 0;0.35;0.35 | 0.2333333 |
| 3 | 0;0.35;0.2875 | 0.2125000 |
| 4 | 0;0.225;0.35 | 0.1916667 |
| 5 | 0;0.225;0.2875 | 0.1708333 |
| 6 | 0;0.35;0.2875 | 0.2125000 |
| 7 | 0.35;0.225;0.35 | 0.3083333 |

| Sample | Ratios | $r_M$ |
|---|---|---|
| 8 | 0.35;0.225;0.2875 | 0.2875000 |
| 9 | 0.35;0.35;0.2875 | 0.3291667 |
| 10 | 0.225;0.35;0.2875 | 0.2875000 |

$$\mu_{r_M} = 0.2425; \quad \sigma^2_{r_M} = 0.0028083$$

$$\frac{\sigma^2_{r_i}}{n}\left(\frac{N-n}{N-1}\right) = \frac{0.01685}{3}\left(\frac{5-3}{5-1}\right) = 0.0028083 = V(r_M) = \sigma^2_{r_M}$$

The variance of the mean of *Y*, estimated using the mean of ratios, is

$$V(\bar{y}_M) = \sigma^2_{\bar{y}_M} = V(r_M\mu_X) \doteq \mu^2_X V(r_M) \tag{19.81}$$

and the variance of the total of *Y* is

$$V(T_{Y_M}) = \sigma^2_{T_{Y_M}} \doteq V(r_M\tau_X) = \tau^2_X V(r_M) = N^2\mu^2_X V(r_M) \tag{19.82}$$

As can be seen from Equations 19.80 through 19.82, if the finite population correction is ignored, the variances of $r_M$, $\bar{y}_M$, and $T_{Y_M}$ are all inversely proportional to the sample size. Sample size also affects the finite population correction, which approaches zero as $n$ approaches $N$. Since the bias of $r_M$ is not a function of $n$, the relative bias of $r_M$ is approximately $1/(1/\sqrt{n}) = \sqrt{n}$, which means that in the case of the mean of ratios the bias relative to the standard error increases as $n$ increases, which is the reverse of the pattern associated with the ratio of means.

The sample-based estimates of the variances are as follows. The variance of the individual ratios, from Equation 19.78, is

$$s^2_{r_i} = \frac{1}{(n-1)}\sum_i^n (r_i - r_M)^2 \to V(r_i) \tag{19.83}$$

the variance of the mean of ratios, from Equation 19.80, is

$$s^2_{r_M} = \frac{s^2_{r_i}}{n}\left(\frac{N-n}{N}\right) \to V(r_M) \tag{19.84}$$

the variance of the estimated mean of *Y*, from Equation 19.81, is

$$s^2_{\bar{y}_M} = \mu^2_x s^2_{r_M} \to V(\bar{y}_M) \tag{19.85}$$

and the variance of the estimated total of *Y*, from Equation 19.82, is

$$s^2_{T_{Y_M}} = N^2\mu^2_x s^2_{r_m} = N^2 s^2_{\bar{y}_M} \to V(T_{Y_M}) \tag{19.86}$$

### 19.6.3 Unbiased Ratio-Type Estimators

Robson (1957), using methods beyond the scope of this book, developed an expression for the exact variance of the estimate of $\mu_Y$ when the Hartley–Ross unbiased ratio-type estimator described in Section 19.8.3 is used. It is assumed that the sampling fraction, $n/N$, is negligible so that the finite population correction can be ignored.

$$V(r_{HR}) = \sigma^2_{r_{HR}} = \frac{1}{\mu^2_x n}[\sigma^2_Y + \mu^2_{r_M}\sigma^2_X - 2\mu_{r_M}\sigma_{XY}] + \frac{1}{\mu^2_X n(n-1)}(\sigma^2_{r_i}\sigma^2_x + \sigma^2_{Xr_i}) \tag{19.87}$$

The sample size must be large so that the correction for the bias of the mean of ratios remains valid.

According to Equations 19.64 and 19.65, the first term in Equation 19.87 is equal to

$$\frac{1}{\mu^2_x nN}\sum_i^N (y_i - Rx_i)^2 \quad \text{or} \quad \sigma^2_{r_R}$$

so that Equation 19.87 can be written

$$V(r_{HR}) = \sigma^2_{r_{HR}} = \sigma^2_{r_R} + \frac{1}{\mu^2_x n(n-1)}(\sigma^2_{r_i}\sigma^2_x + \sigma^2_{Xr_i}) \tag{19.88}$$

Consequently, unless $n$ is so large that the second term essentially vanishes, $V(r_{HR})$ will be larger than $V(r_R)$.

The variance of the estimated mean of $Y$ is then

$$V(\hat{\bar{y}}_{HR}) = \sigma^2_{\hat{\bar{y}}_{HR}} \doteq V(r_{HR}\mu_X) = \mu^2_x V(r_{HR}) \tag{19.89}$$

and the variance of the estimated total of $Y$ is

$$V(T_{Y_{HR}}) = \sigma^2_{T_{HR}} \doteq V(r_{HR}\tau_X) = \tau^2_x V(r_{HR}) = N^2\mu^2_X V(r_{HR}) \tag{19.90}$$

## 19.7 SAMPLE SIZE

When a maximum permitted cost of field operations, $C_F$, has been specified, the number of sampling units that can be located and evaluated, $n$, depends on the cost associated with a single sampling unit, $C_{sv}$. This is true regardless of the type of ratio estimation that is to be used. Thus,

$$n = C_F/C_{sv} \tag{19.91}$$

When a maximum allowable sampling error has been specified, the process of arriving at an appropriate sample size follows that used in the case of simple random sampling.

### 19.7.1 Ratio of Means

According to Equation 19.72,

$$s_{r_R}^2 = \frac{1}{\mu_X^2 n}\left(\frac{N-n}{N}\right)\left(\frac{1}{n-1}\right)\sum_i^n (y_i - r_R x_i)^2$$

Since $r_R$ is an estimate of $R$ and $R$ is the regression coefficient, $\beta_1$, $r_R x_i$ is equivalent to $\hat{y}_i$ in Equation 15.54. Then, $d_i$ is a residual from the regression line:

$$d_i = y_i - r_R x_i$$

and

$$s_d^2 = \frac{1}{(n-1)}\sum_i^n (y_i - R_R x_i)^2 = \frac{1}{(n-1)}\sum_i^n d_i^2 \tag{19.92}$$

Equation 19.72 can then be written, remembering that $\mu_X$ is presumed to be known.

$$s_{r_R}^2 = \frac{1}{\mu_X^2 n}\left(\frac{N-n}{N}\right)s_d^2 \tag{19.93}$$

The sampling error of $r_R$ is

$$\text{SE}_{r_R} = \pm z_c s_{r_R} \tag{19.94}$$

where $z_C$ is the abscissa of the normal distribution which corresponds to $C$, the level of confidence. Because of uncertainties regarding the actual shape of the sampling distribution of $r_R$, the approximate values for $z_c$ will be used. These are $z_{0.67} = 1$ and $z_{0.95} = 2$. Substituting the allowable error, $\text{AE}_{r_R}$, for $\text{SE}_{r_R}$ in Equation 19.94

$$\text{AE}_{r_R} = \pm z_c s_{r_R} \tag{19.95}$$

Squaring both sides,

$$\text{AE}_{r_R}^2 = z_c^2 s_{r_R}^2 = \frac{z_c^2 s_d^2}{\mu_X^2 n}\left(\frac{N-n}{N}\right)$$

$$= \frac{z_c^2 s_d^2 N - z_c^2 s_d^2 n}{\mu_X^2 nN}$$

Then,

$$\text{AE}_{r_R}^2 \mu_X^2 nN = z_c^2 s_d^2 N - z_c^2 s_d^2 n$$

$$n(\text{AE}_{r_R}^2 \mu_X^2 N + z_c^2 s_d^2) = z_c^2 s_d^2 N \tag{19.96}$$

$$n_{\text{Needed}} = \frac{z_c^2 s_d^2 N}{\text{AE}_{r_R}^2 \mu_X^2 N + z_c^2 s_d^2}$$

When the finite population correction can be considered equal to 1, Equation 19.96 reduces to

$$n_{\text{Needed}} = \frac{z_c^2 s_d^2}{\mu_X^2 \text{AE}_{r_R}^2} \tag{19.97}$$

When the allowable sampling error is stated in terms of the total of $Y$, the expression for the needed sample size, when the sampling is without replacement, is

$$n_{\text{Needed}} = \frac{z_c^2 \tau_X^2 s_d^2 N}{\text{AE}_{T_Y}^2 \mu_X^2 N + z_c^2 \tau_X^2 s_d^2} \tag{19.98}$$

and when the finite population correction can be ignored,

$$n_{\text{Needed}} = \frac{z_c^2 \tau_X^2 s_d^2}{\text{AE}_{T_Y}^2 \mu_X^2} \tag{19.99}$$

The equivalent expressions, when the allowable sampling error is in terms of the mean of $Y$, can be obtained by substituting $\mu_X$ for $\tau_X$ in Equations 19.98 and 19.99.

A sound value for $s_d^2$ is essential in these computations. Such would be obtained from a presample or from historical records. One must keep in mind that *the sample must be large enough so that the assumption that $\bar{x}$ is essentially equal to $\mu_X$ is valid.* This sets the minimum sample size at 30 to 50, even if the computation of $n_{\text{Needed}}$ indicates that a sample size smaller than the minimum would produce an acceptable sampling error.

As an example, consider a hypothetical situation where a sawmill owner wishes to buy the logs for the mill by weight instead of by log scale but, for tax purposes, needs to keep records in terms of board feet, Scribner. To do this the ratio of wood volume to wood weight must be established. The allowable error, at the 95% level of confidence, is ±2.5% of the estimate of the ratio. A presample of ten randomly chosen log truck loads was obtained. Each truck load was first weighed, in terms of pounds, and then the load was taken off the truck, spread out, and scaled. Thus, $X$ refers to weight and $Y$ refers to volume. The data from the presample and the sample size computations are as follows.

| Load | $x_i$ (lb) | $y_i$ (bd. ft) | $r_R x_i$ | $d_i$ | $d_i^2$ |
|---|---|---|---|---|---|
| 1 | 32,676 | 2150 | 2,008.1888 | 141.81115 | 20,110.403 |
| 2 | 31,980 | 1830 | 1,965.4144 | −135.41435 | 18,337.046 |
| 3 | 28,210 | 1840 | 1,733.7192 | 106.28084 | 11,295.616 |
| 4 | 41,263 | 2710 | 2,535.9253 | 174.07466 | 30,301.988 |
| 5 | 35,367 | 2060 | 2,173.5713 | −113.57127 | 12,898.434 |
| 6 | 24,619 | 1380 | 1,513.0249 | −133.02489 | 17,695.621 |
| 7 | 23,905 | 1560 | 1,469.1442 | 90.85585 | 8,254.785 |
| 8 | 31,012 | 2040 | 1,905.9234 | 134.07662 | 17,976.539 |
| 9 | 42,448 | 2490 | 2,608.7526 | −118.75261 | 14,102.181 |
| 10 | 30,856 | 1750 | 1,896.3360 | −146.33610 | 21,414.224 |

$$\sum_i^{10} d_i^2 = 172{,}386.840$$

$$\mu_X = \bar{x} = 32{,}233.6 \text{ lb / load; } \bar{y} = 1981.0 \text{ bd. ft / load}$$

$$r_R = 1981.0 / 32{,}233.6 = 0.0614576 \text{ bd. ft / lb}$$

$$s_d^2 = \frac{1}{(10-1)}(172{,}386.840) = 19{,}154.093$$

Since there is no limit to the size of the universe, the finite population correction is equal to 1.

$$s_{r_R}^2 = \frac{19{,}154.093}{(32{,}233.6)^2 10} = 0.0000018$$

$$s_{r_R} = \pm\sqrt{0.0000018} = \pm 0.0013578 \text{ bd. ft / lb}$$

Obviously, a meaningful sampling error cannot be obtained without a sample size that exceeds 30. Consequently, any computed needed sample size that is 30 or less must be discarded and a sample size greater than 30 arbitrarily chosen.

$$\text{AE}_{r_R} = \pm 0.025(0.0614576) = \pm 0.0015364 \text{ bd. ft / lb}$$

$$n_{\text{Needed}} = \frac{(2)^2(19{,}154.093)}{(32{,}233.6)^2(0.0015364)^2} = 31.24 \text{ or } 32 \text{ loads}$$

Iterating will not be necessary since $z_{0.95}$ remains constant at 2. The computed needed sample size is greater than 30 and its use will keep valid the assumptions underlying the estimation of $R$.

In the above example the sawmill owner desired the *ratio* of wood volume to wood weight. Now, consider an example where the *total* of the $Y$ variates is needed and the total of the $X$ variates is known.

The example involves a 1000-acre ownership on which mixed hardwoods occurs as 2- to 5-acre patches scattered through pasturelands. The owner desired an estimate of the volume of timber on the tract and specified that the sampling errors, at the 95% level of confidence, should not exceed 3% of the estimated volume. Using conventional fixed-area plots would have been awkward and could have led to considerable edge effect bias because of the large amount of edge relative to the patch areas. Consequently, the following design, which makes use of the ratio of means, was devised.

Aerial photographic coverage of the tract was available, making it possible to make a stand (patch) map and from it determine the patch areas. In this approach, the individual patches were the sampling units, which were numbered, this forming the sampling frame. There were $N = 53$ patches in the frame, with a total area of $\tau_X = 230$ acres. Thus, $\mu_X = 230/53 = 4.34$ acres/patch.

To determine the sample size needed to meet the specified allowable error, a presample of size $n = 10$ was drawn from the frame using random sampling without replacement. Each of these patches was located in the field and a complete inventory was made of the merchantable timber on it. The data so collected are

| Patch | $x_i$ = Patch Area (acre) | $y_i$ = Patch Volume (ft³) | $r_R x_i$ (ft³) | $d_i = (y_i - r_R x_i)$ | $d_i^2$ |
|---|---|---|---|---|---|
| 1 | 2.6 | 19.3 | 174.7166 | 18.2834 | 334.2820 |
| 2 | 1.4 | 68 | 94.0782 | –26.0782 | 680.0715 |
| 3 | 4.8 | 356 | 322.5538 | 33.4462 | 1,118.6510 |
| 4 | 4.1 | 252 | 275.5147 | –23.5147 | 552.9397 |

| Patch | $x_i$ = Patch Area (acre) | $y_i$ = Patch Volume (ft³) | $r_R x_i$ (ft³) | $d_i = (y_i - r_R x_i)$ | $d_i^2$ |
|---|---|---|---|---|---|
| 5 | 2.8 | 163 | 188.1564 | –25.1564 | 632.8425 |
| 6 | 3.2 | 196 | 215.0358 | –19.0358 | 362.3632 |
| 7 | 3.6 | 273 | 241.9153 | 31.0847 | 966.2573 |
| 8 | 1.9 | 90 | 127.6775 | –37.6775 | 1,419.5963 |
| 9 | 2.8 | 216 | 188.1564 | 27.8436 | 775.2683 |
| 10 | 3.5 | 256 | 235.1955 | 20.8046 | 432.8293 |

$$\sum_i^{10} d_i^2 = 7{,}275.1011$$

$$\bar{x} = 3.07 \text{ acres / patch; } \bar{y} = 206.3 \text{ ft}^3 \text{ / patch}$$

$$r_R = 206.3 / 3.07 = 67.1987 \text{ ft}^3 \text{ / acre}$$

$$T_{Y_R} = 230(67.1987) = 15{,}456 \text{ ft}^3 \text{ on the tract}$$

$$s_d^2 = \frac{1}{(10-1)}(7{,}275.1011) = 808.3446$$

$$s_{r_R}^2 = [1/(4.34)^2 10](53-10)/53]808.3446 = 3.4818469$$

$$s_{r_R} = \pm\sqrt{3.4818469} = \pm 1.8660 \text{ ft}^3 \text{ / acre}$$

$$\text{SE}_{r_R} = \pm z_c s_{r_R} = \pm 2(1.8660) = 3.7320 \text{ ft}^3 \text{ / acre} \quad \text{or}$$

$$\text{SE}_{T_{Y_R}} = \pm \tau_X (3.7320) = \pm 230(3.7320) = \pm 858.4 \text{ ft}^3 \text{ on tract}$$

$$\text{AE}_{r_R} = \pm 0.03 r_R = \pm 0.03(67.1987) = \pm 2.02 \text{ ft}^3 \text{ / acre} \quad \text{or}$$

$$\text{AE}_{T_{Y_R}} = \pm 0.03 T_{Y_R} = \pm 0.03(15{,}456) = \pm 464 \text{ ft}^3 \text{ on tract}$$

Then, using the per acre allowable error in Equation 19.96,

$$n_{\text{Needed}} = \frac{(2)^2(808.3446)53}{(2.02)^2(4.34)^2 53 + (2)^2(808.3446)}$$
$$= 23.5 \text{ or } 24 \text{ patches}$$

The computed needed sample size is less than the required minimum needed to keep the bias of the ratio of means under control. It would have to be arbitrarily increased to 31 or larger.

### 19.7.2 Mean of Ratios

According to Equation 19.84,

$$s_{r_M}^2 = \frac{s_{r_i}^2}{n}\left(\frac{N-n}{(n)}\right)$$

The sampling error of $r_M$, at the $C$ level of confidence, is then

$$\text{SE}_{r_M} = \pm z_C s_{r_M} \tag{19.100}$$

Squaring both sides of the equation,

$$\text{SE}^2_{r_M} = z_c^2 s^2_{r_M} = z_c^2 \frac{s^2_{r_i}}{n}\left(\frac{N-n}{n}\right) = \frac{z_c^2 s^2_{r_i} N - z_C^2 s^2_{r_i} n}{nN}$$

Replacing the sampling error, $\text{SE}_{r_M}$, with the allowable sampling error, $\text{AE}_{r_M}$,

$$\begin{aligned}
\text{AE}^2_{r_M} &= \frac{z_c^2 s^2_{r_i} N - z_c^2 s^2_{r_i} n}{nN} \\
\text{AE}^2_{r_M} nN &= z_c^2 s^2_{r_i} N - z_c^2 s^2_{r_i} n \\
\text{AE}^2_{r_M} nN + z_c^2 s^2_{r_i} n &= z_c^2 s^2_{r_i} N \\
n_{\text{Needed}} &= \frac{z_c^2 s^2_{r_i} N}{\text{AE}^2_{r_M} N + z_c^2 s^2_{r_i}}
\end{aligned} \tag{19.101}$$

When the finite population correction is assumed equal to 1, Equation 19.101 simplifies to

$$n_{\text{Needed}} = \frac{z_c^2 s^2_{r_i}}{\text{AE}^2_{r_M}} \tag{19.102}$$

As was brought out in Section 19.4.2, increasing sample size does not reduce the bias associated with the mean of ratios. If $\bar{x}$ is not exactly equal to $\mu_X$ the estimate will contain the full-sized bias (see Equation 19.47). The variance and the above expressions for the sample size are based on the assumption that $\bar{x}$ is equal to $\mu_X$. For that reason, one cannot justify their use. However, if one must use the mean of ratios, the sample should, as a minimum, exceed 30. If Equation 19.101 or 19.102 indicates a sample size greater than 30, consider that to be a lower bound and make the sample at least that large.

## 19.8 EFFICIENCY OF RATIO ESTIMATORS

The efficiency of any estimator is a function of its variance and its cost. If cost is disregarded, efficiency and precision become synonymous. If the variance of the estimate is small and its cost is low, it is considered to be efficient. If two different estimators of the same parameter are being compared, the one with the smaller variance for a given cost will be considered to be the more efficient. The measure of the difference between the two estimators is their *relative efficiency*, which is the ratio of their variances, which are weighted by cost. In the following discussion, the standard against which the ratio estimators is judged is simple random sampling.

### 19.8.1 Ratio of Means

In the case of simple random sampling the standard error of the mean weighted by the cost of locating and evaluating one sample unit, $c_{\text{Ran}}$, is

$$\sigma_{\bar{y}} c_{\text{Ran}} = \pm\sqrt{\frac{\sigma_Y^2}{n}\left(\frac{N-n}{N-1}\right)} c_{\text{Ran}}$$

The weighted variance of the mean is then

$$V(\bar{y}_{\text{Ran}})_w = \sigma_{\bar{y}}^2 c_{\text{Ran}} = \frac{\sigma_Y^2}{n}\left(\frac{N-n}{N-1}\right) c_{\text{Ran}}^2 \tag{19.103}$$

The standard error of the mean of a ratio estimate, using the ratio of means, is

$$\sigma_{\bar{y}_R} = \frac{1}{\mu_x}\sqrt{\frac{\sigma_Y^2 + R^2\sigma_x^2 - 2R\rho_{XY}\sigma_X\sigma_Y}{n}}$$

This standard error must be weighted by the sum of two costs, one for each of the variables being evaluated. The cost associated with the dependent variable should be the same as that associated with simple random sampling since the work involved is essentially the same. Thus, $c_s = c_{\text{Ran}}$. The predicting variable must also be evaluated, which adds a cost, $c_p$. The weighted standard error is then

$$\sigma_{\bar{y}_R}(c_{\text{Ran}} + c_p)$$

and the weighted variance of the mean is (see Equations 19.64 and 19.70):

$$\begin{aligned} V(\bar{y}_R)_w &= \sigma_{\bar{y}_R}^2 (c_{\text{Ran}} + c_p)^2 \\ &= \frac{\mu_x^2}{\mu_x^2 n}\left(\frac{N-n}{N-1}\right)(\sigma_Y^2 + R^2\sigma_X^2 - 2R\rho_{XY}\sigma_X\sigma_Y)(c_{\text{Ran}} + c_p)^2 \end{aligned} \tag{19.104}$$

The efficiency of ratio estimation, using the ratio of means, is then

$$\begin{aligned} \text{Rel. Eff.} &= V(\bar{y}_{\text{Ran}})_w / V(\bar{y}_R)_w \\ &= \frac{\frac{\sigma_Y^2}{n}\left(\frac{N-n}{N-1}\right)c_{\text{Ran}}^2}{\frac{\mu_X^2}{\mu_X^2 n}\left(\frac{N-n}{N-1}\right)(\sigma_Y^2 + R^2\sigma_X^2 - 2R\rho_{XY}\sigma_X\sigma_Y)(c_{\text{Ran}} + c_p)^2} \\ &= \left(\frac{\sigma_Y^2}{(\sigma_Y^2 + R^2\sigma_X^2 - 2R\rho_{XY}\sigma_X\sigma_Y)}\right)\left(\frac{c_{\text{Ran}}}{c_{\text{Ran}} + c_p}\right)^2 \end{aligned} \tag{19.105}$$

If the relative efficiency is equal to 1, both sampling methods are equally efficient. When the relative efficiency is greater than 1, the ratio estimator is the more efficient.

As can be seen, the efficiency of the ratio estimator depends on the relative magnitude of $R^2\sigma_X^2$ and $2R\rho_{xy}\sigma_x\sigma_y$ and the ratio $c_{\text{Ran}}^2/(c_{\text{Ran}} + c_p)^2$. When $R^2\sigma_X^2 = 2R\rho_{xy}\sigma_x\sigma_y$ the first term within parentheses in Equation 19.105 equals 1, the relative efficiency is equal to $c_{\text{Ran}}^2/(c_{\text{Ran}} + c_p)^2$, and, unless $c_p$ is negligible, the ratio estimator is the less efficient. When $R^2\sigma_X^2 < 2R\rho_{xy}\sigma_x\sigma_y$, ratio estimation will be the more efficient *provided* that $c_{\text{Ran}}^2/(c_{\text{Ran}} + c_p)^2$ does not offset the gain. The key is the correlation coefficient, $\rho_{XY}$. If

$$\rho_{xy} \geq \frac{R^2\sigma_X^2}{2R\sigma_X\sigma_Y}\left[\frac{(c_{\text{Ran}}+c_p)^2}{c_{\text{Ran}}^2}\right] + \frac{R\sigma_X}{2\sigma_Y}\left[\frac{(c_{\text{Ran}}+c_p)^2}{c_{\text{Ran}}^2}\right] \tag{19.106}$$

the ratio estimation will be the more efficient. For example, if $R = 1.2$, $\sigma_X = 100$, $\sigma_Y = 120$, $c_{\text{Ran}} =$ \$10, and $c_p =$ \$1, $\rho_{XY}$ would have to be equal to or greater than 0.605 if the ratio estimate, using the ratio of means, is to be the more efficient.

### 19.8.2 Mean of Ratios

In the case of the mean of ratios, the relative efficiency is

$$\begin{aligned}
\text{Rel. Eff.} &= V(\bar{y}_{\text{Ran}})c_{\text{Ran}}^2 / V(\bar{y}_M)(c_{\text{Ran}}+c_p)^2 \\
&= \frac{\dfrac{\sigma_Y^2}{n}\left(\dfrac{N-n}{N-1}\right)c_{\text{Ran}}^2}{\mu_X^2\dfrac{\sigma_{r_i}^2}{n}\left(\dfrac{N-n}{N-1}\right)(c_{\text{Ran}}+c_p)^2} \\
&= \frac{\sigma_Y^2 c_{\text{Ran}}^2}{\mu_X^2\sigma_{r_i}^2(c_{\text{Ran}}+c_p)^2}
\end{aligned} \tag{19.107}$$

Thus,

$$\text{Rel. Eff.} = \frac{\sigma_Y^2 c_{\text{Ran}}^2}{\sigma_Y^2(c_{\text{Ran}}+c_p)^2} = \frac{c_{\text{Ran}}^2}{(C_{\text{Ran}}+c_p)^2} \tag{19.108}$$

indicating that the efficiency of the mean of ratios estimator is a function of the cost of evaluating the predicting variable. If the cost is negligible, as when the data can be obtained from existing records, the two sampling methods are equally efficient. When $c_p$ is not negligible, the mean of ratios estimator will be less efficient than simple random sampling.

### 19.8.3 The Hartley–Ross Unbiased Ratio Estimator

According to Equation 19.88,

$$V(r_{HR}) = \sigma_{r_R}^2 + \frac{1}{\mu_X^2 n(n-1)}(\sigma_{r_i}^2\sigma_{X_i}^2 + \sigma_{Xr_i}^2)$$

Since $\sigma_{Xr_i} = \rho_{Xr_i}\sigma_X\sigma_{r_i}$, Equation 19.88 can be written

$$V(r_{HR}) = \sigma_{r_R}^2 + \frac{1}{\mu_X^2 n(n-1)}\sigma_{r_i}^2\sigma_X^2(1+\rho_{Xr_i}^2)$$

Since $\rho_{Xr_i}$ is squared, the second term will always be positive. As a result $V(r_{HR})$ cannot be smaller than $V(r_R)$. The difference will be minimized when $n$ is large and/or $\sigma_{r_i}^2$ is small. Thus,

$$\frac{V(\bar{y}_{\text{Ran}})c_{\text{Ran}}^2}{\mu_X^2 V(r_{HR})(c_{\text{Ran}}+c_p)^2} \leq \frac{V(\bar{y}_{\text{Ran}})c_{\text{Ran}}^2}{\mu_X^2 V(r_R)(c_{\text{Ran}}+c_p)^2} \tag{19.109}$$

indicating that the efficiency of the Hartley–Ross estimator, relative to unrestricted random sampling, will usually be less than that of the ratio of means estimator.

## 19.9 RATIO ESTIMATION AND STRATIFIED SAMPLING

The stratification process and related statistical operations were explored in Chapter 16. One can now expand that discussion to include how ratio estimation fits in that context.

In Chapter 16 it was brought out that, *if appropriately done*, stratification can be quite effective in reducing the magnitude of the variance of the estimated mean and/or total. The stratification will be appropriate if it is based on levels of the variable of interest (e.g., in the case of a timber cruise, the stratification should be on the basis of volume classes, not forest types). This remains true when ratio estimation is introduced into the process. The stratification should be based on the dependent variable, $Y$, not on the independent variable, $X$.

According to Equation 16.12, the total of the variates making up the stratified population is

$$\tau_{Y_{ST}} = \sum_i^M \sum_j^{N_i} y_{ij}$$

In the same manner,

$$\tau_{X_{ST}} = \sum_i^M \sum_j^{N_i} x_{ij}$$

The ratio of $Y$ to $X$ is then

$$R_{ST} = \tau_{Y_{ST}} / \tau_{X_{ST}} \tag{19.110}$$

For example, consider the following two-dimensional population which has been divided logically into three strata.

| Stratum | $i$ | $Ni$ | $(Y, X)$ |
|---|---|---|---|
| $A$ | 1 | 4 | (2,10);(3,15);(5,25);(8,40) |
| $B$ | 2 | 6 | (9,36);(11,44);(13,52);(14,56);(16,64);(19,76) |
| $C$ | 3 | 8 | (20,200);(22,220);(23,230);(24,240);(26,260);(28,280);(30,300);(31,310) |

$$\tau_{Y_1} = 18; \quad \tau_{Y_2} = 82; \quad \tau_{Y_3} = 204; \quad \tau_{Y_{ST}} = \sum_i^M \tau_{Y_i} = 304$$

$$\tau_{X_1} = 90; \quad \tau_{X_2} = 328; \quad \tau_{X_3} = 2040; \quad \tau_{X_{ST}} = \sum_i^M \tau_{X_i} = 2458$$

$$R_{ST} = 304 / 2458 = 0.1236778$$

Estimates of $R_{ST}$ can be obtained using the ratio of means within either of two different approaches: (1) the *combined* estimate, in which stratum ratios are not computed; and (2) the *separate* estimate, in which stratum ratios are computed.

### 19.9.1 The Combined Ratio Estimate

In this approach, the data from all the strata are pooled:

$$r_{\text{CST}} = T_{Y_{\text{ST}}} / T_{X_{\text{ST}}} = \sum_{i}^{M} \sum_{j}^{n_i} y_{ij} \Big/ \sum_{i}^{M} \sum_{j}^{n_i} x_{ij} \rightarrow R_{\text{ST}} \tag{19.111}$$

where the subscript CST indicates that the combined estimate approach is being used.

The variance of $r_{\text{CST}}$ is obtained by summing the weighted stratum variances:

$$V(r_{\text{CST}}) = \sum_{i}^{M} \frac{N_i^2}{N^2 \mu_X^2 n_i} \left( \frac{N_i - n_i}{N_i - 1} \right) (\sigma_{Y_i}^2 + R^2 \sigma_{X_i}^2 - 2R\sigma_{XY_i}) \tag{19.112}$$

or, since $N^2 \mu_X^2 = \tau_X^2$,

$$V(r_{\text{CST}}) = \frac{1}{\tau_X^2} \sum_{i}^{M} \frac{N_i^2}{n_i} \left( \frac{N_i - n_i}{N_i - 1} \right) (\sigma_{Y_i}^2 + R^2 \sigma_{X_i}^2 - 2R\sigma_{XY_i}) \tag{19.113}$$

Note that the ratio appearing in this expression is the population ratio, not the stratum ratio. Since the stratum correlation coefficient is

$$\rho_{XY_i} = \sigma_{XY_i} / \sigma_{x_i} \sigma_{Y_i}$$

the variance equation can also be written

$$V(r_{\text{CST}}) = \frac{1}{\tau_X^2} \sum_{i}^{M} \frac{N_i^2}{n_i} \left( \frac{N_i - n_i}{N_i - 1} \right) (\sigma_{Y_i}^2 + R^2 \sigma_{X_i}^2 - 2R\rho_{XY_i} \sigma_{X_i} \sigma_{X_i}) \tag{19.114}$$

which is useful when assessing the effect on the variance of the correlation between the two variables. To simplify further, let

$$\sigma_{ci}^2 = \sigma_{Y_i}^2 + R^2 \sigma_{X_i}^2 - 2R\sigma_{XY_i}$$

Then, with some rearrangement,

$$V(r_{\text{CST}}) = \frac{1}{\tau_X^2} \sum_{i}^{M} \left[ \frac{\sigma_{ci}^2}{n_i} \left( \frac{N_i - n_i}{N_i - 1} \right) N_i^2 \right] \tag{19.115}$$

The estimate of the variance of $r_{\text{CST}}$ is then

$$s_{r_{\text{CST}}}^2 = \frac{1}{N^2 \bar{x}^2} \sum_{i}^{M} \left[ \frac{s_{ci}^2}{n_i} \left( \frac{N_i - n_i}{N_i} \right) N_i^2 \right] \tag{19.116}$$

where $\bar{x} = \left(1 \Big/ \sum_i^M n_i\right) \sum_i^M \sum_j^{n_i} x_{ij}$

$$s_{ci}^2 = s_{Y_i}^2 + r_{\text{CST}}^2 s_{X_i}^2 - 2r_{\text{CST}} s_{XY_i} \tag{19.117}$$

$$r_{\text{CST}} = \sum_i^M \sum_j^{n_i} y_{ij} \Big/ \sum_i^M \sum_j^{n_i} x_{ij} \to V(r_{\text{CSr}}) \tag{19.118}$$

As an example, assume that a proportionally allocated sample of size $n = 18$ was randomly drawn, without replacement, from the two-dimensional population used in the previous example.

| Stratum | $i$ | $N_i$ | $n_i$ | $(y_{ij}, x_{ij})$ |
|---|---|---|---|---|
| A | 1 | 4 | 2 | (3,15);(8,40) |
| B | 2 | 6 | 3 | (13,52);(14,56);(19,76) |
| C | 3 | 8 | 4 | (20,200);(24,240);(26,260);(30,300) |

$$\bar{y}_1 = 5.5; \quad \bar{y}_2 = 15.3333; \quad \bar{y}_3 = 25.0$$

$$T_{Y_i} = 4(5.5) = 22; \quad T_{Y_2} 6(15.3333) = 92; \quad T_{Y_3} = 8(25.0) = 20; \quad T_{Y_{\text{CST}}} = 314$$

$$\bar{x}_1 = 27.5; \quad \bar{x}_2 = 61.3333; \quad \bar{x}_3 = 250.0$$

$$T_{X_i} = 4(27.5) = 110; \quad T_{X_2} = 6(61.3333) = 368; \quad T_{X_3} = 8(250.0) = 2000; \quad T_{X_{\text{CST}}} = 2478$$

$$r_{\text{CST}} = 314 / 2478 = 0.1267151$$

The *true* variance of this estimate is computed as follows.

| Stratum | $i$ | $\mu_{Y_i}$ | $\mu_{X_i}$ | $\sigma_{Y_i}^2$ | $\sigma_{X_i}^2$ | $\sigma_{Y_i}$ | $\sigma_{X_i}$ | $\sigma_{XY_i}$ | $\rho_{XY_i}$ |
|---|---|---|---|---|---|---|---|---|---|
| *A* | 1 | 4.5 | 22.5 | 5.25 | 131.25 | 2.2913 | 11.4564 | 26.2500 | 1.0 |
| *B* | 2 | 13.7 | 54.7 | 10.56 | 168.89 | 3.2489 | 12.9957 | 42.2217 | 1.0 |
| *C* | 3 | 25.5 | 255.0 | 13.50 | 1350.00 | 3.6742 | 36.7423 | 134.9986 | 1.0 |

| Stratum | $i$ | $\sigma_{Y_i}^2 + R_{\text{CST}}^2 \sigma_{X_i}^2 - 2R_{\text{CST}} \sigma_{XY_i}$ | $\sigma_{ci}^2$ |
|---|---|---|---|
| *A* | 1 | $5.25 + (0.1236778)^2 \; 131.25 - 2(0.1236778) \; 26.25 =$ | 0.7645 |
| *B* | 2 | $10.56 + (0.1236778)^2 \; 168.89 - 2(0.1236778) \; 42.2217 =$ | 2.6996 |
| *C* | 3 | $13.50 + (0.1236778)^2 \; 1350.00 - 2(0.1236778) \; 134.9986 =$ | 0.7572 |

Then, using Equation 19.113:

$$V(r_{\text{CST}}) = \frac{1}{(2458)^2}\left[\frac{0.7645}{2}\left(\frac{4-2}{4-1}\right)(4)^2 + \frac{2.6996}{3}\left(\frac{6-3}{6-1}\right)(6)^2 + \frac{0.7572}{4}\left(\frac{8-4}{8-1}\right)(8)^2\right]$$

$$= \frac{1}{6{,}041{,}764}(4.0773 + 19.4371 + 6.9230)$$

$$= 0.000005$$

The estimate of this variance is obtained using the data from the sample.

| Stratum | $i$ | $\bar{y}_i$ | $\bar{x}_i$ | $s^2_{Y_i}$ | $s^2_{X_i}$ | $s_{Y_i}$ | $s_{X_i}$ | $s_{XY_i}$ |
|---|---|---|---|---|---|---|---|---|
| *A* | 1 | 5.5 | 27.5 | 12.50 | 312.50 | 3.5355 | 17.6777 | 62.4995 |
| *B* | 2 | 15.3 | 61.3 | 10.33 | 165.33 | 3.2146 | 12.8582 | 41.3340 |
| *C* | 3 | 25.0 | 250.0 | 17.33 | 1.733.33 | 4.1633 | 41.6333 | 173.3319 |

| Stratum | $i$ | $s^2_{Y_i} + r^2_{CST} s^2_{X_i} - 2r_{CST} s_{XY_i} =$ | $s_{C_i^2}$ |
|---|---|---|---|
| *A* | 1 | 12.50 + (0.1267151)² 312.50 – 2(0.1267151) 62.4995 = | 1.6785 |
| *B* | 2 | 10.33 + (0.1267151)² 165.33 – 2(0.1267151) 41.3340 = | 2.5094 |
| *C* | 3 | 17.33 + (0.1267151)² 1733.33 – 2(0.1267151) 173.3319 = | 1.2341 |

$$s^2_{r_{CST}} = \frac{1}{(18)^2(137.6667)^2}\left[\frac{1.6785}{2}\left(\frac{4-2}{4}\right)(4)^2 + \frac{2.5094}{3}\left(\frac{6-3}{6}\right)(6)^2 + \frac{1.2341}{4}\left(\frac{8-4}{8}\right)(8)^2\right]$$

$$= \frac{1}{44{,}604.011}(6.7140 + 15.0564 + 9.8728)$$

$$= 0.0000052$$

The estimates of $\tau_{Y_{CST}}$ requires a knowledge of $\tau_{X_{ST}}$

$$T_{Y_{CST}} = r_{CST}\tau_{X_{ST}} \rightarrow \tau_{Y_{CST}} \tag{19.119}$$

The variance of the estimated total is then

$$V(T_{Y_{CST}}) = V(r_{CST}\tau_{X_{ST}}) = \tau^2_{X_{ST}} V(r_{CST})$$
$$= \sum_i^M \left[\frac{\sigma^2_{ci}}{n_i}\left(\frac{N_i - n_i}{N_i - 1}\right)N_i^2\right] \tag{19.120}$$

and its estimate is

$$s^2_{T_{Y\,CST}} = \sum_i^M \left[\frac{s^2_{ci}}{n_i}\left(\frac{N_i - n_i}{N_i - 1}\right)N_i^2\right] \tag{19.121}$$

Using the data from the samples

$$T_{Y_{CST}} = 0.1267151(2458) = 311$$

$$V(T_{Y_{CST}}) = 4.0773 + 19.4371 + 6.9230 = 30.4374$$

$$s^2_{T_{Y\,CST}} = 6.7140 + 15.0564 + 9.8728 = 31.6432$$

### 19.9.2 The Separate Ratios Estimate

In contrast to the combined ratio estimate, the separate ratio estimate involves stratum ratios. Thus,

$$r_i = \bar{y}_i / \bar{x}_i \tag{19.122}$$

$$T_{Y_i} = r_i \tau_{X_i} \rightarrow \tau_{Y_i} \tag{19.123}$$

$$T_{Y_{SST}} = \sum_i^M T_{Y_i} \rightarrow \tau_{Y_{SST}} \tag{19.124}$$

$$r_{SST} = T_{Y_{SST}} / \tau_{X_{ST}} \rightarrow R_{SST} \tag{19.125}$$

The subscript SST indicates that the item was obtained using the separate ratios procedure. As an example, using data from the stratified population:

| Stratum | $i$ | $r_i = \bar{y}_i / \bar{x}_i$ | $T_{Y_i} = r_i \tau_{X_i}$ |
|---|---|---|---|
| $A$ | 1 | 5.5/27.5 = 0.20 | 0.20(90) = 18 |
| $B$ | 2 | 15.3/61.3 = 0.25 | 0.25(328) = 82 |
| $C$ | 3 | 25.0/250.0 = 0.10 | 0.10(2.040) = 204 |

$$T_{Y_{SST}} = 18 + 82 + 204 = 304$$

$$r_{SST} = 304 / 2458 = 0.1236778$$

The variance of $r_{SST}$ is then

$$V(r_{SST}) = \frac{1}{N^2 \mu_X^2} \sum_i^M \frac{N_i^2}{n_i} \left( \frac{N_i - n_i}{N_i - 1} \right) (\sigma_{Y_i}^2 + R_i^2 \sigma_{X_i}^2 - 2R_i \sigma_{XY_i}) \tag{19.126}$$

which can be written, since $N^2 \mu_X = \tau_X^2$ and $\rho_{XY_i} = \sigma_{XY_i} / \sigma_{X_i} \sigma_{Y_i}$,

$$V(r_{SST}) = \frac{1}{\tau_X^2} \sum_i^M \frac{N_i}{n_i} \left( \frac{N_i - n_i}{N_i - 1} \right) (\sigma_{Y_i}^2 + R_i^2 \sigma_{X_i}^2 - 2R_i \rho_{XY_i} \sigma_{X_i} \sigma_{Y_i}) \tag{19.127}$$

Letting

$$\sigma_{si}^2 = (\sigma_{Y_i}^2 + R_i^2 \sigma_{X_i}^2 - 2R_i \sigma_{XY_i}) = (\sigma_{Y_i}^2 + R_i^2 \sigma_{X_i} - 2R_i \rho_{XY_i} \sigma_{X_i} \sigma_{y_i})$$

then,

$$V(r_{SST}) = \frac{1}{\tau_X^2} \sum_i^M \left[ \frac{\sigma_{si}^2}{n_i} \left( \frac{N_i - n_i}{N_i - 1} \right) N_i^2 \right] \tag{19.128}$$

and

$$V(T_{Y_{\text{CST}}}) = V(r_{\text{SST}}\tau_X) = \tau_X^2 V(r_{\text{SST}})$$

$$= \sum_i^M \left[ \frac{\sigma_{si}^2}{n_i} \left( \frac{N_i - n_i}{N_i - 1} \right) N_i^2 \right] \tag{19.129}$$

The estimates of these variances are

$$s_{r_{\text{SST}}}^2 = \frac{1}{N^2 \bar{x}^2} \sum_i^M \left[ \frac{s_{si}^2}{n_i} \left( \frac{N_i - n}{N_i} \right) N_i^2 \right] \tag{19.130}$$

and

$$s_{T_{Y_{\text{SST}}}}^2 = \sum_i^M \left[ \frac{s_{si}^2}{n_i} \left( \frac{N_i - n_i}{N_i} \right) N_i^2 \right] \tag{19.131}$$

where

$$s_{si}^2 = s_{Y_i}^2 + r_i s_{X_i}^2 - 2 r_i s_{XY_i} \tag{19.132}$$

Continuing the example,

| Stratum | $i$ | $(\mu_{Y_i} / \mu_{X_i}) = R_i$ | $\sigma_{Y_i}^2 + R_i^2 \sigma_{X_i}^2 - 2R_i \sigma_{XY_i}$ | $\sigma_{si}^2$ |
|---|---|---|---|---|
| $A$ | 1 | 4.5/22.5 = 0.20 | $5.25 + (0.20)^2$ 131.25 – 2(0.20) 26.2500 = | 0 |
| $B$ | 2 | 13.7/54.7 = 0.25 | $10.56 + (0.25)^2$ 168.89 – 2(0.25) 42.2217 = | 0.00478 |
| $C$ | 3 | 25.5/255.0 = 0.10 | $13.50 + (0.10)^2$ 1350.00 – 2(0.10) 134.9986 = | 0.00028 |

$$V(r_{\text{SST}}) = \frac{1}{(2458)^2} \left[ \frac{0}{2} \left( \frac{4-2}{4-1} \right) (4)^2 + \frac{0.00478}{3} \left( \frac{6-3}{6-1} \right) (6)^2 + \frac{0.00028}{4} \left( \frac{8-4}{8-1} \right) (8)^2 \right]$$

$$= 0.00000 \text{ or } 0$$

It should be noted that if it were not for rounding errors, all three $\sigma_{si}^2$ would be equal to zero, making $V(r_{\text{SST}})$ actually equal to zero.

The estimate of $V(r_{\text{SST}})$ is then obtained as follows.

| Stratum | $i$ | $s_{Y_i}^2 + r_i s_{X_i}^2 - 2r_i s_{XY_i}$ | $s_{si}^2$ |
|---|---|---|---|
| $A$ | 1 | $12.50 + (0.20)^2$ 312.50 – 2(0.20) 62.4995 = | 0.00020 |
| $B$ | 2 | $10.33 + (0.25)^2$ 165.33 – 2(0.25) 41.3340 = | –0.00388 |
| $C$ | 3 | $17.33 + (0.10)^2$ 1733.33 – 2(0.10) 173.3319 = | –0.00308 |

$$s^2_{r_{SST}} = \frac{1}{(18)^2(137.6667)^2}\left[\frac{0.00020}{2}\left(\frac{4-2}{4}\right)(4)^2 - \frac{0.00388}{4}\left(\frac{6-3}{6}\right)(6)^2 - \frac{0.00308}{4}\left(\frac{8-4}{8}\right)(8)^2\right]$$

$$= -0.00000 \text{ or } 0$$

Again rounding errors have caused the $s^2_{si}$ to have values other than zero. As a result $S^2_{r_{SST}}$ appears to be negative. These results are due to the fact that the stratum correlation coefficients are all equal to 1, which means that the residuals from the regression lines are all equal to zero. Rounding errors in such cases can cause a computed variance to appear to be negative where in actuality it should be equal to zero.

### 19.9.3 Allocation of the Sample

In Chapter 16 it was brought out that total sample size depended on the requirements laid down by the person requiring the information to be obtained by the sampling operation. That person could specify (1) a certain sampling intensity (see Equation 16.48); (2) that the cost of the operation should not exceed a certain amount (see Equation 16.51); or (3) a certain maximum allowable *standard* error. The first two of these options are independent of the method of allocation chosen, but in the case of the third, the sample size is a function of the method of allocation. This holds true when ratio estimation is combined with stratified sampling.

According to Equation 16.76,

$$V(\bar{y}_o)_{ST} = \sum_i^M \left[\frac{\sigma_i^2}{n_i}\left(\frac{N_i - n_i}{N_i - 1}\right)\frac{N_i^2}{N^2}\right]$$

Since $\tau_X^2 = N^2\mu_X^2$, Equation 19.114 can be written

$$V(r_{CST}) = \sum_i^M \left[\frac{\sigma_{ci}^2}{\mu_X}\left(\frac{1}{n_i}\right)\left(\frac{N_i - n_i}{N_i - 1}\right)\frac{N_i^2}{N^2}\right]$$

As can be seen, these two variance equations differ only in that $(\sigma_{ci}^2/\mu_X)$ is used in the second in place of $\sigma_i^2$. Extrapolating, one can then use the equations in Chapter 16 to compute sample size and the allocation of the samples if one replaces $s_i^2$ by $(s_{ci}^2/\bar{x})$. Thus, in the case of equal allocation (see Equations 16.52, 16.57, and 16.58):

$$n_i = n / M \tag{19.133}$$

When the sampling is with replacement,

$$n = \frac{M}{N^2A^2}\sum_i^M (s_{ci}^2 / \bar{x})N_i \tag{19.134}$$

and when it is without replacement,

$$n = M\sum_i^M (s_{ci}^2 / \bar{x})N_i \Bigg/ \left[N^2A^2 + \sum_i^M (s_{ci}^2 / \bar{x})N_i\right] \tag{19.135}$$

When the allocation is proportional (see Equations 16.59, 16.66, and 16.67):

$$n_i = n(N_i / N) \tag{19.136}$$

When the sampling is with replacement,

$$n = \frac{1}{NA^2} \sum_i^M (s_{ci}^2 / \bar{x}) N_i$$

and when the sampling is without replacement,

$$n = N \sum_i^M (s_{ci}^2 / \bar{x}) N_i \Bigg/ \left[ N^2 A^2 + \sum_i^M (s_{ci}^2 / \bar{x}) N_i \right] \tag{19.137}$$

In the case of optimum allocation (see Equations 16.81, 16.82, 16.84, and 16.86):

$$n_i = \left( n\sqrt{s_{ci}^2 / \bar{x}} N_i / \sqrt{c_i} \right) \Bigg/ \sum_i^M \left( \sqrt{s_{ci}^2 / \bar{x}} N_i / \sqrt{c_i} \right) \tag{19.138}$$

When $c_F$ is specified, and the sampling is either with or without replacement,

$$n = c_F \sum_i^M \left( \sqrt{s_{ci}^2 / \bar{x}} N_i \sqrt{c_i} \right) \Bigg/ \sum_i^M \left( \sqrt{s_{ci}^2 / \bar{x}} N_i / \sqrt{c_i} \right) \tag{19.139}$$

When a maximum allowable standard error is specified and the sampling is with replacement,

$$n = (1 / N^2 A^2) \sum_i^M \left( \sqrt{s_{ci}^2 / \bar{x}} N_i \sqrt{c_i} \right) \sum_i^M \left( \sqrt{s_{ci}^2 / \bar{x}} N_i / \sqrt{c_i} \right) \tag{19.140}$$

When the sampling is without replacement,

$$n = \frac{\sum_i^M \sqrt{s_{ci}^2 / \bar{x}} N_i \sqrt{c_i} \sum_i^M \left( \sqrt{s_{ci}^2 / \bar{x}} N_i / \sqrt{c_i} \right)}{N^2 A^2 + \sum_i^M (s_{ci}^2 / \bar{x}) N_i} \tag{19.141}$$

When the Neyman allocation is being used (see Equations 16.92, 16.97, and 16.98),

$$n_i = n\sqrt{s_{ci}^2 / \bar{x}} N_i \Bigg/ \sum_i^M \left( \sqrt{s_{ci}^2 / \bar{x}} N_i \right) \tag{19.142}$$

When the sampling is with replacement,

$$n = (1/N^2A^2)\left[\sum_i^M\left(\sqrt{s_{ci}^2/\bar{x}N_i}\right)\right]^2 \tag{19.143}$$

and when the sampling is without replacement,

$$n = \left[\sum_i^M\left(\sqrt{s_{ci}^2/\bar{x}N_i}\right)\right]^2 \bigg/ \left[N^2A^2 + \sum_i^M (s_{ci}^2/\bar{x})N_i\right] \tag{19.144}$$

If the separate ratios estimate is being used, one would substitute $s_{si}^2$ for $s_{ci}^2$ in the above expressions.

### 19.9.4 Discussion

Figure 19.4 may make it easier to distinguish between combined and separate ratio estimates. It contains the scatter diagram and associated regression line for each of the three strata in the example and also the regression line based on the combined data. Note that all four regressions pass through the origin. Furthermore, note that the data points associated with each stratum fall on the regression line for that stratum. Because of this the errors of the data points are all equal to zero:

$$d_{ij} = y_{ij} - \hat{y}_{ij} = y_{ij} - R_i x_{ij} = 0$$

Then the sum of squared errors for each stratum is also equal to zero:

$$\sum_i^{N_i} d_{ij}^2 = \sum_j^{N_i} (y_{ij} - R_i x_{ij})^2 = 0$$

Since

$$N_i(\sigma_{Y_i}^2 + R_i^2\sigma_{X_i}^2 - 2R_i\sigma_{XY_i}) = \sum_j^{N_i} (y_{ij} - R_i x_{ij})^2$$

the variance of $r_{SST} = 0$:

$$V(r_{SST}) = \frac{1}{N^2\mu_x^2}\sum_i^M \frac{N_i^2}{n_i}\left(\frac{N_i - n_i}{N_i - 1}\right)(0) = 0$$

as occurred in the example.

If the data points do not fall on their respective regression lines, the sums of squared errors will take on magnitudes greater than zero which, in turn, will cause the variance of $r_{SST}$ to be greater than zero. As the spread of data points increases, the errors become larger and the variance of $r_{SST}$ becomes larger. Thus, as the correlation coefficients for the strata decrease in magnitude, the variance of $r_{SST}$ increases.

Now consider the regression based on the combined data, which is associated with $r_{CST}$. In this case the data points are scattered about the regression and the sum of squared errors will be some positive magnitude. Obviously, in the case shown, the magnitude of $V(r_{CST})$ will be greater than $V(r_{SST})$. However, if the data points for the individual strata do not fall on their respective regressions

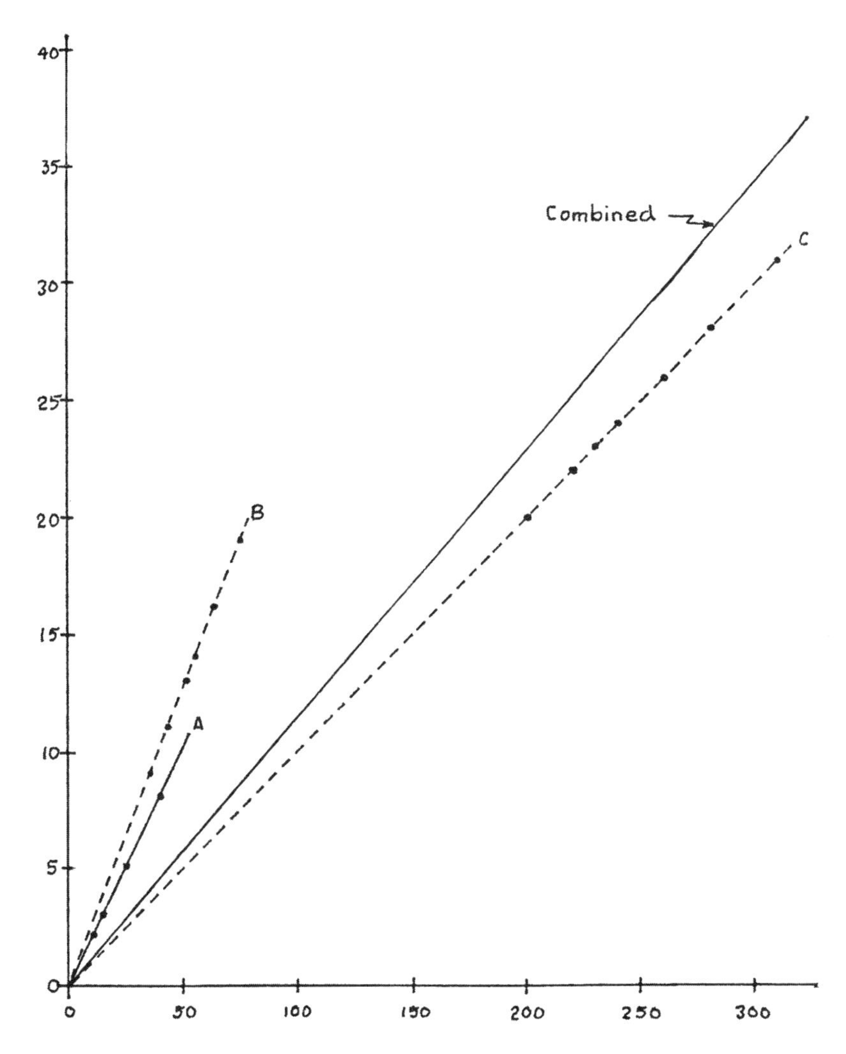


**FIGURE 19.4** Scatter diagrams and regression lines for the three strata along with the regression line for the combined data.

$V(r_{SST})$ will cease being equal to zero and will attain some positive magnitude. As the spread of data points increases, $V(r_{SST})$ will increase and the difference between $V(r_{CST})$ and $V(r_{SST})$ will diminish.

If all the stratum ratios are equal, they will also be equal to the overall ratio. In such a case, the stratum variances would be equal to the overall variance and $V(r_{SST})$ would be equal to $V(r_{CST})$. If the scatter about the stratum regressions remains constant so that the sums of squared errors remain constant and the differences between the stratum ratios, $R_i$, increase (i.e., the differences in the steepness of the stratum regressions increase), the scatter about the overall regression will increase. As a result, $V(r_{SST})$ will remain constant and $V(r_{CST})$ will increase.

The combined ratio estimate,

$$r_{CST} = \bar{y} / \bar{x} = \frac{\sum_i^M \sum_j^{n_i} y_{ij}}{\sum_i^M n_i} \Bigg/ \frac{\sum_i^M \sum_j^{n_i} x_{ij}}{\sum_i^M n_i}$$

Both $\bar{y}$ and $\bar{x}$ are based on $\Sigma_i^M n_i$ observations. In contrast,

$$r_{SST} = \sum_i^M \tau_{X_i} r_i = \sum_i^M \tau_{X_i} (\bar{y}_i / \bar{x}_i) = \sum_i^M \tau_{X_i} \left( \frac{\sum_j^{n_i} y_{ij}}{n_i} \Bigg/ \frac{\sum_j^{n_i} x_{ij}}{n_i} \right)$$

In this case, the $\bar{y}_i$ and $\bar{x}_i$ are *stratum* means which are based on only $n_i$ observations. Since the bias of a ratio is inversely dependent on sample size it can be seen that $r_{CST}$ will be less biased than any one of the $r_i$. When the $r_{SST}$ is computed, these biased $r_i$ are accumulated. Consequently, $r_{CST}$ will be less biased than $r_{SST}$ unless the stratum sample sizes exceed 30 to 50 observations. For this reason, it is recommended that the combined ratio estimate be used if the stratum sample sizes are small. Of course, the separate ratio estimate may be used without regard to the effects of ratio biases if the stratum sample regressions pass through the origin.

## 19.10 RATIO ESTIMATION AND CLUSTER SAMPLING

Ratio estimation can be extended into cluster sampling in much the same manner as was done with stratified sampling. Stratified sampling is essentially a special case of two-stage cluster sampling where the strata are clusters and every cluster (or stratum) is represented in the sample. In contrast, in cluster sampling only a portion of the clusters are chosen in the sampling process. The integration of ratio estimation and cluster sampling in the context of two-stage cluster sampling will be explored briefly.

The population ratio is

$$R_{CL} = \tau_{YCL}/\tau_{XCL} \tag{19.145}$$

and its estimator is

$$r_{CL} = T_{YCL}/T_{XCL} \tag{19.146}$$

where $\tau_{YCL} = \sum_i^M \sum_j^{N_i} y_{ij}$

$$\tau_{XCL} = \sum_i^M \sum_j^{N_i} x_{ij}$$

$$T_{YCL} = \frac{M}{m} \sum_i^m \frac{N_i}{n_i} \sum_j^{n_i} y_{ij} = \frac{M}{m} \sum_i^m N_i \bar{y}_i = \frac{M}{m} T_{Yi} = M\bar{T}_Y$$

$$T_{XCL} = \frac{M}{m} \sum_i^m \frac{N_i}{n_i} \sum_j^{n_i} x_{ij} = \frac{M}{m} \sum_i^m N_i \bar{x}_i = \frac{M}{m} T_{Xi} = M\bar{T}_X$$

$M$ = total number of clusters

$N_i$ = total number of secondary sampling units in stratum $i$

$N = \sum_i^M N_i$ = population size

$m$ = number of clusters in sample

$n_i$ = sample size in cluster $i$

$n = \sum_i^m n_i$ = size of total sample

The variance of $r_{\text{CL}}$ is obtained by modifying Equation 17.40.

$$V(r_{\text{CL}}) = \frac{1}{\tau_{X\text{CL}}^2}\left\{\frac{M}{m}\sum_i^M\left[\frac{\sigma_{ci}^2}{n_i}\left(\frac{N_i - n_i}{N_i - 1}\right)N_i^2\right] + M^2\frac{\sigma_{CB}^2}{m}\left(\frac{M-m}{M-1}\right)\right. \tag{19.147}$$

and its estimator is

$$s_{r_{\text{CL}}}^2 = \frac{1}{T_{X\text{CL}}^2}\left\{\frac{M}{m}\sum_i^m\left[\frac{s_{ci}^2}{n_i}\left(\frac{N_i - n_i}{N_i}\right)N_i^2\right] + M^2\frac{s_{CB}^2}{m}\left(\frac{M-m}{M}\right)\right. \tag{19.148}$$

where

$$\sigma_{ci}^2 = \sigma_{Yi}^2 + R_{\text{CL}}^2\sigma_{Xi}^2 - 2R_{CL}\sigma_{XYi} \quad \text{or} \tag{19.149}$$

$$= \sigma_{Yi}^2 + R_{\text{CL}}^2\sigma_{Xi}^2 - 2R_{\text{CL}}\rho_{XYi}\sigma_{Xi}\sigma_{Yi} \tag{19.150}$$

$$\sigma_{CB}^2 = \sigma_{YB}^2 + R_{\text{CL}}^2\sigma_{XB}^2 - 2R_{\text{CL}}\sigma_{XYB} \quad \text{or} \tag{19.151}$$

$$= \sigma_{YB}^2 + R_{\text{CL}}^2\sigma_{XB}^2 - 2R_{\text{CL}}\rho_B\sigma_{XB}\sigma_{YB} \tag{19.152}$$

$$\sigma_{Yi}^2 = \frac{1}{N_i}\sum_j^{N_i}(y_{ij} - \mu_{Y_i})^2$$

$$\sigma_{Xi}^2 = \frac{1}{N_i}\sum_j^{N_i}(x_{ij} - \mu_{X_i})^2$$

$$\sigma_{XYi} = \frac{1}{N_i}\sum_j^{N_i}(y_{ij} - \mu_{Y_i})(x_{ij} - \mu_{X_i})$$

$$\rho_{XYi} = \sigma_{XY_i} / \sigma_{Xi}\sigma_{Yi}$$

$$\sigma_{YB}^2 = \frac{1}{M}\sum_i^M(\tau_{Yi} - \bar{\tau}_Y)^2$$

$$\sigma_{XB}^2 = \frac{1}{M}\sum_i^M(\tau_{Xi} - \bar{\tau}_X)^2$$

$$\sigma_{XYB} = \frac{1}{M}\sum_i^M(\tau_{Y_i} - \bar{\tau}_Y)(\tau_{X_i} - \bar{\tau}_X)$$

$$\rho_B = \sigma_{XYB} / \sigma_{XB}\sigma_{YB}$$

$$\tau_{Y_i} = \sum_{j}^{N_i} y_{ij}$$

$$\tau_{X_i} = \sum_{j}^{N_i} x_{ij}$$

$$s_{ci}^2 = s_{Yi}^2 + r_{\text{CL}}^2 s_{X_i^2}^2 - 2r_{\text{CL}} s_{YX_i}$$

$$s_{CB}^2 = s_{YB}^2 + r_{\text{CL}}^2 s_{XB}^2 - 2r_{\text{CL}} s_{YXB}$$

$$s_{YB}^2 = \frac{1}{(m-1)} \sum_{i}^{m} (T_{Yi} - \bar{T}_Y)^2$$

$$s_{XB}^2 = \frac{1}{(m-1)} \sum_{i}^{m} (T_{Xi} - \bar{T}_X)^2$$

$$s_{XYB}^2 = \frac{1}{(m-1)} \sum_{k}^{m} (T_{Y_i} - \bar{T}_Y)(T_{X_i} - \bar{T}_X)$$

The size of the sample and its optimum allocation are computed using a modification of the method described in Sections 17.12 and 17.12.1. That method is based on the assumptions that both the cluster size and the sample size within the clusters are constant so that $N_i = A$ and $n_i = a$. As was pointed out in Section 17.12, however, the method can be used with variable cluster and sample sizes if their means are used in place of the indicated constants (i.e., $\bar{N}$ for $A$ and $\bar{n}$ for $a$). The modification of the procedure involves the substitution of

$$\frac{1}{m} \sum_{i}^{m} s_{ci}^2 \quad \text{for} \quad s_w^2$$

and

$$\frac{1}{m} \sum_{i}^{m} s_{CB}^2 \quad \text{for} \quad s_B^2$$

in Equation 17.146 so that

$$a \text{ or } \bar{n} = \sqrt{\frac{c_2}{c_1} \frac{\left(\frac{1}{m} \sum_{i}^{m} s_{ci}^2\right)}{\left(\frac{1}{m} \sum_{i}^{m} s_{CB}^2 - \frac{1}{mA} \sum_{i}^{m} s_{ci}^2\right)}} \tag{19.153}$$

Equation 17.149 remains unmodified:

$$m = \frac{c_F + c_2}{c_1 a + c_2}$$

## 19.11 COMMENTS

Ratio estimation is deceptively simple. In actuality, it is a tricky procedure that is full of traps for the inexperienced sampler. It is based on a series of assumptions that may or may not be correct. As a matter of fact, the probability of some of them being correct is low. Among the assumptions that must be considered when contemplating the use of a ratio estimator are (1) that the relationship between the variables $Y$ and $X$ forms a straight line that passes through the origin; (2) that the scatter of the data points about the regression line is well enough known so that the appropriate estimator, ratio of means or mean of ratios, will be chosen; and (3) that the sample size is large enough to keep the inherent biases within acceptable limits. Associated with assumption 3 is the companion assumption that the sample mean, $\bar{x}$, is equal to the population mean, $\mu_X$. These assumptions leave little room for maneuver. It must therefore be emphasized that inexperienced samplers considering the use of a ratio estimator should seek expert advice before proceeding with the project.

# 20 Double Sampling

## 20.1 INTRODUCTION

Many variables that are of interest to foresters are difficult and/or expensive to measure but are related to other variables that are less difficult or expensive to evaluate. If the relationship between such variables can be established, it becomes possible to use data from the latter to make estimates of the former. This process is known as double sampling. Usually the term *double sampling* is reserved for situations where estimates are made of population means or totals. However, whenever regressions are developed to obtain estimates of difficult-to-measure quantities as, for example, the volume of standing trees using dbh and the height, the procedure could be considered a form of double sampling.

Among many possible applications of double sampling in forestry are as follows:

1. Updating a timber cruise. This can be done by recovering and recruising a portion of the plots used in the original cruise and then finding the relationship between the earlier and the later plot volumes. The relationship can take the form of either a regression or a ratio. Once the relationship is known, the original mean or total volume can be brought up-to-date. In this example the relationship was described using data from a subsample of the original cruise.
2. Improving a photo-based timber cruise. Timber volume estimates using aerial photographs can be made by establishing or ocularly estimating the average total height, average crown diameter, and average stand density of the stands into which the sample plots fall. Such estimates are weak, but they can be made rapidly and with much less cost than ground-based estimates. To improve the photo-based estimates, one can subsample plots on the ground and cruise them in the conventional manner. Obviously, a relationship exists between the photo- and ground-based estimates. That relationship is then described, either as a regression or as a ratio, and the mean or total of the entire photo-based sample is used in the expression to obtain an improved estimate of the timber volume on the tract.
3. Improving a cone count estimate. Cone counts are used to assess an upcoming tree seed crop. Such counts are difficult to make without felling the trees. To keep such destructive sampling to a minimum, counts of cones on standing trees can be made using tripod-mounted binoculars. This can be a slow, expensive, and generally inexact procedure. However, in some cases a relationship between the total count and that on a cone-bearing branch can be developed. When this relationship is known, all one needs to do is to count the number of cone-bearing branches and insert this count into the regression or ratio to arrive at an estimate of the number of cones on the tree.

## 20.2 USING REGRESSION

This discussion will be limited to situations where the relationship is linear.

The large or *primary* sample, of size $n_p$, is used to obtain as strong an estimate as is feasible of the population mean of the predicting variable, $\bar{x}_p \to \mu_X$. The small or *secondary* sample, of size $n_s$, is used to develop the regression equation. This secondary sample, may or may not be a subsample of the primary sample. The subscript $p$ refers to the primary sample, while the subscript $s$ refers to the secondary sample.

### 20.2.1 Estimating $\mu_Y$

According to Equation 15.3, the general two-variable regression equation is

$$\mu_{Y \cdot X} = \beta_0 + \beta_1 X; \quad X_A < X < X_B$$

where $\beta_0 = \mu_Y - \beta_1 \mu_X$

$$\beta_1 = \frac{\sum_i^N (y_i - \mu_Y)(x_i - \mu_X)}{\sum_i^N (x_i - \mu_X)^2}$$

Thus,

$$\mu_{Y \cdot X} = \mu_Y - \beta_1 \mu_X + \beta_1 X = \mu_Y + \beta_1 (X - \mu_X) \tag{20.1}$$

The model of an individual observation is

$$y_i = \mu_Y + \beta_1 (x_i - \mu_X) + \varepsilon_i = \hat{y}_i + \varepsilon_i \tag{20.2}$$

where $\varepsilon_i$ = error of observation i

$$= y_i - \mu_Y - \beta_1 (x_i - \mu_X) \tag{20.3}$$

The sample-based regression equation, generated using data from the secondary sample, is

$$\hat{y}_i = b_0 + b_1 x_i$$

where $b_0 = \bar{y}_s - b_1 \bar{x}_s$

$$b_1 = \frac{\sum_i^{n_s} (y_i - \bar{y}_s)(x_i - \bar{x}_s)}{\sum_i^{n_s} (x_i - \bar{x}_s)^2}$$

or

$$\hat{y}_i = \bar{y}_s - b_1 \bar{x}_s + b_1 x_i = \bar{y}_s + b_1 (x_i - \bar{x}_s) \tag{20.4}$$

An estimate of the population mean, $\mu_Y$, is obtained by substituting the mean of the primary sample, $\bar{x}_p$, for $x_i$ in Equation 20.4

$$\hat{\bar{y}} = \bar{y}_s + b_1(\bar{x}_p - \bar{x}_s) \rightarrow \mu_Y \tag{20.5}$$

The error of this estimate is

$$\begin{aligned}\hat{\bar{y}} - \mu_Y &= \bar{y}_s + b_1(\bar{x}_p - \bar{x}_s) - \mu_Y \\ &= \bar{y}_s - \mu_Y + b_1(\bar{x}_p - \bar{x}_s)\end{aligned}$$

where $\bar{y}_s = \mu_Y + \beta_1(\bar{x}_s - \mu_X) + \bar{\varepsilon}$

Then,

$$\hat{\bar{y}} - \mu_Y = \beta_1(\bar{x}_s - \mu_X) + \bar{\varepsilon} + b_1(\bar{x}_p - \bar{x}_s) \tag{20.6}$$

Since $(y_i - \bar{y}_s) = \mu_Y + \beta_1(x_i - \mu_X) + \varepsilon_i - \bar{y}_s$,

$$b_1 = \frac{\sum_i^{n_s}[\mu_Y + \beta_1(x_i - \mu_X) + \varepsilon_i - \bar{y}_s](x_i - \bar{x}_s)}{\sum_i^{n_s}(x_i - \bar{x}_s)^2}$$

$$= \frac{\mu_Y\sum_i^{n_s}(x_i - \bar{x}_s) + \beta_1\sum_i^{n_s}(x_i - \mu_X)(x_i - \bar{x}_s) + \sum_i^{n_s}\varepsilon_i(x_i - \bar{x}_s) - \bar{y}_s\sum_i^{n_s}(x_i - \bar{x}_s)}{\sum_i^{n_s}(x_i - \bar{x}_s)^2}$$

Since $\Sigma_i^{n_s}(x_i - \bar{x}) = 0$ and $(x_i - \mu_X)(x_i - \bar{x}_s) = x_i^2 - \mu_X x_i - x_i\bar{x}_s + \mu_X\bar{x}_s$,

$$b_1 = \frac{\beta_1\left(\sum_i^{n_s}x_i^2 - \mu_X\sum_i^{n_s}x_i - \sum_i^{n_s}x_i\bar{x}_s + \mu_Y\sum_i^{n_s}\bar{x}_s\right) + \sum_i^{n_s}\varepsilon_i(x_i - \bar{x}_s)}{\sum_i^{n_s}(x_i - \bar{x}_s)^2}$$

Since

$$\sum_i^{n_s}\bar{x}_s = n_s\bar{x}_s + \frac{n_s}{n_s}\sum_i^{n_s}x_i = \sum_i^{n_s}x_i$$

and

$$\sum_{i}^{n_s} x_i \bar{x}_s = \sum_{i}^{n_s} x_i \frac{\sum_{i}^{n_s} x_i}{n_s} = \frac{\left(\sum_{i}^{n_s} x_i\right)^2}{n_s^2}$$

then

$$b_1 = \frac{\beta_1 \left[\sum_{i}^{n_s} x_i^2 - \frac{\left(\sum_{i}^{n_s} x_i\right)^2}{n_s}\right] + \sum_{i}^{n_s} \varepsilon_i (x_i - \bar{x}_s)}{\sum_{i}^{n_s} (x_i - \bar{x}_s)^2}$$

$$= \frac{\beta_1 \sum_{i}^{n_s} (x_i - \bar{x}_s)^2 + \sum_{i}^{n_s} \varepsilon_i (x_i - \bar{x}_s)}{\sum_{i}^{n_s} (x_i - \bar{x}_s)^2} \tag{20.7}$$

$$= \beta_1 + \frac{\sum_{i}^{n_s} \varepsilon_i (x_i - \bar{x}_s)}{\sum_{i}^{n_s} (x_i - \bar{x}_s)^2}$$

Then, returning to Equation 20.6,

$$\hat{\bar{y}} - \mu_Y = \beta_1 (\bar{x}_s - \mu_Y) + \bar{\varepsilon} + \left[\beta_1 + \frac{\sum_{i}^{n_s} \varepsilon_i (x_i - \bar{x}_s)}{\sum_{i}^{n_s} (x_i - \bar{x}_s)^2}\right] (\bar{x}_p - \bar{x}_s)$$

$$= \beta_1 [(\bar{x}_s - \mu_Y) + (\bar{x}_p - \bar{x}_s)] + \bar{\varepsilon} + \left[\frac{\sum_{i}^{n_s} \varepsilon_i (x_i - \bar{x}_s)}{\sum_{i}^{n_s} (x_i - \bar{x}_s)^2}\right] (\bar{x}_p - \bar{x}_s)$$

Since $\beta_1 [(\bar{x}_s - \mu_X) + (\bar{x}_s - \bar{x}_s)] = \beta_1 (\bar{x}_p - \mu_X)$, the error of the estimate is

$$\bar{y} - \mu_Y = \beta_1 (\bar{x}_p - \mu_Y) + \bar{\varepsilon} + \left[\frac{\sum_{i}^{n_s} \varepsilon_i (x_i - \bar{x}_s)}{\sum_{i}^{n_s} (x_i - \bar{x}_s)}\right] (\bar{x}_p - \bar{x}_s) \tag{20.8}$$

To find the expected value of $(\bar{y} - \mu_Y)$ it must be assumed that the $x_i$ values of the primary sample are fixed so that $E(\bar{x}_p) = \bar{x}_p$. Furthermore, as in conventional linear regression operations, the $x_i$ values of the secondary sample must be considered to be fixed. Then,

$$E(\bar{y} - \mu_Y) = \beta_1[E(\bar{x}_p - \mu_X) + E(\bar{\varepsilon}) + E(\bar{x}_p - \bar{x}_s)\left\{\frac{E\left[\sum_i^{n_s} \varepsilon_i(x_i - \bar{x}_s)\right]}{\sum_i^{n_s} E(x_i - \bar{x}_s)^2}\right\}$$

Since $\bar{\varepsilon} = \bar{y}_s - \mu_Y - \beta_1(\bar{x}_s - \mu_X)$,

$$E(\bar{\varepsilon}) = E(\bar{y}_s) - \mu_Y - \beta_1[E(\bar{x}_s) - \mu_X] = \mu_Y - \mu_Y - \beta_1(\mu_X - \mu_X) = 0$$

and

$$E\left[\sum_i^{n_s} \varepsilon_i(x_i - \bar{x}_s)\right] = E\left\{\sum_i^{n_s} [y_i - \mu_Y - \beta_1(x_i - \mu_X)](x_i - \bar{x}_s)\right\}$$

$$= \sum_i^{n_s} \{E(y_i) - \mu_Y - \beta_1[E(x_i) - \mu_X][E(x_i) - E(\bar{x}_s)]\}$$

$$= \sum_i^{n_s} [\mu_Y - \mu_Y - \beta_1(\mu_X - \mu_X)(\mu_X - \mu_X)] = 0$$

Consequently, remembering that $E(\bar{x}_p) = \bar{x}_p$,

$$E(\bar{y} - \mu_Y) = \beta_1(\bar{x}_p - \mu_X) \tag{20.9}$$

which is the bias of the estimate.

As can be seen, if $\bar{x}_p = \mu_X$, there would be no bias. Since the arithmetic mean of a sample is a consistent estimator, its variance decreases as the sample size increases, going to zero as the sample size approaches the size of the population. Consequently, the bias decreases as the size of the primary sample increases. Since $\bar{x}_p$ can be either greater or less than $\mu_X$, the sign of the bias can be either positive or negative.

### 20.2.2 The Variance of the Estimate

The variance of an estimate serves two main purposes: (1) to provide a measure of the repeatability or precision of the estimate and (2) to provide a basis for comparing the precisions of competing sampling schemes.

Consider the following expressions.

$$\frac{1}{M}\sum_i^M (\bar{y}_i - \mu_{\hat{\bar{y}}})^2 \tag{20.10}$$

and

$$\frac{1}{M}\sum_{i}^{M}(\bar{y}_i - \mu_Y)^2 \tag{20.11}$$

where $M$ is equal to total number of samples that can be obtained from the population. Note that in the case of equation 20.10, the differences being squared are from the *mean of the set of* $\bar{y}$, while in the case of Equations 20.11, the differences are from the population mean, $\mu_Y$, the parameter being estimated. If bias is present, as it usually is in the case of $\bar{y}$, the mean of the set of estimates, in this case, $\mu_{\bar{y}}$, will not be equal to the parameter, $\mu_Y$. Consequently, Equation 20.10 will yield a smaller magnitude than will Equation 20.11, because, according to the theory of least squares, the sum of squares will be at a minimum when the differences being squared are from the mean of the set, which in this case is $\mu_{\bar{y}}$.

Equation 20.10 would serve well as a measure of precision or repeatability. However, if the sampling scheme that produced $\bar{y}$ is to be compared with another sampling scheme, such as simple random sampling, there would have to be a common base upon which to make the comparison. In such a case, the comparison should be made using Equation 20.11 for both sampling plans because in both cases the variability would be measured from the same point.

Equation 20.10 yields the conventional variance. Equation 20.11 yields the *mean squared error* or MSE. The MSE is defined as follows:

$$\text{MSE}(\hat{\mu}) = E(\hat{\mu} - \mu)^2$$

where $\hat{\mu}$ is the estimate and $\mu$ is the parameter being estimated. Then, letting $m = E(\hat{\mu})$,

$$\begin{aligned}\text{MSE}(\hat{\mu}) &= E[(\hat{\mu} - m) + (m - \mu)]^2 \\ &= E(\hat{\mu} - m)^2 + 2(m - \mu)E(\hat{\mu} - m) + (m - \mu)^2\end{aligned}$$

Since $E(\hat{\mu}) = m$, the cross-products equal zero and

$$\begin{aligned}\text{MSE}(\hat{\mu}) &= E(\hat{\mu} - m) + (m - \mu)^2 \\ &= \sigma^2_{\hat{\mu}} + (m - \mu)^2 \qquad (20.12) \\ &= \sigma^2_{\hat{\mu}} + \text{BIAS}^2\end{aligned}$$

Note that if $m = \mu$, there is no bias and the variance and the MSE are equal. In the case of $\bar{y}$, the

$$MSE(\hat{\bar{y}}) = \sigma^2_{\wedge/y} + \beta_1^2(\bar{x}_p - \mu_X)^2 \tag{20.13}$$

In the case of a two-variable linear regression,

$$\mu_{Y \cdot X = x_i} = \beta_0 + \beta_1 x_i$$

$$\Delta_i = y_i - \mu_{Y \cdot X = x_i} = y_i - \beta_0 - \beta_1 x_i$$

and, under the assumption that the $x_i$ are fixed,

$$\sum_{i}^{N} \Delta_i^2 = \sum_{i}^{N} (y_i - \mu_Y)^2 - \beta_1 \sum_{i}^{N} (y_i - \mu_Y)(x_i - \mu_X)$$

which, according to Equation 15.113, can be written

$$\sum_{i}^{N} \Delta_i^2 = \sum_{i}^{N} (y_i - \mu_Y)^2 - \beta_1^2 \sum_{i}^{N} (x_i - \mu_X)^2$$

The variance about the regression line is then

$$\sigma_{Y\cdot X}^2 = \frac{1}{N} \sum_{i}^{N} \Delta_i^2$$

$$= \frac{1}{N} \left[ \sum_{i}^{N} (y_i - \mu_Y)^2 - \beta_1^2 \sum_{i}^{N} (x_i - \mu_X)^2 \right] \tag{20.14}$$

$$= \sigma_Y^2 - \beta_1^2 \sigma_X^2 \tag{20.15}$$

Since $\beta_1 = \sigma_{XY} / \sigma_X^2$,

$$\beta_1^2 \sigma_X^2 = \frac{\sigma_{XY}^2}{\sigma_X^2} \tag{20.16}$$

Furthermore, since the correlation coefficient

$$\rho_{XY} = \frac{\sigma_{XY}}{\sigma_X \sigma_Y} \quad \text{and} \quad \rho_{XY}^2 = \frac{\sigma_{XY}^2}{\sigma_X^2 \sigma_Y^2}$$

then,

$$\rho_{XY}^2 = \frac{\beta_1^2 \sigma_X^2}{\sigma_Y^2}$$

$$\rho_{XY}^2 \sigma_Y^2 = \beta_1^2 \sigma_X^2 \quad \text{and} \tag{20.17}$$

$$\sigma_{Y\cdot X}^2 = \sigma_Y^2 - \rho_{XY}^2 \sigma_Y^2 = \sigma_Y^2 (1 - \rho_{XY}^2) \tag{20.18}$$

Switching to the *S* symbology, recognizing that

$$S_{Y\cdot X}^2 = \sigma_{Y\cdot X}^2 \left[ \frac{N}{(N-1)} \right]$$

Equation 20.14 becomes

$$S_{Y\cdot X}^2 = \frac{1}{(N-1)}\left[\sum_i^N (y_i - \mu_Y)^2 - \beta_1^2 \sum_i^N (x_i - \mu_X)^2\right] \quad (20.19)$$

$$= S_Y^2 - \beta_1^2 S_X^2 \quad (20.20)$$

and Equation 20.18 becomes

$$S_{Y\cdot X}^2 = S_Y^2(1 - \rho_{XY}^2) \quad (20.21)$$

According to Equation 15.136, the variance of $\hat{\bar{y}}$, after substituting $S_{Y\cdot X}^2$ for $s_{Y\cdot X}^2$, $n_s$ for $n$, $\bar{x}_P$ for $x_i$, and $n_s\bar{x}_s$ for $\Sigma_i^{n_s} x_i$, is

$$S_{\hat{\bar{y}}}^2 = S_{Y\cdot X}^2\left[\frac{\sum_i^{n_s} x_i^2 + n_s\bar{x}_p^2 - 2n_s\bar{x}_p\bar{x}_s}{n_s\sum_i^{n_s}(x_i - \bar{x}_s)^2}\right] \quad (20.22)$$

Since $2n_s\bar{x}_s^2 = 2\bar{x}_s\Sigma_i^{n_s} x_i$, inserting $n_s\bar{x}_s^2 + n_s\bar{x}^2 - 2\bar{x}_s\Sigma_i^{n_s} x_i$ into Equation 20.22 will have no effect on $S_{\hat{\bar{y}}}^2$.

$$S_{\hat{\bar{y}}}^2 = S_{Y\cdot X}^2\left[\frac{\sum_i^{n_s} x_i^2 - 2\bar{x}_x\sum_i^{n_s} x_i + n_s\bar{x}_s^2 + n_s\bar{x}_p^2 - 2n_s\bar{x}_p\bar{x}_s + n_s\bar{x}_s^2}{n_s\sum_i^{n_s}(x_i - \bar{x}_s)^2}\right]$$

$$= S_{Y\cdot X}^2\left[\frac{\sum_i^{n_s}(\bar{x}_p - \bar{x}_s)^2 + n_s(\bar{x}_p - \bar{x}_s)^2}{n_s\sum_i^{n_s}(x_i - \bar{x}_s)^2}\right] \quad (20.23)$$

$$= S_{Y\cdot X}^2\left[\frac{1}{n_s} + \frac{(\bar{x}_p - \bar{x}_s)^2}{\sum_i^{n_s}(x_i - \bar{x}_s)^2}\right]$$

and the MSE of $\hat{\bar{y}}$ is

$$\text{MSE}(\hat{\bar{y}}) = S_{Y\cdot X}^2\left[\frac{1}{n_s} + \frac{(\bar{x}_p - \bar{x}_s)^2}{\sum_i^{n_s}(x_i - \bar{x}_s)^2}\right] + \beta_1^2(\bar{x}_p - \mu_X)^2 \quad (20.24)$$

This MSE is applicable only when the sets of $x_i$ values making up the primary and secondary samples are fixed, following the rationale for fitting a regression using least squares. To obtain the general variance, it is necessary to replace the

$$\left[\frac{1}{n_s} + \frac{(\bar{x}_p - \bar{x}_s)^2}{\sum_i^{n_s}(x_i - \bar{x}_s)^2}\right] \quad \text{and} \quad (\bar{x}_p - \mu_X)^2$$

terms by the averages of such terms across all possible samples of size $n_p$ and $n_s$. In the case of the $(\bar{x}_p - \mu_X)^2$ term:

$$\frac{1}{\binom{N}{np}} \sum_i^{\binom{N}{np}} (\bar{x}_{p_i} - \bar{x}_{s_i})^2 = S^2_{\bar{x}_p} = \frac{S^2_X}{n_p} \tag{20.25}$$

Finding the average for the

$$\left[\frac{1}{n_s} + \frac{(\bar{x}_p - \bar{x}_s)^2}{\sum_i^{n_s}(x_i - \bar{x}_s)^2}\right]$$

term is not easy. Cochran (1963), working under the assumptions that (1) the primary sample is a random sample, (2) the secondary sample is a random subsample of the primary sample, and (3) the $x_i$ values are normally distributed, found the average to be

$$\frac{1}{n_s} + \left(\frac{1}{n_s} - \frac{1}{n_p}\right)\frac{1}{(n_s - 3)} \quad \text{or} \tag{20.26}$$

$$\frac{1}{n_s} + \left(\frac{n_p - n_s}{n_s n_p}\right)\frac{1}{(n_s - 3)} \tag{20.27}$$

Then the average variance would be

$$V(\hat{\bar{y}}) = S^2_{Y \cdot X}\left[\frac{1}{n_s} + \left(\frac{n_p - n_s}{n_s n_p}\right)\frac{1}{(n_s - 3)}\right] + \frac{\beta_1^2 S_X^2}{n_p} \tag{20.28}$$

which, since $S^2_{Y \cdot X} = S^2_Y(1 - \rho^2_{XY})$ and $\beta_1^2 S_X^2 = S_Y^2 \rho^2_{XY}$, can be written

$$V(\hat{\bar{y}}) = S^2_Y(1 - \rho^2_{XY})\left[\frac{1}{n_s} + \left(\frac{n_p - n_s}{n_s n_p}\right)\frac{1}{(n_s - 3)}\right] + \frac{S_Y^2 \rho^2_{XY}}{n_p} \tag{20.29}$$

Cochran states that only the

$$\left[\frac{1}{(n_s - 3)}\right]$$

term is affected if the secondary sample is deliberately, rather than randomly, selected from the primary sample. Deliberate selection of the secondary sample is often done to obtain as wide a range as possible of the $x_i$ values so that the regression will be as strong as possible. This tends to reduce the

$$\left[\frac{1}{(n_s - 3)}\right]$$

term. If the secondary sample is an independent sample, rather than a subsample, the

$$\left(\frac{1}{n_s} - \frac{1}{n_p}\right)$$

term in Equation 20.26 becomes

$$\left(\frac{1}{n_s} + \frac{1}{n_p}\right)$$

It should be pointed out that, in the case of a given sample, the following expressions would be appropriate.

$$V(\hat{\bar{y}}) = S^2_{Y \cdot X}\left[\frac{1}{n_s} + \frac{(\bar{x}_p - \bar{x}_s)^2}{\sum_i^{n_s}(x_i + \bar{x}_s)^2}\right] + \frac{\beta_1^2 S_X^2}{n_p} \tag{20.30}$$

or

$$V(\hat{\bar{y}}) = S^2_Y(1 - \rho^2_{XY})\left[\frac{1}{n_s} + \frac{(\bar{x}_p - \bar{x}_s)^2}{\sum_i^{n_s}(x_i - \bar{x}_s)^2}\right] + \frac{S_Y^2 \rho_{XY}^2}{n_p} \tag{20.31}$$

## 20.2.3 Estimating the Variance of the Estimate

The estimate of the variance of $\hat{\bar{y}}$ is based on Equation 20.30.

$$s^2_{\hat{\bar{y}}} = s^2_{Y \cdot X}\left[\frac{1}{n_s} + \frac{(\bar{x}_p - \bar{x}_s)^2}{\sum_i^{n_s}(x_i - \bar{x}_s)^2}\right] + \frac{b_1^2 s_X^2}{n_p} \tag{20.32}$$

where
$$s_{Y\cdot X}^2 = \frac{1}{n_s - 2}\left[\sum_i^{n_s}(y_i - \bar{y}_s)^2 - b_1^2\sum_i^{n_s}(x_i - \bar{x}_s)^2\right] \tag{20.33}$$

$$s_X^2 = \frac{1}{n_s - 1}\sum_i^{n_s}(x_i - \bar{x}_s)^2$$

$$b_1 = \frac{s_{XY}}{s_X^2}$$

$$s_{XY} = \frac{1}{n_s - 1}\sum_i^{n_s}(y_i - \bar{y}_s)(x_i - \bar{x}_s)$$

$$= \frac{1}{n_s - 1}\left[\sum_i^{n_s} x_i y_i - \left(\frac{\sum_i^{n_s} y_i \sum_i^{n_s} x_i}{n_s}\right)\right] \tag{20.34}$$

If the sampling fractions $n_s/n_p$ and $n_p/N$ are not negligible, it becomes necessary to include finite population corrections to Equation 20.32.

$$s_{\bar{y}}^2 = s_{Y\cdot X}^2\left[\frac{1}{n_s} + \frac{(\bar{x}_p - \bar{x}_s)^2}{\sum_i^{n_s}(x_i - \bar{x}_s)^2}\right]\left(\frac{n_p - n_s}{n_p}\right) + \frac{b_1^2 s_X^2}{n_p}\left(\frac{N - n_p}{N}\right) \tag{20.35}$$

If it can be assumed that $X$ and $Y$ are perfectly, or nearly perfectly, correlated so that $\rho_{XY} \rightarrow 1$ and $b_1^2 s_x^2 \rightarrow s_y^2$ (see Equation 20.17), Equation 20.35 becomes

$$s_{\bar{y}}^2 = s_{Y\cdot X}^2\left[\frac{1}{n_s} + \frac{(\bar{x}_p - \bar{x}_s)^2}{\sum_i^{n_s}(x_i - \bar{x}_s)}\right]\left(\frac{n_p - n_s}{n_p}\right) + \frac{s_{Y_s}^2}{n_p}\left(\frac{N - n_p}{N}\right) \tag{20.36}$$

Since there is no sense using double sampling with variables that are not strongly correlated, many authors (e.g., Freese, 1962; Cunia, 1963) prefer to use Equation 20.36.

### 20.2.4 Example

Assume a situation where an estimate has been made of the timber volume on a 500-acre tract using 50 1/5-acre plots that were randomly drawn from the sampling frame covering the tract. The sampling was without replacement. The plot centers were located on vertical aerial photographs and estimates of the amounts of timber on each were made with the assistance of aerial stand volume tables which listed gross per acre volumes by average stand height, average crown diameter, and average crown closure. These estimates could be made easily and rapidly when the photographs were viewed stereoscopically.

Since the photo volume tables were based on regional rather than tract-specific data, they could only be considered approximations. In addition, there was a likelihood of photointerpreter bias. Consequently, the cruise was judged to be weak and needed strengthening. To that end, 10 of the 50 photo-plots were randomly chosen using sampling without replacement. These 10 plots were then located in the field and cruised in the conventional manner. The data from these 10 plots were used to develop a regression of ground volume ($Y$) on photo volume ($X$). The data, in terms of board feet per acre, and the computations are as follows:

| Plot | X = Photo Volume | Y = Ground Volume | Plot | X = Photo Volume | Y = Ground Volume | Plot | X = Photo Volume | Y = Ground Volume |
|---|---|---|---|---|---|---|---|---|
| 1 | 12,550 | | 18 | 10,750 | | 35 | 12,550 | |
| 2 | 12,000 | | 19 | 7900 | | 36 | 8750 | |
| 3 | 8,550 | 9100 | 20 | 10,800 | | 37 | 10,800 | |
| 4 | 11,750 | | 21 | 6500 | | 38 | 6500 | |
| 5 | 10,650 | | 22 | 12,550 | | 39 | 9750 | |
| 6 | 11,000 | 11,900 | 23 | 12,000 | 12,800 | 40 | 13,000 | |
| 7 | 10,800 | 11,800 | 24 | 8050 | | 41 | 13,000 | 14,300 |
| 8 | 10,450 | | 25 | 6500 | 6700 | 42 | 7900 | |
| 9 | 11,000 | | 26 | 9500 | 9900 | 43 | 11,700 | |
| 10 | 10,750 | | 27 | 11,750 | 12,500 | 44 | 12,000 | |
| 11 | 10,200 | | 28 | 8750 | | 45 | 6500 | 6700 |
| 12 | 12,750 | | 29 | 8250 | | 46 | 10,650 | |
| 13 | 11,750 | | 30 | 8250 | | 47 | 11,000 | |
| 14 | 11,000 | | 31 | 9750 | | 48 | 12,000 | |
| 15 | 9500 | | 32 | 8550 | | 49 | 12,000 | |
| 16 | 12,500 | 13,200 | 33 | 9150 | | 50 | 12,000 | |
| 17 | 11,000 | | 34 | 10,500 | | | | |

$N = 500$ $\qquad n_s = 10$ $\qquad n_p = 50$

$\bar{x}_s = 10{,}210$ bd. ft/acre $\qquad \bar{y}_s = 10{,}890$ bd. ft/acre $\qquad \bar{x}_p = 10{,}356$ bd. ft/acre

$$\sum_i^{n_s}(x_i - \bar{x}_s)^2 = 50{,}364{,}000 \qquad \sum_i^{n_s}(y_i - \bar{y}_s)^2 = 64{,}349{,}000 \qquad \sum_i^{n_p}(x_i - \bar{x}_p)^2 = 166{,}868{,}200$$

$s^2_{X_s} = 5{,}596{,}000$ $\qquad s^2_{Y_s} = 7{,}149{,}888.9$ $\qquad s^2_{Y_p} = 3{,}405{,}473.5$

$s_{XY} = 6{,}311{,}222.2$ $\qquad s^2_{X_p} = \pm 1845.3925$ bd. ft/acre

$\hat{\rho}_{XY} = 0.9977574$

$b_1 = 1.1278095$

$b_0 = -624.93547$

$$\hat{y} = 1.1278095X - 624.93547 \text{ bd. ft/acre}$$

Substituting $\bar{x}_p$ for $x_i$,

$$\bar{y} = 1.1278095\ (10{,}356) - 624.93547 = 11{,}054.7 \text{ or } 11{,}055 \text{ bd. ft/acre}$$

From Equation 20.32, the estimated variance about the regression is:

$$s_{Y \cdot X}^2 = \frac{1}{(10-2)}[64{,}349{,}000 - (1.1278095)^2(50{,}364{,}000)]$$

$$= \frac{1}{8}(288{,}295.24)$$

$$= 36{,}036.905$$

Then, using Equation 20.35 and assuming $N = 5(500) = 2500$,

$$s_{\hat{\bar{y}}}^2 = 36{,}036.905\left[\frac{1}{10} + \frac{(10{,}356 - 10{,}210)^2}{50{,}364{,}000}\right]\left(\frac{50-10}{50}\right)$$

$$+\left[\frac{(1.1278095)^2(5{,}596{,}000)}{50}\right]\left(\frac{2500-50}{2500}\right)$$

$$= 36{,}036.905(0.1 + 0.0004232)(0.8) + 142{,}357.12(0.98)$$

$$= 2895.1531 + 139{,}509.98$$

$$= 142{,}405.13$$

$$s_{\hat{\bar{y}}} = \pm\sqrt{142{,}405.13} = \pm 377.4 \text{ bd. ft / acre}$$

The final estimate of the timber of the tract, at an approximate confidence level of 0.67, is then 11,055 ± 377.4 bd. ft/acre or 5,527,500 ± 188,683 bd. ft on the tract.

If it is assumed that $\rho_{XY} = 1$ and Equation 20.36 is used,

$$s_{\hat{\bar{y}}}^2 = 36{,}036.905\left[\frac{1}{10} + \frac{(10{,}356 - 10{,}210)^2}{50{,}364{,}000}\right]\left(\frac{50-10}{50}\right)$$

$$+\frac{7{,}149{,}888.9}{50}\left(\frac{2500-50}{2500}\right)$$

$$= 2895.1542 + 140{,}137.82$$

$$= 143{,}032.97$$

$$s_{\hat{\bar{y}}} = \pm\sqrt{143{,}032.97} = \pm 378.2 \text{ bd. ft / acre}$$

Since $\hat{\rho}_{XY} = 0.9977574$, which approaches unity, the two estimates of $V(\hat{\bar{y}})$ are essentially equal.

### 20.2.5 Relative Precision

The precision of the estimate of the population mean obtained using double sampling based on regression compared with that obtained using simple random sampling is essentially a function of the degree of correlation between the variables.

According to Equation 20.31,

$$V(\hat{\bar{y}}) = S_Y^2(1-\rho_{XY}^2)\left[\frac{1}{n_s} + \frac{(\bar{x}_p - \bar{x}_s)^2}{\sum_i^{n_s}(x_i - \bar{x}_s)^2}\right] + \frac{\rho_{XY}^2 S_Y^2}{n_p}$$

The variance of the mean of a simple random sample is

$$V(\bar{y}_0)_{\text{Ran}} = \frac{S_Y^2}{n_{\text{Ran}}}\left(\frac{N - n_{\text{Ran}}}{N}\right) \tag{20.37}$$

Then,

$$S_Y^2 = V(\bar{y}_0)_{\text{Ran}}\, n_{\text{Ran}}\left(\frac{N}{N - n_{\text{Ran}}}\right)$$

Since the mean of a simple random variable is unbiased, its variance is equal to its MSE. Consequently, Equation 20.37 can be compared with Equation 20.31. Substituting

$$V(\bar{y}_0)_{\text{Ran}}\, n_{\text{Ran}}\left(\frac{N}{N - n_{\text{Ran}}}\right)$$

for $S_Y^2$ in Equation 20.31 yields

$$\begin{aligned} V(\hat{\bar{y}}) = {} & V(\bar{y}_0)_{\text{Ran}}\, n_{\text{Ran}}\left(\frac{N}{N - n_{\text{Ran}}}\right)(1-\rho_{XY}^2)\left[\frac{1}{n_s} + \frac{(\bar{x}_p - \bar{x}_x)^2}{\sum_i^{n_s}(x_i - \bar{x}_x)^2}\right] \\ & + \frac{\rho_{XY}^2}{n_p} V(\bar{y}_0)_{\text{Ran}}\, n_{\text{Ran}}\left(\frac{N}{N - n_{\text{Ran}}}\right) \end{aligned} \tag{20.38}$$

The secondary sample involves field operations comparable with those associated with simple random sampling. Thus, $n_{\text{Ran}} = n_s$. Then,

$$\begin{aligned} \frac{V(\hat{\bar{y}})}{V(\bar{y}_0)_{\text{Ran}}} = {} & n_s\left(\frac{N}{N - n_s}\right)(1-\rho_{XY}^2)\left[\frac{1}{n_s} + \frac{(\bar{x}_p - \bar{x}_s)^2}{\sum_i^{n_s}(x_i - \bar{x}_s)^2}\right] \\ & + \rho_{XY}^2\left(\frac{n_s}{n_p}\right)\left(\frac{N}{N - n_s}\right) \end{aligned} \tag{20.39}$$

When the right side of Equation 20.39 equals 1, double sampling using regression and simple random sampling are equally precise. When the ratio is less than 1, double sampling is the more precise, and when it is greater than 1, simple random sampling is the more precise.

Assume that $n_s$ is small relative to $N$ and that

$$\left(\frac{N}{N-n_s}\right) \doteq 1$$

Then Equation 20.39 becomes

$$\frac{V(\hat{\bar{y}})}{V(\bar{y}_0)_{\text{Ran}}} \doteq n_s(1-\rho_{XY}^2)\left[\frac{1}{n_s} + \frac{(\bar{x}_p - \bar{x}_s)^2}{\sum_i^{n_s}(x_i - \bar{x}_s)^2}\right] + \rho_{XY}^2\left(\frac{n_s}{n_p}\right) \quad (20.40)$$

If $n_s$ and $n_p$ are held constant and

$$\rho_{XY} = 1, \quad \frac{V(\hat{\bar{y}})}{V(\bar{y}_0)_{\text{Ran}}} \doteq \frac{n_s}{n_p} \leq 1$$

$$\rho_{XY} = 0, \quad \frac{V(\hat{\bar{y}})}{V(\bar{y}_0)_{\text{Ran}}} \doteq n_s\left[\frac{1}{n_s} + \frac{(\bar{x}_p - \bar{x}_s)^2}{\sum_i^{n_s}(x_i - \bar{x}_s)^2}\right]$$

$$\doteq 1 + \frac{n_s(\bar{x}_p + \bar{x}_s)^2}{\sum_i^{n_s}(x_i - \bar{x}_s)^2} \geq 1$$

The critical value of $\rho_{XY}$ is obtained by setting $V(\hat{\bar{y}})/V(\bar{y}_0)_{\text{Ran}}$ equal to 1 and solving Equation 20.39 for $\rho_{XY}$:

$$\rho_{XY_{\text{crit}}} = \left[\frac{1 + \frac{n_s(\bar{x}_p - \bar{x}_s)^2}{\sum_i^{n_s}(x_i - \bar{x}_s)^2} - \left(\frac{N-n_s}{N}\right)}{1 + \frac{n_s(\bar{x}_p - \bar{x}_s)^2}{\sum_i^{n_s}(x_i - \bar{x}_s)^2} - \frac{n_s}{n_p}}\right]^{1/2} \quad (20.41)$$

As can be seen, the magnitude of $\rho_{XY_{\text{crit}}}$ can be manipulated by changing the ratio $n_s/n_p$. As this ratio is decreased, the magnitude of $\rho_{XY_{\text{crit}}}$ also decreases, indicating that if the ratio $n_s/n_p$ is made sufficiently small, double sampling using regression will yield a more precise estimate of the population mean than will simple random sampling, regardless of the magnitude of $\rho_{XY}$.

As an example, use the data from the previous example where $N = 2500$; $n_p = 50$; $n_s = 10$; $\bar{x}_p = 10{,}356$; $\bar{x}_s = 10{,}210$; $s_{X_s}^2 = 5{,}596{,}000$; and $\hat{\rho}_{XY} = 0.9977574$. Then,

$$\rho_{XY_{\text{crit}}} = \sqrt{\frac{1 + \dfrac{(10{,}346 - 10{,}210)^2}{5{,}596{,}000(10-1)} - \dfrac{(2500-10)}{2500}}{1 + \dfrac{(10{,}356 - 10{,}210)^2}{5{,}596{,}000(10-1)} - \dfrac{10}{50}}} = 0.0743379$$

which is less than $\rho_{XY} = 0.9977574$, indicating that double sampling using regression yields a more precise estimate of $\mu_Y$ than would a simple random sample of size $n_s$.

To determine the effect of the ratio $n_s/n_p$ on the relative precision, hold $\rho_{XY}$ constant at $k$ in Equation 20.40:

$$\frac{V(\hat{\bar{y}})}{V(\bar{y}_0)_{\text{Ran}}} = (n_s - n_s k)\left[\frac{1}{n_s} + \frac{(\bar{x}_p - \bar{x}_s)^2}{\sum_i^{n_s}(x_i - \bar{x}_s)^2}\right] + k\left(\frac{n_s}{n_p}\right)$$

$$= n_s\left[\frac{(\bar{x}_p - \bar{x}_s)^2}{\sum_i^{n_s}(x_i - \bar{x}_s)^2} - \frac{k(\bar{x}_p - \bar{x}_s)^2}{\sum_i^{n_s}(x_i - \bar{x}_s)^2} + \frac{k}{n_p}\right] + (1-k) \quad (20.42)$$

Since $k < 1$, the sum of the first two terms inside the brackets is positive. $k/n_p$ is also positive. $(1 - k)$ is constant. Consequently, as $n_s$ is increased, while $n_p$ is held constant, $V(\hat{\bar{y}})/V(\bar{y}_0)_{\text{Ran}}$ will increase, indicating that the precision of double sampling, relative to that of simple random sampling, is decreasing. In short, when $\rho_{XY}$ is held constant, the precision of the double sampling estimate will be inversely proportional to the magnitude of $n_s/n_p$. To demonstrate this, one can use Equation 20.42 and the data from the earlier example.

$$\frac{V(\hat{\bar{y}})}{V(\bar{y}_0)_{\text{Ran}}} = 10\left\{\frac{(10{,}356 - 10{,}210)^2}{5{,}596{,}000(10-1)} - 0.9977574\left[\frac{(10{,}356 - 10{,}210)^2}{5{,}596{,}000(10-1)}\right] + \frac{0.997754}{50}\right\}$$

$$+ (1 - 0.9977574)$$

$$= 0.2018036$$

indicating that, in this case, the precision of the estimate of $\mu_Y$ obtained using double sampling was 1/0.2018036 or 4.96 times as great as one obtained using simple random sampling.

In summary, the greater the correlation between the dependent and predicting variables and the smaller the ratio of the size of the secondary sample to the size of the primary sample, the greater will be the precision of an estimate of $\mu_Y$ using double sampling with regression relative to that of an equivalent simple random sample.

### 20.2.6 Sample Size

Sample size can be based on either a specified budget or a specified maximum sampling error.

#### 20.2.6.1 Using a Budget Constraint

According to Equation 20.31,

$$V(\hat{\bar{y}}) = S_Y^2(1-\rho_{XY}^2)\left[\frac{1}{n_s} + \frac{(\bar{x}_p - \bar{x}_s)^2}{\sum_i^{n_s}(x_i - \bar{x}_s)^2}\right] + \frac{\rho_{XY}^2 S_Y^2}{n_p}$$

In most cases the difference between $\bar{x}_p$ and $\bar{x}_s$ is small and, consequently, $(\bar{x}_p - \bar{x}_s)^2$ will be small. The secondary sample is usually chosen so as to obtain as wide a spread as is possible of the magnitudes of the predicting variable ($X$). As a result, $\Sigma_i^{n_s}(x_i - \bar{x}_s)^2$ is usually very large relative to $(\bar{x}_p - \bar{x}_s)^2$. For this reason the term

$$\frac{(\bar{x}_p - \bar{x}_s)^2}{\sum_i^{n_s}(x_i - \bar{x}_s)^2}$$

can be considered negligible and Equation 20.31 may be written

$$V(\hat{\bar{y}}) \doteq \frac{S_Y^2(1-\rho_{XY}^2)}{n_s} + \frac{\rho_{XY}^2 S_Y^2}{n_p} \tag{20.43}$$

or

$$\doteq \frac{S_{Y \cdot X}^2}{n_s} + \frac{S_{XY}^2}{n_p S_X^2} \tag{20.44}$$

Then, following a procedure described by Cochran (1963), let

$$V_s = S_Y^2(1-\rho_{XY}^2) = S_{Y \cdot X}^2 \quad \text{and} \quad V_p = \rho_{XY}^2 S_Y^2 = \frac{S_{XY}^2}{S_X^2}$$

so that Equation 20.43 can be written

$$V(\hat{\bar{y}}) \doteq V_s n_s^{-1} + V_p n_p^{-1} \tag{20.45}$$

The cost function, considering only field operations, is

$$c_F = n_s c_s + n_p c_p \tag{20.46}$$

where $c_s$ = cost of locating and evaluating one sampling unit of the secondary sample
$c_p$ = cost of locating and evaluating one sampling unit of the primary sample

Then, using the Lagrangian multiplier,

$$V = V(\hat{\bar{y}}) + \lambda(n_s c_s + n_p c_p - c_F)$$

$$= V_s n_s^{-1} + V_p n_P^{-1} + \lambda(n_s c_s + n_p c_p - c_F)$$

$$\frac{\partial V}{\partial n_s} = -V_s n_s^{-2} + \lambda c_s = 0 \tag{20.47}$$

$$\frac{\partial V}{\partial n_p} = -V_p n_p^{-2} + \lambda c_p = 0 \tag{20.48}$$

Multiplying Equation 20.47 by $n_s^2$ and Equation 20.48 by $n_p^2$ and then simplifying yields

$$-V_s + \lambda c_s n_s^2 = 0 \tag{20.49}$$

and

$$V_p - \lambda c_p n_p^2 = 0 \tag{20.50}$$

Multiplying Equation 20.49 by $V_p$ and Equation 20.50 by $V_s$

$$-V_p V_s + V_p \lambda c_s n_s^2 = 0$$

$$V_p V_s - V_s \lambda c_p n_p^2 = 0$$

Then,

$$V_p c_s n_s^2 = V_s c_p n_p^2$$

$$\frac{n_s^2}{V_s c_p} = \frac{n_p^2}{V_p c_s}$$

$$\frac{n_s}{\sqrt{V_s c_p}} = \frac{n_p}{\sqrt{V_p c_s}} \tag{20.51}$$

$$n_s = n_p \frac{\sqrt{V_s c_p}}{\sqrt{V_p c_s}}$$

Substituting the right side of Equation 20.51 for $n_s$ in the cost function,

$$c_F = n_p \frac{\sqrt{V_s c_p}}{\sqrt{V_p c_s}} c_s + n_p c_p$$

$$= n_p \left( c_s \frac{\sqrt{V_s c_p}}{\sqrt{V_p c_s}} + c_p \right)$$

$$= n_p \left( c_s \frac{\sqrt{V_s c_p} + c_p \sqrt{V_p c_s}}{\sqrt{V_p c_s}} \right)$$

Then, the size of the needed primary sample is

$$n_{p(\text{needed})} = \frac{c_F\sqrt{V_s c_p}}{c_s\sqrt{V_s c_p} + c_p\sqrt{V_p c_s}} \tag{20.52}$$

or

$$n_{p(\text{needed})} = \frac{c_F\sqrt{\dfrac{c_s S_{XY}^2}{S_X^2}}}{c_s\sqrt{c_p\left(\dfrac{S_Y^2 - S_{XY}^2}{S_X^2}\right)} + c_p\sqrt{\dfrac{c_s S_{XY}^2}{S_X^2}}} \tag{20.53}$$

Substituting the right side of Equation 20.52 for $n_p$ in Equation 20.51 yields the size of the secondary sample:

$$\begin{aligned} n_{s(\text{needed})} &= \frac{c_F\sqrt{V_p c_s}}{c_s\sqrt{V_s c_p} + c_p\sqrt{V_p c_s}}\left(\frac{\sqrt{V_s c_p}}{\sqrt{V_p c_s}}\right) \\ &= \frac{c_F\sqrt{V_s c_p}}{c_s\sqrt{V_s c_p} + c_p\sqrt{V_p c_s}} \end{aligned} \tag{20.54}$$

which can be written

$$n_{s(\text{needed})} = \frac{c_F\sqrt{c_p S_{Y \cdot X}^2}}{c_s\sqrt{c_p S_{Y \cdot X}^2} + c_p\sqrt{\dfrac{c_s S_{XY}^2}{S_X^2}}} \tag{20.55}$$

Estimates of $n_s$ and $n_p$ stem from Equations 20.53 and 20.55:

$$n_{p(\text{needed})} = \frac{c_F\sqrt{\dfrac{c_s s_{XY}^2}{s_X^2}}}{c_s\sqrt{c_p\left(\dfrac{s_Y^2 - s_{XY}^2}{s_X^2}\right)} + c_p\sqrt{\dfrac{c_s s_{XY}^2}{s_X^2}}} \tag{20.56}$$

and

$$n_{s(\text{needed})} = \frac{c_F\sqrt{c_p s_{Y \cdot X}^2}}{c_s\sqrt{c_p s_{Y \cdot X}^2} + c_p\sqrt{\dfrac{c_s s_{XY}^2}{s_X^2}}} \tag{20.57}$$

Assume that the example in Section 20.2.4 is for a presample. Then $s^2_{x_s}$ = 5,596,000; $s^2_{Y_s}$ = 7,149,888.9; $s_{XY}$ = 6,311,222.2; and $s^2_{Y \cdot X}$ = 36,036.905. If $c_F$ is set at \$1,000, $c_p$ = \$1, and $c_s$ = \$10, $n_s$, according to Equation 20.57, is

$$n_{s(\text{needed})} = \frac{\$1000\sqrt{\$1(36,036.905)}}{\$10\sqrt{\$1(36,036.905)} + \$1\sqrt{\dfrac{\$10(6,311,222.2)^2}{5,596,000}}}$$

$$= 18.4 \text{ or } 18$$

$n_{s(\text{needed})}$ is rounded downward so that the cost of the operation will not exceed the maximum cost of $c_F$ = \$1000. Then, $n_p$ is found using Equation 20.56

$$n_{p(\text{needed})} = \frac{\$1000\sqrt{\$10\left[\dfrac{(\$6,311,222.2)^2}{5,596,000}\right]}}{\$10\sqrt{\$1[7,149,888.9 - \left[\dfrac{(6,311,222.2)^2}{5,596,000}\right]} + \$1\sqrt{\$10\left[\dfrac{(6,311,222.2)^2}{5,596,000}\right]}}$$

$$= 824.99 \text{ or } 825$$

The cost of the fieldwork would then be

$$c_F = n_s c_s + n_p c_p = 18(\$10) + 825(\$1) = 1005$$

$n_p$ would have to be reduced to 820 to keep the cost within the specified limit of \$1000.

### 20.2.6.2 Using an Allowable Sampling Error Constraint

If an allowable sampling error is specified, rather than a limit on the cost of the fieldwork, the above sample size formulae are not appropriate. According to Equation 13.108, the allowable sampling error is

$$A = t_c(k)s_{\bar{y}}$$

For the sake of simplicity, assume that $n_s$ will be large enough so that the sampling distribution of means can be assumed to be normal. Then $Z_c$ can be substituted for $t_c(k)$. Furthermore, $s_{\bar{y}}$ should be replaced by $S_{\bar{y}}$ since one is generalizing. Then,

$$A = Z_c S_{\bar{y}}$$

Squaring both sides

$$A^2 = Z_c^2 S_{\bar{y}}^2 = Z_c^2 V(\bar{y})$$

Let $V_s = S_Y^2(1 - \rho_{XY}^2)$ and $V_p = \rho_{XY}^2 S_Y^2$. Then, from Equation 20.45

$$V\left(\hat{\bar{y}}\right) = V_s n_s^{-1} + V_p n_p^{-1} = S_{\bar{y}}^2$$

$$A^2 = Z_c^2(V_s n_s^{-1} + V_p n_p^{-1}) \quad \text{and}$$

$$V = \frac{A^2}{Z_c^2} = \frac{V_s}{n_s} + \frac{V_p}{n_p} \tag{20.58}$$

The cost function, as in Equation 20.46, is

$$c_F = n_s c_s + n_p c_p$$

Then, as before (see Equation 20.51),

$$n_s = n_p \frac{\sqrt{V_s c_p}}{\sqrt{V_p c_s}} \tag{20.59}$$

and

$$n_p = n_s \frac{\sqrt{V_p c_s}}{\sqrt{V_s c_p}} \tag{20.60}$$

According to Equation 20.58,

$$\frac{A^2}{Z_c^2} = \frac{V_s}{n_s} + \frac{V_p}{n_p} = \frac{n_p V_s + n_s V_p}{n_s n_p}$$

$$\frac{n_s n_p A^2}{Z_c^2} = n_p V_s + n_s V_p \tag{20.61}$$

$$n_s n_p = \frac{(n_p V_s + n_s V_p) Z_c^2}{A^2}$$

$$n_p = \frac{(n_p V_s + n_s V_p) Z_c^2}{n_s A^2} \tag{20.62}$$

Substituting the right side of Equation 20.60 for $n_p$ on the right side of Equation 20.62,

$$n_{p(\text{needed})} = \frac{\left(V_s n_s \sqrt{\frac{V_p c_s}{V_s c_p}} + n_s V_p\right) Z_c^2}{n_s A^2} = \frac{Z_c^2}{A^2}\left(V_s \sqrt{\frac{V_p c_s}{V_s c_p}} + V_p\right) \tag{20.63}$$

$$= \frac{Z_c^2}{A^2}\left\{S_Y^2(1-\rho_{XY}^2)\sqrt{\frac{c_s}{c_p}\left[\frac{\rho_{XY}^2 S_Y^2}{S_Y^2(1-\rho_{XY})^2}\right]} + \rho_{XY}^2 S_Y^2\right\}$$

$$= \frac{Z_c^2 S_Y^2}{A^2}\left\{(1-\rho_{XY}^2)\sqrt{\frac{c_s}{c_p}\left[\frac{\rho_{XY}^2}{(1-\rho_{XY})^2}\right]} + \rho_{XY}^2\right\} \tag{20.64}$$

which can be written

$$n_{p(\text{needed})} = \frac{Z_c^2}{A^2}\left[S_{Y \cdot X}^2\sqrt{\frac{c_s}{c_p}\left(\frac{S_{XY}^2}{S_{Y \cdot X}^2 S_X^2}\right)} + \frac{S_{XY}^2}{S_Y^2}\right] \tag{20.65}$$

$n_s$ is found in the same manner

$$n_s = \frac{Z_c^2}{A^2}\left(V_s + V_p\sqrt{\frac{V_s c_p}{V_p c_s}}\right) \tag{20.66}$$

$$n_{s(\text{needed})} = \frac{Z_c^2}{A^2}\left\{S_Y^2(1-\rho_{XY}^2) + \rho_{XY}^2 S_Y^2\sqrt{\frac{c_p}{c_s}\left[\frac{S_Y^2(1-\rho_{XY}^2)}{\rho_X^2 S_Y^2}\right]}\right\}$$

$$= \frac{Z_c^2 S_Y^2}{A^2}\left[(1-\rho_{XY}^2) + \rho_{XY}^2\sqrt{\frac{c_p}{c_s}\left(\frac{1-\rho_{XY}^2}{\rho_{XY}^2}\right)}\right] \tag{20.67}$$

which can be written

$$n_{s(\text{needed})} = \frac{Z_c^2}{A^2}\left[S_{Y \cdot X}^2 + \frac{S_{XY}^2}{S_X^2}\sqrt{\frac{c_p}{c_s}\left(\frac{S_{Y \cdot X}^2 S_X^2}{S_{XY}^2}\right)}\right] \tag{20.68}$$

Substituting sample statistics from a presample for the parameters in the above equations,

$$n_{p(\text{needed})} = \frac{t_c^2(k) s_Y^2 s}{A^2}\left\{(1-\hat{\rho}_{XY}^2)\sqrt{\frac{c_s}{c_p}\left(\frac{\hat{\rho}_{XY}^2}{(1-\hat{\rho}_{XY}^2)}\right)} + \hat{\rho}_{XY}^2\right\} \tag{20.69}$$

or

$$n_{p(\text{needed})} = \frac{t_c^2(k)}{A^2}\left[(s_{Y \cdot X}^2)\sqrt{\frac{c_s}{c_p}\left(\frac{s_{XY}^2}{s_{Y \cdot X}^2 s_X^2}\right)} + \frac{s_{XY}^2}{s_X^2}\right] \tag{20.70}$$

$$n_{s(\text{needed})} = \frac{t_c^2(k) s_Y^2}{A^2}\left[(1-\hat{\rho}_{XY}^2) + \hat{\rho}_{XY}^2\sqrt{\frac{c_p}{c_s}\left(\frac{1-\hat{\rho}_{XY}^2}{\hat{\rho}_{XY}^2}\right)}\right] \tag{20.71}$$

or

$$n_{s(\text{needed})} = \frac{t_c^2(k)}{A^2}\left[s_{Y \cdot X}^2 + \frac{s_{XY}^2}{s_X^2}\sqrt{\frac{c_p}{c_s}\left(\frac{s_{Y \cdot X}^2 s_X^2}{s_{XY}^2}\right)}\right] \tag{20.72}$$

As an example, again assume that the data in the example in Section 20.2.4 represents a presample. Further assume that the allowable sampling error has been set at ±1% of the mean volume per acre at a confidence level of 0.68. Then,

$$A = \pm 0.01\bar{x}_s = \pm 102.1 \text{ bd. ft / acre}$$

Since the sampling distribution of means is not known, one can use the approximate value, 1, for $t_c(k)$. Then, using Equations 20.70 and 20.72,

$$n_{p(\text{needed})} = \frac{1^2}{(102.1)^2}\left\{36,036.905\sqrt{\frac{10}{1}\left[\frac{(6,311,222.2)^2}{36,036.905(5,596,000)}\right]} + \frac{(6,311,222.2)^2}{5,596,000}\right\}$$

$$= 836.4 \text{ or } 837 \text{ primary sampling units}$$

$$n_{s(\text{needed})} = \frac{1^2}{(102.1)^2}\left\{36,036.905 + \frac{(6,311,222.2)^2}{5,596,000}\sqrt{\frac{1}{10}\left(\frac{36,036.905(5,596,000)}{(6,311,222.2)^2}\right)}\right\}$$

$$= 18.8 \text{ or } 19 \text{ secondary sampling units}$$

### 20.2.7 Relative Efficiency

The efficiency of double sampling using regression relative to simple random sampling differs from its relative precision because the costs of both the primary and secondary samples must be taken into account. According to Equation 20.45,

$$V(\bar{y}) \doteq V_s n_s^{-1} + V_p n_p^{-1}$$

$$\doteq V_s\left(\frac{c_s\sqrt{V_s c_p} + c_p\sqrt{V_p c_s}}{c_F\sqrt{V_s c_p}}\right) + V_p\left(\frac{c_s\sqrt{V_s c_p} + c_p\sqrt{V_p c_s}}{c_p\sqrt{V_p c_s}}\right)$$

$$\doteq \frac{\left(\sqrt{V_s c_s} + \sqrt{V_p c_p}\right)^2}{c_F} \tag{20.73}$$

$$\doteq \frac{\left[\sqrt{c_s S_Y^2(1-\rho_{XY})^2} + \sqrt{c_p \rho_{XY}^2 s_Y^2}\right]}{c_F}$$

$$\doteq S_Y^2\frac{\left[\sqrt{c_s(1-\rho_{XY}^2)} + \sqrt{c_p\rho_{XY}^2}\right]^2}{c_F}$$

which can be written

$$V(\bar{y}) \doteq \frac{\left[\sqrt{c_s S_{Y\cdot X}^2} + \sqrt{\frac{c_p S_{XY}^2}{S_X^2}}\right]^2}{c_F} \tag{20.74}$$

The cost of locating and evaluating a basic sampling unit, whether when simple random sampling or when obtaining the secondary sample in a double sampling operation, should be the same. Consequently, $c_s = c_{\text{Ran}}$. Then, in the case of simple random sampling, when the sample size is

$$n_{\text{Ran}} = \frac{c_F}{c_{\text{Ran}}}$$

If the finite population correction can be considered to be negligible,

$$V(\bar{y}_0)_{\text{Ran}} = \frac{S_Y^2}{n_{\text{Ran}}} = \frac{S_Y^2 c_{\text{Ran}}}{c_F} \tag{20.75}$$

Then, comparing $V(\bar{y}_0)_{\text{Ran}}$ and $V(\hat{\bar{y}})$,

$$\frac{S_Y^2 c_{\text{Ran}}}{c_F} \overset{?}{=} \frac{S_Y^2\left[\sqrt{c_s(1-1\rho_{XY}^2)} + \sqrt{c_p\rho_{XY}^2}\right]^2}{c_F}$$

and

$$c_{Ran} \overset{?}{=} \left[\sqrt{c_s(1-\rho_{XY}^2)} + \sqrt{c_p\rho_{XY}^2}\right]^2 \tag{20.76}$$

Thus, when $V(\bar{y}_0)_{\text{Ran}} = V(\hat{\bar{y}})$

$$c_{\text{Ran}} = \left[\sqrt{c_s(1-\rho_{XY}^2)} + \sqrt{c_p\rho_{XY}^2}\right]^2$$

and, for a specified cost, both sampling methods will yield equally precise estimates of the population mean. When $V(\bar{y}_0)_{\text{Ran}} > V(\hat{\bar{y}})$,

$$c_{\text{Ran}} > \left[\sqrt{c_s(1-\rho_{XY}^2)} + \sqrt{c_p\rho_{XY}^2}\right]^2 \tag{20.77}$$

double sampling using regression will, for a specified cost, yield a more precise estimate of the population than will simple random sampling and is, therefore, the more efficient sampling method. For example, if $c_{\text{Ran}} = c_s = \$10$, $c_p = \$1$, and $\rho_{XY} = 0.8$, are used in Equation 20.77,

$$\{\sqrt{\$10[1-(0.8)^2]} + \sqrt{\$1(0.8)^2}\}^2 = \$7.28$$

which is less than $c_{\text{Ran}} = \$10$, indicating that double sampling would be more efficient than simple random sampling.

Dividing both sides of Equation 20.77 by $c_p$,

$$\frac{c_{\text{Ran}}}{c_p} > \left[\sqrt{\frac{c_s(1-\rho_{XY}^2)}{c_p}} + \rho_{XY}^2\right]^2$$

$$\sqrt{\frac{c_{\text{Ran}}}{c_p}} > \sqrt{\frac{c_s(1-\rho_{XY}^2)}{c_p}} + \rho_{XY}$$

$$\sqrt{\frac{c_{\text{Ran}}}{c_p}} - \sqrt{\frac{c_s}{c_p}(1-\rho_{XY}^2)} > \rho_{XY}$$

Since $c_{\text{Ran}} = c_s$,

$$\sqrt{\frac{c_s}{c_p}(1-\rho_{XY}^2)} > \rho_{XY}$$

Squaring both sides

$$\frac{c_s}{c_p} > \frac{\rho_{XY}^2}{(1-\sqrt{1-\rho_{XY}^2})^2} \tag{20.78}$$

Thus, where the ratio of $c_s$ to $c_p$ is greater than

$$\frac{\rho_{XY}^2}{(1-\sqrt{1-\rho_{XY}^2})^2}$$

double sampling is more efficient than simple random sampling. Figure 20.1 shows the values of $c_s/c_p$ at which double and simple random sampling are equally efficient.

Whenever $c_s/c_p$ lies above the curve, double sampling is more efficient, and whenever $c_s/c_p$ occurs below the line, double sampling is less efficient than simple random sampling.

## 20.3 USING RATIO ESTIMATORS

The utilization of ratio estimators in double sampling is limited to situations where the regression is linear and passes through the origin. If these conditions are not met, one should use regression rather than ratio estimation. Furthermore, the following discussion is based on the assumption that the variance about the regression is homogeneous or is proportional to the predicting variable so that only the ratio of means will be considered.

### 20.3.1 ESTIMATING $\mu_Y$

In ratio estimation it is assumed that the population mean of the predicting variable, $\mu_X$, is known. If it is not known, it must be estimated with a relatively high degree of precision. This can be done with a large primary sample.

$$\bar{x}_p = \frac{1}{n_p}\sum_{i}^{n_p} x_i \rightarrow \mu_X$$

The secondary sample is used to estimate the ratio.

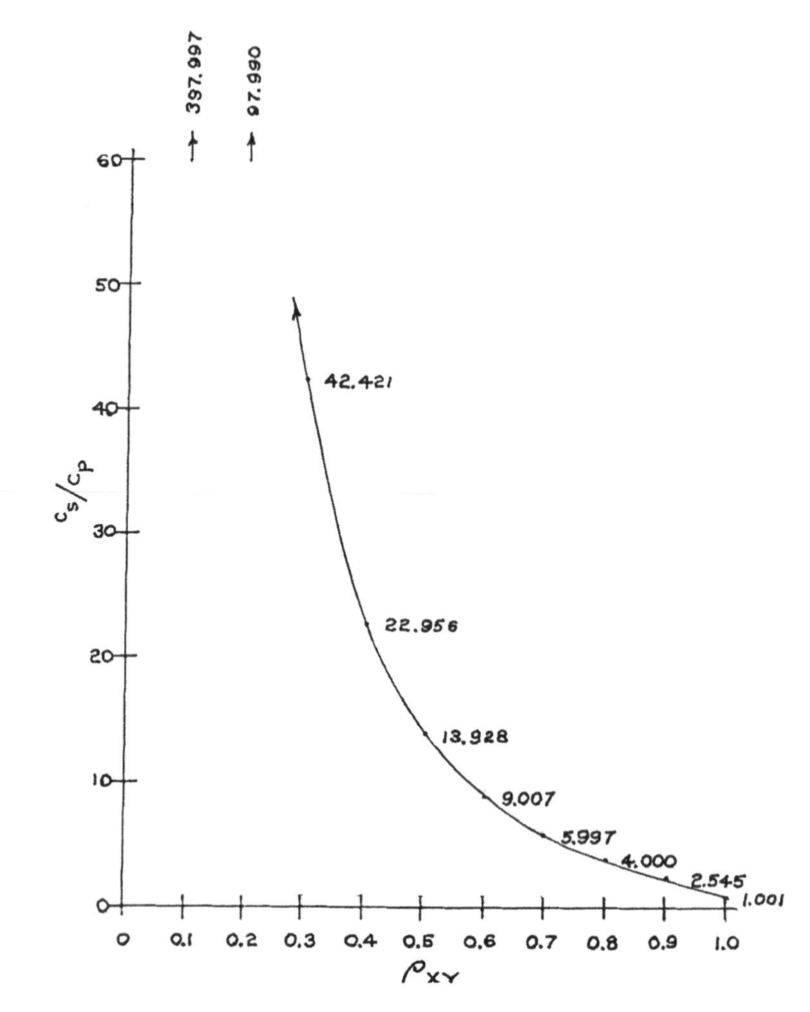


**FIGURE 20.1** Ratio of $c_s$ to $c_p$ at which double sampling using regression is as efficient as simple random sampling. If $c_s/c_p$ is greater than is indicated by the curve, double sampling is more efficient than simple random sampling.

$$\bar{y}_s = \frac{1}{n_s}\sum_{i}^{n_s} y_i$$

$$\bar{x}_s = \frac{1}{n_s}\sum_{i}^{n_s} x_i \quad \text{and}$$

$$\frac{\bar{y}_s}{\bar{x}_s} = r_{\text{DS}} \rightarrow R$$

where the subscript DS indicates double sampling.

According to Equation 19.54,

$$\bar{y}_{\text{DS}} = r_{\text{DS}}\mu_X = \frac{\bar{y}_s}{\bar{x}_s}\mu_X \rightarrow \mu_Y$$

If $\mu_X$ is being estimated by $\bar{x}_p$,

$$\bar{y}_{\text{DS}} = r_{\text{DS}}\bar{x}_p = \frac{\bar{y}_s}{\bar{x}_s}\bar{x}_p \rightarrow \mu_Y \tag{20.79}$$

The estimate of the total is then

$$T_{Y_{\text{DS}}} = N\bar{y}_{\text{DS}} \rightarrow \tau_Y \tag{20.80}$$

According to Equation 9.29,

$$E(\bar{y}_{\text{DS}}) = E(r_{\text{DS}}\bar{x}_p) - E(r_{\text{DS}})E(\bar{x}_p) = \text{COV}(r_{\text{DS}}\bar{x}_p)$$

Then $\mu_{\bar{y}_{\text{DS}}} - \mu_{r_{\text{DS}}}\mu_X = \text{COV}(r_{\text{DS}}\bar{x}_p)$ and

$$\mu_{\bar{y}_{\text{DS}}} = \text{COV}(r_{\text{DS}}\bar{x}_p) + \mu_{r_{\text{DS}}}\mu_X \tag{20.81}$$

According to Equation 19.15,

$$\begin{aligned} \mu_{r_{\text{DS}}} &= \frac{R - \text{COV}(r_{\text{DS}}\bar{x}_s)}{\mu_X} \\ &= \frac{\mu_Y - \text{COV}(r_{\text{DS}}\bar{x}_s)}{\mu_X} \end{aligned} \tag{20.82}$$

Then substituting the right side of Equation 20.82 for $\mu_{r_{\text{DS}}}$ in Equation 20.81,

$$\mu_{\bar{y}_{\text{DS}}} = \text{COV}(r_{\text{DS}}\bar{x}_p) + \mu_Y - \text{COV}(r_{\text{DS}}\bar{x}_s)$$

The bias of the estimate is then

$$\begin{aligned} \text{BIAS} &= \mu_{\bar{y}_{\text{DS}}} - \mu_Y \\ &= \text{COV}(r_{\text{DS}}\bar{x}_p) - \text{COV}(r_{\text{DS}}\bar{x}_x) \end{aligned} \tag{20.83}$$

This is the exact bias, but the expression provides no information regarding the impact of sample size. Since $\text{COV}(r_{\text{DS}}\bar{x}_p) = \rho_{r_{\text{DS}}}\sigma_{r_{\text{DS}}}\sigma_{\bar{x}_p}$ and $\text{COV}(r_{\text{DS}}\bar{x}_s) = \rho_{r_{\text{DS}}}\sigma_{r_{\text{DS}}}\sigma_{\bar{x}_s}$

$$\text{BIAS} = \rho_{r_{\text{DS}}}\sigma_{r_{\text{DS}}}\sigma_{\bar{x}_p} - \rho_{r_{\text{DS}}}\sigma_{r_{\text{DS}}}\sigma_{\bar{x}_s}$$

Assume $\rho_{r_{\text{DS}}}\bar{x}_p = \rho_{r_{\text{DS}}}\bar{x}_s = \rho$. Then, to this level of approximation,

$$\begin{aligned} \text{BIAS} &\doteq \rho\sigma_{r_{\text{DS}}}(\sigma_{\bar{x}_p}\sigma_{\bar{x}_s}) \\ &\doteq \rho\sigma_{r_{\text{DS}}}\left(\frac{\sigma_X}{\sqrt{n_p}} - \frac{\sigma_X}{\sqrt{n_s}}\right) \\ &\doteq \rho\sigma_{r_{\text{DS}}}\sigma_X\left(\frac{1}{\sqrt{n_p}} - \frac{1}{\sqrt{n_s}}\right) \end{aligned} \tag{20.84}$$

Since $n_s < n_p$, the term within the parentheses will be negative. Consequently, Equation 20.84 can be written

$$\text{BIAS} \doteq -\rho\sigma_{r_{DS}}\sigma_X\left(\frac{1}{\sqrt{n_s}} - \frac{1}{\sqrt{n_p}}\right) \tag{20.85}$$

As can be seen, the bias will be negative and its absolute magnitude will decrease as the sample sizes, $n_s$ and $n_p$, increase. Let $\alpha = n_s/n_p$, the proportion of the primary sample which is in the secondary sample. Then,

$$n_p = n_s\alpha$$

$$\left(\frac{1}{\sqrt{n_s}} - \frac{1}{\sqrt{n_p}}\right) = \left(\frac{1-\sqrt{\alpha}}{\sqrt{n_s}}\right)$$

and

$$\text{BIAS} \doteq -\rho\sigma_{r_{DS}}\sigma_X\left(\frac{1-\sqrt{\alpha}}{\sqrt{n_s}}\right) \tag{20.86}$$

The best way to visualize the effect of sample size on the bias is to use an example. One can fix $n_p$ at 100 and vary the size of $n_s$, remembering that $\rho\sigma_{r_{DS}}\sigma_X$ is constant.

| $n_s$ | 4 | 9 | 16 | 25 | 100 |
|---|---|---|---|---|---|
| $\sqrt{n_s}$ | 2 | 3 | 4 | 5 | 10 |
| $\alpha$ | 0.04 | 0.09 | 0.16 | 0.25 | 1 |
| $\sqrt{\alpha}$ | 0.2 | 0.3 | 0.4 | 0.5 | 1 |
| $1 - \sqrt{\alpha}$ | 0.8 | 0.7 | 0.6 | 0.5 | 0 |
| $\left(\frac{1-\sqrt{\alpha}}{\sqrt{n_s}}\right)$ | 0.4 | 0.2333 | 0.15 | 0.1 | 0 |

As can be seen, as the ratio $n_s/n_p$ increases, the bias decreases until, finally, when $n_s/n_p = 1$, the bias vanishes.

### 20.3.2 The Variance of the Estimate

The way the secondary sample is obtained has an effect on the variance of the estimate. Consequently, first consider the situation when the secondary sample is a subsample of the primary sample and then consider the situation when the secondary sample is an independent sample.

#### 20.3.2.1 The Approximate Variance When Subsampling Is Used

Returning to Equation 20.79

$$\bar{y}_{DS} = \frac{\bar{y}_s}{\bar{x}_s}\bar{x}_p \rightarrow \mu_Y$$

The error of this estimate is

$$\begin{aligned}
\bar{y}_{DS} - \mu_Y &= \frac{\bar{y}_s}{\bar{x}_s}\bar{x}_p - \mu_Y \\
&= \frac{\bar{y}_s}{\bar{x}_s}\bar{x}_p - \mu_Y + \frac{\bar{y}_s}{\bar{x}_s}\mu_X - \frac{\bar{y}_s}{\bar{x}_s}\mu_X \\
&= \left(\frac{\bar{y}_s}{\bar{x}_s}\mu_X - \mu_Y\right) + \frac{\bar{y}_s}{\bar{x}_s}(\bar{x}_p - \mu_X) \\
&= \mu_X\left(\frac{\bar{y}_s}{\bar{x}_s} - R\right) + \frac{\bar{y}_s}{\bar{x}_s}(\bar{x}_p - \mu_X) \\
&= \frac{\mu_X}{\bar{x}_s}(\bar{y}_s - R\bar{x}_s) + \frac{\bar{y}_s}{\bar{x}_s}(\bar{x}_p - \mu_X)
\end{aligned} \tag{20.87}$$

Assume that $\mu_X / \bar{x}_s = 1$ and $\bar{y}_s / \bar{x}_s = R = \mu_Y / \mu_X$. Then,

$$\begin{aligned}
\bar{y}_{DS} - \mu_Y &\doteq (\bar{y}_s - R\bar{x}_s) + R(\bar{x}_p - \mu_X) \\
&\doteq \bar{y}_s - R\bar{x}_s + R\bar{x}_p - R\mu_X \\
&\doteq \bar{y}_s - R\bar{x}_s + R\bar{x}_p - \mu_Y \\
&\doteq (\bar{y}_s - \mu_Y) + R(\bar{x}_p - \bar{x}_c)
\end{aligned}$$

According to Item 3 in Section 9.9, the variance of the mean is

$$\begin{aligned}
V(\bar{y}_{DS}) &\doteq V[(\bar{y}_s - \mu_Y) + R(\bar{x}_p - \bar{x}_s)] \\
&\doteq V(\bar{y}_s - \mu_Y) + R^2V(\bar{x}_p - \bar{x}_s) + 2R\,\mathrm{COV}(\bar{y}_s - \mu_Y)(\bar{x}_p - \bar{x}_s)
\end{aligned} \tag{20.88}$$

According to Item 4 in Section 9.9, disregarding the finite population correction, the first term in Equation 20.88 is

$$V(\bar{y}_s - \mu_Y) = V(\bar{y}_s) + V(\mu_Y) - 2\mathrm{COV}(\bar{y}_s\mu_Y)$$

Since $\mu_Y$ is a fixed quantity, $V(\mu_Y) = 0$ and

$$\mathrm{COV}(\bar{y}_s\mu_Y) = \frac{1}{\binom{N}{n}}\sum_{i}^{\binom{N}{n_s}}(\bar{y}_i - \mu_Y)(\mu_Y - \mu_Y) = 0$$

Thus,

$$V(\bar{y}_s - \mu_Y) = \frac{\sigma_Y^2}{n_s} \tag{20.89}$$

Consider now the second term in Equation 20.88.

$$\begin{aligned} V(\bar{x}_p - \bar{x}_s) &= V(\bar{x}_p) + V(\bar{x}_s) - 2\text{COV}(\bar{x}_p \bar{x}_s) \\ &= \frac{\sigma_X^2}{n_p} + \frac{\sigma_X^2}{n_s} - \frac{2\rho_{\bar{x}_p\bar{x}_s}\sigma_X^2}{\sqrt{n_p}\sqrt{n_s}} \end{aligned} \tag{20.90}$$

Assuming $\rho_{\bar{x}_p\bar{x}_s} = 1$ and $\sqrt{n_p}\sqrt{n_s} = n_p$,

$$V(\bar{x}_p - \bar{x}_s) \doteq \frac{\sigma_X^2}{n_p} + \frac{\sigma_X^2}{n_s} - \frac{2\sigma_X^2}{n_p} = \sigma_X^2\left(\frac{1}{n_s} - \frac{1}{n_p}\right)$$

and

$$R^2 V(\bar{x}_p - \bar{x}_s) \doteq R^2\sigma_X^2\left(\frac{1}{n_s} - \frac{1}{n_p}\right) \tag{20.91}$$

Finally, in the third term in Equation 20.88

$$\text{COV}(\bar{y}_s - \mu_Y)(\bar{x}_p - \bar{x}_s) = \rho_{(\bar{y}_s-\mu_Y)(\bar{x}_p-\bar{x}_s)}\sqrt{V(\bar{y}_s - \mu_Y)V(\bar{x}_p - \bar{x}_s)}$$

Referring to Equation 20.90 and assuming $\rho_{\bar{x}_p\bar{x}_s} = 1$,

$$\begin{aligned} V(\bar{x}_p - \bar{x}_s) &\doteq \frac{\sigma_X^2}{n_p} + \frac{\sigma_X^2}{n_s} - \frac{2\sigma_X^2}{\sqrt{n_p}\sqrt{n_s}} \\ &\doteq \sigma_X^2\left(\frac{1}{n_p} + \frac{1}{n_s} - \frac{2}{\sqrt{n_p}\sqrt{n_s}}\right) \end{aligned}$$

Then, assuming that $\rho_{(\bar{y}_s-\mu_Y)(\bar{x}_p-\bar{x}_s)} = \rho_{XY}$,

$$\begin{aligned} \text{COV}(\bar{y}_s - \mu_Y)(\bar{x} - \bar{x}_s) &\doteq \rho_{XY}\sqrt{\frac{\sigma_Y^2}{n_s}\sigma_X^2\left(\frac{1}{n_p} + \frac{1}{n_s} - \frac{2}{\sqrt{n_p}\sqrt{n_s}}\right)} \\ &\doteq \frac{\rho_X\sigma_Y\sigma_X}{\sqrt{n_s}}\sqrt{\frac{1}{n_p} + \frac{1}{n_s} - \frac{2}{\sqrt{n_p}\sqrt{n_s}}} \end{aligned}$$

Since

$$\left(\frac{1}{\sqrt{n_p}} - \frac{1}{\sqrt{n_s}}\right)^2 = \frac{1}{n_p} + \frac{1}{n_s} - \frac{2}{\sqrt{n_p}\sqrt{n_s}}$$

and since $\rho_{XY}\sigma_Y\sigma_X = \sigma_{XY}$,

$$\text{COV}(\bar{y}_s - \mu_Y)(\bar{x}_p - \bar{x}_s) \doteq \frac{\sigma_{XY}}{\sqrt{n_s}}\left(\frac{1}{\sqrt{n_p}} - \frac{1}{\sqrt{n_s}}\right)$$

$$\doteq \sigma_{XY}\left(\frac{1}{\sqrt{n_p}\sqrt{n_s}} - \frac{1}{n_s}\right)$$

Assuming that $\sqrt{n_p}\sqrt{n_s} = n_p$,

$$\text{COV}(\bar{y}_s - \mu_Y)(\bar{x}_p - \bar{x}_s) \doteq \sigma_{XY}\left(\frac{1}{n_p} - \frac{1}{n_s}\right)$$

Since

$$\frac{1}{n_s} > \frac{1}{n_p}$$

the term within the parentheses is negative. Consequently,

$$\text{COV}(\bar{y}_s - \mu_Y)(\bar{x}_p - \bar{x}_s) \doteq -\sigma_{XY}\left(\frac{1}{n_s} - \frac{1}{n_p}\right)$$

The approximate variance of the mean is then

$$V(\bar{y}_{\text{DS}}) \doteq \frac{\sigma_Y^2}{n_s} + R^2\sigma_X^2\left(\frac{1}{n_s} - \frac{1}{n_p}\right) - 2R\sigma_{XY}\left(\frac{1}{n_s} - \frac{1}{n_p}\right)$$
$$\doteq \frac{\sigma_Y^2 + R^2\sigma_X - 2R\sigma_{XY}}{n_s} + \frac{2R\sigma_{XY} - R^2\sigma_X^2}{n_p} \tag{20.92}$$

which can be written

$$V(\bar{y}_{\text{DS}}) \doteq \frac{\sigma_Y^2 + R^2\sigma_X - 2R\rho_{XY}\sigma_X\sigma_Y}{n_s} + \frac{2R\rho_{XY}\sigma_X\sigma_Y - R^2\sigma_X^2}{n_p} \tag{20.93}$$

Let

$$V_s = \frac{\sigma_Y^2 + R^2\sigma_X^2 - 2R\rho_{XY}\sigma_X\sigma_Y}{n_s} \tag{20.94}$$

and

$$V_p = \frac{2R\rho_{XY}\sigma_X\sigma_Y - R^2\sigma_X^2}{n_p} \tag{20.95}$$

If $R^2\sigma_X^2 = 2R\rho_{XY}\sigma_X\sigma_Y$, $V_s$ will be equal to $\sigma_Y^2/n_s$ or $\sigma_{\bar{y}_s}^2$, $V_p$ will be equal to zero, and $V(\bar{y}_{DS})$ will be equal to $\sigma_{\bar{y}_s}^2 = V(\bar{y}_s)$. If $R^2\sigma_X^2 > 2R\rho_{XY}\sigma_X\sigma_Y$, $V_p$ will be negative. By definition, a variance is an average of a set of squared errors or residuals. It cannot be negative. Consequently, in this situation $V_p$ is invalid. If $R^2\sigma_X^2 < 2R\rho_{XY}\sigma_X\sigma_Y$, both $V_s$ and $V_p$ will be valid, provided that $R^2\sigma_X^2 - 2R\rho_{XY}\sigma_{XY} \leq \sigma_Y^2$. The key to these situations is the population correlation coefficient, $\rho_{XY}$. When $R^2\sigma_X^2 = 2R\rho_{XY}\sigma_X\sigma_Y$, $\rho_{XY} = R\sigma_X/2\sigma_Y$; when $R^2\sigma_X^2 > 2R\rho_{XY}\sigma_X\sigma_Y$, $\rho_{XY} < R\sigma_X/2\sigma_Y$; and when $R^2\sigma_X^2 < 2R\rho_{XY}\sigma_X\sigma_Y$, $\rho_{XY} > R\sigma_X/2\sigma_Y$. Furthermore, if $V_s$ is to be valid when $R^2\sigma_X^2 < 2R\rho_{XY}\sigma_X\sigma_Y$, $\sigma_Y^2 + R^2\sigma_X^2 - 2R\rho_{XY}\sigma_X\sigma_Y$ must be equal to or greater than zero:

$$\sigma_Y^2 + R^2\sigma_X^2 \geq 2R\rho_{XY}\sigma_X\sigma_Y$$

$$\rho_{XY} \leq \frac{\sigma_Y^2 + R^2\sigma_X^2}{2R\sigma_X\sigma_Y}$$

Thus, if both $V_s$ and $V_p$ are to be valid, $\rho_{XY}$ must be in the following range:

$$\frac{R\sigma_X}{2\sigma_Y} \leq \rho_{XY} \leq \frac{\sigma_Y^2 + R^2\sigma_X^2}{2R\sigma_X\sigma_Y}$$

This statement is appropriate only when the ratio, $R$, is positive. Squaring it removes this restriction:

$$\frac{R^2\sigma_X^2}{4\sigma_Y^2} \leq \rho_{XY}^2 \leq \left[\frac{\sigma_Y^2 + R^2\sigma_X^2}{2R\sigma_X\sigma_Y}\right]^2 \tag{20.96}$$

As an example, assume a situation where $R = 1.13$; $\sigma_X^2 = 5{,}600{,}000$; and $\sigma_Y^2 = 7{,}150{,}000$. Then $\rho_{XY}^2$ must be in the range

$$\frac{(1.13)^2(5{,}600{,}000)}{4(7{,}150{,}000)} \leq \rho_{XY}^2 \leq \left[\frac{7{,}150{,}000 + (1.13)^2(5{,}600{,}000)}{2(1.13)\sqrt{5{,}600{,}000}\sqrt{7{,}150{,}000}}\right]^2$$

or

$$0.2500 \leq \rho_{XY}^2 \leq 1.0000$$

if both $V_s$ and $V_p$ are to be valid.

Since $R = \beta_1 = \sigma_{XY}/\sigma_X^2 = \rho_{XY}\sigma_X\sigma_Y/\sigma_X^2$, if the correlation between the two variables is sufficiently strong so that $\rho_{XY}$ can be assumed to be equal to 1, $R$ can be assumed to be equal to $\sigma_Y/\sigma_X$ and, from Equation 20.93,

$$V(\bar{y}_{DS}) \doteq \frac{1}{n_s}\left(\sigma_Y^2 + \frac{\sigma_Y^2}{\sigma_X^2}\sigma_X^2 - \frac{2\sigma_Y}{\sigma_X}\sigma_X\sigma_Y\right) + \frac{1}{n_p}\left(\frac{2\sigma_Y}{\sigma_X}\sigma_X\sigma_Y - \frac{\sigma_Y^2}{\sigma_X^2}\sigma_X^2\right)$$

$$\doteq \frac{1}{n_s}(\sigma_Y^2 + \sigma_Y^2 - 2\sigma_Y^2) + \frac{1}{n_p}(2\sigma_Y^2 - \sigma_Y^2) \tag{20.97}$$

$$\doteq \frac{\sigma_Y^2}{n_p} = \sigma_{\bar{y}_p}^2$$

indicating that the primary sample alone provides the needed information and double sampling would be unnecessary.

#### 20.3.2.2 The Approximate Variance When Subsampling Is Not Used

When subsampling is not used, it is assumed that the two samples are independent and that their covariance is equal to zero. Thus,

$$V(\bar{y}_{DS}) = V(\bar{y}_R)_s + V(\bar{y}_R)_p \tag{20.98}$$

According to Equations 19.67 and 19.70

$$V(\bar{y}_R)_s \doteq \frac{1}{n_s}\left(\frac{N - n_s}{N - 1}\right)(\sigma_Y^2 + R^2\sigma_X^2 - 2R\sigma_{XY}) \tag{20.99}$$

According to Equation 19.1,

$$R = \frac{\mu_Y}{\mu_X}$$

Then, $\mu_Y = R\mu_X$.

Substituting $\bar{y}_{R_p}$ for $\mu_Y$ and $\bar{x}_p$ for $\mu_X$ yields

$$\bar{y}_{R_p} \doteq R\bar{x}_p$$

Then,

$$V(\bar{y}_R)_p \doteq V(R\bar{x}_p) = \frac{R^2\sigma_X^2}{n_p}\left(\frac{N - n_p}{N - 1}\right) \tag{20.100}$$

and

$$V(\bar{y}_{DS}) = \sigma_{\bar{y}_{DS}}^2 \doteq \frac{1}{n_s}\left(\frac{N - n_s}{N - 1}\right)(\sigma_Y^2 + R^2\sigma_X^2 - 2R\sigma_{XY}) + \frac{1}{n_p}\left(\frac{N - n_p}{N - 1}\right)R^2\sigma_X^2 \tag{20.101}$$

which can be written

$$V(\bar{y}_{\text{DS}}) \doteq \sigma^2_{\bar{y}_{\text{DS}}} \doteq \frac{1}{n_s}\left(\frac{N-n_s}{N-1}\right)(\sigma^2_Y + R^2\sigma^2_X - 2R\rho_{XY}\sigma_X\sigma_Y) + \frac{1}{n_p}\left(\frac{N-n_p}{N-1}\right)R^2\sigma^2_X \tag{20.102}$$

If $n_s$ and $n_p$ are sufficiently small relative to $N$ so that the finite population correlations can be considered equal to 1, Equation 20.101 becomes

$$V(\bar{y}_{\text{DS}}) = \sigma^2_{\bar{y}_{\text{DS}}} \doteq \frac{\sigma^2_Y + R^2\sigma^2_X - 2R\sigma_{XY}}{n_s} + \frac{R^2\sigma^2_X}{n_p} \tag{20.103}$$

and Equation 20.102 becomes

$$V(\bar{y}_{\text{DS}}) = \sigma^2_{\bar{y}_{\text{DS}}} \doteq \frac{\sigma^2_Y + R^2\sigma^2_X - 2R\rho_{XY}\sigma_X\sigma_Y}{n_s} + \frac{R^2\sigma^2_X}{n_p} \tag{20.104}$$

Since $V(\bar{y}_R)_p$ cannot become negative, it remains valid in all cases. However, $V(\bar{y}_R)_s$ can become negative, and thus invalid, when $2R\sigma_{XY} = 2R\rho_{XY}\sigma_X\sigma_Y > \sigma^2_Y + R^2\sigma^2_X$ or when

$$\rho_{XY} > \frac{\sigma^2_Y + R^2\sigma^2_X}{2R\sigma_X\sigma_Y} \tag{20.105}$$

### 20.3.3 Estimating the Variance of the Estimate

Estimates of the approximate variance of the mean are obtained by substituting the sample statistics $s^2_Y$, $s^2_X$, $s_Y$, $s_X$, $r$, and $s_{XY}$ for, respectively, $\sigma^2_Y$, $\sigma^2_X$, $\sigma_Y$, $\sigma_X$, $R$, and $\sigma_{XY}$ in Equations 20.92, 20.93, 20.95, and 20.101 through 20.105. If the validity requirements cannot be met, a new predicting variable must be found that is more highly correlated with the variable of interest. If such a predicting variable cannot be found, it will be impossible to use double sampling using ratio estimators.

### 20.3.4 Sample Size

As in the case of double sampling using regression, sample size can be controlled by either a budget constraint or by a maximum allowable sampling error.

When the secondary sample is a subsample of the primary sample, the variance of the mean is, according to Equation 20.92,

$$V(\bar{y}_{\text{DS}}) \doteq \frac{\sigma^2_Y + R^2\sigma^2_X - 2R\sigma_{XY}}{n_s} + \frac{2R\sigma_{XY} - R^2\sigma^2_X}{n_p}$$

Let

$$V_s = \sigma^2_Y + R^2\sigma^2_X - 2R\sigma_{XY} \quad \text{and} \quad V_p = 2R\sigma_{XY} - R^2\sigma^2_X$$

Then

$$V(\bar{y}_{DS}) \doteq V_s n_s^{-1} + V_p n_p^{-1}$$

From this point, the development follows that in Section 20.2.6.

When the sample size is constrained by a budgetary limit, from Equation 20.52,

$$\begin{aligned} n_{p(\text{needed})} &= \frac{c_F \sqrt{V_p c_s}}{c_s \sqrt{V_s c_p} + c_p \sqrt{V_p c_s}} \\ &= \frac{c_F \sqrt{(2R\sigma_{DS} - R^2\sigma_X^2)c_s}}{c_s \sqrt{(\sigma_Y^2 + R^2\sigma_X^2 - 2R\sigma_{XY})c_p} + c_p \sqrt{(2R\sigma_{XY} - R^2\sigma_X^2)c_s}} \end{aligned} \quad (20.106)$$

and, from Equation 20.54,

$$\begin{aligned} n_{s(\text{needed})} &= \frac{c_F \sqrt{V_s c_p}}{c_s \sqrt{V_s c_p} + c_p \sqrt{V_p c_s}} \\ &= \frac{c_F \sqrt{(\sigma_Y^2 + R^2\sigma_X^2 - 2R\sigma_{XY})c_p}}{c_s \sqrt{(\sigma_Y^2 + R^2\sigma_X^2 - 2R\sigma_{XY})c_p} + c_p \sqrt{(2R\sigma_{XY} - R^2\sigma_X^2)c_s}} \end{aligned} \quad (20.107)$$

When a maximum allowable sampling error is specified, from Equation 20.63,

$$\begin{aligned} n_{p(\text{needed})} &= \frac{Z_c^2}{A^2}\left(V_s \sqrt{\frac{V_p c_s}{V_s c_p}} + V_p\right) \\ &= \frac{Z_c^2}{A^2}\left[(\sigma_Y^2 + R^2\sigma_X^2 - 2R\sigma_{XY})\sqrt{\frac{c_s}{c_p}\frac{(2R\sigma_{XY} - R^2\sigma_X^2)}{(\sigma_Y^2 + R^2\sigma_X^2 - 2R\sigma_{XY})}} + 2R\sigma_{XY} - R^2\sigma_{XY}\right] \end{aligned} \quad (20.108)$$

and, from Equation 20.66,

$$\begin{aligned} n_{s(\text{needed})} &= \frac{Z_c^2}{A^2}\left(V_s + V_p \sqrt{\frac{V_s c_p}{V_p c_s}}\right) \\ &= \frac{Z_c^2}{A^2}\left[(\sigma_Y^2 + R^2\sigma_X^2 - 2R\sigma_{XY}) + (2R\sigma_{XY} - R^2\sigma_X^2)\sqrt{\frac{c_p}{c_s}\frac{(\sigma_Y^2 + R^2\sigma_X^2 - 2R\sigma_{XY})}{R^2\sigma_X^2}}\right] \end{aligned} \quad (20.109)$$

When the secondary sample is independent of the primary sample (i.e., it is not a subsample), the variance of the mean, according to Equation 20.103, is

$$V(\bar{y}_{DS}) = \sigma_{\bar{y}_{DS}}^2 \doteq \frac{\sigma_Y^2 + R^2\sigma_X^2 - 2R\sigma_{XY}}{n_s} + \frac{R^2\sigma_X^2}{n_p}$$

As can be seen, $V_s$ is as before, but

$$V_p = R^2\sigma_X^2$$

Then, when there is a budgetary constraint,

$$n_{p(\text{needed})} = \frac{c_F\sqrt{(R^2\sigma_X^2)c_s}}{c_s\sqrt{(\sigma_Y^2 + R^2\sigma_X^2 - 2R\sigma_{XY})c_p} + c_p\sqrt{R^2\sigma_X^2 c_s}} \tag{20.110}$$

and

$$n_{s(\text{needed})} = \frac{c_F\sqrt{(\sigma_Y^2 + R^2\sigma_X^2 - 2R\sigma_{XY})c_p}}{c_s\sqrt{(\sigma_Y^2 + R^2\sigma_X^2 - 2R\sigma_{XY})c_p} + c_p\sqrt{R^2\sigma_X^2 c_s}} \tag{20.111}$$

and when there is a specified maximum sampling error,

$$n_{p(\text{needed})} = \frac{Z_c^2}{A^2}\left[(\sigma_Y^2 + R^2\sigma_X^2 - 2R\sigma_{XY})\sqrt{\frac{c_s}{c_p}\frac{R^2\sigma_X^2}{(\sigma_Y^2 + R^2\sigma_X^2 - 2R\sigma_{XY})}} + R^2\sigma_X^2\right] \tag{20.112}$$

$$n_{s(\text{needed})} = \frac{Z_c^2}{A^2}\left[(\sigma_Y^2 + R^2\sigma_X^2 - 2R\sigma_{XY}) + R^2\sigma_X^2\sqrt{\frac{c_p}{c_s}\frac{(\sigma_Y^2 + R^2\sigma_X^2 - 2R\sigma_{XY})}{R^2\sigma_X^2}}\right] \tag{20.113}$$

These sample sizes are estimated using sample statistics in place of the indicated parameters. Since the variance expressions are approximations, one should only substitute 1 for $Z_c$ for an approximate confidence level of 0.68 and 2 for an approximate confidence level of 0.95.

# 21 Sampling with Unequal Probabilities

## 21.1 INTRODUCTION

In the bulk of the material covered up to this point it was implied, if not stated, that the probability of a sampling unit being drawn in a sampling operation was constant for all the sampling units in the sampling frame. It was recognized, of course, that where the sampling frame is finite in size and the sampling is without replacement, the probabilities change from draw to draw but within a draw all the available sampling units have the same probability of being drawn. Now consider the situation where all of the sampling units do not have the same probability of being drawn.

## 21.2 MATHEMATICAL RATIONALE

Probabilities enter into the computation of parameters and statistics as weights associated with the individual variates. Thus, when the probability of being chosen is the same for all the sampling units, and the sampling is with replacement, the weight associated with an individual variate is

$$w_i = p(\text{sampling unit } i) = \frac{1}{M}$$

Each variate is an estimate of the population mean:

$$y_i \rightarrow \mu_y = \left(\frac{1}{M}\right)\sum_i^M y_i$$

The mean of a number of variates is a better estimator of the population mean:

$$\bar{y} = \frac{1}{m}\sum_i^m y_i \rightarrow \mu_Y$$

According to Equation 5.5, probability would be introduced into these expressions as weights. Thus, if the probability is constant at $P(\text{sampling unit } i) = 1/M$, $W_i = 1/M$

$$\mu_Y = \frac{\sum_i^M w_i y_i}{\sum_i^M w_i}$$

$$= \frac{\sum_i^M \left(\frac{1}{M}\right) y_i}{\sum_i^M \left(\frac{1}{M}\right)}$$

$$= \frac{\frac{1}{M} \sum_i^M y_i}{\left(\frac{1}{M}\right) M}$$

$$= \frac{1}{M} \sum_i^M y_i$$

and in the same way,

$$\bar{y} = \frac{\sum_i^m w_i y_i}{\sum_i^m w_i}$$

$$= \frac{\sum_i^m \left(\frac{1}{m}\right) y_i}{\sum_i^m \left(\frac{1}{m}\right)} = \frac{1}{m} \sum_i^m y_i$$

which is the conventional expression.

If, however, the sampling frame contains $M$ sampling units and the probability of a unit being drawn varies from unit to unit so that

$$P(\text{sampling unit } i) = \frac{N_i}{N}$$

where $N = \Sigma_i^M N_i$. The weight assigned to variates associated with sampling unit $i$ would be

$$w_i = \frac{N_i}{N}$$

The weighted population mean is then

$$\mu_{Y_w} = \frac{\sum_i^M w_i y_i}{\sum_i^M w_i} = \frac{\sum_i^M \left(\frac{N_i}{N}\right) y_i}{\sum_i^M \left(\frac{N_i}{N}\right)} = \frac{\sum_i^M N_i y_i}{N} \tag{21.1}$$

Now consider $N_i y_i$. In essence it implies that sampling unit $i$ contains $N_i$ variates, each equal in magnitude to $y_i$. In other words, a sampling unit can be considered a form of cluster. Consequently, many of the mathematical developments associated with cluster sampling can be used in the discussion of variable probability sampling.

Within this context, $y_i$ is equal to the mean of the variates in sampling unit $i$, $y_i = \mu_i$, and $Ny_i = N_i \mu_i = \tau_i$, the total of the variates within the sampling unit. The grand or overall total, across all $M$ sampling units in the frame is, according to Equation 17.6,

$$\tau_o = \sum_i^M \tau_i = \sum_i^M N_i y_i = N \mu_{Y_w} \tag{21.2}$$

Equations 21.1 and 21.2 are general. If the probability of selection is constant across all the sampling units in the sampling frame, $N_i$ will be equal to 1, $N$ will be equal to $M$, and the two expressions reduce to their conventional forms.

In some cases the probability is expressed as a decimal rather than as a fraction. This can simply be because the sampler prefers the decimal format or it can be caused by the fact that the $N_i$ values are imperfectly known and can only be estimated. When the decimal form is used, the mathematical expressions must be modified. For example,

$$\mu_{Y_w} = \sum_i^M P(\text{sampling unit } i) y_i \tag{21.3}$$

is the equivalent of Equation 21.1.

The sample-based estimate of the overall mean is

$$\bar{y}_{o_w} = \frac{M}{Nm} \sum_i^m N_i y_i \rightarrow \mu_{Y_w} \tag{21.4}$$

or

$$\bar{y}_{o_w} = \frac{M}{m}\sum_{i}^{m} P(\text{sampling unit } i)y_i \rightarrow \mu_{Y_w} \tag{21.5}$$

The expected value of $\bar{y}_{o_w}$, assuming $P(\text{sampling unit } i) = N_i/N$, is

$$E(\bar{y}_{o_w}) = \frac{M}{Nm}\sum_{i}^{m} E(N_i y_i)$$

Since

$$N_i y_i = \tau_i \tag{21.6}$$

the total in sampling unit $i$,

$$E(\bar{y}_{o_w}) = \frac{M}{Nm}\sum_{i}^{m} E(\tau_i)$$

$$= \frac{M}{Nm}\sum_{i}^{m} \frac{1}{M}\sum_{i}^{M} \tau_i$$

Since

$$\frac{1}{M}\sum_{i}^{M} \tau_i = \bar{\tau} \tag{21.7}$$

which is the mean sampling unit total,

$$E(\bar{y}_{o_w}) = \frac{M}{Nm}\sum_{i}^{m} \bar{\tau} = \frac{M}{N}\bar{\tau}$$

Since

$$M\bar{\tau} = \tau_0 \tag{21.8}$$

the overall or population total,

$$E(\bar{y}_{o_w}) = \frac{\tau_o}{N} = \mu_{Y_w} \tag{21.9}$$

indicating that $\bar{y}_{o_w}$, is an unbiased estimator of $\mu_{Y_w}$.

The estimate of the overall total is then

$$T_{o_w} = N\bar{y}_{o_w} = \frac{M}{m}\sum_{i}^{m} N_i y_i = \frac{M}{m}\sum_{i}^{m} \tau_i = M\bar{T} \rightarrow \tau_o \tag{21.10}$$

or

$$T_{o_w} = N\bar{y}_{o_w} = \frac{MN}{m}\sum_{i}^{m} P(\text{sampling unit } i)y_i \to \tau_o \tag{21.11}$$

The expected value of $T_{o_w}$ is

$$E(T_{o_w}) = NE(\bar{y}_{o_w}) = N\mu_{Y_w} = \tau_o$$

Thus, $T_{o_w}$ is an unbiased estimator.

The variance of the population is

$$V(Y_w) = \sigma^2_{Y_w} = \frac{\sum_{i}^{M} w_i(y_i - \mu_{Y_w})^2}{\sum_{i}^{M} w_i}$$
$$= \frac{\sum_{i}^{M} N_i(y_i - \mu_{Y_w})^2}{N} \tag{21.12}$$

or

$$V(Y_w) = \sigma^2_{Y_w} = \sum_{i}^{M} P(\text{sampling unit } i)(y_i - \mu_{Y_w})^2 \tag{21.13}$$

The variance of the sampling unit totals is

$$V(\tau) = \sigma^2_\tau = \frac{1}{M}\sum_{i}^{M}(\tau_i - \bar{\tau})^2 \tag{21.14}$$

and the variance of the overall total is

$$V(T_{o_w}) = \sigma^2_{T_{o_w}} = \frac{M^2\sigma^2_\tau}{m} \tag{21.15}$$

It should be noted that Equation 21.5 is the second term in Equation 17.40. The finite population correction has been deleted because the sampling is with replacement. The variance of the overall mean is then

$$V(\bar{y}_{o_w}) = \sigma^2_{\bar{y}_{o_w}} = \frac{\sigma^2_{T_{o_w}}}{N^2} = \frac{M^2\sigma^2_\tau}{N^2_m} \tag{21.16}$$

The estimate of the variance of the sampling unit totals is

$$s_\tau^2 = \frac{1}{m-1}\sum_i^m (\tau_i - \bar{T})^2 \rightarrow \sigma_\tau^2 \tag{21.17}$$

where

$$\bar{T} = \frac{1}{m}\sum_i^m \tau_i \rightarrow \bar{\tau} \tag{21.18}$$

The expected value of $\bar{T}$ is

$$E(\bar{T}) = \frac{1}{m}\sum_i^m E(\tau_i) = \frac{1}{m}\sum_i^m \frac{1}{M}\sum_i^M \tau_i = \bar{\tau}$$

Then,

$$s_{T_{o_w}}^2 = \frac{M^2 s_\tau^2}{m} \rightarrow \sigma_{T_o}^2 \tag{21.19}$$

and

$$s_{\bar{y}_{o_w}}^2 = \frac{s_{T_{o_w}}^2}{N^2} \rightarrow \sigma_{\bar{y}_o}^2 \tag{21.20}$$

Since st² is an unbiased estimator of $\sigma_\tau^2$, $s_{T_{o_w}}^2$ and $s_{\bar{y}_{o_w}}^2$ are also unbiased. An estimate of the population variance cannot be obtained because of the difficulty of determining the appropriate degrees of freedom.

As an example, assume the following population: 2, 4, and 6. The probabilities associated with these variates are as shown below:

| | | | | |
|---|---|---|---|---|
| $i =$ | | **1** | **2** | **3** |
| $y_i =$ | | 2 | 4 | 6 |
| $Pi =$ | $\frac{N_i}{N}$ | $\frac{2}{10}$ | $\frac{5}{10}$ | $\frac{3}{10}$ |
| $N_i =$ | | 2 | 5 | 3 |
| $t_i =$ | $N_i y_i =$ | 2(2) = 4 | 5(4) = 20 | 3(6) = 18 |
| $M =$ | 3 | | | |
| $N =$ | 10 | | | |

$$\mu_{Y_w} = \frac{\sum_i^M N_i y_i}{N} = \frac{[2(2) + 5(4) + 3(6)]}{10} = 4.2$$

$$\sigma_{Y_w}^2 = \frac{\sum_i^M N_i (y_i - \mu_{Y_w})^2}{N}$$

$$= \frac{[2(2-4.2)^2 + 5(4-4.2)^2 + 3(6-4.2)^2]}{10} = 1.96$$

$$\tau_o = \sum_i^M N_i y_i = [2(2) + 5(4) + 3(6)] = 42 \quad \text{or}$$

$$\tau_o = N\mu_{Y_w} = 10(4.2) = 42$$

Now draw samples of size $m = 2$, using sampling with replacement. There are $M^m = 3^2 = 9$ possible samples.

| i | Sample | $\sum_i^m N_i y_i$ | $\bar{y}_w = \frac{M}{Nm}\left[\sum_i^m N_i y_i\right]$ | $\bar{T} = \frac{1}{m}\left[\sum_i^m N_i y_i\right]$ | $T_o = M\bar{T}$ |
|---|---|---|---|---|---|
| 1 | (2,2) | [2(2) + 3(2)] = 8 | 8(0.15) = 1.2 | $\frac{8}{2} = 4$ | 3( 4) = 12 |
| 2 | (2,4) | [2(2) + 5(4)] = 24 | 24(0.15) = 3.6 | $\frac{24}{2} = 12$ | 3(12) = 36 |
| 3 | (2,6) | [2(2) + 3(6)] = 22 | 22(0.15) = 3.3 | $\frac{22}{2} = 11$ | 3(11) = 33 |
| 4 | (4,2) | [5(4) + 2(2)] = 24 | 24(0.15) = 3.6 | $\frac{24}{2} = 12$ | 3(12) = 36 |
| 5 | (4,4) | [5(4) + 5(4)] = 40 | 40(0.15) = 6.0 | $\frac{40}{2} = 20$ | 3(20) = 60 |
| 6 | (4,6) | [5(4) + 3(6)] = 38 | 38(0.15) = 5.7 | $\frac{38}{2} = 19$ | 3(19) = 57 |
| 7 | (6,2) | [3(6) + 2(2)] = 22 | 22(0.15) = 3.3 | $\frac{22}{2} = 11$ | 3(11) = 33 |
| 8 | (6,4) | [3(6) + 5(4)] = 38 | 38(0.15) = 5.7 | $\frac{38}{2} = 19$ | 3(19) = 54 |
| 9 | (6,6) | [3(6) + 3(6)] = 36 | 36(0.15) = 5.4 | $\frac{36}{2} = 18$ | 3(18) = 54 |
| | | | 37.8 | 126 | 378 |

$$\frac{M}{(Nm)} = \frac{3}{[10(2)]} = 0.15$$

$$\mu_{\bar{y}_w} = \frac{37.8}{9} = 4.2 = \mu_{Y_w}$$

$$\bar{\tau} = \frac{126}{9} = 14 \quad \text{or} \quad \frac{[2(2) + 5(4) + 3(6)]}{3} = \frac{42}{3} = 14$$

$$\tau_o = 10(4.2) = 42 \quad \text{or} \quad 3(14) = 42$$

Then,

| i | $s_\tau^2 = \frac{1}{(m-1)}\sum_i^m (\tau_i - \bar{T})^2$ | $s_{T_{o_w}}^2 = M^2 \frac{s_\tau^2}{m}$ | $s_{y_{o_w}}^2 = \frac{s_{T_o}^2}{N^2}$ |
|---|---|---|---|
| 1 | $\left(\frac{1}{1}\right)[(4-4)^2 + (4-4)^2] = 0$ | $\left(\frac{3^2}{2}\right)0 = 0$ | $\frac{0}{10^2} = 0.00$ |
| 2 | $\left(\frac{1}{1}\right)[(4-12)^2 + (20-12)^2] = 128$ | $\left(\frac{3^2}{2}\right)128 = 576$ | $\frac{576}{10^2} = 5.76$ |
| 3 | $\left(\frac{1}{1}\right)[(4-11)^2 + (18-11)^2] = 98$ | $\left(\frac{3^2}{2}\right)98 = 441$ | $\frac{441}{10^2} = 4.41$ |
| 4 | $\left(\frac{1}{1}\right)[(20-12)^2 + (4-12)^2] = 128$ | $\left(\frac{3^2}{2}\right)128 = 576$ | $\frac{576}{10^2} = 5.76$ |
| 5 | $\left(\frac{1}{1}\right)[(20-20)^2 + (20-20)^2] = 0$ | $\left(\frac{3^2}{2}\right)0 = 0$ | $\frac{0}{10^2} = 0.00$ |
| 6 | $\left(\frac{1}{1}\right)[(20-19)^2 + (18-19)^2] = 2$ | $\left(\frac{3^2}{2}\right)2 = 9$ | $\frac{9}{10^2} = 0.09$ |
| 7 | $\left(\frac{1}{1}\right)[(18-11)^2 + (4-11)^2] = 98$ | $\left(\frac{3^2}{2}\right)98 = 441$ | $\frac{441}{10^2} = 4.41$ |
| 8 | $\left(\frac{1}{1}\right)[(18-19)^2 + (20-19)^2] = 2$ | $\left(\frac{3^2}{2}\right)2 = 9$ | $\frac{9}{10^2} = 0.09$ |
| 9 | $\left(\frac{1}{1}\right)[(18-18)^2 + (18-18)^2] = 0$ | $\left(\frac{3^2}{2}\right)0 = 0$ | $\frac{0}{10^2} = 0.00$ |
| | 456 | 2052 | 20.52 |

$$\frac{456}{9} = 50.66667 = \sigma_\tau^2 = \frac{1}{M}\sum_i^M (\tau_i - \bar{\tau})^2 = \frac{1}{3}[(4-14)^2 + (20-14)^2 + (18-14)^2]$$

$$\frac{2052}{9} = 228 = \sigma_{T_{o_w}}^2 = \frac{M^2\sigma_\tau^2}{m} = \frac{3^2(50.66667)}{2}$$

$$\frac{20.52}{9} = 2.28 = \sigma_{\bar{y}_w}^2 = \frac{M^2}{N^2}\frac{\sigma_\tau^2}{m} = \frac{3^2}{10^2}\frac{(50.66667)}{2}$$

As can be seen, these results demonstrate that all the estimators are unbiased.

The overall total can also be estimated using the reciprocal of the probability. Consider first the situation when the probabilities are constant so that $P(\text{cluster } i) = 1/M$. Then $N_i = 1$, $N = M$, and the weight is $N/N_i = M/1 = M$ and

$$\hat{T}_i = My_i \rightarrow M_{\mu_Y} = \tau_o \tag{21.21}$$

The expected value of $\hat{T}_i$ is

$$E(\hat{T}_i) = E(My_i) = ME(y_i) = M\mu_Y = \tau_o$$

indicating that the estimate is unbiased.

When the probabilities are not constant so that $P(\text{cluster } i) = N_i/N$, $\tau_i = N_i y_i$, and $\tau_i/N_i = y_i$, the estimate of $\tau_o$ based on sampling unit $i$ is

$$\hat{T}_i = MP(\text{cluster } i)\left(\frac{\tau_i}{P_i}\right) = \frac{MN_i}{N}\tau_i\left(\frac{N}{N_i}\right) = M\tau_i = MN_i y_i \to \tau_o \tag{21.22}$$

and the expected value of $\hat{T}_i$ is

$$E(\hat{T}_i) = E(M\tau_i) = ME(\tau_i) = M\left(\frac{1}{M}\sum_i^M \tau_i\right) = \sum_i^M \tau_i = \tau_o$$

The mean of $m$ such estimates of the total is

$$\hat{T}_o = \frac{M}{m}\sum_i^m P(\text{cluster } i)\left(\frac{\tau_i}{P_i}\right) \to \tau_o \tag{21.23}$$

or

$$\hat{T}_o = \frac{M}{m}\sum_i^m \left(\frac{N_i}{N}\right)(N_i y_i)\left(\frac{N}{N_i}\right)$$

which reduces to

$$\hat{T}_o = \frac{M}{m}\sum_i^m N_i y_i = \frac{M}{m}\sum_i^m \tau_i = M\bar{T} \to \tau_o \tag{21.24}$$

which is Equation 21.10. The expected value of $\hat{T}_0$ is

$$E(\hat{T}_o) = \frac{M}{m}\sum_i^m E(N_i y_i)$$

$$= \frac{M}{m}\sum_i^m \frac{1}{M}\sum_i^M (N_i y_i)$$

$$= \frac{M}{m}\sum_i^m \frac{1}{M}\sum_i^M \tau_i$$

Since $\Sigma_i^M \tau_i = \tau_o$,

$$E(\hat{T}_o) = \frac{M}{m}\sum_i^m \frac{1}{M}\tau_i = \tau_o$$

To illustrate this process, one can use the data from the previous example. Again, $m = 2$. Then, $M/m = 3/2 = 1.5$.

| i | $\sum_i^m N_i y_i = \sum_i^m \tau_i$ | $\hat{T}_o = \left(\frac{M}{m}\right)\sum_i^m N_i y_i$ |
|---|---|---|
| 1 | [2(2) + 2(2)] = 8 | 1.5( 8) = 12 |
| 2 | [2(2) + 5(4)] = 24 | 1.5(24) = 36 |
| 3 | [2(2) + 3(6)] = 22 | 1.5(22) = 33 |
| 4 | [5(4) + 2(2)] = 24 | 1.5(24) = 36 |
| 5 | [5(4) + 5(4)] = 40 | 1.5(40) = 60 |
| 6 | [5(4) + 3(6)] = 38 | 1.5(38) = 57 |
| 7 | [3(6) + 2(2)] = 22 | 1.5(22) = 33 |
| 8 | [3(6) + 5(4)] = 38 | 1.5(38) = 57 |
| 9 | [3(6) + 3(6)] = 36 | 1.5(36) = 54 |
| | | 378 |

$$\frac{378}{9} = 42 = \tau_o$$

As can be seen, these totals are identical to those in the column headed $T_o = M\bar{T}$ in the previous computations. Consequently, the variance of these totals and the variance of the mean are the same as when computed earlier.

## 21.3 SAMPLING WITH PROBABILITY PROPORTIONAL TO SIZE (PPS)

In essence, in this form of sampling the probability of a sampling unit being drawn is dependent on the size of the sampling unit or on the same-size characteristic associated with the sampling unit, so that sampling units associated with large size are more likely to be chosen than are units associated with small size.

### 21.3.1 List Sampling

In conventional list sampling, the sampling frame consists of a listing of the available sampling units from the specified universe. The sample is generated using simple random sampling, either with or without replacement. When the sampling is without replacement, the probability of a sampling unit being drawn changes from draw to draw, increasing as the sampling progresses, but within a draw all the available sampling units have the same probability of being drawn. If the sampling is with replacement, the probability of a sampling unit being drawn is constant throughout the sampling process. If the sizes of the sampling units are constant or if the variable of interest is not correlated with sampling unit size, the use of conventional simple random sampling from the list is appropriate. If, however, a correlation exists between the variable of interest and the sampling unit size, conventional sample random sampling could produce an unbiased but highly unrepresentative set of statistics. This would be especially troublesome if there were relatively few large sampling units compared with the number of small sampling units. In such a case it would be desirable to make some provision to increase the likelihood of obtaining a representative sample. Such a procedure has been suggested by Hansen and Hurwitz (1943). In this procedure the probability of a sampling unit being chosen is a function of its size. From this beginning came the concept of sampling with probability proportional to size (PPS).

As an example of this procedure, assume a situation where single-stage cluster sampling is to be used and the sampling frame is made up of $M$ clusters, which vary in size. The frame is in the form of a list. The clusters can be in any order. The cluster sizes, $N_i$, are known. The variable of interest is correlated with cluster size, becoming greater as cluster size increases. Proceeding down the list, the cluster sizes are summed so that for each cluster there is a sum of all the cluster sizes preceding and including it.

$$\text{Sum}_h = \sum_{i=1}^{h} N_i$$

where $\text{sum}_h$ is equal to the sum of the cluster sizes when all the cluster sizes up to cluster $h$ have been totaled. When all $M$ cluster sizes have been summed, $h = M$.

When this series of sums has been constructed, there will be a range of sum magnitudes associated with each cluster. This is the "assigned range" of that cluster. Then, using random numbers, a number greater than zero and less than $(\text{Sum}_M + 1)$ is chosen. This number will fall within one of the assigned ranges. The cluster associated with that range becomes one unit included in the sample. This process is repeated, using sampling with replacement, until the sample is of the desired size. In this process the probability of choosing cluster $i$ is

$$P(\text{cluster}_i) = \frac{N_i}{N} \tag{21.25}$$

since the width of the assigned range of cluster $i$ is equal to $N_i$.

As an example of this process, assume that a sample of size $m = 5$ is to be drawn from the set of clusters in Figure 17.10. Those clusters vary in size and one can assume that the variable of interest is correlated with the cluster size.

| h | $N_i$ | $\sum_{i=1}^{h} N_i$ | Assigned Range | $P_i$ | h | $N_i$ | $\sum_{i=1}^{h} N_i$ | Assigned Range | $P_i$ |
|---|---|---|---|---|---|---|---|---|---|
| 1 | 8 | 8 | 1–8 | 8/320 | 15 | 8 | 164 | 157–164 | 3/320 |
| 2 | 12 | 20 | 9–20 | 12/320 | 16 | 12 | 176 | 165–176 | 12/320 |
| 3 | 8 | 28 | 21–28 | 8/320 | 17 | 12 | 188 | 177–188 | 12/320 |
| 4 | 12 | 40 | 29–40 | 12/320 | 18 | 12 | 200 | 189–200 | 12/320 |
| 5 | 16 | 56 | 41–56 | 16/320 | 19 | 8 | 208 | 201–208 | 8/320 |
| 6 | 12 | 68 | 57–68 | 12/320 | 20 | 12 | 220 | 209–220 | 12/320 |
| 7 | 12 | 80 | 69–80 | 12/320 | 21 | 12 | 232 | 221–232 | 12/320 |
| 8 | 8 | 88 | 81–88 | 8/320 | 22 | 16 | 248 | 233–248 | 16/320 |
| 9 | 12 | 100 | 89–100 | 12/320 | 23 | 12 | 260 | 249–260 | 12/320 |
| 10 | 8 | 108 | 101–108 | 8/320 | 24 | 12 | 272 | 261–272 | 12/320 |
| 11 | 16 | 124 | 109–124 | 16/320 | 25 | 16 | 288 | 273–288 | 16/320 |
| 12 | 8 | 132 | 125–132 | 8/320 | 26 | 8 | 296 | 289–296 | 8/320 |
| 13 | 12 | 144 | 133–144 | 12/320 | 27 | 12 | 308 | 297–308 | 12/320 |
| 14 | 12 | 156 | 145–156 | 12/320 | 28 | 12 | 320 | 309–320 | 12/320 |

The random draws yielded the following sample:

| Draw | Random Number | Cluster Chosen |
|---|---|---|
| 1 | 031 | 4 |
| 2 | 293 | 26 |
| 3 | 233 | 22 |
| 4 | 162 | 15 |
| 5 | 129 | 12 |

These clusters would then be located in the field and evaluated in terms of the variable of interest. Since the cluster sizes, $N_i$, are known, the computations would follow the pattern of the examples in Section 21.2.

It should be emphasized that the use of list sampling with probabilities proportional to size is not limited to situations where the sampling units are clusters. The procedure is equally valid when the sampling units are not clusters but still differ in size, provided that the sampling units can be listed and the variable of interest is correlated with sampling unit size: for example, when the sampling units are forestland holdings and the variable of interest is income from timber sales, or when the sampling units are sawmills and the variable of interest is volume of lumber produced in a specified time period.

### 21.3.2 Angle-Count Sampling

Following World War II, Walter Bitterlich, an Austrian forester, devised a new way of estimating basal area per unit of land area which was much faster than the conventional methods then in use (Bitterlich, 1947). His basic idea was seized upon by Lewis Grosenbaugh of the USDA Forest Service, who integrated it with sampling theory and extended it into a general forest inventory procedure (Grosenbaugh, 1952; 1958). Since then a very large literature about the concept and its applications has been generated. It is too extensive for coverage here. See Labau (1967) for a listing of the publications to that date. The concept is of interest here because it makes use of sampling with probability proportional to size.

#### 21.3.2.1 Point Sampling

If the surface of any tract of land, regardless of size, is orthogonally reduced to a horizontal two-dimensional plane, the image on that plane would be a map with a scale of 1:1. Since a plane surface is made up of points with a spacing interval equal to an infinitesimal, the number of points is infinitely large. Each of the points has attributes associated with it. For example, the point might be in a water area, in a road, in a pine plantation, or in a certain soil type or geologic formation, or it could be at a certain elevation above mean sea level, and so forth. If a certain attribute is designated as the variable of interest, the set of variates from the infinite number of points forms the population. Obviously, a complete enumeration would be impossible. If an estimate of some parameter is to be made, a sampling frame, made up of a finite number of points, must be constructed. To do this, the tract map is overlaid with a rectangular grid with an interval that is of a magnitude considered appropriate by the sampler. Each grid intersection defines a point and the set of such grid intersections (points) forms the frame, as shown in Figure 12.2. The frame is actually a cluster in the form of a systematic sample. If the grid is randomly oriented and the start was randomly chosen (e.g., when a dot grid is dropped down on to a map without aligning it with a map edge or specifying that its corner coincide with a tract corner), it can be considered to be one of an infinitely large set of similar grids, each with a different starting point and orientation. When this is the case, all of the points making up this surface have an opportunity to be chosen in a random draw and it can be assumed that the probability of being chosen is consistent across all points.

If the start and orientation of the grid is arbitrary, as when it is deliberately laid down so that the grid lines are aligned with a specific direction and the grid corner is located at a specified point, the probability of the bulk of the points making up the surface being chosen in a random draw is equal to zero. This, as was brought out in Chapters 17 and 18, will result in biased estimators. It can be assumed, with considerable confidence, that those biases will increase as the grid interval increases. In essence, the estimates based on samples from the grid are estimates of the *frame* parameters, not the population parameters.

The sampling of points from a grid is a common practice in forest inventory operations. For example, in line–plot timber cruising, line and plot intervals are specified, as are the initial point and the direction of the line. This creates a grid. Each grid intersection serves as the center of a plot of a specified size and slope. Where areas of land-cover categories are to be estimated using a dot grid or a set of randomly chosen points each dot or point falls into a cover category. That cover category is an attribute of the point. Thus, whenever *points* are chosen in the sampling process, *point sampling* is being used.

In contrast to point sampling, *plot sampling* involves the selection, either randomly or systematically, of plots from a sampling frame made up of *plots*, not points. These plots have a specified size and shape. In the frame they are packed together to minimize unoccupied space, but no overlapping is permitted. If the plots are square or rectangular, there would be no unoccupied space except adjacent to boundaries where full-sized plots cannot be fitted in. This unoccupied space is crosshatched in Figure 21.1. With circular plots, however, there is a definite loss of coverage, disregarding the crosshatched area. If the plots are located as is shown in Figure 21.1, they would cover only 78.54% of the noncrosshatched area.* For this reason, when circular plots are being used, one cannot compute the size of the sampling frame by simply dividing the tract area by the plot area.

### 21.3.2.2 Tree-Centered Plots

According to the rules governing sampling, sampling units must be distinct. They must not overlap. This holds true for both plot and point sampling but it is often apparently violated in the case of point sampling. When random sampling from a grid, it is possible to choose two points that are very close to one another. This is perfectly acceptable. However, when the points are used as the centers of plots, the plots would overlap. This creates the impression of a violation. However, there is no violation since the *points*, not the *plots*, are the sampling units. The timber volume on a plot associated with a point is assigned to the point as an attribute of that point. A point immediately adjacent to the first point might have as its attribute precisely the same volume, based on precisely the same trees, as that of the first point. The two points would yield two distinct data items. This concept may be difficult to comprehend, but if one reverses the tree-choosing process, the situation may be clarified.

Assume that circular plots of a certain size are to be used in a timber cruise. If point sampling is used, a plot is centered on each point, as is shown in Figure 21.2. The trees within the plot are cruised in the usual manner. In the case of Figure 21.2 trees B, D, F, and G would be measured and tallied. Now, reverse this image. Instead of centering the plot on the sampling point, use the bases of the trees as centers of such plots. Figure 21.3 shows the trees in Figure 21.2 with such associated plots. Now, measure and tally the trees *whose circles enclose the sampling point*. Using this procedure, trees B, D, F, and G would be tallied. These are the same trees chosen when the plot was centered on the sampling point. The results are identical. Note that no plot is now associated with the point. Consequently, where point sampling is carried out in this manner, it is sometimes

* If the plot radius is equal to $r$, the squares contain $2r * 2r = 4r^2$ units of area. The circular plots contain $\pi r^2$ units of area or $(\pi/4)100 = 78.54\%$ of the area of the squares.

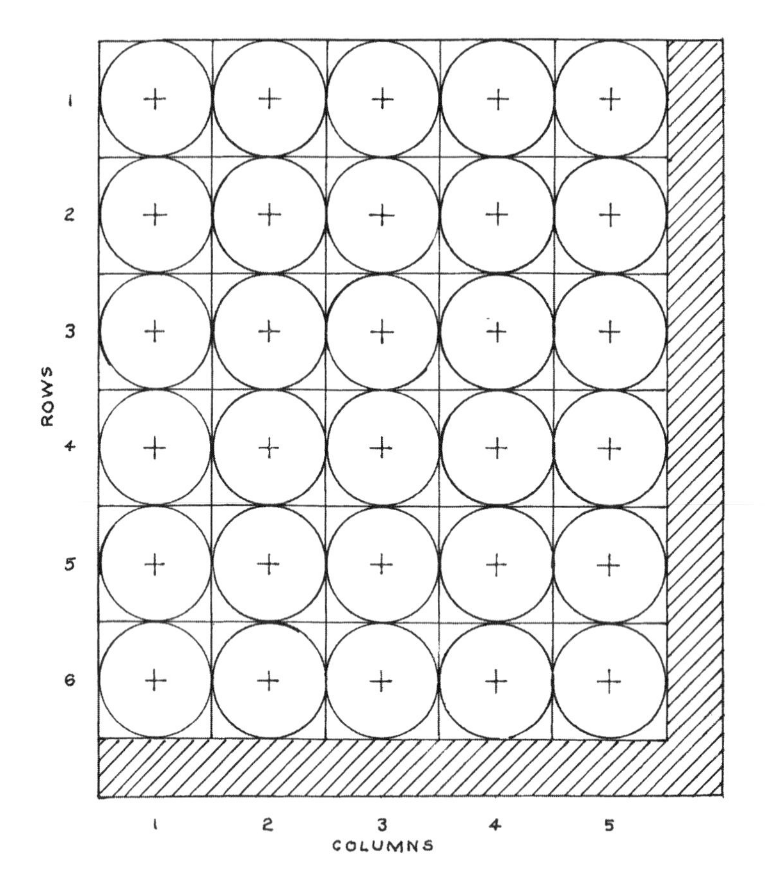


**FIGURE 21.1** Sampling frame for plot sampling. The crosshatched areas cannot be sampled because full-sized plots cannot be fitted into the available space. If circular, rather than square, plots are used, the area between the circles adds to the portion of the tract that cannot be sampled.

referred to as *plotless cruising*. It is admitted that the application of the concept in the field would be awkward, but the idea is valid.

Assume that the plots in Figures 21.2 and 21.3 are $^1/_{10}$ acre in size and the tallied trees have the following volumes: B, 17 ft$^3$; D, 15 ft$^3$; F, 52 ft$^3$; and G, 15 ft$^3$. In the case of the point-centered plot in Figure 21.2, the volume per acre associated with the sampling point is then 10(17 + 15 + 52 + 15) = 990 ft$^3$. The multiplier 10 is the expansion or "blowup" factor and is the *reciprocal of the plot size relative to an acre.* If the tract being cruised covers 100 acres, the estimate of the total volume on the tract, based on the evidence of point A, would be 10(100)(17 + 15 + 52 + 15) = 99,000 ft$^3$. The expansion factor is now 1000, which is the *reciprocal of the plot size relative to the size of the tract.*

In the case of the tree-centered plots in Figure 21.3 the volumes per acre would be obtained tree by tree and then summed: 10(17) + 10(15) + 10(52) + 10(15) = 990 ft$^3$. Thus, both procedures yield identical results.

### 21.3.2.3 Probabilities

In point sampling the universe is the set of points making up the surface of a full-scale orthogonal projection of the surface of the tract. This universe is infinitely large and its constituent points cannot be identified. However, for this discussion it will be assumed that the frame and the universe are coincident. Under this assumption when the tree-centered approach is used, and if the situation near the tract boundaries is ignored, it can be said that the probability of the plot of a tree enclosing

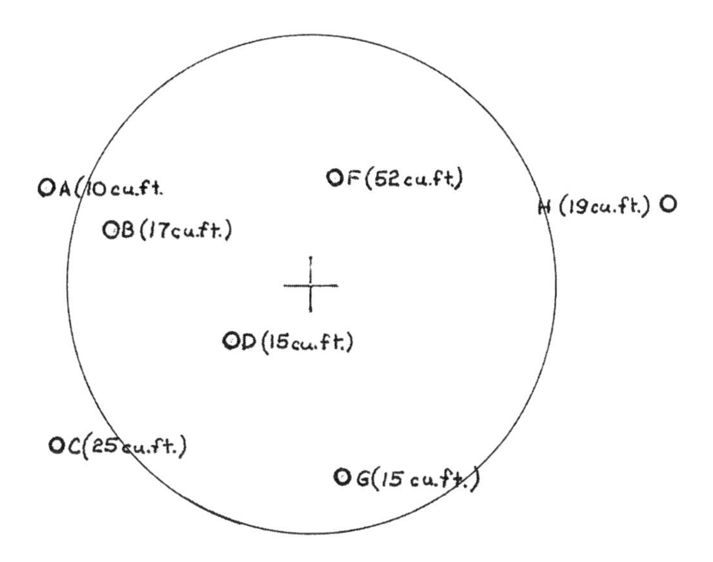


**FIGURE 21.2** Set of trees with a 1/5-acre circular plot centered on the cross. Trees B, D, F, and G are within the plot. The volume per acre, as of this plot, is 990 ft$^3$.

a randomly chosen point is equal to the ratio of plot size to the size of the tract, when both are in terms of the same unit of area.

$$P(\text{tree}_i) = \frac{\text{Plot area}_i}{\text{Tract area}} \tag{21.26}$$

In the case of the trees in Figure 21.3, the plot areas are all the same. Consequently, the probabilities of trees being chosen are constant. If plot area$_i$ = 0.1 acre and the tract area = 100 acres, the probability of the plot associated with tree $j$ enclosing a randomly chosen point is 0.1/100 = 1/1000. On a per acre basis, $P(\text{tree}_i)$ = 0.1/1 = 1/10.

The expansion or blowup of the volume or basal area of the tree to a total on the specified land area (SA; which can refer to either tract, acre, or hectare) is equal to the reciprocal of the probability of selection

$$g_i = \frac{1}{P(\text{tree}_i)} \tag{21.27}$$

Thus, if SA is the tract area and it covers 100 acres, the plot area = 0.1 acre and the tree volume is 17 ft$^3$. The estimated total volume on the tract, based on tree$_i$, is

$$T_{o_i} = 17 \text{ ft}^3\left(\frac{100}{0.1 \text{ acre}}\right) = 17{,}000 \text{ ft}^3$$

On a per acre basis the volume represented by tree$_i$ is

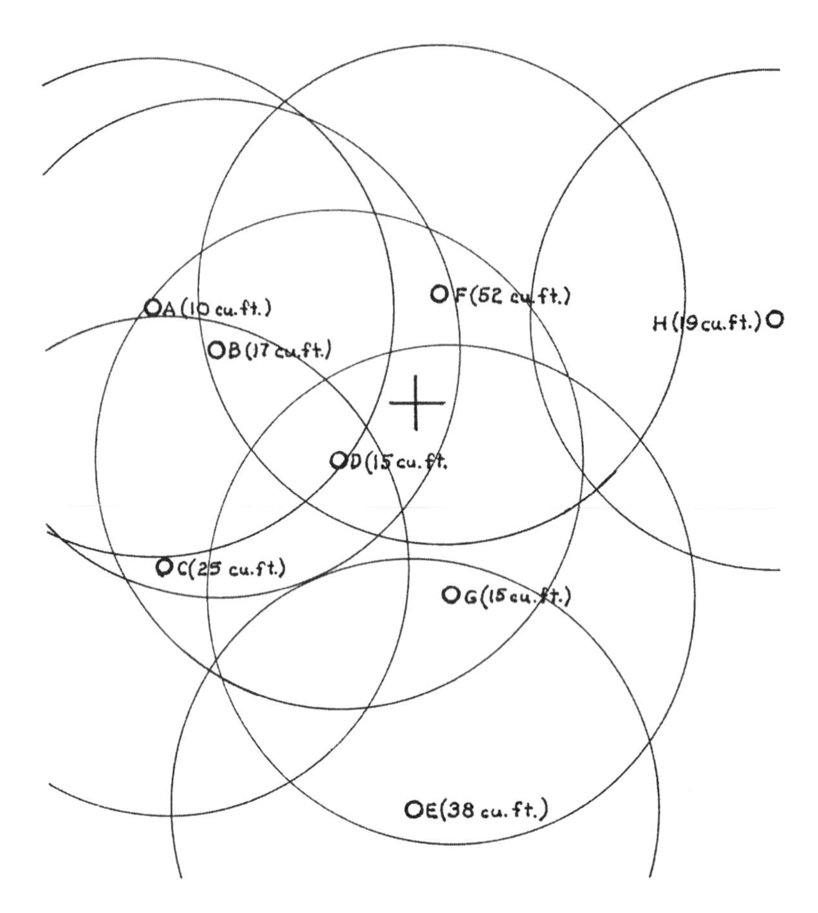


**FIGURE 21.3** The same set of trees as in Figure 21.2. Each is at the center of a 1/10-acre circular plot. The sample point is enclosed by the circles associated with trees B, D, F, and G.

$$T_{A_i} = 17 \text{ ft}^3 \left( \frac{1 \text{ acre}}{0.1 \text{ acre}} \right) = 170 \text{ ft}^3$$

It has long been recognized that the cost of evaluating and tallying a tree in a timber inventory is essentially constant regardless of tree size. Furthermore, small trees are less valuable than large trees of the same species, if one disregards defect. Consequently, the cost per unit of benefit from the trees is greater for small trees than it is for large trees. Compounding the problem, the count of small trees is usually greater than that of large trees. As a result, if all trees are treated alike, a disproportionate amount of time is spent on low-value trees. Timber cruisers have recognized this problem and have attempted to improve the situation by using two concentric plots, one large and one small, at each sampling point, as is shown in Figure 21.4. Small trees would be tallied only if they occurred within the small plot while large trees would be tallied if they occurred anywhere within either the small or large plots.

If this idea of differing plot sizes is carried over into tree-centered point sampling there would be small plots centered on small trees and large plots centered on larger trees. Some arbitrary threshold would be established to distinguish small from large trees. This threshold could be based on the dbh, basal area, volume, or some other measure of the size of the tree. Thus, if one uses the trees in Figure 21.2, a threshold dbh of 12 in., a plot size of 1/10 acre for trees below the threshold and 1/5 acre for trees at or above the threshold, the result is the situation shown in Figure 21.5. Now two trees, D and G, are included in the tally from the 1/10-acre plot while three trees, B, C,

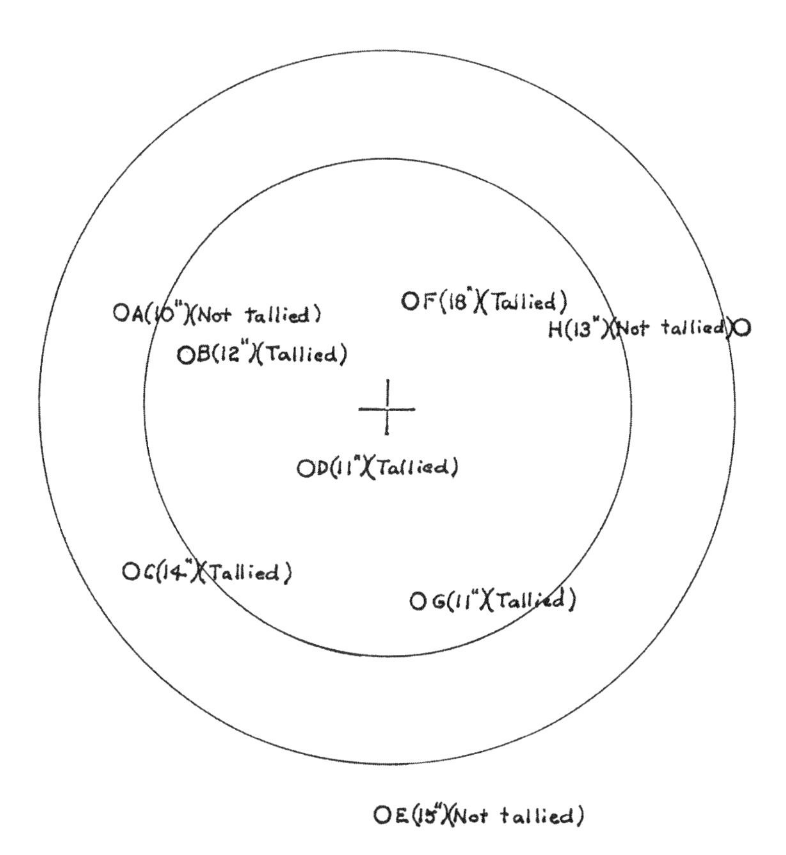


**FIGURE 21.4** Two concentric circular plots ($^1/_{10}$ and $^1/_5$ acre centered on the sampling point). Trees with dbhs less than 12 in. would be tallied only if inside the smaller circle. All trees with dbhs 12 in. or larger would be tallied if inside either the larger or smaller circles. The volumes of trees D and G would be expanded by a factor of 10, whereas the volumes of the rest of the tallied trees would be expanded by a factor of 5.

and F, are from the $^1/_5$-acre plot. The estimated volume, based on the one sampling point, is now $10(15 + 15) + 5(17 + 25 + 52) = 970$ ft$^3$/acre or $1000\ (15 + 15) + 500\ (17 + 25 + 52) = 97{,}000$ ft$^3$ on the tract. This is a crude form of sampling with probability proportional to size.

If more than one sampling point is used, the situation becomes more complicated. One must remember that the *points* are the actual sampling units and can be selected using simple random, stratified random, or even systematic sampling. Sampling with probability proportional to size is not involved in this process. On the other hand, the selection of the trees, upon which the attribute of the point is to be based, makes use of PPS. *The selected trees about a point constitute a cluster.* The cluster can be given a complete enumeration, as was done in the example, or it may be subsampled. The subsampling can involve either constant or variable probability. The possibilities are numerous and a comprehensive coverage of them is beyond the scope of this book.

#### 21.3.2.4 Horizontal Point Sampling

In the previous discussion the use of two different plot sizes was described. The idea is good but the procedure is crude. It would be better if a greater number of plot sizes could be used so that the correlation between tree and plot sizes would be stronger. Bitterlich finally found a way to make this possible.

There is a mathematical theorem that states that, if a circle is inscribed within an acute angle so that the legs of the angle are tangent to the circle, the ratio of the area of the circle to the area of a circle whose radius is equal to the distance from the apex of the angle to the center of the

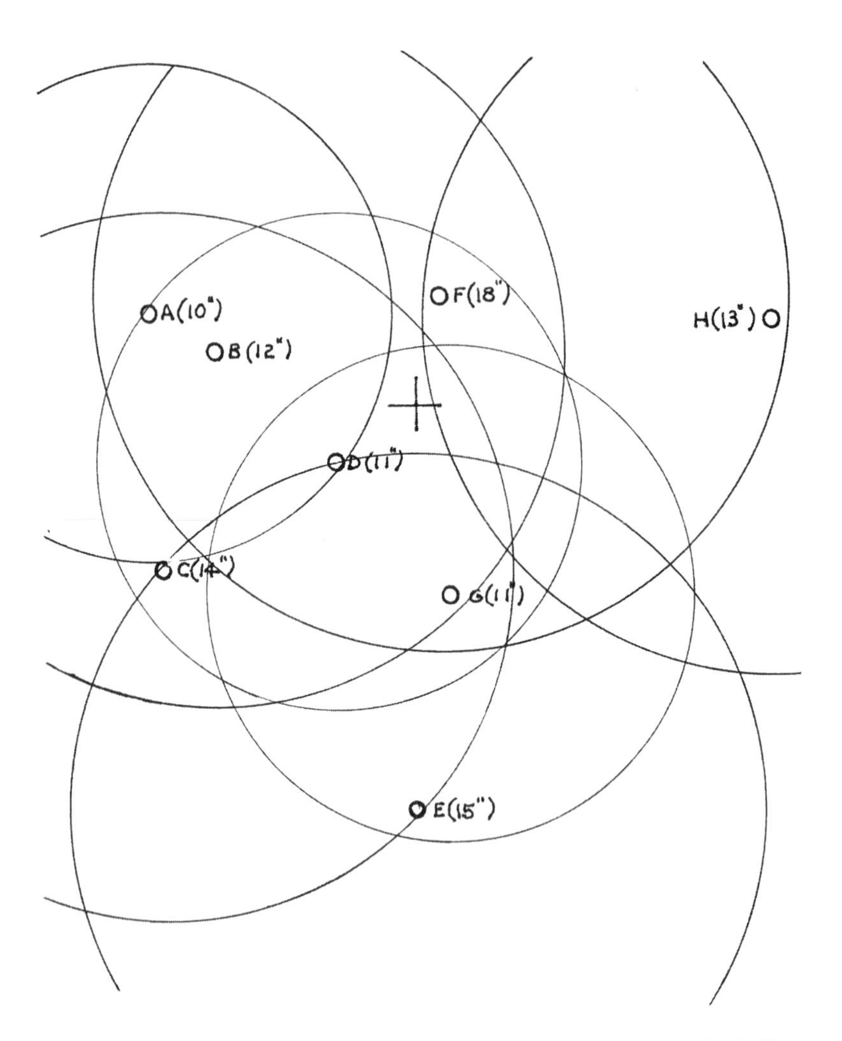


**FIGURE 21.5** Two tree-centered circular plot sizes used with the trees in Figure 21.2, 1/10 acre with trees with dbhs below 12 in. and 1/5 acre plots with trees having dbhs equal to or greater than 12 in.

inscribed circle is a function of the magnitude of the acute angle, as is shown in Figure 21.6. In other words, regardless of the size of the inscribed circle, the ratio of its area to the area of the large circle remains constant, as long as the angle does not change.

Bitterlich made use of this theorem to speed up estimates of basal area per unit area. He reassumed that the small circle could represent the cross section of a tree at breast height and the large circle the size of the plot associated with the tree. To apply the theorem in the field he designed a number of "angle gauges," the simplest of which consists of a stick with a crosspiece at one end and a peep sight at the other. Rays drawn from the peep sight to the ends of the crosspiece define the *critical angle* $\phi$. Then with the peep sight centered over the sampling point, the angle gauge is rotated, in the horizontal plane, about the sampling point. When a tree subtends an angle greater than $\phi$ (i.e., it appears larger than the crosspiece), it indicates that the sampling point is inside the plot of that tree. If the tree subtends an angle equal to $\phi$, it indicates that the sampling point is on the edge of the plot of the tree, and if it subtends an angle smaller than $\phi$, it indicates that the sampling point lies outside the plot of the tree. Figure 21.7 shows this in diagram form. Since the tree plots vary in size depending on the sizes of the trees, the sampling is with probability proportional to size. Furthermore, since the sweep of the angle gauge is in a horizontal plane and about a sampling point the procedure is referred to as *horizontal point sampling*.

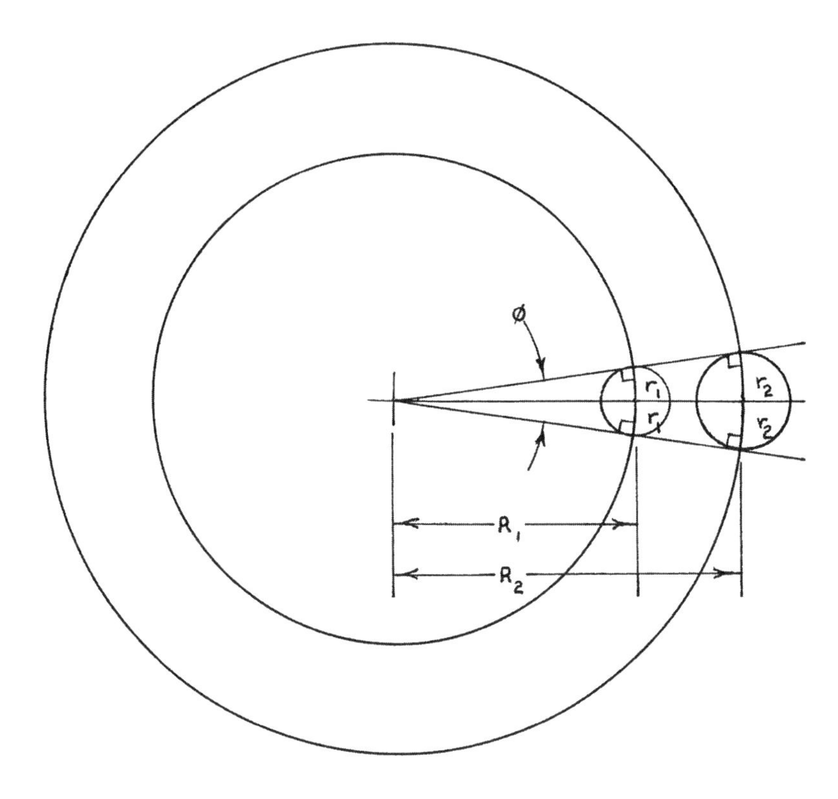


**FIGURE 21.6** The mathematical relationship underlying Bitterlich's contribution.

As can be seen in Figure 21.6, when the basal area (BA) of the tree and its associated plot area (PA) are in the same units

$$\frac{\text{BA}}{\text{PA}} = \frac{\pi r^2}{\pi R^2} = \frac{r^2}{R^2} \tag{21.28}$$

where $r$ = radius of the cross section of the tree at breast height and $R$ = radius of the plot of the tree. Since $r/R = \sin(\phi/2)$,

$$\frac{\text{BA}}{\text{PA}} = \sin^2\left(\frac{\phi}{2}\right) \tag{21.29}$$

When $\phi$ is fixed, the ratio BA/PA is also fixed. As BA changes, PA must also change, to keep the ratio constant. In turn, as BA changes, the probability of a randomly chosen point being enclosed in the plot of the tree also changes, as does the expansion or blowup factor.

Since

$$\frac{\text{BA}}{\text{PA}} = \sin^2\left(\frac{\phi}{2}\right) \quad \text{and} \quad \text{PA} = \frac{\text{BA}}{\sin^2\left(\frac{\phi}{2}\right)}$$

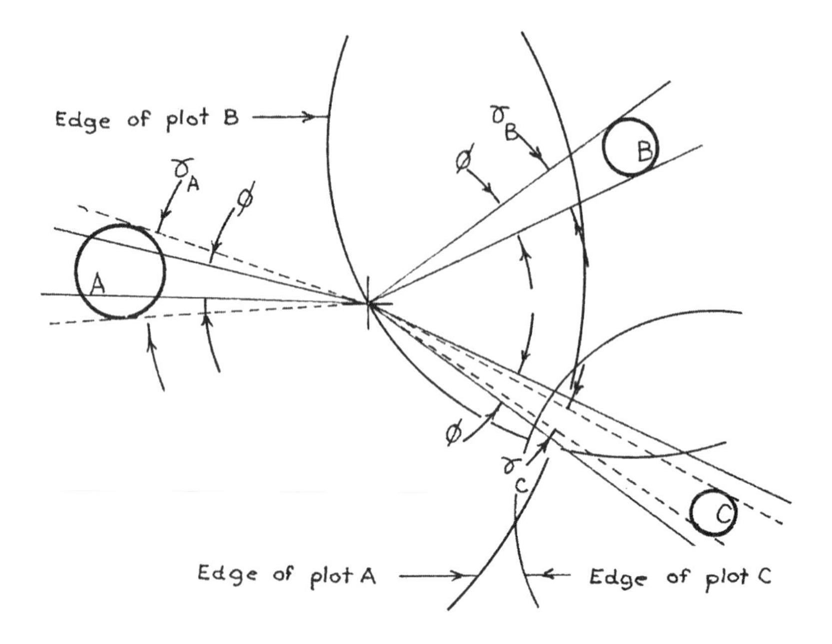


Since:

$\gamma_A > \phi$, A is "In"

$\gamma_B = \phi$, B is "Borderline"

$\gamma_C < \phi$, C is "Out"

**FIGURE 21.7** How an angle gauge is used to select trees.

then

$$P(\text{tree}_i) = \frac{\text{PA}_i}{\text{SA}} = \frac{\text{BA}_i}{\text{SA}\sin^2\left(\frac{\phi}{2}\right)} \tag{21.30}$$

where SA = the specified area (i.e., tract area, acre, or hectare) *in the same units as* $BA_j$ *and* $PA_j$. The expansion or blowup factor, $g_{\text{HP}}$, is then

$$\begin{aligned} g_{\text{HP}_i} &= \frac{1}{P(\text{tree}_i)} = \frac{\text{SA}\sin^2\left(\frac{\phi}{2}\right)}{\text{BA}_i} \\ &= \text{SA}\sin^2\left(\frac{\phi}{2}\right)\left(\frac{4}{\pi D_i^2}\right) \end{aligned} \tag{21.31}$$

where $D$ = dbh. An estimate of the basal area per specified area, $\text{BA}_{\text{SA}}$, based on $\text{tree}_i$, is then

$$\mathrm{BA_{SA}} = \mathrm{BA}_i g_{\mathrm{HP}_i} = \frac{\mathrm{BA}_i\left[\mathrm{SA}\sin^2\left(\frac{\phi}{2}\right)\right]}{\mathrm{BA}_i}$$
$$= \mathrm{SA}\sin^2\left(\frac{\phi}{2}\right) \tag{21.32}$$

Note that $\mathrm{BA_{SA}}$ is dependent only on the specified area and the critical angle, both of which are constants. Consequently, $\mathrm{BA_{SA}}$ is a constant, regardless of the size of the tree, and thus every tree whose plot encloses a given sampling point contributes $\mathrm{BA_{SA}}$ to the total for the point. Because of this, $\mathrm{BA_{SA}}$ is referred to as the *basal area factor*, BAF. Its relationship to the ratio BA/PA is

$$\frac{\mathrm{BA}}{\mathrm{PA}} = \frac{\mathrm{BAF}}{\mathrm{SA}} \tag{21.33}$$

Then,

$$\mathrm{BAF} = \mathrm{SA}\left(\frac{\mathrm{BA}}{\mathrm{PA}}\right) = \mathrm{SA}\sin^2\left(\frac{\phi}{2}\right) \tag{21.34}$$

Thus, if an angle gauge sweep is made about sampling point $j$ and $N_j$ trees have plots enclosing the point, the basal area per specified area, $\tau_j$, as of point $j$, is

$$\tau_j = N_j(\mathrm{BAF}) \tag{21.35}$$

As an example, assume a situation where SA = 1 acre = 43,560 ft$^2$ and $\phi = 2°$. Then,

$$\frac{\phi}{2} = 1°, \quad \sin\left(\frac{\phi}{2}\right) = \sin 1° = 0.017452406, \quad \text{and}$$
$$\sin^2\left(\frac{\phi}{2}\right) = 0.000304586$$

Further assume that the angle gauge sweep about point $j$ yielded two trees whose diameters breast high were 12 and 21 in., respectively. In this case

$$\mathrm{BAF} = 43,560\sin^2(1°)$$
$$= 43,560(0.000304586) = 13.268\ \mathrm{ft}^2/\mathrm{acre}$$

and the estimate of the basal area per acre, as of that sampling point, is equal to $N_j(\mathrm{BAF}) = 2(13.268) = 26.536$ ft$^2$. To confirm this result, make use of Equation 21.31. In the case of the 12-in. tree,

$$r = \frac{\left(\frac{12}{12}\right)}{2} = 0.5\ \mathrm{ft}$$
$$BA = \pi(0.5)^2 = 0.785398163\ \mathrm{ft}^2$$

$$PA = \frac{BA}{\sin^2\left(\frac{\phi}{2}\right)} = \frac{0.785398163}{0.000304586} = 2,578.576046 \text{ ft}^2$$

$$P(12\text{-in. tree}) = \frac{\text{PA}}{\text{SA}} = \frac{2,578.576046}{43,560} = 0.059195961$$

$$K_{\text{HP}} = \frac{1}{P(12\text{-in. tree})} = \frac{1}{0.059195961} = 16.89304455$$

$$\text{BA}_{\text{SA}} = \text{BA}(g_{\text{HP}}) = 0.785398163(16.89304455) = 13.268 \text{ ft}^2/\text{acre}$$

Now the 21-in. tree,

$$r = \frac{\left(\frac{21}{12}\right)}{2} = 0.875 \text{ ft}$$

$$\text{BA} = \pi(0.875)^2 = 2.405281875 \text{ ft}^2$$

$$\text{PA} = \frac{2.405281875}{0.000304586} = 7,896.889138 \text{ ft}^2$$

$$P(12\text{-in. tree}) = \frac{7,896.889138}{43,560} = 0.181287629$$

$$g_{\text{HP}} = \frac{1}{P(12\text{-in. tree})} = \frac{1}{0.181287629} = 5.516096179$$

$$\text{BA}_{\text{SA}} = 2.405281875(5.516096179) = 13.268 \text{ ft}^2/\text{acre}$$

In both cases the basal area per specified area is equal to the BAF and the sum of the two estimates is equal to $N_j$(BAF).

This was Bitterlich's contribution. With the basal area factor known for a given angle gauge, one can obtain an estimate of the basal area per specified land area, as of a given sampling point, simply by counting the tree subtending angles greater than the critical angle, when viewed from the sampling point, and then multiplying that count by the basal area factor. This gives rise to the label "angle count" applied to the procedure.

As an example, consider Figure 21.8 where the angle gauge has a basal area factor of 10 $\text{ft}^2$/acre. Of the eight trees shown, only B, D, F, and G have plots that enclose the sampling point. The estimated basal area per acre, as of that point, is then 4(10) = 40 $\text{ft}^2$.

A given critical angle can be associated with more than one basal area factor. If the specified land area is 1 acre or 43,560 $\text{ft}^2$,

$$\text{BAF}_{\text{A}} = 43,560\sin^2\left(\frac{\phi}{2}\right) \text{ ft}^2/\text{acre} \tag{21.36}$$

if the specified area is 1 ha or 10,000 $\text{m}^2$,

$$\text{BAF}_{\text{H}} = 10,000\sin^2\left(\frac{\phi}{2}\right) \text{ m}^2/\text{ha} \tag{21.37}$$

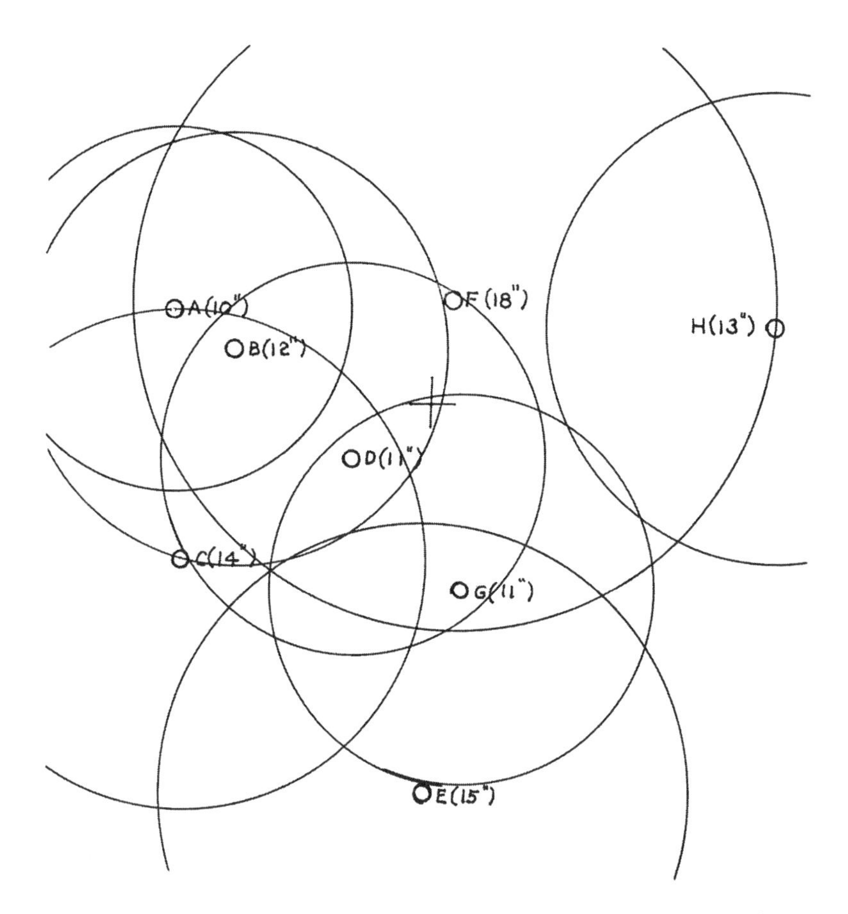


**FIGURE 21.8** The plots associated with the trees in Figure 21.2 when a BAF of 10 ft²/acre is used. Note that the sampling point is enclosed by the circles associated with trees B, D, F, and G.

and if the specified area is the tract size,

$$\text{BAF}_\text{T} = (\text{tract size})\sin^2\left(\frac{\phi}{2}\right) \text{ square units on tract} \tag{21.38}$$

The tract size must be expressed in the same units as the basal area.

### 21.3.2.5 Instrumentation

The instrumentation developed to facilitate the application of angle-count sampling varies from the thumb width observed at arm's length (Carow and Stage, 1953), through the previously described stick and cross-bar, to the wedge prism (Bruce, 1955) and the Relaskop (Bitterlich, 1949). The descriptions of these instruments and their associated operational procedures are too extensive for inclusion here. Reference is made to standard forest mensuration texts such as Husch et al. (1972) and Loetsch et al. (1973).

### 21.3.2.6. The Critical Angle

Once the land area, SA, has been specified, the BAF is dependent on the critical angle or the critical angle is dependent on the BAF. One can specify a convenient BAF and then compute the associated critical angle. From Equation 21.34,

$$\text{BAF} = \text{SA}\sin^2\left(\frac{\phi}{2}\right)$$

Then,

$$\sin^2\left(\frac{\phi}{2}\right) = \frac{\text{BAF}}{\text{SA}}$$

$$\sin\left(\frac{\phi}{2}\right) = \sqrt{\frac{\text{BAF}}{\text{SA}}}$$

and

$$\phi = 2\arcsin\sqrt{\frac{\text{BAF}}{\text{SA}}} \tag{21.39}$$

Thus, if a BAF = 10 ft²/acre is desired,

$$\phi = 2\arcsin\sqrt{\frac{10}{43{,}560}}$$

$$= 2\arcsin 0.015151515$$

$$= 2(0.86815109) = 1.736302181^0 \text{ or } 104.18'$$

The angle gauge would be constructed so that this angle would be generated. In the case of the simple gauge previously mentioned, the stick would be 33 times longer than the crosspiece is wide. If the crosspiece is one unit in width and $\phi = 104.18'$,

$$\tan\left(\frac{\phi}{2}\right) = \frac{\left(\dfrac{\text{Crosspiece width}}{2}\right)}{\text{stick length}}$$

Then,

$$\text{Stick length} = \frac{\left(\dfrac{\text{Crosspiece width}}{2}\right)}{\tan\left(\dfrac{\phi}{2}\right)}$$

$$= \frac{\left(\dfrac{1}{2}\right)}{0.015153255}$$

$$= 32.996 \text{ or } 33 \text{ units}$$

#### 21.3.2.7 Slopover

The portion of a sampling unit that extends beyond the boundary of the sampling frame is referred to as *slopover*. It can occur in virtually any kind of timber inventory. Figure 21.9 illustrates slopover in the context of variable-radius tree-centered plots. It causes a problem since it reduces the effective

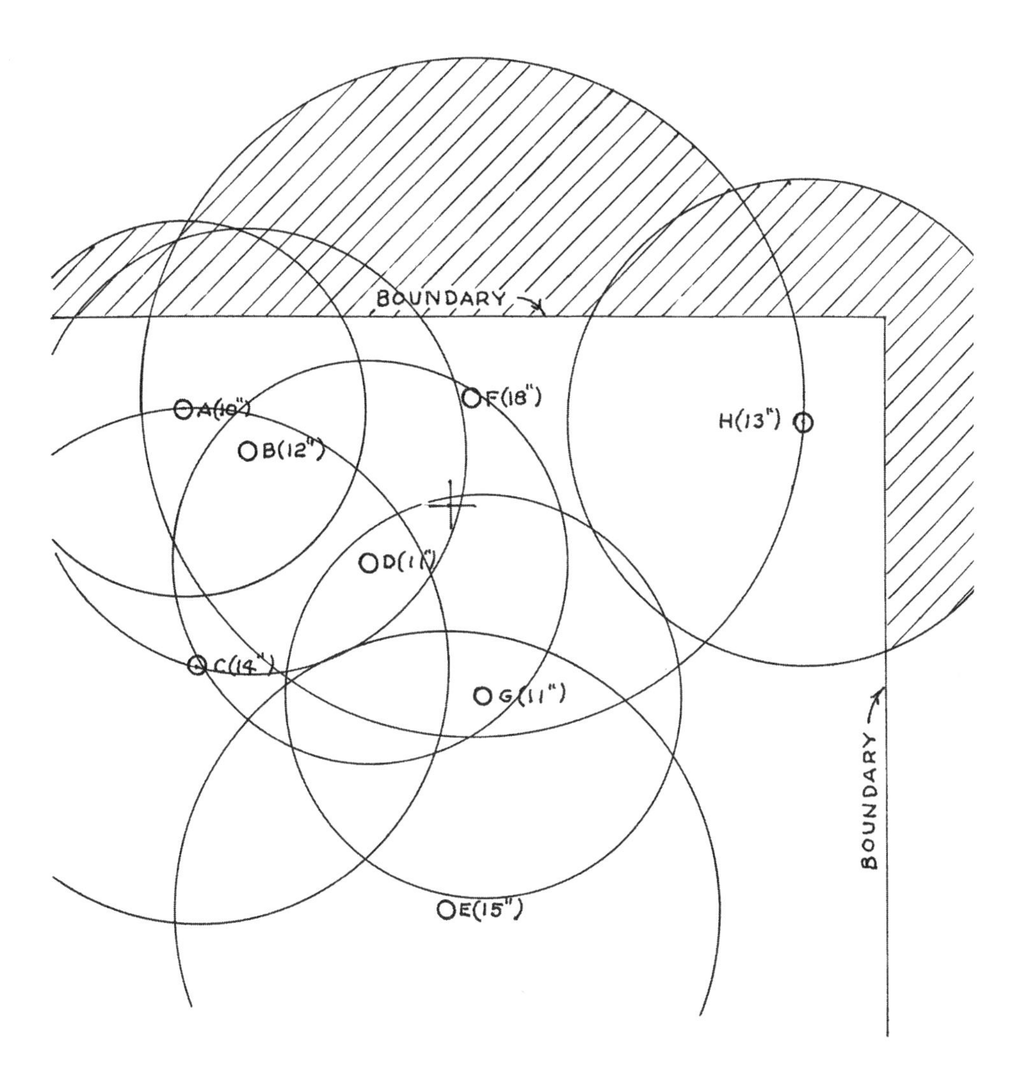


**FIGURE 21.9** Example of slopover. The portions of the tree-centered circles that extend beyond the boundaries constitute slopover.

area of the plot and, in turn, the magnitude of the expansion or blowup factor. The tract boundary is always present but interior boundaries such as those for management units or for stands may also be involved.

In the case of point sampling, the sampling frame extends only to the boundary. Consequently, no points can be chosen beyond the boundary. In turn, this means that trees with plots extending beyond the boundary are less likely to be included in a cluster than similar-sized trees whose plots do not extend beyond the boundary. Since the expansion or blowup factor is the reciprocal of the probability of the tree, this reduction in plot area must be recognized.

Several methods for compensating for slopover have been devised (e.g., Masuyama, 1953; Grosenbaugh, 1955; 1958; Haga and Maezawa, 1959; Barrett, 1964; Beers and Miller, 1964; Beers, 1966; 1977; Schmid, 1969). Probably the simplest of these is that suggested by Grosenbaugh (1955; 1957). He recommended that a zone be established, just inside the boundaries of the sampling frame, that would be a bit deeper than the radius of the plot associated with the largest expected tree, as shown in Figure 21.10. Trees within this zone would have their associated plots truncated to form half or quarter circles, with their straight edges on the side toward the boundary. This is shown in Figure 21.10. This eliminates the slopover. The computation process is unchanged except

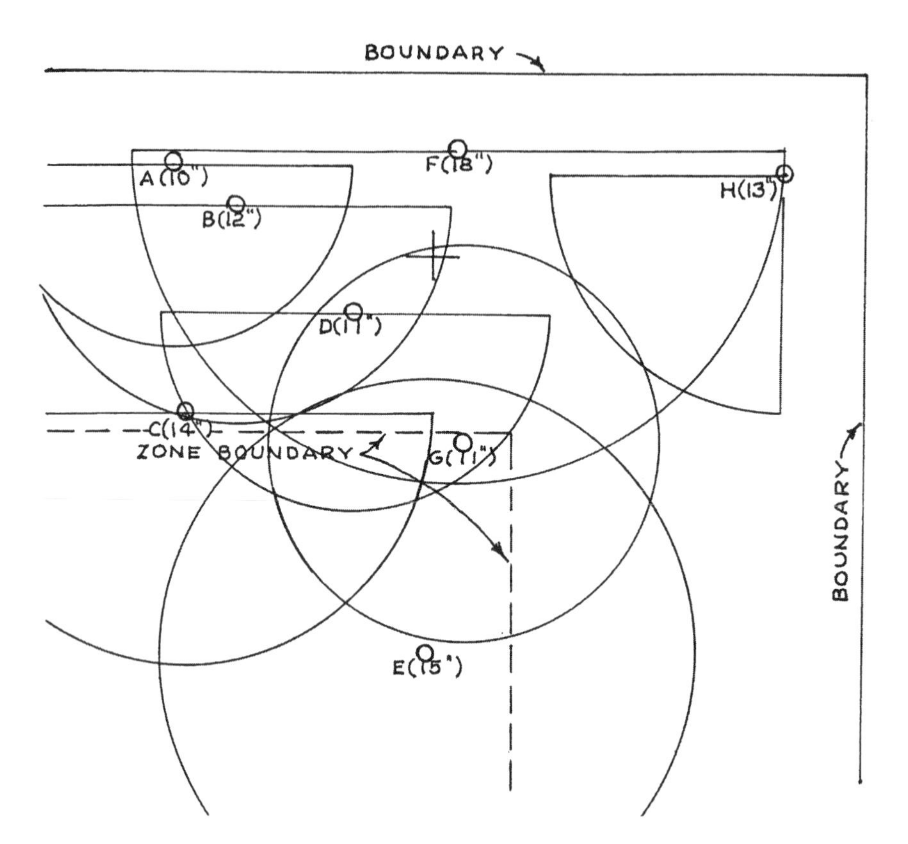


**FIGURE 21.10** Grosenbaugh's solution to the slopover problem.

that the partial, rather than the full-sized, plot areas are used. This causes the blowup factors to be increased. In the case of the trees in Figure 21.10, trees A, B, C, D, and F would have their probabilities halved and their blowup factors doubled. The probability of the plot of tree H enclosing the sampling point would be one fourth of its normal magnitude and its blowup factor would be four times larger than normal. The probabilities and expansion factor associated with trees E and G would be unchanged since they do not lie inside the boundary zone.

In the field, regardless of where the sampling point falls, a 360° sweep is made with the angle gauge. Trees with partial plots will not be included in the tally if the sampling point lies between them and the boundary since in such cases the partial plots cannot enclose the sampling point. Trees C and D in Figure 21.10 are in this category. If a tree in the boundary zone is between the sampling point and the outside boundary, as are trees A, B, and F, and it subtends an angle greater than the critical angle, the sampling point is within the plot of the tree and the tree is tallied, with the notation that its plot is half or one quarter normal size, as the case may be. Trees not in the boundary zone, such as E and G, are treated in the normal manner. As an example, assume that the angle gauge has a BAF of 10 ft²/acre. Trees B and F have half circles which enclose the sampling point. Their blowup factors are equal to 2. Tree G has a full circle and its blowup factor is equal to 1. The estimate of the basal area per acre, as of that sampling point, is then $(2 + 2 + 1)10 = 50$ ft².

Slopover can occur with almost any sampling procedure. In most cases, however, the use of partial plots will eliminate the problem.

### 21.3.2.8 Vertical Point Sampling

The concept of the angle count was extended by Hirata (1955) who suggested the use of a critical angle lying in a *vertical*, rather than a horizontal plane, as is shown in Figure 21.11. Then

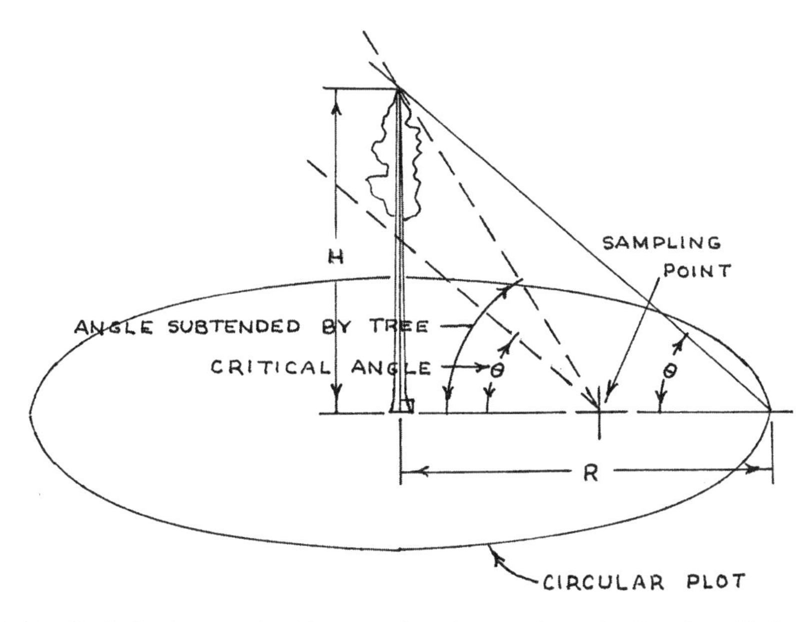


**FIGURE 21.11** Vertical point sampling. The tree subtends an angle greater than the critical angle and thus the plot of the tree encloses the sampling point.

$$\tan \Theta = H / R \tag{21.40}$$

where θ is the vertical critical angle, $H$ is the height (either total or merchantable) of a tree, and $R$ is the radius of the plot, centered on the tree, which is generated in the horizontal plane when the apex of the critical angle occupies the set of positions where the height of the tree subtends an angle exactly equal to the critical angle. Thus, as in the case of horizontal point sampling, each tree is at the center of a circular plot which may or may not enclose a given sampling point. In this case, however, the size of the plot is a function of tree height rather than dbh.

The radius of the plot of the tree is then

$$R = \text{H}/\tan \Theta \tag{21.41}$$

the plot area is

$$\text{PA} = \pi R^2 = \pi H^2/\tan^2 \Theta \tag{21.42}$$

and the probability of the plot enclosing a given sampling point is

$$P(\text{tree}_i) = \text{PA}_i / \text{SA} = \pi H_i^2 / \text{SA} \tan^2 \Theta \tag{21.43}$$

The expansion or blowup factor is

$$g_{\text{VP}_i} = 1 / P(\text{tree}_i) = \text{SA} \tan^2 \theta / \pi H_i^2 \tag{21.44}$$

SA $\tan^2 \Theta/\pi$ is constant, regardless of tree size. Thus, $g_{\text{VP}_i}$ is a function of the square of the tree height.

This type of timber estimation is referred to as *vertical point sampling*.

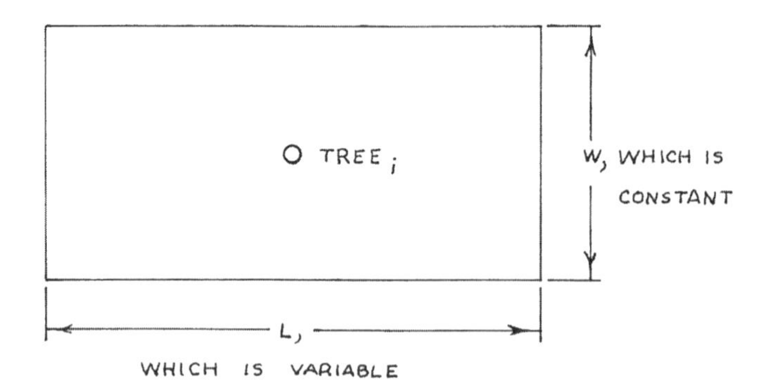


**FIGURE 21.12** Rectangular plot, centered on tree $i$, which varies in area since its length, $L$, is a positive function of some tree dimension such as dbh or height.

### 21.3.2.9 Line Sampling

The basic idea behind point sampling with tree-centered circular plots was extended to rectangular plots by Strand (1957) and was further developed by Beers and Miller (1976). Consider the situation in Figure 21.12 where the tree is at the center of a rectangular plot which has a specified width, $W$, in some specified direction, and a length, $L$, the magnitude of which would be constant, if constant area plots are used, or a positive function of some tree dimension, such as dbh or total height, if variable area plots are used. In the latter case, the plot area, PA, is large where the tree is large and small when the tree is small. The probability of a randomly chosen point being enclosed in such a plot is then, as in the case of the circular plots, equal to the ratio of plot area to tract area, and the blowup factor is equal to the reciprocal of that probability. Consequently, the sampling is with probability proportional to size of the tree dimension which controls the length of the plot.

The field application of this concept makes use of an angle gauge. A line, equal in length to the specified plot width, $W$, and lying in the specified direction, is centered on each of the selected sampling points. This is the *reference line* shown in Figure 21.13. To determine if the sampling point is enclosed in the rectangular plot of a tree, the tree is viewed from the reference line using an angle gauge that is held so that its axis forms a right angle with the reference line. If a tree, such as A in Figure 21.13, subtends an angle greater than the critical angle, the sampling point is within the plot of the tree. If the tree subtends an angle equal to the critical angle, as does C, the sampling point is on the edge of the plot. Finally, if the tree subtends an angle less than the critical angle, as does B, the sampling point is outside the plot of the tree.

If the critical angle lies in the horizontal plane, and trees on both sides of the reference line are candidates for inclusion in the cluster, the variable dimension, $L$, is defined as twice the radius $R$ in equation 21.28.

$$R = r/\sin(\phi/2)$$

and

$$L = 2R = 2r/\sin(\phi/2) \tag{21.45}$$

The plot area is then

$$PA_i = WL_i = 2Wr_i/\sin(\phi/2) \tag{21.46}$$

and since $r_i = D_i/2$, where $D_i = \text{dbh}_i$,

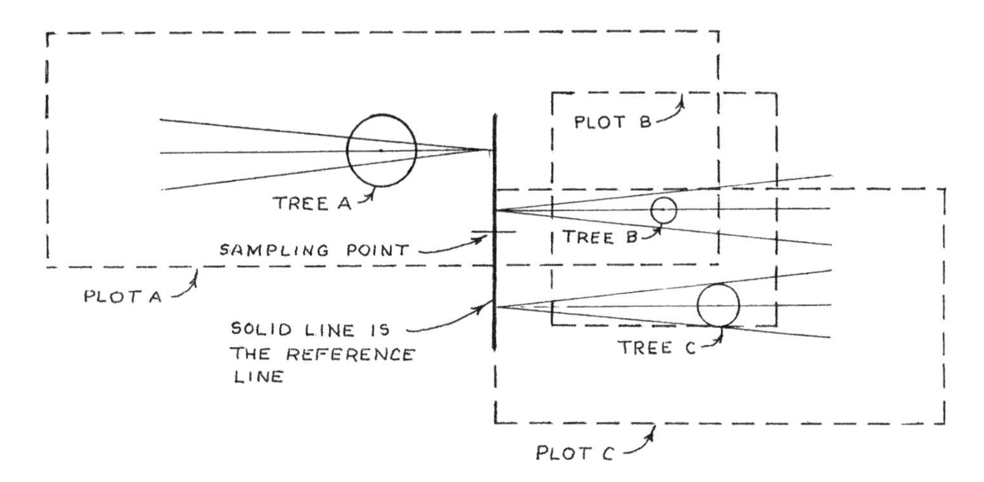


**FIGURE 21.13** Line sampling. Note that the sampling point is inside Plot A, is on the edge of Plot C, and is outside Plot B.

$$\text{PA}_i = WD_i/\sin(\phi/2) \tag{21.47}$$

Then,

$$P(\text{tree}_i) = \text{PA}_i/\text{SA} = WD_i/\text{SA} \sin(\phi/2) \tag{21.48}$$

and the blowup or expansion factor is

$$g_{HL_i} = 1/P(\text{tree}_i) = \text{SA}\sin(\phi/2)/WD_i \tag{21.49}$$

Thus, the expansion factor is a function of dbh, in contrast with horizontal point sampling where the expansion factor, $g_{\text{HP}}$, is a function of basal area or, since $\text{BA} = \pi r^2 = \pi D^2/4$, of the square of the dbh. As might be expected, the procedure is referred to as *horizontal line sampling*.

The relationship between $L$ and BAF is as follows:

$$\text{BAF} = \text{BA(SA)/PA} = \pi r^2(\text{SA})/\pi R^2 = r^2(\text{SA})/R^2$$

$$R^2 = r^2(\text{SA})/\text{BAF}$$

$$R = r\sqrt{\text{SA/BAF}}$$

Then,

$$L = 2R = 2r\sqrt{\text{SA/BAF}} = D\sqrt{\text{SA/BAF}} \tag{21.50}$$

*Vertical line sample* is illustrated in Figure 21.14. Given the vertical angle $\Theta$ and the plot width $W$,

$$\tan\Theta = H/(L/2) \tag{21.51}$$

$$L/2 = H/\tan\Theta \quad \text{and}$$

$$L = 2H/\tan\Theta \tag{21.52}$$

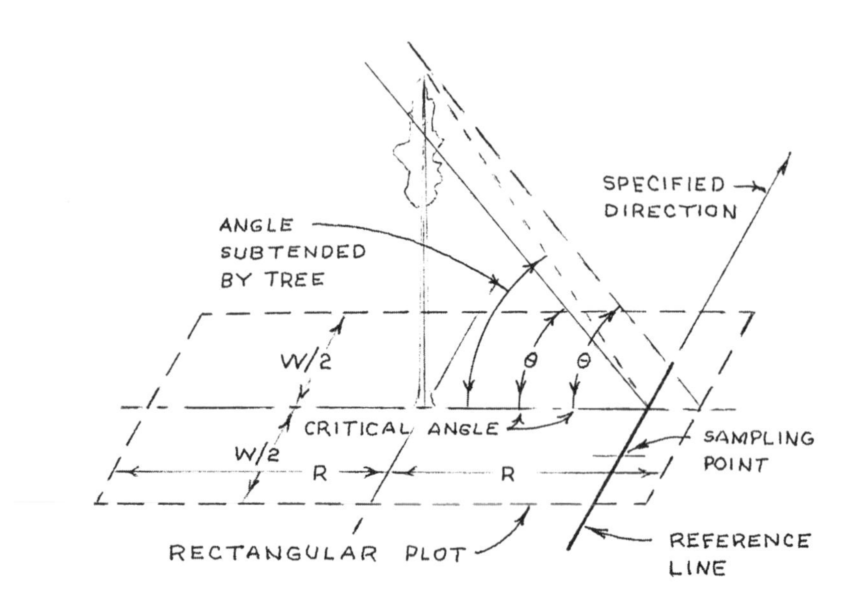


**FIGURE 21.14** Vertical line sampling. The sampling point is enclosed by the plot of the tree because the tree subtends an angle greater than the critical angle.

$$\text{PA} = WL = 2WH/\tan\Theta \tag{21.53}$$

$$P(\text{tree}_i) = \text{PA}_i/\text{SA} = 2WH_i/\text{SA}\tan\Theta \quad \text{and} \tag{21.54}$$

$$g_{\text{VL}_i} = 1/P(\text{tree}_i) = \text{SA}\tan\Theta/2WH_i \tag{21.55}$$

SA tan $\Theta/2W$ is constant, regardless of tree size. Consequently, $g_{\text{VL}}$ is a function of tree height, $H_i$, in contrast to the blowup factor in vertical point sampling, $g_{vp}$, which is a function of the square of the tree height.

### 21.3.2.10 Statistical Properties

Bitterlich's interest lay in a procedure that would rapidly and easily yield basal areas per specified land area so that marking stands for thinning to a desired basal area would be facilitated. He did not apply his idea to timber estimation. This extension was made by Grosenbaugh (1952; 1955; 1958) who with others (e.g., Palley and Horwitz, 1961; Johnson, 1961; Schreuder, 1970) expanded the theory and provided it with a statistical foundation.

Unless otherwise stated, the following discussion will be based on the assumption that the sampling frame grid is a randomly chosen member of an infinitely large set of such grids, each of which has a different initial point and orientation from any of the others. Thus, every point making up the surface of the tract has a probability of having draws in a random draw.

The set of trees having plots enclosing a given sampling point constitutes a cluster. If no subsampling is done, the statistical analysis is carried out in the context of single-stage cluster sampling. Since the selection of the trees is made using variable probability sampling, the raw tree data (dbh, basal area, volume, etc.) must be brought to a common base. This is accomplished by multiplying the raw data, tree by tree, by an appropriate expansion or blowup factor, $g$, to convert the data to a per specified land area basis. Thus, in cluster $j$, letting $z_{ij}$ = tree attribute$_{ij}$,

$$y_{ij} = g_{ij}z_{ij} \text{ units per SA} \tag{21.56}$$

Equation 21.56 can be used with any one of the four angle-count methods and with any tree attribute. However, the process of determining $y_{ij}$ is facilitated if the tree selection procedure is matched with the variable of interest. For example, if the variable of interest is basal area per specified land area, it would be advantageous to use horizontal point sampling since then $y_{ij}$ would be constant for all trees:

$$\begin{aligned} y_{ij} &= g_{\text{HP}_{ij}}(\text{basal area}_{ij}) \\ &= g_{\text{HP}_{ij}} \frac{\pi D_{ij}^2}{4} \\ &= \left[ \text{SA} \sin^2\left(\frac{\phi}{2}\right) \frac{4}{\pi D_{ij}^2} \right] \left[ \frac{\pi D_{ij}^2}{4} \right] \\ &= \text{SA} \sin^2\left(\frac{\phi}{2}\right) = \text{BAF} \end{aligned} \tag{21.57}$$

In the same way, if the variable of interest is the total dbh per SA, one should use horizontal line sampling.

$$\begin{aligned} y_{ij} &= g_{\text{HL}_{ij}} D_{ij} \\ &= \text{SA} \sin\left(\frac{\phi}{2}\right) \frac{D_{ij}}{WD_{ij}} \\ &= \text{SA} \sin \frac{\left(\frac{\phi}{2}\right)}{W} \end{aligned} \tag{21.58}$$

which is constant for all trees. If the variable of interest is the total of the squares of tree heights per SA, vertical point sampling should be used.

$$\begin{aligned} y_{ij} &= g_{\text{VP}_{ij}} H_{ij}^2 \\ &= \left[ \text{SA} \tan^2 \frac{\Theta}{\pi H_{ij}^2} \right] H_{ij}^2 \\ &= \text{SA} \tan^2 \frac{\Theta}{\pi} \end{aligned} \tag{21.59}$$

Finally, if the variable of interest is the total of tree heights per SA, vertical line sampling should be used.

$$\begin{aligned} y_{ij} &= g_{\text{VL}_{ij}} H_{ij} \\ &= \left[ \text{SA} \frac{\tan \Theta}{2WH_{ij}} \right] H_{ij} \\ &= \text{SA} \tan \frac{\Theta}{2W} \end{aligned} \tag{21.60}$$

It must be emphasized that *any* of the sampling procedures can be used to obtain values of $y_{ij}$ for *any* variable of interest (e.g., volume or value). Since horizontal point sampling is the simplest procedure to use in the field, it is the most commonly used of the four.

In the case of single-stage cluster sampling, the statistical analysis is based on cluster totals rather than individual ultimate sampling unit values. Thus, according to Equation 17.2,

$$\tau_j = \sum_i^{N_i} y_{ij} = \sum_i^{N_j} g_{ij} z_{ij} \tag{21.61}$$

where $\tau_j$ is the total of the variable of interest on the specified land area, which may be the whole tract, an acre, or a hectare, as estimated by cluster *j*. $N_j$ is the number of trees making up cluster *j*. If $y_{ij}$ is a constant, as in the cases represented by Equations 21.54 through 21.57, Equation 21.58 is reduced to

$$\tau_j = N_j y_{ij} \tag{21.62}$$

The true total of the variable of interest on the specified land area is

$$\tau_o = \sum_i^{N} z_i \tag{21.63}$$

where $N$ is the number of trees on the specified land area. Each $\tau_j$ is an estimate of $\tau_o$. The estimate of $\tau_o$, based on the $m$ clusters making up the sample, is

$$\bar{T} = \frac{1}{m}\sum_j^{m} \tau_j = \frac{1}{m}\sum_j^{m}\sum_i^{N_j} g_{ij} z_{ij} \rightarrow \tau_o \tag{21.64}$$

This is a general expression, which can be used with any of the four angle-count procedures and with any variable of interest.

The following discussion, leading to an evaluation of the bias of $\bar{T}$, will be in terms of horizontal point sampling using the arguments developed by Palley and Horwitz (1961). It is appropriate only when it can be assumed that every point making up the tract surface has a probability of being chosen in a random draw.

$$\begin{aligned} g_{ij} z_{ij} &= \left[\frac{\text{SA}_T \sin^2\left(\frac{\phi}{2}\right)}{\text{BA}_{ij}}\right] \text{BA}_{ij} \\ &= \text{SA}_T \sin^2\left(\frac{\phi}{2}\right) \text{BAF}_T \end{aligned} \tag{21.65}$$

where $\text{SA}_T$ = tract area and $\text{BAF}_T$ is the corresponding basal area factor. Tract area has been chosen because it can be expressed in any land area measurement unit. If one chooses to set SA equal to an acre or a hectare, the formulae developed will remain the same except for the magnitudes of the constants SA and BAF. However, one must express the acre as 43,560 ft$^2$ or the hectare as 10,000 m$^2$ so as to have a unit of measure that is practical to use to measure basal area. Then,

$$\overline{T}_{BA} = \frac{1}{m}\sum_{j}^{m}\sum_{i}^{N_j} \mathrm{BAF}_T = \frac{\mathrm{BAF}_T}{m}\sum_{j}^{m} N_j = \mathrm{BAF}_T\overline{N} \tag{21.66}$$

where

$$\overline{N} = \frac{1}{m}\sum_{j}^{m} N_j$$

the mean cluster size.

The only variable in Equation 21.66 is the cluster size, $N_j$. Consequently, the expected value of $\overline{T}_{BA}$ is

$$E(\overline{T}_{BA}) = \frac{\mathrm{BAF}_T}{m}\sum_{j}^{m} E(N_j)$$

Let $\alpha$ be an indicator variable that equals 1 when the plot associated with tree $ij$ encloses point $j$ and 0 when the point is not so enclosed. When the plots associated with *all* the trees on the specified land area have been examined, the cluster at point $j$ contains

$$N_j = \sum_{i}^{N} \alpha_{ij} \text{ trees} \tag{21.67}$$

Then,

$$E(N_j) = E\left(\sum_{i}^{N} \alpha_{ij}\right) = \sum_{i}^{N} E(\alpha_{ij})$$

where

$$E(\alpha_{ij}) = P(X_{ij} = 1)(1) + P(X_{ij} = 0)(0) = P(\text{tree}_{ij}) \tag{21.68}$$

which is the probability of tree $i$ having a plot enclosing randomly chosen sampling point $j$. According to Equation 21.30,

$$P(\text{tree}_{ij}) = \frac{\mathrm{PA}_{ij}}{\mathrm{SA}_T} = \frac{\mathrm{BA}_{ij}}{\mathrm{BAF}_T} \tag{21.69}$$

Then

$$E(N_j) = \sum_{i}^{N}\left(\frac{BA_i}{BAF_T}\right)$$

$$E(\overline{T}_{BA}) = \frac{\mathrm{BAF}_T}{m}\sum_{j}^{m}\sum_{j}^{N}\frac{\mathrm{BA}_i}{\mathrm{BAF}_T} \tag{21.70}$$

$$= \sum_{i}^{N} \mathrm{BA}_i = \tau_{o_{BA}}$$

and $\bar{T}_{\text{BA}}$ is an unbiased estimator of $\tau_{o_{\text{BA}}}$.

The variance of $\bar{T}_{\text{BA}}$ is

$$\sigma^2_{\bar{T}_{\text{BA}}} = \text{BAF}_T \sigma^2_{\bar{N}} \tag{21.71}$$

According to Equation 12.4,

$$\sigma^2_{\bar{N}} = E[\bar{N} - E(\bar{N})]^2$$

where $E(\bar{N}) = \mu_{N_j}$.

Then,

$$\sigma^2_{\bar{N}} = E(\bar{N} - \mu_{N_j})^2 = E\left(\frac{1}{m}\sum_j^m N_j - \mu_{N_j}\right)^2$$

$$= \frac{1}{m^2} E\left(\sum_j^m N_j - m\mu_N\right)^2$$

$$= \frac{1}{m^2}\{\sigma_{N_1^2} + \sigma_{N_2^2} + \cdots + \sigma^2_{N_m^2} + 2E[(N_1 - \mu_N)(N_2 - \mu_N)] \tag{21.72}$$

$$+ 2E[(N_1 - \mu_N)(N_3 - \mu_N)] + \cdots + 2E[(N_1 - \mu_N)(N_m - \mu_N)]$$

$$+ \cdots + 2E[(N_{m-1} - \mu_N)(N_m - \mu_N)]\}$$

Since the selection of sampling *points* is done using random sampling, the sum of the cross-product terms equals zero and thus

$$\sigma^2_{\bar{N}} = \frac{\sigma^2_{N_j}}{m} \tag{21.73}$$

Since $N_j = \Sigma_i^N \alpha_{ij}$,

$$\sigma^2_{N_j} = \sigma^2_{\Sigma_i^N \alpha_{ij}} = \sum_i^N \sigma^2_{\alpha_{ij}} + 2\sum_{h=1}^{N-1}\sum_{i=h+1}^{N} \sigma_{\alpha_{hj}\alpha_{ij}} \tag{21.74}$$

where, according to Equation 8.35,

$$\sigma^2_{\alpha_{ij}} = E(\alpha^2_{ij}) - [E(\alpha_{ij})]^2$$

However, since $\alpha_{ij}$ is an indicator variable, taking on only the values 1 and 0, $\alpha^2_{ij} = \alpha_{ij}$, and

$$\sigma^2_{\alpha_{ij}} = \frac{\text{BA}_{ij}}{\text{BAF}_T} - \left(\frac{\text{BA}_{ij}}{\text{BAF}_T}\right)^2 \tag{21.75}$$

According to Equation 9.28,

$$\sigma_{\alpha_{hj}\alpha_{ij}} = E(\alpha_{hj}\alpha_{ij}) - E(\alpha_{hj})E(\alpha_{ij})$$
$$= E(\alpha_{hj}\alpha_{ij}) - \left(\frac{\text{BA}_{hj}}{\text{BAF}_T}\right)\left(\frac{\text{BA}_{ij}}{\text{BAF}_T}\right) \tag{21.76}$$

$$E(\alpha_{hj}\alpha_{ij}) = \frac{0_{hi}}{\text{SA}_T} \tag{21.77}$$

where $0_{hi}$ is the area of overlap of the plots associated with trees $h$ and $i$. $E(\alpha_{hj}\alpha_{ij})$ is then the probability of point $j$ occurring in that overlap. Then,

$$\sigma^2_{N_j} = \sum_j^N \frac{\text{BA}_i}{\text{BAF}_T} - \sum_i^N \frac{\text{BA}_i^2}{\text{BAF}_T^2} + \frac{2}{\text{SA}_T}\sum_{h=1}^{N-1}\sum_{i=h+1}^{N} 0_{hi} + 2\sum_{h=1}^{N-1}\sum_{i=h_1}^{N}\frac{\text{BA}_h\text{BA}_i}{\text{BAF}_T^2} \tag{21.78}$$

Since

$$\left(\sum_i^N \text{BA}_i\right)^2 = \sum_i^N \text{BA}_i^2 - 2\sum_h^{N-1}\sum_{i=h+1}^{N}\text{BA}_h\text{BA}_i$$

then

$$\sigma^2_{N_j} = \frac{1}{\text{BAF}_T}\sum_i^N \text{BA}_i - \frac{1}{\text{BAF}_T^2}\left(\sum_i^N \text{BA}_i\right)^2 + \frac{2}{\text{SA}_T}\sum_{h=1}^{N-1}\sum_{i=h+1}^{N} 0_{hi}$$
$$= \frac{1}{\text{BAF}_T}\tau_{o_{BA}} - \frac{1}{\text{BAF}_T^2}\tau^2_{o_{\text{BA}}} + \frac{2}{\text{SA}_T}\sum_{h=1}^{N-1}\sum_{i=h+1}^{N} 0_{hi} \tag{21.79}$$

$$= \frac{\tau_{o_{\text{BA}}}}{\text{BAF}_T}\left(1 - \frac{\tau_{o_{BA}}}{\text{BAF}_T}\right) + \frac{2}{\text{SA}_T}\sum_{h=1}^{N-1}\sum_{i=h+1}^{N} 0_{hi} \tag{21.80}$$

and

$$\sigma^2_{\bar{T}_{\text{BA}}} = \frac{\text{BAF}_T^2}{m}\left[\frac{\tau_{o_{\text{BA}}}}{\text{BAF}_T}\left(1 - \frac{\tau_{o_{\text{BA}}}}{\text{BAF}_T}\right) + \frac{2}{\text{SA}_T}\sum_{h=1}^{N-1}\sum_{i=h+1}^{N} 0_{hi}\right] \tag{21.81}$$

As can be seen from Equation 21.81, $\sigma^2_{\bar{T}_{\text{BA}}}$ tends to increase as the total basal area on the tract increases and to decrease as the tract size increases. As the critical angle, $\phi$, increases, the plot radii decrease and thus the cluster size decreases. This, in turn, tends to increase $\sigma^2_{\bar{T}_{\text{BA}}}$ because small cluster totals are more likely to be erratic than are large cluster totals. However, as the clusters decrease in size the amount of plot overlap also decreases, affecting to some extent the increase in $\sigma^2_{\bar{T}_{BA}}$ Finally, in common with other variances of the mean, $\sigma^2_{\bar{T}_{BA}}$ is inversely proportional to the sample size. Since $\sigma^2_{\bar{T}_{BA}} = \text{BAF}_T^2 \sigma^2_{\bar{N}}$, its estimator is

$$s^2_{\bar{T}_{BA}} = \text{BAF}_T^2 s^2_{\bar{N}} \tag{21.82}$$

According to Equation 12.3, the estimated variance of the mean cluster size is

$$s^2_{\bar{N}} = \frac{s^2_{N_j}}{m} \tag{21.83}$$

where the estimated variance of the cluster sizes is

$$\begin{aligned} s^2_{N_j} &= \frac{1}{m-1} \sum_j^m (N_j - \bar{N})^2 \\ &= \frac{1}{m-1} \left[ \sum_j^m N_i^2 - \frac{\left( \sum_j^m N_j \right)^2}{m} \right] \end{aligned} \tag{21.84}$$

Then

$$s^2_{\bar{T}_{BA}} = \frac{\text{BAF}_T^2}{m^2(m-1)} \left[ m \sum_j^m N_j^2 - \left( \sum_j^m N_j \right)^2 \right] \tag{21.85}$$

The expected value of $s^2_{\bar{T}_{BA}}$ is

$$\begin{aligned} E(s^2_{\bar{T}_{BA}}) &= \frac{\text{BAF}_T^2}{m(m-1)} E \left[ \sum_j^m N_j^2 - \frac{\left( \sum_j^m N_j \right)^2}{m} \right] \\ &= \frac{\text{BAF}_T^2}{m^2(m-1)} \left[ m \sum_j^m E(N_j^2) - E\left( \sum_j^m N_j \right)^2 \right] \end{aligned} \tag{21.86}$$

Consider the first term inside the brackets in Equation 21.86.

Since $N_j = \Sigma_i^N \alpha_{ij}$,

$$N_j^2 = \left(\sum_i^N \alpha_{ij}\right)^2 = \sum_i^N \alpha_{ij}^2 + 2\sum_{h=1}^{N-1}\sum_{i=h+1}^{N} \alpha_{hj}\alpha_{ij}$$

Then

$$E(N_j^2) = \sum_i^N E(\alpha_{ij}^2) + 2\sum_{h=1}^{N-1}\sum_{i=h+1}^{N} E(\alpha_{hj}\alpha_{ij})$$

Since $\alpha_{ij}$ is an indicator variable, taking on only the values 1 and 0, $\alpha_{ij}^2 = \alpha_{ij}$,

$$E(\alpha_{ij}^2) = E(\alpha_{ij}) = \frac{\text{BA}_{ij}}{\text{BAF}_T}$$

and

$$\sum_i^N E(\alpha_{ij}^2) = \frac{1}{\text{BAF}_T}\sum_i^N \text{BA}_i = \frac{\tau_{o_{\text{BA}}}}{\text{BAF}_T}$$

From Equation 21.77,

$$E(\alpha_{hj}\alpha_{ij}) = \frac{0_{hi}}{\text{SA}_T}$$

Then

$$E(N_j^2) = \frac{\tau_{o_{\text{BA}}}}{\text{BAF}_T} + \frac{2}{\text{SA}_T}\sum_{h=1}^{N-1}\sum_{i=h+1}^{N} 0_{hi}$$

and

$$m\sum_j^m E(N_j^2) = \frac{m^2\tau_{o_{\text{BA}}}}{\text{BAF}_T} + \frac{2m^2}{\text{SA}_T}\sum_{h=1}^{N-1}\sum_{i=h+1}^{N} 0_{hi} \tag{21.87}$$

Consider now the second term within the brackets in Equation 21.86.

$$E\left(\sum_j^m N_j\right)^2 = E\left(\sum_j^m N_j^2 + 2\sum_{j=1}^{m-1}\sum_{k=j+1}^{m} N_j N_k\right)$$

$$= \sum_j^m E(N_j^2) + 2\sum_{j=1}^{m-1}\sum_{k=j+1}^{m} E(N_j N_k)$$

According to Equation 9.30

$$E(N_jN_k) = E(N_j)E(N_k) + \text{COV}(N_jN_k)$$

Since the sampling was random the covariance is equal to zero and thus

$$E(N_jN_k) = E(N_j)E(N_k) = E\left(\sum_h^N \alpha_{hj}\right)E\left(\sum_i^N \alpha_{ik}\right)$$

According to Equations 21.68 and 21.69,

$$E(\alpha_{hj}) = \frac{\text{BA}_{hj}}{\text{BAF}_T} \quad \text{and} \quad E(\alpha_{ik}) = \frac{\text{BA}_{ik}}{\text{BAF}_T}$$

Then

$$\begin{aligned} E(N_jN_k) &= \sum_h^N \left(\frac{\text{BA}_{hj}}{\text{BAF}_T}\right)\sum_i^N \left(\frac{\text{BA}_{ik}}{\text{BAF}_T}\right) \\ &= \frac{\tau^2_{0_{\text{BA}}}}{\text{BAF}_T^2} \end{aligned} \tag{21.88}$$

and

$$\begin{aligned} E\left(\sum_j^m N_j\right)^2 &= \frac{m\tau_{o_{\text{BA}}}}{\text{BAF}_T} + \frac{2m}{\text{SA}_T}\sum_{h=1}^{N-1}\sum_{i=h+1}^{N} 0_{hi} + \frac{2}{\text{BAF}_T^2}\sum_{j=1}^{m-1}\sum_{k=j+1}^{m}\tau^2_{o_{\text{BA}}} \\ &= \frac{m\tau_{o_{\text{BA}}}}{\text{BAF}_T} + \frac{2(m-1)(m)\tau^2_{o_{\text{BA}}}}{\text{BAF}_T^2} + \frac{2m}{\text{SA}_T}\sum_{h=1}^{N-1}\sum_{i=h+1}^{N} 0_{hi} \end{aligned} \tag{21.89}$$

Substituting Equations 21.87 and 21.89 into Equation 21.86,

$$\begin{aligned} E(s^2_{\bar{T}_{\text{BA}}}) &= \frac{\text{BAF}_T^2}{m^2(m-1)}\left[\frac{m^2\tau_{o_{\text{BA}}}}{\text{BAF}_T} + \frac{2m^2}{\text{SA}_T}\sum_{h=1}^{N-1}\sum_{i=h+1}^{N} 0_{hi} - \frac{m\tau_{o_{\text{BA}}}}{\text{BAF}_T}\right. \\ &\quad \left. - \frac{2(m-1)(m)\tau^2_{o_{\text{BA}}}}{\text{BAF}_T^2} - \frac{2m}{\text{SA}_T}\sum_{h=1}^{N-1}\sum_{i=h+1}^{N} 0_{hi}\right] \\ &= \frac{\text{BAF}_T^2}{m}\left[\frac{\tau_{o_{\text{BA}}}}{\text{BAF}_T}\left(1 - \frac{2\tau_{o_{\text{BA}}}}{\text{BAF}_T}\right) + \frac{2}{\text{SA}_T}\sum_{h=1}^{N-1}\sum_{i=h+1}^{N} 0_{hi}\right] \end{aligned} \tag{21.90}$$

Then, $E(s^2_{\bar{T}_{\text{BA}}}) = \sigma^2_{\bar{T}_{\text{BA}}}$ and the estimator is unbiased.

Since the sampling is random, a valid sampling error can be computed:

$$\text{SE}_{\bar{T}_{\text{BA}}} = \pm t_c(k = m-1)s_{\bar{T}_{\text{BA}}}$$

An illustrative numerical example of this process is shown in Section 21.3.2.16.

#### 21.3.2.11 Compensating for Slopover and Borderline Trees

When slopover is present and Grosenbaugh's partial plot procedure is used to correct for the slopover, it becomes necessary to weight the basal area factor.

$$P(\text{tree}_{ij}) = \frac{\text{PA}_{ij}}{\text{SA}_T} = \frac{\text{BA}_{ij}}{\text{BAF}_T}$$

Then

$$g_{ij} = \frac{1}{P(\text{tree}_{ij})} = \frac{\text{BAF}_T}{\text{BA}_{ij}}$$

If the plot area is cut in half as, for example, with trees A, B, C, D, and F in Figure 21.10, $P(\text{tree}_{ij})$ is also cut in half so that

$$P(\text{tree}_{ij}) = \frac{\text{PA}_{ij}}{2\text{SA}_T} = \frac{\text{BA}_{ij}}{2\text{BAF}_T}$$

and

$$g_{ij} = \frac{2\text{BAF}_T}{\text{BA}_{ij}} \quad (21.91)$$

The 2 is then a weight. If $\text{PA}_{ij}$ is full size, $w_{ij} = 1$; if half-size, $w_{ij} = 2$; and if quarter-size, $w_{ij} = 4$. Introducing the weights into Equation 21.61,

$$\tau_{j_s} = \sum_{i}^{N_j} w_{ij}\text{BAF}_T \quad (21.92)$$

Then,

$$\bar{T}_{\text{BA}_s} = \frac{1}{m}\sum_{j}^{m} \tau_i = \frac{\text{BAF}_T}{m}\sum_{j}^{m}\sum_{i}^{N_j} w_{ij} \quad (21.93)$$

which can be written

$$\bar{T}_{\text{BA}_s} = \frac{\text{BAF}_T}{m}\sum_{j}^{m}\sum_{i}^{N} w_{ij}\alpha_{ij} \quad (21.94)$$

The expected value of $\bar{T}_{\text{BA}_s}$ is

$$E(\bar{T}_{\text{BA}_s}) = \frac{\text{BAF}_T}{m}\sum_{j}^{m}\sum_{i}^{N}[E(w_{ij})E(\alpha_{ij})]$$

Since

$$E(w_{ij}) = \frac{1}{N}\sum_{i}^{N} w_{ij} \quad \text{and} \quad E(\alpha_{ij}) = \frac{\text{BA}_{ij}}{w_{ij}\text{BAF}_T}$$

then

$$E(\bar{T}_{\text{BA}_s}) = \frac{1}{m}\sum_{j}^{m}\sum_{i}^{N}\left[\left(\frac{1}{N}\sum_{i}^{N} w_{ij}\right)\left(\frac{\text{BA}_{ij}}{w_{ij}\text{BAF}_T}\right)\right]$$

$$= \frac{\text{BAF}_T}{mN}\sum_{j}^{m}\sum_{i}^{N}\left[\sum_{i}^{N}\left(\frac{w_{ij}\text{BA}_{ij}}{w_{ij}\text{BAF}_T}\right)\right] \tag{21.95}$$

$$= \sum_{i}^{N} \text{BA}_{ij} = \tau_{o_{\text{BA}}}$$

indicating that the estimator is unbiased. It should be recognized that Equation 21.93 is a general expression. If no slopover is present, all the weights are equal to 1 and Equation 21.93 reduces to equation 21.66.

The variance of the estimated mean total when slopover is present is obtained by modifying the probabilities in the development leading to Equation 21.81. According to Equations 21.68 and 21.69,

$$E(\alpha_{ij}) = P(\text{tree}_{ij}) = \frac{\text{BA}_{ij}}{\text{BAF}_T}$$

If slopover is present,

$$E(\alpha_{ij})_s = \frac{\text{BA}_{ij}}{w_{ij}\text{BAF}_T} \tag{21.96}$$

where the subscript $s$ indicates that slopover is being considered. Incorporating this expected value into Equations 21.75,

$$\sigma^2_{\alpha_{ij_s}} = \frac{\text{BA}_{ij}}{w_{ij}\text{BAF}_T} - \left(\frac{\text{BA}_{ij}}{w_{ij}\text{BAF}_T}\right)^2 \tag{21.97}$$

Equation 21.77

$$E\left(\frac{\alpha_{hj}\alpha_{ij}}{w_{ij}w_{ij}}\right) = \frac{0_{hi}}{\text{SA}_T w_{hj} w_{ij}} \tag{21.98}$$

and Equation 21.79

$$\sigma^2_{N_{j_s}} = \frac{1}{\text{BAF}_T}\sum_i^N \frac{\text{BA}_i}{w_i} - \frac{1}{\text{BAF}_T^2}\left(\sum_i^N \frac{\text{BA}_i}{w_i}\right)^2 + \frac{2}{\text{SA}_T}\sum_{h=1}^{N-1}\sum_{i=h+1}^{N}\left(\frac{0_{hi}}{w_{hj}w_{ij}}\right) \tag{21.99}$$

Then

$$\sigma^2_{\bar{T}_{\text{BA}_s}} = \frac{\text{BAF}_T^2}{m}\left[\frac{1}{\text{BAF}_T}\sum_i^N \frac{\text{BA}_i}{w_i} - \frac{1}{\text{BAF}_T^2}\left(\sum_i^N \frac{\text{BA}_i}{w_i}\right)^2 + \frac{2}{\text{SA}_T}\sum_{h=1}^{N-1}\sum_{i=h+1}^{N}\left(\frac{0_{hi}}{w_{hj}w_{ij}}\right)\right] \tag{21.100}$$

The estimator of $\sigma^2_{\bar{T}_{\text{BA}_s}}$ is

$$s^2_{\bar{T}_{\text{BA}_s}} = \frac{\text{BAF}_T^2}{m} s_{N_s^2}$$

When no slopover is present,

$$s_N^2 = \frac{1}{m-1}\left[\sum_j^m N_j^2 - \frac{\left(\sum_j^m N_j\right)^2}{m}\right]$$

However, when slopover is present and some trees have partial plots, the cluster counts must be modified. Thus,

$$s^2_{N_s} = \frac{1}{m-1}\left[\sum_j^m\left(\sum_i^{N_j} w_{ij}\right)^2 - \frac{\left(\sum_j^m\sum_i^{N_j} w_{ij}\right)^2}{m}\right]$$

and

$$s^2_{\bar{T}_{\text{BA}_s}} = \frac{\text{BAF}_T^2}{m^2(m-1)}\left[m\sum_j^m\left(\sum_i^{N_j} w_{ij}\right)^2 - \left(\sum_j^m\sum_i^{N_j} w_{ij}\right)^2\right] \to \sigma_{\bar{T}_{\text{BA}_s}} \tag{21.101}$$

It can be shown, following the pattern used where slopover is not a factor, that this estimate is unbiased. As was the case when slopover was not a factor, a valid sampling error can be computed if the sampling was random.

Borderline trees are trees whose associated plot edges fall *on* the sampling point. If a plot of the same size is centered on the point, rather than the tree, the edge of the circle would pass through the center of the tree. Thus, half of the basal area of the tree would be outside the circle. If the sampler considers the tree to be *in* the cluster, there will be an upward bias in the estimates. If the tree is considered to be outside the circle and, consequently, not a member of the cluster, there will be a negative bias. To remove this source of bias, a weight, similar to those used to correct for slopover, can be introduced into the computations. Since only half of the basal area of the tree is involved, the weight is set at 0.5. Weights of this type can be seen in the example in Section 21.3.2.16.

### 21.3.2.12 Volume Estimation

When horizontal point sampling is being used to estimate the volume of timber on a tract

$$y_{ij} = g_{\mathrm{HP}_{ij}} v_{ij} \tag{21.102}$$

where $y_{ij}$ = volume of timber of the tract based on tree $ij$

$$g_{\mathrm{HP}_{ij}} = \mathrm{SA}_T \sin^2 \frac{\left(\frac{\phi}{2}\right)}{\mathrm{BA}_{ij}} = \frac{\mathrm{BAF}_T}{\mathrm{BA}_{ij}}$$

$v_{ij}$ = volume of tree $ij$ as determined by conventional methods (i.e., using a volume table or equation)

When slopover is present,

$$\begin{aligned} y_{ij} &= w_{ij} g_{\mathrm{HP}_{ij}} v_{ij} \\ &= \mathrm{BAF}_T \left( \frac{w_{ij} v_{ij}}{\mathrm{BA}_{ij}} \right) \end{aligned} \tag{21.103}$$

The ratio of $v_{ij}/\mathrm{BA}_{ij}$ is referred to as the *volume:basal area ratio* or VBAR. In the following discussion it will be assumed that both the volume and the basal area of every tree in cluster $j$ will be determined. Thus, every tree has its own VBAR. It is not a constant. In real life obtaining a VBAR for every tree is rarely done. In some cases a constant VBAR, based on data from previous inventories or from published studies, is used. In other cases several broad dbh classes, within species groups, are recognized and a separate constant VBAR is used with each class.

The estimate of the total volume, based on cluster $j$, is

$$\tau_{v_j} = \sum_{i}^{N_j} y_{ij} = \mathrm{BAF}_T \sum_{i}^{N_j} \left( \frac{w_{ij} v_{ij}}{\mathrm{BA}_{ij}} \right) \tag{21.104}$$

The estimate of the total volume, based on a sample consisting of $m$ clusters, is

$$\bar{T}_v = \frac{\mathrm{BAF}_T}{m} \sum_{j}^{m} \sum_{i}^{N_j} \left( \frac{w_{ij} v_{ij}}{\mathrm{BA}_{ij}} \right) \rightarrow \tau_v \tag{21.105}$$

When slopover is not present, $w_{ij} = 1$, and

$$\bar{T}_v = \frac{\mathrm{BAF}_T}{m} \sum_{j}^{m} \sum_{i}^{N_j} \left( \frac{v_{ij}}{\mathrm{BA}_{ij}} \right) \rightarrow \tau_v \tag{21.106}$$

which was first suggested by Grosenbaugh (1952).

Equation 21.105 can be written

$$\bar{T}_v = \frac{\text{BAF}_T}{m} \sum_j^m \sum_i^{N_j} \left( \frac{w_{ij} \alpha_{ij} v_{ij}}{\text{BA}_{ij}} \right)$$

The expected value of $\bar{T}_v$ is then

$$E(\bar{T}_v) = \frac{\text{BAF}_T}{m} \sum_j^m \sum_i^{N_j} \left[ E(w_{ij}) E(a_{ij}) E(v_{ij}) E\left( \frac{1}{\text{BA}_{ij}} \right) \right]$$

where
$$E(w_{ij}) = \frac{1}{N} \sum_i^N w_i$$

$$E(\alpha_{ij}) = \frac{\text{BA}_{ij}}{w_{ij}\text{BAF}_T}$$

$$E(v_{ij}) = \frac{1}{N} \sum_i^N v_i$$

$$E\left( \frac{1}{\text{BA}_{ij}} \right) = \frac{1}{\frac{1}{N} \sum_i^N \text{BA}_i} = \frac{N}{\sum_i^N \text{BA}_i}$$

Then,

$$E(\bar{T}_{v_s}) = \frac{\text{BAF}_T}{m} \sum_j^m \sum_i^N \left[ \left( \frac{1}{N} \sum_i^N w_i \right) \left( \frac{\text{BA}_i}{w_i \text{BAF}_T} \right) \left( \frac{1}{N} \sum_i^N v_i \right) \left( \frac{N}{\sum_i^N \text{BA}_i} \right) \right]$$

$$= \frac{\text{BAF}_T N}{\text{BAF}_T N^2 m} \sum_j^m \sum_i^N \left[ \sum_i^N \left( \frac{w_i \text{BA}_{ij}}{w_i} \right) \tau_v \left( \frac{1}{\sum_i^N \text{BA}_i} \right) \right]$$

$$= \frac{1}{Nm} \sum_j^m \sum_i^N \tau_v$$

$$= \tau_v$$

Thus, $\bar{T}_{v_s}$ is an unbiased estimator of $\tau_v$, the total volume on the tract.

According to Equation 21.71,

$$\sigma^2_{\bar{T}_{BA}} = BAF_T \sigma^2_{\bar{N}} = \frac{BAF_T^2}{m}\sigma^2_{N_j} = \frac{BAF_T^2}{m}\left[\sum_i^{N_j}\sigma^2_{\alpha_{ij}} + 2\sum_h^{N_j-1}\sum_{k=h+1}^{N_j}\sigma_{\alpha_{hj}\alpha_{kj}}\right]$$

Introducing $(w_{ij}v_{ij}/BA_{ij})$ into the expression,

$$\sigma^2_{\bar{T}_v} = \frac{BAF_T^2}{m}\left[\sum_i^{N_j}\sigma^2_{\alpha_{ij}}\left(\frac{w_{ij}v_{ij}}{BA_{ij}}\right)^2 + 2\sum_h^{N_j-1}\sum_{k=h+1}^{N_j}\sigma_{\alpha_{hj}\alpha_{ij}}\left(\frac{w_{hj}v_{hj}}{BA_{hj}}\right)\left(\frac{w_{kj}v_{kj}}{BA_{kj}}\right)\right] \tag{21.107}$$

where

$$\sigma^2_{\alpha_{ij}}\left(\frac{w_{ij}v_{ij}}{BA_{ij}}\right)^2 = \left[\frac{BA_{ij}}{w_{ij}BAF_T} - \left(\frac{BA_{ij}}{w_{ij}BAF_T}\right)^2\right]\left(\frac{w_{ij}v_{ij}}{BA_{ij}}\right)^2$$

$$= \frac{BA_{ij}\left(\frac{w_{ij}v_{ij}}{BA_{ij}}\right)^2}{w_{ij}BAF_T} - \left(\frac{\frac{BA_{ij}w_{ij}v_{ij}}{BA_{ij}}}{\frac{BA_{ij}}{w_{ij}BAF_T}}\right)^2 \tag{21.108}$$

and

$$\sigma_{\alpha_{ij}\alpha_{ij}}\left(\frac{w_{hj}v_{hj}}{BA_{hj}}\right)\left(\frac{w_{kj}v_{kj}}{BA_{kj}}\right) = 0_{hi}\left(\frac{w_{hj}v_{hj}}{BA_{hj}}\right)\left(\frac{w_{kj}v_{kj}}{BA_{kj}}\right) - \left(\frac{\frac{BA_{hj}w_{hj}v_{hj}}{BA_{hj}}}{w_{hj}BAF_T}\right)\left(\frac{\frac{BA_{kj}w_{kj}v_{kj}}{BA_{kj}}}{w_{kj}BAF_T}\right) \tag{21.109}$$

Consequently,

$$\sigma^2_{\bar{T}_{v_S}} = \frac{BAF_T^2}{m}\left\{\frac{1}{BAF_T}\sum_i^N \frac{BA_i\left(\frac{w_iv_i}{BA_i}\right)^2}{w_i} - \frac{1}{BAF_T^2}\sum_i^N\left(\frac{\frac{BA_iw_iv_i}{BA_i}}{w_i}\right)^2\right.$$

$$+\frac{2}{SA_T}\sum_h^{N-1}\sum_{k=h+1}^N 0_{hk}\left(\frac{w_hv_h}{BA_h}\right)\left(\frac{w_kv_k}{BA_k}\right) \tag{21.110}$$

$$\left. - \frac{2}{BAF_T^2}\sum_h^{N-1}\sum_{k=h+1}^N\left[\frac{\frac{BA_hw_hv_h}{BA_h}}{w_h}\right]\left[\frac{\frac{BA_kw_kv_k}{BA_k}}{w_k}\right]\right\}$$

Since

$$\sum_i^N \left( \frac{\frac{\text{BA}_i w_i v_i}{\text{BA}_i}}{w_i} \right)^2 - 2\sum_h^{N-1} \sum_{k=h+1}^N \left[ \frac{\frac{\text{BA}_h w_h v_h}{\text{BA}_h}}{w_h} \right] \left[ \frac{\frac{\text{BA}_k w_k v_k}{\text{BA}_k}}{w_k} \right]$$

$$= \left( \sum_i^N \frac{\frac{\text{BA}_i w_i v_i}{\text{BA}_i}}{w_i} \right) = \left( \sum_i^N v_i \right)^2 \tag{21.111}$$

and since

$$\frac{\text{BA}_i \left( \frac{w_i v_i}{\text{BA}_i} \right)^2}{w_i} = \frac{w_i v_i^2}{\text{BA}_i} \tag{21.112}$$

then

$$\sigma^2_{\bar{T}_{v_s}} = \frac{\text{BAF}_T^2}{m} \left[ \frac{1}{\text{BAF}_T} \sum_i^N \frac{w_i v_i^2}{\text{BA}_i} - \frac{1}{\text{BAF}_T^2} \left( \sum_i^N v_i \right)^2 \right.$$

$$\left. + \frac{2}{\text{SA}_T} \sum_h^{N-1} \sum_k^N 0_{hk} \left( \frac{w_h v_h}{\text{BA}_h} \right) \left( \frac{w_k v_k}{\text{BA}_k} \right) \right] \tag{21.113}$$

$$= \frac{1}{m} \left[ \text{BAF}_T \sum_i^N \frac{w_i v_i^2}{\text{BA}_i} - \tau_v^2 + \frac{2\text{BAF}_T^2}{\text{SA}_T} \sum_h^{N-1} \sum_{k=h+1}^N 0_{hk} \left( \frac{w_h v_h}{\text{BA}_h} \right) \left( \frac{w_k v_k}{\text{BA}_k} \right) \right]$$

The estimator of $\sigma^2_{\bar{T}_{v_s}}$ is obtained using the same approach followed to obtain the estimator of $\sigma^2_{\bar{T}_{\text{BA}_s}}$.

$$s^2_{\bar{T}_{v_s}} = \frac{\text{BAF}_T^2}{m^2(m-1)} \left\{ m \sum_j^m \left[ \sum_i^{N_j} \left( \frac{w_{ij} v_{ij}}{\text{BA}_{ij}} \right) \right]^2 - \left[ \sum_j^m \sum_i^{n_j} \left( \frac{w_{ij} v_{ij}}{\text{BA}_{ij}} \right) \right]^2 \right\} \tag{21.114}$$

which can be shown to be unbiased. When the sampling is random, one can compute a valid sampling error.

The assumption upon which the preceding discussion is based is that the sampling frame consists of the intersections of a randomly chosen grid from an infinitely large set of similar grids, all of which have the same grid interval but none of which have the same orientation or initial point. Presumably, every one of the infinite number of points making up the surface of the tract would be a potential sampling point if the above sampling frame is superimposed on the tract and the sampling is random. In practice this situation can be assumed to exist if an initial point and an orientation are randomly chosen. This could be done either on the ground or on a map of the tract.

Schreuder (1970) considered the situations where the sampling frame consists of a single grid so that it is finite, containing only $M$ intersections. He based his development on the following assumptions: (1) the number of clusters in which tree $i$ occurs is proportional to its basal area; (2) the number of clusters within which both trees $i$ and $k$ occur is proportional to the area of overlap of their associated plots; (3) the sampling is without replacement; (4) no slopover is present; and (5) the VBAR in the case of an empty cluster is equal to zero. The formulae he developed, in the symbology used here, are

$$\tau_v = \bar{\tau}_v = \frac{\text{BAF}_T}{M} \sum_j^M \sum_i^{N_j} \left( \frac{v_{ij}}{\text{BA}_{ij}} \right) \tag{21.115}$$

$$\bar{T}_v = \frac{\text{BAF}_T}{m} \sum_j^m \sum_i^{N_j} \left( \frac{v_{ij}}{\text{BA}_{ij}} \right) \tag{21.116}$$

$$\sigma^2_{\bar{T}_v} = \frac{\text{BAF}_T^2}{Mm} \left( \frac{M-m}{M-1} \right) \left[ \frac{M}{\text{BAF}_T} \sum_i^N \left( \frac{v_i}{\text{BA}_i} \right)^2 + \frac{2M}{\text{SA}_T} \sum_h^{N-1} \sum_{k=h+1}^{N} 0_{hk} \left( \frac{v_h}{\text{BA}_h} \right) \left( \frac{v_k}{\text{BA}_k} \right) - \frac{M}{\text{BAF}_T^2} \tau_v^2 \right] \tag{21.117}$$

and

$$s^2_{\bar{T}_v} = \frac{\text{BAF}_T}{m^2(m-1)} \left( \frac{M-m}{M} \right) \left\{ m \sum_j^m \left[ \sum_i^{N_j} \left( \frac{v_{ij}}{\text{BA}_{ij}} \right) \right]^2 - \left[ \sum_j^m \sum_i^{N_j} \left( \frac{v_{ij}}{\text{BA}_{ij}} \right) \right]^2 \right\} \tag{21.118}$$

If the assumptions are correct, which seldom occurs, the estimators $\bar{T}_v$ and $s^2_{\bar{T}_v}$ are unbiased. Since bias is usually present, $s^2_{\bar{T}_v}$ is usually an estimate of the mean squared error (MSE) rather than the variance. The bias decreases as the grid density increases and vanishes as $M$ approaches infinity, at which point Equation 21.116 reduces to Equation 21.106 and Equation 21.118 reduces to Equation 21.114.

### 21.3.2.13 Estimating the Number of Trees

When the variable of interest is the number of trees on the tract,

$$y_{ij} = g_{ij} tc_{ij} \tag{21.119}$$

where $y_{ij}$ = the number of trees on the tract, based on tree $ij$

$$g_{ij} = \text{SA}_T \sin^2 \frac{\left( \frac{\phi}{2} \right)}{\text{BA}_{ij}} = \frac{\text{BAF}_T}{\text{BA}_{ij}}$$

$tc_{ij} = 1$

When slopover is present,

$$y_{ij} = w_{ij} g_{ij} tc_{ij}$$
$$= \mathrm{BAF}_T \left( \frac{w_{ij} tc_{ij}}{\mathrm{BA}_{ij}} \right) = \mathrm{BAF}_T \left( \frac{w_{ij}}{\mathrm{BA}_{ij}} \right) \tag{21.120}$$

The estimate of the total number of trees on the tract, based on cluster $j$, is

$$\tau_{\mathrm{trees}_j} = \sum_i^{N_j} y_{ij} = \mathrm{BAF}_T \sum_i^{N} \left( \frac{w_{ij}}{\mathrm{BA}_{ij}} \right) \tag{21.121}$$

The estimated mean number of trees on the tract, based on a sample of $m$ clusters, is

$$\bar{T}_{\mathrm{Trees}} = \frac{\mathrm{BAF}_T}{m} \sum_j^m \sum_i^{N_j} (w_{ij} / \mathrm{BA}_{ij}) \;\rightarrow\; \bar{\tau}_{\mathrm{Trees}} = N \tag{21.122}$$

The expected value of $\bar{T}_{\mathrm{Trees}}$ is then

$$E(\bar{T}_{\mathrm{Trees}}) = \frac{\mathrm{BAF}_T}{m} \sum_j^m E(N_j) E(w_{ij}) E(1/\mathrm{BA}_{ij}) \tag{21.123}$$

According to Equations 21.69 and 21.70,

$$E(N_j) = \frac{1}{\mathrm{BAF}_T} \sum_i^N \mathrm{BA}_i$$

The expected value of $(1/\mathrm{BA}_{ij})$ is

$$E(1/\mathrm{BA}_{ij}) = 1 \Big/ \frac{1}{N} \sum_i^N \mathrm{BA}_i = N \Big/ \sum_i^N \mathrm{BA}_i \tag{21.124}$$

Thus,

$$E(N_j) E(1/\mathrm{BA}_{ij}) = \frac{1}{\mathrm{BAF}_T} \sum_i^N \mathrm{BA}_i \left( N \Big/ \sum_i^N \mathrm{BA}_i \right) = \frac{N}{\mathrm{BAF}_T}$$

and

$$E(\bar{T}_{\mathrm{Trees}}) = \frac{\mathrm{BAF}_T}{m} \sum_j^m \frac{N}{\mathrm{BAF}_T} = N \tag{21.125}$$

indicating that the estimator is unbiased.

Using the same approach that was used to derive the expression for the variance of the mean volume (Equation 21.113), the variance of the mean number of trees on the tract is

$$\sigma^2_{\bar{T}_{\text{Trees}}} = \frac{\text{BAF}_T^2}{m}\left[\frac{1}{BAF_T}\sum_i^N \frac{w_i}{\text{BA}_i} - \frac{N^2}{\text{BAF}_T^2} + \frac{2}{\text{SA}_T}\sum_{h=1}^{N-1}\sum_{k=h+1}^{N} 0_{hk}\left(\frac{w_h}{\text{BA}_h}\right)\left(\frac{w_k}{\text{BA}_k}\right)\right] \tag{21.126}$$

and its unbiased estimator is

$$s^2_{\bar{T}_{\text{Trees}}} = \frac{\text{BAF}_T^2}{m^2(m-1)}\left[m\sum_j^m\left(\sum_i^{N_j}\frac{w_{ij}}{\text{BA}_{ij}}\right)^2 - \left(\sum_j^m\sum_i^{N_j}\frac{w_{ij}}{\text{BA}_{ij}}\right)^2\right] \tag{21.127}$$

#### 21.3.2.14 Estimating the Total Amount of dbh

To estimate the total amount of dbh on a tract using horizontal point sampling, the procedure is essentially the same as with volume or number of trees.

$$y_{ij} = g_{ij}\text{dbh}_{ij}$$

where $y_{ij}$ = the amount of dbh on the tract, based on tree $ij$.

$$g_{ij} = \text{SA}_T \sin^2(\phi/2)/\text{BA}_{ij} = \text{BAF}_T/\text{BA}_{ij}$$

When slopover is present,

$$\begin{aligned} y_{ij} &= w_{ij}g_{ij}\text{dbh}_{ij} \\ &= \text{BAF}_T(w_{ij}\text{dbh}_{ij} / \text{BA}_{ij}) \end{aligned}$$

Then,

$$T_{\text{dbh}_j} = \sum_i^{N_j} y_{ij} = \text{BAF}_T\sum_i^{N_j}(w_{ij}\text{dbh}_{ij} / \text{BA}_{ij}) \tag{21.128}$$

and

$$\bar{T}_{\text{dbh}_j} = \frac{\text{BAF}_T}{m}\sum_j^m\sum_i^{N_j}(w_{ij}\text{dbh}_{ij} / \text{BA}_{ij}) \tag{21.129}$$

The variance of the mean total dbh is

$$\sigma^2_{\bar{T}_{\text{dbh}}} = \frac{\text{BAF}_T^2}{m}\left[\frac{1}{\text{BAF}_T}\sum_i^N\frac{w_i\text{dbh}_i^2}{\text{BA}_i} - \tau^2_{\text{dbh}} + \frac{2\text{BAF}_T}{\text{SA}_T}\sum_{h=1}^{N-1}\sum_{k=h+1}^{N}0_{hk}\left(\frac{w_h\text{dbh}_h}{\text{BA}_h}\right)\left(\frac{w_k\text{dbh}_k}{\text{BA}_k}\right)\right] \tag{21.130}$$

and its estimator is

$$s^2_{\bar{T}_{\text{dbh}}} = \frac{\text{BAF}_T^2}{m^2(m-1)}\left\{m\sum_j^m\left[\sum_i^{N_j}\left(\frac{w_{ij}\text{dbh}_{ij}}{\text{BA}_{ij}}\right)\right]^2 - \left[\sum_j^m\sum_i^{N_j}\left(\frac{w_{ij}\text{dbh}_{ij}}{\text{BA}_{ij}}\right)\right]^2\right\} \tag{21.131}$$

Both $\bar{T}_{\text{dbh}_j}$ and $s^2_{\bar{T}_{\text{dbh}}}$ are unbiased estimators. This can be confirmed using the approaches used in the cases of volume and number of trees.

#### 21.3.2.15 Estimating the Mean dbh

The mean dbh is obtained by dividing the sum of the dbhs on the specified area by the number of trees (see Equation 19.3):

$$\overline{\text{DBH}} = \tau_{\text{dbh}} / \tau_{\text{Trees}} \tag{21.132}$$

where $\overline{\text{DBH}}$ = true mean dhb
$\tau_{\text{dbh}}$ = true sum of the dbhs on the specified area
$\tau_{\text{Trees}}$ = true number of trees contributing to the sum of the dbhs

The estimate of $\overline{\text{DBH}}$, obtained from horizontal point sampling, is

$$\overline{\text{dbh}} = \bar{T}_{\text{dbh}} / \bar{T}_{\text{Trees}} \tag{21.133}$$

This is a ratio of means and, as was discussed in Section 19.3, is a biased estimator unless the correlation between $\overline{\text{dbh}}$ and $\bar{T}_{\text{Trees}}$ is equal to zero (see Equation 19.17). In other words,

$$\bar{T}_{\text{Trees}}(\overline{\text{dbh}}) \xrightarrow{\text{BIASED}} \tau_{\text{dbh}} = \overline{\text{DBH}} \tag{21.134}$$

As can be seen from Equation 19.28, the bias is a function of sample size. As the sample size increases, the bias decreases, vanishing as the sample size approaches population size. As a rule of thumb, one can expect the bias to vary approximately as the reciprocal of the sample size. Another factor affecting the bias is the degree of correlation between the two variables. If the two are perfectly correlated, $\overline{\text{dbh}}$ will be an unbiased estimator of $\overline{\text{DBH}}$ regardless of the sample size.

Because of this ambiguity with respect to bias, the mean squared error of the estimated mean dbh can only be approximated. According to Equation 19.64,

$$\text{MSE}_{\overline{\text{dbh}}} \doteq \frac{1}{\tau^2_{\text{trees}}m}\left(\frac{M-m}{M-1}\right)\left(\frac{1}{M}\right)\sum_j^M(\tau_{\text{dbh}_j} - \overline{\text{DBH}}\tau_{\text{Trees}})^2 \tag{21.135}$$

Its estimator is then

$$s^2_{\overline{\text{DBH}}} = \frac{\text{BAF}^2}{\bar{T}^2_{\text{Trees}}m}\left(\frac{M-m}{M}\right)\left(\frac{1}{m-1}\right)\sum_j^m\left(\sum_i^{N_j}\frac{w_{ij}\text{dbh}_{ij}}{\text{BA}_{ij}} - \overline{\text{dbh}}\sum_i^{N_j}\frac{w_{ij}}{\text{BA}_{ij}}\right)^2 \tag{21.136}$$

Since $M$ is infinitely large, the finite population correction reduces to unity. Consequently,

$$s^2_{\overline{\text{dbh}}} = \frac{\text{BAF}^2}{\overline{T}^2_{\text{Trees}_s} m(m-1)} \sum_j^m \left( \sum_i^{N_j} \frac{w_{ij}\text{dbh}_{ij}}{\text{BA}_{ij}} - \overline{\text{dbh}} \sum_i^{N_j} \frac{w_{ij}}{\text{BA}_{ij}} \right)^2 \quad (21.137)$$

### 21.3.2.16 Estimating the Mean Volume per Tree

Again, the estimate is a ratio of means and is usually biased.

$$\overline{\text{VOL}} = \overline{\tau}_{v_s} / \overline{\tau}_{\text{Tree}_s} \quad (21.138)$$

which is estimated by

$$\overline{\text{vol}} = \overline{T}_{v_s} / \overline{T}_{\text{Tree}_s} \quad (21.139)$$

The estimate of the mean squared error is

$$s^2_{\overline{\text{vol}}_s} = \frac{\text{BAF}^2}{\overline{T}^2_{\text{Trees}_s} m(m-1)} \sum_j^m \left( \sum_i^{N_j} \frac{w_{ij}v_{ij}}{\text{BA}_{ij}} - \overline{\text{vol}} \sum_i^{N_j} \frac{w_{ij}}{\text{BA}_{ij}} \right)^2 \quad (21.140)$$

The derivations of these expressions follow closely the derivation of those in the case of the estimation of the mean dbh.

### 21.3.2.17 Example of the Computations Associated with Horizontal Point Sampling

Assume that the BAF = 10 ft²/acre; $m = 3$; slopover and borderline trees are present; $N_1 = 5$; $N_2 = 4$; and $N_3 = 6$.

| $j$ Point | $i$ Tree | $w_{ij}$ | $\text{dbh}_{ij}$ (in.) | $\text{BA}_{ij}$ (ft²) | $\text{Vol}_{ij}$ (bd. ft) | $g_{ij}$ | $g_{ij}\text{BA}_{ij}$ (ft²/acre) | $g_{ij}\text{Vol}_{ij}$ (bd. ft/acre) | $g_{ij}(1)$ (Trees/acre) |
|---|---|---|---|---|---|---|---|---|---|
| 1 | 1 | 1 | 16 | 1.40 | 173 | 7.1429 | 10 | 1235.7 | 7.1429 |
| | 2 | 1 | 13 | 0.92 | 108 | 10.8696 | 10 | 1173.9 | 10.8696 |
| | 3 | 0.5 | 18 | 1.77 | 268 | 2.8249 | 5 | 757.1 | 2.8249 |
| | 4 | 1 | 10 | 0.55 | 46 | 18.1818 | 10 | 836.4 | 18.1818 |
| | 5 | 1 | 12 | 0.79 | 88 | 12.6582 | 10 | 1113.9 | 12.6582 |
| | | 4.5 | | | | | 45 | 5,117.0 | 51.6774 |
| 2 | 1 | 1 | 13 | 0.92 | 108 | 10.8696 | 10 | 1173.9 | 10.8696 |
| | 2 | 2 | 10 | 0.55 | 46 | 36.3636 | 20 | 1672.7 | 36.3636 |
| | 3 | 1 | 17 | 1.58 | 234 | 6.3291 | 10 | 1481.0 | 6.3291 |
| | 4 | 1 | 21 | 2.40 | 434 | 4.1667 | 10 | 1808.3 | 4.1667 |
| | | 5 | | | | | 50 | 6,135.9 | 57.7290 |
| 3 | 1 | 1 | 17 | 1.58 | 234 | 6.3291 | 10 | 1481.0 | 6.3291 |
| | 2 | 1 | 11 | 0.66 | 58 | 15.1515 | 10 | 878.8 | 15.1515 |
| | 3 | 4 | 15 | 1.23 | 151 | 32.5204 | 40 | 4910.6 | 32.5204 |
| | 4 | 1 | 12 | 0.79 | 88 | 12.6582 | 10 | 1113.9 | 12.6582 |
| | 5 | 2 | 22 | 2.64 | 480 | 7.5758 | 20 | 3636.4 | 7.5758 |
| | 6 | 0.5 | 16 | 1.40 | 173 | 3.5714 | 5 | 617.9 | 3.5714 |
| | | 9.5 | | | | | 95 | 12,638.6 | 77.8064 |

and

| $j$ Point | $i$ Tree | $g_{ij}\mathrm{dbh}_{ij}$ (dbh/acre) | $\frac{w_{ij}}{\mathrm{BA}_{ij}}$ | $\frac{w_{ij}v_{ij}}{\mathrm{BA}_{ij}}$ | $\frac{w_{ij}\mathrm{dbh}_{ij}}{\mathrm{BA}_{ij}}$ | $\frac{\mathrm{dbh}_{ij}}{\mathrm{BA}_{ij}}$ | $1/\mathrm{BA}_{ij}$ |
|---|---|---|---|---|---|---|---|
| 1 | 1 | 114.29 | 0.7143 | 123.57 | 11.4286 | 11.4286 | 0.7143 |
| | 2 | 141.30 | 1.0870 | 117.39 | 14.1304 | 14.1304 | 1.0870 |
| | 3 | 50.85 | 0.2825 | 75.71 | 5.0847 | 10.1695 | 0.5650 |
| | 4 | 181.82 | 1.8182 | 83.64 | 18.1818 | 18.1818 | 1.8182 |
| | 5 | 151.90 | 1.2658 | 111.39 | 15.1899 | 15.1899 | 1.2658 |
| | | 640.16 | 5.1678 | 511.70 | 64.0154 | 69.1002 | 5.4503 |
| 2 | 1 | 141.30 | 108.70 | 117.39 | 14.1304 | 14.1304 | 1.0870 |
| | 2 | 363.64 | 3.6364 | 167.27 | 36.3636 | 18.1818 | 1.8182 |
| | 3 | 107.59 | 0.6329 | 148.10 | 10.7595 | 10.7595 | 0.6329 |
| | 4 | 87.50 | 0.4167 | 180.83 | 8.7500 | 8.7500 | 0.4167 |
| | | 700.03 | 5.7730 | 613.59 | 70.0035 | 51.8217 | 3.9548 |
| 3 | 1 | 107.59 | 0.6329 | 148.10 | 10.7595 | 10.7595 | 0.6329 |
| | 2 | 166.67 | 1.5152 | 87.88 | 16.6667 | 16.6667 | 1.5152 |
| | 3 | 487.80 | 3.2520 | 491.06 | 48.7805 | 12.1951 | 0.8130 |
| | 4 | 151.90 | 1.2658 | 111.39 | 15.1849 | 15.1899 | 1.2658 |
| | 5 | 166.67 | 0.7576 | 363.64 | 16.6667 | 8.3333 | 0.3788 |
| | 6 | 57.14 | 0.3571 | 61.79 | 5.7143 | 11.4286 | 0.7143 |
| | | 1137.77 | 7.7806 | 1263.86 | 113.7776 | 74.5731 | 5.3200 |

To obtain basal area/acre:

$$\bar{T}_{\mathrm{BA}_s} = \frac{\mathrm{BAF}_A}{m}\sum_j^m \sum_i^{N_j} w_{ij} = \frac{10}{3}(4.5+5+9.5) = 63.33 \text{ ft}^2\text{/acre}$$

$$s^2_{\bar{T}_{\mathrm{BA}_s}} = \frac{\mathrm{BAF}_A^2}{m^2(m-1)}\left[m\sum_j^m\left(\sum_i^{N_j} w_{ij}\right)^2 - \left(\sum_j^m\sum_i^{N_j} w_{ij}\right)^2\right]$$

$$= \frac{10^2}{3^2(3-1)}[3(4.5^2+50^2+9.5^2)-(4.5+5.0+9.5)^2]$$

$$= 252.7778$$

$$s_{\bar{T}_{\mathrm{BA}_s}} = \pm\sqrt{252.7778} = \pm 15.90 \text{ ft}^2\text{/acre}$$

$$\mathrm{SE}_{\bar{T}_{\mathrm{BA}_s}} = \pm t_{0.95}(k=2)s_{\bar{T}_{\mathrm{BA}_s}} = \pm 4.303(15.8990) = \pm 68.4134 \text{ ft}^2\text{/acre}$$

To obtain volume/acre:

$$\overline{T}_{v_s} = \frac{\text{BAF}_A}{m} \sum_j^m \sum_i^{N_j} (w_{ij} v_{ij} / \text{BA}_{ij})$$

$$= \frac{10}{3}(511.70 + 613.59 + 1263.86) = 7964 \text{ bd. ft / acre}$$

$$s^2_{\overline{T}_{v_s}} = \frac{\text{BAF}_A^2}{m^2(m-1)} \left[ m \sum_j^m \left( \sum_i^{N_j} \frac{w_{ij} v_{ij}}{\text{BA}_{ij}} \right)^2 - \left( \sum_j^m \sum_i^{N_j} \frac{w_{ij} v_{ij}}{\text{BA}_{ij}} \right)^2 \right]$$

$$= \frac{10^2}{3^2(3-1)} [3(511.70^2 + 613.59^2 + 1263.86^2) - (511.70 + 613.59 + 1263.86)^2]$$

$$= 5{,}549{,}873.948$$

$$s_{\overline{T}_{v_s}} = \sqrt{5{,}549{,}873.948} = \pm 2356 \text{ bd. ft / acre}$$

$$\text{SE}_{\overline{T}_{v_s}} = \pm t_{0.95}(k=2) s_{\overline{T}_{v_s}} - \pm 4.303(2355.8) = \pm 10{,}137 \text{ bd. ft / acre}$$

To obtain the number of trees/acre:

$$\overline{T}_{\text{trees}_s} = \frac{\text{BAF}_A}{m} \sum_j^m \sum_i^{N_j} (w_{ij} / \text{BA}_{ij})$$

$$= \frac{10}{3}(5.1678 + 5.7730 + 7.7806) = 62.4 \text{ trees/acre}$$

$$s^2_{\overline{T}_{\text{Trees}_s}} = \frac{\text{BAF}_A^2}{m^2(m-1)} \left[ m \sum_j^m \left( \sum_i^{N_j} \frac{w_{ij}}{\text{BA}_{ij}} \right)^2 - \left( \sum_j^m \sum_i^{N_j} \frac{w_{ij}}{\text{BA}_{ij}} \right)^2 \right]$$

$$= \frac{10^2}{3^2(3-1)} [3(5.1678^2 + 5.7730^2 + 7.7806^2) - (5.1678 + 5.7730 + 7.7806)^2]$$

$$= 62.35249244$$

$$s_{\overline{T}_{\text{Trees}_s}} = \pm\sqrt{62.35249244} = \pm 7.90 \text{ trees/acre}$$

$$SE_{\overline{T}_{\text{Trees}_s}} = \pm t_{0.95}(k=2) s_{\overline{T}_{\text{Trees}_s}} = \pm 4.303(7.8964) = \pm 33.98 \text{ trees/acre}$$

To obtain the amount of dbh/acre:

$$\overline{T}_{\text{dbh}_s} = \frac{\text{BAF}_A}{m} \sum_j^m \sum_i^{N_j} (w_{ij} \text{dbh}_{ij} / \text{BA}_{ij})$$

$$= \frac{10}{3}(64.0154 + 70.0035 + 113.7776) = 826 \text{ in. of dbh/acre}$$

$$s^2_{\overline{T}_{\text{dbh}_s}} = \frac{\text{BAF}_A^2}{m^2(m-1)}\left[m\sum_j^m\left(\sum_i^{N_j}\frac{w_{ij}\text{dbh}_{ij}}{\text{BA}_{ij}}\right)^2 - \left(\sum_j^m\sum_i^{N_j}\frac{w_{ij}\text{dbh}_{ij}}{\text{BA}_{ij}}\right)^2\right]$$

$$= \frac{10^2}{3^2(3-1)}[3(64.0154^2 + 70.0035^2 + 113.7776^2) - (64.0154 + 70.0035 + 113.7776)^2]$$

$$= 24,601.69846$$

$$s_{\overline{T}_{\text{dbh}_s}} = \pm\sqrt{24,601.69846} = \pm 156.8 \text{ in. of dbh/acre}$$

$$SE_{\overline{T}_{\text{dbh}_s}} = \pm t_{0.95}(k=2)s_{\overline{T}_{\text{dbh}_s}} = \pm 4.303(156.849) = \pm 674.9 \text{ in. of dbh/acre}$$

To obtain the mean dbh/tree:

$$\overline{\text{dbh}} = \overline{T}_{\text{dbh}_s} / \overline{T}_{\text{Trees}_s} = 825.99 / 62.4 = 13.24 \text{ in.}$$

$$s^2_{\overline{\text{dbh}}_s} = \frac{\text{BAF}_A^2}{\overline{T}^2_{\text{Trees}_s} m(m-1)}\sum_j^m\left(\sum_i^{N_j}\frac{w_{ij}\text{dbh}_{ij}}{\text{BA}_{ij}} - \overline{\text{dbh}}\sum_i^{N_j}\frac{w_{ij}}{\text{BA}_{ij}}\right)^2$$

$$= \frac{10^2}{(62.40426667)^2(3)/(3-1)}\{[64.0154 - 13.2361(5.1678)]^2$$

$$+[70.0035 - 13.2361(5.7770)]^2 + [113.7776 - 13.2361(7.7806)]^2\}$$

$$= 0.756626554$$

$$s_{\overline{\text{dbh}}_s} = \pm\sqrt{0.756626554} = \pm 0.8698 \text{ in.}$$

A valid sampling error cannot be computed.

To obtain the mean volume/tree:

$$\overline{\text{vol}}_s = \overline{T}_{v_s} / \overline{T}_{\text{Trees}_s} = 7964 / 62.4 = 127.6 \text{ bd. ft / tree}$$

$$s^2_{\overline{\text{vol}}_s} = \frac{\text{BAF}_A^2}{\overline{T}^2_{\text{Trees}_s} m(m-1)}\sum_j^m\left(\sum_i^{N_j}\frac{w_{ij}v_{ij}}{\text{BA}_{ij}} - \overline{\text{vol}}_s\sum_i^{N_j}\frac{w_{ij}}{\text{BA}_{ij}}\right)^2$$

$$= \frac{10^2}{(62.40426667)^2(3)(3-1)}\{[511.70 - 127.6282(5.1678)]^2$$

$$+[613.59 - 127.6282(5.7770)]^2 + [1263.68 - 127.6282(7.7806)]^2\}$$

$$= 472.4605$$

$$s_{\overline{\text{vol}}_s} = \pm\sqrt{472.4605} = \pm 21.7 \text{ bd. ft/tree}$$

A valid sampling error cannot be computed.

#### 21.3.2.18 Remarks

Angle-count sampling, especially horizontal point sampling, has been widely accepted by foresters, ecologists, and other specialists whose work involves the sampling of vegetation. This acceptance has developed because the method is sound and efficient, usually costing much less than more conventional procedures.

This discussion of angle-count sampling has only skimmed the surface of the topic. The body of literature concerned with all its facets is enormous and even a digest of the material is beyond the scope of this book. Reference is made to the bibliography compiled by Labau (1967) and to those in standard forest mensuration textbooks such as that by Husch et al. (1972). All that has been considered in this book has been in the context of single-stage cluster sampling. However, the procedure can and has been used in many multistage sampling schemes. Furthermore, the arguments presented are in terms of tree-centered plots. Beers and Miller (1964) at Purdue University described the procedure in terms of point-centered plots. This approach is sound. It was not included in this chapter to avoid confusion on the part of the reader.

The instrumentation that has been developed to facilitate angle-count sampling is too extensive for coverage here. Description of these instruments and their use in the field can be found in standard forest mensuration tests.

It must be pointed out that while the angle-count procedures are theoretically sound and efficient, much depends on how well the fieldwork is performed. For example, the apex of the sweeping angle must be kept *at* the sampling point. Failure to do so can result in the inclusion in clusters of trees whose associated circles do not enclose the sampling point. The reverse can occur, where trees that should be included are excluded. Since the number of trees associated with a typical cluster is small, such an error can cause a serious deformation of the resulting estimate. As an example of this type of operator error, consider a tree with a basal area of 0.92 ft$^2$ and a volume of 108 bd. ft. If the BAF = 10 ft$^2$/acre, the tree represents BAF (volume of tree)/basal area of the tree = 10(108)/0.92 = 1174 bd. ft/acre. If the tree had a basal area of 0.55 ft$^2$ and a volume of 46 bd. ft, it would represent 10(46)/0.55 = 836 bd. ft/acre. As can be seen, small trees contribute almost as much as large trees and they cannot be treated lightly. If the work is being done on sloping ground, corrections must be made to compensate for the slope. Several methods have been suggested but probably the most foolproof would be to measure the dbh of any tree that is not clearly in or clearly out of the cluster and then determine if it is to be included in the cluster by measuring, in the *horizontal* plane, the distance from the sampling point to the center of the tree and then comparing that distance to the radius of the circle of the tree.

Since the sampling points are chosen using random sampling, the size of the samples can be determined using one of the methods appropriate for use with simple random sampling.

Finally, unlike conventional fixed-radius plot or strip cruises, one cannot compute or specify a percent cruise when using angle-count sampling. This is because the sampling units are the *points* and they are dimensionless.

## 21.4 3P SAMPLING

3P (*P*robability *P*roportional to *P*rediction) sampling was first described by Grosenbaugh (1963a) with follow-up statements in 1965, 1967a, 1967b, 1971, and 1979. It is essentially equivalent to Poisson sampling (Lahiri, 1951), but is described in the context of a pre-timber-sale inventory. It is part of a package developed by Grosenbaugh, which includes the sampling design, instrumentation suggestions, and computer programs to process the data generated by the procedure. Because of its efficiency, it has been used quite extensively by agencies such as the Bureau of Land Management and the USDA Forest Service.

### 21.4.1 The Context

Consider the situation where an estimate is needed of the volume or value of the trees to be removed from a tract in a thinning or single-tree selection operation. The trees to be removed must be marked. One could measure and otherwise evaluate every tree as it is marked, but this would be time-consuming and expensive. To keep the costs down to acceptable levels some form of sampling must be used. One could complete the marking and then carry out a conventional line–plot, strip, or horizontal point sampling operation but this would require two passes through the tract. The problem would be even worse if simple random sampling or list sampling were to be used because each tree would have to be numbered and its position mapped at the time of marking and then would have to be relocated in the second pass. To be economically desirable the marking and evaluation operations should proceed simultaneously and limit the evaluations to a sample of the marked tree. One possible solution would be to use systematic sampling where the marker would evaluate, for example, every tenth tree that is marked. Such a sampling scheme would possess the flaws inherent in all systematic sampling designs and, in addition, there is the problem that the marker-cruises would be able to choose the trees to be evaluated by simply manipulating the marking sequence. This could introduce a serious bias into the results.

### 21.4.2 An Equal Probability Solution

If the marker-cruiser is equipped with a belt pouch that contains, say, ten marbles, nine of which are white and one is black, and if the marbles are stirred up and then a blind draw made from the pouch, the probability of drawing the black marble is 1 in 10 or 0.1. If this stirring and drawing is done each time a tree is marked and the tree evaluated whenever the black marble is drawn, one would expect approximately 10% of the marked trees would be measured and evaluated. Of course, after every draw the marble would have to be returned to the pouch to keep the probability constant. A record should be kept of the number of trees marked as well as the tree evaluation data. If the stand is a mix of species, the use of value rather than volume may be preferable since species and quality can be incorporated into the value of the tree.

This approach differs fundamentally from the previously described sampling designs in that the sampling frame is developed as the sampling proceeds and that the sample size is not known until the sampling is completed. If the stand contains $N$ marked trees, the sample size could vary from zero, when no trees are chosen, to $N$, when all the marked trees are included in the sample. The sample size is then a random variable with the expected value (see Equation 10.14).

$$E(n) = Np \tag{21.141}$$

and the variance (see Equation 10.15)

$$V(n) = Npq \tag{21.142}$$

where $p = P(\text{black marble})$ and $q = 1 - p = P(\text{white marble})$.

As an example of this procedure, assume that 85 trees have been marked and cruised. The marker-cruiser's pouch contained the ten marbles previously described. The expected sample size is

$$Np = 85(0.1) = 8.5 \text{ trees}$$

Nine trees were actually chosen. The cruise data and computations are

| Tree ($i$) | Value ($y_i$) ($) | |
|---|---|---|
| 1 | 11.50 | $\bar{y} = 50.50/9 = \$5.61/\text{tree}$ |
| 2 | 4.50 | $T_Y = 85(\$5.61) = \$476.94$ |
| 3 | 2.00 | the estimated total value |
| 4 | 7.50 | of the marked trees |
| 5 | 3.00 | $s_Y^2 = 15.73611111$ |
| 6 | 12.50 | $s_Y = \pm\$3.97/\text{tree}$ |
| 7 | 2.50 | C.V. = 3.97/5.61 = 0.708 or 70.8% |
| 8 | 3.00 | $s_{\bar{y}}^2 = \left(\frac{15.73611111}{9}\right)\left(\frac{85-9}{85}\right)$ |
| 9 | 4.00 | $= 1.563326071$ |
| | 50.50 | $s_{\bar{y}} = \pm\$1.25/\text{tree}$ |

These data can be analyzed using a different approach

| Tree (i) | Weight ($w_i$) | Value ($y_i$) | Blowup | Estimate of Total Value ($T_i$) |
|---|---|---|---|---|
| 1 | 1 | 11.50 | 85/1 = 85 | 11.50(85) = 977.50 |
| 2 | 1 | 4.50 | 85/1 = 85 | 4.50(85) = 382.50 |
| 3 | 1 | 2.00 | 85/1 = 85 | 2.00(85) = 170.00 |
| 4 | 1 | 7.50 | 85/1 = 85 | 7.50(85) = 637.50 |
| 5 | 1 | 3.00 | 85/1 = 85 | 3.00(85) = 255.00 |
| 6 | 1 | 12.50 | 85/1 = 85 | 12.50(85) = 1062.50 |
| 7 | 1 | 2.50 | 85/1 = 85 | 2.50(85) = 212.50 |
| 8 | 1 | 3.00 | 85/1 = 85 | 3.00(85) = 255.00 |
| 9 | 1 | 4.00 | 85/1 = 85 | 4.00(85) = 340.00 |
| | | | | 4292.50 |

$$\bar{T}_y = 4292.50/9 = \$476.94 \text{ estimated total value of marked tree}$$

$$\bar{y} = \$476.94/85 = \$5.61/\text{tree}$$

$$s_T^2 = \frac{1}{(n-1)}\left[\sum_i^n T_i^2 - \left(\sum_i^n T_i\right)^2 \Big/ n\right] = 113{,}693.4028$$

$$s_T = \pm\$337.18$$

$$\text{C.V.} = 337.18/476.94 = 0.707 \text{ or } 70.7\%$$

$$s_{\bar{T}}^2 = \left(\frac{113{,}693.4028}{9}\right)\left(\frac{85-9}{85}\right) = 11{,}295.03086$$

$$s_{\bar{T}} = \pm\$106.28$$

Note that in the first procedure the mean of the sample tree values is multiplied or expanded by the total number of marked trees to arrive at the estimate of the total value. In the second, it is recognized that the value of each tree is an estimate of the population mean value and that it is appropriate to expand these estimates individually to obtain estimates of the total value which are then averaged.

When the pouch procedure is used, each marked tree has one chance in ten of being chosen for evaluation. Consequently, the probability of being chosen is constant across all the marked trees and they all have the same weight ($w = 1$). The sum of these weights, $\Sigma_i^N w_i$, is then 85. If a single random draw is made from the set of marked trees, every tree in the set has the same probability of being drawn, 1/85. The reciprocal of this probability, 85, is the expansion or blowup factor. Equal weights for the individual observations is a characteristic of simple random sampling with equal probability.

When this procedure is used it is possible to end up with no trees chosen for evaluation so that $n = 0$. This could be interpreted that the marked trees have no value but this is obviously absurd. If $n - 0$, the cruise should be repeated. The probability of this occurring is slight. For example, in the case of the 85 trees where the probability of drawing a white (rejection) marble is 0.9, the probability of ending up with $n = 0$ is

$$P(n = 0) = P(\text{all white}) = 0.9^{85} = 0.000129007$$

As the number of white marbles in the pouch is increased, while holding the count of black marbles constant at 1, the probability of $n$ equaling zero increases.

In the same manner, it is possible to have $n = N$, so that every marked tree must be evaluated. The likelihood of this occurring is much less than having $n = 0$ because the probability of drawing the black marble is so much less than that of drawing a white marble. In the case of the 85 trees,

$$P(n = N) = P(\text{all black}) = 0.1^{85} = 1^{-85}$$

Both the estimated mean and the estimated variance are unbiased, in spite of the fact that $n$ is not fixed.

$$E(\bar{y}) = \frac{1}{E(n)} \sum_i^{E(n)} E(y_i) = E(y_i) = \mu_Y \tag{21.143}$$

and, following the argument associated with Equation 12.12,

$$s_Y^2 = \frac{1}{n-1} \sum_i^n (y_i - \bar{y})^2 \to \sigma_Y^2 \left( \frac{N}{N-1} \right)$$

$$\begin{aligned} E(s_Y^2) &= \frac{1}{E(n)-1} E\left[ \sum_i^n (y_i - \bar{y})^2 \right] \\ &= \frac{1}{E(n)-1} \left\{ [E(n)-1]\sigma_Y^2 \left( \frac{N}{N-1} \right) \right\} \\ &= \sigma_Y^2 \left( \frac{N}{N-1} \right) \end{aligned} \tag{21.144}$$

According to Equation 12.5,

$$\sigma_{\bar{y}}^2 = \frac{\sigma_Y^2}{n}\left(\frac{N-n}{N-1}\right)$$

Its estimator is

$$s_{\bar{y}}^2 = \frac{s_Y^2}{n}\left(\frac{N-1}{N}\right)\left(\frac{N-n}{N-1}\right) = \frac{s_Y^2}{n}\left(\frac{N-n}{N}\right)$$

When $n$ is variable,

$$\begin{aligned} E(s_{\bar{y}}^2) &= \frac{E(s_Y^2)}{E(n)}\left(\frac{N-1}{N}\right)\left[\frac{N-E(n)}{N-1}\right] \\ &= \frac{1}{E(n)}\left[\frac{N-E(n)}{N-1}\right]\left[\frac{1}{E(n)-1}\right]\left\{[E(n)-1]\sigma_Y^2\left(\frac{N}{N-1}\right)\right\}\left(\frac{N-1}{N}\right) \\ &= \frac{1}{E(n)}\sigma_Y^2\left[\frac{N-E(n)}{N-1}\right] \end{aligned} \tag{21.145}$$

Thus, $s_{\bar{y}}^2$ is also an unbiased estimator.

This procedure is sound and yields unbiased estimates. It is efficient in the sense that only one pass through the stand is needed but low-value trees are just as likely to be chosen for evaluation as are high-value trees.

### 21.4.3 Adjusted 3P Sampling

One can now build on the sampling concept discussed in the previous section. One can define a set of classes or strata into which the marked trees can be assigned on the basis of *ocular* estimates of their value. For example, set up four value classes: class A with trees worth approximately $10; class B with trees worth in the neighborhood of $8; class C, $6; and class D, $3. The marker-cruiser would be equipped with four belt pouches, one for each of the strata. Within each pouch would be 40 marbles. In the pouch associated with class A there would be 4 black and 36 white marbles. In the class B pouch the mix would be 3 black and 37 white, while in the class C pouch the mix would be 2 black and 38 white, and in class D, 1 black and 39 white. Consequently, the high-value trees in class A would be four times as likely to be chosen for evaluation as the low-value trees in class D, and so forth. In this manner the problem of oversampling low-value trees is resolved.

From this point the procedure follows that described in Section 21.4.2. Upon marking a tree, the marker-cruiser makes a quick ocular estimate of the value of the tree and assigns it to a class. The marbles in the pouch associated with that class are then stirred and a blind draw made. If the marble is white, the marker-cruiser records the class of the tree and moves on. The marble must be returned to the proper pouch so that the initial probabilities are maintained. If the marble is black, the tree is evaluated and the class of the tree and its measured value are recorded. This process is continued until all the trees have been marked and subjected to the possibility of evaluation. This is 3P sampling in its purest form.

As an example, using the 85-tree set and the previously described classes, the data resulting from a 3P inventory might be as follows:

| Class | Probability of Selection | Weight ($w$) | Number of Trees Marked ($N$) | Number of Trees Selected | Total Values ($y$) | Total Weight |
|---|---|---|---|---|---|---|
| A | 4:40 | 4 | 15 | 2 | 10.50;11.50 | 4(15) = 60 |
| B | 3:40 | 3 | 20 | 2 | 6.00;8.00 | 3(20) = 60 |
| C | 2:40 | 2 | 20 | 0 | — | 2(20) = 40 |
| D | 1:40 | 1 | 30 | 1 | 2.00 | 1(30) = 30 |
| | | | $N = 85$ | $n = 5$ | | $\sum_i^N w_i = 190$ |

Its analysis is then:

| Sample Tree ($i$) | Weight ($w_i$) | Measured Tree Values ($y_i$, \$) | Blowup Factor $\left(\frac{\sum_i^N w_i}{w_i}\right)$ | Estimate of Total Value ($T_i$, \$) |
|---|---|---|---|---|
| 1 | 4 | 10.50 | $\frac{190}{4} = 47.500$ | 10.50(47.500) = 498.75 |
| 2 | 4 | 11.50 | $\frac{190}{4} = 47.500$ | 11.50(47.500) = 546.25 |
| 3 | 3 | 6.00 | $\frac{190}{3} = 63.333$ | 6.00(63.333) = 380.00 |
| 4 | 3 | 8.00 | $\frac{190}{3} = 63.333$ | 8.00(63.333) = 506.67 |
| 5 | 1 | 2.00 | $\frac{190}{1} = 190.000$ | 2.00(190.000) = 380.00 |
| | | | | $\sum^n T_i = 2311.67$ |

where

$$T_i = \left(\frac{\sum_i^n w_i}{w_i}\right) y_i \quad \text{or} \quad \sum_i^N w_i \left(\frac{y_i}{w_i}\right) \tag{21.146}$$

$$\overline{T}_A = \frac{1}{n}\sum_i^n T_i = \text{estimate of the mean total value of the marked trees}$$

$$= \frac{2371.67}{5} = \$462.33 \tag{21.147}$$

$$s_{T_i}^2 = \frac{1}{(n-1)}\sum_i^n (T_i - \bar{T})^2 = 5975.1343$$

$$s_{T_i} = \pm\$77.30 \text{ per estimate of total value}$$

$$\text{C.V.} = \frac{s_{T_i}}{\bar{T}_A} = \frac{77.30}{462.33} = 0.167 \quad \text{or} \quad 16.7\%$$

$$s_{T_A}^2 = \frac{s_{T_i}^2}{n}\left(\frac{N-n}{N}\right) = \frac{5975.1343}{5}\left(\frac{85-5}{85}\right) = 124.731162$$

$$s_{T_A} = \pm\$33.54 \text{ per estimate}$$

When the analysis is made in this manner the cruise is said to be "adjusted" because it needs no correction to compensate for the probability that $n$ will equal zero.

In most cases, pouches of marbles would not be practical. The only truly practical solution is to use random numbers. However, to set the stage for a description of how the random numbers should be set up and used, one can shift from the pouches to a deck of cards.

The deck would consist of cards bearing two kinds of labeling. First, there would be a card representing each of the classes. Each of these cards would bear an integer representing the class weight. In the case of this example, the deck would have a card with a 4, for class A, a card with a 3 for class B, a card with a 2 for class C, and a card with a 1 for class D. The number of cards bearing integers is equal to $K$. All the remaining cards would be alike. They represent rejection of the tree and are analogous to the white marbles. These cards could be blank or bear some symbol such as an asterisk, a black ball, a zero, or whatever is desired. These cards serve to control the probabilities of trees being chosen. There are $Z$ such cards. Regardless of weight, the greater the value of $Z$, the smaller the probability of any tree being chosen. $Z$, which can be any desired whole number, must equal at least 1 and probability should be greater than 1. In the deck associated with this example, $Z$ would be equal to 36, bringing the deck size up to 40.

The deck is used in the following manner. As before, each tree is visited, examined, judged, and either marked or not marked. If marked, it is ocularly assigned to a class (or given a weight). The deck is then thoroughly shuffled and a card drawn at random from the deck. If the card bears a rejection symbol, the weight of the tree is recorded and the marker-cruiser moves on. If the card bears a number that is larger than the assigned weight, the tree is rejected, the weight is recorded, and the marker-cruiser moves to the next tree. If, however, the card bears a number that is equal to or less than the assigned weight, the tree is evaluated. Table 21.1 shows how the procedure parallels that using the pouches.

### 21.4.4 The Bias in Adjusted 3P Sampling

Consider Equation 21.147, which can be written

$$T_A = \frac{\sum_i^N w_i}{n} \sum_i^n \left(\frac{y_i}{w_i}\right) \tag{21.148}$$

Let $y_i/w_i = r_i$, the ratio of the measured tree value to the assigned weight. Then,

**TABLE 21.1**
**How the Probability of a Tree Being Evaluated Varies According to Its Weight When the 40-Card Deck Is Used**

| Card | Action Taken | Probability of Measurement | Probability of Rejection |
|---|---|---|---|
| **When a tree is assigned $w_i = 1$** | | | |
| * | No measurement | | $\frac{36}{40}$ |
| $4 > w_i = 1$ | No measurement | | $\frac{1}{40}$ |
| $3 > w_i = 1$ | No measurement | | $\frac{1}{40}$ |
| $2 > w_i = 1$ | No measurement | | $\frac{1}{40}$ |
| $1 = w_i = 1$ | Measure | | $\frac{1}{40}$ |
| | | $\frac{1}{40}$ | |
| | | $\frac{1}{40}$ | $\frac{39}{40}$ |

A tree assigned a weight equal to 1 has 1 chance in 40 of being measured.

| Card | Action Taken | Probability of Measurement | Probability of Rejection |
|---|---|---|---|
| **When a tree is assigned $w_i = 2$** | | | |
| * | No measurement | | $\frac{36}{40}$ |
| $4 > w_i = 2$ | No measurement | | $\frac{1}{40}$ |
| $3 > w_i = 2$ | No measurement | | $\frac{1}{40}$ |
| $2 = w_i = 2$ | Measure | $\frac{1}{40}$ | |
| $1 < w_i = 2$ | Measure | $\frac{1}{40}$ | |
| | | $\frac{2}{40}$ | $\frac{38}{40}$ |

A tree assigned a weight equal to 2 has 2 chances in 40 of being measured.

| Card | Action Taken | Probability of Measurement | Probability of Rejection |
|---|---|---|---|
| **When a tree is assigned $w_i = 3$** | | | |
| * | No measurement | | $\frac{36}{40}$ |
| $4 > w_i = 3$ | No measurement | | $\frac{1}{40}$ |

**TABLE 21.1** *(continued)*
**How the Probability of a Tree Being Evaluated Varies According to Its Weight When the 40-Card Deck Is Used**

| Card | Action Taken | Probability of Measurement | Probability of Rejection |
|---|---|---|---|
| $3 = w_i = 3$ | Measure | $\frac{1}{40}$ | |
| $2 < w_i = 3$ | Measure | $\frac{1}{40}$ | |
| $1 < w_i = 3$ | Measure | $\frac{1}{40}$ | |
| | | $\frac{3}{40}$ | $\frac{37}{40}$ |
| A tree assigned a weight equal to 3 has 3 chances in 40 of being measured. | | | |
| **When a tree is assigned $w_i = 4$** | | | |
| * | No measurement | | $\frac{36}{40}$ |
| $4 > w_i = 4$ | Measure | $\frac{1}{40}$ | |
| $3 > w_i = 4$ | Measure | $\frac{1}{40}$ | |
| $2 = w_i = 4$ | Measure | $\frac{1}{40}$ | |
| $1 < w_i = 4$ | Measure | $\frac{1}{40}$ | |
| | | $\frac{4}{40}$ | $\frac{36}{40}$ |
| A tree assigned a weight equal to 4 has 4 chances in 40 of being measured. | | | |

$$\frac{1}{n}\sum_{i}^{n} r_i = r_M \quad \text{or} \quad \bar{r}$$

the estimated mean of ratios. Since $\Sigma_i^N w_i$ is the true total of the covariate $w_i$, it can be used to obtain an estimate of the total of the variable of interest,

$$\sum_{i}^{N} w_i(\bar{r}) = T_A \rightarrow \tau \qquad (21.149)$$

The bias associated with a mean of ratios was discussed in Section 19.3.2. According to Equations 19.31 through 19.33,

$$\begin{aligned}\text{BIAS} &= -\text{COV}(r_i, w_i)\\ &= -\frac{(\mu_Y - \mu_{\bar{r}}\mu_w)}{\mu_w}\\ &= \frac{-P_{r_i w_i}(\sigma_{r_i}\sigma_{w_i})}{\mu_w}\end{aligned}$$

where $\mu_r = \mu_{\bar{r}} = E(r_i) = \frac{1}{N}\sum_r^N r_i$

$$\mu_Y = E(Y) = \frac{1}{N}\sum_i^N Y_i$$

$$\mu_w = E(W) = \frac{1}{N}\sum_i^N W_i$$

$$P_{r_i w_i} = \frac{\text{COV}(Y_i, W_i)}{\sigma_Y \sigma_W}$$

Thus, then the ratio, $r_i$, is independent of $W_i$ (i.e., $r_i$ does not change as $W_i$ changes), $T_A$ is not biased. If a relationship does exist between $r_i$ and $W_i$, $T_A$ will be biased. In other words, if the regression of $Y_i$ on $W_i$ is not a straight line passing through the origin, the estimator is biased. As was shown in Section 19.3.2, this bias is independent of a sample size. Thus, $T_A$ is not a consistent estimator.

As an example where the 3P sample is bias-free, consider the 85-tree population. Assume that the *true* values of the 15 marked trees in class A average \$12.00/tree, the 20 trees in class B average \$8.00/tree, the 20 trees in class C average \$5.00/tree, and the 30 trees in class D average \$3.00/tree. Under these circumstances the actual total value of the marked trees would be \$530.00. Now, instead of four classes, one can establish enough classes so that there would be a weight for every possible tree value when those values are expressed in cents. This would require somewhat more than 1500 classes. If all 85 trees were correctly assigned to these classes, the sum of the weights would be

$$\sum_i^N w_i = 15(1200) + 20(800) + 20(500) + 30(300) = 53{,}000$$

If the same five trees were chosen the computations would be as follows.

| Sample Tree ($i$) | Weight ($w_i$) | Measured Tree Values ($y_i$, \$) | Blowup Factor $\left(\frac{\sum_i^N w_i}{w_i}\right)$ | Estimate of Total Value ($T_i$, \$) |
|---|---|---|---|---|
| 1 | 1050 | 10.50 | 50.476 | 530.00 |
| 2 | 1150 | 11.50 | 46.087 | 530.00 |
| 3 | 600 | 6.00 | 88.333 | 530.00 |
| 4 | 800 | 8.00 | 66.250 | 530.00 |
| 5 | 200 | 2.00 | 265.000 | 530.00 |
| | | | | $\sum_i^N T_i = 2650.00$ |

$$T_A = \frac{1}{5}(2650.00) = \$530.00 \text{ total value}$$

$$s_{T_1} = 0$$

$$s_{T_1} = \pm 0$$

$$\text{C.V.} = 0$$

Note that every estimate of the total value is exactly correct. There is no variability. This result occurred because there was a perfect match between the $w_i$ and $y_i$ values. The population regression of $Y_i$ vs. $W_i$ would be a straight line through the origin. Thus, there would be no bias. There also would be zero variance.

Now, consider the situation where the weights are based on dollars instead of cents and that each tree is assigned its appropriate weight. There would be approximately 15 classes and

$$\sum_i^N w_i = 15(12) + 20(8) + 20(5) + 30(3) = 530$$

Then,

| Sample Tree ($i$) | Weight ($w_i$) | Measured Tree Values ($y_i$, \$) | Blowup Factor $\left(\frac{\sum_i^N w_i}{w_i}\right)$ | Estimate of Total Value ($T_i$, \$) |
|---|---|---|---|---|
| 1 | 11 | 10.50 | 48.182 | 505.91 |
| 2 | 11 | 11.50 | 48.182 | 554.09 |
| 3 | 6 | 6.00 | 88.333 | 530.00 |
| 4 | 8 | 8.00 | 66.250 | 530.00 |
| 5 | 2 | 2.00 | 265.000 | 530.00 |
| | | | | 2650.00 |

$$\bar{T} = \frac{2650.00}{5} = \$530.00 \text{ total value}$$

$$s^2_{T_i} = 290.1641$$

$$s_{T_i} = \pm\$17.03 \text{ per estimate of total value}$$

$$\text{C.V.} = 0.032 \quad \text{or} \quad 3.2\%$$

In this case, the correlation between $Y_i$ and $W_i$ is not perfect (see trees 1 and 2), causing the $T_i$ value to vary. Consequently, the variance of the estimates and the coefficient of variations are not equal to zero. *The greater the agreement between $Y_i$ and $W_i$, the greater the precision of the estimate.* It also should be noted that the C.V. of the $y_i/w_i$ ratios is exactly equal to that of the inventory.

*Ratio*

$$\frac{10.50}{11} = 0.9545$$

$$\frac{11.50}{11} = 1.0455$$

$$\frac{6.00}{6} = 1.0000$$

$$\frac{8.00}{8} = 1.0000$$

$$\frac{2.00}{2} = \frac{1.0000}{5.0000}$$

$$\bar{r} = \frac{5.0000}{5} = 1.0000$$

$$s_r^2 = 0.001033$$

$$s_r = \pm 0.0321$$

$$\text{C.V.} = 0.0321 \quad \text{or} \quad 3.21\%$$

The effect of this lack of perfect correlation on bias depends on the situation. If it causes the population regression of $Y_i$ on $W_i$ to show curvature or to have an intercept other than at the origin, the result would be bias. However, there is no certainty that this would occur.

Faulty inventory design can introduce bias into the estimates. For example, if a maximum weight is specified and trees that should be assigned greater weights are lumped into the class associated with the maximum weight, a bias will be generated. This can be seen if it is assumed that the 85-tree set used earlier contains two additional marked trees worth $20 and $25, respectively, the weights are based on tree values expressed in cents, and the maximum weight is set at 1200. The total value of the marked trees would then be $530 + $20 + $25 = $575. Then, as can be seen below, the bias would be a negative $9.86.

| Class | Mean Weight ($w_i$) | Class Size ($N_i$) | Mean Measured Value ($\bar{y}_i$, $) | $N_i(w_i)$ | Blowup | ($T_i$, $) | $N_i(T_i)$ |
|---|---|---|---|---|---|---|---|
| A | 1200 | 15 | 12.00 | 18,000 | 46.167 | 554.00 | 8310.00 |
| B | 800 | 20 | 8.00 | 16,000 | 69.250 | 554.00 | 11,080.00 |
| C | 500 | 20 | 5.00 | 10,000 | 110,800 | 554.00 | 11,080.00 |
| D | 300 | 30 | 3.00 | 9000 | 184.667 | 554.00 | 16,620.00 |
| A′ | 1200 | 1 | 20.00 | 1200 | 46.167 | 923.34 | 923.34 |
| A″ | 1200 | 1 | 25.00 | 1200 | 46.167 | 1154.17 | 1154.17 |
| | | 87 | $\sum^{N} w_i$ | = 55,400 | | | 49,167.51 |

$$T_A = \frac{49{,}167.51}{87} = \$565.14 \text{ estimated total value}$$

– 575.00 true value
–$ 9.86 BIAS

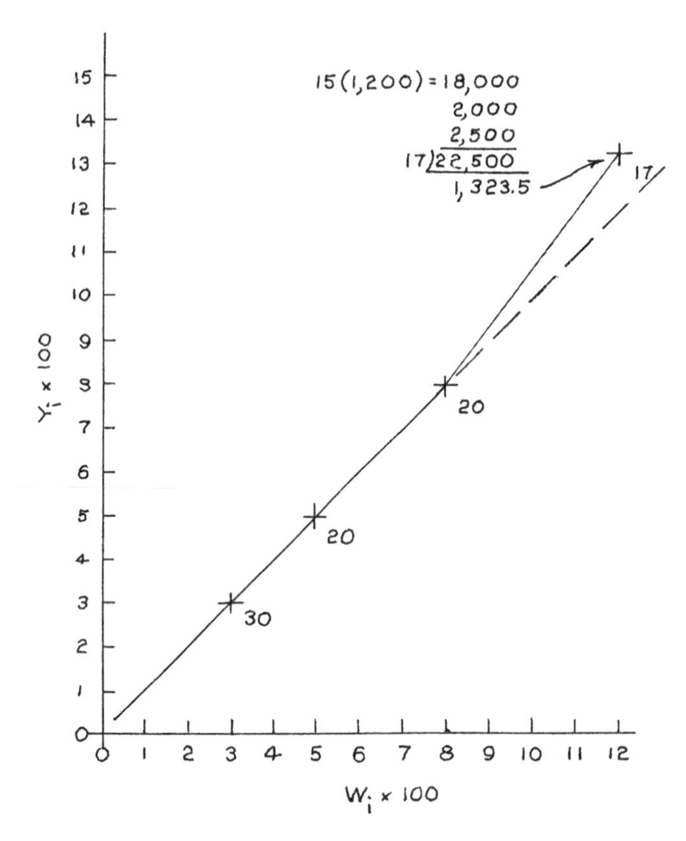


**FIGURE 21.15** Plot of $Y_i$ vs. $W_i$ when the class associated with $W_i$ = 12 is open-ended. The regression is not a straight line, indicating that bias is present.

As can be seen in Figure 21.15, the regression of $Y_i$ on $W_i$ is not straight, which further indicates that the mean of ratio is biased.

Open-ended classes can be avoided in two ways. First, one can eliminate the weight cap so that each tree can be assigned its appropriate weight. Second, one can specify that *all* trees with weights *above the cap* be measured. Then,

$$T_A = \frac{\sum_i^N w_i}{n} \sum_i^n \left(\frac{y_i}{w_i}\right) + \text{sum of the values of trees sure to be measured} \tag{21.150}$$

If the marker-cruiser is biased and consistently assigns weights that are too low or too high, the estimate may or may not be biased. If the marker-cruiser's bias can be described as a ratio the resulting 3P estimate will be unbiased. The cruiser's bias would only affect the slope of the regression line. No curvature would be introduced and the intercept would remain at the origin. The basic blowup or expansion factor is $\Sigma_i^n w_i / w_i$. If the bias can be described as a ratio, the blowup factor would be

$$\frac{\sum_i^n (w_i)(\text{BIAS \%})}{w_i(\text{BIAS \%})} = \frac{\sum_i^N w_i}{w_i}$$

The bias terms cancel out, leaving the bias blowup factor unchanged. As an example, assume that the marker-cruiser of the 85-tree inventory consistently assigned weights which were 10% too high. Then,

| Class | Mean True Weight ($w_i$) | Mean Biased Weight | Class Size ($N_i$) | Mean Measured Value ($\bar{y}_i$, \$) | Blowup | ($T_i$, \$) | $N_i(T_i)$ |
|---|---|---|---|---|---|---|---|
| A | 12 | 13.2 | 15 | 12.00 | 44.167 | 530.00 | 7950.00 |
| B | 8 | 8.8 | 20 | 8.00 | 66.250 | 530.00 | 10,600.00 |
| C | 5 | 5.5 | 20 | 5.00 | 106.000 | 530.00 | 10,600.00 |
| D | 3 | 3.3 | 30 | 3.00 | 176.667 | 530.00 | 15,900.00 |
| | | | 85 | | | | 45,050.00 |

Biased weight: 12 + 12(0.1) = 13.2, etc., which are averages

$$\sum_i^N w_i = 15(13.2) + 20(8.8) + 20(5.5) + 30(3.3) = 583$$

$$T_A = \frac{45,050.00}{85} = \$530.00 \text{ estimated total value}$$

$$- \underline{530.00} \text{ true value}$$

$$\$ \ 0.00 \text{ BIAS}$$

If, on the other hand, the marker-cruiser assigns weights that are consistently a constant number of classes too low or too high, the resulting estimate will be biased. In such a case the blowup factor would be

$$\frac{\sum_i^N (w_i + \text{BIAS})}{w_i + \text{BIAS}}$$

This type of bias cannot be removed. It causes the intercept of the regression $Y_i$ on $W_i$ to move away from the origin. Again using the 85-tree set, assume that the marker-cruiser assigned weights that are one class too high. Then,

| Class | Mean True Weight ($w_i$) | Mean Biased Weight | Class Size ($N_i$) | Mean Measured Value ($\bar{y}_i$, \$) | Blowup | ($T_i$, \$) | $N_i(T_i)$ |
|---|---|---|---|---|---|---|---|
| A | 12 | 13 | 15 | 12.00 | 47.308 | 567.70 | 8515.44 |
| B | 8 | 9 | 20 | 8.00 | 68.333 | 546.66 | 10,933.28 |
| C | 5 | 6 | 20 | 5.00 | 102.500 | 512.50 | 10,250.00 |
| D | 3 | 4 | 30 | 3.00 | 153.750 | 461.25 | 13,837.50 |
| | | | 85 | | | | 43,536.22 |

Biased weight: 12 + 1 = 13, etc., which are averages

$$\sum_i^N w_i = 15(13) + 20(9) + 20(6) + 30(4) = 615$$

$$T_A = \frac{47,536.22}{85} = \$512.19 \text{ estimated total value}$$

$$\begin{array}{rl} - \underline{530.00} & \text{true value} \\ -\$\ 17.81 & \text{BIAS} \end{array}$$

Most cruiser biases are not this clear-cut. Usually they can be characterized as tendencies to be high or low. Mathematically they are akin to the additive type with the magnitudes of the biases relative to the correct weights varying in an essentially random manner. They do not tend to cancel out.

### 21.4.5 Expected Sample Size

Before discussing the variances associated with 3P sampling it will be necessary to consider the expected sample size. As was mentioned earlier, the probability of tree $i$ being chosen for evaluation is equal to the ratio $W_i/K + Z)$. Then, if $\alpha$ is an indicator variable equaling 1 when a tree is chosen and 0 when it is rejected,

$$E(\alpha_i) = \frac{W_i}{(K+Z)} = P(\alpha_i = 1) \tag{21.151}$$

The expected value of $n$ is then

$$\begin{aligned} E(n) &= E\left(\sum_i^N \alpha_i\right) = \sum_i^N E(\alpha_i) = \sum_i^N [P(\alpha_i = 1)] = \sum_i^N \left[\frac{W_i}{(K+Z)}\right] \\ &= \frac{w_i}{(K+Z)} + \frac{w_2}{(K+Z)} + \cdots + \frac{w_N}{(K+Z)} \end{aligned} \tag{21.152}$$

For example, using the 85-tree set with four classes and $K + Z = 40$,

$$\begin{aligned} E(n) &= 15\left(\frac{4}{40}\right) + 20\left(\frac{3}{40}\right) + 20\left(\frac{2}{40}\right) + 30\left(\frac{1}{40}\right) \\ &= 1.50 + 1.50 + 1.00 + 0.75 = 4.75 \end{aligned}$$

The variance of $n$ is also a function of $\alpha$.

$$V(n) = V\left(\sum_i^N \alpha_i\right) = \sum_i^N V(\alpha_i)$$

where

$$V(\alpha_i) = \frac{1}{N}\left[\sum_i^N \alpha_i^2 - \frac{\left(\sum_i^N \alpha_i\right)^2}{N}\right]$$

Since $\alpha_i^2 = \alpha_i$,

$$V(\alpha_i) = \frac{1}{N}\left(\sum_i^N \alpha_i - \frac{\sum_i^N \alpha_i \sum_i^N \alpha_i}{N}\right)$$

$$= \frac{1}{N}\sum_i^N \alpha_i\left(1 - \frac{\sum_i^N \alpha_i}{N}\right) \tag{21.153}$$

$$= E(\alpha_i)[1 - E(\alpha_i)]$$

$$= E(\alpha_i) - [E(\alpha_i)]^2$$

Then

$$V(n) = \sigma_n^2 = \sum_i^N \{E(\alpha_i) - [E(\alpha_i)]^2\}$$

$$= \sum_i^N \left(\frac{W_i}{K+Z}\right) - \sum_i^N \left(\frac{W_i}{K+Z}\right)^2 \tag{21.154}$$

Continuing the example:

| Class ($i$) | Weight ($w_i$) | Class Size ($N_i$) | $N_i w_i$ | $\frac{w_i}{(K+Z)}$ | $\left[\frac{w_i}{(K+Z)}\right]^2$ | $N_i\left[\frac{w_i}{(K+Z)}\right]^2$ |
|---|---|---|---|---|---|---|
| A | 4 | 15 | 60 | 0.1 | 0.01 | 0.15 |
| B | 3 | 20 | 60 | 0.075 | 0.005625 | 0.1125 |
| C | 2 | 20 | 40 | 0.05 | 0.0025 | 0.05 |
| D | 1 | 30 | 30 | 0.025 | 0.000625 | 0.01875 |

$$\sum_i^N N_i w_i = 190 \qquad \sum_i^N N_i\left[\frac{w_i}{(K+Z)}\right]^2 = 0.33125$$

$$V(n) = \left(\frac{190}{40}\right) - 0.33125 = 4.41875$$

$$\sigma_n = \pm 2.1021$$

It can be shown that

$$V(n) = E(n) - \frac{[E(n)]^2}{N}\left[1 + \left(\frac{\sigma_W^2}{\mu_W^2}\right)\right]$$

$$= \sum_i^N \left(\frac{W_i}{K+Z}\right) - \left(\frac{1}{N}\right)\left[\sum_i^N \left(\frac{W_i}{K+Z}\right)\right]^2 (1 + \mathrm{CV}_W^2) \tag{21.155}$$

This indicates that as $CV_W$ increases the second-term increases and the variance of $n$ decreases. It also indicates that if $[E(n)]^2$ is small relative to $N$, the second term will approach zero and $V(n)$ will approach $E(n)$.

Obviously, the expected value and variance of $n$ cannot be determined in the absence of the complete set of weights, which would be available *after* an inventory. However, both can be estimated as is described in Section 21.4.8.

If the set of marked trees is small, the probability of $n = 0$ can be appreciable. If the inventory specifications state that the set be reinventoried if $n = 0$, the expected sample size must be modified to recognize this requirement. According to Grosenbaugh the modified expected sample size is

$$\text{ENZSN} = \frac{E(n)}{p(n \geq 1)} \tag{21.156}$$

Grosenbaugh used very small sets in his examples and it was essential that he use ENZSN in place of $E(n)$. Consequently, the correction appears in his formulae. Usually, however, $P(n \geq 1)$ is so close to unity that the correction can be ignored.

### 21.4.6 Probability of $n = 0$

Since there is always a possibility that $n$ will equal 0 when using 3P sampling, it often is of interest to determine that probability. This must be done class by class and then the class results are combined. The probability of tree $i$ being selected for evaluation is

$$P(\text{tree } i) = \frac{W_i}{(K + Z)}$$

It follows that the probability of tree $i$ not being selected is

$$P(\widetilde{\text{tree}}\ i) = 1 - P(\text{tree } i)$$

According to the multiplication law (see Equation 7.21)

$$\begin{aligned} P(n = 0) &= \prod_i^N P(\widetilde{\text{tree}}\ i) \\ &= \prod_i^N \left[1 - \frac{W_i}{(K + Z)}\right] \\ &= \prod_i^N \left[\frac{(K + Z) - W_i}{(K + Z)}\right] \end{aligned} \tag{21.157}$$

In the case of the 85-tree set when $(K + Z) = 40$

$$P(n = 0) = \left(\frac{40 - 4}{40}\right)^{15}\left(\frac{40 - 3}{40}\right)^{20}\left(\frac{40 - 2}{40}\right)^{20}\left(\frac{40 - 1}{40}\right)^{30} = 0.007262$$

### 21.4.7 Variance of the Adjusted 3P Estimate

According to Equation 19.84, the estimated variance of the mean total obtained using a mean of ratios is

$$s_{\bar{T}_A}^2 = N^2 \mu_\alpha^2 s_{\bar{r}}^2$$

From Equation 19.84,

$$s_{\bar{r}}^2 = \frac{s_{r_i}^2}{n}\left(\frac{N-n}{N}\right)$$

and from Equation 19.83,

$$s_{r_i}^2 = \frac{1}{(n-1)}\sum_i^n (r_i - \bar{r})^2$$

Then, letting $\Sigma_i^N w_i / N = \mu_X$,

$$s_{\bar{T}_A}^2 = N^2 \left(\frac{\sum_i^N w_i}{N}\right)^2 \frac{1}{n(n-1)} \sum_i^n (r_i - \bar{r})^2 \left(\frac{N-n}{N}\right)$$

Since $n$ is usually very small compared with $N$, the finite population correction is usually ignored. Then,

$$s_{\bar{T}_A}^2 = \frac{\left(\sum_i^N w_i\right)}{n(n-1)} \sum_i^n (r_i - \bar{r})^2 \rightarrow V_{\bar{T}_A} \tag{21.158}$$

which is the first of two expressions suggested by Grosenbaugh (1963b). The second (Grosenbaugh, 1974) is

$$s_{\bar{T}_{AE}}^2 = \frac{\left(\sum_i^N w_i\right)^2}{E(n)(n-1)} \sum_i^n (r_i - \bar{r})^2 \rightarrow V_{\bar{T}_{AE}} \tag{21.159}$$

where

$$E(n) = \frac{\sum_i^N w_i}{(K+Z)}$$

In both cases the estimated variance of the individual ratios ($r_i = y_i/w_i$) is expanded to an estimate of the population variance of the mean total. The two expansion terms differ. In the case of $s^2_{\bar{T}_A}$ it is $(\Sigma_i^N w_i)^2 / n$ while with $s^2_{\bar{T}_{AE}}$ it is $(\Sigma_i^N w_i)^2 / E(n)$. Since $n$ is a random variable and $E(n)$ is constant, the mean squared error associated with $s^2_{\bar{T}_A}$ will be greater than that of $s^2_{\bar{T}_{AE}}$. This is borne out in the results of repeated simulations (Schreuder et al., 1968; 1971).

Grosenbaugh (1974) suggested that the expression for $s^2_{\bar{T}_{AE}}$ be modified to compensate for the possibility of $n$ equaling zero. He recommended that $s^2_{\bar{T}_{AE}}$ be multiplied by

$$[1 - P(n = 0)]\left[1 - \left(\frac{E(n)}{N}\right)\right]$$

Van Deusen (1987) stated that simulation studies indicate that the correction should not be used. Typically its magnitude is close to unity. However, if the sample includes a large proportion of the population the correction tends to create a negative bias.

As an example of the use of Equation 21.158, one can use the 85-tree set, the original four classes, weights in terms of whole dollars, and the original five-tree sample. The sum of the weights equals 190.

| **Sample Tree** ($i$) | $w_i$ | $y_i$, $ | $r_i = \frac{y_i}{w_i}$ | $r_i - \bar{r}$ | $(r_i - \bar{r})^2$ |
|---|---|---|---|---|---|
| 1 | 4 | 10.50 | 2.625 | 0.1916 | 0.03671056 |
| 2 | 4 | 11.50 | 2.875 | 0.4416 | 0.19501056 |
| 3 | 3 | 6.00 | 2.000 | –0.4334 | 0.18783556 |
| 4 | 3 | 8.00 | 2.667 | 0.2336 | 0.05456896 |
| 5 | 1 | 2.00 | 2.000 | –0.4334 | 0.18783556 |
| | | | 12.167 | | 0.66196120 |

$$\bar{r} = \frac{12.167}{5} = 2.4334$$

$$T_A = \sum_i^N w_i(\bar{r}) = 190(2.4334) = \$462.33 \text{ total value}$$

$$s^2_{T_A} = \frac{(190)^2}{(5-1)}(0.66196120) = 5974.19983$$

$$s^2_{T_A} = \pm\$77.29 \text{ per estimate}$$

$$s^2_{\bar{T}_A} = \frac{s^2_T}{n} = \frac{5974.19983}{5} = 1194.839966$$

$$s_{\bar{T}_A} = \pm\$34.57 \text{ per estimate}$$

Within rounding errors this agrees with the results in the initial 3P example. The two computation procedures are mathematically equivalent.

### 21.4.8 Control of the Probability of Selection

The rate of sampling is dependent on $K$, the numbers of classes or weight values, and $Z$, the number of rejection cards in the deck. If $K$ is held constant and $Z$ is increased, the likelihood of drawing a rejection card increases. However, the magnitude of $Z$ does not influence the relative probability of choosing a tree of a given class compared with the other classes. For example, in terms of the original four classes, regardless of how large $Z$ is, the trees in class A would be four times as likely to be selected for measurement as the trees in class D. In essence, $Z$ acts as a dilutant.

According to Equation 21.152, the expected sample size is

$$E(n) = \frac{\sum_{i}^{N} W_i}{(K + Z)}$$

In any given situation, the sum of weights would not be known prior to the marking. Consequently, an estimate of its magnitude would have to be obtained from a light presample. For this purpose a set of fixed-radius plots or horizontal point sampling points would be distributed over the area and the trees thereon marked using the same specifications as would be used in the marking proper. Each of the marked trees would be assigned a weight. There would be a sum of such values on each plot. These sums would be averaged and put on a per unit land area basis (acre or hectare). This per land unit value would be multiplied by the number of such units in the tract to arrive at the estimate of $\Sigma_i^N w_i$ needed in the calculations.

$K$ would be specified by the inventory designer so that the desired range in weights would be available, preferably without any open classes.

$Z$ would then control $E(n)$. For example, using the original four classes and weights in terms of whole dollars: $K = 4$; $Z = 36$; $\Sigma_i^N w_i = 190$. Then,

$$E(n) = \frac{190}{(4 + 36)} = 4.75 \quad \text{or} \quad 5 \text{ trees}$$

Had $Z = 96$,

$$E(n) = \frac{190}{(4 + 96)} = 1.90 \quad \text{or} \quad 2 \text{ trees}$$

and if $Z$ had equaled 6,

$$E(n) = \frac{190}{(4 + 6)} = 19 \text{ trees}$$

Because of the problem of bias, valid sampling errors cannot be computed. However, the method of arriving at a sample size sufficient to meet an allowable sampling error can be used to compute a magnitude for $Z$. According to Equation 13.109,

$$n_{\text{Needed}} = \frac{t_C^2(k)s_{T_A}^2}{\text{AE}_{T_A}^2}$$

which can be written

$$n_{\text{Needed}} = \frac{t_C^2(k)(CV_{T_A}^2)}{\text{AE}_{T_A}^2} \tag{21.160}$$

where $\text{CV}_{T_A}$ is the coefficient of variation of the $T_A$ values, $k$ is the number of degrees of freedom, and $\text{AE}_A$ is the allowable sampling error as a proportion of $T_A$. These values would be estimated using the data from the presample. Once computed, the $n_{\text{Needed}}$ would be substituted for $n$ in Equation 21.152 and the equation solved for $Z$.

As an example, assume that an inventory is being designed with 12 classes so that $K = 12$. Further assume that the total number of trees to be marked has been estimated to be 85 and that $\Sigma_i^N w_i$ has been estimated to equal 530. Under these circumstances, $n_{\text{Needed}}$ is found to be equal to 13 trees, using iteration to reconcile $n_{\text{Needed}}$ and the appropriate value of $t$. Then $n_{\text{Needed}}$ is substituted for $E(n)$ in Equation 21.152 and the equation solved for $Z$.

$$n_{\text{Needed}} = \frac{\sum_i^N w_i}{(K+Z)}$$

$$Z = \frac{\sum_i^N w_i - n_{\text{Needed}}K}{n_{\text{Needed}}}$$

$$= \frac{[530 - Z(12)]}{13} = 28.77 \quad \text{or} \quad 29$$

The deck would then contain 12 numbered cards and 29 rejection cards.

There are two constraints that must be observed when designing a 3P inventory. They are

1. $E(n) * (\text{largest } w_i) < \Sigma_i^N w_i$
2. $\left(\frac{Z}{K}\right)^2 > \left(\frac{4}{[E(n)]}\right) - \left(\frac{4}{N}\right)$

Both were met in the example

1. $13(12) = 156 < 530$
2. $\left(\frac{29}{12}\right)^2 = 5.84 > \left(\frac{4}{12}\right) - \left(\frac{4}{85}\right) = 0.2606$

### 21.4.9 Unadjusted 3P Sampling

If $E(n)$ is substituted for $n$ in the demonstration in Equation 21.148, the expression becomes

$$T + \frac{\sum_i^N w_i}{E(n)} \sum_i^N \left(\frac{y_i}{w_i}\right)$$

Since

$$E(n) = \frac{\sum_i^N w_i}{(K+Z)}, \quad \frac{\sum_i^N w_i}{E(n)} = K + Z \quad \text{and} \tag{21.161}$$

$$T_u = (K+Z)\sum_i^N \left(\frac{y_i}{w_i}\right) = \sum_i^N \left(\frac{K+Z}{w_i}\right) y_i$$

If a 3P inventory yields no sample (i.e., $n = 0$), the appropriate procedure is to conduct a new inventory. If Equation 21.161 is to be used in the computations associated with the second inventory, it must be multiplied by the probability of $n \geq 1$. Because of this, Grosenbaugh refers to Equation 21.161 as the *unadjusted* estimator.

As an example of unadjusted 3P sampling, assume that the 85-tree set has been inventoried using weights based on whole dollar values, as was done in Section 21.4.4. Furthermore, assume that $\Sigma_i^N w_i = 530$, $K = 15$, and $Z = 91$, so that $(K + Z) = 106$, and that the same five trees were chosen for evaluation. Then,

| Sample Tree (i) | Weight ($w_i$) | Measured Tree Value ($y_i$, $) | Blowup Factor $\frac{(K+Z)}{w_i}$ | $\frac{y_i(K+Z)}{w_i}$ |
|---|---|---|---|---|
| 1 | 11 | 10.50 | $\frac{106}{11} = 9.6364$ | (9.6364)(10.50) = 101.182 |
| 2 | 11 | 11.50 | $\frac{106}{11} = 9.6364$ | (9.6364)(11.50) = 110.818 |
| 3 | 6 | 6.00 | $\frac{106}{6} = 17.6667$ | (17.6667)(6.00) = 106.000 |
| 4 | 8 | 8.00 | $\frac{106}{8} = 13.2500$ | (13.2500)(8.00) = 106.000 |
| 5 | 2 | 2.00 | $\frac{106}{2} = 53.0000$ | (53.0000)(2.00) = 106.000 |
| | | | | 530.000 |

$$T_u + \frac{\sum_i^N y_i(K+Z)}{w_i} = \$530.00$$

The expected value of $T_u$ can be determined if the indicator variable $\alpha$ is inserted into Equation 21.161.

$$E(T_u) = \sum_i^N E(\alpha_i)(K+Z)\left[\frac{1}{E(W_i)}\right]E(Y_i)$$

Since

$$E(\alpha_i) = \frac{W_i}{(K+Z)}; \quad E(W_i) = \frac{1}{N}\sum_i^N W_i; \quad \text{and} \quad E(Y_i) = \frac{1}{N}\sum_i^N Y_i = \frac{\tau_Y}{N}$$

then

$$E(T_u) = (K+Z)\sum_i^N \left[\left(\frac{W_i}{K+Z}\right)\left(\frac{N}{\sum_i^N W_i}\right)\frac{\tau_Y}{N}\right]$$

$$= \left(\frac{K+Z}{K+Z}\right)\left(\frac{\sum_i^N W_i}{\sum_i^N W_i}\right)\tau_Y$$

$$= \tau_Y$$

Thus, $T_u$ is unbiased.

In his description of unadjusted 3P sampling, Grosenbaugh states that, if the specifications for such an inventory include a revision that the inventory be rerun if $n = 0$, Equation 21.161 should be modified to compensate for the changes in probabilities. The probability of tree $i$ being selected for evaluation is

$$P(\text{tree } i) = \frac{W_i}{(K+Z)}$$

The unmodified blowup factor is the reciprocal of this probability

$$\text{Blowup } i = \frac{(K+Z)}{W_i}$$

The correction involves the multiplication of this blowup factor by the probability that $n$ will not be equal to zero, i.e., $P(N \geq 1)$.

$$P(N \geq 1) = 1 - P(n = 0)$$

where, from Equation 21.156,

$$P(n = 0) = \prod_i^N \left[\frac{(K+Z) - W_i}{(K+Z)}\right]$$

Then

$$
\begin{aligned}
\text{Blowup}_{\text{Mod}_i} &= \frac{(K+Z)}{W_i}[1 - P(n=0)] \\
&= \frac{(K+Z)}{W_i}\left\{1 - \prod_i^N \left[\frac{(K+Z) - W_i}{(K+Z)}\right]\right\}
\end{aligned}
\tag{21.162}
$$

and

$$
T_{u_{\text{Mod}}} = \sum_i^N \left(\frac{K+Z}{W_i}\right)[1 - P(n=0)]Y_i
$$

Using the 85-tree set, when the weights are in terms of whole dollars and $(K + Z) = 106$, as an example,

$$
\begin{aligned}
P(n=0) &= \left(\frac{106-12}{106}\right)^{15}\left(\frac{106-8}{106}\right)^{20}\left(\frac{106-5}{106}\right)^{20}\left(\frac{106-3}{106}\right)^{30} \\
&= 0.005521
\end{aligned}
$$

Then $P(n \geq 1) = 1 - 0.005521 = 0.994429$.

Applying this to the result of the previous example yields

$$
T_{u_{\text{Mod}}} = \$530.00(0.994429) = \$527.07
$$

The probability of $n \geq 1$ approaches 1 as $N$ increases. Consequently, if $N$ is large, it can be assumed that $P(n \geq 1) = 1$ and no correction to Equation 21.161 is needed.

If $N$ is large enough to permit the assumption that Equation 21.161 needs no correction, one can obtain an estimate of the total volume or value without having to keep a record of all assigned weights because the equation does not include the sum of the weights. This estimate will be much greater than that associated with the adjusted estimator.

### 21.4.10 Variance of the Unadjusted 3P Sample

According to Schreuder et al. (1968), the variances of the unadjusted 3P sample are

$$
\begin{aligned}
V(T_u) = \sigma_{T_u}^2 &= \frac{1}{N}\left[\frac{\sum_i^N T_{u_i}^2 - \left(\sum_i^N T_{u_i}\right)^2}{N}\right] \\
&= \frac{1}{N}\sum_i^N T_{u_i}^2 - \frac{1}{N^2}\left(\sum_i^N T_{u_i}\right)^2 \\
&= E(T_u^2) - \tau_Y^2 \\
&= E\left[(K+Z)\sum_i^N \frac{Y_i}{W_i}\right]^2 - \left(\sum_i^N Y_i\right)^2
\end{aligned}
$$

Introducing the indicator variable α which takes on the value 1 when tree$_i$ is chosen for evaluations and 0 when it is not so chosen.

$$V(T_u) = E\left[\frac{(K+Z)\sum_i^N \alpha_i Y_i}{W_i}\right]^2 - \left(\sum_i^N Y_i\right)^2$$

$$= E\left\{\left[\frac{(K+Z)^2 \sum_i^N \alpha_i^2 Y_i^2}{W_i^2}\right] + (K+Z)^2 2\sum_{i \neq j}^N \alpha_i \alpha_j \left(\frac{1}{W_i}\right)\left(\frac{1}{W_j}\right)\right\}$$

$$-\left[\sum_i^N y_i^2 + 2\sum_{i \neq j}^N Y_i Y_j\right]$$

Since $\alpha_i^2 = \alpha_i$, $E(\alpha_i^2) = E(\alpha_i) = W_i/(K+Z)$, and

$$V(T_u) = \frac{(K+Z)^2 \sum_i^N E(\alpha_i) Y_i^2}{W_i^2} + (K+Z)^2 2\sum_{i \neq j}^N E(\alpha_i \alpha_j) Y_i Y_j \left(\frac{1}{W_i}\right)\left(\frac{1}{W_j}\right)$$

$$-\sum_i^N Y_i^2 - 2\sum_{i \neq j}^N Y_i Y_j \tag{21.164}$$

$$= (K+Z)\sum_i^N \frac{Y_i^2}{W_i} - \sum_i^N Y_i^2$$

$$= \sum_i^N Y_i^2 \left(\frac{K+Z}{W_i}\right) - \sum_i^N Y_i^2$$

which also can be written

$$V(T_u) = \sum_i^N Y_i^2 \left[1 - \left(\frac{W_i}{K+Z}\right)\right]\left(\frac{K+Z}{W_i}\right) \tag{21.165}$$

The sample-based estimate of $V(T_u)$ is

$$s_{T_u}^2 = \sum_i^N y_i^2 \left[1 - \left(\frac{w_i}{K+Z}\right)\right]\left(\frac{K+Z}{w_i}\right)^2$$

$$= \sum_i^N y_i^2 \left[\left(\frac{K+Z}{w_i}\right)^2 - \left(\frac{K+Z}{w_i}\right)\right] \tag{21.166}$$

The expected value of $s_{T_u}^2$ is

$$E(s_{T_u}^2) = E\left\{\sum_i^N y_i^2\left[\left(\frac{K+Z}{w_i}\right)^2 - \left(\frac{K+Z}{w_i}\right)\right]\right\}$$

Introducing the indicator variable α into the expression,

$$E(s_{T_u}^2) = E\left\{\sum_i^N \alpha_i y_i\left[\left(\frac{K+Z}{w_i}\right)^2 - \left(\frac{K+Z}{w_i}\right)\right]\right\}$$

$$= \sum_i^N E(\alpha_i) y_i\left[\left(\frac{K+Z}{w_i}\right)^2 - \left(\frac{K+Z}{w_i}\right)\right]$$

Since $E(\alpha_i) = W_i/(K + Z)$,

$$E(s_{T_u}^2) = \sum_i^N \left(\frac{w_i}{K+Z}\right) y_i\left[\left(\frac{K+Z}{w_i}\right)^2 - \left(\frac{K+Z}{w_i}\right)\right]$$

$$= \sum_i^N y_i^2\left(\frac{K+Z}{w_i}\right) - \sum_i^N y_i^2$$

$$= V(T_u)$$

The estimator is unbiased.

Completing the example in Section 21.4.7, where $T_u$ = \$530.00,

| Sample Tree ($i$) | $\left[\frac{y_i(K+Z)}{w_i}\right]^2$ | $\left[\frac{y_i^2(K+Z)}{w_i}\right]$ |
|---|---|---|
| 1 | 10,237.7971 | 1062.411 |
| 2 | 12,280.6291 | 1274.407 |
| 3 | 11,236.0000 | 636.000 |
| 4 | 11,236.0000 | 848.000 |
| 5 | 11,236.0000 | 212.000 |
| | 56,226.4262 | 4032.818 |

$$s_{T_u}^2 = 56,266.4242 - 4032.818 = 52,193.6082$$

$$s_{T_u} = \pm\$228.46$$

As can be seen, the standard error of the unadjusted 3P sample is much larger than that of the equivalent adjusted 3P sample, which was ±\$34.37 (see Section 21.4.5).

### 21.4.11 Interpenetrating Samples

As has been mentioned, it is possible to carry out a 3P inventory and end up with no sample trees (i.e., $n = 0$). Inventory specifications usually require that, if no trees are selected for evaluation, a second inventory be carried out. This is expensive. One way to keep costs down is to use *interpenetrating* samples.

In this process the tract is *simultaneously* cruised two or more times. In such a case, each tree is assigned a weight in the usual manner but then, instead of randomly choosing one card from the deck, several cards are randomly drawn and the weight of the trees compared with each of the card values. *Each of the comparisons represents a separate cruise*. The data of the several cruises must be kept separate and each cruise must be analyzed by itself. The probability is high that at least one of these cruises would yield usable results.

### 21.4.12 Using Random Numbers

A deck of cards such as that used to describe 3P sampling is not a practical tool in the woods. It is better to use a set of random numbers that vary in magnitude in the same manner as the numbered cards in the deck. Diluting these random numbers are symbols that correspond to the rejection cards. Grosenbaugh developed such a program (RN3P) which was first described as "THRP" in 1965 and subsequently revised (Grosenbaugh, 1974). Figure 21.16 shows an example of the printout of this program. The printout can be cut into strips, which are joined together, end-to-end, and rolled onto spools for use in a special random number, generated initially suggested by Grosenbaugh and then developed by Mesavage (1967). This eliminates the problems associated with wet or sticky cards that cannot be drawn with these planned probabilities.

### 21.4.13 Use of Dendrometers

The 3P sampling procedure is designed to select trees to be measured or evaluated so that an efficient estimate of the total volume or value of the whole population can be made. Nowhere in the 3P procedure are any requirements stated regarding how the selected trees are to be evaluated. The evaluation could consist of an ocular estimate; it could be based on a local, standard, or form–class volume table; or it could consist of a series of measurements made along the bole using a dendrometer such as the Spiegelrelaskop (Bitterlich, 1952); the Barr & Stroud FP-15 dendrometer (Grosenbaugh, 1963; Mesavage, 1969a), which is no longer in production; the Zeiss Telemeter Teletop (Mesavage, 1969b); the Wheeler penta-prism dendrometer (Wheeler, 1962); or the McClure mirror caliper (McClure, 1969). These measurements can also be obtained directly if the sample trees are felled, as has been done when a suitable dendrometer was not available.

The dendrometer approach is preferred because 3P samples are characteristically very light, making the probability of compensating errors very small. When the dendrometer is used, or the measurements are made on the felled trees, each tree's *own* volume or value is obtained. They are not estimated using evidence from *other* trees, as would be the case if a volume table or equation is used. Thus, a source of bias is removed and a considerable component of variance is removed.

### 21.4.14 Supporting Computer Programs

3P sampling can be carried out easily and efficiently in the field but the computational phase of the inventory operation, especially if 3P is used as one of the stages in a multistage sampling operation, can be onerous. Grosenbaugh realized this early and began developing computer programs to speed up and simplify the process. These programs were refined and updated over a period of some 10 years (see Grosenbaugh 1964; 1965; 1967a and b; 1971; 1974). Grosenbaugh (1974) refers to the set as GROSENBAUGH * DNDR3P. It consists of four programs (quoting from Grosenbaugh):

```
AGE   62                               STX 3-3-73
================= ILLUSTRATION OF OUTPUT FROM RN3P 1-10-69 ==================

XMPL RANDOM INTEGERS FOR POINT-3P, POINT, AREA-POINT SAMPLES PAGE     1
L=    621489793, LIM=     500, K=      21, KZ=      24                XMPL

      7     19 -99999      4      3      4      1     12 -99999      3
      9      9      8     13     12     18     16      2      6      1
     14      7     20     14     16 -99999      1 -99999 -99999     10
      1      5 -99999     19     11     15     21     19     19     15
     10      7      5     15 -99999     14     17      4     19     10
      8     15     16 -99999     20 -99999      8 -99999      3     19
     19     17     20     14     18     21 -99999 -99999     19 -99999
     20      2      9      2     14     12 -99999     14     10     19
     17     12      1      5     21      2      9      2     12      1
      4     10     19     10     16     11 -99999      6     19     19
      3      7      2     11     11     14     10     16      3 -99999
     13      6      7 -99999 -99999     15 -99999     20     15     17
      5 -99999     12     20     21     16      3     15      8 -99999
      8      7     10 -99999     11      1      4     16      3 -99999
 -99999 -99999 -99999     12      7     18 -99999     19      8     10
     17     13     12     20 -99999      5      5     15      6      1
     11 -99999     20     12      7      4     11      7 -99999      8
      6     20     19      1 -99999      8     16     16     11     19
     21     11     17     17      8      6     13      1     16     20
      7     19      4     19     17     15     17      1     20     10
     13      8     20      7 -99999     21     12      5     20      6
     11     21      2      3      3     12     15     12 -99999     19
      8      9      2     14      1      7      4      2      7      9
      6     16 -99999     11     11     21 -99999      5      3     20
      8      2     17      2 -99999     10      7     19      3      4
     14     16     19     13      4      1 -99999     20      4      9
     20      8     14 -99999     12      2      6     18     19     18
     13      3     17     14     11      7      4      4     15     14
     17     17      1     13     15      1 -99999     15     11 -99999
 -99999     20     18     18      8     11      2 -99999     19     14
     11      3      2      3      6      8 -99999      1      4      8
 -99999     17     14     16      5      6 -99999     20      3      2
      7     12      8 -99999     15     12     10     17     17     11
     13      1      2      1      4      5     17      9     15      8
      8     21     10      1      9     14      9 -99999      1      3
      4     17     12     12     12 -99999      4 -99999     10     17
      2     18      6     12     18      1     20      4      2     20
     19      1     13      4     12     20     16      2     12 -99999
      1     16     15     21 -99999      6     18      8     15     13
     12      9     18      5      4     15     15      2      7 -99999
      7     14      8      8      2     10     17     14     18      4
      4     15     20     11     18 -99999      4     20      7     14
     19 -99999     11     12     10      4      4     13     21     21
      5     20     18      6     17     17 -99999     12      1      3
      7     21      7     15     16      3      5      5      8      2
     10      4      3 -99999     10      7      2      4 -99999     12
     19     14     21      1      8 -99999     15     21     17     12
     18     17      9      2      8      6      1     16      4      3
     10     19 -99999      1      1      2     12      7     17 -99999
     11      1     16      8      8 -99999 -99999     21 -99999      4
                                                                         4679
   XMPL   XMPL   XMPL   XMPL   XMPL   XMPL   XMPL   XMPL   XMPL   XMPL    437
```

**FIGURE 21.16** Example of printout from RN3P showing columns of random numbers. The –99999 entries are rejection symbols. (From Grosenbaugh, L. R., USDA Forest Service Research Paper SE-117, 1974.)

RN3P generates pseudo-random integer … [to be used in the tree selection process].

REX provides capability for a variety of multivariate regression, correlation, or covariance analyses. Its options allow transgeneration, transformation, rearrangement, grouping, weighing, combinatorial screening, fitting best or specific regressions, comparison of observed and predicted values, plots of residuals, etc., for as many as 50 variables, 8 of which can be dependent.

USMT converts STX control cards and data decks whose quantities are expressed in U.S. units of measure to equivalent decks with quantities expressed in SI (metric) units of measure and vice versa.

STX estimates individual tree and aggregate V-S-L by quality-defect category from various types of dendrometry applied to sample trees selected by a wide variety of sample designs. V-S-L by category can be program-converted to any related variable of interest (products, realization values, costs, weights, etc.).

Requests for information concerning the current status of these programs should be referred to the USDA Forest Service Southeastern Forest Experiment Station in Asheville, NC.

### 21.4.15 Units of Measure

In 1967, Grosenbaugh (1967), at a conference on young-growth forest management in California, recommended that forest inventory results be expressed in terms of cubic volume ($V$), surface area ($S$), and length ($L$), supplemented by measures of weight and length obtained while the wood is being hauled. He argued that from these data the seller and/or the buyer could obtain the basic quantities needed by using appropriate conversion factors. These quantities are specific and unambiguous, consistent (i.e., repeated measurements by different people would yield the same results), simple in application, and additive (i.e., capable of being linearly related to a variable of interest). Because of these characteristics, Grosenbaugh (1974) designed a version of STX to yield results in terms of V-S-L.

### 21.4.16 Grosenbaugh's Symbology

To follow the 3P literature one should be familiar with the symbol system used by Grosenbaugh. It was developed in the context of FORTRAN programming, which makes it cumbersome in some situations. Consequently, it was not used, except in a few places, in the preceding development. The basic equivalents are as follows.

| Text | Grosenbaugh |
|---|---|
| $N$ | $M$ |
| $n$ | $N$ |
| $y_i$ | $YI$ |
| $w_i$ | $KPI$ |
| $\sum^N w_i$ | $\sum^N KPI$ |
| $K$ | $K$ |
| $Z$ | $Z$ |
| $K + Z$ | $KZ$ |
| $r_i$ | $TI = \frac{YI}{KPI}$ |
| $E(n)$ | ESN |
| $\frac{E(n)}{p(n \geq 1)}$ | ENZSN |
| $p(n = 0)$ | $PO$ |

### 21.4.17 Remarks

The pure 3P sampling design used as the foundation for this discussion is the simplest and probably the least useful that might be employed. A far more useful design involves the combination of 3P sampling with horizontal point sampling in a two-stage operation, which was initially suggested by Grosenbaugh (1971) and is an option in STX. Van Hooser (1971; 1972) reported on the application of this idea in the USDA Forest Service National Forest Survey. This two-stage procedure can be incorporated in a stratified sampling scheme where the stratification is based on aerial photographic evidence. The possibilities are numerous. A discussion of all these possibilities is beyond the scope of this book. However, it should be noted that STX is capable of handling the data from such complex designs.

3P sampling can be used outside of the context of timber inventory. For example, Johnson et al. (1971) used it in a log scaling context and Wiant and Michael (1978) explored its use in wildlife and range management.

It should be noted that numerous tests of 3P sampling have been made — some in the field and some using computer simulations. A partial listing of the reports of such tests is Ek (1971); Furnival (1979); Furnival et al. (1987); Grosenbaugh (1968; 1976); Schreuder (1976; 1984; 1990); Schreuder et al. (1968; 1971; 1084; 1990; 1992); Space (1974); Van Deuson (1978); Van Hooser (1972; 1974); Wood and Schreuder (1986). A review of this literature will reveal the fact that the emergence of 3P sampling has led to considerable controversy which, in turn, has had a negative impact on its general acceptance.

# Appendix A

**TABLE A.1**
**Cumulative Binomial Probabilities[a]**

$$P = [Y \preceq y_b(n, p) = \sum_{Y=0}^{y_b} \binom{n}{y} p^y q^{n-y}$$

| | | *p* | | | | | | | | | |
|---|---|---|---|---|---|---|---|---|---|---|---|
| *n* | *y* | **0.05** | **0.10** | **0.15** | **0.20** | **0.25** | **0.30** | **0.35** | **0.40** | **0.45** | **0.50** |
| 2 | 0 | 0.9025 | 0.8100 | 0.7225 | 0.6400 | 0.5625 | 0.4900 | 0.4225 | 0.3600 | 0.3025 | 0.2500 |
| | 1 | 0.9975 | 0.9900 | 0.9775 | 0.9600 | 0.9375 | 0.9100 | 0.8755 | 0.8400 | 0.7975 | 0.7500 |
| 3 | 0 | 0.8574 | 0.7290 | 0.6141 | 0.5120 | 0.4219 | 0.3430 | 0.2746 | 0.2160 | 0.1664 | 0.1250 |
| | 1 | 0.9928 | 0.9720 | 0.9392 | 0.8960 | 0.8438 | 0.7840 | 0.7182 | 0.6480 | 0.5748 | 0.5000 |
| | 2 | 0.9999 | 0.9990 | 0.9966 | 0.9920 | 0.9844 | 0.9730 | 0.9571 | 0.9360 | 0.9089 | 0.8750 |
| 4 | 0 | 0.8145 | 0.6561 | 0.5220 | 0.4096 | 0.3164 | 0.2401 | 0.1785 | 0.1296 | 0.0915 | 0.0625 |
| | 1 | 0.9860 | 0.9477 | 0.8905 | 0.8192 | 0.7383 | 0.6517 | 0.5630 | 0.4752 | 0.3910 | 0.3125 |
| | 2 | 0.9995 | 0.9963 | 0.9880 | 0.9728 | 0.9492 | 0.9163 | 0.8735 | 0.8208 | 0.7585 | 0.6875 |
| | 3 | 1.0000 | 0.9999 | 0.9995 | 0.9984 | 0.9961 | 0.9919 | 0.9850 | 0.9744 | 0.9590 | 0.9375 |
| 5 | 0 | 0.7738 | 0.5905 | 0.4437 | 0.3277 | 0.2373 | 0.1681 | 0.1160 | 0.0778 | 0.0503 | 0.0312 |
| | 1 | 0.9774 | 0.9185 | 0.8352 | 0.7373 | 0.6328 | 0.5282 | 0.4284 | 0.3370 | 0.2562 | 0.1875 |
| | 2 | 0.9988 | 0.9914 | 0.9734 | 0.9421 | 0.8965 | 0.8369 | 0.7648 | 0.6826 | 0.5931 | 0.5000 |
| | 3 | 1.0000 | 0.9995 | 0.0078 | 0.9933 | 0.9844 | 0.9692 | 0.9460 | 0.9130 | 0.8688 | 0.8125 |
| | 4 | 1.0000 | 1.0000 | 0.9999 | 0.9997 | 0.9990 | 0.9976 | 0.9947 | 0.9898 | 0.9815 | 0.9688 |
| 6 | 0 | 0.7351 | 0.5314 | 0.3771 | 0.2621 | 0.1780 | 0.1176 | 0.0754 | 0.0467 | 0.0277 | 0.0156 |
| | 1 | 0.9672 | 0.8857 | 0.7765 | 0.6554 | 0.5339 | 0.4202 | 0.3191 | 0.2333 | 0.1636 | 0.1094 |
| | 2 | 0.9978 | 0.9842 | 0.9527 | 0.9011 | 0.8306 | 0.7443 | 0.6471 | 0.5443 | 0.4415 | 0.3438 |
| | 3 | 0.9999 | 0.9987 | 0.9941 | 0.9830 | 0.9624 | 0.9295 | 0.8826 | 0.8208 | 0.7447 | 0.6562 |
| | 4 | 1.0000 | 0.9999 | 0.9996 | 0.9984 | 0.9954 | 0.9891 | 0.9777 | 0.9590 | 0.9308 | 0.8906 |
| | 5 | 1.0000 | 1.0000 | 1.0000 | 0.9999 | 0.9998 | 0.9993 | 0.9982 | 0.9959 | 0.9917 | 0.9844 |
| 7 | 0 | 0.6983 | 0.4783 | 0.3206 | 0.2097 | 0.1335 | 0.0824 | 0.0490 | 0.0280 | 0.0152 | 0.0078 |
| | 1 | 0.9556 | 0.8503 | 0.7166 | 0.5767 | 0.4449 | 0.3294 | 0.2338 | 0.1586 | 0.1024 | 0.0625 |
| | 2 | 0.9962 | 0.9743 | 0.9262 | 0.8520 | 0.7564 | 0.6471 | 0.5323 | 0.4199 | 0.3164 | 0.2266 |
| | 3 | 0.9998 | 0.9973 | 0.9879 | 0.9667 | 0.9294 | 0.8740 | 0.8002 | 0.7102 | 0.6083 | 0.5000 |
| | 4 | 1.0000 | 0.9998 | 0.9988 | 0.9953 | 0.9871 | 0.9712 | 0.9444 | 0.9037 | 0.8471 | 0.7734 |
| | 5 | 1.0000 | 1.0000 | 0.9999 | 0.9996 | 0.9987 | 0.9962 | 0.9910 | 0.9812 | 0.9643 | 0.9375 |
| | 6 | 1.0000 | 1.0000 | 1.0000 | 1.0000 | 0.9999 | 0.9998 | 0.9994 | 0.9984 | 0.9963 | 0.9922 |
| 8 | 0 | 0.6634 | 0.4305 | 0.2725 | 0.1678 | 0.1001 | 0.0576 | 0.0319 | 0.0168 | 0.0084 | 0.0039 |
| | 1 | 0.9428 | 0.8131 | 0.6572 | 0.5033 | 0.3671 | 0.2553 | 0.1691 | 0.1064 | 0.0632 | 0.0352 |

**TABLE A.1** *(continued)*
**Cumulative Binomial Probabilities**[a]

$$P = [Y \preceq y_b(n, p) = \sum_{Y=0}^{y_b} \binom{n}{y} p^y q^{n-y}$$

| n | y | 0.05 | 0.10 | 0.15 | 0.20 | 0.25 | 0.30 | 0.35 | 0.40 | 0.45 | 0.50 |
|---|---|---|---|---|---|---|---|---|---|---|---|
| | 2 | 0.9942 | 0.9619 | 0.8948 | 0.7969 | 0.6785 | 0.5518 | 0.4278 | 0.3154 | 0.2201 | 0.1445 |
| | 3 | 0.9996 | 0.9950 | 0.9786 | 0.9437 | 0.8862 | 0.8059 | 0.7064 | 0.5941 | 0.4770 | 0.3633 |
| | 4 | 1.0000 | 0.9996 | 0.9971 | 0.9896 | 0.9727 | 0.9420 | 0.8939 | 0.8263 | 0.7396 | 0.6367 |
| | 5 | 1.0000 | 1.0000 | 0.9998 | 0.9988 | 0.9958 | 0.9887 | 0.9747 | 0.9502 | 0.9115 | 0.8555 |
| | 6 | 1.0000 | 1.0000 | 1.0000 | 0.9999 | 0.9996 | 0.9987 | 0.9964 | 0.9915 | 0.9819 | 0.9648 |
| | 7 | 1.0000 | 1.0000 | 1.0000 | 1.0000 | 1.0000 | 0.9999 | 0.9998 | 0.9993 | 0.9983 | 0.9961 |
| 9 | 0 | 0.6302 | 0.3874 | 0.2316 | 0.1342 | 0.0751 | 0.0404 | 0.0207 | 0.0101 | 0.0046 | 0.0020 |
| | 1 | 0.9288 | 0.7748 | 0.5995 | 0.4362 | 0.3003 | 0.1960 | 0.1211 | 0.0705 | 0.0385 | 0.0195 |
| | 2 | 0.9916 | 0.9470 | 0.8591 | 0.7382 | 0.6007 | 0.4628 | 0.3373 | 0.2318 | 0.1495 | 0.0898 |
| | 3 | 0.9994 | 0.9917 | 0.9661 | 0.9144 | 0.8343 | 0.7297 | 0.6089 | 0.4826 | 0.3614 | 0.2539 |
| | 4 | 1.0000 | 0.9991 | 0.9944 | 0.9804 | 0.9511 | 0.9012 | 0.8283 | 0.7334 | 0.6214 | 0.5000 |
| | 5 | 1.0000 | 0.9999 | 0.9994 | 0.9969 | 0.9900 | 0.9747 | 0.9464 | 0.9006 | 0.8342 | 0.7461 |
| | 6 | 1.0000 | 1.0000 | 1.0000 | 0.9997 | 0.9987 | 0.9957 | 0.9888 | 0.9750 | 0.9502 | 0.9102 |
| | 7 | 1.0000 | 1.0000 | 1.0000 | 1.0000 | 0.9999 | 0.9996 | 0.9986 | 0.9962 | 0.9909 | 0.9805 |
| | 8 | 1.0000 | 1.0000 | 1.0000 | 1.0000 | 1.0000 | 1.0000 | 0.9999 | 0.9997 | 0.9992 | 0.9980 |
| 10 | 0 | 0.5987 | 0.3487 | 0.1969 | 0.1074 | 0.0563 | 0.0282 | 0.0135 | 0.0060 | 0.0025 | 0.0010 |
| | 1 | 0.9139 | 0.7361 | 0.5443 | 0.3758 | 0.2440 | 0.1493 | 0.0860 | 0.0464 | 0.0232 | 0.0107 |
| | 2 | 0.9885 | 0.9298 | 0.8202 | 0.6778 | 0.5256 | 0.3828 | 0.2616 | 0.1673 | 0.0996 | 0.0547 |
| | 3 | 0.9990 | 0.9872 | 0.9500 | 0.8791 | 0.7759 | 0.6496 | 0.5138 | 0.3823 | 0.2660 | 0.1719 |
| | 4 | 0.9999 | 0.9984 | 0.9901 | 0.9672 | 0.9219 | 0.8497 | 0.7515 | 0.6331 | 0.5044 | 0.3770 |
| | 5 | 1.0000 | 0.9999 | 0.9986 | 0.9936 | 0.9803 | 0.9527 | 0.9051 | 0.8338 | 0.7384 | 0.6230 |
| | 6 | 1.0000 | 1.0000 | 0.9999 | 0.9991 | 0.9965 | 0.9894 | 0.9740 | 0.9452 | 0.8980 | 0.8281 |
| | 7 | 1.0000 | 1.0000 | 1.0000 | 0.9999 | 0.9996 | 0.9984 | 0.9952 | 0.9877 | 0.9726 | 0.9453 |
| | 8 | 1.0000 | 1.0000 | 1.0000 | 1.0000 | 1.0000 | 0.9999 | 0.9995 | 0.9983 | 0.9955 | 0.9893 |
| | 9 | 1.0000 | 1.0000 | 1.0000 | 1.0000 | 1.0000 | 1.0000 | 1.0000 | 0.9999 | 0.9997 | 0.9990 |
| 11 | 0 | 0.5688 | 0.3138 | 0.1673 | 0.0859 | 0.0422 | 0.0198 | 0.0088 | 0.0036 | 0.0014 | 0.0005 |
| | 1 | 0.8981 | 0.6974 | 0.4922 | 0.3221 | 0.1971 | 0.1130 | 0.0606 | 0.0302 | 0.0139 | 0.0059 |
| | 2 | 0.9848 | 0.9104 | 0.7788 | 0.6174 | 0.4552 | 0.3127 | 0.2001 | 0.1189 | 0.0652 | 0.0327 |
| | 3 | 0.9984 | 0.9815 | 0.9306 | 0.8389 | 0.7133 | 0.5696 | 0.4256 | 0.2963 | 0.1911 | 0.1133 |
| | 4 | 0.9999 | 0.9972 | 0.9841 | 0.9496 | 0.8854 | 0.7897 | 0.6683 | 0.5328 | 0.3971 | 0.2744 |
| | 5 | 1.0000 | 0.9997 | 0.9973 | 0.9883 | 0.9657 | 0.9218 | 0.8513 | 0.7535 | 0.6331 | 0.5000 |
| | 6 | 1.0000 | 1.0000 | 0.9997 | 0.9980 | 0.9924 | 0.9784 | 0.9499 | 0.9006 | 0.8262 | 0.7256 |
| | 7 | 1.0000 | 1.0000 | 1.0000 | 0.9998 | 0.9988 | 0.9957 | 0.9878 | 0.9707 | 0.9390 | 0.8867 |
| | 8 | 1.0000 | 1.0000 | 1.0000 | 1.0000 | 0.9999 | 0.9994 | 0.9980 | 0.9941 | 0.9852 | 0.9673 |
| | 9 | 1.0000 | 1.0000 | 1.0000 | 1.0000 | 1.0000 | 1.0000 | 0.9998 | 0.9993 | 0.9978 | 0.9941 |
| | 10 | 1.0000 | 1.0000 | 1.0000 | 1.0000 | 1.0000 | 1.0000 | 1.0000 | 1.0000 | 0.9998 | 0.9995 |
| 12 | 0 | 0.5404 | 0.2824 | 0.1422 | 0.0687 | 0.0317 | 0.0138 | 0.0057 | 0.0022 | 0.0008 | 0.0002 |
| | 1 | 0.8816 | 0.6590 | 0.4435 | 0.2749 | 0.1584 | 0.0850 | 0.0424 | 0.0196 | 0.0083 | 0.0032 |
| | 2 | 0.9804 | 0.8891 | 0.7358 | 0.5583 | 0.3907 | 0.2528 | 0.1513 | 0.0834 | 0.0421 | 0.0193 |

**TABLE A.1** *(continued)*
**Cumulative Binomial Probabilities**[a]

$$P = [Y \preceq y_b(n, p) = \sum_{Y=0}^{y_b} \binom{n}{y} p^y q^{n-y}$$

| n | y | 0.05 | 0.10 | 0.15 | 0.20 | 0.25 | 0.30 | 0.35 | 0.40 | 0.45 | 0.50 |
|---|---|---|---|---|---|---|---|---|---|---|---|
| | 3 | 0.9978 | 0.9744 | 0.9078 | 0.7946 | 0.6488 | 0.4925 | 0.3467 | 0.2253 | 0.1345 | 0.0730 |
| | 4 | 0.9998 | 0.9957 | 0.9761 | 0.9274 | 0.8424 | 0.7237 | 0.5833 | 0.4382 | 0.3044 | 0.1938 |
| | 5 | 1.0000 | 0.9995 | 0.9954 | 0.9806 | 0.9456 | 0.8822 | 0.7873 | 0.6652 | 0.5269 | 0.3872 |
| | 6 | 1.0000 | 0.9999 | 0.9993 | 0.9961 | 0.9857 | 0.9614 | 0.9154 | 0.8418 | 0.7393 | 0.6128 |
| | 7 | 1.0000 | 1.0000 | 0.9999 | 0.9994 | 0.9972 | 0.9905 | 0.9745 | 0.9427 | 0.8883 | 0.8062 |
| | 8 | 1.0000 | 1.0000 | 1.0000 | 0.9999 | 0.9996 | 0.9983 | 0.9944 | 0.9847 | 0.9644 | 0.9270 |
| | 9 | 1.0000 | 1.0000 | 1.0000 | 1.0000 | 1.0000 | 0.9998 | 0.9992 | 0.9972 | 0.9921 | 0.9807 |
| | 10 | 1.0000 | 1.0000 | 1.0000 | 1.0000 | 1.0000 | 1.0000 | 0.9999 | 0.9997 | 0.9989 | 0.9968 |
| | 11 | 1.0000 | 1.0000 | 1.0000 | 1.0000 | 1.0000 | 1.0000 | 1.0000 | 1.0000 | 0.9999 | 0.9998 |
| 13 | 0 | 0.5133 | 0.2542 | 0.1209 | 0.0550 | 0.0238 | 0.0097 | 0.0037 | 0.0013 | 0.0004 | 0.0001 |
| | 1 | 0.8646 | 0.6213 | 0.3983 | 0.2336 | 0.1267 | 0.0637 | 0.0296 | 0.0126 | 0.0049 | 0.0017 |
| | 2 | 0.9755 | 0.8661 | 0.6920 | 0.5017 | 0.3326 | 0.2025 | 0.1132 | 0.0579 | 0.0269 | 0.0112 |
| | 3 | 0.9969 | 0.9658 | 0.8820 | 0.7437 | 0.5843 | 0.4206 | 0.2783 | 0.1686 | 0.0929 | 0.0461 |
| | 4 | 0.9997 | 0.9935 | 0.9658 | 0.9009 | 0.7940 | 0.6543 | 0.5005 | 0.3530 | 0.2279 | 0.1334 |
| | 5 | 1.0000 | 0.9991 | 0.9925 | 0.9700 | 0.9198 | 0.8346 | 0.7159 | 0.5744 | 0.4268 | 0.2905 |
| | 6 | 1.0000 | 0.9999 | 0.9987 | 0.9930 | 0.9757 | 0.9376 | 0.8705 | 0.7712 | 0.6437 | 0.5000 |
| | 7 | 1.0000 | 1.0000 | 0.9998 | 0.9988 | 0.9944 | 0.9818 | 0.9538 | 0.9023 | 0.8212 | 0.7095 |
| | 8 | 1.0000 | 1.0000 | 1.0000 | 0.9998 | 0.9990 | 0.9960 | 0.9874 | 0.9679 | 0.9302 | 0.8666 |
| | 9 | 1.0000 | 1.0000 | 1.0000 | 1.0000 | 0.9999 | 0.9993 | 0.9975 | 0.9922 | 0.9797 | 0.9539 |
| | 10 | 1.0000 | 1.0000 | 1.0000 | 1.0000 | 1.0000 | 0.9999 | 0.9997 | 0.9987 | 0.9959 | 0.9888 |
| | 11 | 1.0000 | 1.0000 | 1.0000 | 1.0000 | 1.0000 | 1.0000 | 1.0000 | 0.9999 | 0.9995 | 0.9983 |
| | 12 | 1.0000 | 1.0000 | 1.0000 | 1.0000 | 1.0000 | 1.0000 | 1.0000 | 1.0000 | 1.0000 | 0.9999 |
| 14 | 0 | 0.4877 | 0.2288 | 0.1028 | 0.0440 | 0.0178 | 0.0068 | 0.0024 | 0.0008 | 0.0002 | 0.0001 |
| | 1 | 0.8470 | 0.5846 | 0.3567 | 0.1979 | 0.1010 | 0.0475 | 0.0205 | 0.0081 | 0.0029 | 0.0009 |
| | 2 | 0.9699 | 0.8416 | 0.6479 | 0.4481 | 0.2811 | 0.1608 | 0.0839 | 0.0398 | 0.0170 | 0.0065 |
| | 3 | 0.9958 | 0.9559 | 0.8535 | 0.6982 | 0.5213 | 0.3552 | 0.2205 | 0.1243 | 0.0632 | 0.0287 |
| | 4 | 0.9996 | 0.9908 | 0.9533 | 0.8702 | 0.7415 | 0.5842 | 0.4227 | 0.2793 | 0.1672 | 0.0898 |
| | 5 | 1.0000 | 0.9985 | 0.9885 | 0.9561 | 0.8883 | 0.7805 | 0.6405 | 0.4859 | 0.3373 | 0.2120 |
| | 6 | 1.0000 | 0.9998 | 0.9978 | 0.9884 | 0.9617 | 0.9067 | 0.8164 | 0.6925 | 0.5461 | 0.3953 |
| | 7 | 1.0000 | 1.0000 | 0.9997 | 0.9976 | 0.9897 | 0.9685 | 0.9247 | 0.8499 | 0.7414 | 0.6074 |
| | 8 | 1.0000 | 1.0000 | 1.0000 | 0.9996 | 0.9978 | 0.9917 | 0.9757 | 0.9417 | 0.8811 | 0.7880 |
| | 9 | 1.0000 | 1.0000 | 1.0000 | 1.0000 | 0.9997 | 0.9983 | 0.9940 | 0.9825 | 0.9574 | 0.9102 |
| | 10 | 1.0000 | 1.0000 | 1.0000 | 1.0000 | 1.0000 | 0.9989 | 0.9989 | 0.9961 | 0.9886 | 0.9713 |
| | 11 | 1.0000 | 1.0000 | 1.0000 | 1.0000 | 1.0000 | 1.0000 | 0.9999 | 0.9994 | 0.9978 | 0.9935 |
| | 12 | 1.0000 | 1.0000 | 1.0000 | 1.0000 | 1.0000 | 1.0000 | 1.0000 | 0.9999 | 0.9997 | 0.9991 |
| | 13 | 1.0000 | 1.0000 | 1.0000 | 1.0000 | 1.0000 | 1.0000 | 1.0000 | 1.0000 | 1.0000 | 0.9999 |
| 15 | 0 | 0.4633 | 0.2059 | 0.0874 | 0.0352 | 0.0134 | 0.0047 | 0.0016 | 0.0005 | 0.0001 | 0.0000 |
| | 1 | 0.8290 | 0.5490 | 0.3186 | 0.1671 | 0.0802 | 0.0353 | 0.0142 | 0.0052 | 0.0017 | 0.0005 |
| | 2 | 0.9638 | 0.8159 | 0.6042 | 0.3980 | 0.2361 | 0.1268 | 0.0617 | 0.0271 | 0.0107 | 0.0037 |

**TABLE A.1** *(continued)*
**Cumulative Binomial Probabilities**[a]

$$P = [Y \preceq y_b(n, p) = \sum_{Y=0}^{y_b} \binom{n}{y} p^y q^{n-y}$$

| n | y | 0.05 | 0.10 | 0.15 | 0.20 | 0.25 | 0.30 | 0.35 | 0.40 | 0.45 | 0.50 |
|---|---|---|---|---|---|---|---|---|---|---|---|
| | 3 | 0.9945 | 0.9444 | 0.8227 | 0.6482 | 0.4613 | 0.2969 | 0.1727 | 0.0905 | 0.0424 | 0.0176 |
| | 4 | 0.9994 | 0.9873 | 0.9383 | 0.8358 | 0.6865 | 0.5155 | 0.3519 | 0.2173 | 0.1204 | 0.0592 |
| | 5 | 0.9999 | 0.9978 | 0.9832 | 0.9389 | 0.8516 | 0.7216 | 0.5643 | 0.4032 | 0.2608 | 0.1509 |
| | 6 | 1.0000 | 0.9997 | 0.9964 | 0.9819 | 0.9434 | 0.8689 | 0.7548 | 0.6098 | 0.4522 | 0.3036 |
| | 7 | 1.0000 | 1.0000 | 0.9996 | 0.9958 | 0.9827 | 0.9500 | 0.8868 | 0.7869 | 0.6535 | 0.5000 |
| | 8 | 1.0000 | 1.0000 | 0.9999 | 0.9992 | 0.9958 | 0.9848 | 0.9578 | 0.9050 | 0.8182 | 0.6964 |
| | 9 | 1.0000 | 1.0000 | 1.0000 | 0.9999 | 0.9992 | 0.9963 | 0.9876 | 0.9662 | 0.9231 | 0.8491 |
| | 10 | 1.0000 | 1.0000 | 1.0000 | 1.0000 | 0.9999 | 0.9993 | 0.9972 | 0.9907 | 0.9745 | 0.9408 |
| | 11 | 1.0000 | 1.0000 | 1.0000 | 1.0000 | 1.0000 | 0.9999 | 0.9995 | 0.9981 | 0.9937 | 0.9824 |
| | 12 | 1.0000 | 1.0000 | 1.0000 | 1.0000 | 1.0000 | 1.0000 | 0.9999 | 0.9997 | 0.9989 | 0.9963 |
| | 13 | 1.0000 | 1.0000 | 1.0000 | 1.0000 | 1.0000 | 1.0000 | 1.0000 | 1.0000 | 0.9999 | 0.9995 |
| | 14 | 1.0000 | 1.0000 | 1.0000 | 1.0000 | 1.0000 | 1.0000 | 1.0000 | 1.0000 | 1.0000 | 1.0000 |
| 16 | 0 | 0.4401 | 0.1853 | 0.0743 | 0.0281 | 0.0100 | 0.0033 | 0.0010 | 0.0003 | 0.0001 | 0.0000 |
| | 1 | 0.8108 | 0.5147 | 0.2839 | 0.1407 | 0.0635 | 0.0261 | 0.0098 | 0.0033 | 0.0010 | 0.0003 |
| | 2 | 0.9571 | 0.7892 | 0.5614 | 0.3518 | 0.1971 | 0.0994 | 0.0451 | 0.0183 | 0.0066 | 0.0021 |
| | 3 | 0.9930 | 0.9316 | 0.7899 | 0.5981 | 0.4050 | 0.2459 | 0.1339 | 0.0651 | 0.0281 | 0.0106 |
| | 4 | 0.9991 | 0.9830 | 0.9209 | 0.7982 | 0.6302 | 0.4499 | 0.2892 | 0.1666 | 0.0853 | 0.0384 |
| | 5 | 0.9999 | 0.9967 | 0.9765 | 0.9183 | 0.8103 | 0.6598 | 0.4900 | 0.3288 | 0.1976 | 0.1051 |
| | 6 | 1.0000 | 0.9995 | 0.9944 | 0.9733 | 0.9204 | 0.8247 | 0.6881 | 0.5272 | 0.3660 | 0.2272 |
| | 7 | 1.0000 | 0.9999 | 0.9989 | 0.9930 | 0.9729 | 0.9256 | 0.8406 | 0.7161 | 0.5629 | 0.4018 |
| | 8 | 1.0000 | 1.0000 | 0.9998 | 0.9985 | 0.9925 | 0.9743 | 0.9329 | 0.8577 | 0.7441 | 0.5982 |
| | 9 | 1.0000 | 1.0000 | 1.0000 | 0.9998 | 0.9984 | 0.9929 | 0.9771 | 0.9417 | 0.8759 | 0.7728 |
| | 10 | 1.0000 | 1.0000 | 1.0000 | 1.0000 | 0.9997 | 0.9984 | 0.9938 | 0.9809 | 0.9514 | 0.8949 |
| | 11 | 1.0000 | 1.0000 | 1.0000 | 1.0000 | 1.0000 | 0.9997 | 0.9987 | 0.9951 | 0.9851 | 0.9616 |
| | 12 | 1.0000 | 1.0000 | 1.0000 | 1.0000 | 1.0000 | 1.0000 | 0.9998 | 0.9991 | 0.9965 | 0.9894 |
| | 13 | 1.0000 | 1.0000 | 1.0000 | 1.0000 | 1.0000 | 1.0000 | 1.0000 | 0.9999 | 0.9994 | 0.9979 |
| | 14 | 1.0000 | 1.0000 | 1.0000 | 1.0000 | 1.0000 | 1.0000 | 1.0000 | 1.0000 | 1.0000 | 0.9997 |
| | 15 | 1.0000 | 1.0000 | 1.0000 | 1.0000 | 1.0000 | 1.0000 | 1.0000 | 1.0000 | 1.0000 | 1.0000 |
| 17 | 0 | 0.4181 | 0.1668 | 0.0631 | 0.0225 | 0.0075 | 0.0023 | 0.0007 | 0.0002 | 0.0000 | 0.0000 |
| | 1 | 0.7922 | 0.4818 | 0.2525 | 0.1182 | 0.0501 | 0.0193 | 0.0067 | 0.0021 | 0.0006 | 0.0001 |
| | 2 | 0.9497 | 0.7618 | 0.5198 | 0.3096 | 0.1637 | 0.0774 | 0.0327 | 0.0123 | 0.0041 | 0.0012 |
| | 3 | 0.9912 | 0.9174 | 0.7556 | 0.5489 | 0.3530 | 0.2019 | 0.1028 | 0.0464 | 0.0184 | 0.0063 |
| | 4 | 0.9988 | 0.9779 | 0.9013 | 0.7582 | 0.5739 | 0.3887 | 0.2348 | 0.1260 | 0.0596 | 0.0245 |
| | 5 | 0.9999 | 0.9953 | 0.9681 | 0.8943 | 0.7653 | 0.5968 | 0.4197 | 0.2639 | 0.1471 | 0.0717 |
| | 6 | 1.0000 | 0.9992 | 0.9917 | 0.9623 | 0.8929 | 0.7752 | 0.6188 | 0.4478 | 0.2902 | 0.1662 |
| | 7 | 1.0000 | 0.9999 | 0.9983 | 0.9891 | 0.9598 | 0.8954 | 0.7872 | 0.6405 | 0.4763 | 0.3145 |
| | 8 | 1.0000 | 1.0000 | 0.9997 | 0.9974 | 0.9876 | 0.9597 | 0.9006 | 0.8011 | 0.6626 | 0.5000 |
| | 9 | 1.0000 | 1.0000 | 1.0000 | 0.9995 | 0.9969 | 0.9873 | 0.9617 | 0.9081 | 0.8166 | 0.6855 |
| | 10 | 1.0000 | 1.0000 | 1.0000 | 0.9999 | 0.9994 | 0.9968 | 0.9880 | 0.9652 | 0.9174 | 0.8338 |

**TABLE A.1** *(continued)*
**Cumulative Binomial Probabilities**[a]

$$P = [Y \preceq y_b(n,p) = \sum_{Y=0}^{y_b} \binom{n}{y} p^y q^{n-y}$$

| n | y | p = 0.05 | 0.10 | 0.15 | 0.20 | 0.25 | 0.30 | 0.35 | 0.40 | 0.45 | 0.50 |
|---|---|---|---|---|---|---|---|---|---|---|---|
| | 11 | 1.0000 | 1.0000 | 1.0000 | 1.0000 | 0.9999 | 0.9993 | 0.9970 | 0.9894 | 0.9699 | 0.9283 |
| | 12 | 1.0000 | 1.0000 | 1.0000 | 1.0000 | 1.0000 | 0.9999 | 0.9994 | 0.9975 | 0.9914 | 0.9755 |
| | 13 | 1.0000 | 1.0000 | 1.0000 | 1.0000 | 1.0000 | 1.0000 | 0.9999 | 0.9995 | 0.9981 | 0.9936 |
| | 14 | 1.0000 | 1.0000 | 1.0000 | 1.0000 | 1.0000 | 1.0000 | 1.0000 | 0.9999 | 0.9997 | 0.9988 |
| | 15 | 1.0000 | 1.0000 | 1.0000 | 1.0000 | 1.0000 | 1.0000 | 1.0000 | 1.0000 | 1.0000 | 0.9999 |
| | 16 | 1.0000 | 1.0000 | 1.0000 | 1.0000 | 1.0000 | 1.0000 | 1.0000 | 1.0000 | 1.0000 | 1.0000 |
| 18 | 0 | 0.3972 | 0.1501 | 0.0536 | 0.0180 | 0.0056 | 0.0016 | 0.0004 | 0.0001 | 0.0000 | 0.0000 |
| | 1 | 0.7735 | 0.4503 | 0.2241 | 0.0991 | 0.0395 | 0.0142 | 0.0046 | 0.0013 | 0.0003 | 0.0001 |
| | 2 | 0.9419 | 0.7338 | 0.4797 | 0.2713 | 0.1353 | 0.0600 | 0.0236 | 0.0082 | 0.0025 | 0.0007 |
| | 3 | 0.9891 | 0.9018 | 0.7202 | 0.5010 | 0.3057 | 0.1646 | 0.0783 | 0.0328 | 0.0120 | 0.0038 |
| | 4 | 0.9985 | 0.9718 | 0.8794 | 0.7164 | 0.5187 | 0.3327 | 0.1886 | 0.0942 | 0.0411 | 0.0154 |
| | 5 | 0.9998 | 0.9936 | 0.9581 | 0.8671 | 0.7175 | 0.5344 | 0.3550 | 0.2088 | 0.1077 | 0.0481 |
| | 6 | 1.0000 | 0.9988 | 0.9882 | 0.9487 | 0.8610 | 0.7217 | 0.5491 | 0.3743 | 0.2258 | 0.1189 |
| | 7 | 1.0000 | 0.9998 | 0.9973 | 0.9837 | 0.9431 | 0.8593 | 0.7283 | 0.5634 | 0.3915 | 0.2403 |
| | 8 | 1.0000 | 1.0000 | 0.9995 | 0.9957 | 0.9807 | 0.9404 | 0.8609 | 0.7368 | 0.5778 | 0.4073 |
| | 9 | 1.0000 | 1.0000 | 0.9999 | 0.9991 | 0.9946 | 0.9790 | 0.9403 | 0.8653 | 0.7473 | 0.5927 |
| | 10 | 1.0000 | 1.0000 | 1.0000 | 0.9998 | 0.9988 | 0.9939 | 0.9788 | 0.9424 | 0.8720 | 0.7597 |
| | 11 | 1.0000 | 1.0000 | 1.0000 | 1.0000 | 0.9998 | 0.9986 | 0.9938 | 0.9797 | 0.9463 | 0.8811 |
| | 12 | 1.0000 | 1.0000 | 1.0000 | 1.0000 | 1.0000 | 0.9997 | 0.9986 | 0.9942 | 0.9817 | 0.9519 |
| | 13 | 1.0000 | 1.0000 | 1.0000 | 1.0000 | 1.0000 | 1.0000 | 0.9997 | 0.9987 | 0.9951 | 0.9846 |
| | 14 | 1.0000 | 1.0000 | 1.0000 | 1.0000 | 1.0000 | 1.0000 | 1.0000 | 0.9998 | 0.9990 | 0.9962 |
| | 15 | 1.0000 | 1.0000 | 1.0000 | 1.0000 | 1.0000 | 1.0000 | 1.0000 | 1.0000 | 0.9999 | 0.9993 |
| | 16 | 1.0000 | 1.0000 | 1.0000 | 1.0000 | 1.0000 | 1.0000 | 1.0000 | 1.0000 | 1.0000 | 0.9999 |
| 19 | 0 | 0.3774 | 0.1351 | 0.0456 | 0.0144 | 0.0042 | 0.0011 | 0.0003 | 0.0001 | 0.0000 | 0.0000 |
| | 1 | 0.7547 | 0.4203 | 0.1985 | 0.0829 | 0.0310 | 0.0104 | 0.0031 | 0.0008 | 0.0002 | 0.0000 |
| | 2 | 0.9335 | 0.7054 | 0.4413 | 0.2369 | 0.1113 | 0.0462 | 0.0170 | 0.0055 | 0.0015 | 0.0004 |
| | 3 | 0.9868 | 0.8850 | 0.6841 | 0.4551 | 0.2630 | 0.1332 | 0.0591 | 0.0230 | 0.0077 | 0.0022 |
| | 4 | 0.9980 | 0.9648 | 0.8556 | 0.6733 | 0.4654 | 0.2822 | 0.1500 | 0.0696 | 0.0280 | 0.0096 |
| | 5 | 0.9998 | 0.9914 | 0.9463 | 0.8369 | 0.6678 | 0.4739 | 0.1968 | 0.1629 | 0.0777 | 0.0318 |
| | 6 | 1.0000 | 0.9983 | 0.9837 | 0.9324 | 0.8251 | 0.6655 | 0.4812 | 0.3081 | 0.1727 | 0.0835 |
| | 7 | 1.0000 | 0.9997 | 0.9959 | 0.9767 | 0.9225 | 0.8180 | 0.6656 | 0.4878 | 0.3169 | 0.1796 |
| | 8 | 1.0000 | 1.0000 | 0.9992 | 0.9933 | 0.9713 | 0.9161 | 0.8145 | 0.6675 | 0.4940 | 0.3238 |
| | 9 | 1.0000 | 1.0000 | 0.9999 | 0.9984 | 0.9911 | 0.9674 | 0.9125 | 0.8139 | 0.6710 | 0.5000 |
| | 10 | 1.0000 | 1.0000 | 1.0000 | 0.9997 | 0.9977 | 0.9895 | 0.9653 | 0.9115 | 0.8159 | 0.6762 |
| | 11 | 1.0000 | 1.0000 | 1.0000 | 1.0000 | 0.9995 | 0.9972 | 0.9886 | 0.9648 | 0.9129 | 0.8204 |
| | 12 | 1.0000 | 1.0000 | 1.0000 | 1.0000 | 0.9999 | 0.9994 | 0.9969 | 0.9884 | 0.9658 | 0.9165 |
| | 13 | 1.0000 | 1.0000 | 1.0000 | 1.0000 | 1.0000 | 0.9999 | 0.9993 | 0.9969 | 0.9891 | 0.9682 |
| | 14 | 1.0000 | 1.0000 | 1.0000 | 1.0000 | 1.0000 | 1.0000 | 0.9999 | 0.9994 | 0.9972 | 0.9904 |
| | 15 | 1.0000 | 1.0000 | 1.0000 | 1.0000 | 1.0000 | 1.0000 | 1.0000 | 0.9999 | 0.9995 | 0.9978 |

**TABLE A.1** *(continued)*
**Cumulative Binomial Probabilities**[a]

$$P = [Y \preceq y_b(n, p) = \sum_{Y=0}^{y_b} \binom{n}{y} p^y q^{n-y}$$

| n | y | 0.05 | 0.10 | 0.15 | 0.20 | 0.25 | 0.30 | 0.35 | 0.40 | 0.45 | 0.50 |
|---|---|---|---|---|---|---|---|---|---|---|---|
| | 16 | 1.0000 | 1.0000 | 1.0000 | 1.0000 | 1.0000 | 1.0000 | 1.0000 | 1.0000 | 0.9999 | 0.9996 |
| | 17 | 1.0000 | 1.0000 | 1.0000 | 1.0000 | 1.0000 | 1.0000 | 1.0000 | 1.0000 | 1.0000 | 1.0000 |
| 20 | 0 | 0.3585 | 0.1216 | 0.0388 | 0.0115 | 0.0032 | 0.0008 | 0.0002 | 0.0000 | 0.0000 | 0.0000 |
| | 1 | 0.7358 | 0.3917 | 0.1756 | 0.0692 | 0.0243 | 0.0076 | 0.0021 | 0.0005 | 0.0001 | 0.0000 |
| | 2 | 0.9245 | 0.6769 | 0.4049 | 0.2061 | 0.0913 | 0.0355 | 0.0121 | 0.0036 | 0.0009 | 0.0002 |
| | 3 | 0.9841 | 0.8670 | 0.6477 | 0.4114 | 0.2252 | 0.1071 | 0.0444 | 0.0160 | 0.0049 | 0.0013 |
| | 4 | 0.9974 | 0.9568 | 0.8298 | 0.6296 | 0.4148 | 0.2375 | 0.1182 | 0.0510 | 0.0189 | 0.0059 |
| | 5 | 0.9997 | 0.9887 | 0.9327 | 0.8042 | 0.6172 | 0.4164 | 0.2454 | 0.1256 | 0.0553 | 0.0207 |
| | 6 | 1.0000 | 0.9976 | 0.9781 | 0.9133 | 0.7858 | 0.6080 | 0.4166 | 0.2500 | 0.1299 | 0.0577 |
| | 7 | 1.0000 | 0.9996 | 0.9941 | 0.9679 | 0.8982 | 0.7723 | 0.6010 | 0.4159 | 0.2520 | 0.1316 |
| | 8 | 1.0000 | 0.9999 | 0.9987 | 0.9900 | 0.9591 | 0.8867 | 0.7624 | 0.5956 | 0.4143 | 0.2517 |
| | 9 | 1.0000 | 1.0000 | 0.9998 | 0.9974 | 0.9861 | 0.9520 | 0.8782 | 0.7553 | 0.5914 | 0.4119 |
| | 10 | 1.0000 | 1.0000 | 1.0000 | 0.9994 | 0.9961 | 0.9829 | 0.9468 | 0.8725 | 0.7507 | 0.5881 |
| | 11 | 1.0000 | 1.0000 | 1.0000 | 0.9999 | 0.9991 | 0.9949 | 0.9804 | 0.9435 | 0.8692 | 0.7483 |
| | 12 | 1.0000 | 1.0000 | 1.0000 | 1.0000 | 0.9998 | 0.9987 | 0.9940 | 0.9790 | 0.9420 | 0.8684 |
| | 13 | 1.0000 | 1.0000 | 1.0000 | 1.0000 | 1.0000 | 0.9997 | 0.9985 | 0.9935 | 0.9786 | 0.9423 |
| | 14 | 1.0000 | 1.0000 | 1.0000 | 1.0000 | 1.0000 | 1.0000 | 0.9997 | 0.9984 | 0.9936 | 0.9793 |
| | 15 | 1.0000 | 1.0000 | 1.0000 | 1.0000 | 1.0000 | 1.0000 | 1.0000 | 0.9997 | 0.9985 | 0.9941 |
| | 16 | 1.0000 | 1.0000 | 1.0000 | 1.0000 | 1.0000 | 1.0000 | 1.0000 | 1.0000 | 0.9997 | 0.9987 |
| | 17 | 1.0000 | 1.0000 | 1.0000 | 1.0000 | 1.0000 | 1.0000 | 1.0000 | 1.0000 | 1.0000 | 0.9998 |
| | 18 | 1.0000 | 1.0000 | 1.0000 | 1.0000 | 1.0000 | 1.0000 | 1.0000 | 1.0000 | 1.0000 | 1.0000 |

[a] When $P > 0.50$, use column $(1 - p)$ and row $(n - y)$; e.g., $P(Y \leq 15 \mid n = 20, p = 0.60) = 0.1256$. The value 0.1256 is from column $(1 - 0.60) = 0.40$ and row $(20 - 15) = 5$.

*Source:* From *Probability and Statistics for Engineers* 2nd ed., by Miller, I. and Freund, J. E., © 1977. Reprinted by permission of Prentice-Hall, Inc., Upper Saddle River, NJ.

**TABLE A.2**
**Cumulative Poisson Probabilities[a]**

$$P = [Y \preceq y_b \mid \lambda] = \sum_{y=0}^{y_b} (\lambda^y / y!e^{\lambda})$$

| y | λ = 0.1 | 0.2 | 0.3 | 0.4 | 0.5 | 0.6 | 0.7 | 0.8 | 0.9 | 1.0 |
|---|---|---|---|---|---|---|---|---|---|---|
| 0 | 0.905 | 0.819 | 0.741 | 0.670 | 0.607 | 0.549 | 0.497 | 0.449 | 0.407 | 0.368 |
| 1 | 0.995 | 0.982 | 0.965 | 0.938 | 0.910 | 0.878 | 0.844 | 0.809 | 0.772 | 0.736 |
| 2 | 1.000 | 0.999 | 0.996 | 0.992 | 0.986 | 0.977 | 0.966 | 0.953 | 0.937 | 0.920 |
| 3 | | 1.000 | 1.000 | 0.999 | 0.998 | 0.997 | 0.994 | 0.991 | 0.987 | 0.981 |
| 4 | | | | 1.000 | 1.000 | 1.000 | 0.999 | 0.999 | 0.998 | 0.996 |
| 5 | | | | | | | 1.000 | 1.000 | 1.000 | 0.999 |
| 6 | | | | | | | | | | 1.000 |

| y | λ = 1.1 | 1.2 | 1.3 | 1.4 | 1.5 | 1.6 | 1.7 | 1.8 | 1.9 | 2.0 |
|---|---|---|---|---|---|---|---|---|---|---|
| 0 | 0.333 | 0.301 | 0.273 | 0.247 | 0.223 | 0.202 | 0.183 | 0.165 | 0.150 | 0.135 |
| 1 | 0.699 | 0.663 | 0.627 | 0.592 | 0.558 | 0.525 | 0.493 | 0.463 | 0.434 | 0.406 |
| 2 | 0.900 | 0.879 | 0.857 | 0.833 | 0.809 | 0.783 | 0.757 | 0.731 | 0.704 | 0.677 |
| 3 | 0.974 | 0.966 | 0.957 | 0.946 | 0.934 | 0.921 | 0.907 | 0.891 | 0.875 | 0.857 |
| 4 | 0.995 | 0.992 | 0.989 | 0.986 | 0.981 | 0.976 | 0.970 | 0.964 | 0.956 | 0.947 |
| 5 | 0.999 | 0.998 | 0.998 | 0.997 | 0.996 | 0.994 | 0.992 | 0.990 | 0.987 | 0.983 |
| 6 | 1.000 | 1.000 | 1.000 | 0.999 | 0.999 | 0.999 | 0.998 | 0.997 | 0.997 | 0.995 |
| 7 | | | | 1.000 | 1.000 | 1.000 | 1.000 | 0.999 | 0.999 | 0.999 |
| 8 | | | | | | | | 1.000 | 1.000 | 1.000 |

| y | λ = 2.1 | 2.2 | 2.3 | 2.4 | 2.5 | 2.6 | 2.7 | 2.8 | 2.9 | 3.0 |
|---|---|---|---|---|---|---|---|---|---|---|
| 0 | 0.123 | 0.111 | 0.100 | 0.091 | 0.082 | 0.074 | 0.067 | 0.061 | 0.055 | 0.050 |
| 1 | 0.380 | 0.335 | 0.331 | 0.308 | 0.287 | 0.267 | 0.249 | 0.231 | 0.215 | 0.199 |
| 2 | 0.650 | 0.623 | 0.596 | 0.570 | 0.544 | 0.518 | 0.494 | 0.469 | 0.446 | 0.423 |
| 3 | 0.839 | 0.819 | 0.799 | 0.779 | 0.758 | 0.736 | 0.714 | 0.692 | 0.670 | 0.647 |
| 4 | 0.938 | 0.928 | 0.916 | 0.904 | 0.891 | 0.877 | 0.863 | 0.848 | 0.832 | 0.815 |
| 5 | 0.980 | 0.975 | 0.970 | 0.964 | 0.958 | 0.951 | 0.943 | 0.935 | 0.926 | 0.916 |
| 6 | 0.994 | 0.993 | 0.991 | 0.988 | 0.986 | 0.983 | 0.980 | 0.976 | 0.971 | 0.966 |
| 7 | 0.999 | 0.998 | 0.998 | 0.997 | 0.996 | 0.995 | 0.994 | 0.992 | 0.990 | 0.988 |
| 8 | 1.000 | 1.000 | 0.999 | 0.999 | 0.999 | 0.999 | 0.998 | 0.998 | 0.997 | 0.996 |
| 9 | | | 1.000 | 1.000 | 1.000 | 1.000 | 1.000 | 0.999 | 0.999 | 0.999 |
| 10 | | | | | | | | 1.000 | 1.000 | 1.000 |

| y | λ = 3.1 | 3.2 | 3.3 | 3.4 | 3.5 | 3.6 | 3.7 | 3.8 | 3.9 | 4.0 |
|---|---|---|---|---|---|---|---|---|---|---|
| 0 | 0.045 | 0.041 | 0.037 | 0.033 | 0.030 | 0.027 | 0.025 | 0.022 | 0.020 | 0.018 |
| 1 | 0.185 | 0.171 | 0.159 | 0.147 | 0.136 | 0.126 | 0.116 | 0.107 | 0.099 | 0.092 |
| 2 | 0.401 | 0.380 | 0.359 | 0.340 | 0.321 | 0.303 | 0.285 | 0.269 | 0.253 | 0.238 |
| 3 | 0.625 | 0.603 | 0.580 | 0.558 | 0.537 | 0.515 | 0.494 | 0.473 | 0.453 | 0.433 |
| 4 | 0.798 | 0.781 | 0.763 | 0.744 | 0.726 | 0.706 | 0.687 | 0.668 | 0.648 | 0.629 |
| 5 | 0.906 | 0.895 | 0.883 | 0.871 | 0.858 | 0.844 | 0.830 | 0.816 | 0.800 | 0.785 |

**TABLE A.2** *(continued)*
**Cumulative Poisson Probabilities**[a]

$$P = [Y \preceq y_b \mid \lambda] = \sum_{y=0}^{y_b} (\lambda^y / y!e^{\lambda})$$

| y | λ 3.1 | 3.2 | 3.3 | 3.4 | 3.5 | 3.6 | 3.7 | 3.8 | 3.9 | 4.0 |
|---|---|---|---|---|---|---|---|---|---|---|
| 6 | 0.961 | 0.955 | 0.949 | 0.942 | 0.935 | 0.927 | 0.918 | 0.909 | 0.899 | 0.889 |
| 7 | 0.986 | 0.983 | 0.980 | 0.977 | 0.973 | 0.969 | 0.965 | 0.960 | 0.954 | 0.949 |
| 8 | 0.995 | 0.994 | 0.993 | 0.992 | 0.990 | 0.988 | 0.986 | 0.984 | 0.981 | 0.979 |
| 9 | 0.999 | 0.998 | 0.998 | 0.997 | 0.997 | 0.996 | 0.995 | 0.994 | 0.993 | 0.992 |
| 10 | 1.000 | 1.000 | 1.000 | 0.999 | 0.999 | 0.999 | 0.999 | 0.998 | 0.997 | 0.997 |
| 11 | | | | 1.000 | 1.000 | 1.000 | 1.000 | 0.999 | 0.999 | 0.999 |
| 12 | | | | | | | | 1.000 | 1.000 | 1.000 |

| y | λ 4.1 | 4.2 | 4.3 | 4.4 | 4.5 | 4.6 | 4.7 | 4.8 | 4.9 | 5.0 |
|---|---|---|---|---|---|---|---|---|---|---|
| 0 | 0.017 | 0.015 | 0.014 | 0.012 | 0.011 | 0.010 | 0.009 | 0.008 | 0.007 | 0.007 |
| 1 | 0.085 | 0.078 | 0.072 | 0.066 | 0.061 | 0.056 | 0.052 | 0.048 | 0.044 | 0.040 |
| 2 | 0.224 | 0.210 | 0.197 | 0.185 | 0.174 | 0.163 | 0.152 | 0.143 | 0.133 | 0.125 |
| 3 | 0.414 | 0.395 | 0.377 | 0.359 | 0.342 | 0.326 | 0.310 | 0.294 | 0.279 | 0.265 |
| 4 | 0.609 | 0.590 | 0.570 | 0.551 | 0.532 | 0.513 | 0.495 | 0.476 | 0.458 | 0.440 |
| 5 | 0.769 | 0.753 | 0.737 | 0.720 | 0.703 | 0.686 | 0.668 | 0.651 | 0.634 | 0.616 |
| 6 | 0.879 | 0.867 | 0.856 | 0.844 | 0.831 | 0.818 | 0.805 | 0.791 | 0.777 | 0.762 |
| 7 | 0.943 | 0.936 | 0.929 | 0.921 | 0.913 | 0.905 | 0.896 | 0.887 | 0.877 | 0.867 |
| 8 | 0.975 | 0.972 | 0.968 | 0.964 | 0.960 | 0.955 | 0.950 | 0.944 | 0.938 | 0.932 |
| 9 | 0.990 | 0.989 | 0.987 | 0.985 | 0.983 | 0.980 | 0.978 | 0.975 | 0.972 | 0.968 |
| 10 | 0.997 | 0.996 | 0.995 | 0.994 | 0.993 | 0.992 | 0.991 | 0.990 | 0.988 | 0.986 |
| 11 | 0.999 | 0.999 | 0.998 | 0.998 | 0.998 | 0.997 | 0.997 | 0.996 | 0.995 | 0.995 |
| 12 | 1.000 | 1.000 | 0.999 | 0.999 | 0.999 | 0.999 | 0.999 | 0.999 | 0.998 | 0.998 |
| 13 | | | 1.000 | 1.000 | 1.000 | 1.000 | 1.000 | 1.000 | 1.000 | 0.999 |
| 14 | | | | | | | | | | 1.000 |

[a] To obtain the probability of obtaining a value of $Y > y$, subtract the tabular value for $(y_b - 1)$ from 1. For example, $(P(Y > 5|\lambda = 3.2) = 0.219$. Since $(y - 1) = 4$, the tabular value of which is 0.718, $1 - 0.781 = 0.219$.

## TABLE A.3
## Ordinates of the Standard Normal Curve

$$y = e^{-z^2/2}$$

| z | 0 | 1 | 2 | 3 | 4 | 5 | 6 | 7 | 8 | 9 |
|---|---|---|---|---|---|---|---|---|---|---|
| 0.0 | 0.3989 | 0.3989 | 0.3989 | 0.3988 | 0.3986 | 0.3984 | 0.3982 | 0.3980 | 0.3977 | 0.3973 |
| 0.1 | 0.3970 | 0.3965 | 0.3961 | 0.3956 | 0.3951 | 0.3945 | 0.3939 | 0.3932 | 0.3925 | 0.3918 |
| 0.2 | 0.3910 | 0.3902 | 0.3894 | 0.3885 | 0.3876 | 0.3867 | 0.3857 | 0.3847 | 0.3836 | 0.3825 |
| 0.3 | 0.3814 | 0.3802 | 0.3790 | 0.3778 | 0.3765 | 0.3752 | 0.3739 | 0.3725 | 0.3712 | 0.3697 |
| 0.4 | 0.3683 | 0.3668 | 0.3653 | 0.3637 | 0.3621 | 0.3605 | 0.3589 | 0.3572 | 0.3555 | 0.3538 |
| 0.5 | 0.3521 | 0.3503 | 0.3485 | 0.3467 | 0.3448 | 0.3429 | 0.3410 | 0.3391 | 0.3372 | 0.3352 |
| 0.6 | 0.3332 | 0.3312 | 0.3292 | 0.3271 | 0.3251 | 0.3230 | 0.3209 | 0.3187 | 0.3166 | 0.3144 |
| 0.7 | 0.3123 | 0.3101 | 0.3079 | 0.3056 | 0.3034 | 0.3011 | 0.2989 | 0.2966 | 0.2943 | 0.2920 |
| 0.8 | 0.2897 | 0.2874 | 0.2850 | 0.2827 | 0.2803 | 0.2780 | 0.2756 | 0.2732 | 0.2709 | 0.2685 |
| 0.9 | 0.2661 | 0.2637 | 0.2613 | 0.2589 | 0.2565 | 0.2541 | 0.2516 | 0.2492 | 0.2468 | 0.2444 |
| 1.0 | 0.2420 | 0.2396 | 0.2371 | 0.2347 | 0.2323 | 0.2299 | 0.2275 | 0.2251 | 0.2227 | 0.2203 |
| 1.1 | 0.2179 | 0.2155 | 0.2131 | 0.2107 | 0.2083 | 0.2059 | 0.2036 | 0.2012 | 0.1989 | 0.1965 |
| 1.2 | 0.1942 | 0.1919 | 0.1895 | 0.1872 | 0.1849 | 0.1826 | 0.1804 | 0.1781 | 0.1758 | 0.1736 |
| 1.3 | 0.1714 | 0.1691 | 0.1669 | 0.1647 | 0.1626 | 0.1604 | 0.1582 | 0.1561 | 0.1539 | 0.1518 |
| 1.4 | 0.1497 | 0.1476 | 0.1456 | 0.1435 | 0.1415 | 0.1394 | 0.1374 | 0.1354 | 0.1334 | 0.1315 |
| 1.5 | 0.1295 | 0.1276 | 0.1257 | 0.1238 | 0.1219 | 0.1200 | 0.1182 | 0.1163 | 0.1145 | 0.1127 |
| 1.6 | 0.1109 | 0.1092 | 0.1074 | 0.1057 | 0.1040 | 0.1023 | 0.1006 | 0.0989 | 0.0973 | 0.0957 |
| 1.7 | 0.0940 | 0.0925 | 0.0909 | 0.0893 | 0.0878 | 0.0863 | 0.0848 | 0.0833 | 0.0818 | 0.0804 |
| 1.8 | 0.0790 | 0.0775 | 0.0761 | 0.0748 | 0.0734 | 0.0721 | 0.0707 | 0.0694 | 0.0681 | 0.0669 |
| 1.9 | 0.0656 | 0.0644 | 0.0632 | 0.0620 | 0.0608 | 0.0596 | 0.0584 | 0.0573 | 0.0562 | 0.0551 |
| 2.0 | 0.0540 | 0.0529 | 0.0519 | 0.0508 | 0.0498 | 0.0488 | 0.0478 | 0.0468 | 0.0459 | 0.0449 |
| 2.1 | 0.0440 | 0.0431 | 0.0422 | 0.0413 | 0.0404 | 0.0396 | 0.0387 | 0.0379 | 0.0371 | 0.0363 |
| 2.2 | 0.0355 | 0.0347 | 0.0339 | 0.0332 | 0.0325 | 0.0317 | 0.0310 | 0.0303 | 0.0297 | 0.0290 |
| 2.3 | 0.0283 | 0.0277 | 0.0270 | 0.0264 | 0.0258 | 0.0252 | 0.0246 | 0.0241 | 0.0235 | 0.0229 |
| 2.4 | 0.0224 | 0.0219 | 0.0213 | 0.0208 | 0.0203 | 0.0198 | 0.0194 | 0.0189 | 0.0184 | 0.0180 |
| 2.5 | 0.0175 | 0.0171 | 0.0167 | 0.0163 | 0.0158 | 0.0154 | 0.0151 | 0.0147 | 0.0143 | 0.0139 |
| 2.6 | 0.0136 | 0.0132 | 0.0129 | 0.0126 | 0.0122 | 0.0119 | 0.0116 | 0.0113 | 0.0110 | 0.0107 |
| 2.7 | 0.0104 | 0.0101 | 0.0099 | 0.0096 | 0.0093 | 0.0091 | 0.0088 | 0.0086 | 0.0084 | 0.0081 |
| 2.8 | 0.0079 | 0.0077 | 0.0075 | 0.0073 | 0.0071 | 0.0069 | 0.0067 | 0.0065 | 0.0063 | 0.0061 |
| 2.9 | 0.0060 | 0.0058 | 0.0056 | 0.0055 | 0.0053 | 0.0051 | 0.0050 | 0.0048 | 0.0047 | 0.0046 |
| 3.0 | 0.0044 | 0.0043 | 0.0042 | 0.0040 | 0.0039 | 0.0038 | 0.0037 | 0.0036 | 0.0035 | 0.0034 |
| 3.1 | 0.0033 | 0.0032 | 0.0031 | 0.0030 | 0.0029 | 0.0028 | 0.0027 | 0.0026 | 0.0025 | 0.0025 |
| 3.2 | 0.0024 | 0.0023 | 0.0022 | 0.0022 | 0.0021 | 0.0020 | 0.0020 | 0.0019 | 0.0018 | 0.0018 |
| 3.3 | 0.0017 | 0.0017 | 0.0016 | 0.0016 | 0.0015 | 0.0015 | 0.0014 | 0.0014 | 0.0013 | 0.0013 |
| 3.4 | 0.0012 | 0.0012 | 0.0012 | 0.0011 | 0.0011 | 0.0010 | 0.0010 | 0.0010 | 0.0009 | 0.0009 |
| 3.5 | 0.0009 | 0.0008 | 0.0008 | 0.0008 | 0.0008 | 0.0007 | 0.0007 | 0.0007 | 0.0007 | 0.0006 |
| 3.6 | 0.0006 | 0.0006 | 0.0006 | 0.0005 | 0.0005 | 0.0005 | 0.0005 | 0.0005 | 0.0005 | 0.0004 |
| 3.7 | 0.0004 | 0.0004 | 0.0004 | 0.0004 | 0.0004 | 0.0004 | 0.0003 | 0.0003 | 0.0003 | 0.0003 |
| 3.8 | 0.0003 | 0.0003 | 0.0003 | 0.0003 | 0.0003 | 0.0002 | 0.0002 | 0.0002 | 0.0002 | 0.0002 |
| 3.9 | 0.0002 | 0.0002 | 0.0002 | 0.0002 | 0.0002 | 0.0002 | 0.0002 | 0.0002 | 0.0001 | 0.0001 |

*Source:* Spiegel, M. R., *Theory and Problems of Statistics*, Schaum, New York, 1961. With permission.

## TABLE A.4
## Cumulative Standard Normal Probability Distribution for $Z \geq 0$

$$P(Z < z_b) = \frac{1}{\sqrt{2\pi}} \int_0^{z_b} e^{-z^2/2} dz$$

| z | 0 | 1 | 2 | 3 | 4 | 5 | 6 | 7 | 8 | 9 |
|---|---|---|---|---|---|---|---|---|---|---|
| 0.0 | 0.0000 | 0.0040 | 0.0080 | 0.0120 | 0.0160 | 0.0199 | 0.0239 | 0.0279 | 0.0319 | 0.0359 |
| 0.1 | 0.0398 | 0.0438 | 0.0478 | 0.0517 | 0.0557 | 0.0596 | 0.0636 | 0.0675 | 0.0714 | 0.0754 |
| 0.2 | 0.0793 | 0.0832 | 0.0871 | 0.0910 | 0.0948 | 0.0987 | 0.1026 | 0.1064 | 0.1103 | 0.1141 |
| 0.3 | 0.1179 | 0.1217 | 0.1255 | 0.1293 | 0.1331 | 0.1368 | 0.1406 | 0.1443 | 0.1480 | 0.1517 |
| 0.4 | 0.1554 | 0.1591 | 0.1628 | 0.1664 | 0.1700 | 0.1736 | 0.1772 | 0.1808 | 0.1844 | 0.1879 |
| 0.5 | 0.1915 | 0.1950 | 0.1985 | 0.2019 | 0.2054 | 0.2088 | 0.2123 | 0.2157 | 0.2190 | 0.2224 |
| 0.6 | 0.2258 | 0.2291 | 0.2324 | 0.2357 | 0.2389 | 0.2422 | 0.2454 | 0.2486 | 0.2518 | 0.2549 |
| 0.7 | 0.2580 | 0.2612 | 0.2642 | 0.2673 | 0.2704 | 0.2734 | 0.2764 | 0.2794 | 0.2823 | 0.2852 |
| 0.8 | 0.2881 | 0.2910 | 0.2939 | 0.2967 | 0.2996 | 0.3023 | 0.3051 | 0.3078 | 0.3106 | 0.3133 |
| 0.9 | 0.3159 | 0.3186 | 0.3212 | 0.3238 | 0.3264 | 0.3289 | 0.3315 | 0.3340 | 0.3365 | 0.3389 |
| 1.0 | 0.3413 | 0.3438 | 0.3461 | 0.3485 | 0.3508 | 0.3531 | 0.3554 | 0.3577 | 0.3599 | 0.3621 |
| 1.1 | 0.3643 | 0.3665 | 0.3686 | 0.3708 | 0.3729 | 0.3749 | 0.3770 | 0.3790 | 0.3810 | 0.3830 |
| 1.2 | 0.3849 | 0.3869 | 0.3888 | 0.3907 | 0.3925 | 0.3944 | 0.3962 | 0.3980 | 0.3997 | 0.4015 |
| 1.3 | 0.4032 | 0.4049 | 0.4066 | 0.4082 | 0.4099 | 0.4115 | 0.4131 | 0.4147 | 0.4162 | 0.4177 |
| 1.4 | 0.4192 | 0.4207 | 0.4222 | 0.4236 | 0.4251 | 0.4265 | 0.4279 | 0.4292 | 0.4306 | 0.4319 |
| 1.5 | 0.4332 | 0.4345 | 0.4357 | 0.4370 | 0.4382 | 0.4394 | 0.4406 | 0.4418 | 0.4429 | 0.4441 |
| 1.6 | 0.4452 | 0.4463 | 0.4474 | 0.4484 | 0.4495 | 0.4505 | 0.4515 | 0.4525 | 0.4535 | 0.4545 |
| 1.7 | 0.4554 | 0.4564 | 0.4573 | 0.4582 | 0.4591 | 0.4599 | 0.4608 | 0.4616 | 0.4625 | 0.4633 |
| 1.8 | 0.4641 | 0.4649 | 0.4656 | 0.4664 | 0.4671 | 0.4678 | 0.4686 | 0.4693 | 0.4699 | 0.4706 |
| 1.9 | 0.4713 | 0.4719 | 0.4726 | 0.4732 | 0.4738 | 0.4744 | 0.4750 | 0.4756 | 0.4761 | 0.4767 |
| 2.0 | 0.4772 | 0.4778 | 0.4783 | 0.4788 | 0.4793 | 0.4798 | 0.4803 | 0.4808 | 0.4812 | 0.4817 |
| 2.1 | 0.4821 | 0.4826 | 0.4830 | 0.4834 | 0.4838 | 0.4842 | 0.4846 | 0.4850 | 0.4854 | 0.4857 |
| 2.2 | 0.4861 | 0.4864 | 0.4868 | 0.4871 | 0.4875 | 0.4878 | 0.4881 | 0.4884 | 0.4887 | 0.4890 |
| 2.3 | 0.4893 | 0.4896 | 0.4898 | 0.4901 | 0.4904 | 0.4906 | 0.4909 | 0.4911 | 0.4913 | 0.4916 |
| 2.4 | 0.4918 | 0.4920 | 0.4922 | 0.4925 | 0.4927 | 0.4929 | 0.4931 | 0.4932 | 0.4934 | 0.4936 |
| 2.5 | 0.4938 | 0.4940 | 0.4941 | 0.4943 | 0.4945 | 0.4946 | 0.4948 | 0.4949 | 0.4951 | 0.4952 |
| 2.6 | 0.4953 | 0.4955 | 0.4956 | 0.4957 | 0.4959 | 0.4960 | 0.4961 | 0.4962 | 0.4963 | 0.4964 |
| 2.7 | 0.4965 | 0.4966 | 0.4967 | 0.4968 | 0.4969 | 0.4970 | 0.4971 | 0.4972 | 0.4973 | 0.4974 |
| 2.8 | 0.4974 | 0.4975 | 0.4976 | 0.4977 | 0.4977 | 0.4978 | 0.4979 | 0.4979 | 0.4980 | 0.4981 |
| 2.9 | 0.4981 | 0.4982 | 0.4982 | 0.4983 | 0.4984 | 0.4984 | 0.4985 | 0.4985 | 0.4986 | 0.4986 |
| 3.0 | 0.4987 | 0.4987 | 0.4987 | 0.4988 | 0.4988 | 0.4989 | 0.4989 | 0.4989 | 0.4990 | 0.4990 |
| 3.1 | 0.4990 | 0.4991 | 0.4991 | 0.4991 | 0.4992 | 0.4992 | 0.4992 | 0.4992 | 0.4993 | 0.4993 |
| 3.2 | 0.4993 | 0.4993 | 0.4994 | 0.4994 | 0.4994 | 0.4994 | 0.4994 | 0.4995 | 0.4995 | 0.4995 |
| 3.3 | 0.4995 | 0.4995 | 0.4995 | 0.4996 | 0.4996 | 0.4996 | 0.4996 | 0.4996 | 0.4996 | 0.4997 |
| 3.4 | 0.4997 | 0.4997 | 0.4997 | 0.4997 | 0.4997 | 0.4997 | 0.4997 | 0.4997 | 0.4997 | 0.4998 |
| 3.5 | 0.4998 | 0.4998 | 0.4998 | 0.4998 | 0.4998 | 0.4998 | 0.4998 | 0.4998 | 0.4998 | 0.4998 |
| 3.6 | 0.4998 | 0.4998 | 0.4999 | 0.4999 | 0.4999 | 0.4999 | 0.4999 | 0.4999 | 0.4999 | 0.4999 |
| 3.7 | 0.4999 | 0.4999 | 0.4999 | 0.4999 | 0.4999 | 0.4999 | 0.4999 | 0.4999 | 0.4999 | 0.4999 |
| 3.8 | 0.4999 | 0.4999 | 0.4999 | 0.4999 | 0.4999 | 0.4999 | 0.4999 | 0.4999 | 0.4999 | 0.4999 |
| 3.9 | 0.5000 | 0.5000 | 0.5000 | 0.5000 | 0.5000 | 0.5000 | 0.5000 | 0.5000 | 0.5000 | 0.5000 |

*Source:* Spiegel, M. R., *Theory and Problems of Statistics*, Schaum, New York, 1961. With permission.

## TABLE A.5
## Values of the Gamma Function

$$\Gamma(r) = \int_0^\infty y^{r-1} e^{-y} dy$$

| $r$ | $\Gamma(r)$ | $r$ | $\Gamma(r)$ | $r$ | $\Gamma(r)$ | $r$ | $\Gamma(r)$ |
|---|---|---|---|---|---|---|---|
| 1.000 | 1.00000 00 | 1.250 | 0.90640 25 | 1.500 | 0.88622 69 | 1.750 | 0.91906 25 |
| 1.005 | 0.99713 85 | 1.255 | 0.90538 58 | 1.505 | 0.88639 90 | 1.755 | 0.92020 92 |
| 1.010 | 0.99432 59 | 1.260 | 0.90439 71 | 1.510 | 0.88659 17 | 1.760 | 0.92137 49 |
| 1.015 | 0.99156 13 | 1.265 | 0.90343 63 | 1.515 | 0.88680 50 | 1.765 | 0.92255 95 |
| 1.020 | 0.98884 42 | 1.270 | 0.90250 31 | 1.520 | 0.88703 88 | 1.770 | 0.92376 31 |
| 1.025 | 0.98617 40 | 1.275 | 0.90159 72 | 1.525 | 0.88729 30 | 1.775 | 0.92498 57 |
| 1.030 | 0.98355 00 | 1.280 | 0.90071 85 | 1.530 | 0.88756 76 | 1.780 | 0.92622 73 |
| 1.035 | 0.98097 16 | 1.285 | 0.89986 67 | 1.535 | 0.88786 25 | 1.785 | 0.92748 79 |
| 1.040 | 0.97843 82 | 1.290 | 0.89904 16 | 1.540 | 0.88817 77 | 1.790 | 0.92876 75 |
| 1.045 | 0.97694 93 | 1.295 | 0.89824 30 | 1.545 | 0.88851 30 | 1.795 | 0.93006 61 |
| 1.050 | 0.97350 43 | 1.300 | 0.89747 07 | 1.550 | 0.88886 83 | 1.800 | 0.93138 38 |
| 1.055 | 0.97110 26 | 1.305 | 0.89672 45 | 1.555 | 0.88924 38 | 1.805 | 0.93272 05 |
| 1.060 | 0.96874 36 | 1.310 | 0.89600 42 | 1.560 | 0.88963 92 | 1.810 | 0.93407 63 |
| 1.065 | 0.96642 70 | 1.315 | 0.89530 96 | 1.565 | 0.89005 45 | 1.815 | 0.93545 11 |
| 1.070 | 0.96415 20 | 1.320 | 0.89464 05 | 1.570 | 0.89048 97 | 1.820 | 0.93684 51 |
| 1.075 | 0.96191 83 | 1.325 | 0.89399 67 | 1.575 | 0.89094 48 | 1.825 | 0.93825 82 |
| 1.080 | 0.95972 53 | 1.330 | 0.89337 81 | 1.580 | 0.89141 96 | 1.830 | 0.93969 04 |
| 1.085 | 0.95757 25 | 1.335 | 0.89278 44 | 1.585 | 0.89191 41 | 1.835 | 0.94114 18 |
| 1.090 | 0.95545 95 | 1.340 | 0.89221 55 | 1.590 | 0.89242 82 | 1.840 | 0.94261 24 |
| 1.095 | 0.95338 57 | 1.345 | 0.89167 12 | 1.595 | 0.89296 20 | 1.845 | 0.94410 22 |
| 1.100 | 0.95135 08 | 1.350 | 0.89115 14 | 1.600 | 0.89351 53 | 1.850 | 0.94561 12 |
| 1.105 | 0.94935 42 | 1.355 | 0.89065 59 | 1.605 | 0.89408 82 | 1.855 | 0.94713 95 |
| 1.110 | 0.94739 55 | 1.360 | 0.89018 45 | 1.610 | 0.89418 06 | 1.860 | 0.94868 70 |
| 1.115 | 0.94547 43 | 1.365 | 0.88973 71 | 1.615 | 0.89529 24 | 1.865 | 0.95025 39 |
| 1.120 | 0.94359 02 | 1.370 | 0.88931 35 | 1.620 | 0.89592 37 | 1.870 | 0.95184 02 |
| 1.125 | 0.94174 27 | 1.375 | 0.88891 36 | 1.625 | 0.89657 43 | 1.875 | 0.95344 58 |
| 1.130 | 0.93993 14 | 1.380 | 0.88853 71 | 1.630 | 0.89724 42 | 1.880 | 0.95507 09 |
| 1.135 | 0.93815 60 | 1.385 | 0.88818 41 | 1.635 | 0.89793 35 | 1.885 | 0.95671 53 |
| 1.140 | 0.93641 61 | 1.390 | 0.88785 43 | 1.640 | 0.89864 20 | 1.890 | 0.95837 93 |
| 1.145 | 0.93471 12 | 1.395 | 0.88754 76 | 1.645 | 0.89936 98 | 1.895 | 0.96006 28 |
| 1.150 | 0.93300 09 | 1.400 | 0.88726 38 | 1.650 | 0.90011 68 | 1.900 | 0.96176 58 |
| 1.155 | 0.93149 50 | 1.405 | 0.88700 29 | 1.655 | 0.90088 30 | 1.905 | 0.96348 85 |
| 1.160 | 0.92980 31 | 1.410 | 0.88676 47 | 1.660 | 0.90166 84 | 1.910 | 0.96523 07 |
| 1.165 | 0.92823 47 | 1.415 | 0.88654 90 | 1.665 | 0.90247 29 | 1.915 | 0.96699 27 |
| 1.170 | 0.92669 96 | 1.420 | 0.88635 58 | 1.670 | 0.90329 65 | 1.920 | 0.96877 43 |
| 1.175 | 0.92519 74 | 1.425 | 0.88618 49 | 1.675 | 0.90413 92 | 1.925 | 0.97057 57 |
| 1.180 | 0.92372 78 | 1.430 | 0.88603 62 | 1.680 | 0.90500 10 | 1.930 | 0.97239 69 |
| 1.185 | 0.92229 05 | 1.435 | 0.88590 97 | 1.685 | 0.90588 19 | 1.935 | 0.97423 80 |
| 1.190 | 0.92088 50 | 1.440 | 0.88580 51 | 1.690 | 0.90678 18 | 1.940 | 0.97609 89 |
| 1.195 | 0.91951 12 | 1.445 | 0.88572 23 | 1.695 | 0.90770 08 | 1.945 | 0.97797 98 |

**TABLE A.5** *(continued)*
**Values of the Gamma Function**

$$\Gamma(r) = \int_0^\infty y^{r-1}e^{-y}dy$$

| $r$ | $\Gamma(r)$ | $r$ | $\Gamma(r)$ | $r$ | $\Gamma(r)$ | $r$ | $\Gamma(r)$ |
|---|---|---|---|---|---|---|---|
| 1.200 | 0.91816 87 | 1.450 | 0.88566 14 | 1.700 | 0.90868 87 | 1.950 | 0.97988 07 |
| 1.205 | 0.91685 73 | 1.455 | 0.88562 21 | 1.705 | 0.90959 57 | 1.955 | 0.98180 16 |
| 1.210 | 0.91557 65 | 1.460 | 0.88560 43 | 1.710 | 0.91057 17 | 1.960 | 0.98374 25 |
| 1.215 | 0.91432 62 | 1.465 | 0.88560 80 | 1.715 | 0.91156 66 | 1.965 | 0.98570 37 |
| 1.220 | 0.91310 59 | 1.470 | 0.88563 31 | 1.720 | 0.91258 06 | 1.970 | 0.98768 50 |
| 1.225 | 0.91191 56 | 1.475 | 0.88567 95 | 1.725 | 0.91361 35 | 1.975 | 0.98968 65 |
| 1.230 | 0.91075 49 | 1.480 | 0.88574 70 | 1.730 | 0.91466 54 | 1.980 | 0.99170 84 |
| 1.235 | 0.90962 34 | 1.485 | 0.88583 56 | 1.735 | 0.91573 62 | 1.985 | 0.99375 06 |
| 1.240 | 0.90852 11 | 1.490 | 0.88594 51 | 1.740 | 0.91682 60 | 1.990 | 0.99581 33 |
| 1.245 | 0.90744 75 | 1.495 | 0.88607 56 | 1.745 | 0.91793 48 | 1.995 | 0.99789 64 |
| | | | | | | 2.000 | 1.00000 00 |

*Source:* Adapted from Weast, R. C. and Selby, S. M., *Handbook of Tables for Mathematics,* 4th ed., CRC Press, Boca Raton, FL, 1975.

**TABLE A.6**
**Fractiles (or Percentage Points) of the Chi-Square Probability Distribution**

$$P[\chi^2 < \chi_b^2(k)] = [1/2^{k/2}\Gamma(k/2)]\int_0^{\chi^2} y^{(k/2)-1}e^{-y/2}dy$$

$$P[\chi^2 < \chi^2(k)]$$

| $k$ | 0.005 | 0.010 | 0.025 | 0.050 | 0.100 | 0.250 | 0.500 | 0.750 | 0.900 | 0.950 | 0.975 | 0.990 | 0.995 |
|---|---|---|---|---|---|---|---|---|---|---|---|---|---|
| 1 | 0.0000393 | 0.000157 | 0.000982 | 0.00393 | 0.0158 | 0.102 | 0.455 | 1.32 | 2.71 | 3.84 | 5.02 | 6.63 | 7.88 |
| 2 | 0.0100 | 0.0201 | 0.0506 | 0.103 | 0.211 | 0.575 | 1.39 | 2.77 | 4.61 | 5.99 | 7.38 | 9.21 | 10.6 |
| 3 | 0.0717 | 0.115 | 0.216 | 0.352 | 0.584 | 1.21 | 2.37 | 4.11 | 6.25 | 7.81 | 9.35 | 11.3 | 12.8 |
| 4 | 0.207 | 0.297 | 0.484 | 0.711 | 1.06 | 1.92 | 3.36 | 5.39 | 7.78 | 9.49 | 11.1 | 13.3 | 14.9 |
| 5 | 0.412 | 0.554 | 0.831 | 1.15 | 1.61 | 2.67 | 4.35 | 6.63 | 9.24 | 11.1 | 12.8 | 15.1 | 16.7 |
| 6 | 0.676 | 0.872 | 1.24 | 1.64 | 2.20 | 3.45 | 5.35 | 7.84 | 10.6 | 12.6 | 14.4 | 16.8 | 18.5 |
| 7 | 0.989 | 1.24 | 1.69 | 2.17 | 2.83 | 4.25 | 6.35 | 9.04 | 12.0 | 14.1 | 16.0 | 18.5 | 20.3 |
| 8 | 1.34 | 1.65 | 2.18 | 2.73 | 3.49 | 5.07 | 7.34 | 10.2 | 13.4 | 15.5 | 17.5 | 20.1 | 22.0 |
| 9 | 1.73 | 2.09 | 2.70 | 3.33 | 4.17 | 5.90 | 8.34 | 11.4 | 14.7 | 16.9 | 19.0 | 21.7 | 23.6 |
| 10 | 2.16 | 2.56 | 3.25 | 3.94 | 4.87 | 6.74 | 9.34 | 12.5 | 16.0 | 18.3 | 20.5 | 23.2 | 25.2 |
| 11 | 2.60 | 3.05 | 3.82 | 4.57 | 5.58 | 7.58 | 10.3 | 13.7 | 17.3 | 19.7 | 21.9 | 24.7 | 26.8 |
| 12 | 3.07 | 3.57 | 4.40 | 5.23 | 6.30 | 8.44 | 11.3 | 14.8 | 18.5 | 21.0 | 23.3 | 26.2 | 28.3 |
| 13 | 3.57 | 4.11 | 5.01 | 5.89 | 7.04 | 9.30 | 12.3 | 16.0 | 19.8 | 22.4 | 24.7 | 27.7 | 29.8 |
| 14 | 4.07 | 4.66 | 5.63 | 6.57 | 7.79 | 10.2 | 13.3 | 17.1 | 21.1 | 23.7 | 26.1 | 29.1 | 31.3 |
| 15 | 4.60 | 5.23 | 6.26 | 7.26 | 8.55 | 11.0 | 14.3 | 18.2 | 22.3 | 25.0 | 27.5 | 30.6 | 32.8 |
| 16 | 5.14 | 5.81 | 6.91 | 7.96 | 9.31 | 11.9 | 15.3 | 19.4 | 23.5 | 26.3 | 28.8 | 32.0 | 34.3 |
| 17 | 5.70 | 6.41 | 7.56 | 8.67 | 10.1 | 12.8 | 16.3 | 20.5 | 24.8 | 27.6 | 30.2 | 33.4 | 35.7 |
| 18 | 6.26 | 7.01 | 8.23 | 9.39 | 10.9 | 13.7 | 17.3 | 21.6 | 26.0 | 28.9 | 31.5 | 34.8 | 37.2 |
| 19 | 6.84 | 7.63 | 8.91 | 10.1 | 11.7 | 14.6 | 18.3 | 22.7 | 27.2 | 30.1 | 32.9 | 36.2 | 38.6 |
| 20 | 7.43 | 8.26 | 9.59 | 10.9 | 12.4 | 15.5 | 19.3 | 23.8 | 28.4 | 31.4 | 34.2 | 37.6 | 40.0 |
| 21 | 8.03 | 8.90 | 10.3 | 11.6 | 13.2 | 16.3 | 20.3 | 24.9 | 29.6 | 32.7 | 35.5 | 38.9 | 41.4 |
| 22 | 8.64 | 9.54 | 11.0 | 12.3 | 14.0 | 17.2 | 21.3 | 26.0 | 30.8 | 33.9 | 36.8 | 40.3 | 42.8 |
| 23 | 9.26 | 10.2 | 11.7 | 13.1 | 14.8 | 18.1 | 22.3 | 27.1 | 32.0 | 35.2 | 38.1 | 41.6 | 44.2 |
| 24 | 9.89 | 10.9 | 12.4 | 13.8 | 15.7 | 19.0 | 23.3 | 28.2 | 33.2 | 36.4 | 39.4 | 43.0 | 45.6 |
| 25 | 10.5 | 11.5 | 13.1 | 14.6 | 16.5 | 19.9 | 24.3 | 29.3 | 34.4 | 37.7 | 40.6 | 44.3 | 46.9 |
| 26 | 11.2 | 12.2 | 13.8 | 15.4 | 17.3 | 20.8 | 25.3 | 30.4 | 35.6 | 38.9 | 41.9 | 45.6 | 48.3 |
| 27 | 11.8 | 12.9 | 14.6 | 16.2 | 18.1 | 21.7 | 26.3 | 31.5 | 36.7 | 40.1 | 43.2 | 47.0 | 49.6 |
| 28 | 12.5 | 13.6 | 15.3 | 16.9 | 18.9 | 22.7 | 27.3 | 32.6 | 37.9 | 41.3 | 44.5 | 48.3 | 51.0 |
| 29 | 13.1 | 14.3 | 16.0 | 17.7 | 19.8 | 23.6 | 28.3 | 33.7 | 39.1 | 42.6 | 45.7 | 49.6 | 52.3 |
| 30 | 13.8 | 15.0 | 16.8 | 18.5 | 20.6 | 24.5 | 29.3 | 34.8 | 40.3 | 43.8 | 47.0 | 50.9 | 53.7 |
| 40 | 20.7 | 22.2 | 24.4 | 26.5 | 29.1 | 33.7 | 39.3 | 45.6 | 51.8 | 55.8 | 59.3 | 63.7 | 66.8 |
| 50 | 28.0 | 29.7 | 32.4 | 34.8 | 37.7 | 42.9 | 49.3 | 56.3 | 63.2 | 67.5 | 71.4 | 76.2 | 79.5 |
| 60 | 35.5 | 37.5 | 40.5 | 43.2 | 46.5 | 52.3 | 59.3 | 67.0 | 74.4 | 79.1 | 83.3 | 88.4 | 92.0 |
| 70 | 43.3 | 45.4 | 48.8 | 51.7 | 55.3 | 61.7 | 69.3 | 77.6 | 85.5 | 90.5 | 95.0 | 100.4 | 104.2 |
| 80 | 51.2 | 53.5 | 57.2 | 60.4 | 64.3 | 71.1 | 79.3 | 88.1 | 96.6 | 101.9 | 106.6 | 112.3 | 116.3 |
| 90 | 59.2 | 61.8 | 65.6 | 69.1 | 73.3 | 80.6 | 89.3 | 98.6 | 107.6 | 113.1 | 118.1 | 124.1 | 128.3 |
| 100 | 67.3 | 70.1 | 74.2 | 77.9 | 82.4 | 90.1 | 99.3 | 109.1 | 118.5 | 124.3 | 129.6 | 135.8 | 140.2 |

*Source:* Adapted from Bayer, W. H., *Handbook of Tables for Probability and Statistics*, 2nd ed., CRC Press, Boca Raton, FL, 1968.

**TABLE A.7**
**Critical Values of the Beta Probability Distribution When the Probability Level ($I_{y_b}$) is 0.05[a,b]**

$$P[Y < y_b(r_1, r_2)] = I_{y_b}(r_1, r_2) = \int_0^{y_b} y^{r_1-1}(1-y)^{r_2-1}dy \Big/ \int_0^1 y^{r_1-1}(1-y)^{r_2-1}dy = 0.05$$

| $r_2$ \ $r_1$ | 1 | 2 | 3 | 4 | 5 | 6 | 7 | 8 | 9 | 10 | 12 | 15 | 20 | 24 | 30 | 40 | 80 | 120 |
|---|---|---|---|---|---|---|---|---|---|---|---|---|---|---|---|---|---|---|
| 1 | 0.0061558 | 0.0025000 | 0.0015429 | 0.0011119 | 0.086820 | 0.071179 | 0.060300 | 0.052300 | 0.046170 | 0.041325 | 0.034155 | 0.027098 | 0.020156 | 0.016727 | 0.013326 | 0.099535 | 0.066082 | 0.032904 |
| 2 | 0.097500 | 0.050000 | 0.033617 | 0.025321 | 0.020308 | 0.016952 | 0.014548 | 0.012741 | 0.011334 | 0.010206 | 0.0085124 | 0.0068158 | 0.0051162 | 0.0042653 | 0.0034137 | 0.0025614 | 0.0017083 | 0.085452 |
| 3 | 0.22852 | 0.13572 | 0.097308 | 0.076010 | 0.062412 | 0.052962 | 0.046007 | 0.040671 | 0.036447 | 0.033020 | 0.027794 | 0.022465 | 0.017026 | 0.014264 | 0.011472 | 0.0086511 | 0.0057991 | 0.0029157 |
| 4 | 0.34163 | 0.22361 | 0.16825 | 0.13535 | 0.11338 | 0.097611 | 0.085728 | 0.076440 | 0.068979 | 0.062850 | 0.053375 | 0.043541 | 0.033319 | 0.028053 | 0.022679 | 0.017191 | 0.011585 | 0.0058567 |
| 5 | 0.43074 | 0.30171 | 0.23553 | 0.19403 | 0.16528 | 0.14408 | 0.12778 | 0.11482 | 0.10427 | 0.095510 | 0.081790 | 0.067312 | 0.051995 | 0.043994 | 0.035747 | 0.027240 | 0.018458 | 0.0093841 |
| 6 | 0.50053 | 0.36840 | 0.29599 | 0.24860 | 0.21477 | 0.18926 | 0.16927 | 0.15316 | 0.13989 | 0.12876 | 0.11111 | 0.092207 | 0.071870 | 0.061103 | 0.049898 | 0.038224 | 0.026043 | 0.013317 |
| 7 | 0.55593 | 0.42489 | 0.34929 | 0.29811 | 0.26063 | 0.23182 | 0.20890 | 0.19019 | 0.17461 | 0.16142 | 0.14029 | 0.11733 | 0.092238 | 0.078783 | 0.064651 | 0.049781 | 0.034103 | 0.017540 |
| 8 | 0.60071 | 0.47287 | 0.39607 | 0.34259 | 0.30260 | 0.27134 | 0.24613 | 0.22532 | 0.20783 | 0.19290 | 0.16875 | 0.14216 | 0.11267 | 0.096658 | 0.079695 | 0.061675 | 0.042481 | 0.021976 |
| 9 | 0.63751 | 0.51390 | 0.43716 | 0.38245 | 0.34080 | 0.30777 | 0.28082 | 0.25835 | 0.23930 | 0.22292 | 0.19618 | 0.16638 | 0.13288 | 0.11449 | 0.094827 | 0.073748 | 0.051068 | 0.026572 |
| 10 | 0.66824 | 0.54928 | 0.47338 | 0.41820 | 0.37553 | 0.34126 | 0.31301 | 0.28924 | 0.26894 | 0.25137 | 0.22244 | 0.18984 | 0.15272 | 0.13211 | 0.10991 | 0.085885 | 0.059785 | 0.031288 |
| 11 | 0.69425 | 0.58003 | 0.50546 | 0.45033 | 0.40712 | 0.37203 | 0.34283 | 0.31807 | 0.29677 | 0.27823 | 0.24746 | 0.21244 | 0.17207 | 0.14943 | 0.12484 | 0.098008 | 0.068575 | 0.036094 |
| 12 | 0.71654 | 0.60696 | 0.53402 | 0.47930 | 0.43590 | 0.40031 | 0.37044 | 0.34494 | 0.32286 | 0.30354 | 0.27125 | 0.23413 | 0.19086 | 0.16636 | 0.13955 | 0.11006 | 0.077394 | 0.040967 |
| 13 | 0.73583 | 0.63073 | 0.55958 | 0.50551 | 0.46219 | 0.42635 | 0.39604 | 0.37000 | 0.34732 | 0.32737 | 0.29383 | 0.25492 | 0.20908 | 0.18288 | 0.15401 | 0.12199 | 0.086209 | 0.045889 |
| 14 | 0.75268 | 0.65184 | 0.58256 | 0.52932 | 0.48626 | 0.45036 | 0.41980 | 0.39338 | 0.37025 | 0.34981 | 0.31524 | 0.27481 | 0.22669 | 0.19895 | 0.16818 | 0.13377 | 0.094994 | 0.050847 |
| 15 | 0.76754 | 0.67070 | 0.60333 | 0.55102 | 0.50836 | 0.47255 | 0.44187 | 0.41521 | 0.39176 | 0.37095 | 0.33554 | 0.29382 | 0.24370 | 0.21457 | 0.18203 | 0.14539 | 0.10373 | 0.055827 |
| 16 | 0.78072 | 0.68766 | 0.62217 | 0.57086 | 0.52872 | 0.49310 | 0.46242 | 0.43563 | 0.41196 | 0.39086 | 0.35480 | 0.31199 | 0.26011 | 0.22972 | 0.19556 | 0.15682 | 0.11240 | 0.060821 |
| 17 | 0.79249 | 0.70297 | 0.63933 | 0.58907 | 0.54750 | 0.51217 | 0.48158 | 0.45474 | 0.43094 | 0.40965 | 0.37307 | 0.32936 | 0.27594 | 0.24441 | 0.20877 | 0.16805 | 0.12099 | 0.065820 |
| 18 | 0.80307 | 0.71687 | 0.65503 | 0.60584 | 0.56490 | 0.52991 | 0.49949 | 0.47267 | 0.44880 | 0.42738 | 0.39041 | 0.34596 | 0.29120 | 0.25865 | 0.22164 | 0.17908 | 0.12950 | 0.070818 |
| 19 | 0.81263 | 0.72954 | 0.66944 | 0.62131 | 0.58103 | 0.54645 | 0.51624 | 0.48951 | 0.46564 | 0.44414 | 0.40689 | 0.36183 | 0.30591 | 0.27244 | 0.23418 | 0.18989 | 0.13791 | 0.075809 |

| | | | | | | | | | | | | | | | | | | |
|---|---|---|---|---|---|---|---|---|---|---|---|---|---|---|---|---|---|---|
| 20 | 0.82131 | 0.74113 | 0.68271 | 0.63564 | 0.59605 | 0.56189 | 0.53194 | 0.50535 | 0.48152 | 0.45999 | 0.42256 | 0.37701 | 0.32009 | 0.28580 | 0.24639 | 0.20050 | 0.14622 | 0.080789 |
| 21 | 0.82923 | 0.75178 | 0.69496 | 0.64894 | 0.61004 | 0.57635 | 0.54669 | 0.52027 | 0.49652 | 0.47501 | 0.43746 | 0.39154 | 0.33375 | 0.29874 | 0.25828 | 0.21088 | 0.15442 | 0.085753 |
| 22 | 0.83647 | 0.76160 | 0.70632 | 0.66132 | 0.62312 | 0.58990 | 0.56056 | 0.53434 | 0.51071 | 0.48925 | 0.45165 | 0.40544 | 0.34693 | 0.31126 | 0.26985 | 0.22106 | 0.16252 | 0.090698 |
| 23 | 0.84313 | 0.77067 | 0.71687 | 0.67287 | 0.63536 | 0.60263 | 0.57363 | 0.54764 | 0.52415 | 0.50276 | 0.46518 | 0.41877 | 0.35964 | 0.32340 | 0.28112 | 0.23102 | 0.17051 | 0.095621 |
| 24 | 0.84927 | 0.77908 | 0.72669 | 0.68366 | 0.64684 | 0.61461 | 0.58506 | 0.56022 | 0.53689 | 0.51560 | 0.47808 | 0.43154 | 0.37190 | 0.33515 | 0.29208 | 0.24078 | 0.17838 | 0.10052 |
| 25 | 0.85494 | 0.78690 | 0.73586 | 0.69377 | 0.65764 | 0.62590 | 0.59761 | 0.57213 | 0.54898 | 0.52782 | 0.49040 | 0.44379 | 0.38373 | 0.34653 | 0.30275 | 0.25032 | 0.18615 | 0.10539 |
| 26 | 0.86021 | 0.79418 | 0.74444 | 0.70327 | 0.66780 | 0.63656 | 0.60864 | 0.58343 | 0.56048 | 0.53945 | 0.50217 | 0.45554 | 0.39516 | 0.35756 | 0.31314 | 0.25966 | 0.19379 | 0.11024 |
| 27 | 0.86511 | 0.80099 | 0.75249 | 0.71219 | 0.67738 | 0.64663 | 0.61909 | 0.59416 | 0.57141 | 0.55054 | 0.51343 | 0.46683 | 0.40619 | 0.36826 | 0.32325 | 0.26880 | 0.20133 | 0.11505 |
| 28 | 0.86967 | 0.80736 | 0.76004 | 0.72060 | 0.68643 | 0.65617 | 0.62900 | 0.60436 | 0.58183 | 0.56112 | 0.52420 | 0.47768 | 0.41685 | 0.37862 | 0.33309 | 0.27775 | 0.20875 | 0.11983 |
| 29 | 0.87394 | 0.81334 | 0.76715 | 0.72854 | 0.60499 | 0.66522 | 0.63842 | 0.61407 | 0.59177 | 0.57122 | 0.53452 | 0.48812 | 0.42715 | 0.38867 | 0.34267 | 0.28650 | 0.21606 | 0.12458 |
| 30 | 0.87794 | 0.81896 | 0.77386 | 0.73604 | 0.70311 | 0.67381 | 0.64738 | 0.62332 | 0.60125 | 0.58088 | 0.54442 | 0.49816 | 0.43711 | 0.39842 | 0.35200 | 0.29507 | 0.22326 | 0.12930 |
| 40 | 0.90734 | 0.86089 | 0.82447 | 0.79327 | 0.76559 | 0.74053 | 0.71758 | 0.69636 | 0.67663 | 0.65819 | 0.62460 | 0.58083 | 0.52099 | 0.48175 | 0.43321 | 0.37136 | 0.28936 | 0.17453 |
| 60 | 0.93748 | 0.90497 | 0.87881 | 0.85591 | 0.83517 | 0.81606 | 0.79824 | 0.78150 | 0.76569 | 0.75070 | 0.72282 | 0.68535 | 0.63185 | 0.59522 | 0.54807 | 0.48477 | 0.39458 | 0.25416 |
| 120 | 0.96837 | 0.95130 | 0.93720 | 0.92458 | 0.91290 | 0.90192 | 0.89148 | 0.88150 | 0.87191 | 0.86266 | 0.84504 | 0.82047 | 0.78342 | 0.75661 | 0.72016 | 0.66738 | 0.58326 | 0.42519 |
| ∞ | 1.00000 | 1.00000 | 1.00000 | 1.00000 | 1.00000 | 1.00000 | 1.00000 | 1.00000 | 1.00000 | 1.00000 | 1.00000 | 1.00000 | 1.00000 | 1.00000 | 1.00000 | 1.00000 | 1.00000 | 1.00000 |

*Note:* When $V_1 = \infty$, $y_b = 0$.

[a] Where $V_1 = 2r_2$, $V_2 = 2$.

[b] See Beyer (1968) for other probability levels.

*Source:* Adapted from Beyer, W. H., *Handbook of Tables for Probability and Statistics*, 2nd ed., CRC Press, Boca Raton, FL, 1968.

## TABLE A.8
## Critical Values of Student's *t* Distribution

$$P[|t|<|t_b|(k)] = 2\left\{\frac{\Gamma[(k+1)/2}{\sqrt{\Pi k}\,\Gamma(k/2)}\int_0^{t_b}[(t^2/k)+1^{-(k+1)/2}dt\right\}$$

$$P[t<t_b(k)] = \frac{\Gamma[(k+1)/2}{\sqrt{\Pi k}\,\Gamma(k/2)}\int_{-\infty}^{t_b}[(t^2/k)+1]^{-(k+1)/2}dt$$

| | $P[\|t\| < \|t_b\|(k)]$ (two-tailed) | | | | | | | | | | | |
|---|---|---|---|---|---|---|---|---|---|---|---|---|
| *k* | 0.1 | 0.2 | 0.3 | 0.4 | 0.5 | 0.6 | 0.7 | 0.8 | 0.9 | 0.95 | 0.975 | 0.99 |
| | | | | | | | $t_b$ | | | | | |
| 1 | 0.158 | 0.325 | 0.510 | 0.727 | 1.000 | 1.376 | 1.963 | 3.078 | 6.314 | 12.706 | 25.452 | 63.657 |
| 2 | 0.142 | 0.209 | 0.445 | 0.617 | 0.816 | 1.061 | 1.386 | 1.886 | 2.920 | 4.303 | 6.205 | 9.925 |
| 3 | 0.137 | 0.277 | 0.424 | 0.584 | 0.765 | 0.978 | 1.250 | 1.638 | 2.353 | 3.182 | 4.176 | 5.841 |
| 4 | 0.134 | 0.271 | 0.414 | 0.569 | 0.741 | 0.941 | 1.190 | 1.533 | 2.132 | 2.776 | 3.495 | 4.604 |
| 5 | 0.132 | 0.267 | 0.408 | 0.559 | 0.727 | 0.920 | 1.150 | 1.476 | 2.015 | 2.571 | 3.163 | 4.032 |
| 6 | 0.131 | 0.265 | 0.404 | 0.553 | 0.718 | 0.906 | 1.134 | 1.440 | 1.943 | 2.447 | 2.969 | 3.707 |
| 7 | 0.130 | 0.268 | 0.402 | 0.349 | 0.711 | 0.896 | 1.119 | 1.435 | 1.895 | 2.365 | 2.841 | 3.499 |
| 8 | 0.130 | 0.261 | 0.389 | 0.546 | 0.706 | 0.889 | 1.108 | 1.397 | 1.860 | 2.306 | 2.752 | 3.355 |
| 9 | 0.129 | 0.260 | 0.398 | 0.543 | 0.703 | 0.883 | 1.100 | 1.383 | 1.833 | 2.262 | 2.685 | 3.250 |
| 10 | 0.129 | 0.260 | 0.397 | 0.542 | 0.700 | 0.879 | 1.093 | 1.872 | 1.812 | 2.228 | 2.634 | 3.169 |
| 11 | 0.129 | 0.259 | 0.396 | 0.540 | 0.697 | 0.876 | 1.088 | 1.363 | 1.796 | 2.201 | 2.593 | 3.106 |
| 12 | 0.128 | 0.259 | 0.395 | 0.539 | 0.695 | 0.873 | 1.083 | 1.356 | 1.782 | 2.179 | 2.560 | 3.055 |
| 13 | 0.128 | 0.258 | 0.394 | 0.538 | 0.694 | 0.870 | 1.079 | 1.350 | 1.771 | 2.160 | 2.533 | 3.012 |
| 14 | 0.128 | 0.258 | 0.393 | 0.537 | 0.692 | 0.868 | 1.076 | 1.345 | 1.761 | 2.145 | 2.510 | 2.977 |
| 15 | 0.128 | 0.258 | 0.393 | 0.536 | 0.691 | 0.866 | 1.074 | 1.341 | 1.753 | 2.131 | 2.490 | 2.947 |
| 16 | 0.128 | 0.257 | 0.392 | 0.535 | 0.690 | 0.865 | 1.071 | 1.337 | 1.746 | 2.120 | 2.473 | 2.921 |
| 17 | 0.128 | 0.257 | 0.392 | 0.534 | 0.689 | 0.863 | 1.069 | 1.333 | 1.740 | 2.110 | 2.458 | 2.898 |
| 18 | 0.127 | 0.257 | 0.392 | 0.534 | 0.688 | 0.862 | 1.067 | 1.330 | 1.734 | 2.101 | 2.445 | 2.878 |
| 19 | 0.127 | 0.257 | 0.391 | 0.533 | 0.688 | 0.861 | 1.066 | 1.328 | 1.729 | 2.093 | 2.433 | 2.861 |
| 20 | 0.127 | 0.257 | 0.391 | 0.533 | 0.687 | 0.860 | 1.064 | 1.325 | 1.725 | 2.086 | 2.423 | 2.845 |
| 21 | 0.127 | 0.257 | 0.391 | 0.532 | 0.686 | 0.859 | 1.063 | 1.323 | 1.721 | 2.080 | 2.414 | 2.831 |
| 22 | 0.127 | 0.256 | 0.390 | 0.532 | 0.686 | 0.858 | 1.061 | 1.321 | 1.717 | 2.074 | 2.406 | 2.819 |
| 23 | 0.127 | 0.256 | 0.390 | 0.532 | 0.685 | 0.858 | 1.060 | 1.319 | 1.714 | 2.069 | 2.398 | 2.807 |
| 24 | 0.127 | 0.256 | 0.390 | 0.531 | 0.685 | 0.857 | 1.059 | 1.318 | 1.711 | 2.064 | 2.391 | 2.797 |
| 25 | 0.127 | 0.256 | 0.390 | 0.531 | 0.684 | 0.856 | 1.058 | 0.316 | 1.708 | 2.060 | 2.385 | 2.787 |
| 26 | 0.127 | 0.256 | 0.390 | 0.531 | 0.684 | 0.856 | 1.058 | 1.315 | 1.706 | 2.056 | 2.379 | 2.779 |
| 27 | 0.127 | 0.256 | 0.389 | 0.531 | 0.684 | 0.855 | 1.057 | 1.314 | 1.703 | 2.052 | 2.373 | 2.771 |
| 28 | 0.127 | 0.256 | 0.289 | 0.530 | 0.683 | 0.855 | 1.056 | 1.313 | 1.701 | 2.048 | 2.368 | 2.763 |
| 29 | 0.127 | 0.256 | 0.389 | 0.530 | 0.683 | 0.854 | 1.055 | 1.311 | 1.699 | 2.045 | 2.364 | 2.756 |
| 30 | 0.127 | 0.256 | 0.389 | 0.530 | 0.683 | 0.854 | 1.055 | 1.310 | 1.697 | 2.042 | 2.360 | 2.750 |
| ∞ | 0.12566 | 0.25335 | 0.38532 | 0.52440 | 0.67449 | 0.84162 | 1.03643 | 1.28155 | 1.64485 | 1.95996 | 2.2414 | 2.57532 |
| *k* | 0.55 | 0.60 | 0.65 | 0.70 | 0.75 | 0.80 | 0.85 | 0.90 | 0.95 | 0.975 | 0.9875 | 0.995 |
| | $P(t < t_b)$ (one-tailed) | | | | | | | | | | | |

*Source:* Adapted from Fisher, R. A. and Yates, F., 1957.

**TABLE A.9A**
**Critical Values of the $F$-Distribution when $\alpha = 0.05$**

$$\alpha = P[F > F_b(k_1, k_2)] = 1 - \left[ \frac{\Gamma(k/2) k_1^{k_1/2} k_2^{k_2/2}}{\Gamma(k_1/2)\Gamma(k_2/2)} \int_0^{F_b} \frac{F^{(k_1/2)-1}}{(k_1 F + k_2)^{k/2}} d_F \right] = 0.05$$

| $k_2$ | $k_1$ | | | | | | | | | | | | | | | | | | |
|---|---|---|---|---|---|---|---|---|---|---|---|---|---|---|---|---|---|---|---|
| | 1 | 2 | 3 | 4 | 5 | 6 | 7 | 8 | 9 | 10 | 12 | 15 | 20 | 24 | 30 | 40 | 60 | 120 | ∞ |
| 1 | 161.4 | 199.5 | 215.7 | 224.6 | 230.2 | 234.0 | 236.8 | 238.9 | 240.5 | 241.9 | 243.9 | 245.9 | 248.0 | 249.1 | 250.1 | 251.1 | 252.2 | 253.3 | 254.3 |
| 2 | 18.51 | 19.00 | 19.16 | 19.25 | 19.30 | 19.33 | 19.35 | 19.37 | 19.38 | 19.40 | 19.41 | 19.43 | 19.45 | 19.45 | 19.46 | 19.47 | 19.48 | 19.49 | 19.50 |
| 3 | 10.13 | 9.55 | 9.28 | 9.12 | 9.01 | 8.94 | 8.89 | 8.85 | 8.81 | 8.79 | 8.74 | 8.70 | 8.66 | 8.64 | 8.62 | 8.59 | 8.57 | 8.55 | 8.53 |
| 4 | 7.71 | 6.94 | 6.59 | 6.39 | 6.26 | 6.16 | 6.09 | 6.04 | 6.00 | 5.96 | 5.91 | 5.86 | 5.80 | 5.77 | 5.75 | 5.72 | 5.69 | 5.66 | 5.63 |
| 5 | 6.61 | 5.79 | 5.41 | 5.19 | 5.05 | 4.95 | 4.88 | 4.82 | 4.77 | 4.74 | 4.68 | 4.62 | 4.56 | 4.53 | 4.50 | 4.46 | 4.43 | 4.40 | 4.36 |
| 6 | 5.99 | 5.14 | 4.76 | 4.53 | 4.39 | 4.28 | 4.21 | 4.15 | 4.10 | 4.06 | 4.00 | 3.94 | 3.87 | 3.84 | 3.81 | 3.77 | 3.74 | 3.70 | 3.67 |
| 7 | 5.59 | 4.74 | 4.35 | 4.12 | 3.97 | 3.87 | 3.79 | 3.73 | 3.68 | 3.64 | 3.57 | 3.51 | 3.44 | 3.41 | 3.38 | 3.34 | 3.30 | 3.27 | 3.23 |
| 8 | 5.32 | 4.46 | 4.07 | 3.84 | 3.69 | 3.58 | 3.50 | 3.44 | 3.39 | 3.35 | 3.28 | 3.22 | 3.15 | 3.12 | 3.08 | 3.04 | 3.01 | 2.97 | 2.93 |
| 9 | 5.12 | 4.26 | 3.86 | 3.63 | 3.48 | 3.37 | 3.29 | 3.23 | 3.18 | 3.14 | 3.07 | 3.01 | 2.94 | 2.90 | 2.86 | 2.83 | 2.79 | 2.75 | 2.71 |
| 10 | 4.96 | 4.10 | 3.71 | 3.48 | 3.33 | 3.22 | 3.14 | 3.07 | 3.02 | 2.98 | 2.91 | 2.85 | 2.77 | 2.74 | 2.70 | 2.66 | 2.62 | 2.58 | 2.54 |
| 11 | 4.84 | 3.98 | 3.59 | 3.36 | 3.20 | 3.09 | 3.01 | 2.95 | 2.90 | 2.85 | 2.79 | 2.72 | 2.65 | 2.61 | 2.57 | 2.53 | 2.49 | 2.45 | 2.40 |
| 12 | 4.75 | 3.89 | 3.49 | 3.26 | 3.11 | 3.00 | 2.91 | 2.85 | 2.80 | 2.75 | 2.69 | 2.62 | 2.54 | 2.51 | 2.47 | 2.43 | 2.38 | 2.34 | 2.30 |
| 13 | 4.67 | 3.81 | 3.41 | 3.18 | 3.03 | 2.92 | 2.83 | 2.77 | 2.71 | 2.67 | 2.60 | 2.53 | 2.46 | 2.42 | 2.38 | 2.34 | 2.30 | 2.25 | 2.21 |
| 14 | 4.60 | 3.74 | 3.34 | 3.11 | 2.96 | 2.85 | 2.76 | 2.70 | 2.65 | 2.60 | 2.53 | 2.46 | 2.39 | 2.35 | 2.31 | 2.27 | 2.22 | 2.18 | 2.13 |
| 15 | 4.54 | 3.68 | 3.29 | 3.06 | 2.90 | 2.79 | 2.71 | 2.64 | 2.59 | 2.54 | 2.48 | 2.40 | 2.33 | 2.29 | 2.25 | 2.20 | 2.16 | 2.11 | 2.07 |
| 16 | 4.49 | 3.63 | 3.24 | 3.01 | 2.85 | 2.74 | 2.66 | 2.59 | 2.54 | 2.49 | 2.42 | 2.35 | 2.28 | 2.24 | 2.19 | 2.15 | 2.11 | 2.06 | 2.01 |
| 17 | 4.45 | 3.59 | 3.20 | 2.96 | 2.81 | 2.70 | 2.61 | 2.55 | 2.49 | 2.45 | 2.38 | 2.31 | 2.23 | 2.19 | 2.15 | 2.10 | 2.06 | 2.01 | 1.96 |
| 18 | 4.41 | 3.55 | 3.16 | 2.93 | 2.77 | 2.66 | 2.58 | 2.51 | 2.46 | 2.41 | 2.34 | 2.27 | 2.19 | 2.15 | 2.11 | 2.06 | 2.02 | 1.97 | 1.92 |
| 19 | 4.38 | 3.52 | 3.13 | 2.90 | 2.74 | 2.63 | 2.54 | 2.48 | 2.42 | 2.38 | 2.31 | 2.23 | 2.16 | 2.11 | 2.07 | 2.03 | 1.98 | 1.93 | 1.88 |

**TABLE A.9A** *(continued)*
**Critical Values of the *F*-Distribution when α = 0.05**

$$\alpha = P[F > F_b(k_1, k_2)] = 1 - \left[\frac{\Gamma(k/2)k_1^{k_1/2}k_2^{k_2/2}}{\Gamma(k_1/2)\Gamma(k_2/2)}\int_0^{F_b}\frac{F^{(k_1/2)-1}}{(k_1F+k_2)^{k/2}}d_F\right] = 0.05$$

| $k_2$ | $k_1$ = 1 | 2 | 3 | 4 | 5 | 6 | 7 | 8 | 9 | 10 | 12 | 15 | 20 | 24 | 30 | 40 | 60 | 120 | ∞ |
|---|---|---|---|---|---|---|---|---|---|---|---|---|---|---|---|---|---|---|---|
| 20 | 4.35 | 3.49 | 3.10 | 2.87 | 2.71 | 2.60 | 2.51 | 2.45 | 2.39 | 2.35 | 2.28 | 2.20 | 2.12 | 2.08 | 2.04 | 1.99 | 1.95 | 1.90 | 1.84 |
| 21 | 4.32 | 3.47 | 3.07 | 2.84 | 2.68 | 2.57 | 2.49 | 2.42 | 2.37 | 2.32 | 2.25 | 2.18 | 2.10 | 2.05 | 2.01 | 1.96 | 1.92 | 1.87 | 1.81 |
| 22 | 4.30 | 3.44 | 3.05 | 2.82 | 2.66 | 2.55 | 2.46 | 2.40 | 2.34 | 2.30 | 2.23 | 2.15 | 2.07 | 2.03 | 1.98 | 1.94 | 1.89 | 1.84 | 1.78 |
| 23 | 4.28 | 3.42 | 3.03 | 2.80 | 2.64 | 2.53 | 2.44 | 2.37 | 2.32 | 2.27 | 2.20 | 2.13 | 2.05 | 2.01 | 1.96 | 1.91 | 1.86 | 1.81 | 1.76 |
| 24 | 4.26 | 3.40 | 3.01 | 2.78 | 2.62 | 2.51 | 2.42 | 2.36 | 2.30 | 2.25 | 2.13 | 2.11 | 2.03 | 1.98 | 1.94 | 1.89 | 1.84 | 1.79 | 1.73 |
| 25 | 4.24 | 3.39 | 2.99 | 2.76 | 2.60 | 2.49 | 2.40 | 2.34 | 2.28 | 2.24 | 2.16 | 2.09 | 2.01 | 1.96 | 1.92 | 1.87 | 1.82 | 1.77 | 1.71 |
| 26 | 4.23 | 3.37 | 2.98 | 2.74 | 2.59 | 2.47 | 2.39 | 2.32 | 2.27 | 2.22 | 2.15 | 2.07 | 1.99 | 1.95 | 1.90 | 1.85 | 1.80 | 1.75 | 1.69 |
| 27 | 4.21 | 3.35 | 2.96 | 2.73 | 2.57 | 2.46 | 2.37 | 2.31 | 2.25 | 2.20 | 2.13 | 2.06 | 1.97 | 1.93 | 1.88 | 1.84 | 1.79 | 1.73 | 1.67 |
| 28 | 4.20 | 3.34 | 2.95 | 2.71 | 2.56 | 2.45 | 2.36 | 2.29 | 2.24 | 2.19 | 2.12 | 2.04 | 1.96 | 1.91 | 1.87 | 1.82 | 1.77 | 1.71 | 1.65 |
| 29 | 4.18 | 3.33 | 2.93 | 2.70 | 2.55 | 2.43 | 2.35 | 2.28 | 2.22 | 2.18 | 2.10 | 2.03 | 1.94 | 1.90 | 1.85 | 1.81 | 1.75 | 1.70 | 1.64 |
| 30 | 4.17 | 3.32 | 2.92 | 2.69 | 2.53 | 2.42 | 2.33 | 2.27 | 2.21 | 2.16 | 2.09 | 2.01 | 1.93 | 1.89 | 1.84 | 1.79 | 1.74 | 1.68 | 1.62 |
| 40 | 4.08 | 3.23 | 2.84 | 2.61 | 2.45 | 2.34 | 2.25 | 2.18 | 2.12 | 2.08 | 2.00 | 1.92 | 1.84 | 1.79 | 1.74 | 1.69 | 1.64 | 1.58 | 1.51 |
| 60 | 4.00 | 3.15 | 2.76 | 2.53 | 2.37 | 2.25 | 2.17 | 2.10 | 2.04 | 1.99 | 1.92 | 1.84 | 1.75 | 1.70 | 1.65 | 1.59 | 1.53 | 1.47 | 1.39 |
| 120 | 3.92 | 3.07 | 2.68 | 2.45 | 2.29 | 2.17 | 2.09 | 2.02 | 1.96 | 1.91 | 1.83 | 1.75 | 1.66 | 1.61 | 1.55 | 1.50 | 1.43 | 1.35 | 1.25 |
| ∞ | 3.84 | 3.00 | 2.60 | 2.37 | 2.21 | 2.10 | 2.01 | 1.94 | 1.88 | 1.83 | 1.75 | 1.67 | 1.57 | 1.52 | 1.46 | 1.39 | 1.32 | 1.22 | 1.00 |

*Note:* Where $k_1 = n_1 - 1$, $k_2 = n_2 - 1$, and $k = k_1 + k_2$.

*Source:* Adapted from Beyer, W. H., *Handbook of Tables for Probability and Statistics*, 2nd ed., CRC Press, Boca Raton, FL, 1968.

**TABLE A.9B**
**Critical Values of the $F$-Distribution When $\alpha = 0.025$**

| $k_2$ | $k_1$ = 1 | 2 | 3 | 4 | 5 | 6 | 7 | 8 | 9 | 10 | 12 | 15 | 20 | 24 | 30 | 40 | 60 | 120 | ∞ |
|---|---|---|---|---|---|---|---|---|---|---|---|---|---|---|---|---|---|---|---|
| 1 | 647.8 | 799.5 | 864.2 | 899.6 | 921.8 | 937.1 | 948.2 | 956.7 | 963.6 | 968.6 | 976.7 | 984.9 | 993.1 | 997.2 | 1001 | 1006 | 1010 | 1014 | 1018 |
| 2 | 38.51 | 39.00 | 39.17 | 39.25 | 39.30 | 39.33 | 39.36 | 39.37 | 39.39 | 39.40 | 39.41 | 39.43 | 39.45 | 39.46 | 39.46 | 39.47 | 39.48 | 39.49 | 39.50 |
| 3 | 17.44 | 16.04 | 15.44 | 15.10 | 14.88 | 14.73 | 14.32 | 14.54 | 14.47 | 14.42 | 14.34 | 14.25 | 14.17 | 14.12 | 14.08 | 14.04 | 13.99 | 13.95 | 13.90 |
| 4 | 12.22 | 10.65 | 9.98 | 9.60 | 9.36 | 9.20 | 9.07 | 8.98 | 8.90 | 8.84 | 8.75 | 8.66 | 8.56 | 8.51 | 8.46 | 8.41 | 8.36 | 8.31 | 8.26 |
| 5 | 10.01 | 8.43 | 7.76 | 7.39 | 7.15 | 6.93 | 6.85 | 6.76 | 6.68 | 6.62 | 6.52 | 6.43 | 6.33 | 6.28 | 6.23 | 6.18 | 6.12 | 6.07 | 6.02 |
| 6 | 8.81 | 7.26 | 6.60 | 6.23 | 5.99 | 5.82 | 5.70 | 5.60 | 5.52 | 5.46 | 5.37 | 5.27 | 5.17 | 5.12 | 5.07 | 5.01 | 4.96 | 4.90 | 4.85 |
| 7 | 8.07 | 6.54 | 5.89 | 5.52 | 5.29 | 5.12 | 4.99 | 4.90 | 4.82 | 4.76 | 4.67 | 4.57 | 4.47 | 4.42 | 4.36 | 4.31 | 4.25 | 4.20 | 4.14 |
| 8 | 7.57 | 6.06 | 5.42 | 5.05 | 4.82 | 4.65 | 4.53 | 4.43 | 4.36 | 4.30 | 4.20 | 4.10 | 4.00 | 3.95 | 3.89 | 3.84 | 3.78 | 3.73 | 3.67 |
| 9 | 7.21 | 5.71 | 5.08 | 4.72 | 4.48 | 4.32 | 4.20 | 4.10 | 4.03 | 3.96 | 3.87 | 3.77 | 3.67 | 3.61 | 3.56 | 3.51 | 3.45 | 3.39 | 3.33 |
| 10 | 6.94 | 5.46 | 4.83 | 4.47 | 4.24 | 4.07 | 3.95 | 3.85 | 3.78 | 3.72 | 3.62 | 3.52 | 3.42 | 3.37 | 3.31 | 3.26 | 3.20 | 3.14 | 3.08 |
| 11 | 6.72 | 5.26 | 4.63 | 4.28 | 4.04 | 3.88 | 3.76 | 3.66 | 3.59 | 3.53 | 3.43 | 3.33 | 3.23 | 3.17 | 3.12 | 3.06 | 3.00 | 2.94 | 2.88 |
| 12 | 6.55 | 5.10 | 4.47 | 4.12 | 3.89 | 3.73 | 3.61 | 3.51 | 3.44 | 3.37 | 3.28 | 3.18 | 3.07 | 3.02 | 2.96 | 2.91 | 2.85 | 2.79 | 2.72 |
| 13 | 6.41 | 4.97 | 4.35 | 4.00 | 3.77 | 3.60 | 3.48 | 3.39 | 3.31 | 3.25 | 3.15 | 3.05 | 2.95 | 2.89 | 2.84 | 2.78 | 2.72 | 2.66 | 2.60 |
| 14 | 6.30 | 4.86 | 4.24 | 3.89 | 3.66 | 3.50 | 3.38 | 3.29 | 3.21 | 3.15 | 3.05 | 2.95 | 2.84 | 2.79 | 2.73 | 2.67 | 2.61 | 2.55 | 2.49 |
| 15 | 6.20 | 4.77 | 4.15 | 3.80 | 3.58 | 3.41 | 3.29 | 3.20 | 3.12 | 3.06 | 2.96 | 2.86 | 2.76 | 2.70 | 2.64 | 2.59 | 2.52 | 2.46 | 2.40 |
| 16 | 6.12 | 4.69 | 4.08 | 3.73 | 3.50 | 3.34 | 3.22 | 3.12 | 3.05 | 2.99 | 2.89 | 2.79 | 2.68 | 2.63 | 2.57 | 2.51 | 2.45 | 2.38 | 2.32 |
| 17 | 6.04 | 4.62 | 4.01 | 3.66 | 3.44 | 3.28 | 3.16 | 3.06 | 2.98 | 2.92 | 2.82 | 2.72 | 2.62 | 2.56 | 2.50 | 2.44 | 2.38 | 2.32 | 2.25 |
| 18 | 5.98 | 4.56 | 3.95 | 3.61 | 3.38 | 3.22 | 3.10 | 3.01 | 2.93 | 2.87 | 2.77 | 2.67 | 2.56 | 2.50 | 2.44 | 2.38 | 2.32 | 2.26 | 2.19 |
| 19 | 5.92 | 4.51 | 3.90 | 3.56 | 3.33 | 3.17 | 3.05 | 2.96 | 2.88 | 2.82 | 2.72 | 2.62 | 2.51 | 2.45 | 2.39 | 2.33 | 2.27 | 2.20 | 2.13 |

**TABLE A.9B** *(continued)*
**Critical Values of the *F*-Distribution When $\alpha = 0.025$**

| $k_2$ | $k_1$ | | | | | | | | | | | | | | | | | | |
|---|---|---|---|---|---|---|---|---|---|---|---|---|---|---|---|---|---|---|---|
| | 1 | 2 | 3 | 4 | 5 | 6 | 7 | 8 | 9 | 10 | 12 | 15 | 20 | 24 | 30 | 40 | 60 | 120 | ∞ |
| 20 | 5.87 | 4.46 | 3.86 | 3.51 | 3.29 | 3.13 | 3.01 | 2.91 | 2.84 | 2.77 | 2.68 | 2.57 | 2.46 | 2.41 | 2.35 | 2.29 | 2.22 | 2.16 | 2.09 |
| 21 | 5.83 | 4.42 | 3.82 | 3.48 | 3.25 | 3.09 | 2.97 | 2.87 | 2.80 | 2.73 | 2.64 | 2.53 | 2.42 | 2.37 | 2.31 | 2.25 | 2.18 | 2.11 | 2.04 |
| 22 | 5.79 | 4.38 | 3.78 | 3.44 | 3.22 | 3.05 | 2.93 | 2.84 | 2.76 | 2.70 | 2.60 | 2.50 | 2.39 | 2.33 | 2.27 | 2.21 | 2.14 | 2.08 | 2.00 |
| 23 | 5.75 | 4.35 | 3.75 | 3.41 | 3.18 | 3.02 | 2.90 | 2.81 | 2.73 | 2.67 | 2.57 | 2.47 | 2.36 | 2.30 | 2.24 | 2.18 | 2.11 | 2.04 | 1.97 |
| 24 | 5.72 | 4.32 | 3.72 | 3.38 | 3.15 | 2.99 | 2.87 | 2.78 | 2.70 | 2.64 | 2.54 | 2.44 | 2.33 | 2.27 | 2.21 | 2.15 | 2.08 | 2.01 | 1.94 |
| 25 | 5.69 | 4.29 | 3.69 | 3.35 | 3.13 | 2.97 | 2.85 | 2.75 | 2.68 | 2.61 | 2.51 | 2.41 | 2.30 | 2.24 | 2.18 | 2.12 | 2.05 | 1.98 | 1.91 |
| 26 | 5.66 | 4.27 | 3.67 | 3.33 | 3.10 | 2.94 | 2.82 | 2.73 | 2.65 | 2.59 | 2.49 | 2.39 | 2.28 | 2.22 | 2.16 | 2.09 | 2.03 | 1.95 | 1.88 |
| 27 | 5.63 | 4.24 | 3.65 | 3.31 | 3.08 | 2.92 | 2.80 | 2.71 | 2.63 | 2.57 | 2.47 | 2.36 | 2.25 | 2.19 | 2.13 | 2.07 | 2.00 | 1.93 | 1.85 |
| 28 | 5.61 | 4.22 | 3.63 | 3.29 | 3.06 | 2.90 | 2.78 | 2.69 | 2.61 | 2.55 | 2.45 | 2.34 | 2.23 | 2.17 | 2.11 | 2.05 | 1.98 | 1.91 | 1.83 |
| 29 | 5.59 | 4.20 | 3.61 | 3.27 | 3.04 | 2.88 | 2.76 | 2.67 | 2.59 | 2.53 | 2.43 | 2.32 | 2.21 | 2.15 | 2.09 | 2.03 | 1.95 | 1.89 | 1.81 |
| 30 | 5.57 | 4.18 | 3.59 | 3.25 | 3.03 | 2.87 | 2.75 | 2.65 | 2.57 | 2.51 | 2.41 | 2.31 | 2.20 | 2.14 | 2.07 | 2.01 | 1.94 | 1.87 | 1.79 |
| 40 | 5.42 | 4.05 | 3.46 | 3.13 | 2.90 | 2.74 | 2.62 | 2.53 | 2.45 | 2.39 | 2.29 | 2.18 | 2.07 | 2.01 | 1.94 | 1.88 | 1.80 | 1.72 | 1.64 |
| 60 | 5.29 | 3.93 | 3.34 | 3.01 | 2.79 | 2.63 | 2.51 | 2.41 | 2.33 | 2.27 | 2.17 | 2.06 | 1.94 | 1.88 | 1.82 | 1.74 | 1.67 | 1.58 | 1.48 |
| 120 | 5.15 | 3.80 | 3.23 | 2.89 | 2.67 | 2.52 | 2.39 | 2.30 | 2.22 | 2.16 | 2.05 | 1.94 | 1.82 | 1.76 | 1.69 | 1.61 | 1.53 | 1.43 | 1.31 |
| ∞ | 5.02 | 3.69 | 3.12 | 2.79 | 2.57 | 2.41 | 2.29 | 2.19 | 2.11 | 2.05 | 1.94 | 1.83 | 1.71 | 1.64 | 1.57 | 1.48 | 1.39 | 1.27 | 1.00 |

**TABLE A.9C**
**Critical Values of the *F*-Distribution When $\alpha = 0.01$**

| $k_2$ | $k_1$: 1 | 2 | 3 | 4 | 5 | 6 | 7 | 8 | 9 | 10 | 12 | 15 | 20 | 24 | 30 | 40 | 60 | 120 | ∞ |
|---|---|---|---|---|---|---|---|---|---|---|---|---|---|---|---|---|---|---|---|
| 1 | 4052 | 4999.5 | 5403 | 5625 | 5764 | 5859 | 5928 | 5892 | 6022 | 6056 | 6106 | 6157 | 6209 | 6235 | 6261 | 6287 | 6313 | 6339 | 6366 |
| 2 | 98.50 | 99.00 | 99.17 | 99.25 | 99.30 | 99.33 | 99.36 | 99.37 | 99.39 | 99.40 | 99.42 | 99.43 | 99.45 | 99.46 | 99.47 | 99.47 | 99.48 | 99.49 | 99.50 |
| 3 | 34.12 | 30.82 | 29.46 | 28.71 | 28.24 | 27.91 | 27.67 | 27.49 | 27.35 | 27.23 | 27.05 | 26.87 | 26.69 | 26.60 | 26.50 | 26.41 | 26.32 | 26.22 | 26.13 |
| 4 | 21.20 | 18.00 | 16.69 | 15.98 | 15.52 | 15.21 | 14.98 | 14.80 | 14.66 | 14.55 | 14.37 | 14.20 | 14.02 | 13.93 | 13.84 | 13.75 | 13.65 | 13.56 | 13.46 |
| 5 | 16.26 | 13.27 | 12.06 | 11.39 | 10.97 | 10.67 | 10.46 | 10.29 | 10.16 | 10.05 | 9.89 | 9.72 | 9.55 | 9.47 | 9.38 | 9.29 | 9.20 | 9.11 | 9.02 |
| 6 | 13.75 | 10.92 | 9.78 | 9.15 | 8.75 | 8.47 | 8.26 | 8.10 | 7.98 | 7.87 | 7.72 | 7.56 | 7.40 | 7.31 | 7.23 | 7.14 | 7.06 | 6.97 | 6.88 |
| 7 | 12.25 | 9.55 | 8.45 | 7.85 | 7.46 | 7.19 | 6.99 | 6.84 | 6.72 | 6.62 | 6.47 | 6.31 | 6.16 | 6.07 | 5.99 | 5.91 | 5.82 | 5.74 | 5.65 |
| 8 | 11.26 | 8.65 | 7.59 | 7.01 | 6.63 | 6.37 | 6.18 | 6.03 | 5.91 | 5.81 | 5.67 | 5.52 | 5.36 | 5.28 | 5.20 | 5.12 | 5.03 | 4.95 | 4.86 |
| 9 | 10.56 | 8.02 | 6.99 | 6.42 | 6.06 | 5.80 | 5.61 | 5.47 | 5.35 | 5.26 | 5.11 | 4.96 | 4.81 | 4.73 | 4.65 | 4.57 | 4.48 | 4.40 | 4.31 |
| 10 | 10.04 | 7.56 | 6.55 | 5.99 | 5.64 | 5.39 | 5.20 | 5.06 | 4.94 | 4.85 | 4.71 | 4.56 | 4.41 | 4.33 | 4.25 | 4.17 | 4.08 | 4.00 | 3.91 |
| 11 | 9.65 | 7.21 | 6.22 | 5.67 | 5.32 | 5.07 | 4.89 | 4.74 | 4.63 | 4.54 | 4.40 | 4.25 | 4.10 | 4.02 | 3.94 | 3.86 | 3.78 | 3.69 | 3.60 |
| 12 | 9.33 | 6.93 | 5.95 | 5.41 | 5.06 | 4.82 | 4.64 | 4.50 | 4.39 | 4.30 | 4.16 | 4.01 | 3.86 | 3.78 | 3.70 | 3.62 | 3.54 | 3.45 | 3.36 |
| 13 | 9.07 | 6.70 | 5.74 | 5.21 | 4.86 | 4.62 | 4.44 | 4.30 | 4.19 | 4.10 | 3.96 | 3.82 | 3.66 | 3.59 | 3.51 | 3.43 | 3.34 | 3.25 | 3.17 |
| 14 | 8.86 | 6.51 | 5.56 | 5.04 | 4.69 | 4.46 | 4.28 | 4.14 | 4.03 | 3.94 | 3.80 | 3.66 | 3.51 | 3.43 | 3.35 | 3.27 | 3.18 | 3.09 | 3.00 |
| 15 | 8.68 | 6.36 | 5.42 | 4.89 | 4.56 | 4.32 | 4.14 | 4.00 | 3.89 | 3.80 | 3.67 | 3.52 | 3.37 | 3.29 | 3.21 | 3.13 | 3.05 | 2.96 | 2.87 |
| 16 | 8.53 | 6.23 | 5.29 | 4.77 | 4.44 | 4.20 | 4.03 | 3.89 | 3.78 | 3.69 | 3.55 | 3.41 | 3.26 | 3.18 | 3.10 | 3.02 | 2.93 | 2.84 | 2.75 |
| 17 | 8.40 | 6.11 | 5.18 | 4.67 | 4.34 | 4.10 | 3.93 | 3.79 | 3.68 | 3.59 | 3.46 | 3.31 | 3.16 | 3.08 | 3.00 | 2.92 | 2.83 | 2.75 | 2.65 |
| 18 | 8.29 | 6.01 | 5.09 | 4.58 | 4.25 | 4.01 | 3.84 | 3.71 | 3.60 | 3.51 | 3.37 | 3.23 | 3.08 | 3.00 | 2.92 | 2.84 | 2.75 | 2.66 | 2.57 |
| 19 | 8.18 | 5.93 | 5.01 | 4.50 | 4.17 | 3.94 | 3.77 | 3.63 | 3.52 | 3.43 | 3.30 | 3.15 | 3.00 | 2.92 | 2.84 | 2.76 | 2.67 | 2.58 | 2.49 |

**TABLE A.9C** *(continued)*
**Critical Values of the *F*-Distribution When $\alpha = 0.01$**

| $k_2$ | $k_1$ | | | | | | | | | | | | | | | | | | |
|---|---|---|---|---|---|---|---|---|---|---|---|---|---|---|---|---|---|---|---|
| | **1** | **2** | **3** | **4** | **5** | **6** | **7** | **8** | **9** | **10** | **12** | **15** | **20** | **24** | **30** | **40** | **60** | **120** | **∞** |
| 20 | 8.10 | 5.85 | 4.94 | 4.43 | 4.10 | 3.87 | 3.70 | 3.56 | 3.46 | 3.37 | 3.23 | 3.09 | 2.94 | 2.86 | 2.78 | 2.69 | 2.61 | 2.52 | 2.42 |
| 21 | 8.02 | 5.78 | 4.87 | 4.37 | 4.04 | 3.81 | 3.64 | 3.51 | 3.40 | 3.31 | 3.17 | 3.03 | 2.88 | 2.80 | 2.72 | 2.64 | 2.55 | 2.46 | 2.36 |
| 22 | 7.95 | 5.72 | 4.82 | 4.31 | 3.99 | 3.76 | 3.59 | 3.45 | 3.35 | 3.26 | 3.12 | 2.98 | 2.83 | 2.75 | 2.67 | 2.58 | 2.50 | 2.40 | 2.31 |
| 23 | 7.88 | 5.66 | 4.76 | 4.26 | 3.94 | 3.71 | 3.54 | 3.41 | 3.30 | 3.21 | 3.07 | 2.93 | 2.78 | 2.70 | 2.62 | 2.54 | 2.45 | 2.35 | 2.26 |
| 24 | 7.82 | 5.61 | 4.72 | 4.22 | 3.90 | 3.67 | 3.50 | 3.36 | 3.26 | 3.17 | 3.03 | 2.89 | 2.74 | 2.66 | 2.58 | 2.49 | 2.40 | 2.31 | 2.21 |
| 25 | 7.77 | 5.57 | 4.08 | 4.18 | 3.85 | 3.63 | 3.46 | 3.32 | 3.22 | 3.13 | 2.99 | 2.85 | 2.70 | 2.62 | 2.54 | 2.45 | 2.36 | 2.27 | 2.17 |
| 26 | 7.72 | 5.53 | 4.64 | 4.14 | 3.82 | 3.59 | 3.42 | 3.29 | 3.18 | 3.09 | 2.96 | 2.81 | 2.66 | 2.58 | 2.50 | 2.42 | 2.33 | 2.23 | 2.13 |
| 27 | 7.68 | 5.49 | 4.60 | 4.11 | 3.78 | 3.56 | 3.39 | 3.26 | 3.15 | 3.06 | 2.93 | 2.78 | 2.63 | 2.55 | 2.47 | 2.38 | 2.29 | 2.20 | 2.10 |
| 28 | 7.64 | 5.45 | 4.57 | 4.07 | 3.75 | 3.53 | 3.36 | 3.23 | 3.12 | 3.03 | 2.90 | 2.75 | 2.60 | 2.52 | 2.44 | 2.35 | 2.26 | 2.17 | 2.06 |
| 29 | 7.60 | 5.42 | 4.54 | 4.04 | 3.73 | 3.50 | 3.33 | 3.20 | 3.09 | 3.00 | 2.87 | 2.73 | 2.57 | 2.49 | 2.41 | 2.33 | 2.23 | 2.14 | 2.03 |
| 30 | 7.56 | 5.39 | 4.51 | 4.02 | 3.70 | 3.47 | 3.30 | 3.17 | 3.07 | 2.98 | 2.84 | 2.70 | 2.55 | 2.47 | 2.39 | 2.30 | 2.21 | 2.11 | 2.01 |
| 40 | 7.31 | 5.18 | 4.31 | 3.83 | 3.51 | 3.29 | 3.12 | 2.99 | 2.89 | 2.80 | 2.66 | 2.52 | 2.37 | 2.29 | 2.20 | 2.11 | 2.02 | 1.92 | 1.80 |
| 60 | 7.08 | 4.98 | 4.13 | 3.65 | 3.34 | 3.12 | 2.95 | 2.82 | 2.72 | 2.63 | 2.50 | 2.35 | 2.20 | 2.12 | 2.03 | 1.94 | 1.84 | 1.73 | 1.60 |
| 120 | 6.85 | 4.79 | 3.95 | 3.48 | 3.17 | 2.96 | 2.79 | 2.66 | 2.56 | 2.47 | 2.34 | 2.19 | 2.03 | 1.95 | 1.86 | 1.76 | 1.66 | 1.53 | 1.38 |
| ∞ | 6.63 | 4.61 | 3.78 | 3.32 | 3.02 | 2.80 | 2.64 | 2.51 | 2.41 | 2.32 | 2.18 | 2.04 | 1.88 | 1.79 | 1.70 | 1.59 | 1.47 | 1.32 | 1.00 |

## TABLE A.10A
## Confidence Belts for Proportions

**$C = 0.95$**

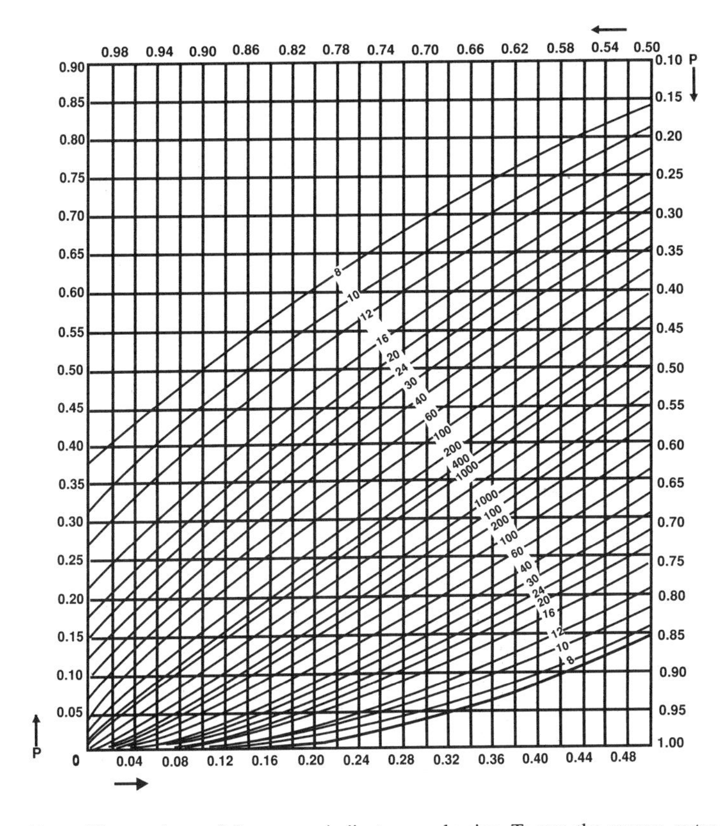


*Note:* The numbers of the curves indicate sample size. To use the curves, enter along the abscissa with $\hat{p}_o = y_o/n$ and read the ordinates $\hat{p}_a$ and $\hat{p}_b$ from the upper and lower curves to obtain the interval $\hat{p}_a \le p \le \hat{p}_b$.

*Source:* Adapted from Beyer, W. H., *Handbook of Tables for Probability and Statistics,* 2nd ed., CRC Press, Boca Raton, FL, 1968.

## TABLE A.10B
## Confidence Belts for Proportions

$C = 0.99$

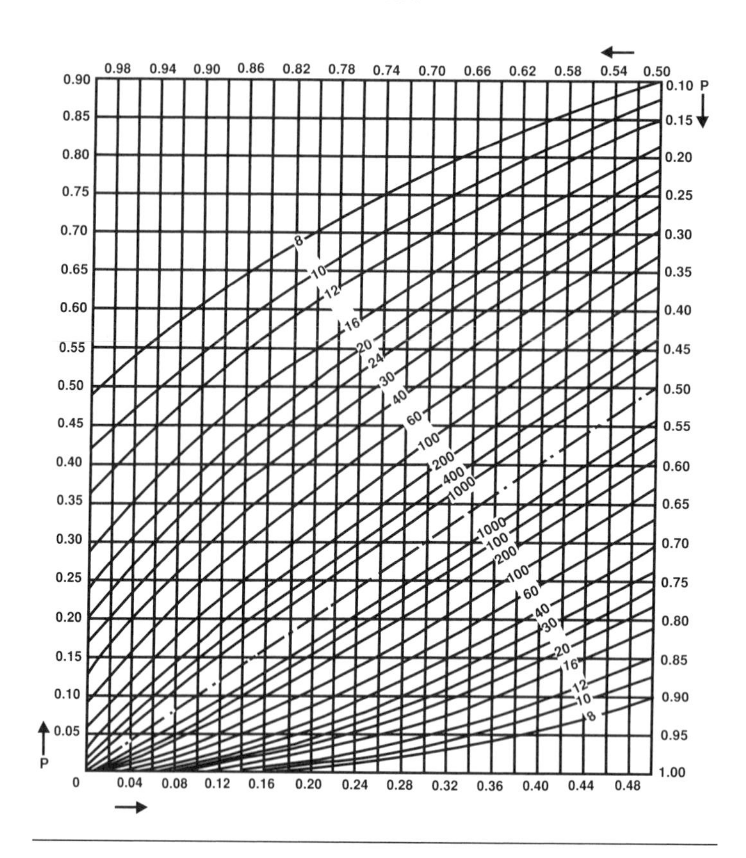

**TABLE A.11**
**Upper 5% and 1% Critical Values of the Studentized Range**

$$q = (y_{max} - y_{min})s_y$$

| Error | | $p$ = Number of Treatment Means | | | | | | | | | | | | | | | | | | | Error |
|---|---|---|---|---|---|---|---|---|---|---|---|---|---|---|---|---|---|---|---|---|---|
| $k$ | α | 2 | 3 | 4 | 5 | 6 | 7 | 8 | 9 | 10 | 11 | 12 | 13 | 14 | 15 | 16 | 17 | 18 | 19 | 20 | α | $k$ |
| 5 | 0.05 | 3.64 | 4.60 | 5.22 | 5.67 | 6.03 | 6.33 | 6.58 | 6.80 | 6.99 | 7.17 | 7.32 | 7.47 | 7.60 | 7.72 | 7.83 | 7.93 | 8.03 | 8.12 | 8.21 | 0.05 | 5 |
| | 0.01 | 5.70 | 6.97 | 7.80 | 8.42 | 8.91 | 9.32 | 9.67 | 9.97 | 10.24 | 10.48 | 10.70 | 10.89 | 11.08 | 11.24 | 11.40 | 11.55 | 11.68 | 11.81 | 11.93 | 0.01 | |
| 6 | 0.05 | 3.46 | 4.34 | 4.90 | 5.31 | 5.63 | 5.89 | 6.12 | 6.32 | 6.49 | 6.65 | 6.79 | 6.92 | 7.03 | 7.14 | 7.24 | 7.34 | 7.43 | 7.51 | 7.59 | 0.05 | 6 |
| | 0.01 | 5.24 | 6.33 | 7.03 | 7.56 | 7.97 | 8.32 | 8.61 | 8.87 | 9.10 | 9.30 | 9.49 | 9.65 | 9.81 | 9.95 | 10.08 | 10.21 | 10.32 | 10.43 | 10.54 | 0.01 | |
| 7 | 0.05 | 3.34 | 4.16 | 4.68 | 5.06 | 5.36 | 5.61 | 5.82 | 6.00 | 6.16 | 6.30 | 6.43 | 6.55 | 6.66 | 6.76 | 6.85 | 6.94 | 7.02 | 7.09 | 7.17 | 0.05 | 7 |
| | 0.01 | 4.95 | 5.92 | 6.54 | 7.01 | 7.37 | 7.68 | 7.94 | 8.17 | 8.37 | 8.55 | 8.71 | 8.86 | 9.00 | 9.12 | 9.24 | 9.35 | 9.46 | 9.55 | 9.65 | 0.01 | |
| 8 | 0.05 | 3.26 | 4.04 | 4.53 | 4.89 | 5.17 | 5.40 | 5.60 | 5.77 | 5.92 | 6.05 | 6.18 | 6.29 | 6.39 | 6.48 | 6.57 | 6.65 | 6.73 | 6.80 | 6.87 | 0.05 | 8 |
| | 0.01 | 4.74 | 5.63 | 6.20 | 6.63 | 6.96 | 7.24 | 7.47 | 7.68 | 7.87 | 8.03 | 8.18 | 8.31 | 8.44 | 8.55 | 8.66 | 8.76 | 8.85 | 8.94 | 9.03 | 0.01 | |
| 9 | 0.05 | 3.20 | 3.95 | 4.42 | 4.76 | 5.02 | 5.24 | 5.43 | 5.60 | 5.74 | 5.87 | 5.98 | 6.09 | 6.19 | 6.28 | 6.36 | 6.44 | 6.51 | 6.58 | 6.64 | 0.05 | 9 |
| | 0.01 | 4.60 | 5.43 | 5.96 | 6.35 | 6.66 | 6.91 | 7.13 | 7.32 | 7.49 | 7.65 | 7.78 | 7.91 | 8.03 | 8.13 | 8.23 | 8.32 | 8.41 | 8.49 | 8.57 | 0.01 | |
| 10 | 0.05 | 3.15 | 3.88 | 4.33 | 4.65 | 4.91 | 5.12 | 5.30 | 5.46 | 5.60 | 5.72 | 5.83 | 5.93 | 6.03 | 6.11 | 6.20 | 6.27 | 6.34 | 6.40 | 6.47 | 0.05 | 10 |
| | 0.01 | 4.48 | 5.27 | 5.77 | 6.14 | 6.43 | 6.67 | 6.87 | 7.05 | 7.21 | 7.36 | 7.48 | 7.60 | 7.71 | 7.81 | 7.91 | 7.99 | 8.07 | 8.15 | 8.22 | 0.01 | |
| 11 | 0.05 | 3.11 | 3.82 | 4.26 | 4.57 | 4.82 | 5.03 | 5.20 | 5.35 | 5.49 | 5.61 | 5.71 | 5.81 | 5.90 | 5.99 | 6.06 | 6.14 | 6.20 | 6.26 | 6.33 | 0.05 | 11 |
| | 0.01 | 4.39 | 5.14 | 5.62 | 5.97 | 6.25 | 6.48 | 6.67 | 6.84 | 6.99 | 7.13 | 7.25 | 7.36 | 7.46 | 7.56 | 7.65 | 7.73 | 7.81 | 7.88 | 7.95 | 0.01 | |
| 12 | 0.05 | 3.08 | 3.77 | 4.20 | 4.51 | 4.75 | 4.95 | 5.12 | 5.27 | 5.40 | 5.51 | 5.62 | 5.71 | 5.80 | 5.88 | 5.95 | 6.03 | 6.09 | 6.15 | 6.21 | 0.05 | 12 |
| | 0.01 | 4.32 | 5.04 | 5.50 | 5.84 | 6.10 | 6.32 | 6.51 | 6.67 | 6.81 | 6.94 | 7.06 | 7.17 | 7.26 | 7.36 | 7.44 | 7.52 | 7.59 | 7.66 | 7.73 | 0.01 | |
| 13 | 0.05 | 3.06 | 3.73 | 4.15 | 4.45 | 4.69 | 4.88 | 5.05 | 5.19 | 5.32 | 5.43 | 5.53 | 5.63 | 5.71 | 5.79 | 5.86 | 5.93 | 6.00 | 6.05 | 6.11 | 0.05 | 13 |
| | 0.01 | 4.26 | 4.96 | 5.40 | 5.73 | 5.98 | 6.19 | 6.37 | 6.53 | 6.67 | 6.79 | 6.90 | 7.01 | 7.10 | 7.19 | 7.27 | 7.34 | 7.42 | 7.48 | 7.55 | 0.01 | |
| 14 | 0.05 | 3.03 | 3.70 | 4.11 | 4.41 | 4.64 | 4.83 | 4.99 | 5.13 | 5.25 | 5.36 | 5.46 | 5.55 | 5.64 | 5.72 | 5.79 | 5.85 | 5.92 | 5.97 | 6.03 | 0.05 | 14 |
| | 0.01 | 4.21 | 4.89 | 5.32 | 5.63 | 5.88 | 6.08 | 6.26 | 6.41 | 6.54 | 6.66 | 6.77 | 6.87 | 6.96 | 7.05 | 7.12 | 7.20 | 7.27 | 7.33 | 7.39 | 0.01 | |
| 15 | 0.05 | 3.01 | 3.67 | 4.08 | 4.37 | 4.60 | 4.78 | 4.94 | 5.08 | 5.20 | 5.31 | 5.40 | 5.49 | 5.58 | 5.65 | 5.72 | 5.79 | 5.85 | 5.90 | 5.96 | 0.05 | 15 |
| | 0.01 | 4.17 | 4.83 | 5.25 | 5.56 | 5.80 | 5.99 | 6.16 | 6.31 | 6.44 | 6.55 | 6.66 | 6.76 | 6.84 | 6.93 | 7.00 | 7.07 | 7.14 | 7.20 | 7.26 | 0.01 | |
| 16 | 0.05 | 3.00 | 3.65 | 4.05 | 4.33 | 4.56 | 4.74 | 4.90 | 5.03 | 5.15 | 5.26 | 5.35 | 5.44 | 5.52 | 5.59 | 5.66 | 5.72 | 5.79 | 5.84 | 5.90 | 0.05 | 16 |
| | 0.01 | 4.13 | 4.78 | 5.19 | 5.49 | 5.72 | 5.92 | 6.08 | 6.22 | 6.35 | 6.46 | 6.56 | 6.66 | 6.74 | 6.82 | 6.90 | 6.97 | 7.03 | 7.09 | 7.15 | 0.01 | |
| 17 | 0.05 | 2.98 | 3.63 | 4.02 | 4.30 | 4.52 | 4.71 | 4.86 | 4.99 | 5.11 | 5.21 | 5.31 | 5.39 | 5.47 | 5.55 | 5.61 | 5.68 | 5.74 | 5.79 | 5.84 | 0.05 | 17 |
| | 0.01 | 4.10 | 4.74 | 5.14 | 5.43 | 5.66 | 5.85 | 6.01 | 6.15 | 6.27 | 6.38 | 6.48 | 6.57 | 6.66 | 6.73 | 6.80 | 6.87 | 6.94 | 7.00 | 7.05 | 0.01 | |

**TABLE A.11** *(continued)*
**Upper 5% and 1% Critical Values of the Studentized Range**

$$q = (y_{max} - y_{min})s_y$$

| Error | | p = Number of Treatment Means | | | | | | | | | | | | | | | | | | | | Error |
|---|---|---|---|---|---|---|---|---|---|---|---|---|---|---|---|---|---|---|---|---|---|---|
| k | α | 2 | 3 | 4 | 5 | 6 | 7 | 8 | 9 | 10 | 11 | 12 | 13 | 14 | 15 | 16 | 17 | 18 | 19 | 20 | α | k |
| 18 | 0.05 | 2.97 | 3.61 | 4.00 | 4.28 | 4.49 | 4.67 | 4.82 | 4.96 | 5.07 | 5.17 | 5.27 | 5.35 | 5.43 | 5.50 | 5.57 | 5.63 | 5.69 | 5.74 | 5.79 | 0.05 | 18 |
| | 0.01 | 4.07 | 4.70 | 5.09 | 5.38 | 5.60 | 5.79 | 5.94 | 6.08 | 6.20 | 6.31 | 6.41 | 6.50 | 6.58 | 6.65 | 6.72 | 6.79 | 6.85 | 6.91 | 6.96 | 0.01 | |
| 19 | 0.05 | 2.96 | 3.59 | 3.98 | 4.25 | 4.47 | 4.65 | 4.79 | 4.92 | 5.04 | 5.14 | 5.23 | 5.32 | 5.39 | 5.46 | 5.53 | 5.59 | 5.65 | 5.70 | 5.75 | 0.05 | 19 |
| | 0.01 | 4.05 | 4.67 | 5.05 | 5.33 | 5.55 | 5.73 | 5.89 | 6.02 | 6.14 | 6.25 | 6.34 | 6.43 | 6.51 | 6.58 | 6.65 | 6.72 | 6.78 | 6.84 | 6.89 | 0.01 | |
| 20 | 0.05 | 2.95 | 3.58 | 3.96 | 4.23 | 4.45 | 4.62 | 4.77 | 4.90 | 5.01 | 5.11 | 5.20 | 5.28 | 5.36 | 5.43 | 5.49 | 5.55 | 5.61 | 5.66 | 5.71 | 0.05 | 20 |
| | 0.01 | 4.02 | 4.64 | 5.02 | 5.29 | 5.51 | 5.69 | 5.84 | 5.97 | 6.09 | 6.19 | 6.29 | 6.37 | 6.45 | 6.52 | 6.59 | 6.65 | 6.71 | 6.76 | 6.82 | 0.01 | |
| 24 | 0.05 | 2.92 | 3.53 | 3.90 | 4.17 | 4.37 | 4.54 | 4.68 | 4.81 | 4.92 | 5.01 | 5.10 | 5.18 | 5.25 | 5.32 | 5.38 | 5.44 | 5.50 | 5.54 | 5.59 | 0.05 | 24 |
| | 0.01 | 3.96 | 4.54 | 4.91 | 5.17 | 5.37 | 5.54 | 5.69 | 5.81 | 5.92 | 6.02 | 6.11 | 6.19 | 6.26 | 6.33 | 6.39 | 6.45 | 6.51 | 6.56 | 6.61 | 0.01 | |
| 30 | 0.05 | 2.89 | 3.49 | 3.84 | 4.10 | 4.30 | 4.46 | 4.60 | 4.72 | 4.83 | 4.92 | 5.00 | 5.08 | 5.15 | 5.21 | 5.27 | 5.33 | 5.38 | 5.43 | 5.48 | 0.05 | 30 |
| | 0.01 | 3.89 | 4.45 | 4.80 | 5.05 | 5.24 | 5.40 | 5.54 | 5.65 | 5.76 | 5.85 | 5.93 | 6.01 | 6.08 | 6.14 | 6.20 | 6.26 | 6.31 | 6.36 | 6.41 | 0.01 | |
| 40 | 0.05 | 2.86 | 3.44 | 3.79 | 4.04 | 4.23 | 4.39 | 4.52 | 4.63 | 4.74 | 4.82 | 4.91 | 4.98 | 5.05 | 5.11 | 5.16 | 5.22 | 5.27 | 5.31 | 5.36 | 0.05 | 40 |
| | 0.01 | 3.82 | 4.37 | 4.70 | 4.93 | 5.11 | 5.27 | 5.39 | 5.50 | 5.60 | 5.69 | 5.77 | 5.84 | 5.90 | 5.96 | 6.02 | 6.07 | 6.12 | 6.17 | 6.21 | 0.01 | |
| 60 | 0.05 | 2.83 | 3.40 | 3.74 | 3.98 | 4.16 | 4.31 | 4.44 | 4.55 | 4.65 | 4.73 | 4.81 | 4.88 | 4.94 | 5.00 | 5.06 | 5.11 | 5.16 | 5.20 | 5.24 | 0.05 | 60 |
| | 0.01 | 3.76 | 4.28 | 4.60 | 4.82 | 4.99 | 5.13 | 5.25 | 5.36 | 5.45 | 5.53 | 5.60 | 5.67 | 5.73 | 5.79 | 5.84 | 5.89 | 5.93 | 5.98 | 6.02 | 0.01 | |
| 120 | 0.05 | 2.80 | 3.36 | 3.69 | 3.92 | 4.10 | 4.24 | 4.36 | 4.48 | 4.56 | 4.64 | 4.72 | 4.78 | 4.84 | 4.90 | 4.95 | 5.00 | 5.05 | 5.09 | 5.13 | 0.05 | 120 |
| | 0.01 | 3.70 | 4.20 | 4.50 | 4.71 | 4.87 | 5.01 | 5.12 | 5.21 | 5.30 | 5.38 | 5.44 | 5.51 | 5.56 | 5.61 | 5.66 | 5.71 | 5.75 | 5.79 | 5.83 | 0.01 | |
| ∞ | 0.05 | 2.77 | 3.31 | 3.63 | 3.86 | 4.03 | 4.17 | 4.29 | 4.39 | 4.47 | 4.55 | 4.62 | 4.68 | 4.74 | 4.80 | 4.85 | 4.89 | 4.93 | 4.97 | 5.01 | 0.05 | ∞ |
| | 0.01 | 3.64 | 4.12 | 4.40 | 4.60 | 4.76 | 4.88 | 4.99 | 5.08 | 5.16 | 5.23 | 5.29 | 5.35 | 5.40 | 5.45 | 5.49 | 5.54 | 5.57 | 5.61 | 5.65 | 0.01 | |

*Source:* May, J. M., *Biometrika*, 39, 192–195, 1952. By permission of Oxford University Press.

## TABLE A.12
## Significant Studentized Ranges for 5% and 1% Level New Multiple Range Tests

| Error | Protection | $p$ = Number of Means for Range Being Tested | | | | | | | | | | | | | |
|---|---|---|---|---|---|---|---|---|---|---|---|---|---|---|---|
| $k$ | Level | 2 | 3 | 4 | 5 | 6 | 7 | 8 | 9 | 10 | 12 | 14 | 16 | 18 | 20 |
| 1 | 0.05 | 18.0 | 18.0 | 18.0 | 18.0 | 18.0 | 18.0 | 18.0 | 18.0 | 18.0 | 18.0 | 18.0 | 18.0 | 18.0 | 18.0 |
| | 0.01 | 90.0 | 90.0 | 90.0 | 90.0 | 90.0 | 90.0 | 90.0 | 90.0 | 90.0 | 90.0 | 90.0 | 90.0 | 90.0 | 90.0 |
| 2 | 0.05 | 6.09 | 6.09 | 6.09 | 6.09 | 6.09 | 6.09 | 6.09 | 6.09 | 6.09 | 6.09 | 6.09 | 6.09 | 6.09 | 6.09 |
| | 0.01 | 14.0 | 14.0 | 14.0 | 14.0 | 14.0 | 14.0 | 14.0 | 14.0 | 14.0 | 14.0 | 14.0 | 14.0 | 14.0 | 14.0 |
| 3 | 0.05 | 4.50 | 4.50 | 4.50 | 4.50 | 4.50 | 4.50 | 4.50 | 4.50 | 4.50 | 4.50 | 4.50 | 4.50 | 4.50 | 4.50 |
| | 0.01 | 8.26 | 8.5 | 8.6 | 8.7 | 8.8 | 8.9 | 8.9 | 9.0 | 9.0 | 9.0 | 9.1 | 9.2 | 9.3 | 9.3 |
| 4 | 0.05 | 3.93 | 4.01 | 4.02 | 4.02 | 4.02 | 4.02 | 4.02 | 4.02 | 4.02 | 4.02 | 4.02 | 4.02 | 4.02 | 4.02 |
| | 0.01 | 6.51 | 6.8 | 6.9 | 7.0 | 7.1 | 7.1 | 7.2 | 7.2 | 7.3 | 7.3 | 7.4 | 7.4 | 7.5 | 7.5 |
| 5 | 0.05 | 3.64 | 3.74 | 3.79 | 3.83 | 3.83 | 3.83 | 3.83 | 3.83 | 3.83 | 3.83 | 3.83 | 3.83 | 3.83 | 3.83 |
| | 0.01 | 5.70 | 5.96 | 6.11 | 6.18 | 6.26 | 6.33 | 6.40 | 6.44 | 6.5 | 6.6 | 6.6 | 6.7 | 6.7 | 6.8 |
| 6 | 0.05 | 3.46 | 3.58 | 3.64 | 3.68 | 3.68 | 3.68 | 3.68 | 3.68 | 3.68 | 3.68 | 3.68 | 3.68 | 3.68 | 3.68 |
| | 0.01 | 5.24 | 5.51 | 5.65 | 5.73 | 5.81 | 5.88 | 5.95 | 6.00 | 6.0 | 6.1 | 6.2 | 6.2 | 6.3 | 6.3 |
| 7 | 0.05 | 3.35 | 3.47 | 3.54 | 3.58 | 3.60 | 3.61 | 3.61 | 3.61 | 3.61 | 3.61 | 3.61 | 3.61 | 3.61 | 3.61 |
| | 0.01 | 4.95 | 5.22 | 5.37 | 5.45 | 5.53 | 5.61 | 5.69 | 5.73 | 5.8 | 5.8 | 5.9 | 5.9 | 6.0 | 6.0 |
| 8 | 0.05 | 3.26 | 3.39 | 3.47 | 3.52 | 3.55 | 3.56 | 3.56 | 3.56 | 3.56 | 3.56 | 3.56 | 3.56 | 3.56 | 3.56 |
| | 0.01 | 4.74 | 5.00 | 5.14 | 5.23 | 5.32 | 5.40 | 5.47 | 5.51 | 5.5 | 5.6 | 5.7 | 5.7 | 5.8 | 5.8 |
| 9 | 0.05 | 3.20 | 3.34 | 3.41 | 3.47 | 3.50 | 3.52 | 3.52 | 3.52 | 3.52 | 3.52 | 3.52 | 3.52 | 3.52 | 3.52 |
| | 0.01 | 4.60 | 4.86 | 4.99 | 5.08 | 5.17 | 5.25 | 5.32 | 5.36 | 5.4 | 5.5 | 5.5 | 5.6 | 5.7 | 5.7 |
| 10 | 0.05 | 3.15 | 3.30 | 3.37 | 3.43 | 3.46 | 3.47 | 3.47 | 3.47 | 3.47 | 3.47 | 3.47 | 3.47 | 3.47 | 3.48 |
| | 0.01 | 4.48 | 4.73 | 4.88 | 4.96 | 5.06 | 5.13 | 5.20 | 5.24 | 5.28 | 5.36 | 5.42 | 5.48 | 5.54 | 5.55 |
| 11 | 0.05 | 3.11 | 3.27 | 3.35 | 3.39 | 3.43 | 3.44 | 3.45 | 3.46 | 4.46 | 3.46 | 3.46 | 3.46 | 3.47 | 3.48 |
| | 0.01 | 4.39 | 4.63 | 4.77 | 4.86 | 4.94 | 5.01 | 5.06 | 5.12 | 5.15 | 5.24 | 5.28 | 5.34 | 5.38 | 5.39 |
| 12 | 0.05 | 3.08 | 3.23 | 3.33 | 3.36 | 3.40 | 3.42 | 3.44 | 3.44 | 3.46 | 3.46 | 3.46 | 3.46 | 3.47 | 3.48 |
| | 0.01 | 4.32 | 4.55 | 4.68 | 4.76 | 4.81 | 4.92 | 4.96 | 5.02 | 5.07 | 5.13 | 5.17 | 5.22 | 5.24 | 5.26 |
| 13 | 0.05 | 3.06 | 3.21 | 3.30 | 3.35 | 3.38 | 3.41 | 3.42 | 3.44 | 3.45 | 3.45 | 3.46 | 3.46 | 3.47 | 3.47 |
| | 0.01 | 4.26 | 4.48 | 4.62 | 4.69 | 4.74 | 4.84 | 4.88 | 4.94 | 4.98 | 5.04 | 5.08 | 5.13 | 5.14 | 5.15 |
| 14 | 0.05 | 3.03 | 3.18 | 3.27 | 3.33 | 3.37 | 3.39 | 3.41 | 3.42 | 3.44 | 3.45 | 3.46 | 3.46 | 3.47 | 3.47 |
| | 0.01 | 4.21 | 4.42 | 4.55 | 4.63 | 4.70 | 4.78 | 4.83 | 4.87 | 4.91 | 4.96 | 5.00 | 5.04 | 5.06 | 5.07 |
| 15 | 0.05 | 3.01 | 3.16 | 3.25 | 3.31 | 3.36 | 3.38 | 3.40 | 3.42 | 3.43 | 3.44 | 3.45 | 3.46 | 3.47 | 3.47 |
| | 0.01 | 4.17 | 4.37 | 4.50 | 4.58 | 4.64 | 4.72 | 4.77 | 4.81 | 4.84 | 4.90 | 4.94 | 4.97 | 4.99 | 5.00 |
| 16 | 0.05 | 3.00 | 3.15 | 3.23 | 3.30 | 3.34 | 3.37 | 3.39 | 3.41 | 3.43 | 3.44 | 3.45 | 3.46 | 3.47 | 3.47 |
| | 0.01 | 4.13 | 4.34 | 4.45 | 4.54 | 4.60 | 4.67 | 4.72 | 4.76 | 4.79 | 4.84 | 4.88 | 4.91 | 4.93 | 4.94 |
| 17 | 0.05 | 2.98 | 3.13 | 3.22 | 3.28 | 3.33 | 3.36 | 3.38 | 3.40 | 3.42 | 3.44 | 3.45 | 3.46 | 3.47 | 3.47 |
| | 0.01 | 4.10 | 4.30 | 4.41 | 4.50 | 4.56 | 4.63 | 4.68 | 4.72 | 4.75 | 4.80 | 4.83 | 4.86 | 4.88 | 4.89 |
| 18 | 0.05 | 2.97 | 3.12 | 3.21 | 3.27 | 3.32 | 3.35 | 3.37 | 3.39 | 3.41 | 3.43 | 3.45 | 3.46 | 3.47 | 3.47 |
| | 0.01 | 4.07 | 4.27 | 4.38 | 4.46 | 4.53 | 4.59 | 4.64 | 4.68 | 4.71 | 4.76 | 4.79 | 4.82 | 4.84 | 4.85 |
| 19 | 0.05 | 2.96 | 3.11 | 3.19 | 3.26 | 3.31 | 3.35 | 3.37 | 3.39 | 3.41 | 3.43 | 3.44 | 3.46 | 3.47 | 3.47 |
| | 0.01 | 4.05 | 4.24 | 4.35 | 4.43 | 4.50 | 4.56 | 4.61 | 4.64 | 4.67 | 4.72 | 4.76 | 4.79 | 4.81 | 4.82 |
| 20 | 0.05 | 2.95 | 3.10 | 3.18 | 3.25 | 3.30 | 3.34 | 3.36 | 3.38 | 3.40 | 3.43 | 3.44 | 3.46 | 3.46 | 3.47 |
| | 0.01 | 4.02 | 4.22 | 4.33 | 4.40 | 4.47 | 4.53 | 4.58 | 4.61 | 4.65 | 4.69 | 4.73 | 4.76 | 4.78 | 4.79 |
| 22 | 0.05 | 2.93 | 3.08 | 3.17 | 3.24 | 3.29 | 3.32 | 3.35 | 3.37 | 3.39 | 3.42 | 3.44 | 3.45 | 3.46 | 3.47 |
| | 0.01 | 3.99 | 4.17 | 4.28 | 4.36 | 4.42 | 4.48 | 4.53 | 4.57 | 4.60 | 4.65 | 4.68 | 4.71 | 4.74 | 4.75 |
| 24 | 0.05 | 2.92 | 3.07 | 3.15 | 3.22 | 3.28 | 3.31 | 3.34 | 3.37 | 3.38 | 3.41 | 3.44 | 3.45 | 3.46 | 3.47 |
| | 0.01 | 3.96 | 4.14 | 4.24 | 4.33 | 4.39 | 4.44 | 4.49 | 4.53 | 4.57 | 4.62 | 4.64 | 4.67 | 4.70 | 4.72 |
| 26 | 0.05 | 2.91 | 3.06 | 3.14 | 3.21 | 3.27 | 3.30 | 3.34 | 3.36 | 3.38 | 3.41 | 3.43 | 3.45 | 3.46 | 3.47 |
| | 0.01 | 3.93 | 4.11 | 4.21 | 4.30 | 4.36 | 4.41 | 4.46 | 4.50 | 4.53 | 4.58 | 4.62 | 4.65 | 4.67 | 4.69 |
| 28 | 0.05 | 2.90 | 3.04 | 3.13 | 3.20 | 3.26 | 3.30 | 3.33 | 3.35 | 3.37 | 3.40 | 3.43 | 3.45 | 3.46 | 3.47 |
| | 0.01 | 3.91 | 4.08 | 4.18 | 4.28 | 4.34 | 4.39 | 4.43 | 4.47 | 4.51 | 4.56 | 4.60 | 4.62 | 4.65 | 4.67 |

**TABLE A.12** *(continued)*
**Significant Studentized Ranges for 5% and 1% Level New Multiple Range Tests**

| Error $k$ | Protection Level | $p$ = Number of Means for Range Being Tested | | | | | | | | | | | | | |
|---|---|---|---|---|---|---|---|---|---|---|---|---|---|---|---|
| | | 2 | 3 | 4 | 5 | 6 | 7 | 8 | 9 | 10 | 12 | 14 | 16 | 18 | 20 |
| 30 | 0.05 | 2.89 | 3.04 | 3.12 | 3.20 | 3.25 | 3.29 | 3.32 | 3.35 | 3.37 | 3.40 | 3.43 | 3.44 | 3.46 | 3.47 |
| | 0.01 | 3.89 | 4.06 | 4.16 | 4.22 | 4.32 | 4.36 | 4.41 | 4.45 | 4.48 | 4.54 | 4.58 | 4.61 | 4.63 | 4.65 |
| 40 | 0.05 | 2.86 | 3.01 | 3.10 | 3.17 | 3.22 | 3.27 | 3.30 | 3.33 | 3.35 | 3.39 | 3.42 | 3.44 | 3.46 | 3.47 |
| | 0.01 | 3.82 | 3.99 | 4.10 | 4.17 | 4.24 | 4.30 | 4.34 | 4.37 | 4.41 | 4.46 | 4.51 | 4.54 | 4.57 | 4.59 |
| 60 | 0.05 | 2.83 | 2.98 | 3.08 | 3.14 | 3.20 | 3.24 | 3.28 | 3.31 | 3.33 | 3.37 | 3.40 | 3.43 | 3.45 | 3.47 |
| | 0.01 | 3.76 | 3.92 | 4.03 | 4.12 | 4.17 | 4.23 | 4.27 | 4.31 | 4.34 | 4.39 | 4.44 | 4.47 | 4.50 | 4.53 |
| 100 | 0.05 | 2.80 | 2.95 | 3.05 | 3.12 | 3.18 | 3.22 | 3.26 | 3.29 | 3.32 | 3.36 | 3.40 | 3.42 | 3.45 | 3.47 |
| | 0.01 | 3.71 | 3.86 | 3.98 | 4.06 | 4.11 | 4.17 | 4.21 | 4.25 | 4.29 | 4.35 | 4.38 | 4.42 | 4.45 | 4.48 |
| $\infty$ | 0.05 | 2.77 | 2.92 | 3.02 | 3.09 | 3.15 | 3.19 | 3.23 | 3.26 | 3.29 | 3.34 | 3.38 | 3.41 | 3.44 | 3.47 |
| | 0.01 | 3.64 | 3.80 | 3.90 | 3.98 | 4.04 | 4.09 | 4.14 | 4.17 | 4.20 | 4.26 | 4.31 | 4.34 | 4.38 | 4.41 |

*Source:* Duncan, D. B., *Biometrics*, 11, 1–42, 1955. With permission.

**TABLE A.13**
**Table of Random Numbers**

| | | | | | | | | | | | | | | | |
|---|---|---|---|---|---|---|---|---|---|---|---|---|---|---|---|
| 2952 | 6641 | 3992 | 9792 | 7979 | 5911 | 3170 | 5624 | 2002 | 3965 | 9830 | 5657 | 0106 | 3920 | 8316 | 9933 |
| 4167 | 9524 | 1545 | 1396 | 7203 | 5356 | 1300 | 2693 | 5558 | 1397 | 1616 | 3421 | 3179 | 0089 | 8577 | 3804 |
| 2730 | 7483 | 3408 | 2762 | 3563 | 1089 | 6913 | 7691 | 0754 | 3099 | 8955 | 5016 | 2501 | 3686 | 2497 | 6478 |
| 0560 | 5246 | 1112 | 6107 | 6008 | 8126 | 4233 | 8776 | 1551 | 0425 | 2703 | 6001 | 8768 | 0535 | 2809 | 5682 |
| 2754 | 9143 | 1405 | 9025 | 7002 | 6111 | 8816 | 6446 | 9887 | 3519 | 7728 | 1338 | 9894 | 4485 | 2318 | 5219 |
| 5870 | 2859 | 4988 | 1658 | 2922 | 6166 | 6069 | 2763 | 4109 | 5807 | 2066 | 4443 | 0552 | 1304 | 5387 | 2851 |
| 9263 | 2466 | 3398 | 5440 | 8738 | 6028 | 5048 | 2683 | 9694 | 5636 | 7209 | 6345 | 1296 | 4774 | 4254 | 3462 |
| 2002 | 7840 | 1690 | 7505 | 0423 | 8430 | 8759 | 7108 | 3077 | 7595 | 7436 | 1091 | 6957 | 6384 | 6631 | 6160 |
| 9568 | 2835 | 9427 | 3668 | 2596 | 8820 | 1955 | 6515 | 2799 | 3153 | 0609 | 5469 | 4965 | 4800 | 8714 | 9360 |
| 8243 | 1579 | 1930 | 5026 | 3426 | 7088 | 3991 | 7151 | 0427 | 9396 | 9498 | 8282 | 8204 | 4597 | 6375 | 9136 |
| 5667 | 3513 | 9270 | 6298 | 6396 | 7306 | 7898 | 7842 | 5748 | 3267 | 5974 | 9794 | 6701 | 6567 | 4691 | 5365 |
| 1018 | 6891 | 1212 | 6563 | 2201 | 5013 | 0730 | 2405 | 0418 | 8561 | 8849 | 8392 | 9816 | 7342 | 9414 | 9071 |
| 6841 | 5111 | 5688 | 3777 | 7354 | 3434 | 8336 | 6424 | 8516 | 4703 | 2340 | 6613 | 8192 | 3256 | 5725 | 7918 |
| 2041 | 2207 | 4889 | 7346 | 2865 | 1550 | 5960 | 5479 | 7801 | 4345 | 7510 | 9733 | 3441 | 8177 | 3886 | 2615 |
| 5565 | 4764 | 2617 | 5281 | 1870 | 6497 | 5744 | 9576 | 4218 | 0099 | 4947 | 0353 | 4201 | 8115 | 8989 | 9270 |
| 4508 | 1808 | 3289 | 3993 | 9485 | 4240 | 2835 | 9955 | 1353 | 2965 | 9832 | 8173 | 5700 | 7361 | 2514 | 2100 |
| 2152 | 6475 | 5692 | 9309 | 7661 | 1668 | 5431 | 7658 | 0297 | 8753 | 7503 | 6788 | 6699 | 3480 | 4938 | 4651 |
| 6917 | 4113 | 7340 | 6853 | 1172 | 7229 | 1279 | 5085 | 1258 | 5569 | 4135 | 7016 | 6081 | 4714 | 6238 | 3650 |
| 8241 | 4124 | 4131 | 9500 | 5657 | 3932 | 5942 | 3317 | 4020 | 9799 | 1473 | 0351 | 5895 | 4711 | 9096 | 5095 |
| 7913 | 3709 | 5944 | 9763 | 2755 | 4211 | 4995 | 8657 | 8374 | 4000 | 2392 | 1352 | 9626 | 9160 | 3161 | 4864 |
| 9385 | 7125 | 3230 | 0737 | 2957 | 1013 | 6369 | 4494 | 8573 | 9047 | 6273 | 5370 | 1692 | 8642 | 4723 | 6395 |
| 3436 | 6293 | 6025 | 9384 | 3343 | 1071 | 1468 | 4801 | 5908 | 6753 | 2968 | 0493 | 6963 | 0932 | 0530 | 6225 |
| 9094 | 1634 | 5070 | 0664 | 6510 | 0918 | 4601 | 4294 | 2323 | 0931 | 1318 | 8502 | 0393 | 7352 | 5943 | 9679 |
| 9226 | 9296 | 2796 | 7097 | 4067 | 2074 | 6297 | 2587 | 2631 | 0929 | 9749 | 9533 | 2543 | 0582 | 2726 | 4243 |
| 7781 | 3760 | 2895 | 7653 | 0091 | 7012 | 1308 | 1946 | 3216 | 6505 | 2356 | 5935 | 3123 | 2939 | 6567 | 7700 |
| 9742 | 9694 | 7347 | 0017 | 9572 | 1850 | 0116 | 1899 | 0971 | 7262 | 6057 | 0544 | 4519 | 1483 | 6533 | 7521 |
| 9420 | 9210 | 8787 | 9375 | 4663 | 0396 | 6717 | 5562 | 9900 | 3867 | 9348 | 0523 | 6698 | 9726 | 4095 | 1889 |
| 1179 | 3571 | 5992 | 3059 | 9015 | 5608 | 2348 | 8144 | 6595 | 1946 | 1833 | 6994 | 2662 | 3287 | 5559 | 2437 |
| 0708 | 4011 | 4057 | 1550 | 1674 | 1376 | 5243 | 4427 | 2135 | 3917 | 6708 | 2659 | 7562 | 2070 | 9136 | 8124 |
| 6350 | 3996 | 3795 | 2176 | 8182 | 4514 | 6349 | 3483 | 7277 | 3541 | 1868 | 6176 | 5605 | 3278 | 4894 | 0898 |
| 1414 | 7152 | 3658 | 1636 | 0638 | 3443 | 4440 | 3086 | 2114 | 9399 | 2242 | 2143 | 8724 | 1212 | 9485 | 3985 |
| 7041 | 8985 | 7011 | 5676 | 7570 | 6685 | 1776 | 3154 | 7280 | 0130 | 7791 | 6272 | 1446 | 8574 | 6895 | 5119 |
| 3243 | 2783 | 0840 | 9054 | 8862 | 5173 | 8433 | 9117 | 3049 | 5947 | 7154 | 2744 | 1040 | 2042 | 8679 | 4307 |
| 7922 | 4931 | 5753 | 6160 | 6566 | 8602 | 3423 | 9074 | 2626 | 7739 | 6694 | 5994 | 7517 | 1339 | 6812 | 4139 |
| 8769 | 3513 | 8976 | 0780 | 6382 | 0029 | 2619 | 5982 | 6938 | 8098 | 6140 | 2013 | 8824 | 1349 | 2077 | 0075 |
| 2510 | 7274 | 8743 | 0000 | 1850 | 2408 | 3602 | 5179 | 9423 | 9305 | 4635 | 7627 | 5754 | 6274 | 1842 | 3913 |
| 0224 | 2404 | 9811 | 6641 | 9732 | 1662 | 9158 | 1404 | 4900 | 5906 | 3203 | 4680 | 8972 | 2175 | 6758 | 1408 |
| 3009 | 8516 | 7245 | 9409 | 2844 | 0717 | 1072 | 3137 | 6736 | 3153 | 8194 | 3403 | 5709 | 3691 | 9809 | 3202 |
| 7489 | 0221 | 7921 | 2351 | 2696 | 4906 | 2484 | 3868 | 2029 | 0157 | 8942 | 9125 | 9147 | 5712 | 8165 | 6909 |
| 5188 | 1825 | 2220 | 9382 | 0532 | 1915 | 1790 | 2081 | 5131 | 8000 | 3060 | 5383 | 2425 | 5712 | 8165 | 6909 |
| 1198 | 2546 | 2482 | 9607 | 0067 | 3744 | 9866 | 5096 | 6196 | 7762 | 4053 | 2338 | 9339 | 4559 | 5769 | 0054 |
| 3908 | 4676 | 7816 | 6517 | 9121 | 3171 | 4119 | 3615 | 7873 | 0152 | 0061 | 8299 | 3340 | 8162 | 0809 | 6769 |
| 1094 | 2223 | 1675 | 2282 | 3712 | 8191 | 1130 | 1454 | 8367 | 7197 | 2065 | 4458 | 2517 | 1874 | 5165 | 2182 |
| 1817 | 7723 | 5582 | 7153 | 9518 | 0231 | 7782 | 5742 | 4472 | 5622 | 6875 | 1626 | 8839 | 5268 | 5193 | 3333 |
| 6208 | 9598 | 9623 | 2114 | 7747 | 2096 | 5027 | 0561 | 3527 | 6466 | 6874 | 0215 | 9079 | 4781 | 8888 | 5622 |

**TABLE A.13** *(continued)*
**Table of Random Numbers**

| | | | | | | | | | | | | | | | |
|---|---|---|---|---|---|---|---|---|---|---|---|---|---|---|---|
| 4752 | 4519 | 2749 | 8020 | 4642 | 1190 | 7302 | 8350 | 9279 | 1166 | 3175 | 9130 | 6135 | 1674 | 3075 | 3674 |
| 0486 | 6993 | 3115 | 5025 | 4887 | 1571 | 9819 | 6804 | 7979 | 4612 | 5231 | 8130 | 8459 | 0139 | 0959 | 7728 |
| 4942 | 3004 | 1442 | 2810 | 1479 | 0970 | 7302 | 3775 | 7026 | 4980 | 1357 | 1297 | 7974 | 6421 | 2094 | 3218 |
| 4930 | 9785 | 7460 | 3996 | 2864 | 0559 | 3985 | 8092 | 3794 | 0065 | 6435 | 5299 | 5271 | 4296 | 2281 | 8933 |
| 2349 | 1594 | 7152 | 0257 | 4041 | 4105 | 3180 | 9806 | 8439 | 9065 | 1010 | 2859 | 9417 | 1004 | 3007 | 1702 |
| 1596 | 3069 | 7906 | 6656 | 5298 | 5090 | 8580 | 6756 | 1254 | 2858 | 7358 | 4024 | 3684 | 8485 | 2617 | 5488 |
| 4951 | 6993 | 5243 | 6375 | 5088 | 2078 | 8339 | 0322 | 5443 | 4911 | 0922 | 7134 | 4798 | 1311 | 8701 | 2210 |
| 9867 | 0399 | 6741 | 2579 | 7552 | 4171 | 1364 | 9390 | 3262 | 2322 | 4112 | 9877 | 4776 | 4512 | 1746 | 2593 |
| 2670 | 7969 | 2909 | 8705 | 0188 | 9808 | 7206 | 8104 | 7809 | 0297 | 8956 | 2158 | 7780 | 0753 | 1232 | 7181 |
| 5808 | 4329 | 6681 | 9852 | 9458 | 0863 | 8491 | 1166 | 6862 | 4194 | 3596 | 5072 | 4473 | 3099 | 0729 | 4950 |
| 8298 | 1125 | 9153 | 9517 | 1481 | 0879 | 2355 | 2615 | 9179 | 3814 | 9153 | 2127 | 6745 | 9646 | 8105 | 3133 |
| 3117 | 0239 | 0770 | 8034 | 9978 | 9120 | 0967 | 0334 | 5317 | 0986 | 0633 | 6480 | 4834 | 8710 | 8829 | 8572 |
| 2385 | 1979 | 1120 | 3109 | 1169 | 0151 | 3869 | 3360 | 0126 | 4777 | 8034 | 9217 | 2128 | 2232 | 5039 | 8637 |
| 0039 | 4461 | 8674 | 5296 | 8111 | 5748 | 7451 | 2141 | 2372 | 7774 | 9446 | 7178 | 8403 | 3971 | 0899 | 5274 |
| 2400 | 4971 | 0464 | 1849 | 7055 | 3093 | 6860 | 7777 | 0357 | 5276 | 3999 | 0261 | 9255 | 5780 | 5728 | 0032 |
| 5989 | 6749 | 8832 | 2760 | 5220 | 3344 | 5704 | 6359 | 7855 | 9707 | 5259 | 4263 | 9878 | 4918 | 0987 | 9118 |
| 0996 | 0851 | 9659 | 7107 | 0121 | 5224 | 1225 | 4881 | 2510 | 4254 | 1543 | 0224 | 0112 | 6523 | 8667 | 4707 |
| 6909 | 4971 | 4087 | 4783 | 9694 | 8725 | 4088 | 4023 | 6639 | 1913 | 3120 | 9149 | 6145 | 5895 | 0726 | 3883 |
| 7841 | 5930 | 8157 | 0130 | 8818 | 5682 | 5231 | 6721 | 6769 | 1435 | 9107 | 4762 | 9902 | 3764 | 7388 | 2729 |
| 4891 | 6223 | 9252 | 9259 | 2174 | 0304 | 0848 | 6429 | 4527 | 8000 | 8648 | 3366 | 7945 | 4847 | 4317 | 9636 |
| 1036 | 9133 | 5407 | 1734 | 5114 | 3940 | 1052 | 6210 | 5699 | 9883 | 2456 | 0893 | 4132 | 6668 | 0799 | 6137 |
| 9140 | 7896 | 0881 | 7064 | 5862 | 2466 | 9764 | 7913 | 4160 | 1445 | 2887 | 0724 | 1294 | 8988 | 1527 | 1467 |
| 1281 | 8613 | 3431 | 0104 | 0689 | 3742 | 4788 | 9887 | 4506 | 2474 | 3590 | 5308 | 7640 | 7128 | 1023 | 2418 |
| 9426 | 3819 | 5761 | 8019 | 3637 | 9217 | 1250 | 1852 | 4645 | 0613 | 9846 | 4458 | 5666 | 7671 | 1184 | 2328 |
| 9156 | 4867 | 9269 | 6149 | 4471 | 8007 | 5156 | 2117 | 6686 | 3544 | 9828 | 9187 | 0506 | 6473 | 5356 | 8940 |
| 7611 | 3508 | 9182 | 1738 | 5772 | 5037 | 9252 | 0914 | 6503 | 0329 | 7899 | 8211 | 0852 | 8066 | 5706 | 1940 |
| 5045 | 3524 | 8250 | 4053 | 4343 | 3222 | 9205 | 4775 | 1658 | 4288 | 1856 | 5319 | 3512 | 8981 | 7468 | 3836 |
| 5302 | 2660 | 1618 | 1595 | 0344 | 6798 | 3884 | 0570 | 1319 | 1204 | 3344 | 8886 | 3946 | 6777 | 1700 | 0323 |
| 3689 | 5028 | 6302 | 4552 | 9513 | 5867 | 5072 | 1420 | 5030 | 4027 | 2077 | 9812 | 8645 | 3290 | 7191 | 6152 |
| 0148 | 2449 | 5491 | 2607 | 5425 | 0310 | 9764 | 1875 | 7866 | 2029 | 5156 | 2003 | 2940 | 0237 | 7670 | 2852 |
| 2881 | 7835 | 1188 | 1705 | 2254 | 4336 | 4869 | 0663 | 1983 | 8992 | 1017 | 7263 | 7699 | 4151 | 8132 | 7271 |
| 7924 | 5505 | 0432 | 3101 | 0995 | 2741 | 1570 | 2507 | 9944 | 0845 | 7468 | 3936 | 8002 | 0857 | 5784 | 4480 |
| 1980 | 8420 | 7225 | 8292 | 2557 | 8286 | 9977 | 4910 | 0330 | 9913 | 4990 | 7790 | 6932 | 0871 | 1988 | 9881 |
| 0987 | 2467 | 1820 | 3694 | 7650 | 2971 | 1982 | 3115 | 9903 | 7914 | 4138 | 6826 | 0230 | 1337 | 7413 | 8840 |
| 7558 | 5563 | 2431 | 0144 | 0099 | 9578 | 9302 | 8922 | 1614 | 7862 | 9500 | 4109 | 1037 | 2978 | 6075 | 0971 |
| 9149 | 1081 | 0426 | 4201 | 1984 | 1069 | 3092 | 2143 | 2096 | 1164 | 3788 | 6257 | 0632 | 0693 | 2263 | 5290 |
| 6830 | 6866 | 0032 | 9777 | 1599 | 8792 | 5895 | 3951 | 0511 | 0229 | 5951 | 6808 | 1409 | 7624 | 4903 | 4692 |
| 3690 | 4634 | 8850 | 7257 | 1523 | 3725 | 8352 | 7958 | 0524 | 4056 | 9140 | 6371 | 2099 | 8290 | 3611 | 6501 |
| 6311 | 0073 | 2706 | 0558 | 1831 | 3718 | 5649 | 2346 | 7381 | 7386 | 6568 | 1568 | 4160 | 0429 | 3488 | 3741 |
| 4500 | 1635 | 9652 | 3850 | 3005 | 1152 | 9703 | 5892 | 3311 | 3733 | 7882 | 6985 | 7874 | 7264 | 4587 | 9591 |
| 2638 | 2413 | 0427 | 1386 | 7036 | 9050 | 0426 | 5529 | 6874 | 2534 | 7485 | 9695 | 9086 | 2701 | 4967 | 1588 |
| 5878 | 1097 | 8222 | 2528 | 4893 | 6114 | 5975 | 8436 | 8987 | 3121 | 3628 | 0372 | 1059 | 6339 | 8973 | 4218 |
| 4228 | 3005 | 7018 | 1359 | 9809 | 4135 | 4390 | 8052 | 9205 | 2358 | 0393 | 3196 | 1612 | 8397 | 4390 | 0187 |
| 1902 | 0565 | 9486 | 3143 | 0381 | 1076 | 4311 | 8009 | 9749 | 7983 | 7076 | 9791 | 1530 | 8127 | 4474 | 1895 |
| 6021 | 0790 | 3778 | 8780 | 4416 | 1079 | 3643 | 5781 | 2183 | 2109 | 2874 | 5733 | 1567 | 7764 | 4939 | 9919 |

**TABLE A.13** *(continued)*
**Table of Random Numbers**

| | | | | | | | | | | | | | | | |
|---|---|---|---|---|---|---|---|---|---|---|---|---|---|---|---|
| 6925 | 7225 | 7044 | 1013 | 8455 | 2282 | 4424 | 2742 | 6926 | 3085 | 2079 | 3330 | 4432 | 9524 | 0327 | 9640 |
| 6216 | 0623 | 7919 | 1403 | 9369 | 1368 | 3535 | 9378 | 3373 | 3567 | 0371 | 5932 | 3923 | 7250 | 8578 | 5869 |
| 3119 | 6623 | 8283 | 1162 | 5702 | 3438 | 2940 | 1399 | 9771 | 5542 | 4715 | 5527 | 3763 | 3167 | 3679 | 4399 |
| 3834 | 5149 | 0358 | 7261 | 9637 | 0449 | 8158 | 2765 | 9353 | 5576 | 5474 | 0190 | 7274 | 6993 | 3920 | 7272 |
| 9894 | 4662 | 2386 | 1148 | 5722 | 3917 | 8587 | 6513 | 5761 | 6301 | 3558 | 6205 | 3012 | 6195 | 8461 | 2046 |
| 8585 | 5029 | 4318 | 0608 | 1135 | 5188 | 9062 | 4824 | 6850 | 8122 | 4455 | 2940 | 9945 | 9688 | 3588 | 8311 |
| 1976 | 7922 | 3906 | 8808 | 2128 | 0110 | 4461 | 2906 | 6616 | 6760 | 4938 | 0066 | 0391 | 8898 | 4753 | 7402 |
| 5657 | 3262 | 3811 | 8320 | 2547 | 9713 | 1175 | 9142 | 2633 | 4255 | 8755 | 8434 | 4334 | 8992 | 1260 | 0547 |
| 5762 | 0776 | 2463 | 1786 | 3847 | 9388 | 4315 | 5001 | 6837 | 4898 | 9527 | 3526 | 6536 | 7703 | 9941 | 8454 |
| 9639 | 0038 | 7097 | 8784 | 9044 | 0376 | 4533 | 6807 | 4523 | 7772 | 3888 | 7243 | 1330 | 7115 | 4897 | 6618 |

*Source:* Tippett, L. H. C., *Random Sampling Numbers*, Cambridge University Press, London, 1959. Reprinted with permission of Cambridge University Press.

# References


Anscombe, F. J. 1949. The statistical analysis of insect counts based on the negative binomial distribution. *Biometrics* 5:165–173.

Asthana, R. S. 1950. *The Size of Sub-Sampling Unit in Area Estimation*. Thesis for Diploma, I.C.A.R., New Delhi. *See* Sukhatmi, P. V. 1954. *Sampling Theory of Surveys with Application*. Iowa State College Press, Ames.

Avery, T. E. and Burkhart, H. E. 1983. *Forest Measurements*, 3rd. ed., McGraw-Hill, New York.

Ayres, F., Jr. 1962. *Theory and Problems of Matrices*. (Schaum's Outline Series), Schaum, New York.

Bailey, R. L. 1974. Weibull model for *Pinus radiata* diameter distributions. *In Statistics in Forestry Research, Proc. IV Conf. of Advisory Group of Forest Statisticians*. I.U.F.R.O. Subj. Group 56.02.

Bailey, R. L. and Dell, T. R. 1973. Quantifying diameter distribution with the Weibull function. *For. Sci.* 19:97–104.

Barrett, J. P. 1964. Correction for edge effect bias in point sampling. *For. Sci.* 10:52–55.

Bartlett, M. S. 1937a. Properties of sufficiency and statistical tests. *Proc. R. Soc.* A 160:268–282.

Bartlett, M. S. 1937b. Some examples of statistical methods of research in agriculture and applied biology. *J. R. Stat. Soc. Suppl.* 4:137–183.

Baskerville, G. L. 1972. Use of logarithmic regression in the estimation of plant biomass. *Can. J. For. Res.* 2:49–53.

Beauchamp, J. J. and Olson, J. S. 1973. Corrections for bias in regression estimates after logarithmic transformation. *Ecology* 54:1,403–1,407.

Beers, T. W. 1966. *The Direct Correction for Boundary-Line Slopover in Horizontal Point Sampling*. Purdue University Agricultural Experiment Station Res. Prog. Rep. 224.

Beers, T. W. 1977. Practical corrections of boundary slopover. *South. J. Appl. For.* 1:16–18.

Beers, T. W. and Miller, C. I. 1964. *Point Sampling: Research Results, Theory, and Applications*. Purdue University Agricultural Experiment Station Res. Bull. 786.

Beers, T. W. and Miller, C. I. 1976. *Line Sampling for Forest Inventory*. Purdue University Agricultural Experiment Station Res. Bull. 934.

Beyer, W. H. 1968. *Handbook of Tables for Probability and Statistics*. 2nd ed., Chemical Rubber Co., Cleveland, OH.

Bitterlich, W. 1947. Die Winkelzähl messung [The angle-count method]. *Allg. Forst Holzwirtsch. Z.* 58:94–96.

Bitterlich, W. 1949. Das Relaskop. *Allg. Forst Holzwirtsch. Z.* 60:41–42.

Bitterlich, W. 1952. Das Spregelrelaskop. *Österr. Forst Holzwirtsch.* 7:3–7.

Bruce, D. 1955. A new way to look at trees. *J. For.* 53:163–167.

Burk, T. E. and Ek, A. R. 1982. Application of empirical Bayes/James Stein procedures to simulate estimation problems in forest inventory. *For. Sci.* 28:753–770.

Carow, J. and Stage, A. 1953. Basal area by the thumb. *J. For.* 51:512.

Chiang, A. C. 1974. *Fundamental Methods of Mathematical Economics*. McGraw-Hill, New York.

Clutter, J. L. and Bennett, F. A. 1965. *Diameter Distributions in Old-Field Slash Pine Plantations*. Georgia Forest Research Council Rep. 13.

Clutter, J. L. and Shephard, M. 1968. *A Diameter Distribution Technique for Predicting Multiple Product Yields in Old-Field Slash Pine Plantations*. University of Georgia, School of Forest Research Final Rep. to USDA Forest Service, S.E. Forestry Experiment Station Contract No. 12-11-008-876.

Clutter, J. L., Fortson, J. C., Pienaar, L. V., Brister, G. H., and Bailey, R. L. 1983. *Timber Management: A Quantitative Approach*. John Wiley & Sons, New York.

Cochran, W. G. 1946. Relative accuracy of systematic and stratified random samples for a certain class of populations. *Ann. Math. Stat.* 17:164–177.

Cochran, W. G. 1963; 1977. *Sampling Techniques*, 2nd ed., 1963; 3rd ed., 1977, John Wiley & Sons, New York.
Cochran, W. G. and Cox, G. M. 1957. *Experimental Designs,* 2nd ed., John Wiley & Sons, New York.
Cunia, T. 1985. More Basic Designs for Survey Sampling: Two-Stage and Two-Phase Sampling, Ratio and Regression Estimates and Sampling on Successive Occasions. S.U.N.Y. College of Environmental Science and Forestry, Forest Biometry Memograph Series No. 4.
Dalenius, T. 1957. *Sampling in Sweden. Contributions to the Methods and Theories of Sampling Survey Practice.* Almqvist & Wicksell, Stockholm.
DeLury, D. B. 1950. *Values and Integrals of the Orthogonal Polynomials up to* n = *26.* University of Toronto Press, Toronto.
Dell, T. R., Feduccia, D. P., Campbell, T. E., Mann, W. F., Jr., and Polmer, B. H. 1979. *Yields of Unthinned Slash Pine Plantations on Cutover Sites in the West Gulf Region.* USDA Forest Service Res. Pap. SO-147.
Draper, N. R. and Smith, H. 1981. *Applied Regression Analysis,* 2nd ed., John Wiley & Sons, New York.
Dubey, S. D. 1967. Some percentile estimates for Weibull parameters. *Technometrics* 9:119–129.
Duncan, A. J. 1959. *Quality Control and Industrial Statistics* (Rev.). Richard D. Irwin, Homewood, IL.
Duncan, D. B. 1951. A significance test for differences between ranked treatments in an analysis of variance. *Va. J. Sci.* 2:171–189.
Duncan, D. B. 1955. Multiple range and multiple F-tests. *Biometrics* 11:1–42.
Ek, A. R. 1971. A comparison of some estimators in forest sampling. *For. Sci.* 17:2–13.
Ek, A. R. 1974. Nonlinear models for stand table projection in northern hardwood stands. *Can. J. For. Res.* 4:23–27.
Ek, A. R. and Issos, J. N. 1978. Bayesian theory and multi-resource inventory. In *Proc. Workshop on Integrated Inventories of Renewable Natural Resources.* Society of American Foresters, Tucson, AZ, Jan. 8–12, 1978. pp. 291-298.
Ezekiel, M. 1941. *Methods of Correlation Analysis*, 2nd ed., John Wiley & Sons, New York.
Feduccia, D. P., Dell, T. R., Mann, W. F., Jr., Campbell, T. E., and Polmer, B. H. 1979. *Yields of Unthinned Loblolly Pine Plantations on Cutover Sites in the West Gulf Region.* USDA Forest Service Res. Pap. SO-147.
Feller, W. 1957. *An Introduction to Probability Theory and Its Applications,* Vol. 1. John Wiley & Sons, New York.
Finney, D. J. 1952. *Probit Analysis* (Rev.). Cambridge University Press, London.
Fisher, R. A. 1921. On the "probable error" of a coefficient of correlation deduced from a small sample. *Metron* 1:3–32.
Fisher, R. A. and Yates, F. 1957. *Statistical Tables for Biological, Agricultural and Medical Research*, 5th ed., Oliver & Boyd, Edinburgh.
Flewelling, J. W. and Pienaar, L. V. 1981. Multiplicative regression with logarithmic errors. *For. Sci.* 27:281–289.
Freese, F. 1962. Elementary Forest Sampling. USDA Forest Service Agricultural Handbook 232.
Freese, F. 1964. Linear Regression Methods for Forest Research. USDA Forest Service Res. Pap. FPL 17.
Furnival, G. M. 1961. An index for comparing equations used in constructing volume tables. *For. Sci.* 7:337–341.
Furnival, G. M. 1964. More on the elusive formula of best fit. In *Proc. Soc. Am. For.*, Denver, CO, 201–207.
Furnival, G. M. 1979. Forest sampling-past performance and future expectations. In *Proc. Workshop in Forest Resource Inventories*, Ft. Collins, CO, 320–326.
Furnival, G. M., Gregoire, T. G., and Grosenbaugh, L. R. 1987. Adjusted inclusion probabilities with 3-P sampling. *For. Sci.* 33:617–631.
Gautschi, W. 1957. Some remarks on systematic sampling. *Ann. Math. Stat.* 28:385.
Grosenbaugh, L. R. 1948. Forest parameters and their statistical estimation. Paper presented at the Alabama Polytech. Inst. Conf. on Statistics Applied to Research, Auburn, AL, Sept. 8, 1948, 10 pp. mimeo.
Grosenbaugh, L. R. 1952. Plotless timber estimates — new, fast, easy. *Jour. For.* 50:32–37.
Grosenbaugh, L. R. 1955. *Better Diagnosis and Prescription in Southern Forest Management.* USDA Forest Service Southern Forestry Experiment Station Occas. Pap. 145.
Grosenbaugh, L. R. 1958a. *Point Sampling and Line Sampling: Probability Theory, Geometic Implications, Synthesis.* USDA Forest Service SE Forestry Experiment Station Occas. Pap. 160.
Grosenbaugh, L. R. 1958b. *The Elusive Formula of Best Fit: A Comprehensive New Machine Program.* USDA Forest Service Southern Forestry Experiment Station Occas. Pap. 158.

Grosenbaugh, L. R. 1963a. *Optical Dendrometers for Out-of-Reach Diameters: A Conspectus and Some New Theory.* Forest Science Monogr. 4.

Grosenbaugh, L. R. 1963b. Some suggestions for better sample tree measurement. In *Proc. Soc. Am. For.*, Denver, CO, 36–42.

Grosenbaugh, L. R. 1964. *STX-FORTRAN-4 Program for Estimates of Tree Populations from 3-P Sample Tree Measurements.* USDA Forest Service Res. Pap. PSW 13.

Grosenbaugh, L. R. 1965. *Three-Pee Sampling Theory and Program THRP.* USDA Forest Service Res. Pap. PSW 21.

Grosenbaugh, L. R. 1967a. The gains from sample tree selection with unequal probabilities. *J. For.* 65:203–206.

Grosenbaugh, L. R. 1967b. *STX-FORTRAN-4 Program for Estimates of Tree Populations from 3-P Sample Tree Measurements.* USDA Forest Service Res. Pap. PSW (Rev.).

Grosenbaugh, L. R. 1967c. *REX-FORTRAN-4 System for Combinational Screening or Conventional Analysis of Multivariate Regressions.* USDA Forest Service Res. Pap. PSW 44 (Updated 9/20/71).

Grosenbaugh, L. R. 1967d. Choosing units of measure. In *Proc. Conf. on Young Growth Forest Management in California.* University of California Agricultural Experiment Service, Berkeley, CA, March 1–3, 1967, 143–146.

Grosenbaugh, L. R. 1968. Comments on "3-P sampling and some alternatives, I" by Schreuder, Sedransk, and Ware. *For. Sci.* 14:453–454.

Grosenbaugh, L. R. 1971. *STX 1-11-71 for Dendrometry of Multistage 3-P Samples.* USDA Forest Service Pub. FS-277.

Grosenbaugh, L. R. 1974. *STX 3-3-73: Tree Content and Value Estimations Using Various Sample Designs, Dendrometry Methods, and V-S-L Conversion Coefficients.* USDA Forest Service Res. Pap. SE 117.

Grosenbaugh, L. R. 1976. Approximate sampling variance of adjusted 3-P estimates. *For. Sci.* 22:173–176.

Grosenbaugh, L. R. 1979. *3P Sampling Theory, Examples, and Rationale.* USDA Bureau of Land Management Tech. Note 331.

Hafley, W. L. and Schreuder, H. T. 1977. Statistical distributions for fitting diameter and height data in even-aged stands. *Can. J. For. Res.* 7:481–487.

Haga, T. and Moezawa, K. 1959. Bias due to the edge effect. *For. Sci.* 5:370–376.

Hald, A. 1952a. *Statistical Theory with Engineering Applications.* John Wiley & Sons, New York.

Hald, A. 1952b. *Statistical Tables and Formulas.* John Wiley & Sons, New York.

Hansen, M. H. and Hurwitz, W. N. 1943. On the theory of sampling from finite populations. *Ann. Math. Stat.* 14:333–362.

Hansen, M. H., Hurwitz, W. N., and Madow, W. G. 1953. *Sample Survey Methods and Theory,* Vol. I, *Methods and Applications.* John Wiley & Sons, New York.

Harter, H. L. 1957. Error ratio and sample sizes for range tests in multiple comparisons. *Biometrika* 8:511–536.

Hartley, H. O. 1955. Some recent developments in analysis of variance. *Comm. Pure Appl. Math.* 8:47–72.

Hartley, H. O. and Ross, A. 1954. Unbiased ratio estimates. *Nature* 174:270–271.

Hasel, A. A. 1942. Estimation of volume in timber stands by strip sampling. *Ann. Math. Stat.* 13:179–206.

Hastings, N. A. J. and Peacock, J. B. 1974. *Statistical Distributions: A Handbook for Students and Practitioners.* John Wiley & Sons, New York.

Hatheway, W. H. and Williams, E. J. 1958. Efficient estimation of the relationship between plot size and the variability of crop yields. *Biometrics* 14:207–222.

Hendricks, W. A. 1944. The relative efficiencies of groups of farms as sampling units. *J. Am. Stat. Assoc.* 39:367–376.

Hirata, T. 1955. Height estimation through Bitterlich's method, vertical angle count sampling. *Jpn. J. For.* 37:479–480.

Hoel, P. G., Post, S. C., and Stone, C. J. 1971. *Introduction to Probability Theory.* Houghton-Mifflin, Boston.

Husch, B., Miller, C. I., and Beers, T. W. 1972. *Forest Mensuration*, 2nd ed., Ronald Press, New York.

Hyink, D. M. and Moser, J. W., Jr. 1979. Application of diameter distribution for yield projection in uneven-aged forests. In *Proc. Forest Resource Inventories Workshop*, Vol. II. Colorado State University, Department of Forestry and Wood Science, Ft. Collins, CO, 906–915.

Hyink, D. M. and Moser, J. W., Jr. 1983. Generalized framework for projecting forest yield and stand structure using diameter distributions. *For. Sci.* 29:85–95.

Issos, J. N., Ek, A. R., and Hansen, M. H. 1977. *Computer Programs for Bayesian Estimation in Forest Sampling.* University of Wisconsin, Department of Forestry, Staff Pap. 8.

Jensen, C. E. 1964. *Algebraic Descriptions of Forms in Space*. USDA Forest Service Central States Forestry Experiment Station.

Jensen, C. E. 1973. *MATCHACURVE-3: Multiple Component Mathematical Models for Natural Resource Studies*. USDA Forest Service Res. Pap. INT-146.

Jensen, C. E. 1976. *MATCHACURVE-4: Segmented Mathematical Descriptors for Asymmetric Curve Forms*. USDA Forest Service Res. Pap. INT-182.

Jensen, C. E. 1979. *$e^{-k}$, A Function for the Modeler*. USDA Forest Service Res. Pap. INT-240.

Jensen, C. E. and Homeyer, J. W. 1970. *MATCHACURVE-1 for Algebraic Transforms to Describe Sigmoid- or Bell-Shaped Curves*. USDA Forest Service Intermountain Forestry & Range Experiment Station.

Jensen, C. E. and Homeyer. J. W. 1971. *MATCHACURVE-2 for Algebraic Transforms to Describe Curves of the Class $X^n$*. USDA Forest Service Res. Pap. INT-106.

Jessen, R. J. 1942. *Statistical Investigation of a Sample Survey for Obtaining Farm Facts*. Iowa Agricultural Experiment Station Res. Bull. 304.

Jessen, R. J. 1978. *Statistical Survey Techniques*. John Wiley & Sons, New York.

Johnson, F. A. 1961. *Standard Error of Estimated Average Timber Volume per Acre under Point Sampling When Trees are Measured for Volume on a Subsample of All Points*. USDA Forest Service Pacific N.W. Forestry & Range Experiment Station Res. Note 201.

Johnson, F. A., Lowrie, J. B., and Gohlke, M. 1971. *3P Sample Log Scaling*. USDA Forest Service Res. Note PNW 162.

Johnson, N. L. and Kotz, S. 1969. *Distributions in Statistics: Discrete Distributions*. Houghton-Mifflin, Boston.

Johnson, N. L. and Kotz, S. 1970. *Distributions in Statistics: Continuous Univariate Distributions*. Vols. I and II. John Wiley & Sons, New York.

Kendall, M. G. and Stuart, A. 1963. *The Advanced Theory of Statistics:* Vol. 1, *Distribution Theory*. Hafner, New York.

Ker, J. W. 1954. Distribution series arising in quadrat sampling of reproduction. *J. For.* 52:838–841.

Keuls, M. 1952. The use of "Studentized Range" in connection with an analysis of variance. *Euphytica* 1:112–122.

Koch, E. J. and Rigney, J. A. 1951. A method of estimating optimum plot size from experimental data. *Agron. J.* 43:17–21.

Kotz, S. and Johnson, N. L. 1986. *Encyclopaedia of Statistical Sciences*. Vols. 1–9. John Wiley & Sons, New York.

Labau, V. J. 1967. Literature on the Bitterlich Method of Forest Cruising. USDA Forest Service Res. Pap. PNW 47.

Lahiri, D. B. 1951. A method of sample selection providing unbiased ratio estimates. *Bull. Int. Stat. Inst.* 33(2):433–440.

Langsaeter, A. 1926. Om beregning av middelfeilen ved regelmessige linjetakseringer. *Medd. Nor. Skogsforsøksves.* 2(7):5.

Langsaeter, A. 1927. Om beregning av middelfeilen vid regelmessige linjetakseringer. *Medd. Nor. Skogsforsøksves.* 8:71.

Langsaeter, A. 1932. Nøiaktigheten vid lingetaksering av. skog. *Medd. Nor. Skogsforsøksves.* 4:432–563; 5:405–445.

Lappi-Seppälä, M. 1924. Linja-arvioimisestaja sen farkkuudesta. *Medd. Forstvetensk Försöksanst.* 7:49.

Lass, H. and Gottlieb, P. 1971. *Probability and Statistics*, Addison-Wesley, Reading, MA.

Leak, W. B. 1965. The J-shaped probability distribution. *For. Sci.* 11:405–409.

Lenhart, J. D. and Clutter, J. L. 1971. *Cubic Foot Yield Tables for Old-Field Loblolly Pine Plantations in the Georgia Piedmont*. Georgia Forest Res. Coun. Rep. 22-Series E.

Lindeberg, J. W. 1923. Über die Berechnung des Mittelfehlers des Resultates einer Linientaxierung. *Acta For. Fenn.* 25.

Loetsch, F. and Haller, K. A. 1964; 1973. *Forest Inventory*. BLV Verlagsgesellschaft, Munich (Rev., with Zöhrer, F., 1973).

Lyle, E. A., Patterson, R. M., Hicks, D. R., and Bradley, E. L. 1975. *Polymorphic Site Index Curves for Natural Sweetgum in Alabama*. Alabama Agricultural Experiment Station Circ. 220.

Mahalanobis, P. C. 1940. A sample survey of the acreage under jute in Bengal. *Sankhya* 4:511–530.

Masuyama, M. 1953. A rapid method of estimating basal area in timber survey — an application of integral geometry to areal sampling problems. *Sankhya* 12:291–302.

Matérn, B. 1947. Metoder att upskatta noggrannheten vid linge-och provytetaxering. *Medd. Statens Skogsförskningsinst.* 36(1):1–138.

Matérn, B. 1960. Stochastic models and their application to some problems in forest surveys and other sampling investigations. *Medd. Statens Skogsförskningsinst.* 49(5):1–144.

Maxim, L. D. and Harrington, L. 1983. The application of pseudo-Bayesian estimators to remote sensing data: ideas and examples. *Photogramm. Eng. Remote Sensing* 49:649–658.

May, J. M. 1952. Extended and corrected table of the upper percentage points of the "Studentized" range. *Biometrika* 39:192–195.

McClure, J. P. 1969. The Mirror Caliper, a New Optical Dendrometer. USDA Forest Service Res. Note SE 112.

McGee, C. E. and Della-Bianca, L. 1967. Diameter Distributions in Natural Yellow-Poplar Stands. USDA Forest Service Res. Pap. SE 25.

Mendenhall, W. and Schaeffer, R. L. 1973. *Mathematical Statistics with Applications.* Duxbury Press, North Scituate, MA.

Mesavage, C. 1967. Random Integer Dispenser. USDA Forest Service Res. Pap. SO 49.

Mesavage, C. 1969a. New Barr and Stroud dendrometer Model FP-15. *J. For.* 67:40–41.

Mesavage, C. 1969b. Modification of a Zeiss Telemeter Teletop for Precision Dendrometry. USDA Forest Service Res. Note SO 95.

Mesavage, C. 1971. STX Timber Estimating with 3P Sampling and Dendrometry. USDA (Forest Service) Agricultural Handbook 415.

Meyer, H. A. 1941. A Correction for a Systematic Error Occurring in the Application of the Logarithmic Volume Equation. Pennsylvania State University Forestry School Res. Pap. 7. (Also Pennsylvania Agricultural Experiment Station J. Ser. Pap. 1058).

Meyer, H. A. 1952. Structure, growth, and drain in balanced uneven aged forests. *J. For.* 50:85–92.

Meyer, H. A. 1953. *Forest Mensuration.* Penns Valley Publishers, State College, PA.

Meyer, H. A. 1956. The calculation of the sampling error of a cruise from the mean square successive difference. *J. For.* 54:341.

Meyer, P. L. 1970. *Introductory Probability and Statistical Applications,* 2nd ed., Addison-Wesley, Reading, MA.

Mickey, M. R. 1959. Some finite population unbiased ratio and regression estimators. *J. Am. Stat. Assoc.* 54:594–612.

Miller, I. and Freund, J. E. 1977. *Probability and Statistics for Engineers,* 2nd ed., Prentice-Hall, Englewood Cliffs, NJ.

Moore, P. G. 1955. The properties of the mean square successive difference in samples from various populations. *J. Am. Stat. Assoc.* 50:434–456.

Morse, A. P. and Grubbs, F. E. 1947. The estimation of dispersion from differences. *Ann. Math. Stat.* 18:194–214.

Moser, J. W., Jr. 1976. Specification of density for the inverse J-shaped diameter distribution. *For. Sci.* 22:177–180.

Näslund, M. 1930. Om medelfelets beräkning vid linjetaxering. *Sven. Skogsvardsfören Tidskr.* 28:309.

Näslund, M. 1939. Om medelfelets härledning vid linje-och provytetaxering. *Medd. Statens Skogsförskningsinst.* 31:301–344.

Nelson, T. C. 1964. Diameter distribution and growth in loblolly pine. *For. Sci.* 10:105–114.

Newman, D. 1939. The distribution of range in samples from a normal population, expressed in terms of an independent estimate of standard deviation. *Biometrika* 31:20–30.

Neyman, J. 1934. On the two different aspects of the representative method; the method of stratified sampling and the method of purposive selection. *J. R. Stat. Soc.* 97:558–606.

Osborne, J. G. 1942. Sampling errors of systematic and random surveys of cover type area. *J. Am. Stat. Assoc.* 37:256–264.

Östlind, J. 1932. Eforderlig taxeringsprocent vid linjetaxering av skog. *Sven. Skogsvardsfören. Tidskr.* 30:417–514.

Owen, D. B. 1962. *Handbook of Statistical Tables,* Addison-Wesley, Reading, MA.

Palley, M. N. and Horwitz, L. G. 1961. Properties of some random and systematic point sampling estimators. *For. Sci.* 7:52–65.

Patel, J. K., Kapadia, C. H., and Owen, D. B. 1976. *Handbook of Statistical Distributions,* Marcel Dekker, New York.

Pearson, E. S. and Hartley, H. O. (Eds.) 1954. *Biometrika Tables for Statisticians,* Vol. I. Cambridge University Press, London.

Pearson, K. 1931. *Tables for Biometricians and Statisticians.* Part II, London.

Quenouille, M. H. 1953. *The Design and Analysis of Experiments.* Hafner, New York.

Quenouille, M. H. 1956. Notes on bias in estimation. *Biometrika* 43:353–360.

Raj, D. 1972. *The Design of Sample Surveys.* McGraw-Hill, New York.

Randolph, J. F. 1967. *Calculus U Analytic Geometry U Vectors.* Dikenson, Belmont, CA.

Robson, D. S. 1957. Applications of multivariate polykays to the theory of unbiased ratio type estimation. *J. Am. Stat. Assoc.* 52:511–522.

Roeder, A. 1979. Bayesian estimation of information value. In *Proc. Workshop on Forest Resources Inventories.* Colorado State University, Ft. Collins, 174–182.

Sassaman, W. J. and Barrett, J. P. 1974. White Pine Yields by Diameter Classes. New Hampshire Agricultural Experiment Station Res. Rep. 38.

Satterthwaite, F. E. 1946. An approximate distribution of estimates of variance components. *Biometrics* 2:110–114.

Sayn-Wittgenstein, L. 1966. A Statistical Test of Systematic Sampling in Forest Surveys. Ph.D. dissertation, Yale University, New Haven, CT. [University Microfilms, Ann Arbor, MI (66-14,998).]

Schmid, P. 1969. Stichproben am Waldrand. *Mitt. Schweiz. Anst. Forstl. Versuchswes.* 45:234–303.

Schreuder, H. T. 1970. Point sampling theory in the framework of equal probability cluster sampling. *For. Sci.* 16:240–246.

Schreuder, H. T. 1976. Comment on note by L. R. Grosenbaugh "Approximate sample variance of adjusted 3P samples." *For. Sci.* 22:176.

Schreuder, H. T. 1984. Poisson sampling and some alternatives in timber sales sampling. *For. Ecol. Manage.* 8:149–160.

Schreuder, H. T. 1990. The gain in efficiency in PPS sampling. *For. Sci.* 36:1146–1152.

Schreuder, H. T., Brink, G. E., and Wilson, R. L. 1984. Alternative estimators for point-Poisson sampling. *For. Sci.* 30:803–812.

Schreuder, H. T., Gregoire, T. G., and Wood, G. B. 1993. *Sampling Methods for Multi-Resource Forest Inventory.* John Wiley & Sons, New York.

Schreuder, H. T., Li, H. G., and Sadooghi-Alvandi, S. M. 1990. Sunter's PPS without Replacement Sampling as an Alternative to Poisson Sampling. USDA Forest Service Res. Pap. RM290.

Schreuder, H. T., Ouyang, Z., and Williams, M. 1992. Point-Poisson, point-PPS, and modified point PPS sampling: efficiency and variance estimation. *Can. J. For. Res.* 22:725–728.

Schreuder, H. T., Sedransk, J., and Ware, K. D. 1968. 3-P sampling and some alternatives, I. *For. Sci.* 14:429–453.

Schreuder, H. T., Sedransk, J., Ware, K. D., and Hamilton, D. A. 1971. 3-P sampling and some alternatives, II. *For. Sci.* 17:103–118.

Schumacher, F. X. and Chapman, R. A. 1948. Sampling Methods in Forestry and Range Management. Duke University School of Forestry Bull. 7.

Sethi, V. K. 1965. On optimum pairing of units. *Sankhya* Ser. B, 27:315–320.

Shifley, S. R., Moser, J. W., Jr., and Brown, K. M. 1982. Growth and Yield Model for the Elm-Ash-Cottonwood Type in Indiana. USDA Forest Service Res. Pap. NC218.

Shiue, C. J. 1960. Systematic sampling with multiple random starts. *For. Sci.* 21:179–203.

Singh, D., Jindal, K. K., and Garg, J. N. 1968. On modified systematic sampling. *Biometrika* 55:341–546.

Smalley, G. W. and Bailey, R. L. 1974a. *Yield Tables and Stand Structure for Loblolly Pine Plantations in Tennessee, Alabama, and Georgia Highlands.* USDA Forest Service Res. Pap. SO 96.

Smalley, G. W. and Bailey, R. L. 1974b. *Yield Tables and Stand Structure for Shortleaf Pine Plantations in Tennessee, Alabama, and Georgia Highlands.* USDA Forest Service Res. Pap. SO 97.

Smith, H. F. 1938. An empirical law describing heterogeneity in the yields of agricultural crops. *J. Agric. Sci.* 28:1–23.

Smith, J. H. G. and Ker, J. W. 1957. Some distributions encountered in sampling forest stands. *For. Sci.* 3:137–144.

Snedecor, G. W. 1934. *Analysis of Variance and Covariance.* Collegiate Press, Ames, IA.

Snedecor, G. W. and Cochran, W. G. 1967. *Statistical Methods*, 6th ed., Iowa State University Press, Ames.

Sokal, R. R. and Rohlf, F. J. 1969. *Biometry: The Principles and Practice of Statistics in Biological Research.* W. H. Freeman & Co., San Francisco, CA.

Som, R. K. 1973. *A Manual of Sampling Techniques,* Heinemann Educational Books, London.

Sommerfeld, R. A. 1974. A Weibull prediction of the tensile strength–volume relationship of snow. *J. Geophys. Res.* 79:3354–3356.

Space, J. C. 1974. 3-P sampling in the South. In *Proc. Workshop on Inventory Design and Analysis.* Colorado State University, Ft. Collins, 38–52.

Spiegel, M. R. 1961. *Theory and Problems of Statistics,* Schaum, New York.

Spurr, S. H. 1952. *Forest Inventory,* Ronald Press, New York.

Steel, R. G. D. and Torrie, J. H. 1960. *Principles and Procedures of Statistics.* McGraw-Hill, New York.

Stephan, F. F. 1941. Stratification in representative sampling. *J. Marketing* 6:38–46.

Sterling, T. D. and Pollack, S. V. 1968. *Introduction to Statistical Data Processing,* Prentice-Hall, Englewood Cliffs, NJ.

Sternitzke, H. S. 1960. Arkansas Forests. USDA Forest Service, Southern Forestry Experiment Station 58 pp.

Sternitzke, H. S. 1960. Tennessee Forests. USDA Forest Service, Southern Forestry Experiment Station 29 pp.

Strand, L. 1957. "Relaskopisk" hoyde og kubikkmasseebestemmelse. *Nor. Skogbruk.* 3:535–538.

Sukhatme, P. V., Sukhatme, B. V., Sukhatme, S., and Asok, C. 1984. *Sampling Theory of Surveys with Applications,* Iowa State University Press, Ames.

Swindel, B. F. 1961. Application and Precision of Permanent Point Inventory Control on Small Forests in the Piedmont. Master's thesis, University of Georgia, School of Forestry. [See Sayn-Wittgenstein, 1966.]

Swindel, B. F. 1972. The Bayesian Controversy. USDA Forest Service Res. Pap. SE 95.

Swindel, B. F. and Bamping, J. H. 1963. A Comparison of Two Methods to Estimate Sampling Error on Small Forest Tracts, unpublished manuscript, University of Georgia, School of Forestry. [See Sayn-Wittgenstein, 1966.]

Tin, M. 1965. Comparison of some ratio estimators. *J. Am. Stat. Assoc.* 60:294–307.

Tippett, L. 1959. *Random Sampling Numbers,* University of London, Department of Statistics, Tracts for Computers No. 15. Cambridge University Press, London.

Tukey, J. W. 1953. *The Problem of Multiple Comparisons.* Princeton University ditto, Princeton, NJ.

U.S. Forest Service. 1929. Volumes, Yield and Stand Tables for Second-Growth Southern Pines. USDA Misc. Pub. 50.

Van Deusen, P. C. 1987. 3-P sampling and design versus model-based estimates. *Can. J. For. Res.* 17:115–117.

Van Hooser, D. D. 1972. *Evaluation of Two-Stage 3P Sampling for Forest Surveys.* USDA Forest Service Res. Pap. SO 77.

Van Hooser, D. D. 1974. Inventory remeasurement with two-stage 3P sampling. In *Proc. Workshop on Inventory Design and Analysis,* Colorado State University, Ft. Collins, 53–62.

van Laar, A. and Geldenhuys, C. J. 1975. Distribution and correlation patterns in an indigenous forest. *S. Afr. For. J.,* 94(Sept.):20–28.

Von Neumann, J., Kent, R. H., Bellinson, H. R., and Hart, B. I. 1941. The mean square successive difference. *Am. Math. Stat.* 12:153–162.

Waters, W. E. 1955. Sequential sampling in forest insect surveys. *For. Sci.* 1:68–79.

Waters, W. E. and Henson, W. R. 1959. Some sampling attributes of the negative binomial distribution with special reference to forest insects. *For. Sci.* 5:397–412.

Weast, R. C. and Selby, S. M. 1975. *Handbook of Tables for Mathematics,* Rev. 4th ed., Chemical Rubber Co., Cleveland, OH.

Welch, B. L. 1949. Further note on Mrs. Aspin's tables and on certain approximations to the tabled function. *Biometrika* 36:293–296.

Wheeler, P. R. 1956. *Point-Sampling on the Forest Survey.* USDA Forest Service, Southern Forestry Experiment Station, Forest Survey Eastern Techniques Meeting.

Wheeler, P. R. 1962. Penta prisen caliper for upper-stem diameter measurement. *J. For.* 60:877–878.

Wiant, H. V. and Michael, E. D. 1978. Potential of 3-P sampling in wildlife and range management. In *Proc. Workshop in Integrated Inventories of Renewable Natural Resources.* Society of American Foresters, Tucson, AZ, Jan. 8–12, 1978, 304–306.

Wishart, J. and Metakides, T. 1953. Orthogonal polynomial fitting. *Biometrika* 40:361–369.

Wood, G. B. and Schreuder, H. T. 1986. Implementing point-Poisson and point-model based sampling in forest inventory. *For. Ecol. Manage.* 14:141–156.

Yamana, T. 1967. *Elementary Sampling Theory.* Prentice-Hall, Englewood Cliffs, NJ.

Yates, F. 1946. A review of recent statistical developments in sampling and sampling surveys. *J. R. Stat. Soc.* 109:12-30.

Yates, F. 1948. Systematic sampling. *R. Soc. London Philos. Trans.* 241(A 834):345–377.

Yates, F. 1949. *Sampling Methods for Censuses and Surveys.* Chas. Griffith & Co., London.

Yates, F. 1981. *Sampling Methods for Censuses and Surveys,* 4th ed., Macmillan, New York.

Zar, J. H. 1968. Calculation and miscalculation of the allometric equation as a model in biological data. *Bio. Sci.* 18:1118–1120.

Zöhrer, F. 1970. *The Beta-Distribution for Best Fit of Stem Diameter Distributions.* IUFRO 3rd Conf. Adv. Group For., Stat. Inst. Natl. Rech. Agr. Publ. 1972–3.

# Index

## A

## B



## C

## D



## E

## F



## G

## H



## I



## J

## K



## L



## M

## N



## O



## P

# Q



# R

## S

## T

## U



## V

## W